Methodicum Chimicum

A Critical Survey of Proven Methods
and Their Application in Chemistry,
Natural Science, and Medicine

Editor-in-Chief

Friedhelm Korte

Volume Editors

H. Aebi
H. Batzer
E. Baumgartner
K.-H. Büchel
J. Falbe
R. Gompper
M. Goto

U. Hasserodt
H. Machleidt
K. Niedenzu
G. Ohloff
H. Zimmer
F. Zymalkowski

Academic Press New York · San Francisco · London   1974

Georg Thieme Publishers Stuttgart

Volume 1

Analytical Methods

Part A

Purification, Wet Processes
Determination of Structure

Edited by
Friedhelm Korte

English Editorial Advisor: Laeticia Kilzer

Contributions from

H. Batzer, Basel
E. Bayer, Tübingen
A. Becker, Basel
H.-D. Beckey, Bonn
H. Beyer, Aachen
B. Briat, Paris
W. Büchler, Basel
G. Buttgereit, Leverkusen
F. Christen, Basel
J. T. Clerc, Zürich
B. Cornils, Oberhausen
F. Coulston, Albany
H. Determann, Frankfurt
H. F. Ebel, Heidelberg
F. Effenberger, Stuttgart
H. Egan, London
G. Erdtmann, Jülich
J. Falbe, Oberhausen
H.-W. Fehlhaber, Bonn
H. Feltkamp, Wuppertal-
  Elberfeld
P. Fischer, Stuttgart
J. Fleischhauer, Aachen
E. Fluck, Stuttgart
J. K. Foreman, London
E. K. Franke, Cincinnati
G. Giesselmann, Hanau-Wolfgang
E. Glotter, Rehovot
J. Haartz, Cincinnati
J. Haase, Karlsruhe
H.-D. Hahn, Oberhausen
U. Hasserodt, Hamburg
K. H. Haussler, Heidelberg
G. Hesse, Erlangen
O. Hockwin, Bonn

W. Hofmann, Basel
G. Hohlneicher, München
A. W. Hubbard, London
T. Ishii, Tokyo
H. Jork, Saarbrücken
H. Kaiser, Dortmund
F. Kaplan, Cincinnati
B. Kastening, Jülich
H. J. Keller, Heidelberg
H. Kienitz, Ludwigshafen
W. Klein, St. Augustin
B. Kolb, Überlingen
L. Kraus, Saarbrücken
K. Lampert, Frankfurt
D. C. Lankin, Cincinnati
D. Lavie, Rehovot
G. Legler, Köln
O. F. H. Lehmann, Hamburg
F. Lohse, Basel
A. Maas, Bonn
K. Maas, Heidelberg
A. Manzetti, Basel
Y. Mazur, Rehovot
E. W. McChesney, Albany
K. A. McCully, Ottawa
J. Meier, Basel
D. Merz, Frankfurt
G. Michal, Tutzing
K. Möbius, Berlin
H. V. Morley, Ottawa
R. Neeb, Mainz
W. Niemitz, Berlin
H. W. Nürnberg, Jülich
H. Pauschmann, Tübingen
G. F. Phillips, London

D. G. Porter, London
A. Prox, München
H. Rink, Bonn
E. Robens, Frankfurt
I. Rosenblum, Albany
H. Runge, Ludwigshafen
R. Sawyer, London
H.-D. Scharf, Aachen
J. Scharschmidt, Krefeld-
  Uerdingen
H. Schildknecht, Heidelberg
K. Schlögl, Wien
R. Schmid, Basel
W. Schöniger †
B. Schrader, Dortmund
E. Schultze-Rhonhof, Bonn
K. E. Schwarzhans, München
J. F. Sievers, Berlin
W. Simon, Zürich
G. Snatzke, Bonn
K. F. Soike, Albany
H. Stiller, Jülich
J. E. Todd, Cincinnati
G. Tölg, Stuttgart
A. Walter, Basel
G. Walter, Frankfurt
H. Wamhoff, Bonn
J. Weber, Oberhausen
H. Weitkamp, Wuppertal
H. Wenger, Basel
G. Will, Bonn
A. Yogev, Rehovot
S. A. Zahir, Basel
W. Zeil, Karlsruhe
H. Zimmer, Cincinnati

288 illustrations in 368 individual presentations, 7 schemes, 146 tables

Academic Press New York · San Francisco · London   1974

Georg Thieme Publishers Stuttgart

Editors

Prof. Dr. *Hugo Aebi*
Direktor des Medizinisch-Chemischen Instituts der
Universität Bern
CH-3000 Bern 9, Bühlstraße 28

Prof. Dr. *Hans Batzer*
Direktor, Ciba-Geigy AG
CH-4000 Basel

Prof. Dr. *Erich Baumgartner*
Direktor des Instituts für Lebensmittelchemie der
Universität Bern
CH-3000 Bern

Dr. *Karl-Heinz Büchel*
Direktor, Bayer AG
5600 Wuppertal 1, Postfach 130105

Dr. *Jürgen Falbe*
Mitglied des Vorstandes, Ruhrchemie AG
4200 Oberhausen-Holten, Bruchstraße

Prof. Dr. *Rudolf Gompper*
Institut für Organische Chemie der Universität
München
8000 München, Karlstraße 23

Prof. Dr. *Miki Goto*
Faculty of Science
Mejiro, Tokyo

Dr. *Ulrich Hasserodt*
Direktor, Deutsche Shell AG
6000 Frankfurt, Shellhaus

Prof. Dr. *Friedhelm Korte*
Direktor des Instituts für ökologische Chemie der
Technischen Universität München, 8050 Weihen-
stephan
Leiter des Instituts für ökologische Chemie der GSF
5205 St. Augustin, Schloß Birlinghoven

Prof. Dr. *Hans Machleidt*
Direktor, Dr. Karl Thomae GmbH
7950 Biberach, Postfach 720

Prof. Dr. *Kurt Niedenzu*
Department of Chemistry University of Kentucky
Lexington, Kentucky 40506/USA

Dr. *Günther Ohloff*
Direktor, Firmenich & Cie.
CH-1211 Genf 8

Prof. Dr. *Hans Zimmer*
Department of Chemistry, University of Cincinnati,
Cincinnati, Ohio 45221/USA

Prof. Dr. *Felix Zymalkowski*
Direktor des Pharmazeutischen Instituts der Universität Bonn
5300 Bonn 1, Kreuzbergweg 26

English Editorial Advisor:

Prof. Dr. *Laeticia Kilzer*
Department of Chemistry, Mount Marty College,
Yankton, South Dakota 57078

© 1974 Academic Press Inc., New York · San Francisco · London. — Typesetting: Filmsatz Stauffer + Cie., Basel. — Text Processing and Computer-controlled Photo-Typesetting of Index: Siemens TEAM Programme System. Register Typesetting: CRT Photo-Typesetting Equipment of Satz AG, Zürich. — Printed in Germany by Grammlich, Pliezhausen.

Library of Congress Catalog Card Number: 74-21580
ISBN 0-12-460701-2 (Academic Press)
ISBN 3 13 504301 0 (Thieme)

Preface of the Series

The METHODICUM CHIMICUM is a short critical description of chemical methods applied in scientific research and practice. It is particularly aimed at chemists as well as scientists working in associated areas including medicine who make use of chemical methods to solve their 'interrelated' problems.

Considering the present development of science and the necessity for concise and unambiguous information, the series provides a guide to rapid and reliable detection of the method suitable for the solution of the problem concerned. Thus, particular emphasis is placed on the description of proved procedure whereby a complete and exhaustive compilation of all reported methods and also a detailed description of experimental techniques have been deliberately omitted. Newer methods as well as those which have not yet been reported in review articles are treated more extensively, whereas conventional methods are dealt with cincisely. Biological procedures which, in specific cases, are more will be discussed in the analytical volume. The interrelated methods and concepts which are constantly gaining importance will be fully discussed in the third 'Specific Part'.

The METHODICUM CHIMICUM is comprised of three parts. The first, the 'General Part' consists of Volumes 1, 2 and 3. Volume 1 (Analytical Methods) is concerned with chemical, physical, and biological analytical methods including those necessary for the elucidation of structures of compounds.

Volume 2 (Planning of Syntheses) contains a review on fundamentals, principles, and models with particular respect to the concepts and applications of theoretical chemistry essential to the practically working scientist.

Volume 3 (Types of Reactions) is designed to illustrate the scope and utility of proved working techniques and syntheses.

The second part (Vols. 4–8), which is particularly devoted to 'Systematic Syntheses', deals with proved methods for syntheses of specific compounds. These procedures are classified according to functional groups linked together in the last step of reaction.

Volume 4 (Syntheses of Skeletons) describes the construction of hydrocarbons and heterocyclic compounds.

Volume 5 the formation of C–O-bonds, Volume 6 the formation of C–N-bonds, Volume 7 the syntheses of compounds containing main group elements, and Volume 8 compounds containing transition metal elements.

The third 'Special Part' (Volume 9–11) is concerned with the chemical aspects connected with the formulation of a question or problem.

Volume 9 deals with nonmetallic synthetic fibers and synthetic materials as well as their additives, Volume 10 with synthetic compounds and Volume 11 with natural products and natural occuring compounds.

All Volumes should not contain more than 900 printed pages. They are intended to give the chemist and any person working in fields related of chemistry a sufficient answer to his problem. Selected review articles or important original works are cited for the sake of detailed information.

We wish to thank the Georg Thieme Verlag, Stuttgart, for making possible the realization of the basic concept of METHODICUM CHIMICUM and for the excellent presentation of the work.

Bonn, September 1974   Friedhelm Korte

Preface of Volume 1

Volume 1 of METHODICUM CHIMICUM is devoted to analytical methods and techniques of compound separation which are of interest not only to the chemist but also to the scientists who works in associated areas, including medicine.

This compilation contains a short discussion of the principles of well-proved procedures so that the experienced chemist can apply them without the use of secondary literature (e.g. Chapter 3). Theory and practical techniques of the newer methods (e.g. Chapter 7) or procedures (e.g. Chapter 5, 6, 9, 14), which are rapid development, are discussed in more detail. Sufficient room is also given to those contributions which present a novel aspect to a problem (e.g. Chapters 1, 3.1, 5.2).

We considered it necessary to include the application of various methods of trace analysis to organic materials (see Chapter 10) and selected methods for the analysis of significant technical products (see Chapter 11). A critical survey of procedures for the determination of carbohydrates, proteins and nucleic acids is also given (see Chapter 12).

It is well known that the investigation of enzymatic and microbiological methods is essential in modern chemical analysis. Furthermore, techniques for the determination of chemicals in animals and higher plants (see Chapter 13.3 and 13.4) as well as methods for the biochemical transformation of compounds of pharmacological and toxicological interest are treated. In the whole volume, the literature has been surveyed up to 1. January 1972.

We are very grateful to all authors and collaborators for their cooperation and their many valuable suggestions. The publication of the first volume has been somewhat delayed. Thanks to the close collaboration of all concerned, it has been possible to include all new significant results. Further volumes will be published in rapid succession.

We hope that this volume will convince the reader that the objective of the METHODICUM CHIMICUM is to provide the chemist and the scientist working in related areas of chemistry with a rapid and reliable source of information on the method applicable to his specific problem. This applies both to the whole chemistry and all the other fields of science including medicine which are closely connected with chemistry. The disposition of METHODICUM CHIMICUM fully considers this interrelated character.

I am particularly grateful to Professor Otto Hockwin who has given me valuable support during the copy editing of the first volume and who will also advise us in the planning and coordination of the future volumes.

Bonn, September 1974   Friedhelm Korte

Table of Contents for Volume A (Chapter 1–8)

6 Fragmentation Methods 479

7 Diffraction Methods 513

8 Equilibrium and Kinetic Methods 583

Table of Contents for Volume B (Chapter 9–14)

9 Special Physical Methods 629

Table of Contents

1 Foundations for the Critical Discussion of Analytical Methods

Contributed by

Heinrich Kaiser, Dortmund

Preliminary Note

In the following chapter the principles pertaining to chemical analysis are critically observed, partially from a new point of view. Therefore, some of the concepts, relationships, and words used may appear unfamiliar to the reader accustomed to the normal chemical literature. However, the adequate presentation of the ideas has required the use of special terms. Their choice was not made haphazardly from a stock of appropriate synonyms, but by careful selection from mathematics, physics, and from the colloquial language. To assist the reader to grasp the rather abstract ideas and to stimulate his imagination, words with somewhat 'plastic' meanings were chosen. However, they have been given a precise meaning in this text. It is best to take them literally in their original sense and to transfer their understanding as far as possible into the field of the new relationships.

1.1 Basic Concepts

1 Introduction

Chemical Analyses are information processes planned — in most cases fairly well — with the aim of obtaining knowledge about the composition of the substance under investigation. For assessment and evaluation of such processes and of the information provided by them, two essentially different aspects must be considered: content and form.

A critical discussion of the content is possible only in the context of a given individual case, rather than as a generality: for instance, the content of a formally simple analysis may be of utmost importance in a trial in a court of law, for commercial negotiations, or for solution of a scientific problem. However, information processes and the information arising from them always have a formal structure, upon which the type, the scope and the reliability of the obtainable information depend; these questions can be treated in a general way within the framework of a theory. When such a theory is applied to an individual case, it may produce the basis for assessment of the content and the importance of the reported results.

The formal structure of an information process also determines the structure of the information in the sense of knowledge which is produced by this process. For chemical analyses, this means: the questions of importance regarding the type, scope, and reliability of analytical information, and the questions of *accuracy, precision, power of detection, sensitivity, specificity* and *selectivity* all of which relate to the analytical procedure as such. All these concepts, used here in their colloquial sense, if exactly defined, lead to *figures of merit* for analytical procedures. Only from these is it possible to arrive at a critical conclusion regarding analytical results which have been produced by a definite analytical procedure (cf. Section 1.2:6).

2 Concept of a 'Complete Analytical Procedure'

Figures of merit in information theory must be capable of being stated in an objective way. They can, therefore, be given only in relation to concrete analytical procedures, not to general analytical principles, such as titrimetry, spectrometry, polarography, neutron activation, etc. Figures of merit for an analysis always relate only to a definite *complete analytical procedure,* which is specified, in every detail, by fixed working directions (order of analysis) and which is used for a particular analytical task.

For a complete analytical procedure, everything must be predetermined: the analytical task, the apparatus, the external conditions, the experimental procedure, the evaluation and the calibration. If any feature is altered, a different analytical procedure results.

If, for instance, quartz vessels are used instead of glass vessels, or if normal distilled water is replaced by doubly-distilled water, or if one moves to another laboratory with air conditioning and clean air, then one may have a new (and probably better) analytical procedure, in spite of the fact that the general mode of operation has remained unchanged.

One of the most effective means of achieving higher precision in chemical analysis is to take the average of the values from a number of repeated analyses. However, even this alteration of the evaluation process constitutes a change in the analytical procedure. The same would apply if the measurement period were extended (equivalent to better time averaging), for example in X-ray fluorescence analysis.

3 Structure of Analytical Procedures, Classes of Measurable Quantities

Chemical analytical procedures are a special group of technical and scientific *measurement procedures.* Very little attention has hitherto been paid to their logical structure, because interest has been concentrated on the chemical reactions and the operations to be carried out.

Any measurement procedure is embodied in the experimental setup, the instruments, substances, etc., and also in the relevant directions, which must specify the procedure in every detail.

If a measurement procedure is employed in a particular case, it produces a *measured value* x (sometimes for short called a *measure*) for the measurable quantity in question. The aim of the measurement is to find the *true value* of the measurable quantity. However, the 'measure' may deviate to some extent from the unknown true value. The difference (measure minus true value) is the *error* of this particular experimental result. It would be possible to determine how large or small the error is in a particular case only if, in addition to the measured value, the true value were also known in some way. Such a case is, however, only of theoretical interest since a measurement is normally made because one wishes to know the so far unknown true value of the measurable quantity.

Chemical analyses are carried out in order to determine the *concentrations* (symbol: c) of the substances under study, or their *absolute quantities (q),* and often the ratios $(c_r$ or $q_r)$ of such quantities. Various units for these quantities are used, depending on circumstances and practice; it is, therefore, essential always to indicate which *units* are used, in order to avoid misunderstandings.

[Symbols in formulae stand for the measurable quantities, and always denote the product of a numerical value and the appropriate unit.]

This class of measurable quantities is the class of the *target quantities* of the analysis. To make the following considerations more concrete, we shall apply the term *content* to all target quantities, and will use the symbol c for them.

Chemical contents cannot be measured directly; their numerical values, the 'results of an analysis', are derived from measurements of other quantities such as weight, volume, current strength, refractive index, absorbance (extinction in German), intensity of a spectral line, pulse rate, photographic blackening etc. For these measurable quantities, which relate to the very core of a procedure, the term *measure* (symbol: x) should be reserved in its specific sense. What may be considered as the 'core' of an analytical procedure is often indicated by the type designation, for instance one speaks of gravimetric, polarographic, chromatographic, spectrochemical, and neutron activation analysis.

The measurable quantities which are of decisive importance in analysis are nowadays almost exclusively physical in character. However, in many cases these measures are not observed directly, but are derived from the 'reading' of an indicating instrument.

Very often this is a geometrical quantity, for instance the position of a needle, or a recorder trace, or the print-out from a digital voltmeter, etc. This is so obvious that one often overlooks the fact that the 'indicated quantities' are necessary intermediate steps, mainly because most instruments are calibrated to give numerical values of the important physical quantities. For this reason, the intermediate steps, such as reading the position of a needle, will not be considered separately in the following text. It must not, however, be forgotten that indicating instruments may limit the power of a measurement procedure by having too low a precision, too restricted a range of measurement, or too high a 'reading threshold'.

Besides the two quantitative classes of decisive importance, considered thus far, the contents c and the measured quantities x (including the readings, etc.), there is a third class of quantities, the *parameters* (symbol: r) of the analytical procedure. The parameters (i.e. temperature, pressure, concentration of chemical reagents, pH value, wavelength in a spectrum, time) determine the conditions under which the measurements are made during the analysis.

The values of the parameters can be measured; they are often given by the external conditions (e.g. room temperature, atmospheric pressure), and in many cases they can be preset (e.g. pH value, reaction temperature, wavelength). Parameters may be adjusted to fixed values, or they may be varied over wide ranges during an analysis. Unnoticed changes of experimental parameters are often the cause of analytical errors which may appear to be accidental, but which in fact are systematic in nature and, therefore, can be corrected for if the appropriate parameter is measured (e.g. fluctuations in temperature, varying background in a spectrum). This is often disregarded; consequently, all directions relating to procedure should state specifically which parameters are important, and to which values they should be adjusted.

The various quantities, contents c, measures x, and parameters r are interdependent through a system of empirical correlations, theoretically derived functions, systems of equations, diagrams, etc.

1.2 Aspects of the Assessment of Analytical Procedures

1 Calibration

The relationship between measures x (weight, volume, current strength, intensity, etc.) and the desired values for the contents c is described by the 'analytical function', which must be regarded as an essential component of an analytical procedure. If this function is very simple, for instance if it is given by a stoichiometric conversion factor, its involvement may go unnoticed. However, many problems of chemical analysis cannot be understood correctly if one forgets that the link between the measure and the content is always the analytical function.

If one wishes to rely upon stoichiometric relationships or upon theoretical derivations, it is essential to have tested whether the reactions, which have been presupposed as the basis of the calculations, proceed completely and undisturbed. This is why, in publications on new analytical procedures, tables are very often given comparing the 'given' chemical contents with those 'found' by analysis. When there are systematic deviations between the 'given' and the 'found' values, these can be corrected by establishing a so-called 'empirical factor'. This procedure is none other than a calibration of the analytical procedure with standard samples of known composition, even if not so named.

Calibration is a necessary operation when building up a quantitative analytical procedure. There are no analytical procedures which are 'absolute' in the true sense of this word, even if some are so called. To analyze absolutely – i.e. without any presuppositions – it would be necessary to identify the atoms and the molecules of the substances to be determined, to sort them out, and to count them individually and completely.

During any analysis, the sample under investigation is compared with the standard samples which were used to calibrate the analytical procedure. This is done by using the relevant 'analytical function' (which sometimes may be given only by a conversion factor). The calibration of many of the common, well proved procedures may have taken place a long time ago; the author and his paper may be forgotten; his results are, however, part of the anonymous treasure of experience in Analytical Chemistry. Every analysis, therefore, contains comparison as a basic operation in its structure. However, comparison of different objects is possible only if they are in some way 'alike' – this general concept of likeness must be specified for each particular case.

The samples submitted for analysis, and the standard samples used for calibration of the analytical procedure, must be of the 'same kind' with respect to all operations, reactions and measurements to which they are submitted during the course of the 'particular complete analytical procedure': they must belong to the same 'family' of analytical samples.

2 Classification of Calibration Methods

The method by which an analytical procedure is calibrated is an important characteristic for its assessment. The accuracy (see Section 1.2:6) of the analytical results, the range of applicability of the procedure, and the instrumental and running costs are partly dependent on the method of calibration which was used. The various methods for calibration of analytical procedures may be classified in order of decreasing effectiveness; in the following text, they are indicated by small Greek letters.

2.1 Complete Calibration Methods

2.1.1 σ-Calibration (with Synthetic Standard Samples)

Standard samples* which have been synthesized from pure substances – together with the calibration functions based on them – are in their entirety the supporting structure of Analytical Chemistry. During the course of a long historical development, these samples have been made more and

more reliable by many cross-checks, controls and corrections. Today, we are able to prepare many substances in a pure state as starting materials for the calibration of analytical procedures; in particular, we have learned to determine quantitatively how pure these substances are.

When it is possible to prepare standard samples in a reliable way, for a particular analytical procedure from pure substances, then an analytical function can be established which has no bias; and subsequent analytical results will, consequently, be free from systematic errors. Since the true contents of the standard samples are known from their composition, they can be directly correlated with the measured quantities; this allows the inevitable accidental errors of the individual calibration measurements to be eliminated by taking the average of a large number of such measurements.

σ-Calibrations are found everywhere in the practice of chemical analysis; many of the stoichiometric conversion factors in the literature have been determined in this way – mostly, however, for ideal analytical conditions. Standard samples for σ-calibrations can be prepared with reasonable expenditure only for relatively simple analytical problems, for which the type of the sample and of the components to be determined is known. Furthermore, only a few components should be involved or if many components are involved lower analytical precision should be regarded as adequate. Examples are: calibration solutions for *flame spectrometric methods*, mixtures of gases, and simple metallic alloys with known and clear phase diagrams.

In analytical organic chemistry, and particularly in biochemistry, it may be difficult to obtain the starting materials. If traces of contaminants are to be determined in substances of high purity, it may be impossible to obtain these substances free from these contaminations; they cannot, therefore, be used to synthesize σ-standard samples.

Even when synthetic samples have the correct composition, it may not be feasible to transfer them into the same physical or chemical state as the samples for analysis; this means that the standard samples and the analytical samples are not of the same kind with respect to the analytical procedure, and cannot, therefore, be directly compared.

In practice, direct σ-calibration of a complex analytical procedure is the exception rather than the rule. 'Accurate analytical' results are, however, to be expected only if they relate directly or indirectly to σ-calibrations. This is achieved with the next class, α-calibration.

---

* The term 'standard sample' follows the present official nomenclature. However, there is a strong movement to restrict the use of the term 'standard' to samples which have been officially analyzed and issued by some authoritative organization. It has been proposed to use the term 'reference sample' or 'calibration sample' for such material coming from other sources without an official certificate of composition.

2.1.2 α-Calibration (with Analyzed Standard Samples)

This type of calibration is especially important for analytical procedures used for complex analyses of large series of similar samples, for instance for production control in steel mills. α-Calibration proceeds in a way converse to synthetic calibration; it starts with the selection of a set of homogeneous samples which cover the whole range of the compositions in question. Tests must be carried out to determine whether the desired range is covered, and whether these selected samples are of the 'same kind' as the other samples: belong to the same 'family'. Such tests can often be carried out by using the measured value given by the uncalibrated procedure, or by performing auxiliary experiments (e.g. investigation of the crystal structure). In other cases a thorough knowledge of the task may be helpful.

The second step of an α-calibration is the analysis of the selected samples by another analytical procedure which has been σ-calibrated. There are two different ways of achieving this goal: either all samples involved must be treated in such a way that ultimately they are of the same kind with respect to the analytical procedure used for the σ-calibration, or the total analytical procedure must be split up into a number of different parts.

The first way is feasible only with relatively simple analytical procedures. For instance, metallic samples which are to be used as standards for rapid spectrochemical analysis of solid cast samples may be dissolved and then compared with σ-standard solutions made from pure salts of the elements to be determined. The second way involves more work, but is more practicable. The sample for analysis is selectively subdivided into portions of different composition by a series of separation operations, in such a way that each of the final preparations contains only one or a few of the components. Their contents can then be determined by relatively simple reactions. At the end of such an analytical procedure, the σ-calibrations for the different components occur independently of each other. (A prototype of such a procedure is the classical inorganic analysis by a sequence of precipitations.) The topological scheme for such a multi-branched method of analysis is the 'tree' (see 1.2:3.1, Fig. 1, A).

It may be very difficult to get reliable results from a highly branched method of analysis. Incomplete separations may lead to systematic errors; and also, the random errors of the results will, in the end, become systematic analytical errors of the subsidiary α-calibrated procedure. This possible transfer of errors to all subsidiary analyses requires the utmost care. Checks by parallel analyses using other procedures and in other laboratories must be made. Analyzed standard samples of good reliability are generally expensive, and are commercially available only for the most important technical analytical purposes.

If standard samples are required for the determination of *chemical compounds* rather than elements, it may be very difficult or even impossible to design separation processes for the analysis of these standard samples such that the analytical information is not partly lost during the course of the operations. This is especially true for sensitive organic compounds and natural products, which are easily destroyed. In order to draw conclusions about what the original substance may have been, one must try to obtain as many relationships as possible between the observed fragments and the measured quantities in the same way as when elucidating a chemical structure. The topological scheme for such an analytical procedure is the 'network' (see 1.2:3.2, Fig. 1, N).

2.1.3 δ-Calibration (by Differential Additions)

In many cases, an analytical procedure can be calibrated by adding small but known amounts of the component to be determined to the sample undergoing analysis. This is like scanning the analytical function in small differential steps. In order to obtain a smooth and simple function, it is necessary to make a suitable choice for the measurable quantity, and to eliminate the influence of interfering parameters by applying corrections (e.g. corrections for blank values in chemical reactions, background in a spectrum, distortions of recorded traces). If the result is a calibration function with an easily apparent form, then it may be possible to extrapolate this function beyond the range which was covered by the additions, and thus to determine the unknown content which was originally present in the analytical sample.

This δ-calibration procedure, sometimes called the *addition method of calibration,* is the only one which allows quantitative determination of very small trace amounts when the basic material of the analytical sample cannot be obtained completely free from impurities[1,2].

The δ-calibration procedure presupposes that the added amount of the component to be determined behaves analytically in the same way as that part of the component which was originally present in the sample (the standard sample and the analytical sample must be of the same kind).

An example of how difficult the task can be is the analysis of hard refractory ceramics, such as Al_2O_3, for traces of alkalis. The alkali atoms which are added do not enter the crystal lattice; the atoms in the lattice do

[1] G. Ehrlich, R. Gerbatsch, Z. Anal. Chem. *209,* 35 (1965).

[2] F. Rosendahl, Spectrochim. Acta *10,* 201 (1957/-58).

not come out[3]. If the sample is dissolved or melted, contamination or losses are probable, and, if the sample is diluted during the course of such operations, the detection limit for traces may become undesirably high.

2.1.4 ϑ-Calibration (on a Theoretical Basis)

With some reservations, it is also possible to include in this group of complete calibration methods those which derive the contents to be determined from the measured quantities by using quite general principles and known data. Such laws, for instance, include the law of mass action, the Lambert-Beer law, the Boltzmann distribution, etc.

General laws are, of course, idealizations; they are valid only if definite assumptions are correct. In most cases they have been found and verified only under special, very pure experimental conditions. 'Pure' conditions are, under the practical conditions of chemical analysis, nearly a *contradictio in adjecto*. Interfering factors of all kinds must be taken into consideration. Factors of yield or activities demonstrate clearly that pure theory is often not sufficient. Even in cases where a theoretically derived relationship has been established and has long been proved, its validity has in most cases been verified by the use of synthesized standard samples (for instance for many analytical reactions which follow the theoretically predicted stoichiometric relationships).

However, ϑ-calibrations in the two meanings of this term are of importance for new analytical procedures. They provide a first approximation, until they are either verified or replaced by σ-calibrations. They often lead to analytical results which are at least almost correct; in this sense, ϑ-calibrations represent a transition to the next group.

2.2 Abridged Calibration Methods

2.2.1 κ-Calibration (by Convention)

Very often the results of chemical analyses provide a basis for decisions in industry, commerce, or in legal affairs. In such cases, the chemical composition of a substance may be of interest only in so far as it determines the required qualities. When technical requirements as to quality and the conditions of delivery are fulfilled, and the prescribed tolerances are observed, then one may be able to manage without knowing the 'true contents'. On the other hand, it may be necessary to state differences or ratios exactly, in order to ensure that quality specifications are fulfilled, or to decide a dispute. In such cases, the partners must agree, by convention, on a procedure which leads to directly comparable results. There are two possibilities for reaching such a procedure: either one agrees upon an 'umpire assay', including fixed calibration factors, or one calibrates the analyses by using '*official* standard samples'.

The convention is then to accept the results of such umpire analyses as correct within the limits set by the inevitable random errors. This assumption may be true in many cases; every effort is made to select optimal procedures for arbitration and standard samples; however, it is essential that one need not check the validity of the assumption for each single case, but rather appeal to the convention.

2.2.2 β-Calibration (Broad Band Calibration)

This term is used for analytical procedures for which it is known in advance that the calibration will not give a high analytical precision.

Instead of a calibration curve (with relatively small scatter), the relationship between measurable quantity x and content c (see 1.1:3) is represented by a broader band.

One may be compelled to forego analytical precision because this is no longer attainable; there are, however, many analytical problems whose solution does not require very high precision. Examples include many procedures for survey analysis, where classification of the contents according to their orders of magnitude may be sufficient. Such analytical procedures are sometimes termed 'semi-quantitative'. The costs of instruments and time are generally much smaller than with highly refined precision procedures. Furthermore, the problem of ensuring that the analytical samples and the standard samples are of the same kind is considerably simplified. In the broad band which represents the calibration function, the systematic errors (see 1.2:6) which are due to the different composition of the samples disappear at least partially. They can now be regarded as random errors caused by accidental differences in the composition of the samples. This opens the way to 'universal analytical procedures'[4]. The smaller the requirements for precision and limits of detection are, the 'more universal' such procedures may be as regards the nature of the samples.

2.3 Calibration with Auxiliary Scales

For all the classes of calibration methods mentioned above, it is, in principle, possible to give the 'target quantities' of the analysis (contents, concentrations, absolute quantities) in SI units, i.e. in kg, m and mol. However, this presupposes that the nature of the substances to be determined is known, i.e. that their density or their molecular weight can be determined. If this is not the case, units or suitable scales must be introduced in an ad hoc manner.

---

[3] *H. Waechter* in *E. Rexer,* Reinststoffprobleme, Bd. II — Reinststoffanalytik, p. 245, Akademie Verlag, Berlin 1966.

[4] *Ch.E. Harvey,* Semiquantitative Spectrochemistry, Applied Research Laboratories Inc., Glendale Cal. 1964.

2.3.1 ω-Calibration (with Agreed Units)

ω-Calibrations are often found in biochemical or pharmacological analyses. The interest may, for example, be in a substance, such as a hormone, an enzyme or a poison, principally as regards its biological effect, which is usually dependent on the quantity of substance present. In such cases, one attempts to find an analytical measure, for instance the color or turbidity of a solution, which can be attributed to the active component of the substance. In order to be able to derive from such an observation a useful value for the quantity present, the procedure must be calibrated with some 'standard sample', whose content has been determined in some arbitrarily fixed unit derived from the observed biological effect. Of this type are, for instance, the *international units* of antibiotics, or the *LD50* values (lethal dose) used in the case of poisons. The logical structure of the ω-calibration, therefore, corresponds to an α-calibration, with the difference that the basic unit is specific and valid only for the substance in question.

2.3.2 τ-Calibration (with Technical Scales)

In calibration with technical scales, no attempt is made to give any value for contents; instead, the analytical measures are directly correlated with characteristics or properties of technical or practical importance. This is possible when raw materials, chemicals, metallic alloys, etc. are to be classified. In general, only a very rough subdivision into different classes is necessary, i.e. a broad-band calibration is involved.

2.4 Stability of Calibration

The question of how long and under what circumstances the calibration values derived for a definite analytical procedure remain valid, has always been of importance for the critical assessment of a procedure and of the analytical results produced by it. Recently this question has reemerged in a hidden form, with the catch-phrase 'inter-laboratory reproducibility' which is used in the statistical treatment of series of analyses. Obviously it has not hitherto been realized that this is not so much a question of analytical precision, but rather a question of successful calibration.

Stability of calibration involves two questions: How 'robust' is the structure of an analytical procedure towards variations of the experimental parameters? To what extent has one control over an analytical procedure, as regards theory and technique, in order to keep the experimental conditions constant or to take into account the influence of changing parameters by suitable measurements and corrections?

Let us consider a system of separated analytical functions, which may also be dependent on various parameters r_h, $h = 1...j$; then a procedure may be termed 'robust' if the partial differential quotients

$$\frac{\partial f_i}{\partial r_h} \quad i = 1, ..., n; h = 1, ..., j$$

are small throughout the whole range of application; at best, they would be zero. This may be achieved most nearly with analytical procedures whose structure is simple, and in which only a few variables, a few parameters, and relatively simple operations are involved.

Complex analytical procedures are in most cases not 'robust', because not all sensitivities towards parameter variations,

$$\frac{\partial f_i}{\partial r_h}$$

can be made small simultaneously and throughout the whole range. For assessment of analytical procedures as regards stability, it is useful to distinguish three degrees of calibration stability.

2.4.1 Perfect Calibration (p)

The highest degree of calibration stability exists if it is possible to describe the analytical procedure sufficiently precisely, i.e. to prescribe every detail, so that it can be reproduced anywhere and at any time in such a way that the calibration function, once established, can be accepted. Analytical procedures with such a generally transferable calibration should be called perfectly or permanently calibrated (symbol: p).

Perfectly calibrated in this sense are many of the classical *precipitation reactions,* as well as analytical procedures using *titrimetry, coulometry* and *spectrophotometry.* Most of these are relatively simple procedures as regards task and operation. A complex analytical procedure which is composed of perfectly calibrated subsidiary procedures may not as a whole be perfectly calibrated, because the transitions between the component parts of the total procedure (for instance preliminary separations) may not be totally reproducible. On the other hand, it should be mentioned that very complex analytical procedures do exist which can be perfectly calibrated.

2.4.2 Fixed Calibration (f)

The second degree of calibration stability exists when the calibration values, once derived, cannot be generally transferred, but remain valid for a definite experimental arrangement, a type of instrument or even an individual instrument. A necessary condition for this degree of stability is that attention is paid to the 'environmental parameters' (pressure, temperature, humidity and cleanness of the air), and also to the identity of chemical reagents originating from different sources. Analytical procedures for which the calibration validity is restricted to definite conditions and experimental arrangements should be termed firmly calibrated (symbol: f) or 'fixed' in calibration. Most analytical procedures used in routine production control are firmly calibrated.

In production control laboratories where many analyses are made with firmly calibrated procedures, it is the custom to check the calibration from time to time, for instance every day, with the aid of analysis control samples. This must be considered as a summary inspection of the procedure and the apparatus, but not as a 'control calibration'. A new calibration requires much more work, and should not be necessary for a procedure with sufficient calibration stability.

2.4.3 Calibration Linked to 'Leader Standard Samples' (l)

For many types of analysis, it is not possible to keep the experimental conditions constant over a relatively long time. To calibrate such procedures of limited stability, it is necessary to include measurements on the standard samples in each particular analysis; these standard samples lead the analysis samples through the whole procedure, and are called leader standard samples, the analytical procedures being designated 'calibrated with leader samples' (symbol: l).

2.4.4 Uncertain Calibration (u?)

Characteristic difficulties may be observed when new types of analytical procedures are being developed; at this stage, the new procedures are not yet 'complete'. The cause of large variations in the calibration function is generally some kind of factor concerned with yield, which depends greatly on the experimental conditions. What is missing, then, is a reference quantity which runs through the whole procedure and which is influenced in the same way as the quantity to be measured.

A classical example of a procedure without a definite calibration function is the 'ultimate line' technique of de Gramont, used in spectrochemistry. It has been superseded by the 'homologous line pair' technique of Gerlach, in which the analysis line is compared with one or more lines of a reference element[5].

---

[5] W. Gerlach, Z. Anorg. Chem. 142, 383 (1925).

3 Topology of Analytical Procedures

The logical interrelation of operations and decisions which, in the course of an analysis, lead from the analytical sample to the analytical result, can be represented in topological diagrams (Fig. 1). These diagrams are called 'topological' because only the general schemes of the interrelations are given, without any metrical aspect.

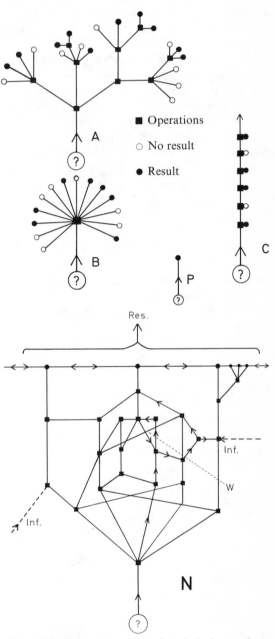

■ Operations
○ No result
● Result

Fig. 1. Topological structure of analytical procedures
A = tree (arbor)
B = bundle (local)
C = chain (temporal)
P = point
N = net; Inf = information; W = loop; Res = total result

3.1 Simple and Branched Structures

The topological tree (A) gives the interrelation of classical analytical procedures, characterized by their sequence of separations. Starting from the stem, the analytical sample, the procedures branch increasingly until finally, at the end of the last twigs, the measured values for the particular components or elements to be determined are found.

The structure of the 'tree' demonstrates clearly how, in this case, the procedure as a whole is composed of individual, mostly simple, analytical operations (see 1.2:4.1). The topological structure of a bundle (B) is characteristic of all analytical procedures in which the different measured values for the various components to be determined are obtained in parallel, practically simultaneously, and mostly by using the same principle. This group includes particularly all *spectroscopic procedures* with several parallel measuring channels. Large numbers of measurement channels require expensive instrumentation, but the advantage is the saving in time during the course of the analyses themselves.

The topological structure of a chain (C) corresponds 'dually' (in the sense of mathematics) to the structure of a bundle (B); the axis of the chain may be considered as representing the sequence in time; and, the individual members of the chain may represent the analytical decisions, which are mostly based on one analytical principle, as for the bundle. Analytical procedures with this structure are 'single channel procedures', usually requiring low instrument expense, but are time-consuming. This group includes procedures involving *electrochemical separation, fractional distillation, fractional extraction, chromatography,* or the recording of spectra in a time sequence, etc.

The topological structure of a point (P) can be regarded as a degenerated form of either B or C; it can be applied to analytical procedures in which the decision is made on the basis of one simple measurement, without analytical (A), spatial (B) or sequential (C) preseparation. Examples are: determination of the concentration of a particular substance in solution by measuring the *density,* the *vapor pressure,* the *optical rotation,* the *melting point* or the *refractive index,* etc. Simply *weighing* a grain of gold in order to determine the quantity of gold also belongs to this group. (In contrast, the determination of gold by a docimastic process, during which the less noble metals in the sample are first removed, obviously has the structure C.)

3.2 Complex Analytical Procedures with the Topological Structure of a Network (Compound Procedures)

The topological structure of a network (N) espe-cially describes procedures by means of which complicated problems in organic analysis, for instance the elucidation of chemical structures[6], are solved. It is a characteristic feature of such procedures that the strategy of the analytical process cannot be fixed in advance, but is developed during the course of the work. At the knots of the net, different partial results which have been gained so far are combined, and together determine the subsequent procedure. It may happen that the same knot of a network is passed several times by going along a loop.

For instance, the first attack on the problem by means of a number of measurements started in parallel may have demonstrated that other types of measurements must be made before sufficient information can be produced to decide how to proceed further with the analysis. It may happen that additional information must be brought in from an external source and must be fed into the network. It is very useful for the analyst himself to write down the course and the interrelations of a complicated analytical procedure in the form of a network; it may be possible to recognize a dead-end at an early stage, and also to detect a shorter way.

A typical analytical problem requiring a compound procedure of this type exists, for example, when one has to determine the nature and structure of a new organic compound.

Methods of purification, such as distillation and crystallization, combined with analytical control measurements, provide a homogeneous and well-defined substance for any further work. An *elemental analysis* or a high resolution *mass spectrum* provides the empirical formula. By spectrochemical *emission analysis,* metals which are present may be detected. *Infrared spectrometry* indicates the functional groups and some features of the molecular skeleton. *Raman spectrometry* particularly reveals highly symmetrical arrangements of atoms, whose vibrations do not appear in the infrared spectrum. The *UV spectrum* indicates the electronic configuration; the *NMR spectrum* gives the numbers and the positions of the protons in the molecule. In a *mass spectrum* the stable fragments from the molecule can be observed. *X-Ray diffraction* provides information about the crystal structure and the texture of the substance. The classical reactions of organic chemistry, preparative or analytical, provide further indications or confirmation.

This whole network of measurement procedures, operations, and decisions constitutes a complex analytical procedure with very high 'informing power'.

4 Optimization

4.1 Aims of Optimization

The task of discussing a particular analytical procedure critically very often includes the question of

---

[6] *W. Simon,* Z. Anal. Chem. *221,* 368 (1966).

what is the best procedure for solving a definite analytical problem. According to the particular needs, the optimum may mean low costs, short times for analyses, high analytical precision, low limits of detection, etc. Furthermore, one must describe the limitations, or the requirements to be fulfilled, for instance which instruments are available or can be obtained, the maximum admissible cost, the number of required analyses per unit time, etc. The range of possibilities may thus be limited to such an extent that the problem of optimization can be tackled. The solution of such problems has already been well treated in the mathematical theories of operations research. In order to make such mathematical investigations feasible, a formalized mathematical model for chemical analysis must be developed.

4.2 Optimization with Respect to Topology

4.2.1 Tree Structure (A)

For example, let us suppose that several single possible operations from which such an analytical procedure is built up are equivalent as regards cost and time. The task of optimization is then reduced to the problem of isolating the individual components by the smallest possible number of separations necessary to determine their contents. To achieve this, the separation processes must be so chosen that the subgroups of separated components produced are nearly equally occupied, thus permitting simultaneous further treatment of as many of the components as possible. Isolation and determination of single components must take place at the end of the whole procedure.

A problem of optimization is also hidden in the question: which analytical tasks can be solved by the use of analytical procedures of a definite topological and calibration structure? The answer to such a question can provide only general indications. Procedures following the scheme of classical separation analysis with the topological structure of a tree (A) are composed of many relatively simple and independent operations. It is, therefore, possible to calibrate them perfectly (p) with synthesized samples (σ); the calibration values can generally be transferred, and may be found very often in the literature as stoichiometric factors. This type is, therefore, designated by the symbol A(σp).

A(σp) procedures are highly versatile, because they are designed according to a building box principle. The bricks of this analytical building box are available ready-made, in large numbers. The number of possible combinations is unimaginably high. When an analyti-

cal method must be found in an ad hoc manner for a new and hitherto untreated analytical task, one should first try to compose this new procedure as one of type A. The instrumental costs will in most cases be relatively small, whereas the required time may be considerable.

In spite of their versatility, the A(σp) procedures are not very suitable for survey analyses and for universal analytical procedures; they are too tedious and too expensive. Here, other procedures are required whose details are all established, and which, therefore, need no check measurements or alterations for particular cases. Complex analytical procedures of type A are also generally not very suitable for automation. Apparatus used for the *automation* of A-procedures (automatic balances; transporting, filling, dividing and tapping devices; optical and electrical sensors) are very often robots that replace the human hand, eye and sense of touch, which formerly had to play an active part in the procedure.

In this article, the concepts 'automated' or 'partly automated' are applied to all analytical procedures which operate wholly or in part without human action. This corresponds to the use of the word 'automated' in colloquial usage. By stating whether the analytical procedures are governed by a program or controlled by feedback, or whether they are partly governed and partly controlled, one has a clear distinction between the various possibilities, particularly if the topological structure with A, B, C and P is also given.

4.2.2 Bundle Structure (B)

Procedures of this type, (B), were in most cases originally determined by modern techniques rather than by 'craftsmanship', and can be automated without fundamental difficulties; however, the technical expenditure may be very great. The main point is that, in the core of such procedures, a single analytical principle operates, not a variety of branched operations and reactions as in the A-procedures. The 'analytical work' of separation is taken over by a variable parameter, such as potential, energy, wavelength, mass number or (sometimes) time. (Regarding the concept of 'resolving power', see Section 1.2:5.2.)

Analytical procedures with the bundle structure (B), i.e. mostly with spatial separation, are suitable for *rapid analyses*. Depending on the analytical task, they must have many parallel measurement channels, and very often the evaluation of the measured quantities is done by an electronic computer. Examples include the direct reading optical emission spectrometer, and the X-ray fluorescence spectrometer, which are nowadays used in the laboratories of steel works, with very extensive but fixed analytical programs. These procedures are firmly calibrated with analyzed standard samples, and, therefore, are of type B (α, f).

4.2.3 Chain Structure (C)

These procedures in general are slower, but in most cases cheaper, because one measurement channel is sufficient. The separating variable parameter is either time itself (e.g. time consumed during chromatography or electrophoresis) or a time-dependent parameter (e.g. the potential applied during a polarographic analysis). These procedures can be automated very easily without great expense. If many samples are to be analyzed, it is possible to have a large number of installations running in parallel. The number of components to be determined with a C-procedure is generally smaller than the number of components with a B-procedure, because constant conditions must be maintained throughout the longer duration of the analyses, and this may be difficult to achieve. However, when time is available, it is possible to use not only the rapid α-calibration but also the more time-consuming σ-calibration; the most frequent types are C (αf) and C (σf).

4.2.4 Network Structure (N)

Compound procedures of this structure must be considered in the same way as procedures with the topological tree structure (A). They may also be composed of subsidiary procedures, as in a building box system; they can be adapted not only to individual problems, but also to whole *fields of problems* in Chemical Analysis, for instance to the determination of structures of organic compounds.

The urgent necessity of finding any suitable solution for a difficult analytical task, may predominate over the question of costs and of the time required for analysis. Furthermore, the question as to whether the procedure as a whole can be automated has little meaning, whereas the subsidiary procedures co-operating in a compound analytical procedure may themselves proceed automatically, either partially or completely. These subsidiary procedures have mostly the structure B or C. It is interesting to consider what happens at the knots of the network where the intermediate results are evaluated together. Very often, the experience and the imagination of the analytical chemist is indispensable for deciding how the work must be continued. However, if the decisions are made according to formal

rules, then logical tools may be employed, for instance correlation tables, punched cards, or a computer.

5 Informing Power

5.1 Definitions

Analytical procedures are information processes (cf. Ref. [7, 8]). Information of all kinds* is transferred and retained by *signals*. Signals are either configurations in space (e.g. printed letters, pictures, punched tapes, or magnetic tapes) or they are processes which occur in space and time (e.g. sound, light, electric currents). All these signals are finite in space and time; in consequence, only a finite number of structural details can be distinguished in each signal. These can be represented by a finite number of numerical values. In order to understand the meaning of this representation, one must know the classification system or code.

Because the binary system of numbers is widely used in electronic data processing, this system has been generally adopted for the theoretical treatment of information processes as well; it will, therefore, be applied for the following considerations.

A binary position (bit) is considered as the unit with dimension 1, by which an 'amount of information' can be measured as a metric quantity. (As all units it is to be written in the singular form. If, in a context, this term is used as an *abbreviation of* 'binary digit' in this concrete sense, the plural should be used.) A single position in a binary system, with its two digits 0 and 1, makes it possible to represent symbolically the two cases of a yes/no decision.

Let us now consider how many binary digits would be required to assign binary numbers to all the different results which might possibly be produced by an analytical procedure, so that one could then look up the corresponding text (for instance values for the concentrations). This number of required binary digits is a metric quantity which indicates how much an analytical procedure can do as an information process, considered from a purely formal point of view. This quantity may be called the 'informing power' P_{inf} of the analytical procedure[9].

---

* The nomenclature in this field is still ambiguous. The word 'information' is used in at least four different meanings, often in the same text. In this article the following nomenclature will be used:
In general sense: Information = process of instruction, Information (s) = (new) knowledge with respect to content (preferably plural).
In the sense of metric quantities: Informing power, P_{inf} (of a method); Amount of information, M_{inf} (required technically to communicate some definite desired knowledge); Information capacity (capability of a technical system to transmit or store an amount of information).

[7] *L.M. Ivancov, P.G. Kuznecov, Ju.I. Stacheev*, in *E. Rexer*[3], p. 31 ff.
[8] General literature about information theory see Ref. [47] (with many references), also articles in [48] vol. II and [50].
[9] *H. Kaiser*[27].

The informing power of an analytical procedure is determined by the number n of the different measurable quantities, and by the number S of distinguishable steps of values for each of these quantities. The formula is

$$P_{inf} = \sum_{i=1}^{n} \log_2 S_i \qquad (1)$$

or

$$P_{inf} = n \log_2 \bar{S} \qquad (1a)$$

if the number of the distinguishable steps for all n measures is of nearly the same order of magnitude, so that the calculation can be made with an average value $\bar{S}$. It is obvious that the relatively coarse measurement (S small) of a second quantity may give much more additional informing power than a considerable increase in the precision of measurement (S large) which can very often only be achieved at high cost and with much effort. If the number of measurable quantities is doubled P_{inf} will be increased by a factor of 2; if the number of distinguishalbe steps is duplicated, P_{inf} rises by one bit only.

5.2 Parameters and Resolution

The n different measures may belong to different values of a variable parameter (frequency, wavelength, mass number, time, location in space, electrical potential etc.). If only a few distinct values for the parameter (measurement positions or 'measurement channels') are allowed, formulae (1) and (1a) are valid. However, if the experimental parameter is varied over a wide range in a continuous way, then the number of distinguishable measurement positions will be very large; in this case a reformulation of formula (1) is most enlightening.

Let the symbol for the parameter be v. The concept of *resolution* R is defined by $R(v) = v/\delta v$, where δv is the smallest distinguishable difference in v which can be recognized for practical purposes. Let the smallest distinguishable difference from basic principles be called $\delta_0 v$; the corresponding quantity $R_0 = v/\delta_0 v$ is called the (theoretical) 'resolving power'. Within a small range of Δv, there are $\Delta v/\delta v$ different measurement positions. Then,

$$\Delta v/\delta v = R(v) \frac{\Delta v}{v}$$

If this is inserted into formula (1), and if summation is replaced by integration, the informing power of an analytical procedure with (spectral) decomposition through a parameter v in the range from v_a to v_b is given as

$$P_{inf} = \int_{v_a}^{v_b} R(v) \log_2 S(v) \cdot \frac{dv}{v} \qquad (2)$$

A hint for a formula for the 'channel capacity' of a spectrograph was given by Wolter, Marburg[10]. When R and S are practically constant throughout the range of application, their average values may be taken, and the formula is then simplified to:

$$P_{inf} = \bar{R} \log_2 \bar{S} \cdot \ln \frac{v_b}{v_a} \qquad (2a)$$

It is obvious that the resolution R is decisive for a high informing power. High *spectral resolution* is, therefore, much more important than a wide spectral range or a large number of steps S (precision) in the measurement.

5.3 Useful Resolution

From formula (1), the informing power of an analytical procedure by which only one component is to be determined, using only one measurable quantity, must be of the order of 10 bit. (One-dimensional, non-dispersive procedure.) Analytical procedures with several measures (multichannel- or polychromatic procedures) give about 100 to 500 bit. In contrast, spectroscopic analytical procedures may have informing powers of 10^4 to 10^6 bit, assuming that the spectral resolving power is used to its full extent.

One restriction must be considered in particular. In formula (2), $R(v)$ stands for the practical resolution taken for the analytical procedure as a whole, and not the theoretical resolving power $R_0(v)$ of the spectroscopic instrument in the core of the procedure. The informing power offered by the instrument very often cannot be fully used, because the observed physical phenomenon does not permit this. For instance, in molecular spectroscopy the absorption bands of liquids and of solids are relatively broad; the practical resolution R which has to be used in the formula is, therefore, entirely determined by the width of the bands.

During discussion of the solution of an analytical problem, the question may arise as to the minimum number of bit required to communicate the desired information. This quantity is called the required amount of information M_{inf}. It can be calculated by a logical analysis; one must ask how many yes/no decisions are necessary to represent completely the desired information. Should the informing power P_{inf} offered by the analytical procedure be smaller than the required amount of information, M_{inf}, three possibilities remain:

a) reduction of the task,
b) use of 'pre-information' or 'joint information' from other sources,
c) combination of several analytical procedures into one compound procedure (usually of topological structure N), which procedure as a whole may yield the required amount of information.

Example: It is required to analyze a sample of unknown composition quantitatively for all 100 elements of the periodic system; the concentrations of the different elements are to be given in a concentration scale having 1000 steps. If such a scale is logarithmic and equally subdivided, then the concentration from step to step would grow by a factor of 1.023; such a scale

---

[10] *H. Wolter*, unpubl., 1964.

would cover the range from 100% down to 10^{-8}%. From equation (1a), the amount of information required for the determination of the 100 elements would be $M_{inf} = 100 \cdot \log_2 1000 \approx 1000$ bit

5.4 Principles with High Resolving Power

Analytical principles (not methods) which provide such a high informing power are *optical emission-, X-ray-* and *mass-spectroscopy*. The universal analytical procedures which are based on these principles offer informing powers from 10 000 to 200 000 bit (see formula 2a). The surplus of informing power (redundancy) may be used to increase the reliability and the precision of the analysis.

Comparison of the amount of information M_{inf}, which is required to solve an analytical problem with the informing power P_{inf}, of an analytical procedure is very useful, but one should not exaggerate such formal considerations. For example, no practical conclusions can be drawn from the statement that about 10^8 bit are necessary merely to represent the numerical values of all possible organic analyses.

6 Figures of Merit

There are two groups of figures of merit in use for analytical procedures: a functional group, and a statistical group. The functional group comprises *sensitivity, selectivity, specificity,* and *accuracy*. The statistical group comprises *precision, detection power, limits of precision*.

6.1 Functional Figures of Merit for Simple Analytical Procedures

In a simple analysis, the content c of the component to be determined is derived from the measure x of the appropriate measurable quantity, with the use of the analytical function. $c = f(x)$. This function is the inverse of another one, $x = g(c)$, which is experimentally determined during the process of calibration of the analytical procedure, performed with standard samples of known composition. There is, therefore, a pair of functions which are inverse to each other. The inversion is possible within a limited range, when the differen-

tial quotient of the function g (c) exists in the whole $c_a...c_b$ range and when it is nowhere zero. This must be expressly stated, because calibration functions are known which cannot be reversed everywhere, and above all not unambiguously.

Where the analytical curve is retrogressive (Fig. 2), the differential quotient is zero,

$$\frac{dx}{dc} = 0$$

Such functions often occur: for instance, the electrical conductivity of a solution sometimes decreases at increasing concentrations; the main spectral lines used in emission spectroscopy sometimes show so strong a self-reversal at increasing concentrations that their intensity decreases with rising concentrations. Such functions can be used only in the restricted range of c in which they can be reversed unambiguously.

The *sensitivity* of a measurement procedure is, in general, defined as the differential quotient of the characteristic function of the procedure[11]. An analytical procedure should, therefore, be termed sensitive when a small variation in c causes a large variation in the measured quantity x.

The 'sensitivity' is constant only in those cases where the calibration function is linear; in general, the sensitivity is a function of c. The 'sensitivity' of an analytical procedure has nothing at all to do with the *power of detection* of the procedure.

The term sensitivity of detection, which used to be employed, is misleading and its use should be abandoned.

Accuracy: If an analytical procedure were 'completely accurate', then its calibration function (and also the inverse analytical function) would have to be free from systematic errors. If the law which such systematic errors follow is known, the calibration function can be corrected. When systematic errors are suspected to be present, but when their size and their functional dependence are not known, then it is necessary to investigate the analytical procedure critically and to compare analytical results which have been obtained for identical samples by different procedures. Sometimes it is possible by such observations to derive an upper and a lower limit for the systematic errors which may be occurring, and to indicate an interval within which the correct calibration curve should lie.

If Δc is the width of this interval, then the smallest value of the ratio $c/\Delta c$ which occurs in the whole range of application may be regarded as a useful measure of the overall accuracy A of the analytical procedure:

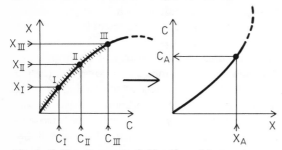

Fig. 2. Calibration curve (left) with ambiguous inverse analytical curve (right).

---

[11] DIN 1319, Abschnitt 6, Grundbegriffe der Meßtechnik.

$$A = \mathrm{Min}\, \frac{c}{\Delta c} \qquad (3)$$

A procedure will have a higher figure of merit, the more 'accurate' it is, and the smaller the uncertainty interval which must be taken into account because of the possible, but unknown, systematic errors. For instance, when it is known that the systematic errors in the whole range are smaller then 0.01 c, the accuracy can be numerically expressed as $A = 100$.

The situation is much simpler when it is known that the systematic errors of the calibration function are only the 'frozen in' random errors of the calibration measurements. In this case, over the interval in which the calibration function may be found, it is possible to establish a probability function for its course. Paradoxically, it is then possible to state the probability with which analytical results might be systematically wrong by a certain amount, if they were derived from a particular calibration function.

These considerations can be applied in a similar way to multicomponent analyses; for calculation of the regions of uncertainty, the new methods of 'interval arithmetics', may be useful.

6.2 Functional Figures of Merit for Complex Analytical Procedures

When n components whose contents are independent of each other are to be determined in one sample, at least n independent measurable quantities are required (cf. Ref. [12]). When there are more measures available, the superfluous ones may be dropped, or several of the original measures may be combined into new ones in such a way that n independent measures remain. The one measurable quantity x used for a simple analysis is now replaced by the set $(x_1, ..., x_n)$. These n figures may be considered as the coordinates of a point in a space of n dimensions. Correspondingly, the contents $(c_1, ..., c_n)$ can be regarded as the coordinates of a point in another space, that of the chemical constitution. Calibration of an analytical procedure experimentally establishes a correlation between the points $(c_1, ..., c_n)$ and the points $(x_1, ..., x_n)$. This leads to a system of 'calibration functions', n in number

$$
\begin{aligned}
x_1 &= g_1(c_1, ..., c_n) \\
x_2 &= g_2(c_1, ..., c_n) \\
&\cdots\cdots\cdots\cdots\cdots \\
x_n &= g_n(c_1, ..., c_n)
\end{aligned}
\qquad (4)
$$

The inverse system of these functions is used for the analysis. This leads to a system of analytical functions

$$
\begin{aligned}
c_1 &= f_1(x_1, ..., x_n) \\
c_2 &= f_2(x_1, ..., x_n) \\
&\cdots\cdots\cdots\cdots\cdots \\
c_n &= f_n(x_1, ..., x_n)
\end{aligned}
\qquad (5)
$$

If the functions in (4) are throughout continuously differentiable, then they can be approximated in the environment of each point by the first linear terms of a Taylor series. The system of calibration functions can, therefore, be locally represented by a system of n linear equations. The coefficients of this system of equations are the partial differential quotients of the measures with respect to the contents,

$$\gamma_{ik} = \frac{\partial x_i}{\partial c_k}.$$

Obviously, these γ_{ik} are none other than the (partial) sensitivities of the individual measures towards variations of the contents of the different components. The relationship between contents and measures in its totality is throughout represented in a sufficiently small region about the points $(c_1, ..., c_n)$ under consideration, by the appropriate matrix of the partial sensitivities.

$$
\begin{matrix}
\gamma_{11} & \gamma_{12} & \cdots & \gamma_{1n} \\
\gamma_{21} & \gamma_{22} & \cdots & \gamma_{2n} \\
\multicolumn{4}{c}{\cdots\cdots\cdots\cdots} \\
\gamma_{n1} & \gamma_{n2} & \cdots & \gamma_{nn}
\end{matrix}
\qquad (6)
$$

The functional figures of merit must be derived from this matrix, which may be termed the 'calibration matrix' (of the analytical procedure).

From the system of calibration functions (4), the system of analytical functions can be derived as its inversion only when the determinant of the matrix (6) is not zero. When this requirement is fulfilled throughout the range of application, then the system of the n analytical functions can always be calculated in linear approximation; however, it must be noted that the solutions of the calibration functions for the contents $c_1, ..., c_n$ are valid only *locally,* and not for the whole range of application. Only when the partial sensitivities γ_{ik} have constant values throughout the whole range, i.e. when the calibration functions are really linear and not merely in local approximation, does the inversion give the corresponding system of n (likewise linear) analytical functions which are valid for the whole range. (The matrix of the analytical functions is the inverted calibration matrix.) In this case the work expended in calibration measurements and their evaluation is relatively small, because then n^2 coefficients γ_{ik} have to be determined only once, and are valid for the whole range.

It is, therefore, an important and positive verdict on an analytical procedure, when it can be stated that the calibration and analytical functions are (practically) linear throughout the whole range of application. Often the system of calibration functions can be linearized, if the variables x_i and c_i are suitably transformed mathematically. An example is Beer's law in *photometric analysis;* it can be written as a linear equation if, instead of the original measured quantity (the *transmittance* of the sample), its negative logarithm, the *absorbance,* is used. In many cases one must find the best way by trial and error: power functions, simple rational functions, and logarithms offer many possibilities.

The value of the determinant of the calibration matrix, det (γ_{ik}), is the transfer factor by means of which a volume element in the n-dimensional space of the constitution $(c_1, ..., c_n)$ is represented in the space of the measurable quantities $(x_1, ..., x_n)$, and, to this extent, it is the generalization of the concept 'sensitivity' from the simple to the complex analytical procedures.

6.2.1 Selectivity

A generally applicable quantitative definition for the concept 'selectivity' can also be derived from the mathematical properties of the calibration matrix. Obviously, an analytical procedure would be called 'fully selective' (in the colloquial language of analytical chemists) when only the elements of the principal diagonal of its calibration matrix, γ_{ii}, are non-zero $(i = 1, ..., n)$. The procedure would then break down into n independent subprocedures — at least as regards calibration and evaluation. Each of the components (i) to be determined would then be measured by its own measurable quantity x_i, which depends only on the content c_i of this one component. Analytical procedures which are fully selective in this sense occur for instance in *emission spectroscopic analysis* and in *mass spectrometry.*

For a procedure of moderate selectivity, there should be at least some correspondence between measures and components, such that the corresponding x and c can be given identical indices, without ambiguity. It must, therefore, be possible to choose the index numbers in such a way that the matrix elements of the principal diagonal are the greatest in their row (regardless of sign). This is presupposed in the following, but is still not sufficient for a definition of 'selectivity'.

One gets further, if one requires that the system of calibration functions shall be soluble, for the contents $c_1, ...c_n$ to be determined, by the use of an 'iteration process'. In such a process, the system of equations is solved in several sequential steps by successively inserting approximate solutions until

the numerical values for the solutions no longer change. The first approximation is obtained by taking into consideration only the matrix elements in the principal diagonal; in this first step, one proceeds as if the analytical procedure were fully selective. This iteration process for solution converges only, when in each row of the calibration matrix the element in the diagonal is larger than all other elements of the same row taken together i.e. if the inequality (7) is valid for all i[12], Ref. [12], p. 159.

$$\frac{|\gamma_{ii}|}{\sum_{k=1}^{n} |\gamma_{ik}| - |\gamma_{ii}|} > 1 \qquad (7)$$

The larger these quotients are, the better the iteration procedure converges. If, therefore, we take the smallest value of these quotients which occurs in the matrix and in the range of application of the procedure, we have a quantitative measure Ξ of selectivity in the following expression:

$$\Xi = \min_{i=1,...n} \frac{|\gamma_{ii}|}{\sum_{k=1}^{n} |\gamma_{ik}| - |\gamma_{ii}|} - 1 \qquad (8)$$

For a 'fully selective' procedure, Ξ becomes very great (formally infinite); when the value of Ξ is only a little above zero, one can hardly speak of selectivity.

Even analytical procedures which are not selective in this sense may be very useful for the determination of several components. The system of calibration functions can always be solved for the contents when the determinant of the calibration matrix is different from zero. It is in no way necessary that measurable quantities and components should be connected in pairs. However, selective procedures have many practical advantages: They are clear, and relatively simple to calibrate and to evaluate.

To avoid misunderstandings, a practical difficulty which may occur during the course of numerical solution of such systems of linear equations must be mentioned: When the unavoidable measurement errors for the main components have too strong an effect in the equations for the minor components, or when the γ_{ik} are not known sufficiently exactly to allow calculation of the determinant, then the calculated results may be nonsensical. For instance, one should not try to determine trace elements in pure iron indirectly, on the basis of the measures obtained for the iron content. However, these are problems of numerical calculation, and have nothing to do with the functional relationships which lead to a definition of selectivity.

[12] *R. Zurmühl,* Praktische Mathematik für Ingenieure und Physiker, 5. Aufl., Springer Verlag, Berlin, Heidelberg, New York 1965.

6.2.2 Specificity

An analytical procedure is generally called specific for a substance if an analytical signal is generated only by that particular component (i) in a multi-component sample. A specific procedure, therefore, is always also selective.

If a procedure is only approximately specific a formal analogy to 'selectivity' can lead to a quantitative definition of the degree of 'specificity' in terms of the following formula:

$$\Psi = \frac{|\gamma_{ii}|}{\sum\limits_{k=1}^{n} |\gamma_{kk}| - |\gamma_{ii}|} - 1 \qquad (9)$$

The most notable feature of these definitions is their freedom from arbitrariness. The calibration matrix alone contains all the information about the sensitivity, selectivity, and specificity of a complete analytical procedure, for each point in the chemical constitution space.

6.3 Statistical Figures of Merit

Methods of mathematical statistics are tools, which can be used with three very different groups of tasks.

a) To describe and to classify large masses óf data.

b) To compress, in a descriptive way, numerous data by the use of statistical characteristic figures.

c) To derive predictions, to establish decisions with indications of the risks.

In the following, only the most important statistical figures of merit will be treated, briefly and with reference to the very extensive literature.

6.3.1 Descriptive Statistics

From extensive observations material can be ordered, subdivided and presented in a clear fashion by 'classification' and by stating the observed frequency distribution[13, 39].

Two statistical quantities in particular are used for data compression: the 'mean' and the 'standard deviation'. The calculation is purely formal. Suppose N measurements (observations, analyses) have been made, and that the measured values are $x_1, x_2, ..., x_N$ (these need not all be different). The 'mean' $\bar{x}$ (often also termed the 'average') is defined as

$$\bar{x} = \frac{x_1 + x_2 + ... x_N}{N} = \frac{1}{N}\sum_{i=1}^{N} x_i \qquad (10)$$

The standard deviation s is defined as

$$s = +\sqrt{\frac{1}{N-1}\sum_{i=1}^{N} (x_i - \bar{x})^2} \qquad (11)$$

If the standard deviation s is divided by the mean $\bar{x}$, the result is called the relative standard deviation, s_r.

$$s_r = s/\bar{x} \qquad (12)$$

In chemical analysis, the relative standard deviation should always be given as a decimal fraction, not as a percentage, in order to avoid confusion with concentration values which also are often given as percentages[14].

In many cases, there may not be enough material for a large number of analyses on the same sample. Alternatively, the time required for an analysis may be so long that for practical reasons it is impossible to make numerous analyses merely for the purpose of checking the procedure. However, it is possible to determine the standard deviation not only from a large number of analyses made on one particular sample, but also from a large number of analyses on slightly different samples, each of which has been analyzed several times. Analyses which are made routinely in many laboratories, very often involving double or multiple determinations, can be used to give the numerical values necessary for investigation of the precision of the procedure in question[15].

In practically all cases, the standard deviation is the best value to characterize the 'scatter' of an analytical procedure. Under no circumstances should one instead use the range between the highest and the lowest measured values, which may by chance have occurred in a particular series of experiments, since these are in fact the two most uncertain values obtained.

6.3.2 Prognostic Statistics

The real problems begin when the frequency of occurrence of future events must be predicted on the basis of past observations. Such considerations are necessary for the estimation of risks connected with decisions to be taken. The bridge between the past and the future is formed by mathematical probability theory, which is nowadays being developed in an axiomatic way following the model established by Komolgoroff[16]. Mathematical models can be established for frequency distributions of the different types observed in practice[9]. However, it is possible to find out whether such a model is suitable for treatment of a particular case only by critical comparison of the model with

---

[13] *J. Pfanzagl*[44], Bd. I, Sammlung Göschen, Bd. 746, 746a, 1967.

[14] *R.W. Fennell, T.S. West*, Recommendations for the Presentation of the Results of Chemical Analysis, IUPAC Commission on Analytical Nomenclature, Pure Appl. Chem. *18*, 439–442 (1969).

[15] *H. Kaiser, H. Specker*[28].

[16] *A.N. Kolmogoroff*, Grundbegriffe der Wahrscheinlichkeitsrechnung, Springer Verlag, Berlin 1933.

empirical findings. The same is true also for critical assessment of analytical procedures.

The widespread opinion that the distribution of errors in chemical analyses can always be adequately described in terms of a Gaussian normal distribution is plainly wrong. The essential problem of mathematical statistics is to find an appropriate probability function which best 'fits' a more or less numerous set of observations. This must be done in such a way that conclusions for successful practical action can be drawn from this probability function.

At present, there is a tendency to give the so-called confidence limits — generally for 95% or 99% confidence (or statistical certainty) — instead of the observed standard deviations. This tendency is dangerous for two reasons. Firstly 'confidence', 'certainty', and 'risk' are concepts of practical ethics; 'how much' confidence one has, i.e. how much certainty is required to make a decision, cannot be fixed once and for all by convention; in every individual case, this must be decided after taking into account the facts and problems of life. Secondly, the conventional figures for the relationship between statistical confidence and standard deviation (for instance 95% confidence corresponds to the range $\pm\, 2\, \sigma$) are all based on the assumption that the population in question has a 'normal' Gaussian distribution. The same is true for cases where the confidence range was calculated using Student's t-distribution, since this too presupposes a Gaussian distribution for the whole population from which a relatively small statistical series was taken.

As Gauss[17] himself first demonstrated, many observations in science can be described, to a good approximation, by assuming a normal distribution. This is especially true for the distribution of random measurement errors. In chemical analysis, however, for practical reasons it is often not possible to make very long series of experiments in order to verify that the distribution of errors about the mean does actually correspond to a normal distribution. For such an investigation, very many measurements, at least 100, would have to be made in order to be sure that the dangerous (but rather rare) large deviations from the mean do not occur too often.

The most important reason for accepting a Gaussian normal distribution is the 'central limit theorem' of probability theory. In order to understand its importance, one must realize that the so often investigated distribution of the measured values about the mean of many measures is by no means the distribution which is of importance for discussion of chemical analyses. A different distribution function, which indicates how the possible true

values are distributed round a given measured value, is needed. This at first sight seems surprising, since it is natural to adhere to the correct view that there is just one true content in the sample for the element being determined.

The meaning of the above can best be clarified by an example. Suppose that in an analytical sample one finds 2.70 μg Fe per ml. The 'true content' is unknown; it might in fact be, say, 2.62 or 2.75 or 2.71 or 2.67 μg. One might select, from a very large series of such analyses, all the observations in which one found exactly 2.70 μg Fe. Now imagine that the true values are subsequently found by a better analytical process, or by looking them up in a list. Then, for the many different samples which gave precisely the same measured analytical value, quite different true Fe contents would have been established, distributed for example in the range 2.51—2.93 μg, probably with most of them near to 2.70 μg. The conclusion is that one particular measured value can arise from different true values.

Obviously, this distribution of the possible true values round a measured value must be dependent on the reserve of true values which are admitted. The type will be different, according to whether a continuous sequence of values is possible with equal probability, or whether the measurable quantity can assume only one or two discrete values. It is fortunate that in many cases of chemical analyses it can be assumed that the possible true values occur in a continuous distribution with equal frequency. The following statements are valid for this case:

Distribution function for the possible true value leading to a particular measure:
If a) the possible true values in the range of application under consideration are distributed continuously and with equal relative frequency and if b) the random errors are produced by the combined action of many independent sources of error of approximately equal magnitude, then a Gaussian normal distribution (bell-shaped curve) is valid for the inference from the measured value to the true value. The position of the maximum of the Gaussian curve is at the measured value.

Distribution function for the possible true values leading to a particular mean of a large series of measures. If the final result is the mean from a *very* large number of single measured values, then the distribution function for the inference from the mean to the true value is always given by a normal Gaussian distribution function, with its maximum at that mean. This statement is independent of the assumption that the total error of a measure is the sum of many individual errors, and is also independent of the distribution of the possible true values. In addition, for this case the special form of the Gaussian function is determined by the 'standard deviation' which can be calculated from the many individual measured values.

Experience indicates that these assumptions, in any combination, correspond to most cases of practical analysis. It is therefore reasonable to base discussion of the size and distribution of random errors first on a Gaussian function. One should not, however, forget the risk connected with such an assumption.

6.3.3 Limit of Detection

For a correct appraisal of an analytical procedure by which very small contents, 'traces', are to be determined, it is necessary to know which is the smallest value the procedure can give for a content. This is a problem both of measurement and of statistics. At low concentrations, there is uncertainty about whether (and to what degree) an observed 'measure' is really due to the amount of the desired substance in the sample, or whether it is caused by uncontrolled chance disturbing influences. This uncertainty of assessment, although limited by the statistical definition of the limit of detection, is not completely removed. Nevertheless, a decision with calculable risk can be made by using a criterion agreed upon by convention [17, 18]: Chance disturbing influences are already operative when the analytical procedure is applied to a blank sample, and leads to measures x_{bl}, which exhibit chance fluctuations whose origin cannot be investigated in detail, nor indeed would one wish to do so. It is important that the cause of the uncertainty in the analytical value is not the size itself of the blank measure, but the size of its fluctuations. A constant blank measure of any size can always be compensated for. The same is also true for the constant component of the fluctuating blank measures which is given by the statistical mean $\bar{x}_{bl}$.

The fact that chance fluctuations set a limit to the measurement of low values has been known since Brownian motion was observed. Recently it has gained wide-spread popularity under the term 'signal to noise ratio' which was coined by electrical engineers. In chemistry it is appropriate to speak of an 'analytical signal'; however, it is hard to speak of (analytical) 'noise', even if the final measurement is made electronically. The disturbing influences which give rise to the chance fluctuations of the blank measurements can be of many different kinds. Examples are: impurities in reagents, losses through adsorption on the walls of the vessels, errors of weighing or titration, secondary reactions, temperature fluctuations of the light source in spectrochemical analysis, etc. The size of the fluctuations due to such causes is usually not predictable from theory. In practice, however, the magnitude of the fluctuations for each analytical procedure can be found numerically by carrying out a sufficiently large number of blank analyses, and then statistically evaluating the measures $\bar{x}_{bl}$ found in the course of them. The man $\bar{x}_{bl}$ and the standard deviation s_{bl} are calculated. For the definite detection of a substance, a requirement is that the difference between the analytical measure x and the mean blank value $\bar{x}_{bl}$ must be greater than a definite multiple k of the standard deviation s_{bl} of the blanks. Smaller measures are discarded as not sound.

Experience has shown that it is appropriate to choose $k = 3$. This compensates for the uncertainty which is due to the fact that only 'estimates' for the mean blank value and the standard deviation can be obtained, and above all that the type of the distribution function is not known. (One should not without further investigation assume a normal distribution). If at least 20 blank analyses are carried out, and k is taken as 3, then the risk of erroneously eliminating a measure as not sound is at most a few percent.

The measure x at the limit of detection is, therefore, defined by the equation

$$\underline{x} = \bar{x}_{bl} + 3\,s_{bl} \tag{13}$$

The content c at the limit of detection also follows from the analytical function, as

$$\underline{c} = f(\underline{x}) \tag{14}$$

This is the smallest value for the content which the analytical procedure in question can ever yield. This, therefore, is a *characteristic feature of the procedure itself*. However, the concept of 'limit of detection' has been formulated with regard to the analytical problem. If one were speaking of the procedure as such a better colloquial term would be 'power of detection'.

Near the limit of detection $\underline{c}$ all quantitative determinations of the content are rather imprecise; it follows from the definition that − depending on the slope of the analysis function − the relative standard deviation for $\underline{c}$ must be about 0.3.

For critical assessment of the efficiency of an analytical procedure at small concentrations, not only is there the question of the possibility of detection of a desired substance, but also that of the 'precision' of the analysis in the region of small concentrations in general, i.e. the question of the extent to which the concentration may be reduced before the noise level peculiar to the particular analytical procedure pushes the standard deviation of the analytical measure over a preselected threshold value. This could be described by a 'limit of precision' (for...), in which term the required standard deviation must be inserted in the brackets. The particular analytical task will determine what is admissible. The limit of detection $\underline{c}$ is a figure of merit for the analytical procedure as such, not for the individual analysis. All analytical results which can be obtained by this procedure must be $\geq \underline{c}$. If the desired substance has not been found, it must never be asserted that the content is below the limit of detection $\underline{c}$. The 'limit of guarantee for purity', which one would wish to indicate

---

[17] *H. Kaiser,* Spectrochim. Acta *3,* 40 (1947).
[18] *H. Kaiser* [21].

Other important concepts, for instance selectivity, specificity, and stability of calibration of analytical procedures, are only conceived 'qualitatively', i.e. by instinct; they are still floating about in the outer periphery of scientific concept formation, and thus belong to the 'colloquial language' of the specific technical field. Definitions have been tried, in most cases, only for a particular problem. This situation is characteristic of a scientific field which is in a state of rapid development. Characteristic also is the inclination to introduce the new intellectual tools − particularly mathematical statistics − in an 'unconsidered way', i.e. without definitions and limitations being given the necessary precision. Alternatively, these may be laden with too much unnecessary erudition. In such an open situation, the reader himself must form his judgment by comparison; sometimes, on the same page he may find both wheat and chaff.

These considerations have compelled the author to present the bibliography differently from the other chapters of this book. The bibliography contains mainly new books and papers, arranged according to general subject. Many of these contain extensive references to the older literature, and indicate literature for more extensive studies. Besides the books describing the basis of mathematical statistics, it seemed appropriate also to add some literature dealing with the theory of information and mathematics in general. This was not merely for the purpose of giving the sources of various terms used in this chapter, but rather, because as Analytical Chemistry develops into an all-embracing field of science, it will need the intellectual approach of modern mathematics to develop its own systematic concepts.

8.1 Trace Determination, Detection limits, General

[19] H. Kaiser, Zum Problem der Nachweisgrenze, Z. Anal. Chem. *209*, 1 (1965).

[20] Analytiker-Tagung, Lindau 1966, Teil 1: Analytik kleinster Substanzmengen, Z. Anal. Chem. *221*, 1 (1966).

[21] H. Kaiser, Zur Definition der Nachweisgrenze, der Garantiegrenze und der dabei benutzten Begriffe, Z. Anal. Chem. *216*, 80 (1966).

[22] Trace Characterization Chemical and Physical, U.S. Department of Commerce, National Bureau of Standards Monograph 100, Washington 1967.

[23] Optical and X-Ray Spectroscopy (Contributed Papers and Discussion) in Ref. [22], p. 149.

[24] V. Svoboda, R. Gerbatsch, Zur Definition von Grenzwerten für das Nachweisvermögen, Z. Anal. Chem. *242*, 1 (1968).

[25] H. Kaiser, A.C. Menzies, The Limit of Detection of a Complete Analytical Procedure, Adam Hilger London 1968.

[26] G. Ehrlich, H. Scholze, R. Gerbatsch, Zur objektiven Bewertung des Nachweisvermögens in der Emissionsspektroskopie IV. Spectrochim. Acta *24B*, 641 (1969).

[27] H. Kaiser, Quantitation in Elemental Analysis, Anal. Chem. *42*, Nr. 2, 24A, Nr. 4, 26A (1970).

8.2 Statistics in Chemistry

[28] H. Kaiser, H. Specker, Bewertung und Vergleich von Analysenverfahren, Z. Anal. Chem. *149*, 46 (1956).

[29] W.J. Youden, Statistical Methods for Chemists, 5th Ed., Wiley, New York 1961.

[30] G. Gottschalk, Statistik in der quantitativen chemischen Analyse, Die chemische Analyse, Bd. 49, Ferdinand Enke Verlag, Stuttgart 1962.

[31] V.V. Nalimov, The Application of Mathematical Statistics to Chemical Analysis, Pergamon Press, Oxford 1963.

[32] K. Doerffel, Beurteilung von Analysenverfahren und -ergebnissen, 2. Aufl., Springer Verlag, Berlin, Heidelberg, New York, J.F. Bergmann, München 1965.

[33] Brookes, Bettely, Loxton, Mathematics and Statistics for Chemists, Wiley, New York 1966.

[34] G. Gottschalk, Einführung in die Grundlagen der chemischen Materialprüfung, S. Hirzel Verlag, Stuttgart 1966.

8.3 Theory and Statistics of Experiments

[35] O.L. Davies, Design and Analysis of Industrial Experiments, Oliver and Boyd, London 1960.

[36] N.L. Johnson, F.C. Leone, Statistics and Experimental Design, Vol. I, II, John Wiley & Sons, New York, London, Sydney 1964.

[37] M.G. Natrella, Experimental Statistics, NBS, Handbook 91, Washington 1966.

[38] V.V. Nalimov, Theory of the Experiment (russ.), Publ. of the Academy, Moscow 1971.

8.4 Mathematical Statistics, Tables, Standards

[39] Deutsche Normen, DIN 55 302, Blatt 1 und Blatt 2

[40] E. Kreyszig, Statistische Methoden und ihre Anwendungen, Vandenhoeck & Ruprecht, Göttingen 1965.

[41] K. Stange, H.-J. Henning, Formeln und Tabellen der mathematischen Statistik, 2. Aufl., Springer Verlag, Berlin, Heidelberg, New York 1966.

[42] G.W. Snedecor, W.G. Cochran, Statistical Methods, Iowa State Univ. Press., Ames, Iowa, 1967.

[43] I.M. Chakravarti, R.G. Laha, J. Roy, Handbook of Methods of Applied Statistics, Vol. I and II, Wiley, New York 1967.

[44] J. Pfanzagl, Allgemeine Methodenlehre der Statistik II, 3. Aufl., Sammlung Göschen, Bd. 747/747a, Walter de Gruyter & Co., Berlin 1968.

[45] S. Koller, Neue graphische Tafeln zur Beurteilung statistischer Zahlen, 4. Aufl., Verlag Dietrich Steinkopff, Darmstadt 1969.

[46] L. Sachs, Statistische Auswertungsmethoden, 2. Aufl., Springer Verlag, Berlin, Heidelberg, New York 1969.

8.5 Information Theory, Mathematics in General

[47] W. Meyer-Eppler, Grundlagen und Anwendungen der Informationstheorie, 2. Aufl., Springer Verlag, Heidelberg, Berlin, New York 1969.

[48] H. Margenau, G.M. Murphy, The Mathematics of Physics and Chemistry, Vol. I, 1964, Vol. II, 1966, D. Van Nostrand Company, Inc., London.

[49] H. Behnke, R. Remmert, H.G. Steiner, H. Tietz, Mathematik 1, Fischer Lexikon 29/1, Fischer Bücherei, Frankfurt/Main, Hamburg 1964.

[50] H. Behnke, H. Tietz, Mathematik 2, Fischer Lexikon 9/2, Fischer Bücherei, Frankfurt/Main, Hamburg 1966.

2 Methods of Separation

Contributed by

H. Batzer, Basel
E. Bayer, Tübingen
B. Cornils, Oberhausen
H. Determann, Frankfurt
J. Falbe, Oberhausen
H. Feltkamp, Wuppertal-Elberfeld
G. Hesse, Erlangen
W. Hofmann, Basel
H. Jork, Saarbrücken
L. Kraus, Saarbrücken
K. Lampert, Frankfurt
G. Legler, Köln
F. Lohse, Basel
K. Maas, Heidelberg
H. Pauschmann, Tübingen
H. Schildknecht, Heidelberg
R. Schmid, Basel
S.A. Zahir, Basel

2.1 Distillation, Centrifuging, Crystallization

Boy Cornils and Jürgen Falbe
Ruhrchemie A.G., D-42 Oberhausen-Holten

1 Distillation and Rectification

1.1 Physical Principles

Distillation and rectification are among the most important separation methods. Moreover, many other separation methods can be simplified if a preliminary distillation is carried out.

In order to avoid the confusion which is frequently found in the literature, the definitions used in DIN 7052 will be employed in this survey. Distillation is defined as the evaporation of a liquid, the vapors being led away and condensed. The general term 'distillation' includes both simple distillation *(batch distillation)* and rectification *(counter-current distillation)*.

In *simple distillation* (also known as single-stage or differential distillation), the separation is effected by (in most cases partial) vaporization of the liquid mixture, which is to be separated, in such a way that the ascending vapors and the condensate move in the same direction.

In *rectification* (also known as multi-stage distillation), separation of a liquid mixture involves repeated evaporation and condensation, such that the vapor moves in the opposite direction to the returning condensate (countercurrent flow); thus both material exchange and heat exchange take place.

The separation efficiency of a single-stage distillation is basically lower than that of rectification (see Section 2.1:1.1.2). However, single-stage distillation is adequate for many separations, e.g., the evaporation of solvents. Therefore, the choice of method depends on the composition of the mixture to be separated and on the problem to be solved.

1.1.1 Single-stage Distillation

In *single-stage distillation* only a minor part of the vapor is condensed as it ascends, and the resulting countercurrent flow effects are generally small. The optimal efficiency of a single-stage distillation thus corresponds to one theoretical plate (see Section 2.1:1.1.2). Thus, in a favorable case, the lower boiling component can reach a vapor concentration which corresponds to the equilibrium concentration of the vapor, determined by the different liquid concentrations at the point where vapor

leaves the liquid[1]. Therefore, single-stage distillation is worthwhile only when there is a sufficiently large difference between the vapor pressures of the lower boiling compound and the residue (e.g. *solvent stripping*), or if the mixtures to be separated contain thermally unstable materials. In this case, repeated single-stage distillation is often preferred to rectification in spite of an unsatisfactory separation since the temperature in the distillation flask can be kept lower (cf. Section 2.1:1.2.1).

The main elements of the apparatus for a single-stage distillaton are: a device for feeding liquid to the distillation flask, an evaporator, a condenser, and a means for removal of the residue and the distillate. In recent years, optimal combination of all these components, especially the evaporator and condenser, has enabled the development of efficient single-stage distillation equipment, permitting the separation of mixtures of materials which are very labile at high temperatures.

Continuously-operating single-stage distillation equipment in laboratory use includes rotary evaporators and film evaporators; discontinuously operating equipment includes the well-known standardized Engler and ASTM distillation apparatus (see Section 2.1:1.2.1).

1.1.2 Rectification

Rectification[2-18] is based on the equilibrium between vapor and condensate. This equilibrium is achieved by a countercurrent flow process occuring mainly through mechanical contact at so-

---

[1] *L. Gemmeker, H. Stage,* Glas-Instrum.-Tech. *8,* 413, 503, 563 (1964).

[2] *K. Sigwart* in *Houben-Weyl,* Methoden der organischen Chemie, Bd. I/1, p. 777, Georg Thieme Verlag, Stuttgart 1958.

[3] *K. Sigwart* in *Ullmann,* Encyclopädie der technischen Chemie, Bd. I, p. 429, Urban & Schwarzenberg, München, Berlin 1951.

[4] *P. Grassmann,* Einführung in die thermische Verfahrenstechnik, Walter de Gruyter Verlag, Berlin 1967.

[5] *K.B. Wiberg,* Laboratory Technique in Organic Chemistry, McGraw-Hill Book Co., New York 1960.

[6] *P.A. Washer,* Distillation, Engineering Extension Service, Texas Agriculture and Mechanical System, 1963.

[7] *E.A. Coulson, E.F.G. Herington,* Laboratory Distillation Practice, G. Newnes, London 1958.

[8] *T.P. Carney,* Laboratory Fractional Distillation, Macmillan, New York 1949.

[9] *M.I. Rosengart,* Die Technik der Destillation und Rektifikation im Laboratorium, VEB Verlag Technik, Berlin 1954.

called plates or trays. The possibility of separating a given mixture and the choice of apparatus depend on the 'equilibrium curve'. For a given liquid composition (abcissa, cf. Fig. 1), the corresponding vapor composition (ordinate) is obtained provided that the boiling liquid and the vapor are in close contact long enough to reach a state of equilibrium. These curves can be determined experimentally[3, 13, 19–23, 37, 400] and represent the behavior of the binary mixture investigated. The slope of a curve (Fig. 1), compared with a line drawn at 45°, indicates whether the mixture is an ideal binary one (1), a mixture with partially miscible components (2), or one with an azeotropic point (3,4). Fig. 3[20] is an illustration of equipment for the determination of phase equilibria.

If a rectification column (Fig. 2) operating in an ideal (i.e. adiabatic) manner is filled with an ideal binary system [Fig. 1, equilibrium curve (1)] and is operated with total reflux ($v = R/D = \infty$), then the material balance during the rectification process is given by the equation:

$$V = R + D \qquad (1)$$

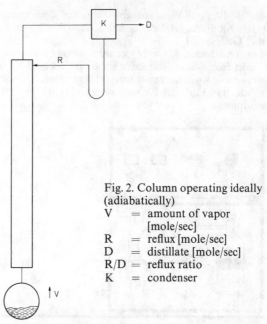

Fig. 2. Column operating ideally (adiabatically)

V = amount of vapor [mole/sec]
R = reflux [mole/sec]
D = distillate [mole/sec]
R/D = reflux ratio
K = condenser

This equation holds for any diameter, and, with plate columns, for any plate. Provided that the upgoing vapors and the refluxing liquid are in perfect equilibrium in each section of the column, (and, in plate-containing columns, on each plate) the following equations are valid:

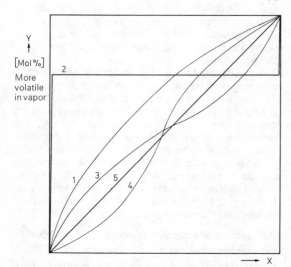

Fig. 1. Equilibrium curves of various binary mixtures
1 Binary mixture without azeotropic point (e.g. methanol-water)
2 Insoluble or partially miscible components (benzene-water)
3 Azeotropic mixture with minimum boiling point (n-butanol-water)
4 Azeotropic mixture with maximum boiling point (nitric acid-water)
5 45° line

[10] A. Rose, E. Rose, Distillation, in Weissberger, Technique of Organic Chemistry, Vol. IV, Interscience Publishers, New York 1951.
[11] R. Jacobs, Destillier-Rektifizier-Anlagen, Verlag R. Oldenbourg, München 1950.
[12] R. Billet, Grundlagen der thermischen Flüssigkeitszerlegung, Hochschultaschenbücher Bd. XX, Bibliographisches Institut, Mannheim 1962.
[13] E. Krell, Handbuch der Laboratoriumsdestillation, 2. Aufl., VEB Deutscher Verlag der Wissenschaften, Berlin 1960.
[14] E. Kirschbaum, Destillier- und Rektifiziertechnik, 4. Aufl., Springer Verlag, Berlin, Heidelberg, New York 1969.
[15] H. Stage, G.R. Schultze, Die Kolonnendestillation im Laboratorium, VDI-Verlag, Berlin 1944 und Dechema Monographie 14, 17 (1950).
[16] M. van Winkle, Distillation, McGraw-Hill Book Co., New York 1967.
[17] H. Stage, E. Müller, L. Gemmeker, Chem.-Ztg. 85, 387 (1961).
[18] H. Röck, Destillation im Laboratorium, Steinkopff-Verlag, Darmstadt 1960.
[19] H. Stage, I.S. Baumgarten, Öl Kohle 40, 126 (1944).
[20] H. Stage, W.G. Fischer, Glas-Instrum.-Tech. 12, 1167 (1968).
[21] W.G. Kogan, W.M. Fridman, Handbuch der Dampf-Flüssigkeits-Gleichgewichte, VEB Deutscher Verlag der Wissenschaften, Berlin 1961.
[22] E. Müller, H. Stage, Verfahrenstechnik in Einzeldarstellungen, Bd. XI, Springer Verlag, Berlin, Göttingen, Heidelberg 1961.
[23] F. Patat, K. Kirchner, Praktikum der technischen Chemie, 2. Aufl., Walter de Gruyter & Co., Berlin 1968.
[24] W.L. McCabe, E.W. Thiele, Ind. Eng. Chem. 17, 605 (1925).
[25] M.R. Fenske, S.O. Tongberg, D. Quiggle, Ind. Eng. Chem. 26, 1169 (1934).

Fig. 3. Illustration of an apparatus for determination of phase equilibria (ex Fischer Labor- und Verfahrenstechnik, Bad Godesberg)
1 heating device, 2 vessel, 3 vacuum jacket, 4 thermocouple, 5 inlet feed, 6 magnetic valve for liquid take-off, 7 receiver, 8 magnetic valve for vapor take-off, 9 receiver, 10 connection to vacuum, 11 device for vacuum stability, 12 control equipment

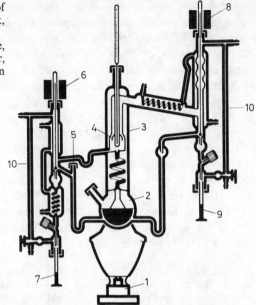

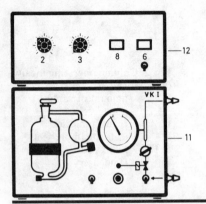

$$V \cdot y = R \cdot x + D \cdot x_D \qquad (2)$$

$$y = \frac{R}{D+R} \cdot x + \frac{D}{D+R} \cdot x_D \qquad (3)$$

$$y = \frac{v}{v+1} \cdot x + \frac{1}{v+1} \cdot x_D \qquad (4)$$

Here, y is the concentration of the more volatile material in the vapor (ordinate in Fig. 1 in Mol-%), x is the concentration of the more volatile component in the liquid (abcissa in Fig. 1 in Mol-%), and x_D is the concentration of the more volatile component in the distillate. Equation (4) corresponds to a line with a slope of $v/v + 1$ and an intercept of $x_D/v + 1$; it gives the composition of the vapor as a function of the composition of the liquid for any horizontal section of column operating in an ideal manner. For $R/D = v = \infty$, the slope becomes 1 and the intercept zero, e.g. this line (known as the *operating line* or *material balance line*) is identical with the 45° line.

The number of theoretical plates for a given two component system [(1) in Fig. 1] containing, for example, 30 mole-% of x, to be concentrated to 80 mole-% of x, is determined from the number of steps drawn between the operating line and the equilibrium curve[24], e.g. between x = 30 and 80 Mol-%, (method of McCabe-Thiele, Fig. 4). This method is less suitable for flat equilibrium curves and total reflux. Alternately, the number of plates necessary may be found by experimental determination of x_D at known reflux ratio (mostly ∞, e.g. $D \ll R$). The number of theoretical plates can also be determined using nomograms[15, 25-28, 300].

All the methods mentioned above give the theoretical number of plates n_{th}, a 'theoretical plate' being a hypothetical ideal section in the column where vapors coming up from the next unit below this section and the liquid coming down from the next section above are in thermodynamic equilibrium. In practice, a correction must be made for the number of theoretical plates; the ratio of a theoretical plate n_{th} to a 'practical' plate n is defined as the plate efficiency S_m.

$$S_m = \frac{n_{th}}{n} \qquad (5)$$

For packed columns, the HETP value (height equivalent to a theoretical plate) is obtained by dividing the height of the packed column by the number of theoretical plates n_{th}.

$$HEPT = \frac{height}{n_{th}} \qquad (6)$$

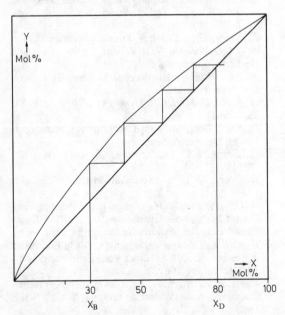

Fig. 4. Determination of the required number of theoretical plates
Graphical determination by drawing a step curve between x = 30 and 80 Mole-% (v = ∞)

Plate-containing columns show measurable concentration differences between the individual plates. In packed columns, in contrast, the vapor does not stay long enough at any one point to reach full thermodynamic equilibrium. The equilibrium curve and the operating line are, therefore, normally not parallel. Consequently, the methods[15, 25-28, 300] mentioned cannot be used. Generally, the required number of plates of a packed column n_a is greater than the number of theoretical plates n_{th} of a plate-containing column if the equilibrium curve and the operating line converge to greater values of y, and vice versa. In the case of total reflux $(v = \infty)$, the determination can be carried out nomographically[29]. The length of the column which corresponds to one theoretical exchange unit is called the HTU value (height of transfer unit).

$$HTU = \frac{height}{n_a}$$

Under laboratory conditions, differences between the HETP and HTU values are small compared with the effects of the load, pressure and pressure drop, or reflux ratio. Moreover, uncertainties in the equilibrium data used cause greater differences in the number of plates than those caused by the different modes of operation of packed and plate-containing columns.

S_m and HETP are characteristic criteria essential for judging the efficiency of the various columns described in Section 2.1:1.2.2[30-32].

However, the number of theoretical plates alone is not sufficient to characterize a column. Under given operating conditions, such as pressure, feed rate and reflux ratio, the nature and composition of the test mixture have an important influence. Test mixtures frequently used are: benzene/carbon tetrachloride, benzene/xylene, n-heptane/methylcyclohexane, methanol/water, or benzene/cyclohexane[33-40].

Various calculation methods have also been introduced during recent years[42-57, 59-63, 266, 320, 401, 406].

1.2 Application

1.2.1 Single-stage Distillation

The technique of batch distillations has been extensively reviewed[2, 7, 10, 13, 18, 64, 132]. The *determination of boiling range* involves vaporization of the contents of the still at constant pressure, determination of the vapor temperature, condensation of the vapor, and determination of the amount of

[26] *M.R. Fenske*, Ind. Eng. Chem. *24*, 482 (1932).

[27] *F.W. Melpolder, C.E. Headington*, Ind. Eng. Chem. *39*, 763 (1947).

[28] *E.H. Smoker*, Ind. Eng. Chem. *34*, 509 (1942).

[29] *H. Stage, J. Juilfs*, Chem.-Ztg. *77*, 511, 538, 575 (1953); *78*, 43, 78, 112, 143, 182, 217 (1954).

[30] *H.V. Adolphi*, Chem.Tech. (Leipzig) *20*, 543 (1968).

[31] *F. Kastanek, G. Standart*, Wirkungsgrad von Destillationsböden, Akademie Verlag, Prag 1966.

[32] *M. Hartmann*, Eine Studie neuer Patente von Böden für Destillationskolonnen, Akademie Verlag, Prag 1966.

[33] *D.G. Peacock*, Chem. Eng. Sci. *22*, 957 (1967).

[34] *C. Haughton*, Brit. Chem. Eng. *10*, 237 (1965).

[35] *E. Kauer*, Chem.Tech. (Leipzig) *20*, 681, 406 (1968).

[36] *H. Lentz, H.G. Wagner*, Ber. Bunsenges. Phys. Chem. *73*, 66 (1969).

[37] *A.J. Brainard, G.B. Williams*, A.I.Ch.E.-Journal *13*, 60 (1967).

[38] *J.R. Maa*, Ind. Eng. Chem., Fundamentals *6*, 504 (1967).

[39] *H. Brandt, H. Röck*, Chem.-Ing.-Tech. *29*, 397 (1957).

[40] *R.J. Hellwag*, Chem.-Ing.-Tech. *37*, 539 (1965).

[41] *K. Dieter, W. Hübner*, Chem.-Ztg. *94*, 319 (1970).

[42] *G. Kortüm, H. Buchholz-Meisenheimer*, Die Theorie der Destillation und Extraktion von Flüssigkeiten, Springer Verlag, Berlin 1952.

[43] *R. Haase*, Thermodynamik der Mischphasen, Springer Verlag, Berlin 1956.

[44] *R.C. Reid, T.K. Sherwood*, The Properties of Gases and Liquids, McGraw-Hill Book Co., New York 1958.

[45] Autorenkollektiv, Berechnung thermodynamischer Stoffwerte von Gasen und Flüssigkeiten, VEB Deutscher Verlag für Grundstoffindustrie, Leipzig 1966.

[46] *R.R. Dreisbach*, Pressure-Volume Relationships of Organic Compounds, 3. Ed., Handbook Publ., Sandusky, Ohio 1952.

[47] *E. Hala, J. Pick, V. Fried, O. Vilim*, Vapor Liquid Equilibrium, Pergamon Press, New York 1958, Deutsche Ausgabe Akademie Verlag, Berlin 1960.

[48] *J.C. Chu*, Distillation Equilibrium Data, Reinhold Publishing Co., New York 1950.

[49] *R.A. Eckart, A. Rose*, Hydrocarbon Proc. *47*, 165 (1968).

[50] *K. Hoppe, H. Künne, H. Bendix*, Chem. Tech. (Leipzig) *20*, 204 (1968).

[51] *S.R.M. Ellis, C. McDermott, C.S. Chiang*, Brit. Chem. Eng. *12*, 727 (1967).

[52] *H.L. Chang*, A.I.Ch.E.-Journal *12*, 11 (1966).

[53] *F.O. Mixon*, Ind. Eng. Chem., Fundamentals *4*, 455 (1965).

[54] *A.S. Wigborow*, Zh. Prikl. Khim. (Leningrad) *40*, 1052 (1967).

[55] *J.H. Prausnitz, F. Fleck*, Chem. Eng. *75*, 157 (1968); Ind. Eng. Chem. *57*, 18 (1965).

[56] *R.N. Finch, M. van Winkle*, A.I.Ch.E.-Journal *8*, 455 (1962).

[57] *A.E. van Arkel*, Trans. Faraday Soc. *42B*, 81 (1946).

[58] *H. Stage*, Glas-Instrum.-Tech. *14*, 337 (1970).

[59] *R.W.H. Sargent, B.A. Murtagh*, Trans. Inst. Chem. Eng. *47*, 85 (1969).

[60] *B.D. Smith*, Design of Equilibrium Stage Processes, McGraw-Hill Book Co., New York 1963.

[61] *M.B. King*, Phase Equilibrium in Mixtures, Pergamon Press, New York 1967.

[62] *L.A. Bouron*, Calcul et Disposition des Appareils de Distillation, Libr. Polytechnique Ch. Béranger, Paris 1959.

[63] *D.N. Hanson, J.H. Duffin, G.F. Somerville*, Computation of Multistage Separation Processes, Reinhold Publishing Co., New York 1962.

condensate. The *true boiling-point curve* (vapor temperature versus amount of condensate) allows conclusions to be drawn about the composition or purity of the liquid being analyzed.

In order to eliminate even small interactions between the rising vapor and the refluxing condensate, equipment for this type of analysis has been standardized with regard to all components and is commercially available.

Using the most common methods – Engler distillation[65] in Germany and ASTM distillation[66, 67] in the U.S.A. – all distillation results are identical, provided that thermometers of the same kind and with the same depth of immersion (stem correction) are used. The specifications of many distillable organic products contain figures for the boiling range (e. g. the vapor temperatures measured when 2% and 97% of the product have been condensed). The ASTM method is increasingly displacing the Engler method even in German-speaking countries.

Steam distillation is a special kind of single-stage distillation. In this method, the total distillation pressure is the sum of the vapor pressures of the steam and of the compound to be distilled[10, 12, 14, 68–73]. In many cases, the boiling temperature of the mixture to be separated can be lowered to a value at which thermal effects (chemical reactions, cracking) are very unlikely[72, 73].

Single-stage distillation is preferred for the separation of liquids with very different boiling points[74, 75], or for the distillation of water from salt solutions[76–79].

Recently designed vaporizers are the flash and film evaporators. In *flash evaporators*, the vaporizer is filled with liquid or with a mixture of liquid and vapor, the latter being separated in a special flash chamber[1, 58, 80, 351]. In *run through evaporators*[1, 81–86], the residue is completely removed from the evaporator after one pass. However, in *thermosyphon evaporators*[1, 86–88], the non-vaporized residue is recycled to the evaporator. Both types can be adapted easily to suit different problems. Foams may interfere greatly with the separation of liquid and vapor in the flash chamber; however, they can be prevented by using a special type of construction[1, 86].

Flash vaporizers are commercially available, but *film evaporators,* which permit vacuum distillation, are often preferred. This may be due to an increasing trend toward vacuum distillation. In all flash evaporators vaporization must overcome the pressure resulting from the liquid and vapor column pressing on the heating device. This results in high temperatures in the residue which may cause troubles with heat-sensitive compounds. With vacuum distillation, however, these can be distilled under mild conditions, especially when the films of the liquid phase are very thin. Such liquid films can be prepared by rotary, spray, twisttube, falling film or shortpath evaporators.

The best known type of film evaporators are the *rotary evaporators*[1, 82, 89–96]. The feed device does not allow continuous operation because the feed rate cannot be regulated to match the heat transfer which depends on the size of the rotating flask, the

[64] *F. J. Zuiderweg,* Laboratory Manual of Batch Distillation, Interscience Publishers, New York 1957.
[65] Deutsche Normblätter DIN 53 169 und 51 751.
[66] American Society for Testing Materials (ASTM) D 1078-63, E 133-58 und D 86-59 (cf. 1967 Book of ASTM-Standards, Part. 20, ASTM, Philadelphia 1967).
[67] British Standard Specification 658 (1962); Standards of the Institute of Petroleum I.P. 24/55.
[68] *J. Reilly, W. N. Rae,* Physico-Chemical Methods, Vol. II, 5th Ed., p. 155, Methuen & Co., London 1954.
[69] *J.M. Coulson, J.F. Richardson,* Chemical Engineering, Vol. II, Pergamon Press, London 1955.
[70] *U. von Weber, E. Gildemeister, F. Hoffmann,* Die ätherischen Öle, p. 309, Akademie Verlag, Berlin 1956.
[71] *W. Wintermeyer,* Chem.-Ztg. *92,* 289 (1968).
[72] *R. Rigamonti, A. Gionetto,* Dechema-Monographie *28,* 75 (1956).
[73] *H. Stage,* Erdöl, Kohle *3,* 478 (1950); Fette Seifen, Anstrichm. *55,* 217, 284, 375, 513, 580 (1953); *58,* 358 (1956); Chem.-Ztg. *88,* 412 (1964).
[74] *G. Guerreri,* Brit. Chem. Eng. *13,* 524 (1968).
[75] *E. Thrun,* Chem.-Ztg. *91,* 198 (1967).
[76] *F. Jakob,* Chem.-Ing.-Tech. *41,* 481 (1969).
[77] *M.R. Bloch,* Naturwissenschaften *56,* 184 (1969).
[78] *H. Germerdonk,* Chem.-Ing.-Tech. *41,* 110 (1969).
[79] *T. Messing,* Chem.-Ztg. *92,* 45 (1968).
[80] *H. Stage, L. Gemmeker,* Glas.-Instrum.-Tech. *7,* 608, 687 (1963); *8,* 413, 503, 563 (1964).
[81] *R.B. Smith, T. Dresser, H.F. Hopp, T.H. Paulsen,* Ind. Eng. Chem. *43,* 766 (1961).
[82] *W. Frank, D. Kutsche,* Die schonende Destillation, Krausskopf Verlag, Mainz 1969.
[83] *R.E. Jentoft, J.F. Johnson,* Ind. Eng. Chem. *51,* 519 (1959).
[84] *W.H. Bartholomew,* Anal. Chem. *21,* 527 (1949).
[85] *A. Hartrampen,* Chem-Ing.-Tech. *41,* 238 (1969).
[86] *R. Billet,* Verdampfer-Technik, Bibliographisches Institut, Mannheim 1965.
[87] *F. Petuely, N. Meixner,* Mikrochim. Acta 613 (1957).
[88] *M. Beroza,* Anal. Chem. *26,* 1251 (1954).
[89] *H. Stegemann,* Chem.-Ztg. *81,* 110 (1957).
[90] *K.C.D. Hickman,* Ind. Eng. Chem. *39,* 686 (1947).
[91] *H. Schwarz* in *Houben-Weyl,* Methoden der organischen Chemie, Bd. I/1, p. 889, Georg Thieme Verlag, Stuttgart 1958.
[92] *D.W. West,* Chem. Ind. [London] *1963,* 1118.
[93] *F.A. Metzsch,* Chem.-Ing.-Tech. *28,* 62 (1956).
[94] *F. Genser, H. Richter,* Glas-Instrum.-Tech. *7,* 620 (1963).
[95] *H.E. Zaugg, J. Shavel,* Anal. Chem. *26,* 1999 (1954).

pressure, and the speed of rotation. Rotary evaporators up to the semi-technical scale (flask volumes up to 100 l) are commercially available. Nearly all rotary evaporators can be used to evaporate to dryness.

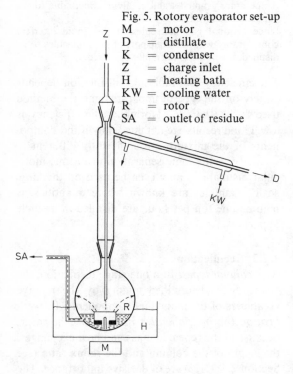

Fig. 5. Rotory evaporator set-up

M   = motor
D   = distillate
K   = condenser
Z   = charge inlet
H   = heating bath
KW = cooling water
R   = rotor
SA = outlet of residue

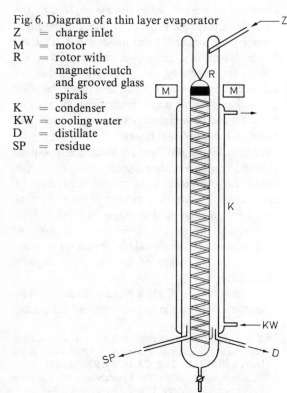

Fig. 6. Diagram of a thin layer evaporator

Z   = charge inlet
M   = motor
R   = rotor with
      magnetic clutch
      and grooved glass
      spirals
K   = condenser
KW = cooling water
D   = distillate
SP = residue

In *spray evaporators* (Fig. 5) the liquid to be evaporated is dispersed by dropping onto rotors, whence it flows as a thin film on the walls of the flask[1, 97]. In other models, the liquid contained in the flask is repeatedly agitated by injecting an inert gas. An advantage of the spray evaporator over the rotary evaporator is the better vacuum which can be reached (magnetic stirring device); moreover, continuous operation is possible with spray evaporators.

Efficient separation under mild conditions is achieved with *falling film evaporators* or *thin layer evaporators*, in which the material to be distilled is fed to the upper part of a heated pipe and then flows down along the wall of the pipe as a thin film. The high boiling compounds are obtained at the end of the pipe as a residue (Fig. 6).

The various models of this type of evaporator[1, 2, 13, 41, 82, 98−106, 226, 351, 403] differ in the type of feed device which must be of a special design to avoid incomplete wetting of the pipe wall. Counter-current or co-current vaporizers differ in the manner in which the condenser is connected[1]. Uniform wetting of the surface of the evaporator may be achieved by use of annular slots[100], perforated plates, spirals[99], or other devices. Another possibility is the use of a rotor which distributes the liquid product on the wall[103−105, 403]. Horizontal or slightly sloping thin layer vaporizers have been described[102]; these are suitable especially for highly viscous materials and may be employed also on the pilot plant scale.

The *short path evaporators*[82, 102, 105, 107−114] operate in a special way and are of a special design in

[96] *K. Egli*, Glas-Instrum.-Tech. *11*, 509, 705, 869 (1967).

[97] *D.J. Trevoy, W.A. Torpey*, Anal. Chem. *26*, 992 (1954).

[98] *M.S. Tendulkar*, Chem. Eng. *67*, 208 (1960).

[99] *R.L. Hopkins*, Ind. Eng. Chem. *43*, 1456 (1951).

[100] *L.H. Walker, D.C. Patterson*, Ind. Eng. Chem. *43*, 934 (1951).

[101] *E. Krell*, Chem. Tech. *2*, 252 (1950); *4*, 443 (1952).

[102] *G.E. Utzinger*, Chem.-Ing.-Tech. *26*, 129 (1954); Chem. Tech. *7/8*, 61 (1963).

[103] *R. Schneider*, Chem.-Ing.-Tech. *27*, 257 (1955).

[104] *E. Keunecke* in *Ullmann*, Encyklopädie der technischen Chemie, Band II/1, p. 72, Verlag Urban & Schwarzenberg, München 1961; Dechema-Monographie *31*, p. 233.

[105] *W. Frank*, Chem.-Ztg. *87*, 365 (1963); Verfahrenstechnik *2*, 259 (1968).

[106] *R.G. Nester*, J. Sci. Instrum. *31*, 1002, (1960).

[107] *R. Jaeckel* in *Houben-Weyl*, Methoden der organischen Chemie, Bd. I/1, p. 901, Georg Thieme Verlag, Stuttgart 1958.

[108] *G.W. Oetjen* in *Ullmann*, Encyklopädie der technischen Chemie, Bd. I, p. 156, Verlag Urban & Schwarzenberg, München 1951.

which the distance between the surface of the vaporizer and the condenser is kept as short as possible. By using high vacuum, this distance can be kept shorter than the free path length of the molecules evaporated. Thus the individual molecules may be condensed before they are able to hit other molecules (including inert gas molecules present in the system). This method of distillation is, therefore, termed *molecular distillation*[102, 109, 110]. However, the distance between the surface of the vaporizer and the condenser cannot be kept shorter than the free path length under all circumstances. Thus, the term *short path distillation* seems to be more appropriate for this type of distillation. The extremely high vacuum enables the required large free path length. Thus, the separation by molecular distillation takes place at a temperature as low as possible, and so this distillation method has the lowest thermal requirements. Therefore, it is widely employed with naturally occurring compounds such as *vitamins*[115], *glycerides*[116], *perfumes*[117, 118], *plant constituents*[104, 107, 108, 118], and in the field of natural and synthetic polymers which, for a long time, were regarded undistillable (separation of monomers from polymers[105, 117–119]).

The various types of molecular distillation apparatus which are common today have been reviewed[41, 82, 104, 117, 226] (rotating, falling film, and centrifugal molecular distillation). Special types of construction permit the four basic steps of molecular distillation to be in optimal sequence: transport of the molecules to the surface of the liquid, evaporation of the molecules from the surface, transport through the vapor phase, and condensation.

As in other types of distillation, in molecular distillation the apparatus must meet the requirements of the specific separation problem. Of course, the residence time of

molecules of a heat-sensitive compound in the apparatus should be minimal. Residence times from 1 to 5×10^{-3} sec can be reached in centrifugal molecular distillation. However, this is not always compatible with the requirement that there should be no impoverishment of the compound to be evaporated in the liquid film. Moreover, care must be taken to avoid local temperature gradients and catalytic effects due to the vaporizer materials[115, 119].

Since residual gas molecules in the liquid interfere considerably with the transport of the molecules to be distilled, the liquid is normally degassed[107, 121].

The efficiency of a molecular distillation depends largely on the type of apparatus and the method used[107, 108, 111, 120, 121]. Normally, the efficiency is low. Good results are obtained only if the components of the mixture are sufficiently different in volatility. With some separation problems, molecular distillations may even be used on the large scale; examples are known where quantities as high as one ton per hour are distilled in a single apparatus.

1.2.2 Rectification

New column types: In a one-stage distillation, the efficiency is determined mainly by the relative volatilities of the individual components. In *rectification* (or *multiplicative distillation*), however, because of the repeated mass and heat exchanges, the length of the column and the reflux ratio (see Section 2.1:1.1.2) are of decisive importance. The efficiency of a rectification column is given by the number of theoretical plates (which can be measured by the methods listed in Section 2.1:1.1.2). In the laboratory, *plate columns* or *packed columns* are normally used. All are designed to achieve a mixing of the liquid and the gas phases as ideal as possible and a minimum resistance to the mass and heat exchange processes which take place on the individual plates.

In *plate columns*, *practical plates* are fixed perpendicular to the axis of the column, thus allowing the liquid and gas phases to be mixed thoroughly. In continuous distillation, a definite volume of liquid is present on each plate through which the vapor rising from the next plate below must pass. A certain amount of the liquid (holdup) returns to the lower plate by means of special reflux devices (descenders).

The various types of glass plate columns used in laboratories may differ in the design of the plates,

[109] *J.N. Brönstedt, G.v. Hevesy*, Z. phys. Chem. *A99*, 189 (1921).

[110] *K.C.D. Hickman, T.J. Trevoy*, Chem. Rev. *34*, 51 (1944), Ind. Eng. Chem. *43*, 68 (1951), Ind. Eng. Chem. Anal. *44*, 1882 (1952), Chem.-Ing.-Tech. *25*, 672 (1953), Ind. Eng. Chem. *39*, 19, 686 (1947), *40*, 16 (1948), *42*, 36 (1950).

[111] *L.W. Masch*, Chem.-Ing.-Tech. *22*, 141 (1950).

[112] *G. Burrows*, Molecular Distillation, Clarendon Press, Oxford 1960.

[113] *P.R. Watt*, Molecular Stills, Chapman & Hall, London 1963.

[114] *G. Kretschmar, J. Pictet*, Chem.-Ing.-Tech. *29*, 16 (1957).

[115] *E.S. Perry*, Vacuum Symposium Transact. p. 151, (1956).

[116] *R. Habendorff*, Génie chim. (Chim. et Ind.) *95*, 2, 54 (1966).

[117] *B. Sehrt*, Chem. Tech. *20*, 137 (1968).

[118] *W. Heimann, K.H. Strackenbrock*, Z. Lebensm.-Unters. *120*, 273 (1963).

[119] *D. Kutsche*, Chem.-Ztg. *90*, 568 (1966).

[120] *K.C.D. Hickman*, Adv. in Vacuum Chemistry, Science in Progress, 4. Series, Yale University, New Haven 1945; Ind. Eng. Chem. *39*, 686 (1947).

[121] *S. Dushmann*, Scientific Foundation of Vacuum Technology, J. Wiley & Sons, New York 1949.

the method of mixing of the phases, and the construction of the descenders. The different constructions have been reviewed by Stage[13, 15, 17, 122, 128]. Designs with external descenders[3, 123] guarantee a sufficient capacity for liquid reflux without the vapor being carried over, but are both difficult to insulate and very fragile. With columns with internal descenders[124–127], however, there is a danger of liquid flowing from the upper plate directly to the descenders of the lower plate without taking part in the mass and heat exchange occurring in this lower plate (Fig. 7).

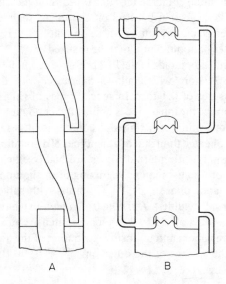

Fig. 7. Diagram of laboratory plate columns
A = With internal descenders (Oldershaw)[126]
B = With external descenders (Bruun)[123]

Consequently, the plate efficiency S_m of these columns is relatively low, whereas columns with external descenders give much better values[218]. However, plate efficiencies of more than 80% may be attained in specially constructed columns with internal descenders (jet trays, perforated plates

and others[128, 131]). Glass columns of this type are commercially available.

All plate columns with descenders suffer from the disadvantage that a considerable amount of the product to be distilled remains on the plates after the distillation is finished (holdup >0.4 ml per plate). The *sieve plate distillation columns*[126–130] which are empty after distillation are an exception. However, these plates require a certain minimum vapor velocity in order to keep the necessary amount of liquid on the plates during distillation. One of the advantages of plate columns – the relatively large load range within which the efficiency of the column is nearly constant – is not present in this type of column.

Packed columns which consist mainly of a tube filled with various types of packing materials are frequently used in the laboratory[122, 132–136, 144]. The efficiency of a packed column depends very much on the packing; of importance also are the surface per unit volume, the packing material, the flow of the phases and their distribution in the packing, the pressure drop, mechanical strength, and the possibilities for cleaning. The various commercially available packings (Raschig rings, steel spirals, spheres, saddles, Dixon rings, Braunschweiger spirals and others[7–16, 134, 137, 218, 225, 316]) generally fulfil only some of the requirements so that compromises must be made. In the laboratory only seldom can allowance be made for effects such as the angle of pouring the packings, foam formation, or the wall effect of the refluxing liquid film[135, 138–144, 154].

[122] *H. Stage, L. Gemmeker,* Chemie Lab. Betr. *15*, 177, 232, 290 (1964).
[123] *J.H. Bruun, W.B. Faulcower, S.D. West,* Ind. Eng. Chem. Anal. *1*, 212 (1929), *8*, 224 (1936) und *9*, 192, 247 (1937).
[124] *W.H. Keesom, H. van Dijk,* Proc. Kon. Acad. Wetensch. *34*, 42 (1931).
[125] *H. Stage, K. Klein, Gg.R. Schultze,* Z. Phys. Chem. *A189*, 163 (1941), Z. Elektro-Chem. u. Angew. Phys. Chem. *47*, 848 (1941), DBP 1.082.229 (1959), FP 1.252.869 (1960), Angew. Chem. *B19*, 182 (1947).
[126] *C.F. Oldershaw,* Ind. Eng. Chem. Anal. *13*, 265 (1941).
[127] *E. Krell,* Silikattechnik *5*, 2 (1954); Chem. Tech. (Leipzig) *7*, 585 (1957).

[128] *H. Stage,* Glas-Instrum.-Tech. *13*, 1285 (1969), *5*, 384, 425, 479, (1961), Z. Instrumententechnik *72*, 176 (1964).
[129] *K.P. Dimick, M.S. Simone,* Ind. Eng. Chem. *44*, 2487 (1952).
[130] *E. Zelfel,* Chem.-Ing.-Tech. *39*, 433 (1967).
[131] *G. Kortüm, A. Bittel,* Chem.-Ing.-Tech. *28*, 40 (1956).
[132] *A. Bittel* in *Ullmann,* Encyklopädie der technischen Chemie, Bd. II/1, p. 50, Verlag Urban & Schwarzenberg, München 1961.
[133] *H. Stage, K. Bose,* Bd. XIII der Verfahrenstechnik in Einzeldarstellungen, Springer Verlag, Berlin 1962.
[134] *R. Billet, L. Raichle,* Chem.-Ing.-Tech. *40*, 43 (1968), *39*, 133 (1967); Chem. Eng. *74*, 145, 195 (1967).
[135] *H. Gelbe,* Chem.-Ing.-Tech. *40*, 528 (1968).
[136] *C.B. Willingham, F.D. Rossini,* J. Res. Bur. Stand. *37*, 15 (1946).
[137] *H. Kolling,* Chem.-Ing.-Tech. *24*, 405 (1952).
[138] *F. Wolf, W. Günther,* Chem. Tech. (Leipzig) *20*, 141 (1968).
[139] *T. Ankel,* Chem.-Ing.-Tech. *40*, 1153 (1968).

The efficiency of packed columns is indicated by the HETP value; this can be determined by one of the methods listed in Section 2.1:1.1.2. The measurement techniques recommended by various authors[3, 13, 134, 137] suffice for approximate results. Packed columns offer a number of advantages for use in the laboratory. They are readily adaptable to different problems (the packings can be changed quickly) and show only low pressure drop under standard conditions. Packed columns have also been used for separations on the micro scale[145].

Good efficiencies at higher distillation rates may be achieved with columns containing fixed devices instead of removable packings. These devices (which should not be confused with the plates of plate columns) may be twisted spirals, sieve spirals, protruded wire gauze, or other types[142, 146–153]. *Inclined film columns* contain metal sieves which are slightly inclined towards each other[154].

The Vigreaux column and the so-called Widmer spiral are intermediates between the types described above and the true *wetted wall columns*. In the Vigreaux column the surface of the pipe is enlarged by constrictions; the main advantage is the low pressure drop. The Widmer spiral has multiple reversals of the vapors and shows low separation efficiency.

Pure *wetted wall columns* consist of empty vertical pipes; their efficiency depends on the ratio of the wall surface to the cross-sectional area and on the uniform distribution of the liquid on the wall[13, 155–161]. Various methods can be used to enlarge the wall surface wetted by the liquid phase: pipes arranged side-by-side or one inside another, spirally wound or grooved pipes, or two concentric pipes[162–167]. These columns, which are termed *ring slot columns*, must be made from precisely cylindrical pipes, a slight rotation of the liquid reflux being caused by spiral etchings or grooves. Distillation equipment of this type is commercially available. It can be used on the micro-scale and under vacuum; under these conditions, HETP values of about 5 mm may be realized[122, 167, 168]. A special advantage is the low pressure drop of 10^{-1} to 10^{-2} torr which allows operation with flask pressures of 0.1 torr. In special types of ring slot columns, the inner pipe is cooled by air. Generally, the operation of a ring slot column is little more complicated than that of other types of distillation.

Columns with internal rotors are highly efficient; the rotor causes intensive mixing of the liquid and the vapor phase[13, 170–174]. Columns with rotating internal cylinders *(turning roll columns)* having slot widths of 0.1 to 2 mm cause intensive mixing of the vapors and uniform distribution of the liquid on the surface of the outer pipe[174–176]; 100 theo-

[140] *G.A. Hughmark*, Ind. Eng. Chem., Fundamentals *6*, 408 (1967).

[141] *K.P. Schönherr*, Chem.-Ing.-Tech. *40*, 889 (1968).

[142] *A. Sperandio*, Chem.-Ing.-Tech. *37*, 22 (1965).

[143] *B. Böhlen, H.U. Bürki, A. Guyer*, Helv. Chim. Acta *48*, 1270 (1965).

[144] *H. Stage*, Großtechnische Kolonnendestillation empfindlicher Substanzen, Krausskopf Verlag, Mainz 1971.

[145] *A.A. Morton, J.F. Mahoney*, Ind. Eng. Chem. Anal. *13*, 494 (1941).

[146] *W.J. Podbielniak*, Petr. Refiner *30*, 85, 145 (1951), Ind. Eng. Chem. Anal. *13*, 639 (1941).

[147] *H.S. Lecky, R.H. Ewell*, Ind. Eng. Chem. Anal. *12*, 544 (1940).

[148] *D.F. Stedman*, Trans. Amer. Inst. Chem. Eng. *33*, 153 (1937).

[149] *S.R.M. Ellis, J. Vargavandi*, Chem. Process Eng. *39*, 239 (1958).

[150] *J.H. de Groot*, Chem.-Ing.-Tech. *39*, 56 (1967).

[151] *K. Nakanishi*, J. Chem. Soc. Jap., Ind. Chem. Sect. *58*, Nr. 9, 635 (1955), *59*, Nr. 1, 122 (1956).

[152] *M. Huber*, Dechema-Monographie *55*, 33 (1965).

[153] *A. Paleni*, Dechema-Monographie *55*, 115 (1955).

[154] *H.J. John, C.E. Rehberg*, Ind. Eng. Chem. *41*, 1056 (1949).

[155] *A. Rose*, Ind. Eng. Chem. *28*, 1210 (1936).

[156] *E. Krell*, Brit. Chem. Eng. *12*, 562 (1967).

[157] *C.T. Everitt, H.P. Hutchison*, Trans. Inst. Chem. Eng. *45*, Nr. 1, 9 (1967).

[158] *H.P. Hutchison, L.R. Lusis*, Trans. Inst. Chem. Eng. *46*, Nr. 5, 158 (1968).

[159] *W. Kuhn, K. Ryffel*, Helv. Chim. Acta *26*, 1693 (1943); *25*, 252 (1942).

[160] *J. Hempel*, Chem.-Ing.-Tech. *39*, 1136 (1967).

[161] *R. Hoffmann*, Chem.-Ing.-Tech. *41*, 442 (1969).

[162] *E. Jantzen*, Angew. Chem. *36*, 529 (1923), *39*, 675 (1926), Chem.-Ztg. *50*, 419 (1926).

[163] *L. Craig*, Ind. Eng. Chem. Anal. *8*, 219 (1936), *9*, 441 (1937).

[164] *M.L. Selker, R.E. Burk, H.P. Lonkelma*, Ind. Eng. Chem. Anal. *12*, 352 (1940).

[165] *B.C.K. Donnel, R.M. Kennedy*, Ind. Eng. Chem. *42*, 2327 (1950).

[166] *J.W. Westhaver*, Ind. Eng. Chem. *34*, 126 (1942).

[167] *H. Stage, L. Gemmeker, G. Fischer, R. Richter*, Glas-Instrum.-Tech. *5*, 138, 204, 301, 347, 370 (1961).

[168] *W.G. Fischer*, Glas-Instrum.-Tech. *13*, 535 (1969), Chem.-Ztg. *94*, 157 (1970).

[169] Anonym, Chem.-Ing.-Tech. *42*, A 295 (1970).

[170] *G.B. Pegram, H.C. Vrey, J.R. Hoffman*, Phys. Rev. *49*, 883 (1936), J. Chem. Phys. *4*, 623 (1936).

[171] *E. Krell*, Chem. Tech. (Leipzig) *9*, 333 (1957).

[172] *F. Wolf, W. Böttger*, Chem. Tech. (Leipzig) *19*, 608 (1967).

[173] *J.H. Taylor*, Chem. Eng. *75*, 109 (1968).

[174] *W. Jost*, Angew. Chem. B *20*, 231 (1948), Dechema-Monographie *22*, 31 (1953), Chem.-Ing.-Tech. *25*, 291 (1953); *25*, 356 (1953).

retical plates per meter of column may be reached. The separation efficiency depends both on the rotation velocity and on the distillation rate[159, 174, 177, 178]. However, according to Jantzen and Wieckhorst[179], a turning roll column is not always better than a ring slot column (turning roll column with fixed rotor).

In most turning roll columns, the outer and inner pipes must fit together precisely; consequently, the manufacturing costs are high and the set-up is very delicate. These disadvantages are avoided with the *spinning band columns,* which are also frequently used in the laboratory and which contain a heat resistant Teflon tape as rotating element[181–183, 185].

This flexible tape may smooth over small unevennesses in the wall of the tube; flat tapes, spirals or rotors with star-shaped cross-sections have also been described. As with the columns with rotating internal cylinders, the efficiency increases with the speed of rotation[183]. An advantage is the small volume of the column which makes it especially useful for micro or semimicro distillations. Plate numbers of 30 to 100 theoretical plates per meter of column length have been quoted[175, 181, 184]. It is recommended that the reflux ratio should not be smaller than the number of theoretical plates; the distillation rate with this type of column is, therefore, relatively small. For vacuum distillation, a considerably higher speed of rotation is necessary since the higher vapor velocities in vacuo require a correspondingly higher distribution energy for the liquid. Apparatuses of lengths up to one meter are commercially available.

Horizontal columns are also used for distillations on the micro scale. If very small amounts are distilled in normal columns, there is not sufficient material to guarantee reflux. Moreover, in these cases the heat loss through the walls of the columns is large compared with the small amount of heat being exchanged in the column[186]. In horizontal distillations, the countercurrent principle is achieved by passing the vapors over a thin liquid film which covers the surface of the column. This principle is especially useful for compounds which are practically non-volatile at room temperature[180]. An external temperature gradient is applied to avoid heat transfer from the vapors. Backmixing against the temperature gradient is prevented by a stream of inert gas. By moving the temperature gradient along the column, high efficiencies may be reached. Variations of this method have also been recommended for packed columns[187–189].

1.2.3 Special Applications

The equilibrium curves of binary mixtures (see Fig. 1) may be influenced by various factors. This is especially important not only for mixtures with narrow boiling ranges, but also for azeotropic mixtures — whose separation possibilities are limited due to the intersection of the equilibrium curve with the 45° line[12, 13, 43, 190, 404]. The easiest means of changing or eliminating the azeotropic point is to alter the pressure. Generally, a reduction in pressure results in a shift of the azeotropic composition in favor of the compound with the lower heat of vaporization[12, 13, 190, 191].

For example, the water/ethanol mixture no longer has an azeotropic point below 70 torr. Pressure reduction is often used for technical purposes. Binary mixtures require two rectification columns operating under different pressures. However, the desired pressure reduction (which also results in a lowering of the temperature in the residue flask) has a limit at which the cooling medium available is no longer able to condense the vapors.

The partial pressures of components in a binary mixture may also be affected by adding salts *(salt effect).* For instance, the addition of 10 g of calcium chloride to 100 ml of the water/ethanol azeotropic mixture eliminates the azeotropic point[13, 192]. For many other azeotropic mixtures,

[175] *C.B. Willingham, V.A. Sedlak, F.D. Rossini, J.V. Westhaver,* Ind. Eng. Chem. *39,* 706 (1947).

[176] *W. Machard, K.J. Matterson,* Chem. Eng. Sci. *10,* 254 (1959).

[177] *N.J. Gelperin, M.S. Chatzenko,* Zh. Prikl. Khim. *25,* 610 (1952).

[178] *V.R. Ruchinsky,* Khim. i neft [USSR] Nr. 3, 20 (1967).

[179] *E. Jantzen, O. Wieckhorst,* Chem.-Ing.-Tech. *26,* 392 (1954).

[180] *E. Jantzen, H. Witgert,* Fette u. Seifen *46,* 563 (1939).

[181] *A.G. Nerheim, R.A. Dinerstein,* Anal. Chem. *28,* 1029 (1956), *29,* 1546 (1957).

[182] *R.G. Nester,* Anal. Chem. *28,* 278 (1965).

[183] *K.E. Murray,* J. Am. Oil Chem. Soc. *28,* 235, 461 (1951).

[184] *A.L. Irlin, B.P. Brun,* J. Anal. Chim. *5,* 44 (1950).

[185] Anonym, Chem.-Ztg. *93,* 194 (1969).

[186] *W. Kuhn, P. Baertschi, M. Thürkauf,* Chimia *8,* 109, 145 (1954).

[187] *B. Hausdörfer,* Chem.-Ing.-Tech. *40,* 370 (1968).

[188] *G. von Elbe, B.B. Scott,* Ind. Eng. Chem. Anal. *10,* 284 (1938).

[189] *M. Blumer,* Anal. Chem. *34,* 704 (1962).

[190] *L.H. Horsley,* Ind. Eng. Chem. Anal. *19,* 508, 602 (1947); *21,* 831 (1949).

[191] *J. Joffe,* Ind. Eng. Chem. *47,* 2533 (1955).

[192] *W. Jost,* Chem.-Ing.-Tech. *23,* 64 (1951), Z. Naturforsch. *1,* 576 (1946).

or mixtures with narrow boiling ranges, corresponding salt additives have been found experimentally[14, 193-197]. However, in continuous distillations (which are generally used on the large scale) the relatively high salt concentrations may cause difficulties (crystallization in the column, corrosion, etc.).

The best known *distillation processes with auxiliary compounds* – besides steam distillation (see Section 2.1:1.2.1) – are azeotropic distillation and extractive distillation. In these methods, the ratio of the activity coefficients of the two components to be separated is changed by adding liquid auxiliary compounds (theory, see[13, 14, 42, 198-201]).

The term *azeotropic distillation* is not always used correctly. Strictly, a distillation is azeotropic if a mixture with a narrow boiling range is split into azeotropes of the individual components having larger boiling point differences, on addition of an auxiliary compound.

Thus the indene/diphenyl mixture (boiling point difference 0.6 °C) on addition of diethylene glycol gives two azeotropes which can easily be separated: one consists of diphenyl/diethylene glycol (b.p. 230.4°), the other of indene/diethylene glycol (b.p. 242.6°). The lower boiling azeotrope separates in the receiver; diethylene glycol is taken off and continuously recycled to the top of the column[202].

The addition of an auxiliary compound to a binary mixture may also result in the formation of an easily separated ternary mixture (e.g. azeotropic drying of a water/ethanol mixture by addition of benzene, followed by distillation).

A different kind of azeotropic distillation is involved in the drying of n-butanol, in which one of the compounds (n-butanol) serves as carrier for the other. In the receiver, the distillate separates into a water-containing butanol layer, which is recycled to the column, and a butanol-containing water phase. This method of azeotropic distillation may be used to advantage if the azeotropic composition lies within the range of partial miscibility of the two components[203]. Both components may be obtained pure if two continuously operating columns are used[12] (cf. Fig. 8).

Suitable auxiliary compounds and separation methods have been described for many mixtures of organic compounds[2, 13, 42]. Moreover, useful rules have been established for choice of suitable additives[190, 200, 204-207].

Azeotropic distillation heads are commercially available in which the layers from the condensate are separated, and the layer which is rich in the auxiliary compound is returned to the column[13, 80, 213, 399].

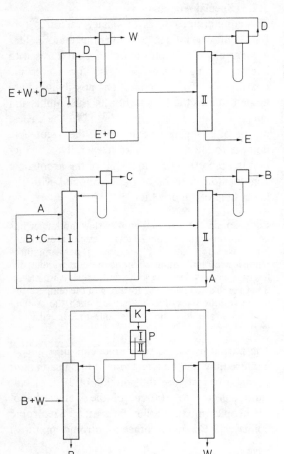

Fig. 8. Principles of azeotropic and extractive distillation

Azeotropic separation of glacial acetic acid (E) and water (W) by means of dichloroethane (D)

Extractive separation of benzene (B) and cyclohexane (C), using aniline (A)

Two-column apparatus for separation of n-butanol (B) and water (W) by means of (B):

K = condenser	I = butanol-rich phase
P = phase separator	II = water-rich phase

[193] *R.R. Tursi, R. Thompson*, Chem. Eng. Progr. *47*, 304 (1951).
[194] *L. Garwin, K.E. Hutchison*, Ind. Eng. Chem. *42*, 727 (1950).
[195] *L. Belck*, Chem.-Ing.-Tech. *23*, 90 (1951).
[196] *A. Guyer, B.K. Johnsen*, Helv. Chim. Acta *38*, 946 (1955).
[197] *W.F. Furter, R.A. Cook*, Heat and Mass Transfer *10*, 23 (1967).
[198] *E.J. Hoffman*, Aceotropic and Extractive Distillation, Interscience Publishers, New York 1964.
[199] *G. Kortüm*, Chem.-Ztg. *74*, 151 (1950).
[200] *R.H. Ewell, J.M. Harrison, L. Berg*, Ind. Eng. Chem. *36*, 871 (1944).
[201] *K. Nakanishi*, Ind. Eng. Chem., Fundamentals *7*, 381 (1968).
[202] *H.G. Franck*, Angew. Chem. *63*, 260 (1951).
[203] *F. Neumann*, Erdöl Kohle *4*, 561 (1951).
[204] *B.J. Mair*, J. Res. Bur. Stand. *27*, 39 (1941).
[205] *O. Nagel, R. Sinn*, Chem.-Ing.-Tech. *39*, 275, 671 (1967).
[206] *R.F. Marschner, W.P. Cropper*, Ind. Eng. Chem. *38*, 262 (1946).
[207] *K.J. Nowikowa, A.G. Natradse*, Chem. Ind. [USSR] *2*, 102 (1958).

Extractive distillation (Distex process) is another method for the separation of mixtures having narrow boiling ranges or azeotropic points. In this type of distillation, it is not necessary for the auxiliary compound to form an azeotrope with one or both of the components of the binary mixture. It may also have a boiling point which is much higher than those of the components of the binary mixture.

Generally, the auxiliary compounds change the relative volatility [13, 14, 192, 205, 208–212]. The best known example is the separation of benzene and cyclohexane in which the azeotropic point can be eliminated by adding aniline as extracting agent [14, 192]. Mixtures are also known in which a minimum azeotrope is changed into a maximum azeotrope by addition of an auxiliary compound. In azeotropic distillation the mixture to be separated and the auxiliary compound are fed simultaneously. In contrast, in extractive distillation it is advantageous to feed in the auxiliary compound at a higher plate than the mixture of components to be separated. In this case, the whole length of the column between the inlet of the mixture and the inlet of the auxiliary compound is available as extraction zone (cf. Fig. 8).

1.2.4 Modern Accessories for Distillation

The distillation head of one-stage distillation is derived from the equipment used in Claisen distillation; the progress during recent years involves mainly details of construction: wide connections to avoid damming-up of liquid caused by the high vapor velocities in the neck of the flask, introduction of new conical joints with fire-polished surfaces which do not require lubricants, introduction of Teflon connections (which allow the construction of flexible apparatus), and the use of Teflon-lined vacuum fittings.

In contrast, for rectification a number of different distillation heads have been described which are designed for optimum accuracy in keeping the reflux ratio constant. In some cases, they allow the reflux ratio and amount of condensation (or dephlegmatisation) of the ascending vapor mixture to be measured. A review and critical discussion have already been published by Stage et al. [213]. Two basically different types are used: the still

head for liquid take-off and the still head for vapor take-off.

In the *still head for liquid take-off*, all of the vapor arriving at the top of the column is condensed; part of it is separated as distillate, and the rest recycled as reflux to the column. In the *still heads for vapor take-off*, only a part of the vapor is separately condensed. Still heads for liquid take-off have a certain minimum liquid hold-up, but their construction and use are simple. Still heads for vapor take-off have a much smaller dead volume. They are especially suitable for micro-distillations and low temperature rectifications. They are less suitable for a high distillation rate, because of the large vapor volumes.

Dephlegmators operate differently. Only sufficient vapor is condensed for reflux; the rest is condensed separately in a descending cooler. The vapor leaving the reflux condenser has a composition different from that entering this condenser (partial condensation). Dephlegmators offer the advantage that the vapor does not come into contact with valves so that dead spaces and contamination by lubricants are eliminated.

Numerous distillation heads for various types of special distillations have been described and reviewed [213], e.g., heads for azeotropic distillations with recycling of one or both heterogenous phases, for low temperature rectification, for distillation of compounds which are high melting or which sublime, and others. Many of these commercially available devices may be used in combination with each other.

Automatic operation of distillation heads is normally achieved by use of magnetically operated valves, swivel funnels, etc. Correct operation is controlled by separate 'head controllers', various models being commercially available [13, 14, 216]. Normally, their functions are: provision of electrical impulses to the magnet controlling the reflux ratio (constant impulse periods and variable pause times, or variable impulse and pause times), control of the heating of the distillation flask, and disconnection if programmed temperatures are reached either at the head or in the residue flask.

All the theoretical principles of distillation or rectification discussed so far are valid only for *adia-*

[208] *E.G. Scheibel*, Chem. Eng. Progr. *44*, 927 (1948).

[209] *W.B. Kogan*, Zh. Fiz. Khim. *29*, 1470 (1953).

[210] *W.W. Kafarow, L.A. Gordijewski*, Zh. Prikl. Khim. (Leningrad) *29*, 176 (1956).

[211] *F.H. Garner, S.R.M. Ellis*, J. Inst. Petrol, London *42*, 148 (1956).

[212] *J.A. Gerster*, Petr. Refiner *39*, Nr. 4, 146 (1960).

[213] *H. Stage, L. Gemmeker*, Glas-Instrum.-Tech. *7*, 18, 606 (1963); *6*, 69, 114, 169, 206, 372, 455, 494, (1962).

[214] *A.T. Williamson*, Proc. Roy. Soc. *A* 195, 97 (1948).

[215] *J.R. Bowman, K. Yu, J. Coull*, J. Inst. Petrol, London 35, 311 (1949).

[216] *G. Löttel*, Zum Stand der Regelung von Destillationskolonnen, Techn. Information, VEB Geräte- und Reglerwerk Teltow, 1968.

batic operation. Strictly adiabatic operation is, however, difficult to achieve. For instance, the energy transfer processes on the exchange plates (interplay between vaporization and condensation) are already affected by heat radiation from the column (concerning *isothermal distillation,* see[214, 215]). If the external temperature is lower than the temperature on the plate in a column which is not thermally insulated, then the proportion of rising vapor condensed is greater than that corresponding to the phase equilibrium *(wild reflux).* If the external temperature is too high, insufficient condensate will be formed. Both effects lower the efficiency of the column. This is a particular disadvantage with micro- and vacuum-distillations.

To secure adiabatic conditions, each section of the column must be supplied with as much heat as is necessary to compensate for the radiation losses; normal insulation is generally inadequate[10, 13, 131, 137, 217–220, 229]. The (often recommended) heating jackets which are heated by heating media cannot be regarded as optimal devices. Internally silver-plated vacuum jackets have proven much better. They allow almost adiabatic operation up to temperatures exceeding 130 °C[137, 146, 220–224]. The pressure in the jacket ranges from 10^{-4} down to 10^{-6} torr. Electrically heated compensation jackets have proven to be the best solution; individual sections of the column are heated electrically, heating being controlled by internal thermocouples. The smaller the individual sections, the more accurate is the compensation for heat losses[13, 131, 136, 218, 223, 224, 227, 402]. According to the literature, the standard length of these sections should be 10 to 50 cm.

The trend in modern laboratory distillation is towards fully automated operation (Fig. 9). Such equipment may be classified according to the mode of operation-batch-wise or continuous – or the degree of automation[228, 229]. The completely automated apparatus shown is suitable for a capacity of 250 to 3,000 ml and for the determination of 'true boiling point curves', which are necessary for control in technical distillation towers.

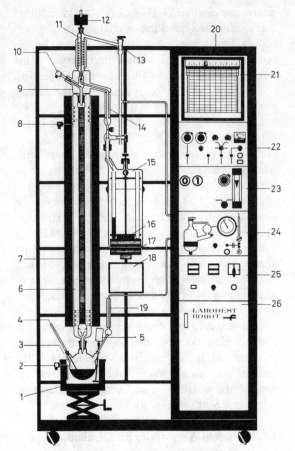

Fig. 9. Diagram of a fully-automatic distillation apparatus (ex Fischer Labor- und Verfahrenstechnik, Bad Godesberg)

1 heating device	11, 12 magnet
2, 4, 8, 10 thermocouples	14 cooler for
3 vessel	distillate
5 stirrer	15–18 fraction sampler
6 column	19 cooler
7 heating jacket	20–26 control devices
9 still head with condenser	

2 Centrifuging

2.1 Physical Principles, Technical Properties and Limitations

Centrifuging is one of the most important methods for solid/liquid separation (for solid/gas or liquid/gas separations in conifuges, see Section 2.1:2.2.2). Whereas in normal sedimentation the only driving force is the earth's gravitational field, in centrifuges considerably higher forces may be

[217] *C. Junge,* Chem. Tech. (Leipzig) *6,* 37 (1954).

[218] *H. Stage,* Angew. Chem. *B 19,* 219 (1947).

[219] *H. Brandt,* Chem.-Ing.-Tech. *29,* 86 (1957).

[220] *H. Kolling, H. Tramm,* Chem.-Ing.-Tech. *21,* 9 (1949).

[221] *U. von Weber,* Chem. Tech. (Leipzig) *1,* 159 (1949).

[222] *L. Zampachova,* Chem. Prum. *7/32,* 408 (1957).

[223] *P.E. Weston,* Ind. Eng. Chem. Anal. *5,* 179 (1933).

[224] *W.J. Podbielniak,* Ind. Eng. Chem. Anal. *5,* 119 (1933).

[225] *A. Neuhaus,* Chimia *18,* 93 (1964), Jahrbuch des Landes NRW, p. 487, Westdeutscher Verlag, Köln/Opladen 1965.

[226] *J. Wiegand,* Chem.-Ztg. *93,* 939 (1969).

[227] *K.D. Uitti,* Petr. Refiner *29,* 130 (1950).

[228] *N. Huppes, J.J. de Jong,* Chem. Anal. *1959,* 436.

[229] *W.G. Fischer,* Glas-Instrum.-Tech. *14,* 23 (1970).

used for separation. Centrifuging is generally defined as sedimentation in an artificial gravitational field; the greater forces reduce considerably the time required for separation.

A solid particle with the distance r from the center z having the mass m and moving in a circular path is influenced by the gravitation, the friction of motion, the buoyancy, and the centrifugal force P_z:

$$P_z = m \cdot r \cdot w^2 \qquad (1)$$

w being the angular velocity $d\alpha/dt$. Normally, w is expressed by the peripheral velocity $v = w \ r$ and w is substituted by the number of revolutions n. This leads to

$$w = \frac{2\pi \cdot n}{60}$$

Combined with Equation (1) the following equation results:

$$P_z = m \cdot r \cdot \frac{v^2}{r^2} = \frac{m \cdot v^2}{r} \qquad (2)$$

Of course, P_z can serve for comparison of various centrifuges only if m, v, and r are separately determined or known. Therefore the *centrifuge coefficient* Z (centrifuge number), giving the ratio of P_z and the weight G, was defined as a universal applicable relation.

$$Z = \frac{P_z}{G} \qquad (3)$$

With $G = m \cdot g$ (g = acceleration of gravity), Z can be expressed as:

$$Z = \frac{r \cdot w^2}{g} = \frac{r \cdot 4\pi^2 \cdot n^2}{g \cdot 3600} \qquad (4)$$

and with $\pi^2 \approx g$

$$Z = \frac{r \cdot n^2}{900} \qquad (5)$$

Z gives not only the ratio of P_z/G but also the ratio of the centrifugal force and the acceleration of gravity because, according to Newton, with $P = m \cdot b$ (P = force and b = acceleration), $r \cdot w^2$ is the centrifugal acceleration and, according to Equation (4), Z may also be given as $r \cdot w^2/g$. With ultracentrifuges the maximum realizeable artificial field of gravity is said to be 100,000 g.

According to Equation (5) the ratio of acceleration Z is determined by the second power of the number of cycles n and by the radius r. This means that the separation efficiency and the separation time required depend much more on the number of cycles than on the radius. Consequently the trend in the construction of modern centrifuges is in the direction of higher numbers of cycles, eventually with reduction of the diameter.

Analogous to the fundamentals of sedimentation[23, 230-232], the sedimentation velocity U_r under the influence of the centrifugal force is given by

$$U_r = w \cdot \sqrt{\frac{4}{3 c_w} \frac{D_F - D_L}{D_L} \cdot r \cdot d} \qquad (6)$$

in which D_F and D_L mean the density of the solid and of the surrounding liquid resp. and C_W the resistance coefficient. In the range of small Reynolds numbers (Re < 2, laminar flow) the law of Stokes is valid. According to this law C_W equals 24/Re (v_L being the kinematic viscosity of the liquid):

$$U_r = \frac{d^2}{18 \cdot v_L} \cdot \frac{D_F - D_L}{D_L} \cdot r \cdot w^2 \qquad (7)$$

The ratio of the radial sedimentation velocity U_r and the stationary sedimentation velocity U_s is given by

$$\frac{U_r}{U_s} = \frac{w \sqrt{\dfrac{r}{c_{wr}}}}{\sqrt{\dfrac{g}{c_{ws}}}} \qquad (8)$$

Proportionality exists between the sedimentation velocity and the coefficient Z of a centrifuge. Since the centrifugal force increases with increasing density $(P = m \cdot b)$, liquids of different density can be separated.

Laboratory and technical centrifuges have moderate revolution rates (Z up to 50,000) and are used mainly for the separation of solids. In contrast, ultracentrifuges (Z up to 10^6 and more) are used also for determinations of molecular weights and sedimentation constants[233-236].

From Equation (1) combined with density $d = m/volume$ V follows:

$$P_z = V \cdot D \cdot r \cdot w^2$$

and

$$P_z = V \cdot (D_F - D_L) \cdot r \cdot w^2 \qquad (9)$$

For a great number of particles, the sum of which corresponds to Avogadro's constant k, m is equal to $k \cdot M$, M being the molecular weight. This leads to

$$P_z = \frac{M}{D_F} (D_F - D_L) \cdot r \cdot w^2$$

and

$$P_z = M \cdot \left(1 - \frac{D_L}{D_F} \right) \cdot r \cdot w^2$$

By defining $1/D_F$ as partial specific volume V, equation (10) follows:

$$P_z = M \cdot (1 - V \cdot D_L) \cdot r \cdot w^2 \qquad (10)$$

The friction force, $f \cdot (dr/dt)$, has a direction opposite to the centrifugal force P_z. In a stationary state both forces are equal

$$M \cdot (1 - V \cdot D_L) = f \cdot \frac{dr/dt}{w^2 \cdot r}$$

For dilute solutions (<0.5%), which are normally used in ultracentrifuges, $f = R \cdot T/D$. Therefore, the Svedberg equation holds:

$$M = \frac{RT}{D(1 - V \cdot D_L)} \cdot \frac{dr/dt}{w^2 \cdot r} \equiv \frac{R \cdot T \cdot S}{D \cdot (1 - V \cdot D_L)} \qquad (11)$$

R is the universal gas constant, T the absolute temperature, and D the diffusion constant.

This equation is valid if sedimentation is much faster than diffusion. This is normally the case with ultracentrifuges. D may be determined separately. Therefore, the sedimentation constant S allows the molecular weights of compounds which are dissolved in the centrifuge liquid (density D_L) to be found by determination of the factor

$$\frac{dr/dt}{w^2 \cdot r}$$

In practice r (distance between the solvent/solution boundary layer and the rotation center) is determined for appropriate experimental times t. The molecular weight is then calculated using the known values of D and T. Optical methods can be used for observation of the solvent/solution boundary layer[233, 234]. The ultracentrifuges must be constructed so that the rotation speed and the temperature can be kept constant if necessary during several days.

Since temperature constancy is difficult to achieve due to the heat of friction of the rotating sample and the surrounding gas molecules, the rotor of ultracentrifuges is operated normally under hydrogen atmosphere or in vacuum[237]. The special problem in the construction of all centrifuges is the avoidance of critical rotational velocities in which the amplitude of vibration of the rotating system corresponds to the number of revolutions. In centrifuges with firmly fixed axes, the torque of the centrifugal force will be small only if the axis of rotation is both the axis of the center of gravity and the axis of inertia. If this is not the case the centrifugal moment will be compensated by a moment of compulsion (moment of bearing) which may destroy the bearings of the axis. Therefore, centrifuges with firmly fixed axes are operated only below such critical rotational speeds. Equal distribution of the mass around the axis of rotation is necessary for unimpeded operation.

Centrifuges with high rotational speeds have no firmly fixed bearings of the axes. Either they are suspended swinging or they have an elastic connection between the rotor and the motor. In this case centrifugal moments which may arise by unsymmetrical distribution of masses (unbalance) must not be compensated by moments of bearing. The alternation between precession and nutation leads to an automatic stabilization of the gyro system. For this reason the region of critical cycles is passed as quickly as possible in order to reach the supercritical and stable range.

Due to the high centrifugal forces which arise in ultracentrifuges, the rotor must be constructed of materials with high tensile strengths (steel, nickel, titanium, or various aluminum alloys). For the same reason the rotors are made mostly from massive material.

Normally centrifuges are used for separation of solids from liquids. Moreover heavy liquids can be separated from light liquids, or solids from gases. In the *separation of solids* from liquids a distinction is usually made between filterable and non-filterable solids[238]. Whereas filterable solids are charged into *screen centrifuges* in which the centrifuge basket is perforated and covered by a filter medium, non-filterable solids are separated in *imperforate bowl centrifuges* (Table 1).

Liquids with density differences <1% can normally be separated from each other only by using special precautions, e.g. devices in the centrifuge basket which prevent the formation of eddies and the running ahead of a liquid phase (*disc centrifuges*). Solid or liquid particles suspended in gases can be separated in *conifuges*[239-241].

[230] *H. Madel* in *A. Eucken/M. Jacob*, Der Chemie-Ingenieur, Bd. I/2, p. 308, Akademische Verlagsgesellschaft, Leipzig 1933.

[231] *H. Robel, V. Clauss, C. Küchler, K. Luckert*, Chem.-Tech. *20*, 529 (1968).

[232] *F.W. Schneider*, Maschinenmarkt *73*, 1733 (1967), *74*, 234 (1968).

[233] *E. Wiedemann* in *Ullmann*, Encyklopädie der technischen Chemie, Bd. II/1, p. 808, Verlag Urban & Schwarzenberg, München 1961.

[234] *H.K. Schachman*, Ultracentrifugation in Biochemistry, Academic Press, New York, London 1959.

[235] *T. Svedberg, K.O. Pedersen*, Die Ultrazentrifuge, Theodor Steinkopff Verlag, Dresden, Leipzig 1940.

[236] *E. Wiedemann*, Dechema-Monographie *26*, 333 (1956).

[237] *E. Rödel*, Chem. Tech. (Leipzig) *20*, 630 (1968).

[238] *H. Mießner* in *Ullmann*, Encyklopädie der technischen Chemie, Bd. I, p. 510, Verlag Urban & Schwarzenberg, München 1951.

[239] *W. Stöber, U. Zessack*, Z. biol. Aerosolforsch. *13*, 263 (1966); Staub, Reinhalt. Luft *29*, 186 (1969).

[240] *K.F. Sawyer, W.H. Walton*, J. Sci. Instrum. *27*, 272 (1950).

[241] *H. Hauck, J.A. Schedling*, Staub, Reinhalt. Luft *28*, 18 (1968).

Table 1. Data for different centrifuge designs

	Centrifuge coefficient Z	Max. performance slurry m^3/h	product t/h	Centrifugate	Sediment
Imperforate bowl centrifuges					
ultracentrifuges	$< 10^6$			clear	pasty-solid
bottle-centrifuges	$< 6 \cdot 10^3$			clear	pasty-solid
tubular-centrifuges	$< 5 \cdot 10^4$	< 3		clear	pasty-solid
chamber-type centrifuges	$< 6 \cdot 10^3$	< 4		clear	pasty-solid
disc-centrifuges	$< 12 \cdot 10^3$	< 20		clear	pasty-solid
scroll-centrifuges			< 20	turbid	granular-dry
Screen centrifuges					
horiz. batch basket-centrif.	$< 2 \cdot 10^3$		< 35	turbid-clear	dry
pusher-centrifuges	$< 2 \cdot 10^3$		< 25	turbid-clear	dry
scroll-centrifuges	200–1,600		< 60	turbid-clear	dry

According to Miessner[238], the various centrifuge designs may be classified as given in Table 1 [232, 242-244, 273, 274].

2.2 Applications and Design

2.2.1 Centrifuges for Preparative Organic and Biochemical Work

For simple organic preparative or analytical laboratory work, *test-tube centrifuges* with freely-swinging containers have proven to be efficient[245, 246]. Their disadvantage resides in the fact that the value of Z is not constant throughout the whole contents of a tube so there is differential sedimentation. More uniform sedimentation is achieved with *angle centrifuges*[246, 247] in which the containers are fixed at angles between 15° to 45° with respect to the rotational axis. Rotation speeds in the supercritical range, and fields of gravity up to 35,000 g, may be achieved by using massive rotors which are also standard equipment in ultracentrifuges. Angle centrifuges whose rpm and gravity field ranges equal those of the ultracentrifuges, but which are merely used for preparative or

analytical purposes, are commercially available as *supercentrifuges* or *fast preparative centrifuges*. As in ultracentrifuges the rotors of supercentrifuges are normally operated under vacuum or in hydrogen at reduced pressure.

Because of the high gravitational fields attained, (centrifuge coefficients up to 10^6) ultracentrifuges must meet special requirements[233-236, 246-252].

The main constructive features are:
Rotors made of materials with high tensive strengths (special alloys of steel, titanium, nickel, or aluminum). Massive rotors with an efficient radius of 6.5 cm (standard design). Normally a balance cell for better balance is contained. The balance cell contains zero marks for the control of sedimentation. Preparative supercentrifuges have fixed angle rotors with angles of 20 up to 30 degrees.
Possibility of the installation of stabilizers allowing a damping of the motion of rapture which arises at low values of rpm. These motions which are typical for selfstabilizing or supported rotors may damage the rotor and cause a whirling of the sediment. Moreover, the optical determination methods require a smooth run of the rotor and rotational speeds as constant as possible.
Constant temperature up to a maximum of 0.2° deviation to avoid any thermal convection of the centrifugate. For special problems the rotor shell must be cooled.
In multipurpose centrifuges which may be used either as preparative supercentrifuges or as ultracentrifuges, the rotor normally can be exchanged. Rotors are used

[242] *R.D. Kearney*, Chem. Eng. Progr. *63*, Nr. 12, 72 (1967).

[243] *H.E. Schultze, E.J.T. Wiedemann*, in *Houben-Weyl*, Methoden der organischen Chemie, Bd. I/1, p. 626, Georg Thieme Verlag, Stuttgart 1958.

[244] *A. Fierstine*, Chem. Eng. Progr. *63*, Nr. 10, 115 (1967).

[245] J. Chem. Educ. *45*, 744 (1968).

[246] *E.G. Pickels*, J. Gen. Physiol. *26*, 341 (1947); Electr. Manufg. *45*, 66 (1950); Machine Design *22*, 102 (1950); Chem. Rev. *30*, 341 (1942).

[247] *H.B. Golding* in *A. Weissberger*, Technique of Organic Chemistry, Vol. III, Interscience Publishers, New York 1951.

[248] *J.B. Nichols, E.D. Bailey* in *A. Weissberger*, Physical Methods of Organic Chemistry, Vol. I/1, p. 621., Interscience Publishers, New York 1949.

[249] *K.O. Pedersen*, J. Phys. Chem. *62*, 1282 (1958).

[250] *E.G. Pickels* in *F.M. Huber*, Biophysical Research Methods, Interscience Publ., New York 1950.

[251] *G. Meyerhoff*, Makromol. Chem. *15*, 68 (1955).

[252] *H. Metzner*, Naturwissenschaften *40*, 388 (1953).

with different maximum values of rpm containing a safety system for occasional superspeeding.

Devices should be included for optical determination and sedimentation control. Common are refractometers, interferometers, or light absorption methods[233-235, 253-257].

Devices are usual for taking samples out of preparative ultracentrifuges without stirring up the sediment[258, 259] and for adjusting the density gradient[234, 251, 253, 260-262]. An efficient break system is necessary to avoid a whirling of the sediment or the density gradient.

Possibility of using test-tubes for smaller amounts of compounds and for special purposes[263-265].

Possibility of changing the test-tubes for various problems. Usually the test-tubes are of polypropylene, polycarbonate, polyamide, cellulose, borosilicate glass, or steel. Furthermore zonal rotors with corrosion resistant installations are available.

Devices for filling and taking samples from zonal rotors during operation. With modern integrators the determination of the integral $\int w^2 dt$ is possible, even for the start-up period and low values of rpm[237].

Preparative ultracentrifuges have made possible the separation of various types of *virus, bacteriophages*, etc. Analytical ultracentrifuges permit the determination of *sedimentation constants* and *molecular weights* in the range from about 20,000 to 3×10^6. This range can now be extended to molecular weights below 1,000 using the density gradient method. *Polymers, mineral oils, fats*, and other technical products are now also open to investigation. Ultracentrifuges have become especially important in biochemistry[243] (see also Sec-

tions 2.7; 9; 12.1; 12.2). In medicine, the determination of *serum proteins* by means of ultracentrifuges has become a standard method.

2.2.2 Semi-Technical and Technical Plants

The transition from laboratory operation to the semi-technical or technical scale presents additional requirements which centrifuges must fulfil:

continuous operation
capacity increase, by some powers of ten
special devices for slurry feed, filtrate, and sediment.

Tubular centrifuges are intermediate between laboratory and technical centrifuges both as regards their rotation speeds (up to 50,000 rpm, corresponding to about 62,000 g) and their maximum capacity (up to 5,000 l/h).

Disc centrifuges and other types of centrifuges (Table 1 and 2) are also used in semi-technical or technical operations[233, 236, 238, 242, 244, 267-274].

3 Crystallization

Crystallization – defined as separation of solids from a solution, a melt, or a gas phase – is normally preceded by supersaturation resulting from cooling or concentration. Separation of the individual components of a multi-component mixture may be achieved by fractional crystallization[69, 275-288, 381, 382, 387, 397].

[253] *K. Strohmaier*, Anal. Biochem. *15*, 109 (1966).

[254] *E. Wiedemann*, Chem.-Ing.-Tech. *28*, 263 (1956), Helv. Chim. Acta. *40*, 2074 (1957), *35*, 23 (1952).

[255] *J.W. Beams, N. Snidow, A. Robeson, H.M. Dixon*, Science *120*, 619 (1954), Rev. Sci. Instrum. *25*, 295 (1954).

[256] *L.G. Longsworth*, Anal. Chem. *23*, 346 (1951), J. Phys. Chem. *58*, 770 (1954).

[257] *V.N. Schumaker, H.K. Schachman*, Biochem. Biophys. Acta *23*, 628 (1957).

[258] *M.L. Randolph, J.R. Snavely, W.H. Goldwater*, Phys. Review. *74*, 118 (1948); Science *112*, 528 (1950).

[259] *G.H. Hogeboom, E.L. Kuff*, J. Biol. Chem. *210*, 733 (1954).

[260] *M. Christ, K. Strohmaier*, DBP 1.171.178 (1960/1964).

[261] *H.K. Schachman, M.A. Lauffer, W.F. Harrington, E.G. Pickels*, J. Am. Chem. Soc. *72*, 4266 (1950), Proc. Nat. Acad. US *38*, 943 (1952); J. Poly. Sci. *12*, 379 (1954).

[262] *C.S. Yannoni*, J. Am. Chem. Soc. *89*, 2833 (1967).

[263] *V. Neuhoff*, Glas-Instrum.-Tech. *13*, 86 (1969); Anal. Biochem. *23*, 359 (1968).

[264] *B. Gelotte, A. Emnéus*, Chem.-Ing.-Tech. *38*, 445 (1966).

[265] *H. Determann*, Chimia *23*, 94 (1969).

[266] *E. Hala, I. Wichterle, J. Pick, T. Boublik*, Vapor-Liquid Equilibrium Data at Normal Pressures, Pergamon Press, Oxford 1968.

[267] *W.J. Podbielniak, C.M. Doyle*, USP 2.004.001 (1935), 2.109.375 (1938), 2.281.796 (1942), 2.670.132 (1954), 2.758.784 (1956), Chem. Eng. Progr. *61*, Nr. 5, 69 (1965) and *64*, Nr. 12, 68 (1968), USP 3.107.218 (1963), 3.292.850 (1966), USP 3.114.706 (1963), 3.116.246 (1963), 3.132.100 (1964), USP 3.217.980 (1965), 3.254.832 (1966), 3.327.939 (1967).

[268] *R.E. Treybal*, Liquid Extraction, 2. Ed., p. 531, Mc Graw-Hill Book Co., New York 1963.

[269] *L. Alders*, Liquid-Liquid Extraction, p. 86, Elsevier, Publ. Co., Amsterdam 1955.

[270] *S. Kießkalt* in *Winnacker-Küchler*, Chemische Technologie, 2. Aufl., Bd. I/1, p. 111. C. Hanser-Verlag, München 1958.

[271] *A.E. Flowers* in *Perry*, Chemical Engineers Handbook, 2. Ed., McGraw-Hill Book Co., New York 1941.

[272] *K. Schneider*, Maschinenmarkt *74*, Nr. 14 234 (1968).

[273] *C. Gösling, G. Mundil*, Chem. Ind. *19*, 762 (1967).

[274] *G. Pahlitzsch, R. Richter*, Maschinenmarkt *74*, Nr. 10, 160 (1968); Chem.-Ing.-Tech. *37*, 848 (1965).

[275] *G. Matz*, Die Kristallisation in der Verfahrenstechnik, p. 40, Springer Verlag, Berlin, Göttingen, Heidelberg 1954; s.[298].

Table 2. Principle set-up and technical data for industrial centrifuges

Group	Imperforate bowl centrifuges	
Type	Chamber-type bowl centrifuge	Scroll centrifuge

Operation	discontinous	continous
Position of axes	vertical	horizontal
Bearing of axes	fixed	fixed
Centr. coefficient Z	up to 6,000	up to 2,000
Capacity	up to 4 m³/h	up to 10 t/h sediment
Conc. of precipitate	up to 1%	up to 30%
Centrifugate	clear	turbid
		granular, dry
Sediment	pasty-solid	(also available as screen centrifuges)
Remarks	1 feed inlet	1,2 drive of basket 8 and of scroll 4
	2 axis of motion	3 feed inlet
	3 sediment	4 scroll
	4 centrifugate outlet	5 sediment outlet
		6 centrifugate outlet
		7 overflow
		8 basket

[276] H.E. Buckley, Crystal Growth, John Wiley & Sons, New York 1951.

[277] A. van Hook, Crystallization, Theory and Practice, John Wiley & Sons, New York 1961.

[278] R.C. Reid, G.D. Botsaris, G. Margolis, D.J. Kirwan, E.G. Denk, G.S. Ersan, J. Tester, F. Wong, Ind. Eng. Chem. 62, 52 (1970).

[279] J.W. Mullin, Crystallization, Butterworths, London 1961; J.W. Mullin in Kirk-Othmer, Encyclopedia of Chemical Technology, 2nd Ed., Vol. VI, p. 482, Interscience Publ., John Wiley & Sons, New York 1965.

[280] N.N. Sirota, F.K. Gòrskii, V.M. Varikash, Crystallization Processes, New York Consultant Bur., 1966; Solid State Transformations, New York Consultant Bur., 1966.

[281] H. Mauser in Ullmann, Encyklopädie der technischen Chemie, Bd. II/1, p. 660., Verlag Urban & Schwarzenberg, München 1961.

[282] A. Lüttringhaus in Houben-Weyl, Methoden der organischen Chemie, Bd. I/1, p. 345, Georg Thieme Verlag, Stuttgart 1958.

[283] R.S. Tipson in A. Weissberger, Technique of Organic Chemistry, p. 395, Interscience Publishers, New York 1956.

[284] E. Hegelmann, C. Beck in Ullmann, Encyklopädie der technischen Chemie, Bd. I, p. 544, Verlag Urban & Schwarzenberg, München 1951.

[285] R. Klockmann in Ullmann[284], p. 687.

[286] H.S. Peiser, Crystal Growth, Pergamon Press, Oxford 1967.

[287] C. Kittel, Einführung in die Festkörperphysik, R. Oldenbourg Verlag, München 1968.

[288] K.T. Wilke, Methoden zur Kristallzüchtung, Verlag Harry Deutsch, Frankfurt/Main, Zürich 1963.

[289] G. Matz, Kristallisation — Grundlagen und Technik, 2. Aufl., Springer Verlag, Berlin, Heidelberg, New York 1969.

Table 2 (cont.)

Screen centrifuges		
Batch basket centrifuge	Screen centrifuge	Pusher centrifuge

| continuous
horizontal
fixed
up to 2,000
up to 35 t/h sediment
up to 40%
turbid to clear
dry
(also available as
imperforate bowl centrifuges) | discontinuous
vertical
mostly suspended swinging
available in various types

Mostly used for
separating mother liquors
or others | continuous
horizontal
fixed
up to 2,000
up to 25 t/h sediment
up to 90%
turbid to clear
dry
(also available as screen centrifuges) |
| 1 drive of basket 2
3,4 conveyor device
5 piston-rod
6 feed inlet
7 centrifugate outlet
8 sediment outlet | 1 axis of motion
2 feed inlet
3 sediment
4 centrifugate outlet
5 wash water | 1 axis of motion
2 piston-rod for pusher 3
4 wash water
5 feed inlet
6 centrifugate outlet
7 sediment outlet |

Crystallization is the best method for purification of very small amounts of crystallizable compounds, especially where structure determination is involved, using methods which form well developed crystals or crystalline powders (cf. Sections 2.2; 2.12; 5.3; 5.4; 7.1; 7.3; 7.4; 11.5; and 12.2). During recent years, materials or classes of compounds which were at one time regarded as non-crystallizable have been obtained crystalline; examples are *polymers, vitamins, proteins,* and *nucleic acids*[290-295, 352, 386, 397].

Undesired crystallization phenomena may occur especially when solvent diagrams having negative curves of solubility are involved (solubility of salts decreasing with increasing temperatures).

3.1 Physical Principles

3.1.1 Crystal Forms

Crystals are homogeneous anisotropic solid materials which are bounded by natural flat surfaces. They differ from amorphous substances in the regular order of their basic units *(space lattice)*. These units are fixed in equilibrium positions; their positions seldom change. In an ideal crystal, the elementary cell always repeats in the same man-

[290] *W. Gerlach,* Z. Angew. Chem. *42,* 693 (1929).
[291] *F. Sanger, E.D.P. Thompson,* Biochem. J. *53,* 353 (1953).
[292] *H. Tuppy,* Biochim. Biophys. Acta *11,* 449 (1953).
[293] *B.F.C. Clark,* Nature *219,* 1222 (1968).
[294] *B.F. Birdwell, F.W. Jessen,* Nature *209,* Nr. 5021 366 (1966).
[295] *F. Gornick, J.D. Hoffman,* Ind. Eng. Chem. *58,* 41 (1966).

ner; hence, the well-known regularity in the external appearance of crystals. The combination of different planes on a crystal is called the *habit*. Because of the anisotropic nature of crystals, their properties and behavior are direction-dependent within the crystal (crystals belonging to the cubic system are an exception).

Real crystals generally contain dislocations, empty sites in the space lattice, impurities, etc. The properties of crystals depend to a considerable extent on these discontinuities[296, 297].

3.1.2 Crystal Nuclei Formation

The presence of *crystal nuclei* is a prerequisite for crystal formation. Crystal growth results from concentration differences[282, 284, 298–306, 313, 390, 405]. Formation of nuclei and crystal growth are mutually independent, being temperature-dependent to different extents.

Nucleation may occur in two ways: homogeneous (or spontaneous) nucleation, which results from the assembly of dissolved molecules or ions, and heterogeneous nucleation, which is initiated by sub-microscopic impurities to which the dissolved particles become attached. If a solution (e.g. a freshly prepared, unfiltered solution) contains many impurity nuclei, preferred precipitation of small-grained crystals occurs[307–309]. Formation of crystals is also favored by nuclei at the walls; hence, scraping with a glass rod may initiate crystallization[307, 310]. Other means of initiating nucleation include: shaking, pressure, magnetic fields,

and ultrasonics[282, 311–314]. Nucleation (which is very difficult to determine experimentally) has been considered theoretically[299, 301, 307, 310, 315]. In the absence of impurities, considerable supersaturation of the solution may be observed.

3.1.3 Crystal Growth

Concentration differences favoring crystal growth can be obtained, both in the laboratory and on the technical scale, by cooling or evaporation, or a combination of both[297, 317–319, 321–323] (cf. Section 2.1:3.2). While the classical theory[301] interpreted crystal growth in terms of the formation of two-dimensional nuclei, the presence of spiral dislocations has also been regarded as responsible for crystal growth[324, 325]. The latter concept avoids the contradiction that crystal growth can be observed even when the formation of multidimensional nuclei would often require an energy of activation higher than that available[326, 327, 390]. According to Heyer[327], examples of both types of growth can be found independent of the kind of mother phase (melt, solution, or gas). Deviations of experimentally observed crystal growth from theory have been discussed in a number of recent papers[279, 321, 328–332]. Determination of crystallization rate, the effect of dissolved additives, and the relationship between the equilibrium form and the surface energy and crystallization kinetics have similarly been discussed[322, 323, 333–343].

[296] *H. Gülzow*, Kristall, Tech. *1*, 411 (1966).

[297] *G.I. Distler, B.B. Zvyagin*, Nature *212*, Nr. 5064 807 (1966).

[298] *D.R. Uhlmann, B. Chalmers*, Ind. Eng. Chem. *57*, Nr. 9, 19 (1965).

[299] *A.G. Walton*, Science *148*, Nr. 3670, 601 (1965).

[300] *S. Bakowski*, Brit. Chem. Eng. *14*, 1213 (1969).

[301] *M. Volmer*, Kinetik der Phasenbildung, Theodor Steinkopff Verlag, Dresden 1939.

[302] *W. Vogel*, Angew. Chem. *77*, 109 (1965).

[303] *A.C. Montefinale, H.M. Papée*, Z. Angew. Math. Phys. *16*, 740 (1965).

[304] *L.F. Evans*, Nature *206*, Nr. 4986, 822 (1965); *213*, Nr. 5074, 384 (1967).

[305] *T.P. Melia*, J. Appl. Chem. *15*, 345 (1965).

[306] *R.A. Kuntze*, Nature *211*, Nr. 5047, 406 (1966).

[307] *A.E. Nielsen*, Kinetics of Precipitation, Pergamon Press, Oxford 1964; Acta Chem. Scand. *11*, 1512 (1957).

[308] *L. Erdey, E. Gere, L. Polos*, Acta Chim. Acad. Sci. Hung. *26*, 43 (1961).

[309] *E.J. Bogon, H.V. Moyer*, Anal. Chem. *28*, 473 (1956).

[310] *M. Kahlweit*, Z. Phys. Chem. (N.F.) *25*, 1 (1960).

[311] *P. Günther, W. Zeil*, Z. Anorg. Chem. *285*, 191 (1956).

[312] *L. Bergmann*, Der Ultraschall, 6. Aufl., Verlag S. Hirzel, Stuttgart 1954.

[313] *A.N. Kirgincev, M. Sokolov*, Kolloid Zh. *27*, 697 (1965).

[314] *G. Schmid, A. Jetter*, Ber. Bunsenges. Phys. Chem. *56*, 760 (1952).

[315] *R. Becker, W. Döring*, Ann. Phys. *24*, 719 (1935).

[316] *V. Brezina*, Glas-Instrum.-Tech. *14*, 493 (1970).

[317] *T. Messing*, Chem.-Ing.-Tech. *40*, 793 (1968): Chem.-Ztg. *91*, 963 (1967).

[318] *G. Matz*, Verfahrenstechnik *3*, 191 (1969); Wärme *75*, 1 (1969).

[319] *A.R. Allnatt, A.V. Chadwick*, Chem. Rev. *67*, 681 (1967).

[320] *W.L. Bolles*, Ind. Eng. Chem. *62*, 81 (1970).

[321] *L.O. Meleshko*, Sov. Phys.-Cryst. *10*, 575 (1966).

[322] *S. Nielsen*, Sci. J. *1*, Nr. 10, 59 (1966).

[323] *L. Graf, L. Scheiner*, Z. Metallkunde *58*, 271 (1967).

[324] *W.K. Burton, N. Cabrera, F.C. Frank*, Nature *163*, 398 (1949).

[325] *F.C. Frank*, Discuss. Faraday Soc. *5*, 48 (1949).

[326] *H. Fischer*, Angew. Chem. *81*, 101 (1969).

[327] *H. Heyer*, Angew. Chem. *78*, 130 (1966).

[328] *G.C. Trigunayat*, Z. Kristallogr. *122*, 463 (1965).

[329] *B. Cleaver*, Nature *207*, Nr. 5003, 1291 (1965).

[330] *N. Cabrera*, Surface Sci. *2*, 320 (1964).

[331] *H. Gatos*, The Surface Chemistry of Metals and Semiconductors, John Wiley & Sons, New York 1960.

3.1.4 Crystal Morphology

Polymorphism: existence of a compound in several different crystal forms.

Epitaxy: oriented growth of a compound on a crystal of a different material. In this case, both crystals possess at least one common crystal face and one common zonal direction.

Pseudomorphism: chemical transformation of the crystalline substance without the external form changing (topochemical reaction). This kind of reaction is often observed during the weathering of minerals [344].

Inclusions: enclosure of impurities (in liquid, gaseous, or crystalline form) in crystals [319, 345]. *Clathrates* are a different kind of inclusion compound in which foreign molecules are enclosed, particularly those which fit sterically into the crystal cavities *(adductive crystallization).* A number of clathrates are of preparative or analytical importance, e.g. those of urea for the separation of *paraffins, rare gases, dextrins, isomers, racemates,* etc. The *deparaffination of petroleum* and the extraction of *naphthalene* and other *aromatics* are of especial commercial importance [346–348, 350, 352, 373, 374].

Crystallization includes *ripening* and *recrystallization.* During ripening, crystals which, as a result of crystal dislocations, are not in thermodynamic equilibrium with the surrounding solution dissolve and build bigger and more regular crystals. A number of smaller crystals may also coalesce to form a bigger crystal (cementation). Recrystallization after crystal formation involves rearrangement of irregular crystals yielding crystals which are in better equilibrium with the solution. In contrast to ripening, however, the number and size of the crystals remain unchanged [390].

However, the most thermodynamically stable final state is the ideal *single crystal.* Single crystals are used extensively as *piëzocrystals,* for *transistors* and *semiconductors, crystal counters* in radiochemistry, as bearings in the watch industry, or for gas turbine blades [353].

Liquid crystals are liquids which show anisotropic behavior under the polarization microscope [349, 354–357] *(mesophases, mesomorphism).* In contrast to isotropic liquids, this is due to a long-range arrangement which transceds shorter range configurations. The various long-range arrangements are classified into nematic, smectic, and cholesteric phases. Compounds of this so-called 'fourth state of aggregation', of which several thousand have become known, are of considerable importance in biological systems, in polymers and detergents, for temperature measurements, as solvents in NMR spectroscopy, or as stationary phases in GLC [262, 354, 357–361].

The preparation of *whiskers,* filamentous single crystals of metals or compounds, has also gained technical importance. They are preferably made by reacting a gaseous compound with a crystal

[332] *J.H.A. Palermo,* Ind. Eng. Chem. *60,* Nr. 4, 65 (1968).

[333] *H. Gülzow, G. Gülzow, W. Lüdke,* Phys. Status Solidi *11,* 205 (1965); Kristall, Technik *1,* 411 (1966).

[334] *R. Kaischew, D. Nenow,* Kristall, Technik *1,* 369 (1966).

[335] *H. Karge, H. Heyer, G.M. Pound,* Z. Phys. Chem. (N.F.) *53,* 294 (1967).

[336] *N.N. Sirota,* Nauka i Tekhnika 460 (1964).

[337] *M. Zief, W.R. Wilcox,* Fractional Solidification, Vol. I, Marcel Dekker, New York 1967.

[338] *E.R. Hauser, L. Sogor, A.G. Walton,* J. Chem. Phys. *45,* 1071 (1966).

[339] *M. Smutek,* Collect. Czech. Chem. Commun. *32,* 922 (1962).

[340] *G. Zinsmeister,* Vacuum *16,* 529 (1966).

[341] *T.F. Canning, A.D. Randolph,* A.I.Ch.E.-J. *13,* 5 (1967).

[342] *U. Steinike,* Kristall, Technik *1,* 113 (1966).

[343] *K. Schneider, R. Adoutte,* Brit. Chem. Eng. *11,* 1217 (1966).

[344] *K. Hauffe, C. Seyferth,* Reaktionen in und an festen Stoffen, in: Anorganische Chemie in Einzeldarstellungen, Bd. II, Springer Verlag, Berlin 1966.

[345] *W.R. Wilcox,* Ind. Eng. Chem. *60,* Nr. 3, 13 (1968).

[346] *W. Schlenk* in *Houben-Weyl,* Methoden der organischen Chemie, Bd. I/1, p. 391, Georg Thieme Verlag, Stuttgart 1958; Experientia *8,* 337 (1952); Liebigs Ann. Chem. *565,* 204 (1949).

[347] *F. Cramer,* Einschlußverbindungen, Springer Verlag, Berlin 1954.

[348] *A. Hoppe, H. Franz,* Erdöl, Kohle *8,* 411 (1955); Petr. Refiner *36,* Nr. 5, 221 (1957).

[349] Anonym, Nachr. Chem. Tech. *18,* 361 (1970).

[350] *A.D. Rogaceva,* Dokl. Akad. Nauk SSSR *165,* 1298 (1965).

[351] *H. Stage,* Glas-Instrum.-Tech. *14,* 213 (1970).

[352] *M. Freund, S. Keszthely, R. Csikos, G. Mozes,* Chem. Tech. (Leipzig) *19,* 688 (1967).

[353] *A. Neuhaus,* Angew. Chem. *28,* 155, 350 (1956).

[354] *A. Mannschreck,* Chem.-Ztg. *92,* 69 (1968).

[355] *G.W. Gray,* Molecular Structure and the Properties of Liquid Crystals, Academic Press, London 1962.

[356] *C. Wiegand* in *Houben-Weyl,* Methoden der organischen Chemie, Bd. III/1, p. 681, Georg Thieme Verlag, Stuttgart 1955.

[357] *R.S. Porter, J.F. Johnson,* Ordered Fluids and Liquid Crystals, Adv. Chem. Vol. 63, American Chemical Society, Washington 1967.

[358] *C. Robinson,* Tetrahedron *13,* 219 (1961).

[359] *H. Athenstaedt,* Naturwissenschaften *49,* 433 (1962).

[360] *H. Kelker, B. Scheurle, H. Winterscheidt,* Anal. Chim. Acta *38,* 17 (1967); Z. Anal. Chem. *198,* 254 (1963).

[361] *J.L. Ferguson, N.N. Goldberg, R.J. Nadalin,* Mol. Cryst. *1,* 293, 309 (1966); Sci. American *211,* 76 (1964).

[362] *R.V. Coleman,* Metallurg. Rev. *9,* Nr. 35, 261 (1964).

[363] *N. Franssen, H. Schladitz, H. Borchers,* Metall *19,* 423 (1965).

nucleus. Their importance is due to their outstanding mechanical and thermal properties[169, 362–365].

3.2 Applications

In principle, the thermal separation method of crystallization may be from a melt (e.g. as zone melting, see Section 2.2), from the gas phase (sublimation, preparation of whiskers), or from a solution. The classical method is crystallization from a solution[277, 282–285, 288]. A number of new procedures have been developed in recent years.

Crystallization under pressure (100,000 atm, 10,000 °K) can be used in the laboratory for the preparation of precious stones (diamonds, rubies, and emeralds) whose properties closely resemble those of naturally occurring crystals. Diamonds can now be produced under considerably less severe conditions[255, 366–372].

Adductive crystallization can be extended to the separation of isomers or racemates by use of special methods (clathration, formation of complexes, or solid solutions[350, 373, 374]). This procedure is of technical importance in the dewaxing of petroleum, and in the desalting of sea water via the hydrate process.

Crystallization from the gas phase is of technical interest for the preparation of silicon and germanium crystals for the electronic industry, for the preparation of thin films on the most varied materials (e.g. for corrosion inhibition[375]), and for the manufacture of whiskers.

Electrolytic crystallization (electro-crystallization) is used for the refining of definite structural types of various metals[376–379]. This is important for the purification of metals and for corrosion inhibition. The *growth of single crystals*[225, 275, 289, 353, 380–383, 405] is also important for the electronic industry.

Crystal growth in gels[384, 385] is important with compounds which (because of low solubility or low dissociation temperature) do not yield single crystals by other methods.

Two types of *precipitation crystallization* methods exist: in one, the compound to be crystallized is precipitated by addition of solvents[282, 386–389], whereas in the other a new insoluble compound is formed by a chemical reaction[307, 390].

Combined rectification and crystallization can be used for the separation of isomer mixtures[391].

Polymerization and crystallization occurring in parallel or consecutively to form *crystalline polymers* have also been investigated intensively[295, 392–398].

For crystallization on the large scale, the following requirements must be fulfilled:

[364] *W.U. Wagner*, Z. Naturforsch. *20a*, 705 (1965).
[365] *E.F. Riebling*, Science *162*, 468 (1968).
[366] *H. Honda, Y. Sanada, K. Inous*, Carbon (Oxford) *1*, 127 (1964).
[367] *F.P. Bundy*, J. Chem. Phys. *38*, 631 (1963).
[368] *W. Wilson*, J. Appl. Phys. *36*, 268 (1965); Chem. Weekbl. *96*, Nr. 17, 116 (1965).
[369] *C.B. Alcock*, Chem. Brit. 216 (1968).
[370] *K.A. Müller*, Naturwiss. Rundschau *19*, 147 (1966).
[371] *J.C. Angus*, Chem. Eng. News *46*, Nr. 30, 44 (1968).
[372] *T. Evans, P.F. James*, Proc. Roy. Soc. *A 227*, 260 (1964).
[373] *C.J. Santhanam*, Chem. Eng. *73*, Nr. 12, 165, 170 (1966).
[374] *P. Sherwood*, Brit. Chem. Eng. *10*, 382 (1965).
[375] *A. Neuhaus*, Coll. Int. Nr. 152, (Nancy 1965), Edit. CNRS, Paris *1965*, p. 1; Werkst. Korrosion *17*, 567 (1966).
[376] *H. Fischer*, Angew. Chem. *81*, 101 (1969); Ber. Bunsenges. Phys. Chem. *70*, 856 (1966); Z. Phys. Chem. (N.F.) *53*, 29 (1967).

[377] *H. Fischer*, Elektrolytische Abscheidung und Elektrokristallisation von Metallen, Springer Verlag, Berlin 1954.
[378] *R. Kaischew, C. Nanev, E. Budewski, W. Bostanoff, T. Witanoff*, Dokl. bolgarskoj Akad. Nauk *2*, 29 (1949); Electrochim. Acta *11*, 1697 (1966); Chem.-Ing.-Tech. *39*, 554 (1967).
[379] *J.O.M. Bockris, G.A. Razumney*, Fundamental Aspects of Electrocrystallization, Plenum-Press, New York 1967.
[380] *V.M. Goldfarb, B.M. Goltsman, A.V. Donskoi, A.V. Stepanov*, Sov. Phys.-Cryst. *10*, 450 (1966).
[381] *E.A.D. White*, Brit. J. Appl. Phys. *16*, 1415 (1965).
[382] *D.B. Wilson*, Chem. Eng. *72*, 119 (1965).
[383] *J.E. Wardill, D.J. Dowling*, Chem. Brit. 226 (1969).
[384] *H.K. Henisch*, J. Phys. Chem. Solids *26*, 493 (1965).
[385] *J.W. McCauley, R. Roy*, Am. Ceram. Soc. Bull. *44*, 635 (1965).
[386] *J.B. Sumner*, J. Biol. Chem. *69*, 435 (1926).
[387] *E.L. Holland-Merten*, Chem. Tech. (Leipzig) *17*, 335 (1965).
[388] *L.N. Matusevic*, Kristall, Technik *1*, 127 (1966).
[389] *E. Tietze* in *Houben-Weyl*, Methoden der organischen Chemie, Bd. I/1, p. 453, Georg Thieme Verlag, Stuttgart 1958.
[390] *K.H. Lieser*, Z. Anorg. Allg. Chem. *292*, 97 (1957); *335*, 225 (1965); Angew. Chem. *81*, 206, Angew. Chem. Internat. Ed. Engl. *8*, 188 (1969).
[391] *H. Clasen*, Chem.-Ing.-Tech. *39*, 1279 (1967).
[392] *R.W. Lenz*, Organic Chemistry of Synthetic High Polymers, Interscience Publishers, New York 1967.
[393] *L. Mandelkern*, Crystallization of Polymers, McGraw-Hill Book Co., New York 1964.
[394] *P.H. Geil*, Polymer Single Crystals, Interscience Publishers, New York 1963.
[395] *F.G.A. Stone, W.A.G. Graham*, Inorganic Polymers, Academic Press, New York 1962.

sufficient crystal nucleus formation (this is seldom a problem);

sufficiently high crystallization velocity (which may be affected either by the temperature or the degree of supersaturation); a too high velocity is not desirable since it may lead to enclosure of impurities;

sufficient supply of the compound to be crystallized from the solution (stirring of the crystallization vessels);

sufficiently large cooling devices for removal of the heat of crystallization.

In crystallization under normal pressure the supersaturation normally is effected by indirect cooling or by evaporation of the solvent.

For better efficiency the crystallization vessels may be shaken (agitated crystallizers). Crystallization by cooling is applied especially if the solubility of the compound to be crystallized is strongly reduced with decreasing temperature (e.g. potassium nitrate in water). Crystallization by evaporation is used if various solvents must be removed. In vacuum crystallization the solution is supersaturated by evaporation of the solvent under reduced pressure. Because of the hydrostatic pressure of the liquid, a small height of liquid is preferred. Vacuum crystallization is of special importance for heat sensitive compounds and non-aqueous solvents.

In special procedures special equipment is used such as scrolls or tapes for crystallization.

Which crystallizer is to be chosen depends on the thermostability of the product, the economics, and sometimes the desired form of the crystals.

For reviews of crystallizers used in industry see Ref. [69, 270, 275, 279, 280, 284, 285, 317, 318, 332].

The authors wish to thank Dr. H. Stage, Köln-Niehl, for advice and discussions, and also Mr. W.G. Fischer, Fischer Labor- und Verfahrenstechnik, Bad Godesberg, for advice and the preparation of additional figures.

[396] *R.G.R. Gimblett*, Inorganic Polymer Chemistry, Butterworths, London 1963.

[397] *A. Sharples*, Introduction to Polymer Crystallization, Edward Arnold Publishers, London 1966.

[398] *B. Wunderlich*, Angew. Chem. *80*, 1009; Angew. Chem. Internat. Ed. Engl. 7, 912 (1968).

[399] *H.D. Martin*, Glas-Instrum.-Tech. *12*, 1293 (1968).

[400] *W. Schäfer, H. Stage*, Chem.-Ing.-Tech. *21*, 416 (1949).

[401] *A. Rose, E. Rose*, Ind. Eng. Chem. *32*, 668 (1940), *33*, 594 (1941), *42*, 2145 (1950).

[402] *Anonym*, Glas-Instrum.-Tech. *12*, 1196 (1968).

[403] *H. Brusset, J.C. Levain*, Chim. Ind. (Milan) *97*, 909 (1967).

[404] *W. Kuhn, P. Massini, H.J. Kuhn*, Helv. Chim. Acta *33*, 737 (1950); *40*, 2433 (1957).

[405] *A. Neuhaus*, Chem.-Ing.-Tech. *28*, 155, 350 (1956); Z. Phys. Chem. (N.F.) *53*, 163 (1967).

[406] *C.J. Liddle*, Chem. Eng. (London) *75*, 137 (1968).

2.2 Zone Melting and Column Crystallization

Klaus Maas and *Hermann Schildknecht*
Organisch-Chemisches Institut der Universität, D-69 Heidelberg

1 Analytical Applications

Zone melting and column crystallization are based on the concentration changes that occur during crystallization of substances from their melts. Passage through many molten zones in zone melting and countercurrent flow of crystals and melt through a temperature gradient during column crystallization multiply the single effect many times (multistage separation), see Figs. 1 and 5.

Repeated crystallization from a given melt thus requires no auxiliary substances (solvents, adsorbents, etc.) and is very suitable for:

Preparing impurity concentrates (for analytical evaluation)

Obtaining *ultrapurity materials*

Applied to substances that are stable at the melting point, and using a protective gas if necessary, these methods are loss-free. By employing the inverse principle, zone melting enables different compounds to be very uniformly distributed in one another (addition of dopants, Section 2.2:1.6).

1.1 Fundamentals of Zone Melting and Column Crystallization

Fusion and crystallization of binary mixtures follow the temperature versus concentration phase diagrams (the relationship is very clearly described in Kofler[1]). Ternary mixtures require a corresponding three-dimensional representation. The two main groups

Systems without solid solution formation (2.2:1.1.1)

Systems with solid solution formation (2.2:1.1.2)

[1] *L. Kofler, A. Kofler*, Thermo-Mikro-Methoden zur Kennzeichnung organischer Stoffe und Stoffgemische, Verlag Chemie, Weinheim/Bergstr. 1954.

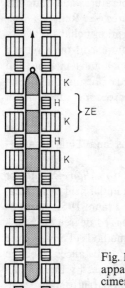

Fig. 1. Schematic zone melting apparatus with zone melting specimen
K   cooling element
H   heating element
ZE 'zone melting unit'

can be subdivided further. Roozeboom[2] divides the latter into 5 types, *viz.*, simple continuous solid solution formation and modifications involving a maximum, minimum, peritectic, or eutectic. Purification of organic compounds usually involves systems forming eutectics. Zone melting and column crystallization merely lead to *higher* and *lower* melting fractions in which the individual components are distributed according to the phase diagram.

1.1.1 Systems without Solid Solution Formation

When a molten binary mixture is cooled under ideal conditions, the pure component present in excess of the eutectic ratio crystallizes until, at the eutectic point, the minimum melting point mixture of crystals of both components deposits. Systems of this type can thus be separated only into the pure 'excess component' and the binary or, in higher systems, the ternary, *etc.*, eutectic.

In theory, this separation takes place in one crystallization step, but in practice generally several are required because the advancing crystal front entrains impurities which do not diffuse quickly enough across the phase boundary. [When a tube full of aqueous solution is immersed in a freezing bath little aggregation occurs; vigorous stirring at the phase boundary is needed to produce a solid

ingot of (almost) pure excess component. Its composition progressively approaches the eutectic ratio of the individual components in the upper part of the tube (principle of 'normal' or 'progressive' freezing[3]).]

1.1.2 Systems with Solid Solution Formation

Substances capable of replacing one another in the lattice (solid solution formers) require several crystallization stages to achieve separation even in theory, since only concentration displacements take place during melting and crystallization. Each stage is expressed quantitatively by the distribution coefficient

$$k_0 = \frac{c_{\text{solid}}}{c_{\text{liquid}}}$$

which may be determined from (ideal) normal freezing or by means of single-stage (microscale) zone melting[4]. k_0 is constant only at high dilution; the closer it lies to unity, the more crystallization stages are necessary to achieve separation (or a greater number of zones n in zone melting). Since theoretical equilibrium is not attained because of slight diffusion, in practice it is necessary to employ the effective distribution coefficient k_{eff} with a value intermediate between k_0 and unity. According to the equation

$$k_{\text{eff}} = \frac{k_0}{k_0 + (1 - k_0)\, e^{-f\delta/D}}$$

established by Burton, Prim and Slichter[5], k_{eff} is dependent on the diffusion coefficient D and the boundary (diffusion) layer thickness δ. For a given system and apparatus the separation process is governed by the rate of solidification f. δ may be reduced in value by providing mechanical mixing of the melt and the crystals (normal freezing[3], oscillatory rotation of the zone melting specimen[6]) and by thermal convection so as to increase the value of f. The theory of zone melting is discussed in Ref.[7-11] and the monographs listed in Section 2.2:4.

[2] *H.W.B. Roozeboom*, Z. Phys. Chem. *10*, 145 (1892).

[3] *H. Schildknecht, G. Rauch, F. Schlegelmilch*, Chem.-Ztg. *83*, 549 (1959); *H. Schildknecht, F. Schlegelmilch*, Chem.-Ing.-Tech. *35*, 637 (1963). *J.S. Matthews, D. Coggeshall*, Anal. Chem. *31*, 1124 (1959).

[4] *H. Vetter, S. Rössler, H. Schildknecht*, Symposium Zonenschmelzen und Kolonnenkristallisieren, p. 57. Karlsruhe 1963.

[5] *J.A. Burton, R.C. Prim, W.P. Slichter*, J. Chem. Phys. *21*, 1987 (1953).

[6] *N.J.G. Bollen, M.J. VanEssen, W.M. Smit*, Anal. Chim. Acta *38*, 279 (1967).

While countercurrent-flow is not possible during zone melting[11], in multistage column crystallization[12] the melt is led past the crystals in an adiabatically operated part of the column to form a temperature and concentration gradient (akin to that obtained during distillation in a packed column; for theory see Ref. [13, 14]; general reviews of fractional crystallization are given by Matz[93], Zief[41], Wilcox[92], and in Ref. [15, 16]).

1.2 Necessary Properties of Substances

Two requirements, stability at the melting point (prolonged if necessary) and ability to crystallize, must be met to the degree demanded by the method employed. By using very slow advancing rates, substances whose crystallization rate is too slow for successful column crystallization can still be zone-melted; details of the (expensive) improvement of nucleation and acceleration of crystal growth by using electric fields and vibration or ultrasonics can be found in Ref. [17]. Conversely, since column crystallization is a 10 to 100 times faster separation process, it does not demand the same high degree of stability. According to a USSR Patent[18], systems of eutectic composition should be chosen for zone melting of labile substances (vitamins).

Highly viscous melts retard both crystal growth and the required diffusion; in column crystallization the movement of crystals through the viscous melt is made more difficult (especially in fats and oils, where modification changes interfere additionally[19, 20]). The temperature gradient between the hot and cool zones promotes the formation of crystal nuclei at lower temperatures and optimum growth at higher temperatures (see Ref. [21]).

A good guide to the separation behavior of unknown substances can be obtained from a related analytical method since analogous melting and crystallizing can be observed on heating or cooling under the melting-point microscope, especially in polarized light[1].

1.3 Results of Separations and Time Requirements

Zone melting of *metals* and *transition elements* for use as semiconductors (the initial field of application of zone melting; see Pfann 1952[22]) yields materials of almost any desired degree of purity. Correspondingly sensitive methods of analysis are not available for organic substances (Section 2.2:1.5); a theoretical limit is set by thermal decomposition to which even apparently stable compounds are prone (decomposition is especially marked where radioactive labeling is used[23]).

Substances forming *solid solutions* may be separated into the pure components very slowly if the value of k_0 is unfavorable (zone melting fractionation[24]). Naturally, with most *eutectic*-forming systems only the pure excess component and a concentrate of the eutectic composition is obtained. However, the latter can be concentrated as a higher eutectic by means of further zone melting after adding an extra substance.

The following factors govern the time required to achieve separation:

Nature of the substance (composition, k_0 or k_{eff} of the individual components, crystallizability),
Purity or enrichment required,
Method employed (normal freezing, zone melting, column crystallization, instrumental modifications such as mechanical mixing at the phase boundary, temperature gradient),
Adaptation of the method to the required flow.

With small amounts of substance microscale and ultramicroscale zone melting (see 2.2:2.1.1) achieve a rate of separation that is comparable to

[7] *W.R. Wilcox, C.R. Wilke*, Am. Inst. Chem. Engrs. J. *10*, 160 (1964).

[8] *G. Matz*, Chem.-Ing.-Tech. *36*, 381 (1964).

[9] *P.J. Jannke, R. Friedenberg*, Talanta *12*, 617 (1965).

[10] *R. Friedenberg, P.J. Jannke, W. Hilding*, Talanta *13*, 245 (1966).

[11] *W.G. Pfann*, Ind. Eng. Chem., Fundamentals *8*, 357 (1969).

[12] *H. Schildknecht, H. Vetter*, Angew. Chem. *73*, 612 (1961).

[13] *J.D. Henry jr., M.D. Danyi, J.E. Powers*, in Ref. [94].

[14] *H. Schildknecht, J. Breiter*, Chem.-Ztg. *94*, 1 (1970); Chem.-Zgt. *94*, 81 (1970).

[15] *R.T. Southin, G.A. Chadwick*, Sci. Progr. [London] *57*, 353 (1969).

[16] *G.D. Botsaris, E.G. Denk, G.S. Ersan, D.J. Kirwan, G. Margolis, M. Ohara, R.C. Reid, J. Tester*, Ind. Eng. Chem. *61*, (10), 86 (1969); Ind. Eng. Chem. *61*, (11), 92 (1969).

[17] *S.L. Hem*, Ultrasonics 5 (Okt.) 202 (1967).

[18] UdSSR-Pat. 186637 (Erf. *R.V. Fedorova, P.I. Fedorov, Yu.P. Shvedov, O.D. Belova*) v. 3.10.1966.

[19] *P. de Bruyne, M. v.d. Tempel*, Symposium Zonenschmelzen und Kolonnenkristallisieren, p. 391. Karlsruhe 1963.

[20] *J. Hannewijk*, Chem. Weekbl. *60*, 309 (1964).

[21] *A. Lüttringhaus* in *Houben-Weyl*, Methoden der Organischen Chemie, Bd. I/1, 4. Aufl., Georg Thieme Verlag, Stuttgart 1958.

[22] *W.G. Pfann*, Trans. Aime *194*, 747 (1952).

[23] *H. Schildknecht*, Kerntechnik 6, 249 (1964).

[24] *H. Schildknecht, H. Vetter*, Angew. Chem. *71*, 723 (1959).

macroscale column crystallization, provided the melting and crystallization zones are close together. However, in general one to several days are required for obtaining a concentrate, several days to a week for separating a solid solution system, and some weeks for very similar compounds (homologous higher paraffins, waxy alcohols, and fatty acids)[24].

Zone refining accelerated 10–100 times (by using rapidly alternating rotation[6]) and continuous zone melting[25, 26] are feasible with more complex apparatus. The effective separation times for microscale column crystallization (see Section 2.2:3.3), macroscale column crystallization, and usual macroscale zone melting are in the approximate ratio $1:10–10^2: 10^3–10^4$; the separation efficiencies of the latter two methods have been compared by Ammon[27] (*nitrates*, solid solutions) and Pouyet[28] (*aromatic amines*).

1.4 Solvents and Primary Standards

Primary standards are usually obtained by microscale or semimicroscale zone melting, physically pure solvents (theoretical degree of purity as in Section 2.2:1.3, impurity content below the limit of chemical detection) either by semimicroscale or macroscale zone melting depending on the amount involved, or by continuous column crystallization (e.g. with the all-glass column referred to in 2.2:3.1.2. Traces of abraded glass or metal can be removed by brief zone melting for stringent requirements, e.g. meteorite extraction[29]). These methods are very suitable for aromatic compounds, but not for glassy solids like hydrogen-bonded aliphatic polyhydroxy or polyamino compounds (for examples see Section 2.2:2.4).

As each zone melting step merely goes to equilibrium, and approaches it in progressively decreasing steps, portions corresponding zones of several zone-melted specimens are combined for the quantitative analysis of solid solution forming compounds. After renewed zone melting the separation process is repeated. Zone melting fractionation follows the scheme shown in Fig. 6 for continuous column crystallization.

1.5 Concentrates for Analytical Evaluation, 'Zone Melting Analysis'

The trace substances sought can be found at almost any dilution; often the limit of detection is set by the zone melting fractionation process. As a result of the mild thermal treatment there is less risk of isolating artifacts than during distillation, and entraining of foreign matter is eliminated. Unlike enrichment, slightly impure solvents and adsorbents have little effect on the subsequent separation of the concentrate. The concentrates obtained from zone-melted high purity substances allow conclusions concerning the degree of purity of thermally stable compounds (*zone melting analysis*).

1.6 Addition of Dopants

Moving the molten zone to and fro along an ingot of the substance or, in a more complicated fashion, in one direction around a ring of the substance inside an annular vessel[22, 30], distributes the compounds present very uniformly within the matrix substance (Fig. 2). This process reverses the separation principle. The demixing or segregation that otherwise occurs during solidification of a solution is suppressed ('zone leveling' or 'zone alloying').

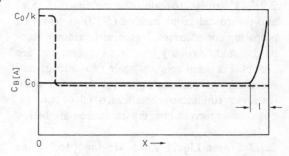

Fig. 2. Concentration distributions before (– – – –) and after (——) zone leveling (idealized); the component previously introduced at the beginning of the ingot is uniformly distributed; some is concentrated at the end at (1).

The technique was developed for adding dopants to *semiconductors;* it is suitable also for producing *scintillator materials* and especially for preparing *organic standards* of *definite activity.*

By performing a subsequent brief zone melting, eutectic-forming impurities can be separated without appreciably altering the $^{12}C/^{14}C$ concentration ratio; more pronounced *isotope* and *exchange effects* appear in D and T labeling[31].

Generally speaking, *tracer* substances should be zone-melted before their incorporation in order to remove radioactive breakdown products formed during storage.

[25] *W.G. Pfann,* Trans. Aime *203,* 297 (1955).
[26] *J.K. Kennedy,* Rev. Sci. Instrum. *33,* 387 (1962).
[27] *R. von Ammon,* Chem.-Ing.-Tech. *39,* 428 (1967).
[28] *B. Pouyet,* Anal. Chim. Acta *38,* 291 (1967).
[29] *H. Schildknecht,* Anal. Chim. Acta *38,* 261 (1967).

[30] *H. Schildknecht,* Z. Naturforsch. *12b,* 23 (1957).
[31] *G.J. Sloan,* J. Am. Chem. Soc. *85,* 3899 (1963).

2 Zone Melting Techniques

2.1 Systematic Classification

Zone melting methods may be classified either according to the weight of substance to be processed or in terms of the characteristics of the apparatus used (method of melting, zone travel, temperature range, *etc.*).

2.1.1 Classification by Weight of Substance

In 1964, Wilcox, *et al.,* published 'histograms' of melt temperatures, melting zone widths, *etc.*[32]. At that time ingot diameters between 5 and 10 mm were in common use; today they range from a few 1/100 mm[35] to about 10 cm[34, 35].

Ultramicroscale zone melting using <0.5 mg requires a special technique (see Sections 2.2:2.3.1 and 2.3.2). The smallest weight of substance handled to date is about 10 μg[33]. Because of short separation times when closely spaced melting zones are employed, microscale zone melting (0.5 mg–0.5 g) is of analytical interest in biological chemistry[24, 33] (see 'Zone Melting Unit', Section 2.2:2.2). Semimicroscale zone melting (0.5–5 g) and macroscale zone melting (5–50 g) serve for preparing concentrates, high purity primary standards, and auxiliary materials (solvents, see above). For semitechnical-scale (50–500 g) and especially technical-scale (>500 g) zone melting of readily crystallizable substances, batch or continuous countercurrent-flow crystallization are better.

2.1.2 Frozen-Liquid Zone Melting ('Ice' Zone Melting)

Ice zone melting[36], whice is zone melting of liquids, is distinguished mainly by the cooling temperatures, the insulation, and the exclusion of atmospheric moisture[37–40]. Interesting variations include ring ice zone melting (*e.g.*, in a desiccator[30]),

movement of the specimen tube in a cooling bath provided with insulated electrical heaters[41], and the use of Peltier elements[42].

2.1.3 Special Methods

In addition to zone melting methods involving continuous flow of substances[25, 26], techniques have been described that utilize sublimation[43] (*vapor zone refining*[44]) or, contrary to true zone melting, addition of solvent (*zone precipitation*[45], fractionation of polystyrene in benzene[46]) or fractionation of substances in very narrow zones along a long solid ingot. The latter uses very many stages analogous to chromatography (*zone chromatography*[47], *cf.* Ref. [48], and Ref. [49]).

2.2 Principles of the Zone Melting Apparatus

In zone melting one or preferably several simultaneous molten or liquid zones move along the length of an ingot of substance and cause the lower melting material to be concentrated in the direction of motion. With organic compounds, which possess a low thermal conductivity and generally a high melt viscosity, closely spaced alternate heating and cooling elements provide the necessary mixing at the phase boundaries (heat convection) by virtue of a steep temperature gradient and also yield a large number of zones per unit time. The 'zone melting unit'[33] (ZE) given by

$$ZE = H + K + 2A$$

(H the width of the heating element,
K the width of the cooling element and
A the distance between the elements),
can be used to compare different apparatuses (see Fig. 1).

On empirical evidence K is approximately 1.5–3 H; A should not exceed a few millimeters to a few tenth of a millimeter. K can be reduced to as little as 1.7 mm for ultramicroscale zone melting, but may reach the decimeter range in large apparatus. Interposed cooling elements should not be eliminated, even with high melting organic compounds.

[32] *W.R. Wilcox, R. Friedenberg, N. Back,* Chem. Rev. *64*, 187 (1964).

[33] *K. Maas, H. Schildknecht,* Anal. Chim. Acta *38*, 299 (1967).

[34] *J. Cremer, H. Kribbe,* Chem.-Ing.-Tech. *36*, 957 (1964).

[35] *H. Plancher, J.C. Morris, W.E. Haines,* Anal. Chem. *40*, 1592 (1968).

[36] *H. Schildknecht, A. Mannl,* Angew. Chem. *69*, 634 (1957).

[37] *E.F.G. Herington, R. Handley, A.J. Cook,* Chem. Ind. [London] *1956*, 292.

[38] *H. Schildknecht, U. Hopf,* Chem.-Ing.-Tech. *33*, 352 (1961).

[39] *M. Zief, H. Ruch, C.H. Schramm,* J. Chem. Educ. *40*, 351 (1963).

[40] *R. Kieffer,* Bull. Soc. Chim. Fr. 3024 (1967), Bull. Soc. Chim. Fr. 3029 (1967).

[41] *J.N. Carides,* Rev. Sci. Instrum. *39*, 1811 (1968).

[42] *J. Schnell, H. Schildknecht,* Symposium Zonenschmelzen und Kolonnenkristallisieren, p. 159, Karlsruhe 1963.

[43] *G.J. Sloan,* Symposium Zonenschmelzen und Kolonnenkristallisieren, p. 277, Karlsruhe 1963.

[44] *L.R. Weisberg, R.D. Rosi,* Rev. Sci. Instrum. *31*, 2061 (1960).

[45] *I.A. Eldib,* Ind. Eng. Chem., Process Design Develop. *1*, 2 (1962).

[46] *J.D. Loconti, J.W. Cahill,* Am. Chem. Soc., Div. Polym. Chem., Preprints *1*, 129 (1960); J. Polym. Sci. *1*, 3163 (1963).

[47] *W.G. Pfann,* Anal. Chem. *36*, 2231 (1964).

[48] *H. Reiss, E. Helfand,* J. Appl. Phys. *32*, 228 (1961).

[49] *H. Plancher, T.E. Cogswell, D.R. Latham* in Ref. [94].

Fig. 3. Ultramicroscale
zone melting apparatus
(example of a 'comb'
apparatus)[33]

K cooling block
(Brass)
H heating block
(invar)
T transport device
(Rubber rollers for
the transport wire)
Magnified diagram:
st   steel capillary
gl_1 guide capillary
(glass)
gl_2 substance capillary
(glass)
subst. = substance

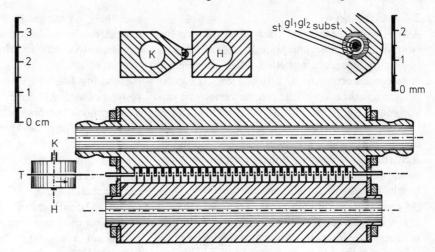

2.2.1 Heating Elements for Zone Melting

During manufacture of high purity metals and semi-conductors melting zones are usually generated by electron bombardment, arc discharge or induction (see monographs referred to in Section 2.2:4). Container-free zone melting of rubies by CO_2 laser beams has also been reported[50]. For organic compounds, which generally melt between -100 and $+300°$, focussed heat radiation[51] and hot-air heating have been used less than the methods that follow.

The power output of *electrical resistance heating* is adjusted via transformers or pulse generators. An accurate temperature controller (thermocouple or thermistor sensors for each heating unit) is usually unnecessary, because the width of the melting zone is controlled visually (glass tubes). Where large differences in melting point exist between the beginning and end of the ingot, the heating elements are controlled separately or the ingot being processed is divided up as in zone-melting fractionation.

Thermostatic heating with water or silicone oil as the heat transfer medium, similar to liquid cooling, is less sensitive to external temperature fluctuations but more expensive than the above (see Ref.[52]).

Indirect *induction heating* in place of the above direct contact methods is particularly suitable for large tubes more than 3 cm in diameter. According to Ref.[34], up to 20 kg of substance can be zone-melted in tubes of up to 10 cm in diameter with inductively heated perforated plates of encased soft iron (see Ref.[35] and the first patent covering this method[53]).

Large-scale apparatus comprise a series of heating and cooling elements fitted either on insulated mountings or on a guide tube; in microscale apparatus the zone melting specimens travel in wire loops[24] or through drilled plates which are in alternating contact with either a heating block or a cooling block ('comb' apparatus, see Fig. 3).

2.2.2 Cooling Elements for Zone Melting

Adequately dimensioned cooling elements ($K = 1.5–3$ H) need no accurate control. Intensive cooling produces a beneficial steep temperature gradient[54], but the substance should not be overheated when narrow heating elements are used in the molten zones.

The coolants (water, water/methanol, or methyl-cyclohexane) flow either through a centrally drilled metal cooler for the zone melting specimen ('Liebig condenser') or through a coiled copper or lead tube which can be soldered to metal spools. Glass or thin-section chromium-nickel steel guide tubes allow very convenient centering of the heating and cooling elements[29]. Especially sharp phase boundaries are obtained if the cooling liquid streams are passed directly over the filled rotating tube ('direct contact cooling technique'[55], see method of Carides[41] in Section 2.2:2.1.2). The two sides of a Peltier element can be employed as a heating or cooling pair for temperatures down to about $-20°$[42, 56].

[50] *K. Eickhoff, K. Guers,* J. Cryst. Growth *6,* 21 (1969).
[51] *R. Handley, E.F.G. Herington,* Chem. Ind. [London] 304 (1956); Chem. Ind. [London] 1184 (1957).
[52] *F. Ordway,* Anal. Chem. *20,* 1178 (1965).
[53] Belg. Pat. 615 104 (Knapsack-Griesheim AG) v. 29.6.1962.
[54] *A. Deluzarche, A. Maillard, J.C. Maire, J. Moritz,* Bull. Soc. Chim. Fr. 89 (1963).
[55] *W.G. Pfann, C.E. Miller, J.D. Hunt,* Rev. Sci. Instrum. *37,* 649 (1966).
[56] *E. Scharrer, K. Böke, J. Schnell, H. Polnitzky, K. Hannig, H. Schildknecht,* Chem.-Ing.-Tech. *37,* 1039 (1965).

2.2.3 Containers

Almost without exception glass tubes are used for zone melting organic compounds (ease of fabrication of the generally eyeletted tubes, see Fig. 1; absence of corrosion or diffusion of, e.g., plasticizer; visualization of the molten zones and observation of segregation). Flexible hollow PTFE inserts have proved successful[57] for large ingot diameters (kg range) and for aqueous solutions. According to Röck[58], a substance can be zone-melted in the annular space between two concentrically placed tubes. Fabricating accurately circular containers is difficult. This fact, added to the expense of the apparatus, explains why the otherwise very elegant method of ring zone melting[22, 30] using a discontinuous ring, or ring zone alloying using a continuous ring of substance, has not been adopted widely. In open 'boats' the separation process can be followed directly in the case of labeled substances[23, 30] (Fig. 4) or by continuous sampling[12]. This is no routine procedure, needs inert gas and is unsuitable for volatile or spreading compounds.

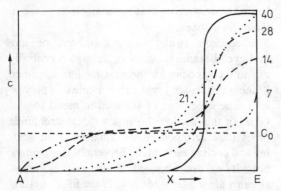

Fig. 4. Separation effect versus zone number of a mixture of ^{14}C cetyl alcohol and stearyl alcohol forming a solid solution series with a minimum melting point.

2.2.4 Zone Travel Mechanisms

The zones travel by virtue of either the movement of the tube containing the substance or the displacement of the heating[41] and/or cooling[55] elements (relative movement). In the first case, which is in widespread use for zone melting organic substances, the specimen is retraced along its path when it begins to leave a long system of zone-melting units (Fig. 1), or else is moved slowly forward and then very quickly backward through one respective zone-melting unit (by automatic means, e.g. limit switches). The ring zone melting method operates without to-and-fro motion; another method involves successive heating of a series of resistance wires without using mechanical aids[59].

Synchronous motors, with a range of reducing gears or stepped pulleys for the filament pulling the specimen tube, provide the control for adjusting the relevant constant speeds for the individual substances (see Sections 2.2:1.2 and 1.3; the usual range covered is 0.5–100 mm/h and exceptionally up to 1000 mm/h according to the method used).

A slow rotation superimposed on the longitudinal movement results in sharply defined melting zones[41, 55]. Very high revolutions combined with rapid reversing of direction (125 alternate clockwise and anticlockwise movements a second) reduce the adverse diffusion zone at the crystal front by shearing the melt and allow zone movement speeds of 10–100 cm an hour for substances with good crystallization characteristics[6].

2.3 Zone Melting Techniques

2.3.1 Sample Preparation

Impure starting material with a very wide melting range that is to be converted into high purity substances should be given a preliminary refining by a more rapid method (recrystallization from solution). The simple techniques described below (Fig. 1, tensile filament) may be replaced by other appropriate methods.

Macroscale zone melting: One end of the glass tube, selected in accord with the type of apparatus and the amount of substance, is sealed with a blowpipe; the glass filament produced is formed into an eyelet; and thoroughly annealed to import resistance to the subsequent heating and cooling stresses. After cooling of the substance, charged either as liquid or as a subsequently melted powder, the other end of the tube is cleaned and carefully sealed by melting (for horizontal zone melting, see Section 2.2:2.3.2; the tube should be only 1/2–2/3 full and have a second eyelet for the return-movement filament or spring).

Microscale zone melting: The powdered substance is packed into the tube in small portions, liquefied, and any air bubbles formed are removed by tapping or centrifuging. Liquid material can be introduced directly with a fine capillary. A thin silicone rubber buffer is useful for substances

[57] *J.C. Maire, M.A. Delmas,* Rev. Trav. Chim. *85,* 268 (1966).

[58] *H. Röck,* Naturwissenschaften *43,* 81 (1956).

[59] *T. Bohl, R.W. Christy,* J. Sci. Instrum. *36,* 98 (1959).

which sublime or display high thermal expansion[60]. Within certain limits, larger equipment with thick-walled guide tubes and/or capillary tubes acting as the substance containers can be used in the microscale region. Microscale apparatus may be used also for working in the ultramicroscale region (see Fig. 3[33]).

Ultramicroscale zone melting: The molten substance is drawn in by capillary action and centered in the capillary tube by tilting. The tiny bulbs produced by fusing with a microburner must be checked under the microscope.

2.3.2 Zone Melting
The zone number n (the number of zones which have passed through the whole ingot including zone fractions) is generally 10–50. Difficult-to-separate solid solution systems like paraffins, higher fatty acids and waxy alcohols (see Section 2.2:2.4) may require several hundred cycles (zone melting fractionation). Intrinsically, wide melting zones produce a more rapid equilibrium (at small n)[8, 61], narrow zones (~ ingot diameter) yield a better final separation at high values of n. The two principles may be combined, but the narrowest possible zones should be used for routine work.

Vertical macroscale zone melting: The substance tube travels upwards in order to utilize the heat convection at the crystallization front. Under these circumstances the compromise between adequate diffusion time and many zones per unit time has the more favorable values of 0.5 to several cm/h.

Horizontal macroscale zone melting: Where there is the risk of fracture of the tube due to ice, thermal expansion, or sublimation, horizontal processing using partially empty glass tubes is employed. Tilting the apparatus backwards slightly prevents a rapid accumulation of substance at the front due to surface tension induced movements counter to the molten zones[38]. The specimen tube moves uniformly only if the filament is kept taut (e.g., by a counteracting spring attached to the rear eyelet).

Microscale and ultramicroscale zone melting: Small microtubes may be drawn through the apparatus with fused-on glass rods[24, 60]. Fig. 3 shows very fine capillaries being pressed through the guide capillary be means of a wire. The small amounts of substance and the small zone melting units require the temperature of the heating zone to be controlled with special care (use of a magnifying glass), and drafts must be absent. Fragments of cracked thin filaments of substance behave like individual molten specimens and are difficult to reunite.

2.3.3 Evaluation of the Samples
Where a detailed study of the molten specimens is not required, the final third and the first quarter are rejected on empirical grounds when high purity material is the objective. Accurate information is obtainable from graphs showing the melting interval (hot-stage microscope) of specimens taken from the ingot plotted against the length along the ingot (see Fig. 7). High degrees of purity may be qualitatively characterized by zone melting analysis (Section 2.2:1.5). Quantitative methods include thermoanalysis[62–64], radiation from labeled compounds (^{14}C and T[23]; [82]Br[65]) or fluorescence analysis (anthracene[66], naphthalene[67]).

2.4 Examples of Applications
Literature surveys and monographs list many examples of organic compounds that have been treated by zone melting[32, 88–92].

Well-known *aromatic compounds* and important derivatives, and also the important crystallizable *heterocyclic* and *aliphatic* compounds were described as long ago as 1963[43, 51, 60, 63, 65, 68–71]. In general, only eutectic-forming impurities were separated. However, use of multistage apparatus permitted the preparation of pure individual C_{20}–C_{30} *waxy alcohols*[72] (which form solid solutions and cannot be separated by recrystallization from solution) and the separation of an eicosanol-hexacosanol mixture[24]. For a mixture of cetyl and stearyl alcohols which form a solid solution series of minimum melting point, Fig. 4 illustrates, how the ^{14}C-labeled cetyl alcohol becomes enriched at

[60] *H. Schildknecht, H. Vetter,* Angew. Chem. *73,* 240 (1961).

[61] *L. Burris, C.H. Stockman, J.G. Dillon,* J. Metals *7,* 1017 (1955).

[62] *W.M. Smit, G. Kateman,* Anal. Chim. Acta *17,* 161 (1957).

[63] *H. Schildknecht, U. Hopf,* Z. Anal. Chem. *193,* 401 (1963).

[64] *J. Moritz, J.C. Maire,* Symposium Zonenschmelzen und Kolonnenkristallisieren, p. 69, Karlsruhe 1963.

[65] *P. Süe, J. Pauly, A. Nouaille,* Bull. Soc. Chim. Fr. *1958,* 593.

[66] *Y. Lupien, D.F. Williams,* Mol. Crystallogr. *5,* 1 (1968).

[67] *K.W. Benz,* Z. Naturforsch. *24a,* 298 (1969).

[68] DBP 1015804 (Erf. *H.C. Wolf*) v. 19.9.1957.

[69] *E.F.G. Herington,* Analyst *84,* 680 (1959).

[70] *J.H. Beynon, R.A. Saunders,* Brit. J. Appl. Phys. *11,* 128 (1960).

[71] *E.A. Wynne,* Microchem. J. *5,* 175 (1961).

[72] *H. Schildknecht, G. Renner, W. Keess,* Fette, Seifen, Anstrichm. *64,* 493 (1962).

the end of the process as a function of the number of zones[23]. The reported fractionation of labeled and nonlabeled organic compounds was, however, denied. A mixture of deuterated (C_6H_5COOD) and nondeuterated benzoic acids did show a separation after almost 300 cycles, but this ran parallel with the enrichment of D_2O and HOD at the end of the treated sample. *Hydrocarbons* displayed practically no change in concentration[31]. An *isotope effect* can be expected only at low molecular weights (*e.g.* with H_2O-D_2O[73]).

Ultrapure *primary standards* have been prepared for spectroscopy (*nitrogen compounds*[74]) and gas chromatography (*lauric, myristic, palmitic,* and *stearic* acids[75]). Others have been investigated as photographic developers[76], scintillators (*anthracene, trans-stilbene, quaterphenyl*[77]), Kerr cell liquids (*nitrobenzene*[78]), and organic semiconductors (*imidazole*[79]).

Tracer substances can be purified by zone melting. The impurity concentrated from tritiated compounds enabled conclusions to be drawn concerning a characteristic decomposition reaction (splitting of the C-N bond in *amino acids*)[23]. Primary standards (conjugated dienes[40]) have been obtained by ice zone melting down to $-140°$. Here again, it is the concentrates such as *ascorbic acid, quinones, vanillin, enzymes, bacteria,* and *bacteriophages* from aqueous solution[36], and higher *aromatic compounds* and *alkanes* extracted with ultrapure benzene from meteorites[29] that are of interest. In the transition region between inorganic and organic chemistry, zone melting can be applied to temperature-sensitive organometallic compounds (*e.g.* Al-alkyls) and *metal complexes* in an excess of chelating agent or organic solvent (*oxinates, benzoylacetonates,* and *acetylacetonates* of Fe, Al, Cu, etc.[80, 81]). Polymers, too, can be fractionated in solvents (polystyrene in benzene[46], polystyrene in naphthalene[82]).

[73] *H.A. Smith, C.O. Thomas,* J. Phys. Chem. *63,* 445 (1959).

[74] *J.S. Ball, C.R. Ferrin, R.V. Helm,* Am. Chem. Soc., Div. Petrol. Chem., Symposium 6, No. 2A, 125 (1961).

[75] *W.H. Schaeppi,* Chimia *16,* 291 (1962).

[76] *W.E. Lee, W.R. Strother,* Photogr. Sci. Eng. *11,* 400 (1967).

[77] *P.T. Perdue, M.D. Brown,* Nucl. Instrum. Methods *71,* 113 (1969).

[78] DBP 1070606 (Askania-Werke AG; Erf. *M. Biermann*) v. 10.12.1959.

[79] *G.P. Brown, S. Aftergut,* J. Chem. Phys. *38,* 1356 (1963).

[80] *K. Ueno, H. Kaneko, Y. Watanbe* in Microchim. J. *10,* 244 (1966).

[81] *K. Ueno, H. Kobayashi, H. Kaneko* in Ref. [94].

3 Column Crystallization Techniques

Column crystallization should be considered here as a relatively simple (and unique) laboratory method using a genuine countercurrent-flow of crystals and melt (for other crystallization methods see Matz[93] and Section 2.1:3).

3.1 Column Crystallization Apparatus

3.1.1 Analogy between Fractional Column Distillation and Column Crystallization

As in fractionating column distillation, three different thermal-state regions are distinguished in column crystallization:

A lower region where heat is supplied for evaporating and melting the substance,
An upper region where heat is abstracted for condensation and crystallization,
An intermediate insulated adiabatic part with a temperature and concentration gradient.

Continuous melting of low melting fractions and crystallization of higher melting fractions within the corresponding temperature range produces separation within a single column analogous to multiple 'one step' distillation or crystallization (see Ref. [14]).

3.1.2 Countercurrent Transport of Crystals and Melt

In contrast to fractionating column distillation, column crystallization requires mechanically aided conveyance because of the small difference in density between the solid and liquid phases. For laboratory-scale work a coil spring rotating in the annular space between two concentric tubes (Fig. 5) and moving the higher melting fractions crystallizing in the cooling zone through the liquid phase[12] is very suitable. Depending on the column diameter, 10–150 rpm are appropriate; about 90 rpm are suitable for a diameter of 2 cm[14]. The aggregation efficiency can be improved by making the conveying spiral pulsate[83, 14] and by providing 'reverse windings' (1–3 oppositely wound turns) at the lower end of the spiral[84]. The profile effectiveness for crystal transport increases in the order circular, square, diamond (see Fig. 5), lenticular; the lenticular shape also gives the greatest useful volume.

[82] *F.W. Peaker, J.C. Robb,* Nature *182,* 1591 (1958).

[83] *H. Schildknecht, J.E. Powers,* Chem.-Ztg. *90,* 135 (1966).

[84] *H. Schildknecht, J. Breiter, K. Maas,* Separ. Sci. *5,* 99 (1970); s.a. *Schildknecht,* Chimia *17,* 145 (1963).

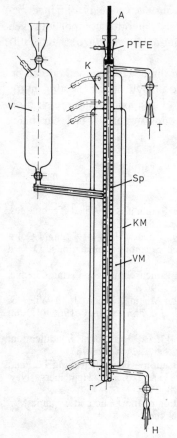

If the impurity becomes concentrated as a eutectic, as is generally the case, and sufficient starting material is available, the separation process is carried out at a large T/H ratio to obtain pure substances and at a small T/H ratio for the production of concentrates (the limits are approximately 10:1 or 1:10 respectively); a further fractionation can follow, as shown in Fig. 6.

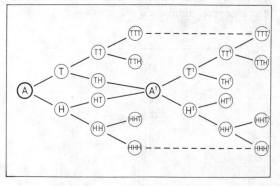

Fig. 6. Schematic representation of multicycle zone melting fractionation and multicolumn crystallization.
A starting substance
H higher melting fraction
T lower melting fraction

Fig. 5. Glass column for continuous column crystallization
A drive shaft
Sp spiral
V vacuum jacket
T top fraction take-off
H bottom fraction take-off

3.1.3 Introduction and Withdrawal of the Substance

The fluid crystal slurry allows continuous operation[83, 84] following laws similar to those of continuous multistage distillation[14]. If possible, the starting substances is introduced in the middle section of the column at the point occupied by identical composition material after equilibrium has been established (generally 2/5–1/2 along the column). This liquid or molten feedstock is stored in a reservoir under nitrogen pressure and is fed into the column at the same rate as fractions are removed. At the foot of the column the higher melting fraction H is obtained through PTFE taps or needle valves, while the lower melting fraction T is withdrawn at the top.

3.2 Operating Technique

The cooler temperature and (electrical) heating are adjusted to produce a dense crystal slurry; non-visual control using thermocouples is feasible[14, 83] but expensive; aggregation is best observed directly (in an all-glass column, Fig. 5). External heating or cooling adjusted to approximately the melt temperature is used to augment the insulation (or vacuum jacket) in the column region proper.

Glass columns have been employed at temperatures between −120 and +250°. An empirical rule states that the hourly flow is to be about half the column capacity. The number of stages (HETP for a pulsating spiral = about 5 cm), the reflux ratio, etc., have been determined for the solid solution-forming azobenzene-stilbene system using a 2.5 cm-column with built-in thermocouples[14]. Possible applications include the preparation of high purity solvents (e.g. benzene[13, 29, 83, 85]), the purification of larger quantities of substances for use in preparative chemistry, and the concentration of traces for analytical evaluation (see also Sections 2.2:1.3, 1.5, 2.4).

3.3 Microscale Column Crystallization

Microscale column crystallization[84, 86, 87] achieves a very high aggregation rate with simple operating techniques. The apparatus consists of a variable-speed geared motor (about 500–2000 rpm) fitted with downward-feeding threaded rods (1–5 mm dia., Cr-Ni steel or similar) and an accurately

[85] *W.D. Betts, J.W. Freeman, D. McNeil,* J. Appl. Chem. *18,* 180 (1968).

[86] *K. Maas, H. Schildknecht,* Z. Anal. Chem. *236,* 451 (1968).

[87] *H. Schildknecht, V. Reimann-Dubbers, K. Maas,* Chem.-Ztg. *94,* 437 (1970).

fitting glass capillary tube as column (of. e.g. 2 mm-dia for 50–100 mg substance). A cooled or heated stream of air[87] is most suitable for coarse temperature control between −10 and +250°; the crystal/melt ratio should be as high as possible.

In 2 mm-columns at between 1000 and 2000 rpm, the aggregation process is complete in a period of 10 seconds to a few minutes. Separation is improved by heating and cooling the transition regions several times. Fig. 7 shows that only a relatively small transition zone is obtained between the completely separated components even with the solid solution-forming azobenzene/stilbene system.

Using different rod diameters, the method can be employed over the analytically interesting mg/g range. An example of its application is preparation in GLC (e.g. detecting impurities in ε-caprolactam[87]).

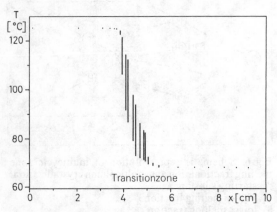

Fig. 7. Melting point diagram along a microcolumn (microscale column crystallization of the system azobenzene-stilbene which forms solid solutions).

4 Bibliography

[88] *N.L. Parr,* Zone Refining and Allied Techniques, George Newnes, London 1960.

[89] *E.F.G. Herington,* Zone Melting of Organic Compounds, Blackwell Scientific Publications, Oxford 1963.

[90] *H. Schildknecht,* Zonenschmelzen, Verlag Chemie, Weinheim/Bergstr. 1964.

[91] *W.G. Pfann,* Zone Melting, 2. Edn. John Wiley & Sons, New York, London, Sydney 1966 (1. Edn. 1958).

[92] *M. Zief, W.R. Wilcox,* Fractional Solidification, Vol. I, Marcel Dekker, New York 1967.

[93] *G. Matz,* Kristallisation — Grundlagen und Technik, 2. Aufl. Springer Verlag, Berlin, Heidelberg, New York 1969 (1. Aufl. 1954).

[94] *M. Zief,* Purification of Inorganic and Organic Materials, Marcel Dekker, New York 1969.

2.3 Countercurrent Distribution

Herbert Feltkamp
Farbenfabriken Bayer, D-56 Wuppertal-Elberfeld

Countercurrent distribution has been used successfully for the separation of numerous organic compounds (Section 2.3:6, Bibliography, and review articles in Analytical Chemistry[39] and Industrial and Engineering Chemistry[40]). However, because of progress in the development of chromatography, the importance of countercurrent distribution for organic analysis has decreased in many areas. In contrast, its use for separation of materials on a preparative scale is still very much open to development.

For numerous classes of substances[1] increasing interest exists in the relationship between distribution behavior and biological activity. The use of countercurrent distribution and extraction in inorganic analysis has grown considerably. By far the majority of current publications concerning countercurrent distribution deal with the extraction and distribution of metals[39].

1 Definition and Use

The terms countercurrent distribution and liquid-liquid partition, or simply distribution or partition, are used for separation processes which involve the use of two incompletely miscible phases.

If a substance is dissolved in the lower phase, and the solution shaken with the appropriate upper phase, then material will pass from the lower into the upper phase until the partition equilibrium is attained. A mixture of two substances can be separated in a phase system if the ratios of the solubilities of the solutes in the upper and lower phases are different.

The most important partition process is simple extraction. This is always used for separation when the differences between the solubilities of the substances to be separated are so large that a single separation stage is sufficient.

Multiple extraction is employed mainly to avoid

---

[1] *C. Hansch, T. Fujita,* J. Am. Chem. Soc. **86,** 1616 (1964).

losses of material, rather than to improve separation.

As a rule, differences in solubilities are so small that numerous separation steps must be carried out successively. This is termed multi-stage distribution. The main use of this technique is the separation of solid substances. Compared with other separation processes, distribution offers the following advantages:

the separation is carried out under very mild conditions;
separation is often quite accurate;
even when small amounts of material are being separated, losses are only small;
the partition processes can also be carried out continuously, in order to obtain large amounts of material;
the separation results can be easily calculated in advance.

This last point is of particular importance for the planning and prediction of a separation. The main disadvantage of partition is that the substances separated are obtained in dilute solution from which they must be isolated.

Numerous types of equipment have been developed for carrying out multi-stage distributions. These can be classified according to the way in which the phases move and the substance is added.

Table 1. Naming of the different partition processes according to their authors

Substance addition	One moving phase		Two moving phases	
	Stepwise flow	Constant flow	Stepwise flow	Constant flow
Continuous	—	—	O'Keeffe[2] Watanabe[4]	Jantzen[3] van Dijck[5]
Single	Craig[6]	Martin, Synge[7]		Cornish[8]

[2] *A.E. O'Keeffe*, J. Am. Chem. Soc. *71*, 1517 (1949).
[3] *E. Jantzen*, Dechema Monographie Nr. 48, Verlag Chemie, Berlin 1932.
[4] *S. Watanabe, K. Morikawa*, J. Soc. Chem. Ind., Jap. *36*, 585 (1933).
[5] *W.J.D. van Dijck, A.H. Ruys*, Perfumery Essent. Oil Record *28*, 91 (1937).
[6] *L.C. Craig*, J. Biol. Chem. *150*, 33 (1943).
L.C. Craig, J. Biol. Chem. *155*, 519 (1944).
L.C. Craig, Fortschr. Chem. Forsch. *1*, 292, 302, 312 (1949/50).
L.C. Craig, O. Post, Anal. Chem. *21*, 500 (1949).
[7] *A.J.M. Martin, R.L.M. Synge*, Biochem. J. *35*, 91 (1941).
[8] *R.E. Cornish*, Ind. Eng. Chem. *26*, 397 (1934).

The Craig and the O'Keeffe distribution processes are particularly suitable for laboratory purposes. The constant flow continuous processes of Jantzen or van Dijck are especially suitable for industrial use. The distinction between field of use is, however, by no means sharp.

2 Parameters

If a substance A is partitioned between two phases, then the ratio between the concentrations of substance A in the light phase ($c_{A, LP}$) and the heavy phase ($c_{A, HP}$) is termed the *partition coefficient* of A (K_A).

$$K_A = \frac{c_{A, LP}}{c_{A, HP}} \qquad (1)$$

In the ideal case, i.e. for sufficiently dilute solutions, K is independent of concentration and of other substances present in the solution, but does depend upon the nature of the solvent system and on the temperature (the Nernst distribution law). For laboratory purposes, it is recommended that operations should be carried out within the region for which the Nernst law holds. For non-dissociated organic molecules, the Nernst law is normally valid up to a concentration of 0.1 Mol/l. For dissociated molecules, the concentrations are considerably lower since dissociation itself is concentration-dependent.

Closely related to the partition coefficient is the *partition number* G, which is defined as the ratio of the amounts of substance in the upper and lower phases.

$$G_A = \frac{p_A}{q_A} \qquad (2)$$

For practical reasons, p and q are usually replaced by relative quantities, so that

$$p_A + q_A = 1. \qquad (3)$$

If one designates the ratio of the volumes of upper and lower phases used as V,

$$V = \frac{Vol_{LP}}{Vol_{HP}} \qquad (4)$$

then the relationship between K and G is

$$G_A = K_A \cdot V \qquad (5)$$

The partition number G is the most important parameter for the performance and the calculation of distributions.

For a given apparatus, the optimal separation of two substances is obtained when $G_A = 1/G_B$[9].

[9] *M.T. Bush, P.M. Densen*, Anal. Chem. *20*, 121 (1948).

By substituting into Equation (5) and solving for V, the most favorable phase ratio emerges as

$$V = \frac{1}{\sqrt{K_A \cdot K_B}} \qquad (6)$$

If one considers two materials simultaneously present in a distribution system, then the ratio of the partition coefficients or of the partition numbers is a measure of the extent to which the substances can be separated.

$$\frac{K_A}{K_B} = \frac{G_A}{G_B} \geq 1 = \beta \qquad (7)$$

(This *separation factor β* is always greater than or equal to 1. If this were not so, two separations of equal difficulty would have two different β values.) If the β value of a pair of substances is 1, then the substances in the given system cannot be separated by partition.

3 Partition Processes and their Calculation

3.1 Extraction
As a model for treatment of extraction, let us consider a separatory funnel. It contains a solution of a substance (LP) which is to be extracted with an equal volume of a heavier extractant (HP). One can either extract the upper phase with the total quantity of lower phase in one operation, or the lower phase can be divided into n parts and the extraction repeated n times. In the first case, the total amount of the extractant is added, the mixture is shaken, and the phases are separated. The upper phase, which now contains the relative amount of substance p, remains in the separatory funnel. The value of p is given in Equations (2) and (3).
If the lower phase is divided into n equal parts, then n extractions can be carried out.

In this case

$$p^n = \frac{1}{\left(1 + \dfrac{1}{K \cdot V \cdot n}\right)^n} \qquad (8)$$

This expression decreases with increasing n and tends towards a limiting value

$$p^n_{\substack{n \to \infty}} = e^{-\frac{1}{K \cdot V}} \qquad (9)$$

Equation (9) shows in mathematical terms that it is more efficient to extract a solution repeatedly with small amounts of the extractant, than to extract once with the entire amount of extractant. By suitable choice of solvent, K can often be changed in a desirable way.

3.2 Craig Distribution

3.2.1 Theoretical Calculation
As a model for calculations on multi-stage processes, let us consider an apparatus consisting of individual glass tubes which is so constructed that the upper phase can be transferred from one tube into the next, whereas the lower phase always remains in the same tube (Craig distribution). The z tubes are numbered from 0 to m. Initially, all tubes contain lower phase and tube 0 also contains upper phase. If the relative amount 1 of a substance of a uniform material is placed in tube 0, and equilibration is effected by shaking, then the following picture is obtained:

And after the transfer of the light phase

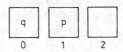

This completes the first partition stage. The second begins with the addition of fresh light phase into tube 0, and is followed by further equilibration. The result is that in both tubes 0 and 1, a p'th of the substance present in each case goes into the upper phase and a q'th part goes into the lower phase.

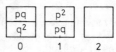

And after the second transfer of the light phase

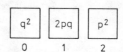

The same process is repeated for each partition stage. The substance in the apparatus distributes itself throughout the individual tubes, as given by the expression

$$(p+q)^n = 1 \qquad (10)$$

Here, the terms of equal dimension of the binomial (10) each represent the relative contents of substance in a tube; n is the number of partition stages carried out. Graphical representation gives a Pasqual triangle. When the basic process of a distribution is completed, the number of distribution stages is given by n = m.
If a mixture of two substances is partitioned, then Equation (10) is valid for both substances. If these

substances have different partition numbers, and thus differ with respect to p and q, then separation occurs during the course of partition and can be calculated using Equation (10).

For a pair of substances with $G_A = 2$ and $G_B = 0.5$, $\beta = 4$. It follows that in tube 0 there is 89% B, and correspondingly for tube three 89% A — provided that equal parts of A and B were present in the initial mixture. The larger the number of partition stages n, and thus the larger the exponent, the greater is the percentage purity of the peak fractions.

The result of a partition is normally represented graphically by plotting the contents of substances in the tubes against the tube number r. This distribution curve is an important aid for calculation and evaluation of distribution processes (cf. Fig. 1).

As an example of the partition of a mixture, the technique of Craig distribution will be explained and methods for calculation and evaluation will be demonstrated.

It is assumed that the mixture contains two substances, nothing being known of their properties. Experiments should, therefore, be carried out to find a suitable solvent system for the mixture, and to determine a total partition coefficient. In the case under consideration let us suppose that this is 0.89. By using Equation (5), one now calculates V for $G = 1$, since this is the requirement for best utilization of the distribution equipment. (This is strictly true only for the case where the two substances each comprise 50% of the mixture.) In our example, $V = 1.12$. Using these experimental conditions, the basic process ($n = 29$) is carried out in a 30 stage Craig apparatus. Since the substances

are non-volatile, the shape of the distribution curve can be determined by evaporating and weighing the contents of the individual tubes. The upper and lower phases from each tube are evaporated and weighed together. A graph is drawn of the weights against the element number r, thus giving the distribution curve (Fig. 1).

One should now attempt to determine the partition numbers of the individual substances from the shape of the distribution curve. These can be approximately determined from the positions of the maxima, provided that the maxima are separated from each other.

$$G = \frac{r_{max} + 0.5}{n - r_{max} + 0.5} \qquad (11)$$

If separate maxima cannot be observed, partition numbers can be calculated from the contents of materials in neighboring tubes ($y_{n,r}$; $y_{n,r-1}$), using the following equation

$$G = \frac{y_{n,r}}{y_{n,r-1}} \cdot \frac{r}{n+1-r} \qquad (12)$$

If this is done for the present example, the following values are obtained:

Table 2. Partition numbers from the contents of adjacent tubes

r	G	r	G	r	G	r	G
5	0.64	10	0.68	15	1.07	20	1.26
6	0.64	11	0.72	16	1.15	21	1.27
7	0.64	12	0.78	17	1.20	22	1.27
8	0.66	13	0.87	18	1.24	23	1.27
9	0.66	14	0.97	19	1.25		

If a uniform substance or a non-separable mixture is involved, the partition numbers must agree within the limits of experimental error. If, however, there is partial separation, as in the present example, then the values on the left hand side of the distribution curve correspond to the substance with the smaller partition coefficient and the substance with the greater partition coefficient are on the right hand side. With partition numbers determined in this way, a first calculation can be made of the theoretical distribution curve.

In the experiment under consideration, the values are $G_A = 0.64$ and $G_B = 1.27$. It is often worthwhile to establish the partition curve by evaluating only alternate tubes. The contents of the remaining tubes can then be used for other investigations. In particular, thin layer chromatography or gas chromatography can provide very important information, and permit better interpretation of the distribution curve. In this case, the partition num-

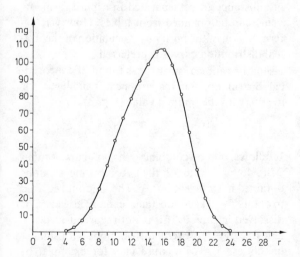

Fig. 1. Experimental Curve; $n = 29$; $K_{total} = 0.89$; $V = 1.12$.

bers can be calculated from the content of the alternate tubes using the following formula:

$$G = \sqrt{\frac{y_{n,r} \cdot r\,(r-1)}{y_{n,r-2} \cdot (n+1-r)\,(n+2-r)}} \qquad (13)$$

If the probable values for G_A and G_B have been determined this way, then the theoretical distribution curves for both values are calculated, using the following equation

$$T_{n,r} = \frac{n! \cdot G^r}{r!\,(n-r)! \cdot (1+G)^n} \qquad (14)$$

This permits the individual members of the binomial $(p+q)^n = 1$ to be calculated. For practical calculations, a different form of this equation is used:

$$T_{n,r} = T_{n,r-1} \cdot G\,\frac{n+1-r}{r} \qquad (15)$$

By this means the relative amount of substance $T_{n,r}$ can be calculated in a simple manner from $T_{n,r-1}$. The first value of $T_{n,r}$ of practical interest is obtained from the logarithmic form of Equation (14)

$$\log T_{n,r} = \log n! + r \log G - \\ [\log r! + \log(n-r)! + n \log(1+G)] \qquad (16)$$

This value is substituted as $T_{n,r-1}$ into Equation (15). Since they are based upon p and q, Equations (14) and (15) yield the relative amounts of materials. It is recommended that these be converted into percentages, by multiplication by 100. The theoretical curves for G_A and G_B calculated in this way in percentages are compared with the experimental curve (Fig. 1 and 2).

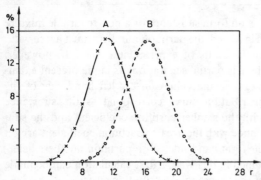

Fig. 2. Theoretical 100 %-curve, n = 29, G_A = 0.64, G_B = 1.27.

Since tube 7 contains a negligible amount of B, the weight of 14.1 mg must be due to A. From the theoretical curve, this corresponds to 4.3% A. The same considerations can be applied to B in tube 21; here, 20.1 mg corresponds to 3.1% B. With a sample weight of 1000 mg, it can be calculated that A will be 35.2% and B 64.8% of

the total quantity. We can now give the factor for each material by which the percentage value for each tube must be multiplied in order to fit the theoretical curve with the experimental one. The sum of the two theoretical curves must then coincide with the experimental one. If this does not happen, one should try to achieve better agreement by changing the partition numbers or the transformation factors. If the agreement is satisfactory, then the partition numbers of the mixture and the percentage of the composition have been determined by analysis of the distribution curve. In the present case, the agreement is good. K_A is found from $K_A = G_A/V = 0.57$, and $K_B = G_B/V = 1.13$. Subsequent manipulations depend upon the apparatus available, requirements for purity, and amounts of the substances to be separated.

In the present example, suppose it is necessary to separate at least 5 g of each of the two matrials in a purity of at least 99%, and that a 100 stage apparatus is available. The largest amount of the mixture of substances which can be dissolved in one tube is 5 g.

Hecker[10] has described separation functions which relate the degree of purity of the substances obtained to the number of separation stages and to the β value. They are based on the approximation that the distribution curves are Gaussian; this is sufficiently exact, for $n > 25$. Assuming that the two substances are present in equal amounts, and that the volume factor $V = 1$, this separation function appears as

$$n = \frac{C_R}{[(\beta-1)/\sqrt{\beta}]-2} \qquad (17)$$

C_R is a constant, which depends upon the quantity of substance to be separated in a pure form. Its value can be obtained from tables. However, determination of the purity by calculation of theoretical distribution curves is preferred.

Using the ratios of substances found, the theoretical distribution curves are next calculated for $n = 99$, with the optimal value (Fig. 3) for

$$V = 1/\sqrt{K_A \cdot K_B} = 1,25$$

It follows, that by combining the contents of tubes 27 to 43, and 55 to 72, the two substances can be obtained in sufficient purity. The amount required to obtain 5 g of each substance cannot, however, be dissolved in tube 0. If it is not regarded as desirable to carry out several distributions with smaller quantities, 5 g of mixture can, for example, be

[10] E. Hecker, Z. Naturforsch. 8b, 77 (1953).

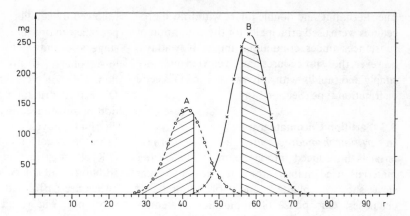

Fig. 3. Distribution curve of two substances; n = 99, $G_A = 0.71$, $G_B = 1.41$ quantity: 5 g mixture

dissolved in each of the first ten tubes, and 99 partition steps can be performed. The resulting total partition curve is then the sum of 10 equal curves which are each staggered by one tube. Although the region in which the two curves overlap does become greater when using this technique, the concentrations in the tubes are smaller.

Combination of the contents of tubes 27 to 45 and tubes 62 to 82 now gives a sufficiently large quantity in the required purity. Exact calculation using Equation (14) and addition of 10 staggered curves is very laborious without use of a suitable calculator. The approximate shape of the distribution curve can be obtained by calculating one single curve for the total quantity, and by assuming that material is added at the center of the set of ten charged tubes. The purity of the materials cannot, however, be determined exactly from such a curve, since the actual curves are broader and, therefore, overlap more.

A further means of utilizing a given apparatus more efficiently is to continue the distribution beyond the basic process.

3.2.2 Recycling Process

The upper phase fraction which leaves the last tube is returned to tube 0. In this way, as many partition stages as required can be carried out in a given apparatus. The method for calculating the theoretical curve is the same as for the normal Craig distribution, except that finally the results in the correct tubes of the apparatus concerned must be added up, since each tube has been used more than once.

The practical usefulness of the recycling process is limited, due to broadening of the distribution curves. If for example a substance with G = 1 in a 30-stage apparatus is partitioned using the basic process, then the curve extends from r = 7 to r = 22. If 60 additioned partition stages are carried out, using the recycling process, the maximum is still

positioned in the same tube of the apparatus, but the curve will have broadened to such an extent that no space is left for a second substance (Fig. 4).

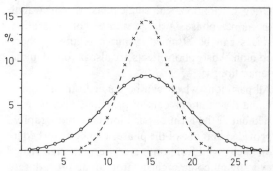

Fig. 4. Basic process n = 29 (– × – × – × –) and recycling process n = 89 (—o—o—o—) in a 30 stage apparatus.

The recycling process can be used for the separation of two substances only when the faster moving substance has not yet reached the substance which moves more slowly. If only one single substance is to be isolated, the position of its distribution curve can be determined or calculated periodically, and the contents of those tubes containing only insignificant amounts of the substance can be replaced by fresh phases.

3.2.3 Single Phase Withdrawal Procedure

After each partition stage, the upper phase fraction which leaves the tube r = m is withdrawn, and (beginning with $\rho = 0$) counted consecutively. Fresh upper phase is added to tube 0 after each partition stage.

It is similarly possible to withdraw the lower phase after each stage, but an additional tube is then needed.

The contents of the withdrawn less dense phase can be calculated, using the following equation:

$$T_{m,\rho} = \frac{(m+\rho)!}{\rho!\, m!} \cdot p^{m+1} q^{\rho} \qquad (18)$$

The alternating and double phase withdrawal procedures were used principally for the separation of substances under continuous input. Nowadays, however, these procedures can be carried out more simply and rapidly with apparatus of the O'Keeffe distribution type (Section 3.4).

3.3 Partition Chromatography[41]

In *partition chromatography*[11], a solid carrier material is introduced into a column. The carrier materials used mainly are: silica gel, molecular sieves such as Sephadex®, and variously-prepared celluloses. One phase of a partition system is adsorbed onto such a carrier; usually it is an aqueous lower phase. The mixture of substances to be separated is placed at the start of the column and is eluted with the second phase. The partition chromatography process can best be regarded as a Craig distribution, with single phase withdrawal of the upper phase. A large number of separation stages can be attained, using a relatively short column. Adsorption effects are also involved in the separation.

All partitions which can be carried out as liquid-liquid distributions can, in principle, also be carried out in the form of partition chromatography, provided that one of the phases can be adsorbed in an unchanged form onto a carrier material. However, by all column chromatographic procedures, only small quantities can be separated. Moreover, the results are difficult to calculate in advance, since the height of a theoretical plate depends to a considerable extent on the load of the column, and on diffusion and other factors. Vink[12] has given theoretical data for calculation of the separation.

3.4 O'Keeffe Distribution

Just as for the Craig distribution, a model based on individual tubes can be used for calculation. The O'Keeffe distribution begins with the addition of material to the middle tube of the model. This is

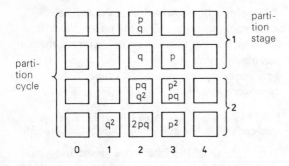

[11] *A.J.P. Martin, R.L.M. Synge,* Biochem. J. *35,* 1358 (1941).

[12] *H. Vink,* J. Chromatogr. *25,* 71 (1966).

followed by equilibration and transport of the upper phase in one direction. After this first partition stage, a second equilibration is carried out, and the lower phase is transferred in the opposite direction. This completes the first partition cycle of an O'Keeffe distribution. Beginning with further addition of substance, any desired number of similar distribution cycles can be carried out.

In order to calculate the separation effect of an O'Keeffe distribution, consider first one single addition of substance, i.e., the subsequent partition cycles are performed without further addition of substance. The result of carrying out the basic process is the same as after performance of the basic process in a Craig distribution. Equation (10) again describes the relative amounts of substance contained in the tubes. The distribution of material in a five stage O'Keeffe distribution sequence after 4 partition steps is:

$$
\boxed{q^4} \quad \boxed{4pq^3} \quad \boxed{6p^2q^2} \quad \boxed{4p^3q} \quad \boxed{p^4} \qquad (19)
$$
$$
\ \ 0 \qquad\ \ 1 \qquad\ \ 2 \qquad\ \ 3 \qquad 4
$$

The difference between the two processes initially consists only in the different direction of movement of the substance with respect to the tube originally charged with material. If the Craig distribution is continued beyond the basic process by single phase withdrawal, then fractions are obtained of different composition, as given by Equation (18). If however an O'Keeffe distribution is continued, by withdrawal of fractions of upper and lower phases, then all fractions have the same percentage composition. This can easily be shown by continuation of the model calculation:

After the fifth step, the relative amount of each substance leaving the apparatus with the upper phase is p^5. The ratio of the substances is p_A^5/p_B^5. After the seventh partition step, a further upper phase fraction leaves the apparatus; it contains an amount $5p^6q$ of each substance. The ratio is now $5p_A^6q_A/5p_B^6q_B$. If the symmetry condition $G_A = 1/G_B$ is fulfilled, then $p_A = q_B$ and $p_B = q_A$, and the ratio is again p_A^5/p_B^5. The same is true for all subsequent fractions. Thus, the separation is dependent only upon the separation factor β and the number of tubes, not on the number of partition steps. In an O'Keeffe distribution one can thus obtain all of the substances in a purity which, in the basic process of a Craig distribution, is attained only in the peak fractions.

If further substance is now added at each partition cycle of an O'Keeffe distribution, then the purity of the fractions is not altered, since each addition can be regarded as the starting point of a new distribution curve which is identical with the preceding one. The total distribution curve of an O'Keeffe distribution is then the sum of the individual curves. Consideration of the mass balance of an O'Keeffe distribution is, therefore, an important part of the calculation. The further addition of material during each cycle causes a considerable rise in the concentration within the apparatus. This increase will be the greater, the more substance is introduced per cycle, the more stages are in the apparatus, and the closer the partition numbers are to 1.

In an O'Keeffe distribution, the concentration approaches an upper limit, which is reached when the amount of substance leaving the apparatus per partition cycle equals the amount of substance added within the same period of time. At this stage, the distribution has reached its steady state, i.e., even when further substance is added, the concentrations in the individual stages of the apparatus no longer change[13].

This emerges on continuation of the calculation from (19). It follows, that

$$\frac{T_{n,LP}^{A} \cdot T_{n,HP}^{B}}{T_{n,HP}^{A} \cdot T_{n,LP}^{B}} = \beta^{z}, \qquad (20)$$

where T_n is the relative amount of substance in simultaneously taken fractions, and z is the number of separation stages. Equation (20) gives the general separation function of the O'Keeffe distribution; it permits the degree of purity of the fractions to be calculated for known partition numbers and a given apparatus.

A variant of the O'Keeffe distribution, known[14] as O'Keeffe distribution of the second type, differs from the O'Keeffe procedure described above in that the upper and lower phases are transferred in opposite directions after each equilibration. The separation function for O'Keeffe distribution of the second type is:

$$\frac{T_{n,LP}^{A} \cdot T_{n,HP}^{B}}{T_{n,HP}^{A} \cdot T_{n,LP}^{B}} = \beta^{\left(\frac{z+1}{2}\right)} \qquad (21)$$

i.e. to achieve the same separation results, approximately twice as many tubes are necessary

as for the procedure of the first type. The loading of the apparatus can, therefore, be twice as high. The tube into which the material should be introduced for O'Keeffe distribution depends upon the composition of the initial mixture, and on the requirements for purity of the products to be obtained[15].

If an O'Keeffe distribution is carried out in such a way that the partition number of one substance equals 1, then this substance is enriched in the apparatus, whereas, all other substances leave the apparatus in the upper and lower phases respectively. This method can be used to enrich and obtain substances which are present only in small quantities in a mixture[16]. The O'Keeffe procedure leads directly to Section 2.3:3.5.

3.5 Constant-Flow Partition Processes

These are generally carried out in columns, in which the two phases move in a counter-current manner. Instead of the phase ratio V of batch-type procedures, one uses the ratio of the phase velocities, φ. This has the same significance as V, and is calculated in a corresponding manner, using Equation (6). The mixture can be introduced at the end of the column or at some intermediate point. The most important difference from the batch-type processes is that the columns have no distinguishable practical separation stages. Rather, one must calculate the number of separation stages from the results of an experimental separation. However, the number of separation stages for a given column is not constant, but depends very much on the experimental conditions. Since constant-flow partition processes correspond to O'Keeffe distribution of the second type — after each equilibration, the two phases are transported in different directions — Equation (21) can be used for calculation of the number of separation stages.

If the number of separation stages of a column has been determined for a definite phase system and for constant experimental conditions, a constant-flow distribution can be calculated by the same methods as for an O'Keeffe distribution of the second type.

4 Practical Performance of Distributions

4.1 Apparatus[17]

Distribution procedures in which the phases move in a batchwise manner can be carried out using

[13] C.R. Bartels, G. Kleimann, Chem. Eng. Progress, 45, 589 (1949).

[14] E. Hecker, Naturwissenschaften 50, 290 (1963).

[15] A. Klinkenberg, Ind. Eng. Chem. 45, 653 (1953).

[16] J.G. Fleetwood, Brit. Med. Bull. 22, 127 (1966).

[17] H. Stage, L. Gemmeker, Chem.-Ztg. 88, 517 (1964).

separatory funnels and beakers. However, in practice, this is possible only when the number of separation stages is very small. For distributions in which a larger number of separation stages are required, an apparatus is used which consists of an assembly of glass tubes mounted on a tilting frame. Each tube performs the function of a separatory funnel. Since the tubes are all mounted on a common frame which can be tilted, the contents of all tubes can be equilibrated simultaneously. The tubes are arranged in such a way that when they are decanted, one phase is transferred into the next tube. In a counter-current apparatus, the two phases can be moved in opposite directions. The apparatus can be operated either manually or completely automatically. Automatic operation is best for large equipment with 50 or more stages.

In practice, the apparatus should fulfil the following minimum requirements: no taps or ground glass joints which must be greased should be present. In order to avoid vapor losses (and thus changes in the phases), the whole battery of partition tubes must be sealed. Only the first and last tubes are left open during the entire distribution, for charging, withdrawal, and avoidance of pressure build-up. With automatic equipment, shaking and settling times – and, if possible, the shaking speed – must be variable. It must be possible to vary the quantity of upper phase. Automatic rinsing is strongly recommended.

Separation columns with various types of cores or fittings (fillers, perforated discs) are used for constant-flow distribution processes. Material exchange is brought about by stirring or by causing the contents of the column to pulsate in a vertical direction[18].

4.2 Solvent Systems[19]

The requirements for the *solvent system* are severe: the difference between the densities of the phases should be sufficiently large for the phases to separate well; the system should have no tendency to form emulsions; the separation factor should be as large as possible; the partition coefficients should be between 0.1 and 10; the substances to be separated must not react with the system; the solvents should not interfere with isolation of the materials and their qualitative and quantitative determination. For safety reasons, the system should be non-toxic, and not too readily flammable.

Systems of four solvents have proved suitable for the separation of a very large number of the most diverse organic compounds. The properties of these systems can be varied within wide limits, by changing their compositions.

Such a mixture, which has also been called a 'standard system', is obtained, for example, by mixing various amounts of ligroin, ethyl acetate, dimethylformamide and water. Suppose that a substance A has a partition coefficient of 1 in a mixture of equal parts of these four solvents.

If the amounts of water and ethyl acetate in the system are reduced compared with dimethylformamide and ligroin, then K_A becomes smaller. If on the other hand the proportions of water and ethyl acetate are increased, K_A becomes larger. In this way, K_A can be varied by a factor of 10^4 to 10^5.

By simultaneously increasing the amounts of ethyl acetate and dimethylformamide, the solubility of a substance can be raised without affecting K. It is clear that compositions based on this single system can be found which are suitable for a very large selection of organic compounds. Moreover, the properties of this system are good, and high separation factors are often obtained.

Mixtures of various quantities of ligroin, ethyl acetate, methanol and water represent a similar system – which is easier to evaporate, but which often gives inferior separation factors.

Even though a large number of organic compounds can be distributed in these and other standard systems, it is often recommended that special systems be employed for particular separation problems.

Buffer systems comprised of aqueous buffers and organic solvents are suitable for substances which dissociate and which have good solubility in water; butanol is frequently used as the organic phase. However, most of these systems cannot be evaporated completely, phase separation is poor, and they have a tendency to form emulsions.

Substances with high solubility in water can be displaced from the aqueous into the organic phase by addition of salts. The solubilities of numerous substances in aqueous phases can be improved by the addition of solution promoters, such as alkali salts of benzoic acid derivatives. Very hydrophilic substances can be partitioned between systems consisting of water, dextrans, and, e.g., polyethylene glycols[20]. Complexing agents are generally used for partitioning of inorganic compounds,

[18] *A. Bittel*, Ref. [33], p. 99.
[19] *F.A. von Metzsch*, Angew. Chem. *65*, 586 (1953).

[20] *L. Rudin*, Biochem. Biophys. Acta *134*, 199 (1967). *P.A. Albertson*, Partition of Cell Particles and Macromolecules, Wiley, New York, Almquist and Wiksell, Stockholm 1960.

since they increase considerably the solubility of the elements in organic phases.

Hecker[21] has published information concerning the systematic choice of solvent systems.

4.3 Determination of Partition Coefficients

If a suitable solvent system has been selected, then a sample of the substance is distributed between equal amounts of the upper and lower phases in a separatory funnel, and the concentrations of the substance in the phases are determined. If the substance is non-volatile, determination is carried out simply by evaporation and weighing. With sensitive biological samples, no consideration is necessary regarding possible decomposition since no substance must be isolated. If acids or bases are involved, direct titration of the phases can be very useful. Any appropriate analytical procedure can be used. If the partition coefficient has a value within the desired range, then a 20 to 30-stage distribution can be carried out, and the partition coefficients and separation factors can be determined exactly, using the procedures described earlier in this chapter.

4.4 Performing a Distribution

To obtain optimal results, it is very important that the phases are in complete equilibrium. They should, therefore, be re-equilibrated immediately before partitioning by shaking or stirring intensively. Since the composition of the phases is very temperature-dependent, the equilibration and partitioning should be carried out at as constant a temperature as possible. The temperature dependence of the partition coefficients is usually of lesser importance. With Craig distributions, it has proved advantageous to allow additional upper and lower phases to run through some tubes, preceding the substance. By this means, change in the phase volume (which otherwise frequently occurs during the course of a distribution) is avoided to a large extent. If the equilibrium is achieved during the distribution by shaking, 30 to 50 oscillations are sufficient. Movement should on no account be violent, since emulsions may be formed and the settling time increased. After the calculated number of distribution steps have been performed, the result of the partition can be checked by analysis of several random samples. The tubes or fractions with similar contents are combined, and the dissolved substances isolated. If isolation cannot be achieved by simple evaporation or removal of solvent by distillation, the substance

can be transferred into a more volatile solvent. With aqueous solutions, freeze-drying can be employed; this technique causes little or no decomposition of the solute. Substances can also be converted into derivatives of low solubility or volatility.

4.5 Aids to Calculation

Hecker[22] has described calculations necessary for evaluation of distributions, with the aid of tables and electric calculators. Such calculations are very time-consuming if one wishes to obtain all the information potentially available. The use of programmable electronic desk calculators has proved very satisfactory for this purpose.

A laboratory which is seriously involved in distribution problems, and which desires to follow separations with calculations, should either use a programmable electronic desk calculator, or should have access to a larger computer.

5 Application Example

For metabolic studies, the substance (**a**) had been labelled with ^{35}S[23].

During the synthesis, approximately 1% of the homolog (**b**) was formed; this by-product lacked the methyl group, but possessed an activity about 50 times that of (**a**). Consequently, about 1/3 of the total activity was due to (**b**). The requirement was that less than 1% of the activity should be due to this impurity; this corresponded to a content of less than 0.04% by weight. 1.75 g of the labelled material were available. The preliminary experiments were carried out using the non-labelled substances which were available in sufficient quantities. After trial experiments using a separatory funnel, a Craig distribution was carried out with $n = 99$ partition steps, and the phase ratio $V = 0.75$ (Fig. 5).

Evaluation of the curves showed K_a to be 2.9, and K_b 1.3. Calculation using Equation (16) shows that the required separation with the optimal value for $V = 1 \sqrt{K_a\ K_b}$ could not be achieved in a 100-stage apparatus. If a single phase withdrawal process is added, then the faster moving substance leaves the apparatus too rapidly, and not enough

---

[21] *E. Hecker*, Chimia *8*, 229 (1954).

[22] *E. Hecker*, Verteilungsverfahren im Laboratorium, Verlag Chemie, Weinheim/Bergstr. 1955.

[23] *W. Maul*, Farbenfabriken Bayer AG, Wuppertal-Elberfeld, unpubl.

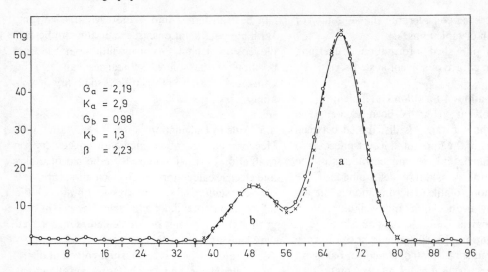

$$G_a = 2{,}19$$
$$K_a = 2{,}9$$
$$G_b = 0{,}98$$
$$K_b = 1{,}3$$
$$\beta = 2{,}23$$

Fig. 5. Distribution curves of (a) and (b). n = 99.

of substance (**a**) is obtained in the required purity. The volume ratio is, therefore, lowered to 0.5 in order to reduce the speed at which (**a**) leaves the apparatus; G_a thus becomes 1.45, and G_b 0.65. Calculation now shows that more than 99% of (**b**) still remains in the apparatus, after 200 distribution steps. The fractions removed with the upper phase must, therefore, contain (**a**) in the desired purity. The distribution was carried out over 232 steps, and the fractions which left the apparatus between the 139th and 191th steps were combined. Using a higher number of partition steps, and having a lower number of combined and used fractions, gave an additional degree of reliability.

The progress of the partition was followed by continually measuring the radioactivity in the individual fractions and tubes. The solvent was evaporated under vacuum, and the residue recrystallized. The yield was 1.28 g, i.e., 73% of the theoretical amount. No (**b**) was detectable in the autoradiogram obtained on TLC of this sample.

6 Bibliography

[24] *L.C. Craig, D. Craig*, Extraction and Distribution, in *A. Weissberger,* Technique of Organic Chemistry Vol. III, p. 171, Interscience Publishers, New York 1950.

[25] *T.K. Sherwood, R.L. Pigford,* Absorption and Extraction, 2. Edn. MacGraw-Hill Book Co., New York 1952.

[26] *L. Alders,* Liquid-Liquid Extraction, Elsevier, Publ. Co., Amsterdam 1955.

[27] *E. Hecker, K. Allemann,* Trennung von Substanzen durch Verteilung zwischen zwei flüssigen Phasen, Angew. Chem. *66,* 557 (1954).

[28] *L.C. Craig, D. Craig, E.G. Scheibel,* Laboratory Extraction and Countercurrent Distribution in *A. Weissberger,* Technique of Organic Chemistry,

2. Edn. Vol. III, p. 149, Interscience Publishers, New York 1956.

[29] *F.A. von Metzsch,* Anwendungsbeispiele multiplikativer Verteilungen, Angew. Chem. *68,* 323 (1956).

[30] *L.C. Craig, T.P. King,* Theory, Choice of Solvent Systems and Experimental Processes. Encycl. Ind. Chem. Anal. *1,* 506 (1966); C.A. *64,* 17051 (1966).

[31] *W.F. Linke,* Solubilities, D. Van Nostrand Co., Princeton, New Jersey 1958.

[32] *O. Jübermann,* Verteilen und Extrahieren, in *Houben-Weyl,* Methoden der organischen Chemie, 4. Aufl. Bd. I/1, p. 223, Georg Thieme Verlag, Stuttgart 1958.

[33] *A. Bittel,* Flüssig-Flüssig-Extraktion in *Ullmanns* Encyklopädie der technischen Chemie, 3. Aufl. Bd. II/1, p. 77, Urban & Schwarzenberg, München, Berlin 1961.

[34] *V.V. Fomin,* Chemistry of Extraction Processes, The Israel Programm for Scientific Translation, Jerusalem 1962.

[35] *R.E. Treybal,* Liquid Extraction, 2. Edn. MacGraw-Hill Book Co., New York 1963.

[36] *A.W. Francis,* Liquid-Liquid-Equilibriums, Interscience Publishers, New York 1963.

[37] *H. Stephen, T. Stephen,* Solubilities of Inorganic and Organic Compounds, Vol. I, II, Pergamon Press, Oxford 1963/64.

[38] *J.K.K. Huber,* Liquid Chromatography in Columns in *C.L. Wilson, D.W. Wilson,* Comprehensive Analytical Chemistry, Vol. IIB, Physical Separation Methods, Elsevier, Publ. Co., London 1968.

[39] *H. Freiser,* Ann. Rev., Extraction, Anal. Chem. *40,* 522R (1968).

[40] *R.B. Beckmann,* Solvent Extraction, Ind. Eng. Chem. *60,* Nr. 11, 43 (1968).

[41] *G. Zweig,* Ann. Rev., Chromatography, Anal. Chem. *40,* 490R (1968).

[42] *P.A. Albertson,* Partition in Polymer Two-Phase Systems, *H. Walter,* Factors in The Partition of Blood Cells in Aqueous Dextrane-Polyethelene Glycol Two-Phase Systems, *D.E. Pettijohn,* Partition of Nucleic Acids in Polymer Two-Phase System in *T. Gerritsen,* Progress in Separation and Purification, Vol. II, Modern Separation Methods of Macromolecules and Particles, Wiley Intersci. Publ., New York 1969.

2.4 Paper and Thin Layer Chromatography

Hellmut Jork and *Ljubomir Kraus*

Institut für Pharmakognosie und Analytische Phyto-
chemie. Institut für Botanik, Abt. Pflanzenphysiologie.
D-66 Saarbrücken 11

1 Introduction

Paper chromatography (PC) and thin layer chromatography (TLC) will be discussed together in
the following text, although they were not developed simultaneously and the processes of separation
are not the same. Borderline cases are: firstly,
different solubilities of substances undergoing
chromatography, between two phases of limited
miscibility (*partition chromatography,* Chapter
2.3) and secondly, selective adsorption on the separating layer (*adsorption chromatography,* Chapter 2.5). Generally, partition and adsorption processes take place simultaneously[364].

Section 2.4:7 lists standard literature, which may
be consulted for more specific questions concern-
ing PC[366–372, 377] and TLC[373, 378, 379, 382, 383, 387–389, 392, 394]. Suggestions are also given, regarding
standardization of TLC[390].

These methods of separation must be differentiated from *paper electrophoresis*[397] and *thin layer
electrophoresis* (*ionophoresis,* Chapter 2.9)[391].
Here, if experimental conditions are suitable, migration of electrically charged particles to the oppositely charged poles takes place in an electric
field. The charge magnitude, hydration, and configuration, etc., determine the speed of migration.
The separation effect also depends to a large extent on the stationary phase (cf. Section 2.4:2.1).

Paper chromatography and thin layer chromatography have numerous common characteristics, in
practical technique and procedure:

The mixture to be examined is applied, usually as a
solution, to the previously marked starting points on
the prepared chromatographic papers or layers (*stationary phase*), using a pipette (2.4:2.4). For development, the zones should be as small and as 'dry' as
possible, and should not be damaged mechanically.

A solvent (*mobile phase*, 2.4:2.5) is applied to effect the
chromatographic process, which is carried out in an
ascending or descending manner in sealed separation
tanks containing high solvent vapor concentrations.
This saturated vapor phase has a marked influence on
the process of separation (2.4:2.3.3).

After development, the chromatograms are dried very
carefully, and the (usually colorless) zones are rendered
visible by the application of chemical reagents (*detection*, 2.4:4). Visible spots appear, whose color and
location on the chromatogram provide specific information about the nature of the substances. Quotations
of hR_f *values*[390] (h$R_f = R_f \times 100$) are meaningful as
standard values only if uniform separating phases are
used and if all external chromatographic conditions are
known.

If a standard substance *st* is chromatographed at the
same time, R_{st}-values can be given as the ratios of the
migration distances of the substances and the standard.

The quantitative evaluation of chromatograms is basically the same for the two separation methods. A
distinction must be made between direct and indirect
methods of estimation (2.4:6).

The most important differences between the two
chromatographic separation methods are summarized in Table 1.

Table 1. Summary of methodical and technical differences between paper chromatography and thin
layer chromatography.

Feature	PC	TLC
adsorbent	organic (cellulose or similar)	inorganic or organic
structure	fibrous net	small-grained ($\sim$ 5–50 μm)
pH value	neutral	acid, neutral, basic
size of carrier	58×60 cm	20×10 or 20×20 cm
development distance	30–50 cm	10–15 cm
direction of development	usually descending	usually ascending
development time	5 to 12 hours	20 to 40 min
spot size	200 to 10000 mm²	30 to 100 mm²
detection	non-corrosive reagents	no limitation, for inorganic adsorbents
usual sample size range	10 to 100 μg	0.5–10 μg

TLC is more universally applicable than PC because it has greater possibilities of variation and is much more sensitive. Because separation is sharper, the TLC method is now preferred in numerous fields of research. Paper chromatography is still employed on a large scale for the separation of *amino acids, nucleic acids, peptides, sugars* and other mainly hydrophilic compounds [377].

2 General Conditions

Selection of conditions for chromatography depends on the nature of the mixture of substances to be separated. If lipophilic solvents are used for extraction, the majority of the compounds extracted will also be lipophilic (non-polar). Chromatographic separation can be successfully achieved by way of the 'triangular' scheme [390, 399], using active stationary phases and non-polar solvents *(adsorption on the stationary phase)*. Polar substances can be separated on inactive layers, with hydrophilic (polar) solvents *(partition between two liquids)*.

2.1 Stationary Phases

A distinction is made between organic and inorganic stationary phases. Their choice depends upon their separation characteristics, and the permissible means of detection (2.4:4). Organic phases are used mostly for the separation of hydrophilic substances. Lipophilic mixtures, on the other hand, can be fractionated more effectively on adsorptive inorganic stationary phases. Here, particular use is made of silica gel and alumina. *Grain size* [μm], *specific surface* [m^2/g], *pore diameter* [Å] and *pore volume* [cm^3/g] are of importance for assessment of separation properties. The migration speed and the time required also depend on these factors.

In TLC, *binders* are frequently added to the stationary phases, in order to provide better adhesion of the powdered adsorbents to the support. The binders may either be inorganic (e.g. Plaster of Paris, colloidal silicic acids, or sodium silicate) or may consist of organic materials such as starch. Their concentration is sufficiently low (<15%) for them not to interfere with the chromatographic process. Abrasion-resistant layers can also be prepared by addition of polymers (e.g. polyvinyl alcohol or polyvinyl chloride). This is essential in the case of factory-made ready-for-use chromatographic plates or foils (2.4:2.3).

Cellulose and its derivatives are the most widely used *organic stationary phases*. Starch, saccharose, mannitol and the dextran gels are used in special cases only.

The most important of the *inorganic carrier materials*, which can also be used for the impregnation of chromatographic papers, are silica gel (silica gel glass fiber paper [1, 2]) and alumina (alumina paper) [3]. Calcium and magnesium silicates, phosphates or sulfates are also employed. Kieselguhr or kieselguhr paper [4] and glass powder are used as non-polar adsorbents.

2.1.1 Cellulose

Cellulose layers have separation properties analogous to those of pure cellulose papers. Frequently the same solvents can be used for development. There are differences, however, in the efficiency of separation and the duration of the chromatographic process.

Apart from the native, fibrous, cellulose powder, a *'micro-crystalline' cellulose* (e.g. Avicel®) is used in TLC. This is prepared from the powder by hydrolysis, and has a lower degree of polymerization [5, 6]. *Acetylated cellulose:* Whereas pure cellulose papers or powders are suitable for the separation of mixtures of hydrophilic substances, acetylated cellulose is preferable for the separation of lipophilic materials (e.g. *fat coloring matters, fatty acid hydroxamates* [7], *aryl halides* [8], *polycyclic aromatics* [9]). The lipophilic nature of the stationary phase increases with increasing degree of esterification. Visualization of the separated zones frequently presents difficulties.

Ion exchange cellulose: Native cellulose can be regarded as a weakly acidic cation exchanger. The

---

[1] *J.W. Dieckert, M. Brown, D.A. Yeadon, L.A. Goldblatt*, Anal. Chem. *29*, 30 (1957).
J.W. Dieckert, N.J. Morris, Anal. Chem. *29*, 31 (1957).
J.W. Dieckert, N.J. Morris, A.F. Mason, Arch. Biochem. *82*, 220 (1959).
J.W. Dieckert, R. Reiser, Science *120*, 678 (1954).
J.W. Dieckert, R. Reiser, J. Am. Oil Chem. Soc. *33*, 123 (1956).

[2] *J.G. Hamilton, J.W. Dieckert*, Arch. Biochem. Biophys. *82*, 203, 212 (1959).

[3] *A. Grüne*, Über Spezialpapiere für die Chromatographie, Druckschrift Fa. C. Schleicher & Schüll.
C. Schleicher & Schüll, Über Spezialpapiere für die Chromatographie und Elektrophorese, Dassel, Firmenschrift.

[4] *A. Jensen, S.L. Jensen*, Acta Chem. Scand. *13*, 1863 (1959).

[5] *M. Baudler, M. Mengel*, Z. Anal. Chem. *211*, 42 (1965).

[6] *P.P. Waring, Z.Z. Ziporin*, J. Chromatogr. *15*, 168 (1964).

[7] *F. Micheel, H. Schweppe*, Mikrochim. Acta 52 (1954).

[8] *A. Kabil, V. Prey*, Monatsh. Chem. *89*, 497 (1958).

[9] *T. Wieland, W. Kracht*, Angew. Chem. *69*, 172 (1957).

exchange capacity [mEquiv/g] can be altered either by chemical modification or by impregnation with exchange substances (Table 2).

The various cellulose products have different swelling capacities. Papers and TLC layers should be used in the moist state; otherwise reproducibility of separation cannot be guaranteed. In spite of their smaller exchange capacity, cellulose products are superior to synthetic resin exchangers (Chapter 2.8) for the separation of big molecules (nucleosides, proteins, etc.). Buffer or salt solutions are used mostly as solvents.

2.1.2 Dextran Gels

Dextran gels (e.g. Sephadex products) can be used also for chromatography of high molecular weight compounds *(peptides, enzymes, hormones,* etc.). Their ability to separate depends essentially on a filtration process. Depending on the degree of crosslinking, the dextrans can have different pore sizes. Separation of equally sized molecules into fractions is achieved in the same way as with ion exchange cellulose products, i.e. with buffer solutions (further details are given in Chapter 2.8).

Table 2. Most important ion exchange chromatographic cellulose products[390].

Abbreviations[a]	Group designation	Active groups	Field of application
AE	aminoethyl-cellulose	$R-O-C_2H_4 \cdot NH_2$	—
CM	carboxymethyl-cellulose	$R-O-CH_2COOH$	—
DEAE	diethylaminoethyl-cellulose	$R-O-C_2H_4 \cdot N(C_2H_5)_2$	nucleotides[10, 388], purines and pyrimidines[11], steroid sulfates[12]
TEAE	triethylaminoethyl-cellulose	stronger base than AE	[13]
Oxicellulose		polyfunctional	arginine[14]
ECTEOLA	reaction product of epichlorohydrin, triethanolamine and alkali-cellulose	weaker base than DEAE	sugar phosphates and nucleotides[15], glucoronosides[12], desoxyribonucleotides[16, 17]
P	phosphorylated cellulose	$R-O-PO_3H_2$	cations[18]
PEI	impregnation with polyethylene imines	$(-CH_2-CH_2-NH-)_X$	mono- and oligo-nucleotides[19, 388]
Poly-P	impregnation with sodium polyphosphate	$(-PO_3Na)_X$	nucleosides, nucleic bases[388]
resin ion exchanger	impregnation e.g. with Dowex 50, IRA 400		carboxylic acids and their esters[20], basic amino acids[21], uranium/bismuth separation[22]

[a] The abbreviations used designate the natures of the stationary phases, and do not pertain to particular firms.

[10] *T.A. Dyer,* J. Chromatogr. *11,* 414 (1963).
[11] *R.G. Coffey, R.W. Newburgh,* J. Chromatogr. *11,* 376 (1963).
[12] *G.W. Oertel, M.C. Tornero, K. Groot,* J. Chromatogr. *14,* 509 (1964).
[13] *J. Porath,* Ark. Kemi *11,* 259 (1957).
[14] *T. Wieland, A. Berg,* Angew. Chem. *64,* 418 (1952).
[15] *C.P. Dietrich, S.M.C. Dietrich, H.G. Pontis,* J. Chromatogr. *15,* 277 (1964).
[16] *R.D. Bauer, K.D. Martin,* J. Chromatogr. *16,* 519 (1964).
[17] *J.E. Meinhard, N.F. Hall,* Anal. Chem. *21,* 185 (1949).

[18] *N.F. Kember, R.A. Wells,* Nature *175,* 512 (1955).
[19] *G. Weimann, K. Randerath,* Experientia *19,* 49 (1963).
[20] *G. Illing,* Arch. Pharm. *290/62,* 581 (1957).
[21] *M.M. Tuckermann, R.A. Osteryoung, F.C. Nachod,* Anal. Chim. Acta *19,* 250 (1958).
[22] *H.T. Peterson,* Anal. Chem. *31,* 1279 (1959).
[23] *V.D. Canić, S.M. Petrović,* Z. Anal. Chem. *211,* 321 (1965).
[24] *B. Colman, W. Vishniac,* Biochem. Biophys. Acta *82,* 616 (1964).
[25] *L.W. Smith, R.W. Breidenbach, D. Rubenstein,* Science *148,* 508 (1965).

2.1.3 Starch, Saccharose, Mannitol
Starch, saccharose and mannitol are highly hydrophilic adsorbents. They are used for the chromatography of *cations* (starch[23]) and *plant pigments* (saccharose[24], mannitol[25]). Lipophilic mobile phases are used exclusively as eluants.

2.1.4 Polyamides (Perlon®, Nylon®)
Because of their cross-linking, polyamide papers or layers swell considerably in alcohol, acetone and dimethyl formamide, but are practically insoluble. Their main application is for the separation of *phenolic compounds (tannins[26], phenolic carboxylic acids,* and *nitro phenols[27]). Quinones* become irreversibly adsorbed on polycaprolactam powder (Perlon®) unless the free amino groups are blocked by acetylation[27].

2.1.5 Silica Gel
The adsorbent used most frequently in TLC is silica gel.

The amorphous, porous powder usually has a particle size of 5 to 50 µm. The *specific surface area* may be as high as 800–1000 m²/g, and is about 400 to 500 m²/g for most commercially available materials. The *pore diameter* is usually between 20 to 150 Å.

Chemical structural elements on the surface of the silica gel particles are silanol (–SiOH) groups and siloxane (–Si–O–Si–) bridges. The silanol groups behave as weak proton donors in aqueous systems[28], whereas the siloxane bridges act as proton acceptors. The more (physically bound) water molecules surrounding the active centers[29, 30], the less active is the layer material[31, 384].

2.1.6 Alumina
As in column chromatography, alumina is also used successfully in TLC[32, 33] and for impregnation of chromatographic papers[3].

The grain size is below 60 µm, and the mean pore diameter is 20–80 Å. The specific surface has been determined as being between 100 and 350 m²/g. The surface is covered with hydroxy groups[34]. In addition, the oxygen atoms of the oxide lattice have an adsorbing effect through hydrogen bonds formation.

Because of its high coordinating power, alumina is very suitable for the separation of substances with various functional groups. Compounds which form intramolecular hydrogen bridges *(multivalent alcohols, phenols* and *carboxylic acids)* are separated better on alumina than on silica gel.

The pH value of the stationary phase frequently provides support. Alumina contains alkali ions, originating from their process of preparation. Consequently they give a basic reaction in aqueous suspension. The commercially available neutral and weakly acidic types are produced[35, 385, 392] by washing with acids.

2.1.7 Kieselguhr (Diatomaceous Earth, Celite®)
The specific surface of this amorphous, porous, purified powder as prepared for chromatography is small (1 to 5 m²/g), and consequently the adsorptive influence of the silanol and siloxane groups is only weak. Kieselguhr is very suitable as carrier material for special types of impregnation.

2.1.8 Calcium and Magnesium Silicates
Calcium and magnesium silicates (e.g. Florisil®) have so far shown advantages only for special separations of *polyhydroxy compounds[36–38], lanatosides[39]* and some *terpene derivatives[40].* In aqueous suspension, these silicates give an alkaline reaction (pH 8–10). For preparation of TLC layers, the silicate powder may be slurried with alcohol, without forming lumps.

2.2 Selection of the Stationary Phase
The selection of a suitable stationary phase depends on the mixture of substances to be separated. The various adsorbents may be classified according to their activity[41, 42].

Tabulations of adsorbent activity can serve only as guides, since orders of activity may change with the external chromatographic conditions. In the selection of an adsorbent, it is, therefore, often preferable to consider the possible types of interaction of the stationary phases (Table 3).

[26] *P. Stadler, H. Endres,* J. Chromatogr. *17,* 587 (1965).
[27] *H. Endres* in *E. Stahl*[390]; *H. Endres* in *K. Macek, I.M. Hais*[369].
[28] *A. Kiselev, V.J. Lygin,* Kolloid-Zh. *21,* 581 (1953).
[29] *S. Heřmánek* in *L. Lábler*[385].
[30] *J. Pitra, J. Štěrba,* Chem. Listy *56,* 544 (1962); *57,* 389 (1963).
[31] *H.W. Kohlschütter, K. Unger* in *E. Stahl*[390].
[32] *E.A. Mistriukov,* Collect. Czech. Chem. Commun. *26,* 2071 (1961).
[33] *M. Mottier, M. Potterat,* Mitt. Gebiete Lebensm. Hyg. *47,* 372 (1956); *49,* 454 (1958).
[34] *H. Rössler* in *E. Stahl*[390].

[35] *Lj. Kraus* in *M. Šaršunová* et al. [392].
[36] *H. Grasshof,* Dtsch. Apoth.-Ztg. *103,* 1396 (1963); J. Chromatogr. *14,* 513 (1964).
[37] *J.P. Tore,* J. Chromatogr. *12,* 413 (1963).
[38] *M.L. Wolfrom, R.M. De Lederkremer, L.E. Anderson,* Anal. Chem. *35,* 1357 (1963).
[39] *J. Zurkowska, A. Ozarowski,* Planta Medica *12,* 222 (1964).
[40] *L.H. Bryant,* Nature *175,* 556 (1955).
[41] *J. Dvořák, I.M. Hais, A. Tockstein,* in *I.M. Hais*[369]
[42] *C. Hesse, I. Daniel, G. Wohlleben,* Angew. Chem. *64,* 103 (1952).

Table 3. Summary of the adsorbents used in TLC, indicating their types of interaction[29]. The number of + signs symbolizes the strength of adsorptive power. + = weak; + + = medium; + + + = strong

Adsorbent	Dipole effect	Hydrogen bonds	Dispersion forces	Coordination bonding	Salt effect ac.	Salt effect bas.	Ion exchange capacity
Silica gel	+	+ + +	+ +	+ +	—	+ +	+
Al$_2$O$_3$, alk.	+ +	+ + +	—	+ + +	+ + +	—	+ +
Al$_2$O$_3$, ac.	+ +	+ + +	—	+ + +	—	+ +	+ +
Al$_2$O$_3$, neutr.	+ +	+ + +	—	+ + +	—	—	+
Ca(OH)$_2$	+ +	+ + +	+	+ +	+ + +	—	+ +
CaCO$_3$	+ +	+ +	+	+ +	+ +	—	+ +
Ca$_3$(PO$_4$)$_2$	+ +	+	+	+ + +	—	—	+ +
CaSO$_4$	+ + +	+	+	+ + +	—	—	+ +
MgO	+ +	+ +	—	+ + +	+	—	+ +
Mg$_3$(PO$_4$)$_2$	+ + +	+	—	+ +	—	—	+ +
Mg(OH)$_2$	+ +	+ +	+	+ + +	+ +	—	+ +
Mg silicate	+ +	+ +	+	+ +	+	—	+ +
Fe$_2$O$_3$ hydrate	+ +	+ +	+	+ + +	+	—	+
glass	+	+	+ +	+ + +	—	—	+
Na$_2$CO$_3$	+ +	+	+	+ +	+ + +	—	+ +
polyamide	+ + +	+ + +	+ +	+ +	—	—	—
polyacrylonitrile	+ + +	+	+ +	—	—	—	—

Table 4. Properties of the most important chromatographic papers (see also Ref. [369]) (used with the Partridge mixture: butanol/acetic acid/water = 40/10/50).

Type of paper	Weight [g/m²]	Thickness [mm]	Capillary adsorption period for 30 cm [min]	Migration time [h]
standard	85–125	0.16–0.26	140–370	15–20
slow-flowing	90–125	0.16–0.17	400–780	20–45
fast-flowing	95–145	0.18–0.20	70–190	7–9
for preparative purposes	180–280 600–700 2400–2500	0.31–0.36 0.51–0.70 3.60–4.20	140–250 270–290 ⎱ 260–320 ⎰	9–11 23–28

Alcohols, ethers or *esters* separate best on adsorbents capable of forming hydrogen bonds or dipole associates.

As eluants, liquids are used whose donor or acceptor qualities, or dipole strengths, correspond to those of the samples under investigation. If the adsorbent has a considerably greater affinity for the compounds being chromatographed than for the mobile phase, the compounds will remain at the starting point. In the converse case, they will migrate near the solvent front.

2.3 Production and Preparation of the Stationary Phase

2.3.1 Papers for Chromatography

Chromatographic papers comprise fibrous networks of pure cellulose, which can be used as stationary phase when in the swollen state, and which can serve as carrier material for special impregnations. The thickness of the paper, and its ability to adsorb or swell, are important factors determining the capacity and duration of chromatography (Table 4). Selection of a paper depends on the type of substances to be separated, and on the solvent and the required degree of separation. Fast-flowing papers usually give less efficient separation[43, 44, 363].

2.3.2 TLC Plates

In TLC, the stationary phase is always supported on a carrier material (e.g. glass, plastic, or metal foil). Sufficient adhesion by the powder is ensured by incorporation of inorganic or organic binders.

[43] *G.N. Kowkabany, H.G. Cassidy*, Anal. Chem. *22*, 817 (1950).
[44] *D.P. Burma*, J. Indian Chem. Soc. *29*, 567 (1952).

Abrasion-resistant ready-for-use layers − *plates or foils* − are commercially available, as are chromatographic papers. Their advantages and disadvantages have already been summarized[45]. A spreading or spraying procedure is generally used for plates smaller than 20×20 cm or 20×10 cm[390].

2.3.3 Drying, Activation, Conditioning and Saturation of Stationary Phases

TLC layers produced according to any of the processes mentioned above usually contain unknown quantities of water which must be largely removed before chromatography. The procedure for drying and activation of the stationary phase depends on the nature of the mixture to be separated and on the external chromatographic conditions (atmospheric moisture, ect.). If *hydrophilic compounds* such as amino acids or sugars are to be chromatographed, it is usually sufficient to dry the layers in air for a few hours, until the initial watery sheen of the surface and the transparency have disappeared, and equilibrium with the surrounding atmosphere has been attained. This process of evaporation (up to 10 ml per 20×20 cm layer) can be accelerated by cautious ventilation.

With the separation of *lipophilic compounds*, more intensive drying (activation) is often advantageous. This is effected by heating to $110-130°$ for at least 30 min in a drying oven with circulating air. Commercial precoated TLC plates can be treated similarly. The activated plates are cooled and stored over silica gel desiccant in vacuum desiccators. However, this is worthwhile only if the subsequent sample application and chromatography are also carried out in rooms with dried air; otherwise, the moisture equilibrium mentioned above will cause deactivation[46, 47] within a few minutes.

Intentional deactivation is sometimes called *conditioning*, or *climatization*. Here, the chromatographic papers or layers are exposed to the appropriate atmosphere of the mobile phase, after spotting with the solution, and before developing. This process causes saturation (2.4:3.1) of the stationary phase with vapor molecules, or swelling of organic layers. Only after this is development carried out (cf. [47−51]).

2.3.4 Impregnation of Stationary Phases

The separation properties of a stationary phase can be changed intentionally by *impregnation*. This can be done by special treatment of the carrier material before the stationary phase is produced, e.g. acetylation of cellulose, silanization of silica gel[52, 390] or flat ready-for-use carriers may be subjected to dipping, spraying or ascending treatment, similar to development. Impregnation by amounts of more than 5 to 10% does not usually improve the separation effect any further (control with gradient TLC: 2.4:3.3).

In TLC, some acids can be used instead of water to prepare the material to be spread (e.g. sulfuric, phosphoric or oxalic acids[53]); basic solutions such as caustic soda of suitable normality can also be used. Buffer systems (boric acid/borate mixtures) or salt solutions can also be employed. Copper or silver salts are used if compounds with different numbers of $C=C$ double bonds are to be separated; complexes are formed between the metal ions and the π-electrons of the unsaturated groups; these have bonds of differing strenghts, thus permitting stepwise separation[54, 381, 390].

Two types of organic impregnating media can be distinguished: temporary and permanent. Both types cause the stationary phase to become hydrophobic and thus lead to phase reversion. Hydrophilic solvents saturated with the impregnation medium are used for development. Permanent *impregnation* can, for example, be done with paraffin oil, silicone oil or squalene.

Temporary impregnations (e.g. with undecane) may usually be removed by heating before detection of the separated zones. Chromatographic separations, however, are naturally less easily reproducible.

2.4 Sample Application

2.4.1 Application of Liquids

In many cases, the nature of the solution to be investigated is determined by an extraction procedure. Frequently, a preliminary separation into lipophilic and hydrophilic compounds is carried out. Should too many contaminants which would interfere with chromatography still be present, appropriate purification steps must be taken (precipitation of proteins, removal of fats or salts). The more volatile the solvents used for application, the smaller the diameter of the starting zone can be

[45] *H. Jork*, Pharma International *4*, 33 (1968).
[46] *E. Dumont*, Dissertation, Universität Saarbrücken, 1968.
[47] *F. Geiss, H. Schlitt, A. Klose*, Z. Anal. Chem. *211*, 37 (1965); *213*, 321, 331, (1965).
[48] *F. Geiss, S. Sandroni*, J. Chromatogr. *33*, 201, 208 (1968).
[49] *J. Pitra*, J. Chromatogr. *33*, 220 (1965).
[50] *H.W. Prinzler, H. Tauchmann*, J. Chromatogr. *29*, 142 (1967).
[51] *R.A. de Zeeuw*, J. Chromatogr. *32*, 43 (1968); *33*, 222 (1968).
[52] *D.K. Halle*, Chem. Ind. [London] 1147 (1955).
[53] *E. Stahl*, Arch. Pharm. *292*, 411 (1959).
[54] *P.J. Schorn*, Z. Anal. Chem. *205*, 298 (1964).

kept; the sharpness of separation depends very much on the size of this zone.

In descending PC, starting points are marked about 15 cm from the paper edge. For ascending chromatography, a distance of 4–5 cm is sufficient, (in TLC, even 1.5 cm is adequate). The distances between the starting spots can be 1–3 cm, depending on the chromatographic technique and the problem involved.

The concentration of the sample solution should be 0.1–1% in TLC, and 1–5% in PC. When application is in the form of spots, 1 to 10 μl should be applied using pipettes or micro-syringes. The following different kinds can be used: pipettes, self-filling capillary pipettes (microcaps), constriction pipettes (Lambda pipettes), bulb pipettes and micro-syringes. For quantitative work, calibrated bulb pipettes or micro-syringes are mostly recommended[55]. Reproducibility of applied dosage is between v=±0.6–0.8%[56, 396].

$$ v = \pm \frac{100 \cdot s}{\bar{x}} = \frac{100 \cdot \sqrt{\dfrac{\Sigma (\Delta x)^2}{n-1}}}{\bar{x}} $$

v = coefficient of variability
s = relative standard deviation
Δx = difference between mean value and measured value
$\bar{x}$ = mean value
n = number of measured values

Application of the solution in streaks rather than spots usually yields better results, because separation is better[46, 57]. This method is especially advantageous in micro-preparative chromatography. Automatic spreaders[57, 58] give the best results.

Occasionally a problem involves powdered substances, tissue sections, or single glandules rather than extracts, and requires direct study or transference of contents onto the separation layer[59–63].

2.4.2 Thermomicro Separation and Application of Substances (TAS Method)

In the TAS method, solid samples are usually inserted directly[64, 65]. Normally, 10 to 25 mg are

sufficient for one analysis. The sample is put into a high-melting glass mandrel, either with a spatula, or in a small aluminium cartridge. One end of the glass mandrel is drawn out to a short capillary tube and the other can be closed with a clip (Fig. 1). This special mandrel is inserted into a hole bored in a preheated aluminium block, so that the end of the capillary pipe protrudes horizontally. Substances that are volatile at the prevailing temperature are evaporated or sublimed onto the layer, which is situated just in front of the oven. This procedure may be carried out as a microscale steam-distillation if moistened propellants (starch, deactivated silica gel) are inserted into the cartridge at the same time.

This qualitative method has proved especially valuable for samples which are hard to handle analytically, such as *ointments, emulsions* and *plasters*[66, 67]. The method also offers advantages in the analysis of *drugs*[66, 399].

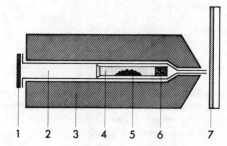

Fig. 1. Diagram showing longitudinal section of TAS heating block (acc.[67])

1 seal	5 sample
2 glass mandrel	6 quartz wool
3 heating block	7 TLC layer
4 aluminum cartridge	

2.5 Mobile Phases (Eluants)

2.5.1 Eluotropic Series

In adsorption chromatography, *elution power* serves as a classification principle. Apart from some irregularities, the following sequence[68] is widely accepted. Here, the elution effect increases in parallel with the dielectric constant:

[55] J.W. Fairbairn, S.J. Relph, J. Chromatogr. 33, 494 (1968).

[56] H. Jork in E.J. Shellard[395].

[57] E. Stahl, E. Dumont, J. Chromatogr. 39, 157 (1969); Talanta 16, 657 (1969); J. Chromatogr. Sci. 7, 517 (1969).

[58] E. Stahl, Z. Anal. Chem. 236, 294 (1968); Qualitas Plantarum et Materiae Vegetabiles 20, XII., 35 (1968).

[59] R. Antoszewski, J. Chromatogr. 2, 220 (1959).

[60] W.M. Doizoki, M. Lziere, Proc. Soc. Exp. Biol. Med. 113, 91 (1963).

[61] J. Herrmann, Pharmazie 10, 259 (1955).

[62] G. Lindner, Arch. Klin. Exptl. Dermatol. 206, 379 (1957).

[63] E. Stahl, Z. Anal. Chem. 221, 3 (1966).

[64] Lj. Kraus, E. Stahl, Česk. Farm. 18, 486 (1969).

[65] E. Stahl, J. Chromatogr. 37, 99 (1968); Z. Anal. Chem. 234, 1 (1968).
E. Stahl, J. Fuchs, Dtsch. Apoth.-Ztg. 108, 1227 (1968).

[66] Lj. Kraus, E. Stahl, Česk. Farm. 18, 535 (1969); Arzneim.-Forsch. 20, 1814 (1970).
Lj. Kraus, E. Stahl, H. Gross, Česk. Farm. 19, 325 (1970).

[67] E. Stahl, Lj. Kraus, Arzneimittel-Forsch. 19, 684 (1969).

[68] W. Trappe, Biochem. Z. 305, 151 (1940).

Table 5. Possible interrelations of mobile phases in chromatographic systems[69]. The number of + signs symbolizes intensity of bonding: + weak, + + medium, + + + strong.

Solvent	H acceptor	H donor	Dipole bond	Dispersion forces	Coordination forces	Ionic bond
aliphatic hydrocarbons	—	—	—	+++	—	—
carbon tetrachloride	—	—	+	+++	+	—
carbon disulfide	—	—	—	+++	+	—
benzene	+	—	—	+++	+	—
chloroform	—	++	++	++	+	—
dichloromethane	—	+	+++	++	+	—
dichloroethylene	—	+	+++	+++	+	—
diethyl ether	+++	—	++	+++	+++	—
isopropyl ether	+++	—	++	+++	++	—
tetrahydrofuran	+++	—	+++	++	+++	—
methanol	++	+++	+++	++	+++	+
n-butanol	+++	++	++	+++	++	+
phenol	++	+++	+++	++	++	++
acetone	+++	+	+++	++	+++	—
acetic acid	++	+++	++	++	++	+++
ethyl acetate	++	+	+++	+++	++	—
water	+++	+++	+++	+	+++	++
formamide	+++	++	+++	+	+++	+
dimethyl formamide	+++	+	+++	++	+++	—
pyridine	+++	—	++	++	+++	—

petroleum ether < cyclohexane < carbon tetrachlorid- < trichloroethylene < toluene < benzene < dichloromethane < chloroform < diethyl ether < ethyl acetate < acetone < n-propanol < ethanol < methanol < water.

This order relates only to systems in which dispersion forces are effective, or where dipolar associations appear. If bonding of a substance depends on the formation of hydrogen bridges, then the donor or acceptor properties of the mobile phase have a decisive influence on elution. Similar effects are recognized for coordination bonds (Table 5).

With development on polyamide phases, eluting powder increases in the following order: water, methanol, acetone, dilute caustic soda, formamide, dimethyl formamide[27].

2.5.2 Selection of Eluants

For a given substance and a selected stationary phase, a suitable solvent can be chosen from Table 5. The so-called *micro-circular technique* may be helpful (2.4:3.2.1) as a rapid means of obtaining guidance on the separation effect.

The mixture to be separated is applied in spots to the paper or separation layer, and is then eluted with the pure mobile phase from the center (pipette). The eluant can be considered optimal if as many well-separated coaxial rings as possible develop. If, however, the substances migrate close together near the solvent front, the mobile phase is too polar; if they remain near the origin, the polarity of the solvent should be increased.

Similar considerations apply to reversed-phase chromatography.

If a mobile phase has been found which provides successful separation, this system may often be replaced by *isopartitive solvents;* variations to suit specific problems are thus possible[69, 70]. By altering the concentrations of polar and non-polar components, physico-chemical interaction effects may be taken into account.

3 Chromatography and Practical Techniques

3.1 Types of Separation Tanks

Well sealed chromatography tanks (troughs) are used, on whose walls filter paper can be fixed in order to accelerate saturation of the atmosphere in the tank with solvent vapor *(chamber saturation).* This precaution provides better reproducibility of the chromatographic results, and suppression of the so-called *edge effect.* It is best to work in rooms of constant humidity and temperature[47, 51, 53]. In PC, this additional measure to produce faster equilibration is less common *(normal saturation).* Here, development is usually a matter of some hours. Usually the paper is first suspended in the tank to attain a degree of swelling dependent

---

[69] S. Heřmánek in Šaršunová [392].
[70] K. Macek, Z. Procházka in I.M. Hais, K. Macek [377].

on the system involved, before the actual chromatographic process is carried out[390].

For the development of *ring chromatograms* in PC, frequently two Petri dishes fitted together so that their edges meet are used as a vessel for development.

The paper or foil is spread between the two dishes. The eluant is fed from below with a paper tongue or cotton wick; elution from above is possible only if a suitable pipette is used.

If coated glass plates are developed radially, the carrier plates can serve as the upper covering. In this case, the stationary phase faces the eluant which is fed from below.

3.2 Methods of development

In *ascending chromatography* (usually used in TLC) a continuous flow development is possible, if the solvent is continually removed from the front of the chromatogram[71-74].

In *descending chromatography* (normally used in PC) the solvent enters the stationary phase from a trough situated at a higher position. This technique can be used for continuous flow development if the solvent is allowed to drip down evenly. *Horizontal chromatography* is frequently coupled with special experimental techniques (2.4:3.2.1).

3.2.1 Special Techniques

A variation of horizontal development is the *circular (radial) technique*. If this is used on rotating stationary phases, the process of separation occurs under the influence of *centrifugal force*[75-79]. The eluant is fed from the center of the stationary phase, the mixtures being applied as spots equidistant from the center.

During chromatography, the substances spread along circular paths, and are concentrated in small areas. This technique is preferred when the concentration differences between the individual substances are relatively large. It may, however, also be used for preliminary selection of solvent systems (cf. 2.4:2.5.2).

The *wedged tip technique* was derived from radial horizontal development. This technique can also be used for vertical chromatography[80-82]. Mixtures of materials are applied on strips cut out[83] from or scratched[82] into TLC layers; the mobile phase flows through these strips. Above the starting zone, widening occurs, usually wedge-shaped; here, too, zones form circular arcs.

3.2.2 Multiple Development, Step Technique

Separation results may be improved occasionally by *multiple development* with the same or different mobile phases[84-87]. Here, the initial development is succeeded by a further chromatographic process. With suitable intermediate drying, this sequence can be repeated several times until the intended separation has been successfully reached. If the migration distance of the mobile phase is not the same every time, the procedure is called a *step technique*.

3.2.3 Two-dimensional Development

The advantage of *two-dimensional development* with the same solvent system is that a preliminary separation of a complicated mixture can be performed in the first dimension, and mutual interaction of different substances during the second development can be prevented. Complicated mixtures of similar substances are often separated by two-dimensional development with different eluants. This can be used to check the homogeneity of a substance.

3.2.4 SRS-Technique, Functional Chromatography

If substances are altered intentionally between two development processes, the procedure is called the SRS-technique (SRS = separation–reaction–separation). By carrying out reactions (hydrolysis[88], coupling[89], or photochemical processes[53]) which are specific for certain groups, this tech-

[71] R.D. Bennett, E. Heftmann, J. Chromatogr. *12*, 245 (1963).

[72] M. Brenner, A. Niederwieser, Experientia *17*, 237 (1961).

[73] R. Hüttenrauch, J. Schulze, Pharmazie *19*, 334 (1964).

[74] E.V. Truter, J. Chromatogr. *14*, 57 (1964).

[75] C.G. Caronna, Chim. Ind. [Milan] *37*, 113 (1955).

[76] Z. Deyl, M. Pavlíček, J. Rosmus, J. Chromatogr. 7, 19 (1962).

[77] H.J. McDonald, E.W. Bermes jr., H.G. Sheperd, Naturwiss. *44*, 9, 616 (1957); Chromatogr. Methods *2*, 1 (1957); Anal. Chem. *31*, 825 (1959).

[78] B.P. Korzum, S. Brody, J. Pharm. Sci. *53*, 454 (1964).

[79] J. Rosmus, M. Pavlíček, Z. Deyl in G.B. Marini-Bettòlo[382].

[80] W. Matthias, Naturwissenschaften *41*, 17 (1954).

[81] L.F. Reindel, H.W. Hoppe, Naturwissenschaften *40*, 245 (1953).

[82] E. Stahl, Parfümerie und Kosm. *39*, 564 (1958).

[83] G.H. Götte, P.D. Pätze, Angew. Chem. *69*, 608 (1957).

[84] E. Stahl, H. Vollmann, Talanta *12*, 525 (1965).

[85] E. Schratz, W. Egels, Planta Medica *6*, 148 (1958).

[86] A. Jeanes, C.S. Wise, R.J. Dimler, Anal. Chem. *23*, 415 (1951).

[87] H.P. Lenk, Z. Anal. Chem. *184*, 107 (1961).

[88] H.P. Kaufmann, Z. Makus, J.H. Khoe, Fette, Seifen, Anstrichm. *64*, 1 (1962).

[89] A. Fono, A.M. Sapse, T.S. Ma, Mikrochim. Acta 1098 (1965).

nique can be used for the characterization of chromatographically separated substances.

In the case of so-called *functional chromatography,* a chemical reaction, e.g. an epoxidation, bromination or esterification, is carried out at the origin[90-92] before development, either in sample application tubes or directly.

3.3 Gradient Techniques

Various gradient techniques exist: vapor phase gradient, gradient elution, and chromatography on gradient layers. In the first two cases, almost without exception, uniform stationary phases are employed. hR_f values can be quoted.

For *chromatography on gradient layers,* the hR_f value is a function of the respective layer property. Curves are obtained, which are substance-specific, and which can often be used to obtain better identification.

A *vapor phase gradient*[47, 51] exists in every separation chamber (cf. also 2.4:2.3.3).

In the *gradient elution technique*[390], the elution effect of an eluant is continuously changed by addition of further components (cf. Chapter 2.5). In these gradient techniques, only ascending and descending development are possible[93]. Special types of apparatus for use with lipids are described elsewhere[94, 95].

More universally applicable than the two gradient techniques mentioned above are the *gradient layers*[96]; so far, these have been developed only for TLC. Besides adsorbent gradients (e.g. silica gel-cellulose), activity gradients, layer thickness and grain size gradients, impregnation gradients have proved most successful. By developing in a direction transverse to the gradient, the concentration which yields optimum separation can be determined quickly[97]. This holds also for pH gradient layers[46, 57, 98-101], which can be prepared with the so-called GM spreader, usually by using the diagonal divider[96].

3.4 Micro-preparative Chromatography

When a mixture has been separated successfully in the μg level, separation on the preparative scale is often required. This is especially so if previous attempts with column chromatography have failed. If differences between the hR_f values are small *(fine separation),* correspondingly many chromatograms will be required on these thin layers which have a low loading capacity. If the zones are further apart (crude separation), papers or layers of 1 mm thickness can be used for chromatography.

Silica gel and alumina layers of thickness up to 5 mm can be used, but thickness of 0.5 to 1 mm are most usual (cellulose layers are generally <1 mm[102]). The solutions chosen for spreading have concentrations of 5 to 20%. In order to make maximum use of the stationary phase, application of the sample in bands is recommended (2.4:2.4.1); this also diminishes the danger of overloading the layer.

Nevertheless, tailing is often observed. Since the solvent evaporates more quickly on the surface, migration is faster on the surface layer than in lower layers. The distribution profile of the compounds being separated is thus oblique. This may explain the fact that neighbouring zones cannot usually be separated completely by micro-preparative methods. This is also true for adsorbents which have been developed especially for micro-preparative TLC.

The wide bands resulting after the first elution may, under certain circumstances, be narrowed by multiple development. On larger layers (20×40 or 20×100 cm)[102] up to 10 g of substance may be separated in this way.

The time required for drying at normal room temperatures depends on the thickness of the preparative layers, and is between 5 to 7 hours per 1 mm layer thickness. Subsequent activation is recommended (e.g. with hot air blowers). To check the success of separation, any methods may be used which make the zones visible without the action of heat (cf. 2.4:4.1), and which do not decompose the compounds on the chromatogram. For micro-preparative extraction of selected substances, the appropriate portions of the chromatogram should be cut out or scraped off immediately after development. Small 'vacuum cleaners' may be used to remove the layer, unless there is a danger that the substance may oxidize (Fig. 2)[390].

[90] *A. Kaess, C. Mathis,* IV. Intern. Symposium Chromatogr. Brüssel 1966, Symposiumsband 1968, p. 525; Ann. Pharm. France *24,* 753 (1966).
[91] *C. Mathis, G. Ourisson,* J. Chromatogr. *12,* 94 (1963); Ann. Pharm. France *23,* 331 (1965).
[92] *G. Ponsinet, G. Ourisson,* Phytochemistry *4,* 799 (1965).
[93] *M. Knedel, Fatech-Moghdam,* Glas-Instrum.-Tech. *9,* 675 (1965).
[94] *A. Niederwieser, C.G. Honegger,* Helv. Chim. Acta *48,* 893 (1965).
      A. Niederwieser, J. Chromatogr. *21,* 326 (1966).
      S.M. Rybicka, Chem. & Ind. [London] 308 (1962).
[95] *T. Wieland, H.D. Determann,* Experientia *18,* 431 (1962).
[96] *E. Stahl,* Chem.-Ing.-Tech. *36,* 941 (1964).

[97] *K.E. Rozumek,* J. Chromatogr. *40,* 97 (1969).
[98] *Lj. Kraus, E. Dumont,* J. Chromatogr. *48,* 96 (1970); Z. Anal. Chem. *252,* 380 (1970).
[99] *E. Dumont, Lj. Kraus,* J. Chromatogr. *48,* 106 (1970).
[100] *J. Poldermann,* Pharm. Weekblad *101,* 421 (1966).
[101] *E.J. Shellard, M.C. Alam, J. Armah,* FIP-Kongress, Symp.-Band, Prag 1965.
[102] *H. Halpaap,* Chem.-Ing.-Tech. *35,* 488 (1963); Chem.-Ztg. *89,* 835 (1965).

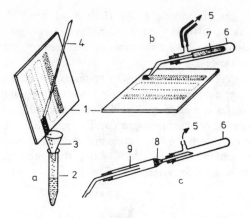

Fig. 2. Alternative methods for collection of zones after separation[390], a) scraping with micro-spatula, b) removal with 'vacuum cleaner', c) modified vacuum cleaner

1 TLC plate with layer	6 vacuum tube
2 test tube with eluant	7 Soxhlet thimble
3 funnel	8 cotton wool plug or
4 micro-spatula	filter plate
5 vacuum connection	9 bent glass tube

Subsequent elution of the substances from the stationary phase is best done in small flasks or by percolation in Allihn tubes of suitable size. Filtration through washed filter paper is thus avoided. If polar solvents are used for extraction, high blank values must be expected, especially with photometric measurements. Materials which interfere may originate from the stationary phase or the eluant. The materials used must, therefore, be of very high purity.

3.5 Chromatography of Radioactively labelled Substances
Radioactively labelled substances are often employed for solution of analytical problems in the fields of biology and medicine, e.g. in the study of *lipids*[103−105, 390], *steroid alcohols*[106, 107, 363], *amino acids*[108], *isoflavones*[108, 109] and [131]*J-insulin*[110]. Special precautions are legally required[111].

[103] *F.M. Ganis*, Anal. Chem. *30*, 2068 (1958).

[104] *H.K. Mangold*, Fette, Seifen, Anstrichm. *61*, 877 (1959).
H.K. Mangold in *E. Stahl*[390].

[105] *H. Schlenk, J.L. Gellermann*, Anal. Chem. *32*, 1412 (1960).

[106] *V.P. Hollander, J. Vinecour*, Anal. Chem. *30*, 1429 (1958).

[107] *B. Kliman, R.E. Peterson*, J. Biol. Chem. *235*, 1639 (1960).

[108] *H. Griesebach, N. Doerr*, Z. Naturforsch. *14b*, 802 (1959); *15b*, 284 (1960).

[109] *H. Griesebach, L. Patschke*, Chem. Ber. *93*, 2326 (1960); Z. Naturforsch. *16b*, 645 (1961).

[110] *A. Massaglia, U. Rosa*, J. Label. Compounds *1*, 141 (1965).

[111] Strahlenschutzverordnung I mit Anlagen vom 24. III. 1964. Bundesgesetzblatt I, S. 233.

4 Methods of Detection
Colorless compounds cannot easily be detected directly on the chromatogram, and in these cases special detection methods must be used, as described below.

4.1 Physico-chemical Detection Methods
These include:

− fluorescent excitation of materials, with suitable light sources; visualization of absorbing compounds by quenching of fluorescence on fluorescent stationary phases. For substances which absorb radiation at wavelengths near $\lambda = 254$ nm, dark zones appear on a yellow-green, blue or red background, depending on the fluorescent indicator used;
− thermal methods, including pyrolysis;
− micro-sublimation with subsequent melting point determination or crystallooptical analysis;
− detection of labelled substances[363, 364, 390] by autoradiography, scintillation procedures or direct estimation (2.4:6.5.4).

4.2 Chemical Detection Methods
In the most popular chemical detection methods, the stationary phases after development are either exposed to a suitable gas atmosphere (ammonia, bromine or iodine), or sprayed with a suitable reagent solution.

Almost all the reagents common in PC are also utilized in TLC. In addition, corrosive spray solutions such as concentrated sulfuric acid, chromic/sulfuric acid, etc. can be used on inorganic separation layers. By adding aliphatic or aromatic aldehydes (formaldehyde, anisaldehyde, furfural, salicylaldehyde, vanillin or 4-dimethylaminobenzaldehyde), spray solutions can be obtained which often react even at low temperatures with the chromatographed substances, forming colored derivatives or complexes. In many cases, heating to temperatures between 100 and 200° is of advantage (drying cabinet, hot-plate, IR heater). The color changes appearing on heating are sometimes characteristic. Further universal reagents and so-called group-specific spray solutions are cited in the bibliography (2.4:7).

4.3 Biological-physiological Detection Methods
Compounds with physiological activity can frequently be detected by biological means. Even when special or color reactions are lacking, or have insufficient sensitivity, bio-autographic methods are often successful. Techniques of this kind have been used for the detection of *vitamins*[112, 113], *saponins*[114−118] and *antibiotics*[119−123, 142].

[112] *L. Cima, R. Mantovan*, Farmaco. Ed. Prat. *17*, 473 (1962).

[113] *W.A. Winsten, E. Eigen*, Proc. Soc. Exp. Biol. Med. *67*, 513 (1948).

[114] *K. Hiller, B. Linzer, S. Pfeiffer*, Pharmazie *21*, 182 (1966); *22*, 321 (1967).

Inactive chromatogram zones do not interfere. It may, however, be a matter of several hours to several days before the final result is available[124]. This is true not only for direct detection methods, but also for the examination of scraped-off or cut-out chromatogram zones, e.g. insecticidal *pyrethrins*[125] or *growth-regulating indole derivatives*[126]. Bitter-tasting materials can sometimes be detected by an organoleptic test[117, 127, 390].

4.4 Coupling with Other Analytical Methods

The hR_f value of a compound is only an approximate value. Comparison of the migration behavior with authentic material is preferable. However, unambiguous identification requires coupling with more definitive analytical methods.

Direct coupling with other analytical methods: UV absorption curves and/or fluorescence spectra may often be taken directly from the chromatogram (PC[128–134], TLC[135–141, 396]). In some cases, it is advisable to perform a preliminary separation

by gas chromatography. Here, both the GLC peaks and the PC or TLC zones are studied to determine whether they can be assigned to single substances.

Indirect coupling, transfer methods: when two analytical methods are coupled indirectly, manual transfer is necessary. The difficulty then arises of isolating the chromatographically-separated compounds from the large excess of stationary phase present, without loss[63, 143]. Extraction processes can be used for separation (Fig. 3, cf. 2.4:3.4).

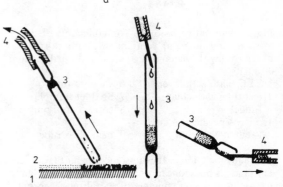

Fig. 3. a) Mechanical transfer of a chromatographed substance zone[390]; operations: transfer of scraped-off chromatogram zone by suction into a micro-percolation tube, addition of a few drops of solvent, suction of extraction fluid (1–4 see Fig. 3 b).

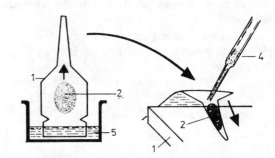

Fig. 3. b) Mechanical transfer of substance from a paper chromatogram to a TLC layer[388]

1 TLC plate or foil, or cut-out paper chromatogram
2 chromatogram spot
3 micro-percolation pipe
4 pipette or micrometer syringe
5 polar eluant.

[115] *D. Vagujfalvi, Gy. Held, P. Tétényi*, Arch. Pharm. *299*, 812 (1966).

[116] *Cl. Wagner*, Dissertation, Universität Saarbrücken 1969.

[117] *E. Weigert, P.J. Schorn*, Tribuna Farmaceut. *30*, 48 (1962).

[118] *W. Winkler*, Kolloid.-Z., Z. Polym. *177*, 63 (1961).

[119] *T.F. Brodasky*, Anal. Chem. *35*, 343 (1963).

[120] *R.R. Goodall, A.A. Levi*, Nature *158*, 675 (1946); Analyst *72*, 277 (1947).

[121] *M. Iglóy, A. Missei*, J. Chromatogr. *28*, 456 (1967).

[122] *E. Meyers, D.A. Smith*, J. Chromatogr. *14*, 129 (1964).

[123] *E. Meyers, R.C. Erickson*, J. Chromatogr. *26*, 531 (1967).

[124] *F. Jungnickel*, J. Chromatogr. *34*, 560 (1968).

[125] *E. Stahl*, Arch. Pharm. *293*, 531 (1960).
E. Stahl, J. Pfeifle, Naturwissenschaften *52*, 620 (1965).

[126] *H. Kaldewey* in *E. Stahl*[390].
H. Kaldewey, E. Stahl, Planta *62*, 22 (1964).
H. Kaldewey, Lj. Kraus in H. Kaldewey et al.[400]

[127] *Y. Kuroiwa, H. Hashimoto*, J. Inst. Brew. *67*, 347, 352 (1961).

[128] *H. Asmis, E. Bächli, H. Schmid, P. Karrer*, Helv. Chim. Acta *37*, 1993 (1954).

[129] *M. Heros, L.M. Anny*, Bull. Soc. Chim. Fr. 367 (1955).

[130] *R.M. Horowitz, L.F. Atkinson*, Anal. Chem. *29*, 1385 (1957).

[131] *K.A. Montagu*, Biochem. J. *63*, 559 (1956).

[132] *M. Nonaka, E.L. Pippen, G.F. Bailey*, Anal. Chem. *31*, 875 (1959).

[133] *J.H.C. Smith, K. Shibata, R.W. Hart*, Arch. Biochem. Biophys. *72*, 457 (1957).

[134] *K. Yamaguchi, K. Fujii, S. Tabata, S. Kato*, J. Pharm. Sci. Jap. *74*, 1322 (1954).

[135] *M.M. Frodyma, T. van Lieu, R.W. Frei*, J. Chromatogr. *17*, 131 (1965); *18*, 520 (1965).

[136] *J. Janák*, J. Gas Chromatogr. *1*, 20 (1963).
J. Janák, I. Klimeš, K. Hana, J. Chromatogr. *18*, 270 (1965).

[137] *R. Kaiser*, Z. Anal. Chem. *205*, 284 (1964); Chimia *21*, 235 (1967).

[138] *E. Sawicki, T.W. Stanley, H. Johnson*, Mikrochem. J. *8*, 257 (1964).
E. Sawicki, T.W. Stanley, W.C. Elbert, J.D. Pfaff, Anal. Chem. *36*, 497 (1964).

Polar compounds adsorbed on the adsorbent can be isolated in a reproducible manner only with difficulty. Deliberate deactivation of the stationary phase[144] or special transfer techniques may provide a solution[145]. The 'wick-stick' method is an example of such a transfer technique[146, 147]. Direct layer transfer of silica gel to potassium bromide has also been reported[148]. With complex-forming substances, the adsorbent must not contain any traces of metal; metals must be removed by suitable purification[149, 150].

If chemical reactions are to be carried out subsequently, suitable reagent solutions can be used for extraction, especially if the reaction products are more soluble than the initial compounds.

The eluates obtained can be studied further, qualitatively and quantitatively, by the methods listed in Table 6. Spectrophotometric analyses in the UV and IR wavelength ranges are often not clear because of residual adsorbent which cannot be removed by centrifuging or filtration (PC[151-154], TLC[155-159]). Mass spectroscopic analyses also become more difficult because of such contaminants[160-163]; backgrounds in the spectra contain stray peaks. Since, in this case, the stationary phase also causes difficulties in the instrument[394], a simple extraction method similar to the 'wick-stick' method, using micro-tubes, is preferred[147]. Interfering effects encountered in polarographic[164], gas chromatographic[165, 166] or radiometric analyses[167, 168] can be dealt with similarly.

5 Standardization and Documentation

For routine application of an analysis method, it is important to record all experimental conditions as exactly as possible[390, 398], especially the following:

Method: development (ascending, descending, horizontal), practical techniques (wedged-tip technique, gradient technique, etc.), type of tank (trough, sandwich, BN- or Vario-S-tank), saturation conditions and method of application (spots or streaks).
Separation phase: adsorbent (silica gel, alumina, cellulose, etc.; manufacturer); size and thickness of the layer; conditions for drying and activation.
Mobile phase: solvent composition (vol.-%) e.g. benzene/methanol (75/25); length of run (TLC: mostly 10 cm); time needed for development.
Detection: UV light ($\lambda = 254$ or 366 nm), spray reagent (composition), exact procedure, result (color, special effects, tailing, etc.)

Deviations from the standard conditions must be recorded.

A worthwhile complement to these data is to reproduce the chromatogram; in many cases, a sketch or a tracing is sufficient. Frequently, original chromatograms (paper chromatograms and TLC foils) can be filed directly.

To permit TLC layers to be detached, the plates must be sprayed with a special plastic solution; after drying an adhesive foil is rolled onto the layer, and the cross-linked stationary phase adheres to the foil.

Photocopies or photographs of chromatograms are often used for documentation. Contact copies can be produced on commercially available

[139] *C. Nigam, M. Sahasrabudhe, L. Levi,* Can. J. Chem. *41,* 1535 (1963).
[140] *H. Seiler,* Helv. Chim. Acta *46,* 2629 (1963).
N. Seiler, G. Werner, M. Wiechmann, Naturwissenschaften *50,* 643 (1963).
N. Seiler, M. Wiechmann, Hoppe-Seyler's Z. Physiol. Chem. *337,* 229 (1964); Z. Anal. Chem. *220,* 109 (1966).
[141] *N. Seiler,* Hoppe-Seyler's Z. Physiol. Chem. *348,* 765, 768, (1967).
[142] *H. Heding* in Ref.[393], p. 569.
[143] *G.B. Marini-Bettòlo* in Ref.[393], p. 74.
[144] *K. Hirayama, K. Inone,* J. Pharm. Sci. *56,* 444 (1967).
[145] *D.D. Rice,* Anal. Chem. *39,* 1907 (1967).
[146] *H.R. Garner, H. Packer,* Appl. Spectrosc. *22,* 122 (1967).
[147] *G. Székely,* J. Chromatogr. *48,* 313 (1970).
[148] *G.W. Goodman* in *E.J. Shellard*[395].
[149] *L.S. de Bohner, E.F. Soto, T. de Cohan,* J. Chromatogr. *17,* 513 (1965).
[150] *J.M. Brand,* J. Chromatogr. *21,* 424 (1966).
[151] *D. Doskočilová,* Chem. Listy *55,* 193 (1960).
[152] *A.L. Hayden,* J. Pharm. Sci. *51,* 176 (1962).
[153] *H. Seligson, B. Kramer, D. Seligson, H. Baltoush,* Anal. Biochem. *6,* 362 (1963).
[154] *T.Y. Toribara, V. di Stefano,* Anal. Chem. *24,* 1519 (1954).
[155] *K. Beyermann,* Z. Anal. Chem. *226,* 16 (1967).
K. Beyermann, E. Röder, Z. Anal. Chem. *230,* 347 (1967).

[156] *V. Černý, J. Joska, L. Lábler,* Collect. Czech. Chem. Commun. *26,* 1658 (1961).
[157] *R.N. McCoy, E.L. Fiebig,* Anal. Chem. *37,* 593 (1965).
[158] *R. Fontanges, P. Heritier, P. Coeur,* Bull. Soc. Chim. Biol. *46,* 1223 (1964).
[159] *M.K. Snavely, J.G. Grasselli,* Develop. in Appl. Spectrosc. *3,* 119 (1963).
[160] *M.A. Schwartz, P. Bommer, F.M. Vane,* Arch. Biochem. Biophys. *121,* 508 (1967).
M.A. Schwartz, F.M. Vane, E. Sostma, Biochem. Pharmacol. *17,* 965 (1968).
[161] *F.T. Deverse, E. Gipstein, L.F. Lesoine,* Perkin-Elmer Instrument News *18,* 16 (1967).
[162] *W.H. McFadden,* Separation Sci. *1,* 723 (1966).
[163] *M. Fetizon* in *G.B. Marini-Bettòlo*[382].
[164] *H. Oelschläger, J. Volke, G.T. Lim,* Arch. Pharm. *298,* 213 (1965); Arzneim.-Forsch. *17,* 637 (1967).
[165] *Y. Tomita, A. Nomori, H. Minato,* Phytochemistry *8,* 2249 (1969).
[166] *H. Vollmann,* Dissertation, Universität Saarbrücken 1967.
[167] *G.V. Vahonny, C.R. Borja, S. Weersing,* Anal. Biochem. *6,* 555 (1963).
[168] *J. Wolleweber, B.A. Kottke, C.A. Owen,* J. Chromatogr. *24,* 99 (1966).

Table 6. Possible techniques for quantitative determination after extraction (indirect evaluation).

Operating range:	1–100 mg	10–1000 µg	10–1000 ng
methods:	gravimetry titrimetry polarimetry	spectroscopy (UV, IR, Mass, NMR) refractometry polarography gas chromatography isotope measurements	fluorescence phosphorescence biological-physiological determinations

copiers, using ordinary copying papers, e.g. Ozalid 200 SS or Diazo 1200 SS papers. When working with UV light, X-ray films have been effective.

If smaller sizes are sufficient, photography (e.g. with a Polaroid camera set-up) is recommended. For UV photographs, skylight filters should be used in order to filter off short wave remission radiation, which would interfere. The 'blue tinge' which appears so frequently in color photographs can be suppressed by employing a yellow-green or yellow-orange filter.

6 Quantitative Evaluation

Both indirect and direct methods of quantitative evaluation exist for determining the quantitative distribution of the individual components.

All procedures in which the substance to be estimated is separated from the stationary phase before measurement are designated as indirect methods. The separation usually involves extraction. *Blank values* and *elution errors* are determined in control experiments.

Both blank values and elution errors can become considerable[169] if the substances to be measured are polar. In such cases, a correspondingly polar and specially purified eluant must be used (2.4:3.4).

The elution error can be estimated, e.g. by means of radiometry[170]. Depending on the sensitivity of the subsequent method of measurement, three fields of application can be differentiated, as shown in Table 6.

6.1 Indirect Determination for the Milligram Level

If sufficient quantities of substance are available, gravimetric, titrimetric or polarimetric methods can be used. Concentration on larger and thicker stationary phases of larger size is usually necessary. This type of concentration is especially suitable for rough separations, i.e. for chromatograms in which the individual substances are separated from each other by at least the breadth of one zone. When differences in hR_f values are smaller, various types of continuous flow chromatography are sometimes preferred[377] unless micro-preparative methods (2.4:3.4) are applicable.

6.1.1 Gravimetric Method

The (mostly lipophilic) substances eluted after chromatography are determined gravimetrically, after the solution has been evaporated[156, 171–174]. Only compounds with a low vapor pressure give sufficiently exact results. If the substances to be estimated are temperature-sensitive, freeze-drying of the extracts is recommended[175]. Easily oxidizable components should be kept in an inert gas atmosphere.

6.1.2 Titrimetric Method

Basic or acidic components can be determined by microtitration (PC[176], TLC[177, 178]).

TLC model analyses[179] give a recovery of 95 to 98% for free *fatty acids*. After addition of boric acid, *polyhydric alcohols* such as *glycerol, glycols* and *sugars*[180] can be titrated acidimetrically or iodometrically with suitable indicators. *Peroxides* in polystyrenes can be determined successfully by means of the iodometric method[181].

Organic halogen compounds may be titrated with an aqueous solution of mercury perchlorate, after combustion in a suitable apparatus[182]; the recovery is between 97 and 99%.

6.1.3 Polarimetric Method

Micropolarimetry yields reliable results only with substances having high rotation values. This meth-

[169] P. Paquin, M. Lepage, J. Chromatogr. 12, 57 (1963).
[170] S. De Witt, S. Goodman, T. Shiratori, J. Lip. Res. 5, 578 (1964).

[171] G. Adam, K. Schreiber, Z. Chem. 3, 100 (1963).
[172] G. Jurriens, B. de Vries, L. Schouten, J. Lipid Res. 5, 267 (1964).
[173] E. Levin, C. Head, Anal. Biochem. 10, 23 (1965).
[174] D.E. Walsh, O.J. Banasik, K.A. Gilles, J. Chromatogr. 17, 278 (1965).
[175] N.S. Radin, J. Am. Oil Chem. Soc. 42, 569 (1965).
[176] P. Flesch, D. Jerchel, Wein-Wiss., Beiheft Fachz. deutsch. Weinbau 9, [3] 1 (1955); Angew. Chem. 67, 682 (1955).
[177] M. Ikram, M.K. Baklish, Anal. Chem. 36, 111 (1964).
[178] G. Schlierf, P. Wood, J. Lipid Res. 6, 317 (1965).
[179] V.P. Skipski, J.J. Good, M. Barcley, H.B. Reggio, Biochem. Biophys. Acta 152, 10 (1968).
[180] G. Pastuska, Z. Anal. Chem. 179, 427 (1961).
[181] J.A. Brammer, S. Frost, V.W. Reid, Analyst (London) 92, 91 (1967).
[182] Y. Imai, A. Yamanchi, S. Terabe, R. Konaka, J. Sugita, J. Chromatogr. 36, 545 (1968).

od, like the titrimetric method, is specific for certain substances and chemical groups, and, in this respect, is superior to gravimetry. The introduction of spectropolarimeters represents a considerable advance in terms of equipment.

6.2 Indirect Determination at the Microgram Level

At the microgram level, usually up to 100 μg, physico-chemical methods, e.g. especially spectroscopy, are used. The choice of the eluant depends on:

- the compound to be eluted,
- the wavelength range used for subsequent measurement.

If necessary, the extracted material must be taken up in a suitable solvent.
Of all indirect quantitative determination methods, spectrophotometric measurements have been found superior.

6.2.1 Measurements in the UV Spectral Region

Quantities of material necessary for determination by UV are usually between 50 and 100 μg; the higher the molar extinction coefficient of a compound, the lower is the required quantity. If the best eluant has been chosen, the recovery is between 90 and 100%. If errors amount to more than 15%, the extraction procedure should be checked.

6.2.2 Measurements in the Visible Spectral Region

Before estimation by densitometric or colorimetric methods[183, 184], colorless substances should be converted into colored compounds using chemical reactions which are as specific as possible[185]. This procedure is of advantage:

- if other substances are present, which, for example, would affect UV measurements, but do not give the same coloration as the compounds to be investigated. This applies also to impurities from the stationary phase or the eluant, which would otherwise yield a very high blank value[186];
- if a *guiding chromatogram* must be provided. The procedure is especially efficient if one uses the same reagent as for the subsequent photometric measurement. 'Quenching' effects, which may be caused by certain stationary phases, are usually negligible[60];

- if compounds can be eluted only imperfectly. If a suitable reagent solution is used as eluant, the distribution equilibrium is often shifted, and more complete separation from the stationary phase is obtained.

Lack of knowledge concerning reproducibility of an analysis is a disadvantage of this method of photometric evaluation. The literature contains only a few data on:

- the course of reaction with time, and the stability of 'auxiliary substances';
- the optimum pH value for reaction;
- the temperature-dependence of reaction;
- color shifts caused by differences in concentration.

If the range of direct proportionality has been found by systematic investigations, random errors are no higher than for measurements in the ultraviolet spectral range.

6.2.3 IR Spectroscopy

Compared with the photometric estimation methods mentioned above, IR spectroscopy is only of secondary importance in quantitative evaluation[187, 188].
For *triglycerides*[189, 190, 376], synthetic *phosphatase inhibitors*[191] and other esters, the band at $\lambda = 5.76$ μm is used for measurement. Substances with *hydroxy groups* (in the absence of water, alcohol and peroxides) are determined by means of the intensity of OH-bands[175]. To determine *nitro compounds,* the absorption at 1659 cm^{-1} is used; *hydrocarbons* can be determined at $\lambda = 3.42$ μm[179].
To obtain recoveries of 95 to 104%, the amount of substance required is usually 10 to 30 mg/ml[189, 192].

6.2.4 Refractometry

Interference refractometers should be used for determination of the refractive indices of eluted chromatogram zones. Errors of only ± 4 to 8% have been reported for model experiments on cellulose carriers[193]. The quantities of chromatographed substances are between 2.5 and 15 μg.

[183] *G. Gorbach*, Mikrochim. Acta *39*, 204 (1952).
[184] *J. Jakubec, M. Zahradníček*, Chem. Listy *50*, 1459 (1956).
[185] *M. Pöhm, L. Fuchs*, Naturwissenschaften *40*, 244 (1953).
[186] *J.C. Morrison, L.G. Chatten*, J. Pharm. Pharmacol. *17*, 655 (1965).

[187] *J.D.S. Goulden*, Nature *173*, 646 (1954).
[188] *D.R. Kalkwarf, A.A. Frost*, Anal. Chem. *26*, 101 (1954).
[189] *K. Krell, S.A. Hashim*, J. Lipid Res. *4*, 407 (1963).
[190] *H.H. Oelert*, Erdöl, Kohle, Erdgas, Petrochem. *18*, 876 (1965).
[191] *A.F. Rosenthal, S. Ching-Hsien Han,* Biochim. Biophys. Acta *152*, 96 (1968).
[192] *G. Blunden, R. Hardman, J.C. Morrison*, J. Pharm. Sci. *56*, 948 (1968).
[193] *N.H. Choulis*, J. Chromatogr. *30*, 618 (1967).

6.2.5 Mass Spectroscopic Determination

The insertion of samples, which may be partly contaminated with the stationary phase[163, 194], is a problem in quantitative mass-spectroscopic determinations. Special micro-probes are sometimes used[161] for scraping off TLC spots and for positioning the sample in front of the ion source. The substance must vaporize completely so that calibration curves showing the relationship between the total ion current and the quantity of substance inserted may be constructed.

6.2.6 Polarographic and Oscillo-polarographic Determination

If substances with a high redox potential are to be estimated, polarographic or oscillo-polarographic methods may be used successfully[195–198]. This is especially true for various *vitamins*[199, 200], as well as for *nitro* and *nitroso compounds*[164, 201]. Determinations of *picric acid* can be carried out with quantities as small as 1 to 5 μg/ml solution if cathode ray polarographs are used[202]. If a certain quantity of substance is added as an internal standard in this measurement, the time required is only slightly greater than that required for the calibration method which is usually used in routine work, or even for the spectrophotometric method.

6.2.7 Gas Chromatographic Determination

Compared with spectrophotometric methods, gas chromatography is less specific for particular substances. In spite of this fact, coupling of these two methods has frequently given good results[203, 396], especially after improvements in methods of transfer and in detector systems[204–207].

To obtain better results when working with fatty acids, it is recommended that these be esterified before injection into the gas chromatograph[208–215]. With steroid hormones, acetylation[216, 217] or tosylation[218, 219] may be helpful.

The addition of an internal standard is recommended[220], especially in cases where combination with isotope-measuring techniques is intended[170].

6.2.8 Determination of Radioactivity

Quantitative estimation of radioactively labelled compounds is concerned largely with medical and biochemical research. From the medical point of view, it is necessary to use isotopes with short halflife times in order to keep the decay time in the organism as short as possible. Calibration measurements usually give straight lines which, in contrast to absorption measurements, are not specific for substances, and are isotope-dependent. Measurements on eluted chromatogram zones can be carried out either directly, e.g. with various modified Geiger-Müller counting tubes or gas flow detectors, or indirectly with scintillation counters[104, 221–226]. *Scintillation measurements* are suit-

[194] *K. Heyns, H.F. Grützmacher*, Angew. Chem. *74*, 387 (1962).

[195] *J.A. Lewis*, Chem. Ind. [London], 284 (1949).

[196] *V. Prey, E. Waldmann, F. Ludwig, H. Berbalk*, Monatsh. Chem. *83*, 1344 (1952).

[197] *J. Proszt, J. Kis*, Acta Chim. Acad. Sci. Hung. *9*, 191 (1956).

[198] *M. Voldan*, Česk. Farm. *4*, 407 (1955).

[199] *K. Černý*, Sborník celostatní pracovní konference chemiků CSSR, p. 279, Praha 1953.

[200] *W.H. Harrison, J.E. Gander, E.R. Blakley, P.D. Bayer*, Biochem. Biophys. Acta *21*, 150 (1956).

[201] *D. Heusser*, Planta Medica *12*, 237 (1964); *D. Heusser, J. Jackwerth*, Dtsch. Apoth.-Ztg. *105*, 107 (1965).

[202] *W. Huber*, J. Chromatogr. *33*, 378 (1968); Chromatographia *1*, 212 (1968).

[203] *E.W. Day jr., T. Golab, J.R. Koons*, Anal. Chem. *38*, 1053 (1966).

[204] *R. Konaka, S. Terabe*, J. Chromatogr. *24*, 236 (1966).

[205] *H.K. Mangold, R. Kammereck*, Chem. Ind. [London] 1032 (1961).

[206] *B.M. Mulhern*, J. Chromatogr. *34*, 556 (1968).

[207] *H.H. Wotiz, S.C. Chattoraj*, Anal. Chem. *36*, 1466 (1964).

[208] *M.L. Blank, O.S. Privett*, J. Lipid Res. *4*, 470 (1963).

[209] *E. Gordis*, J. Clin. Invest. *44*, 1451 (1965).

[210] *F.D. Gunstone, F.B. Padley*, J. Am. Oil Chem. Soc. *42*, 957 (1965).

[211] *G.W. Oërtel, K. Groot*, Hoppe-Seyler's Z. Physiol. Chem. *341*, 1 (1965).

[212] *N. Persmark, B. Töregard*, J. Chromatogr. *37*, 121 (1968).

[213] *M.R. Sahasrabudhe, S.J. Read*, J. Am. Oil Chem. Soc. *42*, 864 (1965).

[214] *H.K. Mangold, R. Kammereck*, J. Am. Oil Chem. Soc. *39*, 201 (1962).

[215] *C. Lichtfield, M. Farguhar, R. Reiser*, J. Am. Oil Chem. Soc. *41*, 588 (1964).

[216] *B.R. Hopper, W.W. Tullner*, Steroids *9*, 517 (1967).

[217] *M. Sparagana*, Steroids *5*, 773 (1965).

[218] *P.B. Raman, R. Avramov, N.L. McNiven, R.J. Dorfman*, Steroids *6*, 177 (1965).

[219] *D.H. Sandberg, N. Ahmad, M. Zachmann, W.W. Cleveland*, Steroids *6*, 777 (1965).

[220] *O.S. Privett, M.L. Blank, D.W. Codding, E.C. Mickell*, J. Am. Oil Chem. Soc. *42*, 381 (1965).

[221] *A.A. Abdell-Latif, F.E. Chang*, J. Chromatogr. *24*, 435 (1966).

[222] *S. Bleecken, G. Kaufmann, K. Kummer*, J. Chromatogr. *19*, 105 (1965).

[223] *H. Danielsson, K. Einarsson*, J. Biol. Chem. *241*, 1449 (1966).

[224] *B. Goldrick, J. Hirsch*, J. Lipid Res. *4*, 482 (1963).

[225] *H.K. Mangold, R. Kammereck, D.C. Malins*, Mikrochem. J. Symposiumsband *2*, 697 (1962).

[226] *H. Schildknecht, O. Volkert*, Naturwissenschaften *50*, 442 (1963).

able for the determination of weak β-emitters. Decrease in the fluorescence yield (quenching) caused by contaminants may occur with moderate concentrations of aldehydes, nitrocompounds, sulfuric acid and silver ions (Chap. 10.5).

If the substance to be measured is not eluted from the stationary phase, suitable suspensions can be measured directly in scintillation solutions[227]. One must, however, determine whether or not a decrease in the count yield occurs due to absorption processes at the stationary phase[228, 167, 168].

Deactivation of the adsorbent[229] or the use of especially fine-grained layer materials often have a beneficial effect on the measurements[230] so that additional evaluation corrections become superfluous[229, 231].

The recovery rate is generally below 95%, i.e. below that of spectrophotometric methods of evaluation. Systematic errors seem to be larger.

6.2.9 Measurements of Gas Volumes

If quantitative determination of labelled substances cannot be carried out by scintillation measurements, the compounds concerned can be converted into $^{14}CO_2$ gas and the activity or volume measured[232, 233]. It is also possible to determine inactive substances, such as α-hydroxy-acids or α-keto-acids: the gas volume obtained by decarboxylation in a Warburg apparatus is between 21 and 68 μl per micromol of substance. The accuracy of measurement is $\pm 6\%$[234].

6.3 Estimation of Nanogram Quantities

6.3.1 Phosphorimetric Measurements

Phosphorimetric determination of eluted substances[138] is not yet a very suitable technique for quantitative investigations. Due to external influencing factors, the reproducibility is often unsatisfactory. Even small temperature variations result in distinct changes in the measured results. The recovery for nornicotine, for example, is $<90\%$[235].

6.3.2 Fluorimetric Measurements

Estimation of fluorescence radiation, and measurement of phosphorescence, are restricted to light-emitting compounds. Since the quantities of substance available are small, complete separation from the stationary phase is sometimes difficult. In such cases, better solubility may be achieved by addition of small amounts of Tween 20[236], or by the influence of ultrasonics. The elution equilibrium can often be shifted by chemical reactions; also, non-fluorescent materials may be converted into fluorescent derivatives.

This technique is frequently made use of in peptide and protein chemistry, e.g. for end-group determinations with immunoglobulins[237]. Reaction with the Procházka reagent[390] has a similar use for the determination of indole derivatives, and for the determination of cardiac glycosides[238]. Weakly or non-fluorescent substances are estimated after formation of complexes. Flavonol glycosides, for example, fluoresce more strongly in a 1% aluminium chloride solution than in methanolic solution[239]. Similarly, Δ^4-3-oxo steroids give considerably more intense emission in the presence of lithium hydroxide[240].

Frequently, fluorescence bands can be shifted by chemical reactions. This is advantageous for 5-hydroxytryptophol because otherwise absorption and emission maxima partly overlap[241].

6.3.3 Biological Methods of Determination

This involves the determination of the critical concentrations at which positive reactions just occur. The scatter of results is relatively large.

As little as 0.2 μg of indolyl-acetic acid was estimated by means of the curvature test on Avena-Coleoptiles[126, 242, 390].

Investigations on insecticidal components from pyrethrum extracts[125] and the enzymatic determination of Parathion® or Trichlorphon®[243, 244] can be carried out in a similar manner. The cut-out or scraped-off chromatogram zones are used directly for measurement.

[227] P.H. Ekdahl, A. Gottfries, T. Scherstein, Scand. J. Chim. Lab. Invest. 17, 103 (1965).

[228] F. Snyder, N. Stephens, Anal. Biochem. 4, 128 (1962).
F. Snyder, Atomlight No. 58, 1 (1967).

[229] F. Snyder, D. Smith, Separation Sci. 1, 709 (1966).
F. Snyder, Anal. Biochem. 9, 183 (1964).

[230] S. Helf in C.G. Bell, F.N. Hayes[365].

[231] P.H. Ekdahl, A. Gottfries, T. Scherstein, Atomlight No. 56, 1 (1966).

[232] F. Drawert, O. Bachmann, K.H. Reuther, J. Chromatogr. 9, 376 (1962).

[233] K.H. Kolb, G. Jänicke, M. Kramer, P.E. Schulze, G. Raspé, Arzneim.-Forsch. 15, 1292 (1965).

[234] M. Trap, S. Grassman, A. Pinsky, J. Chromatogr. 38, 411 (1968).

[235] J.D. Winefordner, H.A. Moye, Anal. Chim. Acta 32, 278 (1965).

[236] E. Ullmann, H. Kassalitzky, Arch. Pharm. 295, 37 (1962).

[237] G. Schmer, Hoppe-Seyler's Z. Physiol. Chem. 348, 199 (1966).

[238] R.W. Jelliffee, J. Chromatogr. 27, 172 (1967).

[239] R.E. Hagen, W.J. Dunlap, J.W. Mizelle, S.H. Wender, Anal. Biochem. 12, 472 (1965).

[240] E. Nowotny, H. Staudinger, Z. Klin. Chem. 4, 203 (1966).

[241] V.E. Davis, J.L. Cashaw, J.A. Hulf, H. Brown, Proc. Soc. Exp. Biol. Med. 122, 890 (1966).

[242] L.C. Luckwill, L.E. Powel jr., Science 123, 225 (1956).

[243] C.E. Mendoza, P.J. Wales, H.A. McLead, W.P. McKinley, Analyst (London) 93, 34, 173 (1968).

[244] H. Ackermann, Nahrung 10, 273 (1966).

The active substance diffuses slowly from the stationary phase into the reaction layer. Substances with antibiotic activity can be determined similarly[369, 371].

6.4 Direct Quantitative Determination without Apparatus

The methods mentioned in Sections 2.4:6.1 to 2.4:6.3 require quantities of substance which can be gained only by micro-preparative procedures. Therefore, methods which can be applied to the chromatogram itself, i.e. in the analytical range of the separation process (2.4:6.4 to 2.4:6.5), have much in their favor.

The following technique can be used for PC as well as for TLC. However, in TLC a concentration of the substance occurs on the surface since the solvent evaporates only from one side.

Separation from the adsorbent is no longer necessary for direct estimation (see Table 7). The determination of blanks is also simpler, so that possible sources of error are reduced, and the time required for determination can be shortened.

Table 7. Possible techniques for direct quantitative determination of substances on the microgram and nanogram scales.

Method	Possible techniques
comparison of zone sizes	visual assessment planimeter
densitometry spectrophotometry }	transmission, reflexion
fluorimetry	self-fluorescence, quenching of fluorescence
isotope technique	scintillation, direct determination

The majority of methods used for direct estimation are based on optical measurements. Essentials are:

– chromatograms on which the various substances are well separated;
– radiation sources which can be used to illuminate the zones;
– detector systems which measure changes of energy caused by partial absorption or emission, in comparison with a set standard.

The requirements of these three differ for the various methods of measurement.

6.4.1 Visual Comparison of Spot Areas

Of the various so-called in-situ determination methods, direct visual comparison of spot areas and color intensities with standard spots is easiest in practice[245, 246, 396]. This procedure involves comparison of two (preferably neighboring) zones. It is usually necessary to apply reagents to produce

colorations. Spraying and heating of the chromatograms should be as even as possible[247, 396]. Zones of high contrast[248] are desirable, and the reagents for detection are chosen accordingly[249]. The contrast can be intensified by using fine grained screen foils[250] or suitable color filters. Evaluation is most sensitive[251–254] near the detection limit since the curve relating spot size and amount of applied substance is steepest here[390].

Errors observed in model studies amounted to $\pm 10\%$ for TLC[255, 256]. However, the deviation from the real value can rise to $\pm 100\%$, if various influential factors are neglected (see above)[257].

The fact that numerical data are lacking for the zone area is sometimes considered to be a disadvantage of this evaluation method. Therefore it is often recommended that the chromatogram spots be traced on to transparent paper in transmitted light before evaluation. Another frequently-used procedure is to take a photograph or photocopy, and then detect the area planimetrically or by counting the area in square millimeters.

6.4.2 Planimetric Evaluation

In order to keep planimetric errors as small as possible, the quantities of substances chromatographed (and thus the zone areas on the chromatogram) are usually bigger for this kind of evaluation than with visual comparative determination[258, 259]. If planimeter measurements are

[245] R.C. Brimly, Nature 163, 215 (1949).
[246] J.C. Giddings, R.A. Keller, J. Chromatogr. 2, 626 (1959).
[247] J.E. Bush in D. Glick[380];
J.E. Bush, J. Chromatogr. 29, 157 (1967);
J.E. Bush, C.P. Hoffman, J. Chromatogr. 30, 164 (1967).
[248] P.J. Curtis, Chem. Ind. [London], 247 (1966).
[249] H. Scheiz, Mikrochim. Acta, 49 (1967).
[250] G. Székely in Ref. [386], p. 325.
[251] D.C. Abbott, J.A. Bunting, J. Thompson, Analyst (London) 90, 356 (1965).
D.C. Abbott, R.B. Harrison, J.O. Tatto, J. Thompson, Nature 211, 259 (1966).
D.C. Abbott, K.W. Blake, K.R. Tarrant, J. Thompson, J. Chromatogr. 30, 136 (1967).
[252] H. Feltkamp, E. Syrbe, Arch. Pharm. 301, 374 (1968).
[253] D.O. Eberle, R.G. Delley, G. Székely, H.U. Stammbach, Agric. and Food Chem. 15, 213 (1967).
[254] Ch. Meythaler, G. Dworak, E. Schmid, Arzneim.-Forsch. 16, 800 (1966).
[255] L.T. Heaysman, E.R. Sawyer, Analyst (London) 89, 529 (1964).
[256] D. Waldi, Mitt. Dtsch. Pharm. Ges. 32, 125 (1962).
[257] C.G. Honegger, Helv. Chim. Acta 45, 2020 (1962).
[258] N. Oswald, H. Flück, Sci. Pharm. 32, 136 (1964).
[259] J. Sliviok, Z. Kwapniewski, Mikrochim. Acta, 1 (1965).

not done on the chromatogram itself[260, 261], the possible copying error should be considered since it can rise considerably in diffusely-bordered zones[262]. For areas of 100 mm^2, it amounts to as much as v$=\pm5$ to 8%. For TLC, ±10 to 15% can be expected[263] with the smaller zones. If graphs are drawn of the measured zone area against the quantity of substance, one obtains curves, which within limits, can be replaced by straight lines or semi-logarithmic functions (PC[264], TLC[258, 265–268]). If a relationship is assumed between the square root of the zone area and the logarithm of the amount of substance, straight calibration lines should result[269, 379].

Many groups of workers have used this relationship[251, 270–272]. It can be used successfully also for measurements with *triglycerides*[178], *amines*, and *amine oxides*[273, 274] as well as for the sequence analysis of *peptides*[275].

Model analyses indicated standard deviations of between ±3 and 5%; errors are likely to be larger in routine analyses. Apart from errors associated with application of the solutions and chromatographic uncertainties, the visual detection of the spot area may be an added source of errors (see also 2.4:6.4.1).

6.5 Methods of Photometric Estimation
The original aim of direct photoelectric estimation methods was the objective determination of spot

areas. Measurements with densitometers, relating measured values to sample amounts[371, 377], were carried out only later.

6.5.1 Densitometry
The simplest densitometers work with white light (e.g. tungsten lamps; for other details, see Ref.[374, 398]). Their sensitivity and specificity are poor, but can be improved by use of suitable filters. Measurement is generally carried out by slitwise scanning of the chromatogram zones. Because part of the incident energy is absorbed with each measurable chromatogram zone, the light striking the photomultiplier is attenuated. This energy change yields absorption curves dependent upon the coordinates of the system. The calibration curves are constructed by plotting the curve areas against the quantities of substance[276, 277]. Straight calibration lines for use in special problems can be obtained by inserting so-called cams in the apparatus. Details of relationships between measured values and sample amounts will be given later (concerning transmission and reflexion measurements).

With colorless compounds, a suitable coloration is a prerequisite for these measurements. The more specific a color reaction is, the more selective is the result for the compound concerned. Apart from the factors mentioned in 2.4:6.4.1, the following conditions must exist:

– the chemical reaction must proceed independently of the concentration, and must yield a reaction product of the same color;
– an excess of the reagent either must not interfere, or it must be removed before the measurement;
– the reaction should proceed quantitatively if possible, or should be reproducible;
– if the adsorbent is changed, one must determine whether the color of the chromatogram zone also changes (choice of filters)[63, 278].

In order to avoid these possible difficulties, carbonization may be advantageous, especially in the TLC of lipids[279]. Oxidizing reagents containing sulfuric acid are generally used. The color intensity of the spots depends on the number of carbon atoms and double bonds[280]. Conversion of carbon into carbon dioxide must be avoided. With olefinic $C=C$ double bonds, this can be effected by first

[260] *F. Jaminet*, Pharm. Acta Helv. *34*, 571 (1959).

[261] *Lj. Kraus*, Pharmazie *12*, 693 (1957).

[262] *J.C. Morrison, L.G. Chatten*, J. Pharm. Sci. *53*, 1205 (1964).

[263] *G. Wagner, J. Wandel*, Pharmazie *21*, 105 (1966).

[264] *R.B. Fischer, D.S. Parsons, R. Holmes*, Nature *164*, 182 (1949).

[265] *J.A. Kohlbeck*, Anal. Chem. *37*, 1282 (1965).

[266] *G. Pastuska, H.J. Petrowitz*, Chem.-Ztg. *86*, 311 (1962).

[267] *R. Wiedemann*, Dissertation, Universität Bern, 1966.

[268] *H.J. Petrowitz*, Mitt. Dtsch. Ges. Holzforschg. *48*, 57 (1962).

[269] *S.J. Purdy, E.V. Truter*, Chem. Ind. [London] 506 (1962); Analyst *87*, 802 (1962); Proc. Roy. Soc. B. *158*, 536 (1963); Laboratory Practice 500 (1964).

[270] *E. Baucher, J. Washüttl, M.Dj. Olfat*, Mikrochim. Acta 773 (1968).

[271] *G. Blunden, R. Hardman*, J. Chromatogr. *34*, 507 (1968).

[272] *J.R. Davies, S.T. Thuraisingham*, J. Chromatogr. *35*, 43, 513 (1968).

[273] *R. Gnehm, H.U. Reich, P. Guyer*, Chimia *19*, 585 (1965).

[274] *J.R. Pelka, L.D. Metcalfe*, Anal. Chem. *37*, 603 (1965).

[275] *G. Pataki*, Helv. Chim. Acta *47*, 1763 (1964).

[276] *E.F. McFarren, K. Brand, H.R. Rutkowski*, Anal. Chem. *23*, 1146 (1951).

[277] *S.V. Vaeck*, Anal. Chim. Acta *12*, 443 (1955).

[278] *Netheler & Hinz GmbH*, Druckschrift No. B 261.

[279] *H. Rasmussen*, J. Chromatogr. *27*, 142 (1967).

[280] *L.J. Nutter, O.S. Privett*, J. Chromatogr. *35*, 519 (1968).

carrying out hydrogenation, thus improving the reproducibility of the evaluation. For quantities between 5 and 50 μg, relative errors below $\pm 10\%$ have been reported.

Densitometric investigations on chromatograms can be carried out as transmission or reflexion measurements.

Transmission measurements: If the radiation strikes the chromatogram vertically, scattering occurs at the stationary phase. The radiation which passes through the adsorbent diverges over a hemispherical space. A Taylor integration sphere should, therefore, be used for the measurement[281]. Hyperchromic effects may render the result incorrect[282].

Often, for reasons connected with the apparatus, only evaluation of chromatogram strips is possible[283-286]. Attempts are being made, however, to develop apparatus which can also be used for measurements with TLC-plates of size 20×20 cm. Measurement apertures are usually slits. Circular measurement apertures are also used[140, 287]. Both single-beam and double-beam instruments are available. Coupling with a monochromator provides advantages similar to those described in 2.4:6.5.2.

For transmission operation, it has been shown that the well-known Bouguer-Lambert-Beer law does not hold in its general form without some correcting factors[245, 288, 289]. The high blank values often have a disturbing effect[290].

In addition, the limit law just mentioned was based on the following requirements (which are not fulfilled in evaluations with PC or TLC chromatograms):

– The sample must be clear, and must not contain scattering particles.
– The distribution of substance throughout the volume must be regular, so that absorption is homogenous.

– The individual molecules of the sample must not influence each other, and must not be affected by the adsorbent.
– The radiation must be monochromatic.
– The layer thickness must be constant.

Of these factors, inhomogeneity of the stationary phase, and the resulting uncontrollable scattering of light, cause the greatest disturbance.

In order to eliminate light scattering, solutions have been used to impregnate the stationary phase[227, 291] (by dipping, spraying etc.). The refractive indices of these liquids are equal to that of the stationary phase (PC: n_D^{20} about 1.51) e.g. α-bromonaphthalene/paraffin solutions[292] or solutions of liquid paraffin[294]. Other impregnation media include mineral oils[296, 297], anisole[298, 299], methyl salicylate[300], glycerol/water mixtures, Krylon clarifier spray solutions[301] and also synthetic resins such as polyurethanes[302].

Although exact results have been claimed, procedures involving the chromatogram transparent must be regarded as variable[303, 304], since some scattering effects still remain after such treatment[293].

Sometimes the results may be even worse[305]. Consequently the procedure is superfluous for routine investigations[306, 307].

Transmission measurement on photographs: To eliminate differences in layer thickness and the

[281] *U. Heber*, Zeiss-Mitt. *2*, 3 (1960).
[282] *T.D. Price*, Federation Proc. *14*, 264 (1955).
[283] *J. Barrolier, J. Heilmann, E. Watzke*, J. Chromatogr. *1*, 434 (1958).
[284] *H. Eberle*, Naturwissenschaften *41*, 479 (1954).
[285] *J.K. Miettinen, T. Moisio*, Acta Chem. Scand. *7*, 1225 (1953).
[286] *H. Zeutner*, Aust. J. Sci. *17*, 131 (1955); C.A. *49*, 1333 (1955).
[287] *J. Tichy*, J. Chromatogr. *32*, 580 (1968).
J. Tichy, S.J. Dencker, J. Chromatogr. *33*, 262 (1968).
[288] *E.M. Crook, H. Harris, H. Fatma, F.L. Warren*, Biochem. J. *56*, 434 (1964).
[289] *R. Mykolajewycz*, Anal. Chem. *29*, 1300 (1957).
[290] *N. Zöllner, G. Wolram, G. Amin*, Klin. Wochenschr. *40*, 273 (1962).

[291] *W. Teichmann, G. Schmidt*, Pharmazie *12*, 80 (1957).
[292] *W. Grassmann, K. Hannig, M. Knedel*, Dtsch. Med. Wochenschr. *76*, 333 (1951).
[293] *H. Röttger*, Experientia *9*, 150 (1953); Klin. Wochenschr. *31*, 85 (1953).
[294] *T.D. Price, P.B. Hudson, D.F. Ashman*, Nature *175*, 45 (1955).
[295] *S. Sakuraba*, Jap. Analyst *4*, 496 (1955).
[296] *W. Grassmann, K. Hannig*, Naturwissenschaften *37*, 496 (1950); *38*, 200 (1951).
[297] *A.E. Thomas III, J.E. Scharaun, H. Ralston*, J. Am. Oil Chem. Soc. *42*, 789 (1965).
[298] *C. Bergamini, W. Versorese*, Anal. Chim. Acta *10*, 328 (1954).
[299] *J. Pieper*, Klin. Wochenschr. *36*, 605 (1958).
[300] *A.L. Latner*, Biochem. J. *52*, 29 (1952).
A.L. Latner, L. Molyneux, J.D. Rose, J. Lab. Medic. *43*, 157 (1954).
[301] *J. Rosenberg, M. Bolgar*, Anal. Chem. *35*, 1559 (1963).
[302] *J. Barrolier*, Naturwissenschaften *42*, 126 (1955).
[303] *J. Brüggemann, K. Drepper*, Naturwissenschaften *39*, 302 (1952).
[304] *R.B. Ingle, E. Minshall*, J. Chromatogr. *8*, 369 (1962).
[305] *J. Büchi, A. Zimmermann*, Pharm. Acta Helv. *40*, 292, 395 (1965).
[306] *J. Mastner, P. Franek, L. Novák*, Collect. Czech. Chem. Commun. *24*, 2959 (1959).
[307] *G. Richter, P. Muscholl*, J. Chromatogr. *15*, 39 (1964).

effects of light scattering, some groups of workers recommend direct measurements of transmitted light with photographic negatives[308]. Positives on X-ray[309-311] or other special films[312-315] can be used equally well for measurement. Evaluation of photostats[316-318] has only limited applicability because of the scattering of radiation. Colored or fluorescent spots[319, 320] or radioactively labelled compounds can be used for development (cf. autoradiography, 2.4:6.5.4). Evaluation of photographs can have the following advantages:

– Background noise can be corrected during photographic development.
– Weakly colored zones which absorb very little between $\lambda = 400$ and 800 nm can be measured more effectively by contrasting development.
– Spots which fade rapidly in air are detected by photography at the same time, so that the measurement time has no effect on the accuracy of the estimation.
– The size of the photograph can be adapted so as to suit best the size of the instrumental measurements. It is then not necessary to cut up thin layer plates for evaluation with 'strip densitometers'.
– The photograph must be taken with great care[310]. Uniform illumination of the chromatogram is of great importance. After choosing a suitable film, care must be taken to develop and fix uniformly. For series of analyses, it is recommended that the same bath be used at a constant temperature[311]. Too high a density should be avoided[308], otherwise direct proportionality to the logarithm of the energy radiated will no longer be obtained. Fig. 4 indicates that linear calibration curves exist only for a medium range of greyness (Curve 3)[279].

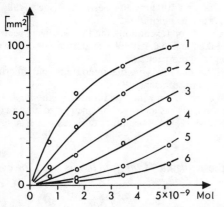

Fig. 4. Dependence of the shape of a calibration curve on the density of a photograph[279]; degrees of transmission 1:68%; 2:46%; 3:31%; 4:24%; 5:15%; 6:12%.

Reflexion measurements: Reflexion measurements have been used for direct evaluation of chromatograms only during the last 20 years[293, 300, 320-323].

Of the few photometers which can be used, two are the self-registering and digitally-integrating Chromoscan Reflexion Densitometer and the Extinction-Registering Instruments ERI 10 or ERI 65. Attachments are also available for the Photovolt and Vitatron densitometers, and the Eppendorf Photometer. These also operate only in the visible spectral region, so that operational problems mentioned above also exist here.
In order to get *straight calibration lines,* changeable control discs–so-called cams–have been proposed. These cams, whose nature depends on the instrument and the problem, can be so arranged that the scale is spread out in the direction of the ordinate. This can increase with rising concentration, and thus lead to a linear relationship between the amount of the compound and the measured value. Further relationships have been treated theoretically[324, 325]. In model studies involving densitometric reflexion measurements, the recovery was about 95%. Under certain circumstances, this can apply also to carbonized chromatogram zones[326-329].

6.5.2 Spectrophotometric Determination

Direct spectrophotometric determination methods have the advantage over densitometric evaluation that they are specific for particular substances. Compared with polychromatic measurements,

[312] P. Darmon, D. Faucquembergue, Ann. Pharm. Fr. 12, 766 (1954); C.A. 49, 9080 (1955).
[313] G.M. Jacobsohn, Anal. Chem. 36, 275, 2030 (1964).
[314] H. Sulser, Mitt. Geb. Lebensm. Hyg. 49, 344 (1958); 50, 275 (1959).
H. Sulser, O. Högl, Mitt. Geb. Lebensm. Hyg. 44, 79 (1953).
[315] A.N. Siakatos, G. Rouser, Anal. Biochem. 14, 162 (1966).
[316] F.W. Hefendehl, Planta Medica 8, 65 (1960).
[317] H.K. Mangold in E. Stahl[390].
[318] E. Zappi, G. Hoppe, Z. Klin. Chem. Biochem. 6, 105 (1968).
[319] E. Gerhards, H. Gibian, G. Raspé, Arzneim.-Forsch. 15, 1295 (1965).
[320] H. Kaiser, L. Wildemann, Intern. Z. Vitaminforsch. 27, 131 (1956).
[321] J. Fellegi, J. Sláma, Chem. Zvesti 10, 315 (1956); J. Fellegi, Zellstoff, Papier (1958), No.3,p.1.
[322] R.S. Hunter, J. Opt. Soc. Am. 50, 44 (1960).
[323] R.H. Müller, Anal. Chem. 22, 72 (1950).
[324] J. Goldmann, R.R. Goodall, J. Chromatogr. 32, 24 (1968).
[325] E.H. Winslow, H.A. Liebhafsky, Anal. Chem. 21, 1338 (1949).
[326] C.B. Barret, M.S.J. Dallas, F.B. Padley, J. Am. Oil Chem. Soc. 40, 580 (1963).
[327] C.M. von Gent, Z. Anal. Chem. 236, 344 (1968).
[328] F. Osman, M.R. Subbaram, K.T. Achaya, J. Chromatogr. 26, 286 (1967).
[329] H.W.H. Schmidt, H. Neukom, Helv. Chim. Acta 49, 510 (1966).

[308] H. Kastien, R. Tomingas, Helv. Chim. Acta 49, 2348 (1966).
[309] G. Jayme, H. Knolle, Z. Anal. Chem. 178, 84 (1960).
[310] R. Miram, S. Pfeifer, Pharm. Zentralh. 96, 457 (1957).
[311] D. Neubauer, K. Mothes, Planta Medica 9, 466 (1961).

Table 8. Methods of measurement with the Chromatogram Spectrophotometer

Characteristics of the object	Measurement problem	Type of measurement[a] aperture	Light source
	Absorption measurement		
light scattering, non-transparent	no interference through fluorescence	M-Pr(R) slit	hydrogen or tungsten lamp
	fluorescence of the indicator	M-Pr(R) slit	mercury lamp
	interference through fluorescence	Pr-M slit	hydrogen or tungsten lamp
transparent, non-scattering	no interference through fluorescence	M-Pr(T) slit	hydrogen or tungsten lamp
	Fluorescence measurement		
	emission spectra	Pr-M iris	mercury lamp
	quantitative emission measurement with selectable wavelength (e.g. λ_{max})	Pr-M iris	mercury lamp
	excitation spectra;	M-Pr iris	xenon lamp with special filters in front of the photomultiplier
	integral measurement with excitation radiation of selectable wavelength	M-Pr iris	xenon lamp

[a] M = monochromator; Pr = sample; R = reflexion; T = transmission

they are considerably more sensitive and accurate. Here, too, special attachments for spectrophotometers, for transmission and reflexion measurements, are available[330, 396]. Kortüm[374, 398] has summarized technical details and physical principles. At present, the most universally equipped instrument is the Zeiss Chromatogram Spectrophotometer. It consists of:

– the Zeiss PMQ II spectrophotometer;
– a coordinate table, on which chromatograms of size up to 200 × 400 mm can be fixed horizontally;
– the remission unit arranged above, which is connected with the monochromator by a special device containing the achromatic lens system for focussing the slit or iris.

Reflexion measurements are usually carried out with this instrument. For transmission measurements on non-scattering objects, the photomultiplier can also be fixed underneath the coordinate table. The chromatogram spectrophotometer can be used for various kinds of measurement (Table 8) since it possesses the following advantages:

[330] H. Jork, Pharma International 3, 11 (1967).

– Compounds which absorb or fluoresce between $\lambda = 200$ and 800 nm can be estimated specifically without previous coloration.
– Absorption (excitation) and fluorescence spectra can be taken independent of the chromatogram without danger of destruction.
– Even quantities in the nanogram and microgram ranges are sufficient.
– In comparison with direct testing methods, therefore, the sensitivity is increased by one to two powers of ten.
– No additional expenditure of time or operating costs is required[396].

If the *Kubelka-Munk function* is used for evaluation,

$$\frac{c \cdot \varepsilon}{s} = \frac{(1 - R_\infty)^2}{2 R_\infty}$$

where
c = quantity applied per spot area
ε = molar decadic extinction coefficient
s = scattering coefficient
R_∞ = absolute degree of remission of the substance

straight line calibration plots result which pass through the origin[396, 398]. If the range of concentration does not exceed two powers of ten, the squares of the recorded peak area also yield straight calibration lines when plotted against amounts of substances.

Error considerations have revealed[331, 332] that in the analysis of model mixtures by TLC, the standard deviation from the theoretical value is between v = ±3 and 4%. In routine work, the reproducibility falls to v = ±4 to 6%. Systematic errors should, of course, be eliminated beforehand.

6.5.3 Measurement of Fluorescence

Either simple filter fluorimeters[333–336, 387] or spectrofluorimeters can be used[201, 297, 337] for in situ determination of fluorescence. The most universally operated instruments are those with two monochromators[140].

In contrast to the absorption measurements mentioned thus far, fluorescence measurements estimate specific changes in the intensity of secondary emission of substances. This method has the following advantages:

- Measurements of uncorrected fluorescence are simple to perform.
- The substance quantities necessary are usually 100 to 1000 times smaller than those needed for absorption measurements.
- Differences in size and form of the chromatogram spots are less important than in photometric absorption measurement.
- If non-fluorescent stationary phases are used for chromatography, one can carry out an integrated operation using circular measuring apertures.
- Scattered light effects can often be neglected.

The emitted light, summed over the whole hemisphere, is linearly proportional to the amount of substance. Measurements of quenching substances on fluorescent TLC plates are erroneously regarded as a special case of quantitative emission determination. Actually, this is an indirect absorption measurement which can be carried out with fluorimeters[338–341].

The incident primary radiation is attenuated by a filter effect caused by the chromatographed substance. Less energy strikes the fluorescence indicator in the layer, so that the light emission also becomes weaker. As a result, one obtains dark zones on a generally yellowish-green background. This is concentration dependent and can, therefore, be evaluated quantitatively.

Not all absorbing substances can be analyzed by fluorescence quenching. If the compounds to be determined have an absorption minimum in the range of wavelengths concerned, the 'quenching effect' is small and evaluation is possible (if at all) only for larger quantities. This method of determination is less sensitive and accurate than direct measurement of fluorescence, or even direct determination of absorption at corresponding wavelengths[342, 396].

6.5.4 Determination of Radioactive Compounds

Chromatograms of radioactively labelled compounds can be estimated most rapidly by direct evaluation. If a radioscanner is not available, photometers adapted for the purpose may be used[343–347]. Geiger-Müller counters or gas flow meters[317] are generally used as detectors. On commercially available strip measurement instruments, the Geiger counters for evaluation of paper chromatograms are placed in a 4π arrangement[337, 377]. For measuring thin-layer chromatograms, a 2π arrangement with an altered transport system for the chromatograms has been recommended, since the carrier plates do not allow downward-directed radiation to pass through. The sensitivity of evaluation in this case is 20% smaller than for measurements of paper chromatograms[348].

If double labelled compounds are to be determined, foil windows with a defined thickness (and thus defined transmittance) can be employed. In this way ^3H, ^{14}C and ^{32}P can be determined separately in the presence of each other[349]. With TLC

[331] H. Jork, Z. Anal. Chem. 221, 17 (1966); 236, 310 (1968); J. Chromatogr. 33, 297 (1968).
[332] E. Mutschler, K. Kraus, H. Rochelmeyer, J. Chromatogr. 40, 244 (1969).
[333] W.M. Connors, W.K. Boak, J. Chromatogr. 16, 243 (1964).
[334] D. Jänchen in E.J. Shellard[395].
[335] R.D. Stubblefield, O.L. Shotwell, C.W. Hesseltine, M.L. Smith, H.H. Hall, Appl. Microbiol. 15, 186 (1967).
[336] L.J. Fischer, S. Riegelman, J. Chromatogr. 21, 268 (1966); J. Pharm. Sci. 56, 469 (1967).
[337] F. Korte, H. Weitkamp, Angew. Chem. 70, 434 (1958).
F. Korte, H. Weitkamp, H.G. Schicke, Biochem. Ztg. 329, 458 (1958).
[338] W. Messerschmidt, J. Chromatogr. 33, 551 (1968).
W. Messerschmidt, W. Weiser, J. Chromatogr. 38, 156 (1968).

[339] G. Pataki, Helv. Chim. Acta 50, 1069 (1967).
G. Pataki, A. Niederwieser, J. Chromatogr. 29, 133 (1967).
G. Pataki. E. Strasky in Ref. [393].
[340] A. Uettwiller, M. Keller, J. Chromatogr. 35, 526 (1968).
[341] R. Klaus, J. Chromatogr. 16, 311 (1964).
[342] D. Jänchen in Ref. [393].
[343] G.S. Boyd, M.A. Eastwood, Biochem. Biophys. Acta 152, 169 (1968).
[344] E.L. Durrum, S.R. Gilford, Rev. Sci. Instrum. 26, 51 (1955).
[345] D. Hofmann-Credner, Naturwissenschaften 40, 623 (1953).
[346] B.M. Johnstone, G.P. Briner, J. Chromatogr. 2, 513 (1959); Nature 183, 681 (1959).
[347] T.D. Price, P.B. Hudson, Anal. Chem. 26, 1127 (1954).
[348] A.S. Csalany, H.H. Draper, Anal. Biochem. 4, 418 (1962).

measurement, however, the count yield (which is normally 20 to 30% for ^{14}C and ^{35}S, 50% for ^{32}P, and 0.6% for ^{3}H) is diminished, and varies according to the thickness of the stationary phase.

For evaluation of bigger chromatograms, special scanners are available. These units have a fixed detector, under which the chromatograms can be moved. Recording is done by writer – and printer-units [349–353, 375].

For indirect determination of labelled compounds with the aid of fluorescent additives, one uses scintillation scanners. The developed chromatograms are sprayed uniformly with suitable scintillation solutions. An excess of reagent solution must not interfere with the results. This determination technique has proved especially successful for weak β-emitters [354, 355]. Self-absorption of β-particles does not usually occur. Anthracene solutions can be used [356–358] with ^{3}H labelled compounds. Weakly emitting isotopes can be determined by *autoradiography*, even in two-dimensional chromatography.

Depending on the isotope, the amount of substance and the photographic material, different contact times are necessary [359]. 'Contact copies' are often prepared, using special casettes [359, 360–362].

Radiographs are usually evaluated densitometrically (cf. 2.4:6.5.1). However, the cost is comparatively high, and the results are rather inaccurate. An important reason for this is an uncontrollable enlargement of zones during the exposure, brought about by scattering within the stationary phase.

7 Bibliography

[363] *M. Calvin, C. Heidelberger, F.C. Reid, B.M. Tolbert, P.E. Yankwich*, Isotopic Carbon, John Wiley & Sons, New York 1949.

[364] *T. Wieland, F. Turba* in *Houben-Weyl*, Methoden der organischen Chemie, 4. Aufl. Bd. II, p. 867, Georg Thieme Verlag, Stuttgart 1953.

[365] *C.G. Bell, F.N. Hayes*, Conference on Liquid Scintillation Counting, Pergamon Press, New York 1957.

[366] *E. Lederer, M. Lederer*, Chromatography. A Review of Principles and Applications. 2. Ed., Elsevier, Publ. Co., Amsterdam 1959.

[367] *D. Waldi*, Chromatographie, 2. Aufl., Merck, Darmstadt 1959.

[368] *H.F. Linskens*, Papierchromatographie in der Botanik, 2. Aufl., Springer Verlag, Berlin 1959.

[369] *I.M. Hais, K. Macek*, Handbuch der Papierchromatographie, 1. Aufl., Bd. II, VEB Gustav Fischer Verlag, Jena 1960.

[370] *H.F. Linskens, L. Stange*, Praktikum der Papierchromatographie, Springer Verlag, Berlin 1961.

[371] *K. Macek, I.M. Hais, J. Gasparič, J. Kopecký, V. Rábek*, Bibliography of Paper Chromatography 1957–1960, Publishing House of the Czechoslovak Academy of Sciences, Prag 1962.

[372] *F. Cramer*, Papierchromatographie, 5. Aufl., Verlag Chemie, Weinheim/Bergstr. 1962.

[373] *Y. Hashimoto*, Thin-Layer Chromatography, Hirokawa, Tokyo 1962.

[374] *G. Kortüm*, Kolorimetrie, Photometrie u. Spektroskopie, 4. Aufl., Springer Verlag, Berlin, Heidelberg, New York 1962.

[375] *M. Wenzel, P.E. Schulze*, Tritium-Markierung, Walter de Gruyter & Co., Berlin 1962.

[376] *N.K. Freeman, F.T. Lindgren, A.V. Nichols* in *R.T. Holman, W.O. Lundberg, T. Malkin*, Progress in the Chemistry of Fats and Other Lipids *6*, 215 (1963).

[377] *I.M. Hais, K. Macek*, Handbuch der Papierchromatographie, 2. Aufl., Bd. I, VEB Gustav Fischer Verlag, Jena 1963.

[378] *J.M. Bobbitt*, Thin-Layer Chromatography, Reinhold Publishing Co., New York 1963.

[379] *E.V. Truter*, Thin-Film Chromatography, Cleaver-Hume Press, London 1963.

[380] *D. Glick*, Methods of Biochemical Analysis. Vol. XI, Interscience Publishers, London 1963.

[381] *A.T. James, L.J. Morris*, New Biochemical Separations, van Nostrand Co., London 1964.

[382] *G.B. Marini-Bettòlo*, Thin-Layer Chromatography, Elsevier, Publ. Co., Amsterdam 1964.

[383] *A.A. Achrem, A.J. Kuznetsova*, Dünnschicht-Chromatographie (russ.), Izdatelstwo Nauk Moskau, 1964.

[384] *K. Macek, I.M. Hais*, Stationary Phase in Paper- and Thin-Layer Chromatography, Elsevier Publ. Co.. Amsterdam 1965.

[385] *L. Lábler, V. Schwarz*, Thin-Layer Chromatography, CSAV, Praha 1965 (Czech.).

[386] III. Intern. Symposium Chromatogr. Brüssel 1964, Symposiumsband, 1968.

[387] *G. Pataki*, Dünnschicht-Chromatographie in der Aminosäure- und Peptidchemie, Walter de Gruyter & Co., Berlin 1966.

[349] *M. Wenzel, P.E. Schulze, H. Wollenberg*, Naturwissenschaften *49*, 515 (1962); Z. Anal. Chem. *201*, 349 (1964); Naturwissenschaften *52*, 129 (1965).

[350] *P. Karlson, R. Maurer, M. Wenzel*, Z. Naturforsch. *18b*, 219 (1963).

[351] *E. Kön493k*, J. Chromatogr. *37*, 128 (1968).

[352] *P.E. Schulze, M. Wenzel*, Angew. Chem. *74*, 777 (1962).

[353] *F. Berthold, M. Wenzel*, Instrum. Nucl. Med. *1*, 251 (1967).

[354] *S. Prydz, H.S. Skammelsrud*, J. Chromatogr. *32*, 732 (1968).

[355] *J.C. Roucayrol, P. Taillandier*, C.R. Acad. Sci., Paris *256*, 4653 (1963).
J.C. Roucayrol, J.A. Bergner, G. Meymel, J. Perrin, Int. J. Appl. Radiat. Isotopes *15*, 871 (1964).

[356] *G. Jolchine*, Physiol. Vegetale *2*, 341 (1964).

[357] *N. Lüthi, P.G. Waser*, Nature *205*, 1190 (1965).

[358] *A.T. Wilson, D.J. Spedding*, J. Chromatogr. *18*, 76 (1965).

[359] *R.M. Fink*, Nature *160*, 801 (1947); Anal. Chem. *28*, 4 (1956).

[360] *M.L. Blank, J.A. Schmitt, O.S. Privett*, J. Am. Oil Chem. Soc. *41*, 371 (1964).

[361] *G.S.J. Richardson, W. Weliky, M. Griffith, L.L. Engel*, J. Chromatogr. *12*, 115 (1963).

[362] *H.S. Sodhi, P.D.S. Wood*, Proc. Soc. Exptl. Biol. Med. *113*, 714 (1963).

[388] *K. Randerath,* Dünnschicht-Chromatographie, 2. Aufl., Verlag Chemie, Weinheim/Bergstr. 1966.

[389] *J.G. Kirchner,* Thin-Layer Chromatography, Interscience Publishers, New York, London, Sydney 1967.

[390] *E. Stahl,* Dünnschicht-Chromatographie, 2. Aufl. Springer Verlag, Berlin, Heidelberg, New York 1967.

[391] *K. Hannig, G. Pascher* in *E. Stahl*[390].

[392] *M. Šaršunová, V. Schwarz, Č. Michalec, Lj. Kraus, F. Perényi,* Thin-Layer Chromatography in Pharmacy and Clinical Biochemistry, Verlag-Obzor, Bratislava 1968 (Czech.).

[393] IV. Intern. Sympos. Chromatogr. Brüssel 1966; Symposiumsband 1968.

[394] *G. Spitteler* in *H. Kienitz,* Massenspektrometrie, Verlag Chemie, Weinheim/Bergstr. 1968.

[395] *E.J. Shellard,* Quantitative Paper- and Thin-Layer Chromatography, Academic Press, New York, London 1968.

[396] *H. Jork,* Qualitative und quantitative Auswertung von Dünnschicht-Chromatogrammen unter besonderer Berücksichtigung photoelektrischer Verfahren (Habilitationsschrift), Universität Saarbrücken 1969.

[397] *W. Preetz,* Fortschritte der chemischen Forschung, Bd. II, Springer Verlag, Berlin, Heidelberg, New York 1969.

[398] *G. Kortüm,* Reflexionsspektroskopie, Springer Verlag. Berlin. Heidelberg, New York 1969.

[399] *E. Stahl, E. Dumont, H. Jork, Lj. Kraus, K.-E. Rozumek, P.J. Schorn,* Mikroskopische und chromatographische Untersuchungen, G. Fischer Verlag, Stuttgart 1970.

[400] *H. Kaldewey, Y. Vardar,* Hormonal Regulation in Plant Growth and Development, Verlag Chemie, Weinheim 1972.

2.5 Column Chromatography from Solutions

Gerhard Hesse

Institut für Organische Chemie der Universität
D-852 Erlangen

The scope of column chromatography includes the separation of pure substances from mixtures on a preparative scale and also quantitative analysis (2.5:3.7). For qualitative purposes and for orientative quantitative analysis, thin-layer and paper chromatography are more suitable (cf. 2.4).

In its commonly employed current procedure the column consists of a vertical glass tube (2.5:3) packed with the sorbent material supported on a porous plate. The latter is impervious to the packing but allows free passage of liquid. The term sorbent includes every stationary phase (cf. 2.5:3.2) which retains the dissolved substances under equilibrium conditions and retards them in such a way that they reach the end of the column at different times. Retardation can occur as a result of the following mechanisms (cf. 2.5:3.7):

Adsorption at the surface or in the pores of the sorbent (cf. 2.5:3.2);

Formation of *hydrogen bonds* with the acceptor sites of the sorbent (cf. 2.5:4.2);

Transfer to a solvent which is dispersed over the grainy material or bound chemically to the latter (*partition,* f. 23);

Transfer to a liquid contained within a gel (cf. 2.6);

Ion exchange with similarly charged ions of the sorbent (ion exchangers, cf. 2.8).

1 Introduction

Column chromatography in the liquid phase (Liquid Column Chromatography, LCC), as used in practice, is based primarily on adsorption or partition. (For ion-exchange see Chapter 2.8; dialysis, Chapter 2.6). The formation of hydrogen bonds between the sample and the adsorbent is considered a special case of adsorption. During a chromatographic separation, often more than one of the cited phenomena are simultaneously effective. This usually results in zone-spreading. It is therefore desirable to find out which principle of separation is predominent, and to change the system accordingly. All the measures aimed towards optimization are dependent on the actual separation principle; hence knowledge of the mechanism has practical significance.

Adsorption chromatography (Liquid Solid Chromatography, LSC) is particularly suitable for distinguishing substances differing in the number and configuration of their functional groups. Whether all or only a few of these groups can contribute towards the sorption of the molecules depends upon their arrangements within the molecules.

Adsorption chromatography is, therefore, highly specific with respect to conformation[1,2].

Partition chromatography (Liquid Liquid Chromatography) has been most widely used with paper and thin layers (2.4) but can also be carried out on a preparative scale in columns. The advantage of this technique is that the substances concerned are handled under nondestructive conditions; its disadvantage is that the column loading capacity is normally ten times smaller, thus requiring the use of large equipment. According to Halász[3], stationary organic phases can be used, which are fixed to porous carriers, such as glass, by means of atomic bonds. An example is Durapak® (Waters). Cellulose esters developed by Stetter[4] may also be used for separations in columns.

---

[1] *G. Hesse,* Chromatographia *1,* 302 (1968).

[2] *K.D. Stscherbakowa,* Gas-Chromatographie, p. 533. Akademie Verlag, Berlin 1968.

[3] *I. Halász, I. Sebestian,* Angew. Chem. *81,* 464 (1969).

[4] *H. Stetter, J. Schroeder,* Angew. Chem. *80,* 1035 (1968).

The use of non-ideal gases[5], such as carbon dioxide or ammonia at high pressures (200–2000 atm.), is comparable to gas chromatography. Under these conditions the gases acquire properties resembling liquids; they dissolve many organic molecules, such as sensitive α-carotene, difficultly volatile sugars, and high polymers like Carbowax 20,000. By suitable arrangements, column chromatographic separations can be developed and, in addition to temperature, the pressure can be utilized as a parameter. A similar development has begun with supercritical liquids[6]. In both these cases the required analysis time is relatively short.

2 Column Chromatography from Solutions
(cf. Houben-Weyl, 4. Ed., I/1, p. 465)

2.1 Definitions
The portion of a column which is occupied by a particular substance is called its *zone*. The narrower the zones, the larger the number of substances which can be separated in a column of a definite length, and the more concentrated are the resulting eluates. At constant conditions, each substance maintains at least its *zone width* with respect to the initial width of the mixed zone. It is, therefore, important to keep the starting zone as narrow as possible. This requirement is satisfied by dissolving, the mixture in a solvent, from which the sorption is very strong. However, for the separation a solvent must be used from which the sorption is incomplete; otherwise no migration of the zones occurs. Sorbent and eluting solvent must be balanced with each other. Together they form the *sorption milieu*.

2.2 Methods
It is often necessary to employ various eluting solvents consecutively, in order to separate the strongly sorbing components from each other.

Another method of reducing the sorption is by *increasing the temperature*[7]. For better reproducibility, the temperature is maintained constant as far as possible during chromatography from solutions. Uneven heating of the separating column, caused by sun light or other sources of heat, must be strictly avoided. Otherwise strong distortion of the zones occurs.

Constant zone width, as in the ideal case, is rarely achieved, principally because of diffusion during

long experiments. Therefore recent trends in the development of liquid chromatography are directed towards reducing the experimental time by the *application of pressure*. Even greater effects are achieved by gradually *changing the solvent* so that its eluting power is increased *(gradient elution)*. Thus the frontal part of each zone is always in a stronger sorption environment than the rear section, and hence the former migrates more slowly. The zones are thus compressed so that they can become even narrower than the original starting zone.

Detection of zones of colorless substances in the separating column presents a problem. UV light is sometimes helpful (cf. 2.5:3.2). Alternately, the components are eluted consecutively from the column and collected in numerous small fractions. The individual fractions are then separately examined.

Thin-layer chromatography is very suitable for this purpose, because many samples can be analyzed in a single experiment. This technique provides criteria of purity since much sharper separations are obtained than by column chromatography.

The trend of further developments is directed towards monitoring the eluate of a column using a *continuously recording detector*. Most of the present liquid detectors are applicable only if the solvent is not changed during the experiment (cf. 2.5:3.7.1).

Frequently, elution of all the components of the mixture from the column is unnecessary; the zones are then separated mechanically and eluted individually (cf. 2.5:3.7.4).

3 Apparatus for Liquid Chromatography

3.1 Columns
A simple straight glass tube, tapered at the bottom and often fitted with a tap is still commonly used. The ratio of length to diameter should be greater than 20:1; a ratio of 40:1 can be taken as standard. Long and narrow tubes are used for difficult separations. Columns fitted with cooling jackets are used when constant temperature conditions are required.

Separating columns with detachable adapters are also commercially available (LKB, Nester-Faust, etc.). The sorbent is packed in these columns between two sintered glass discs, and small-bore teflon tubing is used for liquid input and output.

Such columns have desirable low dead volumes. They can be employed also for the ascending technique, which is advantageous, e.g. when a heavier liquid follows a lighter one. These tubes with 1 cm ∅ can

---

[5] *J.C. Giddings*, J. Chromatogr. Sci. *7*, 276 (1969); Science *162*, 67 (1968).
   D. Bartmann, G.M. Schneider, Chem.-Ing.-Tech. *42*, 702 (1970).
[6] *S.T. Sie, G.W.A. Rijnders*, Anal. Chim. Acta *38*, 31 (1967); Separation Sci. *1*, 459 (1966); *2*, 729, 755 (1967).
[7] *G. Hesse, H. Engelhardt*, J. Chromatogr. *21*, 228 (1966).

withstand pressures up to 20 atm. For higher pressures special metal tubes are required. Stainless steel is preferred to copper tubes. V4A steel is more suitable than V2A for chemical reasons.

3.1.1 Packing of Columns

Special care should be taken in *packing the columns* with sorbent (cf. Houben-Weyl, Vol. I/1, p. 484). Course particles, loose sites in the sorbent bed, and particularly air bubbles in a wet column are the most common causes of distortion of the zones. They impair the separation considerably.

Before filling, the tapered end of the tube is plugged uniformly with glass-wool or cotton (without optical bleach) using moderate pressure. A 0.5 cm layer of pure sand or preferably glass beads (0.2 mm diameter) is then added to the vertically clamped tube so that its surface is flat and even.

Alumina and silica gel can be introduced into the column in a dry state by using a funnel, taking care that the freely flowing powder falls centrally in the cross section of the tube. After filling, the tube is tapped or vibrated from various sides and tapped gently several times on a wooden surface. A uniform packing of the sorbent is thus achieved. The upper layer of the sorbent bed is then levelled and a layer (1 cm) of sand or glass beads is poured on top. The upper surface of this layer is also levelled.

The dissolved substance is best introduced at the center of the layer covering the packing by means of a capillary pipet. Uniform distribution of the substance over the whole surface occurs only with columns of large cross sections. Before use, a dry filled column must be flushed with the eluting solvent (exception: dry column 2.5:3.8) in order to remove air bubbles.

Many sorbents, particularly those which swell in the presence of eluants, must be introduced into the column as suspensions. In such cases the sorbent is mixed with the solvent to obtain a slurry that can be easily poured. To make swelling complete further addition of solvent may be necessary. Often it is desirable to degas the suspension by evacuation in a round bottom flask. The suspension is then poured through a funnel in the column, which is held exactly vertical. The excess liquid is allowed to drain. During this procedure the same precautions must be taken as mentioned earlier. Filling materials which are lighter than sand cannot be covered by the latter. In such cases a disc of filter paper that exactly fits into the cross section of the tube is pressed on the upper surface of the sorbent bed.

3.1.2 Packings with Activity Gradients

For mixtures with widely varying sorptive properties, it is desirable to use a column in which the activity increases steadily from top to bottom. The effects achieved on a *gradient column* are similar to those due to gradient elution in homogeneously packed columns (cf. 2.5:3.1). It should, however, be noted that a large number of detectors currently available do not permit alteration of the eluting solvent during the experiment.

Gradient column with aluminum oxide of activities I–V[8]:
The tube (60 × 1.5 cm) is filled to one half of its length with 75 g aluminum oxide, activity grade I, in a dry state. Instead of sand, a cotton plug, previously soaked in a dilute solution of cobalt(II) chloride and thoroughly dried, is firmly pressed on to the alumina. A second wad of cotton-wool is then placed in the column at a distance of about 15 cm above the other plug. A definite amount of water (4.6 ml, cf. Table 1; 2.5:3.2) which is required by the sorbent to attain the activity stage III is pipetted into the second wad of cotton-wool, and the column is inverted and held in a vertical position. Air, which has been dried thoroughly over P_2O_5 is now sucked into the column from its lower (previously upper) end. The current of air carries water to the sorbent and deactivates the latter, starting from the point of entry. The cotton plug at the head of the packing serves as a humidity indicator. Its blue color turns to pale pink as soon as the humidification begins; the flow of air must be stopped as soon as the color starts changing again to blue. The column is then returned to its original position, both the plugs are removed, and the upper surface of the sorbent bed is levelled and covered with sand.

All the six test dyes of Brockmann and Schodder[9] could be completely separated in a single experiment in such a column using petroleum ether/benzene (4:1) as solvent. These columns are suitable only for chromatography with non-polar solvents, since other solvents might remove the water used for deactivation of the aluminum oxide.

3.2 Sorbents

3.2.1 General Requirements

All solid phases without impregnation which can be used for packing columns are considered here as sorbents. In order to obtain columns with good filtration properties, the particles should be spherical in shape and uniform in size. Their mechanical stability must be great enough to prevent the formation of fine dust which might be deposited in the channels of the packing. Dry-sieving is usually inadequate for the removal of dust already present

---

[8] *G. Hesse, G. Roscher,* Chromatographia *2,* 512 (1969).

[9] *H. Brockmann, H. Schodder,* Ber. *74,* 73 (1941); vgl. *G. Hesse* in *Houben-Weyl,* Methoden der organischen Chemie, 4. Aufl., Bd. I/1, p. 470, Georg Thieme Verlag, Stuttgart 1958.

in the sorbent. Wet-sieving or sedimentation from a solvent, usually water, is more effective.

The sorbent should not react chemically, either with the eluting solvent or with the sample components. It should contain as small amount of soluble components as possible. The material should be catalytically inactive and, as a rule, have a neutral surface. Exceptions are the ion exchangers.

3.2.2 Examination of Sorbents

Various batches of the same material often differ considerably from each other with respect to the cited properties and also with respect to their effective surfaces, which are sometimes *generally* considered as measures of activity. The effective surface is measured statically by gas adsorption, according to the method of Brunauer, Emmett and Teller[10]. However, since this procedure is carried out under vacuum, the volatile substances, such as water, that are present in the sorbent intentionally or not and which play an important part in the chromatographic behavior, are removed. Therefore the modified procedure according to Nelson and Eggertsen[11], which can be carried out in a sorptometer (Perkin-Elmer, Norwalk, Conn.), is more suitable[12]. According to this method, nitrogen is adsorbed from a mixed stream of nitrogen and helium by cooling with liquid nitrogen and is desorbed later by warming to ambient temperature. The gas is continuously analyzed by a thermal conductivity detector. The free surface available for adsorption is obtained from the area of the desorption peak.

However, the above method has a drawback since larger molecules can occupy only part of the active centers, and cannot penetrate into the fine pores and channels of the sorbent. Therefore, the adsorption of dyes from non-polar solvents is generally taken as a measure of activity.

In addition to the standard method of Brockmann[9], a simpler procedure using a single dye has been described and is adequate in most cases[13].

Activity of aluminum oxide and silica gel. A narrow tube (1×75 mm; cf. Houben-Weyl I/1, p. 483), closed at one end, is filled with the sorbent and one drop of benzene is added to prevent the sorbent from falling out if the tube is inverted. The

closed end of the tube is then cut off with a glass file. The other end is immersed for a short time in a 0.5% solution of *p*-amino-azobenzene (for Al_2O_3) or *p*-dimethylamino-azobenzene (for SiO_2), both in benzene, until some of the dyes is drawn into the sorbent. Subsequently the tube is placed vertically in a small test tube containing a little (5 mm high) benzene, and the miniature tube is allowed to develop. The activity and water content above the Brockmann grade I level are determined from the R_f values according to the following tables. The desired activity grade can be adjusted by adding the required amount of water.

Table 1. Activity according to Brockmann and water content of adsorbents, determined from R_F values of *p*-aminoazobenzene in benzene

Additional water [%]	R_F value/Activity grade	
	Aluminum oxide	*Silicagel*
0	0 /I	0.15/I
3	0.12/II	0.22/—
6	0.24/III	0.33/—
8	0.46/IV	—
9	—	0.44/—
10	0.54/V	—
12	—	0,55/II
15	—	0,65/III

A useful kit for rapid determination of the activity of aluminum oxide has been described by Stahl[14] (DESAGA, Heidelberg).

Deactivation can also be effected by strongly adsorbing substances, such as glycol or glycerol[15]. These additives are not readily washed out and withstand relatively high temperatures. They alter the column in such a way that partition as well as adsorption influences the separation.

Nonpolar adsorbents, such as activated charcoal, can be impregnated with substances like stearic acid[16] or other fatty acids and their alkali salts (soaps), which have some similarity to hydrocarbons. The hydrocarbon squalane has proved to be useful in gas chromatography[17].

Precoating is an excellent means for adjusting the activity of adsorbents. Formerly used oxidic adsorbents have no current use because of their low activities, and because adsorbents of low activity can be prepared from aluminum oxide or silica gel by the precoating technique. Partial precoating blocks only the most active centers, and hence the surface of the adsorbent attains more uniform

[10] *S. Brunauer, P.H. Emmett, E. Teller,* J. Am. Chem. Soc. *60,* 309 (1938).

[11] *H. Engelhardt, B.P. Engelbrecht,* Chromatographia *4,* 66 (1971).

[12] *F.M. Nelson, F.T. Eggertsen,* Anal. Chem. *30,* 1387 (1958).

[13] *B. Loev, M.M. Goodman,* Chem. Ind. [London] *1967,* 2026.

[14] *E. Stahl,* Chem.-Ztg. *85,* 371 (1961).

[15] *G. Hesse, G. Roscher,* Z. Anal. Chem. *200,* 3 (1964).

[16] *J. Porath,* Ark. Kemi *7,* 535 (1954).

[17] *F.T. Eggertsen, H.S. Knight, S. Groennings,* Anal. Chem. *28,* 303 (1956).

adsorptive behavior desirable for separations. At the same time decomposition of substances, which is associated with strong adsorption, is eliminated. However, there is always a possibility that the agent used for precoating is rinsed away during the course of a chromatographic separation. This is particularly the case with water. This phenomenon can be circumvented by addition of water to the eluting solvent (cf. 2.5:3.3).

The adsorbent can also be precoated with a fluorescent material[18] (cf. Houben-Weyl I/1, p. 485). This permits the detection of colorless compounds in the column by quenching of fluorescence. The same effect is achieved by incorporation of ca. 0.5% of an inert inorganic fluorescent color to the sorbent (e.g. Leuchtstoff grün, Woelm, D-344 Eschwege), which is also called fluorescent indicator.

In the following discussions the currently most frequently used adsorbents are described and characterized. It is not advisable to prepare one's own adsorbents (cf. Houben-Weyl I/1, p. 483) since it is rarely possible to reproduce properties of small batch materials of this type.

3.2.3 Aluminum Oxides

The commercial products differ mainly in particle size distribution, activity and surface reaction. The surface is mostly alkaline and has cation exchange properties ('cationotropic'). The exchanging ion is sodium. An acidic ('anionotropic') surface is obtained by treatment with strong acids. This surface can be converted to a neutral product without ion exchange properties by thermal hydrolysis. (For differentiation and test procedures cf. DAB 7 and Ref. [19]).

All three products have activity I according to the Brockmann scale after being heated to 300°. By adding definite amounts of water, the other *activity grades* can easily be obtained (cf. Houben-Weyl I/1, p. 470 and Section 2.5:3.2.2).

Recently Engelbrecht[20] has extended the series in both directions and developed modified procedures for measuring the activity. The new activity grade 'Super I' (Woelm) has a specific surface of $200\ m^2/g$, which is approximately twice the capacity of the hitherto known grade I (av. $100\ m^2/g$). The former product is very constant in its properties. Originating from this material, the various amounts of water required to attain the activity I–V are supplied by the manufacturer[21].

Aluminum oxide gradually equilibrates with the humidity of the air (grades II–III). Grade I in particular, should be handled as a hygroscopic material.

Since aluminum oxide is a Lewis acid it can sometimes cause exchange of substances in a column (cf. 2.5:4.3). This is, however, rarely the case with grades beyond II.

3.2.4 Silica Gels

The properties and the degree of purity of commercial silica gels vary much more widely than those of aluminum oxides. The specific surface of silica gels varies from 10^2 to $10^3\ m^2/g$. Various types with broad or narrow pores are known. Compared with aluminum oxides it is more difficult to maintain the activity grades of silica gels since they more readily take up water from the air or from solvents. However, a distinct advantage is that silica gel is not a Lewis acid, and hence substances are seldom altered on it. However, care must be taken that the manufactured product does not contain mineral acids or bases. The different types of 'Supelcosil'[22] are well reproducible as adsorbents or carrier materials.

Other oxidic adsorbents (cf. Houben-Weyl I/1, p. 469) are rarely used at present, because they can be replaced by the deactivated forms of the previously mentioned adsorbents. Specific affinities result from the relationship between the distance of the polar groups in a molecule and the pattern of charge in the crystal lattices of oxides and salts which serve as adsorbents. *Magnesium silicate, zinc carbonate,* and *calcium phosphate* are examples of such agents. Also a definite porous structure, such as that shown by porous glasses or permutites, can cause specific sorption due to size, for example, as in molecular sieving[23]. Suitable sorbents for liquid chromatography are Porasil® (Waters) and Corning®-GPC series supports, especially the sieve fraction between 350–120 mesh.

3.2.5 Activated Charcoal

Graphitic carbon black is one of the few solid hydrophobic supports which effects extensive sorption of organic substances from aqueous solutions. The eluotropic series (cf. 2.5:3.3) is applicable in the reverse order of the solvents. Activated charcoal has been used often in biochemistry (cf. Houben-Weyl I/1, p. 476). However, its general use is limited since no product with properties consistent for chromatography is commercially available. Results cannot even be reproduced on different batches from the same manufacturer. In many instances, polyamides can replace the activated charcoals.

[18] *H. Brockmann, F. Volpers,* Chem. Ber. *80,* 77 (1947).

[19] *H. Böhme,* Arch. Pharm. *299,* 282 (1966).

[20] *B.P. Engelbrecht,* Z. Anal. Chem. *244,* 388 (1969).

[21] *M. Woelm,* D-344 Eschwege, Druckschriften 104–106.

[22] Supelco, Inc., Bellefonte, Pennsylvania 16823.

[23] *A.J. de Vries,* Anal. Chem. *39,* 935 (1967).

3.2.6 Polyamides

Various polyamides, such as Nylon, Perlon, Ultramid BM 228 and powdered polyacryl amide (particle size ca. 0.04 mm) are suitable for the separation of lower *fatty acids*[24], *phenols*[25-29], *flavones*[30], *pharmaceuticals*[31], *quinones*[32] and *synthetic dyes*[33].

Filling of the columns[33]; The polyamide slurry is charged into the chromatographic tube after being allowed to swell for several hours. In order to remove the monomers, about 100 ml of the eluting solvent is passed through the column (100 ml content). The substances to be resolved are added to the column as a 2% solution in the eluting solvent (10 ml) and the chromatogram is developed. The eluotropic series for polyamides is presented in chapter 2.5:3.3.

3.2.7 Other Organic Substances as Sorbents

Powdered sugar, sieved and dried in a desiccator, has been used for the separation of the coloring matters of leaves[34] as well as for *polyphenols*. Use of *milk-sugar* permitted the first chromatographic separation of racemates (cf. 2.5:4.1.3)[35]; a very effective product was obtained by boiling the sorbent with chloroform and subsequently grinding in a ball mill with steel balls. *Starch*, which is also suitable for the separation of racemates, has been prepared as follows[36]:

Potato starch is eluted with water to remove the fine particles, extracted with methanol for 6 hrs in a Soxhlet, dried in air, and sifted (0.05–0.075 mm).

Dextrans treated with epichlorohydrin to various degrees are employed as Sephadex® for gel chromatography (cf. 2.6)[37]. Tautomeric keto and enol forms of *p-hydroxyphenyl pyruvic acid* have been separated on Sephadex G 25, although they do not differ in molecular size[38] (cf. 2.5:4.2). Sephadex LH 20 and the Merckogels® OR[39] (E. Merck, Darmstadt) are also suitable in conjunction with organic solvents. They are used mainly for the fractionation of *high polymers*[40] (cf. 2.7), *lipids*[41] and *oligosaccharides*. For separation according to molecular size, Biogels, i.e. hydrophilic polyacryl amide gels[42], are used. Cellulose powder (Schleicher and Schüll; No. 123a) is employed to a limited extent for the separation of racemates from aqueous solution[43], and as a support for the stationary phase in partition column chromatography[44] (cf. Houben-Weyl I/1, p. 492). Cellulose-2, 5-acetate appears to have the greatest potential for column chromatographic separation of racemates from benzene or alcohol[45] (cf. 2.5:4.1.3). For ion exchangers on cellulose base see Chapter 2.8.

3.3 Solvents and Eluting Agents

The solvents commonly used as mobile phases can be arranged according to their eluting power. Thus *eluotropic series* are obtained which vary with type of adsorbent used. Mixtures of solvents are also included in the term mobile phase.

The series given in Table 2 is valid for oxidic adsorbents and salts. The eluting power of the solvents is practically parallel to their dielectric constants. Their densities are included for practical purposes (2.5:3.7.2).

Table 2. Eluotropic series for hydrophilic adsorbents

Solvent	D.C. (20°)	Density
Petroleum ether	1.9	0.65–0.7
Cyclohexane	2.02	0.78
Carbon tetrachloride	2.24	1.60
Trichloroethylene	3.4	1.47
Benzene	2.28	0.88
Chloroform (free of alcohol)	4.81	1.50
Ether (absolute)	4.34	0.72
Ethyl acetate	6.11	0.90
Pyridine	12.4	0.98
Acetone	21.4	0.79
n-Propanol	21.8	0.80
Ethanol	25.8	0.79
Methanol	33.6	0.79
Water	80.4	1.00
▼Formamide	> 84	1.13

(left margin, vertical: increasing polarity)

[24] *V. Carelli*, Nature *176*, 10 (1955).
[25] *W. Graßmann*, DBP 1 063 833, Priorität v. 26.4. 1957.
[26] *H. Endres, H. Hörmann*, Angew. Chem. *75*, 288 (1963).
[27] *W.N. Martin, R.M. Husband*, Anal. Chem. *33*, 840 (1961).
[28] *W. Graßmann*, Makromol. Chem. *21*, 37 (1956).
[29] *H. Inouye*, J. Chromatogr. *25*, 167 (1966).
[30] *K. Egger, M. Keil*, Z. Anal. Chem. *210*, 201 (1965).
[31] *H. Hörhammer*, Pharm. Ztg. *104*, 783 (1959).
[32] *K. Egger, H. Kleinig*, Z. Anal. Chem. *211*, 187 (1965).
[33] *H. Schweppe*, GAMS-Symposium Lausanne 1969 Säulenchromatographie, Suppl. Chimia *1970*, 56.
[34] *A. Winterstein, G. Stein*, Hoppe-Seyler's Z. Physiol. Chem. *220*, 247 (1933).
[35] *G.M. Henderson, H.G. Rule*, J. Chem. Soc. 1568 (1939).
[36] *H. Musso*, Chem. Ber. *91*, 349 (1958).
[37] Pharmacia, Uppsala (Sweden).

[38] *R. Haavaldsen, T. Norseth*, Anal. Biochem. *15*, 536 (1966).
[39] *W. Heitz*, Makromol. Chem. *127*, 113 (1969).
[40] *W. Heitz*, GAMS-Symposium Lausanne 1969. Säulenchromatographie, Suppl. Chimia *1970*, 126.
[41] *H. Bende*, Fette, Seifen, Anstrichm. *70*, 937 (1968).
[42] *S. Hjertén, R. Mosbach*, Anal. Biochem. *3*, 109 (1962).
[43] *W. Mayer, F. Merger*, Liebigs Ann. Chem. *644*, 65 (1961).
[44] *A. Heesing*, Chem. Ber. *95*, 3008 (1962).
[45] *A. Lüttringhaus, K.C. Peters*, Angew. Chem. *78*, 603 (1966); engl. *5*, 593 (1966); Z. Naturforsch. *22b*, 1296 (1967).

Adsorption of materials on oxidic (polar) adsorbents decreases with increasing polarity of the mobile phase as the eluants react more and more competitively. This is true, however, only if the solvents are very pure and dry. The first eight members of the above eluotropic series are conveniently purified by filtration through silica gel or aluminum oxide[46]; the remainder of the series can be purified by conventional methods[47].

Adsorbents, which have been adjusted to definite activities by addition of water, may release water to the eluting solvent (cf. 2.5:3.2.2). Thus the top of the column becomes more and more active, while the activity decreases steadily towards the lower end of the adsorbent bed. This reverse activity gradient causes broadening and diffusion of the zones. Addition of a certain amount of water to the eluting solvent[15] (according to Table 3) helps to overcome this difficulty.

Table 3. Amounts of water [%] required to equilibrate some solvents for alumina of various activity (20°)

Solvent	activity II	III	IV	V
Petroleum ether	—	—	—	—
Benzene	0.01	0.03	0.04	0.05%
Chloroform	0.04	0.08	0.10	0.11%
Methylene chloride	0.05	0.13	0.15	0.17%
Diethyl ether	0.15	0.55	0.73	0.75%
Ethyl acetate	0.57	2.06	2.4	2.6 %

The eluotropic series for *hydrophobic adsorbents,* such as activated charcoal, acetyl cellulose, polystyrene or polyethylene powder, is in reverse order to that represented in Table 2. The dissolved substances can be adsorbed most strongly from aqueous solutions, whereas adsorption from hydrocarbons is weakest.

For *polyamides,* on which the sorption occurs by hydrogen bonds, an entirely different eluotropic series more or less parallel to the basicity of the solvent is applicable. The order is:

water–methanol–ethanol–butanone–acetone–formamide–dimethyl formamide–aqueous sodium hydroxide.

Planning and operation of chromatographic separations is simplified by a knowledge of these eluotropic series. As far as possible, a strong adsorption is required for a suitable starting zone as well as for frontal and displacement chromatography. In order to achieve this purpose, a combination of the sorbent and a solvent of the eluotropic series is chosen. The process of chromatographic separation begins either when a strongly eluting solvent is taken as such or is added to the

eluting agent (elution chromatography), or, when an otherwise strongly adsorbed substance which operates as a displacer is used.

3.4 Detectors
Detectors are used to monitor and, as far as possible, to determine quantitatively the dissolved substances emerging from the column.

3.4.1 Optical Detectors
In the oldest type of flow analyzer a small cell made from glass or quartz is used for continuous *photometric analysis* with visible or UV light of the appropriate wavelength[48].

A modern instrument of this type is the LKB 8300 Uvicord II, using a low-pressure mercury lamp. Its intensity is strongest at 254 nm, which is the absorption region of *nucleic acids.* Using a fluorescent rod in combination with filters, light sources of longer wave length having a maximum at 280 nm can also be employed. Many organic substances, e.g. proteins, absorb in the latter region. Various wavelengths ranging from 400 to 700 nm can be chosen by using a multi-ray flow photometer LKB-5900, which is provided with a tungsten lamp and interference filters. This procedure is suitable for colored solutions. The photometric accuracy of this apparatus is of the order of $+0.5\%$ of the transmission, the detection limit of the detector being 10^{-7} g. In both the units mentioned above, the results can be registered with a recorder.

If in the cited regions the sample exhibits no appreciable absorption, a reagent can be added to produce a *color reaction* which can be measured photometrically. The latter procedure has been employed successfully in the *amino acid* analyzer according to Moore and Stein[49, 50]. Similarly, *sugars* can be determined and the results continuously recorded by the use of anthrone sulfuric acid[51]. Unfortunately, it is rarely possible and at the same time too elaborate to perform these color reactions quantitatively.

3.4.2 Differential Refractometer
Claesson[52] has employed the refraction of the emerging solution for detection. This method has been improved to a *differential refractometer* having a sensitivity (limit of detection 10^{-6} g) which is comparable to that of thermal conductivity detectors used in gas chromatography.

[46] *G. Hesse,* Z. Anal. Chem. *241,* 91 (1968).
[47] *A. Weissberger, E.S. Proskauer,* Organic Solvents, 2. Edn., Interscience Publishers, New York 1955.
[48] *G.S. Begg,* Anal. Chem. *33,* 1290 (1961).
[49] *St. Moore, W.G. Stein,* J. Biol. Chem. *192,* 663 (1951); *211,* 907 (1954).
[50] *K. Hannig,* Clin. Chim. Acta *4,* 51 (1959).
[51] *P. Jonsson, O. Samuelson,* Science Tools *13,* 17 (1966). The LKB Instrument Journal, Firmenschrift.
[52] *S. Claesson,* Ark. Kemi. *23A,* 1 (1946).

3.4.3 Detectors Based on Heat of Adsorption

Micro-adsorption detectors have a good future potential[53]. In this system, the liquid emerging from a separating column is passed through two cells (2–4 mm), one located on top of the other. The upper cell is either empty or filled with glass beads; the lower cell is filled with an adsorbent, preferably the one used in the column. In the center of the packing of each of the cells, the glass-covered measuring point of a small thermistor is located. As soon as the solvent containing the sample enters the cell, the measuring point indicates a heating (adsorption) followed by a cooling (desorption). The thermistors are in a Wheatstone bridge circuit connected to a recorder via an amplifier. The complete detector is constructed from teflon discs, which are screwed together. The adsorbent can easily be renewed. Up to 10^{-8} g of a sharp zone can be detected in this system without any special thermostatic control. However, the sensitivity is strongly dependent on the solvent, the dissolved substances and the adsorbent used. The total deflection (positive and negative peak) is proportional to the concentration, at least to a range of 10^2.

3.4.4 Other Detectors

Flame ionization detectors according to James[54] and Scott[55] are based on the following principle: An endless metal wire, or a metal wire wound between two spools is passed by the column exit. The wire is thus coated with the solution and is subsequently passed through a drying oven in which the solvent is evaporated. Later, the wire is moved through a pyrolytic chamber. The decomposition products of the substances, which are transported by the wire, are led to a flame ionization or argon detector (Pye). Before starting a new cycle of operation, the wire is washed and cleaned by heating.

This method has been modified frequently[56]. A great advantage of this elaborate technique is its considerable independence of the type of solvent used permitting gradient elution. Its disadvantage resides in the fact that highly volatile substances cannot be detected and also substances having low volatility are partly lost together with the solvent vapor. The response is not quantitative. The boiling point (point of sublimation) must be well above the drying temperature.

Conductivity detectors are suitable for ionized substances in aqueous solutions. The effluent of the column is passed through the measuring cell of the detector which contains two or three platinum electrodes within a Wheatstone bridge circuit and is operated by alternating current.

Some *other detectors* have been used occasionally. However, either these have not been tried extensively or they are designed for specific purposes. Examples are *microwave detector*[57], *ultrasonic detector*[58], *vapor pressure detector*[59-61], *solvent front detector*[62] and *mass detector* operated by continous weighing of the eluate[63]. For continuous measurement of radioactive effluent cf. H.E. Dobbs[64].

3.5 Fraction Collectors

(cf. Houben-Weyl I/1, pp. 486–490)
In many cases, particularly if no detector is available, the column effluent must be collected in numerous small fractions. The receivers (test tubes) can be exchanged either at selected time intervals (LKB) or after collection of a definite volume (Bender und Hobein, D-8 Munich). With some equipment either of these two methods can be chosen (e.g. model linear II, LKB, No. 90 000), for which additional accessories are available. Fraction collectors operating on a volumetric basis can be used for multiple samples of identical size in kinetic studies etc. The number of receivers generally varies between 150 and 500.

3.6 Automatic Apparatus

Originating with the construction of the amino acid analyzer according to Moore and Stein (cf. 2.5:3.4) an increasing trend exists to develop automated units for liquid chromatography, including those for preparative purposes. Important preliminary work in this respect has been published by Scott[65], Stewart[66] and Snyder[67]. The ultimate goal is to monitor results with a recorder and to attain a separation time which is not significantly longer than that in gas chromatography. High pressure

[53] *K.P. Hupe, E. Bayer,* J. Gas Chromatogr. 197 (1967).
[54] *A.T. James* in *A. Goldup,* Gas chromatography, London, Institute of Petroleum, p. 197, 1964.
[55] *R.P.W. Scott,* U.K. Patent 998, 107, July 14, 1963.
[56] *R.J. Maggs, W.G. Pye* and Co., Chromatographia *1,* 1/2, 43 (53), 1968.

[57] *E.H. Adema,* Chem. Weekbl. *61,* 353 (1965).
[58] *F.W. Noble,* Anal. Chem. *36,* 1421 (1964).
[59] *L.R. Snyder,* Anal. Chem. *35,* 1290 (1963); *39,* 648 (1967).
[60] *J.J. Kirkland,* Anal. Chem. *40,* 391 (1968).
[61] *R.E. Poulson, H.B. Jensen,* Anal. Chem. *40,* 1206 (1968).
[62] *N. Harding,* J. Chromatogr. *24,* 482 (1966).
[63] *J.G. Lawrence, R.P.W. Scott,* Anal. Chem. *39,* 830 (1967).
[64] *H.E. Dobbs,* J. Chromatogr. *2,* 572 (1959); Z. Anal. Chem. *175,* 37 (1960).
[65] *R.P.W. Scott,* J. Gas Chromatogr. *1967,* 183.
[66] *H.N.M. Stewart,* J. Chromatogr. *38,* 209 (1968).
[67] *L.R. Snyder,* Anal. Chem. *39,* 698, 705 (1967).

(10–200 atm) is required[68]. In attempts to utilize compressed gas for that purpose, the eluting solvent is saturated with it. At the outlet of the column the gas pressure is reduced and hence the packing is destroyed. In a special case, a U-tube filled with mercury has been placed between the gas and the eluting solvent[69]. Generally high-pressure pumps, which should be practically free from pulsations, are used. Flow rate and temperature must be maintained constant within narrow ranges. In the case of petroleum products, plate heights comparable to those of gas chromatography have been obtained[66] (HETP between 0.1 and 12,5 mm).

3.7 Experimental Procedure

Every chromatographic separation begins with the *selection of suitable separating systems:* sorbent, eluting solvent, column dimension, temperature. Usually comparable cases are considered as a basis for such choice. Essential information regarding the predominating phenomena (adsorption, hydrogen bond, partition, ion exchange, or molecular sieve effects) must be available if a system is to be adopted or modified for a particular purpose. Useful data are available for this (cf. Houben-Weyl I/1, p. 474, or Ref. [70], p. 218 and 221). Frequently, various possibilities exist, e.g. *sugars* can be separated by any of the following means[71] (cf. 12.1):

Adsorption on charcoal[72],
Ion exchange of their borate complexes[73, 74],
Hydrogen bonding with polyamides[1],
Partition columns[75, 76] or capillary columns[77],
As methyl ethers[78] or trimethyl silyl derivatives by gas chromatography[79].

The various methods supplement each other. Possibilities of overcoming the drawbacks of a system are much greater if a completely different system is selected rather than if the original system is modified.

3.7.1 Preliminary Experiments

Preliminary experiments are performed after theoretical planning. *Thin-layer chromatography* (cf. 2.4), particularly the *micro-circular technique*[80] or the technique employing precoated microscopic slides[81], is suitable for this purpose. The sorbent used should be identical to the one used subsequently for the main experiment; TLC sorbents must be of smaller particle size. Due to the humidity of air special attention should be paid to the activity of the sorbent. Results of preliminary tests are reproduced most readily with dry columns (cf. 2.5:3.8).

The starting solution should, as far as possible, be identical with the one used to start the elution. When handling multicomponent mixtures (natural products), it is desirable to separate the material by extraction into basic, acidic and neutral fractions, since often these three groups of compounds must be treated entirely differently during chromatography. The solutions must be thoroughly dried if the sorbent properties are changed by water. Dry sodium chloride or sodium sulfate are generally used for drying. Aqueous solutions must be desalted. This is accomplished preferably by filtration on a gel of narrow pore diameter (cf. 2.6). The starting solution should be as concentrated as possible and should be introduced in the middle of the top of the column at a slow rate. Small amounts are applied conveniently with a syringe. The amount of substance used depends upon the experimental conditions: normally in adsorption[82] and partition columns 1% and 0.1% respectively of a given amount of stationary phase in preparative separations. Ten to a hundred times smaller amounts are used for analytical separations, in order to work within the linear range of adsorption isotherms. Polyamide columns can often be charged to much greater extents.

3.7.2 Elution Chromatography

As a rule, the eluting solvent is added immediately after the disappearance of the original solution in the sorbent bed. At a constant temperature and

[68] *A.B. Littlewood*, J. Gas Chromatogr. *6*, 353 (1968).
[69] *S.G. Perry*, GAMS-Symposium Lausanne 1969, Säulenchromatographie, Suppl. Chimia *1970*, 47.
[70] *G. Hesse*, Chromatographisches Praktikum. Akademische Verlagsgesellschaft, Frankfurt/Main 1967.
[71] *D.J. Bell* in *Paech-Tracey*, Moderne Methoden der Pflanzenanalyse, Bd. II, p. 22. Springer Verlag, Heidelberg 1955.
[72] *R.L. Whistler, D.F. Durso*, J. Am. Chem. Soc. *72*, 677 (1950); *73*, 4189 (1951); *74*, 514 (1952).
[73] *J.X. Khym, L.P. Zill*, J. Am. Chem. Soc. *73*, 2399 (1951).
[74] *S.A. Barker*, J. Chem. Soc. 4276 (1955).
[75] *E.L. Hirst*, J. Chem. Soc. 3145 (1949).
[76] *J. Dahlberg, O. Samuelson*, Acta Chem. Scand. *17*, 2136 (1963).
[77] *E. Bayer*, F.P. 1 403 251, Ger. Appl. 3. 7. 1963; C.A. *65*, P 4670b (1966).
[78] *E. Bayer, R. Widder*, Liebigs Ann. Chem. *686*, 181, 197 (1965).
[79] *C.C. Sweeley*, J. Am. Chem. Soc. *85*, 2497 (1963).

[80] *E. Stahl*, Chem.-Ztg. *82*, 323 (1958).
[81] *J.J. Pfeifer*, Mikrochim. Acta *1962*, 529.
[82] *L.R. Snyder*, J. Chromatogr. *5*, 430 (1960); Anal. Chem. *33*, 1527 (1961).

constant flow rate, the eluting solvent is allowed to flow until the separation is achieved. When working without a detector, the column effluent is collected by a fraction collector and then analyzed. The fractions are combined as required before evaporation of the solvent.

In a technique similar to that of gas chromatography, most separations can be improved by working with increasing temperature[7]. This may have future possibilities in the design of automatic units. A similar effect is achieved by changing the solvent composition during separation in such a manner that the eluting power steadily increases. This can be done by stepwise operation using consecutively those solvents having stronger eluting power according to the eluotropic series (cf. Houben-Weyl I/1, p. 489).

The use of a solvent of appreciably higher density after a lighter solvent, must be avoided since otherwise the heavier solvent penetrates the lighter one and considerably disturbs the zones. The continuous addition of an eluting solvent to the basic solvent by using a mixing chamber is advantageous (cf. Houben-Weyl I/1, p. 481, 490). Various types of mixing chambers for *gradient elution*[83] are commercially available. Such devices can also be assembled from common laboratory equipment[70].

3.7.3 Frontal Chromatography[84]

Consider a solution containing various substances A, B and C, which are all much more strongly adsorbed than the solvent L $(A>B>C\gg L)$. If such a solution is continuously added to an adsorbent column, the effluent from the latter contains only the pure solvent until the column is loaded with the substances A–C. However, the latter are not randomly adsorbed on the colomn; rather, they are located one above another in definite zones without free space between. Actually A displaces B and the latter displaces C from the adsorptive sites. After the lowest zone C has reached the end of the packing, C appears together with the solvent in the effluent, whereas the zone widths of A and B still increase. When zone B arrives at the outlet, the effluent contains L+B+C. Finally, the column is saturated with A and hence the original mixture L+A+B+C emerges as effluent. It is to be noted that this behavior prevails only in isotherms having a saturation point. It does not occur in partition systems.

As seen above, apart from the solvent, only *one* substance that migrats with the front (i.e. C having the least affinity) is obtained in a pure state. The recording is a stairshaped curve, the number of steps representing the number of components of the mixture.

Purification of organic solvents from impurities which can be adsorbed easily, such as water[85], alcohols[86], acids[87] and hydroperoxides[88] as well as unsaturated contaminants, e.g. olefins and aromatics[46] (cf. Houben-Weyl I/1, p. 477): A column provided with cooling jacket is filled with silica gel or aluminum oxide of highest activity grade in the dry state. The technical grade solvent, or the solvent whose quality has become questionable due to storage, is slowly filtered through the column. Optically pure Uvasoles® (Merck) with dielectric constants of up to 12 are obtained in this manner (cf. 2.5:3.3).

A second application involves the enrichment of substances which are present in trace amounts in a large volume of solvent (cf. Houben-Weyl I/1, p. 478). Pure compounds can be isolated from mixtures enriched by 'adsorptive filtration'[89] by applying the rules of elution (cf. 2.5:3.3). In view of the layer arrangement of the column, a preliminary fractionation can be achieved during recovery.

3.7.4 Displacement Chromatography
(cf. Houben-Weyl I/1, p. 480 and 491)
Similar to elution chromatography, a concentrated solution of the mixture is added to the column. Subsequently, the column is developed, not with a pure solvent, but with the solution of a substance which is more strongly adsorbed than all the components of the mixture[90]. Such a substance, which dislodges all the other substances from the starting zone is called the displacer. The displaced substances arrange themselves as various zones in the column. Each substance acts as a displacer for the preceding one. Again, a saturation isotherm is a precondition for the occurrence of this phenomenon. Finally, the solutions of pure components leave the column consecutively, according to their adsorptive affinities, followed finally by the solution of the displacer. This form of chromatogra-

[83] *K. Dorfner*, Chem.-Ztg. *87*, 871 (1963).
[84] *A. Tiselius*, Ark. Kemi, Mineral., Geol. B 14, No. 22 (1940); Endeavour *11*, 5 (1952).

[85] *G. Wohlleben*, Angew. Chem. *67*, 741 (1955).
[86] *G. Wohlleben*, Angew. Chem. *68*, 752 (1956).
[87] *G. Hesse*, Liebigs Ann. Chem. *546*, 233 (1941).
[88] *M. Fichter*, Pharm. Acta Helv. *13*, 123 (1938); *G. Wohlleben*, GAMS-Symposium Lausanne 1969, Säulenchromatographie, Suppl. Chimia *1970*, 277.
[89] *H. Finck*, Ber. *70*, 1477 (1937).
[90] *L. Hagdahl, R.J.P. Williams, A. Tiselius*, Ark. Kemi *4*, 193 (1952).

phy results in very distinct zones, since the usual tailing is compensated for by displacement.

Contrary to frontal chromatography, each of the zones mentioned in the above technique contains only one component. However, since no intermediate gaps occur between zones, this separation technique can seldom be utilized for practical purposes. The procedure can be improved somewhat by addition of easily separable components to the mixture.

The additives must be selected so that their affinities to the adsorbent are intermediate between the affinities of the components to be separated. Such additives are, for example, liquids that can be distilled off easily, or acids that can be removed as salts. Their function is to produce the required distance between the components to be resolved. An example of this procedure has been reported by Holman[91]. A column which has been charged by 'adsorptive filtration' can be used as starting zone for displacement chromatography.

3.7.5 Recovery of Substances from the Column (Elution)

Usually in liquid chromatography not all of the components of a mixture are eluted from the column. Without exception, substances remain bound to the sorbent in the dry column (cf. 2.5:3.8). Usually materials of high R_F values are recovered from the filtrate, whereas slowly migrating components are mechanically separated and subsequently eluted.

Zones remaining in the column are generally detected optically by color, UV fluorescence or fluorescence quenching. However, the latter method can be employed only if fluorescent adsorbents are used (cf. 2.5:3.2).

Elution of distorted column zones:[92] In these cases, the excess liquid in the column is removed by suction and the boundaries of the zones are marked on the wall of the column by visualizing them with a suitable lamp. The moist adsorbent is then loosened to the desired length using a long spatula and carefully poured out of the tube without disturbing the rest of the packing. The remaining zone segments are removed similarly and stored in separate vessels until they are eluted.

Elution of intact column zones: After the solvent ceases to drip from the column, the remainder is driven out by air pressure. While the tube is held horizontally and air pressure is applied from the lower end of the column. The tube is pulled back to obtain a solid cylinder. The solid material emerging from the tube is collected on a suitable glass plate. This material retains its form provided the correct humidity is maintained during this treatment. It can now be viewed under a lamp and divided into various sections. It is also possible to detect the zones by applying a coloring agent on the solid material with a soft brush.

The substances can be extracted easily from the carrier with a suitable solvent (cf. eluotropic series, 2.5:3.3). Often it is adequate to add small amounts of a strong eluting agent to another solvent. The former solvent then acts as a displacer because it is preferentially bound to the adsorbent.

If it is undesireable to change the solvent, the sorption can be weakened by raising the temperature. For substances which are easily soluble, it is sufficient to reflux them in a suitable solvent and filter them hot. Sparingly soluble substances are extracted hot. For quantitative purposes a Soxhlet extractor is used. However it must be borne in mind that the usual extraction thimbles contain soluble impurities, which might be present in the same amount as the materials to be isolated. In some special cases, the sorbent can also be dissolved or chemically altered (cf. Houben-Weyl I/1, p. 486).

3.7.6 Recovery of Pure Materials (Repurification)

In order to remove impurities originating from the sorbent, the eluting solvent, or the elution process itself, a repurification of the materials obtained by chromatography is always necessary prior to elemental analysis. The substances should also be repurified before spectroscopic investigations. The repurification is accomplished by distillation, sublimation, recrystallization or other methods. Incomplete chromatographic separation is usually detected only by very stringent analytical means. Gas chromatography, thin-layer chromatography, and molecular spectroscopy are the most promising tools for this purpose.

3.8 Dry Column

3.8.1 Preparation

The term 'dry column' is used for a special technique described by Loev et al[13, 93]. Originally its purpose was to transfer and scale up the results obtained by thin-layer chromatography directly to column chromatography. Due to its simple and rapid operation, as well as its low solvent requirements, this technique is also useful in other cases.

A separating tube made of glass or quartz and provided with an open stop cock is filled in a dry state with aluminum oxide or silica gel whose activities are exactly adjusted between grades II and III (cf. 2.5:3.2). A tight packing is obtained by tapping the sides or by using a vibrator. It is advantageous to use a sorbent containing fluorescent indicator, e.g. aluminum oxide Woelm 103 or silica gel Woelm 202, which are adjusted to proper activity.

The column packing is levelled at the top and is covered with sand. The substance is dissolved in a volatile solvent such as ether or methylene chloride, is mixed

[91] *R.T. Holman,* J. Am. Chem. Soc. *73,* 1261 (1951).
[92] *J. Turkevich,* Ind. Eng. Chem., Anal. *14,* 792 (1942).

[93] *B. Loev, K.M. Snader,* Chem. & Ind. [London] 15 (1965).

with five times its weight of adsorbent, and the slurry is dried at 30–35 °C. This material is poured as a uniform layer on the top of the column and is again covered with sand. A 3–5 cm layer of solvent is carefully added to the dry column and the liquid level is maintained constant by immersing into it the tip of a dropping funnel which is closed at the top. About 90 ml solvent are required for a tube having 2.5 cm ∅ and 50 cm length.- The experiment is usually stopped after the solvent has reached the lower end of the column (15–30 min.). The edges of the zones are marked under UV light (254 nm) and the zones are separated as described earlier (cf. 2.5:3.7.5).

3.8.2 Choice of Operating Conditions

The R_f values agree closely with those obtained on a thin-layer plate coated with aluminum oxide or silica gel of the same activity and developed in sandwich chambers[94]. In order to attain the same activity, both the thin-layer plate and the column packing material are equilibrated with the humidity of the open air. When the TLC plates are developed in a jar, the adsorbent layer becomes saturated with the vapor of the organic phase. Therefore, good agreement between the TLC and the dry column technique is obtained only if the sorbent for the dry column is also equilibrated previously with the eluting solvent. This condition is achieved by bubbling nitrogen through the eluting solvent in a wash bottle, and subsequently through the packed column. Often it is sufficient to shake the adsorbent thoroughly with 10% by weight of the required amount of eluting solvent. This procedure results in uniform coating after standing for some time.

Nylon tubing can also be employed instead of glass in dry column-techniques. A suitable length of such tubing is closed at one end by welding with a hot iron wire and a small wad of glasswool is pressed into the tube. To enable the air to escape later, a few holes are made in the skin by a needle at a point below the wad of glasswool. The adsorbent is added to the tube in three almost equal portions, and the tube is packed tightly by tapping it on a hard surface. The resulting column is then so hard that it can be held by a clamp. Further treatment of this column is similar to that of a glass column. For separation of the zones after development, the column is simply cut in pieces. Since the thin Nylon skin has good transmittance for UV light of short wave length, the zones can be detected very easily.

Uniform solvents are preferentially used as eluting agents. Dry columns with aluminum oxide have been prepared in lengths to 180 cm. In such a

column up to 50 g of a mixture have been separated in one experiment.

3.9 Capillary Chromatography

E. Bayer[77] has used capillaries made of glass or copper, of 0.05–2 mm internal diameter and 1–20 m length for analytical purposes. The internal surface of the narrow tubing serves as sorbent or support for the liquid phase. However, the surface must be pretreated. Glass capillaries whose internal surface is treated with conc. ammonia at 300° can separate *amino acids* using butanone/pyridine/diluted acetic acid (5:1:1) or *xylose/glucose/maltose* using butanone/acetic acid/water (3:1:6). In a 20 m long capillary which is coated internally with polyethylene glycol, a mixture of *α-pinene/limonene/cedrene* can be separated using hexane. The eluting solvent is forced through the column using compressed nitrogen. Gas and solvent must be free of solid particles. The column effluent is collected in a small fraction collector and is analyzed.

Glass capillaries can be packed with heatstable and absolutely dry adsorbents when being drawn[95, 96]. They are then used as *packed adsorbent columns* or as *partition columns*. In both cases their advantage is high efficiency and considerable saving in the time of analysis.

In addition to packed glass capillaries, Vestergaard and Sayegh[97, 98] have successfully used narrow teflon tubing (1.5–2.5 mm internal diameter, length up to 8 m) packed with aluminum oxide or silica gel (Woelm, for adsorption). An optical detector is used for the analysis of steroids by gradient elution using acetone in chloroform. Pressures up to ca. 60 atm are attained with low pulsation pumps. The separation of seven *urinary steroids*, which requires 36 hrs on a normal column (40 × 0.6 cm) could thus be achieved with better results within 5 hrs.

4 Special Cases

4.1 Separation of Stereoisomers

Differences in molecular weight, and the nature and number of functional groups, which are otherwise considerably significant as factors affecting the separation, are absent in case of stereoisomers.

[94] *S. Hara, K. Mibe,* GAMS-Symposium Lausanne 1969, Säulenchromatographie, Suppl. Chimia *1970,* 39.

[95] *G. Kühlewein,* Diss. Univ. Karlsruhe 1965.
[96] *E. Heine,* Diss. Univ. Frankfurt/Main 1963.
[97] *P. Vestergaard, J.F. Sayegh,* J. Chromatogr. *24,* 422 (1966).
[98] *J.F. Sayegh, P. Vestergaard,* J. Chromatogr. *31,* 213 (1967).

Conformation and the relative position of functional groups are deciding factors. These factors have the most pronounced effect either on adsorption (hydrogen bonding) on a surface or on retention in the swollen matrix of gels containing polar groups. It has been observed that a specific adsorption affinity arises, if the distances of the functional groups of the sample are identical with the distances of the adsorptive sites on the adsorbent[1].

4.1.1 Separation of Geometric Isomers

Winterstein and Stein[34] reported the first chromatographic separation of *cis/trans isomers*. They were able to separate *cis/trans bixin*, and *cis/trans-crocetin dimethyl ester*. Later, Zechmeister[99] demonstrated that many *carotenoids* are reversibly isomerized when their solutions stand for some time. Iodine, in particular, catalyzes this process. The isomers thus formed, can be separated on calcium carbonate, aluminum oxide and other adsorbents. Similarly, *diphenyl octatetraene*[100] can be separated into *all-trans, trans-cis-trans-trans-*, and *trans-cis-cis-trans*-isomers. On brief exposure of the column to light, some additional isomers are formed and even more zones are observed. *Cis/trans isomers of carboxylic acids* have been separated on silica gel and charcoal[101]. *Cis/trans alicyclic compounds*, such as *borneol* and *isoborneol*[102], can be separated also. In all these cases steric factors have a deciding influence. Isomers whose functional groups can approach the surface of the adsorbent more easily are more strongly adsorbed[103, 104].

An example of isomerization at a carbon nitrogen double bond is the separation of *syn/antioximes* of *benzoin* and *anisoin* by adsorption on Neutral Filtrol, during which a slow isomerization of the *syn* form by the adsorbent was observed[105].

A typical example of isomerism at the nitrogen-nitrogen double bond is the separation of *cis/trans-azobenzene*[106, 107]. The polar *cis* configuration is adsorbed more strongly on aluminum oxide and silica gel, whereas the less polar *trans* form is more strongly retained by activated charcoal[108]. The rearrangement is accelerated by the influence of light increasing the amount of *cis* form.

4.1.2 Separation of Diastereomers

Diastereomers can also be successfully separated on various adsorbents[109]. An example is the separation of diastereomeric *7-chloro-7-azobicyclo[4.1.0]-heptane*[110] on silica gel using pentane/diethyl ether as solvent. In those cases where a derivative containing an optically active partner cannot be separated from the latter by crystallization or distillation, the usual racemate separation can be accomplished by chromatography[111].

4.1.3 Separation of Racemates

Resolution of racemic forms into enantiomers is difficult, particularly in the case of materials which cannot be converted to suitable derivatives by means of optically active compounds. Hydrocarbons and nitro compounds are examples of such compounds. Even in all other cases it would be simpler if the steps involving the preparation of intermediates and their resolution could be eliminated.

Chromatographic separation of racemates is possible only if the optically active compounds are involved in the sorption process. Krebs[112] initially studied inorganic octahedron complexes which could often be resolved into their antipodes from aqueous solutions using starch as adsorbent. Krebs also succeeded in achieving some 'activation' in the case of tetrahedral organic racemates, although optically pure compounds could not be obtained[113]. Later, Musso[36, 114] and Krebs[115] succeeded in effecting such separations of certain *biphenyl derivatives* and *3-bromo-DL-camphorsulfonic acid*. The conditions for such separations have been described by Krebs[115] as follows:

[99] *L. Zechmeister*, Ber. *72*, 1340, 1678, 2039 (1939).
[100] *L. Zechmeister, A.L. LeRosen*, J. Am. Chem. Soc. *64*, 2755 (1942).
[101] *H.P. Kaufmann, W. Wolf*, Fette u. Seifen, *50*, 519 (1943).
[102] *G. Vavon, G. Gastambide*, C.R. Acad. Sci., Paris *226*, 1201 (1948).
[103] *G. Vavon, G. Gastambide*, C.R. Acad. Sci., Paris *228*, 1236 (1949); *231*, 1151 (1950).
[104] *G. Vavon, G. Medynski*, C.R. Acad. Sci., Paris *229*, 655 (1949).
[105] *L. Zechmeister*, J. Am. Chem. Soc. *64*, 1922 (1942).
[106] *L. Zechmeister*, Naturwissenschaften *26*, 495 (1938).
[107] *A.H. Cook*, J. Chem. Soc. 876 (1938); 1309 (1939).

[108] *H. Freundlich, H. Heller*, J. Am. Chem. Soc. *61*, 2228 (1939).
[109] *E. Lederer, M. Lederer*[138], p. 426.
[110] *D. Felix, A. Eschenmoser*, Angew. Chem. *80*, 197 (1968).
[111] *A. Stoll, A. Hofmann*, Hoppe-Seyler's Z. Physiol. Chem. *251*, 155 (1938).
[112] *H. Krebs, R. Rasche*, Z. Anorg. Allgem. Chemie *276*, 236 (1954).
[113] *H. Krebs*, Chem. Ber. *89*, 1875 (1956).
[114] *W. Steckelberg*, Chem. Ber. *101*, 1519 (1968).
[115] *H. Krebs, W. Schumacher*, Chem. Ber. *99*, 1341 (1966).

Near the center of asymmetry there must be a polar group through which adsorption on starch occurs, probably due to hydrogen bonds.

The asymmetric center must be linked to groups of different sizes in order that asymmetry has a noticeable effect on the adsorption on starch.

If conditions 1) and 2) are not satisfied, the asymmetry of the molecule can have some effect on the adsorption if the molecule is bound to the adsorbent by several polar groups. These conditions are summarized in the so-called three-point-rule of Dalgliesh[116].

Optically pure antipodes of *racemic catechols* have be separated in good yields on cellulose powder (Schleicher und Schüll No. 123a)[117]. Franck[118, 119] observed a resolution of racemic *morphium alkaloids* on paper chromatograms. It is likely that such separations can be accomplished also on a preparative scale on cellulose columns. The compounds that can be resolved satisfy the rules of Krebs.

Both of the cited adsorbents are most effective in aqueous solutions (buffered if needed) at p_H 7. Lower alcohols or bases which may cleave the hydrogen bonds (cf. 2.5:3.3) interfere considerably. These observations supports the assumption that the bonding occurs via hydrogen bonds.

The first successful separations of racemates using organic solvents[120], were achieved on lactose e.g., the famous separation of enantiomers of *Tröger's base* (1)[121].

1

Prelog and Wieland obtained adsorbents of higher activity by grinding them in a ball mill with heavy steel balls followed by treatment with chloroform. The separation was performed with petroleum ether.

The most widely used adsorbent for racemate separation is cellulose (acetate) 2.5[45]. It is used principally with benzene or ethanol as swelling and eluting agent. This adsorbent must be permitted to swell overnight in the solvent before charging the chromatographic tube. The columns can be re-used several times. Initially the selectivity of the column increases. An optically inactive substance of unknown structure, originating from the packing material, is always found in the eluate to the extent of ca. 0.05 mg/ml. It is apparent from the list of racemates that can be separated by this technique, that a complete resolution of the racemates is rarely achieved[122, 123]. However, the results thus obtained are often sufficient to establish the racemic nature of a material and to determine the stability of the antipodes. Schlögl[124] was able to ascertain *cis-* and *trans*-structures for the *ferrocene derivatives* 2 and 3, since only compound 2 could be further resolved (no molecular symmetry).

2   3

Another principle of separation involves coating a strong adsorbent lacking asymmetric centers, with a sufficiently adhereable layer of an optically active substance. An example is aluminum oxide coated with D-tartaric acid or the sodium salt of L-glutamic acid[125, 126]. The eluting agent must be selected in such a manner that the optically active compound is not dissolved from the carrier. The resolution of DL-*9-sec. butyl phenanthrene* was effected[127] by using a coating of Newman's reagent[126] on silica gel.

The optically active compound can also be chemically bound to the sorbent. Grubhofer and Schleith[128] have converted the carboxylic acid ion exchange resin Amberlite XE 64 with thionyl chloride/pyridine into an acid chloride resin, and subsequently reacted the latter with quinine. The

[116] *C.E. Dalgliesh*, J. Chem. Soc. 3940 (1952).
[117] *W. Mayer, F. Merger*, Liebigs Ann. Chem. *644*, 65 (1961).
[118] *B. Franck, G. Schlingloff*, Liebigs Ann. Chem. *659*, 123 (1962).
[119] *B. Franck, G. Blaschke*, Liebigs Ann. Chem. *695*, 144 (1966).
[120] *G.M. Henderson, H.G. Rule*, J. Chem. Soc. 1568 (1939).
[121] *V. Prelog, P. Wieland*, Helv. Chim. Acta *27*, 1127 (1944).

[122] *M. Woelm*, Eschwege, unpubl.
[123] *W. Steckelberg*, Chem. Ber. *101*, 1519 (1968).
[124] *K. Schlögl*, Fortschr. Chem. Forsch. *6*, 489 (1966).
[125] *G. Karagounis*, J. Chromatogr. *2*, 84 (1959).
[126] *M.S. Newman*, J. Am. Chem. Soc. *77*, 3420 (1955); *78*, 2469 (1956).
[127] *L.H. Klemm*, J. Chromatogr. *14*, 300 (1964).
[128] *N. Grubhofer, L. Schleith*, Naturwissenschaften *40*, 508 (1953).

'quinine resin' contains about 25% quinine: it is an effective ion exchanger having basic optically active centers. Using remarkably short columns (15 cm) these workers obtained, in the initial fractions, nearly pure L (—*mandelic acid* from racemic mandelic acids.

Tailor made adsorbents: Acidification of a solution of sodium silicate containing sodium D-camphorsulfonate forms a silica gel from which the optically active sulfonic acid is removed by extraction with methanol[129]. During chromatography the resultant adsorbent thus obtained retains the D-form of racemic camphorsulfonic acid more strongly. The selectivity increases after using the adsorbent twice. Thereafter the selectivity falls rather rapidly to zero-value.

4.2 Separation of Tautomeric Mixtures

The main problem in the separation of tautomeric compounds is the separation time and the necessity of eliminating catalytic effects on the rate of tautomerization rather than the separation factor itself. Due to the required high temperatures, gas chromatography can rarely be utilized[130]. The keto- and enolforms[131] of *p-hydroxyphenyl pyruvic acid* and *indolyl pyruvic acid* could be separated in the liquid phase.

2 mg *p*-hydroxyphenyl pyruvic acid dissolved in 2 ml of $0.01N$ acetic acid (pH 3.5) were chromatographed on 6 g Sephadex G 25 with $0.01N$ acetic acid. The effluent was monitored at 254 nm using a LKB-Uvicord recording photometer. Between the two pronounced peaks of the tautomeric forms a weak signal of the decarboxylation product, *p-hydroxyphenyl acetaldehyde*, appeared; the amount of the latter compound increased on allowing the solution to stand.

$$HOC_6H_4-CH_2-\underset{\underset{O}{\|}}{C}-COOH \rightleftarrows HOC_6H_4-CH=\underset{\underset{OH}{|}}{C}-COOH$$

$$\longrightarrow HOC_6H_4-CH_2-\underset{\underset{O}{\|}}{C}-H \ + \ CO_2$$

The separation can be accomplished only in weakly acidic medium. The enol-form appears in the eluate before the keto form. However, it is conceivable that the enol form is dimerized to a lactide. In that case the separation occurs on the basis of molecular size, as can be expected usually for gel filtration. Possible *cis/trans* isomers are indicated by a shoulder on the enol peak.

4.3 Sample Changes and other Interferences

Highly active dry adsorbents frequently generate a large amount of heat of adsorption thus destroying heat sensitive substances[131]. Catalytic effects can also occur; these latter are minor with silica gel and are more pronounced with the use of aluminum oxide. Aluminum silicates (bleaching earths, Floridin, Frankonit, etc.) show the strongest catalytic effects[132]. These adsorbents are therefore used only deactivated in dry columns (2.5:3.8).

When the column is already wetted with the eluting solvent, the latter occupies the active sites of the sorbent. The heat of adsorption of the sample is then reduced by an amount corresponding to the heat of desorption of the solvent. Furthermore, the sorption occurs slowly, and hence the generated heat is gradually transferred to the solid support and to the solvent. Nevertheless, dangerous temperatures can be reached on oxidic adsorbents if the dielectric constant of the solvent (cf. 2.5:3.3) is below 10. It is therefore advantageous to use deactivated adsorbents.

In any case, the heat generated by wetting the adsorbent with the mobile phase should have diminished before the column is charged. Surfaces of some adsorbents which are usually considered neutral, actually show strongly alkaline or acidic reactions. The adsorbent samples should, therefore, be tested by suspension in an aqueous solution of an indicator. *Alkaline sorbents* can convert acidic substances into salts and retain the latter particularly as sodium salts. They may also cause *cleavage of esters, aldol condensations* or *elimination reactions*. Acidic surfaces may convert basic substances and retain them in the column; eluting solvents can also be involved in such reactions. *Displacement of double bonds* and *removal of hydroxyl groups* are quite frequent. Structural transformations also occur.

Traces of *heavy metal ions,* particularly iron and copper, which might be present in common preparations of aluminum oxide, silica gel or charcoal, can catalyze the autoxidation of some organic substances as soon as they come in contact with air. Such processes occur mainly during the mechanical separation of various zones of the column and during their elution. Light can promote the autoxidation. The separated zones should be covered with extracting solvent or trans-

[129] *R. Curti, U. Colombo*, J. Am. Chem. Soc. *74*, 3961 (1952).
[130] *G. Hesse*, Z. Anal. Chem. *236*, 192 (1968).
[131] *H. Carlsohn, G. Müller*, Ber. *71*, 863 (1938).

[132] *G. Hesse*, Z. Anal. Chem. *211*, 5 (1965).

ferred to thimbles and the extraction should be started without delay.

An unfavorable ratio between the amounts of material to be separated and the agents employed (sorbent, support, stationary phase and eluting solvents) always increases the chance of contamination of eluates. Hence, resolved compounds must always be repurified (cf. 2.5:3.7.6).

5 Bibliography

5.1 Theory of Column Chromatography

[133] *H.G. Cassidy*, Fundamentals of Chromatography. Interscience Publishers, New York, London 1957.

[134] *J.C. Giddings*, Dynamics of Chromatography, Part I. Principles and Theory. E. Arnold, London, Marcel Dekker, New York 1965.

[135] *L.R. Snyder*, Principles of Adsorption Chromatography. The Separation of Non-Ionic Compounds. Marcel Dekker, New York 1968.

5.2 Apparatus and Procedures

[136] *L. Zechmeister, L. v. Cholnoky*, Die Chromatographische Arbeitsmethode, 2. Aufl., Springer Verlag, Wien 1938.

[137] *L. Zechmeister*, Progress in Chromatography 1938–1947, Chapman & Hall, London 1950.

[138] *E. Lederer, M. Lederer*, Chromatography, A. Review of Principles and Applications, 2. Edn, Elsevier Publ. Co., Amsterdam, London, New York, Princeton 1957.

[139] *R. Stock, C.B.F. Rice*, Chromatographic Methods, Chapman & Hall, London 1963.

[140] *G. Hesse*[70], 1967.

[141] *J.J. Kirkland*, Modern Practice of Liquid Chromatography, Wiley Intersci. Publ., New York, London, Sydney, Toronto 1971.

2.6 Gel Chromatography, Membrane Filtration: Separation by Virtue of Size Differences

Helmut Determann and Klaus Lampert
Institut für Organische Chemie der Universität,
D-6 Frankfurt a.M.

1 Introduction

Of the methods of separation covered in this volume, there are few for which the behavior of molecules can be predicted as accurately as with the techniques described below. When permeating through inert gel networks, molecules become separated solely with regard to size. In the ideal case, which is often realized by these methods, all other properties of the molecules are less important. In biology, separation by means of semipermeable membranes is a method which is both widespread and experimentally well established; only recently have gel beads been used to effect separation according to size differences[1].

Both the membranes used for dialysis or ultrafiltration, and also the separation media employed in gel chromatography, consist of gels: these have macromolecular 'network' structures, which usually swell in solvents (and are always homogeneously permeated by them). On the macroscopic scale, filtration is a process in which particles are held back by the fibrous filter material, and the solvent percolates freely through the filter layer—usually under its own gravity. With microscopic particles (or even molecules), the pores are so small that a high flow resistance to penetration is set up. This resistance can be overcome by application of high pressure. The necessary pressure difference can be brought about either mechanically (in the case of *membrane filtration* 2.6:2) or in consequence of differences in concentration (in the cases of membrane filtration and *gel chromatography* 2.6:3).

In both methods, separation occurs because the biggest molecules (or particles) are bigger than the largest pores of the gel, while the small molecules can penetrate through or into the gel, more or less unhindered. Both dialysis and gel chromatography depend on diffusion for the selective transport of material. In *ultrafiltration*, on the other hand, the solvent is forced through the membrane under high pressure, and the membranes are homogeneous gel layers. In gel chromatography the gel is in bead-form and is packed in a chromatography column through which solvent flows. The entire transport of substances takes place in the eluting solvent. Here, too, selection is caused by diffusion (induced by concentration gradients) into and out of the gel particles.

This chapter is concerned only with separation media (membranes and gel materials) which are generally commercially available. Other interesting alternatives mentioned in the literature will not be considered here.

2. Membrane Filtration

2.1 Materials

Separation with membranes is a time-consuming process. It is well known that the thinner the membrane and the greater the pressure difference, the faster is the penetration of the solute. This is increasingly the case, the smaller the molecules to be retained, and the narrower the pores in the gel of the membrane. Membranes whose function

---

[1] *L.C. Craig, W. Konigsberg*, J. Phys. Chem. *65*, 166 (1961).

Table 1. *Membrane types arranged* in order of their molecular weight "cut-off" limits

MW	Solute[a]	Amicon[b]	Eastman[c]	Millipore[d]	DDS[e]	Sartorius[f]	others
	high MW proteins (e.g. γ-globulin)						
100000		XM—100				$<$ SM—11730 / SM—11530 $<$ SM—11736 / SM—11536	
	Human albumin, hemoglobin						
50000		XM— 50					PEM (Gelman)[g]
	Ovalbumin pepsin			PSDM			
30000		PM — 30 $<$ HF 35 / HT 00		PSED			
	Myoglobin, cytochrome C						
10000		$<$ PM — 10 / UM— 10				$<$ SM—11739 / SM—11539 $<$ SM—12133 / SM—12136	Cellophan[c] (Kalle)[h] $<$ Visking / Cuprophan[c] (Bemberg)[i]
	Peptides, large antibiotics				800		
1000		UM— 2		PSAC	870		
	Sucrose, small antibiotics						
500		UM— 05					
	Glucose, amino acids, urea, $Na^{\oplus}, Cl^{\oplus}, K^{\oplus}$ etc.				875		
20—50		$<$ RO—89 / RO—94 / RO—97			880		

[a] Solutes above the line are retained.
[b] Amicon, Heemskerckstraat 43, Den Haag, Nederlande
[c] Eastman, Chemical Products, Inc., Kingsport, Tenn., USA
[d] Millipore (U.K.) 109 Wembley Hill Rd. Wembley, Middleessex
[e] De Dansk Sukkerfabrikker, 5 Langebrogade, DK-1001 Copenhagen
[f] Sartorius Membranfilter GmbH, D-34 Göttingen, Weender Landstrasse 96
[g] Gelman, Camag, D-1 Berlin 45, Baseler Strasse 65
[h] Kalle AG, D-6202, Wiesbaden-Biebrich
[i] Bemberg AG, D-56 Wuppertal Barmen

is to hold back *particles* only (*bacteria, viruses,* etc.), may be relatively thick (80—150 microns). Sieve plates or tissues are sufficient to support them, since the pressure difference employed is relatively modest (a few atmospheres).

When *molecules* are to be collected, however, membranes with very small pores are necessary; the separating layer must be extremely thin in order to achieve a reasonable flow rate. The mechanical stability necessary results from the structure of these *anisotropic polymers*: These consist of a very thin (0.1—0.5 microns) layer of extremely fine pore texture, supported on one side by a much thicker (50—250 microns) layer of substantially more porous substructure. The surface film is responsible for the actual separation, while the porous substructure serves as the support.

The normal dialysis membranes are intermediate in porosity between the two types mentioned above. Their pores are sufficiently small for molecules with molecular weights exceeding 10,000 generally to be retained. However, the membranes are of such a thickness (30—90 microns) that an efficient flow cannot be maintained, even when pressure is applied. They are consequently suitable only for *dialysis*, i.e. for the exchange of a substance between two solutions. Because of their thickness, the usual membranes can be used without additional support. Their porosity can be altered by stretching and by treatment with zinc chloride[2]; however, these methods have not found wide application.

Table 1 shows the *membrane types*, which are commercially available, arranged in a molecular weight scale according to their 'cut-off' limits, i.e. their abilities to retain molecules larger than those of a given size, and to allow passage of smaller species (as well as solvent). For example, the Amicon-Filter PM-30 allows proteins with molecular weights below 30,000 to pass through, whereas species with larger molecular weights are held back. Naturally these gel layers have a rather broad pore-size distribution, i.e. besides pores of the stated size, decreasing numbers of larger and smaller pores are also present. As regards assessment of the exclusion characteristics of these membranes, it should be emphasized that the arrangement by molecular weight was arbitrarily based upon the results with *globular proteins*. In the case of *random-coil polymers*, it must be remembered that, since they are not so compactly constructed as proteins, they have a greater spatial requirement. Consequently, in ultrafiltration, random-coil polymers can slip through the openings in the membrane and so facilitate passage of complete chains of substances with much higher molecular weight than the proteins. As discussed in section 2.6:4, this is the reverse of the behavior of random-coil polymers in gel chromatography where size is the dominant factor.

Most membranes consist of cellulose gels or their derivatives, except the Amicon products, which are made[3] by interaction of two soluble oppositely charged and strongly ionized polyelectrolytes to form a polyelectrolyte complex or an ionic hydrogel.

2.2 Techniques

As already mentioned, there are three possible ways in which the selectivity of these membranes

can be utilized. In *dialysis* (A in Fig. 1), the solvent (water or salt solution) is on both sides of the membrane. The transport of materials through the membrane takes place by diffusion. For efficient results, the solvent film along the membrane must be kept as thin as possible. Careful mixing in both chambers is, therefore, extremely desirable.

In *ultrafiltration* (B in Fig. 1), the solution is forced through the membrane under high pressure (1–50 atmospheres). The high molecular weight components become concentrated because only the solvent and the low molecular weight substances are able to permeate the membrane. *Proteins* in particular show a tendency to gel when concentrated from aqueous solution. This may lead to clogging of the pores by formation of a 'gelatinous' layer which resists the passage of solvent and eventually brings the filtration to an end.

It is vital to have some means of returning the concentrated macrosolute from the membrane to the bulk of the solution. This can be accomplished effectively in specially developed equipment, such as a rotary magnetic bar stirrer (Amicon), a vibro-mixer (Sartorius) or a forced-flow agitator (Millipore).

It is not necessary in ultrafiltration that the ultrafiltrate be allowed to emerge into the open, as pictured in Figure 1 B; it is quite possible to have liquid on this side of the membrane as well. This arrangement protects the conventional dialysis membranes against the pressure difference, so that these membranes can be used to concentrate solutions of high molecular weight materials by ultrafiltration[4].

Diafiltration (C in Fig. 1) represents a combination of ultrafiltration and dialysis. Repeated or continuous addition of fresh solvent from a reservoir effectively and rapidly washes out or exchanges salts or other micro-species in the ultrafiltration chamber. Such an arrangement is many

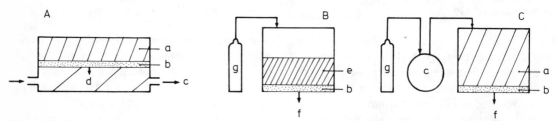

Fig. 1. Illustration of experimental arrangement for dialysis (A), ultrafiltration (B), and diafiltration (C). (a) Retentate, (b) membrane, (c) solvent, (d) dialysate, (e) concentrate, (f) ultra- or diafiltrate, (g) pressure bottle.

[2] *L.C. Craig,* Advan. Anal. Chem. Instrum. *4,* 35 (1965).

[3] *L.L. Markley, H.J. Bixler, R.A. Cross,* J. Biomed. Mater. Res. *2,* 145 (1968).

[4] *B.v. Hofsten, S.O. Falkbring,* Anal. Biochem. *1,* 436 (1960).

times more efficient than a conventional dialysis apparatus. Naturally, with diafiltration the expenditure in equipment is considerably greater.

The selectivity of the membrane is determined by the pore-size distribution. If the substances to be separated differ only slightly in molecular size, it cannot be expected that complete separation can be accomplished with one single membrane. The separation power can be raised considerably by using a cascade system where several ultrafiltration cells, each with the same membrane, or each with an ultrafilter of progressively lower cut-off, are arranged in series. Using the *counter-current principle,* one can also make the dialysate of the first step the retentate of the second step[5, 6].

If the particles which diffuse through the membrane carry an electric charge, the dialysis may be speeded up considerably by applying an electric field. This *electrodialysis*[7] technique is used mainly for the *removal of salt* from neutral high molecular weight substances.

2.3 Applications

Membranes are used most commonly for the removal of micro-solutes from aqueous solutions of macromolecules. The customary *dialysis tubes* are used widely for this purpose. They are simply knotted at the bottom, filled with the crude liquid to be dialysed, knotted at the top, then suspended in the washing solution and kept in constant motion by mechanical means. This technique is applied in most biochemical laboratories, although it is relatively inefficient (cf. Section 2.2). Until recently, membranes with a defined (generally large) pore-size (Table 1) have been used mainly for the removal of *bacteria, viruses or phages,* as well as dust or other *particles,* from aqueous solutions, solvents or gases. Some selected applications in the molecular range are compiled in Table 2.

Table 2. Examples of application of membranes

Area	References
Desalination	8, 9
Waste water	10
Food chemistry	11
Beverages	12
Biochemistry	13
Protein	14
Serum protein	15
Enzymes	16
Lactalbumin	17
Molecular weight	18

[5] R. Signer, H. Hänni, W. Koestler, W. Rottenburg, P.v. Tavel, Helv. Chim. Acta 29, 1984 (1946).

3. Gel Chromatography

3.1 Materials

A gel is suitable for chromatography if it can be prepared in the form of beads which remain stable under the conditions to be used (solvent, pH, temperature, solute), the *pore size* of which can be tailor-made and reproducible, and which have sufficient rigidity under mechanical loading. The materials for gel chromatography (Table 3) currently

Table 3. Materials for gel chromatography

Type	Material	Solvent
Bio-Gel P[a]	Polyacrylamide	aqueous
Bio-Gel A[a]	Agarose	aqueous
Bio-Beads S[a]	Polystyrene (homoporous)	organic
Sephadex G[b]	Dextran	aqueous
Sepharose B[b]	Agarose	aqueous
Sephadex[b] LH–20	Dextran derivative	polar organic
Sagavac SAG[c]	Agarose	aqueous
Styragel[d]	Polystyrene (heteroporous)	organic
Poragel A[d]	Polystyrene	organic

[a] BIO-RAD Laboratories, 32nd Ø Griffin Avenue Richmond, Calif., USA
[b] Pharmacia Fine Chemicals AB, Uppsala, Sweden
[c] Seravac Laboratories (PTY) Limited, Moneyrow Green, Holyport, Maidenhead, Berks., England
[d] Waters Ass. Inc., 61 Fountain St., Framingham, Mass., USA

[6] L.C. Craig, T.P. King, J. Am. Chem. Soc. 77, 6620 (1955).
L.C. Craig, T.P. King, A. Stracher, J. Am. Chem. Soc. 79, 3729 (1957).
[7] E. Manegold, Kapillarsysteme, Bd. II, p. 986, Springer Verlag, Heidelberg 1960.
[8] S. Loeb, J.S. Johnson, Chem. Eng. Progr., Jan. 90 (1967).
[9] K. Popper, R.L. Merson, W.M. Camirand, Science, 159, 1364 (1968).
[10] L.A. Testa, P.F. Bruins, Modern Plastics, May 141 (1968).
[11] C.O. Willits, J.C. Underwood, U. Merten, Food Technol. [Chicago] 21, 24 (1967).
[12] K. Popper, W.M. Camirand, F. Nury, W.L. Stanley, Food Engineering 38, 102 (1966).
[13] W.F. Blatt, S.M. Robinson, H.J. Bixler, Anal. Biochem. 26, (1), 151 (1968).
[14] W.F. Blatt, B.G. Hudson, S.M. Robinson, E.M. Zipilivan, Nature 216, 511 (1967).
[15] V.E. Pollak, M. Gaizutis, J. Rezaian, J. Lab. Clin. Med. 71, (2), 338 (1968).
[16] H.L. Griffin, Y.V. Wu, Biochemistry 7, (9), 3063 (1968).
[17] W.F. Blatt, S.M. Robinson, F.M. Robinson, C.H. Saravis, Anal. Biochem. 18, 81 (1967).
[18] J.R.H. Wake, A.M. Posner, Nature 213, 692 (1967).

on the market more or less fulfil these requirements.

Mostly, the gel structure is formed from primary valence bonds. In the case of Sephadex gels, soluble polymer chains (dextran) are cross-linked (with epichlorohydrin). The Bio-Gel P series are prepared by copolymerization of mono- and bi-functional monomers (acrylamide and methylene-bis-acrylamide). Normally, polymers are obtained which swell in solvents (mostly water), and which shrink on removal of the solvent (Xerogels). The degree of swelling (and thus the porosity) increases as the amount of cross-linking agent decreases. However, the mechanical properties become significantly worse, so that the very soft gels Sephadex G-200 or Bio-Gel P-300 represent the maximum porosity attainable in this way.

The *heteroporous gels* have still larger pores and a considerably more stable structure. Here, the pores are not evenly distributed but are separated from each other by areas of higher polymer concentration.

This concept is realized most thoroughly in the Styragel materials, which are prepared by copolymerization of styrene with a large amount of divinylbenzene, usually with a precipitating agent for the polymer present as a solvent. The fact that the structure is inflexible means that the gels hardly shrink or swell; Therefore, the pore size is almost the same in all solvents. A heteroporous structure of this kind may be postulated also for the agarose gels. Here, the polymer chains are held together only by secondary valence forces (hydrogen bonds), and this results in the formation of regions with a relatively rigid structure. Consequently, because of their large pores, the agarose gels offer the possibility of separating high molecular weight substances in aqueous solutions. However, as is well known, these agarose gels are quite sensitive to environmental influences such as temperature, solvent, and pH values.

3.2 Techniques

Preparation for a gel chromatographic experiment simply requires the gel to be brought into equilibrium with the desired eluant; this takes up to 24 hours for the materials which swell most.

Almost any type of tube can be used as the *chromatographic column*. At the lower end, there should be a device on which the gel bed may settle; a polyamide gauze is best suited for this purpose. Chromatographic columns equipped with this feature are commercially available from several manufacturers. If the gel bed occupies only a small volume, the dead volume (the volume below the gel bed within the column, the volume in the hose connections, etc.) should be kept as small as possible to avoid mixing of the fractions after separation.

For *filling*, the tube should be placed in an exactly vertical position and filled to the halfway point with solvent. The equilibrated gel suspension is then poured in and. as soon as a few cm of the gel packing have settled. the valve of the column is opened gradually. In the case of readily compressible gels, special care should be taken never to allow the hydrostatic pressure on the gel bed to exceed 30–40 cm. This can be accomplished very simply by attaching a hose to the outlet and raising its end to the desired position below the liquid level in the column.

Because the top and bottom fittings are identical on the technically convenient columns mentioned above, 'upwards' elution through the gel bed from bottom to top is possible; this offers certain advantages, especially with soft gels. However, too high an initial rate of flow must not be induced by use of a great difference of levels *(hydrostatic pressure)*, since this may rapidly lead to complete rerouting of the flow of liquid. Depending on the gel type, flow rates of 5–50 ml/h x cm^2 column diameter can be obtained. With heteroporous polystyrene gels, flow rates of 60 ml/h x cm^2 and more are attainable (but with the columns now under considerable pressure).

The manner in which the mixture of substances to be separated is applied to the column has a great influence on the quality of separation. *Sample application* should be either by carefully placing the sample solution beneath the supernatant eluant on the gel bed, or by letting it drain into the gel after removal of the supernatant eluant; in either case, thorough washing on to the bed in the same manner is advisable. The *sample amount* may be between 1/10 and 1/100 of the total column volume; the *concentration* can be varied within wide limits. The upper limit for the sample amount is set by the difference between the viscosities of the sample and of the eluant which affects the degree of separation, while the lower one is naturally fixed by the limit of detection of materials in the eluate. In gel chromatography, separation results are largely independent of sample amounts. This finding, and the fact that the gel packing can be re-used after an experiment without regeneration, distinguish gel chromatography from most other chromatographic separation methods. As soon as the component with the smallest molecular size has left the gel packing, i.e. after an amount of eluant equivalent to the entire volume of the gel bed has passed through, a new run can be started. This is especially important for the purposes of programmed experimentation.

If the flow rate decreases, or if air gets into the gel bed (the column 'runs dry'), the packing should be renewed. The gel itself may be used over and over again.

3.3 Elution Parameters

If a graph is plotted with the concentration of the separated substances in the eluate as the ordinate and the elution volume as the abscissa, an *elution diagram* (Fig. 2) is obtained. The volume in the packing between the gel beads taken up by the eluant is called the *void volume* (V_0). This is also the elution volume of a substance the molecules of which are too big to penetrate even the largest pores of the gel beads (exclusion). No substance

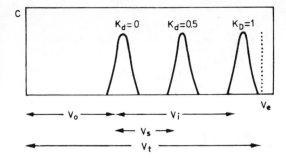

Fig. 2. Illustration of an elution diagram. (V_o) void volume, (V_i) inner volume, (V_s) separation volume, (V_t) total volume of the gel bed, (V_e) elution volume, (C) concentration in the eluate.

can be expected in the eluate before V_o. On the other hand, after the flow of solvent has amounted to one column volume (V_t) (the *total volume* of the gel bed), all substances must have left the column, if the rules of gel filtration are obeyed (separation by size without interaction with the gel). The actual separation — i.e. differentiation according to molecular size — takes place in the *inner volume* of the gel beads (V_i) only. As can be seen in Figure 2, $V_o + V_i$ is somewhat smaller than V_t because of the volume taken up by the impenetrable gel matrix. Substances which can permeate the gel net-work only partly, i.e. materials for which only a part of V_i is available, will leave the column with an *elution volume* (V_e) between V_o and $V_o + V_i$. The distance on the graph between the peaks of two separated substances is called the *separation volume* (V_s).

The elution volume (V_e) is not a very convenient parameter for comparison of the results of different experiments. The *reduced parameters* V_e/V_t and V_e/V_0 are independent of the geometry of the column and may be considered analogous to the well known R_F value of paper chromatography. Besides the K_d *value* which was formerly used

$$(K_d = \frac{V_e - V_o}{V_i})$$

today the K_{av} *value* ($K_{av} = \frac{V_e - V_o}{V_t - V_o}$)

has become especially common in describing gel chromatographic results. As a specific constant, K_d has the properties of a partition coefficient and represents that part of the solvent volume (V_i) within the gel which is accessible to the molecules of the substance in question. The K_{av} value (av = available) contains only variables which are easily measured. This constant differs from the K_d value in that the total volume of the gel phase (matrix + solvent = $V_t - V_o$) is used for its calculation instead

of the inner volume (V_i). Both coefficients always give positive values between 0 and 1 as long as only differences in size are responsible for the elution behavior.

If the K_{av} values of different materials are close together, they can be separated completely only if band-broadening due to diffusion does not predominate over the small separation volume (V_s). The bands will be especially sharp when the *plate count* is high, i.e. if the particle size of the gel is kept small relative to the flow rate. This can be achieved only by using very regular and undeformable particles and a uniformly packed column.

In Table 4 the commercially available gels are listed in terms of their approximate exclusion limits. The exclusion limit is represented by the molecular weight of those substances which are just unable to penetrate the gel. The operative molecular weight ranges (Table 4) for the various gels depend greatly on the pore-size distribution. (Further information is given in the manufacturers' literature.)

3.4 Possibilities for Application
Possibilities for the application of gel chromatography are extremely varied. Table 5 should be regarded as a somewhat arbitrary survey of the uses of the different gels in various fields. It refers to gel chromatography as distinguished from gel filtration because the experimental techniques for the two methods differ greatly. Section 2.6:4 is devoted to the special technique of gel filtration. The possibilities of gel chromatography for molecular weight determination deserve separate discussion and are dealt with in Section 2.6:5. Gel chromatography is still employed most frequently under aqueous conditions for the separation of macromolecules of biological origin (*proteins, nucleic acids*). Nevertheless, gel chromatography has now also achieved an undisputable position for characterization of polymers in organic solvents. The organic chemist is usually faced with mixtures whose components often do not differ greatly in molecular size; here, too gel chromatography is frequently a surprisingly powerful technique (cf. Section 2.6:6).

4 Gel Filtration
In the strict sense of the term, gel filtration is defined here as the separation of small molecules ($K_{av} \simeq 1$) from macromolecular solutes ($K_{av} \simeq 0$). This means that gel filtration competes directly with dialysis (see Section 2.6:1). Separations in which the differences in K_{av} values are very large are much less demanding than with the previously

Table 4. Approximate exclusion limit of various types of gels

MW[a]	Bio-Rad Lab.		Pharmacia	Seravac	Waters Ass.	
	Bio-Gel A—150m			*Sagavac* SAG 2	*Styragel* 10^7 Å	
100 000 000						
	A—50m		*Sepharose* 2B	SAG 3		
50 000 000					10^6 Å	
			4B			
	A—15m			SAG 4		
10 000 000					3×10^5 Å	
				SAG 5		
5 000 000	A—5m		6B	SAG 6	10^5 Å	
				SAG 7		
1 000 000	A—1.5m			SAG 8	3×10^4 Å	
			Sephadex G—200			
				SAG 9		
500 000	A—0,5m			SAG 10	10^4 Å	
	Bio-Gel P—300		G—150			
	P—200					
	P—150		G—100		3×10^3 Å	
100 000	P—100					
			G—75			
	P—60					
50 000					10^3 Å	*Poragel*
	P—30		G—50		500Å	A—25
10 000	P—10				250 Å	
			G—25 *Sephadex*			
5 000						
	P—6	*Bio-Beads*	LH—20 (max.)		100 Å	
	P—4					A—3
	P—2	S—X2			60 Å	
		S—X4	G—15			
1 000		S—X8				A—1
			G—10			

[a]Molecular weight limits from manufacturers' data, (for globular proteins in aqueous systems and for polystyrene standards in organic solvents).

mentioned gel chromatographic separations in which the size differences between the molecules are small. To select the right gel, one should be sure that the desired high molecular weight substance is excluded and also that the properties of the gel packing permit attainment of adequate flow rates. The most common case in practice is the *removal of salts from protein solutions*. For this purpose, Sephadex G—25 or (preferably) G—50, or Bio-Gel P-30 are normally used.

Dilution of the component (caused by diffusion) can be limited. Since the K_{av} values and the corresponding V_s values (cf. Fig. 2) differ greatly, complete separation can still be accomplished even if the sample volume is kept large. For complete separation of high and low molecular weight substances, a rule of thumb is that with Sephadex G—25 or G—50 the sample volume may take up 1/4 to 1/3 of the whole volume of the gel packing (V_t). In considering this rule, a dilution of the separated component by not more than 10–20% must be accepted.

The concentration of the sample solution is limited by

Table 5. Examples of gel chromatographic separations

Solute	Sample volume	Type[a] of gel	Gel bed (ml)	Column (cm)	Solvent	Ref.
Polyethylene glycols	20 mg/1 ml	G—10	136	1.3 × 102	Phosphate buffer, pH 7	[19]
Oligosaccharides from cellulose		G—25	2000	4.5 × 126	dist. H_2O	[20]
Enzyme-substrate complex	2 mg	G—25	12.6	0.4 × 100	Acetate buffer, pH 5.3 + substrate	[21]
Enzyme mixture (lactose synthetase)	900 mg/10 ml	P—30	1130	3 × 160	Tris-HCl/$MgCl_2$, pH 7.4	[22]
Hemoglobin from erythrocytes	3 ml	G—100	163	20 × 52	0,1M NaCl	[23]
Protein solution	micro-MW determination	G—100	10 Gel beads			[24]
DNA/RNA	1 mg/5ml	G—200	172	2.5 × 35	Tris-HCl/NaCl pH 7.2	[25]
Pathological human serum	2 ml	G—200	90.5	1.2 × 80	0.9% NaCl	[26]
Soluble collagen	2 mg/2.5 ml	P—150 A—1.5 m	185	1.5 × 105	Tris-HCl/$CaCl_2$, pH 7.5	[27]
Colostrum (casein dissoc.-assoc.)	5 ml	6B	462	2.5 × 94	Tris-HCl/NaCl a. $CaCl_2$, pH 8.0	[28]
Nucleic acids	2 ml	2B	177	2.1 × 51	Phosphate/$MgCl_2$, pH 6.0	[29]
Triglycerides	40 mg/2 ml	LH-20	170	2.5 × 35	$CHCl_3$	[30]
Polypropylene		Styragel		0.95 × 122	THF	[31]
Polystyrene	13.4 mg/1 ml	Styragel	198	0.84 × 366	Toluene	[32]

[a] see Table 2.

the *viscosity* of the solution. However, by use of a basket centrifuge, it is possible to spin out all of the interstitial water from a gel packing and subsequently to allow a highly viscous solution to percolate under the influence of the high centrifugal force[33]. In this way, relatively large amounts can be de-salted.

The column technique has been expanded also for handling sample volumes up to several hundred liters. Special 'gel filters' by Pharmacia, constructed of V2A-steel and equipped with suitable pumps, valves and switches, allow continuous de-salting of macromolecular solutions[34].

[19] B. Gelotte, J. Porath, Ref. [57], p. 353.
[20] P. Flodin, K. Aspberg, Biological Structure and Function 1, 345 (1961).
[21] J.P. Hummel, W.J. Dreyer, Biochim. Biophys. Acta 63, 530 (1962).
[22] U. Brodbeck, K.E. Ebner, J. Biol. Chem. 241, 762 (1966).
[23] H. Aebi, C.H. Schneider, H. Gang, U. Wiesmann, Experientia 20, 103 (1964).
[24] W. Boguth, R. Repges, Z. Wiss. Mikroskopie 68, 241 (1967).
[25] R. Bartoli, C. Rossi, J. Chromatogr. 28, 30 (1967).
[26] H.E. Müller, Med. Klin. 63, 125 (1968).
[27] K.A. Piez, Anal. Biochem. 26, 305 (1968).
[28] Lundgren, M.K. Joustra in T. Gerritsen[59].
[29] Öberg, Philipson in T. Gerritsen[59].
[30] M.K. Joustra, Protides Biol. Fluids 14, 533 (1966).
[31] G. Meyerhoff, Ber. Bunsenges. Phys. Chem. 69, 866 (1965).
[32] J.C. Moore, J. Polym. Sci., Part A 2, 835 (1964).
[33] N.I.A. Emnéus, J. Chromatogr. 32, 243 (1968).
[34] L. Ek, Process Biochem. 3, 25 (1968).

5 Molecular Weight Determination

Since the elution behavior in gel chromatography is apparently dependent on the molecular size of the macromolecules, it should be possible to draw conclusions from the elution behavior about the *size of the molecule*. It can be said that within a homologous series, for a first approximation, the molecular size—represented by the elution behavior—is proportional to the molecular weight. By plotting a characteristic parameter for the elution behavior (for instance the K_{av} value) against the logarithm of the molecular weight, a *calibration curve* is obtained. With the aid of this curve, the behavior of an unknown substance will provide an idea of its molecular weight[35]. A closer analysis of the calibration curve shows that, within broad limits, it is linear and this renders interpretation of the results quite simple. In the field of high molecular polymers, calibration sometimes becomes difficult because well-defined *calibration substances* showing the extremely narrow molecular weight distribution required are not always available. Nevertheless, some polymer fractions (e.g. of polystyrene) suitable for this purpose are commercially available.

The situation is much more favorable in the case of *proteins*. These macromolecules exist in nature

[35] K.H. Altgelt, J.C. Moore in Ref. [58], p. 123.

as well defined and uniform substances. They can be obtained in a pure state and they give perfectly linear calibration lines[36] (Fig. 3).

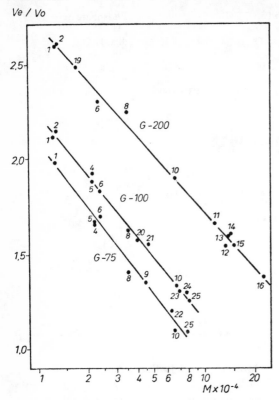

Fig. 3. Dependence of the elution parameter V_e/V_o on the logarithm of the molecular weight of *proteins* for the Sephadex gels G-75, G-100, and G-200. The values were determined in different laboratories[36]. 1 = *cytochrome c* (13,000), 2 = *ribonuclease A* (13,600), 19 = *methemoglobin* (17,000), 4 = *trypsin inhibitor* (soybean) (21,500), 5 = *α-chymotrypsin* (22,500), 6 = *trypsin* (24,000), 8 = *pepsin* (35,500), 20 = *peroxidase-1* (40,000), 9 = *egg albumin* (45,000), 21 = *α-hydroxysteroid dehydrogenase* (47,000), 22 = *phosphoglyceromutase* (64,000), 10 = *serum albumin* (bovine, monomer) (67,000), 23 = *malate dehydrogenase* (79,000), 24 = *enolase* (80,000), 25 = *creatine phosphokinase* (81,000), 11 = *phosphoglyceraldehyde dehydrogenase* (117,000), 12 = *serum albumin* (bovine, dimer) (134,000), 14 = *γ-globulin* (human) (140,000), 13 = *aldolase* (yeast) (147,000), 15 = *alcohol dehydrogenase* (yeast) (150,000), 16 = *catalase* (230,000).

A molecular weight determination is increasingly accurate, the longer the gel bed used and the higher the plate count of the particular packing employed. For determination of the molecular weight distribution of synthetic high polymers in particular, gel chromatography (here often referred to as 'gel permeation chromatography'[35]) has achieved considerable importance (cf. Chapter 9.2).

A further special experimental set-up has proved particularly useful for determination of the *molecular weights of proteins*. This is, in principle, almost an 'open' gel bed where a layer of swollen gel beads is spread out on a glass plate and arranged so that buffer

solution may flow through the gel layer. This *thin-layer gel chromatography*[37] technique also permits calibration with known proteins, but it may be expedient to run a reference protein (e.g. cytochrome C) in each experiment as an internal standard. The proteins on the thin-layer plate can be rendered visible by a color reaction; in this case it is necessary only to measure the migration distances and to compare them with those of the reference protein or with each other.

Because of the close relationship between elution behavior and molecular size, specific conclusions about the *separation mechanism* have often been sought. In this chapter we have tacitly introduced the concept of an exclusion mechanism which has not always been without contradiction. However, fundamental aspects are valid without any restriction.

A gel should be chosen whose expected *separation range* covers the molecular weight range of the mixture under study. Important are not only the numerical values (cf. Table 4) but also the separation characteristics of the gels (Fig. 4).

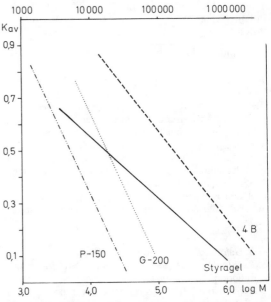

Fig. 4. Separation characteristics of swellable gels (Bio-Gel P-150 and Sephadex G-200) and heteroporous gels (Sepharose 4B and Styragel). Calibration Curves[27, 38, 59], for Random-Coil Polymers (collagen, dextran, polystyrene).

At a given molecular weight difference, very different separation volumes result from various gels. In particular, the lines for the heteroporous materials, for example agarose or Styragel, have especially flat slopes;

[36] *H. Determann, W. Michel*, J. Chromatogr. *25*, 303 (1966).

[37] *H. Determann, W. Michel*, Z. Anal. Chem. *212*, 211 (1965).

[38] *J.C. Moore, J.G. Hendrickson*, J. Polym. Sci., Part C*8*, 233 (1965).

in spite of their better technological behavior, they are inferior in this respect to the conventional gels.

True gel chromatography differs from all other chromatographic techniques in that, to a large extent, elution is independent of the flow rate, the temperature, the ionic strength, and the concentration of the sample. Deviations occur only when sorption forces become involved.

6 Sorption Effects

The fact that, in true gel chromatography, the separation is governed principally by the size and shape of molecules should not lead one to forget that, in addition, all kinds of intermolecular forces may act upon the system and may be superimposed over the mechanism involving separation by size. Among the most important forces are *Coulombic forces, van der Waals forces, hydrogen bonding,* and *hydrophobic interactions.* In molecular weight distribution and molecular weight determinations, the experimental conditions must be chosen in such a manner that all these forces are suppressed. The presence of a small amount of carboxylic groups in Sephadex gels hinders dissociated acids from penetrating into the gel structure; therefore, such acids are eluted too early. Addition of a small amount of electrolyte (0.05 M) to the eluting buffer will normalize their behavior. These apparently disturbing effects can, after study, be used as the basis for new separation methods. Because of the large number of parameters, it is much more difficult to predict the elution volume at which a substance will leave the column than in the case when separation is purely due to size. However, general information useful for practical work can be obtained from some examples selected arbitrarily from various fields (Table 6).

Figure 5 shows the separation of some low molecular weight model substances on a small column (0.65×20.5 cm, $V_t = 6.8$ ml) and illustrates the effectiveness of affinity forces. Substance 1, blue dextran (molecular weight about 2×10^6), indicates the exclusion volume V_0. V_t, the total volume of the column, occurs shortly after the peak due to substances 2/3, benzoic acid and benzene-sulfonic acid (which remain unseparated under these pH conditions). All of the true gel chromatography takes place in the area between these two peaks, 1 and 2/3; all else is *gel adsorption chromatography.*

Table 6. Examples of separation according to gel-solute interactions

Solute	Type of gel[a]	Eluant	Reference
Org. acids and *bases*	G—10	Buffer, pH 2-13	[39]
Purine derivatives	G—10	0.05 M phosphate buffer, pH 7.0	[40]
Keto enol phenyl-pyruvic acid	G—25	0.01 N acetic acid	[41]
Tyrocidines	G—25	10% acetic acid	[42]
Iodotyrosines	G—25	0.02 N NaOH	[43]
Aromatic sulfonic acids	G—25	0.1% KBr	[44]
Conjugated estrogens in urine	G—25	dist. H_2O	[45]
Nucleic acid components	G—25, G—15 and G—10	0.01 M $(NH_4)_2CO_3$ pH 9	[46]
Nucleic acid hydrolysates	P—2	dist. H_2O	[47]
Polycyclic hydrocarbons	LH—20	isopropanol	[48]
Low MW *phenols, alcohols,* and *acids*	LH—20	$CHCl_3$	[49]
Low MW *aromatic* and *heterocyclic compounds*	Styragel	n-butanol	[50]

[a] see Table 2.

[39] *A.J.W. Brook, S. Housley,* J. Chromatogr. *42,* 112 (1969).

[40] *L. Sweetman, W.L. Nyhan,* J. Chromatogr. *32,* 662 (1968).

[41] *R. Haavaldsen, T. Norseth,* Anal. Biochem. *15,* 536 (1966).

[42] *M.A. Ruttenberg, T.P. King, L.C. Craig,* Biochemistry *4,* 11 (1965).

[43] *F. Blasi, R.V. de Masi,* J. Chromatogr. *28,* 33 (1967).

[44] *H. Steuerle,* Z. Anal. Chem. *220,* 413 (1966).

[45] *C.G. Beling,* Nature *192,* 326 (1961).

[46] *G. Gorbach, J. Henke,* J. Chromatogr. *37,* 225 (1968).

[47] *A.N. Schwartz, A.W.G. Yee, B.A. Zabin,* J. Chromatogr. *20,* 154 (1965).

[48] *M. Wilk, J. Rochlitz, H. Bende,* J. Chromatogr. *24,* 414 (1966).

[49] *M.K. Joustra, B. Söderqvist, L. Fischer,* J. Chromatogr. *28,* 21 (1967).

[50] *H. Determann, K. Lampert,* unpubl.

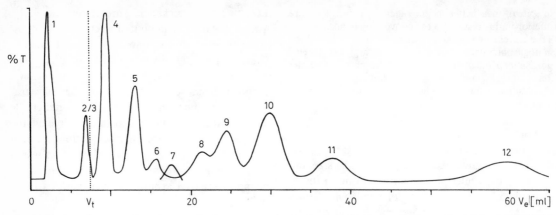

Fig. 5. Elution diagram of a mixture of substances on Sephadex LH-20 (0.65×20.5 cm, V $= 6.8$ ml), 0.006 M Phosphate Buffer containing 0.5 M NaCl, pH 7.2, Flow Rate 5.5 ml/h. Ordinate: percent transmission at 254 nm; abscissa: elution volume. (1) *Blue dextran* 2000 ($=V_0$), (2/3) *benzoic acid* and *benzenesulfonic acid* (about $V_0 + V_i$), (4) *pyridine*, (5) *benzyl alcohol*, (6) *benzaldehyde*, (7) *benzene*, (8) *α-naphthalenesulfonic acid*, (9) *phenol*, (10) *β-naphthalenesulfonic acid*, (11) *resorcinol*, (12) *phloroglucinol* (from data of *Determann* and *Lampert*).

In general, *interactions* with the gel will be strongest for low molecular weight substances resulting in small molecules leaving the column before big ones. It is noteworthy that the elution curves are completely symmetrical and that the positions of the peaks are to a large extent, independent of the concentration of the sample (this does not hold for Styragel). The more polymer contained in the gel packing, the stronger is the retardation. *Aromatic* compounds are particularly prone to retardation. *Phenols* may even be separated according to their number of hydroxyl groups; however, as with the acids mentioned above, the elution is pH dependent[39, 51]. In trying to establish rules for elution behavior or to get some idea of the *separation mechanism*, the remarkable fact must be faced that polymers of such widely different chemical natures as polyacrylamide or dextran ethers undergo essentially similar interactions with low molecular weight substances.

However, on polyacrylamide (Bio-Gel) these are distinctly weaker than on the etherifield dextrans (Sephadex). In the system comprising gel matrix, solute, and eluant, the nature of the solvent exerts a dominating effect on the affinity behavior; consequently, a change of solvent often results in a considerable change in the elution behavior. The closer the solvent is related to the polymer, the less is the solute retarded.

7 Future Prospects

When the development of gel chromatography is examined, a tendency to use gels having better technological properties (which, however, as shown above, are necessarily of lower efficiency) and equipment completely automated and perfected can be observed. Growing requirements for better separations will demand increasingly longer columns and higher plate counts. In the development of more rigid materials, the production of porous glass for gel chromatographic purposes appears to represent the limit of development[52]. Unfortunately, with aqueous systems molecular sieving processes are often surpassed by adsorptive ones.

However, at present this material represents the only possibility for carrying out gel chromatography at high temperatures (above 150°)[53]. For the separation of large molecules (molecular weights exceeding 1,000,000) in aqueous media, cellulose gels[54] will enter into competition with the agarose gels.

Separation of small molecules from mixtures containing large ones can be achieved with both dialysis and gel filtration. Gel filtration provides more definite results in a shorter time, but with a greater requirement of experimental expenditure than dialysis. However, because of the recently developed membranes, ultrafiltration and reverse osmosis are noticeably gaining a wider market.

As far as the separation of molecules whose molecular weight differences are small is considered, gel chromatography today is clearly superior to the membrane techniques. Future development must show whether the advances made recently in

[51] *J.B. Woof*, *J.S. Pierce*, J. Chromatogr. *28*, 94 (1967).

[52] *W.J. Haller*, J. Chem. Phys. *42*, 686 (1965).

[53] *M.J.R. Cantow*, *J.F. Johnson*, J. Appl. Polym. Sci. *11*, 1851 (1967).

[54] *H. Determann*, *H. Rehner*, *Th. Wieland*, *N. Meyer*, *F. Wente*, Makromol. Chem. *114*, 263 (1968).

membrane technology are sufficient for gel chromatography to be displaced in special instances.

8 Bibliography

[55] *H. Determann*, Gelchromatographie, Springer Verlag, Berlin, Heidelberg, New York 1967.
[56] *E. Heftmann*, Chromatography, 2nd Ed., van Norstrand-Reinhold, New York 1967.
[57] *M.J.R. Cantow*, Polymer Fractionation, Academic Press, New York, London 1967.
[58] *T. Gerritsen*, Progress in Separation and Purification, Vol. II, Wiley Intersci., New York 1969.
[59] *L. Fischer* in *T.S. Work, E. Work*, Laboratory Techniques in Biochemistry and Molecular Biology, North Holland Publishing Co., Amsterdam, London 1968.
[60] *W. Heitz*, Gelchromatographie, Angew. Chem. *82*, 675, Angew. Chem. Internat. Ed. Engl. *9*, 689 (1970).

2.7 Methods for the Separation of Macromolecular Substances

Hans Batzer, Walter Hofmann, Friedrich Lohse, Rolf Schmid, Sheik A. Zahir
Ciba-Geigy Limited, CH-4002 Basel

1 Significant Characteristics of Macromolecular Substances

1.1 Structure

In the separation of macromolecules, it is important to consider the factors affecting various properties including solubility. These factors dictate to some extent the choice of separation method and contribute a great deal towards the identification of unknown substances.

The chemical and physical properties of macromolecules are determined both by the nature of their basic structural units and by the chemical bonds linking these units. They are also influenced by the structure of the polymer, by topochemical characteristics, and by the molecular weight (Table 1).

The importance of intermolecular forces, such as hydrogen bonds, and dipole, induction, dispersion and Coulombic forces, must also be stressed.

Unlike cross-linked materials, which sometimes swell in suitable solvents but which are generally insoluble, linear and branched macromolecules (with few exceptions) follow a solubility pattern

Table 1. Characteristics which influence the chemical and physical properties of macromolecules

Characteristics	Example
1. Periodically repeating structural units	
a) structure	Homopolymers (polystyrene, polyoxymethylene, 6-Nylon).
b) bonding between structural units of dissimilar types	Copolymers, terpolymers, quaterpolymers, polyadducts (Perlon U), condensation polymers (66-Nylon).
c) bonding between structural units of similar types	Block polymers, graft polymers.
d) bonding between polymer segments	
2. Nature of the bonds between the structural units in the macromolecule	C–C bonding, ether, ester, amide, ureide groups.
3. Structure of macromolecules	
a) linear	Low pressure polyethylene, polyester, polyamide.
b) slightly branched	High pressure polyethylene, synthetic rubber, starch.
c) strongly branched	Globular macromolecules: glycogen, ovalbumin.
d) cross linked	Phenoplasts, cured epoxy resins and isocyanate resins, polyester resins.
4. Topochemical characteristics	
a) geometrical isomers	Natural rubber, gutta-percha.
b) optical isomers	Polypeptides, polysaccharides.
c) tacticity	Isotactic, syndiotactic, atactic.
d) helix structure	Polypeptides, isotactic polymers.
e) head-tail, head-head or tail-tail bonding	Polyisoprene, polystyrene, polyvinylidene chloride.
5. Molecular weight	

that makes them amenable to physical and chemical separation methods.

1.2 Solubility

1.2.1 Gibbs Free Energy Equation and Solubility Parameter

Solubility of a solute in a solvent will occur if the free energy of mixing, ΔG_m is negative:

$$\Delta G_m = \Delta H_m - T \Delta S_m \qquad (1)$$

where: ΔG_m = free energy of mixing
ΔH_m = heat of mixing
ΔS_m = entropy of mixing

Since the entropy of mixing ΔS_m is always large and positive, the sign of ΔG_m is determined by the magnitude and sign of ΔH_m. If there is positive interaction between polymer and solvent, ΔH_m is negative and solution will occur. When only dispersion forces are involved and ΔH_m is positive, then the magnitude of ΔH_m plays an important role in determining whether ΔG_m is negative or not, i.e. whether solution will occur or not.
The heat of mixing[1] is given by:

$$\Delta H_m = V_m \cdot \Phi_1 \cdot \Phi_2 \left[\left(\frac{\Delta E_1}{V_1} \right)^{\frac{1}{2}} - \left(\frac{\Delta E_2}{V_2} \right)^{\frac{1}{2}} \right]^2 \qquad (2)$$

where: V_m = total volume of mixture
Φ = volume fraction of component 1 or 2 in the mixture
V = molar volume of component 1 or 2
ΔE = energy of vaporization of component 1 or 2

The expression, $\left(\frac{\Delta E}{V} \right)^{\frac{1}{2}}$, is called the solubility parameter δ. Rearranging equation (2), and introducing the solubility parameter,

$$\frac{\Delta H_m}{V_m} = \Phi_1 \cdot \Phi_2 \left(\delta_1 - \delta_2 \right)^2 \qquad (3)$$

For ΔH_m to be small, δ_1 and δ_2 should be similar in magnitude. Thus, in the absence of specific interactions, the most likely solvents for a polymer will be liquids whose solubility parameter is close to that of the polymer. The solubility parameter of a solvent can be calculated readily from the latent heat of vaporization: that of a polymer, however, cannot be determined directly because most polymers cannot be vaporized. It must, therefore, be estimated indirectly. For practical work, classi-

fication of polymers according to the magnitudes of their solubility parameters and their degrees of *crystallinity*[2] (Scheme 1), and of solvents by their tendency to form *hydrogen bonds*[3] (Table 2), can be of great value.

Table 2. Solubility parameter of some solvents[3]

Solvent	Solubility parameter
Poorly hydrogen bonded solvents	
n-Heptane	7.4
Toluene	8.9
o-Dichlorobenzene	10.0
Nitromethane	12.7
Moderately hydrogen bonded solvents	
Diethyl ether	7.4
Di-isobutyl ketone	7.8
Dibutyl phthalate	9.3
Ethylene carbonate	14.7
Strongly hydrogen bonded solvents	
2-Ethylhexanol	9.5
n-Butanol	11.4
Ethanol	12.7
Methanol	14.5

When polymers carrying highly polar groups, which may interact strongly, are considered, the problem of interpreting solubility behavior becomes much more complex.
For crystalline polymers, the free energy of mixing is given by:

$$\Delta G_m = \Delta H_m - T \Delta S_m + \Delta H_f \qquad (4)$$

where ΔH_f is the heat of fusion.

It may be seen that appreciable solubility can be expected at a temperature below the melting point only if the mixing of molten polymer and solvent is sufficiently exothermic to compensate for the heat absorbed during the fusion process.

1.2.2 Flory Parameter

An alternate expression for the effectiveness of a solvent is provided by the Flory parameter χ, which is a measure of the interaction energy between polymer and solvent[4].

χ is defined by the equation:

$$\chi = \frac{\Delta E}{kT} \qquad (5)$$

[1] J.H. Hildebrand, R.L. Scott, The Solubility of Nonelectrolytes, A.C.S.-Monograph No. 17, Reinhold Publishing Co., New York 1950.

[2] H. Burrell, J. Paint. Technol. 40, 197 (1968).
[3] H. Burrell, B. Immergut in Ref. [169], p. IV–341.
[4] Ref. [157].

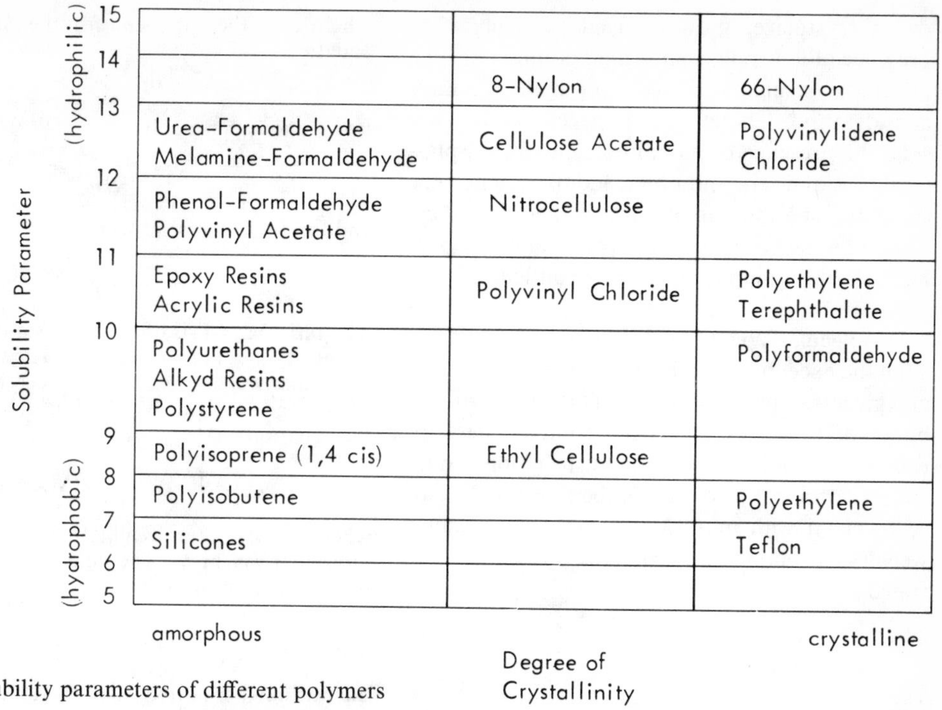

Solubility Parameter	amorphous		crystalline
15 (hydrophilic)			
14			
13		8-Nylon	66-Nylon
	Urea-Formaldehyde Melamine-Formaldehyde	Cellulose Acetate	Polyvinylidene Chloride
12	Phenol-Formaldehyde Polyvinyl Acetate	Nitrocellulose	
11	Epoxy Resins Acrylic Resins	Polyvinyl Chloride	Polyethylene Terephthalate
10	Polyurethanes Alkyd Resins Polystyrene		Polyformaldehyde
9	Polyisoprene (1,4 cis)	Ethyl Cellulose	
8 (hydrophobic)	Polyisobutene		
7			Polyethylene
6	Silicones		Teflon
5			

Degree of Crystallinity

Scheme 1. Solubility parameters of different polymers

where k is Boltzmann's constant, and T the absolute temperature. ΔE represents the difference in energy of a solvent molecule immersed in pure polymer compared with one surrounded by pure solvent.

A solvent for a given polymer should have χ below 0.5. A liquid having a χ value over 0.5 belongs to the category of non-solvents. The critical value χ_c which leads to phase separation depends on the molecular weight of the dissolved polymer molecules. If p is the degree of polymerization, it has been shown[4] that:

$$\chi_c = \frac{1}{2} + \left(\frac{1}{p}\right)^{\frac{1}{2}} \qquad (6)$$

It is thus possible to carry out fractionation, either by fractional solution or by fractional precipitation, by adjusting the χ value of a solvent-nonsolvent mixture.

χ is also a function of temperature[4]:

$$\chi = A + \frac{B}{T} \qquad (7)$$

where A and B are constants, and T is the absolute temperature. Fractional solution can therefore be carried out by increasing the temperature, and so lowering χ. Fractional precipitation by cooling is due to the fact that χ increases as the temperature is lowered.

When a homogeneous polymer solution separates into two liquid phases as a result of a lowering in solvent power due to a change in composition or temperature, a small quantity of a polymer-rich phase will separate in equilibrium with a relatively large liquid phase of lower polymer concentration.

If V_p' is the volume fraction of polymer in the concentrated phase and V_p the volume fraction in the dilute phase, then according to the Flory theory[4],

$$\frac{V_p'}{V_p} = e^{\sigma \cdot p} \qquad (8)$$

where σ is a complex function of the number average molecular weight and χ.

Since V_p'/V_p is proportional to an exponential power of p for a given σ (8), the ratio of concentration of high to low molecular weight polymer molecules is much greater in the precipitate than in the solution phase. If the volume of the more dilute phase is made much larger than that of the more concentrated phase, most of the smaller species will remain in the dilute phase and the higher molecular weight species will collect selectively in the precipitated phase.

Although such studies have proved of great value in the fractionation of known polymers, they cannot replace standard laboratory practice in the search for solvents that might be suitable for the fractionation or separation of macromolecules.

The solubilities of polymers in a large variety of solvents are now available in comprehensive tables[5].

In general, macromolecules with a regular stereochemical structure, or those with a strong tenden-

5 K. Meyersen in Ref. [169], p. IV—185.

cy to crystallize, show reduced solubility. The presence of low molecular weight compounds can affect the solubility behavior of high polymers considerably[6]. Therefore it is necessary to consider the presence of monomers, oligomers, plasticizers, stabilizers and other additives, when the separation of macromolecules is being studied. Their presence can make the preparation of pure high polymer samples much more difficult.

1.3 Molecular Weight Distribution

With the exception of certain biological polymers, any given sample of polymer will contain a distribution of molecular weights. A complete description of the molecular weight distribution would require specification of the number N_i of molecules of species i with mass M_i. A typical molecular weight distribution of a synthetic polymer is illustrated in Fig. 1.

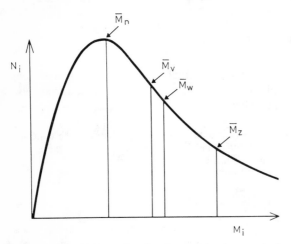

Fig. 1. Distribution of molecular weights in a typical polymer

Such a distribution can be expressed in terms of certain molecular weight averages, $\overline{M}_n$, $\overline{M}_w$, $\overline{M}_z$, which are defined by the equations:

Number average $\overline{M}_n = \dfrac{\Sigma M_i \cdot N_i}{\Sigma N_i}$

Weight average $\overline{M}_w = \dfrac{\Sigma M_i^2 \cdot N_i}{\Sigma M_i \cdot N_i}$

Z-average $\overline{M}_z = \dfrac{\Sigma M_i^3 \cdot N_i}{\Sigma M_i^2 \cdot N_i}$

The viscosity average is defined as follows:

$$\overline{M}_v = \left(\frac{\Sigma M_i^{a+1} \cdot N_i}{\Sigma M_i \cdot N_i}\right)^{\frac{1}{a}}$$

where a is the exponent for the Staudinger-Kuhn equation:

$$[\eta] = K \cdot M^a$$

For monodisperse polymers:

$$\overline{M}_n = \overline{M}_w = \overline{M}_z$$

and for polydisperse polymers:

$$\overline{M}_n < \overline{M}_w < \overline{M}_z$$

The ratio $\overline{M}_w/\overline{M}_n$ is often used as a measure of polydispersity of a polymer sample. The viscosity average molecular weight, $\overline{M}_v$, falls between $\overline{M}_n$ and $\overline{M}_w$. The integral weight distribution, I(M), is defined by the equation:

$$I(M) = \int_0^M W(M)dM$$

where W(M) is the weight fraction of molecules of molecular weight M. The differential weight distribution function is given by:

$$W(M) = \frac{dI(M)}{dM}$$

and the number or frequency distribution by:

$$N(M) = \frac{1}{M} W(M)$$

The distribution of molecular sizes is a direct result of the kinetics of polymerization and polycondensation. From an assumed mechanism of polymerization, analytical functions can be derived from purely statistical or kinetic considerations[7]. The range of values of M_w/M_n observed in synthetic polymers is quite large[8] (Table 3).

Table 3. Polydispersity M_w/M_n of some polymers[8]

Polymer	M_w/M_n
Monodisperse	1.00
Living polymers[9–11]	1.01–1.05
Addition polymers, terminated by combination of two polymer radicals	1.5
Addition polymers, terminated by disproportionation, or condensation polymers	2.0
High conversion vinyl polymers	2–5
Polymers made with autoacceleration (Trommsdorff-effect)	5–10
Coordination polymers	8–30
Branched polymers	20–50

[6] O. Fuchs, Fortschr. Chem. Forsch. 11, 74 (1968).

[7] L.H. Peebles in Ref. [169], p. II–421.
[8] F.W. Billmeyer jr., J. Polym. Sci. C 8, 161 (1965).
[9] M. Szwarc, M. Levy, R. Milkovich, J. Am. Chem. Soc. 78, 2656 (1956).
[10] H.W. McCormick, J. Polym. Sci. 36, 341 (1959).
[11] G. Meyerhoff, M. Cantow, J. Polym. Sci. 34, 503 (1959).

2 Determination of the Molecular Weight Distribution

The molecular weight distribution is of great importance in the quantitative description and characterization of polymers. The determination of the molecular weight distribution from experimental data is illustrated by an actual example[12] (see also[13-15]). In this example, a polyester from sebacic acid and 1,6-hexanediol was separated into several fractions by fractional precipitation. The weight fraction W_n, the molecular weight and the degree of polymerization p'_n of each fraction were determined (Table 4).

Table 4. Fryctionation and determination of the integral weight distribution of a polyester made from sebacic acid and 1,6-hexanediol[12].

Fraction No.	weight [g]	% of total substance $W_n \cdot 100$	$I(p_n) \cdot 10^2$	$\bar{p}_n$
1	0.575	7.50	3.75	36
2	0.469	6.13	10.57	70
3	0.507	6.62	16.94	93.5
4	0.629	8.20	24.35	145.5
5	0.648	8.45	32.68	184.5
6	0.864	11.25	42.53	229
7	1.076	14.05	55.18	320
8	0.631	8.25	66.33	350
9	1.219	15.90	78.40	452
10	1.047	13.65	93.18	561
Total	7.665	100.00		

In order to obtain the integral weight distribution function[16], the weight fractions W_n for 1 gram of polymer from the fractionation data were inserted into the following equation:

$$I(p_n) = \frac{\frac{1}{2}W_n + \sum_{i=1}^{n-1}W_i}{\sum W_n}$$

By constructing a smooth curve through the points of $I(p_n)$ versus $\bar{p}_n$, the integral distribution curve was obtained (Fig. 2, curve I). The differential weight distribution curve, $W(p)$, was obtained by graphical differentiation of the integral curve (Fig. 2, curve II). The maximum of curve II corresponds to the inflection point of curve I.

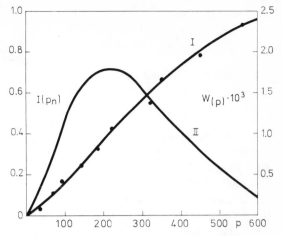

Fig. 2. Molecular weight distributions of a polyester prepared from sebacic acid and 1,6-hexanediol (data of Table 4).
Curve I: Integral distribution curve $I(p_n)$ from experimental data;
Curve II: Differential distribution curve $W(p)$ obtained by graphical differentiation of curve I.

Although this method of treating the data is quite satisfactory for high resolution fractionation data, it cannot completely compensate for the overlapping of fractions. Methods of correcting for overlapping, and for eliminating graphical differentiation, have been proposed by several authors[15, 17-21]. An alternate method of correcting for overlapping, by assuming a *two parameter binomial distribution function*, has also been successfully used[22-24].

At present the ultracentrifuge offers the only direct method for the determination of the molecular weight and the molecular weight distribution of high polymers (see Chapter 2.1). An alternate and rapid method for the determination of the molecular weight distribution is gel permeation chromatography[25-28] (see Chapter 2.7:3.5). The accuracy of the average molecular weights obtained depends on several factors, such as the nature of the distribution and the experimental conditions. Recently, direct determination of the mole-

[12] H. Batzer, F. Wiloth, Makromol. Chem. 8, 41 (1952).
[13] H. Batzer, A. Möschle, Makromol. Chem. 22, 195 (1956).
[14] H. Batzer, A. Nisch, Makromol. Chem. 22, 131 (1956).
[15] L.H. Tung in Ref. [173], p. 379.
[16] G.V. Schulz, Z. Phys. Chem. B47, 155 (1940).

[17] R.W. Hall in Ref. [160], p. 19.
[18] R. Koningsfeld, A.J. Staverman, J. Polym. Sci. A-2 6, 367 (1968).
[19] E.A. Haseley, J. Polym. Sci. 35, 309 (1959).
[20] R.F. Boyer, Ind. Eng. Chem. Anal. 18, 342 (1946).
[21] C. Mussa, J. Polym. Sci. 26, 67 (1957).
[22] G. Beall, J. Polym. Sci. 4, 483 (1949).
[23] C. Booth, L.R. Beason, J. Polym. Sci. 42, 81 (1960).
[24] R. Koningsfeld, C.A.F. Tuijnman, J. Polym. Sci. 39, 445 (1959).
[25] Ref. [174], p. 129.
[26] J.R. Runyon, D.E. Barnes, J.F. Rudd, L.H. Tung, J. Appl. Polym. Sci. 13, 2359 (1969).
[27] L.H. Tung, J. Polym. Sci. 10, 375 (1966).
[28] H.E. Richett, M.J.R. Cantow, J.F. Johnson, J. Appl. Polym. Sci. 10, 917 (1966).

cular weight distribution of some polymers with very high molecular weights has been achieved using the electron microscope[29].

3 Physical Separation Methods[13]

Physical separation methods are preferred to chemical methods as they allow direct isolation of the original components with only slight losses (if any) of the substance being analyzed.

3.1 Fractional Solution

The method consists of extracting fractions of increasing molecular weight by a series of eluants of increasing solvent power. In contrast to fractional precipitation, the lowest molecular weight is obtained first and the highest molecular weight last (Fig. 3). The polymer may be extracted as a finely divided powder, or as a film. It may be deposited on sand in a column and extracted, or a concentrated solution of polymer (coacervate) may be extracted selectively.

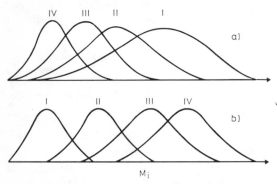

Fig. 3. Theoretical curves for the distribution of molecular weights.
a) Fractional precipitation   b) Fractional solution

Unlike fractional precipitation, much smaller quantities of solvent are required. On theoretical grounds, one would expect the higher fractions obtained by fractional solution to be less contaminated with lower molecular weight polymer than are the corresponding fractions obtained by fractional precipitation. A disadvantage of the method is that it is difficult to establish true equilibrium between the solid polymer and the extracting liquid.

3.1.1 Fractionation by Column Extraction

Of all the fractional solution methods, column elution is by far the most flexible.
Desreux[30-33] pioneered a fractional solution method in which the polymer was deposited in very

thin films on sand particles. These particles were introduced into the top of a column packed with either sand or glass beads. Polymer fractions were then eluted by extraction with solvent/nonsolvent combinations with progressively increasing solvent power until all had been washed through. The solvent power can be increased either by increasing the proportion of solvent in the solvent-nonsolvent combination at a constant temperature, or by progressively increasing the temperature with a single solvent-nonsolvent combination[17, 34-37].

The time required for extraction is of critical importance for the success of the fractionation. If the extraction time is too short, resolution is poor and an inversion of the molecular weight sequence may even occur[38].

Striking improvement in resolution is obtained when the extraction is carried out over a longer period, and also when the sample for extraction is deposited on the carrier material by a special method referred to as selective deposition[36, 39]. In this method, a 1 to 2% solution of the polymer is poured into the packed column which has been heated above the melting point of the polymer. The sample is allowed to deposit as the column is cooled to room temperature. A pre-fractionation of the polymer molecules is thought to occur during the deposition. With decreasing column temperature, the highest molecular weight polymer is presumed to deposit first and then to be covered by layers of progressively lower molecular weight polymer. During extraction, the lower molecular weight polymer in the outermost layer is extracted first.

3.1.2 Baker-Williams Method of Column Fractionation

This method[40] combines the use of a thermal gradient along the column with that of a solvent gradient to achieve a multi-stage fractional preci-

[29] D.V. Quayle, Brit. Polym. J. 1, 15 (1969).

[30] V. Desreux, Rec. Trav. Chim. Pays-Bas 68, 789 (1949).
[31] V. Desreux, M.C. Spiegels, Bull. Soc. Chim. Beles 59, 476 (1950).
[32] V. Desreux, A. Oth, Chem. Weekbl. 48, 247 (1952).
[33] P.S. Francis, R.C. Cooke, J.H. Elliott, J. Polym. Sci. 31, 453 (1958).
[34] J.H. Elliott in Ref. [173], p. 67.
[35] N.S. Schneider, J. Polym. Sci. C8, 179 (1965).
[36] W.L.H. Moll, Kolloid-Z., Z. Polym. 223, 48 (1967).
[37] G.M. Guzmán in Ref. [169], p. IV-235.
[38] S. Shyluk, J. Polym. Sci. 62, 317 (1962).
[39] A.S. Kenyon, J.O. Salyer, J. Polym. Sci. 43, 427 (1960).
[40] C.A. Baker, R.J.P. Williams, J. Chem. Soc. 2352 (1956).

pitation. The polymer-coated glass beads are situated in a constant temperature zone at the top of the column which is also the highest temperature zone. The rest of the column is filled with uncoated beads. The sample is selectively extracted by progressively richer solvent-nonsolvent mixtures. As the extracted polymer moves down the column into the cooler zones, it undergoes a partial reprecipitation. As the solvent-nonsolvent ratio increases, this polymer is redissolved, and moves further down the column, where it is reprecipitated as the solution cools. Thus, each fraction is subjected to precipitation, solution, and reprecipitation etc., as the polymer travels down the column[17, 37, 41].

3.2 Fractional Precipitation

The basic operations for fractional precipitation[17, 42, 43] are as follows. To a stirred solution of a polymer (0.1 to 2%) in a suitable solvent at constant temperature, a nonsolvent is gradually added.

As soon as the solution becomes turbid, it is warmed till it is clear and homogeneous. It is then allowed to cool slowly to the original temperature. The gel phase that separates consists mostly of higher molecular weight polymer. The amount of nonsolvent added, and the resulting turbidity of the mixture, depend to some extent on the number of fractions desired. The experimental conditions for the separation of polymer must be chosen carefully to ensure attainment of equilibrium between the gel and sol phases. The precipitate is separated from the supernatant phase and redissolved in a small amount of solvent. The polymer is reprecipitated by adding it to a large volume of nonsolvent and is then filtered and dried to a constant weight in vacuo. The supernatant liquid is treated with a further volume of nonsolvent and the procedure described above is repeated to obtain the next fraction. Successive precipitations are carried out until the bulk of the polymer has been removed from the solution. At this point, the solution is concentrated under reduced pressure and the fraction with the lowest molecular weight is isolated — either by precipitation or by freeze-drying. Fractional precipitation can be carried out also by solvent evaporation, as well as by progressively cooling a solution of polymer in either a solvent or a solvent/nonsolvent mixture[43].

In the latter method, only one solvent system is required and the volume is fixed. However, there are a number of disadvantages. It is often difficult to find a suitable solvent system that will completely precipitate the polymer on cooling; degradation can occur at the high temperatures necessary to dissolve some poly-

mers. Every fraction isolated will contain an appreciable amount of lower molecular weight components. Some authors[44-47] have proposed methods to avoid the time-consuming nature of fractional precipitation by fractionating quickly without waiting for equilibrium to be attained. This is then followed by refractionating several more times in the same way. Nevertheless, the whole procedure is laborious and, for many polymers, fractional precipitation has been superseded by the more convenient column fractionation methods (e.g. gel chromatography, see 2.7:3.5).

3.3 Crystallization

The crystallizability of polymers is dependent on steric order, the flexibility of the polymer chains, and the nature and magnitude of the secondary valence forces. Therefore, the principles that apply for low molecular weight compounds also hold for the recrystallization of macromolecules. To obtain pure inclusion-free crystals, it is necessary to have a solution with highly mobile chains so that the aggregates of molecules can pass through a stage where the chains are completely separated from each other. For this reason, crystallizations are carried out in dilute solutions by the addition of nonsolvents or at high temperature in solvents in which the polymer is not very soluble.

Single crystals, which are required for X-ray investigations, are also grown by this method. Many examples have been published, e.g. *polyoxymethylene*[48a], *polyoxyethylene*[48b], *polyethylene*[48c, d], *polypropylene*[48e], *polystyrene*[48f], *polyamide*[48g], *polyacrylic acid*[48h], *polyacrylonitrile*[48i] and *cellulose*[48k].
For details of the crystallization of pure samples of single polymer crystals from suitable solvents, appropriate references[36, 49] must be consulted.

41 *R.S. Porter, J.F. Johnson* in Ref.[173], p. 95.
42 *G.V. Schulz* in Ref.[158], Bd. II, p. 726.
43 *A. Kotera* in Ref.[173], p. 44.
44 *G.V. Schulz, A. Dinglinger*, Z. Phys. Chem. B *43*, 47 (1939).

45 *A.M. Meffroy-Biget*, Bull. Soc. Chim. France 465 (1954).
46 *A.M. Meffroy-Biget*, C.R. Acad. Sci. Paris *240*, 1707 (1955).
47 *C.D. Thurmond, B.H. Zimm*, J. Polym. Sci. *8*, 477 (1952).
48a *P.H. Geil, N.K.J. Simons, R.G. Scott*, J. Appl. Phys. *30*, 1516 (1959).
  b *A. Keller*, Polymer *3*, 393 (1962).
  c *P.H. Till*, J. Polym. Sci. *24*, 301 (1957).
  d *E.W. Fischer*, Z. Naturforsch. *12a*, 753 (1957).
  e *B.G. Rånby, F.E. Morehead, N.M. Walter*, J. Polym. Sci. *44*, 349 (1960).
  f *V.A. Kargin*, Vysokomol. Soedin *2*, 1280 (1960).
  g *P.H. Geil*, J. Polym. Sci. *44*, 449 (1960).
  h *M.L. Miller, M.C. Botty, C.E. Ranhut*, J. Colloid Sci. *15*, 83 (1960).
  i *V.F. Holland, S.B. Mitchell, W.L. Hunter, P.H. Lindenmeyer*, J. Polym. Sci. *62*, 145 (1962).
  k *B.G. Rånby, R.W. Noe*, J. Polym. Sci. *51*, 337 (1961).
49 *R.L. Miller* in Ref.[169], p. III—1.

3.4 Distribution and Extraction

The single and multiple distribution methods of separation[50] based on Nernst's distribution law (Chapter 2.3) have occasionally been used with synthetic polymers[36, 51] and with naturally occuring macromolecular substances[52]. — Extraction[36, 50] is used mainly for the separation of substances which have a small distribution coefficient (K < 1.5) and which would otherwise require large quantities of solvent.

3.5 Gel Permeation Chromatography

3.5.1 General

Gel permeation chromatography (GPC)[53–57] is a column chromatographic technique for fractionating materials according to molecular size. It is a modern and highly efficient technique which is of great value in analytical polymer fractionation and in biochemistry. A great advantage of the method is that it is possible to avoid adsorption problems because of the inertness of the gel matrix, a fact which has proved useful in the separation of labile polypeptides. Details of the method and of its applications are given in Chapters 2.6 and 12.2.

Two types of gels are commonly available. Lightly cross-linked soft gels have a flexible network and the pore size is greatly dependent upon the solvent power. Heavily cross-linked gels have a rigid network and swell to a much lower degree; their structure is unaffected by the properties of the solvent and solute.

The distribution coefficient, K_d, for the elution of solute molecules in a column packed with gel particles suspended in a given solvent is defined by the following equation:

$$V_e = V_o + K_d \cdot V_i$$

where V_e = the elution volume of the solute molecule,
V_i = the internal volume in the gel
and V_o = the interstitial volume

When $K_d = 0$, the solute is completely excluded from the column and $V_e = V_o$. This is called the exclusion limit. For very small molecules, $K_d = 1$, and $V_e = V_o + V_i$. A value of K_d greater than 1 indicates that adsorption is occuring between the solute and the gel matrix. K_d is independent of the column size and

geometry but is dependent on any factor that might change the pore size of the gel, i.e. the type of gel, the solvent, and the temperature.

Many eluants may be used; tetrahydrofuran, chloroform, o-dichlorobenzene and dimethyl formamide are generally preferred.

3.5.2 Preparation and Properties (Exclusion Limits) of Porous Gels

The most important gels for fractionating macromolecules in non-aqueous solutions are the polystyrene gels, both heavily and lightly cross-linked types. They are prepared in bead form by suspension polymerization of divinylbenzene/styrene mixtures. Porosity in the highly cross-linked gels is achieved by carrying out the crosslinking in the presence of a poor solvent for the polymer. The proportion of diluent and its solvent power determine the pore size[58–60] (Tables 5 and 6).

Other gels, such as dextran gels, Sephadex®, polyacrylamide gels, porous silica and glass beads have also been used successfully (Chapter 2.6).

Table 5. Exclusion limits of lightly crosslinked polystyrene gels prepared with varying proportions of toluene[58–60].

Styrene (wt. %)	Divinyl benzene (wt. %)	Toluene (wt. %)	Molecular weight exclusion limit
92	8	0	1,000
96	4	0	1,700
99	1	0	3,500
99.9	0.1	0	too soft
79.1	4.2	16.7	2,500
65.7	5.7	28.8	~7,000
30	10	60	7,000

Table 6. Exclusion limits of styrene-divinylbenzene gels prepared with various diluents[58–60].

Diluents (parts per 100 parts of gel)[a]	Molecular weight exclusion limit
60 toluene	$7 \cdot 10^3$
30 toluene, 30 diethylbenzene	$1.5 \cdot 10^4$
60 diethylbenzene	$1.2 \cdot 10^4$
45 toluene, 15 n-dodecane	$1 \cdot 10^5$
30 toluene, 30 n-dodecane	$3 \cdot 10^5$
15 toluene, 45 n-dodecane	$2 \cdot 10^6$
10 toluene, 50 n-dodecane	$< 2 \cdot 10^3$
40 diethylbenzene, 20 isoamyl alcohol	$\sim 3.6 \cdot 10^3$
20 diethylbenzene, 40 isoamyl alcohol	$\sim 8 \cdot 10^6$
13.3 diethylbenzene, 46.7 isoamyl alcohol	$\sim 10^{10}$
60 isoamyl alcohol	extremely high

[a] The gel is made using 21.8% styrene, 8.2% ethylvinylbenzene, 10% divinylbenzene and 60 % diluent.

[50] O. Jübermann in Ref. [159], p. 223.
[51] L.C. Case, Makromol. Chem. 41, 61 (1960).
[52] P.v. Tavel, Chimia 23, 57 (1969).
[53] Ref. [174].
[54] K.H. Altgelt, J.C. Moore in Ref. [173], p. 123.
[55] J. Seidl, J. Malinsky, K. Dusek, W. Hietz, Fortschr. Hochpolym. Forsch. 5, 113 (1967).
[56] D. Kranz, Kolloid-Z. 227, 41 (1968).
[57] J.F. Johnson, R.R. Porter, M.J.R. Cantow, Rev. Macromol. Chem. 1, 393 (1966).

3.5.3 Apparatus and Technique

The apparatus used in gel permeation chromatography is shown diagrammatically in Fig. 4.

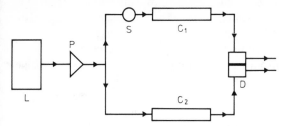

Fig. 4. Schematic diagram of a gel permeation chromatograph

L	Solvent reservoir	D	Differential detector
S	Sample injection	C_1	Sample column
P	Pump	C_2	Reference column

Solvent from a tank passes through a degasser and is pumped through a filter. It is then split into two streams, one of which flows through a reference column. The other flows through an injection device into the sample column. The injection device permits the injection of solute into the columns without interrupting the flow. Differential refractometers, and ultraviolet and infrared spectrometers are used as detectors. The signals from the detector are usually recorded in analog form. They can also be converted into a form suitable for computer evaluation by means of an analog-digital converter and a punched tape. The efficiency of the column is defined by the number of theoretical plates, N, given by the equation:

$$N = \left(\frac{4\,V_e}{w}\right)^2$$

where V_e is the elution volume of a small molecule, and w the baseline width of the curve. Columns commonly used have a theoretical plate count of 20—30 per cm of column length. At flow rates in the region of 0.1 to 1 ml/cm^2/min., the theoretical plates per cm approximately doubles when the flow rate is reduced by a quarter.

Several columns can be coupled in series to increase either the resolution or the upper exclusion limit. The gels are normally characterized by the exclusion limit, given in Angström units, for narrow distribution polystyrene standards.

When a heterogeneous polymer is being eluted through the gel column, its chromatogram is broadened by two processes, i.e. a desirable one

due to the difference in molecular sizes of the component species, and an undesirable one due to longitudinal mixing. The second braodening process has a deleterious effect on the resolution of the column, and methods for correcting it mathematically have been discussed by a number of authors[61-65]. The correction is cumbersome, and is of importance mainly for polymers with a narrow molecular weight distribution.

The gel chromatogram that is finally obtained (Fig. 5)[66] is a curve whose ordinate (recorder sig-

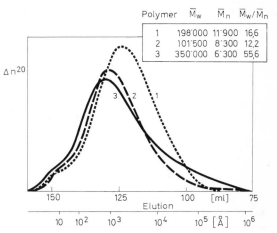

Polymer	$\bar{M}_w$	$\bar{M}_n$	$\bar{M}_w/\bar{M}_n$
1	198'000	11'900	16,6
2	101'500	8'300	12,2
3	350'000	6'300	55,6

Fig. 5. Molecular size distribution of high molecular weight epoxy resins based on 2,2-bis(p-hydroxyphenyl)propane and epichlorohydrin by gel permeation chromatography. Refractive index difference Δn^{20} vs. elution volume [ml] or chain length [Å]. (Styragel columns, exclusion limits 10^4 Å, 10^4 Å, 10^3Å, 60 Å, solvent: tetrahydrofuran).

nal) may be taken as a measure of concentration, and whose abscissa corresponds to elution volume. In order to obtain the desired curve of concentration versus molecular weight, the column must first be calibrated with a series of polymer samples (usually narrow molecular weight fractions of polymer) of known molecular weight.

The assumption is made that the size of the molecule in dilute solution is governed principally by the valency angles and the bond lengths. For each molecular weight M, the total extended length A of

[58] J.C. Moore, J. Polym. Sci. A 2, 835 (1964).

[59] J.C. Moore, J.G. Hendrickson, J. Polym. Sci. C 8, 233 (1965).

[60] J.G. Hendrickson, J.C. Moore, J. Polym. Sci. A-1, 4, 167 (1966).

[61] L.H. Tung, J. Appl. Polym. Sci. 10, 375, 1261, 1271 (1966).

[62] W.N. Smith, J. Appl. Polym. Sci. 11, 639 (1967).

[63] M. Hess, R.F. Kratz, J. Polym. Sci. A-2, 4, 731 (1966).

[64] H.E. Pickett, M.J.R. Cantow, J.F. Johnston, J. Polym. Sci. C 21, 67 (1968).

[65] J.H. Duerksen, A.E. Hamielec, J. Polym. Sci. C 21, 83 (1968).

[66] G.D. Edwards, J. Appl. Polym. Sci. 9, 3845 (1965).

the molecule [in Ångströms] is calculated, M being given by

$$M = A \cdot Q$$

Q is a parameter which depends on the type of molecule.

It is thus possible to prepare a calibration curve of chain length versus elution volume[67-69] (Fig. 6).

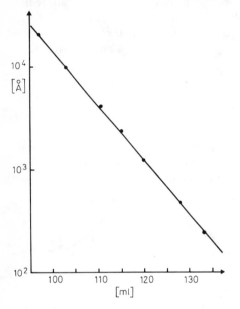

Fig. 6. Calibration curve of chain length vs. elution volume.

The really important parameter in gel chromatography is the hydrodynamic volume of the macromolecule. It has been shown[70, 71] that, if this parameter is plotted against elution volume, a universal calibration curve is obtained.

Detailed studies on the separation of oligomeric compounds have also been published[67-69]. With these low molecular weight polymers, high resolution chromatograms (Fig. 7) can be obtained by use of suitable gels[69].

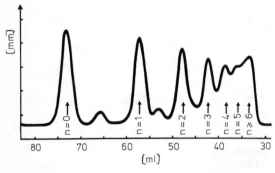

Fig. 7. Gel permeation chromatogram of a solid epoxy resin, based on 2,2-bis-(p-hydroxyphenyl)-propane with an average M.W. of 770, taken with a high resolution column.

3.5.4 Determination of Molecular Weight Distribution

The gel chromatography curve is divided along the abscissa into i segments of equal volume (say 1 ml increments). The heights (ordinates) at each of these volume segments may be taken as a measure of the area under a small segment of the curve. From the calibration curve, it is possible to assign molecular weights (or extended chain lengths, A_i) to every volume V_i. The number average chain length $\overline{A}_n$ and weight average chain length $\overline{A}_w$ of the polymer are then calculated from the equations:

$$\overline{A}_w = \frac{\Sigma\, h_i \cdot A_i}{\Sigma\, h_i} \qquad\qquad \overline{A}_n = \frac{\Sigma\, h_i}{\Sigma\, \dfrac{h_i}{A_i}}$$

where h_i is the height at the ith volume increment and A_i is the corresponding chain length in Angström units.

By multiplying $\overline{A}_n$ and $\overline{A}_w$ by the Q factor for the polymer, the number average and weight average molecular weights may be deduced.

From the same data, it is possible not only to construct the integral distribution curve and the normalized differential distribution curve but also to calculate the Z and Z + 1 molecular weight averages.

3.6 Other Chromatographic Methods

Adsorption chromatography
Precipitation chromatography (Section 2.7:3.1.2)
Distribution chromatography
Thin layer and paper chromatography
Ion exchange chromatography

have met with varying degrees of success for the analysis of macromolecules[36, 37, 41, 54, 72-74]. These chromatographic methods depend for their separating effects not only on the sizes and shapes of molecules, but also on their rheology and on their adsorption behavior on stationary phases in the presence of solvents.

[67] W. Heitz, B. Bömer, H. Ullner, Makromol. Chem. 121, 102 (1969).

[68] J.G. Hendrickson, J. Chromatogr. 32, 543 (1968), Anal. Chem. 40, 49 (1968).

[69] G.D. Edwards, Q.Y. Ng, J. Polym. Sci. C 21, 105 (1968).

[70] Z. Grubisic, P. Rempp, H. Benoit, J. Polym. Sci. B 5, 753 (1967).

[71] J.V. Dawkins, J. Macromol. Sci. (Part B) Phys. 2, 623 (1968).

[72] H.F. Linskens in Ref. [163], p. 97 ff.

[73] M. Florkin, E.H. Stotz, Comprehensive Biochemistry, Vol. IV, Elsevier, Publ. Co., Amsterdam 1962.

[74] E. Lederer, M. Lederer, Chromatography, 2. Ed., Elsevier, Publ. Co., Amsterdam 1957.

Adsorption chromatography (cf. Chapter 2.5) is based on the different molecular migration speeds on polar, stationary phases using eluants of either decreasing or increasing polarity[36, 75]. The adsorption behavior of macromolecules depends on the nature of the solvent and the substrate as well as on the molecular size and the concentration of the polymer to be separated[76]. Only a few reports have been published so far on this topic.

Thin layer[77, 78] *and paper chromatography*[79, 80] (Chapter 2.4) are useful for the identification of polymer mixtures or of additives in polymers. The success of these methods depends to a large extent on finding a highly efficient eluant, and on the available possibilities for rendering separation visible by chemical or physical methods.

Separation by ion exchange[81, 82] (Chapter 2.8) is based primarily on exchange and neutralization processes; adsorption effects due to intermolecular forces are of secondary importance. As ionic reactions are best carried out in aqueous solutions, ion exchangers are used mainly in biochemistry for solving problems of characterization and separation. Unlike low molecular weight compounds, macromolecules can be sufficiently adsorbed onto the ion exchange surface only if the exchangers have correspondingly large specific surface areas. Commercial ion exchangers rarely fulfill these requirements. However, suitable exchangers can be produced by precipitating sulfonated polystyrene on a material with a large specific surface area[82].

Gas chromatography is used for the separation, identification and determination of volatile and other low molecular weight substances present in the polymer[83].

3.7 Diffusion, Dialysis, Ultrafiltration, Electrophoresis

Dialysis[84–86] (Chapter 2.6) is a non-destructive diffusion method which can be used for the separation of low molecular weight materials from high molecular weight substances. In this technique, semi-permeable membranes which are permeable to the solvent and to at least one of the dissolved substances separate the mixture into two fractions according to molecular size.

Non-continuous and continuous apparatus, as well as *extraction dialyzers*, are recommended for this purpose. A new development is *counter-current dialysis*[87–89]. The membranes employed nowadays consist almost exclusively of synthetic materials and the pore size can be chosen to suit any particular problem.

Dialysis is used particularly in biochemical investigations for example, in separation of salts or concentration of solutions.

Non-membrane dialysis has been suggested for the separation of hemoglobin and salts[90] but so far has not achieved great importance.

Electro-dialysis[84, 86, 91] is used for the separation of water-soluble polymers from low molecular electrolytes.

Ultrafiltration[92] (Chapter 2.6) can be used for separation of molecules having a diameter smaller than the pores of the filter from higher molecular weight compounds. However, the separation effect depends not only on the pore size but also on the electrical charge and the affinity of the particles for the membrane wall. Even the fractionation of polydisperse systems[93–95] has been described.

Electrophoresis[96, 97] (Chapter 2.9) has not gained wide acceptance as a technique for the analysis of synthetic macromolecules. It has found greater application for research on natural substances (Chapter 12.2:3).

Isoelectric fractionation or focussing[98–100] is an

[75] *G. Hesse* in Ref. [159], p. 465.

[76] *F. Patat, E. Killmann, C. Schliebener*, Fortschr. Hochpolym.-Forsch. *3*, 332 (1964).

[77] *K. Randerath*, Dünnschichtchromatographie, Verlag Chemie, Weinheim/Bergstr. 1962.

[78] *E. Stahl* in Ref. [163], p. 214.

[79] *F. Cramer*, Papierchromatographie, 4. Aufl., Verlag Chemie, Weinheim/Bergstr. 1958.

[80] *W.J. Langford, D.J. Vanghan*, J. Chromatogr. *2*, 564 (1959).

[81] *F. Helfferich*, Ion Exchange, McGraw Hill Book Co., New York 1962.

[82] *N.K. Boardman* in Ref. [163], p. 190.

[83] *J.G. Cobler* in Ref. [172], p. 1653.

[84] *H.E. Schultze* in Ref. [159], p. 653.

[85] *L.C. Craig, T.P. King* in Ref. [161], p. 175.

[86] *S.B. Tuwiner* in Ref. [171], p. 475.

[87] *L.C. Craig, K. Stewart*, Biochemistry *4*, 2712 (1965).

[88] *J.G. Davis*, Anal. Biochem. *15*, 180 (1966).

[89] *S. Mandeles, E.C. Woods*, Anal. Biochem. *15*, 523 (1966).

[90] *Z. Zasepa, A. Dobry-Duclaus*, J. Chim. Phys. *63*, 675 (1966).

[91] *W.K.W. Chen* in Ref. [166], p. 846.

[92] *W.R. Aehnelt* in Ref. [159], p. 151.

[93] *W.F. Blatt, B.G. Hudson, S.W. Robinson, E.M. Zipilivan*, Nature *216*, 511 (1967).

[94] *W.F. Blatt, S.W. Robinson, F.M. Robbins, C.A. Saravis*, Anal. Biochem. *18*, 81 (1967).

[95] *J.G. Davis, M.B. Wiele*, Anal. Biochem. *20*, 363 (1967).

[96] *W. Grassmann, K. Hannig* in Ref. [159], p. 681.

[97] *J. Conrad, G.F. Sheats* in Ref. [170], p. 738.

[98] *O. Vesterberg, H. Svennson*, Acta Chem. Scand. *20*, 820 (1966).

[99] *T. Flatmark*, Acta Chem. Scand. *20*, 1476 (1966).

[100] *R. Quast, O. Vesterberg*, Acta Chem. Scand. *22*, 1499 (1968).

electrophoretic column separation which uses substrates based on isomeric and homologous poly-amino-polycarboxylic acids. This substrate has a wide isoelectric range (pH 3–10) which makes it possible to separate even high molecular weight polypeptides with a difference of 0.02 pH units between the isoelectric points.

3.8 Distillation, Drying, Thermoanalytical Separation

Low-boiling substances present in precipitated polymers, extracts, eluates, and technical samples are removed by the classical methods of *drying*[101], *distillation*[102], or sublimation. The thermal stability of the macromolecule must be taken into account. In the case of polymer solutions, it is better for practical reasons to give preference to separation by precipitation over concentration (distillation). Precipitation prevents the inclusion[103, 104] of solvents, a fact which cannot be predicted a priori. *Freeze drying* also has become a valuable technique for aqueous[105] and organic[106–108] systems. However, it is possible that the forces occurring during freezing may lead to some degradation of macromolecules[109, 110].

Thermoanalytical separation methods (Chapter 6.4) may also be used for the separation of low molecular weight compounds. *Thermogravimetric analysis* can be employed for quantitative (weight) determination of the low molecular weight portion (cf. 6.4:5).

Most thermobalances are suitable for analysis under high vacuum (10^{-4}–10^{-6} torr) so that pyrolytic decomposition of sensitive substances may largely be avoided. Despite the high vacuum, depolymerization effects and the presence of residual monomer in the distillate cannot be ignored. Thermograms also provide information on processes of evaporation, depolymerization, and decomposition occuring during the experiment.

3.9 Sedimentation and Centrifugation

Macromolecular substances can be separated by *sedimentation*. The greatly increased gravitational field during centrifugation[111] (Chapter 2.1:2) makes it possible to separate not only the usual liquid-solid mixtures but also one-phase solutions of polymer homologs in order of their molecular weights. Extremely high gravitational fields are obtained with *ultra-centrifuges*. The separation effect can be enhanced by adjusting the density of the solution to its optimum or by using a density gradient[112–114]. Besides cane sugar and cesium chloride, the following substances have been found suitable for density gradients: synthetic polyglucose[115], colloidal silicic acid[116], sulfolane, trimethyl phosphate, and urea[117, 118].

Special attention must be given to the removal of the centrifuged substances[111] from the centrifuge cells. Several types of apparatus and techniques are available which help to avoid subsequent remixing in the cells[111, 119–121].

Zonal rotors[114, 122–127] have the advantage that they allow the processing of larger quantities at high speeds of rotation. Separation takes place in a density gradient. The density gradient mixture is introduced into the cell, and the fractions are removed while the rotor is running.

4 Application of Chemical Methods

When separating the components in a mixture of synthetic polymers, it is customary to use physical methods only[128]. In the analytical literature, references concerning chemical conversions with poly-

[101] Ref. [160], p. 16.

[102] *W. Kern, R.C. Schulz* in Ref. [159], Bd. XIV/2, p. 655, 1963.

[103] *H. Staudinger, W. Döhle*, J. Prakt. Chem. *161*, 219 (1942).

[104] *O. Fuchs*, Ber. Bunsenges. Phys. Chem. *60*, 229 (1956).

[105] *R. Jaeckel* in Ref. [159], p. 939.

[106] *F.M. Lewis, P.R. Mayo*, Anal. Chem. *17*, 134 (1945).

[107] *R. Schulz, A. Sabel*, Makromol. Chem. *14*, 115 (1954).

[108] *L. Rey*, Experientia *21*, 241 (1965).

[109] *F. Patat, W. Hägner*, Makromol. Chem. *75*, 85 (1964).

[110] *F. Patat*, Kautschuk Gummi Kunststoffe *20*, 203 (1967).

[111] *H.E. Schulze, E.J.Th. Wiedemann* in Ref. [159], p. 619.

[112] *M.K. Brakke*, J. Am. Chem. Soc. *73*, 1847 (1951).

[113] *P.A. Charlwood*, Brit. Med. Bull. *22*, 121 (1966).

[114] *W. Eichenberger*, Chimia *23*, 85 (1969).

[115] *S.I. Oroszlan, S. Rizvi, T.E. O'Connor, P.T. Mora*, Nature *202*, 780 (1964).

[116] *H. Pertoft*, Biochim. Biophys. Acta *126*, 594 (1966).

[117] *J.H. Parish, J.R.B. Hastings*, Biochim. Biophys. Acta *123*, 202 (1966).

[118] *J.H. Parish, J.R.B. Hastings, K.S. Kirby*, Biochem. J. *99*, 19P (1966).

[119] *T. Oumi, S. Osawa*, Anal. Biochem. *15*, 539 (1966).

[120] *M.L. Randolf, R.R. Ryan*, Science *112*, 528 (1950).

[121] *M. Kalfus, Z. Skupinska*, Polymery *9*, 475 (1964).

[122] *N.G. Anderson*, J. Phys. Chem. *66*, 1984 (1962).

[123] *N.G. Anderson, D.A. Waters, W.D. Fischer, G.B. Cline, C.E. Nunley, L.H. Elrod, C.T. Rankin*, Anal. Biochem. *21*, 235 (1967).

[124] *N.G. Anderson*, Science *154*, 103 (1966).

[125] *H.P. Barringer*, Natl. Cancer Inst. Monograph No. *21*, 77 (1966).

[126] *V.N. Schumaker* in Ref. [168], p. 245.

[127] *N.G. Anderson* in Ref. [161], Vol. 15, p. 271, 1967.

[128] *W. Kupfer*, Z. Anal. Chem. *192*, 219 (1963).

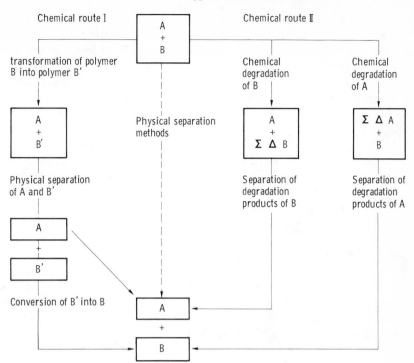

Scheme 2.
Comparison of physical and chemical separation methods

mers deal mainly with identification and only rarely with separation[129].

The principle of chemical separation (cf. Scheme 2) consists in changing at least one component by chemical reaction in such a way that it can be separated out by physical methods on the basis of its newly acquired properties. There are two ways of achieving this:

Formation of complexes, salts, addition compounds or chemically modified macromolecules of identical chain lengths, where the reaction product can be reconverted into the original macromolecule.
Irreversible degradation of a component.

These types of separation require the application of selective reactions and processing methods. Therefore, it is necessary that the components of the mixture or the type of substance present in the mixture are known.

The ability of macromolecules to form complexes must be investigated for each case by a study of the relevant literature. A well-known example is the dissolution of *cellulose* by a series of metal complexes with ammonia, ethylenediamine, biuret or tartaric acid[130, 131].
Salts can, in principle, be obtained from any macromolecule having acidic or basic groups. The solubility

of the salt is a function of the concentration of the salt-forming groups and of the molecular structure.
A typical example of the formation of an *addition compound* is the dissolution of *polyacrolein* in aqueous sodium bisulfite solution[132].

The conversion of macromolecules by chemical means into products with unchanged chain lengths[133] yields products that can be re-converted only with difficulty into their original structures. In principle, all organic reactions can be used provided that the appropriate chemical requirements are fulfilled[134–139].

Separations based on degradation reactions demand the application of selective methods[128] such as solvolysis, ether cleavage, ozonolysis, etc. Thermal degradation methods can be used only in special cases and under closely controlled conditions.

[129] *M. Hofmann, P. Schneider* in Ref. [159], Bd. XIV/2, p. 952, 1963.
[130] *G. Jayme*, Tappi *44*, (4), 299 (1961).
[131] *A.E. Husemann, R. Werner* in Ref. [159], Bd. XIV/2, p. 865, 1963.

[132] *R.C. Schulz*, Kunststoffe *47*, 303 (1957).
[133] *W. Kern, R.C. Schulz*, Angew. Chem. *69*, 153 (1957).
[134] *P. Schneider* in Ref. [159], Bd XIV/2, p. 661, 1963.
[135] *R.C. Schulz*, Angew. Makromol. Chem. *4/5*, 1 (1968).
[136] *D. Braun*, Angew. Makromol. Chem. *4/5*, 91 (1968).
[137] *H. Spoor*, Angew. Makromol. Chem. *4/5*, 142 (1968).
[138] *K. Weissermel, E. Fischer, K.H. Höfner, H. Cherdron*, Angew. Makromol. Chem. *4/5*, 168 (1968).
[139] *H.J. Harwood*, Angew. Makromol. Chem. *4/5*, 279 (1968).

5 Separation of Mixtures
The separation of macromolecules from mixtures depends on the specific properties of the substances in question.

5.1 Separation of Polymers from Reaction Mixtures
For the isolation of products from reaction mixtures[140, 141], it is recommended that an initial rough separation of solvents, starting materials, catalysts, emulsifiers, etc., be carried out prior to further purification. If the polymers cannot be isolated by filtration, the use of suitable solvent-precipitant-combinations may lead to increased purity through repeated precipitation. Attention must be paid to the dissolved portions (oligomers) in the filtrate (cf. Section 2.7:1.2). Drying is not recommended as this will remove only the low molecular weight portions without removing any other trapped products.

Extraction, dialysis, electrodialysis, ultrafiltration and ion exchange may also prove useful.

5.2 Analysis of Plastics
Plastics are either cross-linked (duroplasts) or not cross-linked (thermoplasts) and often contain residual starting material, e.g. monomers or oligomers, as well as catalysts, regulators, etc.

Additives such as plasticizers, stabilizers, lubricants, UV-absorbers, antioxidants, antistatics, emulsifiers, dyes, pigments and fillers are frequently present in plastic materials or fibers. The analysis of soluble plastics and the separation of the polymers from other substances pose little difficulty (for the analysis of cross-linked plastics, cf. Section 2.7:5.4).

Separation scheme: Each case must be treated individually[142]. Solubility tests might indicate an extraction with various solvents and suggest the further course of analysis. The extracts are analyzed for their polymer contents by precipitation tests. Purification or separation of the precipitated macromolecular substances by one or other of the methods described above can then be carried out. All extracts that do not contain polymers, and also the filtrates, are concentrated and subjected to the usual chemical and physical methods of analysis. The insoluble portions can represent inorganic fillers, pigments, or cross-linked poly-

mers (Section 2.7:5.4) which can be characterized by either elemental or spectral analysis.

Selective extraction methods[143] have been described for the separation of additives. Separation methods for various plastics have been described in several papers[144–149] using various methods of *precipitation, column extraction* on suitable substrates, *diffusion* and *electro-dialysis*.

Emulsions must be broken down by coagulation, dialysis, or freezing prior to separation[142].

Gas chromatography[150] is suitable for the separation and simultaneous identification of minute amounts of volatile components in the sample. The product is either injected directly as a solution, or is first subjected to mild pyrolysis in a pyrolysis apparatus[151] coupled to the GLC equipment.

5.3 Separation of Polymer Mixtures
No universal separation methods can be suggested, as differences in solubility behavior may result in a distribution of the components between the phases[152, 153]. The separation of soluble mixtures such as copolymers, co-condensates, co-adducts chemically modified polymers, natural substances, block polymers, graft polymers, and mixtures of various polymer homologs may be accomplished by the methods described above[154, 155]. For separation of unknown mixtures, it is essential to carry out an initial screening by simple preliminary tests, specific identification tests[145, 149, 150, 162, 165, 167, 175], *thin-layer* or *paper chromatography*[80], or by *spectral analysis*[175].

5.4 Cross-linked Polymers
Cross-linked polymeric materials often contain low molecular weight compounds (see Section 2.7:5.2) and sometimes linear or branched poly-

[143] *E. Schröder, E. Hagen,* Plaste Kautschuk *15,* 625 (1968).
[144] *W. Kupfer.* Z. Anal. Chem. *192,* 219 (1963).
[145] *C.A. Lucchesi, J.D. McGuinnes* in Ref. [164], Section 13, p. 210.
[146] Ref. [167], p. 156.
[147] *G.M. Brauer, G.M. Kline* in Ref. [170], Vol. III, p. 634, 1965.
[148] *H. Schweppe,* Paint Technology *27,* (7), 14 (1963).
[149] Ref. [172], Vol II B, p. 2034, 1963.
[150] *J.G. Cobler* in Ref. [172], p. 1579.
[151] *J. Zulaica, G. Guiochon,* Anal. Chem. *35,* 1724 (1963).
[152] *O. Fuchs,* Verhandlungsberichte der Kolloidgesellschaft *18,* 75 (1958).
[153] *H.J. Cantow, O. Fuchs,* Makromol. Chem. *83,* 244 (1965).
[154] *O. Fuchs, W. Schmieder* in Ref. [173], p. 341.
[155] *G.M. Guzmán* in Ref. [169], p. IV–235.

[140] *W. Kern, R.C. Schulz* in Ref. [159], Bd. XIV/1, p. 68, 1961.
[141] *W. Kern, R.C. Schulz* in Ref. [159], Bd. XIV/2, p. 655, 1963.
[142] *G.M. Brauer, E. Horowitz* in Ref. [162], Part III, p. 2.

mers. The investigation, therefore, extends to the separation or isolation of additives from cross-linked insoluble products.

By extraction of finely divided samples, if necessary preceded by swelling, it may be possible to achieve partial separation of the soluble components.

To isolate inorganic components, organic substances are removed as much as possible by chemical degradation or by pyrolysis – if necessary in the presence of oxygen.

Thermogravimetric analyses can yield a great deal of information (Chapter 6.4). Evaporation of additives can be clearly distinguished from degradative reactions. It is often possible to collect the evaporated fractions and to identify them. The thermogram (Fig. 8) of a cured epoxy resin containing plasticizer shows very clearly the evaporation of dibutyl phthalate (first inflection of curve 2) as distinct from the pyrolytic decomposition (second inflection of curve 2)[156]. Using special thermoanalytical measuring devices, it can be shown that, during the entire course of the investigation, no other chemical reaction takes place until pyrolysis begins.

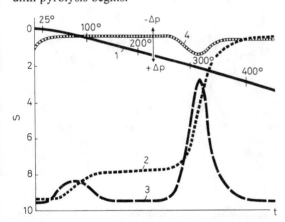

Fig. 8. Thermogravimetric analysis of an epoxy resin plasticized with dibutyl phthalate
curve 1: increase of temperature as function of time. (1.5° per min.)
curve 2: change of weight (1 part of scale = 10 mg)
curve 3: change of weight per time unit
curve 4: recorded pressure (base line: 10^{-2} torr)

---

[156] *H. Batzer, R. Schmid*, unpubl.

6 Bibliography

[157] *P.J. Flory*, Principles of Polymer Chemistry, Cornell University Press, Ithaca, New York 1953.

[158] *H.A. Stuart*, Physik der Hochpolymeren, Bd. I–IV, Springer Verlag, Berlin 1953.

[159] *Houben-Weyl*, Methoden der organischen Chemie, 4. Aufl., Bd. I/1, Georg Thieme Verlag, Stuttgart 1958.

[160] *P.W. Allen*, Techniques of Polymer Characterization, Butterworth Scientific Publications, London 1959.

[161] *D. Glick*, Methods of Biochemical Analysis, Vol. X Interscience Publishers, New York 1962.

[162] *G.M. Kline*, Analytical Chemistry of Polymers, Part III, Interscience Publishers, New York 1962.

[163] *H.F. Linskens, M.V. Tracey*, Moderne Methoden der Pflanzenanalyse, Bd. V, Springer Verlag, Berlin 1962.

[164] *L. Meites*, Handbook of Analytical Chemistry, McGraw-Hill Book Co., New York 1963.

[165] *B. Ke*, Newer Methods of Polymer Characterization, Interscience Publishers, New York 1964.

[166] *Kirk-Othmer*, Encyclopedia of Chemical Technology, 2. Ed., Vol. VII, Interscience Publishers, New York 1965.

[167] *A. Krause, A. Lange*, Kunststoff-Bestimmungsmöglichkeiten, Carl Hanser Verlag, München 1965.

[168] *J.H. Lawrence, J.W. Gofmann*, Advances in Biological and Medical Physics, Vol. XI, Academic Press, New York, London 1967.

[169] *J. Brandrup, E.H. Immergut*, Polymer Handbook, Interscience Publishers, New York 1966.

[170] Encyclopedia of Polymer Science and Technology, Vol. V, Interscience Publishers, New York 1966.

[171] *Snell-Hilton*, Encyclopedia of Industrial Chemical Analysis, Vol. III, Interscience Publishers, New York 1966.

[172] *F.J. Welcher*, Standard Methods of Chemical Analysis 6th Ed., Vol. IIIB, D. van Nostrand Co. Princeton, New Jersey 1966.

[173] *M.J.R. Cantow*, Polymer Fractionation, Academic Press, New York 1967.

[174] *H. Determann*, Gelchromatographie, Springer Verlag, Berlin 1967.

[175] *D. Hummel, F. Scholl*, Atlas der Kunststoff-Analyse, Bd. I, Teile 1 und 2, Carl Hanser Verlag, München, Verlag Chemie, Weinheim/Bergstr. 1968.

2.8 Ion Exchange

Helmut Determann and *Klaus Lampert*
Institut für Organische Chemie der Universität,
D-6 Frankfurt/Main

Because of the particular selectivity forces of the material, separation with ion exchange is used mostly in inorganic chemistry. Organic ions (cf. 2.8:4.1) which form salts with the oppositely charged ions in another phase can be separated also. Occasionally, the properties of the organic grouping outweigh the regular ion behavior (2.8:4.4). The charged group of an organic molecule is not always strong enough to confer water solubility. Almost all ion exchange reactions require aqueous solvents (2.8:4).

Separating mixtures by means of ion exchange has been a widely used method for many years (see 2.8:5). Apart from discussing basic principles (2.8:2), the present article deals exclusively with recent developments in this field, e.g. the increased attention now given to the structure of the charged polymer. Special techniques (2.8:3) and application principles (2.8:4) will be described briefly.

1 Introduction

Similar to gel chromatography, the interaction with the ion exchanger can be predicted largely from simple considerations. However, in the former case, only the size of the molecule need be considered; while, in ion exchange, the pK value of the material to be separated and that of the ion exchanger as well as the pH and ionic strength of the eluant are relevant. The forces that play a role in adsorption chromatography and partition chromatography are even more complex and no more predictable.

2 Ion Exchangers

Most ion exchangers for practical use consist of an insoluble organic polymer into which a charged group has been introduced in some suitable manner (cf. Table 1).

The backbone is usually a styrene-divinylbenzene copolymer. Copolymers of acrylic acid derivatives and divinylbenzene are also frequently employed. For use with biological macromolecules, it is usual to introduce charged groups into cellulose fibers. For several years crosslinked dextran (Sephadex) has also been used as carrier material.

2.1 Functional Groups

Naturally, the polymer framework must be able to carry positively as well as negatively charged groups. The positive groups may be quaternary ammonium ions and, at higher proton concentrations, also lower substituted amines (cf. Table 1). Respectively strongly basic and weakly basic

Table 1. Some Commercially Available ion Exchangers

Trade name[a]	Framework material [functional group]
Cationic exchangers	
$-SO_3H$	
Amberlite IR-120	Styrene/divinyl-
Bio-Rad AG-50W (Xl-16)	benzene copolymers
Dowex 50W (X2-12)	(homoporous)
Duolite C-20 and C-25	
Lewatit S-100	
Wofatit KPS-200	
Zerolit 225	
$-SO_3H$	
Amberlite 200	
Amberlyst 15	(heteroporous)
$-C_2H_4-SO_3H$	
SE Cellulose	cellulose
SE Sephadex C-25 a. C-50	dextran
$-COOH$	
Amberlite IRC-50	Methacrylic acid/
Duolite CS-101	Divinylbenzene
Wofatit CP-300	Copolymers
Zerolit 226	
$-CH_2-COOH$	
CM	Cellulose, fibrous
Whatman CM 22 a. 23	
Whatman CM 32 a. 52	Cellulose, microgranular
CM Sephadex C-25 a. C-50	Dextran
Anionic exchangers	
$-CH_2-N^{\oplus}(CH_3)_3$	
Amberlite IRA-400	*Type I*
Bio-Rad AG 1 (Xl-8)	Styrene/divinyl-
Dowex 1 (Xl-8)	benzene copolymers
Duolite A-42	(homoporous)
Lewatit M-500	
Wofatit L 165	
Zerolit FF-IP	(isoporous)
Amberlite IRA-900 a. 904	(heteroporous)
Amberlyst A 26	
Bio-Rad AG 21 K	
Dowex 21 K	
Duolite A-101 D	
Lewatit MP-500	
$-CH_2-N^{\oplus}-CH_2-CH_2-OH$ $\quad\quad CH_3 \quad CH_3$	
Amberlite IRA-410	*Type II*
Bio-Rad AG 2 (X4 a. 8)	Styrene/divinyl-
Dowex 2 (X4 a. 8)	benzene copolymers
Duolite A-40	(homoporous)
Lewatit M 600	
Wofatit L-150	
Zerolit N-IP	(isoporous)
Amberlite IRA-910 a. 911	(heteroporous)
Amberlyst A-29	

Table 1. cont.

Trade name[a]	Framework material [functional group]
Duolite A-102D Lewatit MP-600	

$$-\overset{|}{\underset{|}{N}}{}^{\oplus}-$$

| QAE Sephadex A-25 a. A-25 | Dextran |

$$-N(R)_2$$

| Amberlite IR-45 Bio-Rad AG 3 X4 Dowex 3 Duolite A-14 Wofatit N Zerolit M-IP | Styrene/divinyl-benzene copolymers (homoporous) (isoporous) |

$$-C_2H_4-N-(C_2H_5)_2$$

Amberlite IRA-93 Amberlyst A-21 Lewatit MP-60	
DE or. DEAE Cellulose Whatman DE 22 a. 23	Cellulose, fibrous
Whatman DE 32 a. 52 DEAE Sephadex A-25 a. A-50	Cellulose, microgranular Dextran

[a] These materials are obtainable in various particle sizes so that not all of them are suitable for analytical purposes.

anionic exchangers are the result. Sulfonic acid and carboxylic acid groups serve as cationic exchange groups.

2.2 Physical Properties

Polystyrene derivatives and acrylic acid derivatives, particularly in bead polymer form, are characterized by mechanical strength combined with uniform particle size and relatively little resistance to flow. For technical purposes a particle size of 0.2–1.5 mm is used, but for laboratory and analytical work particles down to 10 microns are obtainable (for conversion of particle size see Table 2).

Table 2. Particle Size Mesh and mm[a]

mesh	Diameter [mm]
16– 20	1.20–0.85
20– 50	0.85–0.29
50–100	0.29–0.15
100–200	0.15–0.08
200–400	0.08–0.04

[a] The following rough formula also may be used for conversion: $\dfrac{16}{mesh}$ = Diameter of particles [mm].

Naturally, the charged groups are located also inside the resin particles. On contact with water these particles seek to solvate the polar centers in the polymer network. The result is swelling and formation of ion exchange gel. Swelling is the greater the less crosslinked the polymer backbone is and the more charged groups

are attached to it. In these conventional gels the accessibility of the exchange groups is dependent on the molecular size of the oppositely charged ions. For large molecules less crosslinked structures are chosen than for smaller ones. Large pores in these *homoporous gels* can be obtained only at the cost of an appreciable technological disadvantage[1,2]. By using special techniques during polymerization, *heteroporous* structures can be produced in which the pores are preformed, i.e. where it is not swelling that generates them (cf. also Chapter 2.6). Anion exchangers which are not crosslinked by copolymerization with a bifunctional monomer, but only at the substitution stage, are known as *isoporous* and are intended to have a comparatively narrow pore range distribution.

In nonaqueous solvents the homoporous, strongly polar substituted polymers swell only very slightly. For this reason the heteroporous exchangers, also called macroreticular exchangers, are recommended for reactions in organic solvents. The proton catalysis of organic chemical reactions by the insoluble exchanger in acid form is of predominant importance in the organic phase[2].

Naturally, the problem of accessibility of charged centers has a considerable effect on the available *exchange capacity*. The exchange capacity of the materials listed in Table 1 is normally between 2 and 8 meq/g. When comparing the exchangers, it is necessary to consider, wether the values of the capacity are referred to dry material, swollen gel, or bed volume in ml. Due to the spatial ratios, only a certain fraction of the analytically determined nominal capacity is available, which is the smaller, the greater the charged molecules. During ion exchange of macromolecules of biological origin, which because of their sensitive tertiary structure must be separated on hydrophilic polymers such as cellulose or dextran, the difference between the nominal and effective capacity is particularly important. The Sephadex exchangers are the ones most closely related to the conventional resin exchangers having the charged ion groups dispersed in the gel and accessible in accord with the relative size of the molecule and the degree of swelling. These materials swell considerably at low ionic strengths and this specific property restricts their field of application.

The conventional cellulose exchangers occupy an intermediate position, but suffer from the added disadvantage of being composed of fibrous particles with adverse flow-resistance properties. Cellulose exchangers of this type are often made by the

[1] *J.R. Millar, D.G. Smith, W.E. Marr, T.R.E. Kressman*, J. Chem. Soc. 218 (1963).

[2] *J. Seidl, J. Malinsky, K. Dusek, W. Heitz*, Advan. Polym. Sci. *5*, 113 (1967).

user[3, 12]. Removal of the amorphous regions from the cellulose leaves *microgranular exchangers* in which the charged groups are attached exclusively to the surface of the cellulose crystallites.

Recently a method has been described for making cellulose particles in bead form[4] which have been employed as *heteroporous ion exchangers*[5]. Both microgranular and heteroporous cellulose particles provide excellent accessibility of their ion center to the macromolecules, i.e. the nominal and effective capacities are practically identical.

3 Special Techniques

Basically, an ion exchanger represents an insoluble acid (or its salt) or an insoluble amine (or ammonium compound). The former type exchanges cations as ions of opposite charge, the latter distinguish between the different anions.

3.1 Regeneration

Anion exchangers are usually supplied in the form of salts; amines, in particular, are stable only in this form. *Cation exchangers* can be converted into the $H^\oplus$ form by treatment with aqueous acid followed by washing. The ammonium base containing hydroxyl groups is obtained from the strongly basic anion exchanger on treatment with sodium hydroxide, while the weakly basic anion exchangers are converted into free amines. Whatever the form is in which the resin is to be used, it should always be submitted to the full cycle of charging and discharging before use. Carrying out this step enhances its effective capacity.

If, for instance, sulfonic resin is supplied in the sodium salt form, and is also to be used as such, then it is advisable to convert it first into the $H^\oplus$ form with acid and then into a more active sodium salt form by treatment with excess NaOH. This regeneration treatment usually leads to a purification of the material which is beneficial even with 'analytically pure' substances.

The operation is generally carried out in a beaker with the aid of a sintered glass filter plate. The same procedure is employed for batch application. For simple separations the ion-exchange resin can be introduced into the solution first to bind the desired components and the charged resin can then be separated mechanically by filtration.

[3] *E.A. Peterson, H.A. Sober*, J. Am. Chem. Soc. *76*, 1711 (1954), *78*, 751 (1956).

[4] *H. Determann, H. Rehner, Th. Wieland, N. Meyer, F. Wente*, Makromol. Chem. *114*, 263 (1968).

[5] *H. Determann, N. Meyer, Th. Wieland*, Nature *223*, 499 (1969).

3.2 Exchange Chromatography

The ion exchanger can be used as a *filter bed* for removing a component from a percolating liquid under suitable conditions. Such filter devices are used for chromatographic purposes by filling a long tube with the ion exchange bed. The operation then need not be limited to the removal of one or more components from the mixture by reaction with the exchanger; instead, the conditions are selected such that all the components of interest are retained in the upper part of the column. By altering the external conditions the retained ions can be eluted from the column individually in the ideal case.

The eluant is changed either stepwise or continuously. With stepwise elution often only a coarse fractionation is obtained; but this has the advantage that the entire substance is recovered in relatively few fractions. Elution with an eluant of continuously changing composition gives a very much finer separation (for *gradients* cf. Fig. 1).

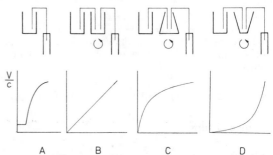

A   B   C   D

Fig. 1. Graphical Representation of the three Possible Gradients (B, C.D) Together with the volume/Concentration Pattern of Each.

In the case of A a gradient occurs merely on account of the buffering effect of the resin filling.

In addition to linear, convex and concave gradients (Fig. 1, Cases B, C and D) a gradient (Case A) is formed in the exchange column due to the buffering effect of the resin. Frequent equilibria are established along the gel packing, resulting in a high degree of separation due to the alternating retention and elution that often suffices for effective separation (2.8:4.1).

The special structure of an ion exchanger (charged groups built into a more or less rigid network) causes the charged groups to repel one another the more, the less is the ionic strength of the swelling medium. This repulsion often causes considerable *swelling*, which is the greater the less the polymer chains are crosslinked, the more groups the ion exchanger contains, with the same charge and the smaller the concentration of counter ions in the surrounding medium. During *regeneration*, washing with distilled water is necessary.

Because of the swelling, it is normally impossible to carry out this operation in the ion exchange column, especially when the crosslinking is comparatively slight. Ion exchange systems based on polysaccharides (Sephadex and conventional cellulose) must be regenerated in a beaker.

4 Selection of Suitable Systems

It is usual to classify ion exchangers according to the pK values of their ion groups (cf. Table 1). Table 3 gives a simplified survey of their working ranges.

Table 3. Operating Range of Commonly used Ion Exchangers

Specification of ion exchanger	Ionic group	Suitable chromatographic medium
Strong acid	$-SO_3^{\ominus}$	acidic and alkaline
Weak acid	$-COO^{\ominus}$	only alkaline
Strong base	$-N^{\oplus}R_3$	acidic and alkaline
Weak base	$-N^{\oplus}HR_2$	only acidic

A more sophisticated idea of the activity of an ion exchanger expressed as a function of the pH of the surrounding solution can be obtained by studying the titration curves for four basic types (Fig. 2).

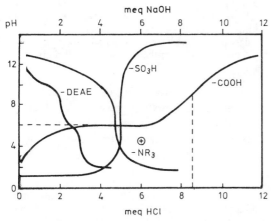

Fig. 2. Titration Curves for Exchange resins. Strongly acidic (—SO₃H), weakly acidic (—COOH), strongly basic (—N⁺R₃), weakly basic (—DEAE). Abscissae: meq NaOH per g cation exchange resin and meq HCl per g anionic exchange resins.

Considering the variation of pH with the addition of hydroxyl ions representative of a carboxylic acid resin, it is seen clearly that the pH initially increases because of the weakly acidic (undissociated) COOH groups. On further addition of alkali and increasing dissociation of the COOH groups, the pH remains substantially constant until, when the capacity limit is reached, it once more increases. The first point of inflection along the horizontal portion of the curve corresponds to the pK of the carboxyl groups (about pH 6), while the second point of inflection projected on the abscissa indicates the exchange capacity (8.5 meq/g). An essential re-

quirement for ion exchange to take place is the presence of charged groups. With carboxylic acid resins these appear initially at about pH 5, while in the case of sulfonic acid resins they begin to form above pH 1.

The strongly basic quaternary ammonium ion exchangers are a direct mirror-image of the sulfonic acid resins, i.e. their working range is below pH 12.5. Two points of inflection (at pH 9.5 and 5.7) are observed in the case of the weakly basic diethylaminoethyl (DEAE) polysaccharides, and the useful exchange range lies between these pH values.

For practical use the dissolved material must be present also in the dissociated form if it is to be retained by the resin. It is most convenient to work at a pH where one of the components will be retained while the other does not interact with the ion exchanger.

A strongly acidic ion exchanger in the $H^{\oplus}$ form can be stirred into a combined aqueous solution of acid and base at high proton concentrations and the residual aqueous acid filtered off. The base is retained by the resin, and can be recovered by eluting with a high concentration of another cation (e.g. $Na^{\oplus}$). If, on the other hand, a weak acidic ion exchanger is to be used, then the mixture must be present in the neutral or alkaline form in order to achieve separation. Elution of the base can then be carried out with aqueous acid (by displacement with protons). The separations described can be carried out in the same fundamental manner with a strongly basic resin (in the alkaline region) and with a weakly basic resin (in the neutral to acidic region). For simple separations widely variable conditions are, therefore, available.

The concentration of the participating ions of opposite charge also plays an important role. The retained ions can be eluted with the aid of a high concentration of cations or anions without changing the pH and, hence, the dissociation relationships. Moreover, there are considerable affinity differences between a given resin and different types of ion. This behavior is called the ion *exchange selectivity*. From a mixture only either cations or anions can be separated individually. However, in water demineralization the cation and anion exchangers are mixed in equivalent amounts and this *mixed bed* is able to separate out both cations and anions in one passage. Regeneration has been solved in an elegant manner by making one resin of lower specific gravity than the other. Flotation can then be used to separate them and each resin is given its specific regeneration treatment.

Greater difficulties are presented in selecting a suitable system for separating a mixture of weak acids (e.g. proteins) on a weakly basic ion exchanger (e.g. DEAE cellulose). Weak acids are dissociated more at higher pH of the solution,

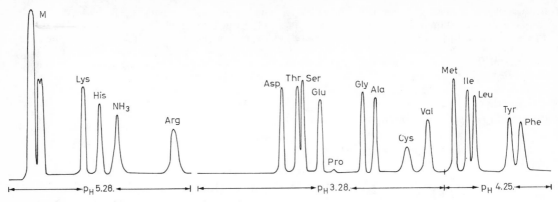

Fig. 3. Mixed Amino Acids in an Amino Acid Analyzer Chromatographed on Special Polystyrene Sulfonic Acid resin.
Ordinate: Extinction of ninhydrin color. Abscissa: Elution volumes. Volume at starting point: 0.5 ml; amount at starting point: $\sim 0.25\ \mu Mol$

pH	5.28	3.28	4.26
Sodium citrate and other component	0.35 N	0.2 N	0.38 N
Amino acids	basic	neutral and acidic	
Column [cm]	6×0.9	69×0.9	

Lys = lysine, His = histidine, NH₃ = ammonia, Arg = arginine, Asp = aspartic acid, Thr = threonine, Ser = serine, Glu = glutamic acid, Pro = proline, Gly = glycine, Ala = alanine, Cys = cystine, Val = valine, Met = methionine, Ile = isoleucine, Leu = leucine, Tyr = tyrosine, Phe = phenylalanine, M = mixture of unseparated amino acids.

while the dissociation of the ion exchanger (and hence its force of retention) is greater if the surrounding medium is more acidic. In order to achieve retention a compromise between these conflicting requirements is necessary; suitably, a pH of 7–8 and a buffer of low ionic strength are chosen. The elution can then be carried out at increasing pH or increasing anion concentration at constant pH. In this very important separation of proteins on DEAE cellulose or DEAE Sephadex, elution is usually performed with a gradient of increasing chloride concentration in order to make maximum use of the very slight pK differences of the protein material.

4.1 Amino Acid Analysis

The complex mixture of 18 *amino acids* obtained by the acid hydrolysis of *proteins* can be split into its components by ion exchange chromatography and determined qualitatively.

There is hardly a method of separation in the entire field of liquid chromatography that has received as much attention as the separation of amino acids on sulfonated polystyrene resins[6]. Complete *amino acid analyzers* which will determine the entire amino acid composition in as little as 0.05–1 mg protein in a few hours are becoming available. The heart of this apparatus is a set of small chromatography tubes filled

with specially prepared resin particles. These particles are complete spheres and have a very small diameter (8–40 μ) and a very narrow size distribution. The resulting uniformity of packing allows a high flow rate at high pressures despite the small interstices. Very rapid establishment of the diffusion equilibrium on the small beads is obtained.

Figure 3 shows a recording of a fully automatic analysis of a test mixture consisting of all amino acids in molar ratios. The area under the peaks is a measure of the amount of material used which can be employed for calibration purposes.

The mixture is first introduced into the very short column at pH 2 and eluted with 0.35 N sodium citrate buffer of pH 5.28, with which the column has been previously equilibrated. Acidic and neutral amino acids at first leave the column unseparated, as is shown by the first peaks (M). They are followed by lysine, histidine, NH₃ and arginine. A second sample is chromatographed in a longer column with 0.2 N sodium citrate buffer at pH 3.28 and subsequently eluted with 0.38 N citrate buffer at pH 4.25.

4.2 Influence on Exchange Chromatography

The order in which the amphoteric ions leave the column does not correspond directly to the pK value sequence. This method thus clearly illustrates that factors other than ionic forces affect the retention. *Hydrophilic ions,* in particular, leave the column at an early stage, whereas *lipophilic ions* do not appear until relatively late. To speed up the process the temperature is increased to 50°. The

[6] *D.H. Spackmann, W.H. Stein, S. Moore,* Anal. Chem. *30,* 1190 (1958).

success of this step demonstrates that retention is influenced also by adsorption forces.

The efficiency of this variously modified procedure is revealed by the fact that up to 50 constituents in physiological fluids can be successfully separated qualitatively and determined quantitatively.

5 Bibliography

[7] *R. Griessbach, G. Naumann*, Ionenaustauscher in *Houben-Weyl*, Methoden der organischen Chemie, Bd. I/1, p. 521, Georg Thieme Verlag, Stuttgart 1958.

[8] *R. Kunin*, Ion Exchange Resins, Wiley Intersci. Publ., New York 1958.

[9] *F. Helfferich*, Ionenaustauscher, Bd. 1, Verlag Chemie, Weinheim/Bergstr. 1959.

[10] *K. Dorfner*, Ionenaustauscher, 2. Aufl., Walter de Gruyter & Co., Berlin 1964.

[11] *J. Inczédy*, Analytische Anwendung von Ionenaustauschern, Verlag der Ungarischen Akademie der Wissenschaften, Budapest 1964.

[12] *M. Rybák, Z. Brada, J.M. Hais*, Säulenchromatographie an Cellulose-Ionenaustauschern, VEB Gustav Fischer Verlag, Jena 1966.

[13] *H.F. Walton*, Principles of Ion Exchange and Techniques and Applications of Ion Exchange Chromatography, in *E. Heftmann*, Chromatography, 2nd Ed., van Norstrand-Reinhold, New York 1967.

[14] *F.C. Saville*, Ion Exchangers, in *C.L. Wilson, D.W. Wilson*, Comprehensive Analytical Chemistry IIb, Elsevier Publ. Co., Amsterdam, London, New York 1968.

[15] *F.J. Wolf*, Separation Methods in Organic Chemistry and Biochemistry, Academic Press, New York, London 1969.

2.9 Electrophoresis

Günter Legler
Institut für Biochemie der Universität
D-5 Köln

1 Theory and General Considerations

The basis for electrophoretic separations is a difference in the migration rate of electrically charged particles in an electric field. This rate is proportional to the field strength and the effective charge of the particles and inversely proportional to the frictional drag depending on their size and shape. The size of the particles to which electrophoretic techniques may be applied ranges from elementary and low molecular weight ions to proteins and even viruses, cell organelles, or whole cells.

In the literature one finds the terms 'electrophoresis' (mostly for high molecular weight compounds and colloidal systems) and 'ionophoresis' (for elementary and low molecular weight ions). There is no basic difference between them.

The theory of electrophoresis has been treated in detail in several recent reviews and monographs[1-3].

Individual ion mobilities which govern the separation problems can be determined with sufficient accuracy only by actual measurements because they are influenced by the degree of solvation, association with other ions, and the formation of complexes. For these reasons mathematical derivations will be omitted, and some general points important for practical performance will be discussed.

Principally, all electrophoretic separations are performed in solutions which contain, in addition to the ionic species to be separated and their counter ions, other positive and negative ions. These are called carrier electrolytes.

If these were absent, the oppositely charged ions would migrate in the electric field only until the field created by the separation of charges equals the external field and cancels it. Furthermore, the field strength at a given voltage is inversely proportional to the conductivity, that is, in the absence of carrier electrolytes the field strength at the site of the ions to be separated and, therefore, their *rate of migration* would be extremely small. For this reason the conductivity of the solution to be separated should not be much larger than that of the carrier electrolytes.

The type of carrier electrolyte used depends on the separation problem. As organic compounds separable by electrophoresis are usually weak acids or bases, a *buffer* is used as the carrier electrolyte. Its pH-value should be close to the pK_s-values of the compounds to be separated. This causes differences in the dissociation constants and, therefore, in

[1] *W. Preetz*, Fortschr. Chem. Forsch. *11*, 375 (1969).

[2] *C.J.U.R. Morris, P. Morris*, Separation Methods in Biochemistry, Pitman & Sons, London 1962.

[3] *K. Dose* in Ullmanns Enzyklopädie der techn. Chemie, Bd. II/1, p. 166, Urban & Schwarzenberg, München 1961.

the degree of ionization to appear as differences in the effective charge, thus facilitating the separation.

The anion of an acid with the dissociation constant K_s and the anion mobility U_A has an effective mobility

$$U_{eff} = \frac{U_A \cdot K_s}{[H^+] + K_s}$$

The differences between individual U_A-values are usually much smaller than those between K_s-values. The selection of the optimal pH of the carrier electrolyte is, therefore, of great importance. With ampholytes (amino acids, peptides, proteins) variation of the pH-value may provide additional possibilities for separation as not only the rate but also the direction of migration may be influenced. Even some non-dissociating substances may be separated electrophoretically (e.g. the separation of *sugars* and *polyalcohols* in the presence of borate or molybdate[4, 5]) by using *complexing agents* as carrier electrolytes.

Another point to consider with preparative separations is the possibility of removing the carrier electrolyte from the separated substances. There is no difficulty with high molecular weight compounds where this can be achieved by dialysis or gel filtration. With materials of low molecular weight one must use volatile buffers. Sometimes it is possible to extract the separated substances with organic solvents after a suitable adjustment of the pH-value.

With separations on a preparative scale the concentration of the carrier electrolyte is governed by two requirements between which a compromise must be made. For separation of large amounts, the reasons mentioned previously make it desirable to use as high a concentration as possible so that the conductivity of the carrier electrolyte is near to that of the sample. The voltage, too, should be as high as possible to reduce the total time of the run (reduction of zone spreading due to diffusion). The increase in the heat generated during electrophoresis is proportional to the conductivity and the square of the voltage. Therefore, an increase of both variables is limited by the cooling capacity of the apparatus (cf. 2.9:3.2.3).

Very often electrophoresis is not carried out in free solution but the electrolyte is distributed in the interstices of a porous supporting medium. This is done in order to stabilize the solution against disturbances by convection caused by local differences in temperature and concentration. The following additional problems arise with this procedure *(carrier electrophoresis)*:

Endosmosis: Usually there is a difference in potential between the electrolyte and the carrier (electrokinetic or zeta potential) which depends on the dielectric constant of the carrier. In an electric field this gives rise to a movement of liquid in the pores and capillaries of the carrier which is directed towards the cathode with aqueous solutions. If necessary its influence can be estimated by measuring the rate of migration of neutral molecules (glucose, dextran, *o*-nitroaniline, thiourea, etc.).

Effect of suction: With paper electrophoresis a capillary movement of liquid from the electrode space to the middle of the carrier may arise if the evaporation of water caused by the electric heat is not prevented.

Chromatographic effect: If there are adsorptive interactions between the supporting medium and the ions to be separated (overlapping with adsorption chromatography), the migration rate of these ions will be decreased. A nonlinear adsorption isotherm may cause tailing. Similar retardation effects (which may be desirable with certain separation problems as they may affect different ions to a different degree) occur when organic solvents are added to the electrolyte (superimposition of *partition chromatography*) or when the supporting medium contains suitable ionized groups *(ion exchange chromatography)*.

Influence of pore size: The ions are not free to move in the direction of the electric field but must follow the bends and turns of the interspaces resulting in a decreased migration rate compared to free electrophoresis. An additional rate decrease will occur if the particle size becomes comparable to that of the pores and capillaries.

The supporting medium will usually not have a uniform pore size but a more or less wide distribution of pore diameters. Thus with increasing size of the particles an increasing part of the pores will become impermeable to them. A practical application of this effect of a separation according to particle size consists in the use of gels from starch or acrylamide as supporting medium which completely fill the separation space (see 2.9:2.2.4).

2 Analytical Separations

2.1 Carrier-Free Electrophoresis[5a]

The technique developed by Tiselius is the oldest method of carrier-free electrophoresis. The mixture to be separated, e.g. different proteins in the anionic state, is placed with the carrier electrolyte

[4] *A.B. Foster*, Advan. Carbohydr. Chem. *12*, 81 (1957).

[5] *E.J. Bourne, D.H. Hutson, H. Weigel*, J. Chem. Soc. 4252 (1960).

in a U-shaped tube. A layer of pure carrier electrolyte is spread on top so as to form a sharp boundary. At the beginning there is a single sharp gradient in the refractive index at this boundary which can be made visible by a schlieren optical system or interferometric technique. Both techniques permit a quantitative evaluation with respect to concentration (for details see Antweiler[6]). The migration of *protein* molecules in the electric field can be followed by the movement of the boundary between solution and pure solvent. If several components of different mobility are present, several boundaries will appear after a sufficient migration. The method is limited to macromolecular ions because only with these the leveling of the gradient at the boundaries by diffusion is sufficiently small to cause no impairment of resolution.

A quantitative determination of the separated compounds is possible by planimetry of the schlieren diagrams, by manual interferometry, or by automatic integration of the light deflection in a Töpler schlieren system[7].

The method has the advantage that there are no disturbances by adsorption on the supporting medium and that systematic errors with all methods based on differences in refractive index are less than 3% even without preliminary calibration. The methods using staining or UV-absorption require a calibration with the pure compound for similar accuracy.

The disadvantages of the method are the relatively large expense of equipment and time, and the necessity of using large amounts of material (50–200 mg). However, Antweiler[8] has described a micro scale apparatus for *proteins* (1–5 mg) which can be analyzed in 20 min. using an automatic scanning and integration device[7]. Another disadvantage is that a complete separation of the components is possible only if a density gradient (mostly with sucrose) is established in the separation tube. The formation of unstable zones can be suppressed if the gradient is sufficiently steep. Otherwise, only those portions of the compounds with fastest and slowest migration rates can be removed for further chemical or serological investigations. The method was first developed for preparative separations (see 2.9:3.1.1).

An apparatus for the automatic analysis of protein mixtures (in the range of 1–5 mg) using UV-absorption for quantitative evaluation has been described by Skeggs and Hochstrasser[9].

2.2 Electrophoresis on Carriers

2.2.1 Electrophoresis on Paper and Cellulose Acetate

Paper electrophoresis[10] has found widespread use because of the simplicity of the equipment needed. At low field strength (below 10 V/cm) no external cooling is needed. A paper strip soaked in the carrier electrolyte is stretched level or roof-like between the two electrode vessels in a moist chamber. The mixture to be separated (0.1–1 mg) is applied in the form of a narrow strip in the center of the paper or at one side if all the components migrate in one direction. After an electrophoresis time of 8–20 hours the substances are located by staining. Suitable staining procedures[2, 10] for *proteins, glycoproteins, lipoproteins, acid polysaccharides, nucleic acids*, and *peptides* sometimes allow detection of different classes of compounds on the same strip or on strips that have been run in parallel. Enzymes may be localized by reaction with chromogenic substrates[11].

A quantitative evaluation is possible by elution of the stained zones or by direct *colorimetry* after the paper has been made transparent. However, reliable results are obtained only after calibration with pure isolated compounds because different proteins may be stained with widely differing intensity by a given method.

A survey for the separation of low molecular weight compounds by paper electrophoresis is given by Thornburg et al.[12]. More than 200 *carboxylic acids, amines, amino acids, peptides, carbohydrates*, and *phosphoric esters* are listed. The authors added 30% formamide to the carrier electrolyte to reduce the evaporation losses at 10 V/cm.

Instead of paper, partially acetylated cellulose is often used as supporting medium. This material shows reduced adsorptive power and greater homogeneity and therefore gives sharper zones and better reproducible results in quantitative analysis.

[5a] *S. Hjertén*, Methods Biochem. Anal. *18*, 55 (1970).
[6] *H.J. Antweiler, Houben-Weyl*, Methoden der organischen Chemie, 4. Aufl., Bd. III/2, p. 211, Georg Thieme Verlag, Stuttgart 1955.
[7] *H.J. Antweiler*, Z. Anal. Chem. *243*, 511 (1968).
[8] *H.J. Antweiler*, Kolloid-Z. *115*, 130 (1949).

[9] *L.T. Skeggs, H. Hochstrasser*, Ann. N. Y. Acad. Sci., *102*, 144 (1962).
[10] *Ch. Wunderly*, Die Papierelektrophorese, 2. Aufl., H.R. Sauerländer & Co., Aarau, Frankfurt/M. 1959.
[11] *H.H.F. Laurell* in Ref.[58], p. 21.
[12] *W.W. Thornburg, L.N. Werum, H.T. Gordon*, J. Chromatogr. *6*, 131 (1961).

In a *micro method* ($<10^{-8}$ g) threads from regenerated cellulose have been described as support for the separation of *nucleotides*[13] and *proteins*[14].

With low molecular weight compounds, blurring of the zones due to diffusion must be reduced. This can be done using shorter electrophoresis times at correspondingly higher field strengths (20—100 V/cm). The heat generated during electrophoresis can be dissipated by cooling the support with a liquid not miscible with water[15] or by cooled plates of glass[16, 17].

The electrophoresis of stronlgy lipophilic substances, e.g. *phosphatides*, presents a special problem. For these substances alcoholic buffers with low water content are recommended[18].

The separation possibilities with complicated mixtures can be increased if unidimensional electrophoresis is combined with *chromatography* at right angles to the direction of electrophoresis. After staining, a two dimensional pattern of spots is obtained which can be used for the characterization of the mixture. With suitable controls the identification of individual compounds is often possible as is their isolation on a micropreparative scale. The main field of application is the investigation of *polypeptide*[19] and *oligonucleotide* mixtures[20].

2.2.2 Thin Layer Electrophoresis

Electrophoresis using thin layers[21] of silica gel, kieselguhr, aluminum oxide, and cellulose powder has the same advantages compared to paper electrophoresis as the corresponding chromatographic techniques, i.e. better separation due to smaller spot size and shorter running time. Using the above-mentioned supporting media, thin layer electrophoresis has been applied to the separation of low molecular weight compounds: e.g. *amino acids* and their derivatives, *peptides*, *phenols* and *phenolic carboxylic acids, dyes, nucleotides*.

As with paper electrophoresis a two dimensional technique combined with chromatography improves considerably the separation of a mixture on the thin layer.

For high molecular weight compounds dextran gels with different degrees of crosslinking have been used instead of cellulose powder[22]. According to Radola[23] these are also suitable for electrophoresis in a stable pH-gradient (see 2.9:3.2.2, isoelectric focussing).

2.2.3 Gel Electrophoresis

With high molecular weight compounds electrophoresis using gels from starch, agar, and polyacryl amide has the advantage of an additional parameter, i.e. the molecular size, which gives further possibilities of separation. According to the pore size of the gels the migration of the larger molecules is impeded more strongly than that of the smaller ones. Because diffusion is impeded too, sharper zones and better resolutions are obtained. As with thin layer electrophoresis the amount of work is somewhat higher than with electrophoresis on paper.

The experimental technique is similar to that for thin layer electrophoresis. The gels are poured in the shape of thin layers or flat blocks. The carrier electrolyte is directly incorporated in the gel medium. The substances to be separated are applied in slits which have been cut into the gels or were formed using suitable spacers. Depending on the amount of material to be separated, they are added directly, spotted on small pieces of filter paper, or mixed with starch or dextran gel. With *starch* gel the amount of material needed is somewhat larger than with paper electrophoresis. With gels from *agar* and *acrylamide* the amount can be considerably smaller ($1-100\mu$ g).

The method has first been described by Smithies[24] with gels from partially hydrolyzed starch which is now commercially available.

The use of agar gels[25] offers the advantage that it is not necessary to make the gels transparent for quantitative evaluations. Furthermore, they are prepared more easily and with smaller thickness.

For the characterization of *protein* mixtures the scope of gel electrophoresis can be expanded by *immuno electrophoresis* according to Grabar and

[13] *J.D. Edstrom*, Biochem. Biophys. Acta *22*, 378 (1956).

[14] *H.G. Noller*, Klin. Wochenschr. *32*, 988 (1964).

[15] *H. Michl*, Monatsh. Chem. *82*, 489 (1951).

[16] *Th. Wieland, G. Pfleiderer*, Angew. Chem. *69*, 257 (1957).

[17] *H. Klamberg, W. Krey, H. Saran*, Z. Anal. Chem. *251*, 365 (1966).

[18] *H. Zipper, M.D. Glantz*, J. Biol. Chem. *230*, 621 (1958).

[19] *J.C. Bennet*, Ref. [58], Vol. XI, p. 330 (1967).

[20] *M. Laskowski jr.* in Ref. [58], Vol. XII, p. 296 (1967).

[21] *K. Hannig, G. Pascher* in *E. Stahl*, Dünnschichtchromatographie, 2. Aufl., Springer Verlag, Berlin, Heidelberg 1967.

[22] *K. Dose, G. Krause*, Naturwissenschaften, *49*, 349 (1962).

[23] *B.J. Radola*, Biochim. Biophys. Acta, *194*, 335 (1969).

[24] *O. Smithies*, Advan. Prot. Chem. *14*, 65 (1959).

[25] *R.J. Wieme*, Agar Gel Elektrophoresis, Elsevier, Amsterdam, London, New York 1965.

Williams[26,60]. After electrophoretic separation an antiserum against the protein mixture is put into slits cut parallel to the direction of electrophoresis. Diffusion of the proteins against the antibodies results in the formation of precipitation zones which allow additional conclusions on the uniformity of the individual zones.

Compared to the above mentioned gel forming media there are some advantages to using polyacrylamide crosslinked with methylenebisacrylamide introduced by Raymond and Weintraub[27]: less adsorption and endosmotic flow, better definition of the synthetic material than the natural gel forming media, controlled pore size[28].

Gels with very large pore size have also been used for the separation of *nucleic acids*[29,30].

With polyarcrylamide as carrier the method of isoelectric focussing (see 2.9:3.2.2) has been extended to analytical problems on a micro scale[31]. The separation of the *protein*-mixture (0.03–0.3 mg) is carried out in a cylindrical gel at a pH-gradient which causes individual proteins to concentrate according to their isoelectric points.

2.2.4 Disc-Electrophoresis

This type of electrophoresis[32] has been developed by Ornstein and Davis[33]. The name is derived from a discontinuity in the carrier electrolyte and the pore size of the gel. This causes the substances being separated to concentrate in a narrow disc (only 50μ wide) before actual separation starts. The degree of resolution obtained by this concentration effect is not reached by any other type of electrophoresis.

The electrophoresis is carried out in glass tubes about 5×70 mm. In these a large-pore concentration-gel (spacer gel) and a small-pore separation-gel (running gel) are formed. pH and mobility of the buffer ions are adjusted so as to cause the above mentioned concentration effect at the interface between the concentration- and the separation-gel. About $10-100 \mu g$ *protein* are needed. In a

micro modification devised by Hyden and Lange[34] less than $1 \mu g$ of protein mixture is used.

A further increase in the resolving power can be achieved by the use of gels with a gradient in pore size[35].

In addition to the separation of proteins, disc-electrophoresis has been applied in the separation of *nucleic acids*[36]. Gels with different pore size, prepared in series in the same tube, have been used.

As with other types of electrophoresis in supporting media, *enzymes* can be localized by reaction with chromogenic substrates. The enzyme zones usually appear wider than the corresponding protein zones because the low molecular weight reaction products show greater diffusion. For enzymes reacting on high molecular weight substrates *(amylases, phosphorylases, pectinases)* Siepmann and Stegemann[37] have developed a procedure based on the incorporation of these substrates into the gels. The electrophoresis is carried out under conditions (low temperature, unfavorable pH) where the activity of the enzymes is rather negligible. After electrophoretic separation the enzymatic reaction is allowed to proceed by an increase of temperature and adjustment of pH. The enzyme zones are then revealed by product specific stains.

The quantitative evaluation of disc-electropherograms is usually done by densitometric determination of dyes adsorbed by the proteins.

According to Weber and Osborne[38] disc-electrophoresis can be used for the estimation of the molecular weights of proteins. The electrophoresis is carried out in the presence of sodium dodecyl-sulfate or urea which denatures the proteins. The procedure is calibrated by running proteins of known molecular weights in parallel.

2.2.5 Isotachophoresis

Under the conditions of isotachophoresis[38a] *all* ions with the same charge migrate in the final state in separate, consecutive zones with the same speed

[26] *P. Grabar, C.A. Williams,* Biochim. Biophys. Acta *17,* 67 (1955).

[27] *S. Raymond, L. Weintraub,* Science *130,* 711 (1959).

[28] *M.L. White,* J. Phys. Chem. *64,* 1563 (1960).

[29] *U.E. Loenig,* Biochem. J., *102,* 251 (1967).

[30] *D.H. Bishop, J.R. Claybrook, S. Spiegelman,* J. Mol. Biol. *26,* 373 (1967).

[31] *C.W. Wrigley,* J. Chromatogr. *36,* 362 (1968).

[32] *H.R. Maurer,* Disk-Elektrophorese, Walter de Gruyter & Co., Berlin 1968

[33] *L. Ornstein, B.J. Davis,* Ann. N. Y. Acad. Sci. *221,* 321, 404 (1964).

[34] *H. Hyden, P.W. Lange,* J. Chromatogr. *35,* 336 (1968).

[35] *J. Margolis, K.G. Kenrick,* Nature *214,* 1334 (1967).

[36] *E.G. Richards, J.A. Coll, W.B. Gratzer,* Anal. Biochem. *12,* 452 (1965).

[37] *R. Siepmann, H. Stegemann,* Naturwissenschaften *54,* 116 (1967); *H. Stegemann,* Z. Physiol. Chem. *348,* 951 (1967).

[38] *K. Weber, M. Osborn,* J. Biol. Chem. *244,* 4406 (1969).

[38a] *H. Haglund,* Sci. Tools *17,* 1 (1970).

toward the anode or cathode respectively. The sequence of the ions corresponds to their mobility; the size of the zones is determined by the amounts of the individual ions. Dilute sample solutions are concentrated during isotachophoresis. According to Kohlrausch a stable boundary between two anions $A^{\ominus}$ and $B^{\ominus}$ (with the common cation $R^{\oplus}$) is obtained when the following ratio of concentrations has been established:

$$\frac{C_A}{C_B} = \frac{U_A}{U_A + U_R} \cdot \frac{U_B + U_R}{U_B} \cdot \frac{L_A}{L_B}$$

(C_A, C_B concentration; U_A, U_B, U_R mobility of the ions, L_A, L_B charge). Spreading of the boundary by diffusion is counteracted by the different field strengths in adjacent zones.

In practice the sample solution is added to the separation compartment between the solution of a leading ion (faster than the fastest component of the sample) and a terminating ion (slower than the slowest component). In the final state the separated zones of sample ions are in immediate contact with each other. To increase the power of resolution and to facilitate their isolation with preparative problems, spacer ions intermediate in mobility between the sample ions may be added to the sample solution.

Isotachophoresis may be applied to the microanalytical separation of low and high molecular weight ions if it is carried out in capillaries. On a preparative scale up to several hundred mg proteins have been separated in columns from acryl amide. The separation power is comparable to that of *isoelectric focussing*. Because all ions migrate with the same speed, this cannot be used to characterize or identify unknown substances. However, mixtures of qualitatively known composition can be analyzed quantitatively after suitable calibration. (10^{-7} to 10^{-8} moles are needed with low molecular weight anions.)

3 Preparative Separations

Almost all methods described for analytical separations can be scaled up without great difficulties for the micro preparative separation of up to a few milligrams. For larger amounts special devices must be found for the dissipation of heat generated during electrophoresis. The temperature gradient across the separation chamber must be kept as small as possible because the ion mobility depends very much on the temperature (about 2% increase per degree centigrade). Higher field strengths (10–50 V/cm) can be used, therefore, only with flat surfaces, (e.g. paper, liquid films, etc.).

3.1 Continuous Methods

3.1.1 Carrier-Free Electrophoresis

In continuous carrier free electrophoresis[39, 40] buffer film (about 0,5 mm thick) flows between two slightly inclined or vertical, cooled glass plates. The mixture to be separated is continuously injected near the upper edge (see Ref.[41] p. 736). The electric field is applied at right angles to the direction of liquid flow (field strenght up to 50 V/cm). The path of the substances in the separation chamber is determined by their migration in the electric field and their flow with the buffer stream. If the differences in the ion mobilities are sufficiently large the individual compounds will leave the separation chamber through separate orifices at the lower edge. With proteins about 0.3 g per hour can be separated whereas with low molecular weight compounds the output amounts to about 0.05–0.1 g per hour.

With a vertical separation chamber the method has been applied to the separation of cellular organelles, viruses, and whole cells[42].

3.1.2 Electrophoresis on Carrier Material

The principle of deflection electrophoresis can be applied to carrier electrophoresis provided the supporting medium permits a continuous flow of buffer (see Ref.[41]). At first flat chambers filled with glass or cellulose powder were used. It was rather difficult to achieve a buffer flow that was constant and even throughout the separation chamber and during the electrophoretic run. Therefore, electrophoresis on free hanging paper or cardboard sheets is preferred, because in these a constant capillary flow can be established if they dip into a trough with constant buffer level. Evaporation losses can be compensated by cooling the sheets with air and condensation of moisture[43]. The separation capacity is comparable to that of carrier free electrophoresis, the degree of resolution is however somewhat smaller.

3.1.3 Electrodecantation (Convection Electrophoresis)

Electrodecantation[44] is a method of electrophore-

[39] *J. Barollier, E. Watzke, H. Gibian,* Z. Naturforsch. *13b,* 754 (1958).
[40] *K. Hannig,* Z. Anal. Chem. *184,* 244 (1961).
[41] *W. Grassmann, K. Hannig* in *Houben-Weyl,* Methoden der organischen Chemie, 4. Aufl., Bd. I/1, p. 685, Georg Thieme Verlag, Stuttgart 1958.
[42] *K. Hannig,* Z. Physiol. Chem. *338,* 211 (1964).
[43] *K. Hannig,* Z. Physiol. Chem. *311,* 63 (1958).
[44] *A. Polson, J.F. Largier* in Ref.[59], p. 161.

sis which permits the separation of compounds with the same isoelectric point from a mixture of high molecular weight ampholytes. Its application, therefore, is limited to *protein* chemistry.

The principle is illustrated in Fig. 1. In a separation chamber (enclosed by semipermeable membranes) the components migrate to the electrodes according to their charge until they are stopped by the semipermeable membranes. They are concentrated here and sink to the bottom of the chamber because the density of the solution increases. Substances with an isoelectric point corresponding to the pH of the carrier electrolyte stay in the center of the separation chamber.

In practice the separation chamber is divided by several membranes to keep the distance covered by electrophoretic migration small; in addition, several cells are connected in series.

For continuous operation the mixture to be separated is fed to the bottom of the first cell, the overflow to the bottom of the second cell and so on. After a certain lag period, the isoelectric compounds appear in the overflow of the last cell; the others are tapped from the bottom as concentrated solutions at a fraction of the feed rate.

The method is especially suited for processing larger amounts of proteins (10 g/h of protein mixture with medium sized equipment), for gaining serumfractions, toxins and enzymes.

For smaller amounts, discontinuous operation is possible. In addition this method can be used for the concentration of dilute protein solutions.

3.2 Discontinuous Methods

With preparative separations the individual compounds must appear in completely separated zones. Different concentrations of material will cause local differences in density. Their levelling by convection currents can be prevented by an additional density gradient or by stabilizing the electrolyte solution by some porous supporting medium.

3.2.1 Carrier-Free Electrophoresis at Constant pH

The electrophoresis is carried out in a vertical, cooled tube which contains in addition to the carrier electrolyte the substances necessary to establish the density gradient[45]. Sucrose is normally used for this purpose. Glycerol, dextran, and ethanol[46] have been used also.

The conditions necessary to establish and maintain stable zones in a density gradient and the influence of gradients of conductivity and viscosity are not discussed here. The isolation of the separated substances is done by slowly emptying the tube into a fraction collector. The low molecular weight substances added for the density gradient can be separated by dialysis or gel filtration and the high molecular weight material (dextran) by ion exchange.

The separation capacity is limited by the steepness of the density gradient and by the cross sectional area of the tube. Because of the difficulty of dissipating the heat generated during electrophoresis without too large a temperature gradient, the latter cannot be increased above 6–8 cm^2. A protein mixture with amounts of 100–300 mg can be separated per run. The time required for electrophoresis depends on the mobility of the compounds and is from 20 to 50 hours.

The degree of resolution as well as the amount of sample is similar to that obtained with column electrophoresis on supporting media. However, the more involved method of carrier free electrophoresis is preferred if interaction with the supporting material prevents effective separations.

The method can be applied to low molecular weight compounds. However, it has not been used except for the separation of dyestuffs for demonstration purposes.

3.2.2 Isoelectric Focussing in a pH-Gradient

This method is based on the principle that in a pH-gradient ampholytes will migrate in the electric field until they reach an area where the pH corre-

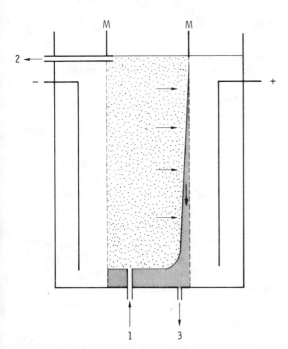

Fig. 1. Model of electrodecantation of a mixture of isoelectric and anionic macromolecules.
M Semipermeable membrane
1 Inflow sample solution
2 Overflow isoelectric compound
3 Outlet anion concentrate

[45] *H. Svensson* in Ref.[59].

[46] *P.A. Charlwood, A.H. Gordon,* Biochem. J. *70,* 433 (1958).

sponds to their isoelectric point. Substances added at some arbitrary location in the separation tube (which must be stabilized by a density gradient or carrier against convection flows) will concentrate in narrow zones according to their isoelectric points.

The method was first used by Kolin[47] for analytical separations. More widespread use and extension to micro preparative separations were possible when Vesterberg[48] introduced synthetic *poly-amino polycarboxylic* acids to build up the pH-gradient. In contrast to previous investigations with a preformed non stable gradient, this is formed during the electrophoretic run and remains constant over the whole time of electrophoresis. It is not necessary to add the material to be separated in a well defined starting zone because the idividual compounds concentrate at their isoelectric points and the zones are sharpened during electrophoresis. When a stationary state has been reached, the separation tube is emptied in a fraction collector and worked up as usual.

The method gives very sharp separations which are comparable to those of disc-electrophoresis. It is used for the characterization of proteins with respect to their isoelectric point and for the micro preparative separation of up to 40 mg. Difficulties may arise with some proteins which have a low solubility at the isoelectric point and which may precipitate during electrophoresis. This can sometimes be overcome by increasing the concentration of carrier electrolytes or by the addition of urea.

The method has been used also to characterize low molecular weight compounds (e.g. *plant pigments*[49]). A preparative application is not feasible due to difficulties encountered in the removal of the carrier ampholytes from the separated compounds because they are of similar molecular weight (300–600).

3.2.3 Electrophoresis on Supporting Media at Constant pH

The following methods described for analytical separations can be applied to micropreparative problems as well: electrophoresis on paper or cardboard, thin layer electrophoresis, electrophoresis in flat blocks of starch gel, polyacryl amide and agar[2, 50]. With high molecular weight compounds and somewhat larger amounts of material (0.1–0.2 g), starch can be used also.

The location of zones is done by staining a control strip or an impression taken on filter paper from the moist support. The elution of the separated substances presents no problem with paper or starch; with gels it is best done by electrophoresis in a special apparatus.

For the separation of larger amounts (up to 20 g in the apparatus described by Porath[51]) the use of vertical columns is preferred. Since the temperature gradient between periphery and center becomes excessively large with column diameters above 3 cm, Porath[51] uses for electrophoresis the annular space between the outer cooling mantle and the central cooling tube.

The most widely used supporting medium for column electrophoresis is cellulose powder which has been treated with ethanol/HCl to reduce the adsorptive power for proteins and to achieve a more even packing. Powders from polyvinyl chloride or from copolymers of vinyl chloride and vinyl acetate have been used also. These show no adsorption of proteins but the disadvantage is their low heat conductivity and large endosmotic flow. Crosslinked dextran gels (Sephadex®) are also suitable as supporting medium. However, at high degrees of crosslinking, only the space between the gel grains is available to large molecules thus giving a reduction in separation capacity. At low degrees of crosslinking gel permeation chromatography (see 2.8) is superimposed on electrophoretic separation. This effect may be an advantage or disadvantage depending on the particular problem. Agarose does not show these drawbacks[52].

3.2.4 Counter Current Electrophoresis

The large distances necessary to separate substances of very similar electrophoretic mobility can be reduced by working with a moving carrier electrolyte having a rate and direction of flow that compensates the electrophoretic migration. This principle has been used mainly for the separation of inorganic ions[1], but there have been some investigations with *proteins*[53] also. The method becomes rather complicated with colorless substances because the correct rate of flow must be determined by preliminary experiments.

An arrangement using a counter flow of buffer to remove the fastest component from a protein mixture has been described by Grubner and Dvorak[54]. The method seems to have found little application.

[47] *A. Kolin*, J. Chem. Phys. *22*, 1628 (1954).
[48] *O. Vesterberg, H. Svensson*, Acta Chem. Scand., *20*, 820 (1966).
[49] *M. Jonsson, E. Petterson*, Sci. Tools *15*, 2 (1968).
[50] *K. Hannig* in Ref. [61], p. 423.

[51] *J. Porath*, Ark. Kemi, *11*, 161 (1957).
[52] *S. Hjerten*, J. Chromatogr. *12*, 510 (1963).
[53] *J. Robbins*, Arch. Biochem. Biophys. *63*, 461 (1956).
[54] *O. Grubner, J. Dvorak*, Collect. Czech. Chem. Commun. *21*, 556 (1956).

3.2.5 Preparative Disc-Electrophoresis

Disc-electrophoresis with its high resolving power can be applied to the separation of amounts up to 30 mg if large cooled gel cylinders are used. Location and elution of the zones may be done by staining a control strip and cutting out the substance containing parts of the gel which may then be extracted with buffer.

However, elution by extending the time of electrophoresis is preferred. The substances migrating out of the gel are recorded by a flow photometer and collected in a fraction collector. Suitable equipment[55, 56] can be used for amounts of up to several 100 mg.

4 Electrodialysis and Related Procedures

Simple electrodialysis is used for rapid desalting of neutral and charged high molecular weight compounds[57]. These are contained in a compartment bounded by two semipermeable membranes and surrounded by running water. The removal of salts is accelerated by the application of an electric field across the dialysis compartment.

In this form electrodialysis is used only for the desalting of large volumes in flow cells. For smaller volumes (1–2 l) it has been replaced by gel filtration (see 2.6). Electrodialysis is of little use with solutions of high salt concentration. Differences in the mobility of anions and cations may cause a shift of pH in the dialysis compartment. This may be suppressed by the use of ion exchange membranes (anion exchange for the anodic and cation exchange for the cathodic side).

The migration of charged macromolecules up to the membranes can be used to effect their concentration and separation. This has already been discussed under 2.9:3.1.3.

5 Bibliography

[58] *S.P. Colowick, N.O. Kaplan,* Methods in Enzymology, Vol. IV, Academic Press, New York 1957.

[59] *P. Alexander, R.J. Block,* Analytical Methods of Protein Chemistry, Pergamon Press, London 1960.

[60] *P. Grabar, P. Burtin,* Immuno-Electrophoretic Analysis, Elsevier Publ. Co., Amsterdam, London, New York 1964.

[61] *M. Bier,* Electrophoresis, Vol. II, p. 423, Academic Press, New York 1967.

[62] *J.R. Corm* in *S.J. Leach,* Physical Principles and Techniques of Protein Chemistry, Academic Press, New York 1969.

---

[55] *T. Jovin, A. Chrambach, M.A. Naughton,* Anal. Biochem. *9,* 351 (1964).

[56] *L. Strauch* in *H. Peeters,* Protides of Biol. Fluids, Proc. 15th Colloquium, Brügge 1967, Elsevier Publ. Co., Amsterdam 1968.

[57] *H.E. Schultze* in *Houben-Weyl,* Methoden der organischen Chemie, 4. Aufl., Bd. I/1, p. 672, Georg Thieme Verlag, Stuttgart 1958.

2.10 Gas Chromatography

Holm Pauschmann and Ernst Bayer
Chemisches Institut der Universität, D-74 Tübingen

1 Principles

Gas chromatography (GC) provides quantitative analytical data on mixtures of substances and on preparative amounts of purified substances. The qualitative information is comparable with that obtained from other measurements of physical properties (e.g. melting point, boiling point, refractive index) but requires additional data (IR, mass spectrum, etc.) for unequivocal identification of a substance. The precision of quantitative analysis corresponds to that of other laboratory methods. It has the advantages of speed, sensitivity and flexibility but the disadvantage that the substance must be volatile without decomposition.

In common with other chromatographic procedures, a separation occurs during GC by distribution of components by absorption or adsorption between two immiscible phases. One of the phases is mobile while the other is stationary. When the mobile phase is liquid, one refers to liquid chromatography (LC) (see Chapter 2.5); when it is gaseous the procedure is termed gas chromatography (GC). The latter is divided according to the state of aggregation of the stationary phase into gas liquid chromatography (GLC) and gas solid chromatography (GSC).

Mixtures of substances of low vapor pressure: Chromatography in which the mobile phase is held near the critical point (e.g. hexane at 60 atmospheres and 250°), where the density is about 0.3 and the viscosity much lower than that of liquids, may be considered as a transition between GC and LC. This process is called dense gas chromatography or fluid chromatography. It is particularly useful for the separation of mixtures of substances which are not amenable to GC treatment because of their very low vapor pressures.

1.1 Variations in Operating Techniques

The usual chromatographic techniques can be applied also to GC, viz:

Elution analysis, in which the components of a mixture are eluted successively by the mobile phase.
Frontal analysis, with displacement of components by the mixture itself.
Displacement analysis, in which components are displaced by an additional substance.

Since the two latter methods do not separate all the components of a mixture as pure substances, elution analysis is used largely at the present time. Several methods are possible according to temperature treatments:

Temperature gradient along the column	Temperature change with time 0	>0 (linear or step-wise)
0	1	2
>0	3	4

Case 1 is the most general, since many separations of mixtures and all retention data are obtained isothermally. Case 2 represents temperature programming, which is used widely at the present time. Case 3 has not been realized since GC operates only with substances which finally leave the separation column. A fixed gradient would be favorable, therefore, only with moving-bed methods[1]. Finally, case 4 represents a temperature gradient which moves along the column and causes the individual components to move at the same speed in a characteristic temperature zone. This method is most suitable for the enrichment of traces of substances and for automation. The programming of the carrier gas flow rate is not as well established as temperature programming[2, 3].

1.2 Scheme of a Gas Chromatographic Separation

A measured amount of sample (see 2.10:4.1), injected via an *injection port* into the *carrier gas stream,* volatilizes and is transported to the *separation column* (see 2.10:3.4).
The separated components leave the column with the carrier gas and reach the *detector.* Here each component of the mixture generates an electrical signal proportional to its amount, which after suitable treatment is recorded *(recorder* or *integrator).*
The recorder gives a *chromatogram* in which each separated component appears as a *peak,* raised above the baseline in the form of a distribution

[1] *W. Kuhn, A. Narten, M. Thürkauf,* Helv. Chim. Acta *41,* 2135 (1958).
[2] *I. Halasz, F.A. Holdinghausen* in Ref. [195], p. 23.
[3] *R.P.W. Scott, A. Goldup* in Gas Chromatography 1964, p. 25, Institute of Petroleum, London 1965.

curve. The baseline is the time axis. The separated components can be collected in fraction collectors.

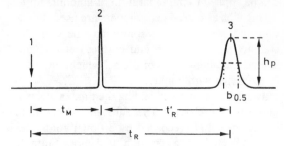

Fig. 1. Gas chromatogram
1 = start; 2 = inert gas peak; 3 = peak of analyzed component; $b_{0.5}$ = peak width at one-half height; h_p = peak height; t_M = retention time of the carrier gas between injector and detector; t'_R = adjusted retention time of an individual component in the stationary phase; t_R = gross retention time

2 Definitions and Relationships of the Chromatographic Parameters

The characteristic data of a column include:

Plate number (2.10:2.1.2)
Selectivity (2.10:2.2.2)
Sample capacity (2.10:2.3)
Pressure drop (2.10:2.4.3)

The most important data for a column packing are:

Plate height (2.10:2.1.2)
Polarity of the stationary phase (2.10:3.4.4)
Coating (2.10:3.4.3)
Permeability (2.10:2.3)

The sample substance can best be characterized gas chromatographically if the retention indices are determined on 3 or 4 separation columns of different polarities (2.10:2.2.2).

2.1 Column Theory

2.1.1 Retention Volume

From the chromatogram the times t_M, t_R, t'_R and $b_{0.5}$ (minutes or seconds) (we prefer $b_{0.5}$, because it is easier to measure than the peak width at base line level) can be taken, as well as the peak height h_p (mV or mA) (cf. Fig. 1). The adjusted retention time can then be determined:

$$t'_R = t_R - t_M \qquad (1)$$

The gross retention volume V_R, the adjusted retention volume V'_R and the gas retention volume V_M are obtained by multiplication of the retention times by the flow rate F_0 at the temperature of the flow meter $\tau_M [^\circ K]$ and the pressure of column outlet p_0:

$$V_R = t_R \cdot F_0$$
$$V'_R = t'_R \cdot F_0 \qquad (2)$$
$$V_M = t_M \cdot F_0$$

Taking into consideration the pressure drop in the column by means of the column pressure gradient correction factor j

$$j = \frac{3\left[\left(\dfrac{p_i}{p_o}\right)^2 - 1\right]}{2\left[\left(\dfrac{p_i}{p_o}\right)^3 - 1\right]} \qquad (3)$$

p_i: column inlet pressure [at, mm Hg]
p_o: column outlet pressure [at, mm Hg]

the (pressure corrected) net retention volume V_N (at the temperature T_M) is obtained:

$$V_N = j \cdot V'_R \qquad (4)$$

and thereby the specific retention volume V_g [ml] (at 273° K and p_o)

$$V_g = \frac{V_N \cdot 273}{w_L \cdot T_C} \qquad (5)$$

w_L: weight of liquid stationary phase [g]
T_C: column temperature [$^\circ$K]

Practical difficulties in the exact determination of V_g limit the value of this theoretically very important parameter.

2.1.2 Partition Ratio and Theoretical Plate Number

From the adjusted retention time and the gas retention time one may calculate the partition ratio (also: capacity ratio) k:

$$k = \frac{t'_R}{t_M} \qquad (6)$$

as the quotient of the retention times in both phases. Multiplying by the phase ratio β (= volume ratio of both phases in the column)

$$\beta = \frac{V_G}{V_L} \qquad (7)$$

V_G: Volume of mobile phase (carrier gas) in column [ml]
V_L: Volume of liquid stationary phase in column at T_C [ml]

the partition coefficient K is obtained. K is independent of the percentage of coating of the support with liquid stationary phase.

$$K = \frac{\text{g substance in 1 ml stationary phase}}{\text{g substance in 1 ml gas phase}} = k \cdot \beta \qquad (8)$$

From t_R and $b_{0.5}$ the theoretical plate number n can be calculated. This term was introduced from the similar concept of theoretical plates used in distillation procedures.

$$n = 5.54 \left(\frac{t_R}{b_{0.5}}\right)^2 \qquad (9)$$

There is no concern here with discrete spaces as with distillation using plate columns or with the Craig's liquid-liquid distribution apparatus. However, an imaginary volume can be defined which assumes a unit concentration of analysis component in both phases, which are instantaneously and totally intermixed and in equilibrium. Its cross section is that of the inside of the column and its height is the plate height. The calculation of the plate height of capillary columns using this equation is subject to large errors on account of their high β values. Therefore, other evaluation methods have been proposed[4].

Owing to somewhat inconstant values of n, the gas retention time is factored out and the effective plate number n_{eff} is defined as

$$n_{eff} = 5.54 \left(\frac{t'_R}{b_{0.5}}\right)^2 \qquad (10)$$

Additional concepts arising from the plate number are: The plate number n^1 per m of column length L

$$n^1 = \frac{n}{L} \qquad (11)$$

The plate height H [mm]

$$H = 1000 \frac{L}{n} \qquad (12)$$

and the effective plate height

$$H_{eff} = 1000 \frac{L}{n_{eff}} \qquad (13)$$

Typical values of H are 0.2 mm (good analytical columns) and 1 mm (preparative columns up to 100 mm i.d.).

2.1.3 Separation Number

The difficulties experienced in measuring constant values of n and n_{eff} led to the recommendation that the latest appearing peak which is still symmetrical should be selected for calculation; and later to the introduction of the separation number n_{sep}[5]. This separation number indicates the number of additional peaks which can be accomodated between the peaks of two adjacent homologs of the n-paraffins (subscripts 1 and 2).

$$n_{sep} = \frac{t_{R2} - t_{R1}}{{}_1 b_{0.5} + {}_2 b_{0.5}} - 1 \qquad (14)$$

---

[4] *R. Kaiser*[199], Bd. II, p. 37, 1968.
[5] Ref.[199], Bd. IV, p. 61.

If the peaks are not completely separated, a quantitative estimate of the degree of separation actually attained is desirable. There are two different expressions for the resolution of two peaks ϑ and R. The resolution ϑ may be obtained graphically by drawing a straight line between the two maxima and dropping a perpendicular from this line to the minimum point in the valley. From the intercepts f and g (Fig. 2)

$$\vartheta = \frac{f}{g} \tag{15}$$

The necessary number of plates for a desired resolution of two substances with known retention times is obtained by using the ratio of the gross retention times as in equations 16 and 17.

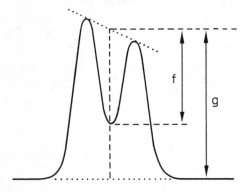

Fig. 2. Calculation of the required plate number

$$\alpha = \frac{t_{R2}}{t_{R1}} \text{ where } \alpha > 1.0 \tag{16}$$

$$n = 2\left(\ln\frac{2}{1-\vartheta}\right)\left(\frac{\alpha+1}{\alpha-1}\right)^2 \tag{17}$$

When a graphical determination of the resolution using Equation (15) is not possible because of unequal peak heights, Equation (18) is used[6]:

$$R_{s2.1} = \frac{(t_{R2}-t_{R1})}{_2 b_{0.5} +_1 b_{0.5}} \tag{18}$$

Poor peak separation naturally introduces difficulties in measuring $b_{0.5}$.

The plate height H is dependent on the nature and method of column packing as well as on the flow rate of the carrier gas. It is given by the Van Deemter equation (19). For each column there is only one optimum gas flow rate. In Equation (19), term A considers the column packing, B the axial diffusion, C the diffusion in the gas phase and D the diffusion in the liquid phase.

$$H = A + \frac{B}{\bar{u}} + (C+D)\bar{u} \tag{19}$$

---

[6] *D. Jentzsch*[198], p. 21.

From the measured values of the average linear velocity of the carrier gas $\bar{u}$ and determined values of h, curves are obtained (Fig. 3) relating to the 4 terms, which have the following significance[6a]:

Parameter		Packed column	Capillary column
Packing	A	$2\lambda d_p$	0
Diffusion: axial	B	$2\gamma D_G$	$2 D_G$
gas phase	C	$\dfrac{1+6k+11k^2}{24(1+k)^2}\cdot\dfrac{d_p^2}{D_G}$	$\dfrac{1+6k+11k^2}{24(1+k)^2}\cdot\dfrac{r^2}{D_G}$
liquid phase	D	$\dfrac{2}{3}\dfrac{k}{(1+k)^2}\dfrac{d_f^2}{D_L}$	

D_G: diffusion coefficient in the gas phase
D_L: diffusion coefficient in the liquid phase
d_p: particle diameter of the solid support
r: capillary radius
k: partition ratio (6)
d_f: the film thickness of the liquid phase
λ: statistical irregularity of the packing
γ: labyrinth factor of the packing pores
$\bar{u}$: average linear gas velocity

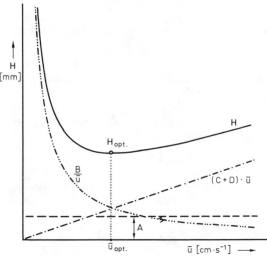

Fig. 3. H (Height equivalent to a theoretical plate) as a function of the carrier gas velocity $\bar{u}$; H_{opt}: smallest possible H.

The separation of this equation into 3 or more terms has been suggested. Also, a large number of modified equations and derivations have been proposed[6b,c].

[6a] Ref. [198], p. 24.
[6b] *R. Kaiser*[199], Bd. I, p. 31.
[6c] *R.S. Juvet, S.D. Nogare*, Anal. Chem. *40*, 33 R (1968).

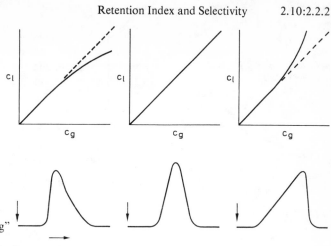

Fig. 4. Partition isotherm and peak form
(Ref.[199])

c_l = concentration of substance in liquid phase

c_g = concentration of substance in gaseous phase

I = c_l/c_g decreases with growing c_g "tailing"

II = c_l/c_g remains constant

III = c_l/c_g increases with growing c_g "heading"

I   II   III

2.2 Evaluation of Chromatograms

2.2.1 Peak Forms

According to thermodynamic theory the peak form results from mass balance of the sample amounts which enter and leave a theoretical plate. This results in a Poisson distribution[8]. From kinetic theory and application of the diffusion laws a Gauss curve[7] is obtained. Both curve forms become similar when the number of plates is high[8].

According to Raoult and Henry's law of ideally dilute solutions, the partial pressure (p_1) of a volatile substance 1 which is dissolved in a nonvolatile substance (3) is proportional to its mole fraction x_1

$$p_1 = p_1^0 x_1 \qquad (20)$$

p_1^0: the vapor pressure of pure substance 1

The conditions which apply in GC require introduction of the activity coefficient $\gamma_{1.3}$ of substance 1 at infinite dilution in liquid phase 3

$$p_1 = p_1^0 \gamma_{1.3}^0 x_1 \qquad (21)$$

When γ remains constant over the observed concentration range (during transit of a peak) a symmetrical peak results (Fig. 4). Failure to obtain a symmetrical peak is usually due to the column temperature being too low or the column being overloaded.

Because substances whose solutions have a low vapor pressure will be transported slower in the liquid phase than those with a high vapor pressure (a greater part of which is found in the gas phase), the following relation between the partial pressure and the relative retention of two substances is applicable:

$$r_{2.1} = \frac{t_{R2}}{t_{R1}} \text{ prop.} \frac{p_1^0}{p_2^0} \frac{\gamma_{1.3}^0}{\gamma_{2.3}^0} \qquad (22)$$

p_1^0, p_2^0: vapor pressure of pure substances 1 and 2

$\gamma_{1.3}^0$, $\gamma_{2.3}^0$: activity coefficients of substances 1 and 2 at infinite dilution in liquid phase 3

This separation equation of Herington[9] indicates that separation of two substances is possible either if they possess identical activity coefficients but different vapor pressures or if they have different activity coefficients but the same vapor pressure. According to the above equation, straightforward separation of azeotropic mixtures is possible.

2.2.2 Retention Index and Selectivity

Retention times relative to a certain standard substance (e.g. pentane, nonane) provide specific data for a particular column packing at a fixed temperature. (cf. equation 22). In principle, the introduction of the retention index provides a more suitable reference system avoiding the disadvantages created by the use of various standard substances, whose retention times sometimes are far removed from those of the substances under investigation[10].

$$I_x = 100 \cdot z + 100 \frac{\log t'_{RX} - \log t'_{RZ}}{\log t'_{R(Z+1)} - \log t'_{RZ}} \qquad (23)$$

The retention index is usually determined graphically[10a] (Fig. 5). The retention times t_R of adjacent n-paraffins with z and z + 1 carbon atoms are plotted on the vertical logarithmic axes and these points are connected by a straight line. The retention time of the substance is plotted on the straight line, and a perpendicular is dropped from this point to the base axis which is divided linearly into 100 units. In the example, the

[7] I. Halasz, J. Gas Chromatogr. 4, 8 (1966).

[8] E. Bayer[187], p. 9.

[9] E.F.G. Herington in D.H. Desty, Vapor Phase Chromatography, p. 5, Butterworth, London 1957.

[10] E.R. Adlard, J. Gas Chromatogr. 3, 298 (1965).

[10a] K.P. Hupe, J. Gas Chromatogr. 3, 12 (1965).

147

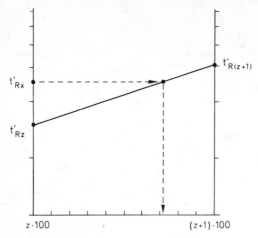

Fig. 5. Graphical determination of the retention index of substance x (t'_{Rx}) between n-paraffins with carbon numbers z and z+1 (t'_{Rz} and t'_{Rz+1})

perpendicular cuts the base at 72 units. The retention index of the substance is accordingly 100 Z + 72.

For two substances possessing the same vapor pressure, only the activity coefficients remain for separation according to the Herington formula. The value of $r_{2.1}$, for the same boiling point, is used to define the selectivity of a column.

$$r_{2.1}(b_{p1} = b_{p2}) = s \qquad (24)$$

By definition both substances must originate from two different homologous series, as shown in tables for instance with [11]

Phase: β,β'-bis (propionitrile)ether
Substances: olefins from paraffins
s: 1.9–4.6
When there are no suitable pairs of substances of the same boiling point available, the retention times of both homologous series can be plotted on a logarithmic scale against the boiling points. Comparison is then made of the retention data of both straight lines at a definite temperature.

In another form, the selectivity is represented as the difference between the retention indices [12]

$$S = I_2 - I_1 \qquad (25)$$

which have been determined for substances of the same boiling point but differing homologous series.
The exact determination of the polarity of sample substances and liquid phases is still not possible. The initially somewhat vague concept of polar alcohols and unpolar paraffins was extended, by a study of paraffins and olefins, to a more precise

classification of binary liquid phases[13]. This was subsequently enlarged to a ternary system including fluoro compounds[14], and finally related to a selectivity S as proposed by Rohrschneider[15].
The retention indices of five characteristic substances (benzene, ethanol, 2-butanone, nitromethane and pyridine) are determined on a column of squalene as well as a column containing the packing under investigation. The differences between the index values of each substance on both columns are divided by 100 giving positive or negative values. These five values provide more meaningful information than the polarity used earlier.

Differences in polarity of GC stationary phases: Variable polar influences account for the large number of column packings (>600), while less than 10 carrier gases are available for mobile phases. This relationship is completely reversed in LC.

Theoretically the total volatilized sample mixture should be located in the gaseous space of the first plate for optimum separation. In practice, this would either result in an excessively large-bore column or the amount of sample would be too small (detector sensitivity problems); the increase in the efficiency of the separation would be small compared with that for the usual sample size. Instead, the plate number n is plotted as a function of the measured sample amount and its maximum value is obtained by extrapolation to zero.

2.3 Sample Capacity and Permeability
The sample capacity of the column is defined as the applied sample amount (mg, g) which still gives 90% of the maximum possible number of theoretical plates. It is proportional to the column cross sectional area. During analytical separations one usually works below this limit, while during preparative work, columns are overloaded. It is desirable to use a preparative column which tolerates considerable overloading without a concomitant loss of separation efficiency. Columns of large cross section, heavily coated with the liquid phase and, if need be, of increased length should be used.

The measurement of sample capacity should be carried out with only one substance, since mixtures are quickly separated decreasing the sample loading. The partition ratio k and the ratio of column length to cross section are also significant[16].

[11] E. Bayer[187], p. 102.
[12] R. Kaiser[199], Bd. II, p. 61.

[13] L. Rohrschneider, Z. Anal. Chem. 170, 16 (1959).
[14] I. Brown, I.L. Chapman, G.J. Nicholson, Australian J. Chem. 21, 1125 (1968).
[15] L. Rohrschneider, J. Chromatogr. 22, 6 (1966).
[16] D. Jentzsch[198], p. 30.

The pressure drop of the carrier gas in the packed column is an unavoidable drawback: optimum carrier gas velocity of the van Deemter curve is realized in only part of the column. The increased inlet pressure p_i interferes with sample volatilization. The usual inlet pressure available suffices in practice only for operation of packed columns up to ca. 20 m and for capillary columns up to 200 m.

The retention data obtained must be corrected by means of the pressure gradient correction factor j, in order to obtain V_g.

Resistance to the gas flow should be kept as minimal as possible, without interference with the necessary properties of the separation column. The term Specific Permeability B_0 (cm^{-2}) [16a] expresses the resistance in the column to gas flow (in a reciprocal form).

$$B_0 = \frac{\eta}{p_i - p_0} L \bar{u} \qquad (26)$$

η [poise, $g \cdot cm^{-1} \cdot s^{-1}$]: viscosity of the carrier gas (for He at $100°$: 229×10^{-6})

L [cm]: column length

$\bar{u}$ $[cm \cdot s^{-1}]$: average linear gas velocity

$p_i - p_0$ $[dyn \cdot cm^{-2}, g \cdot cm^{-1} \cdot s^{-2}]$: pressure drop in the column.

Usual values of B_0 are $1.0 ... 20.0 \times 10^{-7}$ $[2 \times 10^{-7}]$ for packed columns and $100 ... 1000 \times 10^{-7}$ cm^{-2} $[200 \times 10^{-7}]$ for capillary columns. For calculation, the empirical formula (27) is used for packed columns and (28) for capillary columns, cf. Ref. [17, 18]:

$$B_{0 \ pac.} = \frac{d_p^2}{1012} \qquad (27)$$

$$B_{0 \ cap.} = \frac{d^2}{32} \qquad (28)$$

d_p [cm]: the effective particle diameter of the support
d [cm]: the internal diameter of the capillary

2.4 Speed of Analysis

2.4.1 Chromatography at Constant Temperature

The concept of plate number per second is used as a measure of the speed of analysis. Although this is dependent upon the substance under consideration (i.e. its migration rate in the column), comparisons can be made if the substance and its retention data are known:

$$\frac{n}{t} = \frac{L}{H \cdot t_R} \qquad (29)$$

The temperature of the column, operated isothermally, determines the retention data, thus

$$\log V_g = \frac{\Delta H_s}{2,3 \cdot R \, T_C} + C \qquad (30)$$

ΔH_s: partial molar heat of vaporization of the dissolved substance in solution $[cal \cdot mole^{-1}]$
R   gas constant $[1.987 \, cal \cdot °C^{-1} \cdot mole^{-1}]$
T_C: column temperature $[°K]$
C: a constant

Excessive increase of the temperature in an attempt to shorten analysis time is not recommended, because of the reduction in k-values and separation efficiency which occur. Too low a temperature, resulting in an increase in viscosity or solidification of the stationary phase causes peak broadening. There is, therefore, an optimum temperature range between these two cases.

2.4.2 The Dependence of Chromatography on Temperature

The variation of retention time with temperature differs for the various homologous series of substances. This fact may be used for analytical purposes. The alteration in the retention index I is determined for a temperature rise of $10°$, and is designated as $\Delta I / 10$. This value serves as a means of identification [19].

Temperature programming begins by injecting a test sample on the column which first was maintained isothermally at low temperature. Most of the components are already absorbed at the column entrance and only the most volatile substances are transported with the carrier gas. The column temperature is then raised, usually linearly, whereby the vapor pressures of the individual components in the liquid phase increase until they reach a value at which the components begin (one after another) to migrate. The temperature at which this occurs varies for each substance as does the retention time during isothermal separation. Therefore, the separation is not adversely affected.

The advantages of temperature programming are: its time economy, especially when handling mixtures of wide boiling range provided the correct heating rate is employed, and the uniform shape and height of the peaks obtained (no broadening or flattening). However, temperature programming does not improve the separation efficiency of a column beyond that obtained under isothermal conditions. While a component moves through the column, the temperature of the column and therewith the vapor pressure of the component increases. The heating rate must, therefore, be carefully regulated to maintain an efficient separation.

From temperature programming [20] it can be inferred that.

[16a] Ref. [198], p. 43.
[17] S. Dal. Nogare, R.S. Juvet [188].
[18] L.S. Ettre [189].

[19] J.C. Loewenguth, D.A. Tourres, Z. Anal. Chem. 236, 170 (1968).
[20] J.C. Giddings, J. Chem. Educ. 39, 569 (1962).

$$R_r = \frac{HR}{45} t_M \qquad (31)$$

R_r: R_f-value of a component leaving the column (cf. liquid chromatography, $R_r \approx t_M/t_R$)

To maintain efficient separation, R_r should not exceed 0.3; the heating rate $HR[^\circ C \ min^{-1}]$ is then

$$HR \leqslant \frac{13,5}{t_M} \qquad (32)$$

If the programmed heating rate is much too slow, the peaks broaden with increasing retention times similar to isothermal GC.

Since the determination of plate number and retention data is difficult during temperature programming, it is usually determined isothermally.

In thermochromatography with an axial temperature gradient dT/dL [21] moving in the direction of the carrier gas, each substance moves at the exact temperature and at the same speed as the movement of the heater along the column. The temperature gradient not only sharpens each individual peak by accelerating its rear flank and decelerating its fore flank, but can also focus the peaks from a very large volume of sample, i.e. by collecting them into a determined width. Attempts to obtain extremely sharp peaks by the use of very high temperature gradients are counteracted by a poorer separation, since although peak width is proportional to $1/\sqrt{dT/dL}$, the distance between components is proportional to $1/(dT/dL)$.

2.4.3 The Dependence of Chromatography on Flow Rate

Temperature programming is not suitable for substances which are heat sensitive and the isothermal procedure is often too slow. Therefore, occasionally flow programmed gas chromatography is used in which the flow rate is regularly increased during separation. The saving in analysis time is similar to that obtained by temperature programming, but the influence of the increased flow rate on the detector should be noted. For most of the analysis time the carrier gas velocity (u) exceeds the optimum (cf. Fig. 3).

3 Apparatus

3.1 Choice of Carrier Gas

The choice of carrier gas is governed by the detector (2.10:3.6) and column (2.10:3.4). Thermal conductivity detectors usually require that the thermal conductivities of carrier gas (helium or

hydrogen) and sample components be as different as possible. Any permanent gas can be used with flame ionization detectors as long as it is free from organic constituents such as methane. Nitrogen is used mostly.

In order to maintain an optimum flow rate for a given inlet pressure, the viscosity of the carrier gas should not be too high when using long columns. In this respect, hydrogen is better than helium, carbon dioxide, nitrogen or argon [21a]. The gas used must always be purified to remove any residual oxygen which might have a deleterious effect on the stationary phase. The sole disadvantage in the use of hydrogen is the risk of *explosion* which can occur because of defective column fittings etc.

3.2 Control of the Carrier Gas

A two-stage pressure regulator suffices for isothermal columns. In *temperature programming* the gas viscosity increases with temperature and a flow rate regulator is required after the pressure regulator. This is necessary also during switching of columns. In this case the flow rate regulator and the column require a certain time to adjust to the sudden change in pressure and flow rate relationships. Baseline variations can appear during this time interval (ca. 1 min.).

A regulator with a continuously adjustable flow rate is required for flow programming [21b]. It is usually supplied as a separate accessory. All gas regulators should be easily, quickly, and reproducibly adjustable and should be practically unaffected by temperature and inlet pressure variations.

3.3 Injection Systems

3.3.1 Liquid Samples

Liquid samples are usually injected with an injection syringe (analytical: nl [22] to μl [23], preparative: μl to ca. 100 ml). The needle is inserted through an air-cooled rubber septum, which seals the evaporation chamber. This is situated in an electrically heated metal block (made of corrosion resistant metal or with a quartz or glass insert) through which the carrier gas flows. With sensitive substances, the column inlet (particularly for glass columns) leads directly into the evaporation chamber and lies immediately behind the septum).

[21] *A.A. Zuchovitskij, O.V. Zolotareva, V.A. Sokolov, N.M. Turkel'taub,* Dokl. Akad. Nauk SSSR *77*, 435 (1951).

[21a] *H. Farnow* [191], p. 24 ff.
[21b] *S. Clarke,* US Pat. 3, 250, 057 (1966).
[22] Scientific Kit Co., Washington, Pa., P.O. Box 244, USA.
[23] Hamilton Co., Whittier, Cal. 90 608, P.O. Box 307, USA.

It is desirable to have an exact, adjustable heating system, which is independent of the temperature of the apparatus, and which permits the instantaneous evaporation of the sample. The internal volume should not be too great and an exchangeable insert should be available for rapid cleaning. Other devices for introducing liquid samples are: pipet[24], slide valve[25], automatically operating syringes[23, 26, 27], pressure sample injectors (time[27a] or volume[28] dependent).

3.3.2 Gaseous Samples

Gaseous samples are usually introduced with a tube of known volume completely filled with the sample. This tube is inserted through a switching valve into the gas stream so that a plug of sample gas is flushed on to the column[25, 29, 30] (The method requires a relatively large gaseous sample.)

Other possibilities are pistons with gaskets[31] or slide valves[31] and, for small gaseous amounts, gas syringes[23] which can be used in conjunction with liquid injection systems (inexact volumes).

3.3.3 Solid Samples

Injection of *solid samples* is carried out manually[23, 33], semi-automatically[33a] or fully automatically[29, 33b]. Since all these devices are less simple in construction and operation than liquid injectors, solids are injected in solution whenever possible. Solid injection is preferred only when the solvent interferes. Solid sampling is necessary in *pyrolysis-gas chromatography*.

3.3.4 Split System

When using capillary columns, the sample injection component (particularly with a liquid injection system) is followed by a *splitter* which allows only a small part of the carrier gas stream (1/10 to 1/2000) to go through the column, the larger part being vented.

Improper construction of the splitter can give rise to considerable errors[34].

3.4 Columns

3.4.1 Column Size

Columns are made of metal (stainless steel, copper, brass, aluminum), plastics (Teflon, nylon) or glass; those made of stainless steel and glass are preferred.

Inner diamter: preparative columns 8–300 mm; analytical packed columns 1–6 mm; capillary columns[35] 0.1–1 mm. Outer form: straight, bent into V or W shapes, or coiled. The diameter of the coil should be at least 15 times the diameter of the column to avoid a detrimental effect on the separation efficiency. Capillary columns are used exclusively in the coiled form. For the manufacture of glass capillaries cf. Ref. [28, 36, 37].

3.4.2 Support Materials

In principle any solid granular and inert material can be used as a support. However, the surface area should amount to at least $1 \text{ m}^2/\text{g}$[38] and the pores should not be too narrow and deep. Kieselguhr[39] (diatomaceous silica) meets these requirements either in its natural form after heating with acid, water washing and drying[40], or after silanization of the active centers with dimethyldichloro-

[24] *H.M. Tenney, R.J. Harris*, Anal. Chem. *29*, 317 (1957).

[25] Beckmann Instr. Inc., Fullerton, Cal. 92634, 2500 Harbor Blvd., USA.

[26] Hewlett-Packard, Route 41, Avondale, Penn. 19311, USA.

[27] Varian-Aerograph, Walnut Creek, Cal. 94598, P.O. Box 313, USA, D-61 Darmstadt, Hilpertstraße 8.

[27a] *D. Jentzsch* in Ref. [195], p. 116.

[28] Hupe & Busch D-7501 Grötzingen, Gutenbergstr. 6.

[29] Bodenseewerk Perkin-Elmer, D-777 Überlingen, Askaniaweg.

[30] Carle Instr. Inc. 1141 East Ash Ave, Fullerton, Cal. 92631, USA; Techmation, D-4 Düsseldorf.

[31] Loenco, Inc., Mountain View, Cal. 94040, 1062 Linda Vista, USA.

[32] *P. Van Dingenen, D.A. Cramer*, J. Chromatogr. *24*, 167 (1966).
D.A. Podmore, J. Chromatogr. *20*, 131 (1965).
E. von Rudloff, J. Gas Chromatogr. *3*, 390 (1965).
J.A. Hudy, J. Gas Chromatogr. *4*, 350 (1966).
D.R. Roberts, J. Gas Chromatogr. *6*, 126 (1968).
M.E. Yannone, J. Gas Chromatogr. *6*, 465 (1968).
A.O. Lurie, C.A. Villes, J. Gas Chromatogr. *4*, 160 (1966).
R.H. Waltz, Facts & Methods *6*, No. 4, 5 (1965).
R.L. Bowman, A. Karmen, Nature *182*, 1233 (1958).

[33] S.G.E., Biotronik Wissenschaftliche Geräte GmbH, D-6 Frankfurt/M., Borsigallee 22.

[33a] Carlo Erba. I-20159 Milano, Via Imbonati 24.

[33b] Philips Industrie Elektronik GmbH, D-2 Hamburg 63, Röntgenstr. 22.

[34] Ref. [199], Bd. II, p. 182ff.

[35] *M.J.E. Golay* in *V.J. Coates, H.J. Noebels, I.S. Fagerson*, Gas Chromatography, p. 1, Academic Press, New York 1958.

[36] *D.H. Desty, J.N. Haresnape, B.H.F. Whyman*, Anal. Chem. *32*, 302 (1960).

[37] *D.W. Grant*, Z. Anal. Chem. *236*, 118 (1968).

[38] *L.S. Ettre*, J. Chromatogr. *4*, 166 (1960).

[39] *M. De Mets, A. Lagasse*, Chromatographia *2*, 401 (1969).

[40] *G. Hesse*[197], p. 172.

silane or hexamethylsilazane[41, 42], or after previous combustion with added substances[43].

Kieselguhr is the most used support material and can give rise to plate numbers of 3000–5000/m. Smooth, lightly coated glass beads or glass beads with porous surface layers[44] (e.g. for easily decomposed samples), or Teflon powder (for polar or corrosive substances)[45], are also occasionally used and give lower plate numbers /m. The grain size of the support particles lies between 0.1 and 0.4 mm. It should be as uniform as possible (narrow sieve fraction) to ensure that the pressure drop in the column is minimal. Coarser material is used predominately for filling long preparative columns, the finer for short analytical columns.

The coating depends on the nature and application of the column and the support material. It varies from 0.5 − 5 g/100 g for glass beads, ca. 10 g/100 g for Teflon powder, 2 − 15 g/100 g for Kieselguhr in analytical columns and up to 15 − 30 g/100 g for Kieselguhr in preparative columns. High coating leads to smaller plate numbers and higher sample loadings.

3.4.3 Stationary Phase

The support is coated by mixing it with the *stationary* phase[46] dissolved in a suitable solvent, followed by evaporation of the solvent. Coating of capillaries is carried out, after careful preliminary cleaning, by forcing a plug of solution containing the stationary phase through the column under definite and constant conditions of concentration, temperature and pressure[47]. To obtain a uniform, continuous film. This is difficult with polar liquid stationary phases, although the wetability of the glass capillary can be improved, by previous treatment with graphited carbon black[48]. The stationary phase should be liquid and of low viscosity under the column operating conditions and should not react chemically with either the support material or the sample. It should not be altered physically by evaporation of easily volatile components. Its vapor pressure at the operating temperature should lie under 0.5 torr and it should not undergo thermal decomposition. The most important

requirement of the stationary phase is that it should effect the desired separation of the sample into its components. Because of the different forces which are effective between the sample and stationary phase, it is difficult to predict retention data. Although some rules[49] have been formulated, in most cases reference is made to the collected data of retention r[50] or retention indices[51]. A rough division of commonly used stationary phases can be made according to their polarity:

Non-polar phases: paraffins such as squalene and Apiezon® L and M, silicone oil, gum and rubber such as DC 200, DC, SE 30 and SE 52.
Semi-polar phases: carboxylic acid esters such as di-(ethylhexyl) sebacate and di-n-decylphthalate, polyesters like ethylene glycol succinate (EGS), aromatic polyethers like OS 124.
Polar phases: silicone rubber with nitrile groups (XE 60), cyano-ethylated alcohols and glycols, polyethylene and polypropylene glycol such as Carbowax® 1500 and UCON® LB-500 X.
Phases with definite selectivity, including the strongly fluorinated compounds such as fluorosilicone oil FS 1265, which exhibit a particular retention behaviour towards substances containing keto groups.

Column packings in which no stationary liquid phase is used are e.g.

Adsorbents such as active carbon or silica gel
Molecular sieves such as Molsieve 5 Å[52]
Porous organic polymers such as Porapak®[53]
Porous carbon such as carbon − Molsieve B[54]
Gel with chemically modified surface (brush column)[55]
Porous glass beads[56], also with chemically modified surface[55].

3.4.4 Packing

Columns are filled with the prepared *packings* by vibration or by tapping. Coiled metal columns are filled when straight and then coiled. Conditioning of packed columns is usually necessary to remove residual solvent and low boiling impurities in the stationary phase. The column is flushed with a weak stream of carrier gas and its temperature is slowly raised until it is above the intended operating temperature, but below the allowed maximum.

---

[41] Gaschrom. Newsletter Vol. VII, No. 2, April 1966, Fa. Applied Science Laboratories Inc., State College, Pa 16801.

[42] *D.M. Ottenstein*, J. Gas Chromatogr. *6*, 129 (1968).

[43] *D.M. Ottenstein*, J. Gas Chromatogr. *1*, 11 (1963).

[44] *I. Halasz, C. Horvath*, Anal. Chem. *36*, 1179, 2226 (1964).

[45] *I. Halasz*, Z. Anal. Chem. *236*, 15 (1968).

[46] Ref. [199], Bd. III, p. 44.
E. Knapman, GC-Abstracts[207].

[47] *R. Kaiser*[199], Band II, p. 151.

[48] *K. Grob*, Helv. Chim. Acta *48*, 1362 (1965).

[49] *L. Rohrschneider*, Z. Anal. Chem. *236*, 149 (1968).

[50] Ref. [187], 1. Aufl. 1959, p. 102.

[51] Ref. [199], Bd. III, p. 82.

[52] *T.G. Andronikashvili, G.V. Tsitsishili*, Poverkhni Yavleniya na Alyumosilikatakh Akad. Nauk. Gruz. SSR, Sb. Statei 1965, p. 34, C.A. *64*, 16 599 (1966).

[53] *D.L. Hollis*, Anal. Chem. *38*, 1309 (1966); *S.B. Dave*, J. Chromatogr. Sci. *7*, 389 (1969).

[54] *R. Kaiser*, Chromatographia *2*, 453 (1969).

[55] *I. Halasz* in Vth Intern. Symposium on Advances in GC, 20.−23. 1.69 Las Vegas, Nevada, USA.

[56] Ref. [196].

The carrier gas is turned off only after the column has cooled, to avoid contact of the hot packing with air.

In analytical gas chromatography the column is operated discontinuously, so that the largest possible plate number can be achieved by using a small sample.

For preparative purposes, a part of the column separation efficiency must be sacrificed in order to receive as quickly as possible sufficient amounts of separated components for further treatment.

Thus it is possible to use:.

Columns with increased diameter, discontinuous operation, often automatic and cyclic[57].

Moving columns on a rotating table; while sample injection is continuous, components are collected in a number of traps. Advantage: high flow rate, no automatics, no detector, no electronics; Disadvantage: only isothermal operation possible, numerous identical columns necessary, elaborate packing[58].

Movable packing poured into a vertical tube with light vibration. Carrier gas and volatilized sample are continually sampled laterally. Advantage: simple construction, no controls, detectors or electronics. Disadvantage: only two components separable, mechanical abrasion of the support materials is likely[59].

3.5 Column Switching Valves

Since some separations are not practical using only a single packing, a succession of suitable columns must be used and these must be switched at the correct time.

The elimination of the high boiling residues of a mixture by *backflushing*, or use of a preliminary column which can be backflushed, shortens the analysis time. Removal of residues prevents contamination and inactivation of sensitive columns. Column switching is also desirable if temperature programming is dispensed with either by choice or through necessity, e.g. during automated process gas chromatography.

Switching valves correspond in principle to the 6 port valve of the gas injection system (cf. 3.3) and have between 4 and 16 ports. The upper temperature limit is governed by the gas seal and capability function: pneumatic membrane valves are applicable up to about 100°, Teflon sliding switches to about 150°, filled

Teflon material up to about 250°[30]. Most column switches are used for gas analysis[60], e.g. the well-known series − parallel − switching[61].

3.6 Detectors

These transform a characteristic property, dependent on the carrier gas and sample component, into a perceptible signal, usually electric.

3.6.1 Physical Detectors

These yield a signal without any chemical alteration of the sample; the following properties of the sample are employed:

mechanical[62-66]	optical[72]
acoustic[67]	nuclear[73, 74]
thermal[68-70]	electrical[75]
magnetic[71]	electronic[76-86]

[62] *E.C. Creitz*, J. Chromatogr. Sci. *7*, 137 (1969).

[63] *S.C. Bevan, T.A. Gough, S. Thorburn*, J. Chromatogr. *43*, 14 (1969).

[64] *W.H. King*, Belg. Pat. 63 1020 (1963).

[65] *J. Janak*, Collect. Czech. Chem. Commun. *19*, 684 (1954).

[66] Cekoslovenska Akademie Ved., Brit. Pat. 1, 046 162, *M.V. Kulakov, E.F. Shkatov*, Nefte Pererabotka i Nefte Khim. Naukhn.-Tekhn. Sb. 1965, 36; *64*, 14 935 (1966).

[67] *W.H. Grice, D.J. David:* J. Chromatogr. Sci. *7*, 239 (1969).

[68] *A.E. Lawson, J.M. Miller*, J. Gas Chromatogr. *4*, 273 (1966).

[69] *M. Goedert, G. Guiochon*, J. Chromatogr. Sci. *7*, 323 (1969).

[70] Ref. [200], p. 429; *G. Taguet*, Z. Anal. Chem. *236*, 64 (1968).

[71] Ref. [184], p. 11.

[72] *T.J. Klayder*, J. Ass. Offic. Agr. Chemists *47*, 1146 (1964).

[73] *R. Pearson jr., W.C. Fink, A.A. Gordus*, J. Gas Chromatogr. *3*, 381 (1965).

[74] *A.T. James, C. Hitchcock*, Kerntechnik *7*, 5 (1965).

[75] *H.P. Williams, J.D. Winefordner*, J. Gas Chromatogr. *6*, 11 (1968).

[76] *J.E. Lovelock*, Anal. Chem. *33*, 162 (1961).

[77] *A. Karmen* in Ref. [192], Vol. II, p. 293.

[78] *K.P. Dimick, L.A. Rigali*, Principles of GC, Application of four Ionization Detectors, Wilkens Instr. Inc., P.O. Box 313, Walnut Creek, Cal., USA.

[79] *P. Devaux, G. Guiochon*, J. Chromatogr. Sci. *7*, 561 (1969).

[80] *J.E. Lovelock*, US Pat. 3, 247, 375 (1966).

[81] *G.R. Shoemake*, Dissert. Abstr. B 27, 3 (1966) 769 B.

[82] *J.G.W. Price, D.C. Fenimore, P.G. Simmonds, A. Zlatkis*, Anal. Chem. *40*, 541 (1968).

[83] *W.C. Hampton*, J. Gas Chromatogr. *3*, 217 (1965).

[84] *S. Rennhak, C.E. Döhring, G. Schmid, D. Schneller, H. Stürtz, E. Werner*, Chem. Tech. [Leipzig] *17*, 688 (1965).

[85] *A.V. Markevich, S.L. Dobycin*, Gaz Chromatogr. Tr 1-oi (i.e. Pervoi) Vses Konf. Akad. Nauk. SSSR Moscow. Sb. 1964, p. 42; C.A. *65*, 1352 (1966).

[86] *H.J. Arnikar, T.S. Rao, K.H. Karmarkar*, J. Chromatogr. *26*, 30 (1967).

[57] *D. Jentzsch, B. Kolb, B. Kempken, G. Leberecht, D. Gaigalat*, Angew. GC. Heft 6 und Heft 8, Bodenseewerk Perkin-Elmer 1967.

[58] *D. Dinelli, S. Polezzo, M. Taramasso*, J. Chromatogr. *7*, 477 (1962).

[59] *A. Clayer, L. Agneray, G. Vandenbussche, P. Petel*, Z. Anal. Chem. *236*, 240 (1968).

[60] *H. Pauschmann*, Chimia 1970 Suppl., 5th Int. Symp. Sep. Methods, Lausanne 1969, p. 89–90.

[61] *B. Kolb, E. Wiedeking*, Chromatographia *1*, 98 (1968).

Detectors based upon thermal conductivity (TCD; thermistor, hot wire[68, 69]) and electron capture[78–80] (ECD) have widespread application; other properties such as radioactivity[73, 74] (Geiger-counter) and light absorption[72] are often used in combination.

3.6.2 Physicochemical Detectors

These give rise to a signal during chemical reaction of the sample in the detector.

Reaction with gases: Heat of oxidation or hydrogenation (thermistor or hot wire with catalytically active surface[87, 88]), heat of combustion (flame with thermocouple[89]), light emission of the flame (e.g. flame with copper wire and photocell for halogens[90]), light absorption of the flame (atomic absorption spectroscopy[91]), ion formation in the flame (flame ionization detector[77, 93]) and thermionic detectors[94–97].

Reaction with liquids: Color change (optical indicator in titration detector[98]), electrical conductivity (electrical conductivity cell[99]), concentration (E.M.F. or current measurement, coulometric regeneration[100]), current generation (fuel cell or polarographic cell[102, 103]).

Reaction with solids: Current measurement (electrolytic water decomposition on phosphorus pentoxide).

Splitting: Mass number of fragments (mass spectrometer[104] cf. Chapter 12.4)

3.6.3 Physical and Physicochemical Detectors Combined with a Chemical Reaction

The signal in the detector arises not from the sample component itself but from the products of a previous chemical reaction of this component.

Oxidation with a copper oxide reactor to CO_2: Thermal conductivity (thermistor or hot wire, signal increased by all compounds with more than one C-atom[105]), electrical conductivity (dilute NaOH or H_2O[106]), concentration (E.M.F., coulometric regeneration, $BaClO_4$[107]), infra red absorption (non-dispersive photometer[108]), reaction coulometer (the consumption of oxygen is measured by electrolytic supplementation[109]), radioactivity (^{14}C-measurement in a counter[110]). *Compounds with active hydrogen* react with hydrides to form hydrogen which is detected by thermal conductivity cell detectors. Likewise, the reaction with carbide to acetylene is detected with a flame ionization detector.

Also *phosphorus, sulfur, nitrogen* and *halogen* compounds can be detected with a titration cell after previous oxidation or reduction[111].

3.6.4 Biological Detectors

These qualitative detectors use the reaction of the sense organs of living organisms as a signal, e.g. aroma (the experimenter's nose at the end of the column, the behavior of animals towards bait[112]).

3.6.5 Technical Data of Detectors

The expression of the sensitivity depends upon the type of detector. Sensitivity can be conceived as the response a for a definite substance, usually benzene (occasionally also air, methane or n-pentane).

For concentration dependent detectors

$$a_c = \frac{S_c}{C_i} \quad [mV \cdot ml \cdot mg^{-1}],$$

(33)

practically:

$$\frac{peak\ area \cdot carrier\ gas\ flow\ rate}{sample\ weight}$$

[87] *J. Guiliot, H. Bottazzi, S. Guyot, Y. Trambouze*, J. Gas Chromatogr. *6*, 605 (1968).

[88] *J. Sladacek*, Collect. Czech. Chem. Commun. *25*, 636 (1960).

[89] *R.P.W. Scott*, Nature *176*, 422 (1955).

[90] *M.C. Bowman, M. Beroza*, J. Chromatogr. Sci. *7*, 485 (1969).

[91] *R.W. Morrow, J.A. Dean, W.D. Shults, M.R. Guerin*, J. Chromatogr. Sci. *7*, 572 (1969).

[92] *J. Krugers*, Philips Res. Rept. Suppl. *1*, 1 (1965).

[93] *R. Nunnikhoven*, Z. Anal. Chem. *238*, 79 (1968).

[94] *E. Otte*, Chromatographia *1*, 234 (1968).

[95] *M. Dressler, J. Janak*, J. Chromatogr. Sci. *7*, 451 (1969).

[96] *C.H. Hartmann*, J. Chromatogr. Sci. *7*, 163 (1969).

[97] *R.A. Mees, J. Spaans*, Z. Anal. Chem. *247*, 252 (1969).

[98] *A.T. James, A.J.P. Martin*, Biochem. J. *50*, 679 (1952).

[99] *O. Piringer, E. Tataru, M. Pascalu*, J. Gas Chromatogr. *2*, 104 (1964); Hartmann & Braun, D-6 Frankfurt/M.-W 13, Gräfstraße 97.

[100] *G. Burton, A.B. Littlewood, W.A. Wiseman*, Gas Chromatography 1966, p. 193, The Institute of Petroleum, London 1967.
M. Lorant, Chem. Rundschau [Solothurn] *18*, 728 (1968).

[102] *A. Berton*, Chim. Anal. [Paris] *47*, 502 (1965).

[103] *G.E. Hillmann, J. Littlewood*, Anal. Chem. *38*, 1430 (1966).

[104] *K. van Cauwenberghe, M. Vandewalle, M. Verzele*, J. Gas Chromatogr. *6*, 72 (1968).

[105] *J.C. Gourtier* in Ref. [193], p. 185.

[106] *D.M. Coulson* in Ref. [192], p. 197.

[107] Richard Schoeps, Duisburg-Beeck, Arnoldstraße 63–65.

[108] *A.E. Martin, J. Smart*, Nature *175*, 422 (1955).

[109] *G. Burton, A.B. Littlewood, W.A. Wiseman*, Gas Chromatography 1966, p. 193 (Proceedings of the 6th International Symposium on Gas Chromatography and Associates Techniques, Rom, Sept. 1966).

[110] *A.T. James, C. Hitchcock*, Kerntechnik *7*, 5 (1965).

[111] Dohrmann Instr. Co., Mountain View, Calif. 94 040, USA; Techmation, D-4 Düsseldorf.

[112] *A. Butenandt*, Naturw. Rundschau *8*, 457 (1955). Ref. [199], Bd. II, p. 204.

For mass dependent detectors

$$a_m = \frac{S_m}{m_i/t} = \frac{S_m t}{m_i} \qquad [A \cdot sec \cdot g^{-1}], \qquad (34)$$

practically: $\qquad \dfrac{\text{peak area}}{\text{sample weight}}$

It must be known in which ratio the amplifier transforms μA at the input into mV at the output in order to describe the detector (e.g. FID) by means of a_m independent of the remaining system.

S_c: signal of concentration dependent detector e.g. in mV
S_m: signal of mass dependent detector in A
c_i: concentration of sample (mg/ml)
m_i/t: mass flow rate of sample $(g \cdot s^{-1})$

The *correction factor* f is a measure of the altered sensitivity of specific and selective detectors if another substance (x) is to be detected instead of benzene (or other standard substance)

$$f_x = \frac{a_{Bz}}{a_x} \qquad (35)$$

Drift of a detector is a slow, largely continuous variation or fluctuation of the baseline at the measuring region [mV/h or A/h]. It should be as small as possible and, at least during the course of an analysis (up to ca. 2 hours), it should not have an adverse effect on necessary sensitivity ranges. Automatically operated process gas chromatographs have particularly low tolerances in this respect.

Although *sensitivity* should be as high as possible this is not the only important requisite. There is no advantage in using a highly sensitive detector which is particularly noisy.

The *noise* (also static, according to the detector [mV] or [μA]) consists of a very rapid statistical fluctuation of the baseline measured under the normal operating conditions of the detector but without column and sample[114]. The noise should be small (see detection limit). The level is increased on introduction of the column and sample (analytical noise).

The *detection limit* is the smallest amount of substance which is definitely detectable. It is defined as the quantity of sample required to give a signal of double the noise level (c_i or m_i/t)[115].

The *upper saturation limit* is the amount of substance which, when increased, does not cause an increased signal (c_i or m_i/t).

---

[114] Ref. [199], Bd. II, p. 204.
[115] Ref. [199], Bd. II, p. 209.

The *dynamic range* is the quotient of the saturation limit and the detection limit. Depending on the detector it comprises some 10^2 to 10^7 fold.

The *linear range* is that part of the dynamic range over which the signal is directly proportional to the amount of sample within an error of $\pm 2\%$. It should comprise as large a part of the dynamic range as possible, particularly in the case of nonspecific detectors. Frequently a reduced linearity of selective and highly specific detectors must be accepted.

The *time constant* τ is the time which elapses between a stepwise change in the system, and the registration of 63.2% of the change in the signal. It τ is the combination of the time constant of individual parts of the measuring equipment (detector, amplifier, recorder). For the detector, τ should be < 1 second; the value for the FID is as low as milliseconds. Just as each chain is characterized by its weakest link, so the slower potentiometric recorder usually determines the time constant.

The *capacity* of a detector is determined by its volume. Disproportionately large detector volumes (in comparison with the carrier gas flow) or dead volumes in the pneumatic system lead to a slower and less sensitive indication, as well as to peak spreading. The detector volume is particularly important when using capillary columns because of the very low gas flow rates employed. Therefore, microversions of many detectors have been developed (volume: some μl). The FID is particularly suitable for capillary columns.

3.7 Evaluation of the Signal

Concentration detectors give a signal which is proportional to the concentration of sample (e.g. TCD).

Mass detectors generate a signal corresponding to the actual amount of sample per unit time (e.g. FID cf. 2.10:3.6.1).

The assignment of a detector to one of these two groups can be made by measuring the inert peak and then measuring it a second time under the same conditions but with only half the gas flow rate.

With concentration detectors the peak width is doubled and remains about the same height. With mass detectors the peak width is doubled, the height is halved and the area remains constant as long as the detector signal is not disturbed or influenced by the change of carrier gas flow. Treatment of the measured value by the detector:

It is given as a proportional electrical signal directly to the recording apparatus (TCD, FID) and almost all remaining detectors, gauss peaks).

It integrates it with time and passes on the integrated signal (gas volumetric detector[65]; steps, comparable with a polarogram).

It differentiates it and passes on the first derivative (against time) to the recording apparatus (heat of adsorption detector[70]). Signals in the form of S-shaped curves lying above and under the baseline.

Detectors can differ in their *sensitivity* toward various classes of substance

Unspecific detector: Response for all substances is the same (e.g. TCD with helium as carrier gas)
Specific detector: Larger differences in response (e.g. FID with different homologous series of substances containing heteroatoms such as halogens, oxygen, etc.)
Selective detector: The response for certain substances deviates from those of all other homologous series. A response of zero is often encountered, by several powers of ten (e.g. halogen, phosphorus, nitrogen detector).

3.8 Recorders

It is usual to use a potentiometric recorder in the laboratory. Such a recorder should have a recording width (paper width) of at least 120 mm, but preferably of 200 mm. The pen response should be 1—1.5 seconds for normal analysis with packed columns and 0.1—0.5 seconds for rapid analysis or when operating capillary columns. The input range should be between 1 and 5 milli-volts.

The adjustment of amplification and damping should result in only the slightest vibration of the recorder pen; no recognizable deflection in excess of the final deflection of the pen should occur on sudden application of a measuring potential. The recorder should be insensitive to any noise in the a.c. line produced by the oven heater of the gas chromatograph. To avoid a noisy signal which is caused by electrical interference, all grounding wires are connected to one point so that, under certain conditions, the grounded plug of the recorder must be used ungrounded. The ink system should neither dry out nor cause blotting. Ball point pens are usually less suitable than narrow felt peak.

3.9 Integrators

3.9.1 Method of Operation

Integrators are used for the quantitative evaluation of separations of mixtures of many components, as well as for all kinds of routine analyses. They replace the time consuming manual methods of chromatogram evaluation (e.g. height × peak width at one-half height × 1.0632 or planimeter, etc.) and are capable of evaluating the area of very distorted peaks. However, the disadvantage of an integrator is its lack of discrimination so that a small peak trailing a main component may include the tail of the main component. Also the integrator cannot distinguish between genuine peaks of a

sample, and the very broad peaks arising from a previous injection. When using integrators, therefore, it is advisable to have some visual control of the chromatogram on at least a small 100 mm recorder.

Within the framework of these limitations all integrators yield meaningful values of peak area, if:

The peaks are completely or almost completely separated; exception: special circuitry, with superposition of the first derivative can make the chromatogram more suitable for integration[116].
The baseline remains constant; exception: the integrator is drift-compensated by suppressing signal increases below a previously determined threshhold value; disadvantage: shallow peaks at the end of the chromatogram will be suppressed also.
The signals are free from interference; exception: the integrator suppresses signal increases above a specified threshhold; disadvantage: it is difficult to distinguish between substances eluted early and interference impulses during rapid chromatography or when using capillary columns.

When the integrator has the additional advantage of fully automatic printout facilities, the following are obtained:

the printed-out peak area
the printed-out retention time t_R
the peak area multiplied by a correction factor for determination of the percentage weight or volume.

On the other hand problems also result:

The release of the printout signal usually occurs at the beginning of the next peak and requires a signal increase which lies between drift and interference and which is recognized as a peak by the integrator. This means that the first small increase and consequently its area is not included in the correct peak but in the previous one.
Further, the automatic printer requires time to print out, during which the integration is usually stopped. Since this operation takes place just at the beginning of a peak, a part of the front flank of the next peak is not integrated with it. Exception: If the integrator possesses two memory storage units, one may be used to integrate, while the other prints out the peak integral of the previous peak.

3.9.2 Construction

Semi-automatic integrators are usually connected with the potentiometric recorder and are the cheapest systems. However, the inertia of the recorder is also included in the integration:

---

[116] *J.W. Ashley, C.N. Reilley,* Anal. Chem. *37,* 626 (1965).
[117] *A.G. Nerheim,* Anal. Chem. *36,* 1686 (1964).

Ball and disc (mechanical) with counter or second recorder.
Back potential through servomotor and dynamo (electromechanical, with computation of the rotation of the dynamo)
RC-integrator (operation amplifier wired with condenser and resistance, measuring instrument or second recorder)

Fully automatic integrators are connected directly to the gas chromatograph and yield printed data without delay. They possess electronic analog-digital-converters and automatic shift registers for range-switching. The pulse received from the analog-digital-converter can be stored also on magnetic tape and fed later to the printer.
The reproducibility, linearity and precision obtained with an integrator is dependent largely on its construction but is also dependent on the sample and the gas chromatograph[119].

3.10 Collection of the Separated Components
The general problem during collection of the separated components is the removal of a relatively small proportion of the vapor of a condensible substance from an excess of the carrier gas. The partial pressure of the substance above the already condensed material is small so that very strong cooling is necessary. This is made more difficult by the high speed with which the carrier gas flows through the cooling traps particularly in a preparative apparatus. An additional difficulty is mist formation.
According to the amount to be collected one may differentiate between micropreparative (analytical methods), and preparative methods, from which the separated fractions are intended for further processing. For component collection the following are used:

Narrow empty tubes[120, 121] or empty vessels, according to the amount to be collected (at −80 or −180°), special traps agaits mist formation with fiber-glass or stainless steel fabric tubing, or traps with an electrical potential[122].

Subliming substances are collected in traps with wide inlet tubes.
Tubes filled with:
a) active adsorbents, which are also used as column packings in GSC[123]
b) column packing materials from GLC[124] with suitable stationary phases, which are cooled in order to collect the components. The components are later expelled by passing a stream of gas through the warmed material and are condensed afresh in an empty trap or cuvette[125].
c) liquids which, when strongly cooled, solidify to a thick felt of crystals. Sufficient interstices are present, however, to allow passage of the carrier gas (e.g. CCl_4 at −180°)[126]. A solution suitable for IR-spectroscopy, which can be directly filled into the cuvette, is obtained on warming to room temperature.
d) solid, inactive crystals[127] on which the components are precipitated on cooling (e.g. KBr for IR-pressed discs). Disadvantage: low boiling components can evaporate away during evacuation of the disc press.

Thin layer chromatography plates: the plates are strongly cooled on their reverse sides by means of a stream of cold gas, while the carrier gas from the column exit, together with the separated component, impinges on to a small portion of the plate. After passage of a peak, the thin layer plate controlled by the detector of the gas chromatograph is automatically moved on. The components precipitated on the thin layer plate can then be further separated by the various procedures of thin layer chromatography[128].

KBr-discs or chilled IR-cuvettes: if pre-pressed KBr-discs are kept out of contact with air and at low temperature on a strongly cooled metal block (e.g. copper rod, half immersed in a carbon dioxide-acetone bath or liquid nitrogen), substances can be precipitated on their surfaces, which can be analyzed in the IR-spectrometer. The cuvettes are cooled in the same way[129]. KBr-discs are used without cooling for collection of higher boiling substances[130].

3.11 Automatic and Semi-automatic Apparatus
According to the use to which they are put one distinguishes between:
Process gas chromatographs: these analyze the process stream of an industrial plant and their results can be fed back to the plant for control purposes. They consist of the analyzer (gas chromatograph, set up in the plant), programmer (electronic part) and recorder (both stationed in the instrument measuring control panel). An important part of the process chromatograph is the

[119] *R. Kaiser*[199], Bd. IV, p. 130.
D.H. Boege, U. Kohl, B. Kolb, E. Wiedeking, Tip 37 GC, Bodenseewerk Perkin-Elmer, Juni 1968.
B. Kolb, E. Schuhbeck, Tip 39 GC, Bodenseewerk Perkin-Elmer, April 1969.
[120] *M. Beroza*, J. Gas Chromatogr. *2*, 330 (1964).
E. Rudloff, F.M. Couchman, Can. J. Chem. *42*, 1890 (1964).
R.A. Edwards, I.S. Fagerson, Anal. Chem. *37*, 1630 (1965).
[121] *K.R. Burson, C.T. Kenner*, J. Chromatogr. Sci. *7*, 63 (1969).
[122] *W.D. Ross, J.F. Moon, R.L. Evers*, J. Gas Chromatogr. *2*, 340 (1964).

[123] *J.N. Damico, N.P. Wong, J.A. Sphon*, Anal. Chem. *39*, 1045 (1967).
[124] *M.D.D. Howlett, D. Welti*, Analyst (London) *91*, 291 (1966).
[125] *K.H. Kubeczka*, Planta Med. *17*, 294 (1969).
[126] *K. Witte, O. Dissinger*, Z. Anal. Chem. *236*, 119 (1968).
[127] *R.A. Greenstreet*, J. Chromatogr. *33*, 530 (1968).
[128] *R. Kaiser*, Chimia *21*, 235 (1967).
[129] *J.T. Ballinger, T.T. Bartels, J.H. Taylor*, J. Gas Chromatogr. *6*, 295 (1968).
[130] *E.I.M. Bergstedt*, Chromatographia *2*, 545 (1969).

Table 1. Automatic function of various types of apparatus

Function	Process GC	Autom. Lab. GC	Prep. GC
Sampling	+	–	–
Sample preparation	+	–	–
Sample choice	+	(+)	–
Program choice	+	–	–
Baseline adjustment	+	(+)	(+)
Sample injection	+	+	+
Column switching	+	–	–
Temperature programming[20]	–	+	+
Choice of the peaks to be measured or collected	+	(+)	+
Further data utiliation	+	(+)	–
Freezing out of separated substances	–	–	+

+ automatic   – not automatic   (+) often automatic

often elaborate system for sample preparation before the analyzer. This is necessary to remove non-volatile impurities from the process stream.

Automatic laboratory gas chromatographs: there is no separation between analyzer and programmer, no protection against explosion, and no provision for sample preparation. Also the sampling devices are simpler, and usually semi-automatic.

Preparative gas chromatographs: these have an automatically operating program; by frequent repetition of the separation, preparatively usable quantities are obtained. The various operating procedures are automated according to the requirements of the type of apparatus concerned (s. Table 1).

4 Operating Technique

Sampling is discussed in Chapter 1

4.1 Sample Preparation and Injection

Table 2. Sampling of gases

Substance	Example	Method	Sample injection
Permanent gas	pure gas for NH₃ synthesis	branching from process stream	gas sampling valve with sample loops
Liquid gas	propane	from liquid phase through vaporizer	sample loop, heated if necessary
Traces of gas	gases above bacterial cultures	[a] syringes [b] recirculating pump	[a] liquid injection port [b] sample loops
Multiphase systems Condensible components	gases from canned food[131] serum blood alcohol	connecting tube[132] capsules with rubber covers[29] condensate	gas syringe excess pressure thawing out
Damp gases	distillation column residue breath	pre-column <20° from gaseous phase corresp. vapor pressure [a] after warming [b] after drying	desorption >20°[125] gas loop thermostated[71] gas loop
Supersaturated gas	motor exhaust gas	after precipitation	
Liquid gas with polymer content	butadiene	processing system with exchange regeneration	
Traces	rare gases in the air	sample loops with large volume[134]	

[a] with capillary columns, using a well constructed splitter to avoid loss of low boiling components[135].
[b] stored in firmly stoppered vessels until sampling.

[131] *M.F.Mc. Guckin*, US Pat. 3, 347, 678 (1968).
[132] *Mohr*, Verpackungsrundschau *17*, 17 (1966).
[133] *B. Kolb, E. Wiedeking, B. Kempken*, Angew. GC Heft 11, Bodenseewerk Perkin-Elmer 1968.
[134] *B.M. Karlsson*, Anal. Chem. *35*, 1311 (1963).

Table 3. Sampling of liquids

Substance	Example	Method	Sample injection
Solution of narrow boiling range	impurities in $CHCl_3$	—	syringe
Gases in solution	air in hexane	—	syringe
Solution of wide boiling range[a]	diesel fuel	—	syringe
Solution with some non-volatile components	blood alcohol	a) distillation b) pre-column c) see Table 2	syringe syringe head space
Gas containing solution with non-volatile components	blood gas	driving off the gases from the mechanically shaken solution	gas injector[136] (modified)
Solution of low boiling constituents	crude petroleum[b]		destruction of ampoule
Solution of volatile solids	dose of solid substances[c]	—	syringe
Solution not volatilizable without decomposition	fats	derivative formation	syringe
Heterogeneous systems	process control	filtration centrifugation homogenization	syringe

[c] the peak due to the solvent should not overlap the sample peaks; in doubtful cases carry out several separations, with a solvent to each separation.

Table 4. Sampling of solids

Sample	Procedure, sampling	Sample	Procedure, sampling
Volatile without decomposition (biphenyl)	dissolve (cf. Table 3) dissolve and dry[133, 136] cf. 2.10:3.3.3 pressed disc cf. 2.10:3.3.3 indium tubes cf. 2.10:3.3.3	steroids[140] carboxylic acids[140, 141])	rapid pyrolysis with high frequency heating[144] pyrolysis by discharge[145] pyrolysis by laser[146]
		Analytically interesting gases	destruction in gas sampling inlet (modified)
Non-volatile component (tar)	pre-purification by distillation	Dissolved analytically interesting gases	pre-separation after melting in vacuum[147, 148]
Not volatilizable undecomposed (amino acids[138] carbohydrates[139]	formation of derivatives[137–141] slow pyrolysis[142] rapid pyrolysis[143]	Analytically interesting liquids (plasticizer) or monomers) dissolved in solids.	pre-separation by extraction or filtration[143] head space (cf. Table 2)

[135] H. Brudereck, W. Schneider, I. Halasz in Ref. [195], p. 91.

[136] Carlo Erba, Milano, Via Imbonati 24.

[137] K. Hammarstrand, E.J. Bonelli, Derivatives, Varian-Aerograph 1968.

[138] K. Blau in Ref. [201], Vol. 2, p. 1; Anon, Chem. Eng. News 47, 30 (1969).

[139] T. Bhatti, J.R. Clamp, Clin. Chim. Acta 22, 563 (1968).

K. Hammarstrand, Carbohydrates, Varian-Aerograph 1968.

A.L. Larsen, A.W. Engstrom, Am. J. Clin. Pathol. 46, 352 (1966).

4.2 Separation

Table 5. Separation possibilities for various problems

Components	Retention time difference	Example	Temperature	Column
Low boiling components, gas				
Few	large	CO_2, O_2 in air impurities in $CHCl_3$	low	adsorption or partition[149]
Many	small	alkanes, alkenes $C_4 - C_6$	low	high separation efficiency correct polarity[150] if necessary combination of columns
With one column inseparable mixtures		{ coal gas fuel gas }	isothermal	several columns[60] with switching valves
			isothermal	with reaction column[153]
			programmed	molecular sieves
Inseparable at $> 20°$		rare gases	program started at $-180°$	molecular sieves
Traces		aromas	reversion GC[a, 154, 157]	
Sample including high boiling components	large	reaction products	isothermal	back flushing (high boiling substances as summed peak)
Medium to high boiling components				
Few	large	esterification products	medium	polar
Many	small	pre-separated fractions	isothermal	capillary
Many	large	diesel fuel	programmed	capillary
Temperature sensitive[155]		tautomeric substances	isothermal	flow programmed, low coating
Catalyst sensitive		steroids	isothermal or programmed	glass apparatus
Traces		pollutions of water	isothermal or programmed	large loadability[b]
Vapors with mist or smoke		pyrolysis products	isothermal or programmed	protected with glass wool wad
Unknown mixture		odors	isothermal or programmed	2 columns (nonpolar/polar) in series/parallel arrangment
Very high boiling		polyaromatic	isothermal or programmed	salt eutectics[156] or [c]

[a] In the case where the sample has not already been enriched by extraction, preparative GC or LC: reversion gas chromatography with a gradient heater moving along the column, with continuous injection of the sample gas through a cold seal at the column entrance[154].

[b] From the capillary columns only a macro-capillary with porous layer on the walls (100 m long, 1 mm i.d.) is suitable; perhaps also collection of the traces with a preparative column by discarding the main components and collecting all remaining substances for subsequent analytical separation.

[c] Capillary columns with one of the few organic stationary phases which is stable at high temperatures. It can be operated up to about 100° below the boiling point of the sample because of the high phase ratio β of the capillary column.

4.3 Detection

4.3.1 Quantitative Analysis

In principle, any gas chromatographic detector which responds to the component in question may be used, after corresponding calibration. Nevertheless, in practice, the sensitivity should be sufficient to avoid overloading the separation column by large amounts of sample. Further, there are often a number of components present for which calibration is not possible, because they either have not been identified, or are not available in the pure state. Therefore, unspecific detectors are usually preferred for quantitative analysis.

The *thermal conductivity detector* (TCD) meets these requirements to a large extent, since, when using helium as the carrier gas, only hydrogen and neon possess thermal conductivities of the same orders of magnitude. All remaining gases and vapors have much lower thermal conductivities. The dynamic range is about 10^4-fold, while there is no apparent noise when using a 2.5 mV-recorder under normal operating conditions.

Bearing in mind that the *flame ionization detector* (FID) responds primarily to C-H compounds, and that most permanent gases such as nitrogen, oxygen, carbon dioxide, ammonia, water, carbon disulfide, etc., are not registered at all, the F.I.D. may be considered as an almost non specific detector at least within a homologous series of organic compounds.

The order of magnitude of the noise is quite variable and depends to a large extent on the purity of the gas used. The FID tolerates, for a specified jet cross section, only a limited carrier gas flow rate since either the mixture with the burning gas is not ignitable, or at higher hydrogen flow rate, the flame is too large. In general, a carrier gas flow rate of 30—50 ml/ minute should not be exceeded.

The helium detector is highly sensitive for all compounds but it is used predominantly for trace analyses of the permanent gases[157,158]. Its dynamic range is ca. 10^4.

4.3.2 Qualitative Analysis

A specific indication is desired in this case. This is demonstrated most simply by an experiment[197] in which the separated components from the column exit give rise to different colors of a hydrogen flame. In principle, each specific or selective detector yields such information. The disadvantage is that a small signal can be derived from a small amount of substance for which the detector is sensitive equally as well as from a large amount of insensitive substance. Therefore, an unspecific detector must be added in series or parallel in order to obtain meaningful qualitative information. Frequently only one definite class of substances is of interest and is indicated by a suitable special detector, while the remaining components are neglected. Even so, the general properties of detectors

Detector	Volume	Time constant	Sensitivity	Dynamic range
TCD[113]	10–1500 μl	0.3–1.5 sec	3– 26 · 10^3 mV. ml · mg^{-1}	10^5
FID	1 μl	8 msec	5– 50 · 10^{-3} Coulomb/g	10^7
Helium[157,158]			20–200 Coulomb/g	10^4
(see also Ref.[172 a])				

[140] E.C. Horning, W.J.A. Van den Heuvel, Chim. Anal. [Paris] *48*, 12, 79 (1966).

[141] R.T. O'Connor, R.R. Allen, K.M. Brobst, J.R. Chipault, S.F. Herb, C.W. Hoerr, J. Am. Oil Chem. Soc. *46*, 57 (1969).

[142] D.R. Dill, R.L. Kinley, J. Gas Chromatogr. *6*, 68 (1968).

[143] L. Jacque, G. Guiochon, Chim. Anal. [Paris] *49*, 3 (1967).
   R.L. Levy, J. Gas Chromatogr. *5*, 107 (1967).
   R.L. Levy, Chromatogr. Rev. *8*, 49 (1966).

[144] W. Simon, P. Kriemler, J.A. Voellmin, H. Steiner, J. Gas Chromatogr. *5*, 53 (1967).

[145] J.C. Sternberg, R.L. Little, Anal. Chem. *38*, 321 (1966).

[146] O.F. Folmer jr., L.V. Azarraga, J. Chromatogr. Sci. *7*, 680 (1969).

[147] F.R. Bryan, S. Bonfiglio, J. Gas Chromatogr. *2*, 97 (1964).

[148] R.K. Winge, V.A. Fassel, Anal. Chem. *37*, 67 (1965).

[149] Ref. [198], p. 28, 38.

[150] Ref. [198], p. 28, 38.

[151] H. Pauschmann, Z. Anal. Chem. *211*, 32 (1965).

[152] W.F. Wilhite, O.L. Hollis, J. Gas Chromatogr. *6*, 84 (1968).

[153] E.J. Havlena, K.A. Hutchinson, J. Gas Chromatogr. *6*, 419 (1968).

[154] H. Oster, Z. Anal. Chem. *247*, 257 (1969).

[155] G. Hesse, Z. Anal. Chem. *236*, 192 (1968).

[156] F. Geiss, B. Versino, H. Schlitt, Chromatographia *1*, 9 (1968).

[157] D. Gere, J. Shatting, Trace Gas Analyzer Methods, Varian-Aerograph 1968.

[158] Ref. [186], p. 16.

[159] A.A. Zhuchovitskii in Ref. [190], p. 161.

[160] A. Karmen, J. Chromatogr. Sci. *7*, 541 (1969).

[161] D.M. Coulson, J. Gas Chromatogr. *4*, 285 (1966).

Table 6. Application of detectors (examples)

Substance	Column packing	Detector
Higher fatty acid esters	ethyleneglycol succinate	TCD with hot wire FID (also for traces)
Aromatic hydrocarbons	polyphenylether in capillary columns	FID (small measuring volume)
Gases	adsorption or diethylhexyl sebacate	thermistor-TCD
Gases	adsorption or diethylhexyl sebacate	TCD, standard mixture as carrier gas = vacancy chromatography[159]
Main component suppressed:		
Aqueous solutions		FID
Organic solutions		TCD with hot wire[a]

[a] temperatures at which the conductivities of some gases[4] and vapors are identical

	CO_2	N_2
pentane	50°	220°
acetone	200°	—

Table 7. Specific detectors for various heteroatoms

Detector	N	S	P	Halogen	Metal	Radioactive isotope
Thermionic	96, 160	95	160	160	—	—
Microcoulometric	a	163	163	163	—	—
Flame emission	—	162	162	Cu^b, Na^{165}	—	—
Electrochemical solid body cell	—	164	—	—	—	—
ECD	d	e	—	166	166, 167	—
atomic absorption	—	—	—	—	91	—
Geiger counter	—	—	—	—	—	168, 169

[a] After reduction to ammonia[161].

[b] Copper sensitized[90].

[c] Counter with pre-connected arrangement for oxidation to carbon dioxide or reduction to methane[169]

[d] As – NO_2, – ONO_2[170]

[e] *D.M. Oaks, M. Hartmann, K.P. Dimick*, Analysis of Sulfur Compounds with Electron Capture/Hydrogen Flame Dual Channel 6c. Varian-Aerograph, Darmstadt, Hilperstr. 8.

(2.10:3.6.5) should not be overlooked. The sensitivity of such detectors for the substances being registered is often surprisingly high and exceeds that of the paraffins frequently by 3 or 4 powers of ten; considering this fact, the high cost and often a narrower dynamic range are of secondary importance.

[162] *M.C. Bowman, M. Beroza*, Anal. Chem. *40*, 1449 (1968).

[163] *H.P. Burchfield, R.J. Wheeler*, J. Ass. Offic. Anal. Chem. *49*, 651 (1966).
D.F. Adams, G.A. Jensen, J.P. Steadman, R.K. Koppe, T.J. Robertson, Anal. Chem. *38*, (8), 1094 (1966).

[164] *E. Bechthold*, Z. Anal. Chem. *221*, 262 (1966).

[165] *A.V. Nowak, H.V. Malmstedt*, Anal. Chem. *40*, 1108 (1968).

[166] *E.A. Boettner, F.C. Dallos*, J. Gas Chromatogr. *3*, 190 (1965).

[167] *B. Kolb, G. Kemmner, F.H. Schleser, E. Wiedeking*, Z. Anal. Chem. *221*, 166 (1966).

[168] *D.L. Peterson, F. Helfferich, R.J. Carr*, Am. Inst. Chem. Engrs, Paper *12*, 903 (1966).

[169] *F. Drawert, A. Rapps, H. Ullmeyer*, Chem.-Ztg. *88*, 379 (1964).

[170] *C.D. Pearson, R.S. Silas*, Anal. Chem. *39*, 540 (1967).

Also, functional groups can be selectively detected[80], e.g. nitro groups or nitrites with the electron capture detector[170]. Certain properties of a molecule can lead to selective indication, such as its ability to be hydrogenated[171] or oxidized to carbon dioxide and water[172]. The hydrogen (or oxygen) consumed is re-generated electrolytically and is thereby measured. Numerous applications have shown that it is possible to adjust a mass spectrometer to follow a fixed mass number and thus use it as a detector[173]. The biological properties of a substance can be used also for selective detection (see Ref. [112]).

4.4 Information and its Processing

4.4.1 Qualitative Information

Table 8. Characterization of the separated products[174]

Differentiation by	Produced by	Detectable substances
Shift of retention times[a, 175]	reaction column Raney-Ni/H_2 before the separation	olefins
Separation on several columns[176]	diverse column fillings	many
Disappearance of peaks[175]	reaction with CaH_2 after separation	active hydrogen
Physical data[b]	direct coupling with MS, IR etc.	many
Specific detection[177]	parallel operation FID/ECD[b, c]	halogen compounds in hydrocarbons
Subsequent treatment[d]	continuation of the analysis[d]	many

[a] Only undistorted peaks (no fronting or tailing) are suitable for exact determination of retention data[176]

[b] Molecular weights can be determined by the gas density balance[177].

[c] A large difference in the signals obtained from two detectors gives information about the class of substance.

[d] Condensation on thin layer plates and thin layer chromatography (uniformity, R_f-values), microanalytical group reactions or spectroscopic methods (IR, UV/Vis, MS, NMR, EPR) and their evaluation[178].

4.4.2 Quantitative Information

Table 9. Quantitative evaluation

Parameter	Applicability
Peak height [mV]:	slender peaks from capillary columns; examination of the linear region
Peak area [mV · min]: peak height × gross retention time	slender peaks[a]
peak height × peak width (base) (one-half height) (mean of 15 % and 85 % of the maximum)	broad peaks very exact[d] peaks with fronting and tailing rough approximation (cf.[c])
Planimeter	troublesome[178]
Cutting out and weighing[c]	more suitable than planimeter
Integrators	series analyses multicomponent systems

[a] It is necessary to have very constant operating conditions, since fluctuations in the retention time cause proportional errors.

[b] Width of the ink trace is eliminated by consistently measuring from the left edge of the trace. Perfect gauss peaks can be converted into their exact area by multiplication $h_p \cdot h_{0.5}$ 1.0632.

[c] Recommended when the peaks are unsymmetrical and in the absence of an electronic integrator.

[171] A.B. Littlewood, W.A. Wiseman, J. Gas Chromatogr. 5, 334 (1967).

[172] G. Burton, A.B. Littlewood, W.A. Wiseman in Gas Chromatography 1966, p. 193 (Symposium Rom[109]).

[173] E. Gelpi, W.A. König, J. Gibert, J. Oro, J. Chromatogr. Sci. 7, 604 (1969).
D. Joly, Z. Anal. Chem. 236, 259 (1968).
D. Henneberg, G. Schomburg, Z. Anal. Chem. 215, 424 (1966).

[174] G. Schomburg, Z. Anal. Chem. 200, 360 (1964).

[175] M. Beroza, R.A. Coad, J. Gas Chromatogr. 4, 401 (1966).

[176] L. Eek, T. Galceran, Chromatographia 2, 541 (1969).

[177] S.C. Bevan, A. Gough, S. Thorburn, J. Chromatogr. 44, 241 (1969).

[178] Ref. [199], Bd. IV, p. 97.

4.4.3 Evaluation of Quantitative Data[179]

Absolute standardization: Pure standard substances are required. Impurities and sampling errors give rise to proportional errors in the calibration curve obtained. Recommended method for determination of individual substances in mixtures.

100% -method: The area of every peak is multiplied by the specific correction factor, these products are added, and from the relation between this sum and the individually corrected peak areas the proportion of each component concerned is ascertained. The procedure is troublesome since every correction factor should be known. It is recommended for mixtures containing few components, or mixtures of substances which possess similar correction factors (e.g. many homologous series).

Admixture: The components being investigated must be available in the pure state. Two analyses are involved: the sample mixture is investigated first alone and again after addition of a small amount of the pure substance. A neighboring peak is used for comparison. This method works without correction factors. Errors are to be expected if there is curvature in the calibration curve of the detector.

Internal standards: In this case a pure substance (the standard) is required in addition to the component to be determined. The retention times of both component and standard should be close to each other. At first the relation between the correction factors of component and standard is ascertained by sampling a homogeneous mixture containing weighed amounts of both substances. An amount of standard comparable to the amount of component is then added to the sample and the analysis is repeated. This represents a good procedure for traces of substances and samples which are not completely volatilized.

4.4.4 Data Processing

The qualitative and quantitative data may either be supplied in printout form on paper tape from electronic integrators, or it is fed into an *electronic data processor,* which either accepts the results directly from the gas chromatograph or integrator, or it refers them to the memory (e.g. with magnetic tape). The precision requirements for the gas chromatograph are so high, however, that this method is not yet developed. When the demand is not for utmost precision but more for a trend signal, a process gas chromatograph has been useful for control of less complicated operating components (e.g. distillation columns which control a definite percentage of a component in the head fraction).

5 Application

GC as an analyzed method was discovered in 1952[98].

With about 30,000 publications which have appeared to date on GC, the analyst can review his field of work relatively easily despite some short-comings of the abstracts. However, a change in the content of publications has been observed for some years. While at the beginning the emphasis was on methodology and petrochemistry, today applied work is being published in large amounts in the fields of biochemistry, medicine and toxicology.

A literature search in a particular field of works should in each case include the GC abstracts (ed; Knapman)[207] and also Preston card index[208], allowing for a delay of indeterminate length between publication and abstraction.

Indexes of text books[149, 181, 187] can be consulted as to the best column to use for a particular organic chemical separation (2.10.6). Also collections of data and reviews provided by manufactures of apparatus and components are useful[182, 183].

For separation problems which are not exactly defined a survey analysis should be carried out using a silicone rubber column (e.g. OV 17) — programmed from room temperature to the upper temperature limit of the packing. For strongly polar samples it is recommended to use a similar temperature programmed survey analysis with porous polystyrene (e.g. Porapak Q®), with polyethyleneglycol (e.g. Carbowax 1500) as well as with a fluorinated stationary phase (e.g. FS 1265).

---

[181] Ref. [199], Bd. III, p. 45

[182] Bodenseewerk Perkin-Elmer, 777 Überlingen (Bodensee) Tip 9 GC (Retentionsdaten).

[183] Bodenseewerk Perkin-Elmer, 777 Überlingen (Bodensee): Angewandte Gaschromatographie, 1967, Heft 9: Säulendatenblätter.

6 Bibliography

[184] *H. Karthaus,* Die Physikalische Gasanalyse, Hartmann und Braun, Frankfurt/M. 1958.

[185] *S. Dal Nogare, L.W. Safranski,* Gas Chromatography: Organic Analysis, Vol. VI, Interscience Publishers. New York 1960.

[186] *J.E. Lovelock, R.P.W. Scott,* Gas Chromatography 1960, Butterworth Scientific Publ., London 1960.

[187] *E. Bayer,* Gas-Chromatographie, 2. Aufl. Springer Verlag, Berlin 1962.

[188] *S. Dal Nogare, R.S. Juvet,* Gas-Liquid Chromatography — Theory and Practice; Interscience Publishers, New York 1962.

[189] *L.S. Ettre,* Open Tubular Columns in Gas Chromatography, Plenum Press, New York 1965.

[190] *A. Goldup,* Gas Chromatography 1964 (5th Int. Symposium Brighton), Institute of Petroleum, London 1965.

[191] *H. Farnow,* Gas-Chromatographie, Dragoco. Holzminden 1966.

[192] *I.C. Giddings, R.A. Keller,* Advances in Chromatography, Vol. III, Edward Arnold Publishers, New York 1965/66.

---

[179] Ref. [199], Bd. IV, p. 206.

[193] *G. Parissakis*, Chromatography and Methods of Immediate Separation (Journées Hellènes d'Etude des Méthodes de Séparation. Immédiate et de Chromatographie, Athen 1965), Association of Greek Chemists. Athen 1966.

[194] *H.G. Struppe, H. Obst*, Gas-Chromatographie 1965 (5. Symposium in der DDR); Akademie Verlag, Berlin 1966.

[195] *A. Zlatkis*, Advances in Gas Chromatography 1967 (IV. Intern. Symposium New York), Preston Techn. Abstr. Co., Evanston, Ill. 1967.

[196] *K.J. Bombaugh, W.A. Dark, D.F. Horgan, P.W. Farlinger*, Gas Chromatography on Porous Silica Beads, Waters Associates Inc., Framingham, Mass., USA 1968.

[197] *G. Hesse*, Chromatographisches Praktikum, Akademische Verlagsgesellschaft, Frankfurt/M. 1968.

[198] *D. Jentzsch*, Gas-Chromatographie, Frankh'sche Verlagsbuchhandlung, Stuttgart 1968.

[199] *R. Kaiser*, Chromatographie in der Gasphase Bd. I—IV, Hochschultaschenbücher Bibliographisches Institut, Mannheim 1960—68.

[200] *H. Pauschmann* in *H.G. Struppe, D. Obst*, Gas-Chromatographie, Akademie Verlag. Berlin 1968.

[201] *H.A. Szymanski*, Biomedical Applications of Gas Chromatography. Vol. I, II, Plenum Press, New York 1968.

[202] *C.L.A. Harbourn*, Gas Chromatography 1968 (7th Intern. Symposium), Institute of Petroleum, London 1969.

[203] *G.V. Marinetti*, Lipid Chromatographic Analysis, Vol. 2, Marcel Dekker, New York 1969.

[204] *H.M. McNair, E.I. Bonelli*, Basic Gas Chromatography, Consolidated Printers Berkeley, Cal. 1969.

[205] *A. Zlatkis*, Advances in Chromatography (Vth Intern. Symposium Las Vegas 1969), Preston, Evanston 1969.

[206] *D. Jentzsch, E. Otte*, Die Detektoren der Gas-Chromatographie, Akademische Verlagsges. Frankfurt/M. 1970.

[207] *C.E.M. Knapman*, GC-Abstracts, The Instituts of Petroleum (London).

[208] GC-Literature, Preston Technical Abstraction Co., Evanston, Illinois.

2.11 Adsorptive Bubble Separation Methods

Klaus Maas
Organisch-Chemisches Institut der Universität,
D-69 Heidelberg

1 Principles and Analytical Applications

1.1 Changes of Concentration at Interfaces

Dissolved molecules possessing a polar watersoluble part (● = hydrophilic) and an apolar water-repellent part (— = hydrophobic) according to the diagram Fig. 1 prefer existing interfaces between polar liquid and apolar liquid or gaseous phases. The *Gibbs adsorption theorem* (1876) relates the concentration of such surface-active compounds with the reduction of the *surface tension* of their solutions. The quotient obtained from the increased concentration Γ in the interphase layer and the

$$\frac{\Gamma}{c} = -\frac{1}{RT} \cdot \frac{d\gamma}{dc} \quad (1)$$

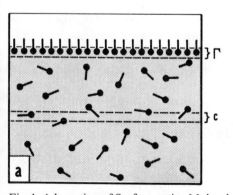

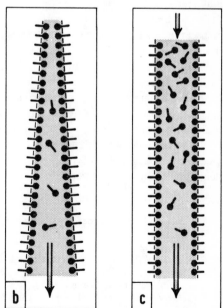

Fig. 1. Adsorption of Surface-active Molecules on the Gaseous: Liquid Phase Interface (diagram)
a concentration comparison with the interior of the solution
b concentration increase by draining of lamellar liquid
c concentration increase by concentrated collabate

concentration c of a similar layer inside the solution must be considered as a distribution coefficient and can be utilized for separation; cf. Fig. 1a and Equation (1).

Where the constants a [dyn/cm] and b (specific capillary activity, [l/mol]) are known, it is possible to calculate Γ e.g. for fatty acids by means of Equation (2) which corresponds to the Langmuir adsorption isotherm[1].

$$\Gamma = \frac{0,4343}{RT} \cdot \frac{abc}{1-bc} \qquad (2)$$

(Limiting areas: bc $\ll$ 1 signifies proportional increase of Γ with c; bc $\approx$ 1 signifies saturation of the surface, i.e. Γ_{max}).

Exact calculations must take activity coefficients of electrolyte solutions into account[2]. Qualitative data are sufficient for practical foam analysis. The conventional methods operate best with 10^{-3} to 10^{-7} molar solutions of surface-active substances[3]. At high concentration, aggregates (micellae) are formed with hydrophile groups directed outwardly. In case of high dilution the boundary layer contains only a few surface-active molecules and surface tension differs little from that of the pure solvent.

Gas bubbles rising in an aqueous medium concentrate the substances adsorbed according to Fig. 1a in the upper part of the liquid column. At a given distribution coefficient Γ/c, this 'extraction' depends on the surface generated, i.e. on the volume and degree of dispersion of the gas phase. Rediffusion and, in particular, mechanical backmixing restrict the concentration factor in foamless 'bubble fractionation' (see 2.11:2.2). As a rule, more or less stable foam lamellae coated on both sides with surface-active molecules form as liquid is discharged. The intermediate stages during '*draining*' between spherical foam ('Kugelschaum'[4]) and dry foam with Plateau surfaces, sharp edges, and angles were described by Kitchener and Cooper[5]. Occurrence and stability of the lamellae are dependent upon the following factors:

surface tension of solution or solvent;
surface activity and concentration of dissolved substances;
orientation of molecules at the phase boundary[6, 7];
viscosity of the solution;
any incorporation of a solid phase (types of flotation).

The lamellae which collapse especially in the upper part of the foam column (Fig. 2) form a concentrated fluid reflux ('collabate'). Consequently, as seen in Fig. 1b, the foam separation methods are able to supply concentrates in that the deplenished liquid drains inside the lamellae and the boundary surface related to the liquid volume is raised. Much more effective, as seen in Fig. 1c, is the exchange of internal laminae liquid by the enriched reflux in a countercurrent process (see 2.11:2.3).

1.2 Surface-active Substances (Tensides)-Function of Hydrophobic and Hydrophilic Groups

In 1960 Götte[8] suggested 'tensides' as a brief and international term for surface active compounds. The discussion below is limited to the aqueous solutions and suspensions which are involved in most cases.

Exceptions are found in flotation, e.g. when NaCl, NH_4Cl, and $CaCl_2$ are concentrated from hydrocarbons using fluorinated carboxylic acids as collector (see 2.11:1.4)[9]. Depending on the solvent, polymers yield fractions of differing degree and type of polymerization[10]. An interesting procedure from the point of view of analysis is the separation of *mucins* from seed pods (e.g. psyllium) by flotation from chloroform[11], because the principle can be applied analogously to other *natural substances*.

The hydrophobic part of surface active molecules may be pure or mixed aliphatic, cycloaliphatic (e.g. in the case of saponins) and aromatic. Fluorination has an increasing effect, whereas the incorporation of nitrogen and oxygen atoms has a descreasing effect. Polar ether, ester, and carbonamide groups are sometimes introduced, forming the transition to the hydrophilic molecule range.

In accordance with the chemical properties of their hydrophilic part, the tensides are classified into

non-ionogenic tensides,
ionogenic tensides (anion tensides, cation tensides, amphotensides)

[1] *R. Brdička,* Grundlagen der physikalischen Chemie, 6. Aufl., p. 555, VEB Deutscher Verlag der Wissenschaften, Berlin 1967.

[2] *E. Rubin, E.L. Gaden* in: New Chemical Engineering Separation Techniques, p. 322, Interscience Publishers, New York, London 1962.

[3] *B.L. Karger, D.G. DeVivo,* Separation Sci. *3,* 393 (1968).

[4] *E. Manegold,* Schaum, Straßenbau-Chemie-Technik, Heidelberg 1953.

[5] *J.A. Kitchener, C.F. Cooper,* Quart. Rev. Chem. Soc. *13,* 71 (1959).

[6] *Wo. Ostwald, A. Siehr,* Chem.-Ztg. *64,* 649 (1937).

[7] *J.W. McBain, G.P. Davies,* J. Am. Chem. Soc. *49,* 2230 (1927).

[8] *E. Götte,* Fette, Seifen, Anstrichm. *62,* 789 (1960).

[9] A.P. 3 186 546 (1962), General Mills, inventor: *J.L. Keen;* C.A. *63,* 3 920c (1965).

[10] *G.L. Gaines, W.G. LeGrand,* J. Polym. Sci. B, Polymer Lett. *6,* 625 (1968).

[11] A.P. 2 132 484, G.D. Searle & Co., inventor: *P.A. Kober;* C.A. *33,* 321[6] (1939).

(cf. e.g. the list in[12]). The various potential foam separation methods correspond to this classification.

In the case of ionogenic tensides, the hydrophilic group may also act as a reagent and absorb dissolved substances on the basis of an ion link. The carboxylic or sulfonic acid groups of the anion tensides carry cations; the positive centers of the cation tensides (amine or ammonium salts, less frequently phosphonium or sulphonium salts) carry anions into the foam. Solubility, size and valence of the constituents dictate concentration shifts; tenside salts which are insoluble or difficult to dissolve are preferentially separated in *ion flotation*[13]. Leaving non-specific salt effects aside, fractionation of the non-ionogenic tensides (hydroxyl, ether, ester, carbonamide groups) correspond essentially to their surface activity. Consequently, they are particularly suitable for testing processes and apparatus, and for use as non-reactive foaming agents in various types of flotation. This group also includes a number of interesting natural substances, containing, for example, a glycoside hydrophilic moiety.

1.3 Systematic Survey

Karger et al. developed a system for distinguishing 'adsorptive bubble separation methods' in 1967[14] (expanded in **Scheme 1**[15]).

The adsorptive bubble separation methods classified by type and concentration of tenside were extended to include auxiliary vapors added to the gas phase[15]. In solvent sublation, a' supernatant immiscible liquid is involved as a third phase; in flotation with a solid third phase, further subdivision must be made in accordance with order of magnitude and mode of formation of the suspended substances.

1.4 Analytical Applications

The methods of interest from the point of view of organic analysis are substantially covered by the term 'Zerschäumungsanalyse' coined by Wo. Ostwald[6, 16]. They are methods for concentrating diluted and highly diluted aqueous solutions (e.g. $> 1:10^{10}$). They can be applied to large quantities of liquid if necessary. The costs are low, because all that is required is the energy for dispersing the air (nitrogen circulation for substances sensitive to oxidation). However, certain variants of flotation require additional tensides and other chemicals. In some cases, e.g. in natural substances and effluents, the surface-activity may be a separation parameter left out by the other methods. As a rule, foam analysis is inferior to conventional chromatography in the fractionation of similar compounds. Further analytical evaluation can be improved by two preceding processes:

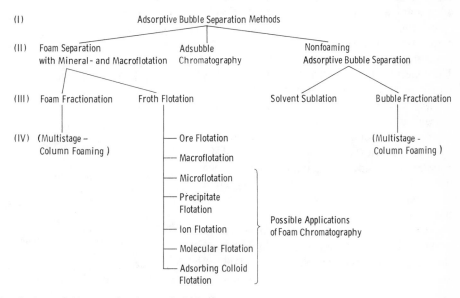

Scheme 1. Methods of adsorptive separation by gas bubbles[15]

---

[12] *K. Matschat* in: Ullmann, Encyklopädie der technischen Chemie, 3. Aufl., Vol. 16, p. 724, Urban & Schwarzenberg, München, Berlin 1965.

[13] *F. Sebba*, Ion Flotation, Elsevier Publ. Co., Amsterdam, New York 1962.

[14] *B.L. Karger*, Separation Sci. *2*, 401 (1967).

[15] *K. Maas*, DGF-Vortragstagung Heidelberg (13. 10. 1969), Fette, Seifen, Anstrichm. *72*, 1032 (1970).

[16] *Wo. Ostwald, A. Siehr*, Kolloid-Z. *76*, 33 (1936).

concentration (particularly in trace analysis) separation of interfering accompanying substances.

In principle, all surface-active substances in the foam can be concentrated, e.g. technical tensides (from *effluents*) and natural substances such as *dextrin, lecithin, urobilin, bilirubin, resin acids* (hop resins), *saponins, bile acid salts, steroid hormones*. In the case of highly diluted solutions without foam formation, separation efficiency can be improved either by adding organic vapors to the gas phase (see 2.11:2.5) or by superimposing an organic solvent which accommodates the molecules adsorbed on the gas bubbles (see 2.11:2.2). Concentration of large *protein* molecules is achieved by flotation processes (molecular flotation of *albumin, γ-globulin, hemoglobin*[17]), particularly since the higher proteins frequently coagulate in the process and pass into the area of micro and macro flotation (use nitrogen for removing albumin in biochemistry). Foreign tenside and precipitation reagents[18] have a favorable effect, as do buffer solutions, because of the reduced solubility of the proteins at the iso-electric point, and because of salt effects.

The literature on concentration of *enzymes* such as *catalase* (E.C. 1.11.1.6), *cholinesterase* (E.C. 3.1.1.8), *diastase, lipase* (E.C. 3.1.1.3) and *urease* (E.C. 3.5.1.5),compiled by Rubin and Gaden[2] was supplemented by Lemlich[19]. Residual activity, after foam analysis, fluctuates considerably depending upon the enzyme used and upon the accompanying substances (alteration of quaternary structure by surface forces).

The concentration of microscopic and sub-microscopic particles from a suspension, e.g. *microorganisms (bacteria* or *bacteriophages*[20]*)* is generally called *micro-flotation;* the term colloid flotation has been proposed for colloids[14]. Flotation of charged particles can best be performed with counter-ion tenside, for example, bacteria (negatively charged surface) with quaternary ammonium compounds; the transition to molecular flotation (see below) is gradual.

In *precipitation flotation*, the solid phase is formed in the liquid – in the simplest case, by introducing the alcoholic or acetonic solution of organic substances into water[21]. Precipitating and complex-

forming[22] reagents are used in particular for inorganic ions (Pt, Au, Ag, Fe, Co, Ni); in organic analysis, consideration should be given to the reverse (e.g. precipitation of $CN^{\ominus}$ with heavy metal ions[23]) or to carbonyl reagents (cf. section 3.4:3). Particularly versatile are *ion flotation*[13] and *molecular flotation* using ionogenic tensides as reagents. Sebba[24] lists, as examples of ion flotation, in addition to anionic and cationic pigments, the alkaloids *strychnine* and *brucin* and the salts of *picric acid, gallic acid* and *tetraphenylborane*. Grieves and Aronica[25] concentrated *phenol* and phenolate. For concentration and decomposition of natural substances and pharmaceuticals according to Hofman, see 2.11:2. On the basis of a number of analogies (ion exchange, adsorption, elution, fractionation), the term *foam chromatography*[26] is suitable for various kinds of 'reaction' flotation.

Absorbing colloid flotation[27], on which very little work has been done, requires neither surface activity nor precipitation of the substance to be separated. The molecules adsorbed on colloids undergo flotation with the latter. From the behavior of solutions during foaming, direct analytical conclusions can be drawn for many systems (cf. Ref.[28]).

2 Methods

2.1 Methods to Produce Liquid/Gas Interfaces
On an industrial scale, electrolytic generation of gas bubbles for flotation appears to be economically promising[29] (effluent purification). *Electroflotation* has been used also for concentrating *vitamin B_{12}* from culture solutions[30]. It has not been used thus far in analytical foaming. Mechanical dispersion of gaseous and liquid phase[31] has

[17] *J. Spurny, B. Jakoubek*, Tenside 5, 1 (1968).
[18] *R. Zahn*, Angew. Chem. *62*, 170 (1950).
[19] *R. Lemlich*, Ind. Eng. Chem. *60*, 16 (1968).
[20] *H. Ruska*, Kolloid-Z. *110*, 175 (1948).
[21] *T.A. Pinfold, E.J. Mahne*, Chem. Ind. [London] *31*, 1917 (1967).

[22] *J.A. Lusher, F. Sebba*, J. Appl. Chem. *16*, 129 (1966).
[23] *R.B. Grieves, D. Bhattacharyya*, Separation Sci. *3*, 185 (1968).
[24] *F. Sebba*, Nature *188*, 736 (1960).
[25] *R.B. Grieves, R.C. Aronica*, Nature *210*, 901 (1966).
[26] *K. Maas*, Chimia, Supplementum «5. Internationale Tagung über Trennmethoden, Lausanne 1969», *1970*, 238.
[27] *S.G. Mokrushin*, Ref. Zh. Khim. *1954*, 30 406; C.A. *49*, 2 149i (1955); cit. in Ref.[14].
[28] *D. Peters*, Angew. Chem. *64*, 586 (1952).
[29] *P. Ellwood*, Chem. Eng. *75*, 82 (1968); s.a. Nachr. Chem. Tech. *16*, 310 (1968).
[30] *B.M. Matov*, Izv. Vyssh. Ucheb. Zavede., Pishch. Tekhnol. (1) 116 (1966); C.A. *65*, I 341e (1966).
[31] *P. Siedel* in *Ullmann*[12], Bd. I, p. 666.

not been used in industrial flotation. A review of suitable gas dispersion devices will be found in Cassidy[32].

Glass or metal[33] frits form fairly uniform gas bubbles on their entire surface area. As a compromise between increasing adsorption surface with decreasing bubble diameter on the one hand, and increasing pressure drop on the other (cf. Ref.[34]) middle coarse pore widths (in Germany G or D 2 or D 3) have much to commend them in the case of glass frits. Dissolved salts[34] and tensides, and many organic vapors in the gas phase, reduce the diameter of the gas bubbles and make it difficult for them to coalesce.

2.2 Concentration on Ascending Gas Bubbles

Since no stable foam lamellae can form if the tenside content is very low, the concentration effects shown in Fig. 1b and 1c do not occur. Dorman and Lemlich[35] described the relatively small concentration shift caused solely by adsorption on the ascending gas bubbles as 'bubble fractionation'[35]. Apart from spherical foam chromatography (2.11:2.5), *solvent sublation ('foamless gas bubble separation with absorption')*[36, 37] should produce better results, in that the organic solvent (anisole or octanol) floating on the aqueous solution dissolves the molecules supplied or adsorbs them at the liquid/liquid phase interface. A typical example of solvent sublation is the fractionation of *methyl orange/rhodamine B* by differential reaction with added cation tenside. The process is related to ion flotation and requires little tenside as reagent because no foam is required above the aqueous phase; conversely, the solvent layer prevents the build-up of interfering stable columns of foam in the case of high tenside concentrations[13].

2.3 Foam Analysis in Columns

The diameter of laboratory columns is, usually between 1 and 10 cm (a figure of about 100 cm is sometimes given for decontamination experiments). Depending on cross-section and separation problem involved, columns with a length of 0.5−5 m produce good results (minimum to date 2 mm ∅, 30 cm length, about 0.1 ml

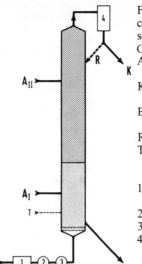

Fig. 2. Diagram showing a column for various foam separation methods

G gas phase

A_I, A_{II} feed points for original solution

K top fraction (concentrated)

B bottom fraction (deconcentrated)

R reflux

T feed point for tenside or reagents

1 saturation of gas phase with water vapor
2 manometer
3 flowmeter
4 foam breaker; mechanical in most cases

solution in *laminae column foaming*[38]). Columns with frits of column diameter show exceptionally uniform ascent of gas bubbles. As a rule, the gas phase is saturated with water vapor. This is not necessary in a closed gas cycle[38] with membrane or hose pump. Moreover it operates practically without loss with inert gas. A manometer 2 and a flowmeter 3 are used for monitoring.

Dilution series of surface-active pigments (e.g. patent blue V) and inactive dyestuffs (red neococcin)[16] are the most suitable substances for testing columns and for initiation into separation processes. Acidic and basic dyestuffs should not be used simultaneously; non-ionogenic dyestuffs should be suitable.

In a 'column process' foam should be formed and destroyed repeatedly over the entire length of the column ('column foaming'[38]) by analogy with column distillation (2.1:1.2.2) and column crystallization (2.2). Previously, 'multi-stage' operation has been used in successive units, e.g. Grieves[39] (theory) and Kishimoto[40] (see Ref.[3]). One possibility is to use a multiple frit column with alternating long foaming and short reflux cycle (change of direction of gas flow) and staggered discharge of fractions or a multichannel-pump to feed back the reflux to each plate (K. Maas, unpublished). Even without a 'column' according to the strict definition of the term, increased concentration is obtained in accordance with Fig. 1b and, in particular, 1c. According to Fig. 3, adsorptive

[32] *H.G. Cassidy* in: Technique of Organic Chemistry, Vol. X, Fundamentals of Chromatography, C. XI, Interscience Publishers, London, New York 1957.

[33] *W. Siemes, E. Borchers*, Chem.-Ing.-Tech. *28*, 783 (1956).

[34] *R. Seeliger*, Naturwissenschaften *36*, 41 (1949).

[35] *D.C. Dorman, R. Lemlich*, Nature *207*, 145 (1965).

[36] *A.B. Caragay, B.L. Karger*, Anal. Chem. *38*, 652 (1966).

[37] *B.L. Karger, A.B. Caragay, S.B. Lee*, Separation Sci. *2*, 39 (1967).

[38] *K. Maas*, Separation Sci. *4*, 69 (1969).

[39] *R.B. Grieves*, Separation Sci. *1*, 395 (1966).

[40] *H. Kishimoto*, Kolloid-Z., Z. Polymere *192*, 66 (1963).

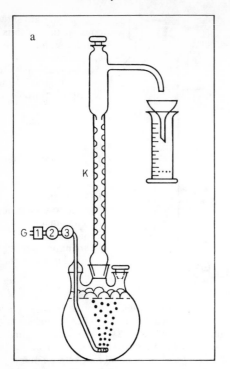

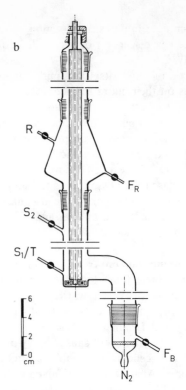

Fig. 3. Apparatus for Foam Chromatography (or for various flotation processes according to scheme 1)

a) 'gas bubble chromatography'[44] with modified apparatus according to [45]

    G gas phase
    K Vigreux column
    1   saturation with water vapor
    2   manometer
    3   flowmeter

b) 'foam chromatography'[46] with stabilized column of foam

    S   starting solution
    F_R eluted fraction
    R   reagent solution
    F_B discharge of deconcentrated solution (bottom fraction flow)
    N_2 gas inlet

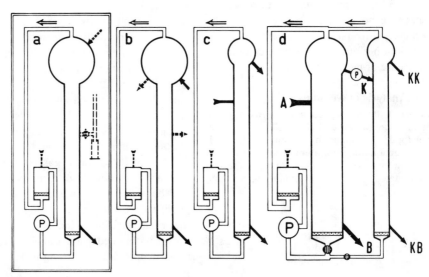

Fig. 4. Possible Apparatus and Methods of Adsubble Chromatography[15] (without auxiliary phase also valid for the other methods of Fig. 2).

P circulating pump (diaphragm or peristaltic pump)
A original solution
K headfraction
B bottom fraction

a batchwise operation
b semicontinuous (concentration)
c continuous
d cascade (concentration)

bubble separation methods offer the following variations (see Lemlich[19]).

In batchwise operation, the reflux consists of normally collapsed foam. As a rule, the equilibrium is reached after 30 minutes – 2 hours, so that the foam fractions can be withdrawn and the non-foaming residue discharged. A considerable advance was made by Harper, Lavi and Lemlich[41,42] who induced integral collapse of the foam using a mechanical 'foam breaker' (4 in Fig. 2) and who recirculated the reflux R in definite quantities. The foam breakers are usually rapidly rotating wire baskets or discs, but heat[43], ultrasonics, and even alpha radiation are also used.

In continuous operation, the feed point of the original solution A depends on the tenside content, the viscosity, and the specifications for the fractions (concentrated pure head fraction K or largely deconcentrated bottom fraction B). A_I corresponds to 'single-stage' bubble fractionation, A_{II} is closer to the conditions of 'foam chromatography'. In accordance with section 2.11:1.4, the tensides or reagent solutions fed in at T extend the scope by the various flotation methods.

2.4 Foam Chromatography

The paper by Hofman[44] which reports fractionated micro-flotation, ion flotation, and molecular flotation of plant material, organ homogenates, *barbiturates, caffeine, amidopyridine,* and *ethylmorphine,* using control by thin layer chromatography, is a good example of foam chromatography (influence of tensides, pH, etc.). Hofman performed 'gas bubble chromatography' in a mod-

ified Karger and Rogers[45] apparatus. (cf. Fig. 3a, p. 170).

A rotating rod stabilizes the foam in a column for separating substances from fairly large quantities of liquid ($=A_{II}$, Fig. 2). The rotating column of foam can be washed in extended chambers, treated with reagents and divided into individual fractions ('eluted'; *foam chromatography*[26], cf. Fig. 3b, p. 170). The adsorption, precipitation and partially colloidal solution processes can be confined in accordance with 2.11:1.4.

2.5 Adsorptive Bubble (Adsubble) Chromatography

If the circulating gas phase is exposed to organic (inorganic) vapor in an interposed wash bottle, differential, temperature-dependent separation effects are obtained in aqueous solutions, depending on the organic 'auxiliary phase'. (For the used example – at the system water/patent blue/neococcin – benzene, pentane, $CHCl_3$ have a positive effect; ether and ethyl acetate have a negative effect)[15,26,46]. In the transition region, where the lamellae show little stability in other respects, foam formation is markedly increased *(booster bubble fractionation*[46]*)*. However, at dilutions greater than $1:10^{10}$ (1 ppb to 0.01 ppb), concentration by several powers of 10 was possible even without foam. A boundary layer of organic phase, tenside, and water must be assumed [Adsubble (adsorptive bubble) chromatography[33]]: moreover, the dense accumulation of gas bubbles *('Kugelschaum'* according to Manegold[4]) produces additional concentration according to Fig. 1b or 1c. The various potentialities of *adsubble chromatography* shown in Fig. 4 are, of course, valid in principle for all foam separation processes using columns.

[41] *R. Lemlich, E. Lavi,* Science *134,* 191 (1961).
[42] *D.O. Harper, R. Lemlich,* Ind. Eng. Proc. Des. Dev. *4,* 13 (1965).
[43] *B.L. Karger,* Anal. Chem. *38,* 764 (1966).
[44] *M. Hofman,* Mitt. Deut. Pharm. Ges. *38,* 54 (1968).
[45] *B.L. Karger, L.B. Rogers,* Anal. Chem. *33,* 1165 (1961).

[46] *K. Maas,* Separation Sci. *4,* 457 (1969); *R. Lemlich,* Adsorptive Bubble Separation Techniques, Academic Press, New York, London 1972.

2.12 Aspects of the Selection of Suitable Separation Processes

Gerhard Hesse
Institut für Organische Chemie der Universität,
D-852 Erlangen

1 Introduction

Chapter 2 deals principally with methods for separation of materials on the preparative scale. This section is intended to facilitate the selection of suitable separation procedures. The isolation of a material in an analytically pure state is essential for the correct determination of an empirical formula which often is a prerequisite for futher work.

Constancy of analytical values after further purification procedures is not an adequate criterion of purity. Elemental analysis cannot reveal contamination by isomers, and especially by tautomers. Other types of impurity frequently can be detected only when the amounts present are as high as several percent. The properties of many substances can, however, be significantly affected even by *traces of contaminants*[1] as low as 1 ppb*. These contaminants are most easily detectable by optical, catalytic or biological methods — frequently, only after concentration by *zone melting* (2.2), *reversion gas chromatography*[1] or other methods of trace analysis. *Mass spectrometry* has a detection limit[2] of 0.01 ppm, and also provides the most accurate values for molecular weights. In the case of macromolecular materials, determination of the average molecular weight provides no information about homogeneity.

2 Treatment of Crystalline Materials

2.1 Criteria of Purity

Many doubts are resolved if a substance is already crystalline, especially if examination under a lens or microscope indicates a crystal of uniform shope; straight edges, sharp and non-rounded angles between crystal faces, and the absence of steps or matt places on the crystal faces have also proved suitable as criteria. Melting points (and perhaps also mixed melting points) should always be determined, since information about the behavior of the material on heating is gained at the same time. Greater purity is likely if the compound is recrystallized from several solvents. Evidence for solvent adducts can also be seen in this way. By analytical chromatography (2.4 and 2.10), it is usually possible to determine if a substance is pure, or to estimate the degree of purity. Substances which are capable of tautomerism may appear to be inhomogeneous in solution even when pure in the crystalline state.

2.2 Separation and Purification

If crystals melt without decomposition, they can be purified further by *normal solidification*[3] of the stirred fused mass, by cooling from the bottom upwards. *Column crystallization* (2.2) is essentially this process constantly repeated. An even more effective technique is *zone melting* (2.2) which can be used with advantage for organic compounds. *Sublimation* is very suitable for the removal of non-volatile impurities; but it is very difficult to separate mixtures of volatile materials by this technique.

If *separation from solutions* is difficult because of limited temperature stability, *chromatographic procedures* (2.5, 2.6, 2.8, 2.12:5) in addition to simple *recrystallization* are especially suitable. Electrophoresis at a definite pH is a convenient method (2.9) for substances which have a natural ionic charge. In very difficult cases, *partition between two liquid phases* (2.3) or *fractional crystallization* using the triangle technique can be used; these cause little or no decomposition of the substance. For growing *single crystals*[4,5] (which are essential, for X-ray structural analysis) only be used. Extremely pure starting materials should be used.

3 Handling of Liquid Mixtures

3.1 Liquids of Comparable Volatility

The purity of liquids is difficult to determine. They are more difficult to purify than crystalline substances. It is good to determine whether they solidify as crystals on stepwise cooling. If so, *freezing out* or *recrystallization* from low boiling solvents is possible; otherwise purification can be achieved by *rectification* (2.1:1.2.2) through a good column, or by *preparative gas chromatography* (2.10).

The latter is the preferred technique for small quantities, especially since the problem of azeotropes does not exist with this technique. In both cases, great care must

---

* (1 ppb $= 10^{-3}$ ppm)

[1] *R. Kaiser*, Naturwissenschaften **57**, 295 (1970).

[2] *H. Kienitz* in Ref. [12], p. 408.

[3] *T.H. Gouw* in Ref. [13], p. 57.

[4] *A. Neuhaus*, Chem.-Ing.-Tech. **28**, 155, 350 (1956).

[5] *H. Schäfer*, Chemische Transportreaktionen, Verlag Chemie, Weinheim/Bergstr. 1962.

be taken to avoid changing the vapors chemically by prolonged contact with the walls and with the packing material. The most inert material is quartz, followed by glass, and then (greatly inferior) metals such as stainless steel, platinum and copper. With heat-sensitive materials, separation from solution is the only possibility, and in this case chromatographic methods and multiple partition are the most important.

3.2 Solutions

Frequently materials consist of mixtures of solids and liquids, e.g. solutions, extracts, pressed juices, and reaction by-products. A fraction of interest can be separated out by *foam fractionation* or *foam chromatography* (2.11). Removal of the foam-forming material can make evaporation considerably easier.

3.2.1 Evaporation

In a *pre-separation* step, volatile materials are separated from the less volatile ones. This is usually done by *evaporation* (2.1:1.2.1), but occasionally by *freezing* out part of the solvent. The latter procedure has the advantage that heat sensitive materials are not decomposed, and readily volatile ones are not lost. The use of a short Vigreux column when evaporating the solvent is often recommended. For simple evaporation, the rotary evaporator (2.1:1.2.2) is used. The separation of solid or resinous precipitates usually does not cause any difficulties, even when working under vacuum. A *spray drier* (Lurgi, Gesellschaft für Wärmetechnik, D-6 Frankfurt) is recommended if there is a chance that partial aggregation of the dissolved material may occur. This avoids the formation of an inhomogeneous final product. In a spray drier, the solution is blown through fine nozzles into an ascending stream of warm air or gas, and the solid is collected as fine granules at the bottom of the drier. It is important that mixtures (for example washing powders, fertilizers, instant tea and coffee) be homogeneous. With solids, it is possible to obtain uniform distribution of additives by *zone levelling*[6] (2.2). Water can be removed from heat-sensitive or readily decomposed organic materials by *freeze drying* (Ref. [11], p. 939). For this *lyophilization* procedure, which can be used not only for proteins and enzymes, but even with living materials (bacteria and viruses), very efficient equipment is commercially available (Leybold-Heraeus, D-5 Köln 51). Following evaporation, chromatographic analysis of both the volatile (2.10) and the non-volatile (2.4) components should be carried out.

3.2.2 Separation of Volatile Materials

This is carried out by rectification on a plate column or spinning band column, or by preparative gas chromatography. Distillation is sometimes performed with auxiliary materials (2.1:1.2.3). *Molecular distillation* (2.1:1.2) is not suitable for the fractionation of volatile components; even a modern high performance apparatus with 10 stages has only seven theoretical plates. The main use of this type of apparatus is the removal of volatile impurities from high boiling oils, such as separation phases employed in gas chromatography, or oils used in high vacuum pumps.

3.2.3 Separation of Non-volatile Materials

Before further separation is possible, these must be redissolved in a suitable homogeneous solvent. By preliminary *digestion* or *extraction* processes, group separations can be carried out in many cases. Extraction with water-immiscible solvents makes possible the extraction of organic acids by shaking with bases, and organic bases (alkaloids, etc.) with acids. By use of buffer solutions, still finer separation can be achieved. One can often obtain a crude crystalline product from a pre-purified solution by crystallization, *salting out*, or mixing with other solvents. Chromatographic or distribution processes can also be used for separation; these too generally require some pre-purification.

Another method increasingly used, especially in macromolecular chemistry and biochemistry, is fractionation according to *particle size*. Examples are: the separation of suspended particles by centrifuging (2.1:2) or by ultrafiltration (2.6:2), and the removal of very small particles by dialysis (2.7:3.7). A generally more simple and powerful technique is *gel filtration* (2.6), in which one successively uses gels with different exclusion limits. In favorable cases, further separation can be achieved by *gel chromatography* (2.6:3), which can also be used for the determination of molecular size distributions in oligopolymers[7]. Molecular weight determinations can be carried out on quite large molecules in this way without requiring the preliminary purification which is necessary in other cases (2.6:5).

Substances which have a natural ionic charge can be separated from aqueous solutions by *ion-exchange chromatography* (2.8) or by *electrophoresis* (2.9). Both processes are very pH dependent.

---

[6] *H. Schildknecht* in Ref. [12], p. 190.

[7] *W. Heitz*, Angew. Chem. *82*, 675; Angew. Chem. Intern., Ed. Engl. *10*, 689 (1970).

The optimum pH is first obtained by using buffers. In the case of large molecules, the buffer can be removed or exchanged if required, by dialysis, electro-dialysis or gel filtration; with small molecules, volatile buffer materials are used. For ampholytes such as proteins, it is sometimes sufficient to adjust the pH of the solution to the isoelectric point of a component so that it separates; to do this, organic solvents or inorganic salts can be added if necessary.

3.3 Residues which do not Dissolve without Decomposition

Investigation of the heat stability of insoluble and nonvolatile *residues* of solution processes is essential. Such residues often contain inorganic oxides or salts of low solubility which can be detected by inorganic elemental analysis. If organic substances are present (usually detectable by charring on heating), information about their structure is often obtained from the pyrolysis products; the most suitable method is *pyrolysis-gas chromatography* (2.10:4.1.3).

4 Gases

Analysis of gas mixtures can be carried out effectively by gas chromatography (2.10), so that classical methods have very little importance. For *preparative separation,* chemical processes such as absorption in acids or alkalis or specific complexing agents are used. Solvent vapors are adsorbed by activated charcoal in the order of their molecular weights, and are then desorbed. In a similar manner, ethanol can be obtained from the waste gases of fermentation plants. Diffusion procedures and gas centrifuging are used for separation in special cases. The most important general method for separation of gases is *fractional distillation* of the liquefied gas mixture[8].

5 Characterization of Chromatographic Separation Methods

The only method of separation which can be used for all molecular sizes and states of aggregation is chromatography in its various forms. The sole prerequisite is that the material should dissolve or vaporize without decomposition. Chromatography is discussed in Chapters 2.3–2.11.

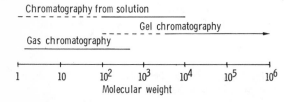

The range of *gas chromatography* extends from molecular weight 2 to about 400. The range of normal *liquid chromatography* is between 100 and ca. 10,000; in the case of ion exchange chromatography it is even as low as 1. Separation by means of *gel chromatography* is feasible from 100 up to the limits of solubility. Its limited specificity can be vastly improved in suitable cases by using ion exchange gels (2.8).

For analytical purposes, *gas chromatography* is used wherever possible. This technique is quick, quantitative, and easy to carry out. For larger and heat-sensitive molecules, its complements are *thin layer chromatography* and *paper chromatography*.

For preparative purposes, *liquid chromatography* is generally simpler and less liable to cause decomposition of the sample; it is also less expensive than gas chromatography. Gel chromatography is used primarily for *separation according to molecular size,* although separations of substances within narrow molecular weight ranges are possible (2.6, Fig. 5).

Liquid chromatography provides the analyst with additional freedom, because of the choice of solvents and the possibility of changing them during the experiment; the technique is thus very versatile. At the same time, however, the choice of the most suitable system for separation, and the optimization of conditions, become more difficult. With the aid of empirically constructed tables[9], planning of experiments can to some extent be made easier.

The foregoing discussion has related to elution chromatography. In special cases, *frontal analysis* 2.5) has attained significance, both for the purification of solvents and gases (removal of trace contaminants), and for their concentration and recovery. This can be carried out only with adsorbents such as silica gel, charcoal, or ion exchangers which have sharply defined absorption limits (saturation isotherm).

6 Separation by Derivative Formation

Another possibility for both solids and liquids is purification or separation by forming derivatives such as salts, esters, oximes (4.2). Here, it should be realized that conversion and regeneration can seldom be carried out quantitatively. However, the non-reacted species can generally be removed from solution (2.5) by adsorption chromatography.

Final purifications is carried out by crystallizing the material or a derivative.

---

[8] *A. Bittel* in Ref.[12], p. 67.

7 Combinations of Separation Procedures

The combination of several separation procedures is always more promising than the repetition of a single one. Materials which have been obtained by preparative gas chromatography should be subjected to a final simple distillation, in order to remove volatile components originating from the stationary phase. Materials obtained by crystallization should, where possible, be sublimed once, in order to remove solid residues (crystallization nuclei) and solvents which had been used for crystallization.

8 Bibliography

[9] *G. Hesse,* Chromatographisches Praktikum. Methoden der Analyse in der Chemie, Bd. VI, p. 218, Akademische Verlagsgesellschaft, Frankfurt/Main 1967.

[10] *H. Musso,* Moderne Trennungsmethoden in der Chemie. Naturwissenschaften *45,* 97 (1958).

[11] *Houben-Weyl,* Methoden der organischen Chemie, 4. Aufl., Bd. I/1, Thieme Verlag, Stuttgart 1958.

[12] Ullmanns Encyclopädie der technischen Chemie, Bd. II/1, Anwendung physikalischer und physikalisch-chemischer Methoden im Laboratorium. Urban & Schwarzenberg, München, Berlin 1961.

[13] *E.S. Perry,* Progress in Separation and Purification, Vol. I. Interscience Publishers, Wiley & Sons, New York, London, Sydney 1968.

[14] *F.J. Wolf,* Separation Methods in Organic Chemistry and Biochemistry, Academic Press, New York, London 1969.

3 Determination of Classes of Compounds and Functional Groups by Chemical Methods

Contributed by

A. Becker, Basel
W. Büchler, Basel
F. Christen, Basel
H.F. Ebel, Heidelberg
A. Manzetti, Basel
J. Meier, Basel
K. Schlögl, Wien
W. Schöniger, Basel
A. Walter, Basel
H. Weitkamp, Wuppertal
H. Wenger, Basel

3.1 Developments in the Automation of CHNO Analyses

Horst Weitkamp

Bayer AG, D-56 Wuppertal

1 Introduction

Well-proven methods and comprehensive instructions[1] are available for the determination of the elements C, H, N, and O in organic substances. However, the conventional procedure for CHNO determinations involves many manual operations.

This presents no problems where analysis times are long, since there is then adequate time for the preparation of the sample, for determination of the final products by weighing or titration, for calculation, for writing up results, and much more.

In many laboratories, however, there is a steadily increasing inflow of substances, which requires a corresponding increase in capacity. Since a steady expansion in terms of space and personnel is frequently impossible, experiments have been carried out in various laboratories with the object of increasing the number of samples that can be ana-

lyzed per apparatus and per person. If the cycle time is reduced to e.g. 10 minutes or less, however, all the operations that are necessary for an analysis can no longer be performed manually, and the various resources of the rapidly developing measurement and control technology have therefore been used in an effort to solve the problem.

The developments have taken various directions, corresponding to the special requirements of the various groups of workers, ranging from regulation of the course of combustion and automatic determination of the combustion products to printing out of the percentage contents.

The development of electronics, particularly in the *data processing* sector, was already felt in mi-

---

[1] *H. Roth* in *Houben-Weyl*, Methoden der organischen Chemie, 4. Aufl., Bd. II, p. 31, Georg Thieme Verlag, Stuttgart 1953.
H. Roth in *Pregl-Roth*, Quantitative organische Mikroanalyse, Springer Verlag, Wien 1958.
A. Steyermark, Quantitative Organic Microanalysis, Academic Press, New York, London 1961.

Table 1. Characteristic Data for Various Analytical Methods. The References are Indicated with the Figures

Sample size[a]	Other elements present	Number of combustions per tube filling	Cycle time	Standard deviation				Fig.	Section 3.1:
				C	H	N	O		
3–5 mg[b]	Any[c]	150– 500	12 min.	0.17	0.18	—	—	1a	2.1
2–5 mg[d]	Any except F	150– 500	11 min.	—	—	—	0.12	1b	2.1
3–5 mg[b]	Any[c]	150– 500	11 min.	—	—	0.15	—	1c	2.1
2–10 mg[e]	Any	300– 400	4 min.	—	—	0.14	—	2	2.1
3–5 mg[e,f]	Any	300– 400	8 min.	<0.25	<0.25	—	—	3	2.1
1–2 mg[g]	Any	1200	10 min.	0.07	0.04	—	—	4	2.2
1–5 mg[h]	Br, Cl, I, N, S	600– 800	20 min.	—	—	0.22	—	5	2.2
0.5–2 mg[i]	Any	—	13 min.	0.20	0.08	0.30	—	6	2.3
1 mg[j]	Br, Cl, I, O, S[k]	1000–5000	12 min.	0.5[l]	0.1[l]	1[l]	—	7	2.3
0.1–0.9 mg[m]	Any	1500–2000	10 min.	0.18	0.05	0.3	—	8	2.3
0.1–0.6 mg[n]	—	—	15 min.	—	—	—	0.15– 0.28	9	2.3
3–5 mg[o]	—	—	—	[p]	[p]	[p]	—	10	2.4
3–50 mg	—	—	5 min.	—	—	—	0.17	11	2.5

[a] In a platinum boat unless otherwise stated.
[b] Liquids in glass capillaries with one end sealed, the open end being covered with WO_3.
[c] Substances containing alkali metals are covered with a layer of WO_3.
[d] Samples containing P in quartz boats. Induline Base RM is added to substances containing alkali metals.
[e] Al boat, liquids sealed in phosphate glass capillaries.
[f] Covered with a layer of 50 mg of MnO_2.
[g] Pt boat in quartz capsule, liquids in glass capillaries.
[h] Volatile liquids in glass capillaries.
[i] In capillaries, liquids sealed in.

[j] Sample size up to 2 mg[52].
[k] Modification when F, P, and metals present[58].
[l] [%] 'Maximum uncertainty'.
[m] Al boat, covered with a layer of Co_3O_4, substances containing alkali metals covered with WO_3[46]; liquids in calibrated capillaries open at both ends.
[n] Ag boat.
In quartz tube together with 1 g of CuO. The quartz tube is evacuated and sealed. The combustion is carried out at 925°.
[p] The authors expect 0.03 after refinement of the computer program.

cro-elementary analysis even in 1964[2]. Nowadays it is possible to obtain programmable desk calculators that can accept programs from suitable storage media (magnetic cards, punched tapes), so that once a program library suited to the field in question has been built up, the results even of complicated calculations are obtained by simply feeding the experimental values into the calculator. Electronic computers are also available which, when coupled directly to a large number of instruments, receive data from these instruments, store it temporarily, and print out the results in response to an inquiry or to instruction in the program.

The numerous publications and the growing range of equipment offered by the instrument manufacturing industry make it increasingly difficult for a laboratory to select the instrument or combination of instruments that best suits its requirements. To illustrate the present position, a selection of different published methods will be presented schematically (cf. Section 3.1:4).

2 Schematic Presentation of Various Analytical Methods

The following schemes outline, without details, various procedures with varying degrees of automation for CHNO determinations. The diagrams are preceded by some essential data (Table 1). Publications on the same topic are listed at the end of each subsection, and publications of a general nature are also included under 3.1:2.1. The published methods generally do not correspond to present-day practice since many changes have not been published, even where they have led to important improvements in practice.

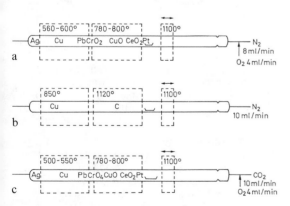

Fig. 1. Microdeterminations in an Automatically Controlled Combustion Apparatus[3].
a) *Carbon hydrogen*
b) *Oxygen* (Leipert titration)
c) *Nitrogen* (The leveling vessel is adjusted manually, and the volume is read automatically by a photoelectric cell and indicated on a digital counter.)

2.1 Gravimetric, Volumetric, or Titrimetric Determination of the Combustion Products

In the following methods (Figs. 1—3), all regulation processes are controlled, where necessary, by a programmed switching mechanism (Figs. 2, 3).

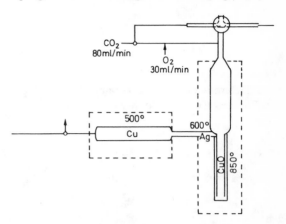

Fig. 2. Automatic Rapid Method for the Determination of *Nitrogen* with an Electronically Controlled Nitrometer Allowing Print-out of the Volume of *Nitrogen*[4].

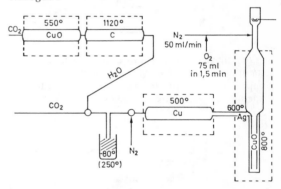

Fig. 3. Automatic Rapid Method for the Determination of *Carbon* and *Hydrogen* (after conversion into equivalent quantities of CO_2).
Absorption of CO_2 with automatic titration and print-out of the experimental values[5].

2.2 Conductimetric Determination of the Combustion Products

The burner and the filling and rinsing of the conductivity cell and also the regulation of the Keidel cell are controlled in this case. The experimental values can be read off from digital counters.

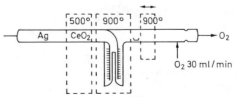

Fig. 4. Rapid combustion apparatus for the determination of *carbon* and *hydrogen*[6].

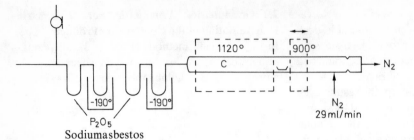

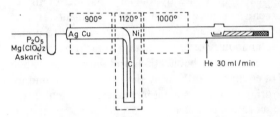

Fig. 5. Rapid combustion apparatus for the determination of *oxygen*[7].

2.3 Determination of the Combustion Products by Thermal Conductivity Measurements

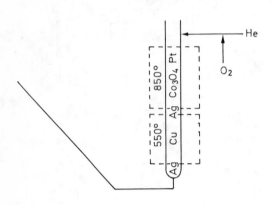

Fig. 9. Determination of *oxygen* by integral evaluation of the katharometer signals[11] (print-out of the experimental values).
All processes from the introduction of the sample to the print-out of the values are controlled.

Fig. 6. Fully automatic apparatus for the simultaneous determination of *carbon, hydrogen*, and *nitrogen* in 16 samples in milligram and submilligram quantities[8]. Experimental values are printed out.

2.4 Manometric Determination of the Combustion Products

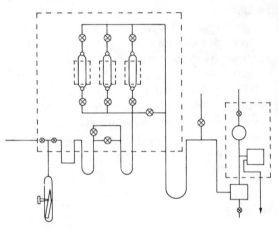

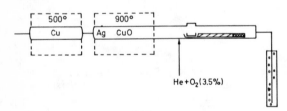

Fig. 7. Rapid ultramicro method for the simultaneous determination of *carbon, hydrogen*, and *nitrogen*[9]. Experimental values for C + N, C and H can be read off from the counter.

Fig. 10. Computer-controlled apparatus for the simultaneous determination of *carbon, hydrogen*, and *nitrogen*[12] (switching to manual operation is possible). The percentages are printed out.

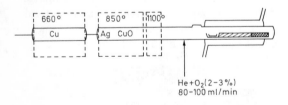

Fig. 8. Printing automatic apparatus for the simultaneous determination of *carbon, hydrogen*, and *nitrogen* as percentage contents[10].
All processes involved, from the introduction of the sample to the print-out of the percentages, are controlled.

[2] *W. Schöniger*, Z. Anal. Chem. *205*, 13 (1964).
[3] *I. Monar*, Mikrochim. Acta 208 (1965); 934 (1966).
[4] *W. Merz*, Z. Anal. Chem. *237*, 272 (1968).
[5] *W. Merz*, Anal. Chim. Acta *48*, 318 (1969).
[6] *F. Salzer*, Z. Anal. Chem. *205*, 66 (1964).
[7] *F. Salzer*, Mikrochim. Acta 835 (1962).
[8] *J.T. Clerc, R. Dohner, W. Sauter, W. Simon*, Helv. Chim. Acta *46*, 2369 (1963).
[9] *W. Walisch*, Chem. Ber. *94*, 2314 (1961).
[10] *H. Weitkamp, F. Korte*, Chem.-Ing.-Tech. *35*, 429 (1963).
     H. Weitkamp, R. Mayntz, F. Korte, Z. Anal. Chem. *205*, 81 (1964).

2.5 Spectrophotometric Determination of the Combustion Products

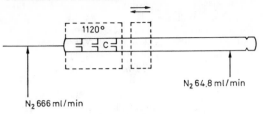

Fig. 11. Determination of *oxygen* in organic substances with the aid of nondispersive infrared absorption. Experimental value can be read off from the integrator[13].

3 Concluding Remarks

Whereas in other fields, such as spectroscopy, the instrument manufacturers carried out extensive development work, much of the work in the field of elementary analysis was carried out in analytical laboratories, whose real function is to carry out analyses. It is not surprising that first consideration was given to the interests of the laboratory in question and not to the development of apparatus suitable for universal use. Many problems have nevertheless been solved in the last decade.

Universal tube fillings are now available. The *sensitivity* has been increased by more than an order of magnitude. Great experience has now been gained in the various methods for the detection and quantitative determination of the combustion products. *Calibration* is necessary in most of the methods. Though this implies a loss of precision according to the law of propagation of error, the errors in the final result (in % absolute) are quite comparable to those of the conventional methods. Many methods have proven themselves in practice over a number of years, and have led to a substantial increase in productivity. It is possible to determine several elements in a single sample. Analyses can be carried out with the same error (in % absolute) in the decimilligram range as in the milligram range. Sample homogeneities that are satisfactory for the milligram range are also adequate for the decimilligram range. These methods are of particular interest for the analysis of substances that are obtainable only in small quantities, e.g. in *biochemistry, metabolite* studies, and investigations on *radioactive preparations*, where the quantities of material required are so small

that even with ten times the activities involved there would be no need for protection against radiation. If the number of samples involved is small, however, the best method for substances with sufficiently high vapor pressures is *high-resolution mass spectroscopy*.

Developments are still progressing rapidly. The value of an analytical result is becoming increasingly dependent on the time that elapses between sampling and receipt of the result. This fact does apply not only for the process-control, but also for scientific work. As analytical methods become faster, however, this delay is largely determined by the time of administration. Analysis systems that are fully integrated with data processing and documentation are becoming necessary, particularly in large laboratories, because of the increasing numbers of samples to be dealt with and the decreasing working time. The means are already at hand. Electronic balances provide the weight in a form suitable for data processing. Instruments with automatic input of a large number of samples have been described. Many instruments already give the final result in digital form. From the point of view of cost, special attention should be paid to the methods (as illustrated in Fig. 3) that do not require expensive instruments for digital measurement (better than 0.1%) of the combustion products. To be able to meet the rapidly changing needs, the development of suitable unit systems and adaptors is particularly important. The methods should be simple from the point of view of measuring technique, and should be capable of being switched to manual operation when necessary.

4 Bibliography

4.1 Gravimetry, Volumetry, Titrimetry

[14] *J. Körbl*, Verwendung des thermischen Zersetzungsproduktes von Silberpermanganat in der organischen Elementaranalyse, Mikrochim. Acta 1705 (1956).

[16] *F. Ehrenberger, S. Gorbach, W. Mann*, Rationalisierung der mikroanalytischen Bestimmung des Sauerstoffes in organischen Substanzen und die Erweiterung der Methode auf die Analyse von fluor- und phosphororganischen Substanzen, Z. Anal. Chem. *198*, 242 (1963).

[17] *K. Hozumi, W.J. Kirsten*, Simultaneous Determination of Carbon, Hydrogen, and Nitrogen at the Decimilligram Level, Anal. Chem. *35*, 1522 (1963).

[18] *G. Kainz, F. Scheidl*, Die Oxidation organischer Verbindungen an Platinkontaktsternen unter den Bedingungen der C-, H-Analyse, Mikrochim. Acta 641 (1964).

[19] *H. Trutnovsky*, Versuche zur Automatisierung der gravimetrischen C-, H-Bestimmung, Z. Anal. Chem. *222*, 254 (1966).

[11] *F. Ehrenberger, O. Weber*, Mikrochim. Acta 513 (1967).

[12] *J.W. Frazer, R. Stump*, Mikrochim. Acta 1968 (1969).

[13] *W. Thürauf, H. Assenmacher*, Z. Anal. Chem. *245*, 26 (1969).

[20] R.C. Rittner, R. Culmo, Simultaneous Microdetermination of Carbon, Hydrogen and Sulfur, Microchem. J. 11, 269 (1966).

[21] H. Malissa, Automation in der analytischen Chemie und mit der analytischen Chemie, Z. Anal. Chem. 222, 100 (1966).

[22] F. Ehrenberger, Bestimmung des Stickstoffs in organischen Substanzen nach Verbrennung im Sauerstoffstrom, Z. Anal. Chem. 228, 106 (1967).

[23] G. Kainz, G. Chromy, Über die quantitative Oxydation von Kohlerückständen, die im Verlaufe der Elementaranalyse auftreten. Mikrochim. Acta 714 (1967).

[24] G. Kainz, K. Zidek, Über die Reduktion der Stickoxyde an Metallen, Beitrag zur C-, H- und N-Bestimmung, Mikrochim. Acta 725 (1967).
Absorption von Stickstoffdioxid bei 20 °C im Rahmen der C-H-Bestimmung. Vergleich verschiedener Absorbentien, Mikrochim. Acta 7 (1967).

[25] Y.A. Gawargious, A.B. Farag, Microdetermination of Carbon and Hydrogen in Steroids, Mikrochim. Acta 585 (1968).

[26] A.D. Campbell, Weighing Hygroscopic Samples, Mikrochim. Acta 833 (1968).

[27] A.D. Campbell, L.S. Harn, R. Monk, D.R. Petrie, Determination of Carbon, Hydrogen, and Nitrogen in Organo-Metallic Perchlorates and Nitrates, Mikrochim. Acta 836 (1968).

[28] O. Hadzija, Absorption Properties of Körbl-Catalyst and Manganese Dioxide. Application to the Simultaneous Determination of Carbon, Hydrogen and Halogen, Mikrochim. Acta 917 (1968).

[29] E. Pella, Beseitigung der Stickoxyde durch Autoreduktion bei der C-H-Mikrobestimmung, Mikrochim. Acta 490 (1969).

[30] F. Sels, P. Demoen, Fast Gravimetric Micro Carbon-Hydrogen Determination, Mikrochim. Acta 530 (1969).

4.2 Conductometry

[31] F.A. Keidel, Determination of Water by Direct Amperometric Measurement, Anal. Chem. 31, 2043 (1959).

[32] L.G. Cole, M. Czula, R.W. Mosley, D.T. Sawyer, Continous Coulometric Determination of Parts per Million of Moisture in Organic Liquids, Anal. Chem. 31, 2048 (1959).

[33] W. Stuck, Eine Mikromethode zur schnellen Bestimmung von Kohlenstoff in organischen Substanzen, Mikrochim. Acta 42 (1960).

[34] N.E. Gelman, Van Ven-Yun, Ein Verfahren zur konduktometrischen Mikrobestimmung von Kohlenstoff und Wasserstoff in organischen Verbindungen, Zh. Anal. Khim. 15, 487 (1960).

[35] W. Schmidts, W. Bartscher, Grundlagen der konduktometrischen Kohlenstoffbestimmung, Z. Anal. Chem. 181, 54 (1961).

[36] H.S. Haber, K.W. Gardiner, A Rapid Instrumental Method for the Microdetermination of Carbon and Hydrogen in Organic Compounds, Microchem. J. 6, 83 (1962).

[37] H. Malissa, W. Schmidts, Relative Conductometric Determination of Carbon, Hydrogen and Sulfur, Microchem. J. 8, 180 (1964).

[38] H.S. Haber, D.A. Bude, R.P. Buck, K.W. Gardiner, Carbon and Hydrogen Analysis by Coulometric Electrolysis of Water, Anal. Chem. 37, 116 (1965).

[39] F. Salzer, Application Range of Rapid Combustion Apparatus with Electric Indication for the Determination of Carbon and Hydrogen, Microchem. J. 10, 27 (1966).

[40] G. Kainz, K. Zidek, G. Chromy, Mikrobestimmung von Kohlenstoff und Wasserstoff in organischen Verbindungen durch konduktometrische Bestimmung der Verbrennungsprodukte, Mikrochim. Acta 235 (1968).

4.3 Conductivity

[41] A.A. Duswalt, W.W. Brandt, Carbon-Hydrogen Determination by Gas Chromatography, Anal. Chem. 32, 272 (1960).

[42] O.E. Sundberg, C. Maresh, Application of Gas Chromatography to Microdetermination of Carbon and Hydrogen, Anal. Chem. 32, 274 (1960).

[43] W. Simon, P.F. Sommer, G.H. Lyssy, Complete Automation of the Microdetermination of Carbon and Hydrogen in Organic Compounds, Microchem. J. 6, 239 (1962).

[44] P.F. Sommer, W. Sauter, J.T. Clerc, W. Simon, Selbstintegrierende Methode zur raschen Bestimmung von Kohlenstoff und Wasserstoff in organischen Verbindungen an Milligramm- und Submilligramm-Mengen unter Verwendung von Katharometern, Helv. Chim. Acta 45, 595 (1962).

[45] C.F. Nightingale, J.M. Walker, Simultaneous Carbon-Hydrogen-Nitrogen Determination by Gas Chromatography, Anal. Chem. 34, 1435 (1962).

[46] J.T. Clerc, W. Simon, Limitations of a Self-Integrating Method for the Simultaneous, C, H, and N Determination by Thermal Conductivity, Microchem. J. 7, 422 (1963).

[47] K. Derge, Anwendung der Gas-Chromatographie in der CHN-Analyse, Chem.-Ing.-Tech. 37, 718 (1965).

[48] F. Ehrenberger, H. Kelker, O. Weber, Simultanbestimmung von CHN unter Verwendung eines Analog-Digital-Umsetzers bei Wägung und Detektion, Z. Anal. Chem. 222, 260 (1966).

[49] G. Kainz, E. Wachberger, Über eine neue Mikro- und Ultramikro-Bestimmung von Kohlenstoff, Wasserstoff und Stickstoff in organischen Verbindungen mit Hilfe der Wärmeleitfähigkeit, Z. Anal. Chem. 222, 271 (1966).

[50] G. Kainz, E. Wachberger, Eine neue Mikro- und Ultramikro-Bestimmung von Sauerstoff in organischen Verbindungen mit Hilfe eines Wärmeleitfähigkeitsdetektors, Z. Anal. Chem. 222, 278 (1966).

[51] K. Derge, Mikroanalyse auf C, H und N, Chem.-Ztg., Chem. App. 90, 283 (1966).

[52] W. Walisch, G. Scheuerbrandt, W. Marks, Additional Techniques for Submicro CHN-Analysis: Extension of Sample Range, Combustion of Liquids and Hygroscopic Compounds, Microchem. J. 11, 315 (1966).

[53] W. Walisch, W. Marks, Eine Ultramikromethode zur Bestimmung des Sauerstoffgehaltes organischer Verbindungen, Mikrochim. Acta 1051 (1967).

[54] A.M.G. Macdonald, Turton, The Automatic Analysis of Highly Fluorinated Organic Materials, Microchem. J. 13, 1 (1968).

[55] *P.B. Olson*, Determination of Carbon and Hydrogen in Highly Fluorinated Substances with a Commercial Carbon Hydrogen Analyzer, Microchim. J. *13*, 75 (1968).

[56] *H. Graham*, Improved Instrumental Carbon, Hydrogen, and Nitrogen Analysis with Electronic Integration, Microchem. J. *13*, 327 (1968).

[57] *K.P. Kunz*, The Perkin Elmer Elemental Analyzer Made Automatic, Microchem. J. *13*, 463 (1968).

[58] *W. Walisch, K. Schäfer*, Zur Verbrennung fluor-, phosphor- und metallhaltiger Verbindungen mit einem CHN-Automaten, Mikrochim. Acta 765 (1968).

[59] *R. Culmo*, Microdetermination of Oxygen in Organic Compounds with an Automatic Elemental Analyzer, Mikrochim. Acta 811 (1968).

[60] *R. Culmo, R. Fyans*, A. New Method for Handling Liquid Samples in Organic Elemental Analysis, Mikrochim. Acta 816 (1968).

[61] *F. Scheidl*, Erfahrungen mit einigen Sauerstoffdonatoren der CHN-Analyse im Dezimilligramm- und Ultramikrobereich, Z. Anal. Chem. *245*, 30 (1969).

4.4 Manometry

[62] *H. Simon, G. Müllhofer*, Manometrisches Verfahren zur C-H-Analyse mit Einwaagen ab 0,3 Milligramm, Z. Anal. Chem. *181*, 85 (1961).

[63] *J.W. Frazer*, Simultaneous Determination of Carbon, Hydrogen, and Nitrogen, Mikrochim. Acta 993 (1962).

[64] *W. Pfab, W. Merz*, Erprobte Verfahren zur Ultramikro-Elementaranalyse, Z. Anal. Chem. *200*, 385 (1964).

4.5 Photometry

[65] *J.A. Kuck, A.J. Andreatch, J.P. Mohns*, Microdetermination of Oxygen in Organic Compounds Using Nondispersive Infrared Absorptiometry, Anal. Chem. *39*, 1249 (1967).

3.2 Detection and Estimation of Heteroatoms

Wolfgang Schöniger

Sandoz A.G., CH-4002 Basel

1 Introduction

The quantitative analysis of heteroatoms in organic compounds with few exceptions still requires mineralization, i.e. decomposition and simultaneous conversion of the heteroelements to specific inorganic molecules. In recent years a number of non-destructive physical methods have been employed which permit the quantitative determination of certain heteroelements. By means of X-ray fluorescence analysis halogens as well as metals can be determined quantitatively in organic molecules.

However, owing to matrix effects it is impossible to make general use of this technique since accurate results can be obtained only by calibrating with compounds having the same matrix or by mineralizing and imbedding the organic compound in the same carrier resulting in an elimination of matrix effects (e.g. Chapter 10.7).

2 Mineralization

2.1 Dry Mineralization

2.1.1 Combustion in Streaming Oxygen

Catalytic combustion in a stream of oxygen is applicable to all types of compounds[95] (3.2:4). Difficulties arise with highly volatile or perfluorinated compounds. Various forms of platinum serve as catalyst (platinum stars, perforated cylinders, asbestos, etc.). Combustion products can be trapped by absorption. This method is very time consuming, since the sample must burn and vaporize under constant supervision[1].

A substantially faster mineralization results from combustion in an empty tube[2] whose combustion chamber is divided by traversing plates to increase catalytic surface area and ensure good mixture of oxygen and substance vapor. Microsamples can be quantitatively burnt in a 50 ml/min gas flow within five minutes. The combustion products are collected in an absorption vessel connected to the combustion tube.

The most effective technique for mineralization is the combustion in an oxygen-hydrogen burner. This method[3] has been used for trace *sulfur* analysis. Nevertheless, its application extends to trace *halogens*, particularly to *fluorinated compounds*, since even perfluorinated derivatives can be mineralized quantitatively. An additional advantage of this method is the fact that samples of practically any size can be analyzed. Milligram amounts of halogens and sulfur have been determined by this procedure[4-6].

[1] *H. Roth*[95], p. 120.

[2] *R. Belcher, G. Ingram*, Anal. Chim. Acta 7, 319 (1952).

[3] *R. Wickbold*, Angew. Chem. *69*, 530 (1957).

[4] *R. Levy*, Mikrochim. Acta 224 (1962).

[5] *F. Martin, A. Floret*, Chim. Anal. [Paris] *40*, 120 (1968); *41*, 181 (1959).

[6] *F. Ehrenberger*, Mikrochim. Acta 192 (1959); 590 (1961).

A universal apparatus with appropriate head-burners which will burn liquids, solutions, or solids is commercially available.

2.1.2 Combustion in Closed Systems

The development of exact microanalytical techniques for the determination of *halogen, phosphorus* and *sulfur*[7-11] has rapidly displaced the usual oxidative decomposition methods. There are several advantages to combustion in an oxygen-filled flask.

The operation is simple. High combustion temperatures (ca. 1200°) ensure complete decomposition of even difficultly combustible types of compounds. Combustion products are collected by a small volume of absorption solution which in most cases, can be employed directly for final determinations. The latter may be modified according to requirements. Duration of combustion is only 20–30 seconds. During the subsequent absorption of products other work may be performed. Thus, the flask combustion technique is excellent for analyses in a routine laboratory.

A 250–300 ml flask equipped with a ground-glass stopper to which a glass rod is attached is used for microanalysis. A platinum net containing the probe is placed inside the glass rod.

Solids are weighed by difference on filter paper (Schleicher and Schüll 589 or 1575; Whatman 44 or 54) and placed after folding inside the platinum net with the ignition strips projecting out underneath.
Liquids may be sealed in thin-walled capillaries[12] of polyethylene or methylcellulose. These are placed in a weighed filter paper and then secured in the probe carrier.
The absorption solution is placed in the flask. Then a fast stream of oxygen is injected filling the flask. The tips of the ignition strips are ignited with a gas flame and the stopper immediately secured in place. The flask is tilted for the absorption solution to seal the ground joint. The stopper must be held fixed to the neck of the flask during the 20 seconds of combustion. Finally the flask is left to stand for 30 to 45 minutes or shaken in order to accelerate absorption of the combustion products. Then the stopper is removed while the sample carrier is washed with absorption solution. The solution is ready for final determination.

Some additional points to observe are:

The volume of the flask should be sufficient to supply both the sample and carrier with abundant oxygen (e.g., 10–*l* round flasks for trace element determination[13]).
Normally glass may be used, but quartz, polyethylene, or polypropylene flasks[14] should be employed for *fluorine* assays.
The sample carrier should have no blank.
The wire strength and mesh width of the platinum net are selected in such a way that, on the one hand, the net doesn't melt, while on the other hand, sufficient circulation of oxygen is guaranted. Appropriate nets are commercially obtainable.

A modified method exists[15] especially for submilligram range samples, for fluorine in highly fluorinated compounds, and for *volatile substances* and is recommended by other authors[16, 17].

2.1.3 Decomposition with Sodium Peroxide

This method, performed in a 'decomposition bomb', is universally applicable except that difficulties may arise in volumetric or electrometric final determinations because of the large excess of sodium and residual acid ions (formed by neutralization of decomposed materials).

2.1.4 Reductive Mineralization

The well-known process of *catalytic hydrogenation*[18] has been applied recently to a greater extent especially to trace *sulfur* analysis[19].
Dry decomposition in a closed tube with sodium, potassium[20], or magnesium[21, 22] represents the most important method for *halogen* and *sulfur* determination. Jenik[23] has shown this method to be particularly effective since *nitrogen, phosphorus, arsenic, antimony* and *silicon* can also be estimated quantitatively using magnesium decomposition.

7   *W. Schöniger*, Mikrochim. Acta 123 (1955); 869 (1956).
8   *W. Hempel*, Angew. Chem. *13*, 393 (1892).
9   *E. Votocek*, Chem. Listy *16*, 248 (1922).
10 *O. Mikl, J. Pech*, Chem. Listy *46*, 382 (1952); *47*, 904 (1953).
11 *W. Schöniger*, Ref.[96], p. 93.
12 *W. Kirsten*, Ref.[97], p. 137.
13 *D.J. Lisk*, J. Agr. Food. Chem. *8*, 119 (1960); *8*, 306 (1960); *9*, 50, 468, 488, 489 (1961).

14 *M. Fernandopulle, A.M.G. MacDonald*, Microchem. J. *11*, 41 (1966).
15 *W. Kirsten*, Microchem. J. *7*, 34 (1963).
16 *H.J. Francis jr., J.H. Deonarine, D.D. Persing*, Microchem. J. *14*, 580 (1969).
17 *J. Pavel, R. Kübler, H. Wagner*, Microchem. J. *15*, 192 (1970).
18 *K. Bürger*, Chem. Fabrik *13*, 218 (1940); Angew. Chem. *54*, 392 (1941).
19 *E.E.H. Pitt, W.E. Rupprecht*, Fuel *43*, 417 (1964).
20 *W. Zimmermann*, Mikrochemie, *31*, 15 (1944); *33*, 122 (1948); *35*, 80 (1930); *40*, 162 (1952).
21 *W. Schöniger*, Mikrochim. Acta 74 (1954); 44 (1955).
22 *P.N. Fedoseev, N.P. Ivasova*, Zh. Anal. Khim. *7*, 112 (1952).
23 *J. Jenik*, Ref.[98], p. 183.

Nickel bomb decomposition with potassium is highly recommended for analysis of very stable *fluorine* compounds[24]. The bomb must be heated in a muffle furnace for a long time (suggested times and temperatures vary from 1 hour and 600° to 2 hours and 950°).

2.2 Wet Mineralization

2.2.1 Oxidizing Mineralization

Decomposition with nitric acid under pressure (Carius method) is one of the oldest oxidation procedures[1]. In laboratory, however, it has no longer been used for a couple of decades and is mentioned here only for the sake of completeness. The same applies to mineralization with concentrated sulfuric acid in the presence of potassium dichromate. This method as well does not apply for the determination of hetero-elements. However, it is sometimes used for the determination of organic carbon, especially for the analysis of solutions.

In addition, metals or non-metals are still determined through decomposition with sulfuric acid/nitric acid or sulfuric acid/hydrogen peroxide in a Kjeldahl flask.

2.2.2 Reducing Mineralization

Reduction of organically bound halogens by alkali metals in organic solvents according to the classical method of Stepanov[25] may not be applied universally although it has undergone many modifications recently[26-28].

3 Determination

Choice of a particular procedure for final determination depends on the mode of decomposition. Dry oxidation is highly recommended as it ensures elimination of foreign ions which is a prerequisite to accurate final determinations. In the case of sodium peroxide decomposition purification of the dissolved reaction products by means of an ion exchanger prior to decomposition is required. Potentiometric titration would otherwise be inaccurate.

3.1 Halogens

3.1.1 Chlorine

Chloride determination by silver ion or, even better, mercury ion has proved to be the most suitable method. Mercury ions form soluble, non-dissociated mercury chloride, while silver ion forms a silver chloride precipitate which adsorbs the indicator during titration. Two recommended indicators are diphenylcarbazide[29] and diphenylcarbazone[30]. An improved end-point is observed in 80% ethanolic solution[31].

Attention should be paid to the Vollhard titration[1] which is still in use mainly for macro analysis beside other titration methods with visual end-point detection. Clearly, chloride can also be determined acidimetrically as long as the combustion produces no acids and the absorption solvent contains no strong electrolytes.

Electrometric end-point[32, 33] and coulometric methods[34, 35] exist; colorimetry may be applied only to a limited[36] extent.

3.1.2 Bromine

In principle, bromine may be determined as described in Section 3.1.1, although the conversion factor is much less favorable. Frequently, it is preferable to analyze bromine specifically[37], free of impurities such as sulfur, phosphorus, or fluorine which would affect bromine determination.

3.1.3 Chlorine and Bromine

If both elements are present in the same sample and are to be simultaneously analyzed independently, then potentiometric titration should be given preference. This well-known technique has some disadvantages. The sensitivity is affected by foreign electrolytes; thus for individual halide estimation it is advisable to use absorption solutions after combustion of the product in the flask. In simultaneous titrations, errors are caused by

[24] *C.L. Wilson, D.W. Wilson*, Ref. [97], p. 557.

[25] *A. Stepanov*, Ber. *39*, 4056 (1906).

[26] *P. Johncock, W.K.R. Musgrave, A. Wiper*, Analyst (London) *84*, 245 (1959).

[27] *R.D. Chambers, W.K.R. Musgrave, J. Savory*, Analyst (London) *86*, 356 (1961).

[28] *R.C. Blinn*, Anal. Chem. *32*, 292 (1960).

[29] *K. Präger, H. Fürst*, Chem. Tech. [Leipzig] *10*, 537 (1958).

[30] *J.V. Dubsky, J. Trtilek*, Mikrochemie *12*, 315 (1933); *15*, 302 (1934).

[31] *F.W. Cheng*, Microchem. J. *3*, 537 (1959).

[32] *D.G. Newman, C. Tomlinson*, Mikrochim. Acta 73 (1961).

[33] *J.P. Dixon*, Ref. [100], p. 140.

[34] *D.F. Ketchum, H.E. Pragle Johnson*, Microchem. J. *11*, 139 (1966).

[35] *D.M. Coulson, L.A. Cavanagh*, Anal. Chem. *32*, 1245 (1960).

[36] *I. Lysij*, Microchem. J. *3*, 529 (1959).

[37] *W. Schöniger*, Mikrochim. Acta 869 (1956).

coprecipitation of mixed crystals or solid solutions. The higher the molar ratio of the halides, the greater the individual results of the estimation will differ. Similar limitations apply to coulometric and amperometric methods. The accuracy of potentiometric titration of chlorides and bromides is increased if first the total halide in an aliquot is estimated, then bromide removed from a second aliquot (oxidation to bromine with acid hydrogen peroxide solution and reaction with oxime) which is subsequently titrated for chloride[38]. Since iodide is supposed to react similarly, three titrations of three appropriately treated aliquots are sufficient for the three estimations:

total halogen potentiometrically,
chloride potentiometrically after bromide and iodide removal, and
iodide by thiosulfate titration.

3.1.4 Iodine

Volumetric determination of iodate (obtained by iodide oxidation)[39] is an alternative to one of the procedures already mentioned. This method permits determination of minute iodide quantities with higher accuracy owing to the chemical multiplication principle.

3.1.5 Fluorine

This element has achieved prominence in all areas of organic chemistry within the last ten years and, consequently, its accurate estimation is important.

Molecules containing only few C–F bonds or few –CF$_3$ groups can be subjected to normal flask combustion (3.2:2.1.2). The upper limit of the fluorine content is found to be 45%. Quartz[40, 41] or propylene[13] flasks must be used since fluorine reacts with glass. For very stable fluorinated compounds containing highly fluorinated rings or long chains, both the method of Kirsten[15] and the combustion in an oxygen-hydrogen burner[42] are recommended. However, the most suitable method of determining fluorine content of stable highly fluorinated compounds is still the reduction by means of potassium.

For fluoride ion determination, no distinct method can be recommended since all methods require time-consuming preparations and both volumetric and colorimetric methods demand calibration against pure fluoride salt solutions.

Gravimetric determination: Precipitation as lead chlorofluoride[43] was found to be very suitable for samples of more than 10 mg. Difficulties arise, however, from cations forming strong fluoride complexes or from anions giving insoluble lead compounds.

Thorium nitrate titration in 50% ethanolic solution using alizarin red-S indicator is a visual end-point *volumetric method* which is used most frequently. Erroneous results are caused by sulfate, phosphate, arsenate, and many metal ions so that fluorosilicic acid must be separated by distillation according to the method of Willard and Winter[44]. Most *colorimetric procedures* are based on the decrease in intensity of color of either thorium, zirconium, or aluminum complexes through the action of fluoride ion. Cerium(III)ions bind alizarin to give a red complex which reacts with fluoride ions forming a water soluble blue complex[45]. Recently cerium ion has been replaced by the more stable and more effective lanthanum ion[46, 47–49]. Aluminum and iron(III)ions interfere. *Use of a fluoride sensitive electrode:* This reliable and routine method is based on the end-point determination in lanthanum or thorium ion titrations[50–52]. Direct determination of fluorine[53] after combustion has also proved useful. It is based on pH-measurements and comparison with standards[53].

3.2 Sulfur

The final analysis is determined mainly by the choice of mineralization. In most cases oxidative mineralization is applied and produces both sulfur dioxide and sulfur trioxide. Particular care is needed to include an oxidizing agent in the absorption solution to convert all *sulfur* of the organic com-

[38] *T.S. Prokopov,* Anal. Chem. *42,* 1096 (1970).
[39] *T. Leipert,* Mikrochemie, Pregl-Festschrift p. 266 (1929).
[40] *E.C. Olson, S.R. Shaw,* Microchem. J. *5,* 101 (1961).
[41] *R. Levy, E. Debal,* Mikrochim. Acta 223 (1962).
[42] *F. Ehrenberger,* Mikrochim. Acta 192 (1959).

[43] *R. Belcher, A.M.G. MacDonald,* Mikrochim. Acta 1111 (1957).
[44] *H.H. Willard, O.B. Winter,* Ind. Eng. Chem. Anal. *5,* 7 (1933).
[45] *R. Belcher, M.A. Leonard, T.S. West,* Talanta *2,* 92 (1959); J. Chem. Soc. 3577 (1959).
[46] *M.E. Fernandopulle, A.M.G. MacDonald,* Mikrochem. J. *11,* 41 (1966).
[47] *R. Greenhalgh, J.P. Riley,* Anal. Chim. Acta *25,* 179 (1961); *35,* 365 (1966).
[48] *A.G. Davies, J.K. Foreman,* Proc. S.A.C. Conf. Nottingham, p. 167, Heffer, Cambridge 1966.
[49] *L.H. Weinstein,* J. Air Pollut. Control Ass. *15,* 222 (1965).
[50] *J.J. Lingane,* Anal. Chem. *39,* 881 (1967).
[51] *C. Harzdorf,* Z. Anal. Chem. *245,* 67 (1969).
[52] *D.A. Shearer, G.F. Morris,* Microchem. J. *15,* 199 (1970).
[53] *J. Pavel, R. Kübler, H. Wagner,* Microchem. J. *15,* 192 (1970).

pound to sulfate. Aqueous hydrogen peroxide solution has proved highly effective for this purpose. If the sample is free of other acid forming elements besides sulfur, then sulfate ions can be determined directly acidimetrically. Chlorine can also be determined as described in Section 3.2:3.1. Metals giving water soluble sulfates can be removed by means of a cation exchanger[1], and then acidimetric end-point determination may be carried out.

However, in most cases preference is given to barium perchlorate titration[54] where *sulfate* ions are titrated in 80% ethanolic solution using thorin as an indicator. Methylene blue[55] is also added; otherwise the end-point (yellow to pale pink) would be difficult to recognize. A more precise end-point is obtained with sulfonazo III (red to blue)[56]. Further volumetric visual end-point methods are of limited value such as the titration with lead nitrate[57] or 4-amino-4′-chlorodiphenyl hydrochloride[58].

For *colorimetric sulfate* quantitative detection in the milli- and microgram range, barium chloranilate[59] is most suitable. With sulfate ions this insoluble compound gives a barium sulfate precipitate accompanied by an equivalent release of chloranilate ions the concentration of which can be measured at 530 and 332 nm. Trace concentrations of sulfate ions may be determined by way of the nephelometric method[60].

The variety of applied methods is of great value. After adsorption of sulfur oxides on silver metal in a combustion tube, the produced silver sulfate can be dissolved in water and titrated potentiometrically[61]. A final conductometric determination with barium chloride or acetate can be carried out profitably only if the total ion concentration is low; otherwise the change in conductivity at the equivalence point is too small.

After reductive decomposition, i.e. after treatment with alkali, sulfide ion is estimated iodometrically[62]. For colorimetric sulfide determination, *p*-amino-*N*,*N*-dimethylaniline in the presence of iron(III) ions can be converted quantitatively to

methylene blue by sulfide ions[63]. This method is best for trace analysis and for biological material. It is not affected by other ions; carefully controlled reaction conditions are absolutely necessary.

3.3 Phosphorus

The best methods involve oxidative decompositions. For series analyses, particularly samples of small *phosphorus* content, decomposition with sulfuric acid/nitric acid in Kjeldahl flasks is still used owing to a wide range of sample size and the possibility of a colorimetric final determination.

However, the conditions of decomposition should be controlled carefully. Sodium peroxide decomposition is recommended for substances containing phosphorus and fluorine[64], although oxygen combustion may be used. In nearly all cases, mineralization by combustion in the flask is completely sufficient. Various authors suggest the addition of sodium carbonate[65], ammonium sulfate[66] or sodium peroxide[67] to the sample. The absorption solution should contain an oxidizing agent to convert small amounts of pyro- and meta-phosphate and low phosphoric oxides to the major oxidation product, orthophosphate. Thus hydrogen peroxide[68], dilute nitric acid[69], or alkaline hypobromite[70] solution are used.

For the final determination, volumetric and colorimetric methods are considered besides classical gravimetric ones. *Colorimetric methods* are based on the well-known molybdate-blue reaction, a process which has been subject to numerous modifications. Since this method generally requires plenty of time, it is used only for routine analysis. *Volumetric methods:* Phosphate can be converted to quinoline molybdophosphate[71]. The resulting precipitate is filtered, washed, and dissolved in excess sodium hydroxide. The phosphorus content is determined by back titration. The advantage of this method lies in a precise end-point and in the fact that foreign ions do not interfere. Thus, it can be used even after sodium peroxide decomposition. Precipitation with excess magnesium chloride at pH 10 in 50% ethanolic solution and back

---

[54] *J.S. Fritz, S.S. Yamamura*, Anal. Chem. 27, 1461 (1955).
[55] *C.N. Reilley*, Ref. [99], p. 91.
[56] *B. Buděšinsky*, Anal. Chem. 37, 1159 (1965).
[57] *E.E. Archer*, Analyst (London) 82, 208 (1957).
[58] *R. Belcher, W.I. Stephen*, Anal. Chim. Acta 20, 197 (1959).
[59] *I. Lysij, J.E. Zarembo*, Microchem. J. 3, 173 (1959).
[60] *Sant Lal Chopra*, Indian J. Chem. 2, 78 (1964).
[61] *J.P. Dixon*, Talanta 4, 221 (1960).
[62] *J.P. Dixon*, Ref. [100], p. 106.

---

[63] *L. Gustafsson*, Talanta 4, 227, 236 (1960).
[64] *J.D. Burton, J.P. Riley*, Analyst (London) 80, 391 (1955).
[65] *S.J. Gedansky, J.E. Bowen, O.I. Milner*, Anal. Chem. 32, 1447 (1960).
[66] *A. Dirscherl, F. Erne*, Mikrochim. Acta 670 (1960).
[67] *A. Steyermark*, Microchem. J. 3, 523 (1959).
[68] *E. Meier*, Mikrochim. Acta 70 (1961).
[69] *R. Bennewitz, I. Tänzer*, Mikrochim. Acta 835 (1959).
[70] *R. Belcher, A.M.G. MacDonald*, Talanta 1, 185 (1958).
[71] *A.M.G. MacDonald, W.I. Stephen*, J. Chem. Educ. 39, 528 (1962).

titration with EDTA solution using eriochrome-black T as indicator ensure a precise end-point[72]. It is necessary, however, to ascertain an empirical calibration factor.

Titration with cerium (III) nitrate[73] using eriochrome-black T as indicator is by far the best method.

3.4 Other Elements

Only a few of the many elements of the periodic system are significant in organic chemistry. While most metals are still determined by conventional methods, the near future will see a variety of improved instrumental techniques. As described earlier, *atomic absorption spectrometry* may be employed routinely to determine elements after wet mineralization. Increasing application of x-ray fluorescence in analysis must be considered since direct measurements, i.e., without prior decomposition, obviously offer advantages. Moreover non-metals may be determined quantitatively. The present lower limit in the periodic system is fluorine.

3.4.1 Arsenic

Only oxidative decomposition is used for arsenic. Wet mineralization is performed with hydrogen peroxide and sulfuric acid in a Kjeldahl flask. Then excess nitric acid is removed and *arsenic* acid is determined iodometrically[74] at fixed pH.

Alternately, reduction and distillation gives arsenic hydride which is dissolved in a pyridine solution of silver diethyldithiocarbamate and measured colorimetrically at 540 nm[75].

With oxygen combustion an oxidizing absorption solution must be used. Absorption is performed either in dilute iodine solution, and arsenate is determined colorimetrically using molybdate-blue[76], or in weakly basic perhydrol solution, titrating arsenate with lead nitrate solution buffered with hexamethylene-tetramine and using pyridylazoresorcinol or Snazoxs as indicator[77]. The following procedure is likewise recommended: precipitation with silver nitrate[74], subsequent dissolution in tetracyanonickelate, and titration of the released nickel ions with ethylene diamine tetraacetic acid (EDTA) with murexide indicator[78].

3.4.2 Borine

Mineralization of *borine* compounds may be very difficult since they have different stability ranging from extremely stable to extremely unstable. In any case, only oxidative decomposition methods are employed. For wet decomposition procedures, mixtures of nitric, perchloric, and sulfuric acids are suggested. In the sodium peroxide decomposition carried out in a nickel bomb, an extremely violent and spontaneous reaction may take place. In streaming oxygen, an incomplete combustion may occur. In this case, addition of glucose[79] or potassium hydroxide[80] is recommended. Boric acid formed in all these experiments can be determined volumetrically by the well-known mannitol procedure[81] using visual or instrumental end-point methods. Various colorimetric methods, especially for trace concentrations, are reported[82, 83].

3.4.3 Mercury

Classical procedures involve combustion of the organic compound in an oxygen stream and adsorption of *mercury* on gold fibers with subsequent determination of the weight increase of the fibers. Combustion in a closed oxygen atmosphere is generally applied using concentrated nitric acid as absorption solution to convert all mercury to mercury(II)ions. The final determination of mercury(II)ions is performed amperometrically with EDTA solution[84] or in a two-phase titration with sodium diethyldithiocarbamate at pH 9–10[85]. A subsequent potentiometric chlorine determination is possible according to this method. Trace quantities are best determined colorimetrically with dithizone[86].

3.4.4 Selenium

Besides wet ashing with either sulfuric acid or a mixture of sulfuric and nitric acid followed by iodometric end titration, the best method is closed

[72] *K.D. Fleischer*, Anal. Chem. *30*, 152 (1958).
[73] *R. Püschel, H. Wittmann*, Mikrochim. Acta 670 (1960).
[74] *O. Wintersteiner*, Mikrochemie *4*, 155 (1929).
[75] *H. Kashiwagi, Y. Tukamoto, M. Kan*, Ann. Rep. Takeda Res. Lab. *22*, 69 (1962).
[76] *W. Merz*, Mikrochim. Acta 640 (1959).
[77] *R. Püschel, Z. Štefanac*, Mikrochim. Acta 1108 (1962).
[78] *Z. Štefanac*, Mikrochim. Acta 1115 (1962).

[79] *S.K. Yasuda, R.N. Rogers*, Microchem. J. *4*, 155 (1960).
[80] *M. Corner*, Analyst (London) *84*, 41 (1959).
[81] *H. Roth, W. Beck*, Z. Anal. Chem. *141*, 404, 414 (1954).
[82] *C.A. Parker, W.J. Barnes*, Analyst (London) *85*, 828 (1960).
[83] *R.D. Srivastava, P.R. Van Buren, H. Gesser*, Anal. Chem. *34*, 209 (1962).
[84] *B.C. Southworth, J.H. Hodecker, K.D. Fleischer*, Anal. Chem. *30*, 1152 (1958).
[85] *P. Gouverneur, W. Hoedeman*, Anal. Chim. Acta *30*, 519 (1964).
[86] *W.H. Gutenmann, D.J. Lisk*, J. Agr. Food Chem. *8*, 306 (1960).

oxygen combustion. As *selenium* attacks platinum, the usual sample carrier is replaced by a quartz vessel[87, 88]. For final iodometric determinations, the absorption solution must act as oxidizing reagent[89]. Trace concentrations are measured fluorometrically with 2, 3-diaminonaphthalene[90] or 3, 3-diaminobenzidine[91].

3.4.5 Other Metals

When determining other metals the combustion residue may be weighed after CH- or CHN-determination. *Silver, gold,* and *platinum* are left as free metals, while *iron, aluminum, copper, silicon, magnesium,* and *chromium* form definite oxides. *Alkali* and *alkaline earth metals* can be determined gravimetrically as sulfates after evaporation with sulfuric acid.

In the last few years, closed oxygen combustion has been applied successfully to determine metal in organometallics. Methods for the determination of *zinc, magnesium, calcium*[92], *barium, manganese, cobalt, copper*[93] and *tin*[94] are reported.

---

[87] *E. Meier, N. Shaltiel,* Mikrochim. Acta 580 (1960).

[88] *W. Ihn, G. Hesse, P. Neuland,* Mikrochim. Acta 628 (1962).
[89] *E.S. Gould,* Anal. Chem. *23,* 1502 (1951).
[90] *W.H. Allaway, E.E. Cary,* Anal. Chem. *36,* 1359 (1964).
[91] *N. Kunimine, H. Ugajin, M. Nakamaru,* J. Pharm. Soc. Jap. *83,* 59 (1963).
[92] *R. Belcher, A.M. G. MacDonald, T.S. West,* Talanta *1,* 408 (1958).
[93] *A.M.G. MacDonald, P. Sirichanya,* Microchem. J. *14,* 199 (1969).
[94] *R. Reverchon,* Chim. Anal. (Paris) *47,* 70 (1965).

4 Bibliography

[95] *Pregel-Roth,* Quantitative organische Mikroanalyse, 7. Aufl., Springer Verlag, Wien 1958.
[96] Proceedings of the International Symposium on Microchemistry 1958, Pergamon Press, London 1960.
[97] *C.L. Wilson, D.W. Wilson,* Comprehensive Analytical Chemistry, Vol. I B, Elsevier Publ. Co., Amsterdam 1960.
[98] *J. Jenik,* Scientific Papers 1961/II, Institute of Chemical Technology, Pardubice, Statni pedagogické nakladadelstvi, Praha 1961.
[99] Advances in Analytical Chemistry and Instrumentation, Vol. IV, Wiley Intersci. Publ., New York 1965.
[100] *J.P. Dixon,* Modern Methods in Organic Microanalysis, D. van Nostrand Co., London 1968.

3.3 Redox Methods

Willy Büchler, Jürg Meier, and Heinz Wenger
CIBA-GEIGY, CH-4002 Basel

1 Reduction

1.1 Reduction with Hydrogen

Catalytic hydrogenation[4] can be used for reduction of double and triple bonds, both between carbon atoms and between hetero atoms. Nitro and nitroso compounds, nitriles and azo compounds are, for example, converted into amines[1]. Determination of unsaturation via hydrogenation is predominantly used for elucidation of the structures of new compounds. Although halogenation is the usual method for determination of simple unsaturated compounds, hydrogenation has the advantages that side-reactions (substitutions) are avoided, and that even sterically hindered double bonds can be determined.

For good results, correct choice and pre-treatment of the catalyst are essential. Palladium on charcoal is widely used[2]. Alcohols and acetic acid are frequently used as solvents.

Hydrogen as a stoichiometric reducing agent: The conventional methods of hydrogenation employ volumetric or manometric gas measurements[3-11] (see also Section 3.7:2).

Some recent methods avoid the inconvenience associated with gasometric determinations of hydrogen. In quite an elegant electrochemical method[12],

---

[1] *W. Huber,* Z. Anal. Chem. *197,* 236 (1963).

[2] *N.D. Cheronis,* Micro and Semimicro Methods, p. 239, Interscience Publishers, New York 1954.
[3] Ref.[98], p. 292, 482.
[4] *A. Polgár, J.L. Jungnickel* in Ref.[99], Vol. III, p. 256.
[5] *S. Tiong, H.F. Waterman,* Chim. Ind. [Milan] *81,* 204, 357 (1959).
[6] *B. Budĕšinský,* Chem. Listy *53,* 997 (1959).
[7] *J. Horáček, V. Pechanec,* Collect. Czech. Chem. Commun. *27,* 1500 (1962).
[8] *C.L. Ogg, F.J. Cooper,* Anal. Chem. *21,* 1400 (1949).
[9] *A. Reuter,* Z. Anal. Chem. *231,* 356 (1967).
[10] *M. Sedlak,* Anal. Chem. *38,* 1503 (1966).
[11] *C.A. Brown,* Anal. Chem. *39,* 1882 (1967).
[12] *J.W. Miller, D.D. DeFord,* Anal. Chem. *30,* 295 (1958).

an automatically controlled recording coulometer produces hydrogen in such a way that the rate of generation equals the rate of consumption. Hydrogen can also be generated from lithium aluminum hydride and from alkali borohydrides. One can add a known amount of the hydride and titrate its excess with Karl Fischer reagent after reacting the unused hydrogen with oxygen to form water[13]. In another method[14], a manometric hydrogenation valve controls the admission of the hydride solution; a highly active catalyst is generated in situ. Catalytic hydrogenation has a very wide field of application. Such methods therefore are of low specificity, and are suitable only for the determination of uniform compounds.

Hydrogen as a preparative reducing agent: Highly specific assay methods can be developed if the reduction products are determined, rather than the amount of hydrogen necessary for the reduction. For example, nitriles can be converted into primary amines, which in turn can be titrated with perchloric acid. A standard deviation of 0.3% can be obtained; amides do not interfere[1].

Good specificity can also be obtained if a simple hydrogenator coupled with a gas chromatograph[15] is used.

1.2 Reduction with Hydrides

1.2.1 Lithium Aluminum Hydride[16]

Lithium aluminum hydride is widely recognized as an extraordinarily powerful basic reagent for the reduction of functional groups and for the determination of active hydrogen atoms (see 3.7:2.4). The most important substances which react with this reagent are: *alcohols*[16-18], *acids*[16, 17], *primary and secondary amines*[16], *aldehydes* and *ketones*[16, 19], *esters*[16, 19], *anhydrides*[16], *acid halides*[16], *primary, secondary and tertiary amides*[16, 19, 20], *nitriles*[16, 19], *aromatic nitro-compounds*[16, 19, 21], *aryl peroxides*[22], and *halides*[21].

Lithium aluminum hydride also reacts with *aliphatic di-* and *poly-sulfides*[23] and with the following inorganic compounds: *water, ammonia, carbon dioxide, oxygen*, and *metal halides*[16].

Stoichiometric use: The sample is treated with a known excess of the reagent. Depending on the nature of the sample, the reduction is conducted at room temperature or under reflux conditions. After a suitable period, which may vary between 1 and 120 minutes, the excess of the reagent is determined by titration with a standard alcohol solution. Alternatively, the residual lithium aluminum hydride may be decomposed, and the hydrogen liberated can be determined by gasometric methods (see 3.7:2). A blank must be run under identical conditions.

Solvents commonly used include tetrahydrofuran, ether, benzene and other aromatic compounds. Potentiometric methods[18] or visual indicators[17] can be used for detection of the end-point. If the volume of the hydrogen gas liberated during the course of the reduction is measured, it may be possible to differentiate between active hydrogen and reduction of functional groups[16].

If protected from atmospheric moisture, oxygen and carbon dioxide, the lithium aluminum hydride solution is stable for several weeks. Under favorable conditions, relative standard deviations in the range of 0.5 to 1% are obtained; the accuracy is usually better than 2%.
A disadvantage of the use of lithium aluminum hydride as a stoichiometric reductant is that the analytical methods are of low selectivity. Even though different functional groups may require differing conditions for their reduction, it is hardly possible to develop truly specific analytical methods.

Samples containing moisture give erroneous results. Because water has a small equivalent weight, even small amounts cause considerable errors.

Preparative use: The sample is treated with an arbitrary excess of the hydride. The reaction products are then quantitatively determined, often after suitable isolation. Amides can be reduced to the corresponding amines, which can be titrated after steam distillation[20]. In many cases, it may be convenient to determine the reduction products by gas chromatography.

The use of suitable separation processes subsequent to reduction permits specific analytical methods to be

[13] *W. Seaman*, Anal. Chem. *30*, 1840 (1958).
[14] *C.A. Brown, S.C. Sethi, H.C. Brown*, Anal. Chem. *39*, 823 (1967).
[15] *M. Beroza, R. Sarmiento*, Anal. Chem. *38*, 1042 (1966).
[16] *C.J. Lintner, D.A. Zuck, T. Higuchi*, J. Am. Pharm. Assoc., Sc. Ed. *39*, 418 (1950); *T. Higuchi* in Ref. [99], Vol. 2, p. 123.
[17] *T. Higuchi, D.A. Zuck*, J. Am. Chem. Soc. *73*, 2676 (1951).
[18] *C.J. Lintner, R.H. Schleif, T. Higuchi*, Anal. Chem. *22*, 534 (1950).
[19] *G. Jander, K. Kraffczyk*, Z. Anorg. Allg. Chem. *283*, 217 (1956).
[20] *S. Siggia, C.R. Stahl*, Anal. Chem. *27*, 550 (1955).
[21] *J.S. Fritz, G.S. Hammond*, Quantitative Organic Analysis, p. 99, 107, John Wiley & Sons, New York 1957.
[22] *A.P. Terent'ev, G.G. Larikova, E.A. Bondarevskaya*, Zh. Anal. Khim. *21*, 355 (1966).
[23] *M. Porter, B. Saville, A.A. Watson*, J. Chem. Soc. 346 (1963).

developed, even though the reduction step lacks selectivity. Moisture in the sample has no bearing on the results, provided that sufficient hydride is available for the reduction.

Lithium aluminum amide: Lithium aluminum amide is a less active and more specific reductant than lithium aluminum hydride. *Alcohols* may be determined by direct titration in the presence of aldehydes, esters, ketones and amines[24]. The amount of lithium aluminum amide consumed by moisture in the sample can be calculated and corrected for after a Karl Fischer titration.

1.2.2 Alkali Borohydrides

Alkali borohydrides are moderately strong reducing agents, which can be used for the determination of *aldehydes, ketones, esters*[25-27], *carbohydrates*[28, 29], *unsaturated compounds*[30], *disulfides*[31] and *quinones*[32]. In some cases, catalysts are required.

Lithium borohydride is a stronger reductant than sodium borohydride. The reducing action of sodium borohydride can be enhanced through addition of lithium halide[25]. The rate of reduction depends considerably on the nature of the solvent and the catalyst.

Stoichiometric use: The sample is reacted for 5—200 minutes at room temperature or reflux temperature with a known amount of the reductant. The excess of the borohydride is then decomposed by addition of acid, and the liberated hydrogen is determined by gasometric methods (see 3.7:2). The residual reductant can be determined also by iodometric titration[26]. Two different functional groups can be determined simultaneously by reaction rate methods, provided that the rates of reduction are sufficiently different[26]. Aldehydes can be determined by direct photometric titration[27].

Since borohydrides react only very slowly with active hydrogen atoms, there is little limitation in the choice of the solvents. Water, alcohols and ether are commonly used. Standard deviations of the order of 0.3 % can be achieved.

Preparative use: The sample is reduced with an arbitrary excess of the reductant. After separation and purification, the reaction products are determined by suitable methods[31]. *Disulfides*, for example, are reduced to mercaptans, which in turn can be determined by argentometric titration with a reproducibility and accuracy of 1 %.

1.3 Reduction with Metals[105]

1.3.1 Zinc

Zinc and zinc amalgam are powerful reductants. They may be used for the reduction of *nitro* and *nitroso compounds*[33-35], *organic nitrates*[36], *quinones*[34, 36], *sulfoxides* and *disulfides*[35] and, sometimes, double bonds.

Stoichiometric use: If reduction with zinc amalgam is carried out in such a way that generation of hydrogen is suppressed and atmospheric oxygen excluded, reduction of the sample causes liberation of an equivalent quantity of zinc ions from the amalgam[34]:

$$ZnHg \rightarrow Zn^{2\oplus} + 2e + Hg$$

The zinc ions can be determined easily by complexometric titration, and quite accurate results may be obtained.

Preparative use: Zinc and zinc amalgam are used more frequently for preparative purposes than for stoichiometric reductions. After suitable separation and purification, the reduction products can often be determined by acidimetric or redox titration.

1.3.2 Other Metals

Milder reducing agents than zinc include (in order of decreasing reduction potential): *cadmium, lead* and *bismuth amalgams*. They can be used for stoichiometric reductions, in the same way as zinc[34]. Alkali metals and their amalgams are very potent reducing agents. Aluminum, iron, nickel and Dewarda's alloy are used only occasionally for analytical purposes.

Metallic tin and hydrochloric acid can be used for the reduction of *aromatic nitro compounds*. Under certain precautions, the amount of tin converted

[24] *D.E. Jordan*, Anal. Chim. Acta *30*, 297 (1964).

[25] *H.C. Brown, E.J. Mead, B.C.S. Rao*, J. Am. Chem. Soc. 77, 6209 (1955).

[26] *E.H. Jensen, W.A. Struck*, Anal. Chem. 27, 271 (1955).

[27] *E. Cochran, C.A. Reynolds*, Anal. Chem. 33, 1893 (1961).

[28] *P.S. Skell, J.G. Crist*, Nature *173*, 401 (1954).

[29] *A.P. Terent'ev, I.S. Novikova*, Zh. Anal. Khim. *20*, 1226 (1965).

[30] *C.A. Brown, S.C. Sethi, H.C. Brown*, Anal. Chem. *39*, 823 (1967).

[31] *C.R. Stahl, S. Siggia*, Anal. Chem. 29, 154 (1957).

[32] Ref. [104], p. 215.

[33] Ref. [98], p. 628—629.

[34] *W.G. Scribner, C.N. Reilley*, Anal. Chem. *30*, 1452 (1958).

[35] Ref. [104], p. 307, 330, 344.

[36] *J. Doležal, J. Zýka*, Chemist-Analyst *53*, 68 (1964).

into stannous ion is related stoichiometrically to the amount of nitro compound; it can be found by determining the weight loss of the granulated tin[37]. Mercury may be used as a reducing agent for *N-nitro* and *N-nitroso compounds* and for *nitrates*[38, 39]:

$$R_2N - NO_2 + 3/2\, Hg \rightarrow R_2NH + NO + 3/2\, Hg^{2\oplus}$$
$$R_2N - NO + 1/2\, Hg \rightarrow R_2NH + NO + 1/2\, Hg^{2\oplus}$$
$$NO_3^{\ominus} + 3/2\, Hg \rightarrow NO + 3/2\, Hg^{2\oplus}$$

The nitric oxide produced is measured by volumetric methods (see 3.7:2).

1.4 Reduction with Inorganic Cations

1.4.1 Divalent Chromium

As is indicated by its standard potential Cr(III)/-Cr(II) of −0.40 V, chromous ion is a very powerful reducing agent[106]. It is often observed that in cases where titanous ion leads to unsatisfactory or non-stoichiometric reactions, divalent chromium can be used to provide fast and quantitative reduction. The following are a few typical reductions: *quinones* to hydroquinones[40–42, 44], *aromatic nitro-componds*[40–42, 44], *aromatic nitroso-compounds*[40, 41, 43, 44], and *azo-compounds*[40–45] to anilines, and *enynes* to dienes[44, 46].

Titrations with chromous ion may be carried out either by direct or by indirect methods. Elevated temperatures are rarely used, and all operations must be carried out in oxygen-free atmospheres. Solvent mixtures are used which usually contain organic components such as acetic acid, ethanol, or dimethyl formamide, together with at least 15% of water. A blank titration is run to check the stability of the solvent to reduction.

Potentiometric methods or visual indicators[41] may be used for determination of the end point. Methods based on reduction with bivalent chromium usually yield results with an accuracy of 0.5—1%.

Preparation of standard chromium(II) solutions: Solutions of bivalent chromium are most simply prepared by reduction of chromic ion with amalgamated zinc[42, 44, 47, 48], and standardized by titration with a standard ferric solution, which has itself been standardized by iodometric titration. The chromium(II) solution is stable for several weeks.

Direct titration: Quinones and various *nitroso* and *nitro compounds* can be determined by direct titration in acidic solution. If titration is carried out potentiometrically, simultaneous determination of different functional groups or even isomers may be possible[40, 42].

Indirect titration: The indirect method is preferred when the rate of reduction near the end point is slow. The sample is reacted with a 10—500% excess of chromium(II) ion, and after 2—30 minutes this is back-titrated with a suitable oxidant.

Iron(III) and chromium(VI) solutions are commonly used for this back-titration. If *anthrahydroquinones* are present, they are re-oxidized to the quinones. Two potential transitions can then be observed, the first one occuring after oxidation of the excess of chromium(II) ion, the second indicating completion of the oxidation of the hydroquinone. On reduction with chromous ion, nitro and azo groups yield amines, which are not re-oxidized in the back-titration; simultaneous determinations of *quinones* and *azo* or *nitro* compounds are therefore feasible[42].

When it is desirable to avoid re-oxidation of a hydroquinone, the excess of chromium(II) can be determined with a standard solution of anthraquinone-2-sulfonic acid[42].

Reductions with chromium(II) ion can be carried out also in alkaline media; precipitation of chromous ion is prevented by addition of EDTA[42]. The solution is acidified before the back-titration.

1.4.2 Vanadium(II) Ion

Vanadium(II) ion is a powerful reducing agent[106], the standard potential for V(III)/V(II) of −0.25 V lying about halfway between Cr(II) and Ti(III).

As with chromium(II) *quinones* and *nitro compounds* may be reduced to hydroquinones[49, 51] and amines[50] respectively. Depending on the conditions, azo compounds react to give either hydrazo compounds or amines. Vanadium(II) ion has also been used for the reduction of *triphenylmethane dyes*[51]:

$$R_3C^{\oplus} + 2e + H^{\oplus} \rightarrow R_3CH$$

[37] *C.E. Vanderzee, W.F. Edgell*, Anal. Chem. 22, 572 (1950).

[38] Ref. [104], p. 311.

[39] *J.S. Fritz, G.S. Hammond*, Quantitative Organic Analysis, p. 112, John Wiley & Sons, New York 1957.

[40] *H. Jucker*, Anal. Chim. Acta 16, 210 (1957).

[41] *J.P. Tandon*, Z. Anal. Chem. 167, 184 (1959).

[42] *W. Büchler, P. Gisske, J. Meier*, Anal. Chim. Acta 49, 309 (1970).

[43] *N.H. Furman, R.S. Bottei*, Anal. Chem. 29, 121 (1957).

[44] *R.S. Bottei, N.H. Furman*, Anal. Chem. 27, 1182 (1955); 29, 119 (1957).

[45] *V.M. Ivanov, D. Hálová, L. Sommer*, Z. Anal. Chem. 230, 422 (1967).

[46] *R.S. Bottei*, Anal. Chim. Acta 30, 6 (1964).

[47] *J.J. Lingane, R.L. Pecsok*, Anal. Chem. 20, 425 (1948).

[48] *H.W. Stone*, Anal. Chem. 20, 747 (1948).

[49] Ref. [104], p. 215.

[50] *C.M. Ellis, A.I. Vogel*, Analyst (London) 81, 693 (1956).

[51] Ref. [106], p. 147, 152.

Carbonyl compounds can be determined with vanadium(II), after conversion into their dinitrophenylhydrazones[51].

The titrations are usually carried out at room temperature, in buffered to strongly acidic aqueous solution; organic solvents may be added. Atmospheric oxygen must be excluded. If an *indirect titration* procedure is used, the excess of the reducing agent is back-titrated with standard Fe(III) solution. The end-point can be detected potentiometrically, amperometrically, or visually using an indicator[50].

The vanadium(II) solution is prepared by reducing metavanadate with zinc or zinc amalgam. Ferric ion or iodate are commonly used for standardization of the solution[50].

1.4.3 Titanium(III) Ion

Titanium(III)[105] is a mild reducing agent, with a standard potential Ti(IV)/Ti(III) of +0.05 V. It can be used to reduce *quinones* to hydroquinones[52, 53], *aromatic nitro* and *nitroso compounds* to amines[54-56], *azo* and *azoxy compounds* also to amines[53, 55, 57, 58], and *sulfoxides* to sulfides[59, 60]. The reagent can also be used to reduce *diazonium salts*[57] and *peroxides*[61].

Reductions are usually performed with a 10—100% excess of Ti(III) ion, the unreacted reagent being back-titrated with Fe(III) ion. In some cases, polarographic determination of the excess of titanium(III) is claimed to be advantageous[56]. Solvents generally used include water, glacial acetic acid, ethanol and dimethyl formamide. Although titanous ion is a mild reductant in acidic medium, it becomes fairly powerful when buffered to a pH value of about 5. A blank determination must always be carried out, using the same procedures and reagents.

Titrations with titanium(III) are performed in an inert atmosphere. The end point can be detected visually[52], but better precision is obtained by potentiometric titration. The accuracy and reproducibility of titrations with titanium(III) are frequently of the order of 0.5—1%.

Reduction in strongly acidic solution: Because of the moderate reducing power of titanium(III) ion in acidic solution, elevated temperatures and extended reaction periods are often required. The mild reducing action allows functional groups of different reactivity to be determined selectively. *Nitroso* or *azo compounds*, for example, can be determined in admixture with nitro com-

pounds[54, 57]. Titanium(III) ion in buffered solution causes total reduction and hence can be used to determine the sum of the components[56, 58].

In hydrochloric acid solution, titanium(III) may promote aromatic chlorination. In strongly acidic solution, azo compounds may undergo rearrangement, and reactions will then not be stoichiometric.

Reduction in buffered solution: In the presence of acetate or citrate buffers, the reducing power of titanium(III) ion is enhanced to such a degree that most reductions go to completion at room temperature within a few minutes.

1.4.4 Tin(II) Ion

The standard potential of the Sn(IV)/Sn(II) system is in the vicinity of +0.15 V, and is strongly affected by the acidity of the solution. Tin(II) ion can be used as a reducing agent[105] for *peroxides*[62], *aromatic nitro* and *azo compounds*[63], *quinones*[64], *sulfoxides*[65] and *amino phenoxazones*[66].

The indirect titration procedure is used predominantly. In view of the only moderate reducing power of the reducing agent, elevated temperatures and extended reaction times are frequently required. The excess of the reductant can be determined by back-titration with iodine, chromate or ferric iron. The end point is detected either potentiometrically or with indicators. Strongly acidic aqeous — organic mixed systems are generally used as solvents. Exclusion of oxygen is advisable.

1.4.5 Iron(II) Ion

In acidic solution, the standard potential of Fe(III)/Fe(II) is +0.75 V; iron(II) ion is thus a weak reducing agent. However, it can be converted into a powerful one by addition of complexing agents such as EDTA. Iron(II) in alkaline solution containing triethanolamine as a complexing agent is a very powerful reducing agent[67].

[52] *T. Gerstein, T.S. Ma*, Mikrochim. Acta 170 (1965).
[53] *Y.A. Gawargious, S.W. Bishara*, Z. Anal. Chem. *245*, 366 (1969).
[54] *T.S. Ma, J.V. Earley*, Mikrochim. Acta 129 (1959).
[55] Ref.[98], p. 574, 625, 704, 705.
[56] *Y.A. Gawargious, G.M. Habasky, S.W. Bishara*, Mikrochim. Acta 1129 (1968).
[57] *J.V. Earley, T.S. Ma*, Mikrochim. Acta 685 (1960).
[58] *G.M. Habasky, Y.A. Gawargious, S.W. Bishara*, Mikrochim. Acta 44 (1969).
[59] *D. Barnard, K.R. Hargrave*, Anal. Chim. Acta 5, 536 (1951).
[60] *R.R. Legault, K. Groves*, Anal. Chem. *29*, 1495 (1957).
[61] *T.S. Ma, T. Gerstein*, Microchem. J. *5*, 163 (1961).
[62] *D. Barnard, K.R. Hargrave*, Anal. Chim. Acta 5, 476 (1951).
[63] Ref.[98], p. 623, 704.
[64] Ref.[104], p. 215.
[65] *E. Glynn*, Analyst (London) *72*, 248 (1947).
[66] *M. Kotoucek, V. Jirásek, E. Ružička*, Mikrochim. Acta 501 (1966).
[67] *J. Zýka, J. Doležal*, Microchem. J. *10*, 554 (1966). Collect. Czech. Chem. Commun. *29*, 2597 (1964).

Reduction in acidic solution: Ferrous ion may be used to analyze *peroxides*[68] and *organic nitrates*[69-71].

Peroxides are determined by indirect titration; the excess of reducing agent is back-titrated amperometrically with dichromate. The method has the advantages of high sensitivity and good reproducibility but, depending on the reaction conditions, considerable systematic errors may occur.
Organic nitrates are reduced using an excess of ferrous ion, at elevated temperature. The resulting ferric ion can be titrated with titanous ion, using thiocyanate as indicator. Nitrates and aromatic nitro groups in mixtures can be reduced consecutively, using iron(II) and titanium(III) as reductants. A relative standard deviation of less than 0.5% can be achieved.

Reduction in alkaline solution: Alkaline solutions of ferrous ion containing triethanolamine have a reducing power comparable to that of chromous ion, and may be used to determine *anthraquinones, nitroso, nitro* and *azo compounds*[67, 72].

Azo compounds require an excess of the reducing agent and elevated temperature, whereas many of the other groups mentioned can be analyzed by direct titration at room temperature. Care must be taken to exclude atmospheric oxygen; the end point is determined amperometrically. Solvents used include mixtures of water with dioxane or dimethyl formamide. Reproducibility is better than 1.5%.
To avoid problems with unstable alkaline solutions of iron(II), acid solutions are used as titrants. Alkaline reaction conditions are generated by having a corresponding excess of base in the titration flask.

1.5 Reduction with Inorganic Non-metallic Compounds

1.5.1 Iodide
Of the inorganic non-metallic reducing agents, iodide is the most important[105]. It is a mild reductant; *peroxides* are reduced to alcohols[73-76], *quinones* to hydroquinones[73], *phosgene* to carbon monoxide and chloride[73], and *aromatic nitroso*

compounds to hydroxylamines[77]; *nitroso-amines*[78] and *chloramines*[79] form secondary amines. *Diazonium salts* give aromatic compounds[77], and *dibromo-alkanes*[80] form alkenes.
Iodine and Prussic acid are formed from *cyanogen iodide*[73], and *S-oxygen*[81, 83] or *S-chloro compounds*[82] split off hetero-elements to give the corresponding disulfides. *Peroxides* and some *quinones* are reduced in a few minutes at room temperature, but the reduction of *diazonium compounds* requires several hours at 100–300° in a sealed tube. The mild reducing action can be used to differentiate between functional groups of different reactivity.

As a rule, the sample is treated with an excess of iodide, and the resulting iodine is titrated with thiosulfate, using starch as indicator. As with the Karl Fischer titration in mixed organic-aqueous solutions, sulfur dioxide in pyridine-methanol can be used as titrant, with electrometric dead stop indication[79].
When, as is usual, atmospheric oxygen is excluded, care must be taken that the flow of inert gas does not carry away volatile iodine compounds. An inherent limitation of the iodometric method is the fact that the solvent and the sample must not react with free iodine. This problem may, in certain cases, be circumvented by the addition of a known excess of thioglycolate to take up the liberated iodine[76].

1.5.2 Sulfite
Sulfite may be used to determine *disulfide groups* in *proteins*[84, 85].

$$R-S-S-R + SO_3^{2\ominus} \rightarrow RS^{\ominus} + R-S-SO_3^{\ominus}$$

The resulting mercaptan is titrated with silver nitrate.

1.5.3 Sulfide
Sulfonyl chlorides may be reduced with sodium sulfide[86]:

$$R-SO_2Cl + S^{2\ominus} \rightarrow R-SO_2^{\ominus} + Cl^{\ominus} + S$$

[68] *I.M. Kolthoff, A.I. Medalia*, Anal. Chem. *23*, 595 (1951).
[69] *J. Grodzinski*, Anal. Chem. *29*, 150 (1957).
[70] *W.W. Becker*, Ind. Eng. Chem. Anal. *5*, 152 (1933).
[71] *W.E. Shaefer, W.W. Becker*, Anal. Chem. *25*, 1226 (1953).
[72] *B. Eremiáš, Z. Přikryl, J. Zýka*, Collect. Czech. Chem. Commun. *32*, 2478 (1967).
[73] Ref. [98], p. 480, 543, 546, 572.
[74] *R.D. Mair, A.J. Graupner*, Anal. Chem. *36*, 194 (1964).
[75] *T.S. Ma, T. Gerstein*, Microchem. J. *5*, 163 (1961).
[76] *L.K. Dahle, R.T. Holman*, Anal. Chem. *33*, 1960 (1961).

[77] Ref. [104], p. 266, 310.
[78] *J. Gal, E.R. Stedronsky, S.I. Miller*, Anal. Chem. *40*, 168 (1968).
[79] *R.W. Freedman*, Anal. Chem. *28*, 247 (1956).
[80] *I.M. Kolthoff, R. Belcher*, Volumetric Analysis, Vol. III, p. 406, Interscience Publishers, New York 1957.
[81] *H. Bretschneider, W. Klötzner*, Monatsh. Chem. *81*, 589 (1950).
[82] *N. Kharasch, M.M. Wald*, Anal. Chem. *27*, 996 (1955).
[83] *D. Barnard, E.R. Cole*, Anal. Chim. Acta *20*, 540 (1959).
[84] *J.R. Carter*, J. Biol. Chem. *234*, 1705 (1959).
[85] *W. Stricks, I.M. Kolthoff*, J. Am. Chem. Soc. *73*, 4569 (1951).
[86] *M.R.F. Ashworth, W. Walisch, G. Kronz-Dienhart*, Anal. Chim. Acta *20*, 96 (1959).

The determination is carried out as a direct titration in acetone-water, using either polarovoltric or visual end point indication.

1.5.4 Arsenic(III) Ion

Arsenic(III) ion is an even milder reducing agent than iodide. It can be used in place of iodide in situations where liberated iodine would undergo reaction with the sample or the solvent. Arsenic(III) ion may be used to determine *peroxides*[87], *hydroperoxides*[88] and *azides*[89, 90].

1.6 Reduction with Organic Reducing Agents

1.6.1 Phenylhydrazine

Phenylhydrazine reduces *nitroso compounds* to azo-derivatives[91]. *Aromatic azo compounds* form hydrazo compounds, and *quinones* give hydroquinones[91, 92].

The sample is treated with an excess of phenylhydrazine or phenylhydrazine carbamate at elevated temperature. Gasometric methods are used to measure the liberated nitrogen (see 3.7:2).
As reductions with phenylhydrazine may lead to side reactions, this reducing agent is preferred to chromium (II) or titanium(III) ion only in instances, when a mild selective reduction is desired.

1.6.2 Mercaptans

Thiosulfonates are reduced by mercaptans[93], forming sulfinic acids which are determined by titration with alkali:

$$R^1-SO_2-SR^2 + R^3SH \rightarrow R^2-S-S-R^3 + R^1-SO_2H$$

1.6.3 Ascorbic Acid

Ascorbic acid reduces dyes derived from *phenothiazine* and *phenoxazine* to the corresponding leuco compounds[94–96]. *Nitro, nitroso* and *azo compounds* undergo partial reduction with ascorbic acid.

---

[87] *S. Siggia*, Quantitative Organic Analysis via Functional Groups, p. 286, John Wiley & Sons, New York 1964.

[88] *I.M. Kolthoff, E.J. Meehan, S. Bruckenstein, H. Minato*, Microchem. J. *4*, 33 (1960).

[89] *A. Gutmann*, Z. Anal. Chem. *66*, 224 (1925).

[90] Ref. [104], p. 255.

[91] *Houben-Weyl*, Ref. [98], p. 479, 617.

[92] Ref. [104], p. 215, 265.

[93] *D. Barnard, E.R. Cole*, Anal. Chim. Acta *20*, 540 (1959).

[94] *G.G. Rao, V.N. Rao*, Z. Anal. Chem. *147*, 338 (1955).

[95] *E. Ružička, M. Kotouček*, Z. Anal. Chem. *180*, 429 (1961).

[96] *E. Ružička*, Z. Anal. Chem. *168*, 414 (1959).

[97] *M.R. Lindbeck, H. Freund*, Anal. Chim. Acta *35*, 74 (1966).

1.7 Cathodic Reduction

Polarography and potentiostatic coulometry are electrochemical reduction techniques[100–103], which permit *selective reduction* of different functional groups in mixtures. These electrochemical methods have high sensitivity and are thus suitable for the determination of by-products and impurities.

Whereas coulometry is an absolute method based on Faraday's law, polarographic methods require calibration with standard compounds. As a rule, coulometric methods have good reproducibility[97]. Polarographic methods seldom reach the precision attainable with a good chemical reducing agent.

Applications:

conjugated multiple bonds	organic nitrates
aldehydes	N-oxides
ketones	peroxides
quinones	sulfones
double bonds C=N	disulfides
halides	thiocyanates
azo compounds	isothiocyanates
diazo compounds	heterocyclic compounds
nitro compounds	organometallic compounds
nitroso compounds	

1.8 Bibliography

[98] *Houben-Weyl*, Methoden der organischen Chemie, 4. Aufl., Vol. II, Georg Thieme Verlag, Stuttgart 1953.

[99] Organic Analysis, Vol. II, III, Interscience Publishers, New York 1954, 1956.

[100] *K. Schwabe*, Polarographie und chemische Konstitution organischer Verbindungen, Akademie Verlag, Berlin 1957.

[101] *K. Abresch, I. Claassen*, Die coulometrische Analyse, Verlag Chemie, Weinheim/Bergstr. 1961.

[102] *H.W. Nürnberg*, Angew. Chem. *72*, 433 (1960).

[103] *P. Zuman*, Organic Polarographic Analysis, Pergamon Press, Oxford 1964.

[104] *N.D. Cheronis, T.S. Ma*, Organic Functional Group Analysis, Interscience Publishers, New York 1964.

[105] *M.R. Ashworth*, Titrimetric Organic Analysis, Part 1, Part 2, Interscience Publishers, New York 1964, 1965.

[106] *A. Berka, J. Vulterin, J. Zýka*, Newer Redox Titrants, Pergamon Press, Oxford 1965.

2 Oxidation[1, 2]

2.1 Oxidation with Halogen Compounds

In this section, frequently the titrant mentioned is not the same as the active agent. For example if mercaptans are titrated with bromate in the presence of iodide, bromate generates iodine, which is a very mild oxidant. This titration will be dealt with in Section 3.3:2.1.2, Oxidation with Bromate.

---

[1] *M.R.F. Ashworth*, Titrimetric Organic Analysis, Part. I, Interscience Publishers, New York, London, Sydney 1964.

[2] *M.R.F. Ashworth*[1], Part II, 1965.

Halogen compounds[3] can be used for three different types of reactions:

for real oxidations, such as the oxidation of α-glycols with periodate, to form aldehydes;
for substitutions combined with oxidation, such as the bromination of phenols or aromatic amines with bromate/bromide;
for addition reactions combined with oxidation, e.g. determination of the iodine number of unsaturated fatty acids with iodine monochloride.

A wide variety of indication methods is available, particularly when halogen appears or disappears at the end-point. Examples are: *indicators* with reversible or irreversible color change[4,5], extraction of iodine into chloroform or carbon tetrachloride[4,5], potentiometric measurement[5], and the very sensitive methods with polarized electrodes[5,6].

2.1.1 Heptavalent Halogen (Periodate)

Aqueous solutions of periodate are obtained by dissolving periodic acid H_5IO_6 or alkali salts of the metaperiodic acid HIO_4. To standardize, iodide is added and the iodine liberated is titrated with arsenite[7,8].

The Malaprade reaction is the basis of oxidations with periodate. *α-Glycols, polyhydroxy compounds, α-amino-alcohols, α-hydroxy-aldehydes, α-hydroxy-ketones, α-diketones, α-keto-aldehydes* and *glyoxal* are oxidized, the bond between the carbon atoms bearing the oxygen being broken. The reaction is usually carried out in weakly acidic to weakly basic solution at $0-25°$. Under these conditions, oxidation is selective; mono-alcohols, aldehydes and ketones do not react.

The following reactions are of analytical interest:

for α-glycols

$$R^1-\underset{\underset{HO}{|}}{\overset{\overset{H}{|}}{C}}-\underset{\underset{OH}{|}}{\overset{\overset{H}{|}}{C}}-R^2 \;+\; H_4JO_6^\ominus \;\longrightarrow\; R^1-C\overset{\nearrow O}{\underset{\searrow H}{}} \;+$$

$$R^2-C\overset{\nearrow O}{\underset{\searrow H}{}} \;+\; JO_3^\ominus \;+\; 3H_2O$$

for polyhydroxy compounds

$$CH_2OH-(CHOH)_n-CH_2OH \;+\; (n+1)H_4JO_6^\ominus \;\longrightarrow$$

$$n\,HCOOH \;+\; 2\,HCHO \;+\; (n+1)JO_3^\ominus \;+\; (2n+3)H_2O$$

for aldoses, e.g. sucrose

$$CHO-(CHOH)_4-CH_2OH \;+\; 5H_4JO_6^\ominus \;\longrightarrow$$

$$5\,HCOOH \;+\; HCHO \;+\; 5\,JO_3^\ominus \;+\; 10\,H_2O$$

for ketoses, e.g. fructose

$$CH_2OH-CO-(CHOH)_3-CH_2OH \;+\; 4H_4JO_6^\ominus \;\longrightarrow$$

$$3\,HCOOH \;+\; HCHO \;+\; CH_2OH-COOH \;+\; 4\,JO_3^\ominus \;+\; 8\,H_2O$$

for α-amino alcoholes

$$R^1-\underset{\underset{R^2-\underset{\underset{H}{|}}{\overset{\overset{|}{}}{C}}-NH_2}{|}}{\overset{\overset{\overset{H}{|}}{}}{C}}-OH \;+\; H_4JO_6^\ominus \;+\; H^\oplus \;\longrightarrow$$

$$R^1CHO \;+\; R^2CHO \;+\; NH_4^\oplus \;+\; JO_3^\ominus \;+\; 2H_2O$$

An excess of the oxidant is usual. If side reactions are avoided, the amount of periodate consumed is stoichiometrically related to the amount of the oxidizable species[9-11]. The excess of periodate is determined by adding bicarbonate buffer, potassium iodide as a catalyst, and a known amount of arsenite solution. The excess of arsenite is back-titrated with iodine. Standardization of the periodate solution is carried out in the same way.
If the oxidation does not give uniform products, the reaction mixture can be assayed for specific oxidation products, such as carboxylic acids, aldehydes, or, in the case of α-amino alcohols, ammonia[7].

2.1.2 Pentavalent Halogen (Bromate, Iodate)

Bromate: As bromate is a slow-acting oxidant, it is not used analytically in neutral or alkaline solution[3]. In acidic solution and in the presence of bromide, bromine is generated. If, instead of bromide, one uses chloride, iodide or mixed halides, elemental chlorine or iodine or inter-halogen compounds are generated, too. This allows control of the reactivity of the oxidant generated in situ[12].

Bromate is frequently used for true oxidation reactions; it is also commonly used for bromination of aromatic amines and phenols, and for determination of the bromine number of unsaturated compounds.

[3] *G. Den Boef, H.L. Polak,* Talanta *9,* 271 (1962).
[4] *I.M. Kolthoff, R. Belcher,* Ref.[143], p. 205, 451, 502, 576.
[5] *A.I. Vogel,* Ref.[102], p. 345, 375, 383, 395, 908, 1035.
[6] *J.T. Stock,* Amperometric Titrations, p. 550, 570, 620, Interscience Publishers, New York, London, Sydney 1965.
[7] *I.M. Kolthoff, R. Belcher,* Ref.[143], p. 475.
[8] *A. Berka, J. Vulterin, J. Zýka,* Ref.[145], p. 66.

[9] *R.J.B. Reddaway,* Analyst (London) *82,* 506 (1957).
[10] *N.D. Cheronis, T.S. Ma,* Ref.[63], *p. 156, 158, 193.*
[11] *R.B. Paulsen, T. Waaler,* Pharm. Acta Helv. *37,* 125 (1962).
[12] *E. Schulek, K. Burger, J. Laszlovszky,* Talanta *7,* 51 (1960).

True oxidation reactions: Bromate/bromide is the most common reagent for the quantitative oxidation of *disulfides*[13, 16], aliphatic[13, 14, 16], aromatic[14], and heterocyclic[14] *sulfides*, and can be used in the presence of iodide ions for the determination of *mercaptans*[16].

These methods can be used also on the microscale[15].

Titration with bromate is of some importance also for the determination of *benzenesulfinic acid*[17] and for determination of the hydrazine group in *phenylhydrazine*[18], *semicarbazide*[18, 19] and various other hydrazides[18, 19].

Substitution Reactions (Koppeschaar Method):
Potassium bromate solutions and a large excess of bromide ion are used for the *quantitative bromination* of *aromatic amines*[20, 21] and *phenols*[20–22]. Bromine attacks the free *ortho-* and *para-* positions.

It should be noted that bromination can lead to quantitative cleavage of carboxylic and sulfonic groups. Thus, salicylic acid, sulfanilic acid and anthranilic acid upon bromination yield tribromophenol or tribromoaniline.

Brominations with bromate/bromide are almost always conducted as indirect titrations: the sample is oxidized with an excess of bromine, and the unused oxidant is back-titrated iodometrically. The reaction mixture must be kept in the dark.

Temperature and reaction time are chosen such that bromination is quantitative and side-reactions are negligible. Optimal conditions must be sought for every compound (*halophenols, dihydroxydiphenylpropane*[23], *alkylphenols*[23, 24], *amino-benzenesulfonamides*[25, 26]).

Addition Reactions: Titration with bromate/bromide is one of the standard methods for the determination of the *bromine number* of olefins, particularly of products from petrochemical processes[27–29]. In older methods[27], an excess of bromine was required. More recent methods[28, 29] involve direct titrations in appropriate solvent mixtures (containing glacial acetic acid, carbon tetrachloride, methanol, sulfuric acid), with electrometric end point detection. Mercuric chloride is frequently used as a catalyst, but can be omitted[30, 31].

Iodate: The most frequent uses of this oxidant are for direct titration of hydrazines, hydrazides, mercaptans and ascorbic acid.

Oxidation of *hydrazines* and *hydrazides* is usually carried out in the presence of high concentrations of hydrochloric acid. Reaction is usually quantitative and stoichiometrically well defined for monosubstituted hydrazines[32]. The oxidations of *alkyl hydrazines*[33] and *isonicotinic hydrazide*[34] have been reported.

Titration with iodate is a commonly used method for the determination of *mercaptans*. Iodine, which occurs as an intermediate oxidant, causes the oxidation to the disulfide[35]:

$$2 \text{ R–SH} + J_2 \rightarrow \text{R–S–S–R} + 2 J^\ominus + 2 H^\oplus$$

The equation above is valid only for primary mercaptans[36].

Iodate/iodide converts *ascorbic acid* to dehydroascorbic acid[37–39].

[13] S. Siggia, R.L. Edsberg, Anal. Chem. 20, 938 (1948); Houben-Weyl, Methoden der org. Chemie, 4. Aufl., Band II, p. 589, Georg Thieme Verlag, Stuttgart 1953.
[14] W.H. Houff, R.D. Schuetz, Anal. Chem. 25, 1258 (1953).
[15] N.D. Cheronis, T.S. Ma, Ref.[63], p. 327.
[16] B. Jaselskis, Anal. Chem. 31, 928 (1959).
[17] B. Fleszar, Chem. Anal. [Warsaw] 10, 49 (1965).
[18] F. Jančik, O. Činková, J. Körbl, Collect. Czech. Chem. Commun. 24, 2695 (1959).
[19] J. Vulterin, Collect. Czech. Chem. Commun. 28, 1391 (1963).
[20] I.M. Kolthoff, R. Belcher, Ref.[143], p. 534.
[21] A.R. Day, W.T. Taggart, Ind. Eng. Chem. 20, 545 (1928).
[22] N.D. Cheronis, T.S. Ma, Ref.[63], p. 447.
[23] M.M. Sprung, Ind. Eng. Chem. Anal. 13, 35 (1941).

[24] L.V. Erichsen, N. Rudolphi, Erdöl, Kohle 8, 16 (1955).
[25] E. Schulek, P. Rózsa, Z. Anal. Chem. 122, 96 (1941).
[26] P.L. De Reeder, Anal. Chim. Acta 9, 314 (1953).
[27] American Society for Testing Materials, Standards, Designation: D 1158 T.
[28] American Society for Testing Materials, Standards, Designation: D 1159 T.
[29] Deutscher Normenausschuß, DIN 51774 (Febr. 1966).
[30] J.C.S. Wood, Anal. Chem. 30, 372 (1958).
[31] E.H. Unger, Anal. Chem. 30, 375 (1958).
[32] W.R. McBride, R.A. Henry, S. Skolnik, Anal. Chem. 25, 1042 (1953).
[33] L.F. Audrieth, L.H. Diamond, J. Am. Chem. Soc. 76, 4869 (1954).
[34] E. Kühni, M. Jacob, H. Grossglauser, Pharm. Acta Helv. 29, 233 (1954).
[35] N.D. Cheronis, T.S. Ma, Ref.[63], p. 324.
[36] I.M. Kolthoff, W.E. Harris, Anal. Chem. 21, 963 (1949).
[37] G.S. Deshmukh, M.G. Bapat, Z. Anal. Chem. 147, 271 (1955).
[38] L. Erdey, L. Káplár, Z. Anal. Chem. 162, 180 (1958).
[39] L. Erdey, E.B. Gere, E. Bányai, Talanta 3, 209 (1959).

Iodate can be used for the titration of *thiourea* and some of its alkyl and aryl derivatives, in sulfuric acid solution[40].

2.1.3 Monovalent Halogen (Hypochlorite, Hypobromite)

Hypohalites are used principally for oxidation in alkaline or neutral solution[41]. Solutions of hypochlorite are more stable but less reactive than those of hypobromite. The use of calcium hypochlorite as a titrant in the presence of bromide has, therefore, been suggested[42]. Urea and alcohols can not be titrated directly[43]. Oxidation of *urea* with an excess of hypobromite leads to satisfactory results only if the titrant has been standardized with pure urea[44]. Similarly, *thioacetamide*[45] and *thiosemicarbazide*[46] can be determined by indirect titration. N-chlorination according to the equation:

$$R-CO-NH_2 + Cl_2 \rightarrow R-CO-NHCl + H^{\oplus} + Cl^{\ominus}$$

can be used to determine several primary aromatic and aliphatic *amides*[47] by direct titration.

Solutions of hypoiodite are unstable; for oxidations in alkaline media, iodine solutions are employed (see Section 3.3:2.1.4).

2.1.4 Elemental Halogen

For titrations in acid media, it is usually advisable to generate bromine or iodine in situ, using bromate and iodate. Standard solutions of elemental halogens are, however, preferable for titrations in non-aqueous solutions and for reaction in alkaline, neutral or weakly acid media.

Bromine: The determination of *formic acid*[48] in glacial acetic acid/pyridine solution through oxidation with a solution of bromine in glacial acetic acid is of some importance.

The same reagent is used for specific bromination at the free *o-* and *p-* positions of *phenol* and some derivatives[49]. In some cases, bromination by direct

titration is possible, if the reaction rate is increased by addition of optimal amounts of pyridine as a catalyst[50].

The most important use of bromine is for the determination of the *bromine number* by addition to C=C double bonds. For petrochemical products, frequent use is made of solutions of bromine in carbon tetrachloride. A solution of bromine in aqueous or methanolic potassium bromide has been recommended for the determination of *α,β-unsaturated aldehydes, ketones, acids* and *esters*[51, 53].

Another common solvent for bromine is glacial acetic acid; in the presence of pyridine, bromine forms addition compounds[52] (C_6H_5N,H_2SO_4,Br_2 or C_6H_5N,HBr,Br_2). Bromination is carried out at 0° or at room temperature, with an excess of bromine. A few preliminary experiments are usually necessary to determine the optimal reaction time and excess of bromine for each kind of sample. In some cases, the components of mixtures of unsaturated compounds can be determined individually by the reaction rate approach[54]; the decrease of the bromine concentration in the reaction mixture can be followed by spectrophotometry.

Iodine: Solutions of free iodine[55] are used for oxidation in acidic to strongly alkaline media. In alkaline solution, hypoiodite is formed, which oxidizes *aldehydes* (simple aldehydes and aldoses) and *methyl ketones (haloform reaction);* this is the most important use of the reagent.

Iodine is generally dissolved in aqueous potassium iodide; alcoholic solutions are less commonly employed.

Aldehydes[56], *acetone*[57] and *aldoses*[58] are oxidized with a known excess of iodine in alkaline solution. After acidification, the residual iodine is titrated with thiosulfate.

For the oxidation of *thiourea* and its derivatives, an excess of iodine in strongly alkaline[59] or bicarbonate-buffered[60] solutions is used. The excess of halogen is back-titrated with arsenite in neutral to weakly alkaline solution.

[40] *B. Singh, B.C. Verma*, Z. Anal. Chem. *194*, 112 (1963).
[41] *I.M. Kolthoff, R. Belcher*, Ref.[143], p. 573.
[42] *I.M. Kolthoff, V.A. Stenger*, Ind. Eng. Chem. Anal. 7, 79 (1935).
[43] *H.A. Laitinen, D.E. Woerner*, Anal. Chem. 27, 215 (1955).
[44] *K. Wölfel*, Z. Anal. Chem. 90, 170 (1932).
[45] *A. Claeys, H. Sion, A. Campe, H. Thun*, Bull. Soc. Chim. Belg. 70, 576 (1961).
[46] *J. Pijck, A. Campe, A. Claeys*, Bull. Soc. Chim. Belg. 73, 898 (1964).
[47] *W.R. Post, C.A. Reynolds*, Anal. Chem. 37, 1171 (1965).
[48] *J.V. Longstaff, K. Singer*, Analyst (London) 78, 491 (1953).
[49] *A.K. Ingbermann*, Anal. Chem. 30, 1003 (1958).

[50] *C.O. Huber, J.M. Gilbert*, Anal. Chem. 34, 247 (1962).
[51] *A. Polgár, J.L. Jungnickel*, Ref.[142] p. 360.
[52] *A. Polgár, J.L. Jungnickel*, Ref.[142] p. 232, 244, 245.
[53] *F.E. Critchfield*, Anal. Chem. 31, 1406 (1959).
[54] *S. Siggia, J.G. Hanna, N.M. Serencha*, Anal. Chem. 35, 362 (1963).
[55] *I.M. Kolthoff, R. Belcher*, Ref.[143], p. 199, 375.
[56] *O.P. Malhotra, V.D. Anand*, Z. Anal. Chem. 160, 10 (1958).
[57] *G.E. Goltz, D.N. Glew*, Anal. Chem. 29, 816 (1957).
[58] *J.R. Collins*, Anal. Chim. Acta 9, 500 (1953).
[59] *M.K. Joshi*, Anal. Chim. Acta 14, 509 (1956).
[60] *P.C. Gupta*, Analyst (London) 88, 896 (1963).

Hydrazines can be determined by direct titration[61, 62]. Careful investigation of the course of the reaction is frequently required with this type of compound[63].

Xanthates[64, 65] can be titrated directly with iodine solution. On reaction with alcoholic potassium hydroxide, carbon disulfide forms xanthate, which can be determined by the same method.

For the titration of mercaptans and ascorbic acid in acid media, iodine is best generated in situ from iodate (see Section 3.3:2.1.2).

Bromine Monochloride: To prepare the reagent[66], a solution containing bromate and bromide in the exact molar ratio 1:2 is used. Before use, hydrochloric acid is added to form the 0.1 N BrCl standard solution. Alternatively, the BrCl reagent may be generated in situ by first adding hydrochloric acid to the titration flask[67].

Bromine chloride can be used for determination of *hydrazine derivatives*[68], such as *phenylhydrazine, semicarbazide,* and *isonicotinic hydrazide,* and of *phenol*[69] and certain of its derivatives. Under strictly controlled conditions, simple primary and secondary *alcohols* (except methanol), are quantitatively converted into the corresponding acids or ketones by an excess of the reagent[67].

Iodine Monochloride[70, 76]: Oxidation of organic compounds with iodine chloride is usually conducted as direct titration in strongly acidic to neutral media, I^{+1} being transformed to I^0 or I^{-1}. The standard solution is prepared by acidifying a 1:2 molar iodate/iodide solution with hydrochloric acid. The redox potential of this solution can be adjusted precisely to the potential of iodine mono chloride[70]. *Ascorbic acid*[71], *hydroquinone*[72, 73], *hydrazine derivatives*[71, 74], and *mercaptans*[75] are the most important compounds which can be determined with this reagent.

Iodine mono-chloride is frequently used for the determination of C=C double bonds, particularly in fatty acids and esters *(Wijs iodine number)*. The reagent is prepared in glacial acetic acid or in carbon tetrachloride by reacting equimolar amounts of iodine and chlorine, or iodine and iodine trichloride[76, 77].

Iodine Monobromide[76, 77]: Like iodine monochloride, the bromide may be employed for the determination of unsaturated compounds *(Hanus iodine number)*. However, in spite of its good stability, the bromide reagent is less used than the chloride.

2.1.5 Organic Halogen Compounds

N-Bromosuccinimide contains bromine in the oxidation state +1, and is both an oxidant and an electrophilic reagent. Depending on the reaction conditions, intermediary formation of bromine or hypobromite is conceivable.

The over-all oxidation process of *N*-bromosuccinimide can be represented by the following equation:

$$\text{NBr} + H^{\oplus} + 2e \longrightarrow \text{NH} + Br^{\ominus}$$

Hydroquinone, quinhydrone, ascorbic acid and *thiourea* may be determined by direct titration in neutral to acidic solution[78]. The course of the oxidation of *aryl hydrazines* is open to discussion[79, 80].

The reagent has also been recommended for the bromination of aromatic *amines* and *phenols*[81].

The reaction goes to completion within 1—10 minutes, an excess of reagent being used. In the case of amines and phenols bearing negative substituent groups, such as *p-aminobenzoic acid*, N-bromosuccinimide seems to be a more effective brominating agent than elementary bromine.

[61] *R.A. Rowe, L.F. Audrieth,* J. Am. Chem. Soc. *78,* 563 (1956).

[62] *W.R. McBride, H.W. Kruse,* J. Am. Chem. Soc. *79,* 572 (1957).

[63] *N.D. Cheronis, T.S. Ma,* Organic Functional Group Analysis, p. 292, 296, Interscience Publishers, New York, London, Sydney 1964.

[64] *H. Roth,* Angew. Chem. *73,* 167 (1961).

[65] *M. Eusuf, M.H. Khundkar,* Anal. Chim. Acta *24,* 419 (1961).

[66] *E. Schulek, K. Burger,* Talanta *1,* 219 (1958).

[67] *K. Konishi, Y. Mori, H. Inoue, M. Nozoe,* Anal. Chem. *40,* 2198 (1968).

[68] *E. Schulek, K. Burger,* Talanta *1,* 344 (1958).

[69] *E. Schulek, K. Burger,* Talanta *1,* 224 (1958).

[70] *A. Berka, J. Vulterin, J. Zýka,* Ref. [145], p. 55.

[71] *B. Singh, G.P. Kashyap, S.S. Sahota,* Z. Anal. Chem. *162,* 357 (1958).

[72] *J. Cihalik, D. Vavrejnová,* Collect. Czech. Chem. Commun. *21,* 192 (1956).

[73] *B. Singh, G.P. Kashyap,* Z. Anal. Chem. *163,* 338 (1958).

[74] *J. Cihalik, K. Terebová,* Collect. Czech. Chem. Commun. *22,* 756 (1957).

[75] *J. Cihalik, K. Terebová,* Collect Czech. Chem. Commun. *23,* 110 (1958).

[76] *A. Polgar, J.L. Jungnickel,* Ref. [142], p. 219.

[77] *S. Siggia,* Ref. [144], p. 313.

[78] *A. Berka, J. Zýka,* Collect. Czech. Chem. Commun. *23,* 402 (1958).

[79] *M.Z. Barakat, M. Shaker,* Analyst (London) *88,* 59, 330 (1963).

[80] *E.V. Egginton, M.J. Graham,* Analyst (London) *89,* 226 (1964).

[81] *R.D. Tiwari, J.P. Sharma, I.C. Shukla,* Talanta *14,* 853 (1967).

Chloramine-T[82]: The oxidative properties of chloramine-T are similar to those of hypochlorite solution, which, in fact, may be formed through hydrolysis:

$$Tos-\overset{\ominus}{N}Cl \; + \; H_2O \; \rightleftharpoons \; Tos-NH_2 \; + \; OCl^{\ominus}$$

With reducing agents, chloramine-T reacts as follows:

$$Tos-\overset{\ominus}{N}Cl \; + \; 2H^{\oplus} \; + \; 2e \; \longrightarrow \; Tos-NH_2 \; + \; Cl^{\ominus}$$

The standard solutions are stable and are prepared by dissolving the trihydrate of the sodium salt of the reagent in water. Standardization is carried out iodometrically or by titration with arsenite[82, 83]. Chloramine-T acts as an oxidant both in acidic and in alkaline media.

Titrations in acid solution are usually conducted in the presence of iodide or bromide. The halide limits the oxidative power of chloramine-T, and serves as an end point indicator. *Hydrazine derivatives*[82, 84], *thiourea*[82, 84] and *ascorbic acid*[38] may be determined in this way.

In alkaline solution, chloramine-T can convert some *aldehydes* to carboxylic acids[82, 84] and thiourea to urea and sulfate[85].

2.2 Oxidation with other Oxidants

This group includes a number of strong but rather nonspecific oxidants: hexavalent chromium, tetravalent cerium, heptavalent manganese, tetravalent lead, and pentavalent vanadium.

In addition, there are a number of milder oxidants exhibiting some degree of selectivity: divalent copper, hexacyanoferrate(III), monovalent silver and divalent mercury.

2.2.1 Hexavalent Chromium

Potassium dichromate is frequently used for the preparation of standard solutions; because it can be obtained in a very pure state, it can serve as a primary standard[86]. Oxidations are conducted in aqueous acid solution at elevated temperature, almost always with an excess of the reagent. Unreacted chromium(VI) is back-titrated with iodide or with iron(II) ion.

Among the few direct titrations, only the determination of *hydroquinone* deserves mention[87].

Oxidation of *alcohols* and *aldehydes*[88] gives reproducible results; however, only in rare instances can simple stoichiometric factors be used. Chromium(VI) is, therefore, used only for special applications, such as the determination of *azo*[89], *amino*[90] and *nitro*[90] compounds.

2.2.2 Tetravalent Cerium

The oxidation potential of cerium(IV) ion depends on the nature and concentration of the complexing anions present. Ceric perchlorate exhibits the highest redox potential: + 1.7 V in $1N$ perchloric acid[91].

The ceric salts used in standard solutions include ceric sulfate[91–93], ceric perchlorate[91, 92] and diammonium hexanitratocerate[91, 92, 94]. Arsenic(III) oxide is used for standardization, and iron(II) salt solutions are used for back-titration of the excess of ceric ion[91, 93].

Oxidations with ceric ion are carried out in strongly acidic solutions, and osmium tetroxide or iodine monochloride are frequently added as catalysts.

For determination of *oxalic acid,* direct or indirect titration with cerium(IV) can be employed[91–93, 95]; oxalic acid has also been used to standardize the oxidant. In general, the reaction between cerium(IV) and carboxylic acids depends on the constitution of the acids and on the reaction conditions[91, 95].

α-Diols are usually oxidized with an excess of ceric ion[95, 96]. A direct titration procedure at 80° in perchloric acid has been reported for *glycerol*[97]. Ceric salts oxidize *aldoses*[95, 98, 99] quantitatively to form formic acid. *Ketoses*[95, 99] are converted into formic acid and carbon dioxide.

[82] A. Berka, J. Vulterin, Ref. [145], p. 37.

[83] E. Bishop, V.J. Jennings, Talanta 1, 197 (1958).

[84] A. Berka, J. Zýka, Česk. Farm. 5, 335 (1956).

[85] G. Aravamudan, V.R. Satyanarayana Rao, Talanta 11, 55 (1964).

[86] Ref. [102], p. 304.

[87] U.A.Th. Brinkmann, H.A.M. Snelders, Z. Anal. Chem. 204, 337 (1964).

[88] J.A. Barnard, N. Karayannis, Anal. Chim. Acta 26, 253 (1962).

[89] M. Jurecek, P. Kozák, Vl. Novák, Zd. Boháckova, Mikrochim. Acta 643 (1963).

[90] P. Kozák, M. Jurecek, Collect. Czech. Chem. Commun. 34, 521 (1969).

[91] W. Petzold, Die Cerimetrie, Verlag Chemie, Weinheim/Bergstr. 1955.

[92] Ref. [143], p. 121.

[93] Ref. [102], p. 314.

[94] G. Gopala Rao, P.V. Krishna Rao, K.S. Murty, Talanta 9, 835 (1962).

[95] G.F. Smith, F.R. Duke, Ind. Eng. Chem. Anal. 15, 120 (1943).

[96] G.G. Guibault, W.H. McCurdy Jr., Anal. Chem. 33, 580 (1961).

[97] J.R. Sand, C.O. Huber, Talanta 14, 1309 (1967).

[98] A.A. Forist, J.C. Speck Jr., Anal. Chem. 27, 1166 (1955).

[99] E. Michalski, K. Czarnecki, M. Ignaczek, Talanta 5, 137 (1960).

Hydroquinone can be determined cerimetrically with excellent accuracy[100].

Since ceric ion is a strong but non-specific oxidant, cerimetric methods have found only limited use.

2.2.3 Heptavalent Manganese (Permanganate)

Potassium permanganate is used almost exclusively in aqueous solution. (For the preparation of the standard solutions see ref. [101, 102]).

In oxidations with permanganate, an excess of the reagent is usually employed, and quite long reaction periods may be required. It should be noted that considerable analytical errors can arise because of decomposition of the permanganate solution. Indirect titration procedures with this oxidant should, therefore, not be used at elevated temperatures[101].

In acidic solution, the permanganate is reduced to the bivalent state; in neutral or alkaline solution, manganese dioxide is formed.

In acidic solution, the excess of the oxidant can be determined iodometrically[101] or by titration with ferrous salt solutions. In alkaline media, the manganese dioxide formed may be determined[103, 104].

Quantitative oxidation with permanganate usually converts organic compounds into carbon dioxide and water. The oxidation of *tartaric acid* for example proceeds as follows:

$$\begin{array}{c} COOH \\ | \\ HCOH \\ | \\ HCOH \\ | \\ COOH \end{array} + 2MnO_4^{\ominus} + 6H^{\oplus} \longrightarrow 2Mn^{2\oplus} + 4CO_2 + 6H_2O$$

Oxalic acid in acid solution may be used for standardization of permanganate solutions[101, 102].

Formic acid[101, 102] undergoes total oxidation in alkaline solution. Many *carboxylic acids* may be treated with an excess of permanganate in alkaline solution at room temperature; subsequent acidification completes the oxidation to carbon dioxide[101].

The method yields satisfactory results for readily oxidizable carboxylic acids, such as hydroxy-acids or unsaturated acids. Oxidation of certain hydroxy-acids

goes to completion in acidic solution, provided that the appropriate pH is precisely maintained[105]. These methods, however, are prone to error and are not recommended.

Methanol is the only saturated aliphatic monoalcohol which can be oxidized quantitatively[106]. Permanganate oxidation can be used for determination of aliphatic or aromatic polyhydroxy compounds, but is not recommended[107].

2.2.4 Tetravalent Lead

Tetravalent lead is used almost exclusively in the form of lead tetra-acetate in glacial acetic acid solution[108]. Its oxidative properties are similar to those of periodate. Oxidation of α-diols is slightly faster with lead tetra-acetate; however, on the other hand, indirect methods suffer from interference due to the hydrolysis of the excess of lead tetra-acetate.

Readily oxidizable compounds, such as *mercaptans* and *hydrazine derivatives*[109, 110] may be determined by direct titration. Direct titration can also be used to effect oxidative cleavage of carbon-carbon bonds in *cis*-glycols, e.g. in *furanosides*[111]. Indirect methods have been reported for *polyhydroxy compounds* and for *α-hydroxy-carboxylic acids*[112].

2.2.5 Pentavalent Vanadium

Pentavalent vanadium undergoes reduction thus:

$$VO_3^{\ominus} + 4H^{\oplus} + e \rightleftharpoons VO^{2\oplus} + 2H_2O$$

Quantitative oxidation of organic compounds is carried out in acidic solution at elevated temperature, using an excess of the oxidant. Unreacted reagent is back-titrated with ferrous ion. Solutions of ammonium or sodium metavanadate in dilute sulfuric acid are used as the reagent[113, 114].

Under carefully controlled conditions, *α-hydroxy-acids* are oxidized stoichiometrically to give carbon dioxide and a carboxylic acid containing one less carbon atom[114]. For a number of *phenols* and

[100] *U.A.Th. Brinkmann, H.A.M. Snelders*, Z. Anal. Chem. *204*, 337 (1964).

[101] Ref.[143], p. 33.

[102] *A.I. Vogel*, Quantitative Inorganic Analysis, 3rd Edition, p. 277, Longmans, Green and Co., London 1961.

[103] *A. Berka*, Collect. Czech. Chem. Commun. *29*, 2844 (1964).

[104] *H. Flaschka, J. Garrett*, Chemist-Analyst *52*, 101 (1963).

[105] *A. Berka, S. Hilgard*, Mikrochim. Acta 164 (1966).

[106] *J. Sharp*, Anal. Chim. Acta *25*, 139 (1961).

[107] *C.J.B. Smit, M.A. Joslyn, A. Lukton*, Anal. Chem. *27*, 1159 (1955).

[108] Ref.[145], p. 76.

[109] *A. Berka, J. Zýka*, Collect. Czech. Chem. Commun. *24*, 105 (1959).

[110] *L. Suchomelová, J. Zýka*, J. Electroanal. Chem. Interfacial Elektrochem. *5*, 57 (1963).

[111] *R.E. Reeves*, Anal. Chem. *21*, 751 (1949).

[112] *A. Berka*, Z. Anal. Chem. *195*, 263 (1963).

[113] Ref.[145], p. 83. *A. Berka, J. Vulterin, J. Zýka*, Chemist-Analyst *51*, 24 (1962).

[114] *D.M. West, D.A. Skoog*, Anal. Chem. *31*, 583 (1959).

for α- and β-*naphthol*, equations for the course of the oxidation reaction have been postulated[115]. With the exception of glycerol[116], no simple stoichiometry is observed for the oxidation of polyalcohols[114]. Even though a few additional uses have been reported[113], the scope of this reagent is limited.

2.2.6 Divalent Copper

Cupric ion is the most commonly used reagent for the determination of reducing sugars. It reacts also with some hydrazines, liberating nitrogen in quantitative amounts. It can be used in anhydrous media, e.g. for the determination of mercaptans. Direct[118, 119] and indirect[117, 120–122] methods may be used for the determination of *reducing sugars*. No simple stoichiometric relations are observed.

For each series of analyses, the reaction conditions and the rate of addition of titrant are kept constant, and empirical calibration of the method is performed with a standard sugar solution. The oxidant used is an alkaline solution of complexed divalent copper, which upon reduction is transformed to cuprous oxide. The constituents of the standard solution, i.e. cupric salt, complexing agent (tartrate[119, 120], citrate, trihydroxyglutaric acid[118, 121, 122]) and alkali are mixed immediately before use. Direct titrations are performed in the inverse manner: the reagent is titrated with the sugar solution at 80–100°. If an indirect method is employed, the excess of reagent is back-titrated with potassium cyanide[120] or iodometrically[117]. Alternatively, the cuprous oxide formed can be determined with ferric salt and permanganate solutions[121, 122] or iodometrically[117].

In a reliable and specific procedure for determination of *hydrazine derivatives*, oxidation with copper(II) liberates a stoichiometric amount of nitrogen, which is determined by gasometric methods[123] (see 3.7:2).

Oxidation of the thiol group with copper(II) in non-aqueous solution is particularly useful for determination of mercaptans[124, 125] which are not soluble in water. Copper(II) in acetonitrile is an analytical oxidant of potential interest[126, 127].

2.2.7 Hexacyanoferrate(III)

In weakly acidic, neutral or alkaline solution, the redox potential of $Fe(CN)_6^{3-}/Fe(CN)_6^{4-}$ is 0.45 V, and is independent of the pH. Zinc ions

$$Fe(CN)_6^{3\ominus} + e \rightleftharpoons Fe(CN)_6^{4\ominus}$$

increase the oxidative power of the reagent by lowering the equilibrium concentration of hexacynoferrate(II). Potentiometry or polarized electrodes are used for end point determination[128, 129]. An aqueous solution of potassium hexacyanoferrate(III) is used as the oxidizing agent[128, 130].

Direct titrations with hexacyanoferrate(III) may be used for the determination of *formaldehyde*[131] in strongly alkaline solution in the presence of osmium tetroxide, of *ascorbic acid* in the presence of zinc acetate[132] and of *cysteine*[129] at pH 7 in the presence of copper sulfate.

The titration of *reducing sugars* is of more general interest.

Usually, an alkaline hexacyanoferrate(III) solution is titrated with a sugar solution at boiling temperature[128, 133, 134]. The method requires empirical calibration against a standard sugar solution and strict control of temperature, pH, ionic strength and titration rate[128]. Addition of zinc salts improves the titration[134].

Assays of *hemoglobin, penicillin* and *reducing sugars* in biological samples are usually performed by indirect titration[130]. An excess of hexacyanoferrate(III) is used for oxidation; the unused reagent is back-titrated iodometrically. Alternatively, the amount of hexacyanoferrate(II) produced may be titrated with ceric solution.

[115] *R.K. Mittal, R.C. Mehrotra*, Z. Anal. Chem. *209*, 337 (1965).

[116] *D.M. West, D.A. Skoog*, Anal. Chem. *31*, 586 (1959).

[117] *I.M. Kolthoff, R. Belcher*, Ref. [143], p. 357.

[118] *A.V. Ablov, D.G. Batyr*, Zh. Anal. Khim. *15*, 734 (1960).

[119] *L.W. Chubb, A.W. Hartley*, Analyst (London) *83*, 311 (1958).

[120] *K.R. Manolov*, Zh. Anal. Khim. *17*, 898 (1962).

[121] *A.V. Ablov, D.G. Batyr*, Zh. Anal. Khim. *12*, 749 (1957).

[122] *A.V. Ablov, D.G. Batyr*, Zh. Anal. Khim. *15*, 112 (1960).

[123] *S. Siggia*, Ref. [144], p. 541; *S. Siggia, L.J. Lohr*, Anal. Chem. *21*, 1202 (1949).

[124] *H. Roth*, Mikrochim. Acta 769 (1958).

[125] *Z. Hladký, J. Vřešťál*, Collect. Czech. Chem. Commun. *34*, 1098 (1969).

[126] *B. Kratochvil, D.A. Zatko, R. Markuszewski*, Anal. Chem. *38*, 770 (1966).

[127] *D.A. Zatko, B. Kratochvil*, Anal. Chem. *40*, 2120 (1968).

[128] *R.N. Adams, C.N. Reilley, N.H. Furman*, Anal. Chem. *24*, 1200 (1952).

[129] *H.G. Waddill, G. Gorin*, Anal. Chem. *30*, 1069 (1958).

[130] *A. Berka, J. Vulterin, J. Zýka*, Ref. [145], p. 18.

[131] *F. Solymosi*, Chemist-Analyst *51*, 71 (1962).

[132] *B.R. Sant*, Chemist-Analyst *47*, 65 (1958).

[133] *E.T. Polubnaya, P.S. Bukharov*, Zh. Anal. Khim. *3*, 131 (1948).

[134] *E.T. Polubnaya, P.S. Bukharov*, Zh. Anal. Khim. *5*, 300 (1950).

Oxidation of *aryl sulfonic hydrazides*[135] liberates an equivalent amount of nitrogen, which is determined by gasometric methods (see 3.7:2).

2.2.8 Monovalent Silver

The oxidation of aldehydes to carboxylic acids is the only instance in which this reagent is of some importance. Ammoniacal solutions of silver salts *(Tollen's reagent)*[136], silver oxide generated in situ[137], and silver complexed with tert-butylamine[138] are the reagents used.

If silver oxide is taken as the oxidant, the reaction proceeds as follows:

$$RCHO + Ag_2O \rightarrow RCOOH + 2\ Ag$$

A known excess of the silver(I) is added to the sample. Residual oxidant may be determined by titration with 0.1 N potassium iodide in ammoniacal solution[136]. If silver oxide is used as an oxidant, the unreacted oxide is dissolved in sulfuric acid and titrated with thiocyanate[137]. In cases where the butylamine complex is used, the reaction mixture is filtered after the oxidation is completed; the filtrate is acidified and titrated with thiocyanate[138]. The consumption of silver(I) is determined by running a blank.

2.2.9 Divalent Mercury

Alkaline solutions of tetraiodomercuric ion oxidize *aldehydes* according to the equation:

---

[135] *N.N. Dykhanov, A.B. Dzhidzhelava*, Zh. Anal. Khim. *21*, 1277 (1966).
[136] *S. Siggia, E. Segal*, Anal. Chem. *25*, 640 (1953).
[137] *H. Siegel, F.T. Weiss*, Anal. Chem. *26*, 917 (1954).
[138] *J.A. Mayes, E.J. Kuchar, S. Siggia*, Anal. Chem. *36*, 934 (1964).

$$RCHO + HgJ_4^{2\ominus} + 3\ OH^{\ominus} \rightarrow$$
$$RCOO^{\ominus} + Hg + 4\ J^{\ominus} + 2\ H_2O$$

The sample is treated with an excess of the reagent. Precipitated elemental mercury is re-oxidized in acidic solution with a known excess of iodine, which, in turn, is back-titrated with thiosulfate[139].

2.3 Anodic Oxidation

Coulometric generation of bromine, iodine and chlorine is of practical importance for the determination of small samples, preferably when oxidation proceeds rapidly[140, 141].

---

[139] *J.E. Ruch, J.B. Johnson,* Anal. Chem. *28*, 69 (1956).
[140] *J.J. Lingane*, Electroanalytical Chemistry, 2nd Ed., p. 536, Interscience Publishers, New York 1958.
[141] *K. Abresch, I. Claassen*, Die coulometrische Analyse, Verlag Chemie, Weinheim/Bergstr. 1961.

2.4 Bibliography

[142] *A. Polgár, J.L. Jungnickel*, Organic Analysis, Vol. III, p. 229, Interscience Publishers, New York 1956.
[143] *I.M. Kolthoff, R. Belcher*, Volumetric Analysis, Vol. III, Interscience Publishers, New York, London 1957.
[144] *S. Siggia*, Quantitative Organic Analysis via Functional Groups, 3rd Ed., John Wiley & Sons, New York, London 1963.
[145] *A. Berka, J. Vulterin, J. Zýka*, Newer Redox Titrants, Pergamon Press, Oxford 1965.

3.4 Addition, Substitution, Condensation and Complex Formation

Willy Büchler, Alfred Becker and *Armin Walter*
CIBA-GEIGY, CH-4002 Basel

1 Addition

1.1 Addition of Water

Addition of water is not utilized very frequently for analytical purposes. Hydrolysis, formally involving addition, is discussed in Section 3.5:1. The K. Fischer titration method can be used to determine the amount of water used, but it is generally preferable to determine the reaction products.

1.1.1 Addition to Alkynes

Alkynes can react with water in acidic solution, with mercuric sulfate as catalyst; ketones are formed.

If precipitation of the ketone as its dinitrophenylhydrazone is desired, mercury must first be removed as the sulfide[1, 2].
If an α-C-atom is substituted, addition is usually slow. Halogen compounds may react with the mercuric sulfate and so render the catalyst ineffective.

1.1.2 Addition to Epoxides

In dilute perchloric acid containing a trace of periodic acid, *epoxides* can be hydrated. The resulting 1,2-glycols are oxidatively cleaved by periodic acid:

$$>\!\!C\!-\!C\!\!<\ \xrightarrow{H_2O}\ >\!\!C\!-\!C\!\!<\ \xrightarrow{H_5JO_6}\ 2\ >\!\!C\!=\!O$$

The excess reagent can be determined with arsenite[3]. This procedure is said to give accurate results, but is rarely used for epoxide determination.

1.2 Addition of Hydrogen Halides (and Hydrogen Cyanide)

1.2.1 Addition to Carbonyl Compounds

The addition of hydrogen cyanide to *carbonyl groups* is of importance for analysis of *aldehydes* and *sugars*.

After alkaline hydrolysis of the cyanohydrin, the ammonia formed can be determined[4]. The carbonyl content of starch[5] may be estimated in this way. The long reaction times necessary (up to 24 hours) are disadvantageous, and may cause loss of reagent.

Addition of potassium cyanide is used principally in the determination of *formaldehyde*[6-8], which can be detected specifically in the presence of acetaldehyde and acetone. Homologous *aldehydes*[9] can also be determined by means of this reaction.

The course of the reaction can be followed by acidimetric, iodometric or complexometric means[6, 7, 10].

1.2.2 Addition to Epoxides

The addition of hydrogen halides to *1,2-epoxides (oxirans)* is of considerable importance.

$$>\!\!C\!-\!C\!\!<\ +\ H\!-\!Hal\ \longrightarrow\ >\!\!C\!-\!C\!\!<$$

Of the many procedures which have been recommended, two have emerged as the most important and up-to-date: one involving a perchloric acid titration[11], and the other involving titration with hydrogen bromide[12].

In the first method, the epoxy compound is dissolved in chloroform; acetic acid and tetraethylammonium bromide are added, and the solution is titrated directly with perchloric acid. With the other procedure, the sample is dissolved in benzene and titrated in a closed vessel with a 0.1 N solution of hydrogen bromide in acetic acid (accuracy $\pm 0.4\%$).

Interference can occur as a result of various side reactions: hydrolysis, or alcoholysis by the solvent; replacement of hydroxy groups by halogen; polymerization, and isomerization to carbonyl compounds. These side reactions can be suppressed by using suitable solvents, temperatures and reaction times.

Epoxides of low reactivity can be estimated by use of hydrochloric acid in pyridine, at elevated temperatures[13].

Methods involving the addition of hydrogen chloride or hydrogen bromide in dioxane have been superseded by the procedure involving titration with perchloric acid, except for the specific determination of diepoxides or of unreactive epoxides[14]. *1,3-epoxides* and *aziridines* can be determined to some extent by the same methods.

1.3 Addition of Sulfite or Hydrogen Sulfite

The most important use of this type of reaction is for the determination of carbonyl groups. Hydrogen sulfite can add also to alkenes and epoxides, but in the latter case, reaction is subject to interference and is not recommended.

1.3.1 Addition to Alkenes

Sulfite can be added to C=C double bonds conjugated with powerfully electron-attracting groups:

$$NaHSO_3\ +\ H_2C\!=\!CH\!-\!X\ \longrightarrow\ NaO_3S\!-\!CH_2\!-\!CH_2X$$

The reaction is irreversible. Alkalimetric methods[15, 16] can be used for quantitative estimation. Possible inter-

[1] H.A. Iddles, C.E. Jackson, Ind. Eng. Chem. Anal. 6, 454 (1934).

[2] S. Siggia, Anal. Chem. 28, 1481 (1956).

[3] A.M. Eastham, G.A. Latremouille, Can. J. Res. 28 B, 264 (1950).

[4] V.L. Frampton, L.P. Foley, L.L. Smith, J.G. Malone, Anal. Chem. 23, 1244 (1951).

[5] J. Schmorak, M. Lewin, Anal. Chem. 33, 1403 (1961).

[6] E. Schulek, Chem. Ber. 58, 732 (1925).

[7] J.I. de Jong, Recl. Trav. Chim. Pays-Bas 72, 356 (1953).

[8] K. Konno, M. Kageyama, T. Ueda, J. Pharm. Soc. Jap. 72, 1153 (1952).

[9] W.J. Svirbely, J.F. Roth, Anal. Chem. 26, 1377 (1954).

[10] C. Berther, K. Kreis, O. Buchmann, Z. Anal. Chem. 169, 184 (1959).

[11] R. Dijkstra, E.A. Dahmen, Anal. Chim. Acta 31, 38 (1964).
R.R. Jay, Anal. Chem. 36, 667 (1964).

[12] A.J. Durbetaki, Anal. Chem. 28, 2000 (1956).

[13] R.T. Keen, Anal. Chem. 29, 1041 (1957).

[14] H. Jahn, H. Raubach, G. Rodekirch, W. Tiege, Plaste Kautschuk 11, 141 (1964).

[15] F.E. Critchfield, J.B. Johnson, Anal. Chem. 28, 73 (1956).

[16] D.H. Whitehurst, J.B. Johnson, Anal. Chem. 30, 1332 (1958).

ference through addition of sulfur dioxide can be suppressed by addition of isopropanol.

1.3.2 Addition to Carbonyl Compounds

The reaction of aldehyde with sulfite or hydrogen sulfite proceeds as shown in the following equations:

$$R-\underset{\underset{O}{\parallel}}{\overset{H}{C}} + Na_2SO_3 + H_2O \rightleftharpoons R-\underset{\underset{OH}{|}}{\overset{H}{\underset{|}{C}}}-SO_3Na + NaOH$$

$$R-\underset{\underset{O}{\parallel}}{\overset{H}{C}} + NaHSO_3 \rightleftharpoons R-\underset{\underset{OH}{|}}{\overset{H}{\underset{|}{C}}}-SO_3Na$$

Addition of sulfite is quantitative only in the case of *formaldehyde;* here the sodium hydroxide formed can be determined by acidimetric titration. Other *aldehydes* are reacted with hydrogen sulfite or sulfite plus sulfuric acid, to shift the equilibrium to the right. Addition of phosphate buffer has the same effect[17]. Besides the alkalimetric determination methods[18], it can also be advantageous to use an iodometric procedure, in which the excess of hydrogen sulfite is titrated. In the iodometric procedure, reverse reaction is avoided by using low pH, low temperature and an excess of reagent.

Aldehydes, acetals, vinyl ethers and some *ketones* can be determined alkalimetrically[21]. In this titration, addition compounds of ketones may often decompose even before the potential end-point (of the aldehyde-bisulfite compound) is reached. Nevertheless, good results can be obtained often by titrating rapidly and at low temperatures.

Aldehydes and *sugar derivatives* (acetals)[20] can be quantitatively determined with Schiff's reagent. By making use of differences in solubility in chloroform or higher alcohols etc., different aldehydes can be separately determined in their mixtures.

Schiff's reagent is prepared by passing sulfur dioxide into a solution of Fuchsine until decolorisation takes place: leuco-Fuchsine is formed[19]. Reaction of leuco-Fuchsine with an aldehyde involves addition of a sulfine group, the corresponding amount of a quinoid dye, being formed.

1.4 Addition of Metallic Salts and Organometallic Compounds

1.4.1 Grignard Reaction

A Grignard reagent is a solution of an alkylmagnesium halide in ether[21]. It reacts with active hydrogen atoms, e.g. in OH-, NH- and COOH-groups *(Zerewitinoff Reaction),* evolving alkanes, and also with reactive groups (of aldehydes, ketones, ester, azomethines, or nitriles), to extend the carbon skeleton.

These reactions can be used for analytical procedures. The Zerewitinoff reaction is used for quantitative determination of the excess of Grignard reagent after reaction with *carbonyl compounds.*

Since all of these reactions proceed quantitatively, the excess Grignard reagent can be determined by decomposition with water or aniline, and measurement of the methane evolved. However Grignard compounds have low stability towards oxygen, carbon dioxide and water, so that handling is often difficult.

Grignard reagents react differently with ketones and with enols, and so can be used for determination of *enol contents,* e.g. in β-diketo-compounds[22].

1.4.2 Addition of Mercury (II) Salts

Addition of *mercury(II) salts* to double bonds is used analytically in the determination of *allyl alcohol, vinyl ethers, cyclohexenes, vinyl acetate* and *styrenes;* acrylonitrile and methacrylic acid will not interfere[23].

Alkynes and *α,β-unsaturated acids, esters* and *nitriles* do not usually add mercuric salts quantitatively; however, if a small amount of perchloric acid is added as accelerator, these compounds also can be determined[24, 25].

Addition of mercury(II) acetate in methanol (methoxymercuration) has general applicability:

$$\underset{H}{\overset{R^1}{>}}C=C\underset{H}{\overset{R^2}{<}} + (CH_3COO)_2\,Hg + CH_3OH$$

$$\longrightarrow \underset{H_3CO}{\overset{R^1}{\underset{|}{H}C}}-\underset{Hg-OOC-CH_3}{\overset{R^2}{\underset{|}{CH}}} + CH_3COOH$$

Determination is carried out either by alkalimetric titration of the acetic acid liberated, or by complexing the mercury. In other alkalimetric procedures[26, 27], the

[17] *I.R. Hunter, E.F. Potter,* Anal. Chem. *30,* 293 (1958).

[18] *J.G. Reynolds, M. Irwin,* Chem. Ind. [London] 419 (1948).

[19] *J. Alexander,* Science *111,* 13 (1950).

[20] *C.L. Hoffpauir,* Ind. Eng. Chem. Anal. *15,* 605 (1943).

[21] *S. Siggia, W. Maxey,* Ind. Eng. Chem. Anal. *19,* 1023 (1947).

[22] *Houben-Weyl,* Methoden der organischen Chemie, 4. Aufl., Bd. II, Georg Thieme Verlag, Stuttgart 1953.

[23] *M. Wronski,* Z. Anal. Chem. *171,* 177 (1959).

[24] *K.L. Mallik, M.N. Das,* Chem. Ind. [London] 162 (1959).
   K.L. Mallik, Anal. Chem. *32,* 1369 (1960).

[25] *M. Koulkes,* Bull. Soc. Chim. Fr. 402 (1953).

[26] *R.P. Marquardt, E.N. Luce,* Anal. Chem. *39,* 1655 (1967).

[27] *J.B. Johnson, J.P. Fletcher,* Anal. Chem. *31,* 1563 (1959).

excess of mercury acetate is first eliminated by addition of halide, and the acetic acid formed during the reaction is then titrated. In order to protect the addition compound against decomposition, it is advisable to extract it from water with carbon tetrachloride.

The mercury addition compound and the excess of reagent are titrated with standardized hydrochloric acid in propylene glycol[28]. In this reaction, the excess of mercuric acetate combines with two halogen equivalents whereas the mercury bonded to the carbon atom in the compound reacts with only one. This titration can be carried out rapidly, with an accuracy of 1 to 3%. The excess mercury reagent can be determined also by backtitration with thioglycolic acid[29]; thiofluorescein is used as indicator.

1.5 Addition of Alcohols

Reactions involving addition of alcohols to nitriles, isocyanates and carbon disulfide are used mostly for the determination of the alcohol group. Mercaptans and to some extent phenols and thiophenols react analogously.

1.5.1 Addition to a,β-Unsaturated Nitriles

Alcohols, mercaptans (thiols) and *thiophenols* can be added quantitatively to C=C bonds which are adjacent to a nitrile group:

$$R-OH + H_2C=CH-CN \longrightarrow RO-CH_2-CH_2-CN$$

The reaction is catalyzed by potassium hydroxide[30].

The excess of unsaturated nitrile can be determined iodometrically, or by treatment with sodium sulfite (section 3.4:1.3.1); in the latter case, alkali is liberated, in an amount equivalent to the excess of unsaturated nitrile:

$$Na_2SO_3 + H_2C=CH-CN + H_2O$$
$$\longrightarrow NaO_3S-CH_2-CH_2-CN + NaOH$$

1.5.2 Addition to Isocyanates

Isocyanates add alcohols, forming esters of carbamic acid:

$$R^1-N=C=O + R^2-OH \longrightarrow R^1-NH-COOR^2$$

Because of the high volatility and toxicity of the lower aliphatic isocyanates, aromatic isocyanates are usually used. This reaction is frequently used for the determination of *hydroxy groups* in polyhydroxyalkanes and in *polyester resins*[31, 32].

The excess of isocyanate can be titrated directly with di-isobutylamine, using bromophenol blue as indicator[32]. Another possibility is to allow the isocyanate to react with an excess of amine, and to titrate the residual amine with perchloric acid[31].

1.5.3 Addition of Carbon Disulfide

Primary and secondary alcohols react in alkaline solution with *carbon disulfide* to form xanthates, which can be titrated iodometrically[33].

$$2 \; S=C\begin{matrix}OR\\ \\SK\end{matrix} + I_2 \longrightarrow S=\overset{\overset{OR}{|}}{C}-S-S-\overset{\overset{OR}{|}}{C}=S + 2KI$$

For determination of *alcohol*, the xanthate formed is precipitated with an excess of nickel salt. The nickel chelate is filtered off, dissolved in ammonia, and the nickel titrated complexometrically with EDTA[34] (indirect chelatometry). This procedure is suitable for the determination of alcohols in aqueous solution, and is preferred to the acetylation and active hydrogen determination techniques.

1.5.4 Addition to Alkynes

Mono- and disubstituted *acetylenes* add alcohols in the presence of mercuric oxide or boron trifluoride as catalyst, ketals being formed:

$$R^1-C\equiv C-R^2 + 2 \; CH_3OH \longrightarrow R^1-\overset{\overset{OCH_3}{|}}{\underset{\underset{OCH_3}{|}}{C}}-CH_2-R^2$$

Dimethyl Ketal

On addition of water, the ketal is decomposed and the ketone formed can be determined by condensation with hydroxylamine[35].

This procedure, however, can be used only under defined conditions, and with application of a correction factor. Because of this and the very complicated reaction procedure, it cannot be generally recommended.

1.6 Additions of Amines

Additions of *amines* to polarized double bonds, and to isocyanates and carbon disulfide, can often be employed for volumetric determination of an amine or of the other reactant. Since such reactions are often specific to some extent, they are used frequently.

1.6.1 Addition to Alkenes

Amines can add to double bonds which are adjacent to electronegative groups. The reactivity decreases in the following sequence

[28] *M.N. Das*, Anal. Chem. *26*, 1086 (1954).
[29] *M. Wronski*, Z. Anal. Chem. *171*, 177 (1959).
[30] *S.I. Obtemperanskaya*, Zh. Anal. Khim. *16*, 372 (1961).
[31] *D.H. Reed*, Anal. Chem. *35*, 571 (1963).
[32] *B. Dreher*, Farbe Lack *67*, 703 (1961).
[33] *M.P. Matuszak*, Ind. Eng. Chem. Anal. *4*, 98 (1932).
[34] *N.D. Cheronis, T.S. Ma*, Organic Functional Group Analysis, p. 191, Interscience Publishers, New York 1964.
[35] *C.D. Wagner*, Anal. Chem. *19*, 103 (1947).

$$\left(R = \ \overset{|}{\underset{}{>}}C = \overset{|}{C}-\right):$$

$$R-C\equiv N \ > \ R-COOR \ > \ R-COOH \ > \ R-CO-R$$

$$> \ R-COONa$$

In acetic acid solution, *a,β-unsaturated esters, amides* or *nitriles* add morpholine or piperidine[36, 37]. After acetylation of the excess secondary amine, the addition product can be determined as a tertiary amine by acidimetry in non-aqueous solution. If the amine is of low basicity, the titration should be carried out conductimetrically.

Good results can be obtained with *acrylamide* and *acrylonitrile*. Long-chain alkenes do not react as well as acrylic compounds, and they require higher reaction temperatures, *α,β*-unsaturated aldehydes or ketones undergo other reactions: for these compounds, addition of bromine is better.

Analogously, *ketenes* can be determined by addition of aniline[38]:

This reaction is carried out in benzene or dioxane solution, the excess aniline being titrated subsequently with perchloric acid in acetic acid.

1.6.2 Addition to Quinones

Primary amines can add to *quinones*[39, 40]. The addition product is reoxidized easily to a quinone, which can add further amine:

The adducts can be determined colorimetrically, at ca. 350 nm. Anthraquinone does not react in this way.

1.6.3 Addition to Isocyanates

Isocyanates and *isothiocyanates* react with primary or secondary amines to form derivatives of urea and thiourea:

$$R^1-N=C=O \ + \ R^2-NH_2 \ \longrightarrow$$
$$R^1-NH-CO-NHR^2$$

If determination of the isocyanate group is required, an excess of secondary amine (dibutylamine in dioxane or in a mixture of chlorobenzene and methanol) is added, and the excess is back-titrated with standardized acid[41]. This procedure can be used as a micromethod. Isothiocyanates react similarly.

With ammonia, isothiocyanates form thiourea, which may be determined argentometrically (silver sulfide being precipitated), or by oxidation with bromine or iodine[42].

Primary and secondary *amines* can likewise be determined in the presence of each other, by reaction with phenyl isocyanate; here, however, kinetic measurements are necessary. In mixtures containing tertiary amines, primary and secondary amines react with phenyl isocyanate to form urea derivatives; consequently, on subsequent titration with acid in anhydrous solution[43, 44], the tertiary amines alone are determined.

Isocyanates react with hydroxylamine forming hydroxamic acids, which with ferric ions give complexes suitable for colorimetric estimation[45] (see Section 3.4:3.1.1).

1.6.4 Addition to Carbon Disulfide

Primary and secondary *amines* react with *carbon disulfide*, forming dithiocarbamic acids:

$$R-NH_2 \ + \ CS_2 \ \longrightarrow \ RNH-C\overset{\displaystyle S}{\underset{\displaystyle SH}{}}$$

Dithiocarbamic acids can be determined by various methods: titration with sodium hydroxide[46]; precipitation of an insoluble nickel complex, followed by complexometric titration of nickel[47];

[36] *F.E. Critchfield, G.L. Funk, J.B. Johnson*, Anal. Chem. *28*, 76 (1956).
[37] *A.P. Terentev*, Zh. Anal. Khim. *14*, 506 (1959), C.A.
[38] *A.M. Potts*, Arch. Biochem. *24*, 329 (1949).
[39] *H. Karius, G.E. Mapstone*, Chem. Ind. (London) 266 (1956).
[40] *R.J. Lacoste*, Anal. Chem. *32*, 990 (1960).

[41] *A.G. Williamson*, Analyst (London) *77*, 372 (1952).
[42] *L.R. Wetter*, Can. J. Biochem. Physiol. *33*, 980 (1955).
[43] *J.G. Hanna, S. Siggia*, Anal. Chem. *34*, 547 (1962).
[44] *M. Miller, D.A. Keyworth*, Talanta *10*, 1131 (1963).
[45] *R.E. Buckles, C.J. Thelen*, Anal. Chem. *22*, 676 (1950).
[46] *F.E. Critchfield, J.B. Johnson*, Anal. Chem. *28*, 430 (1956).
[47] *L. Nebbia, F. Guerrieri*, Chemica Industria *35*, 896 (1953).

titration with coulometrically generated mercuric ions[48]; colorimetry of copper complexes, which can be extracted with chloroform[49].

To determine the content of *carbon disulfide,* the sample is added to morpholine in pyridine solution, and the dithiocarbamic acid formed is titrated with sodium hydroxide. For microanalysis, di-n-butylamine is used as base, in acetone solution[48].

1.7 Addition of Conjugated Dienes (Diels-Alder Reaction)

The Diels-Alder reaction is usually used for synthetic purposes, but can also be successfully employed analytically, for the determination of dienes. In this reaction, an olefin (dissolved in an organic solvent) reacts with a conjugated diene to form an addition product:

Maleic anhydride is used generally as the olefinic component; recently, tetracyanoethylene has also been used[50]. Undesirable side-reactions include polymerization of the diene, and copolymerization of the diene with the olefinic component; terpenes with conjugated systems frequently react in this way. This reaction is an equilibrium one, so an excess of reagent is often necessary for complete conversion to be attained. Five- or six-membered cyclic addition products are formed at room temperature; those with larger rings require higher temperatures. Cumulenes do not react with dienes.

To perform the reaction, the diene is mixed with an excess of maleic anhydride in a suitable solvent, e.g. ether, xylene or carbon tetrachloride. The excess of maleic anhydride is then extracted with water, and titrated with sodium hydroxide[51]. A recent variation uses a system comprising two liquid phases as the reaction medium, e.g. with xylene or carbon tetrachloride as the lower phase and water as the upper phase.

1.8 Miscellaneous Addition Reactions

1.8.1 Addition of Nitrogen Oxides to Olefins

To a certain extent, N_2O_3 adds to double bonds. Thus, *styrene*[52] in hydrocarbon solution can be precipitated by addition of sodium nitrite and sulfuric acid; a nitro-nitroso compound is formed in 82% yield:

1.8.2 Addition of Sulfuric Acid to Olefins

Olefins are converted into sulfuric acid esters by fuming sulfuric acid:

The reaction products formed depend on the concentration of the acid.

This reaction is used especially for removing unsaturated compounds from hydrocarbons or mixtures of gases. Side-reactions include polymerization, oxidation, and esterification. For quantitative estimation a solution of mercuric sulfate in sulfuric acid (20% wt/vol) which is saturated with magnesium sulfate[53] is used. Olefins and aromatic hydrocarbons can be removed quantitatively from hydrocarbon mixtures by the Kattwinkel reagent (30% phosphoric acid plus 70% sulfuric acid[54]). However, for quantitative determination of olefins, this procedure is being superseded by gas chromatography.

1.8.3 Addition of Mercaptans

Mercaptans add to unsaturated compounds, e.g. *acrylonitrile, acrylamide, acrylic esters* and *α,β-unsaturated carbonyl compounds* (esters, aldehydes, and amides). Long-chain aliphatic mercaptans and thioglycolic acid are used as reagents. Other unsaturated compounds, such as styrene, allyl alcohol, and thiophene, do not react with thioglycolic acid.

This reaction is accelarated by hydroxyl ions; the alkene is reacted with an excess of mercaptan in alkaline alcoholic solution, at room temperature and under nitrogen (reaction time, 2 to 15 minutes). After acidification, the excess of mercaptan can be determined by iodometric or argentometric titration[55] (accuracy ± 0.2%). If the mercaptan used is thioglycolic acid, the excess can be titrated with standardized mercury (II) acetate[57].

The excess of mercaptan can also be determined by addition of a known amount of iodine, followed by colorimetric determination[56] of the residual iodine.

[48] *E.P. Przybylowiez, L.B. Rogers,* Anal. Chim. Acta *18,* 596 (1958).
[49] *M. Weiser, K.M.K. Zacherl,* Mikrochim. Acta 577 (1957).
[50] *K. Hafner, J. Schneider,* Liebigs Ann. Chem. *624,* 37 (1959).
    K. Hafner, K.C. Moritz, Angew. Chem. *72,* 918 (1960).
[51] *W. Schmidt,* Brennst.-Chem. *33,* 176 (1952).
[52] *R.P. Marquardt, E.N. Luce,* Anal. Chem. *23,* 629 (1951).

[53] *F.R. Brooks,* Anal. Chem. *21,* 1105 (1949).
[54] *I.W. Mills,* Anal. Chem. *20,* 333 (1948).
[55] *D.W. Blessing,* Anal. Chem. *21,* 1073 (1949).
[56] *J. Haslam, G. Newlands,* Analyst *80,* 50 (1955).
[57] *M. Wronski,* Z. Anal. Chem. *171,* 177 (1959).

Since most other unsaturated compounds do not react with mercaptans, mercaptan addition is a relatively specific and accurate method for determination of compounds of the classes mentioned.

1.9 Bibliography

[58] *J. Mitchell, Jr.,* Organic Analysis, Vol. I, p. 243, Interscience Publishers, New York 1953.

[59] *M.S. Kharash, O. Reinmuth,* The Grignard Reaction, Prentice Hall, New York 1954.

[60] *S. Siggia,* Quantitative Organic Analysis via Functional Groups, 3rd Ed., John Wiley, New York 1963.

[61] *N.D. Cheronis, T.S. Ma,* Organic Functional Group Analysis, Interscience Publishers, New York 1964.

[62] *J.L. Jungnickel* in: Organic Analysis, Vol. I, p. 127, Interscience Publishers, New York 1964.

[63] *B. Dobinson, W. Hofmann, B.P. Stark,* The Determination of Epoxide Groups, Pergamon Press, Oxford 1969.

2 Substitution and Condensation

2.1 Substitution and Condensation with Reactive Halogen Compounds

Several types of halogen-containing nitrobenzene derivatives react quantitatively with amines and certain phenols. 2,4-Dinitrofluorobenzene, for example, is used for the determination of primary and secondary amines[1,2], phenols and amino acids[3].

The crystalline compounds formed generally have low solubility, and can be determined gravimetrically.

It has been reported that picryl chloride (2,4,6-trinitrochlorobenzene) and (in some cases) 2,4-dinitrochlorobenzene react quantitatively with primary amines but not with secondary and tertiary amines. The chloride formed can be titrated argentometrically; alternatively, the condensation product can be extracted and analyzed colorimetrically. In this way, it is possible to determine *primary amines* in the presence of secondary and

tertiary amines[2,4]. Picryl chloride can be used also in the determination of *hydrazine derivatives*[5]. Techniques involving condensation with picryl chloride can be used both for analysis on the normal (macroscopic) scale, and for analysis of traces of amines. 1-Dimethylaminonaphthalene-5-sulfonyl chloride has been employed as an acylating agent for *phenols*[6]; the reaction is carried out in an aqueous acetone solution at pH > 8.

One of the best known colorimetric methods for phenols is that involving reaction with 2,6-dibromoquinone chloroimine[7], 2,6-dichloroquinone chloroimine[7], or *N*-(benzensulfonyl)-quinonimine[8].

The indophenols produced are determined spectrophotometrically. Reaction with N-(benzenesulfonyl)-quinonimine is rapid and sensitive, and can be used for the determination of traces of phenolic compounds.

2.2 Condensation with Methylene Groups

Dimedone[9] (5,5-dimethyl-1,3-cyclohexane-dione, a reagent with active methylene groups), is frequently used for the determination of aldehydes and ketones.

The reaction products are generally estimated gravimetrically; alternatively, the excess of reagent can be titrated with nitrous acid[9a].

Traces of aldehydes in air are determined either by precipitation with dimedone or by the very sensitive fluorimetric method[9b].

[1] *H. Zahn, A. Würz,* Z. Anal. Chem. *134,* 183 (1951).
H. Zahn, R. Kockläuner, Z. Anal. Chem. *141,* 183 (1954).

[2] *F.C. McIntire, L.M. Clements, M. Sproull,* Anal. Chem. *25,* 1757 (1953).

[3] *G. Koch, W. Weidel,* Z. Physiol. Chem. *303,* 213 (1956).

[4] *G. Spencer, J.E. Brimly,* J. Soc. Chem. Ind. *64,* 53 (1945).

[5] *J.P. Riley,* Analyst *79,* 76 (1954).

[6] *N. Seiler, M. Wiechmann,* Z. Anal. Chem. *220,* 109 (1966).

[7] *G.R. Boreham, J.A.P. Cunningham,* Fuel *38,* 489 (1959).
G. Gorbach, O.G. Koch, G. Dedic, Mikrochim. Acta 882 (1955).
J.P. Sweeney, W.L. Hall, J. Ass. Offic. Agr. Chem. *38,* 697 (1955).

[8] *G.G. Guilbault, D.N. Kramer, E. Hackley,* Anal. Chem. *38,* 1897 (1966).

[9] *J.H. Yoe, L.C. Reid,* Ind. Eng. Chem. Anal. *13,* 238 (1941).

[9a] *C. Ericson,* Acta Pharm. Suecica, *5,* 283 (1968).

[9b] *E. Sawicki, R.A. Cornes,* Mikrochim. Acta 148 (1968).

2,4-Pentanedione has been used for the fluorimetric analysis of *aldehydes,* especially formaldehyde[10].

Cyclopentadiene reacts with aldehydes and ketones, forming colored products, which can be estimated colorimetrically[10a].

2.3 Condensation with Alcohols

The most important reaction is the condensation of sodium alkoxide[11] with acid anhydrides and acid halides.

$$R-C\underset{Cl}{\overset{O}{\big|}} \ + \ CH_3ONa \ \longrightarrow \ R-C\underset{OCH_3}{\overset{O}{\big|}} \ + \ NaCl$$

Under anhydrous conditions, acid anhydrides or acid halides can be titrated directly with standard alkali alkoxide solution. To determine acid anhydrides or halides in the presence of acids, the total acid is titrated with alkali hydroxide in aqueous solution, after addition of a few milliliters of pyridine or piperidine to effect hydrolysis.

$$R-C\underset{OH}{\overset{O}{\big|}} \ + \ CH_3ONa \ \longrightarrow \ R-C\underset{ONa}{\overset{O}{\big|}} \ + \ CH_3OH$$

In a mixture of acid halide and free acid, the former reacts with alkoxide as described above; the free acid is simply neutralized.

One equivalent of alkoxide is consumed in the esterification of acid anhydrides or halides. After hydrolysis, two equivalents of alkali hydroxide are required. An equal quantity of hydroxide or alkoxide is used for titrating free acid.

A differential rate technique can be used for determination of the contents of acid anhydrides, acid halides and free acids.

2.4 Condensation with Amines

Condensation of aldehydes or ketones with hydroxylamine hydrochloride[12], although an old method, is still often used successfully, because of the simplicity of this technique.

The oxime formed is a weaker base than hydroxylamine; consequently, the acid which is set free can be titrated potentiometrically or by using the mixed indicator dimethyl yellow/methylene blue. If hydroxylamine itself is used in the condensation, the excess of hydroxylamine is titrated with hydrochloric acid[13]. Hydroxylamine acetate or formate is used to form oximes in non-aqueous solvents; the excess of reagent is titrated with perchloric acid[14-16]. Acetals do not interfere, and organic acids do not react under these conditions. This method is used for determination of *aldehydes* (especially *formaldehyde),* and *ketones. Diacetyl* reacts with hydroxylamine to form dimethylglyoxime, which can be estimated gravimetrically as the nickel salt.

Oximes can be estimated colorimetrically after conversion into nitroso compounds[16a].

If the condensation is carried out in non-aqueous medium (e.g. in methanol), the water produced can be titrated by the Karl Fischer method (see Section 3.7:1).

Hydrazine, asymm. dimethylhydrazine: These amines are also used for determination of carbonyl groups, but less frequently than hydroxylamine. The hydrazones formed often have low solubility and may be determined gravimetrically. After condensation of carbonyl compounds with hydrazine sulfate, the hydrazone is filtered off and the sulfuric acid in the filtrate is titrated alkalimetrically[17, 18].

Aromatic aldehydes can be estimated in the presence of ketones by using asymm. dimethylhydrazine[18].

Phenylhydrazine,   2,4-dinitrophenylhydrazine: These reagents (especially dinitrophenylhydrazine) react with carbonyl compounds, usually to give insoluble precipitates, and are, therefore, useful for gravimetric analysis of *aldehydes* and *ketones*[19]. After filtering off the precipitate, the excess reagent is titrated oxidimetrically, using thallium (III), lead (IV) acetate, periodate iodine or chloroamine T[20-22]. Direct titration of *aldehydes* with

[10] *S. Belman,* Anal. Chim. Acta *29,* 120 (1963).

[10a] *J.S. Powell, K.C. Edson, E.L. Fisher,* Anal. Chem. *20,* 213 (1948).

[11] *A. Patchornik, S.E. Rogozinski,* Anal. Chem. *31,* 985 (1959).

[12] *D.M. Smith, J. Mitchell Jr.,* Anal. Chem. *22,* 750 (1950).

[13] *L.D. Metcalfe, A.A. Schmitz,* Anal. Chem. *27,* 138 (1955).

[14] *M. Pesez,* Bull. Soc. Chim. Fr. 417 (1957).

[15] *T. Higuchi, C.H. Barnstein,* Anal. Chem. *28,* 1022 (1956).

[16] *J.S. Fritz, S.S. Yamamura, E.C. Bradford,* Anal. Chem. *31,* 260 (1959).

[16a] *O. Wichterle, M. Hudlicky,* Ref. Z. Anal. Chem. *138,* 145 (1953).

[17] *L. Fuchs,* Scientia Pharm. *16,* 50 (1948).

[18] *S. Siggia, C.R. Stahl,* Anal. Chem. *27,* 1975 (1955).

[19] *T.S. Ma, J. Logun, P.P. Mazzella,* Microchem. J. *1,* 67 (1957).

[20] *D.J. Barke, E.R. Cole,* J. Appl. Chem. *5,* 477 (1955).

[21] *A. Berka, J. Zýka,* Chem. Listy *50,* 314 (1956).
      A. Berka, Z. Anal. Chem. *193,* 276 (1963).
      A. Berka, J. Zýka, Z. Anal. Chem. *169,* 40 (1959).

[22] *H. Lieb, W. Schöniger, E. Schivizhoffen,* Mikrochim. Acta 35 (1950).

phenylhydrazine, using G-Orange as indicator, is a very simple and rapid method[23]. Phenylhydrazine is used also for the separation of *sugars*, since monosaccharides are usually converted into osazones of low solubility.

Other amines: Cyclohexylamine has been recommended as a reagent for the specific determination of *glyoxal*[24] in the presence of acetaldehyde, since, under suitable conditions, only glyoxal reacts to form an insoluble product.

Aromatic and aliphatic *aldehydes* condense with aniline[25] or with o-toluidine[26]. The water formed can be titrated by the Karl Fischer method[26].

An important method for determination of *phenols, naphthols* and *aromatic dihydroxy-carboxylic acids* involves reaction with 4-aminoantipyrine in the presence of an oxidant. Intensely colored indophenols are formed:

This procedure can be used both for normal analysis and for detection of trace amounts[27].

Aliphatic *carboxylic acids* react with o-phenylenediamine, forming benzimidazoles, which can be titrated with perchloric acid[28]. *Acid anhydrides* react with primary aromatic amines such as p-chloroaniline or m-nitroaniline[29], yielding amides. The excess of amine can be determined acidimetrically[30, 31], oxidimetrically[32] or by diazotization[33].

2.5 Condensation with Carbonyl Groups

This reaction is used mainly for the estimation of amino groups; active methylene groups can be reacted quantitatively also.

2.5.1 Determination of Amino Groups

Primary aliphatic and aromatic *amines* react with aldehydes, producing water and azomethines (Schiff's bases):

$$R^1{-}NH_2 \; + \; R^2{-}\overset{\displaystyle O}{\underset{\displaystyle H}{C}} \; \xrightarrow{-H_2O} \; R^1{-}N{=}CH{-}R^2$$

Gravimetric, titrimetric[34, 35] or colorimetric[36, 37] techniques can be used.

Sulfonamides react readily with aromatic aldehydes in acetic acid solution forming intensely colored condensation products. When an acetic acid solution of *hydrazine*, monomethylhydrazine or 1, 1-dimethylenhydrazine is treated with salicylaldehyde in the presence of perchloric acid, only the unsubstituted hydrazine reacts[38]. The perchloric acid consumed in titration is a measure of the monomethylhydrazine or 1, 1-dimethylhydrazine present, since these do not condense under the conditions given.

2.5.2 Determination of Active Methylene Groups

The reactions with aldehydes and ketones which were described in section 3.4:2.2 may conversely be used for analysis of active *methylene groups*. After achieved condensation, the excess aldehyde is determined[39, 40].

2.6 Condensation with Carboxyl Groups (Acylation)

For acylation of amines and alcohols, acid anhydrides and acid halides are mainly used. The amounts of reagent consumed are determined by titration with sodium alkoxide solution (Section 3.4:2.3) or by condensation with morpholine[41, 42].

[23] R. Meyer, Z. Anal. Chem. *140*, 184 (1953).
[24] R. Simionovici, C. Titei, N. Budisteanu, F.M. Albert, Z. Anal. Chem. *240*, 386 (1968).
[25] A.B. Skvortsova, L.N. Petrova, E.N. Novikowa, Zh. Anal. Khim. *17*, 896 (1962); C.A. *58*, 9633b (1963).
[26] L.N. Petrova, E.N. Novikova, A.B. Skvortsova, Zh. Anal. Khim. *14*, 347 (1959); C.A. *54*, 8469i (1960).
[27] G. Wagner, R. Flotow, Pharm. Zentralh. 101 (1962).
[28] Z. Stransky, V. Stuzka, E. Rucicka, Mikrochim. Acta 77 (1966).
[29] P. Sorensen, Anal. Chem. *28*, 1318 (1956).
[30] S. Siggia, J.G. Hanna, Anal. Chem. *23*, 1717 (1951).
[31] T. Ellerington, J.J. Nichols, Analyst (London) *82*, 233 (1957).
[32] H. Roth, Mikrochim. Acta 767 (1958).
[33] W.S. Calcott, F.L. English, O.C. Wilbur, Ind. Eng. Chem. *17*, 942 (1925).

[34] W. Hawkins, D.M. Smith, J. Mitchell, J. Am. Chem. Soc. *66*, 1662 (1944).
[35] J.B. Johnson, G.L. Funk, Anal. Chem. *28*, 1977 (1956).
[36] A.J. Milun, Anal. Chem. *29*, 1502 (1957).
[37] F.E. Critchfield, J.B. Johnson, Anal. Chem. *28*, 436 (1956).
[38] N.M. Serencha, J.G. Hanna, E.J. Kuchar, Anal. Chem. *37*, 1116 (1965).
[39] K. Uhrig, E. Lynch, H.C. Becker, Ind. Eng. Chem. Anal. *18*, 550 (1946).
[40] J.S. Powell, K.C. Edson, E.L. Fisher, Anal. Chem. *20*, 213 (1948).
[41] J.B. Johnson, G.L. Funk, Anal. Chem. *27*, 1464 (1955).
[42] Cs. Ömböly, E. Derzsi, Z. Anal. Chem. *183*, 272 (1961).

The excess morpholine is determined by titration with perchloric acid. Carboxylic acids do not interfere; neither do hydrogen halides if mercury acetate is present.

2.6.1 Amines

Acetic, phthalic, succinic and pyromellitic anhydrides are used mainly for the analysis of amines. After reaction of the acylating agent with the amine, either the acid liberated is titrated, or the excess anhydride is hydrolyzed and the sum of the liberated and hydrolyzed acid is determined. Alternatively, hydrolysis may be carried out with a known quantity of water and the excess water determined by the Karl Fischer method (cf. Section 3.7:1). In certain cases, the acid amide is formed, then separated and determined spectrophotometrically[43, 44].

Primary and *secondary* amines are acylated with acetic anhydride, sometimes in the presence of pyridine. Aromatic amines react more rapidly than aliphatic amines; primary amines are more reactive than secondary amines.

Succinic[45] and pyromellitic anhydrides[46, 47] are superior to acetic and phthalic anhydrides in that they are more reactive and less subject to interference by carbonyl compounds. Pyridine, tetrahydrofuran and dimethyl sulfoxide are used as solvents.

Hydrazines react in the same manner as amines[48].

2.6.2 Alcohols

As with the acylation of amines, the excess reagent is estimated, or the reaction products such as ester, acids or water are determined[51].

Pyridine, perchloric acid[49] and boron trifluoride[50] are used as catalysts, and dioxane, chloroform, ethyl acetate, 1,2-dichloroethane and pyridine are employed as solvents. Another procedure is to esterify the hydroxyl groups of *alcohols* and *phenols* with stearic anhydride in boiling m-xylene[52].

A method of hydroxyl determination for primary alcohols of low solubility has been described; here 3-nitrophthalic anhydride is used, with dimethylformamide as solvent. The excess reagent is back-titrated with tetrabutylammonium hydroxide solution[53].

Aliphatic alcohols can be distinguished from phenols with pyromellitic dianhydride.

2.7 Acylation with Sulfonic Acid Chlorides

Alcohols also react with aromatic sulfonyl chlorides, especially *p*-toluenesulfonyl chloride or *m*-nitrobenzenesulfonyl chloride in the presence of pyridine. Primary and secondary *alcohols* and *phenols*[54] can be determined by this method, using a procedure similar to that described for carboxylic acid halides.

2.8 Diazotization and Reaction with Nitrite Ion

Amines, hydrazines, phenols, pyrazolones and other nucleophilic compounds react with alkali nitrite in acidic solution.

Nitrosation of the amino group is followed by secondary reactions:

$$R-NH_2 \xrightarrow[-H_2O]{+NO_2^{\ominus} \; +H^{\oplus}} R-NH-NO$$

$$\xrightarrow[-H_2O]{+H^{\oplus}} R-\overset{\oplus}{N}\equiv N$$

Aliphatic primary amines react with nitrous acid at room temperature to liberate nitrogen; aromatic primary amines generally form diazonium ions in the cold. Secondary amines form nitrosamines. Two different types of analytical procedure are available: a gasometric procedure (see Section 3.7:2.2) and the frequently used titrimetric method which involves measurement of the alkali nitrite consumed.

An aqueous solution of sodium nitrite is generally used as reagent; 0.1 M solutions are stable for several weeks, but very dilute reagents oxidize quickly[58]. Alcoholic solutions of ethyl or butyl nitrite[55] are sometimes used for titration. The titrations are carried out in aqueous hydrochloric acid or hydrobromic acid: the reaction rate is greater with the latter[56, 57].

[43] *G. Neurath, E. Doerk,* Chem. Ber. *97*, 172 (1964).
[44] *Wen-Hai Hong, K.A. Connors,* Anal. Chem. *40*, 1273 (1968).
[45] *C.K. Narang, N.K. Mathur,* Indian J. Chem. *3*, 182 (1965).
[46] *R. Harper, S. Siggia, J.G. Hanna,* Anal. Chem. *37*, 600 (1965).
[47] *S. Siggia, J.G. Hanna, R. Culmo,* Anal. Chem. *33*, 900 (1961).
[48] *H.E. Malone, R.A. Biggers,* Anal. Chem. *36*, 1037 (1964).
[49] *G.H. Schenk, J.S. Fritz,* Anal. Chem. *32*, 987 (1960).
[50] *W.M. Bryant, J. Mitchell Jr., D.M. Smith,* J. Am. Chem. Soc. *62*, 1 (1940).
[51] *N.K. Mathur,* Talanta *13*, 1601 (1966).
[52] *B.D. Sully,* Analyst (London) *87*, 940 (1962).

[53] *J.A. Florio, I.W. Dobratz, J.H. McClure,* Anal. Chem. *36*, 2053 (1964).
[54] *P. Mesnard, B. Gibirila, M. Bertucat,* Chim. Anal. (Paris) *45*, 491 (1963).
[55] *N.K. Mathur, S.P. Rao, D. Narain,* Anal. Chim. Acta *23*, 312 (1960).
[56] *J.H. Ridd,* Quart. Rev. Chem. Soc. *15*, 418 (1961).
[57] *H. Zollinger,* Chemie der Azofarbstoffe, p. 30, Birkhäuser Verlag, Basel, Stuttgart 1958.
[58] *A. Berka, J. Vulterin, J. Zýka*[80], p. 185.

Electrometric procedures are increasingly superseding visual end-point determination[59-62]. For direct titrations, electrometric end-point determination is especially advantageous, because it avoids the inconvenient use of an external indicator (potassium iodide/starch paper). If potassium iodide/starch paper is used as indicator, it is preferable to perform the diazotization or nitrozation with a small excess of nitrite reagent; the nitrous acid is then back-titrated with a reactive amine[63] such as *p*-nitroaniline.

When titrating amines which react slowly, the temperature of the solutions used should be kept at 5° or below, to avoid losses of nitrite. When end-points are measured electrometrically, the titrations can often be carried out at room temperature, and temperatures of 50° are sometimes permissible; in this last case, the tip of the burette must dip into the solution to be titrated, in order to avoid losses of nitrite.

The choice of solvents is restricted because of the reactivity of nitrous acid. Commonly used solvents are: glacial acetic acid and dimethylformamide; dispersing agents such as naphthalene-α-sulfonic acid are employed. Alcohols cannot be used as solvents.

When titrating *aromatic amines*[64-66], their different reaction rates[57, 67] must be taken into consideration. Besides aromatic amines, *hydrazine derivatives*[58] *(phenylhydrazine 2,4-dinitrophenylhydrazine, semicarbazide,* and *thiosemicarbazide), sulfonylhydrazides* $(R-SO_2-NH-NH_2)$[68], *resorcinol,*[69] *3-methyl-1-phenyl-pyrazolone* and many of its derivatives[58, 66] can be determined by nitrozation.

2.9 Coupling Reactions

Aryldiazonium ions react with activated aromatic compounds *(aromatic amines* and *phenols,* particularly naphthalene derivatives) as well as with *pyrazolones* and *acetoacetic acid* derivatives (esters and anilides).

The most reactive species in these cases are the free amine, the phenolate ion and the enolates of the pyrazolone and acetoacetic derivatives[70].

Although exact conversion into the monoazo dye is usual, and the aryldiazonium ion causes hardly any oxidation, the coupling method is seldom employed. This is because the course of many azo coupling[70] reactions is still obscure, diazonium ions are unstable towards heat and light[71, 72], and the method of end-point determination (detection of the diazonium and coupling components on filter paper) is cumbersome. The new electrometric procedures[73-75] for end-point determination have so far not been commonly used.

Moreover, this method requires a good knowledge of azocoupling. In carrying out a quantitative coupling reaction, it is desirable to find (pH) conditions suitable for direct titration. Indirect procedures are prone to difficulties, because of the instability of diazonium ions[76] at high pH values. The estimation techniques for coupling reactions[71, 72, 81], therefore, vary from case to case. A few diazonium salts (*p*-nitro-, *m*-nitro-, and *p*-chlorophenyldiazonium salts) suffice for a large number of different titrations. Numerous studies have been made of the reactivity of substituted aryldiazonium ions, and the stability of their acid salts in solutions[70].

[59] *H.G. Scholten, K.G. Stone,* Anal. Chem. *24,* 749 (1952).
[60] *L.T. Butt, H.E. Stagg,* Anal. Chim. Acta *19,* 208 (1958).
[61] *W. Büchler,* Z. Anal. Chem. *186,* 154 (1962).
[62] *J.T. Stock,* Amperometric Titrations, p. 652, Interscience Publishers, New York, London, Sydney 1965.
[63] *R. Goupil,* Chim. Anal. (Paris) *42,* 300 (1960).
[64] *S. Siggia*[79], p. 446.
[65] *Houben-Weyl*[77], Band II, p. 676.
[66] *M. Matrka,* Chemie Prag *10,* 635 (1958).
[67] *K.H. Saunders,* The Aromatic Diazo-Compounds, 2nd Ed., p. 1, Edward Arnold & Co., London 1949.
[68] *O.P. Shvaika, T.R. Mnatsakanova, D.M. Aleksandrova,* Ž h. Anal. Khim. *20,* 273 (1965).
[69] *V. Popovici, A. Schweiger,* Z. Anal. Chem. *244,* 44 (1969).

[70] *H. Zollinger,* Chemie der Azofarbstoffe, p. 73, 111, 131, 157, Birkhäuser Verlag, Basel, Stuttgart 1958.
[71] *Houben-Weyl*[77], Band II, p. 663.
[72] *S. Siggia*[79], p. 55.
[73] *R.M. Elofson, P.A. Mecherly,* Anal. Chem. *21,* 565 (1949).
[74] *W. Büchler,* Helv. Chim. Acta *47,* 639 (1964).
[75] *M. Matrka, E. Verîsová,* Chem. Prumyse *16,* 660 (1966).
[76] *M. Matrka* et al., Collect. Czech. Chem. Commun. *32,* 1462 (1967).

2.10 Bibliography

[77] *Houben-Weyl,* Methoden der organischen Chemie, 4. Aufl., Bd. II, p. 676, Georg Thieme Verlag, Stuttgart 1953.
[78] *J. Mitchell,* Organic Analysis, Vol. III, p. 103, Interscience Publishers, New York 1956.
[79] *S. Siggia,* Quantitative Organic Analysis via Functional Groups, 3rd Ed., p. 446, John Wiley & Sons, New York, London 1963.
[80] *A. Berka, J. Vulterin, J. Zýka,* Newer Redox Titrants, p. 185, Pergamon Press, Oxford 1965.
[81] *M. Matrka, A. Spevák,* Chem. Listy *61,* 883 (1967).

3 Complexes, Precipitates of Low Solubility, Chelates and Molecular Compounds

3.1 Complexes

Reactions in which complexes are formed can be utilized for analytical purposes if colored complexes, products of low solubility, or soluble chelates and molecular compounds capable of being estimated by titration are produced.

3.1.1 Colored Complexes

The ferric chelate complexes of hydroxamic acids are often employed for photometric determinations:

$$R^1-\underset{\underset{O}{\|}}{C}-OR^2 \;+\; NH_2OH \longrightarrow$$

$$R^1-\underset{\underset{O}{\|}}{C}-NHOH \;+\; R^2OH$$

$$3\,R-\underset{\underset{O}{\|}}{C}-NHOH \;+\; Fe^{3\oplus} \longrightarrow$$

$$\left[(R-\underset{\underset{O}{\|}}{C}-NHO)_3Fe \right] \;+\; 3\,H^{\oplus}$$

Amidoximes react analogously. The reaction can be applied to the determination of *carboxylic esters, lactams, acid anhydrides, acid halides, amides, imides* and *nitriles*[1-5]. Anhydrides and lactams can be estimated in the presence of esters[2]. *Ketenes* can be converted into esters, and then estimated[6]. The ferric complexes of hydroxamic acids are normally stable for 10 to 20 hours (standard deviation ca. 2%).

Alcohols (e.g. in blood), *thiols* and *amines* react with vanadium 8-hydroxyquinolinate[7] to form a colored complex, which can be used for estimation.

Potassium tetrathiocyanatocobaltate $K_2[Co(CNS)_4]$ has been recommended as a reagent for the analysis of primary, secondary and tertiary *amines, quaternary ammonium salts, nitrogen-containing heterocyclic compounds*[8], *polyethylene glycols* and *ethylene oxide adducts*[9, 10]. The colored complexes formed are extracted into organic solvents and measured spectrophotometrically.

The dyes bromophenol blue and eosin are used for determining *quaternary ammonium salts* and non-ionic surface-active compounds[11]. Methylene blue, basic fuchsin, azure A, methyl green and a bromocresolpurpuroprotein complex can be used for the estimation of anionic surfactants[12]. Dithizone forms colored complexes with *quaternary ammonium compounds*[13] in alkaline solution. The reaction products are extracted into organic solvents and determined spectrophotometrically.

2,2-Diphenyl-1-picrylhydrazyl, a stable free radical, reacts with *amines;* the reaction rate depends considerably on the type of compound. With suitable kinetic methods, analysis of amine mixtures is possible[14].

Electron donor-acceptor complexes have achieved some importance. Tetracyanoethylene forms colored complexes with electron donors and can be used for determining *aromatic ring systems* and *1,3-dienes*[15-17]. Other electron acceptors have been used for chromogenic detection of aromatic amines and phenols[17a].

Tetra-aminoethylenes are powerful electron donors, and form colored complexes with electron acceptors. They are of potential importance for the estimation of aromatic *nitro compounds, acrylonitrile, chloranil* and *cyanoethylenes*[18].

Certain *steroid hormones* form electron donor-acceptor complexes in anhydrous solvents with aluminum chloride and aluminum bromide. The reaction products are intensely colored, and some are strongly fluorescent[19-20]. This reaction can be employed for selective determinations in steroid mixtures.

[1] E. Bayer, K.H. Reuther, Chem. Ber. 89, 2541 (1956).
[2] R.F. Goddu, N.F. Le Blanc, Anal. Chem. 27, 1251 (1955).
[3] W. Pilz, Z. Anal. Chem. 193, 338 (1963).
[4] E.A. McComb, R.M. McCready, Anal. Chem. 29, 819 (1957).
[5] S. Soloway, A. Lipschitz, Anal. Chem. 24, 898 (1952).
[6] W.M. Diggle, J.C. Gage, Analyst (London) 78, 473 (1953).
[7] D. Eskes, Mikrochim. Acta 1065 (1965).
[8] J. Bosly, J. Pharm. Belg. 18, 162 (1963).
[9] B.M. Milwidsky, Analyst (London) 94, 377 (1969).
[10] M. Ziegler, Z. Anal. Chem. 171, 111 (1959).

[11] L.D. Metcalfe, Anal. Chem. 32, 70 (1960).
    K. Peter, Riechstoffe, Aromen 61, 11 (1959).
[12] D. Hummel, Tenside 1, 116 (1964).
    W.A.M. den Toukelaar, G. Bergshoeff, Water Res. 3, 31 (1969).
[13] H. Deppeler, A. Becker, Z. Anal. Chem. 199, 414 (1964).
[14] G.J. Papariello, M.A.M. Janish, Anal. Chem. 37, 899 (1965).
[15] R.E. Merrifield, W.D. Phillips, J. Am. Chem. Soc. 80, 2778 (1958).
[16] G.H. Schenk, M. Ozolins, Talanta 8, 109 (1961).
[17] M. Ozolins, G.H. Schenk, Anal. Chem. 33, 1035 (1961).
[17a] O. Hutzinger, Anal. Chem. 41, 1662 (1969).
    H.T. Gordon, M.J. Huraux, Anal. Chem. 31, 302 (1959).
[18] N. Wiberg, Angew. Chem. 80, 809 (1968).
[19] A. Becker, F. Ehinger, Z. Anal. Chem. 198, 162 (1963).
[20] J. Meier, A. Becker, Z. Anal. Chem. 204, 427 (1964).

3.1.2 Compounds of Low Solubility

Protonated bases form precipitates of low solubility with Reinecke salt.

$$BH^{\oplus} + \left[Cr(NH_3)_2(SCN)_4 \right]^{\ominus} \longrightarrow$$

$$BH\left[Cr(NH_3)_2(SCN)_4 \right]$$

Formation of the complexes depends on the basicity of the compounds; under suitable pH conditions, certain bases can be separated quantitatively[21] from one another.

Ammonium reineckate can be used for the determination of *amines, alkaloids*[21-24], *heterocyclic compounds containing nitrogen,* and *cationic surfactants*[12]. As an alternative to a gravimetric method, the reineckates may be decomposed with sodium hydroxide, and the liberated thiocyanate titrated[22].

The following reagents have uses similar to those of Reinecke salt: picric acid, picrolonic acid, perchloric acid, oxalic acid[23], potassium tetraiodobismuthate[26], silico-tungstic acid[25], and phosphomolybdic acid.

Heteropolyacids also form products of low solubility with *polyethylene glycols, ethylene oxide adducts*[27] and basic *triphenylmethane dyes*[29].

Sodium tetraphenylborate[28] can be used for the determination of *aromatic amines, quaternary ammonium salts, heterocyclic compounds containing nitrogen, polyethylene glycols, ethylene oxide adducts* and *epoxides*.

Surface-active substances (ionic surfactants) are estimated with *p*-toluidine hydrochloride, sodium lauryl sulfate, benzidine hydrochloride and hydroamine[12, 30-32]. A complexometric method for the determination of *tertiary amines, quaternary ammonium bases* and their salts is based on a precipitation reaction with a solution of bismuth ethylene diaminotetra-acetate and potassium iodide, followed by titration of the liberated ethylene diaminotetra-acetic acid with a thorium(IV) salt[33]. Picric acid produces a stable intensely colored complex with cationic surfactants[30a]. Polyoxyethylene compounds (nonionic surfactants) form blue water-insoluble complexes[32b] with cobalt thiocyanate.

3.1.3 Chelates and Molecular Compounds

Copper(II) can be used for estimation of *sulfonamides*[34], *glycerol*[35] and *thioglycolic acid*[36] by titration.

Thiourea, thiosemicarbazide, thiosemicarbazones[37] and *thiols*[38] can be titrated directly with mercury nitrate or perchlorate.

Nickel(II) salts react with *cyanohydrins;* the excess reagent is titrated with ethylene diamine tetraacetate (EDTA)[39].

Silver nitrate in pyridine reacts with saturated and unsaturated *thiols*, generating an equivalent amount of pyridinium nitrate, which can be titrated alkalimetrically[40]. Silver nitrate reacts similarly with acetylene and its derivatives; the liberated nitric acid is titrated[41-42]. Uranyl and cupric ions form chelate compounds with acetoacetamides[42a]. Aluminum alkyls[44] produce green solutions in oxygen-free toluene containing methyl violet indicator. The colored complex can be titrated with bases such as pyridine.

[21] *L. Kum-Tatt,* Nature *182,* 655 (1958).
L. Kum-Tatt, Ch. G.G. Farmilio, Nature *180,* 1288 (1957).
[22] *F. Kuffner, S. Sattler-Dornbacher,* Monatsh. Chem. *93,* 99 (1962).
[23] *Houben-Weyl,* Methoden der organischen Chemie, Bd. II, p. 656, Georg Thieme Verlag, Stuttgart 1953.
[24] *W. Poethke, P. Gebert, E. Müller,* Pharm. Zentralh. *98,* 389 (1959).
[25] *E. Graf, E. Fiedler,* Naturwissenschaften *39,* 556 (1952).
[26] *W. Poethke, H. Trabert,* Pharm. Zentralh. *94,* 219 (1955).
[27] *D. Hummel,* Tenside *1,* 116 (1964).
[28] *H. Boden,* Chemist-Analyst *52,* 112–113 (1963).
R. Neu, Fette, Seifen, Anstrichm. *59,* 823 (1957).
[29] *A.K. Babko, Yu. F. Shkaravsky, E.M. Ivashkovich,* Ukr. Khim. Zh. *33,* 951 (1967).
Ref. Z. Anal. Chem. *244,* 330 (1969).
[30] *J.C.H. Hwa,* Anal. Chem. *37,* 619 (1965).
[31] *A.M. Uppertown,* Chem. Ind. (London) 162 (1964).
[32a] *J. Sheiham, T.A. Pinfold,* Analyst (London) *94,* 387 (1969).

[32b] *M. Milwidsky,* Analyst (London) *94,* 377 (1969).
[33] *B. Budesinsky, J. Körbl,* Chem. Listy *52,* 1513 (1958), Anal. Abstr. *7,* 3972 (1960).
[34] *H. Abdine, W.S. Abdel Sayed,* J. Pharm. Pharmacol. *14,* 761 (1962).
[35] *J.T. McAloren, G.F. Reynolds,* Anal. Chim. Acta *32,* 170 (1965).
[36] *S.B. Sant, B.R. Sant,* Anal. Chem. *31,* 1879 (1959).
[37] *N.V. Koshkin,* Zh. Anal. Khim. *18,* 1492 (1963).
[38] *J.S. Fritz, T.A. Palmer,* Anal. Chem. *33,* 98 (1961).
[39] *C. Berther, K. Kreis, O. Buchmann,* Z. Anal. Chem. *169,* 184 (1959).
[40] *B. Saville,* Analyst (London) *86,* 29 (1961).
[41] *J. Vitovec, M. Sadek,* Collect. Czech. Chem. Commun. *25,* 1972 (1960); Anal. Abstr. *8,* 1066 (1961).
[42] *S. Prévost, W. Chodkiewicz, P. Cadiot, A. Willemart,* Bull. Soc. Chim. Fr. 1742 (1960).
[42a] *A. Kettrup, H. Specker,* Z. Anal. Chem. *246,* 108 (1969).
[43] *E. Kurz, G. Kober,* Analyst (London) *92,* 391 (1967).
[44] *G.A. Razuvaev, A.J. Graevski,* Anal. Chem. *34,* 333 R (1962); Dokl. Akad. Nauk. SSSR. *128,* 309 (1959); C.A. 54, 6407 i (1960).

Methylene blue can be used as a reagent[43] for rapid and simple titration of *o-polynitrophenols*, especially picrates and picrolonates.

Similarly tensides (Chapter 2.11) are determined[27].

3.2 Salts with Low Solubility

Metal ions give almost insoluble or colored salts with various organic acids; the products can be determined gravimetrically or colorimetrically.

Magnesium, calcium, barium, lead, mercury and silver are most commonly used for precipitation purposes; after precipitation, the excess metal ion can be determined by back-titration, or measured colorimetrically.

Phthalic acid[45] is precipitated with lead acetate, *tartaric acid*[46] with lead(II) nitrate, and *urea* with mercuric ions.

Compounds containing active hydrogen form salts of low solubility with silver or copper(I) ions. *Vinylacetylene*[47] can be quantitatively precipitated with ammoniacal silver nitrate, and the excess silver back-titrated by the Volhard method.

2-Naphthylamine[48] can be determined in the presence of 1-naphthylamine by precipitation with cadmium or zinc sulfate. The excess reagent can be titrated with EDTA, without removing the precipitate.

3.3 Clathrates (Inclusion Compounds)

Clathrates[50, 51] (particularly those of urea) can be used for the analysis of mixtures. Separation depends on the geometry of the molecules concerned; it is possible to separate aliphatic substances with unbranched chains from those having branched chains, or from cyclic products. Quite good selectivity can be attained through formation or decomposition of adducts. In certain cases, separation of optical antipodes has been achieved[49].

[45] *Z. Gregorowicz, J. Ciba*, Mikrochem. Ichnoanalyt. Acta 733 (1965); Anal. Abstr. *13*, 4218 (1966).

[46] *I. Sarudi*, Z. Anal. Chem. *194*, 195 (1963).

[47] *J. Vitovec, M. Sadek*, Collect, Czech. Chem. Commun. *25*, 1972 (1960); Anal. Abstr. *8*, 1066 (1961).

[48] *F.M. Albert, M. Butuceanu, M. Cupfer*, Rev. Roumaine Chim. *9*, 835 (1964) (German), C.A. 63, 14046 d (1965).

[49] *W. Schlenk* in Ullmanns Encyklopädie der technischen Chemie, Bd. VI, p. 253, Urban & Schwarzenberg, München 1955.

[50] *V.M. Bhatnagar*, J. Chem. Educ. *40*, 646 (1963).

[51] *J.R. Marquart, G.B. Dellow, E.R. Freitas*, Anal. Chem. *40*, 1633 (1968).

3.4 Bibliography
c.f. Section 3.4:1.9 and 3.4:2.10.

[52] *D. Hummel*, Analyse der Tenside, Carl Hauser-Verlag, München 1962.

3.5 Hydrolysis

Willy Büchler and *Fritz Christen*
CIBA-GEIGY, CH-4002 Basel

Probably most hydrolyses in organic analysis are carried out in alkaline media (section 3.5:3). However, there are cases in which water is the reagent used (3.5:1) or where acid hydrolysis is preferred (3.5:2). Amines are also suitable, especially for the splitting off of halogens (section 3.5:3.2)

1 Hydrolytic Fission with Water

Water alone can be used for hydrolysis only in the case of easily hydrolyzed substances such as lower aliphatic *carboxylic acid* and *sulfonic acid chlorides*[1, 2].

The analysis is carried out by allowing the weighed substance in a fused ampoule to react with water in a closed flask and titrating the acid hydrolysis products alkalimetrically after absorption in water. The endpoint is determined electrometrically or by means of an indicator (section 3.6:1). Halogens which are split off may also be determined by a silver nitrate titration.

Hydrolysis with water followed by a Karl Fischer titration (section 3.7:1) can be used in the case of anhydrides. The anhydride is mixed with a measured excess of water and, after the reaction, the excess water is backtitrated[3-5].

2 Acid Hydrolysis

For acid hydrolysis of organic compounds (*esters, acid amides, nitriles*) hydrochloric and sulfuric

[1] Ref.[22], p. 232.

[2] Ref.[26], p. 900.

[3] Ref.[26], p. 901.

[4] *D.M. Smith*, J. Am. Chem. Soc. *63*, 1700 (1941).

[5] Ref.[22], p. 511.

acids are used predominantly and occasionally phosphoric acid[6]. With phosphoric acid certain reactions can be carried out more quickly than with sulfuric acid.

Solvents used are alcohol, benzene or dioxane and, less frequently, water. The analysis is carried out using dilute acid and determining the amount of acid used alkalimetrically, after the reaction. In case the acid is in too great an excess either the products are separated by extraction and determined separately or the estimation is done in some other suitable way[7]. In the case of the splitting up of *proteins* into amino acids[8] the hydrolysis is performed mainly with 10–20% hydrochloric acid or up to 70% sulfuric acid and the products determined by chromatographic separation of the amino acids.

3 Alkaline Hydrolysis

The greatest number of hydrolyses of organic compounds *(esters, acid amides, nitriles, alkyl-halides, carboxylic acid halides and sulphonic acid halides*[9, 10]*)* are carried out in an alkaline medium. The methods differ in the use of solvents, reaction time, reagents and catalysts.

3.1 With Alkali Hydroxides

The solvents used vary according to the solubility; in the case of water soluble substances aqueous alkali may be used; alcoholic solutions are, however, preferred.

As in the standard method for the determination of the *saponification number* of *fats, waxes* and *oils,* these are saponified with ethanolic potassium hydroxide[11, 12].

For substances which are difficult to hydrolyze, higher boiling solvents are used such as amyl alcohol[13] or diethylene glycol[14]. Substances which are insoluble in alkaline solution are brought into solution by addition of benzene[13].

Alkaline hydrolyses are performed mainly by excess titration. The excess alkali is at least half of the hydroxide used, and in the case of low concentrations (0.1 N) even two thirds. 25 or 50 mls of approx. 0.5 N ethanolic KOH are boiled with 1–2 g of the substance for 1/2 to 1 hour.

Acid chlorides and *acid anhydrides* may, under certain conditions, be titrated directly[15] (see section 3.5:4).

To perform the saponification a stock solution of the test substance in a particular solvent is prepared and an aliquot of this stock solution is allowed to react with a measured quantity of the alkali hydroxide solution either in the cold or at higher temperatures; often at the boiling point. After completion of the reaction, the excess hydroxide solution is backtitrated with standard acid.

The choice of reaction conditions depends in great part on the ease with which the organic halogen can be split off. Easily removeable halogens such as in *cyanogen chloride,* in *dichlorohydrin* compounds, or in 1, 2 or 1, 4 *chloronitrobenzenes* may be readily removed quantitatively by mild hydrolysis at 20–50° and with short reaction times. Solvents used here are methanol, acetone, and other low-boiling liquids. More strongly bound halogen can, in many cases, also be estimated by hydrolysis. The reaction temperature chosen must be as high as possible and the reaction time correspondingly long. Suitable solvents are diethylene glycol (b.p. 175°) and dimethylsulfoxide (b.p. 183°).

The titration after the alkaline saponification is performed mainly with standard hydrochloric or sulfuric acid. The endpoint can be followed by the use of color indicators, mainly phenolphthalein. Potentiometric titration is recommended since the endpoint is more exact and, especially in the case of colored solutions, determined with greater certainty (see section 3.6:1). If, during the hydrolysis, halogen is split off, titration with silver solutions is preferable to acidimetric titration.

The hydrolyzability of *esters* varies greatly so the solvent must be chosen carefully[16]. Esters and halogen compounds which are difficult to saponify may be hydrolyzed far more quickly in a solution of potassium hydroxide in dimethylsulfoxide than in ethanol or other solvents[17, 18]. Substances which are difficult to saponify are hydrolyzed only when the reaction is carried out under pressure[19].

Generally *acid amides* saponify less quickly than esters. They must be treated under more severe

[6] *M.R.F. Ashworth*[26], p. 420, 767, 827.
[7] *D.E. Johnson, H.B. Nunn, S. Bruckenstein,* Anal. Chem. *40,* 368 (1968).
[8] *R.J. Block, D. Bolling,* The Amino Acid Composition of Proteins and Foods, 2. Ed., Charles L. Thomas, Publisher, Springfield, Illinois 1951.
[9] Ref[26], p. 42.
[10] Ref[25], p. 174.
[11] DGF-Einheitsmethoden (Deutsche Einheitsmethoden zur Untersuchung von Fetten, Fettprodukten und verwandten Stoffen) C-V 3 (53), Wissenschaftliche Verlagsgesellschaft, Stuttgart 1950–1961.
[12] ASTM Standards 1955, D 94-55 T, p. 728, Vol. 6, D 555-54, p. 234, Vol. 4, D 939-54, p. 404, Vol. 5, American Society for Testing Materials, 1916 Race St., Philadelphia 3, Pa.
[13] Ref[24], p. 139.
[14] *C.P.A. Kappellmeier,* Chemical Analysis of Resin-Based Coating Materials, p. 72, Interscience Publishers, New York 1959.

[15] Ref[22], p. 510.
[16] Ref[24], p. 138.
[17] *W.H. Greive, K.F. Sporek, M.K. Stinson,* Anal. Chem. *38,* 1264 (1966).
[18] *J.A. Vinson, J.S. Fritz, C.A. Kingsbury,* Talanta *13,* 1673 (1966).
[19] Ref[23], p. 40.

conditions such as higher temperatures and longer reaction times. Even then, secondary amides are hydrolyzed only with difficulty. The titration is usually not performed in the reaction solvent but the hydrolyzed amide product (an amine) is determined after separation by distillation[16].

3.2 With Amines
Certain amines, in particular pyridine, piperdine, and morpholine, are used in the hydrolysis of organic compounds. The field of application covers predominantly *acid anhydrides* and *halogen* compounds.

3.2.1 Pyridine
The reaction may be performed directly or indirectly. For the indirect titration the test solution, in aqueous or alcoholic solution, is mixed with about one third of its volume of pyridine. To this solution is added a measured quantity of aqueous alkali. The excess is backtitrated using phenolphthalein as indicator or determining the endpoint potentiometrically.

3.2.2 Piperidine
The reaction with piperidine is suited especially for the splitting off of organic halogen[20, 21]. It follows the reaction scheme:

---

[20] *H. Hundsdiecker*, Chem. Ber. 76, 264 (1943).
[21] *P. Voegeli, F. Christen*, Z. Anal. Chem. 233, 175 (1968).

$$R-Hal + HN\bigcirc \longrightarrow R-N\bigcirc + HHal$$
$$\xrightarrow{H_2O} ROH + HN\bigcirc$$

A wide variety of solvents can be used. One can work in an aqueous medium, with piperidine alone or with several organic solvents such as 2-methoxyethanol. The reaction is carried out by adding, to the dissolved substance, 5–20 times as much piperidine as the quantity of the compound to be hydrolyzed. The cleavage goes quickly in many cases so that the split-off halogen can be titrated directly after the addition of piperidine. Otherwise the reaction is conducted at a higher temperature. This reaction is well suited for 1, 2 and 1, 4-halogeno nitroaromatic compounds and for aliphatic halogen, *cyanogen chloride* and *epichlorhydrin*.

4 Bibliography
[22] *Houben-Weyl*, Methoden der organischen Chemie, 4. Aufl., Bd. II, p. 510, Georg Thieme Verlag, Stuttgart 1953.
[23] *J. Mitchell, jr.*, Organic Analysis, Vol. III, Interscience Publishers, New York 1956.
[24] *S. Siggia*, Quantitative Organic Analysis via Functional Groups, 3. Ed., John Wiley & Sons, New York, London 1963.
[25] *N.D. Cheronis, T.S. Ma*, Organic Functional Groups Analysis by Micro and Semimicro Methods, Interscience Publishers, New York 1964.
[26] *M.R.F. Ashworth*, Titrimetric Organic Analysis, Part II, Interscience Publishers, New York 1965.

3.6 Neutralization Titration

Willy Büchler and *Aldo Manzetti*
CIBA–GEIGY, CH-4002 Basel

1 Determination of Acids

1.1 Titration in Aqueous Solution
Water soluble acid organic compounds with an acid dissociation constant K_a of up to 10^{-6} may be determined by titration in aqueous solution. The titrant used mainly is a standard aqueous solution of alkali hydroxide. Another possibility is the electrolytic production of hydroxyl ions (Coulometry[1, 2]).

1.1.1 Determination of the Endpoint
Potentiometric determination[3]: The change in hydrogen ion concentration, as seen by change in potential, can be followed during the titration with the help of suitable electrodes. The measuring electrode used mainly is the glass electrode and, as reference electrode, the calomel or silver/silver chloride electrode is employed.

From the potential curve obtained, the end-point can be determined as follows:

by geometric construction from the recorded or drawn potential curve[4–6];

---

[1] *K. Abresch, I. Claassen*, Die coulometrische Analyse, Verlag Chemie, Weinheim/Bergstr. 1961.
[2] *W. Büchler, P. Gisske, J. Meier*, Z. Anal. Chem. 239, 289 (1968).
[3] *F.L. Hahn*, pH und potentiometrische Titrierungen, Akademische Verlagsgesellschaft, Frankfurt/Main 1964.
[4] *S. Ebel*, Z. Anal. Chem. 245, 108 (1969).
[5] *C.F. Tubbs*, Anal. Chem. 26, 1670 (1954).

in the region of the end-point the titrant is added in exactly equal amounts and the corresponding change in potential noted. From the values obtained the endpoint is determined graphically[8] or by calculation[7].

Determination by means of indicators: If the approximate dissociation constants of acids are known, they may be titrated with the help of acid-base indicators. The range of the color change of the indicator should correspond with the pH at the end-point, which is calculated as follows:

$pH = \frac{1}{2} pK_w + \frac{1}{2} pK_a + \frac{1}{2} \log c$

K_w = Dissociation constant of water
K_a = Dissociation constant of acid
c = Concentration of acid to be titrated in Mol/l

Photometric determination[9]: The change of color in the region of the end-point is followed with the help of a photometer. The substance may have a particular color or UV-absorption which changes as the compound passes from the acid or base form to the salt form. Alternatively, the change of an added color indicator may be used.

Conductometric determination: Following the electrical conductivity of the titration solution as titrant is added for determining the end-point, has advantages over the above-mentioned methods, since the hydrolysis of weak acids does not have such a great influence[10]. Also, mixtures of strong and weak acids can be determined. Strong acids are titrated first, whereby the conductivity falls, then the weak ones, when the conductivity rises. Even phenol with a pKa of 9.9 can be titrated. The end-point is best determined graphically.

High frequency titration, on the other hand, has found little acceptance because, in contrast to the quoted methods, it has no advantages and the shape of the titration curve is rather complex.

1.1.2 Various Influences on Titrations

Solubility: Only organic acids which are soluble in water may be determined.

Amphoteric behavior of water: In water, which as an amphoteric solvent can take up or give up a proton, the protonated form, the hydronium ion $H_3O^\oplus$, is the strongest possible acid and the deprotonated form, the hydroxyl ion $OH^\ominus$, is the strong-

est possible base. On titrating a weak acid, e.g. phenol, the basicity of the hydroxyl ion is too low to deprotonize phenol fully:

$$C_6H_5OH + OH^\ominus \rightleftarrows C_6H_5O^\ominus + H_2O$$

The equilibrium lies on the left. Although addition of more hydroxyl ions increases the hydroxyl ion concentration steadily, a sudden change in the region of the end-point does not occur as is the case with stronger acids. For this reason weak acids such as phenol or weak bases like aromatic amines do not give sharp end-points.

Also, neutralization reactions occur when acids or bases are dissolved in water (proton transmission).

$$HA + HOH \rightleftarrows A^\ominus + H_3O^\oplus$$
$$B + HOH \rightleftarrows BH^\oplus + OH^\ominus$$

HA = dissolved acid
$A^\ominus$ = conjugate base of acid HA
B = dissolved base
$BH^\oplus$ = conjugate acid of base B

If the equilibria lie strongly on the right (in the case of strong acids or strong bases) hydronium ions are produced from the acids and hydroxyl ions from the bases. Thus, if an aqueous solution of a strong acid is titrated with a base, then the hydronium ions, as the strongest acid in water, will be titrated. In the case of mixtures of strong acids or bases, the original acid or base strengths will be altered to give an average strength *(levelling effect)* and only the total mixture of acids or bases can be determined.

On the other hand, strong and weak acids can be determined together if the difference in the dissociation constants is at least 10^4.

Hydrolysis of salts formed: The ions $A^\ominus$ or $BH^\oplus$ produced in the titration of an acid HA or of a base B can react again with water.

$$A^\ominus + HOH \rightleftarrows HA + OH^\ominus$$
$$BH^\oplus + HOH \rightleftarrows B + H_3O^\oplus$$

This reaction with the solvent renders determination of the end-point difficult.

1.2 Titration in Non-aqueous Solution

The use of organic solvents for the determination of organic acids or bases has the following advantages:

The solubility of many organic acids or bases is better in organic solvents than in water.
Disturbing influences, such as water possesses, are present to a much smaller extent or not at all.

1.2.1 Characteristics of Organic Solvents

A differentiation is made between aprotic and amphoteric or amphiprotic solvents.

[6] *F.L. Hahn, G. Weiler,* Z. Anal. Chem. *69,* 417 (1926).
[7] *F.L. Hahn,* Mikrochim. Acta 395 (1958).
[8] *J.R. Cohen,* Anal. Chem. *38,* 158 (1966).
[9] *J.B. Headridge,* Photometric Titrations, Pergamon Press, Oxford, London, New York, Paris 1961.
[10] *A.P. Kreshkov, T.A. Khudyakova,* Zh. Anal. Khim. *20,* 625 (1965), engl. 580; *22,* 1153 (1967), engl. 972.

Aprotic solvents have no ionizable proton and a small dielectric constant. Typical representatives are: benzene, chlorinated hydrocarbons, acetonitrile.

Amphiprotic solvents such as water, can take or give up a proton and are ionized to a small extent. Both groups can be further divided into neutral, acid and basic representatives.

What has been said about the *levelling effect of water* is also valid for amphiprotic solvents. For this reason amphiprotic solvents are suitable for the determination of total acidity. Aprotic solvents which have no levelling effect should be used for determination of acids or bases of various strengths when mixed together.

Solvolysis: The property of an amphiprotic solvent, like water, to react with salts produced during neutralization, makes the determination of the end-point difficult.

Influence of dielectric constant: In a protolytic reaction (transfer of a proton) the charges of particles are changed. The direction of this reaction is dependent on the solvent used; therefore, the acidity of an acid depends on the solvent. Solvents LH with large dielectric constants stabilize charged particles. This rule can be used for the following acids with various charges:

$$HA^\ominus + LH \rightleftarrows A^{2\ominus} + LH_2^\oplus \qquad (1)$$
$$HA + LH \rightleftarrows A^\ominus + LH_2^\oplus \qquad (2)$$
$$HA^\oplus + LH \rightleftarrows A + LH_2^\oplus \qquad (3)$$
$$HA^{2\oplus} + LH \rightleftarrows A^\oplus + LH_2^\oplus \qquad (4)$$

In examples (1) and (2), with increasing dielectric constant of the solvent, the equilibrium shifts to the right, and, as a consequence of the stabilizing effect, the Coulomb force between the ions on the right hand side falls. The strength of an acid is, therefore, increased.

In case (3) a change of dielectric constant has no effect on the acidity of $HA^\oplus$ because the number of charges on both sides is the same.

In case (4) the equilibrium shifts to the left with increasing dielectric constant. In the reaction with solvent LH a double charge is shared between two molecules. This sharing is favored in solvents with smaller dielectric constants because then two separate single charges are more easily stabilized than one double charge.

Thus, a carboxylic acid is less acid in ethanol (D.C. = 24.3) than in water (D.C. = 78.5; case 2). When titrating a dicarboxylic acid, e.g. *oxalic acid*, in water, both carboxyl groups are titrated together. If one uses a solvent with a lower dielectric constant, the two acid groups can be determined separately. After titration of the first carboxyl group, one has a singly charged acid of which the acidity in a solvent of smaller

dielectric constant is less than that of the neutral acid (case 1). In this way the difference in the acid strengths in each of the two stages is made so large that they can be titrated separately.

1.2.2 Solvents

For the determination of acid and weakly acid compounds the following *solvents*[22] or mixtures of two or three solvents have proven useful:

Toluene, chlorobenzene, methyl isobutyl ketone, methyl ethyl ketone, acetone, acetonitrile, ethylenediamine, n-butylamine, pyridine, dimethylformamide, dimethylsulfoxide, diethyl ether, tert. butanol, n-propanol, ethanol, methanol, 2-methoxyethanol (methylcellosolve), propylene glycol.

In addition to these traditional solvents other less usual solvents with good properties may be used in future. Sulfolanes, which are aprotic solvents with a high dielectric constant, have an extremely large potential range and are well suited for differentiating mixtures of acids or bases[11]. 1,1,3,3-tetramethylguanidine[12,13] and N,N-dimethyl fatty acid amides[14] also have advantages over traditional solvents. They are easy to purify, the glass electrode works well, and the potential range is large.

The usable range of some of the solvents given above can be limited by several factors[14]. Basic solvents such as ethylenediamine[15,16] and n-butylamine[17] have, with their strong levelling influence, only a very small potential range which, in potentiometric titrations, strongly limits the potential change. In addition, these solvents are very sensitive to carbon dioxide and also have an unpleasant odor. Acetone and acetonitrile are unstable in the presence of strong acids and the titration curves of carboxylic acids and phenols rise unusually steeply in the buffer zone which makes differential titration difficult. In pyridine[18,19] a steep rise in the buffer region is found

[11] *D.H. Morman, G.A. Harlow*, Anal. Chem. *39*, 1869 (1967).

[12] *T.R. Williams, J. Custer*, Talanta *9*, 175 (1962).

[13] *J.A. Caruso, G.G. Jones, A.I. Popov*, Anal. Chim. Acta *40*, 49 (1968).

[14] *C.A. Reynolds, J. Little, M. Pattengill*, Anal. Chem. *35*, 973 (1963).

[15] *V.Z. Deal, G.E.A. Wyld*, Anal. Chem. *27*, 47 (1955).

[16] *M.L. Moss, J.H. Elliot, R.T. Hall*, Anal. Chem. *20*, 784 (1948).

[17] *J.S. Fritz, N.M. Lisicki*, Anal. Chem. *23*, 589 (1951).

[18] *R.H. Cundiff, P.C. Markunas*, Anal. Chem. *28*, 792 (1956).

[19] *C.A. Streuli*, Anal. Chem. *32*, 407 (1961).

with many acids and phenols. Pyridine, because of its odor and toxicity, is unpleasant. Dimethylformamide is unstable in the presence of excess base[15]. Methanol and ethanol are useful for titration of relatively strong acids; for weak acids they are too acid to give good results[20]. In contrast tert. butanol permits even weak acids to be titrated with good accuracy[21].

1.2.3 Titrants

For the titration of acid compounds alkali methoxide and tetraalkyl ammonium hydroxide are used almost exclusively.

Solvents[22] for alkalimethoxide are methanol and isopropanol mixed with benzene. Sodium and potassium methoxide are most frequently used; cesium and rubidium methoxides are slightly more basic but offer no special advantages[23].

Disadvantages of alkali methoxides:

wrong indication with glass electrodes (alkali error); disturbing precipitates from alkali salts.

These disadvantages can be avoided by the use of tetraalkyl ammonium bases[26, 27], such as tetrabutylammonium hydroxide, tetramethylammonium hydroxide, hexadecyltrimethylammonium hydroxide. They are produced from the corresponding tetraalkyl ammonium halide by the reaction with silver oxide or with the aid of ion exchangers[24-26]. Mixtures of methanol, ethanol or isopropanol with benzene, isopropanol, tert. butanol, pyridine or butylamine may be used as solvents.

For the standardization of basic titrants benzoic acid or sulfamic acid[28] are most suitable.

1.2.4 Determination of the Endpoint

Potentiometric titration: In non-aqueous solution (cf. 3.6:1.1.1) the course of a titration can be followed potentiometrically. Due to the very low conductivity of weakly polar solvents and the high resistance of the glass electrode a potentiometer with a high input resistance (over 10^{12} ohms) must be used as the measuring instrument.

The following electrodes are recommended:

Measuring electrodes: glass, platinum, antimony, gold; Reference electrodes: calomel, silver/silver chloride.

In spite of bad reproducibility with respect to absolute potential and non-linear output in the presence of potassium and sodium ions (alkali error), glass electrodes are successful because of their ease of handling and rapid adjustment of potential. Extremely practical are the combined glass electrodes (glass silver/silver chloride) where the external aqueous electrolyte is replaced by a salt solution in an organic solvent e.g. lithium chloride in methanol. In place of glass electrodes, metal electrodes:

oxidized platinum/calomel electrode[29], anodized platinum/cathodized antimony[30], platinum/polarized platinum[31], antimony/antimony[16], and antimony/hydrogen[16]

may be used, but none have attained great significance.

Indicator methods: In non-aqueous systems as in aqueous systems the end-point can be determined with acid-base indicators. The choice of indicator is much more critical than in water because they behave differently in different solvents. The selection is made empirically and should be checked with the help of a potentiometric end-point determination. For *pyridine*[32], *tert. butanol*[33] and *acetonitrile*[34] the effective ranges of some indicators have been determined.

Useful indicators are: azoviolet, thymolphthalein, thymol blue, neutral red, o-nitraniline.

Other methods for the determination of endpoints: Additional methods include:

[20] R.H. Cundiff, P.C. Markunas, Anal. Chem. 30, 1447 (1958).
[21] J.S. Fritz, L.W. Marple, Anal. Chem. 34, 921 (1962).
[22] H.B. van der Heijde, Anal. Chim. Acta 17, 512 (1957).
[23] S.F. Ting, W.S. Jeffery, E.L. Grove, Talanta 3, 240 (160).
[24] R.H. Cundiff, P.C. Markunas, Anal. Chem. 28, 792 (1956); 30, 1450 (1958); 34, 584 (1962).
[25] H.V. Malmstadt, D.A. Vasallo, Anal. Chem. 31, 862 (1959).
[26] L.W. Marple, J.S. Fritz, Anal. Chem. 34, 796 (1962).
[27] G.A. Harlow, Anal. Chem. 34, 1482, 1487 (1962).
[28] M.M. Caso, M. Cefola, Anal. Chim. Acta 21, 205 (1959).

[29] V.E. Petrakovich, Q.M. Podurovskaya, Y.I. Tur'yan Zh. Anal. Khim. 20, (8) 785 (1965); engl. 863.
[30] D.A. Lee, Anal. Chem. 38, 1168 (1966).
[31] G.A. Harlow, C.M. Noble, G.E.A. Wyld, Anal. Chem. 28, 784 (1956).
[32] J.S. Fritz, F.E. Gainer, Talanta 13, 939 (1966).
[33] L.W. Marple, J.S. Fritz, Anal. Chem. 35, 1305 (1963).
[34] I.M. Kolthoff, M.K. Chantooni, S. Bhowmik, Anal. Chem. 39, 315 (1967).
[35] L.E.I. Hummelstedt, D.N. Hume, Anal. Chem. 32, 1792 (1960).
[36] K.A. Connors, T. Higuchi, Anal. Chem. 32, 93 (1960).

Photometric titrations[9, 35, 36];
Conductometric titrations[37, 38];
Polarovoltric titrations[39, 40];
Thermometric titrations[41];
Amperometric titrations[42];
High frequency titrations[43].

In many cases these methods give good results, but no method is as general in application as potentiometry. The range is limited either by the physical properties of the solvent, such as electrical conductivity and dissolving power, or by the obscure shape of the titration curve.

2 Determination of Bases

2.1 Titration in Aqueous Solution
Water-soluble amines with a dissociation constant K_b of up to 10^{-6} may be determined in aqueous solution. For bases with smaller dissociation constants the determination of the end-point is, as in the case of acids, made difficult or completely impossible on account of the properties of water (described in Section 3.6:1.1.2).
Titrants: Aqueous solutions of strong inorganic acids such as hydrochloric acid and sulfuric acid or electrolytically generated hydrogen ions are recommended for titration of bases[1, 2]. The end-point determination is as for acids.

2.2 Titration in Non-aqueous Solution

2.2.1 Solvents
The following solvents[22] have proved useful for titrating basic to weakly basic compounds: cyclohexane, dioxane, carbon tetrachloride, benzene, chloroform, chlorobenzene, methyl isobutyl ketone, methyl ethyl ketone, acetone, acetonitrile, sulfolanes, formic acid, acetic acid, propionic acid, acetic anhydride, nitromethane, nitrobenzene, ethylene glycol, propylene glycol, 2-methoxyethanol, isopropanol.
The solvent most used is acetic acid. It forms acetates with the bases and consequently increases the base strength. Because of the levelling effect, only the total basicity of a mixture can be determined. To differentiate mixtures and to estimate weakly basic compounds (up to pKb 15–20 in water) aprotic solvents such as acetonitrile, chloroform, sulfolane[11], methyl isobutyl ketone and acetic anhydride are recommended. Acetic anhydride has the disadvantage that it can acetylate; thus substances which are readily acetylated need great care. In certain cases, mixtures of diols (ethylene glycol or 1, 2-propylene glycol) with hydrocarbons, chloroform, or isopropanol (G-H mixtures) have proved useful[44]. Because of their polarity, they have a higher dissolving power than acetic acid, but are somewhat more basic.

2.2.2 Titrants
Among the strong inorganic acids, perchloric acid has proved to be best. Solvents are acetic acid, dioxane, nitroethane/chlorobenzene[45] and sulfolanes[46]. Acetic acid has the disadvantage that, during the simultaneous determination of bases of different strength, it has a *'levelling'* effect. Solutions of perchloric acid, as well as those of sulfonic acids in acetic anhydride[47] and in acetonitrile[45], are not stable.

In addition to perchloric acid, but to a lesser extent, other acids such as fluorosulfonic acid[48], toluene sulfonic acid, *p*-nitrobenzoic acid, 2, 4-dinitrobenzoic acid[49] and other sulfonic acids[50] are used particularly if, during a titration with perchloric acid, a precipitate of perchlorate is formed which impairs the signal from the glass electrode. For calibrating the strength of acid titrants the following substances are recommended: potassium hydrogen phthalate, diphenylguanidine[51] and tris (hydroxy methyl) aminomethane[52].

2.2.3 Determination of the Endpoint
Potentiometric titration: What has been said for acids applies also to bases, the electrodes used being almost exclusively glass-calomel and glass-silver/silver chloride.

[37] *N. Van Meurs, E.A.M.F. Dahmen*, Anal. Chim. Acta, *21*, 443 (1959), J. Elektroanal. Chem. Interfacial Elektrochem. *1*, 458 (1959–1960).
[38] *C. Bertoglio-Riolo, T. Fulle-Soldi*, Ann. Chim. [Rome] *54*, 923 (1964).
[39] *I. Shain, G.R. Svoboda*, Anal. Chem. *31*, 1857 (1959).
[40] *J.E. Dubois*, C.R. Acad. Sic. Paris *260*, 564 (1965).
[41] *G.A. Vaughan, J.J. Swithenbank*, Analyst (London) *90*, 594 (1965).
[42] *W. Ruskul*, Talanta *13*, 1587 (1966).
[43] *R. Hara, P.W. West*, Anal. Chim. Acta *15*, 193 (1956).

[44] *S.R. Palit*, Ind. Eng. Chem. Anal. *18*, 246 (1946).
[45] *W. Huber*, Z. Anal. Chem. *216*, 260 (1966).
[46] *J.F. Coetzel, R.J. Bertozzi*, Anal. Chem. *41*, 861 (1969).
[47] *D.J. Pietrzyk*, Anal. Chem. *39*, 1367 (1967).
[48] *R.C. Paul, S.S. Pahil*, Anal. Chim. Acta *30*, 466 (1964).
[49] *D.J. Pietrzyk, J. Belisle*, Anal. Chem. *38*, 969 (1966).
[50] *M.M. Caso, M. Cefola*, Anal. Chim. Acta *21*, 374 (1959).
[51] *J.S. Fritz*, Anal. Chem. *22*, 578 (1950).
[52] *J.H. Fossum, P.C. Markunas, J.A. Riddick*, Anal. Chem. *23*, 491 (1951).

Indicator methods: The following indicators[53], among others, are suitable for titrating bases: crystal violet (methyl violet), malachite green, α-naphtholbenzeine, neutral red, eosin, and bromophenol blue.

As opposed to potentiometry, polarovoltry[54] and conductometry[55] have not been very successful for the determination of bases in non-aqueous solutions.

3 Determination of Salts

3.1 Salts of Carboxylic Acids
Salts of carboxylic acids with alkaline or alkaline earth metals, ammonia or organic bases can, depending on their solubility in acetic acid[56-58] or in G-H mixtures[44] (diols with hydrocarbons, chloroform, or isopropanol), be titrated like bases with perchloric acid.

In basic solutions such as dimethylformamide, pyridine or ethylenediamine[59], ammonium salts of carboxylic acids can be titrated with sodium methylate or tetra alkyl ammonium hydroxide. A further possibility for the estimation of salts of organic acids is to convert the salts to the free acid with the help of a strong cation exchanger and then titrate with base[60].

Estimation of metal salts[61]: The metal salt is dissolved in a little water, mixed with an excess of a strong acid, e.g. sulfuric acid, and diluted with acetone or pyridine. Thus the metal salt of the strong acid is precipitated and the two acids, the excess of the strong acid and the carboxylic acid produced, can be differentially determined by potentiometric titration with tetrabutylammonium hydroxide.

3.2 Salts of Nitrogenous Bases
The addition of mercury (II) acetate to halides of nitrogenous bases gives undissociated mercury (II) halide and base acetate. This base acetate can be titrated directly with perchloric acid in glacial acetic acid or chloroform[53]. Most salts of nitro-

genous bases with organic acids may be titrated directly with perchloric acid after dissolving in acetic acid or in a mixture of a neutral and an acid solvent.

The base is released from sulfates, sulfonates, phosphates and nitrates by means of an aqueous solution of sodium hydroxide or with an ion exchanger and extracted with an organic solvent. After drying the organic extract the base can be titrated with perchloric acid or *p*-toluenesulfonic acid.

A mixture of phenol/acetonitrile/chloroform (5:20:20) has proved useful as solvent for the direct titration of sulfates with perchloric acid[62]. The end-point is determined potentiometrically or, in certain cases, with an indicator.

4 Neutralization Reactions of the Lewis Type
Although reaction of Lewis bases (electron donors) with Lewis acids (electron acceptors) may be used analytically[63, 64], they seldom are.

Lewis bases which can be determined are tertiary amines of various types (primary and secondary amines cannot be determined owing to side reactions) and oxygen compounds, mainly ethers.

The following Lewis acids are used: halides of groups IIIA and IVB of the Periodic Table, stannic chloride, antimony tri- and pentachloride and iron(III) chloride.

[62] *L.G. Chatten,* J. Pharm. Pharmacol. *7,* 586 (1955).
[63] *C. Bertogli-Riolo, T. Soldi,* Ann. Chim. [Rome] *50,* (1960).
[64] *E.T. Hitchcock, P.J. Elving,* Anal. Chim. Acta *27,* 501 (1962), *28,* 301 (1963).
[65] *M.C. Henry, J.F. Hazel, W.M. McNabb,* Anal. Chim. Acta *15,* 187, 283 (1956).
[66] *R.C. Paul, J. Singh, S.S. Sandhu,* Anal. Chem. *31,* 1495 (1959).

[53] *C.W. Pifer,* J. Am. Pharm. Assoc. Sci. Ed. *42,* 509 (1953).
[54] *G.R. Svoboda,* Anal. Chem. *33,* 1638 (1961).
[55] *N. van Meurs, E.A.M.F. Dahmen,* Anal. Chim. Acta *21,* 193 (1959).
[56] *P.C. Markunas, J.A. Riddick,* Anal. Chem. *23,* 337 (1951).
[57] *P. Sensi, G.G. Gallo,* Ann. Chim. [Rome] *43,* 453 (1953).
[58] *A.H. Beckett, R.M. Camp, H.W. Marti,* J. Pharm. Pharmacol. *4,* 399 (1952).
[59] *J.S. Fritz,* Anal. Chem. *24,* 306 (1952).
[60] *W. Stuck,* Z. Anal. Chem. *177,* 338 (1960).
[61] *R.H. Cundiff, P.C. Markunas,* Anal. Chim. Acta *21,* 68 (1959).

5 Bibliography
[67] *W. Huber,* Titrationen in nichtwäßrigen Lösungsmitteln, Akademische Verlagsgesellschaft, Frankfurt/Main 1954. English: Titrations in Non-Aqueous Solvents, Academic Press, New York 1967.
[68] *I.M. Kolthoff, S. Bruckenstein* in *I.M. Kolthoff, P.J. Elving,* Treatise on Analytical Chemistry, Vol. I/1, Chap. 11—13, Wiley Inc., New York 1959.
[69] *J. Kucharský, L. Šafařík,* Titrations in Non-Aqueous Solvents, Elsevier Publ. Co., Amsterdam, London, New York 1965.
[70] *I. Gyenes,* Titration in Non-Aqueous Media, Iliffe Books, London, D. van Nostrand Co., Princeton, New Jersey 1967.

Solvents used are acid chlorides such as thionyl chloride, sulphuryl chloride and acetyl chloride, acetic anhydride, nitrobenzene, acetonitrile and benzene.

The end-point can be determined potentiometrically[63], by high frequency titration[64] conductometrically[65], and with indicators[66].

3.7 Special Chemical Methods

Willy Büchler and *Aldo Manzetti*
CIBA-GEIGY AG, CH-4002 Basel

1 Determination of Water by the Karl Fischer Method

1.1 Principles
The iodometric estimation of sulfur dioxide whose basis is shown in Equation (1)

$$2\,H_2O + SO_2 + I_2 \rightleftarrows H_2SO_4 + 2\,HI \qquad (1)$$

was adapted to water estimation by Karl Fischer[1]. He used methanol as a solvent and added pyridine as a base to drive the equilibrium to the right. Further investigation of the reaction[2] demonstrated that methanol also takes part. The reaction proceeds in two stages (2) and (3); first, pyridine-sulfur trioxide is formed as an intermediate, which then reacts with methanol to form pyridine methyl sulfate.

$$C_5H_5N \cdot I_2 + C_5H_5N \cdot SO_2 + C_5H_5N + H_2O \rightarrow$$
$$2\,C_5H_5N \cdot HI + C_5H_5N \cdot SO_3 \qquad (2)$$

$$C_5H_5N \cdot SO_3 + CH_3OH \rightarrow C_5H_5N \cdot HSO_4CH_3 \qquad (3)$$

From the equations it can be seen that one mole of iodine is consumed per mole of water.
In the absence of methanol, the pyridine-sulfur trioxide formed in the first stage reacts further with water, producing pyridinium hydrogen sulfate (4).

$$C_5H_5N \cdot SO_3 + H_2O \rightarrow C_5H_5N \cdot H_2SO_4 \qquad (4)$$

Although the efficiency of the reagent solution can be doubled by reaction (4), in practice, methanol is added.
Methanol has two important advantages:

It reacts faster with the pyridine-sulfur trioxide complex than do other hydroxy compounds. It stabilizes the solution, and hinders the formation of pyridine periodide ($C_5H_5N \cdot J_4$).

The Karl Fischer method can be used to determine the water content of liquids, solids, and gases if these do not react with the reagent. Water determination can be used also in the estimation of functional groups by titrating the water liberated or consumed during a reaction[3, 4].

Apparatus for the determination of water: All vessels must be absolutely dry. The reagent is measured from burets connected directly to a reservoir and protected from moisture by means of a desiccant. Piston burets and graduated syringes are also suitable. Rubber or PVC tubes should not be used, as they are neither absolutely impermeable to moisture, nor sufficiently resistant to Karl Fischer reagent.
The titration vessels should be capable of being sealed, e.g. by use of ground joints[5, 6]. Various types of equipment are commercially available.

1.2 Reagent Solutions
From the reaction equations, the solution should contain methanol, pyridine, sulfur dioxide and iodine. Since a mixture of all these components is not very stable, the reagent solution is made from 2 solutions which are mixed before use.

Karl Fischer Solutions

Solution A:	Methanol	460 ml
	Iodine	84 g
Solution B:	Pyridine	270 ml
	Methanol	200 ml
	Sulfur dioxide	64 g

When mixing equal parts of both solutions a value for water of 3 mg per ml reagent is used.

More dilute reagent solutions are sometimes used, with a water-equivalent of 1 mg/ml[7] or 0.3 mg/ml[8].

[3] *J. Mitchell, D.M. Smith*, Aquametry, Interscience Publishers, New York 1948.
[4] *E. Eberius*, Wasserbestimmung mit Karl-Fischer-Lösung, Verlag Chemie, Weinheim/Bergstr. 1954.
[5] *M.G. Yakubik*, J. Chem. Educ. *35*, 5, (1958).
[6] *J.D. Ponting, D.H. Taylor*, Chemist-Analyst *54*, 123 (1965).
[7] *I. Wiberly*, Anal. Chem. *23*, 656 (1951).
[8] *A.M. L'vov, V.A. Klimova, A.I. Palii*, Zh. Anal. Khim. *19*, 1366 (1964), engl. 1272.

[1] *K. Fischer*, Z. Angew. Chem. *48*, 394 (1935).
[2] *D.M. Smith, W.M.D. Bryant, J. Mitchell*, J. Am. Chem. Soc. *61*, 2407 (1939).

Replacement of methanol by 2-methoxyethanol[9] gives a more stable solution (stabilized Karl Fischer reagent).

'Stabilized' Karl Fischer Reagent:

Iodine	135 g
Sulfur dioxide	70 g
Pyridine	425 ml
2-Methoxyethanol	425 ml

The water-equivalent is 6 mg water per ml reagent.

The titer of the reagent solution is expressed as mg water per ml of solution, and is determined by titration of a known amount of water. The water can be added directly, or in the form of a methanolic solution. Substances having a constant amount of water of crystallization, e.g. sodium tartrate ($C_4H_4O_6Na_2 \cdot 2H_2O$), are also suitable for the determination of the titer.

The stabilized reagent can be used for the determination of water in compounds which react with methanol, e.g. ketones. Despite advantages, the normal reagent appears to be used more often.

A coulometric technique can also be used[10, 11]. Iodine for this can be generated either electrolytically from potassium iodide in a mixture of pyridine, sulfur dioxide and methanol, or from a Karl Fischer solution which has previously been titrated with water.

1.3 Determination of the End-Point

With clear and colorless solutions, the end-point can be detected visually, since the colors of the reacted and the unreacted reagent solutions differ (reacted: yellow; unreacted: reddish brown). The color change is not instantaneous but gradual and not easy to recognize. Titration can be carried out photometrically[12] or using an indicator[13], but these are not of great importance.

The most commonly used techniques are the *polarovoltric titration*[14] and the socalled *dead stop method*[15]. The polarovoltric procedure is based on the measurement of the voltage between two inert metal electrodes, polarized with a constant current

or between one of these electrodes and a reference electrode. In the amperometric (dead stop) method, however, the polarizing voltage between two inert electrodes is held constant, and the resulting current is measured. Because these techniques allow the end point to be measured quickly and conveniently, this Karl Fischer titration method is used widely. The amount of apparatus is small: a sensitive voltmeter or amperometer, two platinum electrodes, a voltage source and resistors.

1.4 Microscale Titrations

Many variations for the micro-determination of water by the Karl Fischer method have been described[8, 16−21].

In the microtitration of water, the following difficulties must be overcome:

The reaction between Karl Fischer reagent and water is slow at low concentrations, and, therefore, detection of the end point is difficult and subjective.
The residual moisture in the apparatus (before addition of the sample) must be titrated exactly, and atmospheric moisture rigorously excluded.

The rate of reaction can be increased by addition of *N*-ethylpiperidine[20]. External moisture can be largely excluded by the use of special vessels and techniques[8, 20, 21]. The technique involving electrolytic generation of the Karl Fischer reagent[18, 22] is especially suitable for micro-titrations.

1.5 Fields of Application

Determination of water by the Karl Fischer method can be applied to a variety of substances (organic and inorganic compounds, pharmaceutical, technical products, and foodstuffs). The samples under investigation can be titrated directly in a suitable solvent, such as methanol, ethylene glycol, dioxane, ether, pyridine, etc. To determine water in gases, a known volume of the gas is passed through a solvent which absorbs water efficiently, or directly through Karl Fischer solution. However, this last method is not very accurate. With

[9] *E.D. Peters, J.L. Jungnickel*, Anal. Chem. *27*, 450 (1955).
[10] *A.S. Meyer, C.M. Boyd*, Anal. Chem. *31*, 215 (1959).
[11] *G.A. Rechnitz, K. Srinivasan*, Z. Anal. Chem. *210*, 9 (1965).
[12] *K.R. Connors, T. Higuchi*, Chemist-Analyst *48*, 91 (1959).
[13] *J. Michaelis*, Pharmazie *13*, 740 (1958).
[14] *R. Gauguin*, Anal. Chim. Acta *18*, 29 (1958).
[15] *G. Wernimont, F.J. Hopkinson*, Ind. Eng. Chem. *15*, 272 (1943).

[16] *A. Campiglio*, Farmaco, Ed. Sci. *20*, 570 (1965).
[17] *J.M. Corliss, M.F. Buckless*, Microchem. J. *10*, 218 (1966).
[18] *M.R. Lindbeck, H. Freund*, Anal. Chem. *37*, 1647 (1965).
[19] *A. Discherl, F. Erne*, Mikrochim. Acta, 794 (1962).
[20] *E.E. Archer, H.W. Jeater*, Analyst (London) *90*, 351 (1965).
[21] *E.E. Archer, H.W. Jeater, J. Martin*, Analyst (London) *92*, 524 (1967).
[22] *M. Pribyl, Z. Slovak*, Mikrochim. Ichnoanalyt. Acta, 1097 (1964), Anal. Abstr. *13*, 734 (1966).

insoluble solids, the water must be extracted with a solvent.

In estimation of water by the Karl Fischer method, it is important that the sample under investigation contain none of the following substances which react with Karl Fischer reagent:

alkali hydroxides and carbonates (iodine consumed)
easily esterified acids (methyl esters formed)
aldehydes and ketones (acetals and ketals are formed with methanol)
unsaturated compounds (addition of iodine may occur)

Under certain circumstances, water can also be estimated in the compounds mentioned above, i.e. if the determination is carried out at low temperature, or if the methanol-free 'stabilized' Karl Fischer reagent is used[23, 24]. It may be advantageous to change the solvent.

[23] *L. Barnes, M.S. Pawlak,* Anal. Chem. *31,* 1875 (1959).
[24] *H. Beyer, K. Varga,* Z. Chem. *6,* 470 (1966). Anal. Abstr. *15,* 1427 (1968).

1.6 Bibliography

Mitchell, Smith[3], 1948.
Eberius[4], 1954.
[25] *W. Kerstan,* Pharmazie *12,* 711 (1957).
[26] *J. Tranchant,* Mém. Poudres *46—47,* 119 (1964—65).
[27] *J. Mitchell jr.* in *F.D. Snell, C.L. Hilton,* Encyclopedia of Industrial Chemical Analysis, Vol. I, p. 142, Interscience Publishers, John Wiley & Sons, New York, London, Sidney 1966.

2 Gasometric Methods

2.1 General

If a gas is evolved or consumed during a chemical reaction, measurement of the amount of gas can be used for quantitative determination of a compound, provided that the reaction proceeds completely in one direction. This method is termed 'gasometry'. Because gas volumes depend very much on temperature and pressure and consequently special apparatus is necessary, gasometry is not of great importance and can often be replaced by simpler methods. At present, it is restricted almost entirely to micro-scale measurements.

The quantity of gas evolved or consumed can be measured in two ways: manometrically or volumetrically. In the manometric method, the volume of the gas is kept constant and the change in pressure is measured; in the volumetric procedure, the pressure is kept constant and the difference in volume is determined.

The apparatus[1] needed for gasometric determinations consists essentially of a reaction vessel connected to a manometer or a gas buret. In most cases, gas wash-bottles and thermostats are also necessary. Specific assemblies have been developed for most determinations, e.g. for the determination of gases (carbon dioxide, oxygen, nitrogen) in blood and other fluids[2], or the determination of functional groups[3].

2.2 Determination of Primary Amines

Primary aliphatic amines and *amino acids* react with nitrous acid, liberating nitrogen (Van Slyke determination, see 3.4:2.7).

$$RNH_2 + HNO_2 \rightarrow ROH + H_2O + N_2$$

2.3 Determination of Hydrazines, Diazo and Diazonium Compounds

Aromatic hydrazines are oxidized by copper (II) salts[4], lead oxide (Pb_3O_4) or lead dioxide (PbO_2)[5] to the corresponding diazonium salts; these are decomposed, and the nitrogen evolved is measured.

Aliphatic diazo compounds are decomposed by treatment with dilute sulfuric acid, the resulting nitrogen being determined gasometrically[6, 7].

Aromatic diazonium compounds also evolve nitrogen in the presence of a catalyst, such as copper (I) chloride[8] or potassium iodide[9], or a reducing agent such as titanium (III) chloride[10].

2.4 Determination of Active Hydrogen

Hydrogen atoms which are bound to such atoms as oxygen, nitrogen or sulfur, or to acetylenic atoms, are designated *active hydrogen*. Active

[1] *B. Budesinsky,* Mikrochim. Acta 811 (1961).
[2] *D.D. van Slyke, J.M. Neill,* J. Biol. Chem. *61,* 523 (1924).
   D.D. van Slyke, J. Biol. Chem. *73,* 121 (1927).
[3] *D.D. van Slyke,* Microchem. J., Symposium Series, Vol. II, Microchemical Techniques, p. 31, Interscience Publishers, New York, London 1962.
[4] *S. Siggia, L.J. Lohr,* Anal. Chem. *21,* 1202 (1949).
[5] *A. Berka, E. Smolkova, E. Bocanovski,* Z. Anal. Chem. *204,* 87 (1964).
[6] *H. Staudinger, A. Gaule,* Chem. Ber. *49,* 1897 (1916).
[7] *L. Nicolas, P. Lampei,* Chim. Anal. *36,* 238 (1954).
[8] *S. Siggia,* p. 54, in Ref. [18].
[9] *F. Gasser,* Öst. Chem.-Ztg. *51,* 206 (1950), C.A. *45,* 4176 (1951).
[10] *W.E. Shaefer, W.W. Becker,* Anal. Chem. *19,* 307 (1947).

hydrogen can be determined gasometrically[11]. It reacts with methylmagnesium halide forming methane (Zerewitinoff determination; see Section 3.4:1.4.1), and with lithium aluminum hydride (cf. Section 3.3:1.2) or diborane[12], liberating hydrogen.

2.5 Determination of C=C Double Bonds
One of the most important gasometric methods involves the addition of hydrogen to a *C=C double bond* in the presence of a catalyst (see Section 3.3:1.1).

$$>C=C< + H_2 \xrightarrow{\text{Cat.}} \; \overset{\displaystyle H \;\; H}{\underset{\textstyle >C-C<}{|\;\;\;|}}$$

2.6 Determination of Ketones and Aldehydes
Lithium aluminum hydride and sodium borohydride reduce *ketones* and *aldehydes* to alcohols (see Section 3.3:1.2). After reaction has been completed, hydrogen is liberated from the excess hydride, and measured.

The determination of *carbonyl compounds*[13] with phenylhydrazine is based on another principle.

The excess phenylhydrazine is oxidized to nitrogen, which is measured gasometrically.

2.7 Other Determinations
The gasometric method can be applied also to the determination of α-amino acids[14, 15], urea[16] and various natural products of biochemical importance[17, 20], where the gases are produced by direct chemical reaction or action of enzymes.

---

[14] *D.D. van Slyke, R.T. Dillon, D.A. MacFadyen, P.B. Hamilton,* J. Biol. Chem. *141,* 627 (1941), *150,* 251 (1943), *164,* 249 (1946).
[15] *W. Gerok, H.D. Waller,* Klin. Wochenschr. *34,* 1284 (1956).
[16] *D.D. van Slyke,* J. Biol. Chem. *73,* 695 (1927).

2.8 Bibliography
[17] *W.W. Umbreit, R.H. Burris, J.F. Stauffer,* Manometric Techniques, Burgess Publishing Co., Minneapolis 1959.
[18] *S. Siggia,* Quantitative Organic Analysis via Functional Groups, 3rd Ed., John Wiley and Sons, Inc., New York 1964.
[19] *N.D. Cheronis, T.S. Ma,* Organic Functional Group Analysis by Micro and Semimicro Methods, Interscience Publishers, New York 1964.
[20] *Arnost Kleinzeller,* Manometrische Methoden und ihre Anwendung in der Biologie und Biochemie, VEB Gustav Fischer Verlag, Jena 1965.

---

[11] *E.D. Ollemann,* Anal. Chem. *24,* 1425 (1952).
[12] *F.E. Martin, R.R. Jay,* Anal. Chem. *34,* 1007 (1962).
[13] *Z. Nowak,* Chem. Anal. [Warsaw] *11,* 753 (1966), Z. Anal. Chem. *237,* 301 (1968).

3.8 Comparative Methods for the Determination of Stereochemical Configurations

Karl Schlögl
Organisch-chemisches Institut der Universität,
A-1090 Wien

1 Introduction
A complete (geometrical) description of multiatomic molecules requires the knowledge of their constitution, configuration, and conformation. The last two aspects are of increasing interest. The ever growing importance of stereochemistry is due at least partly to recent methods of configurational and conformational analysis.

In this article comparative methods for the determination of configurations (mainly of carbon compounds) will be discussed briefly. Their importance for establishing absolute configurations is based on the knowledge of the actual molecular geometry of several key compounds (3.8:2.1).

1.1 Relative and Absolute Configuration
The configuration of a molecule (of given constitution) is defined as the spatial arrangement of its atoms without regard to structures that arise from rotations about single bonds (conformations)[1].

The absolute configuration (chirality) of an optically active compound, namely of *(+)-tartaric acid,* was established for the first time in 1951 by anomalous X-ray diffraction[2]; i.e. the actual arrangement of the atoms in space was determined (cf. 7.2). Thereby all relative configurations [which are related to an arbitrarily chosen configuration

---

[1] IUPAC-Rules Section E: Fundamental Stereochemistry, J. Org. Chem. *35,* 2849 (1970).
[2] *J.M. Bijvoet, A.F. Peerdeman, A.J. van Bommel,* Nature *168,* 271 (1951).

of a key substance, such as (+)-glyceraldehyde] known till then became absolute configurations.

Recently however, doubts as to the reliability of the X-ray (Bijvoet) method[2] (and hence on the basis of the absolute configuration) were expressed[2a].

All comparative (mainly empirical) methods treated in this article are based on the knowledge of absolute chiralities. They should be applied and the results interpreted with due care, however attractive these methods may appear. It is advisable to support the results by the application of independent methods.

1.2 Stereochemical Nomenclature[3]

A molecule that cannot be superimposed on its mirror image is chiral. Chirality (handedness, previously: dissymmetry) is, therefore, the necessary condition for the occurence of enantiomers and of optical activity.

There are three *elements of chirality:* center, axis and plane. Symmetry considerations rendered possible a specification of molecular chirality, i.e. an unambiguous notation of a chiral molecule[3]. The basis for this stereochemical nomenclature, which has been generally accepted (cf. Ref.[4]), is the sequence rule[3]. Thereby a symbol – (R) or (S) – can be assigned to every element of chirality; these symbols are independent of ordinary chemical nomenclature (cf. Methodicum Chimicum, Vol. 2).

2 Chemical Methods

2.1 Direct Chemical Correlation

The chemical conversion of optically active compounds into one another provides the most reliable method for configurational correlations. Obviously, the stereochemical courses of the reactions used must be known. In *centrochiral compounds* an exchange of ligands attached to the center should be avoided or at least only reactions of high stereoselectivity should be employed: be it either with retention of configuration, such as the $S_N 1$-*reaction*, the Hofmann, Curtius, Lossen or Schmidt degradations and the Beckmann, Baeyer-Villiger or Stevens rearrangements, or with unequivocal inversion, such as the Walden inversion accompanying the $S_N 2$-*substitution*.

Until 1951 many, mainly centrochiral compounds (such as amino acids, sugars, alkaloids, terpenes and steroids) had been correlated with each other and lastly with *(+)-glyceraldehyde*, to which E. Fischer arbitrarily had assigned the configuration 1 (as reference compound of the series). All these unequivocally correlated relative configurations were then confirmed by the determination of the configuration 2 for *(+)-tartaric acid;* i.e., they became *absolute configurations.* Freudenberg had previously (in 1914) chemically correlated (+)-tartaric acid with (+)-glyceraldehyde[5]. Consequently, the latter has the configuration (+)(R) as shown in 1. All projection formulas used in this article are Fischer type projections, such as shown for (+)-glyceraldehyde (1).

$(+)(R)-1$ $(+)(R)-2$

There is an abundance of recent correlations of centrochiral compounds: e.g. of *(+)-pinacolyl alcohol* with *(+)(S)-lactic acid*[6], of *(+)-α-²H₁-benzylamine* with *(−)(R)-α-amino phenylacetic acid*[7] and of *(−)-physostigmine* and related *alkaloids* with *(−)(S)-α-ethyl-α-methylsuccinic acid*[8].

By chemical correlations of centro- with axial and planar chiral compounds, the absolute configurations of several of these molecular chiral compounds were elucidated (cf. Ref.[9]). Prior to this configurational correlations had been established within such classes (as the chiral biaryls[10] or metallocenes[11]).

The following examples are illustrative: correlations of axial chiral *cyclohexylidene derivatives,* such as of 3 with *(+)(R)-3-methylcyclohexanone*[12] (4) or of 5 with *4-methyl-α-deuterobenzylamine*[13] (6); of chiral *ferrocenes* as of 7 with *(+)(S)-α-phenylglutaric acid*[14] (8) and of the planar chiral *(−)-trans-cyclooctene* (9) with *(+)-tartaric acid* (2) via 1,2-dimethoxy-*cyclooctane*[15] (10).

[5] *K. Freudenberg*, Ber. *47*, 2027 (1914).
[6] *J. Jacobus, Z. Majerski, K. Mislow, P.v.R. Schleyer*, J. Am. Chem. Soc. *91*, 1998 (1969).
[7] *H. Gerlach*, Helv. Chim. Acta *49*, 2481 (1966).
[8] *R.B. Longmore, B. Robinson*, Chem. Ind. [London] 622 (1969).
[9] *H. Falk*, Österr. Chem. Z. *66*, 242 (1965); G. Krow in Ref.[101], Vol. 5, p. 31 (1970).
[10] *K. Mislow*, Angew. Chem. *70*, 683 (1958).
[11] *K. Schlögl*, Fortschr. Chem. Forsch. *6*, 479 (1966); in Ref.[101], Vol. I, p. 39 (1967); *K. Schlögl*, J. Pure Appl. Chem. *23*, 413 (1970).
[12] *J.H. Brewster, J.E. Privett*, J. Am. chem. Soc. *88*, 1419 (1966).
[13] *H. Gerlach*, Helv. Chim. Acta *49*, 1291 (1966).
[14] *H. Falk, K. Schlögl*, Mh. Chem. *96*, 1065 (1965).
[15] *A.C. Cope, A.S. Mehta*, J. Am. Chem. Soc. *86*, 1268, 5626 (1964).

[2a] *J. Tanaka, C. Katayama, F. Ogura, M. Kuritanai, H. Tatemitsu, M. Nakagawa*, Chimia *26*, 471 (1972); Chem. Commun. 21 (1973).
[3] *R.S. Cahn, C.K. Ingold, V. Prelog*, Angew. Chem. *78*, 413; engl. *5*, 385 (1966).
[4] Ref.[98], p. 111, 196.

$(+)(R)-4$ $(+)(S)-3$ $(R)-6$ $(+)(S)-5$

$(-)(1R)-7$ $(+)(S)-8$

$(-)(R)-9$ $(+)-10$ $(+)(R)-2$

The configuration $(-)$-(R) was established recently for **9** by anomalous X-ray diffraction[15a]. It is in contrast to that previously postulated on the basis of theoretical considerations[15b].

Hexa-[15c] and pentahelicene[15d], resp. were correlated with compounds of known configurations ([2.2] paracyclo-phane-4-carboxylic acid **22**[31, 44, 45] and a binaphthyl derivative, resp) thereby establishing their configurations as $(-)$-(M) and $(+)$-(P)[3], resp. For a correlation of chiral 9.10-dihydro-9.10-etheno-anthracenes with chiral triptycenes (whose configuration were established by X-ray) cf. Ref.[15e].

With careful planning of the reaction sequence, for which obviously no general rules can be given — except avoiding steps of uncertain stereochemistry — the direct chemical correlation affords the most reliable results.

For these correlations several key compounds are available for which the configurations have been determined by anomalous X-ray diffraction[16].

2.2 Asymmetric Syntheses

Asymmetric syntheses[17, 105] are reactions in which a new chiral grouping is produced while the corresponding enantiomers are formed in unequal amounts.

With suitable chiral starting materials or chiral catalysts the diastereomeric transition states — because of different geometries (e.g. conformations) — have different energies. This is reflected in the different reaction rates of the stereoisomeric products (R and S) and (on kinetic control) in the product ratio: $k_R/k_S = [R]/[S]$.

The stereoselectivity of an asymmetric synthesis may be defined by the optical yield p which specifies the per cent purity of the predominant stereoisomer (enantiomer); e.g., if R predominates,

$[R]/[S] = (100 + p)/(100 - p)$.

With the usual optical yields varying from 5 to 50% — apart from biological processes higher p-values are rare — the $\Delta\Delta G^{\neq}$-values for the stereoisomeric products lie (at 25°) between ~ 100 and 600 cal/mole. Therefore, small energy differences in the transition states are sufficient to cause easily measurable effects in the products.

An exact interpretation of an asymmetric synthesis requires a precise knowledge of the geometry of the transition state involved. This is, however, not accessible to measurement; therefore, it is neces-

[15a] *P.C. Manor, D.P. Shoemaker, A.S. Parkes,* J. Am. Chem. Soc. *92,* 5260 (1970).

[15b] *A. Moscowitz, K. Mislow,* J. Am. Chem. Soc. *84,* 4606 (1962).

[15c] *J. Tribout, R.H. Martin, M. Doyle, H. Wynberg,* Tetrahedron Letters 2839 (1972).

[15d] *H.J. Bestmann, W. Both,* Angew. Chem. *84,* 293 (1972); engl.: 296.

[15e] *Y. Shimizu, H. Tatemitsu, F. Ogura, M. Nakagawa,* Chem. Commun. 22 (1973).

[16] *F.H. Allen, S. Neidle, D. Rogers,* Chem. Commun. 838 (1966); 308 (1968); 452 (1969); J. Chem. Soc (C) 2340 (1970).

[17] *H. Pracejus,* Fortschr. chem. Forsch. *8,* 493 (1967); *D.R. Boyd, M.A. McKervey,* Quart. Rev. [London] *22,* 95 (1968); *J. Mathieu, J. Weill-Reynal,* Bull. Soc. Chim. France *1968,* 1211.

sary to use model conceptions, such as the 'steric approach control' and the 'product development control'[18]. These, too, are applicable with certainty only to a few cases (cf. 3.8:2.2.3 and 3.8:2.2.4 and Methodicum Chimicum Vol. 2).

Nevertheless for many reactions valuable empirical relations between the sign of rotation of the predominant stereoisomer and its configuration can be deduced if abundant experimental material is available.

For such correlations the knowledge of the absolute configuration of at least one representative of the class of chiral compounds in question is necessary.

The advantage of using asymmetric syntheses for configurational correlations lies mainly in the often simple experimental procedures. In contrast to chemical correlations, usually no complicated reaction sequences are required. If applied with due care, they are presently — together with the methods of optical comparison, cf. 3.8:3.1 — the most valuable methods for determining stereochemical configurations.

From the abundance of available material only those examples shall be presented which either allow a reliable and generally applicable determination of configurations (cf. especially (3.8:2.2.1 and 2.2.2) or are of some interest with regard to their reaction mechanism (3.8:2.2.3).

2.2.1 Acylation of Alcohols

Partial esterification of a racem. carboxylic acid, such as *mandelic acid,* with an optically active alcohol[19], e.g. *(−)-menthol,* yields a mixture of diastereomeric esters which, after quantitative hydrolysis (cf. Ref. [20]), yields an excess of active acid: e.g. *(+)-mandelic acid.*

As can be shown for numerous examples[21, 22], *alcohols* corresponding to the general projection formula **11** on reaction with an excess of *racemic α-phenylbutyric anhydride* (PBA) in pyridine by preferential formation of the ester with (−)-acid liberate (+)(S)-α-phenylbutyric acid (**12**)[22]. Hence, the sign of rotation of the acid indicates the configuration of the carbinol in question. The optical yields are generally high. In the case of rigid

molecules, such as *hydroxysteroids*[22], *homoisoborneol*[23], *caryophyllen derivatives*[24] or *metallocene carbinols*[25] p lies between 40 and 70%.

Table 1. Determination of the configuration of alcohols, amines and carboxylic acids

Reaction of	with	yields excess of
11	racem.PBA	(+)(S) − **12**
13	racem.PBA	(+)(S) − **12**
14	racem.PBA	(+)(S) − **12**
15	racem.**16**/DCC	(+)(S) − **16**
17	racem.**16**/DCC	(+)(S) − **16**

(also for R = H and L = OH or COOR¹)

Racem. Carbinol (Amine)	(+) - PBA	**11** (**15**)
$C_6H_5(CH_2)_n CH(R)NHCH_3$ **18**	(+)(S) − **16**	(S) − **18**
L−CH(M)−COOH **19**	(+)(S) − **18**/DCC (R=CH₃ , n=1)	(S) − **19**

[18] *W.G. Dauben, G.J. Fonken, D.S. Noyce,* J. Am. Chem. Soc. *78,* 2579 (1956).

[19] *W. Marckwald, A. McKenzie,* Ber. dtsch. chem. Ges. *32,* 2130 (1899).

[20] *O. Červinka,* Collect. Czech. Chem. Commun. *30,* 1738 (1965).

[21] *A. Horeau,* Tetrahedron Letters 506 (1961); 965 (1962).

[22] *A. Horeau, H.B. Kagan,* Tetrahedron *20,* 2431 (1964).

[22a] *H. Falk, K. Schlögl,* Mh. Chem. *96,* 276 (1965).

[23] *A. Marquet, A. Horeau,* Bull. Soc. Chim. France 124 (1967).

[24] *A. Horeau, J.K. Sutherland,* J. Chem. Soc. (C) 247 (1966);
T.J. Mabry, W. Renold, H.E. Miller, H.B. Kagan, J. Org. Chem. *31,* 681 (1966).

[25] *K. Schlögl,* in Ref. [101], p. 70 (1967);
O. Hofer, K. Schlögl, J. Organometal. Chem. *13,* 457 (1968).

Reaction of	with	yields excess of

20 (1S)—**21**

(-)(S)-α-Phen-ethylamine

(+)(S)—**22**

---

[a]L, M and S refer to large, medium-sized and small groups. The knowledge of the relative 'kinetic sizes' of ligands is essential for the interpretation of the results of asymmetric syntheses. For some cases this sequence is still uncertain. A sequence L>M>S valid for a certain reaction should be applied to another reaction only with due care (cf. also 3.8:2.2.6).

In this context it should be pointed out, that the sequence L>M>S is not always equivalent to the priorities of the sequence rule. Thus, while ethynyl is obviously 'smaller' than tert-butyl, it has priority over the latter. Consequently, the symbols (R) and (S) can be assigned to the projection formulas (such as **11**) only after careful considerations with regard to L, M and S (usually H).

The high rotation of (**12**) ($[\alpha]_D = 96°$) allows the detection of rather small effects of optical induction ($p > 0.015\%$), as has been shown in the case of deutero alcohols where an excess of (+)-(**12**) ($p = 0.1$ to 0.5%) indicates the configuration (**13**) (=S)[26]. Thus here, too, as in other cases[27], H is larger than D and corresponds to M. [Cf. also the reaction of *(+)-2-propanol-1-d_3* with PBA[28]: (+)-(**12**) indicates the configuration (S); consequently CH_3 (=L) is more bulky than CD_3 (=M). However er according to the sequence rule, CD_3 has priority over CH_3; cf. Formula (**33**) in Section 3.8:2.2.3.] This method was also applied successfully to *steroids*[22], *terpenes*[29], *alkaloids*[22a] (see here for a kinetic interpretation), *visnaganes*[29a], *metallocene carbinols*[25] (**24**), to the *(–)-trans, trans-spiro* [4.4] nonane-1.6-diol (**26**), accessible from the corresponding diketone[30] (**25**) and to [2.2]*paracyclo-*

phane derivatives[31] (**28**), and thereby to the determination of the configurations of the corresponding chiral ketones (**23, 25** and **27**).

All mentioned carbinols (**24, 26** and **28**) on reaction with PBA yield (+)-**12**; consequently, the asymmetric carbon atoms have the configuration (R). For the metallocene derivatives the results were confirmed by X-ray analysis[32].

Horeau's important method was recently modified (as a semimicro method[32a] and for glc-application[32b]) and applied (after conversion into suitable carbinols) to chiral Spiro-bi-ferrocenophane-dione[32c], bicyclo [2.2.2] octanes[32d], 9.10-dihydro-9.10-ethanoanthracenes[32e], [10]paracyclo-phanes[32f] and 1.6-methano-[10]anulenes[32g] as well as to phenylvinylcarbinol[32g] and to chiral indanols. The latter were transformed into active 1,1'- and 2,2'-spiro-biindanones their configurations being (+) (S)[32h]. For a quantitative treatment of Horeau's method cf. [32i].

(+)(1S)—**23** (+)—**24**

(-)(S)—**25** (-)—**26**

(+)(R)—**27** (-)—**28**

---

[26] *A. Horeau, A. Nouaille*, Tetrahedron Letters 3953 (1966).

[27] *K. Mislow, R. Graeve, A.J. Gordon, G.H. Wahl jr.*, J. Am. Chem. Soc. *85*, 1199 (1963); *86*, 1733 (1964).

[28] *A. Horeau, A. Nouaille, K. Mislow*, J. Am. Chem. Soc. *87*, 4957 (1965).

[29] *D. De Keukeleire, M. Verzele*, Tetrahedron *27*, 4939 (1971).

[29a] *H. Bernotat-Wulf, A. Niggli, L. Ulrich, H. Schmid*, Helv. Chim. Acta *52*, 1165 (1969).

[30] *H. Gerlach*, Helv. Chim. Acta *51*, 1587 (1968).

[31] *H. Falk, P. Reich-Rohrwig, K. Schlögl*, Tetrahedron *26*, 511 (1970).

[32] *O.L. Carter, A.T. McPhail, G.A. Sim*, J. Chem. Soc. (A) 365 (1967);
M.A. Bush, T.A. Dullforce, G.A. Sim, Chem. Commun. 1491 (1969);
G. Haller, K. Schlögl, Mh. Chem. *98*, 2044 (1967);
H. Falk, K. Schlögl, Mh. Chem. *102*, 33 (1971).

[32a] *R. Weidmann, A. Horeau*, Tetrahedron Letters 2979 (1973).

[32b] *C.J.W. Brooks, J.D. Gilbert*, Chem. Commun. 194 (1973).

[32c] *H. Falk, W. Fröstl, K. Schlögl*, Mh. Chem. *102*, 1270 (1971).

[32d] *D. Varech, J. Jacques*, Tetrahedron *28*, 5671 (1972).

2.2.2 Acylation of Amines

Acylation of α-*phenyl-ethylamine*[22a] or α-*mesityl-ethylamine*[33] with PBA yields analogous results.*
In the projection formula 11, OH has then to be replaced by NH_2 (15). From *N*-methyl derivatives of identical configurations and PBA, 12 with opposite sign of rotation is obtained[33] (cf. 14).

The asymmetric acylation of chiral amines 15 can also be achieved by an excess of racem. α-*phenyl-propionic acid* (hydratropic acid, 16) in the presence of *N,N'-dicyclohexyl carbodiimide (DDC)*[34]. The results are in agreement with those of the kinetic resolution of PBA.

In the case of *sec.* amines, α-aminoalcohols, and esters of α-amino acids the effect is again reversed: Compounds corresponding to structure 17 (usually *S*) liberate (+)(S)-16 (with *p* up to 50%[34]).

Configurational comparisons are also possible by kinetic resolution of carbinols and amines with the aid of optically active acids. The active carbinol (or primary amine), which is isolated after reaction of the racemate with (+)-PBA, corresponds to the configuration shown in 11(15)[35]. According to the above mentioned results[33], the reaction of *racem.* (n + 1) methylamino-1-phenyl-alkanes (18) with (+)(S)-16 and DDC affords an excess of active (S)-amines[36]. (For the stereoselective acylation of racem. amines with mixed anhydrides from optically active acids cf. Ref.[37].)

The procedure of *asymmetric aminolysis* can also be employed for establishing the configurations of *chiral carboxylic acids*. After reaction of the racem. α-*alkyl-phenylacetic* acids (19) with (+)(S)-18[R = CH₃, n = 1, which is easily accessible

from (−)-ephedrine] and DDC, the isolated active acid (*p* = 3–7%) has the configuration (S)-19, at least in the case of α-alkyl phenylacetic acids[38]. For R = isopropyl the result was questioned[39]; it could be confirmed by chemical correlation, and the sequence of relative sizes was determined as being alkyl (incl. *iso*propyl) <phenyl <benzyl[40].* (For the application to *necic acids* from pyrrolizidine alkaloids cf. Ref.[41].)

Partial reaction of the anhydrides 20 of racem. α-substituted *metallocenecarboxylic acids* with (−)(S)-α-phenethylamine in pyridine yields active acids 21 (p = 3–19%) having the configuration (1S), as was established by independent methods[42]. This method made possible the configurational assignment not only to α-methylbiferrocenyl-α'-carboxylic acid [(+)(1R)][43] but also — because of the equivalent topology — to [2.2] *para-cyclophanecarboxylic acid* as (+)(S)-22[44]. The latter result was supported by theoretical considerations[45] and additional stereochemical investigations[31].** This asymmetric aminolysis of the anhydrides of planarchiral carboxylic acids (cf. Table 1, bottom) was also successfully applied to the chiral carboxylic acids of 9.10-dihydro-9.10-ethano-anthracene[32e], [10]paracyclophane[32f], 1.6-methano[10]anulene[32g] and [2.2]metacyclo-

---

* In the meantime a new correlation of *(−)-α-isopropyl- (and tert.-butyl-) phenylacetic acid* with *(+)-(S)-hydratropic acid* has shown that, in contrast to Ref.[38] and[40] (with regard to R = *iso*propyl), both acids have the configuration (−)-(R)[40]; consequently, the abovementioned residues are larger than phenyl and correspond to the ligand L in 19.

** Recently also by X-ray analysis (G.W. Frank, personal commun.)

* Cf. however: *H. Brockmann jr., J. Bode,* Liebigs Ann. Chem *748,* 20 (1971).

[32e] *J. Paul, K. Schlögl,* Mh. Chem. *104,* 274 (1973); *M.J. Brienne, J. Jacques,* Tetrahedron Letters 1053 (1973).

[32f] *H. Eberhardt, K. Schlögl,* Liebigs Ann. Chem. *760,* 157 (1972).

[32g] *U. Kuffner, K. Schlögl,* Mh. Chem. *103,* 1320 (1972).

[32h] *J.H. Brewster, R.T. Prudence,* J. Am. Chem. Soc. *95,* 1217 (1973); *H. Falk, W. Fröstl, K, Schlögl,* Tetrahedron Letters 217 (1974).

[32i] *H.J. Schneider, R. Haller,* Tetrahedron *29,* 2509 (1973).

[33] *H. Pracejus, H. Ripperger,* Z. Chem. *8,* 268 (1968).

[34] *O. Červinka,* Collect. Czech. Chem. Commun. *31,* 1371 (1966).

[35] *R. Weidmann, A. Horeau,* Bull. Soc. Chim. France 117 (1967).

[36] *O. Červinka, E. Kroupova, O. Belovsky,* Collect. Czech. Chem. Commun. *33,* 3551 (1968).

[37] *H. Herlinger, H. Kleimann, I. Ugi,* Liebigs Ann. Chem. *706,* 37 (1967).

[38] *O. Červinka, L. Hub,* Chem. Commun. 761 (1966); *O. Červinka, L. Hub,* Collect. Czech. Chem. Commun. *32,* 2295 (1967).

[39] *C. Aaron, D. Dull, J.L. Schmiegel, D. Jaeger, Y. Ohashi, H.S. Mosher,* J. Org. Chem. *32,* 2797 (1967).

[40] *O. Červinka, L. Hub,* Collect. Czech. Chem. Commun. *33,* 1911 (1968).

[40a] *D.R. Clark, H.S. Mosher,* J. Org. Chem. *35,* 1114 (1970).

[41] *O. Červinka, L. Hub,* Collect, Czech. Chem. Commun. *33,* 2933 (1968).

[42] *H. Falk, K. Schlögl,* Mh. Chem. *99,* 578 (1968).

[43] *K. Schlögl, M. Walser,* Mh. Chem. *100,* 1515 (1969).

[44] *H. Falk, K. Schlögl,* Angew. Chem. *80,* 405; engl. *7,* 383 (1968).

[45] *O.E. Weigang jr., M.J. Nugent,* J. Am. Chem. Soc. *91,* 4555 (1969).

[45a] *H.J. Bestmann, H. Scholz, E. Kranz,* Angew. Chem. *82,* 808; Engl. *9,* 796 (1970); *B. Kainradl, E. Langer, H. Lehner, K. Schlögl,* Liebigs Ann. Chem. *766,* 16 (1972).

phane[45a], thereby deducing their configurations as (+)-(9R), (−)(S), (+)(S) and (−)(S), resp.

The configurations of centrochiral carboxylic acids may also be determined by reaction of one mole of the acid chloride with two moles of racem. *benzylidene-methyl-phenyl-n-propyl-phosphorane*[45a]. From the sign of $[\alpha]_D$ of the optically active benzyl-methyl-n-propyl-phosphoniumchloride formed (the configuration of which has been established by the X-ray method), by application of the Ruch-Ugi-Model[82] the configuration of the employed carboxylic acid can be established. From the results some relatives sizes of ligands (ligand-constants λ according to[82]) were deduced.

2.2.3 Reduction

The Meerwein-Ponndorf-Verley-reduction is one of the few reactions, where a well founded concept of the transition state[46] permits a reliable conclusion about the configuration of the product[47]. On reduction of a prochiral ketone (L–CO–M) with the aluminum salt of a *sec.* alcohol having the configuration shown in **29** (usually R) this configuration is transferred to the produced carbinol. Of the two cyclic diastereomeric transition states, obviously **29** is preferred because of the interaction L–L′ in **30**; an excess of the carbinol **11** is formed by kinetic control.

29   **11**

30

This asymmetric reduction permitted for the first time a reliable determination of the configuration of a *chiral biphenyl*[10]; for binaphthyl derivatives the results were subsequently confirmed by the X-ray method[48] (for biphenyls see Ref. [48a]).

After partial reduction of the racemic ketone **31** with (+)(S)-2-octanol or -pinacolyl alcohol, the starting material contains an excess of (+)-**31** while the product is mostly the (−)-carbinol **32**. On the basis of the above mentioned model the configuration (S) can be assigned to (+)-**31**.

(+)(S)   **31**   (−)(R)

(−)(R)-**32**
(enriched)

This chirality-transfer-process was also applied to the reduction of aldehydes to active deutero alcohols of known configurations[49]. Use of *isobornyloxy-magnesium bromide* gave optical yields higher than 80%.

In connection with the preparation of *(+)(S)-2-propanol-1-d_3* (**33**) from (+)(S)-lactic acid kinetic studies on the influence of the relative sizes of R in the carbinol (+)(S)-R-CHOM-Me were performed[50].

(+)(S)-**33**

Also in the case of Grignard-reductions (*p* up to 90%!)[51] and reductions with Li-amides[52] the hydride transfer occurs via a cyclic transition state analogous to **29** and **30**, resp.

Within the last years the asymmetric reduction of ketones with 'chiral $LiAlH_4$' (accessible by pretreatment with active carbinols or amines) was

[46] *R.B. Woodward, N.L. Wendler, F.J. Brutschy*, J. Am. Chem. Soc. *67*, 1425 (1945).

[47] *W. v.E. Doering, R.W. Young*, J. Am. Chem. Soc. *72*, 631 (1950); *74*, 2997 (1952).

[48] *H. Akimoto, T. Shiori, Y. Iitaka, S. Yamada*, Tetrahedron Letters 97 (1968).

[48a] *L.H. Pignolet, R.P. Taylor, W. De W. Horrocks jr.*, J. Am. Chem. Soc. *91*, 5457 (1969).

[49] *A. Streitwieser jr., J.R. Wolfe jr., W.D. Schaeffer*, Tetrahedron *6*, 338 (1959).

[50] *K. Mislow, R.E. O'Brien, H. Schaefer*, J. Am. Chem. Soc. *84*, 1940 (1962).

[51] *J.S. Birtwistle, K. Lee, J.D. Morrison, W.A. Sanderson, H.S. Mosher*, J. Org. Chem. *29*, 37 (1964).

[52] *G. Wittig, U. Thiele*, Liebigs Ann. Chem. *726*, 1 (1969).

studied in some detail, mainly with a view towards the configuration of the formed carbinols.

Use of *sugar derivatives* yields (R)-carbinols of up to 70% optical purity[53]. The asymmetric reduction of ketones with the reagent from $LiAlH_4$ and (+)-4-dimethylamino-3-methyl-1. 2-diphenyl-2-butanol gives not only high optical yields (up to 70%) of the carbinols but also permits conclusions as to their configurations[53a].

Table 2. Asymmetric Reductions

Ketone	reduced with[a]	carbinol	p

H_3C–CO–R $\xrightarrow{(1,2,3)}$ H–C(CH_3)(R)–OH   2–14%   (+)(S)

H_3C–CO–Ar $\xrightarrow{(1,2,3)}$ HO–C(CH_3)(Ar)–H   1–48%   (+)(R)

H_5C_6–CO–Ar $\xrightarrow{(1,3)}$ H–C(C_6H_5)(Ar)–OH   (S)

$H_{11}C_6$–CO–Ar $\xrightarrow{(1)}$ HO–C(C_6H_{11})(Ar)–H   (+)(R?)

[a] $LiAlH_4$-(—)-quinine (1), -(—)-cinchonidine (2), -(—)-ephedrine (3)

Reduction of phenylglyoxalates with $LiAlH_4$-cinchonine, – (–)-menthol and – (+)-camphor-complexes afforded *(R)-mandelic acid* (p up to 49%)[54].

The results of asymmetric reductions with modified $LiAlH_4$[55] may be summarized as shown in Table 2.

In spite of apparent advantages, the application of reduction with 'asymmetric hydrides' for determining configurations seems to be problematic. This statement applies also (at least partly) to the asymmetric hydroboration.

Prochiral ketones can be reduced with *diisopino-campheylborane* (cf. 3.8:2.2.4) to give *optically active carbinols*. The results obtained for *methylketones* [R–CO–Me; R = ethyl, isopropyl and phenyl → (R)-carbinol; R = tert. butyl → (S)-carbinol] are ambiguous and, accordingly, the configuration of *2, 2-dimethyl-3-butanol (pinacolyl alcohol)* was doubted; it could be confirmed, however, as being (+)(S) by chemical correlation with lactic acid[6].

An interpretation of these results as well as of the reduction of aldehydes to *deuteroalcohols* seems to be possible on the basis of a recent model of asymmetric hydroboration[57].

2.2.4 Addition

The stereoselective addition of organometallic compounds to the C=O bond has proven very valuable not only for conformational analyses of transition states (Prelog and Cram's rules)[58] but especially in connection with semiempirical methods for configurational correlations.

If **34** is accepted as the preferred conformation of α-*ketoesters* (of chiral carbinols)[58, 59], the reagent (R^2MgX) will approach from the S side (i.e. from 'above'). Consequently, an auxiliary alcohol of the configuration shown leads – after hydrolysis of the diastereomeric esters (cf. **35**) – to an excess of an α-hydroxy acid of a predominant configuration, e.g. (R)-**36**.

Hence, the sign of $[\alpha]_D$ of the *atrolactic acid* (**36**), which is produced by esterification of an optically active alcohol with *phenylglyoxylic acid* (→ **34**, $R^1 = C_6H_5$), reaction with methylmagnesium bromide, and subsequent hydrolysis, is related to the configuration of the alcohol.

[53] *S.R. Landor*, J. Chem. Soc. (C) *1966*, 1822, 2280; 197 (1967);
O. Červinka, A. Fabryova, Tetrahedron Letters 1179 (1967).

[53a] *S. Yamaguchi, H.S. Mosher*, J. Org. Chem. *38*, 1870 (1973).

[54] *A. Horeau, H.B. Kagan, J.P. Vigneron*, Bull. Soc. Chim. France 3795 (1968).

[55] *O. Cervinka, V. Suchan, B. Masar, O. Belovsky*, Collection Czech. Chem. Commun. *30*, 1684, 1693, 2487 (1965).

[56] *H.C. Brown, D.B. Bigley*, J. Am. Chem. Soc. *83*, 3166 (1961).

[57] *K.R. Varma, E. Caspi*, Tetrahedron *24*, 6365 (1968).

[58] *K. Weinges, W. Kaltenhäuser, F. Nader*, Fortschr. Chem. Forsch. *6*, p. 402 (1966).

[59] *V. Prelog*, Bull. Soc. Chim. France 987 (1956). Cf. however: *R. Parthasovaty, J. Ohrt, A. Horeau, J.P. Vigneron, H.B. Kagan*, Tetrahedron *26*, 4705 (1970).

34   **35** (excess)

For
$R^1 = C_6H_5$ $R^2 = CH_3$

excess of

11 $(+)(R)$-**36**

In many cases (e.g. with *atropisomeric biphenyls*)[10, 60] this method gave correct results. Presently it is replaced by the more simple and elegant method of Horeau (3.8:2.2.1).

The results of the addition to a carbonyl group adjacent to chiral centers can be predicted correctly using Cram's rule[61], although the assumed conformation **37** is doubtful. An improved model for the interpretation of this asymm. induction[62] takes into account recent results of conformational analysis (cf. **38**).

37

38

The asymmetric synthesis of sulfoxides via sulfinate esters can also be used for determining the configurations of alcohols: for this purpose the latter are acylated with *p*-toluenesulfinyl chloride[63]. The resulting mixture of diastereomeric esters (**39**) is treated with MeMgI. The sign of rotation of the predominant enantiomer of *methyl p-tolyl sulfoxide* (**40**, $p = 18$–50%) is related to the

configuration of the inducing alcohol. Carbinols **11** give rise to an excess of $(-)(S)$-**40**[64, 65].

Application to 1, 11-dimethyl-dibenzo[a, c] cyclo-heptadiene-6-ol (**41**) – in agreement with results of the atrolactic acid synthesis and Horeau's method – resulted in the configuration $(+)(S)$-**41**. The method was also successfully applied to the alcohol **33**[65].

11   (excess of) **39**

$(-)(S)$-**40**

$(+)(R)$-**40**
$(p = 6\%)$

$(+)(S)$-**41**

$(-)(S)$-**40**
$(p = 0.26\%)$

$(+)(S)$-**33**

The asymmetric reduction of *carbonyl compounds* with *diisopinocampheylborane* (easily accessible from optically active α-pinene, $NaBH_4$ and BF_3) was already briefly mentioned (Section 3.8:2.2.3). This asymmetric addition had been applied previously to *cis*-olefins, where, after oxidation of the intermediate boranes, active alcohols, such as *(–)(R)-2-butanol from cis-2-butene*, were obtained with optical yields up to 90% (!)[66]. The postulated four membered transition state[67] permits – on the basis of model considerations[66] which were subsequently modified[57, 68] – a prediction of the configurations of the alcohol as well as of the active olefins produced by kinetic resolution. Although this method appears rather attractive, it has been applied so far only to a few cases: e.g. to

[60] *J.A. Berson, M.A. Greenbaum*, J. Am. Chem. Soc. *79*, 2340 (1957); *80*, 445, 653 (1958).

[61] *D.J. Cram, F.A. Abd Elhafez*, J. Am. Chem. Soc. *74*, 5828 (1952).

[62] *G.J. Karabatsos*, J. Am. Chem. Soc. *89*, 1367 (1967).

[63] *J.v. Braun, W. Kaiser*, Ber. dtsch. chem. Ges. *56*, 549 (1923).

[64] *M.M. Green, M. Axelrod, K. Mislow*, J. Am. Chem. Soc. *88*, 861 (1966).

[65] *M. Axelrod*, Dissertation, Princeton University, 1966. Diss. Abstracts B27 (8), 2643 (1967).

[66] *H.C. Brown, N.R. Ayyangar, G. Zweifel*, J. Am. Chem. Soc. *86*, 397, 1071 (1964).

[67] *H.C. Brown*, Hydroboration, p. 13, W.A. Benjamin, New York 1962.

[68] *A. Streitwieser jr., L. Verbit, R. Bittman*, J. Org. Chem. *32*, 1530 (1967).

deutero alcohols[68, 69], *cyclic olefins*[70] and *dimethylallene* (**42**)[71].

$$H_3C-CH=C=CH-CH_3 \xrightarrow[\text{(-)-pinene}]{\text{borane from}} (-)(R)-\mathbf{42}$$

42

$$\longrightarrow \quad (+)(S)-\mathbf{42}$$

(+)-**43**

After partial reaction of racem. **42** with the active borane from (−)-α-pinene the recovered allene is levorotatory. From model considerations as well as from a comparison with diphenylallene of known absolute configuration the (R)-configuration was deduced for (−)-**42**.

This result, already postulated on the basis of theoretical considerations[72], was supported by correlation of (+)-**42** with the *cyclopropane derivative* (+)-**43**[73].

A new concept of stereochemical control (in the addition of organometallics to C=O) was developed[73a]. (Cf. also a paper on asymmetric synthesis and conformation[73b]).

2.2.5 Rearrangement

From the standpoint of a systematic survey of asymmetric syntheses generally applicable for configurational correlations, rearrangement reactions are rather poor although some examples might be very interesting.

A reasonable application is obviously feasible only if the stereochemical course of the reaction(s) is known. Here, as with the related direct chemical correlations, (3.8:2.1), the reaction sequence must be carefully planned in every case.

The Stevens-rearrangement with known stereochemistry[74] was employed successfully for the

configurational correlation of axial and centrochiral compounds. The levorotatory spiro compound **44** undergoes rearrangement to give **45**, the configuration of which was shown to be (S) by oxidative degradation to a derivative **46** of (S)-*aspartic acid*[75]. In accord with an empirical rule[76], from these results for (−)-**44** the configuration (R) can be deduced.

For the Stevens-rearrangement of *chiral biphenyls* cf. Ref. [77].

(−)(R)−**44**

(−)−**45**   **46**

Rearrangement reactions were also very useful for establishing the configurations of axial chiral allenes.

Reaction of (−)-**47** of known absolute configuration (S)[78] with thionylchloride affords the chlorosulfite **48**, which in turn by S_N I-rearrangement is converted into the levorotatory allene **49** having (S)-configuration[78, 79].

(−)(S)−**47**

(−)−**48**   (−)(S)−**49**

---

[69] *H. Weber, J. Seibl, D. Arigoni*, Helv. Chim. Acta *49*, 741 (1966).

[70] *D.K. Shumway, J.D. Barnhurst*, J. Org. Chem. *29*, 2320 (1964);
J. Katsuhara, H. Watanabe, K. Hashimoto, M. Kobayashi, Bull. Chem. Soc. Japan *39*, 617 (1966).

[71] *W.L. Waters, M.C. Caserio*, Tetrahedron Letters 5233 (1968).

[72] *J.H. Brewster* in Ref. [101], Vol. 2, p. 35 (1967).

[73] *W.M. Jones, J.M. Walbrick*, Tetrahedron Letters 5229 (1968).

[73a] *J. Laemmle, E.C. Ashby, P.V. Roling*, J. Org. Chem. *38*, 2526 (1973).

[73b] *R.A. Auerbach, C.A. Kingsbury*, Tetrahedron *29*, 1457 (1973).

[74] *D.J. Cram*, Fundamentals of Carbanion Chemistry, p. 223, Academic Press, New York 1965.

[75] *J.H. Brewster, R.S. Jones jr.*, J. Org. Chem. *34*, 354 (1969).

[76] *G. Lowe*, Chem. Commun. 411 (1965); For an extension cf. Ref. [86c].

[77] *H. Joshua, R. Gans, K. Mislow*, J. Am. Chem. Soc. *90*, 4884 (1968).

[78] *R.J.D. Evans, S.R. Landor*, J. Chem. Soc. 2553 (1965).

[79] *E.L. Eliel*, Tetrahedron Letters No. 8, 16 (1960).

The configuration of allenes was also determined by a Claisen-rearrangement[80] and by correlation with *cyclopropane derivatives*[81].

2.2.6 Theoretical Aspects

As has been mentioned repeatedly, a qualitatively correct interpretation of many asymmetric syntheses is possible on the basis of empirical relations or model considerations. A recent profound group-theoretical model permits, (so far) in a few cases, quantitative predictions also and statements as to the relative sizes of the ligands L, M and S[82]. (Cf. Methodicum Chimicum Vol. 2)

The future will reveal to what extent this model will permit a general prediction of the results of new asymmetric syntheses.

A theory of diastereoisomeric transition states deals with the problem of asymmetric induction[82a]. The theory of chirality functions[82b] was successfully applied to chiral methanes[82c] and allenes[82d].

3 Physical-Chemical Methods

3.1 Optical Comparison

A comparison of chiroptical properties (*optical rotation, rotatory dispersion, ORD* and *circular dichroism, CD*) obviously seems to be especially attractive for configurational correlations.

Rules based solely on a comparison of $[\alpha]_D$ and $[M]_D$-values, resp., such as the rules of optical superposition (cf. Ref. [83] for the application to steroids) or Freudenberg's displacement rule (cf. Ref. [39, 84] for recent examples), or recent rules relating signs of rotations and configurations of *allenes* and *spiranes*[76] or *metallocenes*[85] have been successfully employed in many cases. (For an attempt

to deduce relationships between centrochiral compounds of the group IV and V elements cf. Ref. [86]) ORD and CD are, however, much more reliable for configurational correlation and, thus, for determining (absolute) configurations (Chapter 5.7).

A sector rule for correlating the sign of the ^1L_b Cottoneffect with the configuration of chiral carbophanes gives excellent results, and permits some predictions[86a]. An ORD- and CD-comparison of suitable derivatives lead to the configuration (+)(R) for 1,2-cyclononadiene[86b], not in accordance with Lowe's rule[76] (for an extension of. Ref. [86c]). 1-Oxo-[2.2] metacyclophane has the configuration (−)(S) as deduced from its chiroptical properties[86d].

Considerations regarding relations between chirality and sign of rotation (such as the helix model)[87] exceed the scope of this review.

3.2 Method of Quasiracemates

(+)- and (−)-antipodes of a chiral molecule usually form (stoichiometric) compounds (R)-A · (S)-A. Such racemic compounds can be recognized from *phase diagrams, IR-spectra, X-ray powder diagrams* (cf. Chapter 7.1) and by *calorimetry* (9.3). Compounds having closely related constitutions, but opposite configurations, (R)-A · (S)-A', e.g. *(+)-chloro* and *(−)-bromosuccinic acids*, exhibit an analogous behavior.

This principle may be employed for establishing configurational relations[88]. This very reliable method, however, has somewhat severe limitations: the compounds to be compared must be crystalline and chemically very similar; moreover, (R)-A and (S)-A' should not form conglomerates. As mentioned earlier for racemates, the formation of quasiracemates may be deduced not only from *phase diagrams* but also from solid state *IR-spectra*[89], *X-ray powder diagrams*[89] and by *differential calorimetry*[90]. This method was applied to *biphe-*

[80] E.R.H. Jones, J.D. Loder, M.C. Whiting, Proc. Chem. Soc. 180 (1960).

[81] W.M. Jones, J.M. Walbrick, Tetrahedron Letters 5229 (1968).
W.M. Jones, J.W. Wilson jr., Tetrahedron Letters 1587 (1965).

[82] I. Ugi, Z. Naturforsch. *20b*, 405 (1965);
E. Ruch, I. Ugi, Theor. Chim. Acta *4*, 287 (1966); Ref. [98], p. 537; E. Ruch, I. Ugi in Ref. [101], Vol. 4, p. 99 (1969).

[82a] L. Salem, J. Am. Chem. Soc. *95*, 94 (1973).

[82b] E. Ruch, Accounts Chem. Res. *5*, 49 (1972).

[82c] W.J. Richter, B. Richter, E. Ruch, Angew. Chem. *85*, 21 (1973); engl.: 30.

[82d] E. Ruch, W. Runge, G. Kresze, Angew. Chem. *85*, 10 (1973); engl.: 20.

[83] D.H.R. Barton, W. Klyne, Chem. and Ind. 755 (1948).

[84] D.W. Slocum, K. Mislow, J. Org. Chem. *30*, 2152 (1965);
K. Schlögl, in Ref. [101], Vol. 1 p. 84–85 (1967).

[85] H. Falk, K. Schlögl, Tetrahedron *22*, 3047 (1966); K. Schlögl, J. Pure Appl. Chem. *23*, 413 (1970).

[86] O. Cervinka, H. Belovsky, Collect. Czech. Chem. Commun. *30*, 2859 (1965).

[86a] E. Langer, H. Lehner, K. Schlögl, Tetrahedron *29*, 2473 (1973).

[86b] W.R. Moore, H.W. Anderson, St. D. Clark, T.M. Ozretich, J. Am. Chem. Soc. *93*, 4932 (1971).

[86c] P. Crabbé, E. Velarde, H.W. Anderson, S.D. Clark, W.R. Moore, A.F. Drake, S.F. Mason, Chem. Commun. 1261 (1971).

[86d] K. Mislow, M. Brzechffa, H.W. Gschwend, R.T. Puckett, J. Am. Chem. Soc. *95*, 621 (1973).

[87] J.H. Brewster, in Ref. [101]; Vol. 2, p. 1 (1967).

[88] A. Fredga, Tetrahedron *8*, 126 (1960).

[89] Cf. S. Gronowitz, Ark. Kemi *11*, 361 (1957).

[90] C. Fouquey, J. Jacques, Tetrahedron *23*, 4009 (1967).

nyls[91], to *phenoxypropionic acids* and *tetrahydrothiopyran dicarboxylic acids*[92].

3.3 Method of Occlusion Compounds

Urea crystallizes in the form of enantiomeric screw-lattices. Such configurationally uniform crystals, if grown in solutions of racemates (of longchain chiral compounds, such as α-substituted fatty or dicarboxylic acids), preferentially include one enantiomer. From these, configurational relationships may be deduced[93]. However, a general application is not feasible, mainly because of occuring anomalies.

For recent work on this subject cf. Ref.[93a].

3.4 NMR-Method

Finally, NMR-spectroscopy (cf. Chapter 5.5) in optically active solvents − because of the magnetic non-equivalence of diastereometrically solvated enantiomers − seems, at least in some cases, applicable to configurational correlations[94] (for application to amino acids see Ref.[94a]). Thus e.g., the OMe-signal of methyl *(−)-α-hydroxy-α-trifluoromethyl phenylacetate* in *(+)-α-(1-naphthyl)ethylamine* is shifted upfield, as compared with the (+)-enantiomer. From analogous chemical shifts of α-hydroxyesters of known absolute configurations, the configuration (−)-(S)-**50** was deduced for *3,3,3-trifluoro-2-hydroxy-2-phenyl propionic acid*.

$$
\begin{array}{c}
\text{COOH} \\
\text{HO}\!-\!\!\!-\!\!\!-\!\text{CF}_3 \\
\text{C}_6\text{H}_5
\end{array}
$$

(−)(S)-**50**

In the meantime NMR-spectroscopy proved to be very useful not only for determining enantiomeric purities − especially with the aid of chiral shift reagents[94b] − but also for determining absolute configurations e.g. of sec. alcohols[94c] and carboxylic acids[94d] from the spectra of diastereomeric mandelates, O-methylmandelates and α-methoxy-α-trifluoromethylphenyl acetates[94e] and sec. amides[94d], resp. The latter can be separated by liquid liquid chromatography[94d, 94e]. The enantiomeric purities and configurations of α-deuterated primary alcohols can be determined from the NMR-spectra of their esters with (−)-camphanic acid in the presence of Eu(dpm)[94f].

[94c]*J.A. Dale, H.S. Mosher*, J. Am. Chem. Soc. *95*, 512 (1973).

[94d]*G. Helmchen, R. Ott, K. Sauber*, Tetrahedron Letters 3873 (1972); see also *G. Helmchen, V. Prelog*, Helv. Chim. Acta *55*, 2612 (1972).

[94e]*R. Eberhardt, H. Lehner, K. Schlögl*. Mh. Chem. *104*, 1409 (1973).

[94f]*H. Gerlach, B. Zagelak*, Chem. Commun. 274 (1973).

4 Bibliography

[95] *J.A. Mills, W. Klyne* in *W. Klyne*, Progress in Stereochemistry, Vol., p. 177, Butterworth, London 1954.

[96] *W. Schlenk jr.*, Angew. Chem. *77*, 161 (1965).

[97] *H. Falk*, Österr. Chem. Z. *66*, 242 (1965).

[98] *E.L. Eliel*, Stereochemie der Kohlenstoffverbindungen, Verlag Chemie, Weinheim/Bergstr. 1966.

[99] *K. Mislow*, Einführung in die Stereochemie, Kap. 3.3−3.6, Verlag Chemie, Weinheim/Bergstr. 1967.

[100] *H. Pracejus*, Fortschr. Chem. Forsch. *8*, 493 (1967).

[101] *N.L. Allinger, E.L. Eliel*: Topics in Stereochemistry, Vol. I-VI, Interscience, New York 1967−1971.

[102] *D.R. Boyd, M.A. McKervey*, Quart. Rev. [London] *22*, 95 (1968).

[103] *J. Mathieu, J. Weill-Raynal*, Bull. Soc. Chim. France 1211 (1968).

[104] *G. Krow* in Ref.[101], Vol. V, p. 31 (1970).

[105] *J.D. Morrison, H.S. Mosher*, Asymmetric Organic Reactions, Prentice-Hall, Englewood Cliffs, New York 1971.

[106] Synthesis, Absolute Configuration and Optical Purity of Chiral Allenes; *R. Rossi, P. Diversi*, Synthesis *25* (1973).

[107] Asymmetric Syntheses with the Aid of Homogenous Transition Metal Catalysts; *B. Bogdanovic*, Angew. Chem. *85*, 1013 (1973); engl.: 954.

---

[91] *M. Siegel, K. Mislow*, J. Am. Chem. Soc. *80*, 473 (1958).

[92] *J.E. Nemorin, E. Jonsson, A. Fredga, K. Olsson*, Ark. Kemi *30*, 403, 409 (1969).

[93] *W. Schlenk jr.*, Angew. Chem. *77*, 161 (1965).

[93a]*W. Schlenk jr.*, Liebigs Ann. Chem. 1145, 1156, 1179, 1195 (1973).

[94] *W.H. Pirkle, S.D. Beare*, Tetrahedron Letters 2579 (1968).

[94a]*W.H. Pirkle, S.D. Beare*, J. Am. Chem. Soc. *91*, 5150 (1969).

[94b]*C. Kutal* in 'Nuclear Magnetic Resonance Shift Reagents', edited by *R.E. Sievers*, Academic Press, New York 1973.

3.9 Determination of CH-Acids and Other Weak Acids

Hans F. Ebel
Organisch-Chemisches Institut der Universität,
D-69 Heidelberg

1 Introduction

CH-acidity is defined as the ability of organic molecules to liberate hydrogen, in the form of a proton, from its bond to carbon. Being species of exceedingly high energy, protons never exist in the free state in a condensed phase, but rather must be bonded to a suitable acceptor (Brönsted base B), in an equilibrium reaction. The removal of a proton from its bond to carbon leaves the latter atom negatively charged, so that organic molecules form carbanions, whereas cations form dipolar molecules (ylids) or uncharged molecules. For uncharged CH-acids, two types of reaction can be distinguished, which differ in the charge of the base involved:

$$R^1R^2R^3C\text{---}H + |B \rightleftharpoons R^1R^2R^3C|^{\ominus} + H\text{---}\overset{\oplus}{B} \quad (1)$$

CH-acid	neutral base	carbanion	conjugate acid of the neutral base

$$R^1R^2R^3C\text{---}H + |B^{\ominus} \rightleftharpoons R^1R^2R^3C|^{\ominus} + H\text{---}B \quad (2)$$

CH-acid	anionic base	carbanion	conjugate acid of the anionic base

Here, R^1, R^2, and R^3 represent three monovalent groups, one divalent and one monovalent, or one trivalent group. Depending on the nature of these atoms or groups and on the proton affinity of base B, the equilibria may lie far to the left or to the right. Accordingly, CH-acids are classified as weak or strong acids. The scale of CH-acids arranged in order of decreasing strength ranges from the cyano-hydrocarbon acids, which are stronger than mineral acids and completely dissociated even in dilute hydrochloric or sulfuric acid, to the aliphatic hydrocarbons, which can only be deprotonated with difficulty by the strongest bases available[44].

The logarithmic acidity constants ($pK = - \log K$) derived from the equilibria of acid/base reactions range from negative values through many units (orders of magnitude in K) to high positive values, thus representing CH-acids of very different strengths or 'acidities' (the latter term now being understood in a quantitative sense). Accordingly, various methods of identification and determination must be used. CH-acids which can be deprotonated in aqueous media to a measurable extent may be estimated by the normal methods used for the stronger organic and inorganic acids. These methods, along with some investigations in non-aqueous solution, are dealt with in Section 3.6. In the text below, analytical procedures will be discussed which are of particular importance for weak CH-acids, and which originate mostly from carbanion chemistry and organometallic chemistry. Some of these procedures are easily adaptable to the analyses of other weak acids (OH-, NH-, PH-, and other acids).

2 Qualitative Analysis of CH-Acids and Other Weak Acids

2.1 Identification of Weak Acids with Reagents which Cause Gas Formation

2.1.1 Use of Organometallic Reagents

Formally, the organic compounds of electropositive metals may be considered as salts of carbanions ($C^{\delta-}\text{---}M^{\delta+}$). The metal/hydrogen exchange reaction between two organic groups, according to the equation:

$$(R^1)\text{---}H + (R^2)\text{---}M \rightleftharpoons (R^1)\text{---}M + (R^2)\text{---}H \quad (3)$$

thus appears as a special case of the general transprotonation reaction (2). The bigger the difference in stability between the two potential carbanions $(R^1)^{\ominus}$ and $(R^2)^{\ominus}$ and the more electropositive the metal, the greater will be the driving force of such a reaction. If R^2 is a small group (up to C_4), then (R^2)–H represents a volatile hydrocarbon, whose liberation as a gas is evidence for the reaction (3) and thus demonstrates the presence of an acidic compound (R^1)–H.

(R^2)-M may be methylmagnesium iodide, which can be easily synthesized from methyl iodide and magnesium in ether, by Grignard reaction. If the ethereal solution of methylmagnesium iodide is treated with an acidic substance (or the solution of one in an aprotic solvent), then gas is formed at a rate which depends on the rate of the metal/hydrogen exchange (transmetallation). This assay is known as the Zerewitinoff reaction (cf. Section 3.7:2.4, 3.9:3.1), but the reaction was introduced first as a qualitative method for the identification of hydroxy groups by Tschugaeff[1].

---

[1] *L. Tschugaeff*, Ber. *35*, 3912 (1902).

239

The reagent is best used at a concentration of about 0.1 M, and as an alternative addition of the Grignard reagent to the solution to be tested for acidity is often employed. Carefully dried pipets should be used, since moisture also causes formation of gas (H_2O acting as an OH-acid).

The reaction in its classical form[33] is essentially restricted to the identification of OH- and NH-acids, i.e. of acids with pK $\leq$ 20 on the McEwen scale (cf. Ref.[44], p. 66). It is, therefore, applicable to all alcohols, and supplements the reaction with diazomethane, which is confined to phenols (unless catalysts such as boron trifluoride or aluminum trichloride are present[2]). Strong CH-acids within the indicated pK-range will, however, also react with methylmagnesium iodide or bromide. For example, *diphenylbenzyl phosphine oxide* exhibits 'Zerewitinoff activity', and this behavior is closely related to its use in PO-activated olefin-forming reactions[3] under relatively weakly basic conditions. Even *terminal acetylenes* such as phenylacetylene (pK = 21 on the McEwen scale) slowly react with organomagnesium compounds in ethereal solution[4–6], the reactivities[5] towards methylmagnesium bromide and ethylmagnesium bromide being in the ratio 6:100.

In the presence of hexamethylphosphorictriamide, the metallating power of organomagnesium compounds is greatly increased; other donor solvents also cause some increase[6]. Under suitable conditions isopropylmagnesium chloride can even metallate hydrocarbons such as *indene, fluorene, triphenyl methane,* or *diphenyl methane,* causing formation of gas (propane)[7, 8], i.e. up to the limit of pK $\approx$ 35 on the McEwen scale. No analytical use has been made of this effect.
Other organometallic compounds which should be mentioned in this context are methyllithium and ethylsodium, the latter being one of the strongest metallating agents known. Ethylsodium is of little use for analytical purposes, because its metallating power is so high that the compound undergoes nonspecific reaction with practically all kinds of hydrogen-containing organic compounds (except aliphatic hydrocarbons); in addition, it is comparatively difficult to prepare and is completely insoluble. Furthermore, compounds containing functional substituents (ketones, esters, etc.)

interact with this energy-rich reagent, undergoing addition or substitution without liberation of ethane. The same is also true, but to a lesser extent, for methylmagnesium iodide. For example, acetone predominantly undergoes addition to the C=O double bond, rather than metallation, which would lead to a magnesium enolate and methane. Knowledge of the chemistry of organometallic reagents is a prerequisite for their use[9].

2.1.2 Use of Lithium Aluminum Hydride and Other Hydrides and Amides
The range of usefulness of lithium aluminum hydride is similar to that of the Zerewitinoff reagent (methylmagnesium iodide)[34]. The reagent may be employed in its commercial form, i.e. as powder, or in solution (preferentially, in tetrahydrofuran).

With compounds containing hydroxy groups, hydrogen is evolved. This reaction may be used to identify traces of moisture in aprotic media. One simply adds a small amount of the hydride to a sample of the liquid, care being taken to exclude the atmosphere as completely as possible. If more than traces of water (or another hydroxylic substance) are present, there will be a noticeable development of gas. The reaction can also be used for quantitative analysis (cf. Section 3.9:3.2). The compound behaves as a reducing reagent towards some substances without hydrogen being evolved.

Stronger metallating reagents are *lithium hydride, sodium hydride,* and *sodamide* which attack even very weak CH-acids. For example, the weak acidity of dimethyl sulfoxide (pK $\approx$ 35, cf. Ref.[44], p. 64) can be detected with sodium hydride or amide; warming with these reagents leads to the liberation of hydrogen or ammonia, respectively[10]. Possible side reactions in which no gas is formed (e.g. reduction of ketones, amidation of esters) must be taken into consideration.

2.2 Identification of Weak Acids with Colored Reagents
In the case of colored metallating reagents, the occurrence of metallating reactions with acidic compounds may be detected by loss or change of color. Colored metallating reagents are distinguished by the fact that their carbanionic centers are bonded to large groups capable of causing charge delocalization; this is the case with *phenyl-isopropyl potassium* (cumylpotassium, *2-potassium-2-phenyl-propane*) and *triphenylmethyl sodium.* The red solutions of these compounds in diethyl ether are decolorized — practically instantaneously in most cases — by numerous OH-, NH-,

[2] *L.F. Fieser, M. Fieser,* Reagents for Organic Synthesis, p. 193, John Wiley & Sons, New York 1967.
[3] *L. Horner,* Fortschr. Chem. Forsch. *7,* 1, 29 (1966).
[4] *H. Kleinfeller, H. Lohmann,* Ber. *71,* 2608 (1938).
[5] *J.H. Wotiz, C.A. Hollingsworth, R. Dessy,* J. Am. Chem. Soc. *77,* 103 (1955).
[6] *L.I. Zakharkin, O.Y. Okhlobystin, K.A. Bilevitch,* Tetrahedron *21,* 881 (1965).
[7] *T. Cuvigny, H. Normant,* Bull. Soc. Chim. Fr. 2000 (1964).
[8] *H.F. Ebel, B.O. Wagner,* Chem. Ber. *104,* 320 (1971).

[9] *Houben-Weyl,* Methoden der Organischen Chemie, 4. Aufl., Bd. XIII/1 and XIII/2, cf. Ref.[47–49].
[10] *E.J. Corey, M. Chaykovsky,* J. Am. Chem. Soc. *87,* 1345 (1965).

and CH-acids. Particularly in the case of triphenylmethylsodium, competing processes such as addition to the C=O double bond are very much suppressed by steric hindrance, and side reactions which might simulate the presence of active hydrogen need hardly be feared[11, 12]. Less specific in this respect are the deeply colored solutions of naphthalene sodium and other radical-anionic metal adducts, which rapidly undergo condensation reactions with halogen compounds.

2.3 Identification of Weak Acids by Base Catalysis

2.3.1 Base-Catalyzed Halogenation

If bromination of an organic compound is accelerated by basic catalysts, this indicates involvement of substrate deprotonation in the rate determining step of the bromination, and hence the presence of mobile hydrogen atoms. In fact with numerous ketones, esters, nitriles, nitroalkanes, sulfones, etc. the rate of halogenation in aqueous alkaline media is independent of the concentration and nature of the halogen (chlorine, bromine, iodine), thus proving that the reaction proceeds exclusively via removal of a proton by the base. Quantitative comparisons of kinetic acidities of various substrates are based on this behavior (cf. Ref.[44], p. 17). Moreover, it appears possible to use the characteristic rates of halogenation under well defined conditions for purposes of identification. Quantitative titration of CH-acids with bromine (cf. Section 3.9:3.3) is of greater analytical importance.

2.3.2 Base-Catalyzed Autoxidation

With very weak CH-acids such as toluene, diphenylmethane, and triphenylmethane, similar behavior is observed towards oxygen in the presence of strong bases as has been described for halogenation. The rate-determining step is deprotonation, and the energy-rich carbanion thus formed subsequently reacts rapidly with molecular oxygen[39]. In favorable cases, it may be possible to decide whether there are several kinds of C–H bonds of differing kinetic acidity by following the progress of oxygen consumption with time. Moreover, from the amount of oxygen consumed, the amount of autoxidizable compound(s) may be determined. Little analytical use has been made of these possibilities, although the consumption of oxygen may be conveniently measured by gas volumetric means.

2.3.3 Base-Catalyzed Hydrogen Isotope Exchange

Of greater practical importance is the exchange of hydrogen for deuterium or tritium in the presence of alkali, a method widely used for estimating *proton mobilities* and comparing relative kinetic acidities[36, 37, 40, 42] (cf. also Ref.[44], p. 19 f.). Under drastic conditions, such as heating with potassium amide in liquid deuteroammonia at 120 °C in a closed vessel[13], all kinds of hydrogen atoms in organic molecules may be exchanged, including those in paraffinic bonds. This means that even paraffin hydrocarbons possess acidic properties, though these are extremely weak. In contrast, hydrogen bonded to oxygen can be exchanged with its isotopes in aqueous alkaline media. Proton mobilities between these extremes can easily be differentiated by the many available basic media. Direct observation in an IR or NMR spectrometer, or sampling and measurement of deuterium by various methods (combustion and analysis of the water of combustion, e.g. by the falling-drop method or mass spectrometry) allows one to decide whether hydrogen exchange is occurring, how rapidly and to what extent.

2.4 Identification of Weak Acids by Metallation

2.4.1 Direct Metallation by Alkali Metals

Some hydrocarbons, including all acetylenes with terminal triple bonds, are directly metallated by molten or dispersed alkali metals, or their solutions in ammonia. However, the stoichiometric amount of hydrogen expected from the equation below (cf. Ref.[47], p. 265) is not always evolved:

$$R\text{—}H + M \rightarrow R\text{—}M + \tfrac{1}{2} H_2 \qquad (4)$$

With the blue solutions of alkali metals in liquid ammonia, the reaction is observable by the fading of the color. The acidity of cyclopentadiene was demonstrated[14] by observing the rapid evolution of hydrogen when the cyclic diene was dropped into a suspension of potassium in benzene. When mixtures of hydrocarbons are melted with potassium, indene and fluorene may be separated as potassium salts, because of their acidity (cf. Ref.[47], p. 276). Direct metallations of this kind are often combined with reduction processes, and cannot in general be applied to compounds bearing functional groups.

2.4.2 The Metal/Hydrogen Exchange Reaction

Not only colored metallating reagents (cf. Section 3.9:2.2), but indeed all kinds of metallating

---

[11] *W. Schlenk* in *Houben-Weyl*, Methoden der Organischen Chemie, 2. Aufl., Bd. IV, p. 974, Georg Thieme Verlag, Leipzig 1924.

[12] *K. Ziegler*, Angew. Chem. *49*, 455 (1936).

[13] *A.I. Shatenshtein, L.N. Vasileva, N.M. Dykhno, E.A. Izrailevich*, Dokl. Akad. Nauk SSSR *85*, 381 (1952); C.A. *46*, 9954 (1952).

[14] *J. Thiele*, Ber. *34*, 68 (1901).

reagents — particularly those of the organic-alkali metal series — may be used to metallate compounds in the sense of reaction (3). There must, of course, be a sufficient difference in acidities if the reaction is to be practically completed within a short time. The products of CH-metallation may be transformed, using reactions such as carboxylation (cf. Ref.[49], p. 617), into derivatives which can be separated easily, and which are amenable to structural identification and quantitative determination.

3 Quantitative Analysis of CH-Acids and Other Weak Acids

3.1 Zerewitinoff Titration with Methylmagnesium Iodide

The reaction of OH-acids with methylmagnesium iodide described in Section 3.9:2.1.1 may be adapted for the quantitative determination of hydroxy groups, even on the micro scale[15–17, 33, 38]. Since the high vapor pressure of diethyl ether causes difficulties with the volumetric measurement of the methane evolved, Zerewitinoff[15] used diamyl ether as a solvent, adding pyridine in order to improve the solubility of the OH-acids. Besides simple *alcohols* and *phenols,* more complicated compounds such as *borneol, phloroglucinol,* or *sugars* can also be analyzed in this manner. *Polyhydroxy compounds* generate the molar amounts of methane corresponding to their numbers of hydroxy groups. *Oximes* and stronger acids such as *carboxylic* or *sulfinic acids* can be analyzed by the method, as well as CH-acids such as *nitro-alkanes, nitriles, β-ketoesters,* and *β-diketones,* and even some simple ketones, e.g. acetophenone. In these last cases, it is desirable to titrate with a solution of methylmagnesium iodide of known activity, or to determine unconsumed reagent by back-titration, comparing the total amount of titrant consumed with the amount of methane evolved. With these compounds, it will be found that part of the reagent does not metallate (CH_4), but instead is consumed in other ways, e.g. by addition to double bonds[16]. Comparison of the volume of methane evolved with the total consumption of titrant provides information about the nature of the acidic bonds and the functional groups present.

3.2 Titration with Lithium Aluminum Hydride in Tetrahydrofuran

Instead of methylmagnesium iodide, lithium aluminum hydride may be used to determine hydroxy groups[34]. Apart from direct volumetric analysis of the hydrogen produced, according to the equation[18, 19]:

$$4 \, ROH + LiAlH_4 \rightarrow ROLi + (RO)_3Al + 4 \, H_2 \quad (5)$$

back-titration with propanol, and electrometric detection of the end point may also be employed[20, 21]. Most common is the use of a 0.25 M solution of lithium aluminium hydride in tetrahydrofuran (in the case of direct volumetric determination, concentrations up to 2.5 M are used). This method is used mainly for determination of alcohols and phenols. Functionally substituted CH-acids such as β-diketones etc. also fall within the scope of this technique. The kind of acid(s) may be characterized by comparing the total consumption of reagent with the gas evolution (H_2). Some information about the nature of the active bonds may also be derived from the rate of reaction, since O—H bonds and N—H bonds generally react much more rapidly than do C—H bonds.

3.3 Bromine Titration: Meyer Procedure

Enols add bromine to form enol dibromides, which are stable in cold methanolic solution only for short periods. This is the basis of the Meyer method of determination[22]. Since ketones do not react under these conditions, the method may be used for determination of the enolic contents of compounds capable of ketone/enol tautomerism. The titration should be performed as quickly as possible, since the slow elimination of hydrogen bromide from the enol dibromides accelerates enolization, because of acid catalysis, and thus precludes determination of the original enol content.

On the other hand, not only acid catalysis but also base catalysis may be used to achieve conversion of the ketone into a form capable of reaction with bromine. In the case of base catalysis, this form is the enolate anion. The total amount of enolizable substances (OH-acid plus CH-acid)

[15] *T. Zerewitinoff*, Ber. *40*, 2023 (1907).
[16] *A. Soltys*, Mikrochemie *20*, 107 (1936).
[17] *W. Fuchs, N.H. Ishler, A.G. Sandhoff*, Ind. Eng. Chem., Anal. Ed. *12*, 507 (1940).

[18] *F.A. Hochstein*, J. Am. Chem. Soc. *71*, 305 (1949).
[19] *G.A. Stenmark, F.T. Weiss*, Anal. Chem. *28*, 1784 (1956).
[20] *C.J. Lintner, R.H. Schleif, T. Higuchi*, Anal. Chem. *22*, 534 (1950).
[21] *T. Higuchi* in *J. Mitchell, I.M. Kolthoff, E.S. Proskauer, A. Weissberger,* Organic Analysis, Vol. II, p. 123, Interscience Publishers, New York 1954.
[22] *K.H. Meyer*, Liebigs Ann. Chem. *380*, 212 (1911).

may thus be determined. The same is true for *nitro/aci-nitro systems*. Hence, the bromination method mentioned in Section 3.9:2.3.1 is also suitable for quantitative analysis. With active methylene compounds, it must, however, be recognized that more than one mole of bromine per mole of substrate is often consumed, since all of the active C–H bonds may become C–Br bonds. The amount of bromine consumed by a definite compound under the conditions of titration may be determined in separate experiments.

3.4 Titration with Colored Reagents

Strong CH-acids such as acetylenes or ketones, but especially all types of OH- and NH-acids (alcohols, amines, etc.) may be titrated with the colored ethereal solutions of phenylisopropylpotassium or triphenylmethylsodium. The former reagent is conveniently prepared by cleavage of phenylisopropyl methyl ether with potassium/sodium alloy in diethyl ether, and, according to the literature, can be made practically free of potassium methoxide[23]. A saturated ethereal solution is ca. 0.08 M, but freshly prepared supersaturated solutions of considerably higher concentration (up to 0.5 M) can also be used. The disadvantage of a volatile solvent for the titrant is avoided by the use of dibutyl ether[24], but the saturation concentration is then only about half that of the ethereal solution. Triphenylmethylsodium is best prepared from triphenylchloromethane and sodium amalgam (cf. Ref.[47], p. 391) and recrystallized from diethyl ether in a double Schlenk tube prior to use as titrating reagent.

Reactions with the acids mentioned above are usually rapid and complete, the end point being indicated by a faint pink color just persisting in the mixture. Since the organometallic reagents also react with atmospheric humidity and oxygen, the titration must be performed under nitrogen. The simplest procedure is to run the solution of the titrant, which has been standardized against benzoic acid, from a Schlenk tube fitted with a lateral graduated tube (cf. Ref.[25], p. 373); the end of the buret dips (in a stream of nitrogen) into the neck of the flask containing the solution to be titrated. Special equipment has been described (cf. Ref.[25], p. 376) for use when higher accuracy is required. Since no effect is observed other than disappearance of the color, side reactions other than re-

placement of mobile hydrogen by metal cannot be differentiated, i.e. only the total conversion is measured. However, as mentioned earlier (see Section 3.9:2.2), with the colored organometallic reagents such as triphenylmethyl sodium, possible side reactions are negligible in comparison with metal/hydrogen exchange.

If the molecule contains more than one group with mobile hydrogen (e.g. in the case of polyhydroxy compounds), all of them are metallated. Even certain compounds which at first sight do not appear to be dibasic acids (e.g. phenylacetic acid) partially undergo double metallation to yield *dianionic* metallation products[48]. Special experiments such as deuterolysis are needed to clarify the situation in each particular case.

In many metal/hydrogen exchange reactions conducted in ether-like solvents, the product of metallation precipitates during the titration. This precipitation can be avoided by using dimethyl sulfoxide as solvent. Some metallations of weak CH-acids which proceed only slowly in ethereal solutions are greatly accelerated in dimethyl sulfoxide. Advantage may be taken of the color of triphenylmethylsodium in a way which is quite different from direct titration with this reagent, i.e. by titrating with the sodium salt of dimethyl sulfoxide ('dimsyl sodium') in the presence of triphenylmethane as an indicator[26]. OH-, NH-, CH-, and other acids with pK values up to approximately 30 (on the McEwen scale, cf. Ref.[44]) react rapidly and quantitatively with dimsyl sodium in dimethyl sulfoxide, the first excess of titrant converts the indicator hydrocarbon into triphenylmethylsodium; the red color thus produced indicates the end point.

It is interesting to note that the titration cannot be performed with methanesulfinylmethyl potassium or cesium, since with these the point of coloration precedes the equivalence point[27].

3.5 Other Methods

The lack of specificity which was mentioned previously as a drawback in titrations with colored organometallic titrants may be overcome by combining the metallation with a *carboxylation*. The carboxylic acids thus obtained are separated, with as little loss as possible, by acid and base extraction. Provided that the titration has not been continued beyond the equivalence point, the carboxylic acids are solely derived from the CH-acids

[23] K. Ziegler, H. Dislich, Chem. Ber. *90*, 1107 (1957).
[24] K. Ziegler, F. Dersch, Ber. *62*, 1833 (1929).
[25] H. Metzger, Eu. Müller in Houben-Weyl, Methoden der Organischen Chemie, 4. Aufl., Bd. I/2, p. 321, Georg Thieme Verlag, Stuttgart 1959.

[26] G.G. Price, M.C. Whiting, Chem. Ind. [London] 775 (1963).
[27] E.C. Steiner, J.M. Gilbert, J. Am. Chem. Soc. *85*, 3054 (1963).

originally present. A mixture of carboxylic acids can be analyzed by esterification with diazomethane, followed by gas chromatography of the resulting mixture of methyl esters. This supplementary procedure can be employed also after an excess of a colorless metallating reagent such as butyllithium has been used, with which the equivalence point cannot be observed, — i.e. where the carboxylic acids derived from the CH-acids must be distinguished from the product of carboxylation of the metallating reagent. Carboxylation can also be used for investigating equilibrium systems for which the metal/hydrogen exchange between metallating agent and the CH-acid to be determined is appreciably reversible, and thus may be employed for estimating equilibrium acidities of CH-acids[28-31].

On carboxylation, alcohols and amines are converted via their metallation products (alkoxides, amides) into alkali metal salts of carbonic acid monoalkyl esters and of N-substituted carbamic acids, respectively. Both of these products decompose on acidification, regenerating the components. The method can thus be used only for distinguishing OH- and NH-acids from CH-acids. For details of the particular behavior of phenols and enols, reference should be made to the literature. In contrast, esterification of alkoxides with ethyl chloroformate leads to stable diesters of carbonic acid; with the same reagent, ethyl carboxylates are obtained directly from metallated CH-acids.

Electrometric methods (cf. Chapter 3.6) have rarely been used for the analysis of very weak acids, although potentiometry using appropriate glass electrodes is feasible, even in very strongly basic solutions in which CH-acids such as triphenylmethane are reversibly deprotonated[31]. The requirement for special electrodes and electrolytic bridges, highly purified dimethyl sulfoxide, vigorous exclusion of air and moisture (argon atmosphere), and the use of dimsyl cesium as a titrant render the routine application of this procedure difficult, as does the slow attainment of equilibrium potentials. The procedure has, therefore, gained importance only for pK-measurements.

4 Estimation of Acid Strength

4.1 Methods for Estimation of Equilibrium Acidity

In aprotic neutral solvents such as diethyl ether, pairs of acids and their conjugate anions (as alkali metal salts) can be equilibrated practically without limitation as regards pK, since these solvents do not participate in proton transfer, and hence have no 'levelling' effect.

Only towards the strongest metallating reagents of the type of non-stabilized alkalimetal alkanes (e.g. butyllithium, butylsodium) the so-called aprotic solvents no longer are inert; instead, they interfere with the equilibria of metallation, mostly by irreversible participation.

The position of the equilibrium may be determined by chemical derivatization (e.g. carboxylation) or, in the case of colored anions, photometrically. Relative acidity constants thus obtained may be converted into absolute values via reference acids, if possible effects due to the medium are ignored. Alternatively, acidity constants relating to water as a standard medium may be determined photometrically by means of acidity functions H[-46] (cf. also Ref.[44], p. 245). Finally, various supplementary quantities such as the equilibrium constants of metal/halogen exchange reactions, or the half wave potentials of polarographic reductions involving organomercury derivatives of the CH-acids, provide measures of the relative stabilities of carbanions of the relative strengths of the CH-acids (cf. Ref.[44] for more extensive discussion).

4.2 Methods for Estimation of Kinetic Acidity

Proton mobilities, i.e. the rates at which protons are removed from their bonds to carbon, oxygen or any other atom by certain proton acceptors, parallel the equilibrium acidities in many cases (but by no means all!). They are quoted as the logarithms of the rate constants of deprotonation (kinetic acidities), measured in various processes. Most commonly investigated are systems in which the deprotonation step is rate determining within a complex series of elementary processes, such as base-catalyzed bromination, autoxidation, racemization, or hydrogen isotope exchange[36, 37, 40, 42, 44].

5 Localization of Acidity in Molecules

The anions (molecules, ionic dipoles) arising from deprotonation of molecules (cations) generally take up protons, deuterons, or other electrophiles at the same site at which the original proton was lost. Application of suitable methods of structural analysis, in particular of spectrometric techniques (IR, NMR, mass spectra), or comparison with

[28] J.B. Conant, G.W. Wheland, J. Am. Chem. Soc. 54, 1212 (1936).

[29] R.D. Kleene, G.W. Wheland, J. Am. Chem. Soc. 63, 3321 (1941).

[30] K. Issleib, R. Kümmel, J. Organometal. Chem. 3, 84 (1965).

[31] C.D. Ritchie, R.E. Uschold, J. Am. Chem. Soc. 89, 1721, 2752 (1967).

known compounds, indicates the position where the electrophile enters, and thus the site of acidity. A complicating factor is charge delocalization in the anions, i.e. electrophiles are often not bonded at the original site of deprotonation; such a situation is exemplified by the behavior of enolate and nitronate anions on protonation. In these cases, a distinction must be made between kinetically controlled and thermodynamically controlled derivative formation.

6 Bibliography

[33] *M.S. Kharasch, O. Reinmuth,* Grignard Reactions of Nonmetallic Substances, p. 1166, Prentice Hall, New York 1954.

[34] *N.G. Gaylord,* Reduction with Complex Metal Hydrides, p. 76, Interscience Publishers, New York 1956.

[35] *R.P. Bell,* The Proton in Chemistry, Cornell University Press, Ithaca, New York 1959.

[36] *A.I. Shatenshtein,* Hydrogen Isotope Exchange Reactions of Organic Compounds in Liquid Ammonia, Advan. Org. Chem. *1,* 156 (1963).

[37] *A.I. Schatenstein* (übersetzt aus dem Russischen), Isotopenaustausch und Substitution des Wasserstoffs in Organischen Verbindungen, VEB Deutscher Verlag der Wissenschaften, Berlin 1963.

[38] *G.F. Wright,* Determination of Active Hydrogen, in *L. Meites,* Handbook of Analytical Chemistry, p. 12/93–12/99, McGraw-Hill Book Co., New York 1963.

[39] *G.A. Russell, E.G. Janzen, A.G. Bemis, E.J. Geels, A.J. Moye, S. Mak, E.T. Strom,* Oxidation of Hydrocarbons in Basic Solution, in Advances in Chemistry Series No. 51, p. 112, American Chemical Society, Washington 1965.

[40] *A. Streitwieser, J.H. Hammons,* Acidity of Hydrocarbons, Progr. Phys. Org. Chem. *3,* 41 (1965).

[41] *E.J. King,* Acid-Base Equilibria, Pergamon Press, Oxford, New York 1965.

[42] *D.J. Cram,* Fundamentals of Carbanion Chemistry, Chapter 1 (Carbon Acids), Academic Press, New York, London 1965.

[43] *H. Fischer, D. Rewicki,* Acidic Hydrocarbons, Progr. Org. Chem. *7,* 116 (1968).

[44] *H.F. Ebel,* Die Acidität der CH-Säuren, Georg Thieme Verlag, Stuttgart 1969.

[45] *R.P. Bell,* Acids and Bases — Their Quantitative Behaviour, 2nd Ed., Methuen & Co., London 1969.

[46] *C.H. Rochester,* Acidity Functions, Chapter 7 (Acidity Functions for Concentrated Solutions of Bases), Academic Press, London, New York 1970.

[47] *H.F. Ebel,* Herstellung von alkalimetall-organischen Verbindungen, in *Houben-Weyl,* Methoden der Organischen Chemie, 4. Aufl., Bd. XIII/1, p. 261, Georg Thieme Verlag, Stuttgart 1970.

[48] *H.F. Ebel,* Herstellung von alkalimetall-organischen Verbindungen spezieller Natur, in Ref. [47], p. 438.

[49] *H.F. Ebel, A. Lüttringhaus,* Umwandlung alkalimetall-organischer Verbindungen, in Ref. [47], p. 452.

4 Importance of Chemical Transformations for Analytical Purposes

Günter Giesselmann

Degussa, D-6451 Wolfgang bei Hanau

4.1 Introduction

Separating a mixture of organic substances for analytical purposes is often quite complicated and in many instances succeeds only by employing a combination of physical and chemical separation methods. Chemical methods are used where the differences in volatility, solubility, or stability of the individual species are too small. By carrying out chemical transformations, these properties of individual constituents can be so modified that separation by physical means becomes feasible.

The following describes a selection of chemical reactions which alter the volatility, solubility, and stability of individual substances or mixtures of substances in such a way that a separation or testing for purity becomes possible not only with classical methods of distillation, sublimation, extraction, but also by means of *gas chromatography, thin-layer chromatography, gel chromatography, nuclear magnetic resonance* and *mass spectrometry*. This represents an important step since these latter techniques are likely to find increasing use in future (4.2). – Methods for preparing sparingly soluble and difficultly volatile derivatives are described in other sections, e.g. for amino acids in Chapter 12.5.

4.2 Methods for Increasing the Volatility and Solubility of Organic Compounds

The volatility and solubility of an organic compound is dependent on its molecular weight and the substituents participating in it. Three types of substituents can modify these physical properties:

Substituents which do not essentially alter the typical organic behavior such as ether solubility and water insolubility are halogen atoms and ester and ether oxygen.

Substituents which convert homopolar compounds into heteropolar ones, producing a very sharp reduction in volatility and solubility in ether, but, in return, an increase in water solubility, are trialkylammonium halide, sulfo, sulfonate and carboxylate groups.

Substituents which, in respect to physical behavior, produce a midway position between organic and inorganic compounds are hydroxy, carboxy and amino groups. These substituents lead to water solubility and an appreciable increase in boiling point but retention of solubility in ether. If more than one such group is present in a compound the boiling point is increased appreciably and the solubility in ether sharply reduced.

The above survey makes clear what measures must be taken to make a difficultly volatile compound more volatile or more soluble in non-polar solvents. One technique consists in etherifying or esterifying all the hydroxy or carboxy groups in a particular compound. For this purpose the organosilyl compounds, which have received a great deal of attention during recent years, are particularly suitable. Trialkylsilyl compounds have been found to be especially useful for making difficultly volatile compounds more volatile, more soluble, and more stable. Therefore, these so-called silylation agents will receive detailed treatment in the paragraphs that follow. Other volatile derivatives of amino acids are described in Section 12.3:1. Trifluoroacetylation is reported in Chapter 12.4.

The replacement of an *active hydrogen* atom in a molecule by a trialkylsilyl group is called *silylation*. Fundamentally all compounds containing hydroxy, amino or sulfhydryl groups are accessible to silylation (cf. 4.2:2).

When all the active hydrogen atoms have been silylated, no groups are left to produce internal hydrogen bonding and, therefore, the homopolar character predominates. Such typical organic molecules display the same size in the gaseous state, in solution, and in the crystal. The volatility of silylated compounds is dependent on the size of the molecules and the nature of the bonding of the trialkylsilyl group.

1 Silylation Methods

Of the numerous reagents developed during the past few years for silylation of organic compounds, trimethylchlorosilane, hexamethyldisilazane, N-trimethylsilyldiethylamine, N-trimethylsilyl-n(or t)-butylamine, N-trimethylsilylacetamide, N,O-bis(trimethylsilyl)acetamide, N-methyl-N-trimethyl silylacetamide and N-methyl-N-trimethylsilyltrifluoracetamide have given the best results (Table 1).

Table 1. Survey of the principal silylation reagents

Silylation reagent	Structural formula	Molecular weight	bp. [°C]	bp. Torr	mp.	Refractive index n_D^{20}	Density D_4^{20}
Trimethylchlorosilane	$(CH_3)_3$—S—Cl	108.7	57	760		1.3868	0.8580
N-Trimethylsilylacetamide	CH_3—C—NH—$Si(CH_3)_3$, =O	131.3	187 / 82	760 / 12	52–54		
N-Methyl-N-trimethylsilyl-acetamide	CH_3—C—N(—$Si(CH_3)_3$, —CH_3), =O	145.3	156	760		1.4393	0.9009
N-Methyl-N-trimethylsilyl-trifluoracetamide	CF_3—C—N(—$Si(CH_3)_3$, —CH_3), =O	199.4	132 / 78–79	760 / 130		—	1.072
N-Trimethylsilyldiethylamine	$(CH_3)_3$—Si—$N(C_2H_5)_2$	145.3	126	760		1.4110	0.7627
N-Trimethylsilyl-t-butylamine	$(CH_3)_3$—Si—N—$C(CH_3)_3$	147.2	118–119	760		1.4060	—
N-Trimethylsilylimidazole	N=CH, HC=CH / N—$Si(CH_3)_3$	140.3	89–91	12		1.4754	0.9492
Hexamethyldisilazane	$(CH_3)_3$ Si—NH—$Si(CH_3)_3$	161.4	126	760		1.4080	0.7742
N,O-Bis(trimethylsilyl)acetamide	CH_3—C(—O—$Si(CH_3)_3$, =N—$Si(CH_3)_3$)	203.4	62	25		1.4182	0.8321
Bis(trimethylsilyltri)fluoroacetamide	CF_3—C—N(—$Si(CH_3)_3$, —$Si(CH_3)_3$), =O	257.4	45–50	14		—	—

Trimethylchlorosilane is used mainly in combination with other silylation agents. Hexamethyldisilazane is employed preferably admixed with trimethylchlorosilane in the ratio 2:1 by volume, *N*-trimethylsilyldiethylamine and *N*-trimethylsilyl-n(or t)-butylamine are used either alone in excess or in the presence of acid catalysts. *N*-trimethylsilylacetamide also has given very good results. This reagent partially silylates *hydroxy groups* quantitatively even at room temperature. The acetamide obtained as side product can be removed easily by filtration.

N,O-bis(trimethylsilyl)acetamide is not only a powerful silylation agent but at the same time a good solvent. It is often employed in combination with trimethylchlorosilane.

N-methyl-*N*-trimethylsilyltrifluoroacetamide is the most volatile of all trimethylsilylamides and possesses very powerful silyl donor properties. *N*-methyl-*N*-trimethylsilylacetamide is somewhat more polar than *N*-methyl-*N*-trimethylsilyltrifluoroacetamide and as a result enables *solid substances* to be converted more quickly.

1.1 Trimethylchlorosilane (TMCS)

TMCS reacts with hydroxy, amino and mercapto groups in accord with Equation (1)[1].

$$ROH + (CH_3)_3SiCl \rightarrow ROSi(CH_3)_3 + HCl \qquad (1)$$

Using TMCS alone generally does not produce a complete conversion. 80–100% yields are obtained by adding bases. The hydrochloric acid formed during the reaction can be bound as ammonium chloride by passing ammonia gas into the cooled reaction solution[2]. Pyridine[3,4] or triethylamine[5] may be used as auxiliary bases. An excess or equimolar amount of either amine is added to the substance to be silylated and the TMCS is added drop by drop to the well-stirred reaction mixture.

Thiols are best silylated with TMCS in the form of their metal salts.

Henglein et al.[6] describe the conversion of *carbohydrates* with TMCS dissolved in hexane, with pyridine acting as hydrochloric acid acceptor and formamide as catalyst.

[1] *L. Birkofer, A. Ritter*, Chem. Ber. *93*, 424 (1960).
[2] *J. Speier*, J. Am. Chem. Soc. *74*, 1003 (1952).
[3] *W. Gerrad, K. Kilburn*, J. Chem. Soc. 1536 (1956).
[4] *R. Schwarz, E. Baronetzky, K. Schoeller*, Angew. Chem. *68*, 335 (1956).
[5] *P. DeBenneville, M. Hurwitz*, U.S.Pat. 2876209 (to Rohm & Haas Co.); C.A. *53*, 12321 (1959).
[6] *F.A. Henglein, W. Knoch*, Makromol. Chem. *28*, 10 (1958), s.a. Ref. [54].

1.2 Hexamethyldisilazane (HMDS)

HMDS[7] can be used alone or in combination with TMCS (Reaction 2).

$$2\ ROH + (CH_3)_3-Si-NH-Si(CH_3)_3 \rightarrow \quad (2)$$
$$2\ ROSi(CH_3)_3 + NH_3$$

The reaction is subject to *catalytic influences*[8,9]; acids catalyze it, bases inhibit it. In addition to TMCS, ammonium salts and sodium hydrogen sulfate are effective catalysts. Ammonium sulfate has given very good results. However, HMDS is most frequently employed in combination with TMCS and suitable solvents such as acetonitrile, pyridine, tetrahydrofuran, dimethylformamide, carbon disulfide and dimethylacetamide. Optimum results are obtained with a 2:1 by volume mixture of HMDS and TMCS, corresponding to a molar ratio of 6:5, and pyridine as solvent[7,10].

1.3 Silamines

Silamines can be used to silylate *alcohols, amines, amino acids, amides* and *ureas* in accord with Equation (3).

$$ROH + R'-NH-Si(CH_3)_3 \rightarrow \quad (3)$$
$$ROSi(CH_3)_3 + R'NH_2$$

Good results are obtained above all with *N-trimethylsilyldiethylamine* and *trimethylsilyl-n(or t)-butylamine*. Rühlmann[11] and De Benneville[5] used these silamines mainly for silylating *amino acids* and achieved yields of 80–97%. The amino acid is heated with an excess of the silamine until no further diethylamine or butylamine distills over. This method at the same time causes the amino group and the carboxy group to be silylated. Mason and Smith[8] esterified amino acids with a mixture of N-trimethylsilyldiethylamine and an acid catalyst, e.g. trichloroacetic acid.

One of the most effective silylating agents for *hydroxyl groups* is N-trimethylsilylimidazole. Reaction (4) proceeds even at room temperature.

By contrast, no reaction takes place with basic amino groups. The silylation of sterically hindered hydroxy groups in *steroids* is described by Chambaz and Horning[12]. Sugars can be etherified very well with this reagent.

1.4 Silylamides

Of the numerous silylamides known, *N,O*-bis(trimethylsilyl)acetamide (BSA) is not only an effective silylation agent (Equation 5) for all compounds with *active hydrogen* atoms but at the same time is also a good solvent[13].

BSA is used in combination with trimethylchlorosilane[14]. *N,O*-bis(trimethylsilyl)trifluoroacetamide lends itself particularly readily for silylating *amino acids* to be submitted to gas chromatography. *N*-methyl-*N*-trimethyl silylacetamide has given good results as a silylating agent for *carbohydrates*[15]. The most volatile of all trimethylsilylamides is *N*-methyl-*N*-trimethylsilyltrifluoroacetamide[16] (MSTFA). It reacts in accord with Equation (6).

Donike et al.[17] describe the quantitative gas chromatographic determination of saturated and unsaturated *fatty acids* as their trimethylsilyl

[7] C. Sweeley, R. Bentley, M. Makita, W. Wells, J. Am. Chem. Soc. 85, 2497 (1963).

[8] P. Mason, E. Smith, J. Gas Chromatogr. 4, 398 (1966).

[9] R. Fessenden, D. Crowe, J. Org. Chem. 26, 4638 (1962).

[10] S. Langer, S. Conell, I. Wender, J. Org. Chem. 23, 50 (1958).

[11] K. Rühlmann, Prakt. Chemie [Wien] 9, 315 (1959); Chem. Ber. 94, 1876 (1961); s.a. Ref. [101].

[12] E. Chambaz, E. Hornig, Anal. Lett. 1, 201 (1967).

[13] J. Klebe, H. Finkbeiner, D. White, J. Am. Chem. Soc. 88, 3390 (1966).

[14] M. Horning, E. Boncher, A. Moss, J. Gas Chromatogr. 5, 297 (1967).

[15] L. Birkhofer, M. Donike, J. Chromatogr. 26, 270 (1967).

[16] M. Donike, J. Chromatogr. 42, 103 (1969).

[17] M. Donike, W. Hollmann, D. Stratmann, J. Chromatogr. 43, 490 (1969).

esters. Brief heating is sufficient to form the derivatives, optionally with added catalytic amounts of trimethylchlorosilane.

Silylations with *N-trimethylsilyl-N,N'-diphenylurea*[18, 19] proceed very rapidly and are catalyzed by the diphenylurea obtained as an insoluble product during the reaction. Therefore, this silylamide can be used to carry out silylations that have been impossible to achieve previously with other reagents.

1.5 Hexamethyldisiloxane

Vorankov and Shabarova[20] silylate *alcohols* with hexamethyldisiloxane in the presence of basic catalysts and *phenols* with the agent admixed with acid catalysts. The water formed during the reaction is removed by azeotropic distillation. Yields vary between 50 and 80%.

1.6 Hexamethylthiodisilane

Abel[21] synthesized *hexamethylthiodisilane* from trimethylchlorosilane and sodium sulfide and studied the reactivity of this compound towards alcohols, amines and carboxylic acids. Good yields were obtained with alcohols. In contrast, amines and carboxylic acids are silylated only slowly.

2 Classes of Compounds

2.1 Monohydric and Polyhydric Alcohols

Primary and secondary *alcoholic hydroxyl groups* are silylated relatively easily; etherification of tertiary alcoholic groups presents greater difficulty.

According to Langer et al.[10] silylation of *tertiary hydroxyl groups* succeeds in high yield with a mixture of equal parts hexamethyldisilazane and trimethylchlorosilane in petroleum ether as solvent.

The reaction also proceeds with trimethylchlorosilane and pyridine. Friedman and Kaufman[22] were able to etherify tertiary hydroxyl groups with hexamethyldisilazane in the presence of boiling dimethylformamide or dimethyl sulfoxide. Hetzberg and Jensen[23] demonstrated the silylation of tertiary hydroxyls by spectrophotometric and paper chromatographic examination.

Long-chain *aliphatic monohydric alcohols, glycol* and *aliphatic polyols* can be esterified in good yield with trimethylchlorosilane alone or admixed with hexamethyldisilazane. Sprung and Nelson[24] obtained 2-trimethylsiloxyethanol on silylating *glycol* with a deficiency of trimethylchlorosilane. In the case of the C_3–C_5 polymethylene glycols predominantly both hydroxy groups are silylated. Most polyhydroxy compounds can be silylated in good yield with a mixture of hexamethyldisilazane and trimethylchlorosilane or trimethylchlorosilane and pyridine.

The silylation derivatives of cis- and trans-cyclohexanediol[25] have different boiling points.

Several authors[26, 27] report on the preparation and gas chromatographic analysis of the silyl ethers of ethylene oxide condensation products in order to determine the molecular weight distribution. *Pentaerythritol* can be silylated quantitatively by brief boiling with a mixture of hexamethyldisilazane and trimethylchlorosilane in pyridine[28].

Inositols dissolved in dimethylsulfoxide can be etherified with hexamethyldisilazane/trimethylchlorosilane and pyridine. The gas chromatographic analysis of the trimethylsilyl ethers of inositols is described by several authors[29, 30].

It should be noted that *thermal isomerization* during the silylation of *mono-* and *diglycerides* can lead to conversion of 2-monoglyceride into 1-monoglyceride and 1,2-diglyceride into the 1,3-isomers[31]. Therefore, silylation should be carried out carefully at room temperature. Wood et al.[32] observed no isomerization if the sample was treated for only 5 minutes with a mixture of hexamethyldisilazane and trimethylchlorosilane in pyridine, followed by addition of hexane and removal of excess reagent by washing and evaporation.

Silylated monoglycerides containing no more than 18 C atoms in the side-chain can be separated by gas chromatography[32]. Admittedly, the 1,2-isomers are not separated.

[18] *J. Klebe*, J. Am. Chem. Soc. *86*, 3399 (1964).
[19] *U. Wannagat, H. Bürger, C. Krüger, J. Pump*, Z. Anorg. Allgem. Chem. *321*, 208 (1963).
[20] *M. Voranlov, Z. Shabarova*, Zh. Obshch. Khim., *29*, 1528, Engl. 1501 (1959), C.A. *54*, 8601c (1960).
[21] *E. Abel*, J. Chem. Soc. 4933 (1961).
[22] *S. Friedman, M. Kaufman*, Anal. Chem. *38*, 144 (1966).
[23] *S. Hetzberg, S. Jensen*, Acta Chem. Scand. *21*, 15 (1967).

[24] *M. Sprung, L. Nelson*, J. Org. Chem. *20*, 1750 (1955).
[25] *J. Brimacombe*, J. Chem. Soc. 201 (1960).
[26] *G. Kresze, F. Schäuffelhut*, Z. Anal. Chem. *229*, 401 (1967).
[27] *J. Törnqvist*, Acta Chem. Scand. *21*, 2095 (1967).
[28] *R. Suchanee*, Anal. Chem. *37*, 1361 (1965).
[29] *Y. Lee, C. Ballou*, J. Chromatogr. *18*, 147 (1965).
[30] *W. Wells, T. Pittmann, H. Wells*, Anal. Biochem. *10*, 450 (1965).
[31] *G. Kresze, K. Bederke, F. Schäuffelhut*, Z. Anal. Chem. *209*, 329 (1965).
[32] *R. Wood, P. Raju, R. Reiser*, J. Am. Oil Chem. Soc. *42*, 161 (1965).

When a 2:1 mixture of hexamethyldisilazane and trimethylchlorosilane is used in hexane-ether[33], a *rearrangement* of the 1-monoglyceride into the 2-isomer is observed if the derivatives are left in contact with the reagent too long (overnight). Saturated and singly unsaturated C_{14}–C_{18} monoglycerides, glycerol ethers[34] and also technical monoglycerides[35] can be separated by gas chromatographic means in this way.

2.2 Phenols

Fundamentally, the silylation of *phenols* provides no difficulties. Hexamethyldisilazane has given good results when mixed with trimethylchlorosilane in pyridine, tetrahydrofuran or ethyl acetate as solvent, even with complex natural products. Numerous *hydroxy-* and *polyhydroxyanthraquinones*[36] and *tocopherols*[37] could be silylated with a mixture of hexamethyldisilazane and trimethylchlorosilane in pyridine; the derivatives were investigated by gas chromatography.

The compounds *cannabinol* and *cannabidiol* occurring in the *marihuana extract* from Indian hemp and in *Egyptian hemp oil* have been silylated and analyzed by gas chromatography[38, 39]. *Morphine alkaloids*[40, 41], *hydroxyflavones* and *flavonols*[42] and also *catecholamines*[43, 44] could likewise be silylated.

2.3 Aliphatic and Aromatic Carboxylic Acids

The carboxyl group is silylated less readily than the hydroxyl group. For analytical purposes hexamethyldisilazane in combination with basic solvents has given good results. Certain *phenolic carboxylic acids* can be silylated more advantageously with *N, O*-bis(trimethylsilyl)acetamide.

The stability of the silyl esters is ensured only if an adequate excess of silylation reagent is present; otherwise even atmospheric moisture is sufficient to produce hydrolysis. These compounds have a reactivity akin to that of acid chlorides or anhydrides.

Silyl esters of *aliphatic carboxylic acids* are obtained most easily by treating the salts of these acids with trimethylchlorosilane in the presence of an inert solvent[45, 46]. Dibasic acids can be etherified directly with trimethylchlorosilane. Jones and Schmeltz[47] were able to separate 18 aliphatic trimethylsilyl esters of carboxylic acids by means of gas chromatography.

Hydroxycarboxylic acids occurring in the citric acid cycle can be esterified completely with a mixture of hexamethyldisilazane and trimethylchlorosilane[48]. Under the same conditions α-keto acids are converted not only into the corresponding *esters* but also the possible *enol ethers* are obtained. However, if *hydroxylamine* is added first, trimethylchlorosilane-oxime derivatives of the keto-acids are formed. These are stable on gas chromatographic analysis and gave single peaks. *O-Methylhydroxylamine hydrochloride* in pyridine has given good results as an alternative to hydroxylamine; subsequently a reaction with bis-(trimethylsilyl)acetamide is performed.

N, O-bis(trimethylsilyl)acetamide has given good results for esterifying *fatty acids*.

For esterfication, aromatic polycarboxylic acids must be boiled with a mixture of hexamethyldisilazane and trimethylchlorosilane in toluene or xylene until solution is complete.

Silylation of certain *hydroxy carboxyl acids* succeeds with hexamethyldisilazane/trimethylchlorosilane in pyridine, dioxane, or dimethylformamide. Here, the carboxyl group is converted into the trimethylsilyl ester and the hydroxyl group into the corresponding trimethylsilyl ether. In order to avoid hydrolysis of the derivatives, the excess reagent should not be removed. No interference results during gas chromatographic analysis[14, 49].

o-Hydroxybenzoic acid is silylated only incompletely. Most aromatic hydroxy carboxylic acids can be silylated in good yield only if the reagent used is

[33] *R. Watts, R. Dils*, Nature *212*, 458 (1966).
[34] *M. Rumsby*, J. Chromatogr. *34*, 461 (1968).
[35] *J. Legari*, J. Am. Oil Chem. Soc. *44*, 379 (1967).
[36] *T. Furuya, S. Shibata, H. Fizuka*, J. Chromatogr. *21*, 116 (1966).
[37] *H. Slover, L. Shelley, T. Burks*, J. Am. Oil Chem. Soc. *44*, 161 (1967).
[38] *U. Claussen, W. Borger, F. Korte*, Liebigs Ann. Chem. *693*, 158 (1966).
[39] *L. Heagsman, E. Walker, D. Lewis*, Analyst (London) *92*, 450 (1967).
[40] *E. Brochmann-Hanssen, A. Svendsen*, J. Pharm. Sci. *51*, 1095 (1962); *52*, 1134 (1963).
[41] *G. Martin, J. Swinehart*, Anal. Chem. *38*, 1789 (1966).
[42] *T. Furuya*, J. Chromatogr. *19*, 607 (1965).
[43] *P. Capella, E. Horning*, Anal. Chem. *38*, 316 (1966).
[44] *S. Kawai*, Chem. Pharm. Bull. [Tokyo] *14*, 618 (1966); J. Chromatogr. *25*, 471 (1966).

[45] *H. Anderson*, J. Am. Chem. Soc. *74*, 2371 (1952); J. Org. Chem. *19*, 1296 (1954).
[46] *H. Schuyten, J. Weaver, J. Reid*, J. Am. Chem. Soc. *69*, 2110 (1947).
[47] *T. Jones, I. Schmeltz*, Science *166*, 20 (1968).
[48] *Z. Horii, M. Makita, Y. Tamura*, Chem. Ind. [London] *34*, 1494 (1965).
[49] *C. Dalgliesh*, Biochem. J. *101*, 792 (1966).

allowed to react for prolonged periods at elevated temperature. *Dihydroxy carboxylic acids* such as *2,3-dihydroxybenzoic acid* or *gentisinic acid (2,5-dihydroxybenzoic acid)* yield two silylation products, one of which has phenolic character. Mono- and bistrimethylsilyl derivatives of *salicylic acid* have also been prepared[50-52].

2.4 Carbohydrates

Except for certain special compounds *carbohydrates* can be silylated relatively readily. Using trimethylchlorosilane in pyridine gives yields between 75 and 97%. For analytical purposes a 2:1 mixture of hexamethyldisilazane/trimethylchlorosilane has given good results.

In general, the reaction is complete after 5 minutes; gentle warming is necessary if the samples are difficult to dissolve. Direct gas chromatographic examination of the reaction mixture is feasible[7].

Silylation of *starch*[53] succeeds with trimethylchlorosilane and pyridine in formamide and hexane; Henglein et al.[54] use this method for silylating *monosaccharides* and *uronic acids*.

Kim[55] et al. observed incomplete silylation in the case of a number of *sugars* when hexamethyldisilazane/trimethylchlorosilane in pyridine or *N*-trimethylsilylacetamide was used[56]. Using hexamethyldisilazane/trifluoroacetamide (2:1)[57] reliable silylation is achieved even in the presence of water so that the otherwise necessary drying step is eliminated. *Tri- and tetrasaccharides* in cereals and carbohydrates in other natural products can be treated in this way. To obtain high yields, care must be taken to see that the trifluoroacetic acid is added last and in small portions.

The gas chromatographic determination[5, 58-60]

and the thin-layer chromatographic separation[61, 62] of silylated *sugars* on silica gel have also been investigated.

Often carbohydrate-containing compounds can be separated readily from natural products by conversion into the corresponding *glycosides* which can then be silylated[63, 64]. *Uronic acids* also occur widely in nature. They can be silylated completely with a mixture of hexamethyldisilazane and trimethylchlorosilane in pyridine[65, 66]. Vecchi and Kaiser[67] have studied the silylation of *ascorbic acid*. During the reaction of *aminosugars* with hexamethyldisilazane and trimethylchlorosilane in pyridine the amino group is not attacked; nevertheless, the products readily undergo gas chromatography. Whether the phosphoric acid group in *sugar phosphates* or *ribonucleotides* is completely or only partially silylated by the various silylation agents has not yet been resolved unambiguously[14].

2.5 Steroids

This class of important biologically active substances gives very good results with hexamethyldisilazane and trimethylchlorosilane[68] without solvent.

In pyridine[69] the progesterone metabolites *pregnanediol*, *pregnanetriol* and the *17-ketosteroids* can also be determined. *Sterically hindered hydroxyl groups* can be silylated in the same way.

A further very effective silylation reagent is *N,O*-bis(trimethylsilyl)acetamide[70]. With non-sterically hindered hydroxyl groups this reagent reacts even at room temperature, the *equatorial hydroxyl groups* reacting more rapidly than the corresponding axial groups.

An 8–12-hour reaction with the reagent is necessary to achieve a quantitative silylation, including of the axial groups. By raising the temperature to 65° the conver-

[50] *C. Burkhard*, J. Org. Chem. *22*, 592 (1957).

[51] *E. Choby, M. Neuworth*, J. Org. Chem. *31*, 632 (1966).

[52] *M. Rowland, S. Riegelman*, J. Pharm. Sci. *56*, 717

[53] *R. Kerr, K. Hobbs*, Ind. Eng. Chem. *45*, 2542 (1953).

[54] *F.A. Henglein, G. Abelsnes, B. Kösters, K. Scheinost*, Makromol. Chem. *21*, 59 (1956); *24*, 1 (1957); Chem. Ber. *92*, 1638 (1959).

[55] *S. Kim, R. Bentley, C. Sweeley*, Carbohydrate Res. *5*, 373 (1967).

[56] *L. Birkhofer, A. Ritter, F. Bentz*, Chem. Ber. *97*, 2196 (1964).

[57] *C. Lott, K. Brobst*, Anal. Chem. *38*, 1767 (1966).

[58] *R. Alexander, J. Garbutt*, Anal. Chem. *38*, 1767 (1966).

[59] *Y. Halpern, Y. Houminer, S. Patai*, Analyst (London) *92*, 714 (1967).

[60] *J. Sawardeker*, J. Chromatogr. *20*, 260 (1965).

[61] *J. Kärkkäinen, E. Haahti, A. Lehtonen*, Anal. Chem. *38*, 1316 (1966).

[62] *J. Lehrfeld*, J. Chromatogr. *32*, 685 (1968).

[63] *C. Sweeley, B. Walker*, Anal. Chem. *36*, 1461 (1964).

[64] *D. Vance, C. Sweeley*, J. Lipid. Res. *8*, 621 (1967).

[65] *M. Perry, R. Hulyalkar*, Can. J. Biochem. *43*, 573 (1965).

[66] *O. Rannhardt, H. Schmidt, H. Neukom*, Helv. Chim. Acta *50*, 1267 (1967).

[67] *M. Vecchi, K. Kaiser*, J. Chromatogr. *26*, 22 (1967).

[68] *T. Luukkainen, W. Vanden Heuvel, E. Haahti, E. Horning*, Biochim. Biophys. Acta *52*, 599 (1963).

[69] *M. Makita, W. Wells*, Anal. Biochem. *5*, 523 (1963), Z. Anal. Chem. *208*, 155 (1965).

[70] *L. Birkhofer, A. Ritter, W. Giesler*, Angew. Chem. *75*, 93 (1963).

sion goes to completion in 1 hour. In the absence of a catalyst N,O-bis(trimethylsilyl)acetamide reacts only with non-sterically hindered hydroxy groups even at elevated temperature. The 11β-hydroxy group is etherified only after adding trimethylchlorosilane. If the silylated compounds are to be investigated subsequently by gas chromatography, only a small amount of trimethylchlorosilane may be employed as catalyst. This is because ammonium chloride is formed during the reaction of hexamethyldisilazane and trimethylchlorosilane with the steroid to be silylated, and is retained by the column and causes decomposition. Samples free of ammonium chloride can be obtained only by distilling off the excess silylation reagent and dissolving the residue in hexane or other nonpolar solvents.

N,O-bis(trimethylsilyl)acetamide offers a further advantage. Using this reagent it is possible sometimes to silylate hydroxyl groups in *ketonic hydroxy steroids* even at room temperature after making derivatives of the keto group with O-methylhydroxylamine or N,N'-dimethylhydrazine[71]. Similarly, the dihydroxyaceto group in *glucocorticoids* can be converted into the bismethylenedioxy compound with formaldehyde; subsequently, the 3-hydroxyl group can be silylated with N,O-bis(trimethylsilyl)acetamide[72].

Gas chromatographic[73,74] and *thin-layer chromatographic*[75] separations of steroid trimethylsilyl ethers have been investigated extensively. On neutral silica gel these compounds are relatively resistant to hydrolysis and can be eluted in 85–95% yield.

Sterols[76], for instance *cholesterol*, can be silylated quantitatively with hexamethylenedisilazane mixed with trimethylchlorosilane in tetrahydrofuran as solvent. Of the *plant steroids*[77-79], 15 plant sterols[80] and the corresponding silyl ethers have been investigated by gas chromatography and mass spectrometry.

Bile acids are silylated best by first converting them into their methyl esters. The silylation reaction of the different hydroxyl groups in the bile acids does not proceed uniformly.

Equatorial hydroxyl groups in the 3α, 6α, 7β and 12β or the axial 3β positions are silylated easily by hexamethyldisilazane in acetone[81] or dimethylformamide[82], or with trimethylchlorosilane in dioxane as solvent[83]. By contrast, axial hydroxyl groups in the 7α and 12α positions do not react under these conditions.

Silylation of all the hydroxyl groups in bile acids[69] succeeds with a mixture of hexamethyldisilazane, trimethylchlorosilane, and pyridine. The reaction is complete after 10 minutes at room temperature; in tetrahydrofuran at least 2 hours are required.

Silylation of the *estrogens*[84,85] likewise succeeds with a mixture of hexamethyldisilazane and trimethylchlorosilane in chloroform, tetrahydrofuran or pyridine as solvent (12 hours at 20° or several hours at 60—70°). Numerous estrogens and *metabolites*[86] occur in *urine* during *pregnancy*. A specific method of analysis was sought for *estriol*, because this hormone is formed in increasing amounts during advanced pregnancy[87].

The analysis of the *17-ketosteroids* has become very much simpler since silylation of these compounds has been achieved. The principal 17-ketosteroids occurring in human urine have been subjected to chromatographic analysis[88-90].

Progesterones are silylated by the same technique as that used for the 17-ketosteroids. Since the content of *pregnanediol* and *pregnanetriol* in urine must generally be determined in the urine, these substances, like the 17-ketosteroids, must first be obtained from the urine by hydrolysis and extraction and then purified before the actual silylation can be carried out.

The hormones of the adrenal cortex are not so simple to silylate. Many *corticosteroids* contain the sterically hindered 11β-hydroxyl group, which is more difficult to etherify than the 3α-hydroxyl

[71] H. Fales, T. Luukkainen, Anal. Chem. 37, 955 (1965); Z. Anal. Chem. 227, 230 (1967).

[72] M. Kirschner, H. Fales, Anal. Chem. 34, 1548 (1962); Z. Anal. Chem. 199, 462 (1964).

[73] A. Kuksis in D. Glick, Methods of Biochemical Analysis, Vol. XIV, p. 325, Interscience Publishers, New York 1966.

[74] E. Horning, J. Gas Chromatogr. 5, 283 (1967).

[75] C. Brooks, Biochem. J. 99, 47 P (1966); J. Chromatogr. 31, 396 (1967); Ref. 91.

[76] W. Wells, M. Makita, Anal. Biochem. 4, 204 (1962).

[77] M. McKillican, J. Am. Oil Chem. Soc. 41, 554 (1964).

[78] T. Miettineu, Acta Chem. Scand. 21, 286 (1967).

[79] A. Rozanski, Anal. Chem. 38, 36 (1966).

[80] B. Knights, J. Gas Chromatogr. 5, 273 (1967).

[81] D. Sandberg, J. Lipid Res. 6, 182 (1965).

[82] P. Eneroth, J. Lipid Res. 7, 511 (1966); J. Lipid Res. 7, 524 (1966).

[83] T. Briggs, S. Lipsky, Biochim. Biophys. Acta 97, 579 (1965).

[84] T. Luukkainen, Biochim. Biophys. Acta 52, 599 (1961); 52, 153 (1962).

[85] H. Adlercrentz, T. Luukkainen, Biochim. Biophys. Acta 97, 134 (1965).

[86] H. Adlercrentz, A. Salokangas, T. Luukkainen, Mem. Soc. Endocrinol. 16, 89 (1967).

[87] A. Schindler, Gynaecologia 161, 446 (1966); s.a. in M.B. Lipsett, Gas Chromatography of Steroids in Biological Fluids, p. 237, Plenum Press, New York 1965.

[88] G. Roversi, A. Ferrari, J. Chromatogr. 24, 407 (1966).

[89] I. Hartmann, H. Wotiz, Biochim. Biophys. Acta 90, 334 (1964).

[90] W. Vanden Heuvel, B. Creech, E. Horning, Anal. Biochem. 4, 191 (1962).

group. *5β-androstane-3α,11β-diol-17one* was completely etherified in 12–15 hours with a 5:1 mixture of hexamethyldisilazane/trimethylchlorosilane[91] in pyridine at room temperature and in 20 hours using *N*-trimethylsilylimidazole[12].

During gas chromatography of untreated *cortisone* and related compounds with a 17α-hydroxyl group, the 17β side-chain is partly split off and the corresponding 17-ketosteroid is formed[92]. Even after reacting the 20-keto group with *O*-methyl-hydroxylamine, these compounds are not yet stable enough; silylation of the corticosteroids previously treated with *O*-methylhydroxylamine is necessary for analysis by gas chromatography to succeed[93].

Using *N,O*-bis(trimethylsilyl)acetamide mixed with basic solvents, the sterically hindered 11β and 17α hydroxy groups can be silylated selectively[74]. With *N,O*-bis(trimethylsilyl)acetamide alone complete conversion is not obtained. All hydroxyl groups are silylated also when a 3:3:2 mixture of trimethylsilylimidazole/*N,O*-bis(trimethylsilyl)-acetamide/trimethylchlorosilane is used at 60°. Pyridine may be added to aid in solution.

The selective silylation of *digitalis glycosides*[94] with hexamethyldisilazane, trimethylchlorosilane and pyridine attacks the 3β, 12β and 16β-hydroxy groups. *N,O*-bis(trimethylsilyl)acetamide, pyridine or *N,O*-bis(trimethylsilyl)acetamide, trimethylchlorosilane and pyridine react with these groups and the enolic hydroxyl group on the 23-C atom. A mixture of trimethylsilylimidazole, *N,*-bis(trimethylsilyl)acetamide and trimethylchlorosilane in pyridine produces silylation of all hydroxyl groups at 60°, including the 14β and the enolic hydroxyl groups.

2.6 Amines

The amino group is more difficult to silylate than a hydroxyl group, but an effective silylation mixture can be found in the majority of cases. Trimethyl-chlorosilane or hexamethyldisilazane mixed with bases, e.g. pyridine and triethylamine, have given good results. In many instances an excess of the amine to be silylated also acts as a basic catalyst. Sometimes substitution of two hydrogen atoms in one amino group by two trimethylsilyl groups

succeeds with *N,O*-bis(trimethylsilyl)acetamide and trimethylchlorosilane[95] or a mixture of *N*-methyl-*N*-trimethylsilyldiethylamine and ammonium sulfate[96].

During silylation of hydrazines with trimethyl-chlorosilane[97], 1-methylhydrazine, for example, is converted into either 1-methyl-2-trimethylsilyl-hydrazine or 1-methyl-1,2-bis(trimethylsilyl)-hydrazine depending on the amount of trimethyl-chlorosilane used. After the silylation of phenolic amines, including *catecholamines*, with hexa-methyldisilazane in dimethylformamide[98] a good separation is obtained with gas chromatography[43]. By combining gas chromatography and mass spectrometry it was demonstrated[43, 99] that primary but not secondary amino groups of catechol-amines can be silylated with hexamethyldisilazane and dimethylformamide.

With trimethylsilylimidazole in acetonitrile only the hydroxy groups in the catecholamines can be etherified; even at 60° no *N*-silylation takes place[99]. Using *N,O*-bis(trimethylsilyl)acetamide and trimethylchlorosilane as catalyst, it is possible to deliberately silylate only the primary amino groups without attacking the secondary amino groups.

If, during the analysis of the *sphingosines*, only the hydroxyl groups in the sphingose molecule are to be etherified, the free base must be reacted with hexamethyldisilazane and pyridine[100]. Sphingosine hydrochlorides do not react under these conditions. When *N,O*-bis(trimethylsilyl)acetamide is employed all hydroxyl and amino groups of the free bases are silylated.

2.7 Amino Acids

Amino acids can be silylated in two ways. The carboxyl group invariably reacts more easily than the amino group. All amino acids except cystine and arginine can be silylated relatively easily, including *asparagine* and *glutamine*. Esterification succeeds with hexamethyldisilazane in boiling toluene[101]. Acid catalysts accelerate the reaction

[91] *C. Brooks, E. Chambaz, E. Horning,* Anal. Biochem. *19,* 234 (1967).

[92] *W. Vanden Heuvel, E. Horning,* Biochem. Biophys. Res. Commun. *3,* 356 (1960).

[93] *W. Gardiner, E. Horning,* Biochim. Biophys. Acta *115,* 524 (1966).

[94] *B. Maume, W. Wilson, E. Horning,* Anal. Lett. *1,* 401 (1968).

[95] *L. Birkhofer, D. Brokmeier,* Tetrahedron Lett. *1968,* 1325.

[96] *J. Hils, v. Hagen, H. Ludwig, K. Rühlmann,* Chem. Ber. *99,* 776 (1966).

[97] *U. Wannagat, W. Liehr,* Z. Anorg. Chem. *299,* 341 (1959).

[98] *S. Lindstedt,* Clin. Chim. Acta *9,* 309 (1964).

[99] *M. Horning, A. Moss, E. Horning,* Biochem. Biophys. Acta *148,* 597 (1967).

[100] *R. Gaver, C. Sweeley,* J. Am. Oil Chem. Soc. *42,* 294 (1965).

[101] *K. Rühlmann,* Liebigs Ann. Chem. *683,* 211 (1965); J. Prakt. Chem. *32,* 37 (1966).

which is complete when the amino acid has dissolved completely. Prolonged treatment with hexamethyldisilazane yields *N*-silylated esters as sideproduct. However, by passing ammonia gas into the reaction mixture the trimethylsilyl group is easily split off from the amino group again.

A series of *dipeptides* also could be silylated with trimethylsilyldiethylamine in good yield[101]. Using trimethylchlorosilane, amino acids are converted into the silyl ester hydrochlorides in high yield[102]. In combination with an acid catalyst trimethylsilyldiethylamine[103] is the best reagent for completely silylating an amino acid.

Good results are obtained also with *N,O*-bis(trimethylsilyl)acetamide or with bis(trimethylsilyl)-trifluoroacetamide[104] in acetonitrile. Near the boiling point the compounds dissolve completely. Aminobenzoic acid[105] is silylated easily and amenable to analysis by gas chromatography. *Tyrosine, thyronine* and *thyroxin* can be silylated quantitatively with *N,O*-bis(trimethylsilyl)acetamide[106].

2.8 Aliphatic and Aromatic Amides, Ureas and Carbamates

In aliphatic and aromatic *amides* one hydrogen atom of the amide group is substituted relatively easily by a trimethylsilyl group, but the introduction of a second silyl group is much more difficult[13, 106].

Chloramphenicol and related compounds[107, 108] have been silylated with a mixture of hexamethyldisilazane, trimethylchlorosilane and pyridine.

Ureas and *carbamates*[18] can be silylated very well with *N*-trimethylsilyl-*N,N'*-diphenylurea. The diphenylurea obtained as a by-product during the silylation can be filtered off readily.

2.9 Miscellaneous Compounds

2.9.1 Peroxides

Trialkylorganoperoxysilanes are stable, distillable liquids[109-111]. They have been prepared from the

corresponding alkyl or aralkyl hydroperoxides by reacting with trichlorosilane and pyridine. These compounds can act as *polymerization* initiators.

2.9.2 Enols

Klebe et al.[13] described the silylation of *cyclic β-keto acids* with *N,O*-bis(trimethylsilyl)acetamide in ether or benzene. On gas chromatographic examination of α-keto acids[112] only a single peak was obtained for each acid, even when the α-keto acid had been methylated previously in dimethylformamide and treated only afterwards with hexamethyldisilazane. Krüger et al.[113] initially convert *enol compounds* into the corresponding sodium enolates and allow trimethylchlorosilane to act on the freshly prepared solutions. The sodium enolate can be obtained by reacting the enol with anthracene sodium in tetrahydrofuran.

2.9.3 Epoxides

Epoxides react very readily with trimethylchlorosilane by ring opening. Opening of the oxide ring occurs almost exclusively at the primary C atom. The derivatives obtained in this manner are less volatile than the epoxides and hence of little use for gas chromatographic analysis.

2.9.4 Thiols

Saturated thiols do not react with the usual silylation agents. These need to be converted into the sodium or lead salts before trimethylchlorosilane will react. Unsaturated thiols react more readily. *Thioacetic acid* can be silylated with a mixture of trimethylchlorosilane and triethylamine.

3 Properties of Silylated Compounds

3.1 Physical Properties

The trimethylsilyl group displays the same geometry as the tert-butyl group, but occupies more space and is less crowded within itself; in return, it can produce a greater *steric effect*. The trimethylsilyl groups transmit a positive *inductive* (+I) *effect* on *neighboring groups,* because the silicon key atom is less electronegative than carbon. Bonds with a trimethylsilyl group are often thermodynamically stronger than corresponding bonds with a tert-butyl group.

[102] *J. Hils, K. Rühlmann*, Chem. Ber. *100*, 1638 (1967).
[103] *P. Mason, E. Smith*, J. Gas Chromatogr. *4*, 398 (1966); *E. Smith, H. Sheppard*, Nature *208*, 878 (1965).
[104] *C. Gehrke, R. Zumwalt*, Biochem. Biophys. Res. Commun. *31*, 616 (1968).
[105] *K. Rühlmann*, J. Prakt. Chem. *4*, 86 (1959).
[106] *L. Birkhofer, A. Ritter, H. Dickopp*, Chem. Ber. *96*, 1473 (1963).
[107] *G. Resnick, D. Corbin, D. Sandberg*, Anal. Chem. *38*, 582 (1966).
[108] *P. Shaw*, Anal. Chem. *35*, 1580 (1963).
[109] *E. Buncel, A.G. Davies*, J. Chem. Soc. *1958*, 1550.

[110] *W. Hahn, L. Metzinger*, Makromol. Chem. *21*, 113 (1956).
[111] *A. Simon, H. Arnold*, J. Prakt. Chem. *8*, 241 (1959).
[112] *C. Dalgliesh*, Biochem. J. *101*, 792 (1966).
[113] *C. Krüger, E. Rochow*, J. Organometal. Chem. *1*, 476 (1964).

The exchange of an active hydrogen atom in a compound with a trimethylsilyl group reduces the polarity and, hence, also the tendency to exhibit association. At the same time the volatility, stability, and solubility in nonpolar liquids are increased. In respect to the increased volatility of silylated compounds, this is especially true of the poly-hydroxy compounds, i.e. substances that have a particular tendency to display association.

Substance:	Glycerol	Pentaerythritol	Benzoic acid
Boiling point depression:	60°	112°	30°

3.2 Chemical Properties

Most silyl derivatives are hydrolyzed easily and must be protected even against *atmospheric moisture*. The most stable silyl derivatives are the tri-methylsilyl ethers. *N*-trimethylsilyl compounds and trimethylsilyl esters are very easily hydrolyzed. Thiotrimethylsilyl ethers[114] are split more easily than the *O*-trimethylsilyl ethers. Acid catalysts[91] accelerate the hydrolysis. The conditions under which a silylated compound is completely hydrolyzed are dependent also on *steric arrangement*. Tertiary pentyltrimethylsilyl ether, for example, is stable in boiling aqueous alcohol and pyridine, and is not split unless a trace of hydrochloric acid is added[115, 116].

However, as a rule, most silyl derivatives are hydrolyzed by heating with aqueous alcohol[117] (for a description of other properties see Ref. [118]). Silylated amino acids lend themselves to *peptide syntheses*, and silylated sugars to *saccharide syntheses*. *O*-silylated lactim ethers are converted into *N*-alkyl derivatives by alkyl halides. *Halogen fatty acid silyl esters* give *lactones* with silver cyanate. The thermal very stable *trimethylsilyl azide* reacts in the same way as organic azides.

[114] *M. Rimpler*, Chem. Ber. *99*, 1523 (1966).
[115] *S. Friedman, M. Kaufman*, Anal. Chem. *38*, 144 (1966).
[116] *G. Illuminati, F. Tarli*, Ric. Sci., Rend. Sez. A, *3*, 329 (1963); C.A. *59*, 10 105c (1963).
[117] *R. Martin*, J. Am. Chem. Soc. *74*, 3024 (1952).
[118] *L. Birkhofer, A. Ritter*, Angew. Chem. *77*, 414 (1965).

5 Spectroscopic and Photometric Methods

Contributed by

B. Briat, Paris
E. Fluck, Stuttgart
J.K. Foreman, London
E. Glotter, Rehovot
K.H. Hausser, Heidelberg
G. Hohlneicher, München
H.J. Keller, Heidelberg
D.C. Lankin, Cincinnati
D. Lavie, Rehovot
Y. Mazur, Rehovot
K. Möbius, Berlin
B. Schrader, Dortmund
K.E. Schwarzhans, München
G. Snatzke, Bonn
H. Stiller, Jülich
J.E. Todd, Cincinnati
H. Weitkamp, Wuppertal
A. Yogev, Rehovot
W. Zeil, Karlsruhe
H. Zimmer, Cincinnati

5.1 Electron Spectroscopy

Günther Snatzke

Organisch-chemisches Institut der Universität,
D-53 Bonn

1 Theory

According to the Einstein-Bohr relation $\Delta E = h\nu$, light of wavelength 150–900 nm has just the right energy, if absorbed, to promote an electron of the outermost shell from the ground state (g) to the electronically excited state (a). The inverse process (emission), also induced by light of the same frequency, is observed almost only with atoms. The excitation of loosely bound electrons, such as those of π-bonds or lone (n-) electron pairs, is possible with light of wavelength longer than 185 nm. These electrons can, therefore, be investigated in solution. In contrast, σ-bond electrons can be excited only in the vacuum-UV region (below 185 nm).

In spectroscopy, these electron transitions are designated mostly either by symbols 'borrowed' from group theory ($^1A_1$, $^1A_2$, ...), or by citing the two electron configurations between which the electron jumps. In general, not all orbitals are indicated, but only those between which the transition takes place. For example, the two bands of the oxo group of 2-butanone which are measurable in solution have been recognized[1] as:

$\{(1s_C)^2 (1s_O)^2 \ldots (2p_{y,O})^2 (2\pi_{C=O})^2\} \rightarrow \{(1s_C)^2 (1s_O)^2 \ldots (2p_{y,O}) \ (2\pi_{C=O})^2 (2\pi^*_{C=O})\}$ (ca. 280 nm) and $\{g\} \rightarrow \{(1s_C)^2 (1s_O)^2 \ldots (2p_{y,O}) (2\pi_{C=O})^2 (2\sigma^*_{C=O})\}$ (ca. 190 nm)

For short, these are written as n→π^* and n→σ^*, respectively. The π→π^* – transition appears at still shorter wavelengths (ca. 160 nm). For graphical representation, one uses either an energy diagram (Fig. 1, left), in which the various orbitals are drawn separately, or a Jablonski diagram (Fig. 1, right), which shows more clearly that the electron configurations for excitation into a π^* – orbital (here, a_1 and a_3) differ, according to whether an n- or a π-electron was promoted from the ground state g.

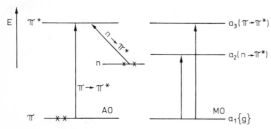

Fig. 1. Absorption spectra, energy level diagram
Left: Atom orbitals (AO), right: Molecular orbitals (MO) (*Jablonski* Diagram). The arrows correspond to the electron transitions during absorption.

[1] *H.H. Perkampus*, Angew. Chem. *80*, 613 (1968).

The energy terms for molecules (particularly those in solution) are never as distinct as for atoms, but are split into multiplets (rotations, vibrations, collision broadening, etc.), so that only overall, broad bands can be measured; in general, fine structure is observable only in nonpolar solvents.

In measurement of absorption (for details of instruments, see Ref.[5]), either the percent *transmittance* (T) or the *extinction* (E, also termed optical density D) is recorded as a function of wavelength or wave number.

If I_0 is the intensity of the incident light beam on a solution, and I the intensity of the emerging light, then the transmittance is defined as:

$$T = 100 \cdot \frac{I}{I_0} [\%] \tag{1}$$

and the extinction as:

$$E = \log \frac{I_0}{I} \tag{2}$$

The following relation exists between these two quantities:

$$E = 2 - \log T \tag{3}$$

According to the Bouguer-Lambert-Beer law, for a solution of concentration c[g/l] of a compound of known molecular weight M, the extinction for an optical path length of d [cm] is given by:

$$E = \varepsilon \cdot \frac{cd}{M} \tag{4}$$

The *molar absorptivity*, ϵ, is (in dilute solutions) a constant (of dimensions [$cm^{-1}\,Mol^{-1}l$]) characteristic of the substance and independent of concentration. As the value of ϵ for various bands of a given compound may vary between 10 and 10^5, log ϵ is often plotted instead of ϵ itself. If the molecular weight is unknown, $E^{1\%}_{cm}$ is used for characterization; this is the extinction which would be obtained with a path length of 1 cm and a concentration of 1%. It is related to ϵ thus:

$$E^{1\%}_{1\,cm} = \frac{\varepsilon \cdot M}{10} \tag{5}$$

The *integrated intensity* $\int\epsilon d\nu$ or $\int\epsilon d\lambda$ is a better measure for quantitative comparison, because curve shapes are not always similar.

According to theory, the integrated intensity or the oscillator strength, f, defined as

$$f = 4{,}315 \cdot 10^{-9} \cdot \int \varepsilon \cdot d\nu \tag{6}$$

can be correlated with the *dipole strength*, which is the square of the electrical transition moment μ_{el}. This quantity, which has vector properties, can be calculated from the equation:

$$\mu_{el} = \int \Psi^*_g \cdot \mu_{Op} \cdot \Psi_a \cdot d\tau \qquad (7)$$

if the wave functions for the ground state (Ψ_g) and the excited state (Ψ_a) are known, and $\mu_{op} = e \cdot \vec{r}$ is the operator corresponding to μ. The correlation between the dipole strength and the integrated intensity is given by:

$$\mu^2_{el} = 9{,}138 \cdot \frac{10^{-39}}{\tilde{\nu}_m} \int \varepsilon \cdot d\nu \qquad (8)$$

where $\tilde{\nu}_m$ is the mean wave number of the absorption band concerned.

By combination of Equations (7) and (8), it follows that the dipole strength (and thus ϵ) may be zero for certain symmetry properties of Ψ_g and Ψ_a. The following types can be distinguished:

Spin-forbidden: The multiplicities of the ground and excited states must be identical, otherwise ϵ is less than 1; for example, singlet-triplet transitions are forbidden.
Parity-forbidden: (only for molecules with a center of symmetry): The ground and excited state must have opposite parities (even/odd). Parity-forbidden bands have $\epsilon < 3$;
Symmetry-forbidden: The integral in (7) becomes zero because of other symmetry properties of Ψ_g, Ψ_a and μ_{Op}. Symmetry-forbidden transitions give $\epsilon < 300$. Only the local symmetry of the chromophore must be taken into account, not the symmetry of the whole molecule.

In contrast, fully allowed transitions have an ϵ of the order of magnitude 10^4 to 10^5. In spite of this, forbidden bands (especially symmetry-forbidden) may appear in the spectrum, particularly since the symmetry is somewhat reduced by vibrations. If only one of the three components μ_x, μ_y, μ_z of the vector $\vec{\mu}$ is non-zero, there is said to be a *polarization* of the band in this direction. On investigation of single crystals with linearly polarized light, one finds appreciable absorption only if the direction of this polarization coincides with that of the transition moment.

2 Simple Chromophores

Atoms or groups of atoms in which the electrons participating in excitation are localized are called *chromophores,* in the nomenclature introduced by Witt. For technical reasons (spectra in solution, no vacuum-UV), such chromophores must always contain loosely bound (n- or π-) electrons (Table 1). Band shifts due to substitution or to change of solvent allow the various transitions to be identi-

Table 1. UV-maxima of some simple chromophores[a]

Chromophore	λ_{max} [nm]	ε	Type of excitation
≡C—O—	185	1000	n→σ*
≡C—N—	200	3000	n→σ*
≡C—S—	200	2000	n→σ*
≡C—Br	200	300	n→σ*
—C—J	260	500	n→σ*
≥N—Cl	270	300	n→σ*
≥N—Br	300	400	n→σ*
—O—O—	200		n→σ*
—S—S—	250–330	1000	n→σ*
≥C—C≤ \S/	265	50	n→σ*
≥C=C≤	190	9000	π→π* (or π→σ*?)
≥C=O	280	20	n→π*
	190	2000	n→σ*
	160		π→π*
—COOR	205	50	n→π*
	165	4000	π→π*
≡C=N—	250	200	n→π*
≥C=NOH	193	2000	n→π*
≥C=S	500	10	n→π*
	240	9000	
≥C=N₂	350	5	n→π*
—N=N—	340	10	n→π*
	240		n→π*
≥S=O	210	2000	
—N=O	675	20	n→π*
	300	100	n→π*
—NNO	350	100	n→π*
	240	8000	
—ONO	310–390	30	n→π*
	220	1000	
—NO₂	330	10	
	280	20	n→π*
—ONO₂	260	20	n→π*
—SCN	245	100	n→π*
—NCS	250	1000	
≡C—N₃	280	30	n→π*
	220	150	
—C≡C—	175	8000	π→π*

[a] Band positions and intensities often depend very much upon substitution. The values cited are mean values. The assignment is not known for every band.

fied. The following terms are used in connection with shifts of maxima: a *bathochromic shift (red shift)* is a shift to longer wavelengths; a *hypsochromic shift (blue shift)* is one to shorter wavelengths. *Hyperchromy* is an increase of extinction; *hypochromy* is a decrease of extinction. In general, on changing from a nonpolar to a polar solvent, an n→π* band will be shifted hypsochromically, and a π→π* band bathochromically.

3 Conjugated Chromophores

Conjugation of two (or more) chromophores always leads to a bathochromic shift, and mainly also to a hyperchromic change.

3.1 Polyolefins

The position of the maximum is governed by rules originally advanced by Woodward, and subsequently modified by Fieser and Scott (Table 2).

Table 2. Rules for determining the positions of the absorption maxima of conjugated olefins

Grouping	Increment [nm]
heteroannular diene	214
homoannular diene	253
every C = C extending the conjugation	30
alkyl or bond in ring	5
exocyclic position of C = C	5
O-acyl substituent	0
O-alkyl substituent	6
S-alkyl substituent	30
Cl-, Br-	5
N(alkyl)$_2$ substituent	60
solvent correction	0

In applying these rules one must differentiate between heteroannular and homoannular positions for the diene grouping: in the former case, the double bonds are distributed over two rings, whereas in the latter, they are both situated in one single ring. For polyenes, one must choose the basic increment of a homoannular diene, if homo- and heteroannular diene groupings are present in the same chromophore.

Cross-conjugated systems (=C⟨)

do not follow these rules; such chromophores often give several bands, which originate from the individual *partial chromophores*.

In example 1, the 'substituents' are marked by thick bonds.

Example 1:

1

Calc.:   253 (basic increment)
        + 30 (extension of conjugation)
        + 15 (3 exocyclic C=C)
        + 25 (5 alkyl substituents)
        ——
        323 nm

Found: Main maximum at 324 nm (11 800); side maxima at 312 (10 400) and 339 (7400) nm.

With compounds which contain many conjugated double bonds, for example carotenoids, the electronic spectra contain several bands. These are ascribed to partial chromophores (half, quarter, ... of the length of the polyene), as well as to the whole system. The lycopenes (**2**) are typical examples.

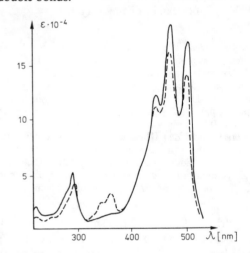

In Fig. 2 the principal maximum at 473 nm (186 000) of the all-*trans* compound is shifted hypsochromically only weakly (λ_{max}=465 nm) by isomerization around the central double bond 15, 15', but its intensity drops appreciably (ϵ= 122 000). In contrast, the first overtone band (the *cis*-band) at 363 nm is symmetry-forbidden (ϵ=14 000) in the all-trans compound, whereas it becomes allowed (λ_{max}=361 nm, ϵ=68 000) on isomerization to the 15, 15'-*cis*-isomer. According to an empirical rule proposed by Dale, λ_{max} of the p-th band of a polyene with n double bonds corresponds with λ_{max} of the absorption band at longest wavelength of a polyene with n/p double bonds.

Fig. 2. Electron spectra of all-trans-lycopene (**2**) (——) and 15.15'-*cis*-lycopene (– – – –). The band at 361 nm is the *cis*-band.

3.2 Polyeneketones

Rules have been formulated for the position of the maximum of the $\pi \rightarrow \pi^*$ band in polyeneketones (Table 3) similar to those developed by Woodward for polyolefines.

Table 3. Rules for determining the position of the absorption maxima (K-band) of polyene ketones of types

(in ethanol solution).

Table 3 (cont.)

Grouping		Increment [nm]
acyclic or six-membered ring-enone		215
five-membered ring-enone		202
each C=C extending the conjugation		30
exocyclic position of C=C		5
homoannular diene		39
alkyl or bond in ring:	α	10
alkyl or bond in ring:	β	12
alkyl or bond in ring:	γ and higher	18
HO-	α	35
	β	30
	δ	50
O-acyl	α, β or δ	6
O-alkyl	α	35
O-alkyl	β	30
O-alkyl	γ	17
O-alkyl	δ	31
S-alkyl	β	85
Cl-	α	15
Cl-	β	12
Br-	α	25
Br-	β	30
N(alkyl)₂	β	95
Solvent correction		
water		− 8
chloroform		+ 1
dioxane		+ 5
ether		+ 7
saturated hydrocarbon		+ 11

The n→π*-band (*R-band*, from the German *Radikal*) and the π→π*-band (*K-band*, from the German *Konjugation*)[2,3] of an oxo group are both red-shifted by conjugation with an olefinic double bond. The former band is, however, shifted less than the latter; if conjugation is sufficiently long, therefore, the K-band may 'overtake' the R-band. Also the n→σ* band should undergo only a weak bathochromic shift (cf. Chapter 5.6).

Example 3: (measured in ethanol)

Calc.:   215 (basic increment)
         + 60 (double extension of conjugation)
         + 5 (exocyclic C=C)
         + 39 (homoannular diene)
         + 12 (β-alkyl)
         + 18 (γ-alkyl)
         ———
         349 nm

Found: Maximum at longest wavelength at 348 nm (11 000). Additional maxima at shorter wavelengths: 278 (3720) and 230 nm (18 600).

[2] *A. Burawoy*, Ber. *63*, 3155 (1930).
[3] *A. Burawoy*, J. Chem. Soc. 1177 (1939); 20 (1914).

Bands of other C=X chromophores are shifted bathochromically by conjugation with double bonds (cf. Table 4).

Table 4. UV-maxima of some conjugated chromophores

Chromophore		λ_{max} [nm]
$>C=C—C=NOH$		235
$>C=C—C=N\cdot NH\cdot CC\cdot NH_2$		265
$>C=C—C=N\cdot NH\cdot CS\cdot NH_2$		300
$>C=C—C=N—$	in neutral solution	220
	in acidic solution	275
$>C=C—NO_2$	n→π*	340
	π→π*	260
$>C=C—C≡N$		213
$>C=C—COOR$	n→π*	260
Rules for π→π* bands:		
β-monosubstituted		208
α, β- or β, β-disubstituted		217
α, β, β-trisubstituted		225
each C=C extending the conjugation		+ 30
exocyclic position of C=C		+ 5

3.3 Cyanine Dyes

Dyes of the *cyanine* type

and the related *merocyanines*

have been extensively investigated, both experimentally and theoretically. The λ_{max} of the longest wavelength absorption band increases by 100 nm with each newly added C=C. With nonsymmetrical derivatives (especially with merocyanines), the position of the maximum depends very much on the polarity of the solvent, as the ground and excited states have quite different electron distributions. For example: the nonpolar ground state of structure 2, which is essentially described by formula **A**, will be less stabilized in polar solvents than the very polar excited state (considerable involvement of structure **B**), so that a bathochromic shift (+20 nm) is found on changing from pyridine solvent to water.

For structure 3, however, the contrary is found ($\Delta\lambda = -50$ nm).

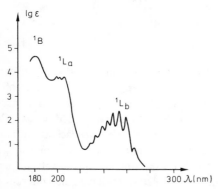

5

This indicates a greater contribution of **B** to the ground state than to the excited state.

3.4 Polyacetylenes

Conjugated polyacetylenes with n C_2-units give an intense band ($\epsilon \simeq 80\,000 \cdot n$) between 205 and 350 nm, which shows pronounced fine structure with a spacing of $\Delta\tilde{\nu} = 2600$ cm^{-1}. In addition, a series of less intense maxima ($\epsilon \simeq$ 200–300) is observed at longer wavelengths, whose separation is $\Delta\tilde{\nu} = 2300$ cm^{-1}. With short-chain compounds, this latter group of bands is red-shifted more strongly with growing n than in the case with long-chain compounds. The λ_{max} of the intense band increases approximately linearly with n. Polyenpolyines follow similar rules.

4 Aromatic Compounds

4.1 Benzene and its Derivatives

Benzene and its simple derivatives have 3 bands at 260, 200 and 185 nm (Fig. 3, Table 5).

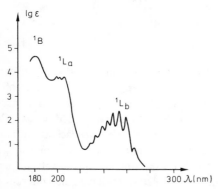

Fig. 3. Electron spectrum of benzene. Band notation of PLATT (cf. Table 5).

Table 5. Different notations for the 3 absorption bands of benzene

λ_{max} [nm]	ϵ	Notation according to				
		Braude	Clar	Platt	Group theory	Doub and Vandenbelt
260	200	I	α	^1L_b	$^1B_{2u}$	secondary
200	7000	II	p	^1L_a	$^1B_{1u}$	primary
185	50000	III	β	^1B	$^1E_{1u}$	second primary

The forbidden band ^1L_b at the longest wavelength, has a pronounced fine structure with a progression of $\Delta\tilde{\nu} = 923$ cm^{-1}; the ^1L_a-band, which is often seen only as a shoulder, has a less well developed one with a progression $\Delta\tilde{\nu} = 1400$ cm^{-1}, whereas the ^1B-band is not split.

Because of their inductive and mesomeric effects, substituent groups generally influence both ϵ and λ_{max} of these bands; in particular, the ^1L_b-absorption is intensified by mixing with the ^1B-transition. All three benzene bands are usually red-shifted by substitution, but the ratio $\lambda(^1L_b)/\lambda(^1L_a)$ remains almost constant (about 1.25). The smaller the ionization energy of the substituent group in question, the larger is this shift of the maximum. For substituents of order I, the sequence of these $\Delta\lambda$ values is:

Alkyl $<$ Cl $<$ Br $<$ OH $<$ OCH$_3$ $<$ NH$_2$ $<$ O$^\ominus$ $<$ NHCOCH$_3$ $<$ N(CH$_3$)$_2$

and for those of order II:

NH$_3^\oplus$ $<$ SO$_2$NH$_2$ $<$ CN = CO$_2^\ominus$ $<$ CO$_2$H $<$ COCH$_3$ $<$ CHO $<$ NO$_2$

Thus, for example, the ^1L_a band of nitrobenzene is at 268.5 nm. The effect of alkyl groups is attributed to hyperconjugation; consequently, the methyl group causes greater shifts than the isopropyl or t-butyl groups.

Polysubstitution leads to an especially pronounced bathochromic shift of the ^1L_a-band if a substituent group of order I is situated *p*- to a group of order II; with o- and m- polysubstitution, still greater red-shifts have been observed.

For conjugated aryl ketones and acids and their derivatives, one might expect an n$\rightarrow\pi^*$ band and

Table 6. Rules for determining the position of the absorption maxima of compounds of general formula Aryl-CO-R (conjugation band) in ethanol.

Grouping			Increment [nm]
Parent compound C$_6$H$_5$CO $\cdot$ R			
R = Alkyl			246
R = H			250
R = OH or O-alkyl			230
Substituent groups	*o-*	*m-*	*p-*
alkyl	3	3	10
OH, O-alkyl	7	7	25
O$^-$	11	20	78
Cl	0	0	10
Br	2	2	15
NH$_2$	13	13	58
NHCOCH$_3$	20	20	45
NHCH$_3$	—	—	73
N(CH$_3$)$_2$	20	20	85

a $\pi\to\pi^*$ band of the carbonyl group (in addition to the benzene bands). The $n\to\pi^*$ band cannot, in general, be observed, as its intensity is low. Scott has described rules (Table 6), for the $\pi\to\pi$ band, which, according to Nagakura, has intramolecular charge transfer character.

Example 6: (measured in ethanol)

Calc.: 246 (basic increment)
+ 3 (o-alkyl)
+ 7 (o-OH)
+ 0 (m-Cl)
256 nm

Found: 257 nm

Example 7: (measured in ethanol)

Calc.: 230 basic increment
+ 7 (o-OH)
+ 25 (p-O-alkyl)
262 nm

Found: 254 nm

The color change of the usual *pH indicators* originates from the fact that neutral and ionized groups cause shifts of different magnitudes.

4.2 Condensed Aromatic Compounds

According to Clar, bands appear for condensed aromatic compounds similar to those for benzene itself. The α-, p- and β-bands (Table 5) shift bathochromically on annulation, the ratio $\tilde{\nu}_\alpha/\tilde{\nu}_\beta$ remaining constant at 1:1.35. Frequently the shift of the p-band changes so that $\tilde{\nu}_{max}$ is not proportional to the other values. For purposes of identification, use can be made of the fact that, on lowering the temperature, the α-band is shifted hypsochromically, whereas the p- and β-bands are shifted bathochromically. Furthermore, all three are shifted bathochromically by going from the vaporized form to solution form: this shift is approximately 250 cm^{-1} for the α-band, and 950 cm^{-1} for p- and β-bands.

5 Heterocyclic Compounds

Replacement of $-CH=$ in benzene by $-N=$, $-O^{\oplus}=$, or $-S^{\oplus}=$ generally causes a reduction of symmetry, and, thus, a hyperchromy of symmetry-forbidden bands. Additional new $n\to\pi^*$ bands may appear in the spectrum. For example, the ^1L_b band of pyridine at about 250 nm has an ϵ of 2000; a new $n\to\pi^*$ band (with fine structure in nonpolar solvents) can be detected as a shoulder on the tailing area of the first portion of the long-wavelength band.

The more nitrogen atoms there are present in the ring, the more is the $n\to\pi^*$ band shifted bathochromically; in contrast, $\pi\to\pi^*$ bands are hardly affected. Electron-donors shift the $n\to\pi^*$ band of an aza-aromatic compound towards the blue, electron-acceptors towards the red. This is especially true if the aromatic orbital has no nodal plane through the C-atom which is substituted (other than the plane of the ring). On the other hand, $\pi\to\pi^*$ bands are red-shifted by any substituent group. Five-membered heterocyclic compounds give only one or two $\pi\to\pi^*$ bands but no $n\to\pi^*$ band (Table 7).

Table 7. UV spectra of some heterocyclic compounds

Compound	$\lambda_{max}(n\to\pi^*)$	ε	$\lambda_{max}(\pi\to\pi^*)$	ε
furan			200	10000
thiophene			235.2	4300
pyrrole			210	15000
pyrazole			211	4100
1,2,3-triazole			210	5000
1,2,4-triazole			187	3300
pyridine	270	450	251	2000
pyrazine	328	1040	260	5600
pyridazine	340	315	246	1300
pyrimidine	298	326	243	2030
sym-triazine	272	890	222	150
sym-tetrazine	542	829	252	2150

The spectra of indole alkaloids, porphines etc., although very complex, can be employed usefully for structure determination or location of double bonds in the skeleton.

6 Interaction of Non-conjugated Chromophores

If two chromophores with strong electric transition moments are so close to each other that electron excitation can proceed only in a cooperative manner, even though the groups are not conjugated (or at least not strongly conjugated, e.g., twisted biphenyls), *band splitting (Davidov splitting)* occurs. If Φ_g and Φ_g' describe the ground states of the two chromophores, the ground state of the whole molecule is given by:

$$\psi_g = \Phi_g \cdot \Phi_g'$$

In the excited state, an electron may be promoted from either of the ground states, so that there are two wave-functions:

$$\Psi_a^\pm = \sqrt{\tfrac{1}{2}}\,(\Phi_a\cdot\Phi_g' \pm \Phi_g\cdot\Phi_a')$$

Instead of a single band, two are generally found, located at longer and shorter wavelengths than

265

the bands of the isolated chromophores. The excitation energy is not localized completely in either of the individual chromophores; the term *exciton delocalization* is, therefore, used. The intensity of the two bands depends on the direction of the transition moments. Thus, bicyclo [2.2.1] heptene (**8**) has a strong $\pi \rightarrow \pi^*$ band near 195 nm. For bicyclo[2.2.1]heptadiene (**9**), theory predicts a weak band at 213 nm, and a strong one near 180 nm. Indeed, a weak absorption band ($\epsilon = 2100$) has been found at about 205 nm, and the absorption curve rises very steeply below 198 nm.

Non-coplanar β, γ-unsaturated ketones in which the π-orbitals of the C=C and C=O groups are directed one towards the other, e.g. bicyclo[2.2.2]octenone (**10**) (cf. Table 8), give a very intense $n \rightarrow \pi^*$ band (ϵ up to 10^3). This cannot be explained solely by exciton interaction, since the C=O chromophore has too small an electric transition moment. The increased intensity of the $n \rightarrow \pi^*$ band can, however, be explained by charge-transfer or by overlap of localized orbitals of the C=C and C=O chromophores, which also causes the appearance of another band at 202 nm ($\epsilon = 3000$). Some examples of

Table 8. Interaction of non-conjugated chromophores

Compound	λ_{max} [nm]	ϵ
	214 284	1500 30
	214,5 270–80	2300 20
	221	5610
10	202 290	3000 110
	287	178
	300	1170

such interactions are given in Table 8. If a new bond is formed in the excited state by such a *transannular conjugation*, the band is termed *photodesmotic*, e.g.

Davidov splitting is not restricted to molecules with two chromophores. It can occur also in the crystalline state or in solution, if identical chromophores of different molecules come close to one another in the lattice or in aggregates. Thus, in concentrated solutions of *cyanine dyes* or of *Rhodamine B* etc., additional new bands appear which are not found in dilute solutions. The same effect also produces the hypochromy of the strong absorption band in *polynucleotides* and *nucleic acids* which is observed during base-stacking.

7 Influence of Stereochemistry

Steric factors may influence λ_{max} as well as ϵ. The ground state of conjugated systems has the lowest energy for the coplanar arrangement; any twisting of bonds which are not cut by nodal planes of the MO will, consequently, increase this energy. In contrast, twisting of a bond which is cut into halves by a nodal plane lowers the energy. The same is true for the excited state, which always contains more nodal planes than the ground state. Depending on which bond of a chromophore is twisted, a hypsochromic or a bathochromic shift may result. ϵ decreases on twisting, following a cos (or $\cos^2$) function (Tables 9–11); the torsion angles θ have been calculated with the help of the equation: $\epsilon/\epsilon_o = \cos^2\theta$.

Table 9. Spectra of hindered acetyl cyclohexenes

Compound	λ_{max} [nm]	ϵ	ϵ/ϵ_0	θ calc.[°]
	232	12 500	1,0	0
	245	6 500	0,52	44
	239	1 300	0,10	71
	243	1 400	0,11	71

Table 10. Spectra of various twisted biphenyl derivatives

X	λ_{max} [nm]	θ calc. [°]
—CH$_2$—	261.5	0
—HH—	251	23
—(CH$_2$)$_2$—	263.5	20
—CH$_2$OCH$_2$—	250	44.5
$-CH_2-\left(\overset{COO-CH_3}{\underset{CH}{\overset{|}{\underset{}{CH}}}}\right)_2 CH_2-$	236.5	61.5

Table 11. Absorption spectra of some substituted N,N-dimethylanilines (in 2,2,4-trimethylpentane)

Substituents	λ_{max} [nm]	ε	$\varepsilon/\varepsilon_\theta$	θ calc. [°]
H	251	15500	(1)	(0)
2-Me	248	6360	0,41	50
2-t-Bu	250	630	0,04	78
2,6-Me$_2$	262	2240	0,14	68
2,4,6-Me$_3$	257	2500	0,16	66
2,4,6-(i-Pr)$_3$	250	1540	0,10	72
2,4-Me$_2$-6-t-Bu	250	800	0,05	77

The absorption maximum of *disulfides* at long wavelengths depends very much on the torsion angle of the S—S-bond; this is shown, for example, by the 1,2-dithiacycloalkane series of homologs:

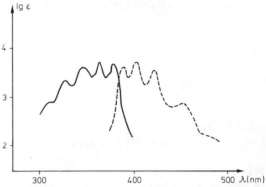

λ max [nm]   334   300   250

Assuming that an n(3p)→σ* band is involved, the excited state is, of course, independent of the torsion angle. The n-orbital is split, however, $\psi_u^\pm = \sqrt{1/2}(\Phi_n \pm \Phi_n')$, and, depending on the torsion angle, either ψ_n^+ or ψ_n^- has the higher energy. This is related [4] to the torsion angle θ, to the simplest approximation, by:

$$E^\pm = \frac{a \pm \beta \cdot \cos\theta}{1 \pm S \cdot \cos\theta}$$

The considerable dependence of the band position on the torsion angle θ follows from this equation.

8 Fluorescence and Phosphorescence

An electronically excited molecule may dissipate its energy with or without emission of a photon.

The first process can proceed spontaneously or may be induced by light. Radiationless transitions between different energy levels of the same multiplicity (singlet-singlet or triplet-triplet) are termed *internal conversions;* such transitions between levels of different multiplicity are called *intersystem crossings.* They always occur iso-energetically, i.e. from a low vibrational state of an electronically higher excited level to a high vibrational state of an electronically lower level.

Fig. 4. Absorption (———) and emission (fluorescence) (————) spectra of anthracene in dioxane solution. The 0–0 transitions do not coincide exactly because the ground and excited states have different solvation.

Emission from the first excited singlet state is called *fluorescence;* whereas, emission from the first excited triplet state is termed *phosphorescence.* The decay time of phosphorescence is about 1 sec; for fluorescence, it is about 10^{-6} to 10^{-8} sec. With a few exceptions, both types of emissions are observed only in a rigid matrix (at −196°). With light *absorption* essentially only the vibrational 0-level in the electronic *ground* state is populated. With *emission* the vibrational 0-level in the electronically *excited* state is populated Therefore, according to the Franck-Condon principle, the 0–0 transition appears as the fine structure band at longest wavelengths in absorption spectra, and at shortest wavelengths in emission spectra. Absorption and emission bands are, therefore, approximately symmetrical with respect to the 0–0 transition (Fig. 4), and both provide similar information. Knowledge of emission spectra is especially important for the mechanistic interpretation of photochemical processes.

9 Bibliography

[4] *G. Bergson,* Ark. Kemi *12,* 233 (1958); *18,* 409 (1962).
J. Lindberg, J.J. Michl, J. Am. Chem. Soc. *92,* 2619 (1970).

[5] *E. Müller,* Houben-Weyl, Methoden der organischen Chemie, Band III/2, p. 593, Georg Thieme Verlag, Stuttgart 1955.

[6] *A.B.F. Duncan, F.A. Matsen*, Electronic Spectra in the Visible and Ultraviolet, in *A. Weissberger*, Technique of Organic Chemistry, Vol IX, p. 581, Interscience Publishers, London 1956.

[7] *A.E. Gillam, E.S. Stern*, Electric Absorption Spectroscopy, 2nd Ed., Arnold, London 1957.

[8] *C.N.R. Rao*, Ultraviolet and Visible Spectroscopy, Butterworth, London 1961.

[9] *H.H. Jaffé, M. Orchin*, Theory and Application of Ultraviolet Spectroscopy, Wiley, New York 1962.

[10] *S.F. Mason*, The Electronic Absorption Spectra of Heterocyclic Compounds, in *A. Katritzky*, Physical Methods in Heterocyclic Chemistry, Vol. II, Academic Press, New York 1963.

[11] *J.N. Murrell*, The Theory of the Electronic Spectra of Organic Molecules, Methuen & Co., London 1963.

[12] *A.I. Scott*, Interpretation of the Ultraviolet Spectra of Natural Products, Pergamon Press, Oxford 1964.

[13] *H.A. Staab*, Einführung in die theoretische organische Chemie, Verlag Chemie, Weinheim/Bergstr. 1966.

[14] *H. Suzuki*, Electronic Absorption Spectra and Geometry of Organic Molecules, Academic Press, New York 1967.

[15] *D.H. Williams, I. Fleming*, Spektroskopische Methoden in der organischen Chemie, Georg Thieme Verlag, Stuttgart 1968.

5.2 IR Spectroscopy

Horst Weitkamp
Farbenfabriken Bayer AG, D-56 Wuppertal

1 Introduction

IR spectroscopy is presently a standard method of analysis in many fields of research and technology. The range of commercial instruments and accessories is so large that suitable equipment is available for the most diverse measurement problems. Regular reviews on advances in methods and equipment have appeared since 1949[1]. The basic principles and many applications have been described in numerous monographs (cf. Section 5.2:5).

2 Identity Checking

If the IR spectra of two substances are recorded under identical conditions, even very small differences indicate that the substances are not identical.

The bands of spectra recorded on different instruments may differ in shape as a result of differences in the recording method, the use of linear wave number or wavelength scales, or differences in resolution between grating and prism monochromators. Greater differences in the spectra of the same substance can result from differences in sample preparation, i.e. from the use of different solvents or different concentrations, or by measurement as a liquid film, KBr disk, etc.

Nevertheless, the identity of a substance can often be established by comparison with a large number of spectra even from different sources. Large collections of spectra for comparison purposes are available on cards, in loose-leaf form, and even on magnetic tapes for electronic data processing[2]. A

---

[1] *R.B. Barnes, R.C. Gore*, Anal. Chem. *21*, 7 (1949); *R.C. Gore*, Anal. Chem. *22*, 7 (1950); *23*, 7 (1951); *24*, 8 (1952); *26*, 11 (1954); *28*, 577 (1956); *30*, 570 (1958); *32*, 238 R (1960).
J.C. Evans, Anal. Chem. *34*, 225 R (1962); *36*, 240 R (1964); *38*, 311 R (1966).
R.O. Crisler, Anal. Chem. *40*, 246 R (1968).

[2] DMS-Dokumentation der Molekularspektroskopie, Verlag Chemie, Weinheim — Documentation of Molecular Spectroscopy — Butterworths, London. Sadtler Research Laboratories Spectra, Sadtler, Philadelphia.
API American Petroleum Institute.
Wyandotte-System, Wyandotte Chemical Company, Detroit.
L.E. Kuentzel, Anal. Chem. *23*, 1413 (1951).
S.S. Stimler, R.E. Kagarise, Infrared Spectra of Plastics and Resins, Part 2, ASTIA AD 634427, US Dep. Commerce, Washington, D.C. 1966.
D.S. Cain, S.S. Stimler, Infrared Spectra of Plastics and Resins, Part 3, ASTIA AD 649004, US Dep. Commerce, Washington D.C. 1967.
K. Dobringer, E.R. Katzenellenbogen, R.N. Jones, G. Robert, B.S. Gallagher, Infrared Absorption Spectra of Steroids — An Atlas, Vol. I, II, Interscience Publishers, New York, London 1958.
W. Neudert, H. Röpke, Steroid-Spektren-Atlas, Atlas of Steroid Spectra, Springer Verlag, Berlin, Göttingen, Heidelberg 1964.
R. Mecke, F. Langenbücher, IR-Spektren ausgewählter chemischer Verbindungen, Heyden & Son, London 1965.
H. Moenke, Mineralspektren, Vol. I, II, Akademie-Verlag, Berlin 1962, 1965.
R.N. Jones, The Infrared Spectra of n-Paraffin Hydrocarbons, N.R.C. Bull. No. 5.
J. Holubek, O. Strouf, Spectral Data and Physical Constants of Alkaloids, Heyden & Son, London 1965.
H.A. Szymanski, Interpreted Infrared Spectra, Vol. I, II, III, Plenum Press, New York 1964, 1966, 1967.

collection of journal references to articles containing IR-spectroscopic data is also obtainable[3].

3 Determination of Structures

IR spectroscopy is an important aid in the determination of structures, since many groups have characteristic frequencies. Excellent standard works[4] exist on structure-band relations. Lists are arranged on the basis of chemical class. To use these data for the interpretation of spectra, one must first decide what information is being sought. In the following list, an attempt is made, for a number of selected structural types, to enable the structural characteristics which are in accordance with the spectra to be found with the aid of key bands.

Complications for the identification of many structural types, arise mainly from the use of the fingerprint region, which generally contains a large number of bands. If strong bands exist for a given structural type, the reliability of the assignment is relatively high. If, on the other hand, weak bands exist, there is always a possibility that these may be present but masked by stronger bands. Thus, if any bands at all are found in the region in question, the corresponding structural type must be considered. Further complications result from the displacement of characteristic frequencies as a result of conformational factors or of the presence of certain substituents. This difficulty can be avoided by widening the band ranges, but this leads to a simultaneous decrease in specificity. A suitable compromise must, therefore, be found to enable the correct assignment to be made with the greatest possible certainty. The degree of uncertainty is reduced if the molecular formula is known. Further selection is possible by the use of other methods of structure analysis.

Table 1: List of the structural units included in Table 2. The numbers in parentheses refer to the answers in column C of Table 2

Substance	Answer No.
Acid anhydride	(53)
Carboxylic acid halide	(56)
Alcohol, primary	(6)
, secondary	(7)
, tertiary	(8)
Aldehyde, saturated	(76)
, $a\beta$-unsaturated	(87)
, $a\beta$, $\gamma\delta$-diunsaturated	(97)
, $a\beta$-unsaturated, β-hydroxy	(103)
, a-aryl	(85)
Alkene, monosubstituted (vinyl)	(25)
, monosubstituted ($H_2C=CH—C=O$)	(26)
, monosubstituted ($H_2C=CH—N$)	(29)
, 1,2-disubstituted (trans)	(30)
, 1,2-disubstituted (cis)	(33)
, 2,2-disubstituted (vinylidene)	(31)
, trisubstituted	(32)
Alkyne, monosubstituted	(15)
, disubstituted	(45)
Allene	(52)
Amide, primary	(96)
, secondary	(89)
, tertiary	(100)
Amine, primary	(11)
, primary aromatic	(12)
, secondary	(13)
, secondary aromatic	(14)
, tertiary	(127)
Anilide	(83)
Aromatic compound	(16)
Benzene, monosubstituted	(17)
, 1,2-disubstituted	(18)
, 1,3-disubstituted	(19)
, 1,4-disubstituted	(22)
, 1,2,3-trisubstituted	(20)
, 1,2,4-trisubstituted	(23)
, 1,3,5-trisubstituted	(21)
Butyl, tertiary	(36)
Carboxylic acid, saturated	(84)
, saturated (intramolecular hydrogen bonding)	(101)
, $a\beta$-unsaturated	(86)
, a-aryl	(88)
, a-halogeno	(77)

[3] *H.M. Hershenson*, Infrared Absorption Spectra Index, Index for 1945—1957 (1959), Index for 1958—1962 (1964), Academic Press, New York, London.

[4] *N.B. Colthups*, J. Opt. Soc. Am. *40*, 397 (1950).
L.J. Bellamy, The Infrared Spectra of Complex Molecules, Methuen & Co., London 1962 and John Wiley & Son, New York 1962.
W. Brügel, Ultrarot-Spektren und chemische Konstitution, Dr. Dietrich Steinkopff Verlag, Darmstadt 1964.
C.N.R. Rao, Chemical Applications of IR-Spectroscopy, Academic Press, New York, London 1964.
R.G.J. Miller, H.A. Willis, H.J. Hediger, IRSCOT-Infrared Structural Correlation Tables, 9 Tables, Heyden & Son, London, Tables 1, 2, 3 1964; 4, 5 1965; 6, 7, 8 1966; 9 1969.
W. Otting, Spektrale Zuordnungstafel der IR-Absorptionsbanden, Springer Verlag, Berlin, Göttingen, Heidelberg 1963.
H.A. Szymanski, Infrared Band Handbook, Plenum Press, New York 1963, 1, 2 1964; 3, 4 1965.

Substance	Answer No.	Substance	Answer No.
Dialkyl sulfite	**(136)**	Monothiocarboxylic acid	**(43)**
1,2-Diketone	**(78)**	Nitric ester	**(104)**
1,3-Diketone (enol form)	**(108)**		
1,4-Diketone	**(79)**	Nitrile, (R—CN)	**(46)**
		Nitrile, (R—C=C—CN, —N=C—CN, Ar—CN)	**(48)**
Dithiocarboxylic acid	**(41)**		
Dithiocarboxylate	**(137)**	Nitrite	**(94)**
		, *trans*	**(95)**
Dixanthogen	**(133)**	Nitroamine	**(109)**
Ester, a-aryl	**(74)**	Nitro compound	**(110)**
, a-halogeno	**(62)**	, $a\beta$-unsaturated	**(111)**
, a-keto	**(69)**		
, β-keto (enol form)	**(102)**	Nitrosoamine, monomeric	**(115)**
, γ-keto	**(72)**	, dimeric	**(116)**
, saturated	**(66)**		
$a\beta$-unsaturated	**(73)**	Nitroso compound, alkyl (monomeric)	**(112)**
, vinyl	**(59)**	, alkyl (dimeric, *cis*)	**(120)**
		, alkyl (dimeric, *trans*)	**(119)**
Ether, vinyl	**(27)**	, aryl (monomeric)	**(114)**
		, aryl (dimeric, *trans*)	**(117)**
Hydroxyl groups			
-OH (free vibration)	**(1)**	Phenol	**(9)**
-OH (intermolecular hydrogen bonds)	**(2)**		
-OH (intramolecular hydrogen bonds)	**(3)**	Phosphorus compounds P—H	**(44)**
-OH (intermolecular polymeric association)	**(4)**	P—NH$_2$	**(10)**
-OH (strong intramolecular bonds, chelate)	**(5)**	P=O	**(132)**
		P—O—C aliphatic	**(143)**
Imide, open-chain	**(82)**	P—O—C aromatic	**(134)**
, cyclic	**(61)**	P—O—P	**(144)**
		P—OH	**(38)**
Isocyanate	**(47)**	Quinone (two C=O in one ring)	**(93)**
		(two C=O in two rings)	**(106)**
Isonitrile	**(49)**		
		Sulfuric ester	**(122)**
Isopropyl	**(35)**		
		Sulfinic acid	**(37)**
Isothiocyanate	**(50)**		
		Sulfinic acid anhydride	**(131)**
Ketone, open-chain or six-membered ring	**(80)**		
, four-membered ring	**(57)**	Sulfinic ester	**(140)**
, five-membered ring	**(65)**		
, a-halogeno	**(71)**	Sulfinyl chloride	**(139)**
, $a\beta$-unsaturated	**(91)**		
, $a\beta, a'\beta'$-diunsaturated	**(92)**	Sulfonamide	**(128)**
, diaryl	**(98)**		
, a-(2-hydroxyaryl)	**(105)**	Sulfone	**(129)**
, a,a'-dihalogeno	**(63)**		
, aryl	**(90)**	Sulfonic acid (ionized)	**(39)**
		(unionized)	**(34)**
β-Lactam	**(68)**		
γ-Lactam	**(75)**	Sulfonic ester	**(126)**
δ-Lactam	**(99)**		
		Sulfonyl chloride (aliphatic)	**(124)**
β-Lactone	**(54)**	(aromatic)	**(123)**
γ-Lactone	**(58)**		
, $a\beta$-unsaturated	**(60)**	Sulfonyl fluoride	**(121)**
, $\beta\gamma$-unsaturated	**(55)**		
δ-Lactone	**(67)**	Sulfoxide	**(142)**
, $a\beta$-unsaturated	**(70)**		
, $a\beta, \gamma\delta$-diunsaturated	**(64)**	Sulfone	**(125)**
Mercaptan	**(42)**		

Substance	Answer No.
Thioamide	**(113)**
Thiocyanate	**(51)**
Thiophenol	**(40)**
Thioketone	**(138)**
Thiocarbonate	**(135)**
Trithiocarbonate	**(141)**
Thiosulfonate	**(130)**
Urea derivative	**(107)**
Urethane	**(81)**

Table 1 contains an alphabetic list of 144 structural features. The list is based mainly on tables of bands and individual data from cited publications[4]. By application of these data to our own spectra as well as to spectra from the DMS cards, some band ranges have been widened and intensity data in particular have been altered[5].

To use Table 2, it is desirable to have a spectrum recorded in the absence of solvent (KBr disk or liquid film) as well as the spectrum of the substance in solution. The spectra should be recorded in such a way that the positions of all the bands

[5] C.W. Joung, R.B. Du Vall, N. Wright, Anal. Chem. **23**, 709 (1951).

Table 2. Structure assignments

Condition A	Condition B	Condition C	Answer No.
3760–2500 M[a] (2.66–4.00)		3650–3570 M[a] (2.74–2.80)	
		–OH (free vibration)	**(1)**
	3570–3450 M[a] (2.79–2.90)	The relative intensity is concentration dependent –OH (dimeric, intermolecular hydrogen bonds)	**(2)**
		The relative intensity is not concentration dependent –OH (intramolecular hydrogen bonds)	**(3)**
		3400–3200 M broad (2.94–3.13) –OH (intermolecular, polymeric association)	**(4)**
		3200–3100 W broad (3.13–3.25) or 2850–2500 W broad (3.51–4.00) –OH (strong, intramolecular bonds, chelate)	**(5)**
	Answer **1, 2, 3, 4** *or* **5** positive	1080–1010 S (9.26–9.90) primary alcohol	**(6)**
		1120–1090 S (8.93–9.17) secondary alcohol	**(7)**
		1160–1140 M (8.62–8.77) tertiary alcohol	**(8)**

[a] Intensities: S = strong, M = medium, W = weak

Condition A	Condition B	Condition C	Answer No.
		1205–1170 M (8.30–8.55)	
		phenol	**(9)**
3500–3100 W (2.86–3.23)		1570–1535 S broad (6.37–6.51)	
two bands		P—NH$_2$	**(10)**
	(position of the second band) = (position of the first band). 0.876 $+345\pm50$ cm^{-1} 1625–1580 M (6.15–6.33)	primary amine	**(11)**
		1345–1245 S (7.43–8.03)	
		primary aromatic amine	**(12)**
3500–3100 M (2.86–3.23)		secondary amine	**(13)**
		1360–1270 S (7.35–7.87)	
		secondary aromatic amine	**(14)**
3390–3200 S (2.95–3.13)		2160–2080 W (4.63–4.81) 700–570 M (14.29–17.54)	
		alkyne monosubstituted	**(15)**
3100–2990 W (3.23–3.34) 1625–1570 W (6.15–6.37) 1525–1465 W (6.56–6.83)		aromatic compound	**(16)**
	790–720 M (12.66–13.89)	710–680 M (14.08–14.71) Absorption range 2000–1650 cm^{-1} (5.00–6.06) similar to Fig. 2	
		benzene, monosubstituted	**(17)**
		Absorption range 2000–1650 cm^{-1} (5.00–6.06) similar to Fig. 3	
		benzene, 1,2-disubstituted	**(18)**
	810–740 M (12.35–13.51)	Absorption range 2000–1650 cm^{-1} (5.00–6.06) similar to Fig. 4	
		benzene, 1,3-disubstituted	**(19)**

Condition A	Condition B	Condition C	Answer No.
		Absorption range 2000–1650 cm⁻¹ (5.00–6.06)	
		similar to Fig. 5	
		benzene, 1, 2, 3-trisubstituted	**(20)**
	860–800 M (11.63–12.50)	730–675 M (13.70–14.81)	
		Absorption range 2000–1650 cm⁻¹ (5.00–6.06) similar to Fig. 6	
		benzene, 1, 3, 5-trisubstituted	**(21)**
		Absorption range 2000–1650 cm⁻¹ (5.00–6.06) similar to Fig. 7	
		benzene, 1, 4-disubstituted	**(22)**
	890–860 M (11.24–11.63) 820–805 M (12.20–12.42)	Absorption range 2000–1650 cm⁻¹ (5.00–6.06) similar to Fig. 8	
		benzene,-1, 2, 4-trisubstituted	**(23)**
3095–3010 W (3.23–3.32)	1690–1610 W (5.92–6.21)	C=C—double bond	**(24)**
	1655–1610 W (6.04–6.21)	1005–980 W (9.95–10.20) 925–905 W (10.81–11.05)	
		alkene monosubstituted (vinyl)	**(25)**
		995–980 W (10.05–10.20) 965–955 W (10.36–10.47)	
		alkene monosubstituted (H₂C=CH—C=O)	**(26)**
		960–940 W (10.42–10.64) 850–810 W (11.76–12.35)	
		vinyl ether	**(27)**
		960–940 W (10.42–10.64) 875–855 W (11.43–11.70)	
		vinyl ester of carboxylic acid	**(28)**

Condition A	Condition B	Condition C	Answer No.
		985–965 W (10.15–10.36) 850–830 W (11.76–12.05) alkene, monosubstituted (H_2C=CH—N)	**(29)**
	1680–1655 W (5.95–6.04)	990–960 W (10.10–10.42) alkene, disubstituted *(trans)*	**(30)**
	1657–1645 W (6.04–6.08)	900–880 W (11.11–11.36) alkene, 2,2-disubstituted (vinylidene)	**(31)**
	1680–1670 W (5.95–5.99)	840–790 W (11.90–12.66) alkene, trisubstituted	**(32)**
	1665–1645 W (6.01–6.08)	740–675 W (13.51–14.81) alkene, disubstituted *(cis)*	**(33)**
3030–2800 S broad (3.30–3.57)		2420–2380 W broad (4.13–4.20) 1355–1340 S (7.38–7.40) 1200–1100 S (8.33–9.09) 910–980 S broad (10.99–11.24) sulfonic acid (unionized)	**(34)**
3000–2840 M (3.33–3.52)	1485–1430 W (6.73–6.99)	1385–1360 W doublet (7.22–7.35) The two bands have equal intensities isopropyl	**(35)**
		1395–1360 W doublet (7.17–7.35) The band at the lower wave number is more intense tertiary butyl	**(36)**
2900–2700 S broad (3.45–3.70)		2500–2350 W broad (4.00–4.26) 1100–1070 S (9.09–9.35) 850–800 S (11.76–12.50) sulfinic acid	**(37)**
2700–2560 W broad (3.70–3.91)		1040–909 S (9.62–11.00) P—OH	**(38)**

Condition A	Condition B	Condition C	Answer No.
2650–1650	W broad		
(3.77–6.06)			
		(3.77–6.06)	
		1220–1150 S	
		(8.20–8.70)	
		1120–1030 S	
		(8.93–9.71)	
		sulfonic acid (ionized)	**(39)**
2600–2550	W	2575–2450 W broad	
(3.85–3.92)		(3.88–4.08)	
		Answer **(16)** positive	
		thiophenol	**(40)**
2600–2500	W	2500–2400 W broad	
(3.85–4.00)		(4.00–4.17)	
		1230–1210 S	
		(8.13–8.26)	
		870–850 S broad	
		(11.49–11.76)	
		dithiocarboxylic acid	**(41)**
2600–2400	W	mercaptan	**(42)**
(3.85–4.17)			
2570–2470	W	1730–1690 M broad	
(3.89–4.05)		(5.78–5.92)	
		970–940 S	
		(10.31–10.64)	
		870–850 M	
		(11.49–11.76)	
		monothiocarboxylic acid	**(43)**
2440–2288	W broad	1150–965 M	
(4.10–4.37)		(8.70–10.36)	
		P—H	**(44)**
2310–2080	W		
(4.33–4.81)			
		alkyne, disubstituted	**(45)**
2280–2240	W	nitrile (R—CN)	**(46)**
(4.39–4.46)			
2280–2230	S	isocyanate	**(47)**
(4.39–4.48)			
2239–2210	W	nitrile (R—C=C—CN, —N=C—CN,	
(4.47–4.52)		AR—CN)	**(48)**
2200–2120	W	isonitrile	**(49)**
(4.55–4.72)			
2200–2080	S broad	isothiocyanate	**(50)**
(4.55–4.81); two bands			
2180–2120	M	thiocyanate	**(51)**
(4.59–4.72)			

Condition A	Condition B	Condition C	Answer No.
1980–1930 M (5.05–5.18)		1075–1060 W (9.30–9.43) allene	**(52)**
1900–1800 S (5.26–5.56)		1800–1740 S (5.56–5.75) acid anhydride	**(53)**
1845–1805 S (5.42–5.54)		β-lactone	**(54)**
1825–1785 S (5.48–5.60)		γ-lactone, $\beta\gamma$-unsaturated	**(55)**
1820–1760 S (5.49–5.68)		acyl halide	**(56)**
1800–1763 S (5.56–5.67)		ketone (four-membered ring)	**(57)**
1800–1760 S (5.56–5.68)		γ-lactone	**(58)**
1800–1760 S (5.56–5.68)		Answer **(28)** positive vinyl ester	**(59)**
1795–1735 S two bands (5.57–5.76)		γ-lactone, $\alpha\beta$-unsaturated	**(60)**
1790–1700 M two bands (5.59–5.88)		imide, cyclic	**(61)**
1780–1740 S (5.62–5.75)		ester, α-halogeno	**(62)**
1775–1735 S (5.63–5.76)		ketone, α, α'-dihalogeno	**(63)**
1772–1732 S (5.64–5.77)		1731–1715 S (5.78–5.83) δ-lactone, $\alpha\beta$, $\gamma\delta$-diunsaturated	**(64)**
1765–1725 S (5.67–5.80		ketone (five-membered ring)	**(65)**
1762–1718 S (5.68–5.82)		ester, saturated	**(66)**
		δ-lactone	**(67)**
1760–1730 S (5.68–5.78)		β-lactam	**(68)**
1760–1715 S (5.70–5.83)		ester, α-keto	**(69)**
1755–1715 S (5.70–5.83)		One of the answers **(25)** to **(29)** or **(33)** positive δ-lactone, $\alpha\beta$-unsaturated	**(70)**
		ketone, α-halogeno	**(71)**

Condition A	Condition B	Condition C	Answer No.
1755–1720 S (5.70–5.81)		1720–1660 S (5.81–6.02)	
		ester, γ-keto	**(72)**
1750–1700 S (5.71–5.88)		One of the answers **(25)** to **(29)** or **(33)** positive ester, αβ-unsaturated	**(73)**
		Answer **(16)** positive	
		ester, α-aryl	**(74)**
		γ-lactam	**(75)**
1750–1710 S (5.71–5.85)		2900–2800 W (3.45–3.57) 2770–2650 W (3.61–3.77)	
		aldehyde (saturated)	**(76)**
1750–1710 S broad (5.71–5.85)		Answer **(5)** positive	
		carboxylic acid, α-halogeno	**(77)**
1737–1700 S (5.76–5.88)		1,2-diketone	**(78)**
		1,4-diketone	**(79)**
1735–1700 S (5.76–5.88)		ketone, (chain or six-membered ring)	**(80)**
		urethane	**(81)**
1735–1700 M (5.76–5.88) two bands		imide, open chain	**(82)**
1730–1690 S (5.78–5.92)		1590–1490 S (6.29–6.71)	
		Answers **(13)** and **(16)** positive anilide	**(83)**
1730–1690 S broad (5.78–5.92)		Answer **(5)** positive	
		carboxylic acid	**(84)**
1725–1675 S (5.80–5.97)		2900–2800 W (3.45–3.57) 2770–2650 W (3.61–3.77) Answer **(16)** positive	
		aldehyde, α-aryl	**(85)**
1715–1675 S broad (5.83–5.97)		Answer **(5)** positive	
		carboxylic acid, αβ-unsaturated	**(86)**
1713–1673 S (5.84–5.98)		2900–2800 W (3.45–3.57) 2700–2650 W W (3.61–3.77)	
		aldehyde, αβ-unsaturated	**(87)**

Condition A	Condition B	Condition C	Answer No.
1710–1670 S broad (5.85–5.99)		Answers **(5)** and **(16)** positive carboxylic acid, α-aryl	**(88)**
1700–1625 S broad (5.88–6.15)		1570–1510 S (6.37–6.62) Answer **(13)** positive amide, secondary	**(89)**
1700–1670 S (5.88–5.99)		Answer **(16)** positive ketone aryl	**(90)**
1965–1645 S (5.90–6.08)		ketone, αβ-unsaturated	**(91)**
		ketone, αβ, α′β′-diunsaturated	**(92)**
1695–1640 S (5.90–6.10)		quinone (2 C=O in one ring)	**(93)**
1690–1610 S (5.92–6.21)	1639–1600 W (6.10–6.25)	nitrite	**(94)**
		815–750 S (12.27–13.33) nitrite *(trans)*	**(95)**
1690–1650 S (5.92–6.06)		Answer **(11)** positive amide, primary	**(96)**
		2900–2800 W (3.45–3.57) 2770–2650 W (3.61–3.77) aldehyde, αβ, γδ-diunsaturated	**(97)**
1690–1660 S (5.92–6.02)		Answer **(16)** positive ketone, diaryl	**(98)**
1685–1630 S (5.93–6.13)		δ-lactam	**(99)**
		amide, tertiary	**(100)**
1680–1640 S broad (5.95–6.10)		Answer **(5)** positive carboxylic acid (intramolecular hydrogen bonding)	**(101)**
1680–1640 S (5.95–6.10)		One of answers **(1)** to **(5)** positive ester, β-keto (enol form)	**(102)**
1677–1637 S (5.96–6.11)		2900–2800 W (3.45–3.57) 2700–2650 W (3.61–3.77) One of answers **(1)** to **(5)** positive aldehyde, αβ-unsaturated, β-hydroxy	**(103)**

Condition A	Condition B	Condition C	Answer No.
1670–1610 S (5.99–6.21)		1300–1250 ˙ S (7.69–8.00) nitric ester	**(104)**
1670–1630 S (5.99–6.13)		Answer **(16)** and one of answers **(1)** to **(5)** positive ketone, a′-(2-hydroxyaryl)	**(105)**
1665–1625 S (6.01–6.15)		quinone (two C=O in two rings)	**(106)**
1660–1610 S (6.02–6.21)		urea derivative	**(107)**
1640–1540 S (6.10–6.49)		One of answers **(2)** to **(5)** positive 1, 3-diketone (enol form)	**(108)**
1635–1530 S (6.12–6.54)		1315–1260 S (7.60–7.94) nitroamine	**(109)**
1600–1490 S broad (6.25–6.71)		1400–1300 S broad (7.14–7.69) nitro compound	**(110)**
		1525–1510 S broad (6.56–6.62) 1360–1335 S broad (7.35–7.49) probably —C=CH—NO$_2$	**(111)**
1585–1540 S (6.31–6.49)		alkyl nitroso compound (monomeric)	**(112)**
1570–1400 S broad (6.37–7.14)		1400–1260 W (7.14–7.94) 1140–940 W (8.77–10.64) thioamide	**(113)**
1515–1480 S (6.60–6.76)		answer **(16)** positive aryl nitroso compound (monomeric)	**(114)**
1510–1300 S (6.62–7.69)	1100–1000 S broad (9.09–10.00)	1510–1440 S (6.62–6.94) nitrosoamine (monomeric)	**(115)**
		1320–1300 S (7.58–7.69) nitrosoamine (dimeric)	**(116)**
1425–1175 S (7.02–8.51)		1300–1290 S (7.69–7.75) answer **(16)** positive aryl nitroso (dimeric, *trans)*	**(117)**

Condition A	Condition B	Condition C	Answer No.
		1289–1241 S (7.76–8.06) aliphatic or aromatic nitroso compound (dimeric, *trans)*	**(118)**
		1240–1175 S (8.06–8.51) aliphatic nitroso compound (dimeric, *trans)*	**(119)**
1425–1385 S (7.02–7.22)		1345–1320 S (7.43–7.58) nitroso compound (dimeric, *cis)*	**(120)**
1420–1400 S (7.04–7.14)		1220–1200 S (8.20–8.33) sulfonyl fluoride	**(121)**
1410–1380 S (7.09–7.25)		1200–1180 S (8.33–8.47) sulfuric ester	**(122)**
1410–1385 S (7.09–7.22)		1205–1175 S (8.30–8.51) Answer **(16)** positive aromatic sulfonyl chloride	**(123)**
1390–1370 S (7.19–7.30)		1190–1160 S (8.40–8.62) aliphatic sulfonyl chloride	**(124)**
1385–1345 S (7.22–7.43)		1175–1160 S (8.51–8.62) sultone	**(125)**
1375–1350 S (7.27–7.41)		1185–1165 S (8.44–8.58) sulfonic ester	**(126)**
1370–1310 S (7.30–7.63)		amine, tertiary	**(127)**
1365–1310 M (7.33–7.63)		1180–1150 S (8.47–8.70) sulfonamide	**(128)**
1355–1275 S (7.38–7.84)		1164–1110 S (8.59–9.01) sulfone	**(129)**
1355–1325 S (7.38–7.55)		1165–1130 S (8.58–8.85) thiosulfonate	**(130)**

Condition A	Condition B	Condition C	Answer No.
1340–1300 S (7.46–7.69)		1150–1130 S (8.70–8.85) 1110–1080 S (9.01–9.26) sulfinic anhydride	**(131)**
1300–1150 S broad (7.69–8.70)		P=O	**(132)**
1275–1250 S (7.84–8.00)		1030–1000 S (9.71–10.00) dixanthogen	**(133)**
1240–1190 M (8.06–8.40)		994–914 S broad (10.06–10.94) Answer **(16)** positive P—O—C aromatic	**(134)**
1240–1200 S (8.06–8.33)		thiocarbonate	**(135)**
1230–1180 S (8.13–8.47)		dialkyl sulfite	**(136)**
1210–1190 S (8.26–8.40)		dithiocarboxylic ester	**(137)**
1180–1140 S (8.47–8.77)		thioketone	**(138)**
1160–1100 S (8.62–9.09)		sulfinyl chloride	**(139)**
1160–1130 S (8.62–8.85)		sulfinic ester	**(140)**
1080–1050 M (9.26–9.52)		trithiocarbonate	**(141)**
1060–990 S (9.43–10.10)		sulfoxide	**(142)**
1055–930 S broad (9.48–10.75)		P—O—C aliphatic	**(143)**
980–900 S broad (10.20–11.11)		P—O—P	**(144)**

and their intensity relations are clearly recognizable. To avoid incorrect conclusions due to strong solvent absorptions, the spectra of some common solvents are reproduced in Fig. 9 to 14. If there is any doubt as to the intensity rating (strong, medium, weak), the higher intensity should be used to avoid loss of information.

The questions in the table are arranged in a block structure with three parts A, B, and C. The ranges of validity of the various blocks are indicated by lines. The layout is illustrated in Fig. 1

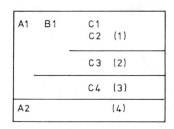

Fig. 1. Illustration of the block structure
Answers (1) (2) (3) (4)

Figures 2–8 give the frequencies of various substituted benzene derivatives in the range of 2000 to 1650 cm⁻¹.

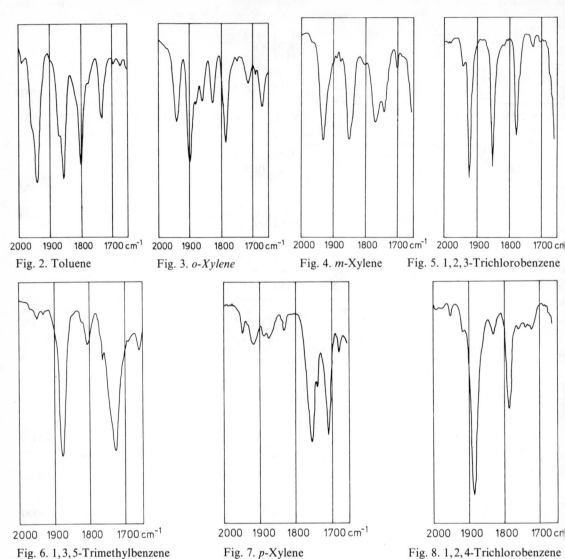

Fig. 2. Toluene   Fig. 3. *o-Xylene*   Fig. 4. *m*-Xylene   Fig. 5. 1,2,3-Trichlorobenzene

Fig. 6. 1,3,5-Trimethylbenzene   Fig. 7. *p*-Xylene   Fig. 8. 1,2,4-Trichlorobenzene

See Ref. 5 for higher substituted derivatives as well as for changes in spectra depending on the nature of the substituent.

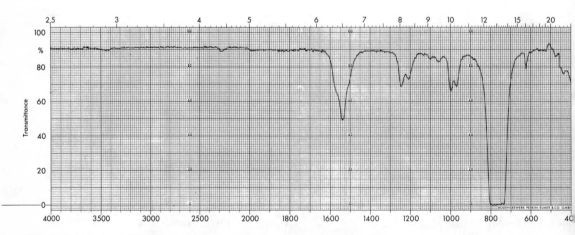

Fig. 9. Spectrum of CCl₄, layer thickness 0.01 cm, measured against air.

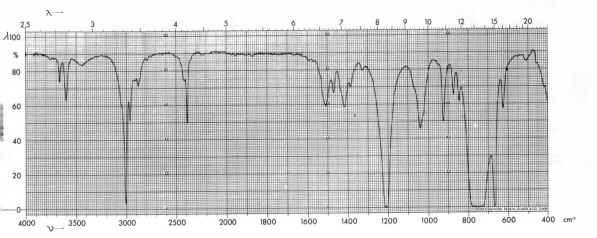

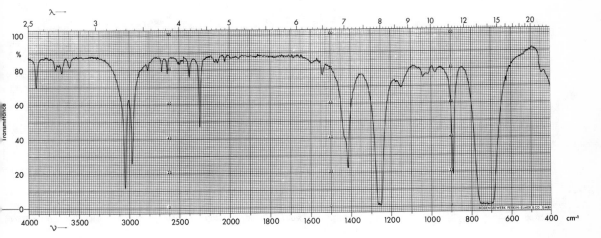

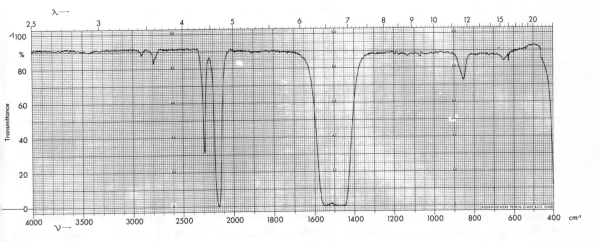

Fig. 10. Spectrum of CHCl₃, layer thickness 0.01 cm, measured against air.

Fig. 11. Spectrum of CH₂Cl₂, layer thickness 0.01 cm, measured against air

Fig. 12. Spectrum of CS₂, layer thickness 0.01 cm, measured against air.

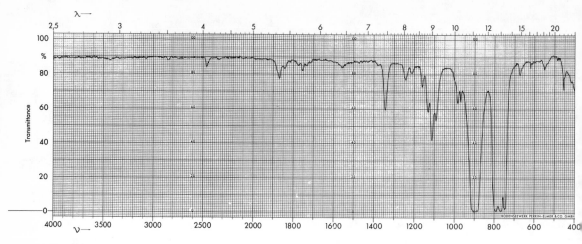

Fig. 13. Spectrum of $Cl_2C = CCl_2$, layer thickness 0.01 cm, measured against air.

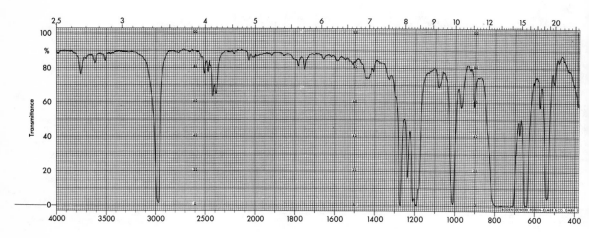

Fig. 14. Spectrum of $Cl_2HC-CHCl_2$, layer thickness 0.01 cm, measured against air.

The conditions A1, B1, C1, and C2 must be satisfied to obtain the answer (1), conditions A1, B1, and C3 for the answer (2), A1 and C4 for the answer (3), and A2 for the answer (4). If a band of medium intensity is required, the condition is satisfied if a band of medium or strong intensity is present in the appropriate region. If only a weak band is required, the presence of any band in the region in question is sufficient. If a band of normal width is required, this condition is satisfied if only a broad band is present in the region in question.

Band positions which have been requested previously will occasionally be required again. In these cases, for the sake of simplicity, the answers obtained earlier will be requested instead of the band positions. Numbers of those answers are enclosed in brackets. The wavelengths $[\mu]$ are given under the wave numbers $[-cm^{-1}]$.

4 Quantitative Analysis

4.1 General

Since small changes in structure lead to different IR spectra, IR spectroscopy is an excellent tool for the quantitative analysis of mixtures. Particularly in cases where the substances to be analyzed cannot be volatilized without decomposion, it is a valuable addition to gas chromatography and mass spectrometry. (A detailed description of the principles and methods of quantitative infrared analysis is given in Ref. [6, 7].) Contents >90% are successfully determined in technical products in many laboratories. The procedure in Section 5.2:4.2 for the analysis of substances in solution minimizes the random error.

4.2 Selection of a Band

In order to select a band for measurement, one must have the fullest possible knowledge of the spectra of all the components of the product to be analyzed. In many cases neither literature spectra nor pure samples will be available. However, often the required reference spectra can be obtained by a combination of *separation methods* and IR spectroscopy. Spectra of satisfactory quality can be recorded with samples as small as 50 μg if *microcells* or *microdisks* are used in conjunction with a microilluminator. After *gas chromatography,* various fractions can be condensed directly in small AgCl cells. A method of recording IR spectra of substances isolated by gas chromatography in the range 1–10 μg using KBr disks has been described by Curry et al[8].

If the components are separated by *thin layer chromatography,* the substances from the scraped-off zones can be concentrated in the vertices of small KBr triangles[9]. The vertices are broken off and formed into microdisks (diameter e.g. 1.5 or 0.5 mm). The spectra are then recorded.

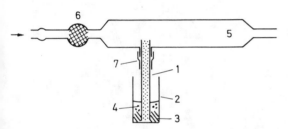

Fig. 15. Concentration of substances isolated by thin layer chromatography in small quantities of KBr

1 Glass tube, 2–3 mm, length 50 mm	4 Solvent
2 Test tube	5 Tube for operation under inert gas
3 Support material from thin layer	6 Glass wool filter
	7 Rubber sleeve

Another simple method[10] utilizes glass tubes (see Fig. 15), which are packed with KBr, after the manner of a chromatography column, to remove substances from the support of a thin layer plate

and to then concentrate them in a small quantity of KBr. Tubes of this nature have an excellent suction power. The substances are deposited in the topmost layer of the KBr. The operation may be carried out under an inert gas.

When comparing the spectra, one chooses a band of the substance to be determined that does not coincide with the bands of the other substances. It is preferable to use a solvent that does not absorb in the chosen region of the spectrum. If an isolated band cannot be found, the other components that absorb at the position of the band to be measured must be determined by means of other determination equations. Other IR as well as suitable experimental values from elementary analysis, thin layer chromatography, or UV absorption may also be used.

4.3 Preparation of Samples

A relatively large sample (100–200 g) is mixed thoroughly, solid samples being ground in a mortar or melted. Depending on the absorption coefficient of the substance, 100–500 mg of this material are weighed into a 10–50 ml volumetric flask. The flask is filled with the chosen solvent just below the mark and shaken until all the soluble components of the sample have dissolved. The flask is now thermostated at its calibration temperature and filled to the mark.

On dissolution of technical products, an insoluble residue sometimes remains. In such cases it is necessary to filter the solution without allowing any evaporation of the solvent. This is achieved by inserting a filter[10] between the syringe and the cell when filling the cell as shown in Fig. 16.

4.4 Measurement

4.4.1 General Remarks

The instrument settings are first adjusted:

slit program
wavelength scale
filter
amplification
balance
0–100% transmission

Careful setting and monitoring of the «balance» is particularly important.

This setting is difficult in some instruments, since the adjusting potentiometers are coarse. It is often helpful, in such cases, to replace the original potentiometer with a multiple-turn version. It is equally important to be able to switch off the filter changer (a facility frequently not provided by the manufac-

[6] *I. Kössler,* Methoden der Infrarot-Spektroskopie in der chemischen Analyse, p. 85, Akademische Verlagsgesellschaft Geest & Portig KG, Leipzig 1961. *C.N.R. Rao*[4].

[7] *M.M. Rochkind,* Infrared Analysis of Multicomponent Gas Mixtures, Anal. Chem. *39,* 567 (1967). *J. Dechant,* Z. Anal. Chem. *5,* 114 (1965).

[8] *A.S. Curry, J.F. Read, C. Brown, R.W. Jenkins,* J. Chromatogr. Sci. *38,* 200 (1968).

[9] «Wick Stick», Harshaw Chemie, Anal. Chem. *39,* 91 A (1967).

[10] *H. Weitkamp, R. Barth,* unpubl.

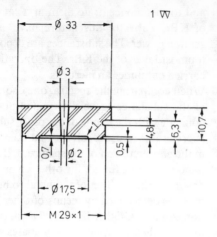

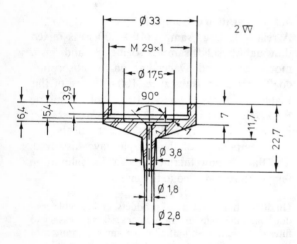

Fig. 16. Microfilter. 1. Top, 2. Bottom. Material: V4A.

turer) to avoid a change of filter in the region of the band to be measured.

A matched pair of cells is used for the measurement. The cell length is so chosen as to minimize the error in the determination of the absorption; this error is smallest[11] for absorptions between 0.2 and 0.6. Since one rarely has two cells with exactly the same length, the cell having the shorter distance is chosen as the reference cell. The absorption of the empty sample cell is then recorded against air in the selected wave-number range to check that the 100% line is satisfactory.

The cell length [in cm] can be calculated at the same time with the aid of the formula

$$d = \frac{n}{2(v_1 - v_2)} \quad (\text{n = number of waves from } v_1 \text{ to } v_2)$$

after measurement of the interference bands. The reference cell is filled with pure solvent. The measurements on all samples in the same series are carried out in the same cell. At the beginning and end of a series of measurements, the spectrum of a very pure sample of the substance to be determined is recorded in the wave-number range in question (standard measurement).

If the sample concentration is so chosen that the difference between the absorption of the standard and that of the sample is small ($< 10\%$), the linearity error of the comb aperture becomes negligible and applicability of the Lambert-Bouguer-Beer law[12] is achieved. Errors in the determination of the cell length are eliminated by the use of the same cell for the reference substance and for the sample. The standard measurement also eliminates uncertainties in the determination of the extinction coefficients; while the standard measurement at the beginning and at the end of a series of measurements provides a check on the stability of the instrument setting.

4.4.2 Evaluation

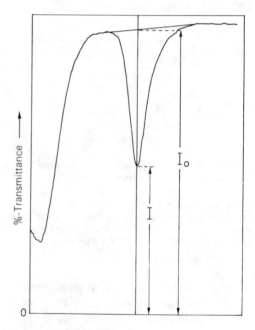

Fig. 17. Evaluation.

The spectra are evaluated by means of Beer's law in its general form

$$A = a \cdot c \cdot d \qquad (1)$$

[11] *H. Hoyer* in *Houben-Weyl,* Methoden der organischen Chemie 4. Aufl., Bd. III/2, p. 870, Georg Thieme Verlag, Stuttgart 1955.

[12] *A. Beer,* Ann. Phys. (2), *86,* 78 (1852);
   H.H. Jaffe, M. Orchin, Theory and Applications of Ultraviolet Spectroscopy, p. 8, J. Wiley & Sons, New York, London 1966.

where A is the absorbance, a is the absorption coefficient $\left[\dfrac{1000\ cm^2}{g}\right]$, c is the concentration of the substance in question [g/l], and d is the cell length [cm]. The absorbance

$$A = \log \frac{I_0}{I} \qquad (2)$$

is first determined from each measurement (cf. Fig. 17). The absorption coefficient

$$a = \frac{As \cdot Vs}{Es \cdot d} \qquad (3)$$

is calculated from the standard measurements by rewriting eq. (1); As is the absorption of a solution of grams of standard in Vs liters of solution. It follows that the average absorption coefficient for n standard measurements is

$$\overline{a} = \frac{Vs}{n \cdot d} \sum_1^n \frac{As_i}{Es_i} \qquad (4)$$

and the percentage content of this substance in the samples is

$$\frac{Ap}{Ep} \cdot \frac{100 \cdot Vp}{a \cdot d} = \text{percentage content} \qquad (5)$$

if Ep grams of sample in Vp liters of solution give the absorption Ap.

4.4.3 Errors

The standard errors obtained by this procedure are between 0.5 and 1% absolute (number of degrees of freedom greater than 100, in some cases greater than 1000). Since both the standard measurements and the measurements of the samples are subject to statistical errors, we find e.g. for standard and sample measurements in triplicate and a standard deviation of 0.7 with 120 degrees of freedom, that the confidence intervals V in a two-sided question for various values of the probability P are:

P	0,75	0,90	0,95	0,98	0,99
V	0,66	0,95	1,13	1,35	1,50

For a one-sided question, such as occurs in the case of a guaranteed minimum content, we find:

P	0,75	0,875	0,93	0,975	0,99
V	0,39	0,66	0,95	1,13	1,35

These values for the accuracy obtainable in routine work are just as good as those from other methods[13].

[13] R. Püschel, Mikrochim. Acta 82, 783 (1968).

If the standard measurements are omitted and the extinction coefficients as found by a single determination are used for the evaluation, much greater errors must be expected, since the determination of the extinction coefficients is strongly dependent on instrument parameters, among other things. Substantial changes occur in particular after an instrument servicing involving extensive adjustments, cleaning of mirrors, etc. For example, the average value found for one substance for the year 1968 (mean) was:

$$a = 1.7209, \sigma = 0.0396, f = 57$$

Further errors must be expected when α values are transferred from one instrument to another, even if the instrument setting and measurement conditions are very carefully fixed[14].

If several components of a mixture absorb in a given measured band, the absorption in general is given by

$$a_1 c_1 d + a_2 c_2 d + \ldots + a_n c_n d = A \qquad (6)$$

Since the concentration (c) and the percentage content (p) are related as follows

$$c = \frac{p}{100} \cdot \frac{E}{V} \qquad (7)$$

equation (6) can also be written in the form

$$a_1 p_1 + a_2 p_2 + \ldots + a_n p_n = \frac{100 \cdot V}{E \cdot d} \cdot A \qquad (8)$$

If some components are present in mixtures to only a few per cent, several methods must be used for the analysis instead of only one. For example, for a five component mixture the following set of equations were obtained for the evalution of IR, UV, and thin layer chromatographic measurements (Tables 3 and 4).

The equations are solved[15] by the method of least squares, and give the values:

$P_1 = 90.1\%$, $P_2 = 5.0\%$, $P_3 = 1.8\%$,
$P_4 = 6.1\%$, $P_5 = 0.3\%$.

[14] H.A. Szymanski, D.W. Teloh, Anal. Chem. 33, 814 (1961).

[15] D.F. Faddejew, W.N. Faddejewa, Numerische Methoden der linearen Algebra, R. Oldenbourg Verlag, München, Wien 1965.
H. Böse, Einführung in die Ausgleichsrechnung, R. Oldenbourg Verlag, München, Wien 1965.

Table 3. Evaluation of various analytical criteria

Criterion	Absorption coefficient					% content eq. (8)		
IR-Band 1160 cm^{-1}	a_{11}	a_{12}	a_{13}	a_{14}	a_{15}	p_1		$\dfrac{100 \cdot V_1}{d_1 \cdot E_1} \cdot A_1$
IR-Band 1074 cm^{-1}	a_{21}	a_{22}	a_{23}	a_{24}	a_{25}	p_2		$\dfrac{100 \cdot V_2}{d_2 \cdot E_2} \cdot A_2$
IR-Band 1040 cm^{-1}	a_{31}	a_{32}	a_{33}	a_{34}	a_{35}	p_3	$=$	$\dfrac{100 \cdot V_3}{d_3 \cdot E_3} \cdot A_3$
IR-Band 890 cm^{-1}	a_{41}	a_{42}	a_{43}	a_{44}	a_{45}	p_4		$\dfrac{100 \cdot V_4}{d_4 \cdot E_4} \cdot A_4$
UV-Band 247 mμ	a_{51}	a_{52}	a_{53}	a_{54}	a_{55}	p_5		$\dfrac{100 \cdot V_5}{d_5 \cdot E_5} \cdot A_5$
TLC value	a_{61}	a_{62}	a_{63}	a_{64}	a_{65}			% according to TLC

Table 4. Numerical Values for Table 3.

IR-Band	1160 cm^{-1}	0,09	0	0	0,18	0,11	p_1			8,88
IR-Band	1074 cm^{-1}	0	0,13	0,02	0	0,35	p_2			0,25
IR-Band	1040 cm^{-1}	0,55	0	0,28	0,32	0,18	p_3	$=$		51,48
IR-Band	890 cm^{-1}	0,12	0	0,24	0	0,13	p_4			11,20
UV-Band	247 mμ	20,70	79,80	1,37	24,70	0,25	p_5			2389,36
TLC value		0	0	0	0	1				0,50

5 Bibliography

R.B. Barnes, R.C. Gore, U. Liddel, V.Z. Williams, Infrared Spectroscopy, Industrial Applications and Bibliography, Reinhold, New York 1944.

H.M. Randall, N. Fuson, R.G. Fowler, J.R. Dangl, Infrared Determination of Organic Structure, Van Nostrand, Toronto, New York, London 1949.

R.A. Sawyer, Experimental Spectroscopy, Prentice-Hall, New York 1951.

S. Mizushima, Structure of Molecules and Internal Rotation, Academic Press, New York 1954.

E.A. Braude, F.C. Nachod, Determination of Organic Structures by Physical Methods, Academic Press, New York, London 1955.

H. Hoyer, Infrarotspektroskopie in Houben-Weyl, Methoden der organischen Chemie, 4. Aufl., Bd. III/2, Georg Thieme Verlag, Stuttgart 1955.

E.B. Wilson, J.C. Delius, P.C. Cross, Molecular Vibrations, McGraw-Hill Book Co., New York 1955.

D. Hummel, Kunststoff-, Lack- und Gummianalyse-Chemische und infrarotspektroskopische Methoden, Carl Hanser Verlag, München 1958.

P.J. Wheatley, The Determination of Molecular Structure, Clarendon Press, Oxford 1959.

G.L. Clark, The Encyclopedia of Spectroscopy, Reinhold Publ. Co., New York, Chapman & Hall, London 1960.

A.D. Cross, An Introduction to Practical IR-Spectroscopy, Butterworths, London 1960.

P. Barchewitz, Spectroscopie Infrarouge. I. Vibration Moléculaires, Gauthier-Villars, Paris 1961.

W. Brügel, Physik und Technik der Ultrarot-Strahlung, Curt R. Vincentz Verlag, Hannover 1961.

K.E. Lawson, Infrared Absorption of Inorganic Substances, Reinhold Publ. Co., New York, Chapman & Hall, London 1961.

B. Bak, Elementary Introduction to Molecular Spectra, John Wiley & Sons, New York 1962.

G.M. Barrow, Introduction to Molecular Spectroscopy, McGraw-Hill Book Co., New York 1962.

R.B. Baumann, Absorption Spectroscopy, John Wiley & Sons, New York 1962.

G.H. Beaven, E.A. Johnson, H.A. Willis, R.G. Miller, Molecular Spectroscopy: Methods and Application in Chemistry, MacMillan Co., New York 1962.

R.E. Dodd, Chemical Spectroscopy, Elsevier Publ. Co., Amsterdam, New York 1962.

D. Hummel, Analyse der Tenside — Infrarotspektroskopische und chemische Methoden, Carl Hanser Verlag, München 1962.

F.C. Nachod, W.D. Philips, Determination of Organic Structures by Physical Methods, Academic Press, New York, London 1962.

K. Nakanishi, Infrared Absorption Spectroscopy — Practical, Holden-Day, San Francisco 1962.

W.A. Shurcliff, Polarized Light, Production and Use, Harvard Univ. Press, Cambridge 1962.

M. Davies, Infrared Spectroscopy and Molecular Structure, Elsevier Publ. Co., New York 1963.

M.St.C. Flett, Characteristic Frequencies of Chemical Groups in the Infrared, Elsevier Publ. Co., New York 1963.

K. Nakamoto, Infrared Spectra of Inorganic and Coordination Compounds, John Wiley & Sons, New York 1963.

H.C. Allen Jr., P.C. Cross, Molecular Vib-Rotors — The Theory and Interpretation of High Resolution Infrared Spectra, John Wiley & Sons, New York 1963.

C.E. Meloan, Elementary Infrared Spectroscopy, MacMillan, New York, London 1963.

W.J. Potts Jr., Chemical Infrared Spectroscopy. Vol. I, Techniques, John Wiley & Sons, New York 1963.

R.M. Silverstein, G.C. Bassler, Spectrometric Identification of Organic Compounds, John Wiley & Sons, New York 1964.

N.B. Colthup, L.H. Daly, Introduction to Infrared and Raman Spectroscopy, Academic Press, New York 1964.

G.W. King, Spectroscopy and Molecular Structure, Holt, Rinehart, Winston, New York 1964.

J.P. Phillips, Spectra-Structure, Correlations, Academic Press, New York 1964.

L.A. Gribov, Intensity Theory of Infrared Spectra of Polyatomic Molecules, Plenum Press, New York 1964.

R.A. White, Handbook of Industrial Infrared Analysis, Plenum Press, New York 1964.

R. Zbinden, Infrared Spectroscopy of High Polymers, Academic Press, New York 1964.

H.A. Szymanski, IR-Theory and Practice of Infrared Spectroscopy, Plenum Press, New York 1964.

A.J. Baker, T. Cairns, Spectroscopy in Education Vol. II. Spectroscopic Techniques in Organic Chemistry, Heyden & Son, London 1965.

T. Cairns, Spectroscopy in Education. Vol. I, Spectroscopic Problems in Organic Chemistry, Heyden & Son, London 1965.

J. Haslam, H.A. Willis, Identification and Analysis of Plastics, Van Nostrand, Princeton 1965.

R.G.J. Miller, Laboratory Methods in Infrared Spectroscopy, Heyden & Son, London 1965.

R.N. Dixon, Spectroscopy and Structure, John Wiley & Sons, New York 1965.

J.R. Dyer, Applications of Absorption Spectroscopy of Organic Compounds, Prentice Hall, Englewood Cliffs, New Jersey 1965.

D.W. Mathieson, Interpretation of Organic Spectra, Academic Press, New York 1965.

C.N. Banwell, Fundamentals of Molecular Spectroscopy, MacGraw-Hill Book Co., New York 1966.

J.C.D. Brand, G. Eglington, Applications of Spectroscopy to Organic Chemistry, Daniel Davey and Co., New York 1966.

R.T. Conley, Infrared Spectroscopy, Allyn and Bacon, Boston 1966.

J.T. Houghton, S.D. Smith, Infrared Physics, Oxford University Press, London 1966.

D.O. Hummel, Infrared Spectra of Polymers in the Medium Long Wavelength Regions, Interscience Publishers Inc., New York 1966.

International Union of Pure and Applied Chemistry, Multilingual Dictionary of Important Terms in Molecular Spectroscopy, National Research of Canada, Ottawa 1966.

D.N. Kendall, Applied Infrared Spectroscopy, Reinhold Publishing Co., New York 1966.

I.N. Kolthoff, P.J. Elving, Treatise on Analytical Chemistry, Theory and Practice, Vol. VI/1. Interscience Publishers Inc., New York 1966.

L.H. Little, Infrared Spectra of Adsorbed Species, Academic Press, New York 1966.

A.E. Martin, Infra-Red Instrumentation and Techniques, Elsevier Publ. Co., New York 1966.

K.N. Rao, C.J. Humphreys, D.H. Rank, Wavelength Standards in the Infrared, Academic Press, New York 1966.

W.W. Wendlandt, N.G. Hecht, Reflectance Spectroscopy, Interscience Publishers Inc., New York 1966.

R.G. Zhbankov, Infrared Spectra of Cellulose and its Derivatives, Consultants Bureau, New York 1966.

N.J. Harrick, Internal Reflection Spectroscopy, Interscience Publishers Inc., New York 1967.

J.C. Henniker, Infrared Spectrometry of Industrial Polymers, Academic Press, New York 1967.

L. May, Spectroscopic Tricks, Plenum Press, New York 1967.

R.M. Silverstein, G.C. Bassler, Spectrometric Identification of Organic Compounds, John Wiley & Sons, New York 1967.

W. Simon, Th. Clerc, Strukturaufklärung organischer Verbindungen mit spektroskopischen Methoden, Akademische Verlagsgesellschaft, Frankfurt/Main 1967.

H.A. Szymanski, Progress in Infrared Spectroscopy, Vol. 3, Plenum Press, New York 1967.

H.A. Szymanski, A Systematic Approach to the Interpretation of Infrared Spectra, Hertillon Press, Buffalo 1967.

P. Kirchmer, IR-Spektroskopie mit polarisiertem Licht, Analysentechnische Berichte, Heft 14, Bodenseewerk Perkin-Elmer, Überlingen 1968.

R. Borsdorf, M. Scholz, Spektroskopische Methoden in der organischen Chemie, Wissenschaftl. Taschenbücher Bd. XXI, Akademie-Verlag, Berlin 1968.

D.O. Hummel, F. Scholl, Atlas der Kunststoffanalyse, Bd. I, Teil 1: Text, Teil 2: Spektren, Verlag Chemie, Weinheim/Bergstr. 1968.

L.J. Bellamy, Advances in Infra-Red Group Frequencies, Methuen & Co., London, Barnes & Nobel, New York 1968.

G. Kemmner, Infrarot-Spektroskopie, Grundlagen, Anwendung, Methoden, Franck'sche Verlagshandlung, Stuttgart 1969.

R.D.B. Fraser, E. Suzuki, Resolution of Overlapping Bands: Functions for Simulating Band Shapes, Anal. Chem. *41,* 37 (1969).

H. Weitkamp, R. Barth, IR-Strukturanalyse — ein dualistisches Interpretationsschema —, Georg Thieme Verlag, Stuttgart 1972.

5.3 Raman Spectroscopy

Bernhard Schrader

Institut für Spektrochemie und angewandte Spektroskopie, Abteilung für theoretische organische Chemie,
Universität D-46 Dortmund

Raman spectroscopy and infrared spectroscopy provide complementary descriptions of the vibrations in
molecules and crystals (5.3:4). An infrared spectrum
reflects changes of dipole moment during vibrations,
whereas a Raman spectrum indicates the change of
electron polarizability. General conditions for Raman
spectroscopy are:

Frequency region: 10–4000 cm^{-1}.

Amount of sample: 1 mg in all states, preferably liquids
and crystal powders. Raman spectra of aqueous solutions and air-sensitive samples are obtained easily.

Information made available: Detection of nonpolar
groups: C=C, C≡C, C–C, N=N, O–O, S–S, etc.
(5.3:7.1 and 7.2); analysis of molecular structure
(5.3:7.3); analysis of conformation equilibrium (5.3:7.2);
symmetry of molecules. Spectra of crystals also provide information about static and dynamic intermolecular interactions and about mechanisms of transitions
between modifications. Spectra of polymers can be
analyzed in terms of chain length, tacticity, and crystallinity (5.3:7.6).

By combining Raman and infrared data, bond orders
may be calculated. Structures of molecules and crystals
can be elucidated and thermodynamic functions computed (5.3:7.5) using frequency calculations.

Application of Raman spectroscopy is strongly recommended for many structural elucidations (5.3:7). In
addition to the other spectroscopic methods (IR, UV,
Mass and NMR spectroscopy), Raman spectroscopy
may provide details about substituent groups and
about the molecular skeleton and symmetry.

As a 'fingerprint' of a molecule, the Raman spectrum is
valuable for identification purposes and is often more
useful than an infrared spectrum, particularly as the
vibrations of the skeleton are more clearly visible
(5.3:4; 5.3:8.2).

1 Introduction

In 1928 Raman[1] and his collaborators observed
that colorless organic substances, illuminated by a
mercury arc in the visible region, emit a spectrum
with new lines in addition to the mercury lines. The
frequencies of the new lines differ from those of the
mercury lines by values equal to the molecular
vibrational frequencies.

This effect had already been predicted by Smekal[2].
Later a comprehensive theoretical interpretation
was developed[3,4]. After 1945 the interest of molecular spectroscopists shifted towards infrared
spectroscopy since simple infrared spectrometers
had been developed requiring only about 1 milligram sample. In contrast, the amount of sample
required for Raman spectroscopy was more than
one gram, and spectra of solid samples were difficult to obtain. Raman spectra are of very low
intensity; only 10^{-5} and 10^{-9} of the exciting light
flux is converted to Raman radiation and is collected by the Raman spectrometer. With the use of
conventional light sources, large sample sizes and
long measuring times were essential.

The discovery of visible light emission by ruby
and helium-neon lasers[5,6] introduced ideal light
sources for Raman spectroscopy:

They emit monochromatic radiation with a spectral
radiance (Power/solid angle × area × bandwidth) several
orders of magnitude higher than that of conventional
light sources.

Laser radiation can be concentrated into very small
volumes whose dimensions are of the same order of
magnitude as the wavelengths of light. Intense Raman
spectra are, therefore, available from very small
samples.

The laser radiation is completely polarized.

Nowadays Raman spectra are obtained from milligrams or micrograms of sample. Spectra of crystal powders, single crystals, polymers, and colored
substances may be recorded. Only few sample
types are unsuitable: samples contaminated with
strongly fluorescent impurities and samples which
absorb in the region of the exciting radiation.
Furthermore, Raman spectra of aqueous solutions
and of air-sensitive and corrosive substances can
be obtained more easily than infrared spectra.

2 Origin of the Raman Effect

2.1 Principle

When a molecule is exposed to light of frequency
ν_o, the electron system responds with on induced
frequency. A dipole moment vibrating with frequency ν_o, is induced with an amplitude proportional to the electron polarizability of the molecule.
Consequently, the molecule emits light of frequency ν_o, the Rayleigh radiation.

The electron polarizability of a molecule is a function of
the size and shape of the molecule, its absolute value

[1] *C.V. Raman,* Nature *121,* 501 619, 711 (1928).

[2] *A. Smekal,* Naturwissenschaften *11,* 873 (1923).

[3] *G. Placzek,* Rayleigh-Streuung und Raman-Effekt,
Handbuch der Radiologie, Band IV/2, Akademische Verlagsgesellschaft Geest & Portig KG, Leipzig 1934.

[4] *K.W.F. Kohlrausch*[72].

[5] *T.H. Maiman,* Nature *187,* 493 (1960).

[6] *A.D. White, J.D. Rigden,* Proceeding of Institut of
Radio Engineer *50,* 1967 (1962).

corresponds approximately to the volume of the molecule. For sodium D-line radiation, the electron polarizability of benzene is about 10 A^3. The polarizability is dependent on the direction, it can be represented by the polarizability ellipsoid. The polarizability has its highest value along the direction of the greatest dimension of the molecule. Polarizability is larger for multiple bonds than for single bonds[7,8].

Vibrations of the atomic nuclei may modify the shape of the molecule and hence the electron polarizability ellipsoid. Consequently the radiation emitted by the molecule will be modulated[9]. The molecule therefore emits radiation containing not only the exciting frequency v_0, but also the sum and difference of the exciting and the vibrational frequency v_s, i.e. $v_R^+ = v_0 + v_s$ and $v_R^- = v_0 - v_s$.

The 'unshifted' line of frequency v_0 is called the Rayleigh line, and the lines with the sum and difference frequencies are called Raman lines (or bands). The same vibrational frequency v_s can be observed directly also in the infrared absorption spectrum (cf. Section 5.3:3).

$h v_R^+ = h v_0 + h v_s$ 'Anti-Stokes' Raman scattering

$h v_R^- = h v_0 - h v_s$ 'Stokes' Raman scattering

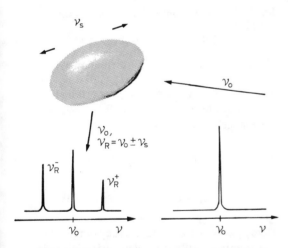

Fig. 1. Modulation of the exciting radiation, frequency v_0, by a molecule vibrating with frequency v_s. The frequency v_0 and sidebands of frequency $v_0 + v_s$ and $v_0 - v_s$ are emitted.

---

[7] G. Kresse, Physikalische Methoden in der Organischen Chemie, 2 Bände, Walter de Gruyter & Co., Berlin 1962.

[8] H.A. Stuart, Molekülstruktur, 3. Aufl. Springer Verlag, Berlin 1967.

[9] H.A. Kramers, W. Heisenberg, Z. Physik 31, 681 (1925).

[10] A. Smekal, Naturwissenschaften 11, 873 (1923).

In addition, the Raman effect can be considered as a process involving collisions between light quanta, $h v_0$, and molecules[10]. Besides elastic collisions, inelastic impacts occur during which quanta of vibrational energy, $h v_s$, are transferred from or to the molecule.

From the Boltzmann law, it follows that there are more molecules in the vibrational ground state than in the excited state, so that energy-deficient light quanta are emitted from the molecule more frequently than energy-rich quanta. Consequently, the Stokes Raman radiation is of higher intensity than the anti-Stokes radiation. From the intensity ratio of the two types of line, one can calculate the vibrational temperature of the molecule. Generally, only the intense Stokes Raman spectrum is recorded.

2.2 Intensity of Raman Lines

The intensity of a Raman line is proportional to the square of the change of the polarizability during a vibration. The induced dipole moment $\vec{\mu}$ of the molecule under the influence of the electric field $\vec{\ell}$ is given by

$$\vec{\mu} = \mathbf{a}\vec{\ell}$$

where $\vec{\mu}$ and $\vec{\ell}$ are vectors and α is the polarizability tensor. Explicity, this formula means:

$$\mu_x = a_{xx}\ell_x + a_{xy}\ell_y + a_{xz}\ell_z$$
$$\mu_y = a_{yx}\ell_x + a_{yy}\ell_y + a_{yz}\ell_z$$
$$\mu_z = a_{zx}\ell_x + a_{zy}\ell_y + a_{zz}\ell_z$$

The terms a_{ij} with $i, j = x, y, z$ are the components of the polarizability tensor, which in our case is symmetrical ($\alpha_{yz} = \alpha_{zy}$, etc.). The polarizability ellipsoid is described by the following formula:

$$a_{yx}x^2 + a_{yy}y^2 + a_{zz}z^2 + 2a_{xy}xy + 2a_{yz}yz + 2a_{zx}zx = 1$$

The axes of the coordinate system can be chosen in such a way that α_{zy}, α_{yz}, and $\alpha_{zx} = 0$ (transformation on main axis). When the molecule is vibrating, all elements of the polarizability tensor can change independently exhibiting the vibrational frequency v_s.
Each element of the polarizability tensor α_{ij} is represented by

$$a_{ij} = a_{ij}^0 + \left(\frac{\partial a_{ij}}{\partial Q}\right) dQ$$

where $\left(\frac{\partial a_{ij}}{\partial Q}\right)$ is the change of the element of the polarizability tensor α_{ij} during a vibration of the molecule, abbreviated to a_{ij}'. Q is the so-called normal coordinate, which describes the displacement of all atomic nuclei during the vibration.

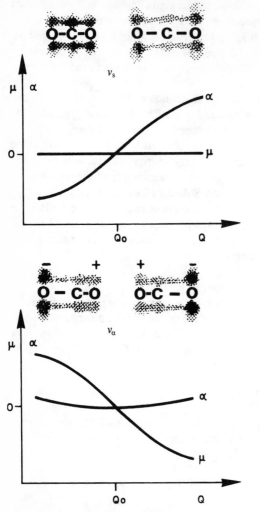

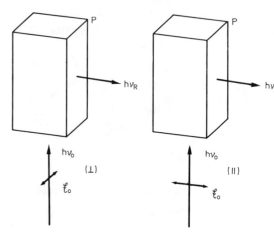

and by the anisotropy of the polarizability change, γ', given by:

$$\gamma'^2 = \frac{1}{2}\,[(a'_{xx}-a'_{yy})^2 + (a'_{yy}-a'_{zz})^2 + (a'_{zz}-a'_{xx})^2 + 6(a'^2_{xy}+a'^2_{xz}+a'^2_{yz})]$$

If the Raman radiation of a sample is examined perpendicular to the direction of the exciting radiation, the intensities of the Raman lines differ, according to whether the electric vector of the exciting radiation is directed parallel ($\parallel$) or perpendicular ($\perp$) to the direction of observation (Fig. 3).

Fig. 3. Determination of depolarization ratio ρ_n of a sample P.
Direction of illumination $h\nu_0$; the electric vector of the exciting radiation is perpendicular ($\perp$), or parallel ($\parallel$), to the direction of observation $h\nu_R$.

Fig. 2. Change of polarizability a and dipole moment μ during stretching vibrations of the CO_2 molecule
For symmetrical stretching, there is no change of dipole moment and therefore no infrared activity. The polarizability increases with elongation, hence there is Raman activity. For the antisymmetric vibration, there is a dipole change but no change of polarizability for small displacements; this vibration is therefore only infrared-active.

Only the changes in the components of the polarizability tensor α_{ij} during a vibration are responsible for the intensity of the Raman line.

If the molecular axes are fixed relative to the axes of the spectrometer, for instance by inclusion in an orientated polymer or a crystal lattice, the changes of the various elements of the polarizability tensor can be observed separately (5.3:7.5). When Raman spectra of molecules in the liquid or gaseous phase are measured, the spectrum observed is the summation of the spectra from all the differently oriented molecules. The intensity of the spectral line is determined by the average change of the diagonal elements of the tensor during the vibration:

$$\bar{a}' = \frac{1}{3}\,(a'_{xx}+a'_{yy}+a'_{zz})$$

The observed intensity is proportional to the Raman scattering coefficients $S\perp$ or $S\parallel$, given by the Placzek theory[3]:

$$S_\perp = \frac{2\pi^2 h}{c}\cdot\frac{(\nu_0-\nu_s)^4}{\nu_s[1-\exp\,(-hc\nu_s/kT)]}\cdot g_s\!\left(a'^2 + \frac{7}{45}\gamma'^2\right)$$

$$S_{\parallel} = \frac{2\pi^2 h}{c}\cdot\frac{(\nu_0-\nu_s)^4}{\nu_s[1-\exp\,(-hc\nu_s/kT)]}\cdot g_s\cdot\frac{6}{45}\gamma'^2$$

Here g_s is the degree of degeneracy of the vibration. The quotient $S\parallel/S\perp$ is the depolarization ratio ρ_n:

$$\rho_n = \frac{6\gamma'^2}{45\bar{a}'^2+7\gamma'^2}$$

These formulae are strictly valid only for Raman radiation perpendicular to the exciting radiation. However, since one observes the radiation within a finite aperture, there are small deviations. Some authors use the depolarization ratio, observed in the arrangement ($\perp$) (Fig. 3) with an analyzer for the Raman radiation perpendicular to and parallel to the direction of excitation. They obtain a different depolarization ratio:

$$\rho_s = 3\gamma'^2/(45\bar{a}'^2+4\gamma'^2).$$

3 Frequencies of Molecular Vibrations

Every atom has three independent degrees of freedom of motion; an n-atomic molecule, therefore, possesses 3n degrees of freedom. Three of these are translational, three relate to rotations of the molecule as a whole, and 3n–6 to vibrations of the atomic nuclei. A linear molecule has only two different rotational degrees of freedom and, therefore, 3n–5 vibrations.

The vibrations are divided into stretching and bending modes[11]. In the stretching modes, the displacements of the atoms are in the direction of the bonds; in the bending modes, only the bond angles are changed. The vibration frequency of a di-atomic molecule is given by:

$$\nu = 1303 \; \sqrt{f\left(\frac{1}{m_1} + \frac{1}{m_2}\right)}$$

where ν is the frequency in wave number units (cm^{-1}), f is the force constant in mdyn/Å, and m_1 and m_2 are the atomic masses in atomic mass units.

Thus, the vibration frequency increases with increasing force constant and decreases with increasing atomic masses. The force constants are in the range 0.1–20 mdyn/Å. The force constants of double and triple bonds are approximately two and three times those of single bonds ($f_{c\equiv c} \sim 15$, $f_{c=c} \sim 10$, $f_{c-c} \sim 5$ mdyn/Å). The force constants of bending vibrations are only about one tenth of those of stretching vibrations[12, 13]. The values of intermolecular force constants are still smaller; these determine the frequencies of the 'frozen' rotations and translations of a molecule in a crystalline lattice – the 'librations' (Latin Libra: the Balance) and the 'translational' vibrations[14].

For hydrogen bonds, they amount to about 0.25 mdyn/Å; for other nonbonded atoms the values are between 0.001 and 0.1 mdyn/Å.

The vibrational spectrum is therefore approximately divided into the following regions:

$\nu[cm^{-1}]$	Vibrations[a]
10– 200	'Librations', 'translational vibrations'
200– 800	Stretching vibrations of $X' - Y$ groups, Bending vibrations of $X - Y$ groups
800–1300	Stretching vibrations of $X - Y$ groups
1000–1600	Stretching vibrations of aromatic compounds, symmetric stretching of $X = Y = Z$ groups Bending vibrations of $X - H$ groups
1500–1900	Stretching vibrations of $X = Y$ groups
1900–2300	Antisymmetric stretching vibrations of $X = Y = Z$ groups Stretching vibrations of $X \equiv Y$ groups
2300–4000	Stretching vibrations of $X - H$ groups

[a] X, Y and Z are elements of the first period, X' denotes elements of the other periods in the Periodic Table. H represents a hydrogen atom.

For an exact description of the molecular vibration frequency it must be taken into account that the change of one bond length or angle will change the elastic properties of the surrounding bonds. Therefore, one must introduce interaction force constants. Finally, also the anharmonicity of the potential functions should be considered.

Mathematical molecular models exist which permit the calculation of frequencies and amplitudes of molecular vibrations when the force constants, the geometry, and the atomic masses are given.

Sometimes the reverse procedure is used: the force constants (a) are calculated from the (observed) frequencies. Up to $a(a+1)/2$ force constants are needed for the calculation of m vibrational frequencies. Therefore, as many as 78 force constants may be needed for calculation of the 12 vibrations of a molecule consisting of 6 atoms. In such cases, an unequivocal solution may be obtained by simultaneous evaluation of the spectra of a number of similar, but isotopically substituted, molecules[15, 16].

Various rules concerning vibrational frequencies exist:

If n oscillators (e.g. di-atomic molecules) of the same frequency are coupled, there will be n vibrations of different frequencies, which are higher and lower than the frequency of the uncoupled oscillators. All oscillators participate in all vibrations but with different phases and amplitudes.

If two oscillators of different frequencies are coupled, then the higher frequency is raised and the lower is reduced. The difference between the frequencies is largest when the frequencies of the isolated oscillators are most similar and the elongations are most colinear.

A molecule containing n atoms with b bonds shows b stretching vibrations and 3n–6–b bending vibrations. For example, *dichloroethylene* contains n=6 atoms and b=5 bonds, and so shows 5 stretching and 7 bending vibrations (Fig. 4a).

If a molecule is changed so that a force constant is reduced or a mass is increased, either all frequencies will be reduced, or some remain constant – none will be raised. If the substituted atom remains at rest during a

[11] R. Mecke, Z. Phys. Chem. Abt. B, 16, 409 (1932); 17, 1 (1932).

[12] J. Goubeau, Angew. Chem. 69, 77 (1957); 73, 305 (1961); 78, 565 (1966).

[13] H. Siebert, Anwendungen der Schwingungsspektroskopie in der anorganischen Chemie, Springer Verlag, Berlin 1966.

[14] B. Schrader[56].

[15] H.J. Becher, Fortschr. chem. Forsch. 10, 156 (1968).

[16] J.H. Schachtschneider, R.G. Snyder, Spectrochim. Acta; Part B, 19, 117 (1963).

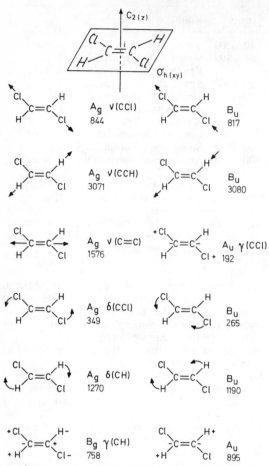

Fig. 4. a) Symmetry elements and vibrations; b) Raman spectrum of *trans*-dichloroethylene, measured with a 'Moritz'

vibration, no change in the frequency will occur (Rayleigh's rule).

If the frequency of a two-atom unit X–Y in a molecule is substantially different from the frequency of adjacent bonds, a vibration will always occur which corresponds principally to the X–Y bond, i.e. it will have a frequency similar to that of the isolated X–Y unit. This frequency is called the characteristic frequency of the group X–Y.

4 Raman and Infrared Activity

The symmetry of the molecule dictates which of the 3n–6 vibrations change the polarizability or the dipole moment of the whole molecule, i.e. whether they are observable in the Raman spectrum or the infrared spectrum. Molecules can possess the following symmetry elements: axes of symmetry, planes of symmetry and centers of symmetry. All symmetry operations possible in a molecule form a point-group.

The *trans-dichloroethylene* molecule belongs to the point-group C_{2h}. This means that a twofold axis of symmetry C_2 and a plane of symmetry σ_h perpendicular to it exist, as well as a center of symmetry i (Fig. 4a). Every vibration may be symmetric or antisymmetric with respect to the symmetry elements, i.e. the symmetry operation either does not change the elongated molecule or it shifts the phase by 180°. The possible vibrations of the molecule may be classified into vibrational species with regard to their symmetry. The point group C_{2h} contains the vibrational species A_g, B_g, A_u and B_u. A and B mean symmetry or antisymmetry with respect to the symmetry axis C_2, g and u

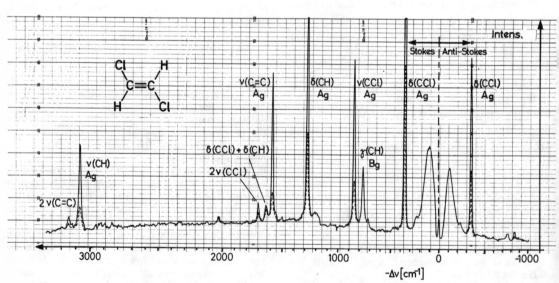

v = Valence vibration δ = Deformation vibration γ = Out of plane vibration

Polarization direction ——— perpendicular ------- parallel to the direction of observation
The exciting line $\Delta v = 0$ is in reality 50,000 times stronger (an interference filter SF was used, *cf.* Fig. 5). Allocation according to Ref.[89].

symmetry and antisymmetry with respect to the center of symmetry. Socalled character-tables exist for the different point groups (e.g. Table 1) which show the properties of the vibrational species.

Table 1. Character table of the point group C_{2h} (cf. 5.3:9). E denotes identity, and 1 and –1 mean symmetry and antisymmetry.

C_{2h}	E	C_2	i	σ_h		
A_g	1	1	1	1	R_z	$\alpha_{xx}, \alpha_{yy}, \alpha_{zz}, \alpha_{xy}$
B_g	1	–1	1	–1	R_x, R_y	α_{yz}, α_{zx}
A_u	1	1	–1	–1	T_z	
B_u	1	–1	–1	1	T_x, T_y	

Infrared activity of the vibrational species (in this case B_u and A_u) is expressed by the symbols T_x, T_y and T_z. They also mean that the translations in the x-, y- and z-directions belong to the corresponding species and that a change of dipole moment occurs in the direction of the x, y and z axes on vibrations of these species. Raman activity is symbolized by the elements α_{ij} of the polarizability tensor, which change during a vibration of the respective species. R_z, R_x and R_y indicate the rotations of the molecule about the axes z, x and y. The table shows that the totally symmetrical vibrations (i.e. those symmetrical with respect to all symmetry elements) – those of species A_g, in the present case – are Raman-active like those of species B_g. The vibrations of the species A_u and B_u are infrared-active but not Raman-active.

There are some general rules about activities and intensities of vibration in Raman and IR spectra:

In the Raman spectrum, vibrations of not more than four species are allowed; in the infrared spectrum not more than three species can be active.

The vibrations belonging to the totally symmetrical species are always permitted to give Raman lines. They usually show the highest intensity.

If the molecule contains a center of symmetry, there is a rule of mutual exclusion: all vibrations which are symmetric with regard to the center of symmetry are infrared-inactive, all antisymmetric vibrations are Raman-inactive.

All Raman lines relating to totally symmetric vibrations of liquid substances are *polarized:* the value of the depolarization ratio ρ_n is between zero and 6/7; for cubic point groups ρ_n is equal to zero. All other Raman-active species show *depolarized* lines: the depolarization ratio ρ_n is equal to 6/7. If one uses ρ_s from Section 5.3:2.1, 6/7 should be replaced by 3/4.

In addition, the following empirical rule holds showing the dependence of the polarizability change from the electron density of a bond.

In Raman spectra, the vibrations of multiple bonds give stronger bands than those of single bonds; those of non-polar bonds are stronger than those of polar bonds

(this is the converse of the situation with infrared spectra).

5 Raman Spectrometers

5.1 Types of Spectrometers

Because of the advantages of laser light sources compared with conventional radiation sources such as the mercury arc, most future Raman spectrometers will use laser light sources. Therefore, we shall mention only the properties of laser Raman spectrometers[17]. A laser Raman spectrometer consists of the following elements:

Laser light source
Primary filter
Sample
Secondary filter or monochromator to suppress 'false' light
Monochromator for spectral dispersion
Radiation detector
Amplifier
Recorder or digital read-out

Some properties of laser light sources used nowadays are given in Table 2.

Table 2. Laser light sources for Raman spectroscopy.

Active substance	λ [Å]	Light flux [mW]	Quantum yield of a photo-cathode (S20 type)
He – Ne	6328	50–200	0.06
Ac$^+$	4880	1000	0.16
Ar$^+$	5145	1000	0.14
Kr$^+$	6471	150	0.05
Ruby[18, 19]	6943	1000^a	0.04

^{a} Quasicontinuous

The following considerations are important when choosing the light source. The intensity of the Raman spectrum increases with the fourth power of the exciting frequency. If the sample is illuminated by the same light flux of an $Ar^\oplus$ or a He-Ne laser, then the $Ar^\oplus$ line ($\lambda_0 = 4880$ Å) yields about three times the Raman intensity, compared with the He-Ne line ($\lambda_0 = 6328$ Å). The quantum yield of the photo-cathode in a red-sensitive radiation detector is nearly three times as great in the first case as in the second. These reasons indicate the superiority of the Argon laser.

The quantum yield of the Raman effect is about 3 to 8 orders of magnitude lower than that of *fluo-*

[17] *B. Schrader*, Chem.-Ing.-Tech. *39*, 1008 (1967).
[18] *D. Roess*, J. Quantum Electronics *2*, 208 (1966).
[19] *W. Kiefer, H.W. Schrötter*, Z. Angew. Physik, *25*, 236 (1968).

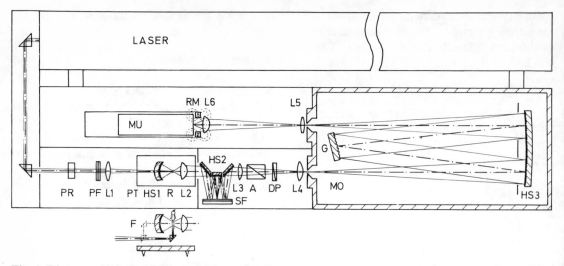

Fig. 5. Diagram of 'Moritz' Raman spectrometer
HS 1, 2 and 3 are concave mirrors, G the grating, L 1–6 are lenses.

rescence. Consequently the fluorescence of impurities – even if they are present in traces of less than 1% – can mask the Raman spectrum completely. The perturbation of the Raman spectrum by fluorescence is less likely if one uses exciting radiation from the red region of the spectrum instead of the blue. In this respect the $Kr^{\oplus}$- or the He-Ne laser is to be preferred over the argon laser. In addition to this, Raman spectra of yellow or orange substances can be obtained with red laser sources. Therefore, the He-Ne laser has emerged as a universal light source for Raman spectroscopy, especially for organic compounds. The argon laser is recommended for excitation of the Raman spectra of gases – which show extremely low intensity. If a light source for exciting lines from the blue-green as well as from the red range of the spectrum is required, this can be achieved by means of an argon-krypton laser.

5.2 Construction Principle

The principles of a simple Raman spectrometer, suitable for application in organic chemistry[17], are shown in Fig. 5.

The (plane-polarized) laser beam passes a polarization rotator PR for optional adjustment of the polarization plane and a 'primary filter' PF, suppressing the undesired lines from the plasma discharge of the laser. The lens, L1, focusses the beam into the sample. Easily exchangeable mounts for the different sample cells allow examination of liquid samples in melting point capillaries (1μl) [Fig. 6a] or of bigger volumes (1–50 μl) in cuvettes with standard ground joints. Other mounts permit the investigation of crystalline powder samples in the forward scattering arrangement[20]; the

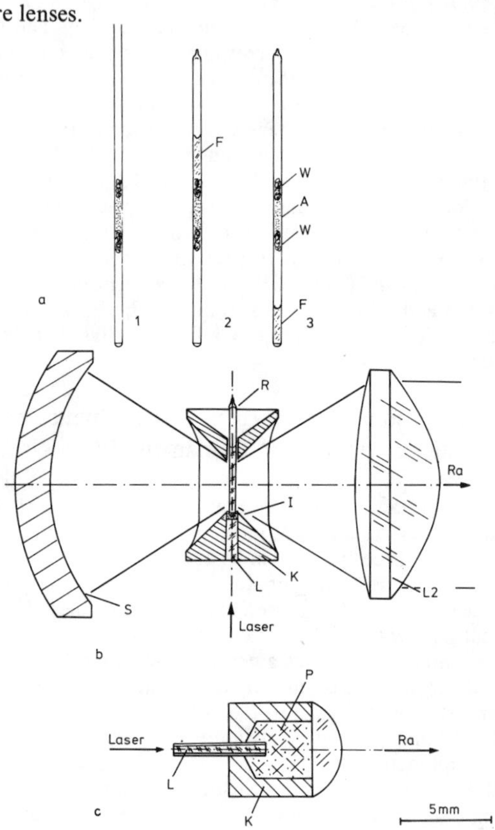

Fig. 6. a) Purification of small liquid samples in the Raman cuvette
(1) tube before use, (2) tube with sample after sealing, (3) after centrifugation, ready for use. F: liquid sample, W: glass wool or cotton wool, A: adsorbent (aluminum oxide)
b) Arrangement for liquids
c) Cuvette for crystal powders
I: immersion liquid; K: cuvette body of stainless steel; L: light guide; S: mirror; L2: lens; P: crystal powder;
LASER: direction of Laser-beam; Ra: direction of observation.

[20] *B. Schrader, F. Nerdel, G. Kresze*, Z. Phys. Chem. N.F., *12*, 132 (1957).

thickness can be varied by means of a small movable light tube (Fig. 6c). A goniometer head is used for adjustment of single crystals. There is also a low (or high) temperature cell for liquids and solids.

The radiation emitted by the sample is collected by a high power aspherical condenser lens L2, and is reflected thrice upon the same interference filter SF. This filter [21] transmits the exciting light and reflects the Raman radiation. The exciting line is attenuated in total by a factor of 2×10^{-5}, relative to the Raman lines, which pass this system almost unchanged. This filter system, with a higher penetration degree for the Raman lines, is equivalent, as regards attenuation of the false light, to one of the monochromators of a double monochromator spectrometer. A drawback of the filter is the attenuation of Raman lines very near the exciting line. Strong lines higher than $50 \, cm^{-1}$ are still recorded; those higher than $100 \, cm^{-1}$ are slightly weakened, and those higher than $200 \, cm^{-1}$ remain practically unweakened. The Raman radiation passes an analyzer A and a polarization scrambler DP in front of the entrance slit of the monochromator MO: The scrambler converts linearly polarized light into normal light and, thereby, avoids the influence of the different reflective power of the grating for different types of linearly polarized light. The exit slit of the monochromator is reduced to 1/8 size and projected on to the photocathode of the photomultiplier MU by the aspheric lens L6. A permanent ring magnet RM with radial magnetization reduces the effective photocathode to the size of the slit image. By doing this, the background in the spectrum is reduced to 1/100 of the value without the magnet, the same signal intensity being retained. Amplification is performed by a photon counting device [22].

The principle of the Raman spectrometer discussed here is similar to that of all commercial Raman spectrometers. Most of the instruments, however, use a double or triple monochromator allowing the registration of lines in the immediate neighborhood of the exciting line.

6 Sample Preparation

All substances to be analyzed by Raman spectroscopy should be free from impurities likely to fluoresce or absorb in the range of the Raman spectrum. Many substances, even those stated to be 'analytically pure', are contaminated by stopcock grease or oxidation products, often strongly fluorescent. Sometimes sample purification can be avoided simply by exposing the sample to strong light fluxes [23, 24]. Substances capable of fluorescing often undergo photochemical transformations

(photolysis, dimerization, or photooxidation), so that fluorescence is extinguished. This process requires between 10 and 60 minutes when a He-Ne laser is used, but only a few minutes with a strong argon laser [25, 26].

Gas chromatography or vacuum distillation can be used for purification of liquids. Filtration through charcoal or alumina may be suitable for purification of liquids or solutions. If only small sample volumes (a few microliters) are available, this filtration can be done by centrifuging the Raman cuvette [26] (Fig. 6a).

Crystalline substances can be examined in solution, as a melt, in powdered form, or as single crystals. The spectra of dissolved samples may be partially masked by the spectrum of the solvent. Water, carbon disulfide, and carbon tetrachloride are normally suitable *solvents*. Substances whose spectra contain many lines (chloroform, benzene, cyclohexane, acetone, etc.) are used less frequently. Solvents are purified by distillation in greasefree apparatus, or by filtration through alumina.

Schrader and Bergmann [27] have made a study of the sample form most suitable for powdered materials.

Coarsely crystalline samples give more intense spectra than do finely powdered samples. Polycrystalline pressed pellets are especially suitable. The forward scattering arrangement (Fig. 6c) gives a spectrum containing less exciting light mixed with Raman radiation than does the back scattering arrangement. The 'optimum' sample thickness is generally between 0.2 and 2 mm, but can be up to 5 mm. Mixing the samples with immersion media such as potassium bromide or polyethylene normally does not result in more intense spectra.

Substances which can be obtained as crystals of at least about 0.1 mm diameter and several mm in length can be adjusted in the laser beam by means of the goniometer head; several different Raman spectra (up to six) may be obtained, and this permits assignment of the observed lines to the vibrational species in the molecule and the crystal (cf. Section 5.3:7.4.). Larger single crystals should be cut in such a way that the crystal surfaces are perpendicular to the axes of the indicatrix. This eliminates undesirable double refraction effects on the spectrum, and the best information on the vibration of the crystal is obtained.

7 Information Provided by Raman Spectroscopy

In principle, Raman spectra and infrared spectra provide identical information about vibrational

[21] W. Meier, Dissertation, Münster 1969; W. Meier, B. Schrader, M. Disarčik, Z. Intrumentenk. 5, 119 (1972).

[22] Y.M. Pao, R.N. Zitter, J.E. Griffiths, J. Opt. Soc. Amer. 56, 1133 (1966).

[23] R.N. Jones, J. Org. Chem. 30, 1822 (1965).

[24] E. Steigner, Dissertation, Münster 1969.

[25] B. Schrader, E. Steigner, unpubl.

[26] B. Schrader, W. Meier, E. Steigner, F. Zöhrer, Z. Anal. Chem. 254, 257 (1971).

[27] B. Schrader, G. Bergmann, Z. Anal. Chem. 225, 230 (1967).

frequencies. Vibrations associated with a large change of dipole moment give strong bands in the infrared; those associated with a large change of polarizability give strong Raman lines. If there are symmetry elements in a molecule, certain vibrations may be observed only in the Raman spectrum and others only in infrared spectra.

If the aim is to obtain a complete assignment of all frequencies to the possible vibrations, or to calculate force constants or thermodynamic properties, information on the vibrational spectra should be as complete as possible. Both Raman and infrared spectra of the substance should, therefore, be measured.

Some frequencies, which are both Raman- and infrared-inactive, can be determined by analysis of all combinations and overtones, or sometimes directly from *neutron spectra* (cf. Chapter 5.4).

Simultaneous analysis of both spectra is also useful when information about the presence of definite groups, or about the geometry of the molecule, is desired.

7.1 Characteristic Frequencies

A chemical group yields characteristic spectral frequencies for those mechanical vibrations which are localized mainly within this group (Section 5.3:3). Specific constituent groups may, therefore, be recognized in different molecules by observation of the same (or slightly shifted) characteristic frequency. The vibrations will also give characteristic intensities in infrared and Raman spectra. The vibrations of multiple bonds and nonpolar bonds give strong bands in Raman spectra, whereas those of polar bonds give strong bands in infrared spectra. Symmetric vibrations (in-phase vibrations) of groups are usually stronger in Raman spectra, and antisymmetric vibrations (out-of-phase vibrations) are often stronger in infrared spectra.

Table 3 lists examples of frequencies and relative intensities of characteristic vibrations which are recognizable either from Raman spectra alone, or together with infrared spectra.

Raman spectra show vibrations particularly well which are due to the following bonds of low or zero polarity: C=C, C≡C (C≡N), C=N, N=N, C–C, O–O, S–S, C–S, as well as the symmetrical skeletal vibrations which are weak or absent in infrared spectra. In contrast, the vibrations of the OH–, C=O and C–O groups are strongly infrared-absorbing but appear only weakly in Raman spectra (Table 3).

Table 3. Relative Raman and infrared intensities and frequency regions of some characteristic vibrations ν stretching, δ bending vibrations, $_s$ symmetric (in-phase), $_a$ antimetric (out-of-phase), ss very strong, s strong, m medium, w weak, o absent.

Vibration	Region	Intensity Raman	Intensity Infrared
ν(O—H)	3650–3000	w	s
ν(N—H)	3500–3300	m	m
ν(≡C—H)	3300	w	s
ν(=C—H)	3100–3000	s	m
ν(—C—H)	3000–2800	s	s
ν(—S—H)	2600–2550	s	w
ν(C≡N)	2255–2220	m–s	s–o
ν(C≡C)	2250–2100	ss	w–o
ν_a(>C=C=C<)	2000–1900	o	s
ν_s(>C=C=C<)	1140–1060	s	o
ν_a(>C=C=O)	2150	w	s
ν_s(>C=C=O)	1120	s	s
ν_a(—N=C=O)	2290–2250	w	ss
ν_s(—N=C=O)	1400–1350	s	w
ν_a(—N=C=S)	2170–2040	m	s
ν_s(—N=C=S)	1090–1045	s	m
ν_a(—N=C=N—)	2145–2115	o	s
ν_s(—N=C=N—)	1040	s	o
ν_a(—N=N=N)	2150–2090	m	s
ν_s(—N=N=N)	1290–1250	s	w
ν(C=O)	1820–1680	s–w	ss
ν(C=C)	1900–1500[a]	ss–m	o–w
ν(C=N)	1680–1610	s	m
ν(N=N), aliph.	1580–1550	m	o
ν(N=N), arom.	1440–1410	m	o
ν_a([C—]NO$_2$)	1590–1530	m	s
ν_s([C—]NO$_2$)	1380–1340	s	m
ν(C—N[O$_2$])	920–820	s	w
δ_s([C—]NO$_2$)	620–600	s	w
ν_a([C—]SO$_2$[—C])	1350–1310	w–o	s
ν_s([C—]SO$_2$[—C])	1160–1120	s	s
ν([C—]SO[—C])	1070–1020	m	s
ν(C=S)	1250–1000	s	w
δ(CH$_2$), δ_a(CH$_3$)	1470–1400	m	m
δ_s(CH$_3$)	1380	m–w,s[b]	s–m
ν(CC), Arom.	1600, 1580	s–m	m–s
	1500, 1450	m–w	m–s
	1000	s[c]	o–w
ν(CC), Alicyclic	1300–550	s–m	m–w
ν_a(CC) ⎱ aliph. chains ⎰	1130–1030	s–w	o–m
ν_s(CC) ⎰	900–650	s–m	o–w
ν_a(C—O—C)	1150–1060	w	s
ν_s(C—O—C)	970–800	s–m	w–o
ν_a(Si—O—Si)	1110–1000	w–o	ss
ν_s(Si—O—Si)	550–450	ss	w–o
ν(O—O)	900–845	s	o–w
ν(N—N)	875–800	s	o–w
ν(S—S)	550–430	s	o–w
ν(Se—Se)	330–290	s	o–w
ν(C$_{arom.}$—S)	1100–1080	s	s–m
ν(C$_{aliph.}$—S)	790–630	s	s–m
ν(C—Cl)[d]	800–550	s	s
ν(C—Br)[d]	700–500	s	s
ν(C—J)[d]	660–480	s	s
δ_s(CC), aliph. chains C$_n$, n=3 ... 12	400–250	s–m	w–o

Vibration	Region	Raman	Infrared
		Intensity	
n > 12	2495/n		
Librations[e]	200–20	f	g
Translation			
vibrations[e]	200–20	s–o	m–o

[a] See Fig. 7
[b] If at C=C
[c] For mono-; meta-; 1,3,5-tri-derivatives
[d] See table 4
[e] Lattice vibrations in molecular crystals, activity determined by crystal symmetry
[f] Proportional to anisotropy
[g] Proportional to dipole momentum

Fig. 7. C = C stretching frequencies of isolated double bonds influenced by various neighboring bonds.

7.2 Influence of Neighboring Bonds on Bond Frequencies

The influence of angles and the frequencies of bonds which are coupled directly to the $C=C$ bond are shown very clearly in Figure 7.

The large range of the values of $\nu(C=C)$ (i.e. 1525–1877 cm^{-1} in *cyclopropenes*[28]) is attributable mainly to the influence of the frequency of the adjacent bonds (cf. Section 5.3:3). The frequencies of coupled groups 'repel' each other. If the adjacent bond has a high frequency, then $\nu(C=C)$ is shifted to smaller values. The effect of the C–D-bond ($\nu(CD)\sim 2250$ cm^{-1}) is larger than that of the C–H-bond ($\nu(CH)\sim 3100$ cm^{-1}). If the hydrogen atoms are replaced by alkyl groups ($\nu(CC)\sim 1000$ cm^{-1}), $\nu(C=C)$ moves to higher values.

Methyl groups raise this frequency more than alkyl chains because of the fact that the symmetric deformation vibration of the CH$_3$ groups [$\delta_s(CH_3)=1380$ cm^{-1}] couples strongly with the $\nu(C=C)$ vibration. In Raman spectra this coupling is also demonstrated by the fact that the methyl vibration near 1380 cm^{-1} is strong only when the methyl group is attached directly to the $C=C$ bond or to the α carbon atom[29].

A significant change in the $C=C$ force constant in Fig. 7 is observed only with *tetrachloroethylene*. The $\nu(C=C)$ of this compound should occur at higher frequencies than that of *ethylene* (1621 cm^{-1}), but the vibration is actually at a lower value (1571 cm^{-1}). This can be attributed to the weakening of the $C=C$ bond due to mesomeric interactions with the Cl atoms[30].

Often the Raman spectrum supplies the only spectroscopic evidence regarding central multiple bonds in a molecule – for instance, in the dimer of a nucleophilic carbene, the *bis-[1,3-diphenylimidazolidinylidene-(2)][31]*(**1**)$\nu(C=C)=1640$cm^{-1}, or the crossed triple in *9,10,19,20-tetrahydrotetrabenzo[a,c;g,i]cyclodecene[32]* (**2**), $\nu(C=C)=2220$ cm^{-1}.

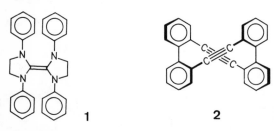

1   **2**

7.3 Interaction between Several Similar Groups

If two identical groups are present in a single molecule, two different vibrational frequencies can be observed. Both groups participate in both vibrations – i.e. the in-phase and out-of-phase vibrations. If the two identical groups are connected through a common atom, the two frequencies (in cm^{-1} to a rough approximation) are given by the formulas:

$$\nu \text{ in phase} = 1303\sqrt{f\left(\frac{1}{m_1}+\frac{1+\cos\alpha}{m_2}\right)}\quad [\text{cm}^{-1}]$$

$$\nu \text{ out of phase} = 1303\sqrt{f\left(\frac{1}{m_1}+\frac{1-\cos\alpha}{m_2}\right)}\quad [\text{cm}^{-1}]$$

In the above formulas, f is the force constant of the two bonds, in mdyn/Å, m_1 is the mass of the outer atoms, m_2 the mass of the common atom in atomic mass units, and α is the bond angle.

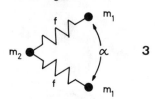

The formulas show that when α is between 0 and 90°, the frequency of the in-phase vibration is higher than that of the out-of-phase vibration. For α between 90 and 180°, which is the case with most groups, the converse holds.

Aliphatic CH$_2$-group	$\nu_a(CH_2)$ 2925	$\nu_s(CH_2)$ 2855
NO$_2$-group in nitrobenzene	$\nu_a(NO_2)$ 1523	$\nu_s(NO_2)$ 1345
Cyclopropane	$\nu_a(\Delta)$ 865	$\nu_s(\Delta)$ 1186
Triphenylcyclopropenyliumbromide[33]	$\nu_a(\Delta)$ 1411	$\nu_s(\Delta)$ 1845

If α is exactly 90°, there is no coupling, and one can observe exactly the frequency of a di-atomic molecule, $(m_1\text{-}m_2)$. The in-phase vibrations are recognized by their greater intensity in the Raman spectrum; the out-of-phase vibrations are usually stronger in the infrared spectrum.

The frequencies of C-halogen bonds are a sensitive indicator of the environment of the C-atom (Table 4)[34].

[28] *G.L. Closs*, Cyclopropenes, Advances in Alicyclic Chemistry, Vol. 1, Academy Press, New York 1966.

[29] *D.G. Rea*, Anal. Chem. *32*, 1638 (1960).

[30] *K.W.F. Kohlrausch*[72].

[31] *H.W. Wanzlick*, Angew. Chem. *74*, 129 (1962).

[32] *H.A. Staab, E. Wehinger*, Angew. Chem. *80*, 240 (1968).

[33] *F. Höfler, B. Schrader, A. Krebs*, Z. Naturforsch. *24a*, 1617 (1969).

[34] *C. Altona, H.J. Hageman, E. Havinga*, Spectrochim. Acta *24A*, 633 (1968).

C. Altona, Tetrahedron Lett. 2325 (1968).

C. Altona, H.J. Hageman, Recl. Trav. Chim. Pays-Bas *88*, 33 (1969).

Table 4. C−Cl-, C−Br-, and C−I stretching parameters [cm^{-1}]
P: primary, S: secondary, T: tertiary. Subscripts: atoms *anti* to X (X = Cl, Br, I); p: lone pair on oxygen, or π-orbital; q: C−X bond flanked by one quaternary carbon atom; ': eclipsed; ^{6}: on cyclohexanoid ring; 4R: shift of frequency due to substitution in position 4 of cyclohexyl halide.

Type[a]	C−Cl	C−Br	C−I
P_H	653	562	503
P_H'	683	614	580
P_C	726	646	594
P_X	726	624	576
P_O	~750	~665	~617
S_{HH}	611	535	486
S_{HH}'	632	578	548
S_{HH}^6	685	660	—
$S_{XH} = S_{CH}$	666	613	577
$S_{XH}^6 = S_{CH}^6$	650	596	—
S_{CC}	742	686	—
S_{HO}	700	660	—
S_{CO}	767	728	—
S_{Hp}	575	534	—
S_{Xp}	708	632	—
T_{HHH}	560	505	487
$T_{XHH} = T_{CHH}$	612	588	573
qT_{CHH}	~607	~548	—
qT_{CHH}^6	~585	~530	—
T_{CHH}	~650	~640	—
4R	+ 18	+ 6	—

[a] Nomenclature as given by *I. Shimanouchi, M. Tasumi*, Spectrochim. Acta *17*, 755 (1961).

Characteristic frequency parameters occur for cyclohexyl chloride with the Cl atom axially or equatorially situated (s. Fig. 4):

$$S_{HH}^6 = 685 \text{ cm}^{-1} \qquad S_{CC} = 742 \text{ cm}^{-1}$$
observed: $\nu(\text{C—Cl}) = 685 \text{ cm}^{-1}$ $\nu(\text{CCl}) = 740 \text{ cm}^{-1}$

The C−Cl vibration gives a strong band in the Raman as well as in the infrared spectrum. *Trans-1,2-dichlorocyclohexane* shows in-phase and out-of-phase vibrations for both conformers, which can be assigned from their typical intensities in the Raman and infrared spectra.

	cm^{-1} Raman	IR		cm^{-1} Raman	IR
γ_s = 690	SS	W	γ_a = 746	m	S
ν_a = 610	W	SS	γ_s = 738	S	m
S_{XH}^6 = 650	—	—	S_{CC} = 742	—	—

The average frequency of the two bands corresponds to the frequency parameter in Table 4. The frequency difference between in-phase and out-of-phase vibrations resulting from coupling is significantly larger for the di-axial conformer (80 cm^{-1}) than for the di-equatorial compound (8 cm^{-1}).

By simultaneous analysis of the Raman and infrared spectra, the structures and conformations of open-chain and cyclic halides can be determined[35, 36]. In addition, the position of a conformational equilibrium can be estimated from intensity measurements[37].

An interesting vibrational coupling has been reported for *cyclopropenone derivatives* 8.

A number of authors have assigned the two very strong infrared bands at 1640 and 1840 cm^{-1} to the C=C and C=O bonds in 8a. Some assigned the higher frequency, and others the lower value, to the C=O group[38].

In the Raman spectrum, only the vibration at 1640 cm^{-1} is very strong. It can be expected that both the isolated C=O and C=C groups would absorb in the 1700–1800 cm^{-1} region (cf. *cyclopropanone*, $\nu(\text{C=O})$ 1820 cm^{-1}, and substituted *cyclopropenes*, $\nu(\text{C=C})$ 1870 cm^{-1}) if a significant participation by the structure 8b is assumed. The bonds in the three-membered ring bring about strong coupling of the two groups, so that they form a vibrating unit; consequently, there should be an in-phase and an out-of-phase vibration. The former is likely to be very strong in the Raman spectrum because the polarizability of the total molecule changes considerably. Both vibrations should show similar intensities in the infrared spectrum because the dipole change is in the C=O direction, perpendicular to the C=C bond. Therefore, the band at 1640 cm^{-1} is assigned to the inphase vibration, and that at 1840 to the out-of-

[35] *C. Altona, H.R. Buys, H.J. Hageman, E. Havinga*, Tetrahedron *23*, 2265 (1967).

[36] *H.R. Buys, C. Altona, E. Havinga*, Tetrahedron *24*, 3019 (1968).

[37] *K.W.F. Kohlrausch*[72].

[38] *A. Krebs, B. Schrader*, Liebigs Ann. Chem. *709*, 46 (1967);
A. Krebs, B. Schrader, F. Höfler, Tetrahedron Lett., 5935 (1968).
E.M. Brigs, I.W. Bird, A.F. Harmer, Spectrochim. Acta *25A*, 1319 (1969).

phase vibration[39]. This assumption is confirmed by [18]O substitution (both bands show similar isotope shift) and by a frequency calculation[40].

7.4 Skeletal Vibrations of Organic Substances

CC-skeletal bonds of organic compounds are mostly nonpolar. Consequently, their vibrations are weak in infrared spectra but strongly Raman-active. The strong 'ring-breathing' vibrations, in which equivalent bonds vibrate in-phase, are of particular significance. These vibrations can be used to characterize the ring system, but they are susceptible to shifts depending on the nature and position of the substituent groups.

The ring-breathing vibration of the benzene nucleus occurs near 1000 cm^{-1} in the case of *benzene* and its mono-, *meta*-di-, and 1,3,5-trisubstituted derivatives. Vibrations with constant fre-

quency, and substituent-sensitive vibrations, have both been studied extensively. By combining data from Raman and infrared spectra, determination of the structure of unknown compounds is made possible[41-43].

With saturated rings, the following ring-breathing frequencies are observed: *cyclohexane,* 802; *cyclopentane,* 886; *cyclobutane,* 960; *cyclopropane,* 1184 cm^{-1}. For substituted and condensed rings, the skeletal vibrations are a sensitive index of the ring skeleton. The bands of the three isomeric compounds *hexamethylbenzene, hexamethyl Dewar benzene* and *hexamethylprismane*[43a, 44] in the region 300–1300 cm^{-1} are due mainly to skeletal vibrations. Therefore, the differences in the Raman spectra are much larger than in the infrared, NMR, UV or mass spectra which, for the most part, provide only indirect information about ring skeletons[45] (Fig. 8).

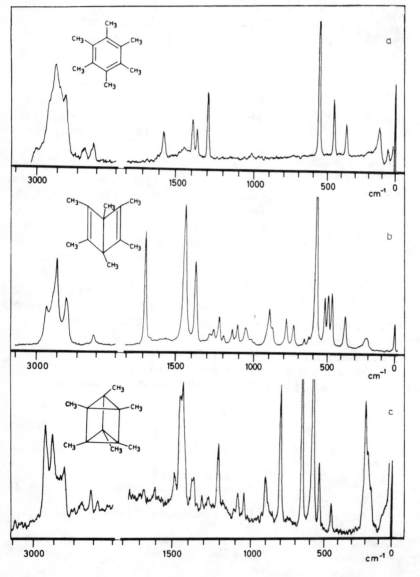

Fig. 8: Raman spectra of
(a) hexamethylbenzene,
(b) hexamethyl Dewar benzene,
(c) hexamethylprismane
Samples (a) and (c) were crystalline, 2 mg;
(b) was liquid, 0.5 μl.

From these 'fingerprint' vibrations, the skeletons of unknown substances can be identified; comparative spectra from spectral collections are used. In the future model calculations may be developed. Recently Olah[46] was able to demonstrate that the *nortricyclonium ion* shows the symmetric structure **9**. In the region 800–1000 cm^{-1}, it gave a Raman spectrum similar to that of the nortricyclene skeleton **10**, but differed strongly from that of *norbornane* **11**. This information could not be obtained from the NMR spectrum, since a rapid valence tautomerism had to be considered.

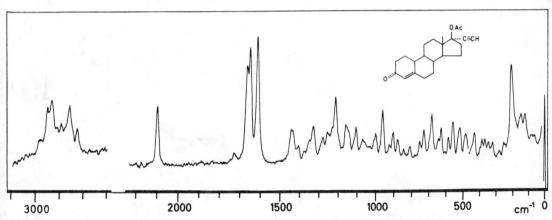

In principle, molecules without symmetry elements – such as the *steroids* – should show all vibrations in the Raman spectra, as well as the infrared spectra. However, here, too, the vibrations of the polar substituents dominate in the infrared spectra and those of the skeleton in the Raman spectra [24, 47, 48] (Fig. 9).

The Raman spectrum of *21-methyl-Δ⁴-19-nor-17α-pregnen-20-yne-17-ol-3-one* (Fig. 9) shows the strong vibrations of the multiple bonds at 2234 cm^{-1}, $\nu(C\equiv C)$, and 1664/1655 and 1613 cm^{-1}, $\nu(C=C-C=O)$[48]. The bonds at 1664/1655 cm^{-1} are probably due to vibrational coupling of the strong infrared-active inphase vibration of the C=C–C=O system of neighboring molecules in the crystal lattice.

By computer evaluation of 80 steroid spectra, it was possible to develop a scheme suitable for the characterization of the ring skeleton from Raman spectra[48–50].

Finally, mention should be made of the application of laser Raman spectroscopy to colored substances such as *ferrocene*[51] and the *metallo-porphyrins*[52].

7.5 Vibrational Spectra of Molecular Crystals

In molecular crystals, all molecules have the same environments, i.e. identical systems with identical frequencies. Consequently, strong intermolecular vibrational coupling should occur. The unit cell is

Fig. 9. Raman spectrum of 21-methyl-Δ⁴-19-nor-pregnen-20-yne-17-ol-3-one (2 mg; 'Moritz' Raman spectrometer).

[39] A. Krebs, B. Schrader, Z. Naturforsch. *21b*, 194 (1966).

[40] F. Höfler, B. Schrader, A. Krebs, Z. Naturforsch. *24a*, 1617 (1969).

[41] K.W.F. Kohlrausch[72]; N.B. Colthup, L.M. Daly, S.E. Wiberley[82]; B. Schrader, W. Meier[127b].

[42] G. Nonnenmacher, Dissertation, Freiburg i. Breisgau 1961.
E.W. Schmid, J. Brandmüller, G. Nonnenmacher, Ber. Bunsenges. Phys. Chem. *24*, 724, 940 (1960).

[43] G. Varsányi, Vibrational Spectra of Benzene Derivatives, Academic Press, London 1969.

[43a] D. Bougeard, Dissertation, Dortmund 1972; D. Bougeard, P. Bleckmann, B. Schrader, unpubl.

[44] W. Schäfer, H. Hellmann, Angew. Chem. *79*, 518, 566 (1967).

[45] B. Schrader, unpubl.

[46] G.A. Olah, A. Commegras, C.Y. Lui, J. Am. Chem. Soc. *90*, 3882 (1968).

[47] R.N. Jones, P.J. Krueger, K. Noack, J.J. Elliot, R.A. Ripley, G.A.A. Nonnenmacher, J.B. DiGorgio, Proceedings of the 8th Colloquium Spectroscopium Internationale, Washington D.C. 1963.

[48] E. Steigner, B. Schrader, Z. Anal. Chem. *254*, 177 (1971).

[49] B. Schrader, E. Steigner, Liebigs Ann. Chem. *735*, 6 (1970).

[50] E. Steigner, B. Schrader, Liebigs Ann. Chem. *735*, 15 (1970).

[51] T.V. Long jun., F.R. Huege, Chem. Comm. 1239 (1968).

[52] H. Bürger, K. Burzyk, J.W. Büchler, J.M. Fulerhop, F. Höfler, B. Schrader, Inorg. Nucl. Chem. Lett. *6*, 171 (1970).

the smallest unit of the crystal lattice having the possible dynamic interactions; it is the vibrating unit of the lattice.

On incorporation of a molecule into the crystal lattice, some of its symmetry properties may be lost; the molecular symmetry is reduced to the symmetry of the site in the lattice. The static crystal field may lead to shifts of the molecular frequencies observed in the gaseous state, as well as to annulment of selection rules and of degeneracy. Because of the dynamic interactions of the molecules within the unit cell, every molecular vibration is split into as many components as there are molecules present in the unit cell. The resulting vibrations now each belong to different vibrational species of the unit cell[53-56].

On inclusion of the molecules into the lattice, the rotational and translational motions will become 'frozen in'. They become 'librations' (hindered rotations of whole molecules about the main inertial axes) and 'translational vibrations' (vibrations of whole molecules in the direction of the crystal axes). If one is fortunate enough to have available

single crystals of the substances to be analyzed, one may observe – using polarized radiation and with the crystal in different orientations – up to six different Raman spectra and three different infrared spectra. In combination, these make possible the assignment of all vibrations. Conclusions about the symmetry of the molecules, the crystal lattice, and the intermolecular interactions are also possible.

Fig. 10 shows three different 'views' of the totally symmetric vibrations (species A_g) of orthorhombic *thiourea*. It also shows the spectra of the other Raman active species of the unit cell B_{1g}, B_{2g}, and B_{3g}. The orientation of the crystal in the spectrometer is shown by a symbol: b(ac)a indicates that the exciting radiation was parallel to the axis b and was polarized in the a-direction; Raman radiation with the electrical vector parallel to the c-axis was observed in the direction a[57]. The symbol (ac) gives the indices of the components of the polarizability tensor under observation.

The spectra show that the symmetrical CN stretching vibration v_s (N–C–N) at 734 cm⁻¹ indicates a large change of polarizability in the

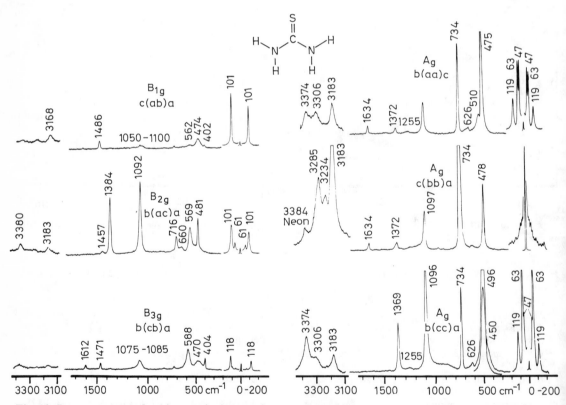

Fig. 10. Raman spectra of a thiourea single crystal at room temperature
For explanations, see text.

[53] *R.S. Halford*, J. Chem. Phys. *14*, 8 (1966).
[54] *D.F. Hornig*, J. Chem. Phys. *16*, 1063 (1948).
[55] *W. Bruhn*, Dissertation Freiburg i. Br. 1959.
[56] *B. Schrader*, Habilitationsschrift Münster 1968.
[57] *S.P.S. Porto, J.A. Giordmaine, T.Z. Damen*, Phys. Rev. *147*, 608 (1966).

direction of the b-axis of the crystal [orientation c(bb)a]. In this orientation the totally symmetrical NH_2 stretching is especially strong. 'Librations' and translational motions which appear in the other spectra in the region of $47–119 \text{ cm}^{-1}$ are missing here (in highly amplified spectra, only 'ghosts' can be observed[58]).

The ν_s (N–C–N) may be found also in the species B_{2g} [orientation b(ac)a] near 716 cm^{-1}. In the infrared spectrum, the same molecular vibration emerges in the species B_{1u} and B_{2u} at 730 and 731 cm^{-1}[59].

On the basis of a frequency calculation, the vibrations at 47 and 63 cm^{-1} in the b(aa)c and b(cc)a spectra have been assigned to translational motions (in the a- and c-directions respectively) and that at 119 cm^{-1} to a 'libration' about an axis of the molecule parallel to the N–N-direction. This calculation also shows that strong hydrogen bonds NH...S are present in the crystal lattice[60].

Shimanouchi and co-workers have developed a method for calculation of all vibrations in molecular crystals; this was applied to many substances[58, 61]. Qualitative predictions of the intensities of the molecular vibrations in the different species of the unit can be made by means of the 'oriented gas model'[62, 63]. Quantitative studies of the intensities of the lattice vibrations in molecular crystals allow determination of the main electron polarizabilities of the molecules as well as of their mutual interactions[64, 65].

7.6 Raman Spectroscopy of Polymers

In their spectra, polymers often exhibit properties of liquids and of crystals simultaneously. This can be related to the existence of amorphous and crystalline regions in the polymers. Stretched polymers strongly resemble molecular crystals. The spectra can be interpreted in terms of unit cells which may contain a few elements of different polymer chains[66]. Like single crystals, stretched polymers may be examined with polarized radiation[67]. Details of the structures of technical polymers and of copolymers can be worked out by frequency calculations using extensive computer programs[68]. Schaufele has demonstrated that the length of a polymer chain can be determined from the totally symmetric deformation motion of the chain ('accordion-motion') in the Raman spectrum[69]. Raman spectroscopy also allows a number of empirical conclusions to be reached about the structural units of macromolecules[70, 71], i.e. $C=C$ bonds (1660 cm^{-1}), benzene nuclei (1000, 1600 and 3060 cm^{-1}), the symmetric Si–O–Si vibration (490 cm^{-1}) in silicones, and the C–S and S–S vibrations (650 and 510 cm^{-1}) in polysulfides. All these vibrations give very intense Raman bands but are weak in (or absent from) infrared spectra. These bands can be used for quantitative analyses. However, serious difficulties may sometimes occur in recording Raman spectra of polymers. The samples often contain fluorescent impurities which cannot be removed by purification processes. Occasionally, photochemical purification might be useful.

[66] M. Tasumi, S. Krimm, J. Chem. Phys. 46, 755 (1967).

[67] P.J. Hendra, H.A. Willis, Chem. Comm. 225 (1968).

[68] G. Zerbi, J. Chem. Phys. 48, 3561, 3813 (1968).

[69] R.F. Schaufele, T. Shimanouchi, J. Chem. Phys. 47, 3605 (1967).

[70] R.F. Schaufele, Macromol. Rev. 3 (1968).

[71] D.S. Cain, A.B. Harvey, Raman Spectra of Polymeric Matrices, NRL Report No. 6792, Naval Research Laboratory, Washington D.C. 1968.

8 Bibliography

8.1 Monographs

[72] K.W.F. Kohlrausch, Raman-Spektren in: Hand- und Jahrbuch der chemischen Physik, Bd. IX/6, Akademische Verlagsgesellschaft Geest & Portig KG, Leipzig 1943.

[73] G. Herzberg, Molecular Spectra and Molecular Structure: I. Spectra of Diatomic Molecules; Van Nostrand, New York 1950; II. Infrared and Raman Spectra of Polyatomic Molecules; Van Nostrand, New York 1945.

[74] J.P. Mathieu, Spectres de Vibration et Symetrie des Molecules et des Cristaux; Hermann, Paris 1945.

[75] W. Otting, Der Raman-Effekt und seine analytische Anwendung, Springer Verlag, Berlin 1952.

[76] E.B. Wilson, jr., J.C. Decius, P.C. Cross, Molecular Vibrations, The Theory of Infrared and Raman Vibrational Spectra; McGraw-Hill Book Co., New York 1955.

[58] B. Schrader, W. Meier, K. Gottlieb, H. Agatha, H. Barentzen, P. Bleckmann, Z. Elektrochem. 75, 1263 (1971).
P. Bleckmann, B. Schrader, W. Meier, H. Takahashi, Z. Elektrochem. 75, 1279 (1971).

[59] K. Gottlieb, Dissertation, Münster 1968.

[60] H. Takahashi, B. Schrader, W. Meier, K. Gottlieb, J. Chem. Phys. 47, 3842 (1967).

[61] T. Shimanouchi, J. Chem. Phys. 35, 1597 (1961); 41, 2651 (1964); 43, 1245 (1965); 44, 2061 (1966); 46, 2708 (1967).

[62] G.C. Pimentel, A.C. McClellan, J. Chem. Phys. 20, 270 (1952).

[63] W.B. Person, J. Chem. Phys. 23, 230, 234 (1955).

[64] A. Kastler, A. Rousset, J. Phys. Rad. VIII/II 49 (1941).

[65] H. Barentzen, B. Schrader, Phys. Stat. Sol. 45, 505 (1971).

[77] *R.N. Jones, C. Sandorfy*, Application of Infrared and Raman Spectroscopy to the Elucidation of Molecular Structure in: Weissberger, Technique of Organic Chemistry, Vol. IX, Interscience Publishers Co., New York 1956.

[78] *S.J. Mizushima*, Raman-Effect in: Handbuch der Physik XVI, Springer Verlag, Berlin 1958.

[79] *J. Brandmüller, H. Moser*, Einführung in die Raman-Spektroskopie; Steinkopff, Darmstadt 1962.

[80] *G.M. Barrow*, Introduction to Molecular Spectroscopy, McGraw-Hill Book Co., New York 1962.

[81] Intensitätsmessungen in der Infrarot- und Raman-Spektroskopie. Abhandlungen der Deutschen Akademie der Wissenschaften zu Berlin, Akademie-Verlag Berlin 1964.

[82] *N.B. Colthup, L.H. Daly, S.E. Wiberley*, Introduction to Infrared and Raman Spectroscopy. Academic Press, New York 1964.

[83] *H. Siebert*, Anwendungen der Schwingungsspektroskopie in der anorganischen Chemie, Springer Verlag, Berlin 1966.

[84] *J. Behringer*, Theorie der molekularen Lichtstreuung. Sektion Physik der Universität München, Lehrstuhl Prof. Brandmüller 1967.

[85] *H.A. Szymanski*, Raman Spectroscopy, Theory and Practice, Plenum Press, New York, Vol. I, 1967; Vol. II, 1970.

[86] *H.A. Stuart*, Molekülstruktur, Springer Verlag, Berlin 1967.

[87] *L.J. Bellamy*, The Infrared Spectra of Complex Molecules, Methuen & Co., London 1958.
L.J. Bellamy, Advances in Infrared Group Frequencies, Methuen & Co., London 1968.

[88] *A.B. Zahlan*, Excitons, Magnons and Phonons in Molecular Crystals, Cambridge University Press 1968.

[88a] *T.R. Gilson, P.J. Hendra*, Laser Raman Spectroscopy, John Wiley & Sons, London 1970.

8.2 Spectra and Data Compilations

[89] *Landolt-Börnstein*, Zahlenwerte und Funktionen, 6. Aufl., Band I/2 und I/3, Molekeln I und II, Springer Verlag, Berlin 1951.

[90] *Landolt-Börnstein*, Zahlenwerte und Funktionen, 6. Aufl., Bd. I/4, Kristalle, Springer Verlag, Berlin 1955.

[91] American Petroleum Institute Research Project 44, Catalog of Raman, Infrared, Ultraviolet and Mass Spectral Data, API Data Distribution Office, A & M Press, College Station, Texas, USA.

[92] DMS fortlaufende Titellisten, Current Literature Lists, Infrarot, Raman, Mikrowellen, mit Junior- und Generalindex, I, 1963; II, 1965; III, 1967; IV, 1969; V, 1971; Verlag Chemie, Weinheim/Bergstr.; Butterworths & Co., London.

[93] Schnellinformation über Infrarot- und Raman-Spektroskopie, Deutsche Akademie der Wissenschaften, 1199 Berlin-Adlershof, Rudower Chaussee 5.

[94] DMS Raman-Atlas, Verlag Chemie, Weinheim; Butterworths & Co., London, unpubl.

8.3 Review Articles

[95] *J. Goubeau*, Raman-Spektroskopie, in *Houben-Weyl*, Methoden der organischen Chemie, 4. Aufl., Band III/2, Georg Thieme Verlag, Stuttgart 1955.

[96] *H. Kienitz*, Ultrarot- und Raman-Spektroskopie, Ullmanns Enzyklopädie der technischen Chemie, 3. Aufl., Bd. II/1, p. 236, Urban & Schwarzenberg, München, Berlin 1961.

[97] *A. Weber*, Raman Spectroscopy Revisited, Spex Speaker *11*, 4 (1966).

[98] *S. Califano*, Applicatione dei Laser in Spettroscopia Molecolare, Alta Frequenza *15*, 54 (1966).

[99] *R.N. Jones, M.K. Jones*, Raman-Spectrometry, Anal. Chem. Reviews *38*, 393 R (1966); *R.E. Hester*, Raman Spectrometry, Anal. Chem. Reviews *42*, 235 R (1970); *44*, 490 R (1972).

[100] *E. Mollwo, W. Kaule*, Maser und Laser, Hochschultaschenbuch 79/79A, Bibliogr. Inst. Mannheim (1966).

[101] *J. Brandmüller*, Raman-Spektren mit Lasern I. Linearer Raman-Effekt, Naturwissenschaften *54*, 293 (1967).
H.W. Schrötter, Raman-Spektren mit Lasern II. Nichtlinearer Raman-Effekt, Naturwissenschaften *54*, 607 (1967).

[102] *E. Ziegler, E.G. Hoffmann*, Fortschritte in der Raman-Spektroskopie, Öst. Chem.-Ztg. *68*, 319 (1967).

[103] *B. Schrader*, Fortschritte in der Technik der Raman-Spektroskopie, Chem.-Ing.-Tech. *39*, 1008 (1967).

[104] *P. Schorygin, L. Kruschinskij*, Zur Theorie des Raman-Effektes, Ber. Bunsenges. Phys. Chem. *72*, 495 (1968).

[105] *S.P.S. Porto*, Light Scattering with Laser Sources, I., II., Spex Speaker *13*, 2 (1968); *14*, 2 (1969).

[106] *M. Delhaye*, Rapid Scanning Raman-Spectroscopy, Applied Optics 7, 2195 (1968).

[107] *H.J. Becher*, Kraftkonstantenberechnungen aus den Schwingungsspektren einfacher organischer Moleküle, Fortschr. chem. Forsch. *10*, 156 (1968).

[108] *P.J. Hendra, P.M. Stratton*, Laser Raman Spectroscopy, Chem. Rev. *69*, 325 (1969).

8.4 Intensities

[109] *D.G. Rea*, Study of the Experimental Factors Affecting Raman Band Intensities in Liquids, J. Opt. Soc. Am. *49*, 90 (1959).

[110] *H.W. Schrötter, H.J. Bernstein*, Absolute Intensities for Some Gases and Vapors, J. Mol. Spectrosc. *13*, 430 (1964).

[111] *G. Eckhardt, W.G. Wagner*, On the Calculation of Absolute Raman Scattering Cross Section from Raman Scattering Coefficients, J. Mol. Spectrosc. *19*, 407 (1966).

[112] *H.J. Höfert*, Die «Lichtstärke» von Monochromatoren, Sonderdruck p. 50/659d, Carl Zeiss, Oberkochen 1966.

[113] *B. Schrader, G. Bergmann*[27].

[114] *F. Gandini, H. Moser, F. Perzl*, Absolute Intensitätsmessungen an Raman-Kristallspektren, Z. ang. Phys. *23*, 461 (1967).

[115] *M.C. Tobin*, Laser Raman Spectroscopy of Crystal Powder, J. Opt. Soc. Am. *58*, 1057 (1968).

[116] *J.G. Skinner, W.G. Nilsen*, Absolute Raman Scattering Cross Section Measurement of the 992 cm^{-1} Line of Benzene, J. Opt. Soc. Am. *58*, 113 (1968).

[117] *J. Tang. A.C. Albrecht*, Studies in Raman Intensity Theory, J. Chem. Phys. *49*, 1144 (1968).

[118] W.R. Hess, H. Hacker, H.W. Schrötter, J. Brandmüller, Zur Bestimmung von Intensität und Depolarisationsgrad der Raman-Linien bei Laser-Erregung, Z. Ang. Physik 27, 233 (1969).

[119] W.F. Murphy, W. Holzer, H.J. Bernstein, Gas Phase Intensities: A Review of Pre-Laser Data, Appl. Spectroscopy 23, 211 (1969).

8.5 Characteristic Bands

[120] D.G. Rea, Some Refinements in the Correlations of the Raman Spectra with Molecular Structures for Acyclic Monoolefine Hydrocarbons, Anal. Chem. 32, 1638 (1960).

[121] G. Kresze, E. Ropte, B. Schrader, IR- und Raman-Spektren von Methylarsulfiden, -sulfoxiden und -sulfonen, Spectrochim. Acta 21, 1633 (1965).

[122] H. Kriegsmann, R. Hess, P. Reich, O. Nillins, Aussagen aus Intensitätsmessungen in der Infrarot- und Raman-Spektroskopie zu Konstitutions- und Bindungsfragen, Z. Chem. 7, 449 (1967).

[123] D.S. Cain, A.B. Harvey, Raman Spectroscopy of Polymeric Materials, NRL Report 6792, Naval Research Laboratory, Washington D.C. 1968.

[124] G. Zerbi, G.J. Hendra, Laser Excited Raman Spectra of Polymers, Hexagonal and Orthorhombic Polyoxymethylene, J. Mol. Spectrosc. 27, 17 (1968).

[125] J.L. Koenig, F.J. Boerio, Raman Scattering and Band Assignments in Polytetrafluoroethylene, J. Chem. Phys. 50, 2823 (1969).

[126] J.R. Durig, W.H. Green, Vibrational Spectra and Structure of Small Ring Compounds, Spectrochim. Acta 25A, 849 (1969).

[127] P.J. Park, E. Wyn-Jones, Infrared and Raman Studies on Meso and (+) 2,3-Dichloro and 2,3-Dibromobutanes, J. Chem. Soc. A, 422 (1969).

[127a] H.A. Szymanski, Correlation of Infrared and Raman Spectra of Organic Compounds, Hertillon Press, Cambridge Springs, Penn. 1969; B. Schrader, W. Meier, Z. Anal. Chem. 260, 248 (1972).

8.6 Crystals

[128] R.S. Halford, Motions of Molecules in Condensed Systems; I. Selection Rules, Relative Intensities, and Orientation Effects for Raman and Infrared Spectra, J. Chem. Phys. 14, 8 (1946).

[129] D.F. Hornig, The Vibrational Spectra of Molecules and Complex Motions in Crystals: I. General Theory, J. Chem. Phys. 16, 1063 (1948).

[130] A.C. Menzies, Raman-Effect in Crystals, Rep. Progr. Physics 16, 83 (1952).

[131] W. Vedder, D.F. Hornig, Adv. Spectrosc. 2, 189 (1961).

[132] D.A. Dows, Infrared Spectra of Molecular Crystals, Chapter 11, in: Physics and Chemistry of the Organic Solid State, Interscience Publishers, New York 1963.

[133] R. Loudon, The Raman Effect in Crystals, Adv. Physics 13, 423 (1964).

[134] S.S. Mitra, P.J. Gielisse, Infrared Spectra of Crystals, in: Progress in Infrared Spectroscopy, Vol. 2, Plenum Press, New York 1964.

[135] J.C. Decius, Dipolar Coupling and Molecular Vibrations in Crystals, J. Chem. Phys. 49, 1387 (1968).

[136] B. Schrader[56].

[137] C.A. Frenzel et al., Raman Spectrum of a Crystalline Polyethylene, J. Chem. Phys. 49, 3789 (1968).

[138] L. Colombo, Low Frequency Raman Spectrum of Imidazole Single Crystals, J. Chem. Phys. 49.

[139] M. Suzuki, M. Ito, Polarized Raman Spectra of p-Di-Chlorobenzene and p-Dibromobenzene Single Crystals, Spectrochim. Acta, A, 25, 1017 (1969).

5.4 Neutron Spectroscopy

Hans Stiller

Institut für Festkörperforschung, Kernforschungsanlage D-517 Jülich

1 Introduction

The scattering of neutron radiation differs from that of electromagnetic radiation in two respects. Firstly, neutrons are scattered predominantly by atomic nuclei, seldom by electron clouds. Scattering in electron clouds occurs only as a result of interaction with the magnetic moment of the neutron, i.e. it takes place solely with atoms which also have a magnetic moment, and then only by those electrons which contribute to this moment. This 'magnetic scattering' will not be considered here. Nuclear scattering takes place with all atoms. Its observation becomes difficult only when the probability of neutron absorption by a nucleus

is high compared with the probality of scattering. This applies to neutrons with energies less than 0.3 eV with the elements Li, B, Rh, Ag, Cd, In, Pm, Sm, Eu, Gd, Dy, Ir, Au, Hg and Ac. As scattering takes place at the nuclei, and the range of the nuclear forces is small compared with the wavelength of low energy neutrons, scattering does not involve an atomic form factor. For the same reason, there are no forbidden transitions.

Secondly, for neutrons the relationship between their energy E and wavelength λ differs from that which holds for electromagnetic radiation, i.e.

$$\lambda = \frac{h}{mv} = \frac{h}{\sqrt{2mE}} \tag{1}$$

where m is the mass of the neutron and v its velocity. From this relationship, neutrons having energies between 0.001 and 0.3 eV (8.04 to

2412 cm^{-1}) – which can be obtained as a reactor beam – have wavelengths between 9 and 0.5 Å. Their energies are comparable to the kinetic energies of atoms in matter. Energy changes which occur as a result of exchanges of energy with atomic motions can be measured easily, essentially as in Raman spectroscopy.

However, in contrast to Raman spectroscopy, the wavelengths are of the order of 1 Å, rather than 10^4 Å. With these short wavelengths, it is possible to use neutron spectroscopy to gain information on the scattering system beyond that obtainable by optical spectroscopy.

If we characterize a wave incident upon a sample by a wave vector $\underline{k}_0$ ($|\underline{k}_0| = 2\pi/\lambda_0$) and a frequency ω_0, and the scattered waves by wave vectors k_1 ($|\underline{k}_1| = 2\pi/\lambda_1$) and frequencies ω_1, then the scattering is described by scattering vectors

$$\kappa = \underline{k}_0 - \underline{k}_1 ,$$
$$\kappa^2 = k_0^2 + k_1^2 - 2\,k_0\,k_1 \cos 2\,\theta =$$
$$= 4\,\pi^2 \left[\left(\frac{1}{\lambda_0} - \frac{1}{\lambda_1} \right)^2 + \frac{4}{\lambda_0 \lambda_1} \sin^2\theta \right] \tag{2}$$

where 2θ is the scattering angle (cf. Fig. 1), and by frequency changes

$$\pm\, \omega = \omega_0 - \omega_1 = \frac{E_0 - E_1}{\hbar}$$

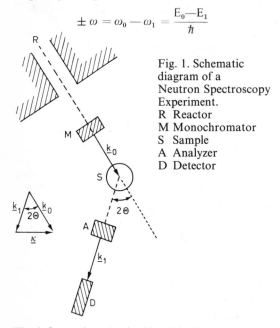

Fig. 1. Schematic diagram of a Neutron Spectroscopy Experiment.
R Reactor
M Monochromator
S Sample
A Analyzer
D Detector

The information obtainable with electromagnetic radiation is limited: With X-rays, only the distribution of scattering vectors, $S(\kappa)$, can be obtained since the frequencies are too high to measure frequency changes. With light, information can be obtained only on the distribution $S(\omega)$ of frequency changes or absorptions at very small κ ($\kappa \lesssim 10^{-3}$ Å^{-1}), because of the very long wave-

lengths of light. In contrast, with neutrons we can measure a distribution $S(\kappa, \omega)$, since (from Equation 1):

$$\pm\, \hbar\, \omega = \frac{\hbar^2}{2\,\mathrm{m}} (k_0^2 - k_1^2) , \tag{3}$$

Fig. 3 shows the ranges of κ and ω accessible with different experimental techniques (cf. Fig. 2).

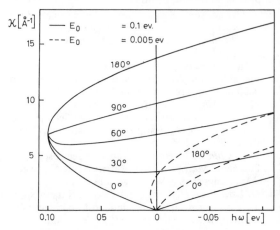

Fig. 2. The crossing experimental paths in the κ–ω plane for fixed incident energy E_0 and fixed scattering angle 2θ. The curves represent equations (2) and (3) after elimination of κ_1. The curves $2\theta = 0°$ and $2\theta = 180°$ limit the area measurable by the given incident energy.

Information about a scattering system is, in principle, always obtained by *Fourier-transformations* of measured distributions S. The corresponding Fourier variables are, on the one hand, the scattering vectors $\underline{\kappa}$ and the particle separations, $\underline{r}$, and, on the other hand, the frequency changes ω and the time intervals t. Thus, a measured distribution $S(\kappa)$ yields a space correlation function g(r). g(r)dr is the probability that in an element of volume dr, another particle will be found within a distance $\underline{r}$ from a given particle. However, if (as with X-ray diffraction) the measurements are integrated over all frequency changes ω, then this space correlation relates to a single moment in time: $g(r)/_{t=0}$: nothing is learned about the motions of the particles. Conversely, a measured distribution $S(\omega)$ provides a function g(t) which describes the temporal behaviour of the scattering particles. However if (as with light scattering) measurement is limited to $\kappa \leq 10^{-3}$ Å^{-1}, then observation is spatially averaged over regions of linear dimension 10^3Å: $\langle g(t) \rangle_r >_{10^3}$Å; for smaller distances, nothing is learned about the spatial behaviour of the particles. In contrast, a distribution $S(\kappa, \omega)$ which can be measured by neutron spectroscopy yields a function G(r,t). Classically, $G(\underline{r}-\underline{r}_0, t-t_0)$ represents the probability that an atomic nucleus will be found at time t at $\underline{r}$ if at time t_0 there was a nucleus at $\underline{r}_0$. The cross-section for single scattering into the solid angle $d\Omega$ with a frequency change between ω and $\omega + d\omega$ is[1] (with $\underline{r}_0 = t_0 = 0$):

$$\frac{d\sigma}{d\,\Omega\,d\omega} = \frac{1}{4\pi}\frac{k_1}{k_0}\left(\sigma_{\mathrm{inc}}\,S_{\mathrm{inc}}\,(\underline{\kappa}, \omega) + \sigma_{\mathrm{coh}}\,S_{\mathrm{coh}}\,(\underline{\kappa}, \omega) \right)$$

$$S_{\mathrm{inc}}\,(\underline{\kappa}, \omega) = \frac{1}{N} \int G_s\,(\mathrm{r}, t)\, e^{i(\underline{\kappa}\underline{r} - \omega t)}\, d\,\underline{r}\,d\,t \tag{4}$$

$$S_{coh}(\kappa, \omega) = \frac{1}{N} \int \{G_s(\underline{r}, t) + G_d(r, t)\} \, e^{i(Kr - \omega t)} \, d\underline{r}\, dt$$

σ_{inc} is the scattering cross-section for incoherent scattering,

σ_{coh} is the cross section for coherent scattering:

$$\sigma_{inc} = 4\pi (\langle a^2 \rangle - \langle a \rangle^2), \quad \sigma_{coh} = 4\pi \langle a \rangle^2$$

$$\langle a^2 \rangle = \frac{1}{N} \sum_\nu N_\nu a_\nu^2, \quad \langle a \rangle^2 = \frac{1}{N^2} (\sum_\nu N_\nu a_\nu)^2 \qquad (5)$$

where a_ν is the scattering amplitude, N_ν the number of nuclei of type ν, with $\sum N_\nu = N$.

$G = G_g + G_d$, where classically $G(\underline{r}, t)$ is the probability that the same nucleus which at time 0 was at $\underline{r} = 0$ will at time t be at r;

$G_d(r, t)$ is the probability that a nucleus will at time t be at r, if a different nucleus was at 0 at time 0.

Neutron spectroscopy thus yields information about the space-time behaviour of the scattering particles (See Sections 5.4:3 and 5.4:4).

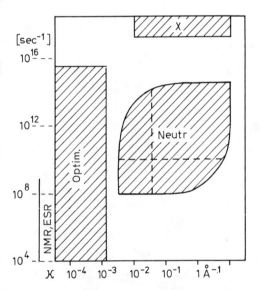

Fig. 3. The areas of κ and ω obtained by various methods. The marked neutron regions ($\kappa < 5.10^{-2}\,\text{Å}^{-1}$, $\omega < 10^{10}\,\text{sec}^{-1}$) are obtained only by special techniques and generally with very low intensities.

X   = Roentgen Region
Neutr. = Neutron Region
Opt. = Optical Spectroscopy

2 Experimental Methods

Equation (4) shows schematically the method of cross-section measurement[4]. Neutrons of a desired wave vector κ_0 are selected from a reactor beam, by means of a monochromator (Fig. 1). The angular distribution (2θ) and the distribution of wavelengths $\lambda_1 = 2\pi/k_1$ in the scattered radiation are measured. The distribution $S(\underline{\kappa}, \omega)$ is then derived from Equations (2), (3) and (4).

[1] L. Van Hove, Phys. Rev. 95, 249 (1954).

2.1 Analyzers and Monochromators

The distribution of wavelengths is generally analyzed either by reflecting the scattered beam from a single crystal, i.e, by variation of a Bragg angle θ_1 (crystal spectrometer): $n\lambda_1 = 2d_1 \sin\theta_1$ ($n = 0, 1, 2 \ldots$; d_1 is the separation between the reflecting planes of the analyzer crystal), or by measuring the velocities v_1 of the scattered neutrons (and using Equation 1). The velocities are found by measuring the times t_1 which the neutrons require to travel along a given flight path L (time-of-flight spectrometer): $v_1 = L/t_1$. For such measurements, the neutron beam must be pulsed, in order to define the starting time. The neutron detectors are connected to an electronic time-of-flight analyzer, which stores the charge pulses created by the neutrons in a number of channels which open consecutively.

Selection of a defined wavelength λ_0 from the primary reactor beam is likewise generally achieved either by Bragg reflection from a single crystal: $n\lambda_0 = 2d_0 \sin\theta_0$, or by selection of a velocity v_0. In order to effect velocity selection, rotating curved slits can be introduced into the beam. If the slits rotate with the surface of a cylinder whose axis is parallel to the beam, a continuous monochromatic beam is obtained. If the slit path is that of a cylinder with diameter perpendicular to the beam (a chopper), the beam is pulsed at the same time. Clearly, the beam may also be pulsed by rotation of a single crystal, which comes into Bragg position only periodically. All of these monochromators and analyzers can be used in various combinations[2-4]. For measurements in a continuous beam, the most common combination is of two crystals (triple axis spectrometer; cf. Fig. 4); for time-of-flight measurements, either a rotating crystal (Fig. 5) or a beam chopper system is used.

2.2 Intensity and Resolution

If absorption is neglected, the number of scattered neutrons detected per unit time is:

$$\frac{dn_1}{dt} = \frac{dI_0(E_0)}{dE_0} \cdot F \cdot \vartheta_0(E_0) \cdot dE_0 \cdot d\Omega_0 \cdot N \cdot D \cdot$$
$$\frac{d\sigma}{d\Omega\, d(\hbar\omega)} \cdot \vartheta_1(E_1) \cdot dE_1 \cdot d\Omega_1 \cdot \varepsilon(E_1) \; s \cdot p \cdot \frac{d\tau}{\tau},$$

[2] B. Brockhouse, S. Hautecler, H. Stiller in Ref. [41], p. 580–642.
[3] R. Brugger in Ref. [42], Chapter 2.
  P. Iyengar in Ref. [42], Chapter 3.
[4] H. Maier-Leibnitz, T. Springer, Nucl. Sci. Eng. 16, 207 (1966).

where $I_0(E_0)$ is the primary intensity, F the effective beam cross-section, $\theta_0(E_0)$ the monochromator transmission, $\theta_1(E_1)$ the analyzer transmission, dE_0 the energy width transmitted by the monochromator, dE_1 that transmitted for E_1 by the analyzer, $d\Omega_0$ the solid-angle seen by the monochromator, $d\Omega_1$ the solid-angle seen by the analyzer, N the density of the scattering nuclei, D the effective sample thickness, (E_1) the detector sensitivity, $d\tau/\tau$ the relative interference time when the beam is pulsed, p the number of energy changes measurable between two interferences, and s the number of scattering angles measured simultaneously (dE_1, $d\Omega_1$ and D are assumed to be the same for all scattering angles).

The instrumental resolutions in κ and ω depend upon dE_0, dE_1, $d\Omega_0$ and $d\Omega_1$. If we designate the beam directions by Z, $d\Omega = dk_x dk_y/k^2$. With selection and analysis of velocities, the resolution in ω depends primarily on dk_z:$dE/E = 2dk_z/\kappa$. The $d\Omega$ are thus only important for the resolution in ω. With crystal spectrometers[5,6], the resolutions in Ω and ω are interdependent: $dk_v/k_v = \mathrm{cotg}\,\theta_v \cdot d\theta_v$, where $d_v = dk/_v$. With time-of-flight spectrometers, dk_1 depends also on d_v and on uncertainties in the length of the flight path (e.g. the sample thickness). With rotating crystals, velocity changes through Doppler shifts in Bragg reflection may become important.

Primary intensities available with research reactors are very low, compared with X-ray tubes and with the light sources used in optical spectroscopy. In general, the velocity spectrum in a reactor beam is approximately a Maxwell distribution:

$$\frac{d\,I_0(E_0)}{d\,E_0} = \frac{\Phi}{2\pi}\left(\frac{E_0}{(k_B T)^3}\right)^{\frac{1}{2}} \exp\left[-\frac{E_0}{k_B T}\right]$$

where Φ is the thermal neutron flux, and T the temperature of the neutron-moderating material in the reactor. With most of the reactors which are at present used for neutron spectroscopy, Φ is of the order of 10^{14} cm^{-2} sec^{-1}. Only in very-high flux reactors values of ca 10^{15} cm^{-2} sec^{-1} have been reached. This low primary intensity is the main reason for limitations in the applicability of neutron spectroscopy which must simultaneously resolve more variables than X-ray diffraction or lightmicroscopy.

Most measurements are very time-consuming, and require considerable effort to reduce background, as well as large samples (F.D) and high densities (N). Useful values for N.D are limited, if multiple scattering is to be kept low. This means that there is a lower limit for concentrations of solutions. The low primary intensity

also has a limiting effect upon the resolution obtainable. The limits of the κ-ω region accessible in neutron spectroscopy (as given in Fig. 3) are intensity-determined.

2.3 Comparison of Methods

Of the two instruments represented in Fig. 4 and 5, the *triple axis spectrometers* have the advantage that $|k_0|$, $|k_1|$, the scattering angle 2θ and (with single crystal or fibrous samples) the sample orientation relative to k_0 can be varied simultaneously between consecutively measured points; whereas, with *time-of-flight spectrometers* data are always gathered as functions of $|k_1|$ (or, if the time-of-flight path is placed in front of the sample with fixed crystals behind it, as functions of $|k_0|$). With triple axis spectrometers, it is consequently possible[2] to conduct experiments along any desired

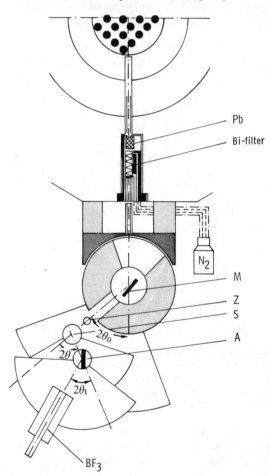

Fig. 4. Triple Axis Spectrometer of Reactor FRJ-2 in Jülich.

The angles θ_0, 2θ and θ_1 and the sample orientation relative to the entering radiation can be varied simultaneously and automatically.

Pb Lead Enclosure	Z Monitor Counter
Bi Bismuth filter cooled	A Analyzer
with liquid Nitrogen	BF$_3$ Borontrifluoride
M Monochromator	Counter

---

[5] *M. Cooper, R. Nathans*, Acta Crystallogr. *23*, 357 (1967).

[6] *B. Dorner, H. Stiller* in Ref.[43], p. 19.

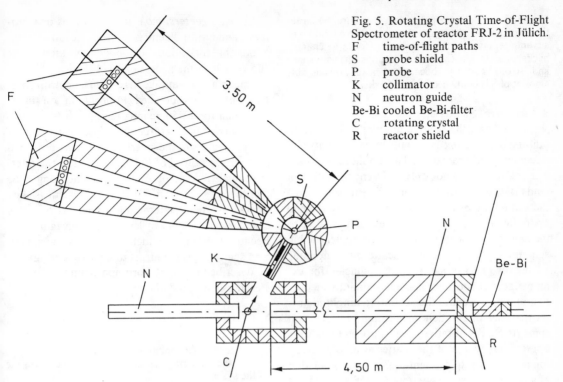

Fig. 5. Rotating Crystal Time-of-Flight Spectrometer of reactor FRJ-2 in Jülich.
F   time-of-flight paths
S   probe shield
P   probe
K   collimator
N   neutron guide
Be-Bi cooled Be-Bi-filter
C   rotating crystal
R   reactor shield

path in κ-ω; this is of special importance with regard to coherent scattering (Section 5.4:4). The propagation of collective atomic motions can be investigated along preselected directions, for example along the directions of symmetry of a crystal, or along the axis of a polymer. Furthermore, the combined instrumental spread in k_0 and k_1 may be orientated in a preferred way, so as to obtain high resolution primarily in directions which are important for interference scattering[5, 6]. For this purpose advanced instruments are equipped with curved crystals and permit a variation not only of angles, but also of the distances between monochromator, sample and analyzer[7].

Time-of-flight spectrometers have the advantage that many scattering angles can be studied simultaneously (e.g. s = 20), although usually not with equal resolution; whereas with a triple axis spectrometer s ≤ 4 in practice. This disadvantage of the triple axis spectrometer is partially compensated by $\rho \cdot d\tau/\tau = 1$ (continuous beam), for a periodically-pulsed time-of-flight instrument, $d\tau/\tau$ is generally $\lesssim 5.10^{-3}$, because $d\tau/\tau$ must be such that the fastest neutrons in one interference group do not reach the detectors earlier than the slowest neutrons from the preceding one.

This loss of intensity can be compensated for by high p only if very many values are of interest (e.g. p = 200). In order to improve utilization of time (up to $d\tau/\tau = 0.5$), so-called 'correlated choppers' have recently been developed[8]. With these, the beam is not opened periodically, but in a preselected statistical sequence $S_c(t)$. The distribution determined in the time-of-flight analyzer is then

$$S_m(t) = \int dt' \, S_c(t-t') \, S(t') + U$$

where $S(t)$ is the scattering function on a time scale, and U the background per channel. If, for instance, $S_c(t)$ is a sequence of N functions $F(i)$, defined as +1 for open-beam transmission and as −1 for zero transmission, then with a time-of-flight analyzer (also of N channels) the count rate in channel i is:

$$S_m(i) = \sum_{i=0}^{N-1} \tfrac{1}{2} \, [F(i-t)+1] \, S(t) + U$$

and one obtains:

$$\sum_{i=0}^{N-1} F(i-t) \, S_m(i) = \frac{N+1}{2} \, S(t)$$

when the series $F(i)$ is statistical: $\sum_{i=0}^{N-1} F(i) \, F(i+j) =$

$= N$ for $j = 0$ and $= -1$ otherwise.

However, the total scattering $S = \sum_{t=0}^{N-1} S(t)$ now appears in each channel.

The relative error for $S(t)$ is

$$\frac{\Delta S(t)}{S(t)} = \frac{\left(\dfrac{N+1}{2} S + N \, U \right)^{\frac{1}{2}}}{\dfrac{N+1}{2} S(t)} ;$$

[7] *H. Maier-Leibnitz*, unpubl.
[8] Inelastic Scattering of Neutrons, International Atomic Energy Agency, Vol. II, Vienna 1968.

The total number of counts must, therefore, be larger than with periodic beam interferences. The method is of advantage in two particular cases: (a) for large channel background U, and (b) for distributions S(t) with sharp and strong maxima, because then a considerable amount of S is contained in such peaks.

3 Incoherent Scattering

With neutrons, the relative amount of incoherent ('diffuse') scattering is usually greater than with electromagnetic radiation. The amplitudes of nuclear scattering are not only different for different kinds of atoms, but also for different isotopes of the same type of atom. In addition, they depend upon the orientation of the nuclear spin relative to the spin of the neutron, so that for nuclei having spin there may be several scattering amplitude values for one isotope. Hydrogen nuclei, for example, scatter neutrons in a predominantly incoherent manner, because the scattering amplitude a_+ for parallel orientation of proton and neutron spins (triplet) is negative, and its absolute value is approximately 1/3 of the scattering amplitude, a_-, for antiparallel spin orientation (singlet), so that $N_- a_- \simeq -N_+ a_+$ (see Equation 5). Of the natural elements, hydrogen and vanadium scatter almost entirely incoherently; some others (for example, carbon and oxygen) scatter almost completely coherently. For most elements, scattering is partly coherent and partly incoherent (cf. Ref.[9]). The incoherent scattering may be observed separately, if it is possible to avoid fulfilling the interference conditions which govern coherent scattering (cf. Section 5.4:4).

Incoherent inelastic scattering of neutrons can be regarded as directly complementing optical spectroscopy. Measurements are extended to optically-forbidden transitions and to larger values of κ (Fig. 3). If data from neutron spectroscopy are compared with optical data, assignment of observed transitions may be facilitated because of the absence of selection rules for the neutron results — particularly if use is also made of the large proton-neutron scattering cross-section. The high magnitude of this cross-section may permit separate observation of proton motions.

For these reasons neutron spectroscopy has recently been used also for investigations of internal molecular modes of motion[10]. In general, however, the main field of application is the investigation of external modes of motion: of translations of molecular mass-centers, molecular rotations or vibrations, and intramolecular or intermolecular fluctuations or migrations of individual atoms. As shown below, the accessibility of large κ values is an important advantage here. Thus, neutron spectroscopy yields detailed information on intermolecular forces: ionic interactions, *multipole forces, Van der Waals forces,* and *hydrogen bonds.* If the frequencies of the external modes are not small compared to those of the internal ones, the intermolecular forces, of course, can no longer be considered separately. The entire system is then treated as one of interacting atoms. Also in this case the accessibility of large κ is important.

Because the range of nuclear forces is short, neutron scattering (in contrast to light scattering) can be described in a first Born approximation with a Fermi pseudo-potential[11]:

$$S_{inc}(\underline{\kappa}, \omega) = \sum_{n_0} \rho_{n_0} \sum_n \sum_j |\langle n | e^{i\underline{\kappa}\underline{r}_j} | n_0 \rangle|^2 \delta(\omega + 2\pi \nu_{n_0 n})$$

(6)

ρ_{n_0} is the thermal condition of the initial state $|n_0\rangle$ of the scattering system, and $|n\rangle$ of the final state of the system;

$\underline{r}_j$ is the position of the jth nucleus.

As with Raman scattering, the δ-function gives the energy state:

$$\pm \hbar\omega = \frac{\hbar^2}{2m}(k_0^2 - k_1^2) = \varepsilon_n - \varepsilon_{n_0} = h\nu_{nn_0}$$

(7)

when ε_{n_0} is the energy of $|n_0\rangle$, ε_n is the energy of $|n\rangle$. From equation (6) follows:

$$S_{inc}(\underline{\kappa}, \omega) = \sum_{n_0} \rho_{n_0} \sum_n \sum_j \frac{1}{2\pi} \int dt\, e^{-i\omega t} \langle n | e^{i\underline{\kappa}\underline{r}_j(0)} | n_0 \rangle$$
$$\langle n_0 | e^{-i\underline{\kappa}\underline{r}_j(t)} | n \rangle$$
$$= \sum_j \frac{1}{2\pi} \int dt\, e^{-i\omega t} \langle e^{i\underline{\kappa}\underline{r}_j(0)} \bar{e}^{-i\underline{\kappa}\underline{r}_j(t)} \rangle .$$

The second of the Equations (4) follows from this.

3.1 Bound Motions

If all nuclei in the system have fixed equilibrium positions, and their movements consist of displacements $\underline{u}_j$ from these positions (as with a crystal in which all motions are mutually coupled): $\underline{r}_j(t) = \underline{r}_j^0 + \underline{u}_j(t)$, then by development in these displacements, we obtain:

$$S_{inc}(\underline{\kappa}, \omega) = \frac{1}{2\pi} \sum_j \int dt\, e^{-i\omega t} e^{-\langle (\underline{\kappa}\underline{u}_j(0))^2 \rangle} \times$$
$$\times (1 + \langle \underline{\kappa}\,\underline{u}_j(0)\underline{\kappa}\underline{u}_j(t) \rangle + \cdots) .$$

(8)

[9] Brookhaven National Laboratory Rep. Nr. 325, 1964–66, Supplement Nr. 2.
[10] P. Reynolds, J. White, Disc. Faraday Soc. 48, 131 (1969).
[11] V. Turchin, Slow Neutrons, Israel Program for Sci. Transl. 1965.

In harmonic approximation, the deviations are:[12]

$$u_j(t) = \sum_{s,q} \frac{c_{js}}{\sqrt{2\pi v_s(\underline{q})}} \, e_{js}(\underline{q})$$

$$\left\{ A_{s\underline{q}} \, e^{i(\underline{q}\underline{r}_j^0 - 2\pi v_s(q)t)} \quad - A_{s\underline{q}}^+ \, e^{i(\underline{q}\underline{r}_j^0 + 2\pi v_s(q)t)} \right\}$$

Here, q is the wave-vector of a collective motion; the index s specifies the mode, C_{js} is a normalization factor, and e (q) is a unit-vector in the direction of polarization of the motion. In the approximation (8), we then obtain:

$$S_{inc}(\underline{\kappa},\omega) = \sum_j e^{-2W_j(\underline{\kappa})} \left\{ \delta(\omega) + \sum_{\underline{q},s} C_{sj} \frac{|\underline{\kappa} \cdot e_{sj}(\underline{q})|^2}{2\pi v_s(\underline{q})} \times \right.$$

$$\left. \times \sum_{\gamma = \pm 1} \left(\coth \frac{h v_s(q)}{2 k_B T} + \gamma \right) \delta(\omega + \gamma 2\pi v_s(\underline{q})) \right\}, \qquad (9)$$

where for the brackets ≪...≫ the thermal condition of the initial state is given explicitly as:

$$2 W_j(\kappa) = \frac{1}{2\pi} \sum_{\underline{q},s} C_{sj}^2 \frac{|\underline{\kappa} \cdot e_{sj}(\underline{q})|^2}{v_s(\underline{q})} \coth \frac{h v_s(q)}{2 k_B T} .$$

We thus have:
the *elastic scattering* $\delta(\omega)$, and
a *spectral distribution* given by summation over all wave vectors *q*.

This distribution will show maxima whenever a particularly large number of *q* fall into an interval of frequencies $v_s(q)$. Maxima are found mostly at the zone boundaries, i.e. in the region of maximum *q*; for acoustic modes, for example, at the Debye frequency. Neutron spectroscopy thus yields frequency distributions averaged over all wavelengths especially for short wavelength modes. This is in contrast to the frequency distributions measurable by optical spectroscopy which, for single excitations, are determined by the long wavelength modes (q→0) alone (because $\kappa \simeq 0$). Truly normalised densities g(*v*) are directly obtainable only when the system has cubic or higher symmetry, because then, averaged over all $v_s(\underline{q})$, $|\kappa \cdot e(q)|^2 = 1/3 \kappa^2$. In this case,

$$S_{inc}(\underline{\kappa},\omega) = e^{-2W(\underline{\kappa})} \left\{ \delta(\omega) + \right.$$

$$\left. + \frac{\hbar}{2M} \kappa^2 \frac{g(\omega)}{\omega} \sum_{\gamma = \pm 1} \left(\coth \frac{h\omega}{2 k_B T} + \gamma \right) \right\}, \qquad (10)$$

M is the Mass of the scattered Atoms 2 $W_{(\kappa)} = \kappa^2 \langle u^2 \rangle$.

Sometimes, it is possible to derive at least an approximate g(*v*) for other structures also. Such possibilities must be considered in individual cases in terms of the particular *lattice* and *molecular*

symmetries involved[13, 14]. If the symmetries do not permit any such direct determination, in order to obtain a quantitative interpretation of the measured $S_{inc}(\kappa,\omega)$, there remains only the possibility of comparison with theoretical models for the interatomic forces operating in the scattering system — calculating the scattering function (9) from the model[15]. As the neutron-proton scattering is incoherent and of particularly high cross-section, investigations have been carried out mostly with molecular systems which contain hydrogen[16].

There is an interesting possibility, which has not yet been investigated, that measurements might be carried out with spin-polarized neutrons[17]. This technique may permit studies of the degeneracies of hindered rotational states, owing to their occupations of different nuclear spin states. A neutron may flip a nuclear spin 1/2, if it has an orientation opposite to that of the neutron spin.

3.2 Unbound Motions

If nuclei are able to move away from their equilibrium positions, or if these sites are themselves moving, then the elastic line and the spectrum of inelastic scattering become broadened. These broadenings are best discussed in terms of the self correlation function, $G_s(\underline{r},t)$, defined by Equation (4). For example, if the molecules are diffusing, the elastic line becomes a Lorentzian,

$$S_{inc}(\underline{k},\omega) = \frac{\kappa^2 D}{\omega^2 + (\kappa^2 D)^2}$$

(where D is the self-diffusion constant) if the diffusion is continuous, i.e. assuming G_s to obey a simple diffusion law[18]. In the case where (for short distances and times) the diffusion is assumed to take place by discontinuous jumps between equilibrium sites[19], a different expression holds:

$$S_{inc}(\kappa,\omega) = \frac{\tau_0 e^{-2W(\kappa)} [1 - e^{-2W(\kappa)} (1 + \kappa^2 D \tau_0)^{-1}]}{\omega^2 \tau_0^2 + [1 - e^{-2W(\kappa)} (1 + \kappa^2 D \tau_0)^{-1}]^2} .$$

Here, τ_0 is the mean duration time of a molecule between jumps, and $W(\underline{\kappa})$ is the Debye-Waller exponent describing the molecular thermal cloud at the equilibrium positions.

[12] *G. Leibfried*, Handb. Phys., Bd. VII/1, Springer, Heidelberg 1955.

[13] *H. Hahn, W. Biem*, Physica status solidi *3*, 1911 (1963).
W. Cochran, G. Pawley, Proc. Roy Soc. *280*, 1 (1964).
[14] *R. Stockmeyer, H. Stiller*, Physica status solidi *27*, 269 (1968).
[15] *S.G. Venkataraman, V. Sahni*, Rev. Mod. Phys. *42*, 409 (1970).
[16] *J. Janik, A. Kowaslka* in Ref. [42], Chapter 10.
[17] *R. Misenta*, Rep. EUR 3910d, EURATOM, Brüssel 1968.
[18] *G. Vineyard*, Phys. Rev. *110*, 999 (1958).
[19] *K. Singwi, A. Sjölander*, Phys. Rev. *119*, 863 (1960).

In such a motion combining vibrations and diffusion (Fig. 6): the peak intensity decreases with increasing κ as $\exp[-2W(\kappa)]$, and at the same time the width is observed to increase[20]. In solids, D, τ_0 and $W(\kappa)$ may be direction-dependent. This dependence provides information about the spatial nature of the diffusion[21].

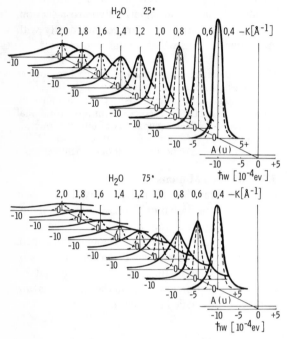

Fig. 6. The broadened elastic line in $S_{inc}(\kappa, \omega)$ for water at 25° and 75°[20]. Measurements on ice are shown in the marked curves.

For large ω, far removed from the broadened elastic scattering, Equation (10) may be applicable in a more generalized sense, if the scattering system (e.g. a liquid) can be considered isotropic. If $\hbar\omega < k_B T$, then $\omega^2 S_{inc}(\kappa, \omega)$ is directly proportional to the density of state $g(\omega)$. Several models for unbound motions in non-crystalline systems have been developed[23] for interpretation of such observations[22] on line broadening and densities of state. Very interesting insight into simple liquids has been gained recently from computations by the methods of molecular dynamics for systems of a few hundred particles[24]. By comparison of such

calculations with neutron spectroscopic data, assumed interaction potentials may be tested. The liquid state actually appears to be even better suited to such tests than the solid state, presumably because in liquids the molecules can experience intermolecular interactions over a larger range.

Comparisons are difficult, however, because the computations do not yet yield scattering functions with sufficient accuracy, and because of uncertainties regarding the correction of experimental data for the effects of multiple scattering. If only a certain percentage of the particles can move freely, or if unbound motions are possible only in certain degrees of freedom (for instance, for rotational motions only), the quasi-elastic scattering may occur in the form of a superimposed sharp elastic line and a broad base. For this case, too, model theories have been developed – in particular, for rotational diffusion[14, 25].

The microscopic observation of such unbound motions permits investigations on complex reactions, for example, in solutions. Neutron spectroscopy may make it possible to observe the fastest reaction steps, those controlled by translational or rotational diffusion. It has, for instance, been suggested that quasi-elastic *incoherent neutron scattering* could be used for investigations on solutions of polymers. The observations could be compared with calculations[26, 27] based on models which regard the polymers as chains of rigid monomer units of mean square length σ^2, and having high orientational freedom. For very dilute solutions, the models predict:

Only for the motions of the monomers coupling with the nearest neighbor the half width of an elastic line is given by: $\Delta\omega = 10^{-2} \cdot \tau^{-1} \cdot \sigma^4 \cdot \kappa^{4'}$; for motions of the monomers which are coupled with a solvent having viscosity η: $\Delta\omega = 7,5 \cdot 10^{-2} \cdot \bar{\tau}^{-1} \cdot \sigma^3 \cdot \kappa^3$. When τ and $\bar{\tau}$ are the relaxation times for the rotation of the monomers relative to each other:

$$\frac{1}{\tau} = \frac{B}{\sigma^2} 3 k_B T; \quad \frac{1}{\bar{\tau}} = \frac{\sqrt{3}}{\pi} \frac{k_B T}{\eta \sigma^3}.$$

B = coupling constant between neighboring monomers.

[20] M. Sakamoto, B. Brockhouse, R. Johnson, N. Pope, J. Phys. Soc. Japan, Suppl. B-II, 370 (1962).
[21] W. Gissler, G. Alefeld, T. Springer, J. Phys. Chem. Solids 31, 2361 (1970).
[22] S. Cocking, AERE Harwell Rep. R-5867, London 1968; J. Phys. C 2 (1969).
[23] A. Sjölander, Ref. 42, Chapter 7.
[24] A. Rahman, Phys. Rev. 136, A405 (1965).
B. Nijboer, A. Rahman, Physica 32, 415 (1966).

[25] K. Larsson, L. Bergstedt, Phys. Rev. 151, 117 (1966).
K. Larsson, Phys. Rev. 167, 171 (1968), 1971.
V. Sears, Can. J. Phys. 44, 1279 (1966), 45, 237 (1967).
P. Egelstaff, J. Chem. Phys. 53, 2590 (1970).
[26] P. De Gennes, Physics 3, 37 (1967).
E. Dubois-Violette, P. De Gennes, Physics 3, 181 (1967).
[27] B. Powell, in Ref. 8, Vol. II, p. 185.

4 Coherent Scattering

Coherent inelastic neutron scattering yields information beyond that obtainable by optical methods. The interference effects which occur in this type of scattering can be used to pick out individual dynamic states for observation. Coherent inelastic neutron scattering is a combination of X-ray diffraction and Raman spectroscopy.

4.1 Many Particle Coherence

Examples of motions in which many atoms are co-related include: sound waves, coupled optical vibrations, skeletal vibrations of macromolecules, and coupled intramolecular or intermolecular atomic movements. Each periodic collective dynamic state is characterized by a wave vector q ($|\underline{q}| = 2\pi/\Lambda$), by a frequency

$$\nu = \nu_s(\underline{q}) \tag{11}$$

and by a polarization $\underline{e}_{js}(\underline{q})$; the index s specifies the mode. In a periodic system with n atoms per unit cell, there are 3n different modes. The index j specifies the nucleus affected by the state.

In crystals, the interference condition for elastic scattering is: $\underline{\kappa} = 2\pi\underline{g}$, where $2\pi\underline{g}$ is a reciprocal lattice vector (cf. Chapter 7.1 and 7.3). This interference condition may also be interpreted in terms of conservation of momentum: $\hbar\underline{\kappa}$ is the momentum transferred from the neutron to the scattering system, and $\hbar\underline{g}$ the momentum taken up by the system as a whole, i.e. without energy change. If the scattering is inelastic, owing to excitation or de-excitation of a collective dynamic state, then a part $\hbar\underline{q}$ of the momentum is taken up with energy change. The interference condition then becomes:

$$\underline{\kappa} = 2\pi\underline{g} + \underline{q} . \tag{12}$$

The inelastic coherent scattering function for single excitation or deexcitation is, in harmonic approximation[10] (cf. Equation 9):

$$S_{coh}^{(1)}(\underline{\kappa},\omega) = \sum_{q,s} \sum_{\gamma=\pm 1} | F_{sq}(\underline{\kappa}) |^2$$

$$\frac{1}{2\pi \nu_s(\underline{q})} \left(\coth \frac{h\nu_s(\underline{q})}{2k_BT} + \gamma \right) \times$$

$$\times \delta(\omega + \gamma 2\pi \nu_s(\underline{q})) \times \sum_g \delta(\underline{\kappa} + \underline{q} + 2\pi \underline{g})_i . \tag{13}$$

$$F_{sq}(\underline{\kappa}) = \sum_j a_j C_{sj} e^{-W_{sj}(\underline{\kappa})} e^{i\underline{\kappa}r_j^0} \underline{\kappa} \cdot e_{sj}(\underline{q})|$$

For the sake of generality we have here redefined the scattering function by including the scattering amplitudes a_j.

With conservation of energy (7), the interference condition (12) and the dispersion relationship (11),

we have three equations for three variables, i.e. ν, q and one experimental quantity, e.g. $|\mathbf{k}_1|$. Hence, we find maxima in $S_{coh}(\underline{\kappa},\omega)$ at those values of these variables which simultaneously satisfy the three equations. In each of these maxima, one collective dynamic state is singled out for observation, with regard to ν_s and q.

4.1.1 Measurements on Single Crystals

In order to pick out a dynamic state, we have varied only one experimental quantity (e.g. $|\mathbf{k}_1|$). If we repeat the measurement with one of the other variables changed, for example, with a different $|\mathbf{k}_0|$, then Equations (7), (11) and (12) will be satisfied for other values of ν, q and $|\mathbf{k}_1|$. In this way, we can measure the entire dispersion relation $\nu = \nu_s(q)$, and thus obtain direct information about interatomic forces. In harmonic approximation, $\nu_s(q)$ is, in fact, the solution of the dynamic matrix, i.e. of the equations of motion[27]:

$$M_a \ddot{u}{}_a^{\overset{m}{\mu}}{}_i = - \sum_{n,\gamma,\beta,j} \Phi{}_a^{\overset{m}{\mu}}{}_i{}_j^{\overset{n}{\gamma}}{}_\beta u_\beta^{\overset{n}{\gamma}}{}_j,$$

where m designates the unit cell, μ the molecule in the unit cell, j the atom in the molecule, and i a component of the displacement u:

$$u{}_a^{\overset{m}{\mu}}{}_i \sim e{}_a^{\mu} \exp [i(\underline{q}\underline{r}{}^0{}_\mu^m{}_a - 2\pi\nu t)] .$$

The coupling parameters Φ are the second space derivatives of the interaction potentials at the equilibrium positions. With measured $\nu_s(q)$, we can thus test assumptions regarding the interactions, and possibly determine the potential parameters. Fig. 7 shows the dispersion of some collective motions in several symmetry directions of a single crystal of *hexamethylenetetramine,* as an example of this type of measurement. Similar measurements have been made for other molecular crystals, and with fibers of *deuterated polethylene*[30] and of *Teflon*[31].

The following additional observations are possible, with the individual dynamic state selected through an interference maximum.

[28] *W. Ludwig,* Recent Developments in Lattice Theory, Springer Tracts in Modern Physics, Vol. 43, Heidelberg 1967.

[29] *B. Powell* in Ref.[8], Vol. II, p. 185.

[30] *L. Feldkamp, G. Venkataraman, J. King,* Ref.[8], Vol. II, p. 159.

[31] *V. La Garde, H. Prask, S. Trevino,* Disc. Faraday Soc. *48,* 15 (1969).

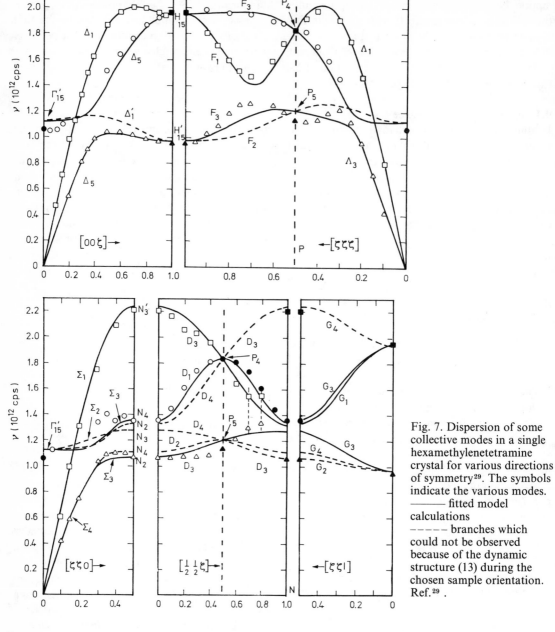

Fig. 7. Dispersion of some collective modes in a single hexamethylenetetramine crystal for various directions of symmetry[29]. The symbols indicate the various modes.
——— fitted model calculations
––––– branches which could not be observed because of the dynamic structure (13) during the chosen sample orientation. Ref.[29] .

Because of the factor $|\kappa \cdot e_{js}(q)e^{ikr°j}|^2$ in Equation (13), the polarization of the state may be derived from the intensity of the maximum. The width of the maximum in ω is determined by the life-time of the state. This lifetime can be measured, and conclusions drawn about the scattering of the state (for example, at defects), or about its interactions with other states — for instance with other sound waves or librations, with electron states, magnetic excitations, etc. Detailed information about such interactions can be obtained, if the widths of the maxima as well as the frequency shifts $v_s(q)$ are observed as functions of temperature and pressure, or as functions of isomorphous replacements which alter the local electron density[32]. Such observations are of special interest in the vicinity of phase transitions, and

may provide detailed information about transition mechanisms[33].

4.1.2 Measurements on Other Systems

For systems which do not consist of single crystals, observation of interference scattering becomes much more difficult. In poly-crystals, the reciprocal lattice vectors g are not defined with respect to direction, and in amorphous and liquid

[32] S. Ng, B. Brockhouse in Ref.[8], Vol. I, p. 253.
[33] J. Skalyo, B. Frazer, G. Shirane, Phys. Rev. B1, 278 (1970).

systems also their absolute values are ill-defined or not defined at all. An expedient is to measure with such a small κ that $g=0$ in the region of reciprocal space involved; the origin of reciprocal space is always defined as the end-point of κ_0. The interference condition is then

$$\underline{\kappa} = \underline{q} , \qquad (14)$$

just as in Brillouin scattering of light.

However, even when condition (14) holds, the values of κ attainable with neutrons are still about 500 times larger than those obtained with light scattering. The main disadvantages of measurements under condition (14) instead of condition (12) are that intensities are small and transverse modes become unexcitable, both because of the factor $|\kappa \cdot e_{sj}(q)|^2$. Dispersion relationships determined in this way with poly-crystals are averaged over all lattice directions[34].

In liquids, it has been possible[35] to observe a Brillouin triplet with neutrons up to frequencies of $\nu = 1.2 \times 10^{12}$ sec^{-1} and scattering vectors $\kappa = 0.6$ Å^{-1}. Just as with results on incoherent scattering from liquids (Section 5.4:3.3), such observations may be interpreted by comparison either with dynamic models for the liquid state, or with molecular dynamics computations. In simple cases, interpretation may also be possible by means of the equations of hydrodynamics[36]. The widths of the maxima in $S_{coh}(\kappa,\omega)$ are then described as originating from *heat conduction* and *viscosity*. These transport coefficients may become frequency-dependent, if relaxation processes (e.g. chemical reactions) take place in the liquid[37]. Because of their short wavelengths, neutrons allow much higher acoustic frequencies to be reached than with light; it, therefore, seems conceivable that with Brillouin scattering of neutrons from liquid solutions the relaxation times of very fast chemical reactions (10^{-10} to 10^{-13} sec) might be determined e.g. those of proton transfer processes.

4.2 Single Particle Coherence

The coherent scattering considered thus far gives rise to interference between waves scattered from different particles, and, consequently, provides information about correlations in the motions of pairs of particles. The term 'coherent' was defined in the Introduction with regard to such interference effects. In addition, however, interference may occur between waves scattered from one and the same particle, independent of whether the scattering is coherent or incoherent in the above sense. The Debye-Waller factor, $\exp[-W_j(\kappa)]$, represents interference of this kind; it results from an assembly average over the distribution of the nuclei in the range of their thermal amplitudes. As can be seen from the explicit form of $W_0(\kappa)$ given with Equation (9), these interference effects may be used for identification of the modes.

For example, with transitions between free rotational states, the Debye-Waller factor takes the form

$$\sum_{l=|I_i-I_f|}^{I_i+I_f} j_l^2(\kappa D_j)$$

where I_i is the angular momentum of the initial state, I_f the angular momentum of the final state, $j_l(x)$ the Bessel function of order l, and D_j the diameter of the sphere on which nucleus j rotates around the molecular center of mass.

The observation of interference of this type is of even greater interest if it results from a time average, and if this time average is not equal to the assembly average, e.g., in the case of a nucleus fluctuating between several equilibrium sites. If the fluctuation frequency is greater than the frequency resolution of the measurement, then the nucleus is seen simultaneously in more than one position; interference may occur between waves scattered at these different sites[38]. With the aid of interference patterns of this type, the tunnelling of protons at hydrogen bonds in paraelectric *potassium dihydrogen phosphate* was recently observed[39]. The time-averaged density distribution of a hydrogen-bonded proton (i.e. the square of its wave function) and its dependence on the polarization of the surrounding medium, and on external electric fields etc., may be determined. Similarly, rotational jumps between equilibrium orientations of the molecules were observed in the plastic phase of solid *adamantane*[40] with the aid of such interference effects. With these interference pheno-

[34] *K. Mika*, Phys. Stat. Sol. *21*, 279 (1967).
[35] *B. Dorner, Th. Plesser, H. Stiller*, Disc. Faraday Soc. *43*, 160 (1967).
K. Mika, B. Dorner, H. Stiller in Ref.[8], Vol. I, p. 599.
[36] *L. Kadanoff, P. Martin*, Ann. Phys. (New York) *24*, 419 (1963).
R. Mountain, Rev. Mod. Phys. *38*, 205 (1966).
[37] *M. Eigen, L. De Maeyer*, Technique Org. Chem. VII/2, 895 (1963).

[38] *H. Stiller*, Ber. Bunsenges. Phys. Chem. *72*, 94 (1968).
[39] *Th. Plesser, H. Stiller*, Solid State Commun. 7, 323 (1969).
H. Grimm, Th. Plesser, H. Stiller, Phys. Stat. Sol. *42*, 207 (1970).
[40] *R. Stockmeyer*, Disc. Faraday Soc. *48*, 156 (1969).

mena, motions of this kind can be studied in much more detail than with relaxation and resonance or optical measurements. Information is obtained, not only on frequencies, but also on the range and directions of motions in space.

5 Final Remarks

Information about the space-time behaviour of scattering molecules usually cannot be obtained from measured $S(\kappa,\omega)$ by the direct Fourier transform relations (4). For such a procedure, the κ-ω region which is accessible with neutron spectroscopy is too small, and experimental errors are too large with the intensities available at present (Fig. 2) Equations (9), (10) and (13) are specialized formualtions for the general Equations (4). They permit evaluation of experimental data without Fourier transforms, and additionally (in the cases of Equations (10) and (13) without comparison with theoretical assumptions. They are valid for bound harmonic motions. In addition, Equation

(10) is specialized for systems of cubic symmetry, and Equation (13) for single crystal samples − or, if averaged with respect to the dynamic structure factor $|F_{sq}(\kappa)|^2$, for κ-values which are smaller than about half the value at which a first diffraction peak appears. If the requirements of these equations are not fulfilled (see e.g. Sections 5.4:3.2 and 5.4:4.2), experimental observations can be evaluated only by comparison of a measured scattering function with theoretically calculated scattering functions. Such comparisons constitute tests for the interaction models which are the bases of the calculations.

6 Bibliography

[41] *R. Strumane*, The Interaction of Radiation with Solids. North Holland Publ. Co., 1964.
[42] *P. Egelstaff*, Thermal Neutron Scattering, Academic Press, New York 1965.
[43] Instrumentation for Neutron Inelastic Scattering Research, IAEA, Vienna 1970.

5.5 Magnetic Resonance Methods

H. Glotter, K.H. Hausser, H.J. Keller, D.C. Lankin, D. Lavie, K. Möbius, K.H. Schwarzhans, J. Todd, H. Zimmer

5.5.1 Principles of Magnetic Resonance

Karl H. Hausser

Max-Planck-Institut für medizinische Forschung, Abt. für Molekulare Physik, D-69 Heidelberg

1.1 Basic Principle

Magnetic resonance is due to the fact that electrons and many nuclei possess an angular momentum $\vec{J}$ and a magnetic moment $\vec{\mu}$ related to each other by the equation $\vec{\mu} = \gamma \vec{J}$, where γ is termed *'magnetogyric ratio'*. While electrons always have a spin $S = 1/2$, there are nuclei where the nuclear spin quantum number $I = 1/2$, such as ^1H, ^{13}C, ^{15}N, ^{19}F, ^{31}P, as well as those with the spin $I \geqslant 1$ (^2H and ^{14}N with $I = 1$, or ^{35}Cl with $I = 3/2$, or ^{17}O with $I = 5/2$). For discussion of the principle of magnetic resonance we shall restrict ourselves to particles where $I = 1/2$, but it should be kept in mind that some nuclei of special importance for chemistry such as ^{12}C and ^{16}O possess the spin $I = 0$, and hence exhibit no magnetic resonance.

1.1.1 Classical Description

Magnetic resonance can be described classically as follows: A homogeneous magnetic field $\vec{H}_0$ exerts a torque on the magnetic moment $\vec{\mu}$ which tries to turn $\vec{\mu}$ in the direction of the field $\vec{H}_0$. In the same manner as a mechanical top moving in the gravitational field of the earth, such a 'magnetic top' also deviates perpendicular to the magnetic force and precesses around the axis of the magnetic field $\vec{H}_0$. The angular frequency ω_L of this *Larmor precession* is proportional to the magnetogyric ratio γ and to the magnetic field $\vec{H}_0$.

$$\omega_L = \gamma \cdot H_0 \tag{1}$$

If this system is irradiated with a radio-frequency rotating magnetic field $\vec{H}_1$ perpendicular to $\vec{H}_0$, resonance occurs provided that the angular frequency ω_1 of this rotating field is equal to the Larmor frequency ω_L.

1.1.2 Quantum-Mechanical Description

Because of spatial quantization, a particle with the spin $1/2$ can orient itself only 'parallel' or 'antiparallel' with respect to H_0, corresponding to the magnetic quantum numbers $m_I = +1/2$ and $-1/2$ (Fig. 1). These two orientations correspond to two discrete energy levels with an energy differ-

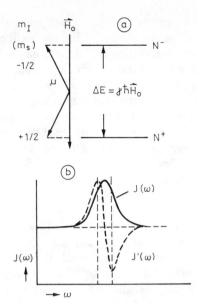

Fig. 1. Magnetic resonance for spin 1/2

a Resonance frequency
$h\nu = \hbar\omega = \Delta E = \gamma\hbar H_0$
Population of energy levels
$N^- = N^+ \exp(-\Delta E/kT)$

b Absorption function $J(\omega)$
and first derivative
$dJ(\omega)/d(\omega)$

ence $\Delta E = C H_0$, i.e. the energy difference is proportional to the magnetic field $\vec{H}_0$; without a magnetic field $\Delta E = 0$. As in other areas of spectroscopy, transitions can be induced by applying a radio frequency field H_1. In our case these transitions are magnetic dipole transitions; they occur if the frequency of the H_1 field obeys the fundamental equation of magnetic resonance:

$$h\nu = \Delta E = C H_0 \qquad (2)$$

The proportionality constant C is the basic quantity which is measured in magnetic resonance. There is a quantitative difference between electrons and nuclei. For nuclei the constant C is equal to the *nuclear magneton* $\mu_N = e\hbar/2m_p c$ multiplied by the nuclear g-value g. Hence (2) becomes for nuclei:

$$h\nu = \Delta E = g_I\mu_N H_0 = \frac{\mu_I}{I}H_0, \qquad (2a)$$

since the relation between the nuclear moment μ_I and the nuclear g-value is $\mu_I/I = g\,\mu_N$. This basic equation holds also for spins $I \geqq 1$. Since the selection rule for nuclear magnetic resonance (NMR) is $\Delta m_I = \pm 1$ and since the energy difference between the $(2I + 1)$ energy levels is equal, one obtains only one resonance frequency.

Analogous to the nuclear case, the constant C for electrons is $C = g_e\,\mu_B$ where g_e is the g value of the electron and μ_B is the Bohr magneton.

With the selection rules $\Delta m_s = \pm 1$ of the electron paramagnetic resonance (EPR), the basic equation becomes:

$$h\nu = \Delta E = g_e\mu_B H_0 \qquad (2b)$$

However, since $\mu_B = e\hbar/2\,m_e c$ is larger than μ_N by the factor M_P/m_e, the resonance frequency of ESR is about 3 orders of magnitude higher than that of NMR. Hence in magnetic fields of $\sim 10^4$ G obtained with ordinary laboratory magnets, the resonance frequency is in the microwave region at about 1 cm wave length corresponding to approximately 30 GHz, while the NMR resonance frequency lies in the region of 10–100 MHz.

Comparison of the quantum-mechanically derived resonance Equation (2) with the classical Equation (1) gives:

$$\omega\;\hbar = \gamma\,\hbar H_0 = \Delta E = C H_0, \text{ i.e. } \gamma\;\hbar = C$$
$$\gamma_I\,\hbar = g_I\mu_N = \mu_I/I \quad \text{or} \quad \gamma_e\,\hbar = g_e\mu_B.$$

1.1.3 Nuclear Quadrupole Resonance

A resonance phenomenon related to NMR is the nuclear quadrupole resonance (NQR, see Section 5.5:6). All nuclei with a spin quantum number $I \geqq 1$ possess, in addition to the magnetic moment, an electric quadrupole moment eQ which measures the deviation of the distribution of the positive charge in the nucleus from spherical symmetry. While e is the elementary charge, $Q = \int r^2(3\cos^2\theta - 1)\varrho d\tau$ with the dimension cm^2 and is defined as positive for a cigar-shaped nucleus and negative for a disc-shaped nucleus. (ϱ is the charge density per volume, r is the distance of the volume element $d\tau$ from origin, and θ is the angle between the radius vector and the axis of quantization of the spins.)

In non-cubic crystals and in almost all molecules the nuclei are situated in an inhomogeneous electric field q which is described by the second derivative of the electric potential in a given direction. Thus, the field gradient q in the direction z of a chemical bond is given by $q = \delta^2 v/\delta z^2$. A nucleus with a quadrupole moment can orient itself only in certain discrete angles with respect to the field gradient, each of which corresponds to a discrete energy value. The energy of these levels is proportional to the product of eQq; hence, this product is the quantity which can be directly measured.

In NQR magnetic dipole transitions are induced, in analogy to NMR, by a magnetic radio frequency field which fulfils the resonance condition $H\nu = \Delta E_Q$ between energy levels E_Q, the energy of which is determined by electric interactions. If the

319

nuclear quadrupole moment eQ is known, a measurement of NQR gives the value of the field gradient q, and, hence, information on quantities like ionic character and hybridization of a bond, which are correlated to q. The measurement by means of NQR is facilitated by the fact that it can be performed in a poly-crystalline solid. On the other hand, the localization of the absorption lines in a large frequency region can be tiresome, since, in many instances, the frequency to be expected cannot be predicted even approximately.

1.2 Relaxation and Linewidth

1.2.1 Population of the Energy Levels

The quantity to be measured is the absorbed radio frequency energy. A measurable absorption can occur only as a consequence of a difference of the population of the energy levels involved, since transitions from the higher to the lower level (induced emission) are induced with the same a priori probability as those from the lower to the higher level (absorption).

The spontaneous emission which is most important in optical spectroscopy can be neglected here because of the low transition probability for magnetic dipole radiation and because of the very low energy difference between the terms involved.

In fact, the number of spins in the higher level N_- differs from that in the lower level N_+ by the *Boltzmann-factor* (see Fig. 1):

$$N_- = N_+ \exp. ^{-\Delta E/kT} \tag{3}$$

The intensity of the observed absorption signal is proportional to the population difference and increases following (3) with decreasing temperature.

1.2.2 Longitudinal or Spin-Lattice Relaxation

If an isolated spin system interacts with the radio frequency field, more transitions are induced from the lower, more highly populated energy level to the higher than vice versa, until the population of the two levels becomes equal. Hence a continuous absorption of radio frequency energy can take place only if the spin system is coupled to its surroundings in such a manner that the absorbed energy can be transferred to these surroundings. Consequently, the spin temperature defined by (3) must remain approximately equal to the temperature of its surroundings which is usually referred to as 'lattice', even for liquids. The process by which temperature equilibrium is reached is termed 'spin-lattice relaxation'. The time constant

of the exponential function which measures the return of the population of the energy levels to Boltzmann-equilibrium after a perturbation is called 'spin-lattice relaxation time', T_1, or, alternatively, 'longitudinal relaxation time' since it describes the building up of the longitudinal magnetization. If the intensity of the radio frequency field is very high or if the efficiency of the relaxation mechanisms is low, i.e. if T_1 is long, the resulting steady state may deviate considerably from Boltzmann-equilibrium; this phenomenon is denoted as 'saturation'.

1.2.3 Transversal Relaxation and Linewidth

In addition to the relaxation time T_1 a second relaxation time has been defined. While the magnetization in a static magnetic field H_0 possesses only one component in the direction of the field, an additional radio frequency field can induce a transversal component. This transversal component disappears again after turning off the radio frequency field, due to interactions between neighboring spins which preserve the energy of the spin system. The time constant of the decay of the transversal component of the magnetization is termed 'transversal relaxation time' T_2.

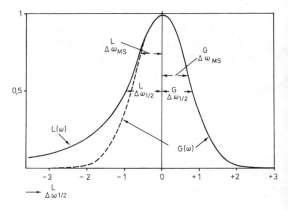

Fig. 2. Lineshape and linewidth $\Delta\omega$ of a Lorentz-function $L(\omega) = 1/(1 + \Delta\omega^2)$ and of a Gauss-function $G(\omega) = e^{-\Delta\omega^2}$; ($\Delta\omega = \omega - \omega_{res.}$) $\Delta\omega_{1/2}$= half linewidth at half height, $\Delta\omega_{MS}$ = half linewidth between points of maximum slope.
$\Delta\omega_{1/2}^L : \Delta\omega_{1/2}^G : \Delta\omega_{MS}^G : \Delta\omega_{MS}^L = 1 : \sqrt{\ln 2} : 1/\sqrt{2} : 1/\sqrt{3}$

T_2 is closely related to the linewidth. The curveshape of a homogeneous, magnetic resonance absorption line is given by a Lorentzian function $L(\omega) = 1/(1 + \Delta\omega^2)$ (Fig. 2). Other curveshapes can also occur, in particular the shape of an inhomogeneously broadened absorption line which is given by a Gaussian function $G(\omega) = e^{-\Delta\omega^2}$ (see 5.5:4).

A particularly simple relation exists between T_2 and half the width at half the maximum $\Delta\omega^L{}_{1/2}$ of a Lorentzian line measured in units of the angular frequency. The proportionality constant becomes 1 and we get:

$$2\pi\,\Delta\overset{L}{v}{}_{1/2} = \Delta\overset{L}{\omega}{}_{1/2} = 1/T_2 \qquad (4)$$

Whereas in high resolution NMR the absorption line is usually measured directly, in wide line NMR and in ESR the first derivative of the absorption line is usually plotted as a function of frequency or of the magnetic field because of the differing experimental technique (field modulation and phase sensitive detection). Hence linewidths are usually given as the distance between points of maximum slope; for a Lorentzian line one obtains $\Delta v_{m\cdot s} = 1/\sqrt{(3}\pi T_2)$ or $\Delta H = 2/(\sqrt{3}\gamma T_2)$

1.3 Information obtained from Magnetic Resonance

1.3.1 The Chemical Shift
The proportionality constant C in Equation (2), and hence the resonance frequency in a given applied magnetic field H_0,varies slightly under the influence of the molecular or crystalline surroundings in which the particle is located. This phenomenon is termed 'chemical shift' in NMR (see 5.5:2), while in ESR it is expressed as deviation Δg of the measured g-factor from the value of the free electron $g_{f.e.} = 2.0023$ (see 5.5:4 and 5).

1.3.2 Splitting Phenomena
The interaction with the magnetic moments of neighboring nuclei results in a splitting of the resonance line into several components. This phenomenon is termed 'spin-spin coupling' in NMR (see 5.5:2), and 'hyperfine structure' in ESR (see 5.5:4).
When observing NMR of nuclei with a spin $I \geqq 1$, an additional quadrupole splitting is obtained if the quadrupole energy ΔE_Q is small compared with the magnetic energy ΔE_{mag}. In the converse case, $\Delta E_Q \gg \Delta E_{mag}$, it is more conveniently described as a Zeeman-splitting of the NQR.

1.3.3 Resolution
As a consequence of these two effects a complicated multiline spectrum is generaly obtained with characteristic frequency differences and intensity ratios between the individual lines. These permit conclusions concerning the structure of the molecules.

As in other fields of spectroscopy a magnetic resonance spectrum provides more information if it has better resolution. In high resolution (HR) magnetic resonance, there is an essential difference between NMR and ESR:
The HR-NMR is essentially a technical problem, i.e., the linewidth, and hence the possible resolution, is determined by the homogeneity and stability of the static magnetic field H_0 and by the stability of the radio frequency transmitters, the attainable limit being about $\Delta v/v = 10^{-9}$. In order to obtain better resolution of resonance lines, the frequency difference of which is due to the chemical shift, one uses the highest possible magnetic fields because the chemical shift is proportional to the field H_0. However, in doing so one does not obtain a better resolution of multiplets which are due to the field independent spin-spin coupling.
Additional information on a complicated HR-NMR spectrum can be obtained by means of the nuclear-nuclear double resonance. One distinguishes between different types of double resonance experiments, namely 'spin decoupling' (high power) and 'spin tickling' (low power) depending on the power of the second radio frequency field, as well as between 'homonuclear' and 'heteronuclear' double resonance depending on the participating nuclei (see 5.5:2 and 3).
In most cases of NMR of nuclei with $I = 1/2$ in diamagnetic molecules, the linewidth resulting from relaxation processes within the sample is narrower than the linewidth due to technical causes and can, therefore, be neglected. This does not hold for solids. Hence HR-NMR is restricted to liquids and solutions.
In contrast to NMR the linewidth of ESR is determined by the physical relaxation processes in the sample; the maximum possible resolution approaches $\Delta v/v \geqq 10^{-6}$ in the most favorable cases of free radicals. When dealing with inhomogeneously broadened lines, the resolution obtainable by ESR can be considerably surpassed by electron nuclear double resonance (ENDOR), because ENDOR spectra possess a much smaller number of lines than ESR spectra (see 5.5:4.5).

1.4 Pulse Technique
Magnetic resonance experiments can be performed with pulsed radio frequency fields as well as with continuous irradiation. With this method the spins which precess with statistical phase are forced into a precession with coherent phase by a short radio frequency pulse; the free precession of the magnetization after the pulse is measured. The main field of application of this method is the

investigation of the physical relaxation processes; formerly its application in chemistry concerned primarily kinetic measurements (see 5.5:2). More recently, two additional important applications are being developed.

Firstly, it is possible to measure in HR-NMR the free precession after a radio-frequency pulse. The modulated decay curve obtained in this manner is the Fourier-transformation of the HR-NMR spectrum. It can be transformed into the latter by means of a computer. The advantage lies in the fact that much less time is needed for running a spectrum many times and taking the time average. The time saved for obtaining the same signal/noise ratio is of the order of 100 or, alternatively, the improvement in the signal/noise ratio for a like period of time is about 10.

Furthermore, it has been shown that a multiple pulse program can considerably reduce the linebroadening in solids due to magnetic dipole interaction. Hence the method of HR-NMR can be extended to solids by means of this technique. The method is, however, still in an early stage of development and considerable experimental difficulties will have to be surmounted before it becomes generally applicable.

5.5.2 Application of Proton Magnetic Resonance Methods to Structural and Stereochemical Problems

Erwin Glotter and *David Lavie*
The Weizmann Institute of Science, Rehovot, Israel

2.1 General Considerations

Nuclear magnetic resonance (NMR) spectroscopy is most useful for the elucidation of general structural and stereochemical problems. For an adequate interpretation of the results, a sound understanding of the theoretical basis of the nuclear resonance phenomenon is desirable (see 5.5:1).

The following chapter is not comprehensive, but rather indicates possible applications and uses. Sections 5.5:2.1–2.3 are extremely condensed since they have been covered by books and literature reviews[1].

2.2 The Chemical Shift

Within a molecule, a magnetic nucleus experiences an effective field slightly different from the external magnetic field H_0. This difference is due to the influence of the molecular environment creating a local field opposed to H_0, and having a screening effect on the given nucleus.

The *screening,* or *shielding* of the nuclei have different origins which are divided, according to Pople[2], into three categories: local diamagnetic effects, *diamagnetic* and *paramagnetic* effects from neighboring atoms, and effects of interatomic currents flowing in close circuits around a molecular path. Since all the shielding effects are proportional to H_0, their total contribution may be condensed in a constant σ called the *shielding constant,* the effective field experienced by the nucleus N being given by the relation

$$H_N = H_0(1-\sigma_N) \qquad (a)$$

The average of $-H_0\sigma_N$ over all the motions of the molecule containing the nucleus N gives the average shielding field acting at N. The differences in $(-H_0\sigma_N)_{av.}$ for the different electronic environments are the chemical shifts, expressed in gauss.

For convenience, chemical shifts are not given in absolute values but as differences between the resonance position of nucleus N in a particular surrounding, and its resonance position in a reference molecule. Chemical shifts can be expressed in Hertz (cycles per second) at a given field intensity or, as usually done, in dimensionless units δ [parts per million, p.p.m.]

$$\delta = \frac{H_{ref} - H_{sample}}{H_{ref}} \qquad (b)$$

The most frequently used reference for protons is tetramethylsilane (TMS)$(CH_3)_4Si$, possessing 12 equivalent protons which give rise to a sharp and intense signal at higher field than most protons in organic molecules. The position of this reference signal is arbitrarily assigned as $\delta = 0$ p.p.m. (or 10, τ scale).

[1a] *J.A. Pople, W.G. Schneider, H.J. Bernstein,* High Resolution Nuclear Magnetic Resonance, McGraw-Hill Book Co., New York 1959.

[b] *L.M. Jackman,* Applications of Nuclear Magnetic Resonance Spectroscopy in Organic Chemistry, Pergamon Press, London 1959

[c] *L.M. Jackman, S. Sternhell,* Applications of NMR Spectroscopy in Organic Chemistry, Pergamon Press, London 1968.

[d] *J.W. Emsley, J. Feeney, L.H. Sutcliffe,* High Resolution Nuclear Magnetic Resonance Spectroscopy, Pergamon Press, London, Vol. I, 1965, Vol. II, 1966.

[e] *E.F. Mooney,* Annual Review of NMR Spectroscopy, Academic Press London, Vol. I, 1968; Vol. II, 1969.

[f] *R.H. Bible,* Interpretation of NMR Spectra, Plenum Press, New York 1965.

[g] *N.S. Bhacca, D.H. Williams,* Applications of NMR Spectroscopy in Organic Chemistry, Holden-Day, San Francisco 1964.

Most measurements are done with the reference compound dissolved in the sample. Other advantages of TMS as internal reference are its magnetically isotropic molecule, its high volatility and miscibility with many organic solvents, and its chemical inertness. For aqueous solutions the sodium salt of trimethylsilylpropane sulfonic acid is used.

For most nuclei the magnitude of the chemical shifts is determined mainly by the diamagnetic effects due to circulation of the electrons around the given nucleus. The total electron density on the hydrogen atom is relatively small, so that its chemical shift will be significantly influenced by contributions due to electron circulations in other atoms within the molecule.

The electronegativity of substituents in simple *aliphatic molecules* influences the electron density on the neighboring protons and affects correspondingly their chemical shifts. The resonance positions of methyl protons in *methyl halides* are in agreement with the *electronegativity* of the halogen: CH_3I, $\delta = 2.16$; CH_3Br, 2.68; CH_3Cl, 3.05; CH_3F, 4.26. The substituents are not, however, the only factors influencing the chemical shift as can be seen from the resonance positions of the protons in *ethane, ethylene* and *acetylene* ($\delta = 0.96$, 5.84 and 2.88, respectively).

The paramagnetic currents may be neglected in the case of hydrogen atoms; an important contribution to the shielding of the proton may, however, arise from such currents on neighboring atoms. If their paramagnetic susceptibility is isotropic, the contribution is zero; if it is anisotropic, the corresponding contribution may be comparable with the local diamagnetic term[2].

2.2.1 Anisotropic Effects

The general quantitative theory of nuclear shielding is due to Ramsey[3a]. Although the application of the theory to the various practical structural problems is almost impossible, many quantitative and semiquantitative relationships have been found between chemical shift and molecular structure.

The contribution to the chemical shift of nucleus N of some specific group G of electrons within the same molecule has been treated theoretically by McConnell[3b] with a direct application to the resonance position of acetylenic protons. The following equation has been developed by McConnell[3b] for groups G possessing an axial symmetry, measuring the contribution of the magnetic anisotropy of the latter to the chemical shift of N.

$$\sigma_N^G = \frac{\Delta\chi^G}{3R^3}(1 - 3\cos^2\gamma) \qquad (c)$$

σ_N^G is the shielding tensor of nucleus N due to the group G, $\Delta\chi^G$ is the anisotropy of G (the difference between the magnetic susceptibilities in two directions, parallel and perpendicular to the symmetry axis of the latter), R is the distance from N to the electrical center of gravity of G and γ is the angle between the symmetry axis of G and R. The mean value will be inserted for ($1 - 3\cos^2\gamma$).

The implication of this equation is that the resonance line of N will depend on the *anisotropy* of the axially symmetric group G and will be at higher or lower field according to its relative position (determined by R and γ) to this group. Such a group is for instance, a *carbon carbon triple bond* which will shield a proton located in a conical region at the ends of the $C \equiv C$ bond, but will deshield a proton above or below its plane (Fig. 1). Similar calculations for a *single bond* in a rigid system lead to the conclusion that an *equatorial proton* should be less shielded than its *axial* counterpart. Indeed, differences of $\sim 0.3-0.5$ p.p.m. have been noted between such protons in otherwise similar systems[4].

If group G contains a *double bond* ($C=O$, $C=C$) the description of its anisotropy ($\Delta\chi^G$) requires the consideration of three reciprocally perpendicular susceptibilities. The McConnell equation can be used, however, for qualitative predictions, leading to the conclusion that protons situated above or below the plane of the double bond will be shielded, whereas, protons in conical regions at the end of the double bond will be deshielded (Fig. 1).

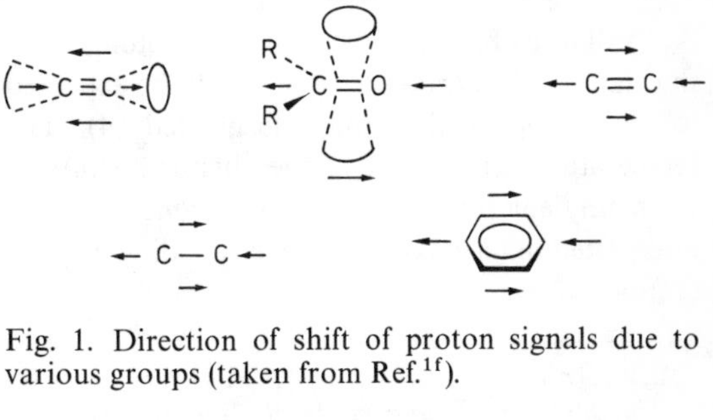

Fig. 1. Direction of shift of proton signals due to various groups (taken from Ref.[1f]).

[2] *J.A. Pople*, Proc. Roy. Soc. A *239*, 541, 550 (1957).
[3a] *N.F. Ramsey*, Phys. Rev. *78*, 699 (1950); *86*, 243 (1958).
[b] *H.M. McConnell*, J. Chem. Phys. *27*, 226 (1957).
s.a. *J.I. Musher* in *J.S. Waugh*, Advances in Magnetic Resonance p. 177, Vol. II. Academic Press, New York 1966.

[4] *A.A. Bothner-By, C. Naar-Colin*, J. Am. Chem. Soc. *80*, 1728 (1958).

The comparison of the gem-dimethyl resonances in *pinane*, *α-pinene* and *β-pinene* (1–3) clearly shows the contribution of the double bond to the shielding of these *methyl groups*. In each of the two pinenes, the *bridge methyl* group, which is closer to the double bond and placed above it, absorbs at significantly higher field than the corresponding methyl in pinane itself (by 0.16 and 0.29 p.p.m. respectively).

2.2.2 Additivity of Substituent Effects

One of the most important consequences of these equations [(c), see also Ref.[5]] is the concept of additivity of substituent effects. The chemical shift produced by the introduction of a substituent into a molecule can be calculated by considering the shielding effects of all bonds which have been displaced, as well as of all those which have been introduced.

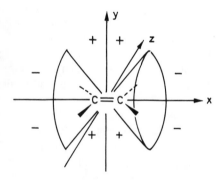

The absorption at $\delta = 10.0\text{-}10.6$ for *aldehydes* is due to the fact that such protons are in the deshielding cone of the carbonyl group (acetaldehyde, $\delta = 10.32$).

The usual picture of the shielding cones near *double bonds* ($C=C$ as well as $C=O$) should be understood in the sense that deshielding of a proton occurs only within a restricted region at the ends of the double bond[5]. Outside this region a proton is shielded whether it lies in the plane of the double bond or above it (Fig. 2).

For instance the difference between the shielding constants of the C_1-H in compounds 5 and 6 ($\Delta\sigma$) (i.e. following substitution of a 4-H in 5 by a methyl group in 6) can be calculated according to a general equation:

$$\Delta\sigma = \Sigma_1^n\sigma(6) - \Sigma_1^n\sigma(5) \qquad (d)$$

where $\sum_1^n\sigma$ is the sum of the shielding effects of all n bonds in the two molecules respectively. However, since the net shift between 5 and 6 is produced only by those bonds which differ in 5 as compared to 6, the problem is simplified to the contribution of these bonds alone. Such calculations give good agreement between calculated and observed chemical shifts, with the exception of the cases when the frame of the molecule is distorted due to nonbonded interactions between the groups. Discrepancies are observed for instance if the hydroxyl and the methyl are in a 1,3-diaxial relationship, as in 7.

The problem of the additivity of substituent effects has been considerably developed in the *steroid* field[7]. From an analysis of a large number of steroids, incremental values produced by different functional groups located in various positions in the steroid nucleus have been calculated for the chemical shift of the angular methyl group (10-CH_3 and 13-CH_3). These data have been tabulated and allow an easy analysis of a proposed structure. The reference compounds are the four stereoisomeric *androstanes* (5α, 5β, 14α, 14β; 8 − 11).

Fig. 2. The shape of the deshielding cones according to Ref.[5].

According to Karabatsos et al.[6], a proton H_A in the plane of the carbonyl group might be expected to be shielded rather than deshielded (4). The temperature dependence of the chemical shifts of the methylene protons in *diethyl ketone* has been interpreted in terms of the above shielding considerations.

[5] *J.W. ApSimon, W.G. Craig, P.V. Demarco, D.W. Mathieson, L. Saunders, W.B. Whalley*, Tetrahedron, *23*, 2357 (1967).

[6] *G.J. Karabatsos, G.C. Sonnichsen, N. Hsi, D.J. Fenoglio*, J. Am. Chem. Soc. *89*, 5067 (1967).

[7] *R.F. Zürcher*, Helv. Chim. Acta, *44*, 1380 (1961); *46*, 2054 (1963);

J.N. Shoolery, M.T. Rogers, J. Am. Chem. Soc. *80*, 5121 (1958); s.a. Ref. [1f]

Additivity can not always be obtained through the sum of the contributions of the groups which are involved: an α,β-*unsaturated ketone*, for example, should be considered as one system and its contribution on an *angular methyl group* as one entity.

	8	9	10	11
10–CH$_3$	0.792	0.767	0.925	0.900
13–CH$_3$	0.692	0.992	0.692	0.992

Dipole-dipole interactions between substituents may also cause deviations from additivity. The resonance position of the 13-CH$_3$ in *12-oxo-5β-ethianic acid (12-oxo-5β, 14α-androstan-17-carboxylic acid)* methyl ester (**12**) is located 5.5 Hz at lower field than calculated from the contributions of the two separate substituents. The spatial orientation of the methoxycarbonyl substituent in compounds of type **12** is not the same as in **13** and the resonance position of the 13-CH$_3$ will be correspondingly affected. Deviations from additivity are also observed if a substituent is hydrogen bonded.

[8] J.C. Jacquesy, R. Jacquesy, J. Levisalles, J.P. Pete, H. Rudler, Bull. Soc. Chim. Fr. 2224 (1964).
[9] J.M. Lehn, G. Ourisson, Bull. Soc. Chim. Fr. 1137 (1962); M. Fétizon, M. Golfier, P. Laszlo, 3205 (1965).
B. Tursch, R. Savoir, G. Chiurdoglu, Bull. Soc. Chim. Belg. *75*, 107 (1966).
[10] L. Salem, Molecular Orbital Theory of Conjugated Systems, Chapter 4, W.A. Benjamin, New York 1966;
F. Sondheimer in Aromaticity, an International Symposium, Sheffield, 1966, p. 75, the Chemical Society 1967; H.C. Longuet-Higgins, p. 109; J.A. Pople, K.G. Untch, J. Am. Chem. Soc. *88*, 4811 (1966).

Table 1. Substituent effects on the chemical shifts of 10-CH$_3$ and 13-CH$_3$ in steroids (δ values)

Substituents	10-CH$_3$	13-CH$_3$
In 5 α-Steroids		
1-oxo	+0.375	+0.017
Δ^1	+0.050	+0.017
Δ^1-3-oxo	+0.250	+0.050
2-oxo	−0.025	+0.008
3-oxo	+0.242	+0.042
3β-OH	+0.033	+0.008
Δ^4-3-oxo	+0.417	+0.075
Δ^4	+0.250	+0.042
6-oxo	−0.050	+0.017
7-oxo	+0.275	+0.008
Δ^7	−0.008	−0.117
Δ^8	+0.125	−0.083
$\Delta^{8(14)}$	−0.117	+0.175
11-oxo	+0.217	−0.033
12-oxo	+0.100	+0.375
in 14 α-Steroids		
Δ^{16}-17-COCH$_3$	+0.017	+0.175
17-oxo	+0.017	+0.167
17αOH-17β-COCH$_2$OAc	−0.008	−0.042
17β-COCH$_3$	−0.008	−0.083
in 14 β-Steroids		
17-oxo	+0.017	+0.083
14β-OH	+0.017	−0.025
14β-OH-17β-COOCH$_3$	+0.025	−0.017
15-oxo	−0.042	+0.192

Values are from the compilation of N.S. Bhacca and D.H. Williams [1g]. The chemical shift of the corresponding methyl group is obtained by addition of the increment due to the substituent to the chemical shift in the corresponding unsubstituted compound given above (**8—11**). A positive value denotes a downfield shift; a negative value − an upfield shift.

One of the basic concepts in the calculation of the contribution of substituents to the chemical shift of an *angular methyl group* is that of equivalent positions, i.e. positions in which substituents have similar orientations with respect to a given methyl. It has been observed, for example, that β-axial hydroxy substituents in positions 2, 4, 6, 8 and 11 of *5α, 14α-androstane* occupy formally equivalent positions with respect to the 10-CH$_3$ and, consequently, will induce similar shifts to this proton: 15.0, 16.0, 13.5, 11.0 and 15.5 Hz, respectively.

The procedure[8] for the calculation of the chemical shifts of *methyl groups* in different substituted derivatives distinguishes between primary, secondary, tertiary etc, contributions, i.e. the contribution of a given substituent in mono, di, tri or more substituted compounds. The most reliable data are, of course, the primary shieldings obtained from monosubstituted derivatives, the contribution of the substituent being obtained by subtraction of the chemical shifts of the angular methyl groups in the unsubstituted compound from the monosubstituted one. Correspondingly, the contribution of the same substituent in a disubstituted compound is obtained by subtraction of the chemical shift of the monosubstituted from that of the disubstituted derivative. The secondary contribution obtained thereby should be the same as the primary contribution provided that no distortions were introduced in the molecule.

Similar measurements have been tabulated in other compounds such as *diterpenes, tetracyclic* and *pentacyclic triterpenes, 4,4-dimethylsteroids*[9].

2.2.3 Aromatic Protons[10]

The contribution of interatomic ring currents in the shielding of protons manifests itself in the chemical shifts of *aromatic protons* or those which, due to stereochemical reasons, are in proximity to the aromatic ring. The sign of the effect depends on the position of the proton relative to the ring, while its intensity is related to the ring current field. The shifts induced are large enough to determine the aromaticity of a system. Nuclear magnetic resonance is, therefore, a powerful tool in the study of the theories of aromaticity. In the case of the *annulenes* and *dehydroannulenes* possessing $4n+2$ out of plane π electrons (n = 3, [14]-annulene, n = 4, [18]-annulene), the outer protons appear at low field ($\delta = 10.36$ to 13.4) while the inner protons resonate at high field ($\delta = 2.34$ to 5.48), as required from the considerations of the diamagnetic ring current sustained by the delocalized π electrons in an external applied field. The NMR spectrum of *(14)-annulene* **(14)** at $-60°$ shows a band at $\delta = 10.0$ for the outer protons and a band at $\delta = 2.4$ for the inner protons (for a detailed discussion of the NMR of [18]-annulene see 5.5:2.10). With the [12]-, [16]-, [20]- and [24]-annulenes, which contain 4n out of plane π electrons (n = 3, 4, 5, 6), the situation is reversed; the inner protons appear at unusually low field ($\delta = 8.4–11.6$) whereas the outer protons resonate at significant higher field values ($\delta = 4.3–6.1$). These observations provide an excellent confirmation of Hückels'rule and are explained by a paramagnetic ring current existing in the latter type of compounds.

The existence of a large ring current of π electrons was first observed in the NMR spectra of the

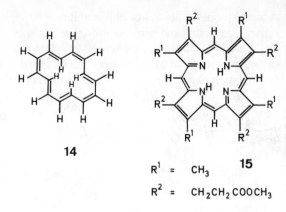

14

R^1 = CH_3

R^2 = $CH_2CH_2COOCH_3$

15

porphyrins[11]. The outer protons resonate at very low field values, whereas the inner protons (NH) are at considerably higher field. In the methyl ester of *coproporphyrin-1* (**15**) there is a $4 \times H$ signal at $\delta = 9.96$ assigned to the protons attached to the double bonds and a broad $2 \times H$ signal at $\delta = 3.89$ for the NH protons. The lowest signal is located at 2.7 p.p.m. lower field than the signal of benzene; the NH signal is 13 p.p.m. higher field than the corresponding signal of this group in pyrrole.

2.2.4 Hydrogen Bonding

The chemical shifts of protons bound to oxygen, nitrogen, or sulfur are usually dependent on temperature, concentration, and solvent, according to the degree of hydrogen bonding. In *ethanol,* for instance, the chemical shift of the hydroxylic proton is shifted upfield by increasing the temperature or by dilution, both lowering the degree of hydrogen bonding. The actual position of the hydroxylic proton in the absence of intermolecular hydrogen bonding can be obtained by extrapolation of the results to infinite dilution.

The formation of a *hydrogen bond* A—H...B modifies the chemical shift of the respective proton; this modification can be evaluated by considering two contributions[1c]: 1) the direct effect of the currents induced in B on the magnetic field of nucleus H, and 2) the disturbance of the electronic structure of the A—H bond, manifesting itself in a change in the shielding of H.

For *intramolecular hydrogen bonds* (e.g. in stable enols) the chemical shift of the hydroxylic proton is concentration independent and is at very low field. The position of acidic protons in *carboxylic acids,* which are known to form stable dimers, is also concentration independent in nonpolar solvents.

[11] *E.D. Becker, R.B. Bradley, C.J. Watson,* J. Am. Chem. Soc. *83,* 3743 (1961).

[12] *J.R. Dyer,* Applications of Absorption Spectroscopy of Organic Compounds, Prentice-Hall, Englewood Cliffs, New Jersey 1965.

[13] *P.L. Corio,* Chem. Rev. *60,* 363 (1960).

Table 2. Chemical shifts of protons in various structural environments

Structural type	δ Value (range)
CH_4	0.233
CH_3—C— saturated	0.85– 0.95
CH_3—C—C—X (X=Cl, Br, I, OH, OR C=O, N)	0.90– 1.10
CH_3—C—X (X=F, Cl, Br, I, OH, OR OAr, N)	1.2 – 1.9
CH_3—C=C<	1.6 – 1.9
CH_3—C=O	2.1 – 2.6
CH_3—Ar	2.25– 2.5
CH_3—S	2.1 – 2.8
CH_3—N<	2.1 – 3.0
CH_3—O—	3.5 – 3.8
—CH_2— cyclopropane	0.2 – 0.3
—CH_2— saturated	1.20– 1.35
CH_2=C< nonconjugated	4.6 – 5.0
CH_2=C< conjugated	5.3 – 5.7
—C—H saturated	1.40– 1.65
—C≡C—H nonconjugated	2.45– 2.65
—C≡C—H conjugated	2.8 – 3.1
—C=C< H acyclic nonconjugated	5.2 – 5.7
—C=C< H cyclic nonconjugated	5.2 – 5.7
—C=C< H conjugated	6.7 – 7.7
ArH	6.6 – 8.0
R—C<H,O aliphatic α, β-unsaturated	10.35–10.50
R—C<H,O aliphatic	10.20–10.30

Structural type	δ Value (range)
Ar—C<H,O Concentration dependent	10.0 –10.3
R_2NH (mole fraction 0.1–0.9 in inert solvent)	0.4 – 1.6
RSH	1.1 – 1.5
RNH_2 (mole fraction 0.1–0.9 in inert solvent)	1.1 – 1.5
ArSH	3.0 – 4.0
$ArNH_2$, ArNHR, Ar_2NH	3.4 – 4.0
ROH (mole fraction 0.1–0.9 in inert solvent)	3.0 – 5.2
>C=N<OH	10.2 –11.2
Enols	15 –16

*According to the compilation of J. R. Dyer, ref. [12]

2.3 Spin-Spin Coupling[1a, b, c, 13]

Compounds possessing neighboring magnetic nuclei exhibit fine structure, i.e., the number of signals is larger than the number of nonequivalent nuclei. This phenomenon was assigned to the reciprocal interactions between the nuclear spins. It could be demonstrated that the energy of this interaction is dependent on the corresponding nuclear spins ($I_N I_{N'}$) and on a coupling constant between the spins ($J_{NN'}$).

$$E_{NN'} = hJ_{NN'} I_N I_{N'} \tag{e}$$

h = Planck's constant

In contrast to the direct interaction between magnetic dipoles, the energy of the spin-spin interaction is not averaged to zero by rapid rotation of the molecule in the liquid; it is also independent of the applied external field H_0 and of the temperature. According to Ramsey and Purcell a nuclear spin induces a small polarization in the nearby electron which is transmitted through the second electron involved in the covalent bond to the next magnetic nucleus. Due to such an interaction one nucleus can 'see' the individual spin states of another neighboring nucleus. In *ethyl bromide*, for instance, the unperturbed signal of the three equivalent protons of the methyl group is split into a triplet, since there are three possible orientations of the spins of the methylene group; the unperturbed signal of the methylene protons is split into a quartet, owing to the four possible orientations of the spins of the three methyl protons, Fig. 3.

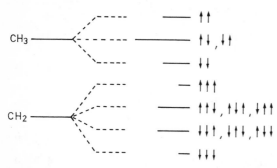

Fig. 3. Splitting scheme of the methyl and methylene signals, in ethyl bromide.

The total intensity of each multiplet is proportional to the number of equivalent nuclei in the corresponding group. The separation of the peaks within the multiplet is the coupling constant J and is expressed in Hz (cycles per second). When the separation of the chemical shifts (Δv) of the two groups of nuclei is large compared to J, the spectral pattern of the interacting nuclei can be predicted according to the following simple rules:
Equivalent nuclei give rise to only one peak (singlet); this does not mean, however, that the nuclei are not reciprocally coupled.
The signal of a group of equivalent nuclei A_m (with spin $I = 1/2$) is split by a neighboring group of equivalent nuclei X_n ($I = 1/2$) into an equally spaced symmetrical multiplet possessing $n + 1$ peaks, their intensities corresponding to the coefficients of the binomial expansion $(n + 1)^n$. When the group A_m is split by two neighboring groups of nuclei X_n and M_p, the multiplicity of the signal is given by $(n + 1)(p + 1)$.

2.3.1 Spin-Spin Coupling Patterns

The analysis of the spin-spin interactions is generally done by dividing the magnetic nuclei into groups according to their chemical shifts[1a,c]. *Nonequivalent nuclei,* in which the differences between the chemical shifts (Δv) are of the same order of magnitude as the corresponding coupling constants (J), are designated as A, B, C, etc. Such nuclei belong necessarily to the same isotopic species. Other groups of magnetic nuclei within the same molecule, which are separated by large Δv from the first group, are called X, Y, Z. Within this latter group the Δv values are comparable to the respective J constants, in the same manner as the members of the A, B, C group. Sometimes a third group of nuclei P, Q, R (or M, N, O) are characterized, possessing chemical shifts

largely different from those of the nuclei belonging to the first two groups.
Nuclei which are magnetically equivalent are assigned the same letter with a subscript indicating their number (for instance the ethyl group $CH_3 \cdot CH_2R$ forms an A_3B_2 system). Nuclei with equivalent chemical shifts but with *unequivalent spin-spin couplings* are primed. An AA'XX' system (sometimes referred as $A_2'X_2'$) consists of two sets of nuclei possessing the same chemical shifts respectively, but differently coupled to each of the nuclei of the second group ($J_{AX} \neq J_{AX'}$).
The magnetic nuclei within a given molecule are differentiated by the A, B... or X, Y... notation in a somewhat arbitrary way. In general, a system will be characterized as A_mX_n if the spectrum can be interpreted by application of the simple splitting rules (first order analysis); in such a case $\Delta v \gg J$. When the coupling constants are comparable in magnitude to the chemical shifts, the problem can be resolved by a quantum mechanical treatment (higher order analysis). Nevertheless, comparison of spectra of appropriate known molecules can give much information as to the structure of the analyzed compound. In some simple cases (AB, ABX) a graphical analysis can be used[1e].
Spectra which are characterized by only one coupling constant and one chemical shift difference are of the A_mB_n type. These spectra are dependent only on the ratio $J/\Delta v$. (see Fig. 4). Spectra in which three coupling constants are involved are of the $A_m B_n C_p$ type (for instance ABX, ABC, etc.). Four coupling constants are encountered in the AA'XX', AA'BB' spectra.

2.3.2 A_mB_n Systems

The AB system is formed by two mutually coupled nuclei: if $\Delta v \gg J$ it will appear as a pair of doublets with the lines of equal intensity. In this case the chemical shift of each nucleus is recorded at the center of the corresponding doublet; such a spectrum is designated as AX, and is the extreme case of the general AB system. The opposite extreme is represented by two nuclei possessing the same chemical shift ($\Delta v = 0$). The system can then be described as A_2 giving rise to only one band. In all intermediate cases two doublets are obtained in which the intensity of the outer bands decreases while that of the inner bands increases with larger values of the ratio $J/\Delta v$ (Fig. 4). The chemical shifts of the interacting nuclei can be calculated from the relative intensities of the absorption lines and from their separation. Examples are the vinylic protons in Δ^1-3-keto-steroids and in *diethyl fumarate* (A_2); see Fig. 5.

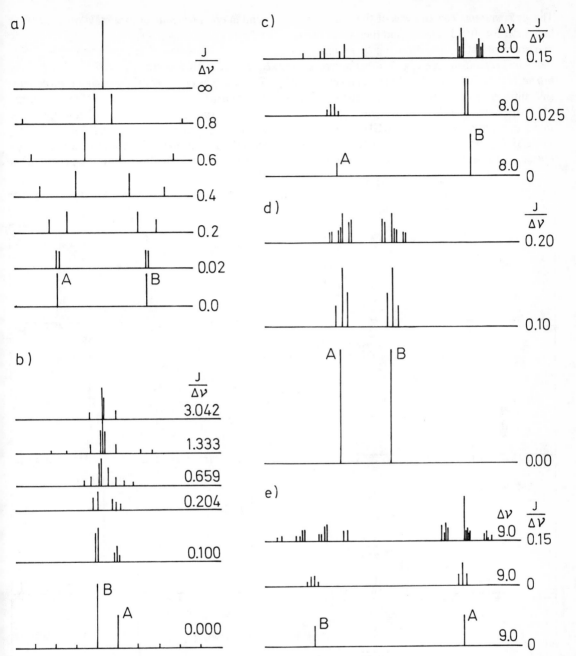

Fig. 4. Some calculated spectra for nuclei with spin I = 1/2:
a) AB system, Δv = 10 Hz
b) AB₂ system, Δv = 10 Hz
c) AB₃ system, Δv = 8.0 Hz
d) A₂B₂ system, Δv = 10 Hz
e) A₃B₂ system, Δv = 9 Hz
   (Adapted from Ref. [1c] Vol. I)

The *AB₂ system* involves the coupling of two magnetically equivalent nuclei with a third nucleus. If $\Delta v \gg J$ ($J/\Delta v$=0.10 in Fig. 4b) the system can be designated as AX₂ and will appear accordingly as a 1:2:1 triplet for A and a 1:1 doublet for X₂. An example is *1,1,2-trichloroethane* (see Fig. 5).

The *AB₃ system:* This pattern can be recognized easily in all CH₃—CH groups. Due to rapid rotation around the C—C bond there is no difference between the three coupling constants J_{AB}; for $\Delta v \gg J$ ($J/\Delta v$=0.025 in Fig. 4c) the system becomes AX₃ and appears as a 1:3:3:1 quartet for A, and a 1:1 doublet for X₃. An example is *acetaldehyde*.

The *A₂B₂ system:* The analysis of the spectrum is first order for $\Delta v \gg J$. The system then gives rise to two symmetrical triplets, for instance, in *difluomethane* or *propadiene* (see Fig. 5 β-methoxypropionitrile, $NCCH_2CH_2OCH_3$). If there is no chemical shift difference between the two groups of nuclei ($\Delta v = 0$), only one line is obtained (e.g. *methyl β-cyanopropionate*, $NCCH_2CH_2COOCH_3$).

The *A₃B₂ system:* This system is particularly important since it appears in all *ethyl groups*. It is found in *ethyl bromide* or in *diethyl fumarate* (see Fig. 5).

2.3.3 The ABX System

The ABX system is characterized by three coupling constants, J_{AB}, J_{AX} and J_{BX}. If Δv_{AB} is large in comparison to J_{AB} a normal 12 line spectrum appears. The spectrum of methyl 2-furoate illustrates the appearance of such a pattern. The system can be considered as due to the perturbation

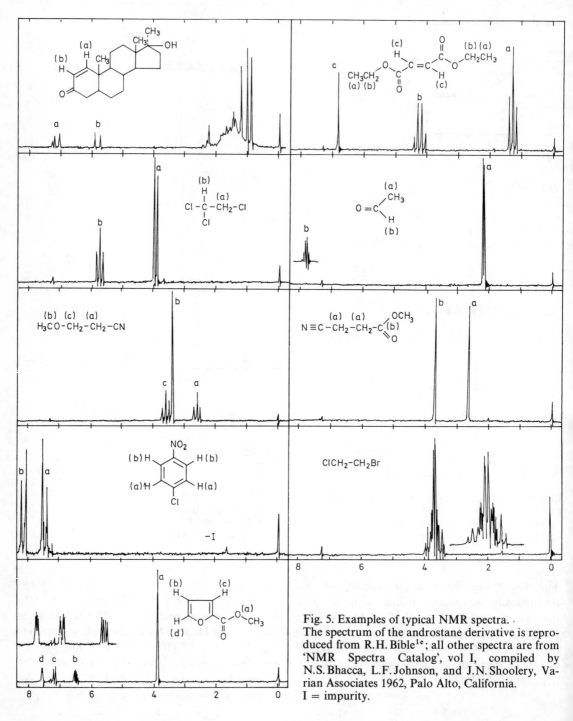

Fig. 5. Examples of typical NMR spectra.
The spectrum of the androstane derivative is reproduced from R.H. Bible[1e]; all other spectra are from 'NMR Spectra Catalog', vol I, compiled by N.S. Bhacca, L.F. Johnson, and J.N. Shoolery, Varian Associates 1962, Palo Alto, California.
I = impurity.

introduced by a third nucleus X into the two doublet pattern of an AB system. Each of the four lines of the AB system will be split into a doublet according to J_{AX} and J_{BX} while the X part of the spectrum will appear as four lines of almost equal intensity.

16

The ABX system present in *2-furanol* (**16**) gives rise to a very simple spectrum as pure liquid. In this case Δv_{AB} and $1/2\ J_{AX} - J_{BX}/J_{AB}$ are small compared to J_{AB}. The X part of the spectrum is a triplet and the AB part a doublet. However, in benzene and acetone solution the spectrum looks quite different due to the changes in the chemical shifts of the interacting protons[14].

2.3.4 The AA'BB' System

The AA'BB' system is characterized by four coupling constants and by two chemical shifts (Fig. 5, *4-chloronitrobenzene* and *1-chloro-2-bromo-ethane*). If $\Delta v = 0$, i.e. the four interacting nuclei have the same chemical shift, only one resonance line is obtained (e.g. *p*-xylene).

2.3.5 Geminal Coupling $J_{HH'}^{gem}$[15, 16, 17]

Geminal protons are attached to the same carbon atom. The problem has been treated theoretically by the valence bond or the molecular orbital method, but neither of these treatments is in good agreement with the negative signs of the geminal coupling constants found experimentally.

Coupling constants are positive or negative, according to the energy of the coupled nuclear spins, which may be oriented antiparallel or parallel. If the antiparallel arrangement has a lower energy than the parallel, the corresponding coupling is positive; conversely, if the parallel arrangement is of lower energy than the antiparallel, the coupling is negative.

The geminal coupling constants become more positive for larger H—C—H angles (from −12 to + 2 Hz), and are larger in absolute magnitude if the intervening carbon atom is sp³ hybridized rather than sp². These couplings are also strongly influenced by the nature and the orientation of the substituents (Table 3).

[14] *R.J. Abraham, H.J. Bernstein,* Can. J. Chem. *39,* 216 (1961).

[15] *N. Barfield, D.M. Grant* in *J.S. Waugh* Advances in Magnetic Resonance, Vol. I, p. 149, Academic Press, New York 1965.

[16] *A.A. Bothner-By,* Ref. [15], p. 195.

[17] *S. Sternhell,* Quart. Rev. *23,* 236 (1969).

Table 3. Geminal coupling constants $J_{HH'}^{gem}$ in some compounds; the protons reported are underlined.

Compound	$J_{HH'}^{gem}$ [Hz]
	± 12.4[a]
	± 14.0[a]
	$+ 12.2$
	± 14.9[a]

[a] The sign of the coupling has not been determined.

Electronegative substituents make positive contributions to $J_{HH'}^{gem}$ in the order[18] I > F > Br > Cl ∼ -OH; for instance $J_{HH'}^{gem}$ in methanol is -10.8 Hz and in CH₃F is -9.6 Hz. For CH₂ groups having an adjacent sp² hybridized carbon atom the magnitude of $J_{HH'}^{gem}$ is dependent on its relative position towards the nodal plane of the π orbital; the effect is largest when the internuclear axis of the two hydrogens is perpendicular on this nodal plane[19]: in *cyclopentenedione* (**17**) $J_{HH'}^{gem}$ is -21.5 Hz.

17   **18**   **19**

The coupling constants between equivalent protons which give rise to one resonance line are calculated in an indirect way. For example, substitution of one hydrogen in a methyl group by a deuterium (CH₃ → CH₂D) is accompanied by the splitting of the resonance line of the two remaining protons into a 1:1:1 triplet (through coupling with the deuterium possessing a spin I = 1); the $J_{HH'}^{gem}$ now is calculated from J_{HD}^{gem} by multiplying the latter with the factor 6.514 (the ratio between the gyromagnetic constants of the proton and the deuteron).

In general, geminal coupling constants do not differ significantly in four, five, or six membered

[18] *H.J. Bernstein, N. Sheppard,* J. Chem. Phys. *37,* 3012 (1962).

[19] *M. Barfield, D.M. Grant,* J. Am. Chem. Soc. *83,* 4726 (1961); *85,* 1901 (1963).

rings from the values determined in acyclic compounds. In small heterocyclic rings there is a trend towards more positive couplings. In *epoxyethylbenzene* (**18**) and *aziridine* (**19**) values of $+5.65$ Hz and ± 2.0 Hz are recorded.

For compounds in which the CH_2 carbon is sp^2 hybridized, the geminal coupling constants are dependent to a large extent on the nature of the β substituent. Large values are quoted for formaldehyde ($+42$ Hz) and ketene (-15.8 Hz).

Table 4. $J_{HH'}^{gem}$ in some ethylenes

$$\begin{array}{c} H \\ \diagdown \\ H \diagup \end{array} C=C \begin{array}{c} X \\ \diagup \\ \diagdown Y \end{array}$$

X	Y	$J_{HH'}^{gem}$ [Hz]
H	H	$+2.5$
H	COOH	$+1.3$
H	NR_2	0
H	Cl	-1.3
F	F	-4.6

2.3.6 Vicinal Coupling $J_{HH'}^{vic}$ [15, 16, 17]

Two protons separated by two carbon atoms are called vicinal; the discussion will deal with the coupling across a formal single bond H—C—C—H' and a formal double bond H—C=C—H'. Since four nuclei are involved in the interaction between the two protons, the mathematical aspect of the problem is by far more complex than that of the geminal coupling.

Valence bond calculations performed by Karplus[20] for the interaction of two protons connected through a C—C fragment in ethane derivatives led to the formulation of a relation between the dihedral angle ϕ (Fig. 6) and the magnitude of the coupling constant:

$$J_{HH'}^{vic} = A\,\cos^2\varphi + B \quad 0° \leqslant \varphi \leqslant 90°$$
$$J_{HH'}^{vic} = A'\cos^2\varphi + B \quad 90° \leqslant \varphi \leqslant 180° \qquad A' > A \quad (f)$$

Although the values predicted according to this equation are usually smaller than the observed coupling constants, the relationship is very useful in correlating the vicinal coupling values. The coefficients are dependent to a significant degree on the electronegativity of the substituents in the investigated molecules, on the ring size, if the fragment is part of a cyclic structure, etc. Therefore, the relationship cannot be used uncritically for the estimation of dihedral angles within a few degrees from experimentally measured coupling

constants. Improvements of the equation have been proposed in which the coefficients A and A' have been determined empirically in order to accomodate experimental coupling constants[21].

A more accurate equation due also to Karplus is the following:

$$J_{HH'}^{vic} = A + B\cos\varphi + C\cos^2\varphi. \qquad (g)$$

For an sp^3 hybridized carbon, with a bond length (C—C) of 1.543 Å and an average energy of 9 eV the coefficients in the last equation are: A = 4.22, B = -0.5 and C = 4.5 Hz. With this set of calculated coefficients, the coupling constants are found to be small for $\phi \sim 90°$, and large for $\phi = 0°$ or $180°$; however, larger for $180°$, than for $0°$.

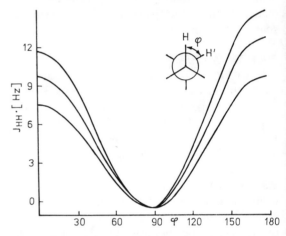

Fig. 6. Plot of the Karplus equation (f) for selected values of A and A' (see Ref.[17]).

In *ethane* the reported vicinal coupling is 8.0 Hz; in a large number of substituted ethyl derivatives the values of $J_{CH_3\,CH_2}$ are in the range of 6.9 to 8.9 Hz. It is obvious that in acyclic compounds the spin-spin couplings are the average values for the various rotational isomers.

Table 5. Some $J_{HH'}^{vic}$ coupling constants [Hz] in ethyl derivatives (according to data compiled by Bothner-By[16])

Compound	$J_{HH'}^{vic}$ [Hz]
CH_3CH_3	8.0
CH_3CH_2MgBr	8.96
$CH_3CH_2CH_3$	7.26
CH_3CH_2Cl	7.23
CH_3CH_2Br	7.33
CH_3CH_2I	7.45
CH_3CH_2CN	7.60
$(CH_3CH_2)_2O$	6.97

[20] *M. Karplus*, J. Chem. Phys. *30*, 11 (1959); J. Am. Chem. Soc. *85*, 2870 (1963).

[21] *K.L. Williamson*, *W.S. Johnson*, J. Am. Chem. Soc. *83*, 4623 (1961).

In *six membered rings*, for *trans* diaxial protons ($\phi = 180°$) $J_{HH'}^{vic}$ is in the range 10.5 to 12.4 Hz and for *cis* couplings ($\phi = 60°$) between 2.7 and 5.4 Hz. In *acetylated sugars* the reported values are somewhat smaller, 5 to 8 Hz for *trans* diaxial, 3 to 4 Hz for *trans* diequatorial, and 2 to 3 Hz for *cis* equatorial-axial couplings. Although the dihedral angle is 60° in the last two cases, there is a significant difference between the corresponding constants.

Anomalies in the magnitude of the vicinal couplings have been reported. In cyclopentenone for instance[22] J_{cis}^{vic} and J_{trans}^{vic} for the fragment —CH_2—CH_2— are 7.2 and 2.2 Hz respectively, in rather good agreement with the corresponding dihedral angles (0° for *cis* and 120° for *trans*). In dihydrothiophene, however, $J_{cis}^{vic} = 10.0$ Hz and $J_{trans}^{vic} = 7.5$ Hz; close values have also been reported for dihydrofuran (10.7 and 8.3 Hz respectively). These values are not only in disagreement with the equation of angular dependence but they increase with the electronegativity of the heteroatom. The possible effect of the lone pair of electrons of the heteroatom is discussed by the authors[22].

In *steroid epoxides* the coupling constants are smaller than the values calculated with the Karplus equation[23].

The vicinal couplings across a *carbon carbon double bond* are in the range +5.7 to +7.0 Hz for *cis* protons and +13.2 to +16.2 Hz for *trans* protons. These experimental values are in rather good agreement with calculations based on the angular dependence relationship of Karplus (g), which gives $J_{cis} = 6.1$ Hz and $J_{trans} = 11.9$ Hz (the calculations are based on a fragment with sp^2 hybridization, a bond length C=C of 1.353Å and an average energy of 9 eV). In cyclic olefins vicinal couplings between the olefinic protons increase with the size of the ring[24], as shown in Table 6.

Table 6. Magnitude of $J_{HH'}^{vic}$ for olefinic protons in cycloalkenes

Ring size	$J_{HH'}^{vic}$ values [Hz]
4	2.5– 3.7
5	5.4– 7.0
6	9.9–10.5
7	9.7–12.5
8	11.8–12.8

[22] *R.J. Abraham, W.A. Thomas*, Chem. Comm. 431 (1965).
[23] *A.D. Cross*, J. Am. Chem. Soc. *84*, 3206 (1962); *K. Tori, T. Komeno, T. Nakagawa*, J. Org. Chem. *29*, 1136 (1964).
[24] *O.L. Chapman*, J. Am. Chem. Soc. *85*, 2014 (1963).

Some coupling constants in aromatic, carbocyclic and heterocyclic rings are given in Table 7.

Table 7. Coupling constants in some aromatic and heterocyclic rings [Hz]

Substance	I values		
	J_o	J_m	J_p
	6–9	1–3	<1
	$J_{\alpha\beta}$	$J_{\beta\gamma}$	$J_{\beta\beta'}$
	4–5	7–8	
X = O:	1.7–1.9	3.1–3.8	
X = NH:	~2.5	~3.7	
X = S:	4.7–5.5	~3.3	

2.4 Long Range Spin-Spin Coupling

Long-range coupling has been defined as that occurring through more than three bonds; such a coupling becomes apparent when the signal due to a proton (or a group of protons) is split more than can be postulated as arising from coupling across two or three bonds. This splitting is usually small and is more effective in unsaturated systems.

The systems which are involved in spin-spin long-range coupling can be arranged in certain groups:
(H–C=C–C–H), allylic
(H–C–C=C–C–H), homoallylic
(H–C=C=C–H), allenic
[H–(C≡C)ₙ–H], acetylenic
conjugated double bonds, coupling through *carbonyl* and through *4σ-bonds* (saturated systems). Long range coupling has also been observed in *aromatic systems* and across *heteroatoms*. We will be concerned with interactions which may contribute to the stereochemical elucidation of a system, restricting ourselves deliberately, since the subject has been thoroughly treated in two extensive reviews[17, 25].

2.4.1 Allylic Coupling

The allylic coupling has probably been most studied. The interaction is transmitted mainly through the π-electron system; the coupling constants are 0–3 Hz. Experimental results show that J_{cisoid} (H^2—H^4) is slightly higher than $J_{transoid}$ (H^1—H^4) by 0.5 Hz.

The steric restrictions on allylic coupling were investigated systematically on rigid systems with well established stereochemistry (mainly natural products). Allylic coupling constants are of maximum values when the angle ϕ is 90° (see formula

[25] *S. Sternhell*, Rev. Pure Appl. Chem. *14*, 15 (1964).

20) and of minimum values when ϕ is 0° or 180°. These results which are compatible with theoretical calculations may be rationalized by considering the $\sigma-\pi$ overlap which is greater when $\varnothing$ is 90° (20).

20

Substitution by *electronegative substituents* (Br, Cl, OH) causes slight variations in the magnitude of the allylic coupling constants showing that the angular dependence is not the only factor influencing the size of this coupling.

Knowing the stereochemical requirements of allylic coupling, the magnitude of $J_{allylic}$ can be of assistance in determining the configuration of substituents.

Allylic bromination e.g. with *N*-bromosuccinimide of a Δ^2-*1-keto steroid*[26] yielded a mixture of two bromo-enones **21** and **22** separable by chromatography.

21   **22**

In compound **21** with the 4-bromo-substituent being axial, the dihedral angle between the plane of the double bond and the 4α-H is very small; the signal of the vinylic 2-H appears like a doublet (J = 10.2 Hz) due to coupling with the vicinal 3-H, no further splitting due to 4α-H being observed. In compound **22** however, the 4-Br is equatorial and consequently the dihedral angle between the double bond and the 4β-H is about 90°; the 2-H in this compound gives rise to a double doublet (J = 10.2 and 1.8 Hz), the small splitting being due to allylic coupling with 4β-H.

2.4.2 Homoallylic Coupling

This coupling deals with protons separated by five bonds located on both sides of a double bond; it lies in the range 0–1.6 Hz, and, comparing *cisoid* and *transoid* systems, the latter class is usually larger by up to 0.5 Hz for pairs of identical substi-

[26] H. Izawa, M. Morisaki, K. Tsuda, Chem. Pharm. Bull. [Tokyo] *14*, 873 (1966).

tuents. As in the allylic coupling constant, the magnitude of J is dependent on the angles ϕ and ϕ' formed by the plane of the double bond and the two C—H bonds involved in the interaction (**23**).

23

Here again the $\sigma-\pi$ overlap can be considered. Somewhat larger coupling constants of 1.9–3.0 Hz are observed in cyclic systems probably due to more favorable angular arrangement of the bonds. For instance (Ref. [1f]) the signal of 6-CH$_3$ in *6-methylpregna-4,6-diene-3,20-dione* (**24**) appears like a quartet of lines. This pattern has been interpreted as arising through allylic coupling with the vinylic 7-H and homoallylic coupling with the 8β-H (axial). In certain cases the homoallylic may be even larger than the corresponding allylic coupling constant.

24   **25**

In *4-hydroxy-3-pentenoic acid lactone* (**25**) the allylic coupling of the methyl group with the olefinic proton is 1.6 Hz, whereas the homoallylic coupling between the same methyl and the methylene protons is 2.7 Hz[25].

2.4.3 Coupling Through Allenic, Acetylenic and Conjugated Systems

In allenic systems the coupling across four bonds is ~ 6 Hz and across five bonds ~ 2–3 Hz. In *2,2-dimethyl-3,4-octadienal* (**26**) the coupling between the two allenic protons 3-H and 5-H is 6.3 Hz and that of 3-H with the C$_6$ methylene is 3.3 Hz. The magnitude of the coupling constants in such a system is explained by effects due to *hyperconjugation*. Since in allenes there is no possibility for *cis* and *trans* isomerism, the coupling constants are equal for both substituents at the end of such a system. In *2-methyl-5-ethyl-5-formyl-2,3-nonadiene* (**27**), $J_{4H-CH_3} = 3$ Hz the pattern of the signal being a regular septet.

26

27   **28**

In *acetylene* itself the vicinal coupling is 9.1 Hz. The coupling is transmitted even through nine bonds, for instance, 0.4 Hz between the methyl and methylene protons in *2,4,6-octatriynol* (**28**). In conjugated systems coupling has been observed between protons five bonds apart. Splitting by 1–2 Hz[25] has been observed in the *cucurbitacins* **29** and **30**.

29   **30**

2.4.4 Coupling Across Four Single Bonds

Usually, protons separated by more than three saturated bonds do not show appreciable coupling. However, this can be observed in cases when the interacting protons are in a definite stereochemical arrangement described as the zig-zag, W or M path.

Comparison of a series of substituted *5-bicyclo-(2,1,1), hexanols*[27] (**31, 32**) shows that in the *exo* series the H_a is coupled to H_a' to an extent of 6.8–8 Hz, while in the *endo* series no coupling is observed.

31   **32**   **33**   **34**

In *endo 6-tricyclo (3,2,1,0^{2,4})-octene* (**33**)[28] a coupling of 2.3 Hz is observed between H_a and H_a' (across five bonds) while in the corresponding *exo* compound (**34**) such a coupling was not seen.

---

[27] *K.B. Wiberg, B.R. Lowry, B.J. Nist*, J. Am. Chem. Soc. *84*, 1594 (1962).

[28] *K. Tori, M. Ohtsuru*, Chem. Comm. 886 (1966).

Measurements of the half height width ($W^{1/2}$) of the methyl group in *9-methyl-decalins* and of the 10-CH_3 in steroids have shown that $W^{1/2}$ is larger in the *trans* than in the corresponding *cis* compounds. This fact is attributed to long range coupling across a W path[29].

In the *5α-steroids (trans)* the axial protons at positions 1,5 and 9 are properly oriented for such a coupling, while in a *5β-steroid (cis)* only the 9-H is in the required geometrical arrangement.

35   **36**

2.4.5 Through Space Coupling

Some unexpected long range couplings, encountered in the NMR spectra of fluorine containing compounds (J_{FF} and J_{HF}), have been accounted for by assuming that the major contribution to the coupling constants is due to a direct through space coupling rather than to a usual through bond mechanism[30]. Such a transmission of interaction may well take place by orbital overlap, when the interacting nuclei are properly oriented along the connecting bonds.

In *4,4-dichloro-3,3-difluoro-1-methyl-1-phenyl-2-deutero-cyclobutane*[31] (**37**), the methyl protons are coupled to the fluorine atom (5σ bonds apart) on the same side of the cyclobutane ring, but no coupling is observed with the second fluorine atom, also separated by 5σ bonds.

37   **38**   **39**

The $J_{2,4}$ couplings between the corresponding fluorine atoms in the two rotamers **38** and **39** of *perfluoroacroleine*[32] cannot be interpreted by a through bond mechanism (see Table 8).

---

[29] *K.L. Williamson, T. Howell, T.A. Spencer*, J. Am. Chem. Soc. *88*, 325 (1966).

[30] *D.R. Davis, R.P. Lutz, J.D. Roberts*, J. Am. Chem. Soc. *83*, 246 (1961).
L. Petrakis, C.H. Sederholm, J. Chem. Phys. *35*, 1243 (1961).

[31] *M. Takahashi, D.R. Davis, J.D. Roberts*, J. Am. Chem. Soc. *84*, 2953 (1962).

[32] *K.C. Ramey, W.S. Brey Jr.*, J. Chem. Phys. *40*, 2349 (1964).

Table 8. Through space coupling

Coupling nuclei		number of σ-bonds (distance)	chemical shifts coupling constants [Hz]				Ref.
			with through space coupling		without through space coupling		
F	F	5 (5)	—/84.5	(38)	—/<2	(39)	32
10—CH$_3$	6—F	5 (5)	—/1.8–5.2	(40)	—/—	(41)	33
13—CH$_3$	11—F	5 (5)	—/3–5.3	(42)			33
N—CH$_3$	F	4 (6)	2.78/1.2	(43)	3.03/—	(43)	34

The long range coupling between H and F has been studied[33] in many fluorinated steroids (e.g. **40–42**) in which the rigid framework of the molecule allows unequivocal stereochemical assignments (Table 8).

40   **41**   **42**

In order to rationalize these observations, the authors[33] have advanced the empirical 'converging vector rule' accounting for the stereochemical requirements of such couplings.

Long range coupling between angular methyl protons and fluorine, five or more σ bonds apart, may occur only when a vector directed along the C—F bond and originating at the carbon atom, converges upon and intersects a vector drawn along an angular methyl C—H bond in the direction of the proton and originating at the methyl carbon. Since the methyl group is rotating, the C—H vector sweeps a cone with the nodal surface intersecting the C—F vector. This rule, however, does not refute unequivocally the through bond mechanism.

A compound in which the involved nuclei are separated by six bonds (4σ, one aromatic, and one with 'partial' double bond character) is *o-fluoro-N,N-dimethyl-benzamide*[34] (**43**); the nonequivalence of the two *N*-methyl groups is due to restricted rotation around the C—N bond.

The coupling occurs only with the higher field methyl group (Table 8) which is closer to the

fluorine but is held outside the plane of the aromatic ring (explaining its high field resonance position).

Inconsistency with the predominance of a through space mechanism is found[35] by comparing the NMR spectra of *1,1,4,4-tetrafluoro-1,3-butadiene* (**44**, s-*trans* according to the dipole moment) and *bis-(4,5-difluoromethylene)-cyclohexene* (**45**).

Though the conformation of the latter is obviously s-*cis* the J$_{FF}$ constants are identical.

The study[36] of long range coupling (J$_{HF}$) in *5,6,7,8-tetrafluoro-1-tert.-butyl-1,4-ethano-naphthalene* (**46**) has revealed the restricted rotation of the tert.-butyl group.

At room temperature two different types of methyl groups are found, giving rise to a 3 × H singlet and 6 × H doublet (J$_{HF}$ 2.9 Hz), the latter being due to coupling with the nearest fluorine atom. At 200° however, enhanced rotation of the tert-butyl substituent leads to equivalence of all three methyl groups appearing as a 9 × H doublet (J$_{HF}$ 2.2 Hz). This coupling is thought to take place 'through space', although a through bond mechanism is not excluded.

A very interesting example[37] of coupling between two hydrogen atoms is given by the NMR spectrum of *1,8,9,10,11,11-hexachloro-4-hydroxy-pentacyclo [6.2.1.1^{3,6}.0^{2,7}.0^{5,9}] dodecane* (**47**).

43

44   **45**

46   **47**

33 *A.D. Cross, P.W. Landis*, J. Am. Chem. Soc. *86*, 4005 (1964); *A.D. Cross*, *86*, 4011 (1964).
34 *A.H. Lewin*, J. Am. Chem. Soc. *86*, 2303 (1964).
35 *K.L. Servis, J.D. Roberts*, J. Am. Chem. Soc. *87*, 1339 (1965).
36 *J.P.N. Brewer, H. Heaney, B.A. Marples*, Chem. Comm. *1967*, 27.
37 *F.A.L. Anet, A.J.R. Bourn, P. Carter, S. Winstein*, J. Am. Chem. Soc. *87*, 5249 (1965).

H_a appears as a well resolved double doublet with splittings of 1.1 and 4.8 Hz, whereas H_b shows a doublet with a splitting of 1.1 Hz; moreover, irradiation of H_c leads to a sharpening of this doublet, thus indicating the existence of a small coupling between H_b and H_c. It is suggested that the coupling between H_a and H_b takes place through space, by way of the unshared electrons of the oxygen. A weak hydrogen bond between H_b and the oxygen atom can also be considered.

2.5 NMR in Conformational Analysis

Conformational analysis concerns the physical and chemical properties of a compound, in terms of the conformations of the pertinent ground states, transition states, and (in the case of spectra) excited states[38]. Nuclear magnetic resonance is one of the best methods for the analysis of the conformation of molecules containing atoms with spin $I \neq 0$; it is based on two kinds of measurements, the difference existing between the chemical shifts of differently oriented magnetic nuclei, and the magnitude of their coupling constants with neighboring protons. In the following discussion we shall confine our attention to some aspects of the magnetic resonance of protons in conformational mobile systems.

2.5.1 Cyclohexane

The interconversion of cyclohexane from one *chair form* to the other results in an equatorial orientation of the previously *axial*, and correspondingly an axial orientation of the *equatorial* hydrogens. In comparison with the rate of inversion for **48** at room temperature, the process leading to the NMR signal (the transition of an excited spin from the parallel to the antiparallel state) is slow.

Consequently, the magnetic environment experienced by a certain cyclohexane proton will be the average of the environments of the given proton in its two spatial orientations (axial and equatorial) corresponding to the two possible chair conformations.

In general, the average chemical shift of a proton in such a mobile system is given by the relation

$$\delta = N_a \delta_a + N_e \delta_e \qquad \text{(h)}$$

where N_a and N_e are the mole fractions of the conformers possessing the given proton in the axial and equatorial orientations respectively, and δ_a and δ_e are the corresponding chemical shifts of the proton in each of the two conformations.

In the case of *cyclohexane,* the rate of interconversion of the two chairs is considerably reduced at low temperature, so that the signals of the axial and equatorial protons (δ_a and δ_e) appear separately, the axial at higher field than the equatorial (see Section 5.5:2.2). Contrary to the very complex spectrum of undeuterated cyclohexane at low temperature, the spectrum of *cyclohexane-d_{11}* at $-100°$ consists of two broadened lines which become very sharp when the deuterons are decoupled from the proton by irradiation at the frequency of the deuterium resonance (Fig. 7). By raising the temperature, these two sharp lines broaden and merge to a single broad band at the coalescence temperature; the band then narrows and becomes extremely sharp at room temperature.

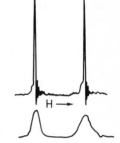

Fig. 7. Proton NMR spectrum of cyclohexane-d_{11} at $-100°$; lower curve, without deuterium decoupling; upper curve, with deuterium decoupling. The separation between the peaks is 28.7 Hz (cf. Ref.[39]).

2.5.2 Substituted Cyclohexanes

A molecule can be fixed in only one conformation by means of the bulky tert-butyl substituent which will force itself into equatorial conformation. The chemical shift of the proton α to the bromine substituent in *cis* and *trans 4-tert-butylbromocyclohexane* is at $\delta = 2.67$ and 3.30 respectively[40]. The chemical shift at room temperature of the proton α to bromine in *bromocyclohexane* is at an intermediate value ($\delta = 3.19$) corresponding to 83% of the conformer possessing the bromine equatorial.

This method relies on the assumption that the chemical shifts of the protons analyzed are not affected by the remote tert-butyl group. Indeed[41], the methoxy resonances in *1,1-dimethoxycyclo-*

[38] *E.L. Eliel, N.L. Allinger, S.J. Angyal, G.A. Morisson,* Conformational Analysis, Interscience Publishers, New York, 1965; *E.L. Eliel,* Angew. Chem. 77, 784, Engl. 4, 761 (1965); *N.C. Franklin, H. Feltkamp,* Angew. Chem. 77, 798, Engl. 4, 774 (1965).

[39] *F.A.L. Anet, A.J.R. Bourn,* J. Am. Chem. Soc. 89, 760 (1967).

[40] *E.L. Eliel,* Chem. Ind. [London] 568 (1959).

[41] *E.L. Eliel, R.J.L. Martin,* J. Am. Chem. Soc. 90, 682 (1968).

hexane (49) give rise to a six proton singlet, since the compound exists as an equilibrium mixture of two indistinguishable conformers.

49

50

4-tert-butyl-1,1-dimethoxycyclohexane (50) the chair-chair interconversion is restricted, and, consequently, the NMR spectrum of this compound shows two signals for the methoxy protons; the location of the six proton signal in the unsubstituted *1,1-dimethoxycyclohexane* is half way between the two signals in the tert-butyl substituted analog. The situation is, however, different in the isomeric *3-tert-butyl-1,1-dimethoxycyclohexane* where the resonances of the equatorial and axial methoxy groups, although resolved, are not equally spaced from the signal in the compound without the tert-butyl group.

2.5.3 Condensed Cyclohexanes

trans-Decalin[42] shows a broad, unresolved and temperature independent band (down to −115°), consistent with a substantial difference between the chemical shifts of the axial and equatorial protons, whereas *cis-decalin* shows a single, relatively narrow absorption.

Since this behavior could well be due to the interconversion between the two possible chair-chair forms of *cis*-decalin, the low temperature NMR spectrum of *2,2-ethylenedioxy-9-methyldecalin* was studied. The spatial relationship between the angular methyl and the dioxolane substituent are different in the two chair-chair conformations 51 and 52; indeed. Below −42°, the signal of the methyl group is split into two lines of nearly equal intensity, separated by 4.1 Hz. The free energy of activation of this interconversion, calculated from the coalescence temperature is 12.5 kcal/mole (for an account of the calculation of the rate of interconversion and of the free energy of activation, see: W.A. Thomas in Ref.[1d], Vol. I, p. 44).

51

52

2.5.4 Heterocyclic Six Membered Rings

Of the studies concerned with the inversion of heterocyclic six membered rings, two examples will be discussed.

53

In *hexahydrotriazines*[43] (53) the methylene protons give a sharp singlet at room temperature which broadens on lowering the temperature and ultimately forms a well defined AB system. The signals of the N-alkylsubstituents are not, however, temperature dependent. These observations have been rationalized as a proof of the interconversion of the two equivalent chair conformations in which fast nitrogen inversion leads to an averaged signal for the N-alkyl groups.

The rate of inversion of the nitrogen relative to the rate of inversion of the ring has been studied[44] by an analysis of the fluorine resonances in *4,4-difluoropiperidine*. At 35°, the F signal is a relatively sharp binomial quintet; the axial and equatorial fluorines show no chemical shift difference due to rapid conformational equilibrium, the multiplicity being due to coupling with the adjacent protons in 54. Below −40° a typical distorted AB pattern of axial and equatorial gem-fluorines appears in chloroform, methanol or acetone; in the last solvent, however, two unequal AB patterns in a ratio of 76:24 are obtained below −88°. These results constitute a proof for the slow inversion of the nitrogen in 55. The lowering of the rate of inversion of nitrogen only in acetone is attributed to a rapid intermolecular NH exchange in the other solvents.

54

55

[42] F.G. Riddell, M.J.T. Robinson, Chem. Comm. 227 (1965).

[43] J.M. Lehn, F.G. Riddell, B.J. Price, I.O. Sutherland, J. Chem. Soc. (B), 384 (1967).

[44] G.A. Yousif, J.D. Roberts, J. Am. Chem. Soc. 90, 6428 (1928).

2.5.5 Rigid Ring Systems

The distortions which can be introduced in a six membered ring, fused in a rigid system like the steroid framework, can be accounted for by the analysis of the signal of the 6α-H (equatorial) in some *6β-substituted* Δ⁴-*3-oxosteroids*[1f] (**56**). If X is relatively small (OH, OCOCH₃) the 6α-H is approximately equally split by the two neighboring 7-H giving rise to a triplet. For a bulkier X, however (Cl, Br), the 6α-H gives rise to a double doublet, i.e. different coupling constants with the 7α and β protons. This signal is interpreted in terms of the repulsion between the bulky 6β substituent and the 10-CH₃ distorting the normal chair conformaion of ring B.

The distortion introduced by a 5,6-epoxide ring in the A/B ring system of steroids can be accounted for by analyzing the pattern of the signal of the 4α-H in *4β-substituted 5α, 6α-(or 5β, 6β)-epoxy-cholestanes*[45] (**57** or **58**).

56   **57**   **58**

In the absence of the epoxide ring, the 4α-H will be equatorial in a 5α-steroid and axial in the 5β counterpart. In the 5α, 6α-epoxy derivative **57** the 4α-H preserves the equatorial orientation, whereas in the 5β, 6β-epoxy derivative **58** the conformational distortion induced in rings A/B imposes the same equatorial configuration for the 4α-H. Consequently, 4α-H will be equally split by the neighboring protons in both compounds giving rise to narrow triplets ($\delta = 4.30$ in **57** and 4.57 in **58**; X = OAc).

2.5.6 Large Rings

The conformational mobility of *larger rings* has been investigated by NMR; stereostructure **a** is assigned to *2,6,6,9-tetramethyl-all-trans-1,4,8-cycloundecatriene* (*humulene*, **59**) by an X ray analysis of its silver nitrate adduct.

a   **59**   **b**

The geminal methyl groups as well as the pairs of protons on carbons 3, 7, 10, 11 are expected to be chemically different[46]. Flipping the ring round the *trans* substituted 8–9-double bond and crossing over the given hydrogens, one obtains conformation **b** which is the mirror image of **a**. The interconversion **a** ⇌ **b** renders identical the substituents on each of the saturated carbon atoms.

The most remarkable temperature dependent changes in such a system have been observed with the 3-oxoderivative of **59** *(zerumbone)*. At 37° the *gem* dimethyl group appears as a broad singlet ($W\frac{1}{2} = 3.3$ Hz); at 75° this signal is a sharp singlet ($W\frac{1}{2} = 0.5$ Hz) whereas below 28° the signal is split into a 1:1 doublet, the separation reaching 10 Hz at −10°.

2.6 Restricted Rotation

The activation energy for the isomerization about a double bond is significantly lowered if the π electron system of the double bond is extensively delocalized in the ground state of the molecule e.g. **60**. The sequence **61** describes the complete equilibrium in such a process[47].

60

61

The shape of the signal of the substituents X and R are temperature dependent, thus indicating their involvement in a kinetic process in which they exchange their identities. Furthermore, it has been established that rotation about the carbon-carbon double bond and the sp² carbon-nitrogen bond are two independent processes occuring in the same molecule. The same applies for enamines, enol

[45] D. Lavie, Y. Kashman, E. Glotter, Tetrahedron, **22**, 1103 (1966).

[46] S. Dev, J.E. Anderson, V. Cormier, N.P. Damodaran, J.D. Roberts, J. Am. Chem. Soc. **90**, 1246 (1968).

[47] Y. Shvo, E.C. Taylor, J. Bartulin, Tetrahedron Lett. 3259 (1967);
Y. Shvo, H. Shanan-Atidi, J. Am. Chem. Soc. **91**, 6683 (1969);
Y. Shvo, J. Belsky, Tetrahedron **25**, 4649 (1969).

ethers, and thioenol ethers. The energy barriers associated with the rotation about the C=C bond are very sensitive to the nature of the substituents. Interesting studies on the partial double bond character of the carbon-nitrogen bond in amides have shown that the two N—CH$_3$ groups in *N,N-dimethylacetamide* appear as two singlets separated by 4.8 Hz (according to the different position of each methyl towards the carbonyl group). The rate of rotation around the C—N bond increases with temperature and, consequently, the signals broaden and finally collapse to one sharp singlet at higher temperature.

In *N,N-dimethyl-2,4,6-trimethylbenzamide*[49], the two N—CH$_3$ groups give rise to two singlets at δ = 2.66 and 3.00 (in hexachlorobutadiene) while at 180° one signal with the intensity of six protons for both N—CH$_3$ groups is obtained (δ 2.77). The free energy of activation calculated from the separation of the signal at 39° (20 Hz) and the coalescence temperature (160°) is 22.5 Kcal/mole.

Restricted rotation is also studied by equilibration methods: *N-methyl-N-benzyl-2,4,6-trimethyl-benzamide,* immediately after dissolution in CCl$_4$, shows singlets for the N-methyl and benzyl protons in **62a**. On standing in the solvent, the intensity of these signals is reduced and concomitantly new signals appear for **63a** corresponding to an isomerization around the C—N bond. At 39° the equilibrium (74:26) is reached after 100 min.

62 a : R = CH$_3$ **b** : R = C(CH$_3$)$_3$ · **63**

	methyl-H	benzyl-H	equi-librium		methyl-H	benzyl-H
62a	2.57	4.67	74%	**62b**	2.55	4.65
63a	2.93	4.20	26%	**63b**	2.97	4.09

The rate of rotation around the C—N bond is significantly lowered in sterically hindered amides. At room temperature *N-methyl-N-benzyl-2,4,6-tri-tert-butyl-benzamide*[50] has two isomers (**62b** and **63b**) which can be separated by fractional crystallization and preparative thin layer chromatography. The assignments of structures **62b** and **63b** respectively are based on the

expected shielding of the N—CH$_3$ by the tri-tert-butyl-phenyl group in isomer **62b** and the corresponding shielding of the N—CH$_2$—C$_6$H$_5$ in **63b**. No isomerization of **62b** into **63b** or vice versa can be detected in solution at 80°. The same equilibrium mixture containing predominantly **62b** is obtained, however, from each of the two isomers at 160–180°. It is noteworthy that the two signals of the N—CH$_3$ groups in the last compound do not coalesce and even do not broaden until 170°.

Also the restricted rotation in substituted *benzophenones, benzils*[51], and *hydrazones*[52] has been investigated.

Table 9. Activation energies [47,48] for rotation about C=C bonds in compounds

X$_1$	X$_2$	R$_1$	R$_2$	ΔG≠ (Kcal/mole)
CO$_2$CH$_3$	CO$_2$CH$_3$	H	N(CH$_3$)$_2$	15.6
CO$_2$CH$_3$	CO$_2$CH$_3$	CH$_3$	N(CH$_3$)$_2$	<8.9
CO$_2$CH$_3$	CO$_2$CH$_3$	H	N(CH$_3$)-⟨⟩-NO$_2$	22.1
CO$_2$CH$_3$	CO$_2$CH$_3$	H	O—CH$_3$	27.7
CO$_2$CH$_3$	CO$_2$CH$_3$	C—(CH$_3$)$_3$	O—CH$_3$	18.3
CN	CO$_2$CH$_3$	S—CH$_3$	S—CH$_3$	24.6
NO$_2$	H	S—CH$_3$	S—CH$_3$	14.8

2.7 Solvent Effects[53]

The solvents which are customarily used in NMR spectroscopy are nonaromatic, including water, and aromatic. In order to avoid overlapping of proton signals of the solvent and of the solute, the solvents are usually deuterated.

2.7.1 Solvent Shifts in Methyl Protons

Protons in *nonpolar molecules,* or remote from a polar group within the molecule show very small variation of their chemical shifts in different solvents of various dielectric constants. Such variations are, however, more pronounced when the corresponding proton or group of equivalent protons are situated in the proximity of the polar substituent. The chemical shifts of three methyl groups attached to the tert-C-atoms 8, 10 and 13 of *8,13-epoxy-14-labdene (epimanoyl oxide)* (**64**) have been measured in a series of solvents of increasing dielectric constant[54]; whereas the position of the 10-CH$_3$ is almost independent of the

[48] G. Isaksson, J. Sandström, I. Wennerbeck, Tetrahedron Lett. 2233 (1967).

[49] A. Mannschreck, Tetrahedron Lett. 1341 (1965).

[50] H.A. Staab, D. Lauer, Tetrahedron Lett. 4593 (1966); Chem. Ber. *101*, 864 (1968).

[51] D. Lauer, H.A. Staab, Tetrahedron Lett. 177 (1968); Chem. Ber. *102*, 1631 (1969).

[52] A. Mannschreck, U. Koelle, Tetrahedron Lett. 863 (1967).

[53] J. Ronayne, D.H. Williams in Ref. [1d], Vol. II, p. 83.

[54] P. Laszlo, Bull. Soc. Chim. Fr. 85 (1964); s.a. Ref. [1f], Chapter 7.

solvent, the chemical shifts of 8-CH$_3$ and 13-CH$_3$, which flank the etheric oxygen in ring C, show significant variations.

64

In his theoretical studies on the long range dipolar shielding of protons, McConnell[3b] has demonstrated that the average shielding of a nucleus in a molecule is influenced by the magnetization induced in the electrons of a neighboring molecule, provided that the latter is magnetically anisotropic; such conditions are most pronounced in aromatic liquids.

The differential shielding of protons in benzene and in chloroform or carbon tetrachloride has been successfully applied in the study of the chemical shifts of protons which are in the proximity of a ketone group.

The difference between the chemical shifts of a given proton in two different solvents is called solvent shift and is defined by the relation:

$$\Delta^{S_1}_{S_2} = \delta s_1 - \delta s_2 \tag{i}$$

Whenever aromatic and nonaromatic solvents are used, the chemical shift measured in the former is subtracted from that of the latter. The signals of the solvent shifts are negative for downfield, and positive for upfield shifts induced by the aromatic solvent. The additivity rules found for the chemical shifts (see 5.5:2.2.2) apply as well to solvent shifts[55].

2.7.2 Solvent Shifts in Keto Compounds

Extensive studies have been performed[56] on the solvent shifts ($\Delta^{CDCl_3}_{C_6H_6}$) of the angular methyl groups in steroidal ketones with the carbonyl group at various positions. In an unsubstituted *steroid* (**8–11**), the resonances of the 10-CH$_3$ and 13-CH$_3$ occur at about the same positions in chloroform and benzene, suggesting that in the absence of polar functional groups, no preferred geometrical relationship exists between the solute molecule and the solvent.

The magnitude and the sign of the solvent shifts experienced by the methyl groups in steroidal ketones reflect the spatial relationship between the given methyl and the carbonyl function (see Table 10).-

Table 10. Solvent shifts of the angular methyl groups in some steroidal ketones (5α-androstane derivatives)[56] in dependence of the position of the oxo group [δ units]

Position	$\Delta^{CDCl_3}_{C_6H_6}$		Position	$\Delta^{CDCl_3}_{C_6H_6}$	
	10–CH$_3$	13–CH$_3$		10–CH$_3$	13–CH$_3$
1	+0.30	0.00	11	−0.14	+0.11
2	+0.16	+0.06	12	+0.24	+0.26
3	+0.37	+0.10	15	+0.10	+0.22
6	+0.12	+0.06	16	+0.15	+0.32
7	+0.32	+0.08	17	+0.12	+0.22

The solvent shifts have been interpreted[56, 57] in terms of an empirical rule which has good predictive value: if a reference plane P is drawn through the carbon of the carbonyl group at right angles to the carbon oxygen bond, then protons close to P show very small solvent shifts; protons in front of P, i.e., on the same side as the oxygen of the carbonyl, are deshielded, while protons behind it are shielded.

Inspection of a Dreiding model of *5α-androstan-11-one* for instance, shows clearly that the 10-CH$_3$ is in front and parallel to the hypothetical plane P, whereas the 13-CH$_3$ is similarly oriented but behind this plane; indeed, the solvent shift experienced by the former methyl group is negative while that of the latter is positive (Table 10).

In the case of 1-oxosteroids the spatial relationship between the 10-CH$_3$ and the carbonyl group depends on the stereochemistry of the junction between rings A and B. If the rings are *trans* fused (5α-H) the 10-CH$_3$ is behind and parallel to the plane P, and this is reflected in a positive $\Delta^{CDCl_3}_{C_6H_6}$ (Table 10); if the rings are, however, *cis* fused (5β-H), the 10-CH$_3$ protons are close to P, i.e., the C=O and the 10-CH$_3$ are almost in the same plane; indeed the solvent shift in a 1-oxo-5β-steroid is slightly negative (−0.11 p.p.m.)[58].

In *2,2,6-trimethyl-cyclohexanone*[59] (**65**), the solvent shift ($\Delta^{CCl_4}_{C_6H_6}$) of the axial methyl group is positive (+0.26 p.p.m.), whereas the two equatorial methyls experience slight negative shifts of −0.07 and −0.09 p.p.m.

65   **66**

[55] M. Fétizon, J.C. Gramain, Bull. Soc. Chim. Fr. 3444 (1966).

[56] D.H. Williams, N.S. Bhacca, Tetrahedron *21*, 2021 (165); s.a. Ref. [53].

[57] J.D. Connolly, R. McCrindle, Chem. Ind. 379 (1965).

[58] E. Glotter, D. Lavie, J. Chem. Soc. (C) 2298 (1967).

[59] M. Fétizon, J. Goré, P. Laszlo, B. Waegell, J. Org. Chem. *31*, 4047 (1966).

Solvent shifts are used efficiently in conformational problems. This method enabled[58] the assignment of a boat conformation to ring A in a derivative **66** of the naturally occurring steroidal lactone *whithaferin A*.

The chemical shifts induced by benzene in ketones have been mechanistically[60] interpreted in terms of a transient 1:1 collision complex between solute and solvent in which the π electrons of the benzene ring interact with the partial positive charge on the carbon atom of the carbonyl. The molecule of benzene probably will be nonplanarly oriented towards the electron deficient site in the solute molecule with the benzene ring as far as possible from the negative end of the dipole. Solvent effects must be measured for dilute solutions only when maximum *complexing* occurs[61].

Johnson-Bovey calculations for the magnetic field around a benzene ring have been used to indicate if the solvent shifts induced by benzene in *5α-androstan-11-one* are due to an association between the benzene and the carbonyl group[62].

An aromatic solvent molecule can interact with a polar site in the solute molecule and assume a preferential orientation. The diamagnetic anisotropy of benzene is such that shielding or deshielding of a proton in the solute will depend on the geometry of the solvent-solute complex. Calculations made for the hypothesis, that the collision complex between benzene and the ketone is such that the π electrons of the benzene and the positive end of the C=O dipole interact in an approximately planar association, cannot account for the deshielding of an equatorial proton adjacent to the carbonyl and the shielding of the corresponding axial proton. A collision complex in which the plane of the benzene ring is almost at right angles to the overall plane of the steroid molecule may, however, well account for the observed shifts.

Solvent shifts ($\Delta_{C_6H_6}^{CCl_4}$) of protons in cycloalkanones of different ring sizes have been studied[63].

For the evaluation of the solvent-solute interaction, a variable temperature study of *5α-androstane-11-one* in toluene as solvent has been performed[65]. At low temperature the equilibrium solvent + ketone ⇌ complex is shifted towards complex formation.

α,β-unsaturated ketones may exist in two approximately coplanar forms known as s-*cis* (**67**) and s-*trans* (**68**).

s-cis   s-trans

The solvent shifts observed[66] with a series of such ketones can best be explained by the same model as proposed for cyclic saturated ketones, in which a reference plane is drawn at right angle to the carbonyl bond.

The main differences are found in the solvent shifts of the Rβ^{cis} substituents which are strongly positive in the s-*trans* series and slightly negative in the s-*cis* ketones.

The solvent shifts of protons in α,β-*unsaturated ketones* are temperature dependent (in toluene-d$_8$)[67], thus substantiating the assumption that on lowering the temperature the equilibrium solvent + solute ⇌ complex is displaced towards complex formation. In *isophrone* (**68a** in table 11), the β cis CH$_2$ moves upfield by 0.6 p.p.m. upon decreasing the temperature from +80° to −80°. In *trans-2-ethylidenecyclohexanone* (**69**, s-*cis*) the β cis H shifts downfield on cooling while the β-trans methyl moves upfield.

The variable temperature study can be applied to the conformational analysis of mobile systems, which can exist in the s-*cis* as well as s-*trans* conformation. *Mesityl oxide* (**70**) was thus shown to exist predominantly in the s-*cis* conformation while the Δ^{16}-20-oxo group in Δ^{16}-*pregnen-3β-ol-20-one acetate* (**71**) prefers the s-*trans* conformation.

2.7.3 Solvent Shifts in Other Oxygen Compounds

In six membered ring lactones, methyl groups on carbon atoms β to the lactonic carbonyl show larger solvent shifts ($\Delta_{C_6H_6}^{CDCl_3}$) for equatorial than for their axial counterparts; in both cases the shifts are positive. All results have been interpreted using the collision complex model[64].

[60] *J. Ronayne, D.H. Williams,* Chem. Comm. 712 (1966).

[61] *K.M. Baker, B.R. Davis,* J. Chem. Soc. *(B)* 261 (1968).

[62] *D.H. Williams, D.A. Wilson,* J. Chem. Soc. *(B)* 144 (1966).

[63] *T. Ledaal,* Tetrahedron Lett. 651 (1968).

[64] *D. Lavie, I. Kirson, E. Glotter, G. Snatzke,* Tetrahedron **26**, 2221 (1970).

[65] *P. Laszlo, D.H. Williams,* J. Am. Chem. Soc. **88**, 2799 (1966).

[66] *C.J. Timmons,* Chem. Comm. 576 (1965).

[67] *J. Ronayne, M.V. Sargent, D.H. Williams,* J. Am. Chem. Soc. **88**, 5288 (1966).

Table 11. Solvent shifts experienced by different protons in unsaturated ketones[66] **67** and **68** [δ-units]

Compound		$\Delta_{C_6H_6}^{CCl_4}$ for different protons			
		R	R^a	R^β *trans*	R^β *cis*
68a	s-*trans*,	CH$_2$ +0.03	H —0.10	CH$_3$ +0.32	CH$_2$ +0.39
b	s-*trans*,	CH$_3$ +0.19	H +0.07	CH$_3$ +0.38	H +0.27
67a				CH$_3$ +0.26	CH$_3$ —0.11
b	s-*cis*,			CH$_3$ +0.32	CH$_3$ —0.01
c		CH$_3$ +0.17	H +0.19	CH$_3$ +0.31	CH$_3$ +0.03

Also protons attached directly to the *epoxide* ring undergo an upfield solvent shift (Table 12), the corresponding shifts decreasing for larger cyclic ethers.

Axial and equatorial acids possessing a methyl group on the adjacent carbon atom can be distinguished according to the solvent shift of this methyl, as shown with the 4-CH$_3$ of *voucapenic acid* (**72**, equatorial) and *vinhaticoic acid* (**73**, axial).

The sign of the solvent shifts is reversed in this case, as compared to those found in ketones; no explanation, however, is advanced by the authors.

Similar observations have been made with the pseudoequatorial or pseudoaxial methyl groups adjacent to tte lactone carbonyl group in γ-lactones[70] and with methoxyl groups[71] in different methoxybenzenes. It seems that the solvent shift of the methoxy group increases with increasing electron withdrawal of the *para* substituent.

72   **73**   **74**

Table 12. Solvent shifts of protons and methyl groups.

Substance	Group	$\Delta_{C_6H_6}^{CCl_4}$	eq. ax (pseudo-eq. ax.)	Ref.
Epoxides	a-H	+0.13–0.38		[68]
72, 73	a-CH$_3$		+0.08–0.03	[69]
γ-Lactone	a-CH$_3$		($\sim$+0.23) ($\sim$+0.46)	[70]
Methoxy benzenes	OCH$_3$	+0.37–0.45		[71]
p-Nitroanisol	OCH$_3$	+0.89		[71]

[68] D.H. Williams, J. Ronayne, H.W. Moore, H.R. Shelden, J. Org. Chem. *33*, 998 (1968).
[69] C.R. Narayanan, N.K. Venkatasubramabian, Tetrahedron Lett. 3639 (1965).
[70] C.R. Narayanan, N.K. Venkatasubramabian, J. Org. Chem. *33*, 3156 (1968).
[71] J.H. Bowie, J. Ronayne, D.H. Williams, J. Chem. Soc. *(B)* 785 (1966).

Enhancement has been observed in certain cases upon addition of a few drops of trifluoroacetic acid[72]; the $\Delta_{C_6H_6}^{CDCl_3}$ shift of the 3-CH_3O group in *3-methoxy-flavone* (**74**) is enhanced from $+0.18$ p.p.m. to $+0.45$ p.p.m. by addition of CF_3COOH.

2.7.4 Solvent Shifts in *N*-Methyl Groups

N-Methyl groups can be readily distinguished from other methyls by the downfield shifts induced on adding some CF_3COOH to their chloroform solution[73].

2.7.5 Influence of Pyridine

The pyridine induced solvent shifts in protons which are in the proximity of ketone groups are in the same direction as the shifts induced by benzene, however, much smaller. These shifts are useful for stereochemical assignments in hydroxyl containing compounds[74]. In rigid systems, protons or methyl groups bearing a 1, 3-diaxial relationship to a hydroxyl group experience negative shifts of -0.2 to -0.4 p.p.m. In the case of 1, 2-relationship, the dihedral angle ϕ between the proton (or methyl and the hydroxyl group seems to play an important role in the size of the shift.

Dihedral angle φ	160°	180°	$\sim 85°$	$\sim 60°$
$\Delta_{C_5H_5N}^{CDCl_3}$	$-0,03$	$-0,05$	$-0,13$	$-0,20$ to $-0,27$

2.7.6 Influence of Dimethylsulfoxide

Dimethylsulfoxide enables classification of alcohols according to the multiplicity of the OH signal[75]. In this solvent the rate of proton exchange is reduced enough to allow the observation of the OH splitting by the neighboring protons. However, the method seems to be unreliable for alcohols possessing strong electron withdrawing substituents in the proximity of the hydroxyl group e.g. *2, 2, 2-trichloroethanol*[76].

An excellent example illustrating the advantages of dimethylsulfoxide is *7-oxo-6-phthalimido-2, 2-dimethyl-2, 3, 4, 7-tetrahydro-1, 4-thiazepine-3-carboxylic acid methyl ester*[77] (**75**). This structure could be assigned by analysis of the signal exhibited by 3-H, 5-H and NH in dimethylsulfoxide solution. 3-H and 5-H appear as doublets at $\delta = 4.58$ ($J = 6$ Hz) and 7.32 ($J = 9$ Hz) respectively. These splittings are undoubtedly due to coupling with the NH proton which gives rise to a double doublet at $\delta = 8.68$ ($J = 9$; 6 Hz). The coupling with the NH proton could not be observed in acetone solution[78].

75   **76**

2.8 Magnetic Nonequivalence Due to Molecular Asymmetry

The protons of a methylene group attached to an asymmetric center are magnetically nonequivalent in chemical shift, in spin-spin coupling or in both. This phenomenon has been attributed to the unequal population of the corresponding conformers giving rise to a difference in the chemical shifts of the methylene protons, and to the 'intrinsic asymmetry' of the molecule. Both criteria have been considered by Gutowsky in the mathematical analysis of the problem[79].

In *isobutyl bromide* $BrCH_2CH(CH_3)_2$[80] the methylene protons are equivalent since they always experience the same electronic environment due to rapid rotation around the C—C bond and give rise to a doublet (vicinal coupling with the methine proton). However, the situation is different when the CH_2 group is attached to an asymmetric carbon atom such as in *methyl 2,3-dibromo-2-methylpropanoate* (**76**). The methylene protons experience different electronic environments in each of the conformers and, consequently, possess different chemical shifts resulting in an AB type spectrum.

In *diethyl sulfite* $O = S (OCH_2CH_3)_2$ the methylene protons give rise to two slightly displaced signals interpreted as the AB part of an ABC_3 system instead of the binomial quartet (1:3:3:1) expected

[72] *R.G. Wilson, D.H. Williams,* J. Chem. Soc. *(C)* 2477 (1968).

[73] *J.B. Davis,* Chem. Ind. 1094 (1968).
J.C.N. Ma, E.W. Warnhoff, Can. J. Chem. **43**, 1849 (1965).

[74] *P.V. Demarco, E. Farkas, D. Doddrell, B.L. Mylard, E. Wenkert,* J. Am. Chem. Soc. **90**, 5480 (1968).

[75] *O.L. Chapman, R.W. King,* J. Am. Chem. Soc. **86**, 1256 (1964).

[76] *J.G. Traynham, G.A. Knesel,* J. Am. Chem. Soc. **87**, 4220 (1965).

[77] *O.K.J. Kovacs, B. Ekström, B. Sjöberg,* Tetrahedron Lett. 1863 (1969).

[78] *S. Wolfe,* unpubl.

[79] *H.S. Gutowsky,* J. Chem. Phys. **37**, 2196 (1962).

[80] *M.L. Martin, G.J. Martin,* Bull. Soc. Chim. Fr. 2117 (1966).

[81] *J.S. Waugh, F.A. Cotton,* J. Phys. Chem. **65**, 562 (1961).

for the case when a group of equivalent methylene protons couples with the equivalent methyl protons.

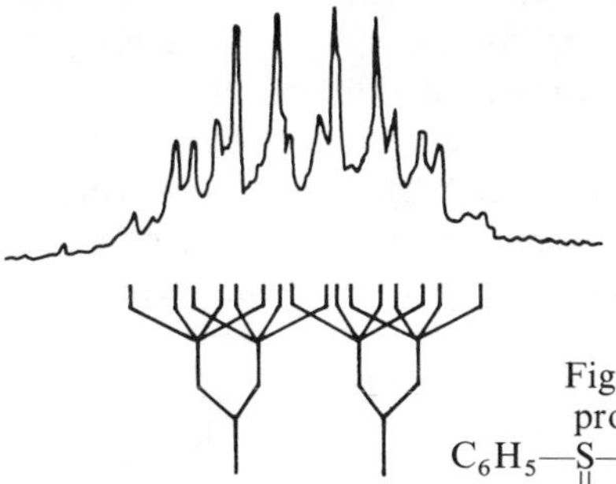

Fig. 8. Methylene protons signal in $C_6H_5-S-OCH_2CH_3$ [81]
(with O double bonded below S)

This spectrum has been explained reasonably by the nonequivalence of the two protons in each methylene group, due to the lack of symmetry of the substituted sulfur atom (double pyramidal configuration of the sulfite group). This assumption was confirmed by the NMR spectrum of $C_6H_5-(SO)-OCH_2CH_3$.

Its methylene group shows also an ABC_3 type spectrum, characteristic for a pair of nonequivalent protoins split by the methyl group (Fig. 8). The nonequivalence of the methylene protons is thus explained only by symmetry arguments, without recourse to conformational isomerism.

2.8.1 Benzylic Compounds

The magnetic nonequivalence of a methylene group may be observed even when the asymmetric center is several bonds apart, or a heteroatom has been inserted between the methylene and the asymmetric center. Such long range effects have been shown in a series of N,N-dimethylbenzylamines[82] substituted in position ortho (77) or meta (78) with an asymmetric group. Whereas the benzylic protons in 77 are nonequivalent and appear as an AB system, the corresponding protons in a symmetric compound (e.g. C_6H_5 instead of OH) give rise to an A_2 type spectrum.

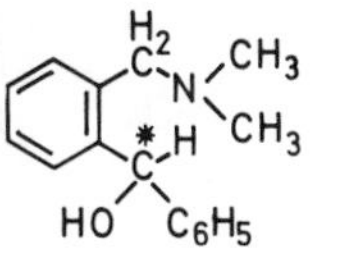

77 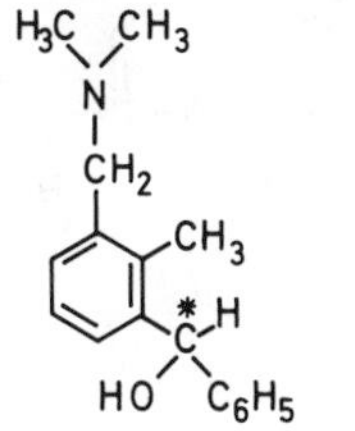   **78**

In order to know the relative contributions to the observed magnetic nonequivalence of the conformational factor (relative populations of the con-

formers), and of the intrinsic asymmetry factor, several compounds of this type have been studied at different temperatures. The data clearly showed that the degree of nonequivalence is reduced but not eliminated at high temperature. Similar results have been obtained with benzylamine 78 and a comparable symmetric compound (C_6H_5 instead of OH).

In benzyl ethers[83], with an asymmetric C-atom next to the oxygen, the magnitude of the observed magnetic nonequivalence of the benzylic protons increases with the size of the R substituent (Table 13).

Table 13. Magnetic nonequivalence of the benzylic protons in a series of compounds of structure:

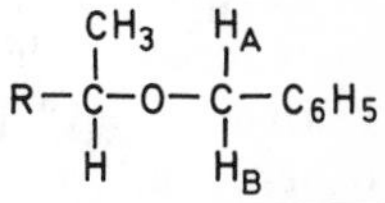

$$R-\underset{\underset{H}{|}}{\overset{\overset{CH_3}{|}}{C}}-O-\underset{\underset{H_B}{|}}{\overset{\overset{H_A}{|}}{C}}-C_6H_5$$

R	$\Delta v = (v_A - v_B)$ [Hz] at 60 MHz		
	in CCl_4	in C_6H_6	in $(CH_3)_2CO$
Ethyl	5.8	6.6	5.8
Isopropyl	8.8	9.3	8.6
Cyclohexyl	9.7	10.2	9.6
tert.-Butyl	14.8	15.7	14.7

2.8.2 Isopropyl Compounds

A similar behavior has been observed with isopropyl groups in the proximity of an asymmetric center. Instead of one doublet for the two equivalent secondary methyl groups, the spectrum of such a compound (79) displays two distinct doublets.

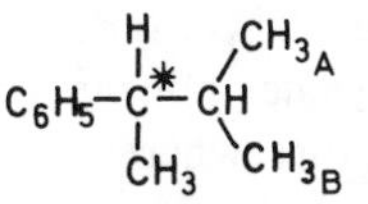

79 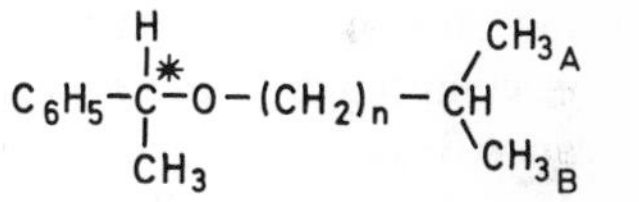   **80**

It is interesting that in compounds of type 80 the nonequivalence is more pronounced when $n = 2$ than $n = 1$. This fact is explained in terms of the conformational preference of the isopropyl group with respect to the asymmetric center, the phenyl ring shielding differently the two isopropyl methyls. The largest effect is observed when the isopropyl group is directly linked to the asymmetric center (Table 14).

[82] J.C. Randall, J.J. McLeskey, P. Smith, M.E. Hobbs, J. Am. Chem. Soc. 86, 3229 (1964).
[83] G.M. Whitesides, D. Holtz, J.D. Roberts, J. Am. Chem. Soc. 86, 2628 (1964).
[84] E. Glotter, M. Bachi, Israel. J. Chem. 8, 633 (1970).

Table 14. Magnetic nonequivalence of the isopropyl protons in compounds **79** and **80**[83]

		$\Delta v (v_{CH_{3-A}} - v_{CH_{3-B}})$ [Hz] at 60 MHz solvent:			
		$(CH_3)_2CO$	C_6H_6	CCl_4	C_5H_5N
79		11.7	8.0	10.9	8.9
80	n=0	5.8	0.8	4.0	3.0
80	n=1	0.0	0.5	0.3	0.0
80	n=2	2.2	1.8	2.5	1.8
80	n=3	0.0	0.8	0.0	0.0
80	n=4	0.0	0.0	0.0	0.0

In a substituted *4-isopropylthiazolidone* (**81a**) the two methyl groups are nonequivalent (two doublets $\delta = 0.92$ and 1.25, $J = 6.5$ Hz; in $CDCl_3$). The degree of nonequivalence of these protons is slightly reduced at higher temperatures (in bromobenzene $\Delta v = 18.3$ Hz at 20°, 13.8 Hz at 140°). In dimethylsulfoxide, however, the compound exists largely in the corresponding thiazolinol form **81b** and the NMR spectrum accordingly shows a six proton doublet due to the now equivalent isopropyl protons (the C_4 is not asymmetric in **81b**)[84].

2.8.3 Biphenyls and Terpenes

In *o,o'*-tetrasubstituted biphenyls[85], the two aromatic rings are not coplanar. In *2,2'-bis-hydroxymethyl-6,6'-diethylbiphenyl* **82** the methylene protons of the CH_2OH groups are magnetically nonequivalent, and give rise to an AB system centered at δ 4.05 and 4.20 with a coupling constant of 12 Hz. In order to observe such nonequivalence, the rotation around the biphenyl bond should be sufficiently restricted; enhancement of the rotation around this bond should, therefore, lead to a lowering of the degree of nonequivalence. Indeed, in compound **83** the *ortho* methylenes appear as an AB quartet at room temperature whereas at 127° the pattern collapses to give an A_2 system.

81a   **81b**

82   **83**

An interesting application of this phenomenon has been made in the study of the stereochemistry of some *terpenes*[86]. In many compounds of type **84** (**a** and **b**) the 4-C-methylene appears as an AB quartet with coupling constants of 10.5–12 Hz. The axial (**84a**) or equatorial (**84b**) orientation of this group can be assigned according to the position of the center of the CH_2 signal; when the group is equatorial, the signal appears at higher field than in the axial counterpart, the difference being ~20 Hz. The nonequivalence of the methylene protons is strongly enhanced if the asymmetric center (C4) is flanked by a carbonyl or a double bond, the difference in the chemical shifts between the two methylenic protons being of the order of 30–40 Hz.

84a   **84b**

2.9 Analysis of Optical Purity[87]

Enantiomers of optically active compounds give superimposable NMR spectra in the usual solvents. However, the optical purity can be determined by converting the enantiomers into diastereomers with an appropriate optically pure chiral reagent, under conditions which exclude possible racemization or epimerization[88]. The ratio of diastereomers must be equal to the ratio of the initial enantiomers.

2.9.1 Formation of Diastereomers

Each diastereomer in the mixture shows a different NMR spectrum and the optical purity can be analyzed without separation of the diastereomers. For instance[89], *o-fluorophenyl-1-ethyl-2-phenylpropionate* (**85**) occurs as a mixture of two diastereomers which can be differentiated in CCl_4 solution according to the signals of the two methyl groups.

85   **86**

[86] *A. Gaudemer, J. Polonsky, E. Wenkert*, Bull. Soc. Chim. Fr. 407 (1964).

[87] *M. Raban, K. Mislow* in *N.L. Allinger, E.L. Eliel*, Topics in Stereochemistry, Vol. II, p. 199, Interscience Publishers, New York 1967.

[85] *W.L. Meyer, R.B. Meyer*, J. Am. Chem. Soc. 85, 2170 (1963).

2.9.2 Application of Optically Active Solvents

The enantiomers of an optically active compound can be differentiated by NMR spectroscopy without having to convert them into a mixture of diastereomers, but only by measuring their spectrum in an optically active solvent[89], since the diastereomeric interactions between solute and solvent are strong enough.

The three equivalent fluorine atoms of racemic *2,2,2-trifluoro-1-phenyl-ethanol* (**86**) in CCl_4 solution give rise to a doublet (by coupling with the neighboring 1-H); however, in optically active (−) α-*phenylethyl amine*, the fluorine signals appear as two equally intense doublets $J_{HF} = 6.9$ Hz, each of the peaks being further split into a narrow triplet due to long range coupling[90]. These doublets will show different intensities if they belong to a partially resolved mixture of enantiomers. If the racemic form of the solvent is used, the fluorines will give one signal (as a doublet of triplets). The carbinyl proton produces one quartet in the racemic solvent and two partially overlapped quartets in the optically active solvent.

The differential shielding of nuclei in enantiomeric compounds dissolved in optically active solvents has been interpreted[91] in terms of the formation of short lived diastereomeric hydrogen bonded carbinol-amine solvates, accounting thereby for the relative positions of the signals in each enantiomer. If partially resolved *2,2,2-trifluorophenyl-ethanol* is used, the more intense higher field carbinyl quartet is assigned to the *dextrorotatory enantiomer*, whereas, the less intense, lower field quartet, is assigned to the *levorotatory enantiomer*. The reverse situation is observed with the fluorine resonances. Conversely, the optical purity of the amine can be determined[92] by running its spectrum in an optically active carbinol.

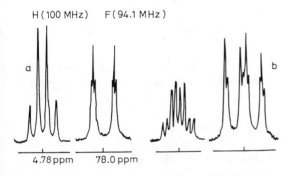

H (100 MHz)   F (94.1 MHz)

a   b

4.78 ppm   78.0 ppm

Fig. 9. Portions of NMR spectra of partially resolved trifluorophenyl ethanol in: a) racemic, and b) optically active (+)α-(1-naphthyl)-ethylamine[91].

These observations can be explained with the aid of a simplified but useful model (**87**). In the dextrorotatory enantiomer, the carbinyl proton is held above the aromatic ring and consequently, its signal is shielded; whereas the CF_3 group, which is further away, is subjected to comparatively less shielding. In the levorotatory enantiomer these two groups change places and, therefore, the CF_3 group will be more shielded.

87   **88**

The same approach[93] has been used for the direct determination of optical purity in sulfoxides. For instance, the spectrum of partially resolved *methyl tert-butylsulfoxide* (**88**) dissolved in an achiral solvent shows two sharp singlets (3:1 ratio) for the tert-butyl and methyl resonances respectively. In (−)R-trifluorophenylethanol as solvent, the spectrum consists of two unequal singlets for the $C-(CH_3)_3$ resonances, and two other unequal singlets for the $S-CH_3$ resonances.

Since the optically active solvent shields the various groups differently in the enantiomers, but does not affect their relative itensities, this method is suitable for direct determination of the proportion of enantiomers in the mixture.

The generality of the method has been proven by showing that such a nonequivalence exists even with methyl-methyl-d_3-sulfoxide, whose asymmetry is due only to isotopic substitution. The fact that the sets of signals collapse by replacement of the optically active solvent with its racemic form, indicates that the molecules of the solute alcohol are equally solvated by the two enantiomers of the amine, due to fast exchange between solute and solvent. Therefore, to observe, a spectral nonequivalence of the solute, the solvent must be optically pure, or diluted with an achiral solvent, or partially resolved. The largest differences between the chemical shifts are obtained in the first case; problems encountered due to the viscosity of such a medium can be overcome by dilution[94]. An NMR study[95] of a

88 *M. Raban, K. Mislow*, Tetrahedron Lett. 3961 (1966).

89 *M. Raban, K. Mislow*, Tetrahedron Lett. 4249 (1965).

90 *W.H. Pirkle*, J. Am. Chem. Soc. **88**, 1837 (1966).

91 *W.H. Pirkle, S.D. Beare*, J. Am. Chem. Soc. **89**, 5485 (1967).

92 *W.H. Pirkle, T.G. Burlingame, S.D. Beare*, Tetrahedron Lett. 5849 (1968).

93 *W.H. Pirkle, S.D. Beare*, J. Am. Chem. Soc. **90**, 6250 (1968).

94 *T.G. Burlingame, W.H. Pirkle*, J. Am. Chem. Soc. **88**, 4294 (1966).

95 *D.A.L. Anet, L.M. Sweeting, T.A. Whitney, D.J. Cram*, Tetrahedron Lett. 2617 (1968).

racemic carbinol (such as **89**) in each of the optically pure diastereomeric sulfoxides **90** and **91**, has shown that hydrogen bonding is a necessary condition to observe diasteromeric chemical shift differences.

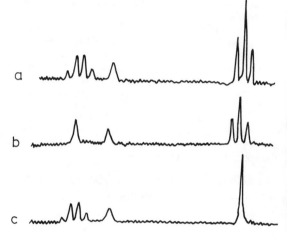

89   **90**   **91**

These decrease at higher temperatures and increase upon dilution. The maximum separation is, however, reached at a molar ratio of 2:1 between solvent and solute. Similar results have been obtained by Pirkle[93].

2.10 Double Resonance and Exchange Processes in Stereochemical Problems[96]

In a single resonance experiment each magnetic nucleus gives rise, as shown in section 5.5:2.2, to a signal which is characterized by its position and multiplicity. Let us consider an AX system of two nuclei with different chemical shifts, reciprocally split, giving rise to two doublets. When a strong oscillatory field, H_2, is applied at the resonance frequency of one of these nuclei (say X) while the second nucleus (A) is irradiated with the usual oscillatory field H_1, the latter being much weaker than H_2 the signal of A will collapse to a singlet. This is due to the rapid transitions induced in X by the strong field H_2 which prevent the transmission of information on its spin states to A. Conversely, irradiation of nucleus A will result in the collapse of the doublet given by X. The amplitude of both the weak and strong oscillatory fields H_1 and H_2 are, however, small in comparison with the static magnetic field H_0.

This type of spectroscopy involving simultaneous irradiation at two different frequencies is called double resonance. It is known as homonuclear or heteronuclear double resonance, depending on whether the nuclei belong to the same, or to different isotopic species. Hereby, it is possible to measure the exact chemical shift of a proton, even among other signals in the same area, and to confirm that the multiplicity of a given proton is due to coupling with the irradiated proton.

To obtain a good decoupling the separation of the chemical shifts of the protons must be at least five times as large as their coupling constant ($\Delta \geqslant 5J$) in order to minimize the effect of the strong field H_2 on the analyzed proton. For example, the signal of

the methyl group in ethanol is a triplet, due to coupling with the two equivalent protons of the methylene group; the latter appears as a quartet by coupling with the three equivalent methyl protons. Irradiation at the resonance position of the methyl group (near $\delta = 0.9$) induces the collapse of the quartet to a singlet, while irradiation at the methylene position (near $\delta = 3.5$) converts the triplet to a singlet (Fig. 10).

Fig. 10. a) Normal spectrum of ethanol
b) Irradiated at the frequency of the methyl group
c) Irradiated at the frequency of the methylene group

In *1-methyl-2-pyridone*[97] (**92**) the protons at C_3 and C_5 give rise to a doublet ($\delta = 6.47$) and a triplet ($\delta = 6.15$) respectively, whereas the protons at C_4 and C_6 show a complex band near $\delta = 7.3$.

92   **93**   **94**

The signals of the 3-H and 5-H can be completely decoupled by irradiation at the position of the 4-H and 6-H, while the C_4 proton will be decoupled only partially by irradiation of the 3-H, since it is also split by the 5-H.

The assignment of the signal due to the 4-H in α-*bromo-camphor*[98] (**93**) was done by irradiation at the resonance position of the 3-H, the chemical shift of which was unambiguously established.

The aromatic protons in the NMR spectrum of *triptycene* (**94**) in CS_2 could be completely anal-

[96] *J.D. Baldeschwieler, E.W. Randall*, Chem. Rev. *63*, 81 (1963).

[97] *D.W. Turner*, J. Chem. Soc. 847 (1962).
[98] *S.L. Manatt, D.D. Elleman*, J. Am. Chem. Soc. *83*, 4095 (1961).

yzed[99] as an AA'BB' system after irradiation at the frequency of the methine proton. The broadening of the low field portion of the spectrum, due to coupling with the methine protons, disappears after irradiation, and splittings of less than 0.25 Hz are clearly resolved.

The three protons flanking the 22-H in a derivative of a natural steroidal lactone (*withaferin A* **95**) were also identified by double resonance[100]. The 22-H appears as a double triplet near $\delta = 4.4$.

95   **96**

Following irradiation at the frequency of the 23-protons the double triplet pattern of 22-H collapsed to a doublet due to coupling with the 20-H only. Similar results were obtained by deuterium exchange[100] of the allylic protons and the vinylic 24-CH$_3$ (**96**). The signal of the 22-H appeared now as a slightly broadened doublet ($J = 4$ Hz), the broadening being due to coupling with the 23-deuterons.

Proton exchange: A common case of collapse of spin-spin multiplet occurs by exchange of acidic protons between different molecules. The hydroxylic proton of ethanol, for example, can change its location from one molecule to another, the process being sharply accelerated by a trace of acid or base. Consequently, the hydroxylic proton of ethanol will usually appear as a singlet, the triplet expected from coupling with the two methylenic protons being observed only in a highly purified sample or at low temperatures.

Double resonance is useful in the study of reversible intramolecular processes[101]. The NMR spectrum of [18]-annulene (**97**) in perdeuteriotoluene is temperature dependent: it shows a single band ($\delta = 5.5$) at 100° which broadens on cooling until it disappears at ~ 50°, then reappears at 20° as two broad signals ($\delta = -3.0$ and $+8.8$); on further cooling to −60° these are resolved into a high field quintet ($\delta = -4.22$) for the six inner protons, and a low field quartet ($\delta = 9.25$) for the outer 12 protons. Upon irradiation of the high field absorption band (at −60°), the low field quartet collapses to a

singlet, and conversely, irradiation of the low field signal causes the collapse of the high field signal to a broad singlet. The irradiation at 20° leads, however, to the complete disappearance of the signal. These effects are due to the transfer of an irradiated outer proton to the position of an inner proton before spin relaxation has completely taken place.

97

Spin decoupling experiments cannot be used when the chemical shift difference is of the same magnitude as the coupling constant between the given protons. However, information concerning the ordering of energy levels of the various transitions and the relative signs of the coupling constants can be obtained in such instances by applying a weak perturbing oscillatory field H$_2$ at each of the resonance positions of one of the protons of the system.

Consider an AB spectrum in which A$_1$, A$_2$ and B$_1$, B$_2$ are the four lines of the system. When the perturbing field H$_2$ is applied at the frequency of A$_1$, then B$_1$ and B$_2$ will be split, the former into a well resolved and the latter into a poorly resolved doublet; A$_2$ remains unchanged. When H$_2$ is applied on A$_2$, then again B$_1$ and B$_2$ will be split into doublets, B$_2$ now giving rise to the well resolved and B$_1$ to the poorly resolved doublet.

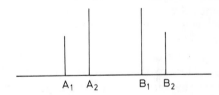

This kind of double resonance experiment is called 'tickling'. A presentation of the phenomenon and reference to the original works of Freeman and Anderson are found in Ref. [1c], vol. I, p. 471.

2.11 The Nuclear Overhauser Effect

The Overhauser effect[102] is an expression of the electronic-nuclear cross-relaxation mechanisms, and its nuclear-nuclear analog is called the nuclear Overhauser effect[103] (NOE). In a double resonance experiment the population of the states which are directly affected, or unaffected, by the strong irradiating field H$_2$ may be perturbed by this field,

[99] *K.G. Kidd, G. Kotowycz, T. Schaefer,* Can. J. Chem. **45**, 2155 (1967).

[100] *D. Lavie, E. Glotter, Y. Shvo,* J. Org. Chem. **30**, 1774 (1965).

[101] *I.C. Calder, P.J. Garratt, F. Sondheimer,* Chem. Comm. 41 (1967).

[102] *A.W. Oberhauser,* Phys. Rev. **89**, 689 (1953); **92**, 411 (1953).

[103] *I. Solomon,* Phys. Rev. **99**, 559 (1955);
I. Solomon, N. Bloembergen, J. Chem. Phys. **25**, 261 (1956).

provided that the corresponding nuclei are in a close spatial proximity to the irradiated nucleus. This effect represents a readjustment of the populations of the states to the new stationary values in presence of the field H_2, and is due to the intramolecular spin-lattice relaxation processes which are strong between close and weak between remote nuclei[96].

Saturation of a nucleus A by irradiation with the field H_2 should lead to an enhancement (theoretically up to 50%) of the integrated intensity of a proton B, provided that the latter is responsible for the relaxation of A, as discussed for β,β-dimethyl-acrylic acid[104] (98).

98   **99**   **100**

The *trans*-methyl group is expected to contribute appreciably to the relaxation of the vinylic proton, whereas the contribution of the *cis*-methyl should be less important. Irradiation at the frequency of the *cis*-methyl group results in a corresponding decoupling of the signal of the vinylic proton (from a septet to a quartet) without practically influencing its integrated area; conversely, irradiation at the frequency of the *trans*-methyl group leads to the same decoupling of the vinylic proton; however, its intensity is now enhanced by 17%.

The geometry of the two isomeric *3-ethylidene-1-azabicyclo (2,2,2) octanes* (99, 100) has been determined using the NOE[105]. The vinylic methyl group is close to the 4-H in compound 99 but remote in 100; therefore, the intramolecular relaxation of 4-H should be strongly affected by the vinylic CH_3 only in compound 99. Indeed, irradiation at the center of the signal of vinylic CH_3 in 100 does not change the integrated intensity of 4-H. However, a similar irradiation of 99 results in an enhancement of 31% of the intensity of 4-H.

In order to ensure good conditions for the observation of a NOE, one must minimize the intermolecular relaxation effects by degasing the sample and by using solvents without nuclei with high magnetic moments. However, the spectrum of 99 has been run in a mixture of $CDCl_3$ and CF_3COOH, the results not being affected by the presence of fluorine nuclei.

The *cis* stereochemistry of the juncture between the two carbocyclic rings in *bakkenolide B* (a member of a new group of sesquiterpene lactones[106]) has been determined by irradiating a derivative 101 at the frequency of the 5-CH_3. Thus the intensity of the integrated area of 6-H was enhanced by 10% without affecting the corresponding band width.

101   **102**

With polycyclic aromatic compounds, the NOE is useful for the assignment of the NMR signals of the various methyl substituents[107]. For instance, the signals of the 1- and 4-methyl groups in *1,2,3,4-tetramethyl-phenanthrene* (102) can be differentiated from the signals of the other two methyl groups by the enhancement of the integrated areas of 5-H and 10-H following irradiation of the frequencies of the 4- and 1-methyls respectively.

The usefulness of this effect has been well demonstrated[108] in the elucidation of the relative stereochemistry and the conformation of the diterpenes of the taxane group. Of all the possible stereostructures of taxinine B (103), only structure 103a can reasonably account for the five nuclear Overhauser effects observed in this compound (with the possible exception at C_5).

103 a

103

---

[104] *F.A.L. Anet, A.J.R. Bourn*, J. Am. Chem. Soc. 87, 5250 (1965).

[105] *J.C. Nouls, G. van Binst, R.H. Martin*, Tetrahedron Lett. 4065 (1967).

[106] *N. Abe*, Tetrahedron Lett. 1993 (1968).

[107] *R.H. Martin, J.C. Nouls*, Tetrahedron Lett. 2727 (1968).

[108] *M.C. Woods, H.C. Chiang, Y. Nakadaira, K. Nakanishi*, J. Am. Chem. Soc. 90, 522 (1968).

Irradiation of the 15-α methyl enhances the areas of the signals of 2-H and 9-H; similarly, irradiation of the olefinic 12-methyl enhances the areas of the 21- and 10-H signals finally, irradiation of 7-H increases the area of the signal due to 10-H.

2.12 Polymers

NMR spectroscopy is being increasingly used during the last years in the study of the stereochemical configuration of polymer chains[109]. This technique provides information on the composition of polymers, sequence distribution in copolymers, and stereoregularity. While most other techniques used heretofore gave rather qualitative information, the NMR method often affords quantitative results[109].

---

[109] *P.R. Sewell*, in Ref. [1d] Vol. I, p. 165.

2.13 Bibliography

J.A. Pople, W.G. Schneider, H.J. Bernstein[1a], 1959.
N.S. Bhacca, D.H. Williams[1g], 1964.
R.H. Bible[1f], 1965.
J.S. Waugh, Vol. I, 1965[15], Vol. II, 1966[3b].
J.W. Emsley, J. Feeney, L.H. Sutcliffe[1d], Vol. I, 1965, Vol. II, 1966.
L.M. Jackman, S. Sternhell[1c], 1968.
E.F. Mooney[1e], Vol. I, 1968; Vol. II, 1969.

5.5:3 Nuclear Magnetic Resonance of Nuclei other than ¹H

Hans Zimmer and *David C. Lankin*

Department of Chemistry, University of Cincinnati, Cincinnati, Ohio 45221, U.S.A.

3.1 Introduction

Nuclear magnetic resonance spectra can be obtained, at least in principle, by all isotopes having nuclei with a nuclear spin number, I, different from zero. Though many isotopes fulfill this requirement, only a relatively small number of such isotopes have been investigated (Table 1). The NMR studies dealing with the solution of purely inorganic problems will be discussed here only briefly or not at all.

The study of the NMR spectra of nuclei other than ¹H has been focused mainly on

the measurement of chemical shift values of NMR-active isotopes and the attempt to recognize the factors causing these shifts;
the investigation and interpretation of the spin-spin coupling constants with the ¹H nucleus.

The latter area is of great importance for structure determinations of organic compounds whereas the former one gives information on electron density around the isotope under investigation.

Hammett relationships have been found to exist between substituents and chemical shifts of various nuclei. These studies have been done extensively with ¹³C and ¹⁵N isotopes (see section 5.5:3.3 and 5.5:3.4). There is a very large body of data published on ¹⁹F, ³¹P, (5.5:3.6 and 5.5:3.7) ¹³C and ¹¹B-NMR-spectroscopy[1-8]. As a consequence, in this chapter, only the main trends in these areas will be presented.

To permit a successful study of the NMR of nuclei other than hydrogen, the nuclei should exhibit the following properties:

Their *natural abundance* must be sufficiently great — or facile methods for enrichment should be available.
The nucleus should have a *large magnetic moment* since the relative sensitivity of nuclei to NMR-detection at constant field strength is proportional to the cubes of their magnetic moments (for theory see Chapter 5.5:1). The nuclei should have a *spin number* I = 1/2; nuclear quadrupole moments are associated with higher values for I and almost always lead to considerable broadening of the lines. This usually leads to the distortion of spin-spin coupling patterns.
The isotopes should have short *spin-lattice relaxation times* to avoid saturation phenomena.
The nuclei under investigation should not interact with unpaired electrons.

The relative intensities of signals of different nuclei (with that of ¹H arbitrarily set at 10,000) are given in Table 1.

With the instrumentation and techniques presently available, a relative strength of about 10^{-5} is, for all practical purposes, the limit of usefulness.
For *low sensitivity measurements* of some of the more important nuclei pertinent to organic chemistry, e.g. ¹³C, ¹⁵N, these samples are run in

---

[1] *J.A. Pople, W.G. Schneider, H.J. Bernstein*, High-Resolution Nuclear Magnetic Resonance, McGraw Hill Book Co., New York 1959.
[2] *P.C. Lauterbur*, Chapter 7 in Ref. [256].
[3] *H. Suhr*, Anwendung der kernmagnetischen Resonanz in der organischen Chemie, Kapitel VI, Springer Verlag, Berlin 1965.
[4] *J.W. Emsley, J. Feeney, L.H. Sutcliffee*, High Resolution Nuclear Magnetic Resonance Spectroscopy, Vol. II, Chapter 12, Pergamon Press, Oxford 1966.
[5] *J. Feeney*, Chapter 9 in Ref. [260].
[6] *E.F. Mooney*, Annual Review of NMR-Spectroscopy, Vol. 2, Academic Press, London, New York 1969.
[7] *M.M. Crotchfield, C.H. Duncan, J.H. Letcher, V. Mark, J.R. Van Wazer* in Ref. [261].
[8] *T.P. Onak, H. Landesman, R.E. Williams, I. Shapiro*, J. Phys. Chem. 63, 1533 (1959).

Table 1. Nuclear Magnetic Resonance from Varian

Isotope	NMR Frequency [MHz] at 14,092G	NMR Frequency [MHz] at 23,487G	Natural abundance [%]	Relative sensitivity for equal number of nuclei at constant field	Spin[a] I in multiples of $h/2\pi$	Electric quadrupole moment Q in multiples of 10^{-24} cm²
H^1	60.000	100.00	99.985	1.00	1/2	—
H^2	9.2104	15.351	1.5×10^{-2}	9.65×10^{-3}	1	2.73×10^{-3}
Li^7	23.317	38.862	92.58	0.293	3/2	-3×10^{-2}
Be^9	8.4321	14.054	100.	1.39×10^{-2}	(—)3/2	5.2×10^{-2}
B^{10}	6.4479	10.746	19.58	1.99×10^{-2}	3	7.4×10^{-2}
B^{11}	19.250	32.084	80.42	0.165	3/2	3.55×10^{-2}
C^{13}	15.087	25.144	1.108	1.59×10^{-2}	1/2	—
N^{14}	4.3343	7.2238	99.63	1.01×10^{-3}	1	7.1×10^{-2}
N^{15}	6.0798	10.133	0.37	1.04×10^{-3}	(—) 1/2	—
O^{17}	8.134	13.56	3.7×10^{-2}	2.91×10^{-2}	(—) 5/2	-2.6×10^{-2}
F^{19}	56.446	94.077	100.	0.833	1/2	—
Na^{23}	15.871	26.452	100.	9.25×10^{-2}	3/2	0.14–0.15
Al^{27}	15.634	26.057	100.	0.206	5/2	0.149
Si^{29}	11.919	19.865	4.70	7.84×10^{-3}	(—) 1/2	—
P^{31}	24.288	40.481	100.	6.63×10^{-2}	1/2	—
S^{33}	4.6018	7.6696	0.76	2.26×10^{-3}	3/2	-6.4×10^{-2}
Cl^{35}	5.8790	9.7983	75.53	4.70×10^{-3}	3/2	-7.89×10^{-2}
Cl^{37}	4.893	8.155	24.47	2.71×10^{-3}	3/2	-6.21×10^{-2}
V^{51}	15.77	26.28	99.76	0.382	7/2	-4×10^{-2}
Mn^{55}	14.798	24.664	100.	0.175	5/2	0.55
Co^{59}	14.168	23.614	100.	0.277	7/2	0.40
Cu^{63}	15.903	26.596	69.09	9.31×10^{-2}	3/2	—0.16
Cu^{65}	17.036	28.394	30.91	0.114	3/2	—0.15
As^{75}	10.276	17.127	100.	2.51×10^{-2}	3/2	0.3
Se^{77}	11.44	19.07	7.58	6.93×10^{-3}	1/2	—
Br^{79}	15.032	25.054	50.54	7.86×10^{-2}	3/2	0.33
Br^{81}	16.204	27.006	49.46	9.85×10^{-2}	3/2	0.28
Rb^{87}	19.632	32.720	27.85	0.175	3/2	0.13
Nb^{93}	14.666	24.443	100.	0.482	9/2	—0.2
Sn^{117}	21.376	35.626	7.61	4.52×10^{-2}	(—) 1/2	—
Sn^{119}	22.363	37.272	8.58	5.18×10^{-2}	(—) 1/2	—
Sb^{121}	14.359	23.931	57.25	0.160	5/2	—0.5
I^{127}	12.004	20.007	100.	9.34×10^{-2}	5/2	—0.69
Cs^{133}	7.8702	13.117	100.	4.74×10^{-2}	7/2	-3×10^{-3}
Pt^{195}	12.90	21.50	33.8	9.94×10^{-3}	1/2	—
Hg^{199}	10.696	17.827	16.84	5.67×10^{-3}	1/2	—
Hg^{201}	3.9598	6.5998	13.22	1.44×10^{-3}	(—) 3/2	0.50
Tl^{203}	34.290	57.150	29.50	0.187	1/2	—
Tl^{205}	34.625	57.709	70.50	0.192	1/2	—
Pb^{207}	12.553	20.922	22.6	9.16×10^{-3}	1/2	—
Bi^{209}	9.6418	16.070	100.	0.137	9/2	—0.4

[a] (—) signifies that the magnetogyric ratio and hence magnetic moment is negative. To calculate the magnetic moment of an isotope, in multiples of the nuclear magneton, multiply the product of columns 2 and 6 by 0.093089; or multiply the product of columns 3 and 6 by 0.955854.

higher concentrations than normal using large tubes. Large tubes usually preclude the spinning of the sample (because homogeneity of the field cannot be maintained over such samples) and thus a broadening of the lines invariably occurs. To overcome this difficulty *computer averaging techniques* (CAT) are used. By repeatedly scanning the spectrum, the computer picks out the true signals appearing at a definite location in the spectrum whereas it neglects peaks due to noise which appear at random locations.

Another factor for broadening of the peaks is the coupling with adjacent and distant hydrogen atoms in the molecules. However, if carefully taken into consideration, the 'long-range' coupling phenomena can be very useful for structure determination (e.g. in the case of *fluorine, phosphorus,* and *tin* compounds).

9 *T.P. Onak, H. Landesman, R.E. Williams, I. Shapiro,* J. Phys. Chem. *63,* 1533 (1959).

3.2 Boron-11

3.2.1 Chemical Shifts

The two naturally occuring *boron* isotopes ^{10}B and ^{11}B (abundance 81.7%) possess nuclear spins of 3 and 3/2 respectively. The ^{11}B chemical shifts are spread over a range of about 220 ppm. These shifts are measured using BCl_3, $B(OCH_3)_3$, $O(CH_3)_2:BF_3$, and $O(C_2H_5)_2:BF_3$ as external standards.

The following approximate adjustments are commonly used:

$\delta B(OCH_3)_3 = \delta BCl_3 - 29{,}6$ ppm
$\delta O(C_2H_5)_2:BF_3 = \delta B(OCH_3)_3 - 18{,}3$ ppm
$[\delta O(C_2H_5)_2:BF_3 = \delta B(OCH_3)_3 - 17{,}4$ ppm$]$
$\delta O(C_2H_5)_2:BF_3 = \delta O(CH_3)_2:BF_3 + 0{,}3$ ppm

$\delta O(C_2H_5)_2:BF_3 = \delta BCl_3 - 47{,}5$ ppm
$\delta O(C_2H_5)_2:BF_3 = \delta NaBH_4 + 42{,}6$ ppm

The chemical shifts of boron-containing compounds range from the low field resonance of $B(CH_3)_3$ at -86 ppm to the high field resonance of $+128$ ppm observed for $[(n\text{-}C_3H_7)_4N]^{\oplus}BI_4^{\ominus}$ (Tabl. 2–6). The shielding provided by the three halide ligands is not proportional to their electronegativities. An explanation can be found in the differences in the double bond character of this bond[9]. It has been postulated that the double bond character of the B-X bond is most significant when $X = F$. The fluorine can 'back-donate' a pair of electrons from its filled p-orbital into the empty p_z orbital of the boron most effectively. However, this explanation of the chemical shifts as applied to the

Table 2. ^{11}B-Chemical Shifts of trisubstituted compounds δ [ppm] from $(C_2H_5)_2O:BF_3$.

Compound	δ [ppm]	Ref.	Compound	δ [ppm]	Ref.
$B(CH_3)_3$	—86	a	$B(C_4H_9)[N(CH_3)_2]_2$	—34.2	f
$B(C_2H_5)_3$	—85.5	a	$B(C_4H_5)[N(CH_3)_2]_2$	—39.3	g
$B(C_6H_5)_3$	—60	b	BF_3	— 9.4	h
$B(CH_3)_2OCH_3$	—53	a	BCl_3	—47.5	h
$B(C_2H_5)_2OCH_3$	—53.6	a	BBr_3	—40.1	h
$B(CH_3)_2F$	—59; —60.3	c, d	BJ_3	+ 5.5	i++
$B(CH_3)_2Cl$	—77.2; —75.7	a, c	$B(OH)_3$	—18.8	h
$B(CH_3)_2Br$	—78.8	c	$B(OCH_3)_3$	—18.3	a
$B(CH_3)_2$	—79.1	c	$B(OC_4H_9)_3$	—18.1	a
$B(C_2H_5)_2F$	—59.6	a, c	$B[O—C(CH_3)_3]_3$	—15.5	a
$B(C_2H_5)_2Cl$	—78	a, c	$B(2—ClC_6H_4O)_3$	—14.0	h
$B(C_2H_5)_2Br$	—81.9	c	$B(2—CH_3C_6H_5O)_3$	—15.0	h
$B(C_2H_5)_2$	—84.4	c	$[(CH_3)_3Si]_2N—B[NHSi(CH_3)_3]_2$	—28	j
$B(C_6H_5)_2F$	—47.4	c			
$B(C_6H_5)_2Cl$	—61.0	a			
$B(C_6H_5)_2Br$	—66.7	c		—27.7	k
$B(C_6H_5)_2$	—69.1	c			
$B(CH_3)F_2$	—28.2	c, d			
$B(CH_3)Cl_2$	—62.3; —61.6	a, c, d		—11	l
$B(CH_3)Br_2$	—62.5	c			
$B(CH_3)_2$	—50.5	c			
$B(C_3H_7)Cl_2$	—63.1	c	$[(CH_3O)_2B]_2$	—30.5	a
$B(C_4H_9)Cl_2$	—63.4	c	$\{[(CH_3)_2N]_2B\}_2$	—36.5	a
$B(n—C_8H_{17})Cl_2$	—63.6	c	$[(CH_3COO)_2B]_2O$	— 1	h
$B(C_6H_5)Cl_2$	—54.8	a, c	$[(C_2H_5)_2B]_2O$	—52.6	a
$B(CH_3)_2N(CH_3)_2$	—44.6	a	$[(C_2H_5)_2B]_2NH$	—57.4	a
$B(C_2H_5)_2N(CH_3)_2$	—45.7	a	$[(C_6H_5)_2B]_2O$	—28.5	a
$B(C_4H_9)_2N(CH_3)_2$	—45.5	a	$[(C_6H_5)_2B]_2NH$	—40.8	a
$[B(CH_3)(N(CH_3)_2)]_2$	—33.5	a	$R^1R^2N—BXY$		m
$B(C_2H_5)[N(CH_3)_2]_2$	—34.5	a	$B(Hal)_n(Hal')_{3-n}$		n

a *H. Nöth, H. Vahrenkamp*, Chem. Ber. *99*, 1049 (1966).

b *H. Landesman, R. E. Williams*, J. Amer. Chem. Soc. *83*, 2663 (1961).

c *H. Nöth, H. Vahrenkamp*, J. Organometallic Chem. *11*, 399 (1968).

d *J. E. DeMoor, G. P. Vander Kelen*, J. Organometallic Chem. *6*, 235 (1966).

e *W. Gerrard, E. F. Mooney, R. G. Rees*, J. Chem. Soc. 1964, 740.

f *J. K. Ruff*, J. Org. Chem. *27*, 1020 (1962).

g *M. F. Hawthorne*, J. Amer. Chem. Soc. *83*, 2671 (1961).

h *T. P. Onak, H. Landesman, R. E. Williams, N. Shapiro*, J. Phys. Chem. *63*, 1533 (1959).

i *T. P. Onak*, J. Amer. Chem. Soc. *83*, 2584 (1961).

j *R. L. Wells, A. L. Collins*, Inorg. Chem. *7*, 419 (1968).

k *F. A. Davis, M. J. S. Dewar, R. Jones*, J. Amer. Chem. Soc. *90*, 706 (1968).

l *R. H. Cragg*, J. Inorg. Nuclear Chem. *30*, 395 (1968).

m *H. Nöth, H. Vahrenkamp*, Chem. Ber. *99*, 2757 (1966) and *100*, 3353 (1967).

n *T. D. Coyle, F. G. A. Stone*, J. Chem. Phys. *32*, 1892 (1960); *P. N. Gates, E. F. Mooney and D. C. Smith*, J. Chem. Soc. 1964, 3511.

boron halides has been modified considerably[10]. From investigations of the ^{11}B-shifts of the hetero-aromatic compounds **1**−**8**, which are isoconjugate with normal aromatic systems, it is concluded that this factor manifests itself in the presence of additional pairs of unshared electrons on atoms adjacent to the boron atom and beyond those needed for π-bonding. The shifts are given in ppm from $O(C_2H_5)_2 : BF_3$ as external standard.

1 R = CH$_3$, −38.4[a]

2 R = OCH$_3$, −30.6[a]

3 R = CH$_3$, −39.9[b]

4 R = OC$_2$H$_5$, −30.2[b]

5 R = C$_6$H$_5$, −37.4[b]

6 R = Cl, −34.6[b]

7 R = OC$_2$H$_5$, −29.7[c]

8 R = Cl, −37.8[c]

[a] *M. J. S. Dewar, R. Dietz*, J. Chem. Soc., 2728 (1959).
[b] *M. J. S. Dewar, V. P. Kubba, R. Pettit*, J. Chem. Soc., 3073 (1958).
[c] *M. J. S. Dewar, R. Dietz*, J. Chem. Soc., 1344 (1960).

For boron halides, ^{11}B chemical shifts[11] are calculated from the equation

$$\delta = \delta\sigma + \delta\pi$$

which readily accounts for their chemical shifts; $\delta\sigma$ represents the linear function of valence state electronegativity and $\delta\pi$ is proportional to π-electron delocalization. Additional calculations are applicable for sp^2 hybridized boron compounds[12].

Table 3. ^{11}B–chemical shifts of monovalent boron-containing anions from $(C_2H_5)_2O : BF_3$.

Anion	δ [ppm]	Cation	Ref.
$[BH_4]^\ominus$	+36.7, +42.6	Al$^{3\oplus}$, Na$^\oplus$	a
$[B(OCH_3)_4]^\ominus$	−3.2, −2.9	Al$^{3\oplus}$, Na$^\oplus$	b,c
$[B(CH^3)_4]^\ominus$	+20.2		d
$[B(C_2H_5)_4]^\ominus$	+17.5, +16.6		d
$BF_4^\ominus$	+2.2, +2.3, +1.81	Ag$^\oplus$, Na$^\oplus$, K$^\oplus$	a, d
$BCl_4^\ominus$	−6.6, −6.7	$(CH_3)_4N^\oplus$ $(C_6H_5)_3C^\oplus$	d
$BBr_4^\ominus$	+24.1, −23.8, +23.9	C$_5$H$_5$NH$^\oplus$ (H$_3$C)$_4$N$^\oplus$ (C$_6$H$_5$)$_3$C$^\oplus$	a, d, d
$BI_4^\ominus$	+128.0	(n-C$_3$H$_7$)$_4$N$^\oplus$	d

[a] *W. D. Philips, H. C. Miller, E. L. Muetterties*, J. Amer. Chem. Soc. *81*, 4496 (1959).
[b] *H. S. Turner, R. J. Warne*, Proc. Chem. Soc. *1962*, 69.
[c] *T. P. Onak, H. Landesman, R. E. Williams, I. Shapiro*, J. phys. Chem. *63*, 1533 (1959).
[d] *R. J. Thompson, J. C. Davis*, Inorg. Chem. *4*, 1464 (1965).

Table 4. ^{11}B–chemical shifts of boron-acceptor-complexes from $(C_2H_5)_2O : BF_3$.

Complex	δ [ppm]	Ref.
Acid Derivatives		
$CH_3COCl : BCl_3$	−45.8	a
$CH_3CO_2C_2H_5 : BF_3$	− 0.3	a
$CH_3CO_2C_2H_5 : BCl_3$	−12.6	a
$CH_3CO_2C_2H_5 : BBr_3$	+ 9.6	a
$CCl_3CO_2C_2H_5 : BF_3$	− 2.4	a
$CCl_3CO_2C_2H_5 : BCl_3$	−47.1	a
$CCl_3CO_2C_2H_5 : BBr_3$	−38.2	a
$C_6H_5C\equiv N : BBr_3$	+10.8	b
Amines		
$CH_3NH_2 : BH_3$	+18.2	c
$(CH_3)_3N : B(CH_3)_3$	− 0.1	d
$(CH_3)_3N : B(C_2H_5)_3$	− 4.3	c
$(CH_3)_3N : B(CH_3)_2OCH_3$	−32.9	c
$(CH_3)_3N : B(C_2H_5)_2OCH_3$	−50.7	c
$(C_2H_5)_3N : BH_3$	+13.3	e
$(C_2H_5)_3N : BF_3$	+ 0.1	f
$(C_2H_5)_3N : BCl_3$	− 9.7	f
$(C_2H_5)_3NBBr_3$	+ 5.4	f
$(C_2H_5)_3N : BF_3$	+60.1	f
$C_5H_5N : BH_3$	+13.2	g
$C_5H_5N : BF_3$	+ 0.6	f
$C_5H_5N : BCl_3$	− 7.7	f
$C_5H_5N : BBr_3$	+ 7.4	f
$C_5H_5N : BI_3$	+60	h
Aromatic Nitrocompounds		
$C_6H_5NO_2 : BF_3$	− 3	i
$C_6H_5NO_2 : BCl_3$	−32.7	i
$C_6H_5NO_2 : BBr_3$	+10.3	i
$C_6F_5NO_2 : BF_3$	+ 0.6	i
$C_6F_3NO_2 : BCl_3$	−46.1	i
$C_6F_5NO_2 : BBr_3$	+39.5	i
Phosphine — BX_3 Compounds		
$(CH_3)_3P : BH_3$	+36.2	j
$(CH_3)_3P : BF_3$	− 1.0	j
$(CH_3)_3P : BCl_3$	− 3.0	j
$(CH_3)_3P : BBr_3$	+14.5	j

[a] *P. G. Davies, E. F. Mooney*, Spectrochim. Acta *22*, 953 (1966).
[b] *J. Chatt, R. L. Richards, B. J. Newman*, J. chem. Soc. A, *1968*, 128.
[c] *H. Nöth, H. Vahrenkamp*, Chem. Ber. *99*, 1049 (1966).
[d] *C. W. Heitsch*, Inorg. Chem. *4*, 1019 (1965).
[e] *J. N. G. Faulks, N. N. Greenwood, J. H. Morris*, J. Inorg. Nuc. Chem. *29*, 329 (1967).
[f] *P. N. Cates, E. J. McLauchlan, E. F. Mooney*, Spectrochim. Acta *21*, 1445 (1965).
[g] *E. F. Mooney, M. A. Qaseem*, J. Inorg. Nuc. Chem. *30*, 1439 (1968).
[h] *H. Landesman, R. E. Williams*, J. Amer. chem. Soc. *83*, 2663 (1961).
[i] *E. F. Mooney, M. A. Qaseem, P. H. Wilson*, J. chem. Soc. B, *1968*, 224.
[j] *D. E. Young, G. E. McAchran, S. G. Shore*, J. Amer. chem. Soc. *88*, 4390 (1966).

For the $[BX_4]^{\ominus}$ species, equations for ^{11}B-chemical shift calculations have been introduced[13-15]. ^{11}B-shifts of *aminoboranes* can be related to ^{13}C-shifts of alkanes using the equation: $1.71 (\delta ^{11}B) = (\delta ^{13}C) -81.0$, with the ^{13}C-shifts taken from benzene[16]. ^{11}B-shifts have been used extensively to elucidate the structure of carboranes[17-50].

Table 5. ^{11}B–chemical shifts of compounds with a boron-metal-bond from $(C_2H_5)_2O:BF_3$.

Compound	δ [ppm]	Ref.
$[(CH_3)_2N]_2B—Si(CH_3)_3$	—36.1	a
$[(CH_3)_2N]_2B—Sn(CH_3)_3$	—39.0	b
$[(CH_3)_2N]_2B—P(C_2H_5)_2$	—35.9	a
$[(CH_3)_2N]_2B—Mn(CO)_5$	—27.1	b
$(C_6H_5)_2B—Fe(CO)_2\pi—C_5H_5$	—43.5	b
$(C_6H_5)_2B—Co(CO)_3PPh_3$	—21.6	b
$(C_6H_5)_2B—PtCl[P(C_2H_5)_3]_2$	—29.2	b
$[(C_6H_5)_2](B—PtP(C_6H_5)_3$	—30	b

a H. Nöth, G. Höllerer, Chem. Ber. 99, 2197 (1966).
b H. Nöth, G. Schmidt, J. prakt. Chem. 1966, 613.

3.2.2 ^{11}B—^{1}H Spin-Spin Coupling Constants

It has been predicted that the spin coupling between a hydrogen nucleus and another NMR active species depends upon the s-character in the bond connecting the two nuclei[51]. In general, the coupling constants for sp^3-hybridized boron are smaller than for the sp^2-hybridized one. Values given in Table 7 appear to support this theory.

Table 7. Spin coupling constants in boron-compounds.

Compound	^{11}B—^{1}H	Ref.
$NaBH_4$	81	a
B_2H_6	137	a
B_4H_{10}		
BH_2 units	132	a
BH units	156	a
B_5H_9		
base	168	a
apex	176	a
$(CH_3)_3N:BH_3$	97 (101)	a, b
$(CH_2)_4O:BH_3$	103	a
$HB[O(CH_3)_3]_2$	141	a
$(CH_3)_2PH:BH_3$	93	c
$C_5H_5N:BH_3$	90	a
$(CH_3)_2NH:BH_3$	91	b
Borazol	136	d

a *W. D. Philips, H. C. Miller, E. L. Muetterties*, J. Amer. chem. Soc. *81*, 4496 (1959).
b *T. P. Onak, H. Landesman, R. E. Williams, I. Shapiro*, J. phys. Chem. *63*, 1533 (1959).
c *J. N. Shoolery*, Discuss. Faraday Soc. *19*, 215 (1955).
d *K. Ito, H. Watanabe, M. Kubo*, J. chem. Phys., *34*, 1034 (1961).

Table 6. ^{11}B-Chemical shifts and ^{11}B—^{1}H coupling constants of alkyl diboranes from $(C_2H_5)_2O:BF_3$.

Compound	δBH_2	δBHR	δBR_2	$J_{B—H_T}$ (terminal)	$J_{B—H_B}$ (bridged)
	—8.8	—26.7	—	127	48
	—10.0	—29.5	—	133	45
	—3.6	—	—36.4	125.5	36 50
	—3.9	—	—40.8	123	37 48
	—	—20.5	—	131.2	47.5
	—	—22.7	—	125.5	42

a *H. H. Lindner, T. Onak*, J. Amer. chem. Soc. *88*, 1890 (1966).

[10] *F. A. Davis, M. J. S. Dewar, R. Jones*, J. Am. Chem. Soc. *90*, 706 (1968).
[11] *C. D. Good, D. M. Ritter*, J. Am. Chem. Soc. *84*, 1162 (1962).
[12] *H. Nöth, H. Vahrenkamp*, Chem. Ber. *99*, 1049 (1966).
[13] *R. J. Thompson, J. C. Davis*, Inorg. Chem. *4*, 1464 (1965).

3.3 Carbon-13

3.3.1 Experimental Techniques

Among the naturally occurring carbon isotopes only ^{13}C has a magnetic moment ($I = 1/2$). Despite its low natural abundance (1.1 percent), its poor sensitivity to NMR detection (only 1.59% of that of ^1H at the same field strength), and its long relaxation times, many useful studies involving this isotope have been reported[52, 53].

These studies involve special techniques for observing spectra before meaningful results can be obtained. Developments in *internuclear double resonance techniques (INDOR)* have proven very useful. In this method, at least two nuclei are irradiated simultaneously by employing two radio frequencies. Thus, while observing the ^{13}C spectrum in the usual way, the protons, which are spin-coupled to the carbon atoms, are irradiated simultaneously at their respective frequencies eliminating ^{13}C—^1H interactions, whereupon the typical multiplets collapse to form singlets. Through the occurence of a positive *Overhauser effect*[54], the singlets (due to ^{13}C) are increased in intensity up to two-fold or more over that expected for the collapse of the multiplet, greatly simplifying the study of ^{13}C-NMR spectra of naturally occuring ^{13}C-compounds. This can be achieved by addition of free radicals to the sample tube[55–57]. The NMR study of naturally occuring ^{13}C-compounds based on refinements of the decoupling method has been improved[58, 59]. To avoid saturation difficulties, use of a continuous flow technique has been recommended[60].

Another new technique consists of taking the spectrum while operating with an extremely slow scan. The signal-to-noise ratio improves as the square root of the ratio of the long sweep time to that of the normal spectrum sweep time *(SUPER SNAIL = Simple Uncomputerized Procedure for Experimental Recording Signals above Noise by Accumulation of Integral Lines)*[61].

Efficient spinning of large sample tubes, in conjunction with such techniques, permitted a detailed study of the ^{13}C-NMR spectrum of *benzene*[62]. Other indirect methods have been used[63–65].

3.3.2 ^{13}C-Chemical Shifts

For a specific nucleus, A, the chemical shift depends on the field H_A experienced at the nucleus and is expressed by $H_A = H_0 (1-\delta_A)$, H_0 being the applied field. Accordingly, the screening constant δ_A can be written as the sum of three parameters[66]:

a diamagnetic term, δ_d, due to local electron currents, a paramagnetic term, δ_p, involving the mixing of electronic states by the applied field H_0, and a term, δ', which represents the contribution from circulation of electrons on neighboring atoms within the molecule as well as the effects of adjacent anisotropic atoms, bonds or ring structures.

[14] D.R. Armstrong, P.G. Perkins, Chem. Comm. 337 (1965).

[15] M.R. Chakrabarty, C.C. Thompson, W.S. Brey, Inorg. Chem. 6, 518 (1967).

[16] B.F. Spielvogel, J.M. Purser, J. Am. Chem. Soc. 89, 5294 (1967).

[17] R.N. Grimes, J. Am. Chem. Soc. 88, 1070 (1966).

[18] I. Shapiro, C.D. Good, R.E. Williams, J. Am. Chem. Soc. 84, 3837 (1962).

[19] R.N. Grimes, J. Am. Chem. Soc. 88, 1895 (1966).

[20] I. Shapiro, B. Keilin, R.E. Williams, C.D. Good, J. Am. Chem. Soc. 85, 3167 (1963).

[21] M.A. Grassberger, E.G. Hoffmann, G. Schomburg, R. Köster, J. Am. Chem. Soc. 90, 56 (1968).

[22] T.P. Onak, G.B. Dunks, J.R. Spielman, F.J. Gerhart, R.E. Williams, J. Am. Chem. Soc. 88, 2061 (1966).

[23] C.L. Bramlett, R.N. Grimes, J. Am. Chem. Soc. 88, 4269 (1966).

[24] P. Binger, Tetrahedron Lett. 2675 (1966).

[25] T. Onak, G.B. Dunks, Inorg. Chem. 5, 439 (1966).

[26] J.R. Spielman, R. Warren, G.B. Dunks, J.E. Scott, T. Onak, Inorg. Chem. 7, 216 (1968).

[27] T. Onak, G.B. Dunks, R.A. Beaudet, R.L. Poynter, J. Am. Chem. Soc. 88, 4622 (1966).

[28] R. Köster, M.A. Grassberger, E.G. Hoffmann, G.W. Rotermund, Tetrahedron Lett. 905 (1966).

[29] F.N. Tebbe, P.M. Garrett, D.C. Young, M.F. Hawthorne, J. Am. Chem. Soc. 88, 609 (1966).

[30] R.E. Williams, F.J. Gerhart, J. Am. Chem. Soc. 87, 3513 (1965).

[31] F.N. Tebbe, P.M. Garrett, M.F. Hawthorne, J. Am. Chem. Soc. 90, 869 (1968).

[32] H.V. Hart, W.N. Lipscomb, J. Am. Chem. Soc. 89, 4220 (1967).

[33] F.N. Tebbe, P.M. Garrett, M.F. Hawthorne, J. Am. Chem. Soc. 88, 607 (1966).

[34] W.H. Knoth, J. Am. Chem. Soc. 89, 1274 (1967).

[35] C.C. Tsai, W.E. Streib, J. Am. Chem. Soc. 88, 4513 (1966).

[36] D.E. Hyatt, D.A. Owen, L.J. Todd, Inorg. Chem. 5, 1749 (1966).

[37] D.E. Hyatt, F.R. Scholer, L.J. Todd, J.L. Warner, Inorg. Chem. 6, 2229 (1967).

[38] F.P. Olsen, M.F. Hawthorne, Inorg. Chem. 4, 1839 (1965).

[39] M.F. Hawthorne, P.A. Wegner, J. Am. Chem. Soc. 90, 896 (1968).

[40] L.I. Zakharkin, V.N. Kalinin, Tetrahedron Lett. 407 (1965).

[41] M.F. Hawthorne, P.A. Wegner, R.C. Stafford, Inorg. Chem. 4, 1675 (1965).

[42] G.D. Vickers, H. Agahigian, E.A. Pier, H. Schroeder, Inorg. Chem. 5, 693 (1966).

[43] J.A. Potenza, W.N. Lipscomb, G.D. Vickers, H. Schroeder, J. Am. Chem. Soc. 88, 628 (1966).

Since δ' is relatively unimportant for nuclei other than ^1H, and the diamagnetic term has been shown to contribute not more than about 50 ppm to the total range of ^{13}C-chemical shifts, it appears that the paramagnetic term, δ_p, is the dominant factor for the ^{13}C-shielding effects. Consequently, considerable efforts have been made to evaluate this term by quantum mechanical treatments[67-69]. A more qualitative approach to the theory of chemical shifts based on molecular orbital theory was also developed[70-72].

The subject of correlation of ^{13}C-chemical shifts with electronegativity has been exhaustively reviewed[1,4]. ^{13}C-chemical shifts range over an area of at least 350 ppm (Fig. 1).

Acetic acid ($CH_3-^{13}COOH$), benzene, carbon disulfide, and aqueous potassium carbonate solution are commonly employed as either external or internal references. Equations used to extrapolate from one reference to the other are:

$$\delta^{ext*}_{CH_3COOH} = \delta^{ext}_{C_6H_6} + 50 \text{ ppm};$$
$$\delta^{int}_{CS_2} = \delta^{ext*}_{CH_3COOH} + 15.6 \text{ ppm};$$
$$\delta_{CS_2} = \delta^{ext}_{K_2CO_3} \text{ aqu.} + 23.2 \text{ ppm}$$

Though acetic acid is an excellent external reference, it cannot be used as an internal standard since the chemical shift of the carbonyl group carbon is solvent dependent. Also, compounds such as dioxane and dimethyl carbonate occasionally have been used as internal or external references. If non-^{13}C enriched CS_2 is used as internal standard, it must be present in about 25 percent by volume. CS_2, as internal standard, is virtually unaffected by a variety of solvents, such as nitrobenzene, acetone, aniline, cyclohexane, carbon disulfide, and bromoform[73] (Fig. 1)[53].

3.3.3 Applications

Aliphatic hydrocarbons: These compounds absorb at high-field in the range of about 172–200 ppm (from CS_2), with methane absorbing at the highest field position of 195.8 ppm. It has been shown that the ^{13}C chemical shift of the kth C-atom of an unbranched alkane can be expressed by the equation[74,75]:

$$^{13}C \text{ chemical shift } \delta(k) = 131.26 + \Sigma A_{lnkl}$$

Thus, δ_c for the kth atom is given by the sum of the products of the number of C-atoms in the 1th position, a

Sheme 1

[44] H. Beall, W.N. Lipscomb, Inorg. Chem. 6, 874 (1967).

[45] M.F. Hawthorne, D.C. Young, P.M. Garrett, D.A. Owen, S.G. Schwerin, F.N. Tebbe, P.A. Wegner, J. Am. Chem. Soc. 90, 862 (1968).

[46] A.B. Harmon, K.M. Harmon, J. Am. Chem. Soc. 88, 4093 (1966).

[47] M.F. Hawthorne, T.A. George, J. Am. Chem. Soc. 89, 7114 (1967).

[48] M.F. Hawthorne, D.C. Young, T.D. Andrews, D.V. Howe, R.L. Pilling, A.D. Pitts, M. Reintjes, L.F. Warren, P.A. Wegner, J. Am. Chem. Soc. 90, 879 (1968).

[49] D.E. Hyatt, J.L. Little, J.T. Moran, F.R. Scholer, L.J. Todd, J. Am. Chem. Soc. 89, 3342 (1967).

[50] J.L. Little, J.T. Moran, L.J. Todd, J. Am. Chem. Soc. 89, 5495 (1967).

factor A, which is an additive parameter for the δ position, plus a constant 131.26, which almost equals the shift difference between benzene and methane.

To calculate the ^{13}C shifts of certain substituted saturated hydrocarbons, several equations have been derived[76-78]. Solvent dependency of ^{13}C-chemical shifts has been investigated using methyl jodide and acetonitrile in a variety of solvents[79].

Cycloalkanes: These absorb in the area of the acyclic analogs[80]. One striking exception is the ^{13}C-chemical shift of *cyclopropane* (196.3 ppm.) which is almost at the same position as that of *methane*[81]. This abnormally high-field position has been attributed to the occurrence of a diamagnetic ring current[82]. ^{13}C-chemical shifts have been used to gain information regarding the conformation of substituted cycloalkanes[83, 84].

Olefins: Carbon atoms of olefinic and aromatic compounds generally absorb in the range of 40–90 ppm (from CS_2). The influence of methyl substituents on the olefinic (aromatic) carbon, and positions α- and β- to it has been investigated[85]. ^{13}C-chemical shifts were shown to be insensitive to conjugation in dienes[86]. The cumulenic carbon atoms, however, exhibited a rather pronounced downfield shift of about 50 ppm. Strain effects in cyclic olefins can be detected by ^{13}C-NMR spectroscopy[87].

Acetylenes: The ^{13}C-shifts for some simple alkynes were observed in the range between 104 and 122 ppm (from CS_2)[1]. In *l-substituted hexynes* bromo- and iodo-substitution at that carbon atom caused a large upfield acetylenic carbon ^{13}C-shift (155.3 and 197 ppm respectively). Since no other ^{13}C-shifts are observed in the region of the —C≡C-group, these can easily serve to identify triple bonds[88].

Aromatic Compounds: Ever since Lauterbur[89] reported that the ^{13}C-chemical shifts show a dependency on the polarity of the substituent, attention has been paid to ^{13}C-shieldings in aromatic systems[90]; a linear correlation with the σ constant of the Hammett relationship has been noted[91]. Since the ^{13}C-chemical shifts of the *para*-carbon showed the best linear correlation, it was concluded[92] that the ^{13}C-shieldings presented a more accurate assessment of the ground-state electron distribution in the aromatic ring than the chemical shifts for either ^{19}F or ^1H. Also it was found[93] that, in *substituted styrenes*, the ^{13}C-shift of the β-carbon atom of the side-chain is much more dependent on electronic effects imposed by the substituents if they are in the *para* position. The higher stability of hybrid structures **9a** and **9b** are thought to be responsible for this behavior:

9   **9a**   **9b**

(X = electron donating or electron withdrawing).

[51] N. Muller, D.E. Pritchard, J. Chem. Phys. *31*, 768 (1959).

[52] P.C. Lauterbur, J. Chem. Phys. *26*, 217 (1957).

[53] J.B. Stothers, Quart. Rev. Chem. Soc. *19*, 144 (1965).

[54] J.D. Baldeschwieler, E.W. Randall, Chem. Rev. Chem. Soc. *63*, 81 (1963).

[55] J.P. Imband, C.R. Acad. Sci. *261*, 5442 (1965).

[56] D.J. Parker, G.A. McLaren, J.J. Conradi, J. Chem. Phys. *33*, 269 (1960).

[57] D.F.S. Natusch, R.E. Richards, Chem. Comm. 579 (1966).

[58] E.B. Baker, J. Chem. Phys. *37*, 911 (1962).

[59] E.G. Paul, D.M. Grant, J. Am. Chem. Soc. *86*, 2977 (1964).

[60] S. Forsen, A. Rupprecht, J. Chem. Phys. *33*, 1888 (1960).

[61] D.S. Dehlsen, A.V. Robertson, Aust. J. Chem. *19*, 269 (1966).

[62] F.J. Weigert, J.D. Roberts, J. Am. Chem. Soc. *89*, 2967 (1967).

[63] J.K. Becconsall, P. Hampson, Mol. Phys. *10*, 21 (1965).

[64] A. Mathias, V.M.S. Gil, Tetrahedron Lett. 3162 (1965).

[65] A. Mathias, Tetrahedron *22*, 217 (1966).

[66] A. Saika, C.P. Slichter, J. Chem. Phys. *22*, 26 (1954).

[67] T. Yonezawa, I. Monishirua, H. Kato, Bull. Chem. Soc. Japan *39*, 1398 (1966).

[68] J.M. Sichel, M.A. Whitehead, Theor. Chim. Acta *5*, 35 (1966).

[69] B.V. Cheney, D.M. Grant, J. Am. Chem. Soc. *89*, 5319 (1967).

[70] J.A. Pople, Discuss. Faraday Soc. *34*, 7 (1962).

[71] M. Karplus, J.A. Pople, J. Chem. Phys. *38*, 2803 (1963).

[72] D.T. Clark, Chem. Comm. 390 (1966).

[73] P.C. Lauterbur, J. Am. Chem. Soc. *83*, 1838 (1961).

[74] E.G. Paul, D.M. Grant, J. Am. Chem. Soc. *85*, 1701 (1963).

[75] D.M. Grant, E.G. Paul, J. Am. Chem. Soc. *86*, 2984 (1964).

[76] G.B. Savitsky, R.M. Pearson, K. Namikawa, J. Phys. Chem. *69*, 1425 (1965).

[77] E.R. Malinowski, T. Vladimiroff, R.F. Tavares, J. Phys. Chem. *70*, 2046 (1966).

[78] W.M. Litchman, D.M. Grant, J. Am. Chem. Soc. *90*, 1400 (1968).

[79] J.K. Becconsall, P. Hampson, Mol. Phys. *10*, 21 (1965).

[80] J.J. Burke, P.C. Lauterbur, J. Am. Chem. Soc. *86*, 1870 (1964).

The irregular trend of ^{13}C-chemical shifts of the β-atom in *ortho-substituted styrenes* was suggested to be due to steric hindrance of resonance[94].

In *substituted benzenes* a reasonable relationship was found to exist between the ^{13}C-shifts and the electronegativities of the substituents[95, 96].

Heterocyclic Compounds: ^{13}C-chemical shifts were used to confirm the aromaticity of heterocycles such as *furan, thiophene* and *pyrrole*[97]. ^{13}C-chemical shifts have been measured for *2- and 4-substituted pyridines*[98, 99] and it was concluded that the shielding mechanisms for carbon atoms *para* to substituents are essentially the same for both the pyridine and benzene systems.

Carbonyl Compounds: The C-atoms of carbonyl groups absorb in the low field region where almost no other signals appear and can be used for identification of such groups. The peaks appear as singlets except for the formyl group, —CH=O, (J^{13}C—^{1}H a doublet, 165–205 cps).

Hydrogen bonding, conjugation, substitution and ring size in cyclic systems are the major factors responsible for carbonyl shielding. Correlations between ^{13}C-chemical shifts of *m-substituted benzaldehydes* and Hammett's σ_n parameters were found to exist[100]. ^{13}C-chemical shift of the carbonyl group of *ortho*-substituted acetophenones strongly indicate the steric inhibition of resonance[101]. These ^{13}C-shifts also reflect the degree of steric hindrance in such similarly substituted ketones and methyl benzoates[102]. The solvent dependency of the ^{13}C-shifts of the carbonyl group has been investigated by a study of ^{13}C carbonyl shifts of acetic acid in a variety of *solvents*[103].

The ^{13}C-chemical shift[104] for $CH_3CO^\oplus$ $SbF_6^\ominus$ at −45.4 ppm (relative to CH_3COF) was taken as evidence that a partial positive charge is located at the carbonyl carbon atom. Similarly, large downfield shifts have been reported for $(CH_3)_3C^\oplus$ $SbF_6^\ominus$, −273 ppm from $(CH_3)_3CCl$ and (C_6H_5) $C^\oplus SbF_6^\ominus$, −129.66 ppm from $(C_6H_5)_3COH$[105, 106] and were taken as evidence for the cationic character of the carbon atoms involved.

3.3.4 ^{13}C-Spin Coupling

Much information is available on ^{13}C—^{1}H coupling[107–112]. In natural abundance spectra, ^{13}C-X coupling interactions, though well resolved, suffer from a lack of precision in their measurement (about ±3 cps). As early as 1959, ^{13}C satellite spectra were used to obtain coupling constants; these are accurate within about 1 cps[113]. Theories which are used to predict the sign and magnitude of coupling involving ^{13}C are well developed[114–116]. Long range ^{13}C-coupling constants have also been reported[117–119]. Direct ^{13}C—^{13}C-couplings can be investigated only in compounds doubly tagged with ^{13}C[119]. In the few reported examples these constants were found to be fairly large[120];

C_6H_5—CH_2—CH_3	C_6H_5—$C\equiv C$—$CO_2C_2H_5$	C_6H_5—$C\equiv CH$
34 Hz	128 Hz	175.9 Hz

The ^{13}C—^{1}H coupling constants are also large[121–128] (Table 8).

Alkane	Olefin and aromat. cpds.	Acetylene
120–130 Hz	150–170 Hz	245–255 Hz

[81] H. Spiesecke, W.G. Schneider, J. Chem. Phys. 35, 722 (1961).
[82] G.E. Maciel, G.B. Savitsky, J. Phys. Chem. 69, 3925 (1965).
[83] G.W. Buchanan, D.A. Ross, J.B. Stothers, J. Am. Chem. Soc. 88, 4301 (1966).
[84] D.K. Dalling, D.M. Grant, J. Am. Chem. Soc. 89, 6612 (1967).
[85] R.A. Friedel, H.L. Retcofsky, J. Am. Chem. Soc. 85, 1300 (1963).
[86] K.S. Dhami, J.B. Stothers, Can. J. Chem. 43, 510 (1965).
[87] T. Lippmaa, V.I. Sokolov, A.I. Olivson, Y.O. Past, Dokl. Akad. Nauk. SSSR 173, 358 (1967).
[88] D.D. Traficante, G.E. Maciel, J. Phys. Chem. 69, 1348 (1965).
[89] P.C. Lauterbur, Ann. N.Y. Acad. Sci. 70, 841 (1958).
[90] P.C. Lauterbur, Tetrahedron Lett. 1961, 274.
[91] H. Spiesecke, W.G. Schneider, J. Chem. Phys. 35, 722, 731 (1961).
[92] R.R. Fraser, Can. J. Chem. 38, 2226 (1960).
[93] K.S. Damai, J.B. Stothers, Can. J. Chem. 3, 510 (1965).
[94] H.D. Marr, J.B. Stothers, Can. J. Chem. 43, 596 (1965).
[95] H. Spiesecke, W.G. Schneider, J. Chem. Phys. 35, 722, 731 (1961).
[96] G.E. Maciel, J. Natterstad, J. Chem. Phys. 42, 2752 (1965).
[97] T.F. Page, T. Alger, D.M. Grant, J. Am. Chem. Soc. 87, 5333 (1965).
[98] H.L. Retcofsky, F.R. McDonald, Tetrahedron Lett. 2575 (1968).
[99] H.L. Retcofsky, R.A. Friedel, J. Phys. Chem. 71, 3592 (1967).
[100] A. Mathias, Tetrahedron 22, 217 (1966).
[101] K.S. Dhami, J.B. Stothers, Can. J. Chem. 43, 479 (1965).
[102] K.S. Dhami, J.B. Stothers, Can. J. Chem. 43, 498 (1965); 45, 233 (1967).
[103] G.E. Maciel, D.D. Traficante, J. Am. Chem. Soc. 88, 220 (1966).
[104] G.A. Olah, W.S. Tolgyesi, S.J. Kuhn, M.E. Moffatt, I.J. Bastien, E.B. Baker, J. Am. Chem. Soc. 85, 1328 (1963).

Table 8. $^{13}C-^1H$ coupling constants.

Compound	J[Hz]	Ref.	Compound	J [Hz]	Ref.
2,2-Dimethylpropane	124	a	Jodomethane	151	d
Tetramethylsilane	119. 120	b, c	Dichloromethane	178	d
Tetramethylgermanium	124	c	Diiodomethane	173	d
Tetramethyltin	126. 128	b, c	Chloroform	209	d
Tetramethyllead	134	c	Acetaldehyde	173	e
Cyclopentane	128	a	Benzaldehyde	174	e
Cyclopropane	161	a	Dimethylformamide	192	e
Cyclohexane	123. 140	a, b	Bromoform	206	d
2-Butyne	131	a	Chloral	207	e
Propyne	248	a	Formyl fluoride	207	e
Ethynylbenzene	251	a	Ethylene	157	f
1,3,5-Trimethylbenzene	126	a, b	Benzene	159	f
Fluoromethane	149	d	Thiophene	187	f
Chloromethane	150	d	Cyclopentadiene	180	f
Bromomethane	152	d	Propadiene	168.2	f

a *N. Mueller, D. E. Pritchard*, J. chem. Phys. *31*, 768 (1959).

b *P. C. Lauterbur*, J. chem. Phys. *26*, 217 (1957).

c *R. S. Drago, N. A. Matwiyoff*, J. Organometal, Chem. *4*, 62 (1965).

d *N. Mueller, D. E. Pritchard*, J. chem. Phys. *31*, 1471 (1959).

e *N. Mueller*, J. chem. Phys. *36*, 359 (1962).

f *J. H. Goldstein, G. S. Reddy*, J. chem. Phys. *36*, 2544 (1962).

For molecules of the type HCXYZ the $^{13}C-^1H$ coupling constant can be predicted[129] using the expression $J^{13}C-^1H = \xi X + \xi Y + \xi Z$, where $\xi X, \xi Y$, and ξZ are contributions to the coupling from the substituents X, Y and Z. The agreement between measured and observed values is satisfactory. Deviations from strong electronegative groups, like F and O, can be corrected using special parameters[130].

3.3.5 $^{13}C-^1H$ Coupling Constants of Group IVA Organometal Compounds

These constants (Table 9) proved useful for structure determination.

A few $^{13}C-^{19}F$ coupling constants are compiled in Table 9.

Table 9. $^{13}C-^{19}F$ coupling constant

Compound	J[Hz]	Ref.
CF_4	259.2	a
CF_3CCl_3	282.5	b
CF_3CO_2H	283.2	b
$(CF_3)_2PCl$	320.2	c
$(CF_3S)_2$	313.8	c

a *S. G. Frankiss*, J. phys. Chem. *67*, 752 (1963).

b *G. V. D. Tiers*, J. phys. Soc. Japan *15*, 354 (1960).

c *R. K. Harris*, J. phys. Chem. *66*, 768 (1962).

[105] *G.A. Olah, E.B. Baker, J.C. Evans, W.S. Tolgyesi, J.S. McIntyre, I.J. Bastien*, J. Am. Chem. Soc. *86*, 1360 (1964).

[106] *G.A. Olah, E.B. Baker, M.B. Comisarow*, J. Am. Chem. Soc. *86*, 1265 (1964).

[107] *P.C. Lauterbur*, J. Chem. Phys. *26*, 217 (1957).

[108] *C.H. Holm*, J. Chem. Phys. *26*, 707 (1957).

[109] *N. Mueller, D.E. Pritchard*, J. Chem. Phys. *31*, 1471 (1959).

[110] *J.N. Shoolery*, J. Chem. Phys. *31*, 1427 (1959).

[111] *M. Karplus, D.M. Grant*, Proc. Nat. Acad. Sci. *45*, 1269 (1959).

[112] *D.M. Grant*, Ann. Rev. Phys. Chem. *15*, 489 (1964).

[113] *N. Sheppard, J.J. Tunier*, Proc. Roy. Soc. A *252*, 506 (1959).

[114] *E. Hiroike*, J. Phys. Soc. Jap. *23*, 1079 (1967).

[115] *T. Yonezawa, I. Morishima, M. Fujii, K. Fukeui*, Bull. Chem. Soc. Jap. *38*, 1224 (1965).

[116] *T. Yonezawa, I. Morishima, M. Fujii*, Bull. Chem. Soc. Jap. *39*, 2110 (1966).

[117] *G.J. Karabatsos*, J. Am. Chem. Soc. *83*, 1230 (1961).

[118] *J.N. Shoolery, L.F. Johnson, W.A. Anderson*, J. Mol. Spectrosc. *5*, 110 (1960).

[119] *G.J. Karabatsos, J.K.D. Graham, F.M. Vane*, J. Am. Chem. Soc. *84*, 37 (1962).

[120] *D.M. Graham, C. Holloway*, Can. J. Chem. *41*, 2114 (1963).

[120a] *K. Frei, H.J. Bernstein*, J. Chem. Phys. *38*, 1216 (1963).

[121] *N. Mueller, D.E. Pritchard*, J. Chem. Phys. *31*, 768, 1471 (1959).

[122] *G.W. Smith*, J. Chem. Phys. *39*, 2031 (1959).

[123] *J.H. Goldstein, G.S. Reddy*, J. Chem. Phys. *36*, 2544 (1962).

[124] *G.P. van der Kelen, Z. Eeckhaut*, J. Mol. Spectrosc. *10*, 141 (1963).

[125] *H. Spiesecke, W.G. Schneider*, J. Chem. Phys. *35*, 722, 731 (1961).

[126] *P.C. Lauterbur*, J. Chem. Phys. *26*, 217 (1957).

[127] *H.M. Huton, W.F. Reynolds, T. Schaefer*, Can. J. Chem. *40*, 1758 (1962).

[128] *R.S. Drago, N.A. Matwiyoff*, J. Organometal. Chem. *4*, 62 (1965).

[129] *E.R. Malinowski*, J. Am. Chem. Soc. *84*, 2649 (1962).

Table.10.$^{13}C-^{1}H$ coupling constants of *Tin and Silicon Compounds* without solvent[a–d,g,l,n–r] or in tetrachloromethane[f, h–k, m, n, s]. For other solvents see ref.[d,e]

Compound	δ	J[Hz]	Ref.
$(CH_3)_4Sn$	— 6.0	127.4	a
	— 4.4	127	b, c
	— 6.0	128±1	f
$(CH_3)_3SnH$	— 4.8	128.8	a
$[(CH_3)_3Sn]_2$	—12.0	128.9	g
$(C\underline{H}_3)_3SnSn(C_2H_5)_3$	—18.9	130.2	a
$(C\underline{H}_3)_3SnCH_2Si(CH_3)_3$	— 5.1	127.4	h
$(CH_3)_3SnC\underline{H}_2Si(CH_3)_3$	+15.5	117.2	h
$(CH_3)_3SnCF_3$	—18.0	131.7	a
$(CH_3)_3SnC \equiv CCH_3$	—13.8	128.5	i
$(CH_3)_3SnC \equiv CC_6H_5$	—21.0	129.5	i
$[(CH_3)_3Sn]_2O$	—14.2	128.8	j
$[(CH_3)_3Sn]_2S$	—23.8	130.8	j
$[(CH_3)_3Sn]_2Se$	—34.5	132.0	j
$(C\underline{H}_3)_3SnOSi(CH_3)_3$	—20.5	128.5	k
$(CH_3)_3SnSSi(CH_3)_3$	—26.6	131.0	j
$(CH_3)_3SnN[Si(CH_3)_3]_2$	—18.8	130.0	k
$(CH_3)_3SnCl$	—37.6	133.0	j
$(CH_3)_3SnBr$	—44.4	132.6	a
	—44.2	134.0	j
$(CH_3)_3SnJ$	—53.0	134.0	j
$(CH_3)_2SnH_2$	—18.0	129.9	l
$(C\underline{H}_3)_2C_2H_5SnH$	— 4.8	126.4	a
$(C\underline{H}_3)_2(HCF_2CF_2)SnH$	—18.6	132.1	l
$(C\underline{H}_3)_2(HCF_2CF_2)_2Sn$	—37.8	137.3	l
$(CH_3)_2(C_6H_5C \equiv C)_2Sn$	—36.6	131.3	i

CH₃-Groups Removed by One or More C from Sn

Compound	δ	J[Hz]	Ref.
$[(CH_3)_3CCH_2]_4Sn$	—62.2	124.4	m
$[C_6H_5(CH_3)_2CCH_2]Sn$	—66.3	128.4	m
$[((CH_3)_3CCH_2)_3Sn]_2$	—62.2	123.6	n
$[((C\underline{H}_3)_3CCH_2)_3Sn]_2O$	—62.4	124.8	n
$[(CH_3)_3CCH_2]_3SnO_2CCH_3$	—63.3	124.5	m
$[(CH_3)_3CCH_2]_2SnCl_2$	—68.0	124.5	m

$$CH-CH \text{ or } C{\overset{\diagup H}{\underset{\diagdown H}{}}}$$

Compound		δ	J[Hz]	Ref.
$(CH_3CH_2)_4Sn$	$(\beta-\alpha)$	—23.4	8.2	n,o
		—	9.3	b
$(CH_2=CH)_4Sn$	*(gem)*	—26.3	2.6	p
	(cis)	+11.6	16.1	p
	(trans)	+37.9	20.7	p
$(CH_2=CH)_3SnCl$	*(gem)*	—20.5	2.6	p
	(cis)	+11.3	16.1	p
	(trans)	+31.8	20.7	p
$CH_2=CH(C_4H_9)_2SnCl$	*(gem)*	—21.7	2.4	h
	(cis)	+15.3	15.0	p
	(trans)	+37.0	21.4	p
$(CH_3CH_2CH_2)_4Sn$	$(\gamma-\beta)$	43.9	~7.8	q
	$(\beta-\alpha)$	—32.2	7.5	q

Miscellaneous

Compound	δ	J[Hz]	Ref.
$(CH_3)_{12}Sn_6$	—33.0	1,3 $(SnSnSnCH_3)$	r
$(CH_3)_2Si\underline{H}CH_2SN(CH_3)_3$	—24.7	31 $(HSi-C-Sn^{117, 119})$	s
$CH_3\overset{O}{\overset{\|}{C}}OSn[CH_2C(CH_3)_3]_3$	—113.4	—	m

a) *H. C. Clark, J. T. Kwon, L. W. Reeves, E. J. Wells,* Inorg. Chem. *3*, 907 (1964).
b) *G. Klose,* Ann. Phys. *8*, 220 (1961).
c) *H. Spiesecke, W. G. Schneider.* J. Chem. Phys. *35*, 722 (1961).
d) *N. Flitcroft, H. D. Kaesz,* J. Am. Chem. Soc. *85*, 1377 (1963).
e) *G. P. Vander Kelen,* Nature [London] *193*, 1069 (1962).
f) *R. S. Drago, N. A. Matwiyoff,* J. Organometal. Chem. *4*, 62 (1965).
g) *H. C. Clark, J. T. Kwon, L. W. Reeves, E. J. Wells,* Can. J. Chem. *42*, 941 (1964).
h) *H. Schmidbaur,* Chem. Ber. *9v*, 270 (1964).
i) *M. LeQuan, P. Cadiot,* Bull. Soc. Chim. Fr., 35 (1965).
j) *H. Schmidbaur, I. Rudisch,* Inorg. Chem. *3*, 599 (1964).

k) *H. Schmidbaur,* J. Am. Chem. Soc. *85*, 233 (1963).
l) *H. C. Clark, J. T. Kwon, L. W. Reeves, E. J. Wells,* Can. J. Chem. *41*, 3005 (1963).
m) *H. Zimmer, O. Homberg,* unpubl.
n) *J. Lorberth, M. R. Kula,* Chem. Ber. *97*, 3444 (1964).
o) *P. T. Narasimhan, M. T. Rogers,* J. Chem. Phys. *34*, 1049 (1961).
p) *W. Brugel, T. Ankel, F. Kruckenberg,* Z. Elektrochem. Ber. Bunsenges. *64*, 1121 (1960).
q) *G. Klose,* Ann. Phys. *10*, 391 (1963).
r) *T. L. Brown, G. L. Morgan,* Inorg. Chem. *2*, 736 (1963).
s) *H. Schmidbaur, S. Waldmann,* Chem. Ber. *97*, 3381 (1964).

3.3.6 Applications

[13]C-shieldings readily lend themselves to studies aimed at examining the factors responsible for nuclear shielding[131, 132]. Furthermore, ^{13}C-shielding studies seem to reveal good correlations of these with π-electron density. Direct evidence can be gained on ground state electron distribution. Studies related to stereochemical problems, e.g., steric hindrance of resonance, strain in *cyclic olefins* and *bicyclic ketones,* can be aided in their solution[133, 134] with the help of ^{13}C-shifts.

Hammett effects have been studied using methyl [13]C—[1]H coupling constants for *substituted toluenes, N,N-dimethylanilines, anisoles* and *tert-butylbenzenes*[135].

A linear relationship between coupling constants and Hammett σ-constants of substituents was found to exist, and the relative slopes of these lines in the given decreasing order (toluene > dimethylaniline > anisole > tert. butylbencene) were interpreted in terms of the mode of transmission of substituent effects to the methyl site. Methyl chemical shifts also exhibit a linear correlation to the σ-constants, with a slightly different order of relative slopes (anisoles $\geqslant$ dimethylanilines). The use of σ^* values for those substituents capable of direct resonance interactions in the dimethylanilines and anisoles greatly improved the linearity of the correlations for these series.

3.4 Nitrogen-14 and Nitrogen-15

The ^{14}N nucleus (abundance 99.635%) with a nuclear spin of 1 and the ^{15}N nucleus (abundance 0.365%) with a nuclear spin of 1/2 are NMR-active nitrogen nuclei.

Even though the ^{14}N nucleus has a high natural abundance, its low magnetic moment makes it very insensitive to NMR detection. Furthermore, because it possesses a nuclear quadrupole, extensive line broadening occurs which precludes detection of spin-spin coupling. It is, therefore, of limited use for structure determination of organic compounds.

Because of low natural abundance of ^{15}N, suitable NMR detection is precluded. However, isotopic ^{15}N enrichment of organic compounds has permitted the study of ^{15}N-NMR. Because ^{15}N has no quadrupole moment, the spin-spin coupling provides useful information on molecular structure.

3.4.1 ^{14}N-Chemical Shifts

In general, a problem arises with the choice of a suitable reference signal for ^{14}N-NMR[4, 8, 136]. Witanowski[137] has proposed a unified scale for the measurement of ^{14}N-chemical shifts (Table 11). ^{14}N-chemical shifts (Table 12) vary linearly with the π-electron densities calculated by the LCAO-MO theory. [14]N chemical shifts, like ^{13}C-shifts, depend on hybridization, symmetry, and steric factors.

Table 11. Internal standards for ^{14}N measurements

	$_\delta$N,[ppm]	Application
CH_3NO_2	0	*Primary:* for general use with organic solvents and concentrated mineral acids.
$NO_3^\ominus$	0	*Primary:* for general use in aqueous solution.
$C(NO_2)_4$	+48	*Secondary:* for nitro-compounds.
$(CH_3)_2NCHO$	+276	*Secondary:* for aromatic nitro compounds and nitrogen heterocycles.

Table 12[8]. [14]N chemical shifts of aromatic nitro compounds

Compound	[14]N chemical shifts[a]		
	o	m	p
Nitrobenzene	6.1		
Nitrophenol	7	9	7
Nitroanisole	6	8	7
Dinitrobenzene	12	13	13
Nitrobenzaldehyde	11	12	12.5
1,3,5-Trinitrobenzene	18.5		
2,4,6-Trinitrotoluene	14		
2,4,6-Trinitrophenol	18		

[a] ppm from nitrate resonance.

3.4.2 ^{14}N-Spin Coupling Constants

Because ^{14}N possesses a quadrupole moment (Spin I = 1), spin-spin coupling to nitrogen is almost impossible to observe. In general [14]N—[1]H and [14]N—C—C—[1]H coupling can be observed indirectly by measuring the proton resonance spectrum[4, 8, 136].

3.4.3 ^{15}N-Chemical Shifts

The ^{15}N nucleus has not been studied by NMR as extensively[8] as the ^{14}N isotope.

Unlike ^{14}N-NMR, no unified standards for the measurement of ^{15}N-NMR have been agreed

[130] *A.W. Douglas,* J. Chem. Phys. *40,* 2413 (1964).

[131] *J.A. Pople,* Discuss. Faraday Soc. *34,* 7 (1962).

[132] *M. Karplus, J.A. Pople,* J. Chem. Phys. *38,* 2803 (1963).

[133] *T. Lippmaa, V.I. Sokolov, A.I. Olovson, Y.U. Past,* Dokl. Akad. Nauk. SSSR *173,* 358 (1967).

[134] *G.B. Savitsky, K. Namikawa, G. Zweifel,* J. Phys. Chem. *69,* 3105 (1965).

[135] *C.H. Yoder, R.H. Tuck, R.E. Hess,* J. Am. Chem. Soc. *91,* 539 (1969).

[136] *D.W. Matthieson*[260].

[137] *M. Witanowski, H. Janus Zewski,* J. Chem. Soc. B 1063 (1967).

upon. The choice of a reference is partially controlled by the availability of suitable materials. Nitrobenzene-^{15}N (Table 13) and H-^{15}NO$_3$ have been used. The chemical shifts of ^{15}N in ATP and ATP-metal ion solutions (Table 14) indicate that Zn$^{2\oplus}$ ion complexes to the NH$_2$ at C—6 and at N—7, whereas Mg$^{2\oplus}$ does not coordinate to the ATP ring at all.

Table 13. ^{15}N chemical shifts of nitrobenzene

Nitrobenzene	^{15}N [a]
H	0.00
4-Hydroxy-	4.38
4-Fluoro-	3.57
4-Chloro-	2.09
4-Bromo-	1.32
4-Nitro-	—3.85

[a]   ppm from nitrobenzene – ^{15}N

Table 14. ^{15}N-Shifts[a] of the nitrogen atoms in ATP and in equimolar metal ion-ATP solutions[138]

Compound	N-1	N-3	N-7	N-9	NH^2-6
ATP	144.7	135.6	129.5	191.6	282.8
Mg$^{2\oplus}$-ATP	144.9	135.0	129.5	191.6	282.3
Zn$^{2\oplus}$-ATP	144.9	135.0	132.5	190.1	279.7

[a]   ppm from H^{15}NO$_3$

Table 15. ^{15}N–H coupling constants. The average deviation of the coupling constants is—if not especially mentioned— ± 0.2 Hz.

Compound	Solvent	$J_{^{15}N-H}$[Hz]	Ref.
(C$_6$H$_5$)$_2$C=NH	Pentane	51.2$\pm$0.4	b
NH$_3$	—	61.2$\pm$0.9	a
NH$_4$Cl	water	73.2	b
CH$_3$NH$_3$Cl	water	75.6	b
HO$_2$CCH$_2$NH$_3$Cl	water	77.0	b
H$_2$N—CONHCONH$_2$	DMSO	88.4$\pm$1.0	b
H$_2$NCONH$_2$	DMSO	89 $\pm$1	b
CH$_3$CONH$_2$	water	89 $\pm$2	b
H$_2$N—C$\equiv$N	DMSO	89.4$\pm$1.0	b
C$_6$H$_5$NH—NHC$_6$H$_5$	DMF	90.5	b
Pyridine Ion	H$_2$SO$_4$	90.5$\pm$1.0	b
C$_6$H$_5$NH COCH$_2$	Acetone	90.9	b
C$_6$H$_5$NH(CS)NHCH$_3$	Ethanol	91.2	b
1-Methylcytosine	DMSO	92.0	d
HCONH$_2$(trans-H)[f]	—	92.0	c
(cis-H)[f]	—	88.0	c
(C$_6$H$_5$)$_2$C=NH$_2$Cl	SO$_2$	92.6$\pm$0.4	b
Phthalimide	DMSO	93.0$\pm$0.8	b
CH$_3$CONHCH$_2$COOH	DMSO	94.5	b
C$_6$H$_5$NH(CS)NHC$_6$H$_5$	DMSO—d$_6$	89.8$\pm$2	e
C$_6$H$_5$NH(CO)NHC$_6$H$_5$	DMSO—d$_6$	89.7$\pm$2	e
C$_6$H$_5$NH—CH=C(COOC$_2$H$_5$)$_2$	CDCl$_3$	91.4$\pm$0.2	e
C$_6$H$_5$CH=N—NH—C$_6$H$_5$	DMSO—d$_6$	92.7$\pm$0.2	e
C$_6$H$_5$NH—N=C(COOCH$_3$)$_2$	CCl$_4$	94.7$\pm$0.3	e
		96.1$\pm$0.3	e
	CCl$_4$	91.2$\pm$0.4 [f]	e
	CCl$_4$	92.8$\pm$0.3 [g]	e
H$_3$C—CO—NH—C$_6$H$_5$	CDCl$_3$	89.9	e

[a]   R. A. Bernheim, H. Bartz-Hernandez, J. chem. Phys, 40, 3446 (1965).
[b]   G. Binsch, J. B. Lambert, B. W. Roberts, J. D. Roberts, J. Amer. chem. Soc., 86, 5564 (1964). Unless otherwise indicated, the average deviation in the coupling constants is 0.2 cps.
[c]   B. Sunners, L. H. Piette, W. G. Schneider, Can. J. Chem., 38, 681 (1960).
[d]   H. T. Miles, R. B. Gradley, E. D. Becker, Science, 142, 1569 (1963).
[e]   A. K. Rose, I. Kugajevsky, Tetrahedron, 23, 1489 (1967).
[f]   Coupling between the carbonyl group and the cis-proton.
[g]   Coupling between the carbonyl group and the trans-proton.

3.4.4 ^{15}N-Spin Coupling Constants

The coupling of ^{15}N with directly bonded protons[139,140], protons separated by two or three bonds (long-range ^{15}N—^{1}H coupling), and ^{15}N—^{13}C couplings[2] have been studied (Tables 15 and 16).

bond instead of the expected 33% s-character, to conform to this correlation, has been attributed to contributions to the nuclear spin-spin coupling from electron orbital motion; the same holds true for the ^{13}C—^{15}N coupling constant in *benzylidenmethylamine* and *acetonitrile* (Tab. 16).

Table 16. Long-Range ^{15}N-Proton coupling constants J

Compound	J(^{15}NCH)	J(^{15}NCCH)	Ref. (s. table 15)
^{13}CH$_3$^{15}NH$_2$	1.0	—	a
^{13}CH$_3$^{15}NH$_3$Cl	0.8	—	a
C$_6$H$_5$^{15}N(CH$_2$CH$_3$)$_2$	0.0	—	a
HCO^{15}NH$_2$	19.0	—	b
	1.4	—	a
ONH(CS)^{15}NH^{13}CH$_3$	0.9	—	a
C$_6$H$_5$CH$^{\alpha}$—^{15}N—^{13}CH$_3$$^{\beta}$	3.9(α)	—	a
	0.6(β)		a
C$_6$H$_5$^{13}CH$^{\alpha}$=^{15}N—CH$_3$$^{\beta}$	3.9(α)	—	a
	0.6(β)		a
^{13}CH$_3$—^{13}N=C=S	3.3	—	a
CH$_3$—^{13}C≡^{15}N	—	1.7	a
	1.2(α)	—	a
	1.1(β)	—	a
	15.6 ( )	—	a
	19.0		c
	3.8±0.1		e
	2.4±0.1		e
	8.4±0.1 (α)		e
	0.0 (β)		e
	8.3±0.1(α)		e
	0.0 (β)		e

Analogous to ^{13}C—^{1}H coupling constants, an empirical relationship of the ^{15}N—^{1}H coupling constants and the %s-character of the nitrogen has been developed[139]. The hybridization predicted from ^{15}N—^{1}H coupling constants supports the theory that the hybridization around nitrogen is a function of bond angle rather than of complete p-hybridizations as suggested by Pauling[141]. The failure of *diphenylketimine* (Table 15), which gives a low value of 15.6% s-character of the ^{15}N—^{1}H

The ^{15}N—C—^{1}H coupling through an sp^3-hybridized carbon atom is small ranging from 0.6 cps to 1.4 cps (Table 16). The ^{15}N=C—^{1}H coupling is somewhat larger. The ^{15}N—(C—^{1}H)X coupling constants are fairly large ranging from 7.5 cps to 19.0 cps[3].

Attempts[142–144] to correlate the ^{15}N—^{1}H coupling constants[143,144] and chemical shifts[143,145] with Hammett σ-constants in *aniline*[140,144] and *nitro-*

[138] *J.A. Happe, M. Morales*, J. Am. Chem. Soc. *88*, 2077 (1966).

[139] *G. Binsch, J.B. Lambert, B.W. Roberts, J.D. Roberts*, J. Am. Chem. Soc. *86*, 5564 (1964).

[140] *A.K. Bose, I. Kugajevsky*, Tetrahedron *23*, 1489 (1967).

[141] *A.L. Pauling*, The Nature of the Chemical Bond, p. 120, Cornell University Press, Ithaca, New York 1960.

[142a] *D.T. Clark, J.D. Roberts*, J. Am. Chem. Soc. *88*, 745 (1966).

[142b] *W. Bremser, J.I. Kroschwitz, J.D. Roberts*, J. Am. Chem. Soc. *91*, 6190 (1969).

benzene derivatives[142] have been reported. In the case of the substituted anilines, a linear correlation of ^{15}N-^1H coupling constants (Table 17) has been observed in DMSO-d_6[143, 144] and CDCl$_3$[143]. The ^{15}N—^1H coupling constant decreases as the electron donating ability of the substituent is increased and is greatest with electron withdrawing groups on the ring. This variation in the coupling constant is attributed to delocalization of the nitrogen lone pair by overlap with the π-system of the aromatic ring, resulting in an increased s-character of the N–H bond compared to a normal sp^3 hybridized ammonium ion[143]. In the case of substituted ^{15}N-nitrobenzenes[142] a fairly good correlation between ^{15}N-chemical shift and Hammett constants was obtained[142b].

Table 17. ^{15}N–^1H coupling constants [Hz] in aniline derivatives

C$_6$H$_5$ ^{15}NH$_2$	DMSO–d$_6$	CDCl$_3$
4-NO$_2$	89,4	86.4
3,5-CH$_3$-4-NO$_2$	87.0	83.2
3-Br	85.3	80.5
3-Cl	85.1	80.9
3-J	84.4	80.4
4-J	84.0	79.7
4-Br	84.0	79.6
4-Cl	83.7	78.9
3-CH$_3$O	83.0	79.4
H	82.6	78.6
3-CH$_3$	82.0	78.2
3,5-(CH$_3$)$_2$	82.1	77.5
4-F	81.6	77.8
4-CH$_3$	81.4	76.5
4-CH$_3$O	79.4	75.6

A plot of *para*—^{13}C-chemical shifts in monosubstituted benzenes and ^{19}F-chemical shifts in *para*-substituted fluorobenzenes vs. ^{15}N-chemical shifts of nitro groups in *para*-substituted nitrobenzenes gave a good linear correlation[142a]. A plot of ^{15}N-chemical shifts vs. π-electron density from *MO calculations* gave considerable deviation from linearity[142b] suggesting that the ^{15}N-chemical shift is probably influenced more by inductive effects than resonance effects. The observed configurational dependence on the ^{15}N—N—C—^1H coupling constants has been useful in determining the geometric configurations of *N-nitrosamines*[145], *N-nitrosohydrazines*[146], *N-nitrosohydroxyl-*

amines[147], and *alkylnitrites*[148]. *Pyrimidine-$^{15}N_2$* derivatives[149] also have been studied by ^{15}N-NMR. The ^{15}N—^1H coupling constants in *quinoline-^{15}N, quinoline-^{15} N-ethiodide* and *quinoline-^{15}N-oxide* have been measured indirectly from the proton magnetic resonance spectrum and the relative signs of the coupling constants assigned[150].

3.5 Oxygen-17

^{17}O is the only oxygen isotope ($I = 5/2$, natural abundance 0.037%) which has a magnetic moment. It is very difficult to observe ^{17}O-NMR signals[1–5, 151].

Studies related to kinetics of keto-enol tautomerism[152] in organic molecules and correlation of the lowest energy electron transitions for compounds containing oxygen bonded to transition metals have been noted. The measurement of ^{17}O-chemical shifts is generally done with reference to H$_2$^{17}O (Table 18).

Table 18. Some characteristic ^{17}O-chemical shifts[a]

Chemical shifts measured in ppm from an H$_2$O external reference

Compounds	Range of chemical shifts [ppm]	
	—O—	=O
Alcohols and ethers	+40 to –40	—
Methyl esters	–130 to –150	–350 to –370
Ethyl and propyl esters	–160 to –180	–350 to –370
Acids	–240 to –260	
Acid anhydrides	–240 to –260	–390 to –393
Ketones	—	–550 to –570
Aldehydes	—	–580 to –600

[a] *H.A. Chust*, Helv. phys. Acta *33*, 572 (1960).

3.6 Fluorine-19

After ^1H the ^{19}F nucleus is the next most studied nucleus[4]. It is the only naturally occurring isotope of fluorine. Its chemical shifts and coupling cons-

[143] *T. Axenrod, P.S. Pregosin, M.J. Wieder, G.W.A. Milne*, J. Am. Chem. Soc. *91*, 3681 (1969).

[144] *M.R. Bramwell, E.W. Randall*, Chem. Comm. 250 (1969).

[145] *T. Axenrod, P.S. Pregosin, G.W.A. Milne*, Chem. Comm. 702 (1968).

[146] *T. Axenrod, P.S. Pregosin, G.W.A. Milne*, Tetrahedron Lett. 5293 (1968).

[147] *T. Axenrod, M.J. Wieder, G.W.A. Milne*, Tetrahedron Lett. 401 (1969).

[148] *T. Axenrod, M.J. Wieder, G.W.A. Milne*, Tetrahedron Lett. 1397 (1969).

[149] *B.W. Roberts, J.B. Lambert, J.D. Roberts*, J. Am. Chem. Soc. *87*, 5439 (1965).

[150] *K. Tori, M. Ohtsuru, K. Aono, Y. Kawazoe, M. Ohnishi*, J. Am. Chem. Soc. *89*, 2765 (1967).

[151] *B.L. Silver, Z. Luz*, Quart. Rev., Chem. Soc. *21*, 458 (1967).

tants are larger than those of ^{1}H and thus render more information about molecular structures and chemical processes for fluorinated molecules than for the hydrogen analogs.

3.6.1 ^{19}F-Chemical Shifts

The ^{19}F-NMR signals occur over a range of about 700 ppm (including binary fluorides); this is in contrast to the range of about 15 ppm for ^{1}H-chemical shifts. The diamagnetic shielding effect of the electron cloud, which is responsible for practically all the shielding experienced by the hydrogen nucleus, contributes only a small amount (about 1%) of the total observed range of ^{19}F-shifts. In ^{19}F the paramagnetic term, σ_p, of the Ramsey[152] equation for nuclear shielding, has been shown to be responsible for most of the observed large chemical shifts. A relationship between the paramagnetic term, which is zero for the spherically symmetrical $F^\ominus$ ion, and the covalent character of the F—X bond has been shown[153].

For ^{19}F-NMR spectroscopy, both external and internal references are used. The reference situation here is not as well agreed upon as is the case in ^{1}H-NMR spectroscopy. Commonly used references[155] are summarized in Table 19. It has been suggested that ^{19}F data should be reported from solutions of less than 10% $CFCl_3$[154] which can be regarded as infinitely dilute.

Table 19. Common references in ^{19}F-NMR-spectroscopy[a]

Compound	C_4F_8	C_6H_5F	CF_3COOH	$CFCl_3$
C_4F_8	0	24.9	61.5	130
C_6H_5F	− 24.9	0	36.6	113.1
CF_4	− 61.3	− 36.4	0.2	76.7
CF_3COOH	− 61.5	− 36.6	0	76.5
$C_6H_5CF_3$	− 74.3	− 49.4	− 12.8	63.75
$CFCl_3$	−138.0	−113.1	− 76.5	0
$C_6H_5SO_2F$	−203.5	−178.6	−142.0	− 65.5

[a] *D.F. Evans*, J. Chem. Soc. *1960*, 877.

3.6.2 Aliphatic Fluorine Compounds

Successive fluorination of methane decreases the ionic character of the C—F bonds; thus a corresponding decrease in the shielding of the ^{19}F

nucleus is observed. Analogously, it is expected that successive fluorine substitution in carbon tetrachloride should be accompanied by a decrease in ^{19}F-chemical shifts (since chlorine like hydrogen is less electronegative than fluorine); however, the observed ^{19}F-chemical shifts are in the opposite direction (Table 20).

To explain these experimental facts, it is assumed that, because of the greater tendency of fluorine to be involved in back-donation (see Section 5.5:3.2) as compared with chlorine, hyperconjugative structures like **10a** contribute significantly more than **10b**.

10 a   **10 b**

Since the F-atom in **10a** is less shielded than in **10b**, its NMR signal will appear at a lower field. Chemical shifts of additional aliphatic fluorine compounds are summarized in tables 21–23.

Table 21. ^{19}F-chemical shifts of selected F-containing aliphatic compounds[a]

All shifts are reported from CF_3COOH using the approximate conversion $\delta CF_3COOH = \delta C_4H_8 + 59$ when necessary.

Compound	δF_1	δF_2	δF_3
CH_2FCH_3	+76.95		
CHF_2CH_3	−25.91		
CHF_2CHF_2	+ 2.21	+ 2.21	
CHF_3CH_3	−73.47		
CF_3CF_2H	−47.76	+ 3.88	
CF_3CFH_2	−56.61	+105.4	
$CHF_2CF_2CHF_2$	+ 3.27	− 0.20	+ 3.27
$CF_3CH_2CF_3$	−71.69		− 71.69
$CF_2CH_2CH_3$	−66.21		
$CHF_2CF_2CH_2F$	+ 3.24	− 7.54	+109.01
$CF_3CF_2CH_2F$	−51.	− 7.97	+108.17
$CF_3CF_2CHF_2$	−56.22	− 1.98	+ 2.86

[a] *D.D. Elleman, L.C. Brown, D. Williams*, J. Mol. Spectr. 7, 322 (1961).

Table 20. ^{19}F-chemical shifts in halomethanes[156].

Compound	δF [ppm]	$J(^1H-^{19}F)$	$J(^{13}C-^{19}F)$	Compound	δF[ppm]
CH_3F	+210[155]	46.4	157.5[156]		
CH_2F_2	+ 80.9	50.2	235	$CFCl_3$	−76.7
CHF_3	+ 18.2	79.7	274.3	CF_2Cl_2	−60.4
CF_4^a	0.0		259.2	CF_3Cl	−36.8

[a]Reference compound

Table 22. ^{19}F-chemical shifts of nitrogen compounds

All shifts are reported from CF_3COOH using the approximate conversion $\delta CF_3COOH = \delta C_4H_8 + 59$ when necessary.

Compound	Group	Chem. Shift	Ref.
(b)F₂ F₂(c) / F₂ NF / (a)F₂ F₂	$CF_2{}^{(a)}$	56.7	a
	$CF_2{}^{(b)}$	54.5	
	$CF_2{}^{(c)}$	32.7	
	NF	36.6	
(b)F₂ F₂(c) / (a)F₂ N— / F₂ F₂ /₂	$CF_2{}^{(a)}$	56.2	a
	$CF_2{}^{(b)}$	55.3	
	$CF_2{}^{(c)}$	19.1	
(b)F₂ F₂(c) / (a)F₂ N—CF₃ / F₂ F₂	$CF_2{}^{(a)}$	57.8	a
	$CF_2{}^{(b)}$	55.2	
	$CF_2{}^{(c)}$	16.8	
	CF_3	—25.3	
(a)F₂ F₂(b) / O N—CF₃ / F₂ F₂	$CF_2{}^{(a)}$	9.6	a
	$CF_2{}^{(b)}$	17.0	
	CF_3	—24.0	
$CF_3CF_2CF_2NF_2$ (a) (b)	CF_3	6.4	a
	$CF_2{}^{(a)}$	27.1	
	$CF_2{}^{(b)}$	51.1	
	NF_2	—92.0	
$(CF_3)_2NC(O)F$	CF_3	—20	a
	CF	—81	
$(CF_3)_2CHC\equiv N$	CF_3	—10.6	a
$CF_2=NCF_3$	CF_3	—18.8	a
	CF	—44.6	
	CF	—25.2	
$FClC=CH—CF_3$	$=CF$	—15.2	b
	CF_3	—15.2	
$CCl_2=CH—CF_3$	CF_3	—14.8	b
$CF_3—CCl=CCl—CF_3$	CF_3	—11.6	
$CF_3—CF=CCl—CF_3$	$CF_3{}^{(a)}$	— 6.2	b
	$CF_3{}^{(b)}$	— 9.7	
F F / F F (a) / F F(b)	$—CF=$	39.2	b
	$CF_2{}^{(a)}$	44.2	
	$CF_2{}^{(b)}$	56.1	
	$=CF—$	77.5	
$HC\equiv C—CF_3$	CF_3	—28.5	b

a N. Mueller, P.C. Lauterbur, G.F. Svatos, J. Amer. Chem. Soc., 79, 1807 (1957); J.A.Young, W.S. Durrell, R.D. Dresdner, J. Amer. Chem. Soc., 84, 2105 (1962).
b H.M. Beisner, L.C. Brown, D. Williams, J. Mol. Spectr., 7, 385 (1961).

3.6.3 Aromatic Fluorine Compounds

^{19}F-chemical shifts have been used repeatedly to obtain information about the electron distribution in aromatic systems. Thus, attempts to correlate the Hammett substituent constants with ^{19}F-che-

[152] N.F. Ramsey, Phys. Rev. 78, 699 (1962).
[153] A. Saika, C.P. Slichter, J. Chem. Phys. 22, 26 (1954).
[154] G. Filipovich, G.V.D. Tiers, J. Phys. Chem. 63, 761 (1959).
[155] L.H. Meyer, H.S. Gutowsky, J. Phys. Chem. 57, 481 (1953).
[156] S.G. Frankiss, J. Phys. Chem. 67, 752 (1965).

Table 23. ^{19}F-chemical shifts of some organometals

All shifts, unless otherwise mentioned, are reported from CF_3COOH using the approximate conversion $\delta CF_3COOH = \delta C_4H_8 + 59$ when necessary.

Compound	Group	Chemical Shift	Ref.
$CF_3Mn(CO)_5$	CF_3	— 85.8 [1]	a
$(CF_3)_2CFMn(CO)_5$	CF_3	— 8.9 [1]	a
(a) (b)	CF	+ 87.1 [1]	
$CF_3C_0(CO)_4$	CF_3	— 87	a
$F_3CCF_2CF_2C_0(CO)_4$	CF_3	+ 2	a
(a) (b)	$CF_2{}^{(a)}$	— 25.5	
	$CF_2{}^{(b)}$	+ 18.4	a
$(F_3CCF_2)_2Hg$	CF_3	+ 6	b
	CF_2	+ 31.9	
$(F_3CCH_2)_2Hg$	CF_3	+ 28.9	b
$(FECCHF)_2Hg$	CF_3	— 0.6	b
	CF	+146.3	
$F_3CCF_2Sn(C_2H_5)_3$	CF_3	+ 4.5	c 2
	CF_2	+ 40.4	
$F_3CCF_2Sn(C_4H_3)_3$	CF_3	4.0	c 2
	CF_3	40.5	
$F_3CCF_2CF_2Sn(C_4H_9)_3$	CF_3	0.4	c 2
(a) (b)	$CF_2{}^{(a)}$	42.8	
	$CF_2{}^{(b)}$	38.3	

a E. Pitcher, A.D.Buckingham, F.G.A. Stone, J. chem. Phys., 36, 124 (1962).
b C.G. Kriespan, J. org. Chem., 25, 105 (1960).
c W.R. McClellan, J. Am. chem. Soc., 83, 1958 (1961).
[1] Fine structure not resolved.
[2] Measured with CCl_3F as external standard; converted into CF_3COOH as external standard by following equation CF_3COOH ext = CCl_3F est + 79.9; s, t, and q refer to singlet, triplet and quartet.

mical shifts for *meta*- and *para*-substituted fluorobenzenes were successful[157] (Table 24). This treatment has been refined and very close correlations have been obtained[158]. Attempts to explain the abnormal behavior of halogen substituents have been reported[159–163].

Table 24. ^{19}F-chemical shifts [ppm] in C_6F_5X compounds from $CFCl_3$ as internal reference[a]

C_6F_5X	ortho	para	meta
X = F	162.28	162.28	162.28
Cl	140.61	156.11	161.48
Br	132.54	154.65	160.60
J	119.18	152.53	159.65
H	138.89	153.50	162.06
$NHCH_3$	161.89	173.07	165.21
C_6	138.25	150.27	160.76
$Sn(CH_3)_3$	122.30	152.75	160.67
$HgCH_3$	121.91	153.52	160.05

[a] $CFCl_3$ present at 5 mole % in a 5% solution of the compounds in CCl_4.

[157] H.S. Gutowsky, D.W. McCall, B.R. McGarvey, L.H. Meyer, J. Am. Chem. Soc. 74, 4809 (1952).
H.W. Johnson, Y. Iwata, J. Org. Chem. 35, 2822 (1970).

Table 25. ^{19}F-chemical shifts and Hammett constants for monosubstituted perfluorinated benzenes

Chemical shifts were measured in ppm from C_6H_5F as internal reference: $\delta CF_3COOH_{ext} = \delta C_6H_5F_{int} +$ 35.6 %

Subst.	ortho δ [ppm]	meta δ [ppm]	Hammett σ const.	para δ [ppm]	Hammett σ const.	Taft σ_I const.	Taft σ_R const.
NO$_2$	+ 5.6	—3.3	+.710	—10.8	+.778	+.63	+.15
CN	— 5.2	—3.0	+.608	— 9.6	+.656	+.59	+.07
J	—19.3	—2.6	+.352	+ 1.2	+.276	+.38	—.10
Br	— 5.5	—2.4	+.391	+ 2.3	+.232	+.45	—.22
Cl	+ 2.7	—2.1	+.373	+ 2.4	+.227	+.47	—.24
F	+25.9	—3.1	+.337	+ 6.4	+.064	+.50	—.44
CH$_3$	+ 5.0	+ .9	—.069	+ 5.5	—1.70	—.05	—.13
OH	+ 2.5	— .9	+.10	+10.6	— .30	+.25	—.61
NH$_2$	+23.1	+ .2	—.161	+14.6	—.660	+.10	—.76

^{19}F-chemical shifts of perfluorinated benzene derivatives (Table 25) have also been reported[164–166]. In several cases the shifts are additive and predictions of ^{19}F-chemical shifts can be made by summation of known chemical shift contributions from the substituents[167, 168]. ^{19}F-chemical shifts for 6- and 7-CH$_3$M-substituted β-fluornaphthalines (M=Si, Ge, Sn, Pb) have been reported[168a]. dπ-pπ-effects are seen to be responsible for differences in ^{19}F-chemical shifts of 7-CH$_3$M-substituted β-fluornaphthalines (M=Si, Ge, Sn, Pb)[168a].

3.6.4 ^{19}F-Spin-Coupling

The ^{1}H—^{19}F coupling constants were first obtained by an investigation of a variety of deuterofluorobenzenes[169]. The measured values were $J^{ortho}_{FH} = 9.4$ Hz. $J^{meta}_{FH} = 5.8$ Hz, and $J^{para}_{FH} = 0.0 \pm 5$ Hz. From these values a full analysis of the ^{1}H-NMR spectrum of monofluorobenzene as belonging to an ABB′CC′X system was reported[170]. A large number of ^{19}F—^{19}F and ^{1}H—^{19}F coupling constants of aromatic-compounds were measured, giving the following results[165–171]:

	^{1}H – ^{19}F	^{19}F – ^{19}F
ortho	6–10	20.2–20.8
meta	6– 8.5	0–7
para	~2	0–15

The relative signs of spin-coupling constants have been reported[172].

$$\pm J^{ortho}_{FF} \qquad \pm J^{meta}_{FF} \qquad \pm J^{para}_{FF}$$

$$\pm J^{ortho}_{HF} \qquad \pm J^{meta}_{HF} \qquad \pm J^{para}_{HF}$$

Many observations exist in relation to ^{1}H—^{19}F (Table 26) and ^{19}F—^{19}F (Table 27) coupling constants in fluoroalkenes[173–187].

Generally the long-range coupling constants decrease with increased number of bonds through which the effect is transmitted. However, there are exceptions in fluoroalkenes where the following constants were observed:

J_{FF} for F—C—C—F ~ O,
J_{FF} for F—C–C–C–F ~ 7–10 Hz.
If there is a nitrogen atom in the chain, the constants usually increase, thus F—C—N—C—F ~ 10–17 Hz, F—C—C—N—C—F ~ 5–7 Hz.

[158] R.W. Taft, J. Am. Chem. Soc. 79, 1045 (1957).
[159] N. Boden, J.W. Emsley, J. Feeney, L.H. Sutcliffe, Mol. Phys. 8, 133 (1964).
[160] M. Karplus, T.P. Das, J. Chem. Phys. 34, 1683 (1961).
[161] F. Prosser, L. Goodman, J. Chem. Phys. 38, 374 (1963).
[162] R.W. Taft, F. Prosser, L. Goodman, G.T. Davis, J. Chem. Phys. 38 (1963).

[163] R.W. Taft, E. Price, I.R. Fox, I.C. Lewis, K.K. Andersen, G.T. Davis, J. Am. Chem. Soc. 85, 3146 (1963).
[164] A.J.R. Bourn, D.G. Gillies, E.W. Randall, Proc. Chem. Soc. 200 (1963).
[165] N. Boden, J.W. Emsley, J. Feeney, L.H. Sutcliffe, Mol. Phys. 8, 133 (1964).
[166] D.E. Fenton, A.G. Massey, J. Inorg. Nucl. Chem. 27, 329 (1965).
[167] H.S. Gutowsky, D.W. McCall, B.R. McGarvey, L.H. Meyer, J. Am. Chem. Soc. 74, 4809 (1952).
[168] G.M. Brooke, J. Burdon, J.C. Tatlow, J. Chem. Soc. 802 (1961).
[168a] W. Adcook, W. Kitching, Chem. Commun. 1163 (1970).
[169] B. Bak, J.N. Shoolery, G.A. Williams, J. Mol. Spectrosc. 2, 525 (1958).
[170] S. Fujiwara, H. Shimizu, J. Chem. Phys. 32, 1636 (1960).
[171] M.S. Gutowsky, C.H. Holm, A. Saika, G.A. Williams, J. Am. Chem. Soc. 79, 4596 (1957).
[172] D.F. Evans, Mol. Phys. 6, 179 (1963).

Table 26. ^{1}H–^{19}F coupling constants of some fluorinated alkenes [Hz]

Compound	J_{HF}^{gem}	J_{HF}^{cis}	J_{HF}^{trans}	Ref.
HFC=CH$_2$	+84.7	+20.1	+55.4	a
F$_2$C=CH$_2$	—	~ 1	34	b
F$_2$C=CHF	72	< 3	12	b
CHF=CHF cis	+72.7	—	+20.4	c
trans	+74.3	+ 4.4	—	c
CHFC=CClH	—	10.8	24.2	d

a C.N. Bannwell, N. Sheppard, Proc. Roy. Soc. A226, 136 (1961).
b H.M. McConnell, C.A. Reilly, A.D. McLean, J. chem. Phys. 24, 479 (1956).
c G.W. Flynn, M. Matsushima, N.C. Craig, J.D. Baldeschwieler, J. chem. Phys. 38, 2295 (1963).
d E. Pitcher, F.G.A. Stone, Spect. Chim. Acta. 17, 1244 (1961).

Table 27. ^{19}F–^{19}F coupling constants[a] of $\begin{smallmatrix}A\\C_6H_5\end{smallmatrix}C=C\begin{smallmatrix}B\\C\end{smallmatrix}$

A	B	C	J_{AC}	J_{AD}	J_{BC}
F	CF$_3$	F	131	23	10
F	F	CF$_2$	13	9	9
CF$_3$	F	CF$_3$	1.5	28	7
CF$_3$	CF$_3$	F	12	12	7

a S. Andreades, J. Amer. chem. Soc., 84, 864 (1962)

173 T.D. Coyle, S.L. Stafford, F.G.A. Stone, Spectrochim. Acta 17, 968 (1961).
174 T.D. Coyle, S.L. Stafford, F.G.A. Stone, Spectrochim. Acta 17, 1244 (1961).
175 C.A. Reilly, J.D. Swalen, J. Chem. Phys. 34, 2122 (1961).
176 H.M. McConnell, C.A. Reilly, A.D. McLean, J. Chem. Phys. 24, 279 (1956).
177 H.M. McConnell, J. Chem. Phys. 24, 460 (1956).
178 T.D. Coyle, S.L. Stafford, F.G.A. Stone, J. Chem. Soc. 743 (1961).
179 T.D. Coyle, S.L. Stafford, F.G.A. Stone, Spectrochim. Acta 17, 1244 (1961).
180 R.A. Beaudet, J.D. Baldeschwieler, J. Mol. Spectrosc. 9, 30 (1962).
181 G.V.D. Tiers, J. Phys. Chem. 66, 1192 (1962).
182 N. Boden, J.W. Emsley, J. Feeney, L.H. Sutcliffe[4], p. 149.
183 N. Boden, J.W. Emsley, J. Feeney, L.H. Sutcliffe, Proc. Roy. Soc. A 282, 559 (1964).
184 G.V.D. Tiers, P.C. Lauterbur, J. Chem. Phys. 36, 1110 (1962).
185 C.A. Reilly, J. Chem. Phys. 37, 456 (1962).
186 G.V.D. Tiers, J. Chem. Phys. 35, 2263 (1961).
187 D.D. Elleman, L. Manatt, J. Chem. Phys. 36, 1945 (1962).
188 L. Petrakis, C.H. Sederholm, J. Chem. Phys. 35, 1243 (1961).
189 S. Ng, C.H. Sederholm, J. Chem. Phys. 40, 2090 (1964).
190 G.V.D. Tiers, J. Phys. Chem. 66, 764 (1962).

Thus it was concluded that the effects responsible for F—F long-range coupling are mostly transmitted through space[188–189], ^{13}C—^{19}F-coupling constants are large (Table 28).

Table 28. ^{13}C–^{19}F Coupling Constants

Compound	$J_{^{13}C-^{19}F}$ [Hz]	Ref.
CF$_4$	259.2	a
CF$_3$CCl$_3$	282.5	b
CF$_3$CO$_2$H	283.2	b
(CF$_3$)$_2$PCl	320.2	c
(CF$_3$S)$_2$	313.8	c

a S.G. Frankiss, J. phys. Chem. 67, 752 (1963).
b G.V.D. Tiers, J. phys. Soc. Japan 15, 354 (1960).
c R.K. Harris, J. phys. Chem. 66, 768 (1962).

3.6.5 Applications of ^{19}F-NMR-Spectroscopy

Similar to ^{13}C-NMR-^{19}F-NMR-spectroscopy lends itself to the investigation of electron distribution in chemical bonds and to studies of the influence of substituent reactivity (Hammett-Taft relationships); a special area of application is conformational studies. The larger variation in ^{19}F shielding over ^{1}H shielding and larger chemical shift differences in different conformations are two major advantages in using ^{19}F-NMR studies over ^{1}H-NMR conformational studies. Many ^{19}F conformational studies of fluorocyclohexanes and similar compounds have been reported[190–198]. Studies involving chemical exchange might also be facilitated by using ^{19}F-NMR-spectroscopy.

3.7 Phosphorus-31

3.7.1 Chemical Shifts

The only naturally occurring isotope of phosphorus, ^{31}P, has a spin quantum number $I=1/2$ ($\mu=1.1305$ nuclear magnetons). ^{31}P-chemical shifts occupy a range of about 700 ppm from -227 ppm for PBr$_3$ to $+488$ ppm for P$_4$. Though ^{31}P is easier to saturate than either ^{1}H or

191 G.V.D. Tiers, Proc. Chem. Soc. 389 (1960).
192 W.D. Phillips, J. Chem. Phys. 25, 949 (1956).
193 J. Feeney, L.H. Sutcliffe, Trans. Faraday Soc. 56, 1559 (1960).
194 J. Feeney, L.H. Sutcliffe, J. Phys. Chem. 65, 1894 (1961).
195 J. Homer, L.F. Thomas, Trans. Faraday Soc. 59, 2431 (1963).
196 L.F. Thomas, J. Homer, Proc. Chem. Soc. 139 (1961).
197 N. Boden, J. Feeney, L.H. Sutcliffe[4].
198 H.H. Hoehn, L. Pratt, K.F. Watterson, G. Wilkinson, J. Chem. Soc. 2738 (1961).

^{19}F, its relaxation time is still short enough to permit measurement of the NMR-spectra using reasonably rapid sweep rates. Due to the low sensitivity of phosphorus (see Table 1), the use of large samples and receiver coils is necessary.

Phosphoric acid (85%) is used almost exclusively as external reference. Recently P_4O_6 is being recommended as a standard. Its signal appears at -112.5 ± 0.1 ppm relative to 85% H_3PO_4. All peaks appearing down-field from 85% H_3PO_4 are said to be shifted negatively; whereas, those peaks found above the reference are regarded as positively shifted. Internal referencing, due to the reactive nature of H_3PO_4, is hardly ever done. The ^{31}P-chemical shift is only very slightly dependent upon the solvent used (^{31}P-chemical shift of $O=P(OCH_3)_3$ shows a change of only 2 ppm in different solvents[199]). A special feature of ^{31}P-NMR spectroscopy is the large ^{31}P—^{1}H coupling constants (about 500–700 Hz) of compounds in which H is directly bonded to P. The question regarding which tautomer 11 or 12, predominates in the equilibrium, can easily be decided in favor of 12, because of the very large coupling constant which is observed.

$$11 \quad (RO)_2\,P(OH) \leftrightarrows (RO)_2\,P(O)H \quad 12$$

For the structure of phosphorous acid 14 was finally established as the stable predominant species, ruling out species 13, by observing a large ^{31}P–^{1}H coupling constant.

$$13 \quad P(OH)_3 \leftrightarrows HP(O)(OH)_2 \quad 14$$

This was confirmed further by observing the appropriate ^{31}P-chemical shifts; $P(OR)_3$ has a chemical shift of about -125 to 140 ppm (depending on R); whereas *phosphorous acid* shows a shift of -4.5 ppm[200].

^{31}P-NMR studies were also applied successfully to solve the problem concerning the structure of the compound from the reaction phosphite esters with α,β-diketones[201].

15

16

In 15, phosphorus would have the coordination number of four; whereas, in 16, a coordination number of five would be present. ^{31}P-chemical shifts for $R_4P^\oplus$ compounds are generally found in the low field range of the ^{31}P-shifts. *Penta-coordinated P-compounds* absorb at much higher fields, e.g., $PCl_5 +78$, and $PF_5 +35$. For the reaction product between phenanthrene quinone and phosphite esters, ^{31}P-chemical shifts with high positive values were observed.

17   18

$R=CH_3$, $+44.7$ ppm; $R=CH(CH_3)_2$, $+48.9$ ppm; $R=C_6H_5$, $+58.9$ ppm.

Thus it was concluded that 17, and analogously 16, represent the correct structures. Additional evidence to support structure 17 was obtained from the study of the ^{31}P splitting pattern and the ^{1}H-NMR spectrum[202-205].

Empirically, the following generalizations concerning ^{31}P-chemical shifts can be made. *Tervalent phosphorus* shows absorptions practically over the entire range of known shifts; *tetravalent phosphorus* covers a much smaller range, ~ -130 to $+120$ ppm. *Pentavalent phosphorus* gives rise to chemical shifts in the range of about $+20$ to about $+100$ ppm. Phosphorus with the coordination number of six seems to exhibit its resonance peak at much higher fields, $+130$ to about $+200$ ppm and even higher ($+281$ for $PCl_6^\ominus$).

There is no simple or direct correlation of the ^{31}P shifts with the electronegativity of the atoms or groups bonded to phosphorus; e.g., the shifts in the trihalide series follow the order $PF_3 > PI_3 > PCl_3 > PBr_3$. A simple and general correlation also does not exist between ^{31}P shifts and Taft's σ^* or Hammett's σ values. This behavior leads to the conclusion that there are at least two

[199] *R.A.Y. Jones, A.R. Katritzky*, Angew. Chem. 74, 60, Engl. 1, 32 (1962).

[200] *J.R. van Wazer, C.F. Callis, J.N. Shoolery, R.C. Jones*, J. Am. Chem. Soc. 78, 5715 (1956).

[201] *V.A. Kukhtin*, Dokl. Akad. Nauk. SSSR 121, 466 (1958).

[202] *F. Ramirez, A.V. Patwaadhan, S.R. Heller*, J. Am. Chem. Soc. 86, 514 (1964).

different factors, e.g. the electronegativity of the ligand (in terms of Pauling electronegativity) and the back donating ability of π-bonds, which are responsible for the phosphorus shielding effects. The latter is not expected to be a linear function of either the electronegativity or of Taft or Hammett functions. Attempts to interpret ^{31}P-NMR shifts on the basis of quantum mechanics have been made[206, 207]. ^{31}P-shifts were used to examine substituent effects in P-organic compounds in the ground state as well as photo-excited state[208].
Data exists on ^{31}P-chemical shifts (Table 29) for a large number of compounds[209-213].

Table 29. ^{31}P-Chemical Shifts

Compound	δ from H_3PO_4 ppm	Ref.
PF_3	-97	a
PCl_3	-219.4	b
PBr_3	-227.4	b
PJ_3	-178	b
PH_3	238	a
	488	b
$P(C_2H_5)Cl_2$	-196.3	c
$P(C_6H_5)Cl_2$	-161.6	d
$P(C_2H_5)_2Cl$	-119	i
$P(C_6H_5)_2Cl$	-80.5	c
$P(CH_3)_3$	62	c
$P(C_6H_5)_3$	5.9	d
$P(OC_2H_5)Cl_2$	-177	f
$P(OC_6H_5)Cl_2$		g
$P(OC_2H_5)_2Cl$	-165	h
$P(OC_6H_5)_2Cl$	-159	g
$P(OC_2H_5)_3$	-136.9	d
$P(OC_6H_5)_3$	-126.8	d
$(C_2H_5O)P\big\langle{}^O_O$	-131	h
$P(SC_2H_5)Cl_2$	-210.7	c
$P(SC_6H_5)Cl_2$	-204.2	c
$P(SC_2H_5)_2Cl$	-186.2	c

Compound	δ from H_3PO_4 ppm	Ref.
$P(SC_6H_5)_2Cl$	-186.7	c
$P(SC_2H_5)_3$	-125.6	c
$P(SC_6H_5)_3$	-115.6	d
$P[N(C_2H_5)_2]_2Cl$	-154	c
$P[N(C_2H_5)]Cl_2$	-162	c
$P[N(C_6H_5)_2]Cl_2$	-151.3	i
$P[N(C_2H_5)_2]_3$	-118.2	d
OPF_3	35.5	c
$OPCl_3$	-1.9	b
$POBr_3$	102.9	d
$OP(C_2H_5)Cl_2$	-53	j
$OP(C_6H_5)Cl_2$	-34.5	d
$OP(C_2H_5)_2Cl$	-75.9	c
$OP(C_6H_5)_2Cl$	-42.7	c
$OP(C_2H_5)_3$	-48.3	d
$OP(C_6H_5)_3$	-23	c
$OP(OC_2H_5)Cl_2$	-3.4	c
$OP(C_6H_5)Cl_2$	-1.5	k
$OP(OC_2H_5)_2Cl$	-2.8	d
$OP(OC_6H_5)_2Cl$	6.2	k
$OP(C_2H_5)(OC_2H_5)_2$	-32.8	d
$OP(C_6H_5)(OC_2H_5)_2$	-16.9	c
$OP[N(CH_3)_2]_3$	-23.4	d
$OP[N(C_6H_5)_2]_3$	-1.7	i
$SP(C_2H_5)Cl_2$	-94.3	d
$SP(C_6H_5)Cl_2$	-74.8	d
$SP(C_2H_5)_2Cl$	-109.8	c
$SP(C_6H_5)_2Cl$	-79.1	c
$SP(C_2H_5)_3$	-54.5	d
$SP(C_6H_5)_3$	-39.9	c
$SP(OC_2H_5)_3$	-68.1	d
$SP(OC_6H_5)_3$	-53.4	d
$SeP(C_2H_5)_3$	-45.8	d
$[P(CH_3)_4]^{\oplus}Cl^{\ominus}$	2.8	c
$[P(CH_3)_4]^{\oplus}Br^{\ominus}$	-25.1	l
$[P(C_6H_5)_4]^{\oplus}I^{\ominus}$	-20.8	m
$[P(C_6H_5)_3(C_2H_5)]^{\oplus}Br^{\ominus}$	-26.2	l
	-19.8	o
	6.8	o
$[PNCl_2]_5$	17	p
$[PNCl_2]_7$	18	p
	-9 ± 2	q
$P(C_6H_5)F_4$	51.7	r
$P(C_6H_5)_2F_3$	34.8	r

[206] J.H. Letcher, J.R. van Wazer, J. Chem. Phys. 45, 2926 (1966); s.a. Ref. [261], Chapter 2.
[207] H.S. Gutowsky, J. Larmann, J. Am. Chem. Soc. 87, 3815 (1965).
[208] H. Goetz, H. Hadamik, H. Juds, Liebigs Ann. Chem. 742, 59 (1970).
[209] H.S. Gutowsky, D.W. McCall, J. Chem. Phys. 22, 162 (1954).
[210] N. Muller, P.C. Lauterbur, J. Goldenson, J. Am. Chem. Soc. 78, 3557 (1956).
[211] J.R. van Wazer, C.F. Callis, J.N. Shoolery, R.C. Jones, J. Am. Chem. Soc. 78, 5715 (1956).
[212] H. Finegold, Ann. N.Y. Acad. Sci. 70, 875 (1958).
[213] V. Mark, C.H. Dungan, M.M. Crutchfield, J.R. van Wazer in Ref. [261], Chapter 4.

Compound	δ from H_3PO_4 ppm	Ref.
	181 ± 2	m
$(C_6H_5)_3P{=}CH_2$	-20.3	n
$(C_6H_5)_3P{=}CHCH_3$	-14.6	n
$(C_6H_5)_3P{=}C(C_2H_5)_2$	-10.7	n
	4.8	n

a H.S. Gutowsky, D.W. McCall, C.P. Slichter, J. chem. Phys. 21, 275 (1953).

b H.S. Gutowsky, D.W. McDall, J. chem. Phys. 22, 162 (1954).

c K. Moedritner, L. Maier, L.C.D. Groenweghe, J. chem. Eng. Data 7, 307 (1962).

d N. Muller, P.C. Lauterbur, J. Goldenson, J. Amer. chem. Soc. 78, 3557 (1956).

e J.R. van Wazer, C.F. Callis, J.N. Shoolery, R.C. Jones, J. Am. chem. Soc. 78, 5715 (1956).

f E. Fluck, J.R. van Wazer, Z. anorg. allg. Chem. 307, 113 (1961).

g E. Fluck, J.R. van Wazer, L.C.D. Groenweghe, J. Am. chem. Soc. 81, 6363 (1959).

h R.A.Y. Jones, A.R. Katritzky, J. chem. Soc. 4376 (1960).

i M.L. Nielsen, J.V. Pustinger, J. Strobel, J. Chem. Eng. Data 9, 167 (1964).

j F.W. Hoffmann, T.C. Simmons, L.J. Glunz, J. Amer. chem. Soc. 79, 3570 (1957).

k E. Schwarzmann, J.R. van Wazer, J. Amer. chem. Soc. 81, 6366 (1959).

l S.O. Grim, W. McFarlane, E.F. Davidoff, T.J. Marks, J. phys. Chem. 70, 581 (1966).

m D. Hellwinkel, Chem. Ber. 98, 576 (1965).

n S.O. Grim, W. McFarlane, T.J. Marks, Chem. Comm. 1191 (1967).

o M. Becke-Goehring, G. Koch, Chem. Ber. 92, 1188 (1959).

p L.G. Lund, N.L. Paddock, J.E. Procter, H.T. Searle, J. chem. Soc. 2542 (1960).

q G.S. Reddy, C.D. Weis, J. org. Chem. 28, 1822 (1963).

r E.L. Muetterties, W. Mahler, R. Schmutzler, Inorg. Chem., 2, 613 (1963).

3.7.2 ^{31}P-Spin-Coupling

Spin coupling interactions between phosphorus and atoms directly bonded to it (Table 30) become increasingly important for structure determination.

[214] F.A. Cotton, R.A. Schunn, J. Am. Chem. Soc. 85, 2394 (1963).

[215] F. Kaplan, G. Singh, H. Zimmer, J. Phys. Chem. 67, 2509 (1963).

Table 30. Some ^{31}P–^{1}H spin coupling constants

Compound	$J_{^{31}P-^{1}H}$	Ref.
H_3P	179	a
H_2PCH_3	205	b
$HP(CH_3)_2$	205	b
$HP(C_6H_5)_2$	218	c
$H_2PC_6H_5$	201	c
$H(CH_3O)PO$	710	b
$H(C_2H_5)_2PO$	701	d
$H(C_2H_5O)_2PS$	645	d

a H.S. Gutowsky, D.W. McCall, C.P. Slichter, J. chem. Phys. 21, 279 (1953).

b C.F. Callis, J.R. van Wazer₁, J.N. Shoolery, W.A. Anderson, J. Am. chem. Soc. 79, 2719 (1957).

c E. Fluck, H. Binder, Z. Naturforsch. 22b, 1001 (1967).

d H. Finegold, Ann. N.Y. Acad. Sci. 70, 875 (1958).

Table 31[215]. Long-range ^{31}P–^{1}H spin-spin coupling data[a]

Compounds	δCH_3[b] [Hz]	J (three bonds)	J (four bonds)
$(C_6H_5)_3PNCH_3$	180.8	24.7	
$[(C_6H_5)_3PNHCH_3]^{\oplus}Br^{\ominus}$	166.5	13.8	
$[(C_6H_5)_3PN(CH_3)_2]^{\oplus}Br^{\ominus}$	184.5	10.4	
$(C_6H_5)_3PNC(CH_3)_3$	71.3		1.21
$[(C_6H_5)_3PNHC(CH_3)_3]^{\oplus}J^{\ominus}$	81.5		0.58
$[(C_6H_5)_3PNCH_3C(CH_3)_3I^{\oplus}J^{\ominus}$	85.0		c
	187.1	10.4	
$(C_6H_5)_3PNN(CH_3)_2$	192.5		c
$[(C_6H_5)_3PNHN(CH_3)_2]^{\oplus}Br^{\ominus}$	166.6		c
$[(C_6H_5)_3PNCH_3N(CH_3)_2]^{\oplus}J^{\ominus}$	150.9		c
	187.0	8.8	
$[(C_6H_5)_3PNCH_3NCH_3Ph]^{\oplus}J^{\ominus}$	187.8		c
	193.6	7.8	
$(C_2H_5O)_2P(O)NHC(CH_3)_3$	72.6		0.73[d]
$(C_2H_5O)_2P(O)NHN(CH_3)_2$	148.5		c
$ROP(S)(OCH_3)NHC(CH_3)_3$	82.0		0.77

a Spectra were obtained in carbontetrachloride or deuterochloroform solutions on a Varian Associate Model A-60 spectrometer.

b From tetramethylsilane.

c ^{31}P–^{1}H Coupling was not observed; this sets an upper limit of J = 4 Hz for such compounds.

d The spacing did not change in a 100 Mc/sec spectrum.

^{31}P—^{1}H long-range coupling (Table 31)[214, 215] has been observed with systems containing P—N—C—C—H or P—N—C—H; whereas, compounds with a P—N—N—C—H system do not show long-range coupling. These results were used to ascertain the location of alkylation of triphenyl-phosphine-aminoimines[216]. Since there was no spin-spin coupling observed, structures of type 20 could be ruled out.

$$(C_6H_5)_3P=N-NXY + RJ \longrightarrow \begin{array}{c} [(C_6H_5)_3\overset{\oplus}{P}-\underset{\underset{R}{|}}{N}-NXY]J^{\ominus} \\ \textbf{19} \\ \\ [(C_6H_5)_3P=\overset{\oplus}{\underset{\underset{R}{|}}{N}}-NXY]J^{\ominus} \\ \textbf{20} \end{array}$$

The influence of the phosphorus lone pair on the ^{31}P—N—C—^{1}H coupling constants was investigated[216a]. Long-range coupling also occurs through bonds containing the P—O—C—C—H and the P—O—C—C—C—H systems. The reported coupling constants are $J=1.0$ and < 0.5 Hz respectively[217]. A linear correlation is observed for the increase in the ^{31}P—O—C—^{1}H coupling constant for the protons on the carbon β to the phosphorus atom with the downfield chemical shifts of protons on Cβ, Cγ (axial), and Cδ in a series of derivatives of **22** and **23** including the corresponding salts[218].

22   **23**

3.8 Tin-115, Tin-117 and Tin-119

3.8.1 ^{119}Sn-Chemical Shifts

Three of the stable naturally occuring tin isotopes, ^{115}Sn (abundance 0.35%), ^{117}Sn (7.67%) and ^{119}Sn (8.68%), have sizable magnetic moments and possess magnetic spin quantum numbers of 1/2. Since ^{119}Sn gives stronger signals than the other two isotopes, it is used almost exclusively in NMR-spectroscopy involving this element. For a reference, the central peak of the *tetramethyltin* multiplet has been chosen for practically all ^{119}Sn-NMR spectroscopy. It is used either as external or internal standard. In some instances an aqueous solution of $SnCl_2 \cdot 2H_2O$ has been recommended as reference. However, it was found that the chemical shift of this 'standard' varies considerably (several 150 ppm) with a change in the pH or the concentration of the solutions. The range of chemical shifts[219] for ^{119}Sn is about 2000 ppm (Table 32).

Table 32. ^{119}Sn chemical shifts from $Sn(CH_3)_4$ with[a–d] or without a solvent [in ppm for a field of 60 Mc using TMS as internal standard].

Compound	δ [ppm]
SnJ_4	1968
$SnBr_4$	638
$SnCl_2 \cdot 2H_2O$	420[a]
$SnCl_2 \cdot 2H_2O$	303[b]
$SnCl_2 \cdot 2H_2O$	282[c]
$SnCl_4$	150
$(n\text{-}C_4H_9)_2Sn(C_6H_5)_2$	44
$(n\text{-}C_4H_9)_3SnCl_3$	14
$(n\text{-}C_4H_9)_4$	8
$(CH_3)_2SnCl_2$	− 34
$(C_2H_5)_2SnCl_2$	− 50
$(n\text{-}C_4H_9)_2SnCl_2$	− 56[b]
$(n\text{-}C_4H_9)_2SnCl_2$	−112[d]
$(n\text{-}C_4H_9)_2SnS$	−124
$(n\text{-}C_4H_9)_3SnCl$	−139
$(C_2H_5)_3SnCl$	−148

[a]H_2O;   [b]$(CH_3)_2 CO$;   [c]C_2H_5OH;   [d]CS_2.

The observed solvent effects point to the well known fact that tin compounds easily form complexes. There seems to be little difference in the ^{119}Sn-chemical shifts between divalent or tetravalent tin compounds.

^{119}Sn-NMR spectroscopy proved of great value in studying the redistribution of halogen substitution in *tin halide* mixtures. Studies on binary mixtures ($SnCl_4/SnBr_4$) and tertiary mixtures ($SnCl_4$, $SnBr_4$ and SnJ_4) showed the halogen substituents to occur in almost a random distribution, indicating a very rapid chemical exchange of the halide groups.

3.8.2 ^{117}Sn—^{1}H and ^{119}Sn—^{1}H Spin-Coupling Phenomena

Of greater importance for structural work are the ^{119}Sn—(C)$_n$—^{1}H and/or ^{117}Sn—(C)$_n$—^{1}H coupling constants which are fairly large. Usually the coupling constant with the H at an α-carbon is smaller than with the one located at a β-carbon.

Thus, in the compound

the coupling constant equals ~ 32 Hz for Sn — CH_2 and ~ 72 Hz for the coupling between the Sn-atom and the CH_3 group[220].

[216] H. Zimmer, G. Singh, J. Org. Chem. *29*, 5179 (1964).

[216a] B.J. Walker, J. Nelson, R. Spratt, J. Chem. Soc. D 1509 (1970).

[217] G.S. Reddy, R. Schmutzler, Z. Naturforsch. *20b*, 104 (1965).

[218] J.G. Verkade, T.J. Huttemann, M.K. Fung, R.W. King, Inorg. Chem. *4*, 83 (1965).

[219] J.J. Burke, P.C. Lauterbur, J. Am. Chem. Soc. *83*, 326 (1961).

[220] P.T. Narasimhan, M.T. Rogers, J. Chem. Phys. *34*, 1049 (1961).

[221] H. Zimmer, O.A. Homberg, unpubl.

[222] S. Boué, M. Gielen, J. Nasielski, Bull. Soc. Chim. Belg. *76*, 559 (1967).

Table 33. ^{117}Sn (^{119}Sn)–^{1}H coupling constants [in Hz for a field of 60 MHz using TMS as internal standard].

Compound	Solvent	δ[Hz]	J^{117}Sn/J^{119}Sn [Hz]	Ref.
(CH$_3$)$_4$Sn	gas phase	— 4.4	51.9/54.2	a
		— 6.0	52/54	b
	—	— 4.2	51.1/53.4	c
(CH$_3$)$_3$SnH	—	— 4.8	54.9/56.9	b
[(CH$_3$)$_3$Sn]$_2$	—	—11.4	46.2/48.3	c
(CH$_3$)$_3$SnCH$_2$Sn(CH$_3$)$_3$	—	— 3.6	50.7/52.9	c
(CH$_3$)$_3$SnN(CH$_3$)$_2$	CCl$_4$	—11.3	53.0/55.2	d
[(CH$_3$)$_3$Sn]$_2$O	CCl$_4$	—14.2	53.6/56.0	e
(CH$_3$)$_3$SnCl	—	—	57.4/59.7	f
	H$_2$O	—37.2	67.7/70.7	g
	CCl	—	56.0/58.5	h
(CH$_3$)$_3$SnBr	—	—44.4	56/58.6	b
	CCl$_4$	—44.2	55.8/58.4	c
(CH$_3$)$_3$SnBr-Py	CDCl$_3$	—46.3	59.8/62.1	g
(CH$_3$)$_3$Sn	CCl$_4$	—53	55.8/58.	e
(CH$_3$)$_2$SnH$_2$	—	—18	57.6/60.2	i
(CH$_3$)$_8$Sn$_3$ (int.)	—	—24	42.8/44.3	j
(CH$_3$)$_2$SnCl$_2$	—	—	68/71	f
	H$_2$O	—	97.4/101.9	f
	D$_2$O	—64.9	102.5/107.5	g
	CDCl$_3$	—70.2	67/69.5	g
(CH$_3$)$_2$SnBr	CCl$_4$	—79.8	63.2/66.2	h
CH$_3$SnH$_3$	—	—16.2	62	k
CH$_3$SnCl$_3$	—	—	95.7/100	f
	D$_2$O	—74.4	122.5/127.5	g
	H$_2$O	—	125.4/131.1	f
	CCl$_4$	—	95.3/99.7	h
	CCl$_4$	—98.8	— —	l
(CH$_3$CH$_2$)$_4$Sn	—	—47.0	49.0/51.5	m
		—	30.8/32.2	n
(CH$_3$CH$_2$CH$_2$)$_4$Sn		49.1	49.1	a
(CH$_3$)$_3$CCH$_2$)$_4$Sn	CCl$_4$	—71.8	46.2/49.2	o
{[(CH$_3$)$_3$CCH$_2$]$_3$S}$_2$	CCl$_4$	—81.0	40.4	o
[C$_6$H$_5$C(CH$_3$)$_2$CH$_2$]$_4$Sn	CDCl$_3$	—47.6	98.6	o
[C$_6$H$_5$C(CH$_3$)$_2$CH$_2$]$_3$	CDCl$_3$	—68.2	46.9/48.9	o
[(CH$_3$)$_3$C(CH$_2$)]	CCl$_4$	—116.2	55.1	o
[C$_6$H$_5$C(CH$_3$)$_2$CH$_2$)]SnCl$_2$	CDCl$_3$	—110.0	51.3	o
(CH$_3$)$_3$SnCH$_2$Sn(CH$_3$)$_3$	—	+16.8	57.7/60.3	c
(CH$_3$CH$_2$)$_4$Sn	—	—80.2	66.8/69.8	a
	CCl$_4$	—71	67.0/69.2	m
(CH$_3$CH$_2$CH$_2$)$_4$Sn	—		47.2	p
(CH$_3$CH$_2$)$_2$SnCl$_2$	CCl$_4$		131.7/137.8	q
[(CH$_3$)$_3$CCH$_2$]$_4$Sn	CCl$_4$	—62.2	105.8/107.9	o
[C$_6$H$_5$C(CH$_3$)$_2$]$_4$Sn	CCl$_4$	—66.3	85.7/91.2	o
	CDCl$_3$	—68.4	98.1/101.7	o
[(CH$_3$)$_3$CCH$_2$]$_3$Sn $_2$	CCl$_4$	—62.2	112.5/115	o
[C$_6$H$_5$C(CH$_3$)$_2$CH$_2$]$_3$Sn $_2$	CDCl$_3$	—76.2	78.5/75.5	p
(CH$_2$=CH)$_4$Sn (trans)	—	—372.2	274.6	r
	a	—360		
	b	—113		
	c	—360	90	s
	a	—360		
	b	—397		
	c	—109	90	s
CH$_3$(C≡C)$_2$Sn(C$_6$H$_5$)$_2$	CCl$_4$	—116.4	5.2	h
(C$_6$H$_5$)$_2$Sn	—	—358.5	31.1/32.5	t
(C$_6$H$_5$)$_4$Sn	CS$_2$	—351.3	25.9/26.9	u

Table 33 (cont.)

compound	solvent	δ[Hz]	$J^{117}Sn/J^{119}Sn$ [Hz]	Lit.
$[C_6H_5C(CH_3)_2CH_2]_4Sn$	$CDCl_3$	—427		o
$[C_6H_5C(CH_3)_2CH_2]_3SnCl$	$CDCl_3$	—430.1		o
$[C_6H_5C(CH_3)_2CH_2]_2SnCl_2$	$CDCl_3$	—435.3		o
SnH_4	—	—231	1846/1931	v
CH_3SnH_3		—228.4	1770/1852	k
$C_6H_5SnH_3$	—	—285.8	1836.7/1921.5	v
$(CH_3)_2SnH_2$	—	—263.4	1717.4/1797.1	i
$(CH_3)_3SnH$	—	—283.8	1640.0/1744.0	w
$(n\text{-}C_8H_{17})_3SnH$	—	—290.3	1534.4/1604.7	w

a *G. Klose*, Ann. Physik 8, 220 (1961).
b *H. C. Clark, J. T. Kwon, L. W. Reeves, E. J. Wells*, Inorg. Chem. 3, 907 (1964).
c *H. D. Kaesz*, J. Amer. chem. Soc. 83, 1514 (1961).
d *J. Lorberth, M. R. Kula*, Chem. Ber. 97, 3444 (1964).
e *H. Schmidbaur, I. Ruidisch*, Inorg. Chem. 3, 599 (1964).
f *J. R. Holmes, H. D. Kaesz*, J. Amer. chem. Soc. 83, 3903 (1961).
g *G. P. Van der Kelen*, Nature [London] 193, 1069 (1962).
h *M. LeQuan, P. Cadiot*, Bull. Soc. Chim. France 1965, 35.
i *H. C. Clark, J. T. Kwon, L. W. Reeves, E. J. Wells*, Canad. Chem. 42, 941 (1964).
j *T. L. Brown, G. L. Morgan*, Inorg. Chem. 2, 736 (1963).
k *N. Flitcroft, H. D. Kaesz*, J. chem. Soc. 85, 1377 (1963).
l *M. P. Brown, E. Webster*, J. phys. Chem. 64, 698 (1960).

A large number of $^{119}Sn(^{117}Sn)$—C—^1H coupling constants (Table 33) have been measured[221].

In a limited series of *alkyltrimethyltin* compounds, it was shown that the values of the coupling constants J(Sn—C—H) follow satisfactorily Taft's σ^* constants but *benzyltrimethyltin* compounds show a deviation[222].

It was also shown that the coupling constants in the *aryltrimethyltin* series decrease as the steric requirements increase. It is suggested that the s-character in the Sn-orbitals of the Sn—C bonds of methylsubstituted *tetraalkyltins* are influenced by steric requirements of the R-groups[223]. A linear relationship between J(^{117}Sn—C—^1H) and the percentage s-character in the tin hybrid orbital was also suggested[224].

The ^{117}Sn—^{119}Sn coupling constant possesses a very large value of $J = 4400 \pm 70$ cps[225].

223 *M. Gielen, M. DeClercq, J. Nasielski*, Bull. Soc. Chim. Belg. 78, 237 (1969).
224 *G.P. Vander Kelen, E.V. Vanden Berghe*, J. Organometal. Chem. 11, 479 (1968).
225 *H.C. Clark, J.T. Kwon, L.W. Reeves, E.J. Wells*, Can. J. Chem. 42, 941 (1964).
226 *R. Brankey, B.N. Figgis, R.S. Nyholm*, Trans. Faraday Soc. 58, 1893 (1962).
227 *D.B. Powell, N. Sheppard*, J. Chem. Soc. 2519 (1960).
228 *M. Orchin, P.J. Schmidt*, Inorg. Chim. Acta 2, 123 (1968).
229 *M. Orchin, P.J. Schmidt*, Coord. Chem. Rev. 3, 345 (1968).
230 *R.E. Dessy, T.J. Flautt, H.H. Jaffé, G.F. Reynolds*, J. Chem. Phys. 30, 1422.

3.9 Other Elements

3.9.1 Platinum-195

Coupling between $^{195}Pt(I = 1/2)$ and ^1H has been observed. In $[(C_2H_5)_3P]_2Pt \cdot HCl$, J ^{195}Pt—^1H of the directly bonded proton has been found to be 1276 Hz. Coupling between ^{195}Pt and hydrogen through two bonds[226] as in $[(CH_3)_3PtJ]_4$ has a value of $J = 70$–80 Hz[226]. π-bonded ligands of platinum show smaller couplings than the σ-bonded ones. For example, in *Zeise's salt*, $K[C_2H_4PtCl_3]$, J ^{195}Pt—^1H = 34. More recently, an interesting NMR-technique involving ^{195}Pt—^1H coupling for assessing the degree of ligand (L) lability in *pyridine*[228], *pyridine N-oxide*[229], *aniline* and *cyclohexylamine*[230] complexes of the type *trans* $(PtCl_2) (C_2H_4)L$ has appeared.

3.9.2 Mercury-199 and Mercury-201

^{199}Hg-*chemical shifts* of a series of *dialkyl mercury* compounds have been determined[231] in order to investigate hyperconjugative effects occurring in the Hg—C-bond (Table 34). As is generally observed in the *coupling phenomena* of metal—^1H-bonds, the α-coupling constants are smaller than the β-coupling constants in organomercury compounds (Table 35). Long-range coupling between ^{199}Hg and ^1H through 5–7 bonds has also been observed; the coupling constants[231] are 26, 14, 38, 28, and 15 for

(*m*- and *p*-isomers)

Table 34. ^{199}Hg chemical shifts in dialkyl-Hg compounds

Compound	δ
$(CH_3)_2Hg$	0
$(C_2H_5)_2Hg$	+330
$(n-C_3H_7)_2Hg$	+240
$[(CH_3)_2CH]_2Hg$	+640

In organomercury compounds, it was also found, that ^{199}Hg—^{1}H-α-coupling is smaller than the ^{199}Hg—^{1}H-β-coupling (Table 35).

such compounds to about $-100°$, fine structure due to ^{205}Tl—^{1}H spin-spin interaction appears in the spectrum. This has been interpreted as a rapid chemical exchange of alkyl groups at room temperature[235-237].

In ^{205}Tl-NMR spectroscopy, the β-coupling, similar to the case of other elements, e.g., ^{117}Sn and ^{119}Sn, ^{199}Hg and ^{201}Hg, is larger than the α-coupling (Table 36).

In an aqueous solution the coupling constants J^{205}Tl—^{1}H of $(CH_3CH_2)_2$ TlX compounds are virtually independent of the anion indicating that

Table 35. Some ^{199}Hg—^{1}H Coupling Constants

Compound	J_{CH_3-Hg}	$J_{\alpha CH_2-Hg}$	$J_{\beta CH_2-Hg}$	$H_{\gamma CH_2-Hg}$	Ref.
CH_3HgCH_3	102				a
$C_2H_5HgC_2H_5$	120	91			a
$C_3H_6HgC_3H_7$		90(α)			a
		108(β)			a
$(CH_3)C_2HHgCH(CH_3)_2$		78(α)			a
		126(β)			a
CH_3HgCN	178				b
CH_3HgJ	200				b
CH_3HgClO_4	233.2				b
$CH_3OCH_2CH_2HgOC(O)CH_3$		230±2	217±2		c
$CH_3OCH_2CH_2HgCl$		225±7	245±2		c
$CH_3OCH_2CH_2HgBr$		207±2	249±2		c
$CH_3CH(OCH_3)CH_2HgOC(O)CH_3$		222±4	288±8	0	c
$(CH_3)_2C(OCH_3)CH_2HgOC(O)CH_3$		215±2		20.8±1	c
$(CH_3)_2C(OCH_3CH_2HgCl$		203±2		21.1±	c
$(CH_3)_2C(OCH_3)CH_2HgBr$		204±2		21.9±—	c

a *H. H. Jaffe', G. F. Reynolds, T.J. Flautt, R. E. Dessy*, J. Chem. Phys. *30*, 1422 (1959).
b *J. V. Hatton, W. G. Schneider, W. Siebrand*, J. Chem. Phys. *39*, 133 (1963).
c *S. Brownstein*, Discuss. Faraday Soc., *34*, 25 (1962).

3.9.3 ^{205}Tl-chemical Shifts

Two thallium isotopes ^{203}Tl and ^{205}Tl (abundancy, 70,78%) occur in nature having $I = 1/2$. A number of ^{205}Tl-chemical shifts for a series of inorganic salts have been reported[232-234].

The observed shifts cover a range of over about 34,000 ppm from 0.3 M TlNO$_3$.

3.9.4 ^{205}Tl—^{1}H Spin Coupling

At ambient temperature, the ^{1}H resonance of *trialkyl thallium* compounds does not show any fine structure. However, upon $-100°$, a solution of

Table 36. Some ^{205}Tl—^{1}H Spin-Spin Coupling Constants [Hz]

Compound	CH$_3$ Group J_{205Tl-^1H}	C$_2$H$_5$ Group $J_{205Tl-CH_2-CH_3}$	$J_{205Tl-CH_2-CH_3}$
Tl(CH$_3$)$_3$	250.8	—	—
Tl(CH$_3$)$_2$C$_2$H$_5$	223	242.4	472.7
Tl(CH$_3$)(C$_2$H$_5$)$_2$	186.9	218.8	441.5
Tl(C$_2$H$_5$)$_3$	—	198.2	396.1

R$_2$TlX compounds are almost completely dissociated in aqueous solution. In pyridine solution, a dependency of the coupling constants on the anion is found due to complexation (Table 37).

[231] *V.S. Petrosyan, O.A. Reutov*, Jzv. Akad. Nauk. Kaz. SSR, Ser. Khim. *6*, 1403 (1970); C.A. *73*, 130368u (1970).
[232] *R. Freemann, R.P.H. Gasser, R.E. Richards*, Mol. Phys. *2*, 301 (1959).
[233] *T.J. Rowland, J.P. Bromberg*, J. Chem. Phys. *29*, 626 (1958).
[234] *B.N. Figgis*, Trans. Faraday Soc. *55*, 1075 (1959).

[235] *J.P. Maher, D.F. Evans*, Proc. Chem. Soc. 208 (1961).
[236] *J.P. Maher, D.F. Evans*, J. Chem. Soc. 5534 (1963).
[237] *J.P. Maher, D.F. Evans*, J. Chem. Soc. 637 (1965).

Table 37[a]. ^{205}Tl—^{1}H Coupling Constants in D$_2$O and Pyridine [Hz]

Anion	$J_{205Tl-^1H(CH_2)}$[cps]	$J_{205Tl-^1H(CH_3)}$[cps]	Solvent
I$^\ominus$	b	—	D$_2$O
I$^\ominus$	332	—	Py
CO$_3^\ominus$	338	623	D$_2$O
CO$_3^\ominus$	358	—	Py
CH$_3$CHOHCO$_2^\ominus$	338	623	D$_2$O
CH$_3$CH$_2$OHCO$_2^\ominus$	359	620	Py
SCN$^\ominus$	337	623	D$_2$O
SCN$^\ominus$	361	628	Py
ClO$_4^\ominus$	337.5	623	D$_2$O
ClO$_4^\ominus$	336	638	Py
NO$_2^\ominus$	338	623	D$_2$O
NO$_3^\ominus$	373	630	Py

[a] *J. V. Hatton*, J. chem. Phys. *40*, 933 (1964).
[b] Not sufficiently soluble

The observed ^{205}Tl—^{1}H coupling constants for olefinic and aromatic organo-thallium compounds[238] are much larger than ^{1}H—^{1}H constants. As expected, the coupling constants are again larger for the ions Tl(C$_6$H$_5$)$_2^\oplus$ and Tl(C$_6$H$_5$)$^{\oplus\oplus}$ than for the *trialkyl thallium* compounds (Table 38).

Investigations concerned with the structure of liquid Tl(OC$_2$H$_5$) indicated that it is tetrameric since there was no ^{205}Tl—^{1}H-coupling observable. The ^{203}Tl—^{205}Tl coupling constant was determined to be 2560 Hz[239].

3.9.5 Lead-207

^{207}Pb (I = 1/2 and natural abundance 21%) possesses a resonance frequency of 8899 MHz/sec at 10,000 Gauss[240, 241]. Chemical shifts show large differences extending over a range of about 16,000 ppm and appear upfield from metallic lead; e.g. Pb = 0 ppm, Pb(CH$_2$CH$_3$)$_4$ = 11,600 ppm (calculated)[3].

^{207}Pb shows fairly large coupling constants with ^{1}H. As is the case with other element —^{1}H coupling constants, the β-^{1}H—^{207}Pb-coupling constants[221] are larger than the corresponding α-coupling constants (Table 39).

Table 38. ^{205}Tl—^{1}H Coupling Constants

Compound	*ortho*		*meta*		*para*	
C$_6$H$_5$Tl$^{\oplus\oplus}$	±948		±365		±123	
(C$_6$H$_5$)$_2$Tl$^\oplus$	451		139		51	
(C$_6$H$_5$)$_3$Tl	±259		±80	±5	±35	±5
	^{205}Tl—^{1}H*(trans)*		^{205}Tl—^{1}H*(cis)*		^{205}Tl—^{1}H*(gem)*	
(CH$_2$=CH)Tl$^{\oplus\oplus}$	3750		1806		2004	
(CH$_2$=CH)$_2$Tl$^\oplus$	1618		805		842	

Table 39. ^{207}Pb—^{1}H Coupling Constants, CH$_3$—Pb

Compound	Solvent	[Hz]	J^{207Pb-^1H}	Ref.
(CH$_3$)$_4$Pb	—	—	55.3	a
	—	—	61.0	b
	—	—	61.2	c
	—	43.8	61.0	d
	CCl$_4$	— 43.8	61.0	d
	C$_6$H$_5$CH$_3$	— 39.0	62.0	e
(CH$_3$)$_3$PbH	none, C$_5$H$_{10}$, C$_5$H$_{12}$	— 51.0	66.7	f
(CH$_3$)$_3$PbC$_5$H$_5$	C$_6$H$_6$	— 36.0	58.0	e
(CH$_3$)$_3$PbOSi(CH$_3$)$_3$	CCl$_4$	— 73.0	69.5	d
(CH$_3$)$_3$PbOH	CHCl$_3$	— 92.0	76.0	e
(CH$_3$)$_3$PbF	CHCl$_3$	— 80.0	81.0	e
(CH$_3$)$_3$PbCl	CHCl$_3$	— 97.5	70.0	e
(CH$_3$)$_3$PbBr	CHCl$_3$	—105.0	68.0	e
(CH$_3$)$_3$PbI	CHCl$_3$	—110.5	63.0	e
(CH$_3$)$_3$Pb(C$_5$H$_5$)$_2$	C$_6$H$_6$	— 25.0	54.0	e

[238] *J.P. Maher, D.F. Evans*, Proc. Chem. Soc. 176 (1973).
[239] Ref. 238, p. 175.

[240] *L.H. Piette, H.E. Weaver*, J. Chem. Phys. *28*, 735 (1958).
[241] *J.M. Rocard, M. Bloom, L.B. Robinson*, Can. J. Phys. *37*, 522 (1959).

Table 39 (cont.)

compound	solvent	[Hz]	$J^{207}Pb—^1H$	Lit.
	$X-CH_2—Pb$			
$(CH_3CH_2)_4Pb$	TMS	— 88.5	41.0	e
$(CH_3CH_2CH_2)_4Pb$			40.5	g
$(CH_3CH_2)_3PbC_5H_5$	C_6H_6	— 94.5	37.0	e
$[C_6H_5C(CH_3)_2CH_2]_4Pb$	$CDCl_3$	— 68.8	20.4	h
$[CH_3C(CH_3)_2CH_2]_2Pb\ _2$	$CDCl_3$	—121.6	19.6	h
$[C_6H_5C(CH_3)_2CH_2]_3Pb\ _2$	$CDCl_3$	—112.2	33.1	h
	$CH_2—C—Pb$			
$(CH_3CH_2)_4Pb$	—		125.0	i
	TMS	— 88.5	125.0	e
$(CH_3CH_2CH_2)_4Pb$			102.4	g
$(CH_3CH_2)_3PbC_5H_5$	C_6H_6	— 94.5	154.0	e
	C_5H_5Pb			
$(C_5H_5)_2Pb(CH_3)_2$	C_6H_6	—368	27	e
$C_5H_5Pb(CH_3)_3$	C_6H_6	—364	25	e
$C_5H_5Pb(C_2H_5)_3$	C_6H_6	—369	17	e
	$Pb—Pb—CH_2—X$			
$[CH_3C(CH_3)_2CH_2]_3Pb\ _2$	$CDCl_3$	—121.6	12.1	h
$[C_6H_5C(CH_3)_2CH_2]_3Pb\ _2$	$CDCl_3$	—112.2	9.4	h
	$Miscellaneous$			
$HPb(CH_3)_3$	—	—460.8	237.8	f
	C_5H_{10}			
	C_5H_{12}			

a G. Klose, Ann. Phys. 9, 262 (1962).
b H. Dreeskamp, Z. Phys. Chem. 38, 121 (1963).
c W.G. Schneider, A.D. Buckingham, Discuss. Faraday Soc. 34, 147 (1962).
d H. Schmidbaur, H. Hussek, J. Organometal. Chem. 1, 257 (1964).
e H.P. Fritz, K.E. Schwarzhans, J. Organometal. Chem. 1, 297 (1963).
f N. Fliteroft, H.D. Kaesz, J. Amer. chem. Soc. 85, 1377 (1963).
g G. Klose, Ann. Phys., 10, 391 (1963).
h H. Zimmer, O.A. Homberg, J. Org. Chem. 31, 947 (1966).
i P.T. Harasimhan, M.T. Rogers, J. Chem. Phys. 34, 1049 (1961).

3.9.6 Less Important Nuclei

Most other nuclei are either of minor importance to organic chemistry or they have not been studied extensively because commercial instruments cannot be used without costly accessories, as is the case with $^{29}Si^4$. Moreover, it is easier to obtain structural information by studying the ^1H-NMR spectra. Some of the less important nuclei as far as structural information in organic chemistry is concerned are $^{59}Co^{242, 243}$ and the halogens $^{35}Cl^{244-248}$, ^{37}Cl, ^{79}Br, ^{81}Br, and ^{127}J; all possess nuclear spin greater than one and, as a consequence, the signals are broadened considerably by electric quadrupole relaxation[249]. The chemical shifts of some aluminum organics have been reported[250]. NMR spectral studies of the alkali metals have been reported and chemical shifts for rubidium and ceasium have been published[251].

[242] W.G. Procter, F.C. Yu, Phys. Rev. 81, 20 (1951).
[243] R. Freeman, G.R. Murray, R.E. Richard, Proc. Roy. Soc. A 242, 455 (1957).
[244] C.H. Laugford, T.R. Stengle, J. Am. Chem. Soc. 91, 4014 (1969).
[245] J.A. Happe, R.L. Ward, J. Am. Chem. Soc. 91, 4906 (1969).
[246] R.G. Bryant, J. Am. Chem. Soc. 91, 976 (1967).
[247] D. Gill, M.P. Klein, G. Kotowycz, J. Am. Chem. Soc. 90, 6870 (1968).
[248] R.G. Bryant, J. Am. Chem. Soc. 89, 2496 (1967).
[249] Y. Masuba, J. Phys. Soc. Jap., Suppl. 11, 670 (1956).
[250] D.E. O'Reilly, J. Chem. Phys. 32, 1007 (1960).

Table 40. ^{7}Li NMR-Chemical Shifts and ^{7}Li—^{1}H Coupling Constants. [Hz][a]

J_{28}	J_{24}	J_{25}	J_{26}	J_{34}	J_{35}	J_{36}	J_{45}	J_{46}	J_{56}	τ_2	τ_3	τ_4	τ_5	τ_6
Phenyllithium[b]														
6.90	1.62	0.52	0.82	7.55	1.66	0.52	7.55	1.62	6.90	1.994	2.956	3.041	2.956	1.994
6.52	1.62	0.84	0.65	7.20	1.17	0.84	7.20	1.62	6.52	1.976	2.975	3.042	2.975	1.976
6.85	1.53	0.51	0.73	7.26	1.65	0.51	7.26	1.53	6.85	1.989	2.950	3.030	2.950	1.989
	1.65			7.14			7.14	1.65						
m-Chlorophenyllithium[b]														
6.50	1.14		0.16	6.51		0.84		1.84		2.095	3.032	3.064		2.187
p-Chlorophenyllithium[b]														
7.40		0.40	0.94		1.74	0.40			7.40	2.112	2.983		2.983	2.112
8.10		0.73	2.20		2.20	0.73			8.10	2.545	2.744		2.744	2.545
m-Tolyllithium[b]														
0.78	1.40		0.29	7.33		0.73		1.83		2.145	3.037	3.200		2.197
p-Tolyllithium[b]														
6.98		0.34	0.82		1.57	0.34			6.98	2.170	3.081		3.081	2.170
7.04		0.57	0.63		1.68	0.57			7.04	2.090	3.080		3.080	2.090

[a] Errors in J values range from 0.05 to 0.10 Hz, the average error being 0.07 Hz.
[b] *J. Parker, J. A. Ladd*, J. Organometal. Chem. *19*, (1969) 1.

3.9.7 Lithium-7

^{7}Li-chemical shifts and ^{7}Li—^{1}H coupling constants (Table 40) for *phenyllithium, m- and p-chlorophenyllithium* as well as *m- and p-tolyllithium* have been measured in ethereal solution[252]. It was observed that all organolithium compounds exhibit a single resonance line downfield from the LiBr reference. Introduction of the mentioned substituents resulted in a downfield shift relative to phenyllithiums. These values seem to be more accurate than those reported previously[253, 254].

[251] *H.S. Gutowsky, B.R. McGarvey*, J. Chem. Phys. *21*, 1423 (1953).
[252] *J. Parker, J.A. Ladd*, J. Organometal. Chem. *19*, 1 (1969).
[253] *G. Fraenkel, D.G. Adams, R.R. Dean*, J. Phys. Chem. *72*, 944 (1968).
[254] *G. Fraenkel, S. Dakachi, S. Kobayashi*, J. Phys. Chem. *72*, 953 (1968).

3.10 Bibliography
[255] *J.A. Pople, W.G. Schneider, H.J. Bernstein*[1],1959.
[256] *F.C. Nachod, W.D. Phillips*[2], 1962.
[257] *B. Pesce*, Nuclear Magnetic Resonance in Chemistry, Academic Press, New York, London 1965.
[258] *G. Suhr*[6], 1965.
[259] *J.W. Emsley, J. Feeney, L.H. Sutcliffe*[4], 1966.
[260] *D.W. Mathieson*, NMR Spectroscopy for Organic Chemists, Academic Press, New York, London 1966.

[261] *M. Grayson, E.J. Griffith*, ^{31}P Nuclear Magnetic Resonance, Interscience Publishers Inc., New York 1967.
[262] *E.F. Mooney*[8], 1969.

5.5:4 Detection of Radicals

Karl H. Hausser and *Klaus Möbius*
Max-Planck-Institut für Medizinische Forschung
Abteilung für Molekulare Physik, D-69 Heidelberg
and II. Physikalisches Institut der Freien Universität
D-1 Berlin

The most important method for detecting the presence of *paramagnetic radicals* and for determining the coupling constants a_i between the unpaired electron and nuclei with a spin $I \neq 0$ is ESR (Section 5.5:4.2). From an ESR spectrum with well-resolved HFS (5.5:4.3.2), any particular radical can be identified. Moreover, ESR is a very sensitive method and, under favorable conditions, 1/10 cm^3 of a solution with a concentration of 1 micromol/l suffices to obtain a spectrum.

The ENDOR method (5.5:4.5) enables a spectrum to be obtained with far fewer single lines. This is especially suitable for determining the coupling constants a_i if the HFS of the ESR-spectrum cannot be resolved further because of too many lines. The main drawback of ENDOR, apart from the high cost involved, is the reduced sensitivity which is about two orders of magnitude. Furthermore, ENDOR does not directly provide the number of equivalent nuclei which contribute to an ENDOR line, since the line-intensity depends on relaxation processes and is not proportional only to the number of nuclei.

NMR spectroscopy (5.5:4.6) is suitable for measuring small coupling constants which cannot be resolved with the other two methods. In addition, NMR provides directly the sign of a_i. This is of particular significance for comparison with theoretical predictions, especially for small a_i where the theoretical sign is far less reliable than with a larger a_i. The chief drawback of NMR is the reduced sensitivity, which in favourable instances is two orders of magnitude lower than with ENDOR, and, in less favourable cases, where large a_i values are involved, is practically zero.

4.1 Introduction

The discussion of methods is restricted to magnetic resonance for two reasons:

1. During the last 15 years magnetic resonance has been developed to become the most important method. In addition to its high sensitivity, it is distinguished by its extremely high specificity for the detection of paramagnetic radicals.
2. The older methods, in particular the magnetochemical methods, have already been described in detail[1].

Electron spin resonance or electron paramagnetic resonance (ESR) is by far the most important method and, hence, is dealt with most comprehensively. In recent years electron-nuclear double resonance (ENDOR) and nuclear magnetic resonance (NMR) have also proved useful for the investigation of radicals (Sections 5.5:4.5 and 5.5:4.6).

Among the more recently published literature, the book by Carrington and McLachlan[90] is particularly recommended for a deeper understanding of applications which are dealt with only briefly here. The monograph by Poole[93] contains a comprehensive description of experimental methods.

4.2 The Parameters of the ESR Spectrum

4.2.1 g-Value
The basic equation of magnetic resonance was introduced in the general section (5.5:1.1) in the form applicable to ESR (Eq. [2c]). The proportionality constant relating the external magnetic field H_0 to the resonance frequency ν, referred to as the g-value, is with radicals mostly a little above the value of the free electron ($g_0 = 2.0023$). The theory of the g-value is dealt with in Section 5.5:4.4.

[1] E. Müller, Magnetochemische Methoden, Houben-Weyl, Methoden der organischen Chemie, 4. Aufl., Bd. III/2, p. 917, Georg Thieme Verlag, Stuttgart 1955.

4.2.2 Hyperfine Structure (HFS)
Most information concerning a radical is provided by the hyperfine splitting of the resonance line due to the interaction between the unpaired electron and the magnetic moments of nuclei of the radical. In order to understand the basic principle of hyperfine structure (HFS) we shall restrict ourselves to the simplest possible system consisting of an unpaired electron spin $S = 1/2$ and a nuclear spin $I = 1/2$ with the magnetogyric ratios γ_S and γ_I respectively. The interaction between these two spins in an external magnetic field H_0 is described by the spin Hamiltonian H:

$$H = \gamma_S \hbar (\vec{S} \cdot \vec{H}_0) - \gamma_I \hbar (\vec{I} \cdot \vec{H}_0) + $$

$$+ \gamma_S \gamma_I \hbar^2 \left[\underbrace{\frac{8\pi}{3} |\Psi(I)|^2 (\vec{I} \cdot \vec{S})} + \underbrace{\left\{ \frac{3(\vec{I} \cdot \vec{r})(\vec{S} \cdot \vec{r})}{\vec{r}^5} - \frac{(\vec{I} \cdot \vec{S})}{\vec{r}^3} \right\}} \right] \quad (1)$$

$$\text{scalar } a(\vec{I} \cdot \vec{S}) \qquad \text{dipolar} \langle (3\cos^2\theta - 1)/\vec{r}^3 \rangle (\vec{I} \cdot \vec{S})$$

The first two Zeeman terms describe the interaction of the electronic spin S and the nuclear spin I with the external field. The third term (Fermi contact term) describes the scalar hyperfine coupling between the spin S and I. This is proportional to the square of the wave function $|\Psi(I)|^2$ of the unpaired electron at the nucleus. The last term describes the classical dipolar coupling between S and I. This term depends explicitly on the angle θ between the connecting vector $\vec{r}$ and the external magnetic field; hence, in a single crystal, it is a function of the orientation of the crystal axis with respect to the external field H_0. In an atom or molecule, $\vec{r}$ and θ are not constant because of the orbital motion of the electron. Therefore, the orientation-dependent part must be averaged over the orbital of the unpaired electron which is expressed by the brackets $\langle \ \rangle$. As a consequence of the orientation dependence, the dipolar term is averaged to zero in liquids of low viscosity by the fast molecular motion. Hence, in such systems, this term does not contribute to the time-independent quantities such as energy levels and transition frequencies, but must be taken into account when considering time-dependent quantities such as relaxation time and line width (see 5.5:4.2.3).

Let us first consider the left side of Fig. 1, in which the electronic Zeeman-levels are dealt with, each of which is split into two HFS components due to the coupling with a spin $I = 1/2$. These HFS terms are, in addition, shifted by the nuclear Zeeman term which, however, does not affect the frequency of the ESR transitions (Fig. 1).

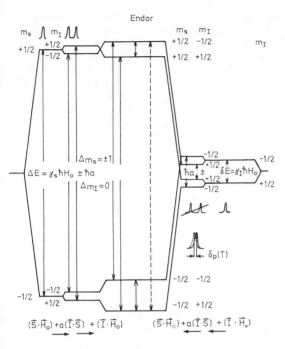

Fig. 1. Energy levels and transitions for a system consisting of an electronic spin S = 1/2 and a nuclear spin I = 1/2 (schematic).

Additional methods for measuring the HFS constant a are described in Sections 5.5:4.5 (ENDOR) and 5.5:4.6 (NMR).

Most radicals contain many nuclei with a spin I > 0; hence, one observes many HFS components. Coupling with one nucleus I = 1 (e.g. ^{14}N) yields three HFS lines with an intensity ratio 1:1:1; for two ^{1}H nuclei one obtains $2^2 = 4$; for n nuclei with I = 1/2 in general 2^n HFS lines result. However, if two H atoms are 'equivalent', i.e. if they have the same coupling constant a, they produce only three HFS lines with an intensity ratio of 1:2:1. In general, interaction with n equivalent nuclei with spin I = 1/2 produces (2n + 1) HFS lines with an intensity ratio of the binomial coefficients. For nuclei with I > 1/2 each case must be calculated separately. The more frequently occurring cases are compiled in a table[2].

In this way complex many-line spectra characteristic of the structure of each radical are obtained under high resolution conditions. They form an excellent means of 'fingerprinting' any specific radical.

An ESR spectrum with well-resolved HFS is also of great interest from the point of view of theoretical chemistry, because it measures directly the

[2] *K.H. Hausser,* Magnetische Elektronen- und Kernresonanz, in Ullmanns Encyklopädie der technischen Chemie, 3. Aufl., Bd. II/1, p. 387.

distribution of the unpaired electron spin-density within the radical. Comparison with the calculated distribution permits a check of the methods of calculation, as will be explained in Section 5.5:4.3.

4.2.3 Relaxation and Linewidth

Concerning the relaxation times T_1 and T_2 and their relation to the linewidth (cf. Section 5.5:1.2) it will suffice here to give a few important methods for obtaining ESR measurements.

The most important relaxation mechanism in *transition metal compounds* is the spin-orbit interaction, in which the electronic spin is coupled with time-dependent processes such as the lattice vibrations or the Brownian motion in liquids. This mechanism is comparatively less important for radicals, since the spin-orbit coupling parameter ζ decreases rapidly with decreasing atomic number Z. The relaxation times of radicals consisting of atoms of the first row are, in general, relatively long as compared to those of metal chelate compounds (order of magnitude: 1/10 to 10 microseconds). This relaxation mechanism is independent of the concentration. In nitrogen- or oxygen-containing radicals it is the sole mechanism below a concentration of 10^{-4} to 10^{-5} mol/l, and determines the linewidth $\Delta \nu$. When investigating such radicals, it is better to work at higher fields and frequencies since the g-value anisotropy, which increases with increasing spin-orbit coupling, causes an increasing linewidth with increasing magnetic field H_0. In contrast to this behavior, relaxation through spin-orbit coupling is negligible for radicals consisting solely of carbon and hydrogen.

There are two other main relaxation mechanisms which affect the linewidth: the exchange interaction and the dipole-dipole coupling.

The exchange interaction, which, in contrast to the chemical exchange (see 5.5:4.6.3), is usually termed physical exchange, arises from an exchange of the unpaired electrons of two radicals if their wave functions overlap during a collision. The magnitude of this effect is directly proportional to the number of collisions and thus to the concentration c. It is also proportional to the temperature T and inversely proportional to the viscosity η. Since the quotient T/η should not be reduced greatly lest the contribution of the electron-nuclear dipole coupling to the linewidth increases, the concentration c must be as low as possible for high resolution; the lower limit is of the order 10^{-6} mol/l because of sensitivity limitations. In the case of dipole-dipole coupling it is necessary to distinguish the concentration-dependent interaction between unpaired electron spins of different radicals from the interaction between electron spins and nuclei in the same radical and in the solvent. Below a radical concentration of about 10^{-4} mol/l the electron-electron dipole coupling can be neglected since it decreases with the sixth power of the distance. (The relaxation time is proportional to the square of the interaction energy, which decreases with r^3.) The remaining concentration-independent relaxation mechanism arises from the last term in Equation (1), which becomes time-dependent in liquids due to molecular motion. Its intermolecular part can be eliminated by using solvents not containing nuclear magnetic moments (CS_2), while its intramolecular part remains effective even with CH-radicals in highest dilutions. For the maximum resolution obtainable when observing these considerations, see Section 5.5:4.2.

4.2.4 Experimental Technique

According to the basic equation of ESR (Eq. 2b in 5.5:1.1), it is possible, when recording a spectrum, to vary continuously either the resonance frequency or the magnetic field H_0. For technical reasons H_0 is usually varied and, in addition, it is modulated by radio frequency. In order to increase the sensitivity, phase-sensitive detection of this radio frequency is used. One obtains not the absorption curve but its first derivative (see Fig. 1 in 5.5:1.2). To measure the HFS, the field is usually calibrated with a Hall probe; however, its accuracy is only about 100 ppm and, hence, inadequate for g-value measurements. For higher accuracy the H_0-field is measured by NMR (see Eq. 9).

Since H_0 is usually varied, the HFS coupling constant a is frequently given in gauss. We prefer to give a in frequency units (MHz), which directly measure the energy. This is particularly suitable when comparing ENDOR, NMR and ESR measurements. While HFS constants given in frequency units are valid in general, those given in gauss must be converted because of the variations in γ_I and γ_S.

Another quantity which can be determined by ESR measurements is the concentration of radicals in a given sample. This concentration is proportional to the area under the absorption curve and is obtained by integrating the first-derivative ESR spectrum twice. The conditions necessary to reduce the error below 10% have been discussed recently[3].

Monographs are available for experimental techniques[93]. Most of the spectrometers commercially available operate in the X-Band region at about 9 GHz $\sim$ 3,3 kG. In theory, a considerable increase in sensitivity can be obtained by using higher frequencies; for example the gain for the Q-band would be $(\nu_Q/\nu_X)^2 \approx 15$ ($\nu_Q \approx 35$ GHz); in practice one achieves an increase in sensitivity of a factor of ten. However, this holds good only for samples of equal size; i.e. if the volume of the sample is restricted to several mm^3. This results in a good filling factor with the small dimensions of the Q-band resonator. When investigating radicals in solution, however, the limiting quantity is usually the concentration, which must be c $<10^{-4}$ mol/l (see 5.5:4.4.2). However, 1 cm^3 of this dilute solution required for a good filling factor at X-band usually is available. Hence, in this case and in most others, measurements at higher frequencies are not advantageous, but they can produce greater linewidths if the linewidth is determined by spin-orbit coupling (see 5.5:4.2.3).

A further important aspect of ESR is the saturation of the transitions. The relevant quantity for the saturation of a Lorentzian line (see 5.5:1.2) is the product $\gamma_S^2 H_1^2 T_1 T_2$ (H_1=radio frequency magnetic field). Since the signal amplitude of an absorption line is proportional to H_1, it is necessary to compromise between sensitivity and resolution. Maximum signal-to-noise ratio is obtained for $\gamma_S^2 H_1^2 T_1 T_2 = 1$; under this condition the linewidth is already increased by saturation broadening by a factor of $\sqrt{2}$ as compared to the unsaturated Lorentzian line. If saturation broadening is to be obviated, the condition $\gamma_S^2 H_1^2 T_1 T_2 \ll 1$ must be fulfilled.

4.3 Theory

4.3.1 Theory of the Hyperfine Structure (HFS)

If the anisotropic dipole-dipole coupling is not averaged out by means of rapid molecular rotation, the HFS interaction must be described by a tensor. The determination of the tensor components and their interpretation represents an important task of ESR spectroscopy. Generally

$$HFS = A \underline{S}_z \underline{I}_z + B \underline{S}_x \underline{I}_x + C \underline{S}_y \underline{I}_y \qquad (2)$$

where $S_{x,y,z}$ and $I_{x,y,z}$ are the components of the spin vector operators in the direction of a cartesian coordinate system fixed in the molecular framework. A, B, and C are the principal values of the HFS tensor. These values have been calculated[4]

[3] G. Casteleijn, J.J. ten Bosch, J. Appl. Phys. 39, 4375 (1968).

[4] H.M. McConnell, J. Strathdee, Mol. Phys. 2, 129 (1959);
O.H. Griffith, D.W. Cornell, H.M. McConnell, J. Chem. Phys. 43, 2909 (1965).

[5] D.H. Whiffen, Pure Appl. Chem. 4, 185 (1962).

for some commonly occurring nuclei, e.g. ^{1}H and ^{14}N. For radicals oriented in a rigid matrix the principal values can be determined directly by means of the angular dependence of the HFS splittings. For this purpose the single crystal must be rotated about its principal axis in the magnetic field[5].

However, when the radicals are incorporated in a glassy or polycrystalline matrix the ESR spectra are very complicated and it is difficult to analyze them.

In fluid solutions of low viscosity, the Brownian motion of the dissolved radicals occurs at a frequency which is high compared with the anisotropies of the HFS interaction. As a result only the isotropic part of the HFS coupling constant can be observed:

$$a = 1/3 \, (A + B + C) \qquad (2a)$$

Therefore the HFS Hamiltonian (2) is commonly divided into an isotropic and an anisotropic part, as has been done already in Equation (1). For the frequently occurring case in which the charge density of the unpaired electron possesses more than a twofold symmetry axis through the nucleus, ESR-lines appear to first order at the following resonance fields:

$$H_{res} = H_0 + \sum_i a_i m_{I_i} + \sum_i b_i (3\cos^2\theta_i - 1)\, m_{I_i} \qquad (3)$$

The third term vanishes in isotropic solutions. In (3) the resonance field $H_0 = h\nu/g\,\mu_B$, due to the interaction H_g is already included (see 5.5:4.3.4). The summation is over all the nuclei in the molecule; a is the isotropic and b the anisotropic HFS coupling constant. θ is the angle between a principal axis of the HFS tensor and the external magnetic field. In case of axial symmetry the principal values are commonly written $A_\parallel \equiv A$, $A_\perp \equiv B = C$ so that

$$a = 1/3(2\,A_\perp + A\|); \quad b = 1/3(A\| - A_\perp) \qquad (2b)$$

According to Equation (1) the isotropic HFS constant a measures directly the s-character of the orbital containing the unpaired electron. The anisotropic HFS constant b, however, vanishes for s-electrons when averaging. Therefore b measures directly the p- or d-character of the unpaired elec-

tron. In this way complementary information concerning the orbital of the unpaired electron is obtained by analyzing the isotropic and anisotropic HFS splittings[6].

4.3.2 Isotropic HFS in π Radicals

In the following we shall confine ourselves to a discussion of organic radicals in solution for which anisotropy effects are averaged out to a large extent.

Although some papers have appeared recently dealing with σ radicals[7], in the vast majority of cases the unpaired electrons occupy π orbitals which are delocalized over the molecule. Aromatic radical ions serve as an example for these π radicals. For σ electrons a nonvanishing probability density $\rho\,(I)$ exists at the nucleus, causing directly a HFS splitting. π orbitals, however, formed by overlap of p_z atomic orbitals, have a node in the plane of the molecule. Therefore, the direct HFS interaction between the unpaired π electron and the ring protons must vanish. The question is, how can an unpaired spin density $\rho\,(I)$ exist in the 1 s hydrogen orbital which is responsible for the observed HFS splitting. Taking the C–H fragment as an example, McConnell and Chesnut[8] have demonstrated that the spin-dependent exchange forces between π and σ electrons result in a slight polarization of the spins in the C–H σ bond. For the spins of the two electrons forming the σ bond, the two structures a) and b) can be drawn:

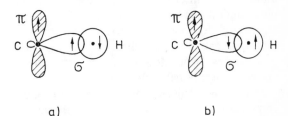

a)   b)

Without considering the π-σ interaction, both structures would be equally important. Because of the exchange interaction, however, there is a smaller repulsion between the unpaired π electron and the σ electron with the same spin direction compared with the σ electron having its spin anti-

[6] *J.R. Morton*, Chem. Rev. *64*, 453 (1964).
[7] *R.W. Fessenden, R.H. Schuler*, J. Chem. Phys. *39*, 2147 (1963);
R.O.C. Norman, B.C. Gilbert, J. Phys. Chem. *71*, 14 (1967);
R. Hoffmann, J. Chem. Phys. *39*, 1397 (1963);
J.A. Pople, D.P. Santry, G.A. Segal, J. Chem. Phys. *43*, 3129 (1965);

J.A. Pople, G.A. Segal, J. Chem. Phys. *44*, 3289 (1966);
G.A. Petersson, A.D. McLachlan, J. Chem. Phys. *45*, 628 (1966);
N.M. Atherton, A. Hinchcliffe, Mol. Phys. *12*, 349 (1967);
R.S. Drago, H. Petersen, J. Am. Chem. Soc. *89*, 3978 (1967).

parallel. As a consequence, structure a) is energetically slightly more favorable, that is, the unpaired π electron polarizes the spins of the σ bond causing a (negative) spin density in the 1 s hydrogen orbital. The extent to which the polarization occurs is proportional to the spin density ρ^π of the unpaired electron in the carbon $2p_z$ atomic orbital. (ρ^π measures the fraction of time the unpaired π electron spends there.) For the C—H fragment $\rho^\pi = 1$, for a molecule with conjugated double bonds $\rho_r^\pi < 1$, whereby for planar molecules the normalization condition

$$\sum_{r=1}^{N} \rho_r^\pi = 1$$

is valid (r indicates the position of the n carbon atoms in the molecule). In this way the isotropic proton HFS constants a_r^H are related to the π spin densities by means of the equation[8]:

$$a_r^H = Q\, \rho_r^\pi \qquad (4)$$

Aromatic radicals have a π-σ parameter Q of the order of -27 G $\simeq -75$ MHz.[9, 10]

Where HFS is due to hetero-atoms, e.g. ^{13}C, spin densities on adjacent carbon atoms also contribute to the spin polarization of the π bonds, resulting in the appearance of several π-σ parameters Q[11].

4.3.3 Spin Density Theories

From a strictly theoretical point of view, ab initio calculations of spin densities are, of course, the ultimate goal. In organic molecules, however, an upper limit is soon reached because of the large number of electrons involved. Thus, in practice, semiempirical molecular theories are generally used. In these, inherent quantities of the formulas are not explicitly calculated but are replaced by experimentally adjusted parameters. The semiempirical methods are distinguished by the degree to which interelectronic interaction is taken into account.

Considerable success in predicting π spin densities is achieved by the LCAO-MO theories of Hückel[12], McLachlan[13], and Amos-Snyder[14]. In all three methods the molecular orbital (MO) Ψ_j is formed by a linear combination of carbon $2p_z$ atomic orbitals (LCAO) ϕ_r:

$$\Psi_j = \sum_{r=1}^{N} c_{jr}\, \varphi_r, \quad j = 1, 2, \ldots n \qquad (5)$$

The AO coefficients c_{jr} are determined by the variational theorem resulting in minimal energy of the π system.

In the Hückel MO theory (HMO) the interaction between the π electrons is explicitly neglected. Implicitly this drastic simplification is partly compensated for by introducing effective molecular parameters, the Coulomb and resonance integrals. As long as we are dealing with conjugated hydrocarbons, numerical values of these parameters are not required for the calculation of spin densities and relative magnitudes of π energy levels. The HMO spin densities ρ_r^π (HMO) are given by the squared LCAO coefficients of the singly occupied π orbital:

$$\rho_r^\pi(\text{HMO}) = c_{or}^2 \qquad (6)$$

that is, ρ_r^π (HMO) $\geqslant 0$. This simple theory has already yielded excellent results, for many radical ions of alternate hydrocarbons. The HMO method has also stood the test for conjugated molecules with heteroatoms. For many molecular classes there are well-established Coulomb and resonance integrals cited in the literature[12]. Even in cases of aliphatic radicals of the type

$$
\begin{array}{c}
X_1 \\
| \\
\cdot\, C\!-\!X_2 \\
| \\
X_3
\end{array}
$$

the HMO theory can explain the experimental result that the π spin density at the α carbon atom is given by[15].

$$\rho_\alpha^\pi = H \overset{3}{\underset{i=1}{\frac{}{\pi}}} \left\{ 1 - \Delta(X_i) \right\} \qquad (7)$$

In this equation the $\Delta(X_i)$ stands for the spin densities which are migrating into the substitutents X_i. Numerical values of the $\Delta(X_i)$ are known for a great variety of substituents[15].

Because of the total splitting which is observed, for instance, for allyl or pyrene$^\ominus$, it can be concluded that positions in the molecule with zero HMO spin density have in fact negative spin densities resulting in

$$\sum_{r=1}^{N} |\rho_r^\pi| > 1.$$

[8] H.M. McConnell, D.B. Chesnut, J. Chem. Phys. 28, 107 (1958).
[9] J.R. Bolton, G.K. Fraenkel, J. Chem. Phys. 40, 3307 (1964).
[10] K. Möbius, Z. Naturforsch. 20a, 1102 (1965).
[11] M. Karplus, G.K. Fraenkel, J. Chem. Phys. 35, 1312 (1961).
[12] A. Streitwieser, Jr. Molecular Orbital Theory, John Wiley & Sons, New York 1961.
[13] A.D. McLachlan, Mol. Phys. 3, 233 (1960).
[14] A.T. Amos, L.C. Snyder, J. Chem. Phys. 41, 1773 (1964); 42, 3670 (1965).

[15] H. Fischer, Z. Naturforsch. 20a, 428 (1965).

This conclusion is supported by NMR measurements (see 5.5:4.6), studies in liquid crystals (see 5.5:4.4.4), and investigations of linewidth variations[16]. Negative spin densities can be predicted by improved MO theories which explicitly consider the interaction between π electrons. Because of the spin polarization in radicals, electrons with different spins occupy spatially different orbitals.

The MO theories of McLachlan and Amos-Snyder start with such a wave function. The spin density ρ_r^π is then no longer given just by the density $(c_{or}^\alpha)^2$ of the unpaired electron, but by the excess of α spin density over β spin density:

$$\rho_r^\pi = \sum_j^{occ.} (c_{jr}^\alpha)^2 - \sum_j^{occ.} (c_{jr}^\beta)^2 \qquad (8)$$

The summation is over all occupied orbitals. In contrast to the HMO method the results of the improved MO theories are critically dependent on the chosen set of interaction parameters.

The self-consistent field (SCF) theories of McLachlan and Amos-Snyder completely neglect differential overlap (ZDO approximation[17]), and, therefore, are unable to account for σ-π interaction. This restriction was removed by the recently developed INDO method (INDO = intermediate neglect of differential overlap[18]). This method no longer distinguishes between π and σ electrons but includes all valence electrons in the basis of the LCAO expansion. Therefore, σ and π radicals are treated within the frame of the same formulae. The INDO method should become especially important for the theory of strongly twisted molecules in which there is considerable overlap between π and σ orbitals[19].

A comparison between the efficiency of the different MO models is presented in Section 5.5:4.4.3. The present situation with regard to the π electron theory is marked by the difficulty of testing experimentally further improvement of spin density calculations as long as the σ-π parameter Q in Equation (4) is not better known. Since both the spin density ρ_r^π and the spin density $\rho(I)$ in the s orbital of the nucleus are calculated in the course of an INDO-MO calculation, additional information concerning the Q factor may be expected from this method.

Advanced MO calculations can, of course, be performed only with the aid of fast electronic digital computers. For all the methods discussed here, appropriate programs are already available from program libraries (for instance from: Quantum Chemistry Program Exchange, Indiana University, Bloomington, Indiana).

So far we have mentioned only HFS splittings due to α protons. However, the HFS splittings due to β protons can also be observed. (β protons are attached to sp^3 carbons which in turn are bonded to π systems.) The methyl protons in radicals of the type $RR'\dot{C}CH_3$ serve as an example. Freely rotating methyl groups exhibit a β splitting of roughly the same magnitude as the splitting from α protons. By analogy to Equation (4) there is a proportionality factor between ρ_α^π and a_{CH_3} with $Q_{CH3} \approx +29$ G $\simeq +81$ MHz[15,20]. The HFS of β protons is explained in terms of hyperconjugation[21].

4.3.4 Theory of the g-Value

In addition to its HFS, a radical is characterized by its g-value. Because of strong electric ligand fields in organic radicals, the orbital angular momentum of the π electrons is largely quenched. For this reason the g-values of these radicals are within 1% of the free electron g-value, $g_o = 2.002319$. Within a molecular class g-values frequently differ only in the fourth and fifth decimal places. To measure any regularities, a commensurate degree of accuracy is required.

According to the resonance condition (3), absolute g-value measurements can be performed by determining the microwave frequency ν and the magnetic resonance field strength H_0 at the center of an ESR spectrum[22]. Because of the required accuracy, H_0 must be measured by means of the nuclear magnetic resonance frequency ν_p. The g-value can then be determined by:

$$g = C' \nu/\nu_p$$

with [22,23] $\qquad (9)$

$$C' = g_o \cdot \gamma_p/\gamma_S = 3{,}0419845 \cdot 10^{-3} \pm 0{,}5 \text{ ppm}$$

(γ_s and γ_p are the gyromagnetic ratios of the electron and proton, respectively)

For a relative measurement of the g-value according to

$$g = g_s(1 - \delta\, H/H_{os}) \qquad (10)$$

[16] E. de Boer, L.L. Mackor, J. Chem. Phys. 38, 1450 (1963).

[17] J.A. Pople, Proc. Phys. Soc. London 68, 81 (1955).

[18] J.A. Pople, D.L. Beveridge, P.A. Dobosh, J. Chem. Phys. 47, 2026 (1967); J. Am. Chem. Soc. 90, 4201 (1968).

[19] J.A. Pople, D.L. Beveridge, J. Chem. Phys. 49, 4725 (1968).

[20] R.W. Fessenden, R.H. Schuler, J. Chem. Phys. 39, 2147 (1963).

[21] J.P. Colpa, E. de Boer, Mol. Phys. 7, 333 (1964).

[22] E. Klein, K. Möbius, H. Winterhoff, Z. Naturforsch. 22a, 1704 (1967).

[23] K. Möbius, H. Haustein, M. Plato, Z. Naturforsch. 23a, 1626 (1968).

it is necessary to select an appropriate standard radical with known g-value g_s (δH is the difference between the resonance field strengths H_o and H_{os} of the standard).

The shift Δ_g between the observed g-value and g_o is due to residual orbital Zeeman and spin-orbit interactions. This follows from the second-order perturbation theory from which it can be shown that the two Hamiltonians are equivalent.

$$Hg = g_0 \mu_B \vec{S}\vec{H} + \mu_B \vec{L}\vec{H} + \zeta \vec{L}\vec{S} \qquad (11a)$$

and

$$Hg = (g_0 + \Delta g)\,\mu_B \vec{S}\vec{H} \equiv g\mu_B \vec{S}\vec{H} \qquad (11b)$$

($\vec{L}$ is the orbital angular momentum operator, ζ is the spin-orbit coupling parameter).

Stone[24] has shown that for planar aromatic hydrocarbon radicals the shift Δg is a linear function of the HMO energy coefficient λ_o of the lowest half-filled π orbital:

$$\Delta g = b + \lambda_0 c \qquad (12)$$

The constants are[25,26] $b = (31.2 \pm 0.4) \times 10^{-5}$ and $c = (-15.1 \pm 0.7) \times 10^{-5}$. Radicals with orbitally degenerate ground states, e.g. *benzene*$^-$ and *coronene*$^-$, have somewhat larger g-values than those predicted by Stone's theory[25].

In strongly twisted phenyl-substituted radicals, such as the radical ions of *diphenylanthracene* and *rubrene,* there is a migration of spin density into the σ system of the phenyl groups. As a result, the g-values decrease drastically[26]. The deviation $\delta \Delta g$ from Stone's straight line can be described by:

$$\delta \Delta g = c\,\Delta\lambda - b'\,\Delta\rho \qquad (13)$$

with $b' \approx 0.5\,b$[27]. According to a hyperconjugative model of strongly twisted phenyl groups, the energy change $\Delta\lambda$ and the spin density $\Delta\rho$ which is migrating into the σ system have been calculated[27]. Provided the twist angles are known, the anomaly of the g-value $\delta \Delta g$ can be predicted. On the other hand, measured Δg-values can give some indication of the stereochemistry of radicals.

Quite often π electron systems contain *heteroatoms* with relatively large ζ values, e.g. *oxygen* or *chlorine.* In these radicals the main contribution to the g shift is due to the promotion of an electron from a non-bonding p orbital of the heteroatom to the half-filled π orbital lying higher by an energy amount of $\Delta E_{n \to \pi}$. Δg is then simply given by[28]:

$$\Delta g = \frac{2}{3}\,\frac{\zeta_{\text{het}}}{\Delta E_{n \to \pi}}\,\Sigma_{\text{het}}\,\rho_{\text{het}}^{\pi} \qquad (14)$$

the index 'het' indicating the heteroatoms. In this way Δg measures the p character of the orbital containing the unpaired electron at the heteroatom. Since, with growing atomic number, the ζ-values increase rapidly[29], Δg also increases in a characteristic way (see Table 1), and is helpful for identifying radicals[30–32]. In the case of *semidione radicals* even *cis-trans stereoisomers* have been distinguished by their g-values; g_{trans} is about 0.01% greater than g_{cis}[33].

Table 1. Spin-orbit coupling parameter ζ[29] and experimental g values[30] of *p-benzo-semiquinone* radicals.

p-Benzo-semiquinone radicals	ζ [cm^{-1}]	g Factor
Unsubstituted	152	2.00468
Tetrachloro	587	2.00568
Tetrabromo	2460	2.00875
Tetrajodo	5060	2.01217

When comparing absolute g-value measurements, solvent and/or counter-ion effects become important. The solvent methylene chloride, for instance, drastically increases the g-values of aromatic radical ions[26], whereas ion pairing with heavy alkali ions may result in a considerable decrease of the g-values[34] (ζ changes sign in going from left to right in the periodic system).

4.4 High-Resolution ESR in Solution (HR-ESR)

4.4.1 Generation of Radicals

Radicals can be generated easily by abstraction of hydrogen atoms from aliphatic molecules in aqueous or alcoholic solution with the aid of OH

[24] *A.J. Stone,* Proc. Roy. Soc. *A271,* 424 (1963); Mol. Phys. *6,* 509 (1963); *7,* 311 (1964).

[25] *B.G. Segal, M. Kaplan, G.K. Fraenkel,* J. Chem. Phys. *43,* 4191 (1965).

[26] *K. Möbius, M. Plato,* Z. Naturforsch. *24a,* 1078 (1969).

[27] *M. Plato, K. Möbius,* Z. Naturforsch. *24a,* 1083 (1969).

[28] *H.W. Brown* in *E. Low,* Paramag. Res., Academic Press, New York 1963.

[29] *D.S. McClure,* J. Chem. Phys. *17,* 905 (1949).

[30] *M.S. Blois, Jr., H.W. Brown, J.E. Maling* in *M.S. Blois,* Free Radicals in Biological Systems, Academic Press, New York 1961.

[31] *K. Möbius, F. Schneider,* Z. Naturforsch. *18a,* 428 (1963).

[32] *K. Möbius, M. Plato,* J. Phys. Chem. *72,* 1830 (1968).

[33] *C. Corvaja, P.L. Nordio, G. Giacometti,* J. Am. Chem. Soc. *89,* 1751 (1967).

[34] *C.L. Dodson, A.H. Reddoch,* J. Chem. Phys. *48,* 3226 (1968).

radicals. The OH radical itself can be prepared in a flow system[35] from $TiCl_3 + H_2O_2$ or photolytically from added H_2O_2[36]. The advantage of the photolytic method is twofold: no paramagnetic side products are formed which might cause line broadening, and there is no restriction to aqueous solutions.

Paramagnetic radical ions can be generated from aromatic molecules by chemical reduction with alkali metals (in dimethoxyethane or tetrahydrofuran)[37] or by electrolytic ionization (in acetonitrile or dimethylformamide)[38]. The main advantage of generating radicals electrolytically is the possibility of using solvents with high dielectric constants. Both this and the use of bulky tetraalkyl-ammonium counter ions results in stably solvated radical ions, thus minimizing obscurity of the spectra due to ion pair formation between radical and counter ions[34]. On the other hand, dynamic interactions with the counter ions may result in interesting linewidth effects which give some information concerning the structure of the complex[39]. Small molecules can often be ionized electrolytically in ammonia at low temperatures[40] or in tetrahydrofuran[41].

4.4.2 Conditions for HR-ESR

As in every other type of spectroscopy, ESR provides a maximum amount of information if a spectrum is completely resolved. In principle this is technically possible, as was mentioned in 5.5:1.3.3, although the necessary conditions are not fulfilled with all spectrometers. The most important conditions for HR-ESR are: frequency stability of the microwave generator better than

1 ppm, and stability and homogeneity of the magnetic field (over the sample volume) likewise better than 1 ppm. The 100 kHz modulation generally used is problematic for highest resolution because of the side bands. While superheterodyne receivers were used in the past, the latest microwave diodes permit a decrease in the modulation frequency by about one order of magnitude without a noticeable loss in the signal-to-noise ratio.

The connection between linewidth and relaxation processes is described in 5.5:4.2.3. The transverse relaxation time T_2 is the relevant quantity for the resolution because of the relation $\Delta v \approx 1/T_2$. The longitudinal relaxation time T_1 is by definition $T_1 \geqslant T_2$; in the case of radicals in solution discussed here the difference is not more than one order of magnitude. These considerations lead to the following conditions for highest resolution.

Radicals with *heteroatoms* always give a lower resolution the higher the atomic number Z and the larger the spin density at the heteroatom; the highest resolution is attainable for C–H radicals.

All paramagnetic impurities must be carefully removed, in particular the oxygen of the air[42]. This is easy to understand since the O_2-concentration is of the order of 10^{-3} mol/l in most organic solvents and on the other hand, a dilution of the radical solution below 10^{-5} mol/l is necessary because of the relaxation mechanisms discussed in 5.5:4.2.3.

The concentration of the solution must be kept below 10^{-5} mol/l mentioned above and both the H_1 field and the modulation amplitude must be reduced sufficiently so as not to cause any line broadening. Since all these measures reduce the amplitude of the signal, the highest resolution always requires the highest sensitivity. The signal-to-noise ratio can be improved, if necessary, by using time-averaging computers.

When recording the spectrum of *1,3-bis-biphenylene-allyl* (Fig. 2)[43], all precautions mentioned above were observed. The spectrum is divided into two halves because of the unusually large coupling constant $a_1 = 37$ MHz of the central H-atom; in the figure the two halves are plotted below each other. The sizes of the other coupling constants are marked by arrows; their values are $a_2 = 5.37$ MHz, $a_3 = 5.20$ MHz, $a_4 = 1.35$ MHz, and $a_5 = 1.01$ MHz. The structure of the molecule with one central proton and four groups of four equivalent protons theoretically leads to 1250 HFS lines.

[35] *W.T. Dixon, R.O.C. Norman*, J. Chem. Soc. 3119 (1963); 3625 (1964).

[36] *J.F. Gibson, D.J.E. Ingram, M.C.R. Symons, M.C. Townsend*, Trans. Faraday Soc. *53*, 914 (1957);
D.J.E. Ingram, Free Radicals, Butterworth & Co., London 1958;
R. Livingston, H. Zeldes, J. Chem. Phys. *44*, 1245 (1966); *45*, 1946 (1966); J. Am. Chem. Soc. *88*, 4333 (1966); J. Chem. Phys. *47*, 1465 (1967); *47*, 4173 (1967); J. Am. Chem. Soc. *90*, 4540 (1968).

[37] *G.J. Hoijtink, P.J. Zandstra*, Mol. Phys. *3*, 371 (1960);
I.C. Lewis, L.S. Singer, J. Chem. Phys. *44*, 2082 (1966).

[38] *D.H. Geske, A.H. Maki*, J. Am. Chem. Soc. *82*, 2671 (1960);
K. Möbius, Z. Naturforsch. *20a*, 1093 (1965);
B. Kastening, Electrochim. Acta *9*, 241 (1964);
R.N. Adams, J. Electroanal. Chem. *8*, 151 (1964);
A.H. Reddoch, J. Chem. Phys. *41*, 444 (1964).

[39] *E. de Boer, E.L. Mackor*, J. Am. Chem. Soc. *86*, 1513 (1964).

[40] *D.H. Levy, R.J. Myers*, J. Chem. Phys. *41*, 1062 (1964).

[41] *W.M. Tolles, D.W. Moore*, J. Chem. Phys. *46*, 2101 (1967).

[42] *K.H. Hausser*, Naturwissenschaften *47*, 251 (1960); Bull. Ampère Fasc. Spéc. 239 (1960).

[43] *K.H. Hausser*, Z. Naturforsch. *17a*, 158 (1962); Proc. Xth Coll. Spectroscop. Internat., p. 707.

This number is reduced by a factor of approximately two, due to accidental coincidence of lines. About 400 of these are visible in the spectrum, the rest disappear in the noise. This is not surprising when one considers that the intensity ratio of the strongest to the weakest HFS component in 1296:1.

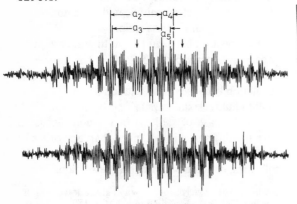

Fig. 2. ESR-spectrum of *1,3-bisdiphenylallyl* in CS_2, Concentration about 10^{-5} mol/l[43].
In the spots marked by an arrow the distances between the HFS components clearly deviate from the equi-distance to be expected. The reason for this is that the magnetic field was not sufficiently constant over the long period of about 15 hours necessary for recording the spectrum.

The number of lines in the ESR spectra of large molecules with low symmetry becomes so large that a complete resolution is no longer possible. In a spectrum the recorded linewidth can become much larger than the homogeneous linewidth of the individual components determined by T_2 due to unresolved substructures; such a spectrum is termed 'inhomogeneously broadened'. Although a reconstruction by lines is impossible, the spectrum can frequently be analyzed if it is simulated by superposition of the first derivative of Lorentzian lines. For this purpose the splitting constant and the linewidth are varied by means of a computer until the best agreement with the experimental spectrum is achieved. A probable set of HFS splitting constants is either obtained by a MO-calculation of the molecule or is taken from the experimental spectrum by means of a correlation calculation[44]. The analysis of a spectrum by digital computers will increasingly become a standard technique of ESR spectroscopy. The necessary programs are already available – like the MO programs. The efficiency of the simulation of spectra is demonstrated in Fig. 3[45] for *chrysene⁻* (6

HFS splitting constants of 2 equivalent protons each, i.e. $3^6 = 729$ HFS components to be expected theoretically, about 90 of which are resolved).

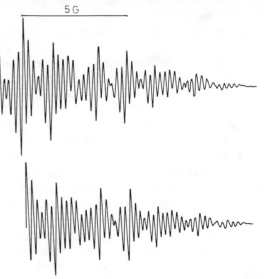

Fig. 3. Experimental and simulated ESR half-spectra of the negative radical ions of *chrysene*.
The HFS splitting constants obtained by simulation have the values[45]: 0.778; 1.426; 3.88; 4.05; 5.92 and 13.75 MHz.

The set of HFS splitting constants obtained is not always unambiguous if the lines in the outer part of the spectrum cannot be recorded because of their low intensity. The amount of information obtained is even less if the spectrum is inhomogeneously broadened so that only the envelope of the HFS components is observed. In this case it is necessary to use the ENDOR method (see 5.5:4.5) by which the number of lines is drastically reduced.

4.4.3 Test of MO Models

Since local properties of wave functions of unpaired electrons can be measured by ESR, this method is especially suitable for testing the quality of different models in quantum chemistry. In the case of silicon-substituted π radicals, strong evidence of d orbital effects in the Si ← C delocalization was found by comparing theoretical and experimental HFS splittings[46].

A convincing and direct proof of the pairing theorem for alternate hydrocarbons has been achieved by ESR. According to this theorem, in alternate hydrocarbons a pair of bonding and antibonding π levels are placed symmetrically about the nonbonding one, the LCAO coefficients of

[44] *E. Ziegler, E.G. Hoffmann*, Z. Anal. Chem. *240*, 145 (1968).
[45] *M. Plato*, Z. Naturforsch. *22a*, 119 (1967).

[46] *F. Gerson, J. Heinzer, H. Bock, H. Alt, H. Seidl*, Helv. Chim. Acta *51*, 707 (1968).

their molecular orbitals being of the same absolute value[12, 18]. Consequently, the squares of these coefficients, i.e. the spin density distribution, should be the same for positive and negative radical ions e.g. for *perylene (Fig. 4)*. The remaining differences in the HFS spectra may be explained by the charge dependence of the Q factor[47]. The validity of the pairing theorem has also been supported by the ^{13}C-HFS[9]. Non-alternate systems do not have pairing properties and, in fact, the ESR spectra of their corresponding positive and negative radical ions are very dissimilar[48].

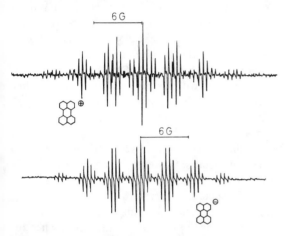

Fig. 4. ESR spectra of the positive and negative radical ions of the alternate hydrocarbon *perylene*, demonstrating the validity of the pairing theorem[10].

For ion radicals, therefore, Equation (4) should be extended to consider the excess charge ϵ_r^π at the rth carbon atom[47]:

$$a_r^H = (Q_0 + K \varepsilon_r^\pi)\, \rho_r^\pi \qquad (15)$$

From Table 2 one can get an idea of the accuracy of the MO theories of Hückel, McLachlan, and Amos-Snyder in calculating spin densities for alternate and non-alternate hydrocarbons[48]. As can be seen, even the simple HMO theory predicts very well the spin density distribution in alternate radical ions. The spin density distribution of non-alternate hydrocarbons, however, is described better if more rigorous π electron interaction is taken into account.

Table 2. σ—π-Parameter Q_0 and K [see eq. (15)] for alternate and non-alternate *hydrocarbon radical* ions[48]. The derivative D is determined by the mean relative deviation between 30 theoretical and experimental HFS coupling constants.

MO Model	Hydrocarbon radical ions					
	alternate			non-alternate		
	Q_0[G]	K[G]	D[%]	Q_0[G]	K[G]	D[%]
Hückel	31.4	17	± 19	34.8	28	± 60
McLachlan	24.3	17	± 23	26.3	30	± 43
Amos-Snyder	24.9	10	± 21	32.1	41	± 35

Quite often information concerning the stereochemistry of a radical can be obtained by interpreting the experimental data, i.e. HFS splittings and g-values, in terms of a resonance integral which depends on the twist angle[12, 49].

For a series of alternate hydrocarbons a correlation between HFS splitting constants and chemical reaction rates at different molecular sites has been observed and interpreted in the frame of HMO theory[50]. Reaction rates are correlated more strongly with HFS constants or spin densities than with the frequently used reaction indices.

4.4.4 Radicals in Liquid Crystals

Additional information may be gained if ESR studies are performed on radicals dissolved in liquid crystals. In a magnetic field of more than 2000 G the molecules of a liquid crystal with a nematic mesophase are aligned in the direction of the magnetic field[51]. A commonly used liquid crystal is *p*-azoxyanisole. Solute molecules are also aligned; in general, however, the degree of ordering of the solvent is different from that of the solute. High resolution ESR spectra show that all the advantages of a liquid are preserved. In addition, shifts of the HFS splitting constants and g-values may occur when passing from the nematic to the isotropic phase. This is accomplished by raising the temperature. From these shifts conclusions can be drawn concerning the magnitude of the tensor components and the sign of spin densities[23, 52]. The ESR studies in liquid crystals have been largely restricted to stable neutral radicals[23, 52, 53], e.g.

[47] *J.P. Colpa, J.R. Bolton*, Mol. Phys. *6*, 273 (1963).
[48] *K. Möbius, M. Plato*, Z. Naturforsch. *22a*, 929 (1967).
[49] *D.H. Geske, J.L. Ragle, M.A. Bambenek, A.L. Balch*, J. Am. Chem. Soc. *86*, 987 (1964);
K. Möbius, Z. Naturforsch. *20a*, 1117 (1964);
M.D. Sevilla, G. Vincow, J. Phys. Chem. *72*, 3641 (1968).

[50] *C.P. Poole, Jr, O.F. Griffith*, J. Phys. Chem. *71*, 3672 (1967).
[51] *A. Saupe*, Angew. Chem. *80*, 99, Engl. *7*; 97 (1968).
[52] *S.H. Glarum, J.H. Marshall*, J. Chem. Phys. *44*, 2884 (1966); *46*, 55 (1967);
H.R. Falle, G.R. Luckhurst, Mol. Phys. *11*, 299 (1966).
[53] *A. Carrington, G.R. Luckhurst*, Mol. Phys. *8*, 401 (1964);
D.H. Chen, G.R. Luckhurst, Trans. Faraday Soc. *65*, 656 (1969).

perinaphthenyl or *vanadyl acetylacetonate*. Recently, however, it has been shown that this technique can be extended to electrolytically generated aromatic radical ions[54].

The question as to whether a radical contains more than one unpaired electron can be decided unequivocally by performing an ESR experiment in a liquid crystal. In a biradical, for instance, which is dissolved in an isotropic liquid, the zero-field splitting D is averaged out by the isotropic Brownian motion. If there is only a small scalar interaction (that is, a small exchange integral, J) between the two electrons, the biradical character is hard to determine. In nematic liquids, however, the radical motion is no longer isotropic. For this reason D does not vanish but splits certain lines into doublets. The magnitude of this splitting depends upon D and the degree of ordering[55]. For a radical containing n unpaired electrons, certain lines are split into n components when passing from the isotropic to the nematic phase.

4.4.5 Intramolecular Motions

As has been shown in the last few years, intramolecular motion, e.g. hindered rotation, can be studied by ESR with a degree of success similar to that achieved with NMR (see 5.5:2). Since in ESR the lifetime of a certain spin state is shorter by several orders of magnitude than in NMR, much faster molecular motion can be investigated by ESR spectroscopy.

When studying hindered rotation by ESR, it is necessary to determine whether the motion simply modifies the magnitude of coupling constants or whether it lowers the symmetry of the radical during the ESR time-scale. In the latter case, the number of HFS constants is influenced.

The Stone-Maki model[56] belongs to the first category. In this model the magnitude of the β proton splitting constant is related to the height of the potential barrier. The model has been tested for a series of nitroalkane anion radicals $RR'CH-NO_2^-$ and its limits of validity have been discussed. In radicals of the type $R-\dot{C}H_2$ (R = alkyl), Fessenden[57] has determined potential barriers from the magnitude of the β splitting. In this case he was able to present a mathematically rigorous solution of the problem.

Rate processes belong to the second category. The exchange rate between different conformations determines the difference between two HFS constants and, therefore, the lineshape. The exchange rate is temperature-dependent. Temperature variations cause very pronounced changes in the spectrum when the exchange rate is comparable in magnitude to the separation of the two lines, i.e. in the order of some MHz. During the exchange process the outer lines of the spectrum remain sharp, whereas in the middle of the spectrum line broadening occurs. The reason for this is explained in Fig. 5. In the two-jump model the jumping occurs between (A) and (B). The splitting constants of the two protons are interchanged while their nuclear spin orientation remains unchanged. Thus the unpaired electron in radicals whose proton spins are parallel – the ones generating the outer lines – feels no change in HFS field when an exchange occurs. However, in radicals whose proton spins are antiparallel, the exchange alters the hyperfine field and, therefore, broadens the central lines. A correlation diagram (Fig. 5) is useful in predicting the linewidth alternation even when the HFS is more complicated. By using a modified Bloch theory[58], which has already proved its value in explaining exchange effects in NMR, potential barriers have been determined from the temperature dependence of the ESR line shapes[59].

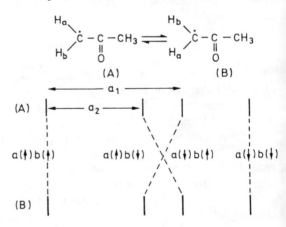

Fig. 5. The two non-equivalent conformations of the acetone radicals $H_2\dot{C}-COCH_3$ and the correlation diagram of the chemical exchange of the H_a and H_b protons.

[54] *H. Haustein, K. Möbius, K.P. Dinse,* Z. Naturforsch. *24a,* 1764, 1768 (1969).

[55] *R. Falle, G.R. Luckhurst, H. Lemaire, Y. Marechal, A. Rassat, P. Rey,* Mol. Phys. *11,* 49 (1966).

[56] *E.W. Stone, A.H. Maki,* J. Chem. Phys. *37,* 1326 (1962);
T.M. McKinney, D.H. Geske, J. Am. Chem. Soc. *89,* 2806 (1967);
D.H. Geske, Progr. Phys. Org. Chem. *4,* 125 (1967).

[57] *R.W. Fessenden,* J. Chim. Phys. *61,* 1570 (1964).

[58] *H.S. Gutowsky, C.H. Holm,* J. Chem. Phys. *25,* 1228 (1956).

4.5 Electron-Nuclear-Double-Resonance (ENDOR)

4.5.1 ENDOR in Solution

In order to understand the ENDOR experiment[60] it is necessary to consider the nuclear Zeeman splitting which could be neglected in ESR (see Fig. 1). In the course of an ENDOR experiment an ESR transition $\Delta m_s = \pm 1$ is saturated and, keeping the magnetic field at its resonance value, the sample is irradiated with a second high-frequency field, which is in the NMR region. This frequency is varied until an NMR transition $\Delta m_I = \pm 1$ occurs. As a result, there is partial desaturation of the ESR and an ENDOR signal appears. Therefore, ENDOR can be considered as a special variety of NMR with the unpaired electron serving as the detector. Because of quantum transformation, ENDOR is much more sensitive than NMR but attains only a few per cent of the sensitivity of ESR.

To first order each group of equivalent nuclei – no matter how many nuclei are present – contributes only two ENDOR lines. These are symmetrically placed about the resonance frequency of the free nucleus ν_0, their distance from this frequency being half the hyperfine coupling constant:

$$\nu_{ENDOR} = \|\nu_0 \pm a/2\| \tag{16}$$

with $\nu_0 = g_I \mu_K H_0/h$. The advantage in resolution of ENDOR over ESR can be demonstrated by comparing the spectral densities Ω, i.e. the number of lines per total width of the spectra:

$$\Omega_{ESR} = \prod_{k=1}^{K} (2 n_k I_k + 1)/2 \sum_{k=1}^{K} a_k n_k I_k \tag{17}$$

$$\Omega_{ENDOR} = 2 \, K/a_{max} \tag{18}$$

Here K indicates the number of groups each containing n_k equivalent nuclei. For organic radicals the total width of the ESR spectra is nearly constant. With increasing K, Ω_{ESR} rises in a multiplicative way, Ω_{ENDOR}, however, only in an additive way. The advantage of ENDOR is especially pronounced for radicals of low symmetry. In the case of *triphenylphenoxyl*, for example, K = 7 and $\Omega_{ESR}/\Omega_{ENDOR} \approx 40$; that is, per unit of energy the ESR spectrum contains 40 times as many lines as the ENDOR spectrum.

This gain in resolution is achieved by a considerable amount of experimental effort. Because of the short electronic relaxation times of radicals in solution, the NMR field strength must be much greater than is required for ENDOR in the solid state[61]. Frequently the intensity of the ENDOR lines is critically temperature-dependent, opening possibilities for studies of relaxation mechanisms[62] and internal molecular motions[63]. Theoretical aspects of ENDOR in solution[62, 64] and experimental results[65, 66] have been compared for the HFS structure of substituted *triarylmethyl radicals*[66]. In case of the *triphenylphenoxyl radical*[67] it has been demonstrated that, even with several almost identical HFS splittings, an ENDOR analysis can be performed with the aid of deuteration. Since $\gamma_P/\gamma_D \approx 6.5$, lines in the ESR spectra are eliminated only if sites of small spin density are deuterated. Deuteration of any site, however, eliminates that site from the proton ENDOR display, since the resonance frequencies of protons and deuterons differ by this factor 6.5.

4.5.2 ENDOR of Organic Radicals in the Solid State

ENDOR studies of organic single crystals[68, 69] and powders[70] have also been performed. Cook and Whiffen[68] have demonstrated that the HFS coupling constant of β protons in π radicals is positive. In the case of the triplet groundstate molecule *fluorenylidene*, Hutchison and Pearson[69] were able to resolve all the proton HFS couplings, thus determining the spin density distribution.

[59] *T.P. Das*, J. Chem. Phys. *27*, 763 (1957);
R.W. Fessenden, R.H. Schuler, J. Chem. Phys. *39*, 2147 (1963);
C. Corvaja, J. Chem. Phys. *44*, 1958 (1966);
G. Golde, K. Möbius, W. Kaminski, Z. Naturforsch. *24a*, 1214 (1969).

[60] *G. Feher*, Phys. Rev. *103*, 834 (1956).

[61] *J.S. Hyde, A.H. Maki*, J. Chem. Phys. *40*, 3117 (1964).

[62] *J.H. Freed*, J. Phys. Chem. *71*, 38 (1967).

[63] *L.D. Kispert, J.S. Hyde, C. de Boer, D. La Follette, R. Breslow*, J. Phys. Chem. *72*, 4276 (1968).

[64] *J.H. Freed*, J. Chem. Phys. *43*, 2312 (1965); *47*, 2761 (1967); *50*, 2271 (1969).

[65] *J.S. Hyde*, J. Chem. Phys. *43*, 1806 (1965);
U. Ranon, J.S. Hyde, Phys. Rev. *141*, 259 (1966);
C. Steelink, J.D. Fitzpatrick, L.D. Kispert, J.S. Hyde, J. Am. Chem. Soc. *90*, 4354 (1968).

[66] *A.H. Maki, R.D. Allendoerfer, J.C. Danner, R.T. Keys*, J. Am. Chem. Soc. *90*, 4225 (1968);
R.D. Allendoerfer, A.H. Maki, J. Am. Chem. Soc. *91*, 1088 (1969).

[67] *J.S. Hyde*, J. Phys. Chem. *71*, 68 (1967).

[68] *R.J. Cook, D.H. Whiffen*, J. Chem. Phys. *43*, 2908 (1965).

[69] *C.A. Hutchison, Jr.*, J. Phys. Chem. *71*, 203 (1967);
C.A. Hutchison, Jr., G.A. Pearson, J. Chem. Phys. *47*, 520 (1967).

[70] *G.H. Rist, J.S. Hyde*, J. Chem. Phys. *50*, 4532 (1969);
J.S. Hyde, G.H. Rist, L.E. Göran Eriksson, J. Phys. Chem. *72*, 4269 (1968).

4.5.3 Electron-Electron Double Resonance

Electron-electron double resonance is analogous to nuclear-nuclear double resonance (see 5.5:2) and has been applied to organic radicals in solution[71]. The method seems promising as an important tool for studies of relaxation processes and determination of HFS constants from complex spectra. In this technique the sample is irradiated with a strong and weak microwave field. The ESR signals are detected by the weak microwave field and decrease in intensity as much as 40% when the microwave frequencies differ by a multiple of a HFS splitting.

4.6 NMR of Radicals in Solution

4.6.1 Paramagnetic Shift

To understand the investigation of radicals by NMR the right side of Fig. 1 is referred to. The nuclear Zeeman term plotted there should be split into a doublet with the same coupling constant a measured in frequency units as in ESR. However, the splitting cannot be observed since it is averaged out because of the fast time dependences of the electronic spins. These time dependencies are the electronic exchange time with the time constant τ_e and the electronic spin-lattice relaxation with the time constant τ_1. If these fulfill Equation (19), which is usually the case,

$$\tau_e \ll 1/2\pi a \quad \text{and/or} \quad \tau_1 \ll 1/2\pi a \qquad (19)$$

one obtains a shift instead of a splitting

$$\delta v = v_p - v_d = a \cdot h v_S / 4kT \qquad (20)$$

Here v_p and v_d are the resonance frequencies of the nuclei in the paramagnetic radicals and in a diamagnetic molecule of a structure as similar as possible, and v_s is the resonance frequency of the electron spins[72].

It is convenient to define a paramagnetic shift

$$\delta_p(T) = \frac{v_p - v_d}{v_d} = \frac{a h v_S}{4kT v_d} = \frac{a h \gamma_S}{4kT \gamma_1} \qquad (21)$$

δ_p depends on the temperature because of the Boltzmann factor; at room temperature (295°K) one obtains an equation suitable for practical measurements:

$$a = C \cdot \delta_p \quad \text{with} \quad C = 3 \cdot 73 \cdot 10^{-2} \text{ MHz/ppm} \qquad (21a)$$

4.6.2 Neutral Radicals

From the NMR spectrum of *3,6-dimethyl-1,5-diphenyl-verdazyl* in di-tert.-butyl-nitroxide (Fig. 6)[73] the following conclusions can be drawn.

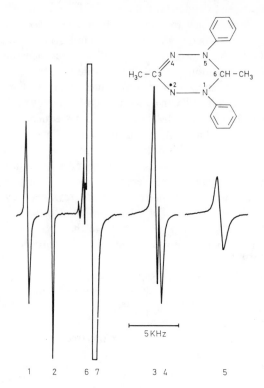

Fig. 6. NMR spectrum of *3,6-dimethyl-1,5-diphenyl-verdazyl* in di-tert.-butyl-nitroxide. Concentration $c \approx 1\ m$. Position of the lines:
$a_1(6—CH_3) = 2.34$ MHz, $a_2(m-C_6H_5) = 1.12$ MHz, $a_3(o-C_6H_5) = -3.17$ MHz, $a_4(p-C_6H_5) = -3.28$ MHz and $a_5(3—CH_3) = -5.58$ MHz.

In contrast to ESR and ENDOR the paramagnetic shift δ_p of the NMR provides directly the sign of a.
It is possible to measure very small coupling constants down to a few kHz.
The relation $\Delta v \alpha a^2$ for the linewidth is valid to a first approximation for coupling constants which are not too small. Because of this proportionality, the widths of the lines originating from larger coupling constants become so large that the limit detectability is reached. In this case the sensitivity can be increased considerably by using time averaging computers and Fourier spectroscopy (see 5.5:1.4). In principle, the situation should be more favorable for hetero nuclei with smaller magnetic moments (^{2}D, ^{13}C); however, application to these is just beginning.

[71] *J.S. Hyde, J.C.W. Chien, J.H. Freed,* J. Chem. Phys. *48,* 4211 (1968).

[72] *K.H. Hausser, H. Brunner, J.C. Jochims,* Mol. Phys. *10,* 253 (1966).

[73] *F.A. Neugebauer, H. Trischmann, G. Taigel,* Monatsh. Chem. *98,* 713 (1967).

[74] *E. de Boer, C. MacLean,* Mol. Phys. *9,* 191 (1965).

[75] *J. Bargon, H. Fischer,* Z. Naturforsch. *22a,* 1556 (1967;
R. Kapstein, J.L. Oosterhoff, Chem. Phys. Lett. *4,* 195 (1969);
H. Fischer, Chem. Phys. Lett. *4,* 611 (1970).

4.6.3 Radical Ions

Somewhat different conditions than with neutral radicals are encountered with radical ions which are produced primarily from aromatic molecules by reduction with alkali metals or by electrolysis. If the reduction is conducted only partially, and if, consequently, neutral molecules M as well as negative radical ions $\cdot M^-$ are present in solution, a chemical exchange takes place according to the equilibrium $M + \cdot M^- \rightleftharpoons \cdot M^- + M$. If we restrict ourselves to the simplest case of fast exchange, the conditions for which are $(a\pi\tau_p)^2 \ll 1$ and $\tau_p \ll \tau_1$ with $\tau_p =$ average lifetime of the paramagnetic ion, we obtain:

$$\delta_p(T, f_p) = \frac{a\,h\,\gamma_S}{4kT\gamma_1} \cdot f_P \qquad (22)$$

i.e. δ_p from Equation (21) must be multiplied by the fraction of the paramagnetic radical ions $f_p = \cdot M^- / \cdot M^- + \cdot M$ [74]. The linewidth is also proportional to f_p in this case

$$\Delta v = \pi \tau_p f_p a^2 \qquad (23)$$

Since both δ_p and Δv are proportional to f_p it is convenient to increase f_p continuously, starting from the NMR spectrum of the diamagnetic molecule. The result is shown in Fig. 7 with the example of *l-n-propyl-naphthalene* in tetrahydrofuran d_8, concentration $c = 1.15$ mol/l [74]. A very small radical concentration $f_p = 10^{-7}$ is sufficient to broaden considerably the NMR lines of the ring protons and of the CH_2-protons in α-position. If the radical concentration is markedly increased to $f_p = 8 \times 10^{-3}$, these lines disappear and one observes a broadening of the absorption lines of the β- and γ-protons as well as a measurable shift of the β-protons. With $f_p = 1/3$, finally, the line of the β-protons is very much broadened and strongly shifted to higher field (δ_p negative), while the line of the γ-protons is also markedly broadened and significantly shifted to low field (δ_p positive).

Such measurements of radical ions as well as those of neutral radicals provide directly the sign of a and also a relative size of the coupling constants of the different nuclei. However, for an absolute measurement of the value of a, f_p must be determined by an independent method, e.g. by the shift of the NMR line of the diamagnetic standard with in-creasing radical concentration due to a variation in the bulk susceptibility [74].

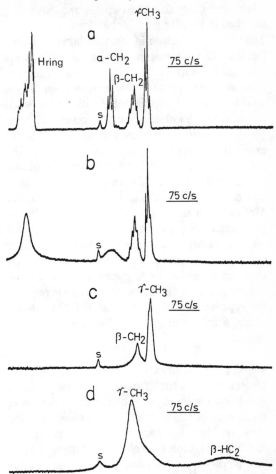

Fig. 7. NMR spectra of 1-n-propylnaphthalene in tetrahydrofuran-d_8 at 56.4 MHz and 20°. Conc. 1,15 mole/l, radical conc. a) O, b) $\leqslant 10^{-7}$ mole/l, c) $9 \cdot 10^{-3}$ mole/l, d) $3.8 \cdot 10^{-1}$ mole/l.

4.6.4 Detection of Radicals by Chemical Nuclear Polarization

During reactions with radical intermediates the nuclei are polarized by a dynamic process. This is detected by observing NMR emission lines of the diamagnetic reaction products [75]. The theory of this 'chemical nuclear polarization' is still being developed.

4.7 Radicals in Biological Systems

Because of its power as a specific and sensitive radical detector, ESR has become a well-established tool for studying molecules of biological interest [76]. The recently published monograph by Ingram [97] gives a critical survey of observations and interpretations so far.

[76] *G. Schoffa*, Chimia *20,* 165 (1966);
H. Beinert in *E.C. Slater*, Flavins and Flavoproteins, Elsevier Publ. Co., Amsterdam 1966;
A. San Pietro, A Symposium on Non-Haem Iron Proteins, Antioch Press, Yellow Springs, Ohio 1965; *P. Hemmerich*, Proc. Roy. Soc. *A302*, 335 (1968).

The technique of 'spin labeling' has been developed to enable ESR studies of large biomolecules. Spin labels often carry an NO group and are chemically attached to reactive sites of the biomolecule[77].

The unpaired electron in the spin label is mainly localized at the nitrogen atom of the NO group. The main feature of the ESR spectrum is, therefore, a triplet splitting. The environment of the label is of special interest, information of which is obtained by its influence on the triplet lineshape. The label performs hindered motions and, to the extent to which this motion does not average out anisotropies of the HFS and the g-value, information is obtained concerning the nature and geometry of the reactive site of the biomolecule. McConnell et al.[78] and Piette et al.[79] have already performed ESR studies of spin labeled *nucleic acids, proteins,* and *enzymes* in solution and in single crystals. ENDOR is very promising for studying biomolecules[80]. Because of its high resolving power, it should be possible to investigate the reactive site of a spin labeled biomolecule within a sphere of 10 Å radius about the NO-radical.

4.8 Triplets, Biradicals

Molecules with a triplet ground state are rare. The best-known is the O_2-molecule. Others have been discovered in recent years, e.g. *diphenylcarbene.* However, there are many molecules possessing an excited, metastable phosphorescent triplet state[81]. The dipole-dipole coupling between the two unpaired electrons of a triplet state causes a spectral splitting without an external magnetic field H_0. This is described by the spin Hamiltonian H_t (without nuclear interaction):

$$H_T = \gamma_S \hbar (\bar{S} \cdot \bar{H}_0) + D \cdot S_z^2 + E \cdot (S_x - S_y)^2 \quad (24)$$

For molecules with at least a threefold axis of symmetry $E = O$. Most D and E values measured in aromatic and heterocyclic molecules are about $0.1 \text{ cm}^{-1} \pm 1$ order of magnitude. In a magnetic field, the frequency of the ESR absorption line depends strongly on the orientation of the molecular axes with respect to the field. This causes a marked broadening of the ESR absorption lines in the case of statistical orientation in rigid solution. Hence the detection of triplet states by ESR spectroscopy is facilitated by doping the host lattice with oriented guest molecules; the first success was obtained with *naphthalene* in *durene*[82]. The less anisotropic $\Delta m = \pm 2$ transitions can also be detected easily in statistical orientation[83]. In the meantime ESR measurements of $\Delta m = \pm 1$ transitions of molecules in statistical orientation in glassy matrices have also been successful[84]. More recently, the ESR of freely mobile triplet excitons in pure molecular crystals was detected[85].

Molecules having two unpaired electrons with only little interaction between them are termed biradicals. It should be emphasized that the difference between triplet states and biradicals is not a qualitative but rather a quantitative one. If the D and E terms in the spin Hamiltonian (24) are only a small perturbation on the Zeeman terms, the compound behaves like a double monoradical. Consequently, the same molecule with small but not negligible D and E terms can behave in a low magnetic field like a triplet state, while in high field it behaves very approximately like a biradical; i.e. its ESR spectrum differs from the spectrum of a monoradical only by a somewhat larger linewidth, due to intramolecular dipole-dipole coupling.

Systems with more than two unpaired electrons also exist, namely *triradicals*[86], and *quintet states*[87].

[77] *O.H. Griffith, D.W. Cornell, H.M. McConnell,* J. Chem. Phys. *43,* 2909 (1965);
O.H. Griffith, H.M. McConnell, Proc. Nat. Acad. Sci. U.S. *55,* 8 (1966).

[78] *J.C.A. Boeyens, H.M. McConnell,* Proc. Nat. Acad. Sci. U.S. *56,* 22 (1966);
S. Ohnishi, J.C.A. Boeyens, H.M. McConnell, Proc. Nat. Acad. Sci. U.S. *56,* 809 (1966);
T.J. Stone et al., Proc. Nat. Acad. U.S. *54,* 1010 (1965);
H.M. McConnell, J.C.A. Boeyens, J. Phys. Chem. *71,* 12 (1967).

[79] *J.C. Hsia, L.H. Piette,* Arch. Biochem. Biophys. *129,* 269 (1969);
J.C. Hsia, D.J. Kosman, L.H. Piette, Biochem. Biophys. Res. Commun. *36,* 75 (1969).

[80] *J.S. Hyde* in *A. Ehrenberg, B.G. Malmström, T. Vänngard,* Magnetic Resonance in Biological Systems, Pergamon Press, Oxford 1967.

[81] *S.P. McGlynn, T. Azumi, M. Kinoshita,* The Triplet State, Prentice Hall, New Jersey 1969.

[82] *C. Hutchison, B.W. Mangum,* J. Chem. Phys. *29,* 952 (1958).

[83] *J.H. van der Waals, M.S. de Groot,* Mol. Phys. *2,* 333 (1959).

[84] *E. Wasserman, L.C. Snyder, W.A. Yager,* J. Chem. Phys. *41,* 1763 (1964).

[85] *M. Schwoerer, H.C. Wolf,* Mol. Cryst. *3,* 177 (1967);
D. Haarer, D. Schmid, H.C. Wolf, Phys. Status Solidi *23,* 633 (1967).

[86] *R. Kuhn, F.A. Neugebauer, H. Trischmann,* Monatsh. Chem. *97,* 525 (1966).

[87] *A.M. Trozzolo, R.W. Murray, G. Smolinsky, W.A. Yager, E. Wasserman,* J. Am. Chem. Soc. *85,* 2526 (1963); *89,* 5076 (1967).

4.9 Bibliography

[88] *M.T. Jones, W.D. Philips*, Ann. Rev. phys. Chem. 323 (1966).

[89] *A.H. Maki*, Ann. Rev. phys. Chem. 9 (1967).

[90] *A. Carrington, A.D. McLachlan*, Introduction to Magnetic Resonance, Harper and Row, New York 1967.

[91] *P.B. Ayscough*, Electron Spin Resonance in Chemistry, Methuen & Co., London 1967.

[92] *F. Gerson*, Hochauflösende ESR-Spektroskopie, Verlag Chemie, Weinheim/Bergstr. 1967.

[93] *Ch. Poole*, Electron Spin Resonance, A Comprehensive Treatise on Experimental Technique, Wiley Inc., New York 1967.

[94] *G. Vincow* in *E.T. Kaiser, L. Kevan*, Radical Ions, Interscience Publishers Inc., New York 1968.

[95] *A. Carrington, G.R. Luckhurst*, Ann. Rev. phys. Chem. 31 (1968).

[96] *M.C.R. Symons*, Ann. Rev. phys. Chem. 219 (1969).

[97] *D.J.E. Ingram*, Biological and Biochemical Applications of Electron Spin Resonance, Adam Hilger, London 1969.

[98] *K. Scheffler, H.B. Stegmann*, Elektronenspinresonanz, Springer Verlag, Berlin 1970.

5.5:5 Metal Complexes

Heimo Jürgen Keller and *Karl Eberhard Schwarzhans*

Anorganisch-Chemisches Institut der Universität, D-69 Heidelberg, and
Anorganisch-Chemisches Laboratorium der Technischen Universität, D-8000 München

This chapter gives an account of the special problems involved in investigating transition metal complexes by NMR (5.5:5.2–5.3) and ESR (5.5:5.4) respectively. Section 5.5:5.2 describes resonance experiments concerning the composition and structure of complex compounds without considering time dependent inter- and intramolecular interactions like conformation changes and exchange reactions, which are discussed in Section 5.5:5.3. The numerous publications dealing with magnetic resonance of diamagnetic (5.5:5.2.1) and paramagnetic (5.5:5.2.2) metal complexes are concerned with structure determination or analytical and kinetic problems.

5.1 Magnetic Resonance in Solutions

The major part of published NMR data (cf. Ref. [1, 2]) on metal complexes deals with experiments in solution. The strong dipole-dipole interactions between the magnetically active particles is averaged out by this procedure. The observed lines are sharp. Very small differences in chemical shifts or coupling constants can be resolved easily in the NMR spectra as well as small g-value variations and weak hyperfine couplings in ESR.

5.2 NMR Experiments of Time-independent Systems

5.2.1 Diamagnetic Metal Complexes

The magnetic properties and hence the positions of lines in the NMR spectra of these complexes are determined mainly by the ligands acting on the metal electrons. Depending on the strength of the metal ligand interactions, more or less of the metal electron density is delocalized over the ligand atoms. Some approximation rules for estimating ^{1}H-NMR parameters of complex compounds are given, since almost every characterization of a new complex is expressed in terms of its NMR parameters[3-7].

In a rough but very useful simplification it can be stated that protons surrounded by the highest electron density show the largest highfield shifts. For this reason ^{1}H-NMR absorptions of hydric protons usually appear in the range between $\tau = 12$ and 45 for transition metal hydrides[8, 9].

With decreasing metal to proton distance, the electron density in hydrogen functions usually increases. The transition metal hydride complexes are typical for this effect, which is, nonetheless, observed generally in transition metal compounds. The aforementioned highfield shift of protons near the metal decreases with increasing positive charge on the complex molecule. The converse is also true.

In the *monomethyl-nickel(II)-corrole*[10] (**1**) the methyl group signal of the unprotonated complex is observed at $\tau = 12.65$.

1

[1] Spectroscopic Properties of Inorganic and Organometallic Compounds, Vol. 1, The Chemical Society, London 1968.

[2] Anal. Chem. Annual. Rev. 1968.

[3] *H.J. Keller*, NMR-Basic Principles and Progress *2*, Springer Verlag, Berlin, Heidelberg, New York 1970.

[4] *D.R. Eaton* in Physical Methods in Advanced Inorganic Chemistry, Interscience Publishers, New York 1968.

[5] *E.F. Mooney, P.H. Winson*, Ann. Rev. NMR Spectr. *2*, 125, 153 (1969).

A drastic shift to $\tau = 7.6$ follows the protonation giving the compound just one positive charge. Besides the increased positive charge an attenuation of the ring current after protonation might be effective in lowering the chemical shift of the methyl protons.

The highfield shifts, usually observed on the approach of a proton to a transition metal ion, are markedly increased if paramagnetic contributions in the free ligand are quenched by complexation. This effect must be considered in π-complexes of aromatic compounds[11] or olefins[12] with a transition metal[11-14]. As a consequence of ring current attenuation and decreasing π-electron contribution the NMR signal of the ring protons in *dibenzenechromium* and related compounds are shifted 1.5–2.2 ppm to higher fields relative to the free ligand absorptions[11].

Appropriate simplifying but useful rules are also helpful in obtaining chemical information from the NMR coupling constants. The indirect proton-proton coupling in organic molecules generally increases with increasing s-character of the electrons forming the bond between the interacting nuclei[15, 16]. Based on this relation it is quite reasonable that internuclear couplings between protons bound to olefinic carbon atoms which are sp^2-hybridized are much larger than between protons coupled via sp^3-hybridized aliphatic carbon atoms.

The theoretical treatment of *proton proton interactions* within a coordinated ligand is almost the same as for the uncomplexed organic molecules[17, 18] and needs no further consideration. The most convenient way of obtaining chemically significant information is by comparison of the couplings of the free and coordinated ligand respectively[1, 13, 14].

The variations in coupling constants usually observed are as follows:

For *olefins* the coupling constants between protons in *cis-* or in *trans*-positions decrease remarkably. The same is true for vicinal hydrogen interactions in *oligoolefins*. Compounds of the *Zeise's Salt* type, e.g., show *cis-couplings* in the range of 8–9 Hz as compared with about 11–12 Hz in the free ligand[17, 18]. The *trans-coupling* constants are even more reduced, that is from 18 Hz in the uncoordinated to 13 Hz in the complexed olefin[12, 19].

The *vic-coupling* constants in coordinated *substituted 1,3-butadienes* fall within the range of 4–5 Hz, compared to the much larger values of about 10 Hz in the free ligand. By these means a reasonable estimate of the degree of π-electron withdrawal caused by the complexation can be obtained[1, 13, 14, 17, 18]. Further information might be gathered by analyzing the *vic-couplings* of oligoolefin metal compounds.

Conjugated oligoolefinic ligands can be considered either as aromatic or as multidentate olefinic ligands. In the latter model alternating *vic-*constants are to be expected while identical *vic-*couplings for all ligand protons support the former arrangement. In *cycloheptatriene-molybdenum-tricarbonyl* (2) the differing *vic-*coupling constants point to an oligoolefinic ligand configuration[20].

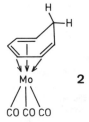

2

Protons of small π-bonded cyclic olefins usually show very small *vic-*couplings even for the free molecule. Instead of the expected values between 9 and 10 Hz only 2.7 Hz for *cyclobutene* and about 1 Hz for *3,3-dimethylcyclopropane* are observed[21]. More insight on the electronic structure can be gained by studying the couplings between ligand and metal nuclei which are directly bonded to each other[22-25, 24a].

[6] *W.G. Henderson, E.F. Mooney,* Ann. Rev. NMR Spectr. *2,* 219 (1969).

[7] *J.F. Nixon, A. Pidcock,* Ann. Rev. NMR Spectr. *2,* 346 (1969).

[8] *M.H.L. Green, D.J. Jones,* Adv. Inorg. Chem. Radiochem. *7,* 115 (1965).

[9] *A.P. Ginsberg,* Trans. Metal Chem. *1,* 111 (1965).

[10] *R. Grigg, R.W. Johnson, G. Shelton,* Chem. Comm. 1151 (1968).

[11] *C.G. Kreiter,* Dissertation, Universität München, 1964.

[12] *H.P. Fritz, K.E. Schwarzhans, D. Sellmann,* J. Organometal. Chem. *6,* 551 (1966).

[13] *M.L. Maddox, S.L. Stafford, H.D. Kaesz,* Adv. Organometal. Chem. *3,* 1 (1965).

[14] *E.O. Fischer, H. Werner,* Metall-π-Komplexe mit di- und oligoolefinischen Liganden, Verlag Chemie, Weinheim/Bergstr. 1963.

[15] *J.A. Pople,* Mol. Phys. *8,* 1 (1964).

[16] *M. Barfield, D.M. Grant,* Adv. Magn. Res. *1,* 149 (1965).

[17] *J.W. Emsley, J. Feeney, L.H. Sutcliffe,* High Resolution NMR-Spectroscopy, Pergamon Press, Oxford 1965.

[18] *A.A. Bothner—By,* Adv. Magn. Res. *1,* 195 (1965).

[19] *T. Kinugasa, M. Nakamura, H. Yamada, R. Saika,* Inorg. Chem. *7,* 2649 (1968).

[20] *H. Günther, W. Grimme,* Angew. Chem. *78,* 1063, Engl. *5,* 1043 (1966).

[21] *H.G. Preston, J.C. Davis,* J. Am. Chem. Soc. *88,* 1585 (1966).

A very remarkable example of couplings between hydrido protons and metal nuclei of the central atom and other ligand atoms respectively is given in Fig. 1 which shows the ^{1}H-NMR spectrum of the complex *hydrido-tetrakis(trifluorophosphino)rhodium(I)*[23].

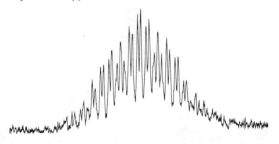

Fig. 1. ^{1}H–NMR-spectrum of complex $RhH(PF_3)_4$

The spectrum consists of 130 symmetric lines caused by couplings between the ^{103}Rh nucleus with $I = 1/2[J(^{103}Rh—^1H) = 5.8$ Hz], the phosphorus nuclei with $I = 1/2[^2J(^{31}P—^1H) = 57.4$ Hz], and finally the fluorine nuclei having also $I = 1/2[^3J(^{19}F—^1H) = 16.6$ Hz]. This gives evidence for the equivalence of the four trifluorophosphine ligands since all the couplings with the hydrido proton are equivalent.

A particularly clear and instructive example is the discussion of ^{31}P—^{1}H-interactions in phosphino compounds, e.g., *bis(phosphino)-chromium tetracarbonyl*[24] or a variety of *methyldiphenylphosphino-metal complexes*[26]. In hydrido complexes of lower symmetry the relative position of phosphino and hydrido ligands respectively can be evaluated. Usually the ^{31}P—^{1}H-couplings in *trans* complexes fall between 80 and 100 Hz while couplings in *cis* complexes are between 10 and 25 Hz. The significance of ^{13}C—^{1}H-constants in metalorganic compounds was pointed out[5]. In some favorable cases even the spin-spin interaction between ligand protons and magnetically active nuclei of the central metal atom (e.g. ^{103}rhodium and ^{195}platinum each with $I=1/2$) can be resolved[12,25].

To complete the list of feasible couplings in metal complexes the interactions between heteronuclei of the same and of different ligands (e.g. ^{19}F—^{31}P-couplings) are quoted[1].

5.2.2 Paramagnetic Metal Complexes

The resolution of NMR-spectra of paramagnetic compounds is usually much less satisfactory than in comparable diamagnetic substances[27]. In only a very few cases, e.g. the *bis(1-amino-1,3,5-heptatriene-7-iminato)nickel(II)* compounds, the spin-spin couplings between ligand protons can be resolved[28,29] (cf. 5.5:1.1.3). For many other paramagnetic complexes NMR-spectra are unobservable, because of the enormous line broadening caused by the action of the unpaired electron spins. The line width at half height of the ^{1}H-NMR-absorptions in solutions of paramagnetic complexes ranges between 2 Hz[28] and several 1000 Hertz.

Besides line broadening, one observes chemical shifts exceeding those of diamagnetic compounds by several ranges of magnitude[29]. These are usually referred to as 'Knight' or 'contact' shifts indicating the relation to the large chemical shifts observed in pure metals. The large differences between chemical shifts of diamagnetic and paramagnetic compounds correspond to the changes in susceptibilities.

Table 1. ^{1}H–NMR shifts of paramagnetic *bis(π-cyclopentadienyl)* metal complexes relative to the standard 1.1′–dimethylferrocene. Positive signs indicate shifts to higher fields.

π-complex	shifts[ppm]: ring	—CH$_3$
V $(π—C_5H_5)_2$	—307.5	
V $(π—C_5H_4—CH_3)_2$[33]	—337	—120
Cr $(π—C_5H_5)_2$	—314.1	
Cr $(π—C_5H_4—CH_3)_2$	—364 —310	— 36
Co $(π—C_5H_5)_2$	+ 53.8	
Co $(π—C_5H_4—CH_3)_2$	+ 72.3 + 49	— 10.4
Ni $(π—C_5H_5)_2$	+254.8	
Ni $(π—C_5H_4—CH_3)_2$	+253.8	—200.5

[22] *H.P. Fritz, K.E. Schwarzhans*, J. Organometal. Chem. *5*, 283 (1966).

[23] *Th. Kruck*, Angew. Chem. *79*, 27, Engl. *6*, 53 (1967).

[24] *E. Moser, E.O. Fischer*, J. Organometal. Chem. *15*, 147 (1968).

[24a] *A.J. Cheney, B.E. Mann, B.L. Shaw*, Chem. Comm. 431 (1971);
L.F. Farnell, E.W. Randall, E. Rosenberg, Chem. Comm. 1078 (1971).

[25] *P.D. Kaplan, M. Orchin*, Inorg. Chem. *6*, 1096 (1967).

[26] *J.M. Jenkins, B.L. Shaw*, J. Chem. Soc. (A), 1711 (1966).

[27] *H.J. Keller, K.E. Schwarzhans*, Angew. Chem. *82*, 227, Engl. *9*, 196 (1970).

[28] *R.E. Benson, D.R. Eaton, A.D. Josey, W.D. Phillips*, J. Am. Chem. Soc. *83*, 3714 (1961).

[29] *D.R. Eaton, A.D. Josey, W.D. Phillips, R.E. Benson*, Discuss. Faraday Soc. *34*, 77 (1962).

[30] *H.M. McConnell, C.H. Holm*, J. Chem. Phys. *28*, 749 (1958).

The paramagnetic complexes to be studied first by NMR methods in 1957, were the *biscyclopentadienyl metal complexes* (Table 1)[30]. Most surprising and formerly not well understood was the fact that some of the signals, e.g. for $(\pi\text{-}C_5H_5)_2Ni$ and $(\pi\text{-}C_5H_5)_2Co$, appear at the high field side of the ferrocene protons while some others, e.g. $(\pi\text{-}C_5H_5)_2Cr$ and $(\pi\text{-}C_5H_5)_2V$, are at lower fields[31]. In regard to the relation between spin densities and contact shifts this is an evident proof that the ring protons of *bis(cyclopentadienyl)-vanadium* and *-chromium* 'feel' positive spin densities, while the corresponding nuclei in *bis(cyclopentadienyl)-cobalt(II)* and *bis(cyclopentadienyl)-nickel(II)* experience negative spin densities (definitions see Section 5.5:1.1). This result caused discussion concerning the electronic structure of these species and especially the nature of the metal to ligand bond.

Using the familiar model of spin delocalization and configurational interaction, overlap of metal and ligand orbitals should lead to a positive spin density in p-carbon orbitals. This is carried over to hydrogen functions by configurational interaction, but the transferred spin density is polarized strongly by the bonding electrons in the C—H[31] bond giving negative spin density at the proton site. Based on this concept the ^{1}H-NMR signals are to be expected at higher fields relative to the diamagnetic ferrocene.

Though much work was done to explain the discrepancy of experimental and theoretical results including investigations of several substituted *biscyclopentadienylmetal compounds,* an unambiguous answer has not been given in our opinion[32-36]. Serious objections can be raised against each of the numerous interpretations of ^{1}H-NMR spectra. The problem is, however, inherent in all discussions of contact shifts. Nevertheless, it seems reasonable to analyze at least the most important aspects.

Even if the Fermi contact interaction (definition see Section 5.5:1.1.3) is assumed to be solely responsible for the observed shifts- an assumption which might be valid for *biscyclopentadienyl-metal complexes* – there are at least three different mechanisms by which positive or negative spin

density in metal orbitals may reach the ligand nuclei.

Unpaired metal d-electrons fill up antibonding and empty ligand orbitals having either σ or π-character $(Ti^{3\oplus}, V^{3\oplus})$[37].

The electrons in the highest occupied energy orbitals of the ligand are transferred totally or partially to the empty metal-d-orbitals.

Metal d-orbitals overlap with ligand orbitals of sterically preferred atoms which are not bonded to the central metal ion in the 'classical' sense of an electron pair bond *(direct transfer).*

The first question to be considered concerns the relationship connecting the experimentally observed shifts with one of these three spin transfer mechanisms. Spin transfer from metal to ligand orbitals takes place. This is expected mainly in complexes of *titanium(III)* and *vanadium(III)*[37] whose d-states have relatively high energies.

The direction of observed shifts depends above all on the character of the antibonding ligand orbital. Positive spin density over the ligand will result if the lowest antibonding orbital has σ-character. The spin density decreases rapidly with increasing distance from the metal ion. The NMR spectra are characterized by remarkably large contact shifts to lower field for nuclei next to the metal while the more distant protons show only small downfield shifts. This type of spectra was observed for several *nickel(II)* and *cobalt(II) complexes* with the ligands *pyridine, pyridine-N-oxide* and other *substituted pyridine derivatives* giving strong evidence for σ-delocalization in these compounds[38-40].

If, on the other hand, the lowest antibonding ligand orbital has π-character, positive spin density is found in carbon p-orbitals of even-numbered conjugated π-systems, while alternating positive and negative spin density is observed in carbon orbitals of odd hydrocarbon ligands[41]. As pointed out before, the spin density in these p-carbon orbitals is transferred to the hydrogen atoms but strongly polarized, leading to negative spin density for protons in even numbered aromatic ligands and to alternating negative and positive spin density for protons in odd aromatic hydrocarbons. The expected alternating spin density has been observed recently in a great number of transition metal complexes with aromatic ligands, e.g. in *bis(triarylphosphino)nickel(II)-dihalogenides* and *bis(amino-1,3,5-heptatriene-7-iminato)nickel(II) compounds*[41].

With increasing atomic number of the central metal ion the energy of the d-functions decreases. This is why the latter ions of the first transition series preferably take over electrons from bonding ligand molecular orbitals.

[31] *H.M. McConnell, D.B. Chesnut,* J. Chem. Phys. *28,* 107 (1958).

[32] *H.P. Fritz, H.J. Keller, K.E. Schwarzhans,* J. Organometal. Chem. *13,* 505 (1968).

[33] *M.F. Rettig, R.S. Drago,* Chem. Comm. *1966,* 891.

[34] *D.A. Levy, L.E. Orgel,* Mol. Phys. *3,* 583 (1961).

[35] *M.R. Rettig, R.S. Drago,* J. Am. Chem. Soc. *91,* 1361, 3432 (1969).

[36] *S.E. Anderson, R.S. Drago,* J. Am. Chem. Soc. *91,* 3656 (1969).

[37] *D.R. Eaton, W.R. McClellan, J.F. Weiher,* Inorg. Chem. *7,* 2040 (1968).

[38] *J. Happe, R.C. Ward,* J. Chem. Phys. *39,* 1211 (1963).

[39] *R.H. Holm, G.W. Everett, W.DeW. Horrocks,* J. Am. Chem. Soc. *88,* 1071 (1966).

[40] *G.N. La Mar, G.R. Van Hecke,* J. Am. Chem. Soc. *91,* 3442, (1969).

[41] *D.R. Eaton, W.D. Phillips,* Adv. Magn. Res. *1,* 103 (1965).

If the unpaired spins in the partially filled metal orbitals align parallel to the external magnetic field, the transfer of ligand spin density with negative polarization is energetically preferred, leaving positive spin density in ligand functions. Evidently both mechanisms lead to positive spin densities in ligand orbitals, differing only in the amount of spin density at different ligand sites and not in their sign. A decision as to which mechanism is operating can be made by calculating theoretical spin densities and comparing the results with the experimental data.

The mechanism of 'direct transfer' results in the characteristics of σ-delocalization. The protons next to the metal ion show large low field shifts caused by the preferred overlap between metal and hydrogen orbitals. On the other hand, the more distant hydrogens are influenced only slightly by the unpaired metal electrons, discernible by small low field shifts.

Without certainty regarding the mechanism by which[35, 37, 42–46] the spin is transferred from metal to ligand orbitals and vice versa, information concerning the character of the metal-ligand bond cannot be obtained.

The interpretation of NMR results in paramagnetic compounds is more difficult, since the shifts are influenced by intramolecular dipole-dipole interactions (pseudo contact interactions) which are not averaged to zero by the tumbling motion in solution. Direction and extent of chemical shifts are influenced in a rather obscure way by this effect. There seems to be only one method to assess the exact amount of pseudo contact contribution; this is to record ESR spectra at low temperature[47].

5.3 Investigation of Time-dependent Systems

Magnetic resonance methods are especially well adapted to investigation of time-dependent processes. Thus, the activation energy and the velocity of the rotational motion involving the C—N-bond in *dimethylformamide*[48] was determined. Rapid changes of the conformation in organic radicals were elucidated by ESR[49–52], in

some diamagnetic and paramagnetic metal complexes by NMR spectroscopy[53, 54]. Finally, the kinetics of ligand substitution processes were studied[55, 56].

5.3.1 Intramolecular Reorientations

Rotations of distinct molecular groups around a common bond axis fall among these reactions, e.g., the already-mentioned motion of the dimethylamine group around the C—N-bond in *dimethylformamide* or *dimethylacetamide*. Rotations involving σ-bonds are easily excited, so that almost free rotation around those axes occurs even at room temperature. The activation energy for the rotation is enhanced considerably if additional π-bonding is taken into consideration. The favorable energetic position of p- and d-orbitals in transition metal complexes makes it highly probable that, apart from the essential σ-bonds, additional π-bonding between the ligands and the metal ion occurs. These contributions prevent free rotation about the metal ligand axis. Metal complexes bearing ligands of lower symmetry can even be isolated as geometric isomers if the activation energy for free rotation appreciably surmounts the lattice energy at room temperature. For all compounds having activation energies in the temperature range which is accessible to NMR experiments, the exact activation energy as well as the kinetic pathway of the reaction can be ascertained by following the temperature dependence of the NMR spectra (cf. 5.5:2.6).

By this method the activation energy for rotation around the metal olefin axis, e.g., in π-bonded *(acetylacetonato)(olefin)platinum(II)halogenide complexes* and in corresponding *ethylene-rhodium compounds*, was in the range of 10 to 15 kcal/mole[57, 57a].

The described method was, furthermore, used to elucidate the structure of coordination com-

[42] *H.-P. Fritz, H.J.Keller, K.E.Schwarzhans*, Z. Naturforsch. *23b*, 298 (1968).

[43] *G.N. La Mar, L.Sacconi*, J. Am. Chem. Soc. *90*, 7216 (1968).

[44] *J.R.Hutchison, G.N.La Mar, W.DeW.Horrocks*, Inorg. Chem. *8*, 126 (1969).

[45] *F.Röhrscheid, R.E.Ernst, R.H.Holm*, J. Am. Chem. Soc. *89*, 6472 (1967).

[46] *H.P.Fritz, W.Gretner, H.J.Keller, K.E.Schwarzhans*, Z. Naturforsch. *23b*, 906 (1968).

[47] *J.P.Jesson*, J. Chem. Phys. *47*, 579, 582 (1967).

[48] *W.D.Phillips*, J. Chem. Phys. *23*, 1363 (1955).

[49] *E. de Boer, A.D.Praat*, Mol. Phys. *8*, 291 (1964).

[50] *E. de Boer, E.L.Mackor*, Proc. Chem. Soc. *23* (1963); J. Am. Chem. Soc. *86*, 1513 (1964).

[51] *J.H. Freed*, J. Chem. Phys. *37*, 1156 (1962); *37*, 1881 (1962); *41*,699 (1964).

[52] *A. Carrington*, Mol. Phys. *5*, 425 (1962).

[53] *D.R. Eaton, A.D. Josey, W.D. Phillips, R.E. Benson*, Discuss. Faraday Soc. *34*, 77 (1962).

[54] *D.R.Eaton, A.D.Josey, W.D.Phillips, R.E.Benson*, J. Chem. Phys. *37*, 347 (1962).

[55] *H.H. Glaeser, H.W. Dodgen, J.P. Hunt*, Inorg. Chem. *4*, 1061 (1965); *T.R. Stengle, C.H. Langford*, Coord. Chem. Rev. *2*, 349 (1967).

[56] *J. Charvolin, P. Rigny*, J. Magn. Res. *4*, 40 (1971).

[57] *C.E. Holloway, G. Hulley, B.F.G. Johnson, J. Lewis*, J. Chem. Soc. (A), 53 (1969).

[57a] *K. Vrieze, P.W.N.M. van Leeuwen*, Progr. Inorg. Chem. *14*, 1 (1971).

[58] *F. Lux, G. Wirth, K.W. Bagnall, D. Brown*, Z. Naturforsch. *24b*, 214 (1969).

pounds. One fairly typical example is *dimethyl-acetamide* with two donor atoms per molecule. It can interact with metal ions of *uranium tetrabromide*[58] by coordinating either the nitrogen or the oxygen atom.

In this adduct the dimethylamino group rotates freely at about 50° compared to a value of 49.8° in the free ligand[59]. The same energy barrier to rotation of about 12 kcal/mole in the free and complexed ligands makes a coordination of the nitrogen as the donor atom very unlikely.

3

The effectiveness of NMR experiments in elucidating complex conformation was impressively demonstrated by the excellent studies on *bis-(amino-1,3,5-heptatriene-7-iminato)nickel(II)complexes*[53, 54] (**3**). They show ^{1}H-NMR spectra with exceptionally small line widths, though in solution a paramagnetic, tetrahedrally coordinated conformer is present in the equilibrium. The related tetrahedrally coordinated *bis(triarylphosphino)nickel(II)-dihalogenide* compounds, with no change in conformation, have comparatively large linewidth, suggesting that the extremely narrow lines of the compounds (**3**) are caused mainly by the participation of a diamagnetic conformer in the rearrangement equilibrium which proceeds at least with a conversion frequency of 10^9 Hz.

These are the only known paramagnetic transition metal complexes for which a pure π-delocalization mechanism has been proved, enabling an accurate characterization of the metal to ligand bond[37, 41].

5.3.2 Intermolecular Reactions

Though the NMR spectra in pure diamagnetic systems are not changed by time-dependent processes to the same extent as those of paramagnetic systems, numerous investigations on *solvation reactions* have been carried out[60-70]. Should the exchange between coordinated and free solvent molecules be slow compared to the NMR time scale (Section 5.5:1.2), different signals for each are observed. By weighing the intensities of these NMR absorptions a quantitative determination of the amount of coordinated solvent molecules can be obtained. The stability of the identified more or less stable complexes depends on the coordinating ability of the solvent molecules and on the charge and size of dissolved metal ions. By this method a scale of the coordinating power of ligands as well as of the acceptor capacity of metal ions may be attained.

If the exchange between coordinated and free solvent molecules is very fast on the NMR time scale, a single time averaged signal shows up. Kinetic data may then be obtained by following the temperature dependence of the spectra especially the line width of absorptions.

^{1}H and ^{17}O investigations have been carried out with great success. While ^{17}O-*resonances* have the advantage of very large chemical shifts between coordinated and free nuclei resulting in separated signals even in rapidly exchanging systems[71], the ^{1}H-method is much easier to handle and much cheaper. Despite the small differences in chemical shifts between free and complexed protons, most investigations use protons as probes for this reason.

The range of reasonable ^{1}H-*NMR experiments* can be expanded by lowering the chemical exchange rate. This can be achieved either by decreasing the sample temperature or by diluting the donor solvent with other less coordinating liquids. The method of low temperature investigations was used with excellent results during studies of aqueous solutions containing the ions *gallium(III), aluminium(III)* and *magnesium(II)*[69, 70].

The second procedure was applied by using acetone or dioxane as diluting agents. Recent findings even suggest that these molecules coordinate extensively only when insufficient amounts of water are available to complex all of the metal ions[69].

In solutions containing stronger donors a competition reaction may occur. In solutions of aluminum(III) ions in *water/dimethylacetamide* four

[59] *H.S. Gutowsky, C.H. Holm*, J. Chem. Phys. *25*, 1228 (1956).

[60] *J.A. Jackson*, J. Chem. Phys. *32*, 553 (1960).

[61] *R.E. Connick, D.N. Fiat*, J. Chem. Phys. *39*, 1349.

[62] *J.H. Swinehart, H. Taube*, J. Chem. Phys. *37*, 1579.

[63] *S. Nakamura, S. Meibomm*, J. Am. Chem. Soc. 89, 1765 (1967).

[64] *A. Fratiello, R.E. Lee, V.M. Nishida, R.E. Schuster*, Chem. Comm. 173 (1968).

[65] *R.E. Schuster, A. Fratiello*, J. Chem. Phys. *47*, 1554 (1967).

[66] *A. Fratiello, D.P. Miller, R.E. Schuster*, Mol. Phys. *12*, 111 (1967).

[67] *A. Takahashi*, J. Phys. Soc. Jap. *24*, 657 (1968).

[68] *P.A. Tennussi, F. Quadrifoglio*, Chem. Comm. 844 (1968).

[69] *A. Fratiello, R.E. Lee, R.E. Schuster*, Chem. Comm. 37 (1969).

[70] *T.D. Alger*, J. Am. Chem. Soc. *91*, 2220 (1969).

[71] *G.E. Glass, W.B. Schwabacher, R.S. Tobias*, Inorg. Chem. 7, 2471 (1968).

signals belonging to complexed and free donor molecules can be observed. It must be assumed that both molecules are involved in exchange reactions[72].

The type of information on the kinetics of exchange reactions, which can be gained by analyzing the temperature and concentration dependence of the ^{1}H-NMR spectra, is demonstrated by the study of some *zirconium, hafnium* and *thorium chelates*[73] and of the reactions between paramagnetic complexes like *bis(salicyclaldehydato)cobalt(II), -nickel(II)* and *-iron(II)* with pyridine or various picolines as donors[74]. The results of ^{1}H- and ^{14}N-NMR experiments using the latter species prove that the reaction rate decreases remarkably when changing from the iron(II) to the nickel(II) complexes. This trend can be observed by studying rates of substitution reactions in octahedrally coordinated compounds[75, 76].

5.4 ESR Investigations

The ESR method has lost much of its significance since the advent of the NMR method at least so far as metal complexes are concerned. ESR-Spectroscopy is restricted to complexes which can not be dissolved sufficiently and/or do not have adequate electronic relaxation times (cf. Section 5.5:2.3). The alkali metal complexes possessing varying types of aromatic molecules as electron acceptors have been the subject of intensive study by ESR[77-79]. In these compounds free spin density is transferred from the metal atom into antibonding molecular orbitals of the organic system. The observed hyperfine splittings reflect the delocalization of the unpaired spin over the aromatic molecule and give a hint of the probability distribution of the lowest lying antibonding molecular orbitals.

The substitution of the alkali metal atoms by transition metal ions results in a marked decrease of the electron spin relaxation time, brought about by the inherent orbital angular momentum of transition metal ions, and leading finally to an increased line width. Particularly for protons having small spin densities and hence only small hyperfine interactions with the unpaired electron spin, the resolution of ESR splittings is not achieved further. Information on the spin distribution is lost for this reason. There are only a few favored complexes in which the spin density at the central metal nucleus and at the nearest directly bonded ligand atoms may be evaluated, e.g. the ^{95}Mo-, ^{97}Mo- and ^{13}C-hyperfine coupling constants in the *octacyanomolybdenum(V)* ion[80] or the ^{53}Cr- and ^{14}N-splittings of *nitroso-pentammin-chromium-(III)-dichloride*[81].

Only a few paramagnetic organometallics are exceptions to this. They possess practically no orbital momentum contribution and exhibit narrow lines in the spectra of dilute solutions, e.g. the sandwich type compounds *bis(π-cyclopentadienyl)vanadium*[82], or *bis(π-benzene)chromium cation*[83, 84] which show 11 and 13 lines respectively caused by the hyperfine interaction of the unpaired electron with 10 and 12 ring protons respectively. Exceedingly large HFS splittings stemming from ligand protons are observed in the ESR spectra of the anion *dihydrido-bis(π-cyclopentadienyl)-titanium(III)* (4). This compound can coordinate molecular nitrogen and contains two hydridic protons directly bonded to the metal atom[85].

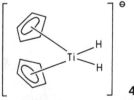

4

Provided that both the starting material and the product of a chemical reaction are paramagnetic and supposing further that their ESR spectra differ markedly from each other, i.e. their g-values and

[72] *R.E. Lee, A. Fratiello, V.M. Nishida, R.E. Schuster*, Inorg. Chem. *8*, 69 (1969);
S.F. Lincoln, Coord. Chem. Rev. *6*, 309 (1917);
J.K.M. Sanders, D.H. Williams, J. Am. Chem. Soc. *93*, 641 (1971);
P.V. Demarco, T.K. Elzey, R.B. Lewis, E. Wenkert, J. Am. Chem. Soc. *92*, 5734, 5737 (1970).
[73] *A.C. Adams, E.M. Larsen*, Inorg. Chem. *5*, 814 (1966).
[74] *W. Gretner*, Dissertation, TH München 1969.
[75] *F. Basolo, R.G. Pearson*, Mechanisms of Inorganic Reactions, J. Wiley and Sons, New York 1967.
[76] *M. Eigen, K. Tamm*, Z. Elektrochem. *66*, 107 (1962).
[77] *A. Carrington, A.D. McLachlan*, Introduction to Magnetic Resonance, Harper & Row, New York 1967.
[78] *P.B. Ayscough*, ESR in Chemistry, Methuen & Co., London 1967.
[79] *K.W. Bowers*, Adv. Magn. Res. *1*, 317 (1965).

[80] *B.R. McGarvey*, Trans. Metal. Chem. *3*, 89 (1966).
[81] *P.T. Manoharan*, J. Am. Chem. Soc. *89*, 4564 (1967).
[82] *R. Prins, J.D.W. van Voorst*, J. Chem. Phys. *46*, 1216 (1967).
[83] *K.H. Hausser*, Naturwissenschaften *48*, 426, 666 (1961).
[84] *R. Prins, F.J. Reinders*, Chem. Phys. Lett. *3*, 45 (1969).
[85] *G. Henrici-Olive, S. Olive*, Angew. Chem. *80*, 398, Engl. *7*, 386 (1968).

hyperfine splittings are quite different, the ESR technique can be used to study the kinetics of this reaction. In such cases the appearance and enhancement of new signals provide a direct and quantitative measure of the increasing concentration of the reaction product, as has be observed with some *tetrachloro-dialkoxy-molybdenum(V)* compounds[86] or in reactions of some 'Ziegler's type' catalysts[87].

Finally data on reaction kinetics are provided by another parameter, namely, the line width of the signals. This technique has been applied to the study of the reduction of aqueous *manganese(II)* solutions mixed with varying amounts of *dithionite*[88].

5.5 Magnetic Resonance in the Solid State

The NMR and ESR spectra of solids differ in two aspects from those in solutions discussed thus far.

They provide information concerning possible *anisotropic magnetic properties* expressed in parameters like g-value, chemical shifts, or hyperfine tensors and throw light on the electronic as well as the geometric structure of complexes.

Resonance spectra are characterized by *dipole-dipole interactions* between the sterically fixed magnetic particles of the sample. The strength of this interaction depends on the relative position of the dipoles giving valuable structural information[89, 90].

ESR investigations of solids are more significant than NMR experiments which give more physically interesting results. Because of the more elaborate instrumentation required for NMR, ist use is less practical.

5.5.1 ESR Measurements

ESR investigations ar usually carried out on magnetically dilute samples. This is done because of the strong intermolecular exchange interaction between adjacent electron spins. For crystals the corresponding diamagnetic compounds crystallizing isomorphic with the investigated paramagnetic complex are especially well suited as diluting agents. Planar, four coordinated *copper(II) complexes* containing the paramagnetic copper(II) ion, e.g., are favorably substituted into the lattice of the related diamagnetic nickel(II) compound. No sub-

stantial change in crystallographic structure is noted. Maki, using this technique, succeeded in resolving even the hyperfine structure caused by the remote aldehydic protons in *bis(salicylaldiminato)copper(II)* mixed by a ratio of 1:100 with the diamagnetic nickel(II) compound[91].

Based on these results, a calculation of spin densities at various points of the organic ligand was accomplished. Since, depending on the strength of the metal ligand bond, more or less of the metal electron spin is transferred into ligand orbitals, the covalency of the metal to ligand bond can be estimated using these results.

The character of the metal to ligand bond can possibly also be determined by analyzing the anisotropy of the g-value, but the necessary calculations are very tedious and extensive[92].

5.5.2 NMR Spectra of Solids

The dipole-dipole interaction between the two protons of crystal water leads to a splitting of the otherwise single ^{1}H-absorption. The amount of splitting depends on the orientation of the proton-proton vector (p—p) relative to the direction of the external field H_o. It can be evaluated from Equation (1)

$$\Delta H = 3\mu r^{-3}(3\cos^2\theta - 1) \tag{1}$$

r is the distance between the interacting protons and θ the angle between their connecting line and the direction of the external field H_0[93-95]. The relative and absolute positions of protons in the crystal lattice may be obtained by this procedure. The results are of special importance, since the X-ray method works unsatisfactorily in evaluating proton-proton distances in crystallized transition metal compounds.

Furthermore, the complex symmetry can be elucidated by using solid state NMR. Exceptionally well suited for this purpose are nuclei of heavier elements like ^{19}F, since for these nuclei only small differences in environment show up in large chemical shift differences. This implies resolved NMR lines for each kind of chemically different nucleus. Hence, this technique allows the determination of even small departures from regular

[86] *D.A. McClung*, Inorg. Chem. *5*, 1958 (1966).

[87] *G. Henrici-Olive, S. Olive*, Angew. Chem. *80*, 796, Engl. *7*, 822 (1968).

[88] *L. Burlamacchi, E. Tiezzi*, J. Mol. Structure *2*, 261 (1968).

[89] *E.R. Andrew*, Nuclear Magnetic Resonance Cambridge University Press 1958.

[90] *W.Low*, Paramagnetic Resonance in Solids, Academic Press, New York 1960.

[91] *A.H. Maki, B.R. McGarvey*, J. Chem. Phys. *29*, 31, 35 (1958).

[92] *B.R. McGarvey*, Trans. Metal. Chem. *3*, 113 (1966).

[93] *G.E. Pake*, J. Chem. Phys. *16*, 327 (1948).

[94] *L.W. Reeves*, Progr. Nucl. Magn. Res. Spectrosc. *4*, 193 (1969).

[95] *H. Kiriyama*, Bull. Chem. Soc. Jap. *35*, 1205 (1962).

[96] *R. Blinc, E. Perkmajer, I. Zupanicic, P. Rigny*, J. Chem. Phys. *43*, 3417, (1965).

ligand arrangements using polycrystalline samples. The splittings in the ^{19}F-NMR-spectra of some hexafluoro compounds were taken as evidence of a tetragonal complex distortion resulting in unequivalent fluoride ligands in *molybdenum, tungsten* and *osmiumhexafluoride*[96, 97]. The same deviation of octahedral symmetry was found in the paramagnetic *dipotassium-sodium-hexafluorotitanate*[98].

The most reliable data on the electronic structure of complexes should be obtainable from resonances of the central metal nuclei. Only a few magnetically active metal nuclei are suited for such experiments, e.g., the ^{59}Co isotope in *cobalt(III) complexes*. The angular dependance of chemical shifts reflects the symmetry of ligand arrangement as well as the strength of the metalligand interaction characteristic of this bond[99].

Some kinetically significant data show that time-dependent reactions, like diffusion processes or hindered rotation of molecular subgroups around a common bond-axis, occur in the solid as well. As for solutions the free rotation about bonds is often hindered heavily by additional π-and δ-bonding. The observed temperature-dependence of the ^{1}H-NMR line widths in solid compounds of the type *bis(cyclopentadienyl)-metal(II)* seems to be an exception to this rule. The rotation about the metalring-axis, which is frozen out at about 80 °K, can be prevented further by substituting bulky groups to the ring. The strong effect on the rotational degrees of freedom shows up in the NMR spectra, but it seems to be a packing effect rather than an effect of various π- and δ-contributions to the metal-ligand bond.

[97] R. Blinc, E. Shomik, I. Zupanicic, J. Chem. Phys. 45, 1488 (1966).

[98] P. Burkert, H. P. Fritz, G. Stefaniak, Z. Naturforsch. 23b, 872 (1968).

[99] H. W. Spiess, H. Haas, H. Hartmann, J. Chem. Phys. 50, 3057 (1969).

5.6 Bibliography

J. W. Emsley, J. Feeney, L. H. Sutcliffe[17], 1965.

A. Carrington, A. D. MacLachlan[77], 1967.

H. J. Keller[3], 1970.

R. R. Ernst, J. Magn. Res. 3, 10 (1970).

P. J. Banney, D. C. McWilliam, P. R. Wells, J. Magn. Res. 2, 235 (1970).

5.5: 6   Direct Nuclear Quadrupole Resonance Spectroscopy

J. E. Todd

Department of Chemistry, Cincinnati/Ohio, 45221/USA

Isotopes of elements whose nuclear spins, I, are unity or greater have electric quadrupole moments which interact with bonding electrons to produce a finite number of preferred orientations of the nuclei with respect to fixed crystalline axes. The energies of the nuclei are quantized and directly related to both the covalent characters of bonds with neighbors and (to a lesser extent) the deviation of the bond from cylindrical symmetry.

Transitions between orientational states may be observed indirectly by nuclear magnetic resonance in solids, and by microwave spectroscopy in gases. Direct measurement of the transition frequencies may be observed in the solid state by means of apparatus closely resembling nuclear magnetic resonance spectrometers, though a magnet is not required.

The theory of nuclear quadrupole interactions with bonding electrons and the temperature dependence of the observed transition frequencies is discussed (5.5:6.1). A description of the experimental apparatus (5.5:6.2) follows with some discussion of conditions required for observation of spectra. Experimental results of spectra are reproduced in tables 1—3 with empirical interpretations (5.5:6.3).

6.1 Principles

As discussed in chapter 5.5:1 all nuclei with spin - $I \geqq 1$ may possess a quadrupole moment which can interact with the electrons in a molecule to produce nuclear energy levels characteristic of the molecule. Transitions between these levels are observed in the radiofrequency region of the electromagnetic spectrum. Isotopes of many elements may be studied by this technique including *chlorine, bromine, iodine, nitrogen, arsenic, antimony, tin*. As in the case of nuclear magnetic resonance ^{12}C and ^{16}O cannot be studied. While ^{2}H quadrupole effects may be detected indirectly by other methods, experimental problems have prevented their observation by the methods discussed here.

6.1.1 Discussion of the Classical Quadrupole

An arrangement of charges having a quadrupole moment is found in the classical linear quadrupole +−−+. Since it has no net charge, the linear quadrupole experiences no forces or torques in a homogeneous electric field but is subjected to a torque in an inhomogeneous field tending to align it in a preferred orientation[1].

In contrast to the classical quadrupole, the nuclear

[1] H. G. Dehmelt, Am. J. Phys. 22, 110 (1954).

quadrupole has an intrinsic spin angular momentum which couples with the torque to produce a precession whose frequency is determined by the quadrupole moment, eQ, and the gradient of the electric field, q.

It is useful to compare this result with the Zeeman effect in which the torque is produced by an external magnetic field interacting with the nuclear magnetic moment. This torque couples with the angular momentum to produce the Larmor precession of nuclear magnetic resonance (see also section 5.5:1.1). The electric field gradient in nuclear quadrupole resonance (NQR) arises from the electrons and nuclei external to the nucleus under investigation and is usually much larger than any field which could be produced by the investigator.

6.1.2 The Nature of the Electric Field Gradient

The electric field gradient q is a second-degree symmetric tensor whose components may be derived from the nine partial derivatives of the form

$$q_{yz} = \frac{\partial^2 V}{\partial y \partial z}, \; q_{xy} = \frac{\partial^2 V}{\partial x \partial y}, \text{ etc.}$$

V is the electrical potential at the nucleus arising from all charges external to the nucleus in question. In a suitably chosen cartesian coordinate system, whose axes are called the principal axes of the tensor, all the components except q_{xx}, q_{yy}, and q_{zz} have zero magnitude.

Since the charges producing the potential are all external to the nucleus, Laplace's equation is obeyed:

$$q_{xx} + q_{yy} + q_{zz} = O \tag{1}$$

Since V is of the form $1/r$, the q components are of the form $1/r^3$ so that it is only necessary to consider charges quite close to the nucleus in evaluating the field gradient components. In practice, the major contribution is made by the electrons 'attached' to the nucleus under observation[2], that is, the inner shell and valence electrons. Since s electrons have a spherically symmetric distribution around the nucleus, they do not contribute to any electric field inhomogeneity. Similarly, filled atomic orbitals are assumed to have no net effect on the field gradient for reasons of symmetry.

In molecules, large field gradients can be produced by the electrons involved in covalent bonding since they appear to be localized between adjacent atoms. In many cases it is satisfactory to assume that the charge distribution is symmetric about the bond axis z. This implies that $q_{xx} = q_{yy} = -1/2 \, q_{zz}$, where equation (1) has been used. When accurate wave functions are available, the potential V, and hence q_{zz}, may be determined theoretically.

Finally, if a nucleus is located in a site of tetrahedral or higher symmetry, the net field gradient will be zero.

6.1.3 Energy Levels and Selection Rule

When the field gradient is axially symmetric the electric quadrupole Hamiltonian is[3]

$$H = eQq_{zz}(I_z^2 - I^2)/4I(2I-1) \tag{2}$$

Here, eQ is called the nuclear quadrupole moment; e is the elctronic charge and Q may be regarded as a property characteristic of the nucleus of interest. Q may be found in tables of nuclear constants and is known for approximately 150 isotopes. I_z is the z-component of the nuclear spin operator and I^2 is the square of the total nuclear spin operator given by $I^2 = I_x^2 + I_y^2 + I_z^2$. I is the total nuclear spin quantum number, simply called the nuclear spin. It may also be found in nuclear tables, e.g., for ^{35}Cl it is 3/2 and for ^{10}B it is 3. q_{zz} has been discussed in section 5.5:6.1.2.

The energy levels E_m are:

$$E_m = eQq_{zz}[3m^2 - I(I+1)]/4I(2I-1) \tag{3}$$

where m may be assigned values of I, I−1 to −I. (If I = 3/2, m may have values 3/2, 1/2, −1/2 and −3/2). The selection rule is m↔m ± 1, that is transitions may occur between adjacent levels only. The quantity eQq_{zz} is called the quadrupole coupling constant and is the quantity usually reported in the literature.

6.1.4 Interaction of Nuclear Quadrupole with Electromagnetic Radiation

The detailed mechanism by which the radiofrequency field produces transitions between quadrupolar energy levels is the same as in nuclear magnetic resonance (see section 5.5:1). Hence, most of the factors affecting line-widths, intensities and saturation effects are operable. The major difference between NMR and NQR lies in the fact that an external magnetic field is used to establish the spacing of the Zeeman levels in NMR while in NQR the spacing arises through the molecular or crystalline electric field and may not be controlled by the observer. It is for this reason that no magnet is needed in NQR spectrometers.

[2] *C.H. Townes, B.P. Dailey,* J. Chem. Phys. *17,* 782 (1949);
W. Gordy, Discuss. Faraday. Soc. *19,* 14 (1955).

[3] *Das, Hahn,* Solid State Physics Supplement 1, Nuclear Quadrupole Resonance p. 5 Academic Press, Inc., New York (1958).

6.2 Experimental Techniques

In crystalline solids, quadrupole splittings may be observed in NMR spectra provided that the coupling constant is not too large, that is when $eQq_{zz} \gtrsim 1$ MHz, approximately. Much larger splittings may be observed in dilute gases by *microwave spectroscopy*. In addition to these methods quadrupole coupling constants have been determined from molecular beam, Mössbauer, and electronic spectra.

When the coupling constant has a value ranging from about 3 MHz to about 500 MHz, the most accurate measurements are made by the methods of direct NQR spectroscopy discussed in the following sections.

6.2.1 Apparatus for NOR Spectroscopy

Many of the single coil NMR spectrometers may be used, without a magnet, to detect NQR signals. The requirements of a good NQR spectrometer are high sensitivity and the capacity for tuning through a wide range of frequencies.

In general, the spectrometers used are oscillating detectors consisting of a tank circuit and a vacuum tube (or transistor) amplifier. The tank circuit contains a coil of a few turns of copper wire in parallel with a tuning condenser. The sample is placed inside the coil and absorbs energy when the circuit is tuned through its quadrupole resonance frequency. This causes a reduction in the level of

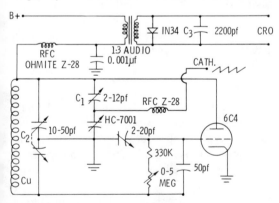

Figure 1. Diagram of a Quadrupole Spectrometer:
Cu   = ca. 10 turns No. 12 wire
CRO = 00000000000
Cath = Cathode follower
C_2 tuning condenser.
HC-7001 = voltage sensitive capacitor used to produce frequency modulation.
The sample coil is to the left of C_2.

[4] *K.E. Weber, J.E. Todd*, Rev. Sci. Instrum. *3*, 390 (1962).

oscillation which is converted into a voltage spike at the output of the amplifier. For oscilloscope display it is necessary to modulate the frequency of the tank circuit over a very narrow range. This allows a repetitive display on the cathode ray tube which can be seen easily.

A simple NQR spectrometer of the superregenerative type (fig. 1) is quite sensitive compared to the continuous wave devices commonly used in NMR. It has the disadvantage of producing a complicated resonance signal and poor line shape and is consequently not a good instrument for NMR, but the additional sensitivity is so desirable that most investigators prefer it when searching for NQR signals.

6.2.2 Sample Requirements

In order to produce observable resonances, the sample must be a crystalline solid. It is desirable, and frequently necessary, to cool the sample to 77° K.

Thus, *hexachloroethane* resonances are so broad as to be unobservable at 25° but are quite intense at liquid nitrogen temperatures. (Liquid sample resonances may not be observed since the strong electrical interactions during collisions cause too wide a variation in the field gradient and the lines are too broad and weak for observation.)

The sample size required depends on the strength of the resonance. A 50 milligram sample of *p-dichlorobenzene* produces observable ^{35}Cl resonance signals whereas several grams of *hexamethylenetetramine* are required to produce a detectable ^{14}N resonance. It is usual practice to completely surround the sample coil with sample when searching for a new resonance.

6.3 Experimental Results and Interpretation

A striking feature of quadrupole spectra is the wide range over which they are observed for a particular nucleus in different molecular environments. The extreme cases for ^{35}Cl are KCl, observed at 20 KHz, and CIF, observed at about 70 MHz[5]. A more interesting example is *chloro acetyl chloride* in which the methylene chlorine resonance is at 37.517 MHz and the acetyl chlorine resonance appears at 30.437 MHz.

Several types of information may be obtained from NQR spectra in regard to:

the field gradient and the chemical bond (5.5:6.3.1)
structure (5.5:6.3.2)
the nucleus.

[5] Ref. [13], p. 174.

6.3.1 Field Gradient and Chemical Bond

Though quantitative determinations of the field gradient tensor components have been made in a few cases, the present discussion will be restricted to a qualitative picture of the field gradient and its relationship to bonding concepts. While the symbol q is used to denote 'the' field gradient, the following considerations apply to the calculation of each of the field gradient components.

If it is assumed that the superposition theorem of potential theory is applicable, then the field gradient is $q = \sum_i q_i$ where the q_i are given by the expression

$$q_i = e\varsigma \, \psi_i^* \, \frac{(3\cos^2\theta - 1)}{r_i^3} \, \psi_i dr \qquad (4)$$

where ψ_i is the wave function of the ith electron
  θ is the 'altitude' angle in polar coordinates
  r_i is the radius vector magnitude, and
  e is the electronic charge.

This equation may be interpreted as meaning that only charges which are quite close to the quadrupolar nucleus contribute appreciably to the field gradient, and hence the quadrupole frequency (since the effect goes as the inverse cube of distance, r). Of these charges, the largest individual contributions to q are made by electrons.

If hydrogenlike wave functions are used, s-electrons should contribute nothing to q because of their spherically symmetric distribution, and d-electrons will contribute very little because of their small probability of being near the nucleus. Moreover, completely filled orbitals (s, p, d, etc.) are spherically symmetric and do not make a net contribution to q. However, atomic electrons in partially vacant p-states may make a significant contribution to q. From the foregoing it may be deduced that an ion, having an inert gas electron configuration, should have q=o, while an atom which has donated p-electrons for the formation of a covalent bond should have a large q.

This statement requires some explanation since a free atom, such as a *halogen*, may have a significent field gradient. On the other hand, when bound, the q of this atom will increase when it donates a p-electron to form a covalent bond and will decrease when the atom accepts an electron and approaches the filled-shell configuration (cf. table 1 [35] chlorine in various types of combination). As might be expected, the field gradient tends to decrease with ionic character and hence with electronegativity difference between adjacent atoms. It follows that adjacent groups with electron donating, or withdrawing, powers relative to the quadrupolar nucleus will have a significant effect on the resonance frequency.

Table 1. Quadrupole Coupling Constants in Compounds of Varying Ionic Character

Molecule	eQq [Mhz]	Bonding
Cl (atomic)	109.74	unbound
I Cl	82.5	covalent
K Cl	0.041	ionic

Hybridization involving p-orbitals with s or d atomic orbitals can produce an indirect effect on the field gradient. In general, the major effect arises from increasing the electron deficiency in the atomic p-orbitals. Thus, an increase in s-character reduces the field gradient, since it increases the spherical symmetry of the electron distribution. The effects of d-hybridization are more complex, but generally tend to decrease the p population and hence increase the quadrupole frequency. The relationship which summarizes these statements is[6]

$$N_z = (l \pm i)(l + s - d) \qquad (5)$$

where N_z is the p_z electron population
  i is the ionic character
  s is the s-character and
  d is the d-character.

Considering the molecule A-B, where A is the observed nucleus, the positive sign applies when the bond has the polarity $A^{\ominus}B^{\oplus}$.

Finally in making quantitative calculations, it should be observed that wave functions (not necessarily LCAO) which yield good results for molecular properties, such as bond energies and dipole moments (dependent on the charge distribution in the valence bond region) will not always be useful in calculating field gradients.

6.3.2 Structural Information

It may be inferred that NQR spectra are equally capable of quantitative prediction, as the other common types of spectra. In practice, the empirical approach is usually taken by the non-specialist and will be followed here with examples of halogenated compounds.

The coupling[7]constants for ^{35}Cl in various chloromethanes show that increasing the number of chlorine atoms increases the coupling constant, èQq/h by ~ 3.5 MHz per chlorine (table 2). This extraordinary chemical shift may be attributed to an effective increase of electronegativity of the entity to which a chlorine atom is bonded, through an inductive effect.

---

[6] Ref. [13], p. 180.
[7] *R. Livingston*, J. Phys. Chem. *57*, 496 (1953).

Table 2. Coupling constants in chloromethanes[a] and chloroacetyl chlorides (eqQ/h[MHz])

Molecule	R = H	R = CO Cl
R CH$_2$Cl	68.0	60.8[8]
R CH$_2$Cl$_2$	72.0	65.0[8]
R CHCl$_3$	76.6	67.4[9]
CCl$_4$	81.2	—

[a] Average of several resonances which arise because of nonequivalence of the nuclei in the unit cell of the crystal.

Also in *chloroacetyl chlorides* there is an increase of the coupling constant with the effective electronegativity of the group bonded to the carbonyl chlorine (R =COCl). Though not included in the table, the chlorines bonded to the non-carbonyl carbons show coupling constants about 10 MHz higher than the carbonyl chlorines.

The average coupling constants for ^{35}Cl in *dichlorobenzenes* depend on the structure (table 3). When there are more than two chlorine atoms, a linear increase of the coupling constant is observed[3].

Table 3. Coupling Constants in Disubstituted Benzenes[10]

Molecule	EqQ/h [Mhz]
o–Cl$_2$C$_6$H$_4$	71.4
m–Cl$_2$C$_6$H$_4$	69.8
p–Cl$_2$C$_6$H$_4$	69.6

The data have been rounded off to one decimal place. The constants for *m*- and *p*-dichlorobenzene appear to be close, but they are still separated by 200 kHz and can be easily distinguished.

While the data given here are necessarily quite limited, they do show the sensitivity of the coupling constant to molecular environment (for intermolecular effects see ref.[3]).

The same sensitivity to surroundings is shown by *bromine* and *iodine* as well as most other quadrupolar nuclei studied. In all cases, atoms in similar environments have coupling constants in the same region of radiofrequency spectrum. A great limitation is the requirement for a moderately large sample. Improved instrumentation may however solve this problem.

6.4 Bibliography

[8] *P. Bray*, J. Chem. Phys. *23*, 704 (1955).
[9] *H. Allen jr.*, J. Am. Chem. Soc. *74*, 6074 (1952).
[10] *P. Bray*, J. Chem. Phys. *25*, 813 (1956).
[11] *E.R. Andrew*, Nuclear Magnetic Resonance, University Press, Cambridge England, 1955.
[12] *R.G. Barnes, S. Segal*, Catalogues of Nuclear Quadrupole Resonance Spectra, United States Atomic Energy Commission, Washington, D.C., U.S.A.
[13] *W.J. Orville-Thomas*, Quart. Rev., Chem. Soc. XI, 162 (1957).

5.6 Microwave Spectroscopy

Werner Zeil
Abteilung für Physikalische Chemie der Universität Ulm, D-75 Karlsruhe

The following topics of chemical interest can be examined with the aid of microwave spectroscopy:
Charge distribution, and its symmetry within the bond (5.6:3);
Molecular structure, i.e. the geometry of molecules (5.6:4);
Dipole moment, and its orientation within the molecule;
Height of the *potential barrier* which restricts free rotation about a bond (5.6:5.5);
Analysis of mixtures (5.6:6).

1 Introduction

With few exceptions, microwave spectra are *rotational spectra,* in the cm and mm wavelength region. The region accessible to microwave analysis without great experimental difficulty ranges from 6000 to 100,000 MHz i.e. 0.2 to 3.33 cm^{-1}. Extension to lower or higher frequencies is possible but serious measuring problems are then encountered. Therefore, work in such extended regions is, at present, limited to laboratories with special equipment.

2 Microwave Spectrometers

Up to the present, practically all microwave spectrometers have been bench-built in the laboratory[1-4]. Only recently has commercial apparatus

[1] *R. Hughes, E.B. Wilson*, Jr., Phys. Rev. *71*, 562 (1947).
[2] *M.W. Strandberg, H.R. Johnson, J.R. Eschbach*, Rev. Sci. Instrum. *25*, 776 (1954).

become available. With further developments this equipment should be applicable also to routine problems[5]. All microwave spectrometers have monochromatic ultra-high frequency tubes as radiation sources. These tubes are tuned while recording a spectrum. Previously, only *reflex klystrons* were used, but, during the last few years, *backward wave oscillators* (Carcinotron, BWO, RWO) have been preferred because of their larger tuning range. For the most commonly used spectral range (wave-length approximately 7.5 mm to 5 cm), microwave spectroscopy groups in Germany use only backward wave oscillators. The recording of a microwave spectrum requires a specific technique because the absorption coefficients are very small (generally of the order of 10^{-7} cm^{-1} to 10^{-10} cm^{-1}). For this reason, waveguide sections of several meters length are used as absorption cells. These cells contain the gas under investigation at a partial pressure between 10^{-2} and 10^{-3} mm Hg. In addition, Stark-effect modulation is used.

An insulated electrode is placed in the absorption cell. (Fig. 1).

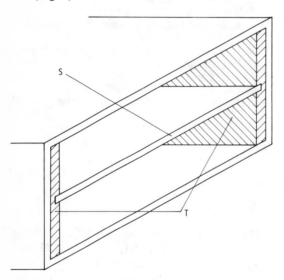

Fig. 1. Part of an absorption cell with Stark electrode (schematic)
S = Stark electrode, T = Teflon insolation tape

A DC voltage from several hundred to approximately 2,000 V is periodically applied to this electrode (square wave voltage). In the particular case of symmetric-top molecules, several volts will often suffice. As a consequence of the Stark-effect, the absorption signals appear or disappear with the same frequency as the square wave of the Stark field. Whenever the radiation frequency corresponds to the frequency of a molecular absorption, the microwave is modulated with the frequency of the Stark square wave. This minute modulation can be separated from the high level of the incident microwave power by extremely selective and sensitive radio frequency amplifiers. The signal is then amplified, and, after rectification, is recorded, giving the microwave spectrum.

3 Microwave Spectra and Molecular Symmetry

Disregarding, for the moment, fine structure effects, which will be discussed later, the character of a microwave spectrum is determined by the symmetry of the molecule under examination. Molecules which, because of their symmetry, have no electrical dipole moment, are not accessible to microwave spectroscopy. Microwave spectroscopy requires the existence of a permanent electric dipole moment.

3.1 Symmetric-Top Molecules

Molecules with C_3- or higher symmetry, i.e. with at least a three-fold symmetry axis, belong to the class of symmetric-top molecules and have a relatively simple microwave spectrum if a permanent electric dipole moment exists. The microwave spectrum consists of equidistant lines, whose frequencies are given by the following equation:

$$\nu = 2\,B\,(J + 1) \quad \text{[MHz]}.$$

In this equation, J stands for the angular momentum quantum number [J = 0, 1, 2,...] and B represents the rotational constant:

$$B = \frac{h}{8\,\pi^2\,I_B} \quad \text{[MHz]}.$$

I_B represents the principal moment of inertia of the molecule $X_pYZ...$ perpendicular to the symmetry axis. The rotation axis must be equal to or higher than three-fold symmetry, $p \geqslant 3$.

The moment of inertia is defined by:

$$I_B = \frac{I_A}{2} + p\,m_X\,r_X^2 + \sum_i m_i\,r_i^2$$

The condition for the center of mass is:

$$p\,m_X\,r_X + \sum_i m_i\,r_i = 0.$$

The sum extends only over the atoms on the major axis. Further, $I_A = p\,m_X\,S_X^2$; r_i = the coordinate of the i-th atom on the axis of inertia; r_X is the distance of the center of

[3] *H.G. Fitzky*, Z. Angew. Physik *10*, 489 (1958).
[4] *H.D. Rudolph*, Z. Angew. Physik *13*, 401 (1961).
[5] *D.J. Millen*, Chem. Brit. *4*, 202 (1968).

mass of the X_p-plane from the molecular center of mass; and S_X is the distance of the X-atoms from the major axis.

From the microwave spectrum of a symmetric-top molecule, *one* principal moment of inertia can be determined.

For diatomic molecules, which are a special case of symmetric-top molecules, the moment of inertia is given by:

$$I = \sum_i m_i r_i^2$$

(m_i = mass of the i-th atom, r_i = the distance of the i-th atom from the center of mass)

Hence, the distance between the two atoms can be determined accurately from the rotational constant which can be derived unambiguously from the spectrum[6].

Linear molecules with more than two atoms, and symmetric-top molecules, have more than one structural parameter. The various isotopic combinations for the molecule must be included in the investigation. Methyl chloride, for example, has several isotopic species in natural abundance; from their moments of inertia, obtained from the respective microwave spectra, the pertinent structural parameters can be determined. Since the resolving power of microwave spectrometers is very high, spectral lines shifted by isotopic substitution can be separated easily. The spectra of the different species, occurring in natural abundance in the present case $^{35}Cl^{12}CH_3$, $^{35}Cl^{13}CH_3$, $^{37}Cl^{12}CH_3$, and $^{37}Cl^{13}CH_3$, can be identified simultaneously. A set of four moments of inertia can be obtained, from which the structure can be determined, assuming that the molecular geometry is independent of isotopic substitution. In the above example, substitution of hydrogen by deuterium would yield additional information.

3.2 Asymmetric-Top Molecules

Molecules which do not possess a three-fold or higher symmetry axis give rise to microwave spectra of the asymmetric-top type. The rotational spectra of these molecules are characterized by three rotational constants which are – like those of symmetric-top molecules – proportional to the respective reciprocal moments of inertia.

The energy levels are given by the following equation:

$$E_{J,\tau} = \frac{A+C}{2} J(J+1) + \frac{A-C}{2} E_{J,\tau}^{(\kappa)}$$

where

$$A = \frac{h}{8\pi^2 I_A}, \quad B = \frac{h}{8\pi^2 I_B}, \quad C = \frac{h}{8\pi^2 I_C}$$

$E_{J\tau}^{(\kappa)}$ is the 'reduced energy'. This energy is a function of the quantum numbers J and τ, as well as of Ray's asymmetry parameter κ which depends on all three rotational constants:

$$\kappa = \frac{2B-A-C}{A-C}.$$

The nature of the microwave spectrum is thus determined by the three rotational constants and the two quantum numbers J and τ[7-10].

The selection rules, and hence the spectra, depend largely on whether all three components of the electric dipole moment are non-vanishing or whether the molecule possesses a plane of symmetry. In the latter case, the component of the electric dipole moment perpendicular to the plane of symmetry equals zero. Each component of the dipole moment along a principal axis of inertia gives rise to a separate microwave spectrum, whose intensity is – to a first approximation – proportional to the square of its dipole moment component.

In the case of asymmetric-top molecules, three principal moments of inertia – and, hence, three pieces of structural information – can be obtained from the spectrum of one isotopic species of a molecule. Exceptions are planar molecules, for which the following relation holds:

$$I_A + I_B - I_C = 0.$$

In this case, a linear dependence exists. Consequently, no more than two structural parameters can be derived from the spectrum of one isotopic combination. For example, the structure of *sulfur dioxide* follows from the spectrum of only one isotopic species assuming C_{2v}-symmetry. For *formaldehyde* with three structural parameters, (C_{2v}-symmetry and planar structure assumed),

[6] *T. Törring*, Z. Naturforsch. *23a*, 777 (1968); *J. Hoeft, E. Tiemann*, Z. Naturforsch. *23a*, 1034 (1968).

[7] *T.M. Sugden, C.N. Kenney*, Microwave Spectroscopy of Gases, Van Nostrand, London 1965.

[8] *H.C. Allen*, Jr., *P.C. Cross*, Molecular Vib-Rotors, John Wiley & Sons, Inc., New York, London 1963.

[9] *J.A. Wollrab*, Rotational Spectra and Molecular Structure, Academic Press, New York, London 1967.

[10] *W. Gordy, R.L. Cook*, Microwave Molecular Spectra, Wiley Intersci. Publ., New York, London, Sydney, Toronto 1970.

two isotopic species must be measured for a complete structural determination.

In general, the following applies for both symmetric- and asymmetric-top molecules: There must be at least as many principal moments of inertia of isotopic species as there are independent structural parameters.

From the microwave spectrum alone, or from the calculated principal moments of inertia, it is possible to determine whether a molecule is of the symmetric-top type, has three fold or higher symmetry (linear molecules are also of this type), is planar, has a plane of symmetry, or has no symmetry element at all.

4 Information on Structure

4.1 Isotopic Combination

Barring accidental degeneracies or very small dipole moment components, the rules concerning information on structure apply to questions of symmetry without exception. A detailed analysis of the bond distances and angles shows that the structure obtained depends on the *choice of isotopic combinations* used. This results from the fact that molecules cannot be regarded as rigid bodies. Even in their vibrational ground state − and only this is under consideration − they are subject to the 'zero point vibration'.

The term «zero-point vibration» should not be taken literally. Rather, it is a metaphorical description of the impossibility of precisely locating an atom in a molecular system, as stated by the Heisenberg Uncertainty principle:

$$\Delta x \ \Delta p \approx \frac{h}{2\pi}$$

Table 1. Bond lengths in carbonoxysulfide calculated by the r_0- and r_s-methods[11]

r_0 structure (5.6:4.1.1.)		r_{co} [Å]	r_{cs} [Å]
Isotopic combinations used for the calculation of the distances			
$^{16}O \ ^{12}C \ ^{32}S$	$^{16}O \ ^{12}C \ ^{34}S$	1.1647	1.5576
$^{16}O \ ^{12}C \ ^{32}S$	$^{16}O \ ^{13}C \ ^{32}S$	1.1629	1.5591
$^{16}O \ ^{12}C \ ^{34}S$	$^{16}O \ ^{13}C \ ^{34}S$	1.1625	1.5594
$^{16}O \ ^{12}C \ ^{32}S$	$^{18}O \ ^{12}C \ ^{32}S$	1.1552	1.5653
Average:		1.1613	1.5604
Scatter:		0.0095	0.0077

r_s structure (5.6:4.1.2.)		
Parent molecule		
$^{16}O \ ^{12}C \ ^{32}S$	1.16012	1.56020
$^{18}O \ ^{12}C \ ^{32}S$	1.15979	1.56063
$^{16}O \ ^{13}C \ ^{32}S$	1.16017	1.56008
$^{16}O \ ^{12}C \ ^{34}S$	1.16075	1.55963
Average:	1.16021	1.56014
Scatter:	0.00096	0.00100

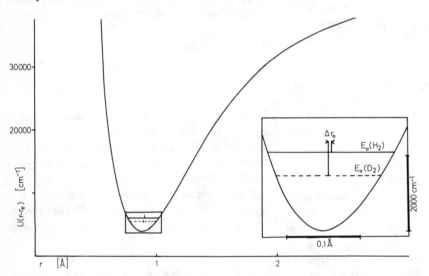

Fig. 2. Morse potential for H_2 and D_2.

$$U(r-r_e) = D \ [1 - \exp(-\beta(r-r_e))]^2$$

$\beta_{H_2} = 18962 \cdot 10^7 \ cm^{-1}$ $\beta_{D_2} = 18762 \cdot 10^7 \ cm^{-1}$

$D_{H_2} = 36103 \ cm^{-1}$ $D_{D_2} = 36724 \ cm^{-1}$

$\omega_{H_2} = \ 4160 \ cm^{-1}$ $\omega_{D_2} = \ 2942 \ cm^{-1}$

For β and D, averages were used. In the insert, Δr_0 gives the difference of the average bond lenghts for H_2 and D_2. E_0 is zero-point energy.

4.1.1 Analysis by the r_0-Method

Because the zero-point vibration in an anharmonic potential − to a first approximation, a MORSE potential − is asymmetric, there is a maximum probability of finding, for example, a particular H- or a D-atom in a molecule at slightly different locations. Different bond distances necessarily result in the two cases (Fig. 2).

It is realized that the rule, by which n independent principal moments of inertia are required for calculation of n structural parameters, will yield results of limited accuracy. Errors up to 0.01 Å may occur (Table 1).

4.1.2 Determination by the r_s-Method

An improvement over the uncertainty of structure obtained by the r_0-method is obtained by using the so-called r_s-method[11]. In this method, only the *differences of the moments of inertia* of isotopically substituted molecules enter into the calculation of the structural parameters. This results in a significant improvement of the space coordinates. The following rule exists: In order to locate n atoms in a molecule (we are concerned here with center of mass coordinates, i.e. coordinates with center of mass as zero), it is necessary to measure the spectra of $n + 1$ different isotopic species of the molecule under investigation, and each atom whose center of mass coordinate is to be determined must be substituted once. For example, for the coordinates of a linear molecule:

$$|z| = \{\mu^{-1} (I'_Y - I_Y)\}^{1/2}$$

I_Y is the moment of inertia of the parent molecule, and I'_Y is the moment of inertia of the isotopically substituted molecule.

$$\mu = \frac{M \Delta M}{M + \Delta M}$$

M = the mass of the parent molecule, and ΔM = the difference between the masses of the parent and of the isotopic substitued molecule. Analogous equations for more complex molecules can be found in publications by Kraitchman[11], Sugden[12], Wollrab[13], and Chutjian[14].

Results obtained by the r_s-method have smaller deviations than those obtained by the r_0-method (Table 1).

[11] *C.C. Costain*, J. Chem. Phys. *29*, 864 (1958);
   J. Kraitchman, Am. J. Phys. *21*, 17 (1953).
[12] *T.M. Sugden, C.N. Kenney*[7], p. 80.
[13] *J.A. Wollrab*[9], p. 88.
[14] *A. Chutjian*, J. Mol. Spectrosc. *14*, 361 (1964).
[15] *C.H. Townes, A.L. Schawlow*, Microwave Spectroscopy, McGraw-Hill, New York, Toronto, London 1955.
[16] *J.A. Wollrab*[9], p. 244.

In cases where the natural abundance of the isotope being investigated is very small, an isotopically substituted molecular species must be specifically prepared. With this method, the atomic coordinates, and from them the bond lengths and bond angles, can be determined within a precision of 0.001 Å and about 1/2 degree.

4.2 Determination of Structures of Molecules with Isotope-free Elements

Using the r_s-method, molecular coordinates of atoms without stable isotopes, e.g. F or P, can be determined only by using additional relationships, such as the center-of-mass relation[12]. The limits of error for the distances of the atoms from the center of mass are greater than those determined directly by isotopic substitution.

5 Fine Structure of Microwave Spectra

5.1 The Stark Effect

The absolute values of the three components of an electric dipole moment along the principal axis of inertia can be determined by measuring the Stark effect. Rotating molecules show a Stark effect just as atoms do. This means that the energy levels in an external electric field split according to the magnetic quantum number M. Each absorption line will thus be split into a great many lines.

For the observed line shift with respect to the zero field line, the following formula holds for linear-top molecules:

$$\Delta v = \frac{\mu^2 |\mathfrak{E}|^2 \{3 M^2 (8 J^2 + 16 J + 5) - 4 J (J+1)^2 (J+2)\}}{h^2 B J (J+1) (J+2) (2J-1) (2J+1) (2J+3) (2J+5)}$$

For symmetric-top molecules:

$$\Delta v = \frac{2 \mu |\mathfrak{E}| K M}{h J (J+1) (J+2)} .$$

In the case of asymmetric-top molecules, a more complex equation is valid[7, 9, 15]. Only the equation for the splitting of an energy level is, therefore, given:

$$\Delta E = \mu_g^2 |\mathfrak{E}|^2 (A' + B'M).$$

μ_g	= electric dipole moment component along the principal axis of inertia g of the molecule
$|\mathfrak{E}|$	= absolute value of the electric field strength
J	= rotational quantum number
K	= angular-momentum quantum number about the figure-axis of the molecule
M	= 'magnetic' quantum number
B	= rotational constant
h	= Planck's constant
A',B'	= functions of the molecular parameters and the quantum numbers in the case of an asymmetric-top molecule

From these relations, it can be seen that symmetric-top molecules show a linear Stark effect, i.e. the frequency-shift is proportional to the field strength applied. This contrasts with linear and asymmetric-top molecules, which have a quadratic Stark effect. Here, the frequency shift depends upon the square of the electric field strength.

These simple statements apply only to molecules whose spectra contain no fine structure effects. Specialized literature should be consulted[16] where such effects exist.

5.2 Results from the Observation of the Stark Effect

From the Stark effect, evidence can be obtained regarding the symmetry of a molecule. On the other hand, the relationships show that the observed frequency shift depends on that component of the electric dipole moment which caused the corresponding transition. The absolute value of the components of the electric dipole moment along the principal axis of inertia can be determined when a sufficient number of Stark-shifts have been measured. However, a statement concerning the direction of the electric dipole moment (direction of the 'arrowhead') is possible only after carrying out additional measurements in a magnetic field[17]. Microwave spectroscopy is the first method which makes it possible to determine not only the magnitude of the total dipole moment — which can be obtained also from dielectric measurements — but also the orientation of the total dipole moment relative to the molecular framework. The value of the total dipole moment can be calculated by vector addition of the separate components corresponding to the main inertia axis.

5.3 Nuclear Quadrupole Coupling

Molecules which contain atomic nuclei with an electric quadrupole moment have microwave spectra with a fine structure. This fine structure results from coupling of the angular momentum (spin) of the quadrupolar nucleus and the total angular momentum of the molecular rotation via the electric field gradient. This electric field gradient is itself caused by the charge distribution created by neighbouring electrons.

In the simplest case, a quadrupole moment arises when two dipoles are oriented antiparallel to each other (Fig. 3).

---

[17] *C.H. Townes, G.C. Douismanis, R.L. White, R.F. Schwarz,* Discuss. Faraday Soc. *19,* 56 (1955).

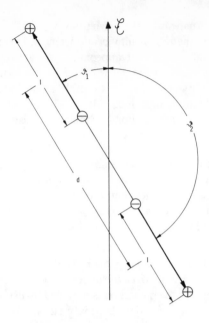

Fig. 3. Simplest representation of an electric quadrupole in an electrical field

The quadrupole moment of the charge distribution shown is:

$$Q = 4 \, e \, d \, l \; [cm^2]$$

The energy of this charge distribution in an inhomogeneous electric field is given by:

$$E = -\frac{1}{8} \, Q \left(\frac{\partial^2 V}{\partial z^2} \right) (3 \cos^2 \vartheta - 1).$$

As the angular dependence shows, this is an energy of orientation. In quantum mechanics, this means that an atomic nucleus with an electric dipole moment is subject to spatial quantization. A finite nuclear quadrupole moment exists whenever the nuclear distribution of positive charge deviates from spherical geometry. It is possible to distinguish between prolate (cigarshaped) and oblate (saucer-shaped) charge distribution, because the two types have different signs for the quadrupole moment. Charge distributions with a finite quadrupole moment are observed for nuclei with a spin ≥ 1 (^{14}N, ^{35}Cl, ^{37}Cl, ^{79}Br, ^{81}Br, etc.).

The above-mentioned coupling between the angular momentum of the atomic nuclei and the total angular momentum of the molecule causes a splitting of the energy levels and — as a consequence of the selection rules — a corresponding multiplet splitting of the microwave absorption lines.

Fig. 4 shows the multiplet expected for the hydrogen cyanide molecule due to the quadrupole moment of the nitrogen nucleus (^{14}N).

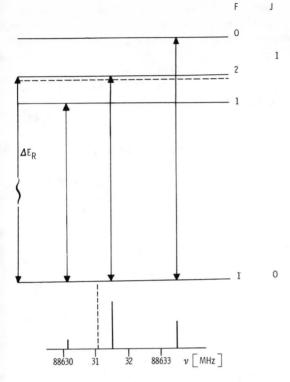

Fig. 4. Quadrupole fine structure of the rotational transition J = 0→1 of hydrogen cyanide.
The dotted line in the schematic spectrum shows the hypothetical position of the non-split line.

The following equations hold for the quadrupole energies:

Symmetric-top

$$E_Q = eQ \left(\frac{\partial^2 V}{\partial z^2}\right) \frac{2J}{2J+3} \left(\frac{3K^2}{J(J+1)}\right)$$

$$\frac{3/4\,C(C+1) - I(I+1)\,J(J+1)}{2J(2J-1)\,2I(2I-1)}.$$

Asymmetric-top

$$E_Q = eQ \left\{ \left(\frac{\partial^2 V}{\partial a^2}\right) \left[J(J+1) + E(\kappa) - (\kappa+1) \left(\frac{\partial E(\kappa)}{\partial \kappa}\right) \right] + \right.$$

$$+ 2 \left(\frac{\partial^2 V}{\partial b^2}\right) \left(\frac{\partial E(\kappa)}{\partial \kappa}\right) +$$

$$+ \left. \left(\frac{\partial^2 V}{\partial c^2}\right) \left[J(J+1) - E(\kappa) + (\kappa-1) \left(\frac{\partial E(\kappa)}{\partial \kappa}\right) \right] \right\}$$

$$\left(\frac{1}{J(J+1)}\right) Y(F).$$

$$Y(F) = \frac{3/4\,C(C+1) - J(J+1)\,I(I+1)}{2(2J-1)\,(2J+3)\,I(2I-1)}$$

$$C = F(F+1) - I(I+1) - J(J+1)$$

$$F = J+I, (J+I-1) \ldots |J-I|.$$

The quantum number F, which arises from a vector addition of the rotational quantum number J and the spin quantum number I, is of decisive importance for the microwave spectrum of a molecule with a quadrupole nucleus. From observation of the splitting pattern of the rotational line, and a knowledge of the selection rules,

$$\Delta J = 0 \pm 1, \quad \Delta F = 0 \pm 1, \quad \Delta K = 0, \quad \Delta I = 0$$

it is possible to find the three diagonal elements of the field gradient tensor. These are the tensor component along the direction of the bond axis which joins the quadrupole nucleus to a neighboring atom, and the two components perpendicular to this axis. They can be interpreted in terms of valency theory, by applying the theory of Townes and Dailey[18, 19].

5.4 Bond Theory Interpretation of Nuclear Quadrupole Fine Structure

When the bond axis component of the field gradient is known, the percentage ionic character of the bond can be calculated from the following equation:

$$i = \left(1 - s^2 - \frac{eQq_{mol}}{eQq_{atom}}\right)$$

q_{atom} represents the field gradient component at the position of the nucleus of the free atom which is known from atomic beam experiments, and q_{mol} represents the measured value of the field gradient component (both components with respect to the bond axis direction). Q is the quadrupole moment of the nucleus, s represents the s-character of the atomic orbital emanating from the quadrupole atom and participating in the bond, and i is the ionic character of the bond.

Similarly, the π-character of a bond can be estimated from nuclear quadrupole coupling constants; frequently, however, additional assumpions[20, 21] must be made.

5.5 Splitting due to Restricted Internal Rotation

Molecules having a methyl or an analogous group show splitting of spectral lines, caused by the hindered rotation of this group around its own bond axis. The width of the multiplets depends on the height of the potential barrier. It depends also on the moment of inertia of the rotating group

[18] C.H. Townes, B.P. Dailey, J. Chem. Phys. 20, 35 (1952).
[19] M. Kaplansky, M.A. Whitehead, Mol. Phys. 16, 481 (1969).
[20] J.H. Goldstein, J.K. Bragg, Phys. Rev. 75, 1453 (1949).
[21] R. Bersohn, J. Chem. Phys. 22, 2078 (1954).

under investigation. The splitting is largest for small barriers and small moments of inertia. In many cases, it is possible to find the barrier potential from the widths of the multiplets by relatively simple methods, though a knowledge of the geometry of the molecule is required.

So far, detailed research work has been carried out on molecules which have either one internally rotating group which is a symmetric-top, for example a CH_3-group[22, 23], or two such tops[24-26]. The theory for three tops has been developed but is still awaiting application[27]; the same is true for molecules containing groups (e.g. CH_2D or CH_2F) which behave as asymmetric tops[28].

6 Analytical Applications of Microwave Spectroscopy

Only a few examples of analytical application have been reported thus far. Qualitative (and in some cases quantitative) analysis of multi-component systems is conceivable, because each microwave spectrum is so highly characteristic for its molecule[29-31].

While it was believed previously that one or two measured lines would suffice for positive identification of a molecule[32], efforts are now being undertaken to make available catalogs which can be adapted to a computerized analysis[31] for wider frequency regions.

7 Examples of Structural Interest[33]

On the basis of chemical and physical studies, it was at one time thought that the compound with the stoichiometric composition FNS should have the structure S-N-F[34, 35]. Microwave spetroscopic experiments, however, provided evidence[36] for a non-linear molecule with the atomic order N-S-F and an angle of 117°.

In the same way, the sequence of atoms in *fulminic acid* was found to be HCNO, thus solving a chemical structural problem of long standing[37].

One of the most noteworthy studies in the field of structural organic chemistry was the exact determination of the geometry of *benzonitrile*[38]. Here, it is interesting to note the deformation of the benzene ring from the regular hexagonal structure (Fig. 5).

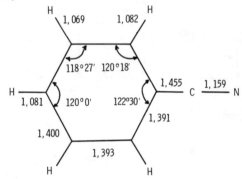

Fig. 5. Structure of benzonitrile

An example of the determination of a molecular dipole moment is that of *vinyl chloride*[39], for which molecule both the magnitude and the orientation of the dipole moment were found. (Fig. 6).

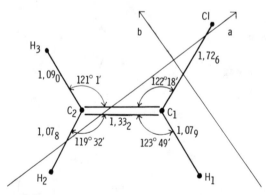

Fig. 6. Molecular structure of vinyl chloride, and orientation of the dipole moment

22 *E.B. Wilson*, Jr., *C.C. Lin, D.R. Lide*, Jr., J. Chem. Phys. *23*, 136 (1955).

23 *D.R. Herschbach*, Tables for Internal Rotation Problem, Dep. of Chemistry, Harvard University, Cambridge 1957.

24 *J.D. Swalen, C.C. Costain*, J. Chem. Phys. *31*, 1562 (1959).

25 *R.J. Myers, E.B. Wilson*, Jr., J. Chem. Phys. *33*, 186 (1960).

26 *H. Dreizler*, Z. Naturforsch. *16a*, 477 (1961).

27 *R.C. Woods*, J. Mol. Spectrosc. *21*, 4 (1966); *22*, 49 (1967).

28 *C.R. Quade, C.C. Lin*, J. Chem. Phys. *38*, 540 (1963);
C.R. Quade, J. Chem. Phys. *44*, 2512 (1966).

29 *W. Zeil*, Z. Anal. Chem. *170*, 19 (1959).
H.W. Harrington, J. Chem. Phys. *46*, 3698 (1967); *49*, 3023 (1968).
R.J. Volpicelli, O.L. Stiefvater, G.W. Flynn, NASA C.R. 967 (1967).
W.F. White, W.C. Easley, NASA T.N. D-5726 (1970).

30 *D.J. Millen*, Chem. Brit. *4*, 202 (1968).

31 *H.G. Fitzky*, Glas-Instrum.-Tech. 884, 1023 (1968).

32 *E.B. Wilson*, Jr., in Molecular Structure and Spectroscopy, p. 23, IUPAC Special Lectures, Butterworths, London 1963.

33 *B. Starck*, Molekelkonstanten aus mikrowellenspektroskopischen Messungen.
Landolt-Börnstein, Gruppe II, Bd. IV. Springer Verlag, Berlin, Heidelberg, New York 1967.

34 *O. Glemser, H. Richert*, Z. Anorg. Chem. *307*, 328 (1961); *O. Glemser, H. Richert, F. Rogowsky*, Naturwissenschaften *47*, 94 (1960).

Acetaldehyde is an example of a molecule for which the sign of the dipole moment has been identified. The microwave spectrum was measured in an external magnetic field[40].

While the structure of these molecules was elucidated purely by microwave spectroscopy methods, this is not always possible for larger molecules. This is the case especially when the atoms lie very near to the principal axis of inertia. Research on *tert-butyl-chloroacetylene* is an example of structure determination involving microwave spectroscopy combined with electron diffraction of gases[41, 42].

For *methylchlorosilane*, it was possible, by analysis of the quadrupole fine structure of the microwave spectrum, to demonstrate that the charge distribution about the Si-Cl bond does not show any deviation from cylindrical symmetry[43]. The asymmetry parameter of the nuclear quadrupole coupling constant (or rather that of the field gradient tensor) was determined as 0.01 — which is zero, within the limits of error. The current hypothesis about a pd-π-bond in this molecule, in the sense of a mesomeric formula such as:

$$H_3CH_2Si\text{-}Cl \qquad\qquad H_3CH_2\overset{\ominus}{Si}=\overset{\oplus}{Cl}$$

requires re-examination.

[35] *F. Rogowsky*, Z. Phys. Chem. NF. *27*, 277 (1961).
[36] *W.H. Kirchhoff, E.B. Wilson*, Jr., J. Am. Chem. Soc. *85*, 1726 (1963).
[37] *M. Winnewisser, H.K. Bodenseh*, Z. Naturforsch. *22a*, 1724 (1967).
[38] *B. Bak, D. Christensen, W.B. Dixon, L. Hansen-Nygard, J. Rastrup-Andersen*, J. Chem. Phys. *37*, 2027 (1962).
[39] *D. Kivelson, E.B. Wilson*, Jr., J. Chem. Phys. *32*, 205 (1960).
[40] *W. Hüttner, M.K. Lo, W.H. Flygare*, J. Chem. Phys. *48*, 1206 (1968).
[41] *H.K. Bodenseh, R. Gegenheimer, J. Mennicke, W. Zeil*, Z.Naturforsch. *22a*, 523 (1967).
[42] *J. Haase, W. Steingross, W. Zeil*, Z. Naturforsch. *22a*, 195 (1967).
[43] *R. Gegenheimer, R. Ronchi, W. Zeil*, unpubl. (Vortrag W. Zeil, Karlsruher Chem. Ges., 30. Okt. 1969).

8 Bibliography

C.H. Townes, A.L. Schawlow[15], 1955.
D.R. Herschbach[23], 1957.
H.C. Allen, Jr., P.C. Cross[8], 1963.
T.M. Sugden, C.N. Kenney[7], 1965.
J.A. Wollrab[9], 1967.
B. Starck[33], 1967.
W. Gordy, R.L. Cook[10], 1970.

5.7 Application of Circular Dichroism, Optical Rotatory Dispersion and Polarimetry in Organic Stereochemistry

Günther Snatzke
Organisch-chemisches Institut der Universität,
D-53 Bonn

1 Basic Principles of ORD and CD

Optical Rotatory Dispersion (ORD) and Circular Dichroism (CD) are optical methods depending on the chirality, and are thus called 'chiroptical methods'[1].

1.1 Optical Rotation

Optical activity is the property of a substance to rotate the plane of linearly polarized light (nowadays usually identical with the plane of the electric field vector) within a greater wavelength range. This can be caused by the chirality[2] of the molecules or by the chiral arrangement (e.g. in a crystal) of achiral molecules or ions. The measured angle of rotation (positive or negative, corresponding to a clockwise or anticlockwise rotation by looking against the ray) is proportional to the length, 1, (in dm) of the tube and to the density, ϱ (for liquids), or the concentration, c$'$ (in g/ml) (for solutions):

$$[a]^1_\lambda = \frac{a}{1\cdot\rho}\ \text{(liquids), or}\ [a]^1_\lambda = \frac{a}{1\cdot c'}\ \text{(solutions)} \tag{1}$$

The proportionality constant (a) is called *specific rotation*. It depends on the temperature, the wavelength, and, in some cases, on the concentration.

To compare compounds of different molecular weights, the *molar (or molecular) rotation* is used

$$[M]\ \text{or}\ [\varPhi]:\quad \frac{[\alpha]\cdot\text{molecular weight}}{100} \tag{2}$$

---

[1] *V. Prelog*, Koninkl. Nederl. Akad. Wetensch. *71B*. 108 (1968).

According to Fresnel[3] linearly polarized light can be thought to be composed of two circularly polarized light rays of equal amplitudes and phases, but opposite senses of helicity. If one faces a circularly polarized light beam, the tip of the light vector (time dependent) describes a circle in a clockwise (right circular) or anticlockwise (left circular) direction. The magnitude of the sum vector of both changes like a cosine-function; the direction of vibration remains constant. In an optically active medium the two oppositely directed circularly polarized light beams travel with different speeds. From this a phase lag between the two rays results growing linearly with the length of the light path, thus leading to a rotation of the plane of vibration of the linearly polarized light:

$$\alpha = \frac{1800 \cdot 1 \cdot \Delta n}{\lambda_{vac}} \text{ [in degrees]} \tag{3}$$

$\Delta n = n_l - n_r$; $\lambda_{vac} =$ wavelength in vacuum; n_l and n_r are the indices of refraction for left and right circularly polarized light of wavelength λ_{vac}.

The absolute value of the rotation increases with decreasing wavelengths. The corresponding curve is called *normal optical rotatory dispersion curve* (ORD; Fig. 1). α can be described by a one-term or multiterm *Drude equation*[4]

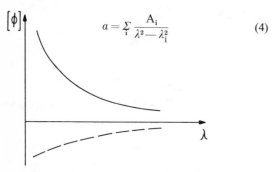

$$a = \sum_i \frac{A_i}{\lambda^2 - \lambda_i^2} \tag{4}$$

Fig. 1. Positive (——) and negative (------) normal ORD-curve

1.2 Circular Dichroism, Anomalous ORD and Cotton Effect

In the range of absorption, in general, the two circularly polarized light rays of opposite direction are absorbed to different degrees in an optically active medium. This property is called *Circular Dichroism* (CD).

$$\Delta\varepsilon = \varepsilon_l - \varepsilon_r \tag{5}$$

(ε_l, ε_r = molar absorptivities for left and right circularly polarized light).

Because of this differential absorption, the two light vectors are no longer of equal length when leaving the cell. The tip of the sum vector, therefore, describes an ellipse, which is traversed clockwise in case of a positive CD, anticlockwise in case of a negative. The *ellipticity,* ψ (defined, as usual, as the arctangent of the ratio of short to long axis) is correlated with $\Delta\varepsilon$ by the equation

$$\psi = 33 \cdot c \cdot d \cdot \Delta\varepsilon \tag{6}$$

d = path length (in cm), c = concentration (in g/l).

In analogy to specific and molar rotation, *specific ellipticity* $[\psi]$ and *molar ellipticity* $[\theta]$ are defined as

$$[\psi] = \frac{\psi}{1 \cdot c'} \text{ and } [\theta] = \frac{[\psi] \cdot \text{molecular weight}}{100} \tag{7}$$

The CD can, therefore, also be described by $[\theta]$; the correlation with $\Delta\varepsilon$ is given by

$$[\theta] \simeq 3300 \cdot \Delta\varepsilon \tag{8}$$

Theory shows that within CD-bands the normal ORD-curve has superimposed on it an S-shaped curve, whose extremes are called *peak* and *trough* (Fig. 2). Such an ORD-curve is called 'anomalous', and both effects, anomalous ORD and CD, are named *Cotton effect*. Proceeding from long to short wavelengths, in case of a positive CD the peak appears before the trough, in case of negative CD the trough comes before the peak. The point of intersection of the anomalous ORD-curve with the normal 'background rotation' curve coincides with the maximum of the CD. If $[\Phi_1]$ is the molar rotation at the longest wavelength of the Cotton effect and $[\Phi_2]$ that at the shortest wavelength, then the *amplitude a* of an ORD curve is used for its quantitative description:

$$a = \frac{[\Phi_1] - [\Phi_2]}{100} \tag{9}$$

For Gaussian shaped CD-curves the following relation holds:

$$a \simeq 40 \cdot \Delta\varepsilon_{max} \tag{10}$$

The positive or negative maximum of a CD-curve need not coincide exactly with the maximum of the isotropic absorption [which is defined as the average value $\varepsilon = \frac{1}{2}(\varepsilon_l + \varepsilon_r)$]. However, it is always determined by the chromophore. For the

[2] *R.S. Cahn, Sir C. Ingold, V. Prelog*, Angew. Chem. *78*, 413 (1966).
[3] s. *T.M. Lowry*[51].

[4] *P. Drude*, Lehrbuch der Optik, p. 379, Hirzel Verlag, Leipzig 1900.

determination of structure one uses mainly the sign of the Cotton effect, its magnitude, and, sometimes, the type of its fine structure.

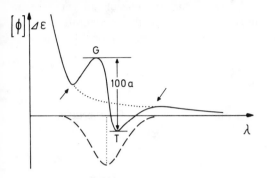

Fig. 2. Cotton effect (here e.g. negative)
------- CD-curve ($\Delta\varepsilon$)
——— anomalous ORD-curve ($[\Phi]$)
.... corresponding plain curve (background rotation)
G = peak   T = trough
a = 'amplitude' of the Cotton effect. For exact determination one must correct for the change of the background rotation within the Cotton effect.
Arrows indicate pseudo-extremes. Inflexion point of ORD- and maximum of CD-curve coincide (λ_{max})

For comparison with theoretical calculations, the Cotton effect is quantitatively described by the *rotational strength* (rotatory strength) R:

$$R = \frac{3 \cdot h \cdot c \cdot 10^3 \cdot \ln 10}{32 \cdot \pi^3 \cdot N_L} \int \frac{\Delta\varepsilon}{v} \, dv \qquad (11)$$

or the *reduced rotational strength* [R]:

$$[R] = \frac{100 \cdot R}{\mu_B \cdot D} = 1.08 \cdot 10^{40} \cdot R \qquad (12)$$

(μ_B = Bohr magneton, D = Debye)

ORD- and CD-curves (within the kth absorption band) are correlated to each other by the *Kronig-Kramers-Relation:*

$$[\Phi_\kappa(\lambda)] = \frac{2}{\pi} \int\limits_0^\infty [\Theta_\kappa(\lambda')] \cdot \frac{\lambda'}{\lambda^2 - \lambda'^2} \, d\lambda' \qquad (13)$$

$$[\Theta_\kappa(\lambda)] = -\frac{2}{\pi\lambda} \int\limits_0^\infty [\Phi_\kappa(\lambda') \cdot \frac{\lambda'^2}{\lambda^2 - \lambda'^2} \, d\lambda' \qquad (14)$$

The ratio of CD to electron absorption has been named dissymmetry factor. Two definitions are used:

$$g = \frac{4R}{D} \qquad (15)$$

and a wavelength dependent one

$$g'(\lambda) = \frac{\Delta\varepsilon}{\varepsilon} \qquad (16)$$

Hitherto g or g' have been used only seldom to solve stereochemical problems (e.g. with lactones[5]).

1.3 Theory
In principle, according to Rosenfeld, the rotational strength can be calculated theoretically. It is the imaginary part of the scalar product of the electric ($\vec{\mu}$) and the magnetic ($\vec{m}$) transition moment vectors, which are correlated with the electron excitation in question: $\Psi_g \to \Psi_a$.

$$R = \text{Im}\{(\vec{\mu} \cdot \vec{m})\} \qquad (17)$$

or in another form:

$$R = \mu \cdot m \cdot \cos(\vec{\mu}, \vec{m}) \qquad (18)$$

$\vec{\mu}$ and $\vec{m}$ are integrals of the form

$$\vec{\mu} \equiv \langle \Psi_g \mid e \sum_i r_i \mid \Psi_a \rangle \,;$$

$$\vec{m} \equiv \langle \Psi_e \mid -\frac{e\hbar i}{2m_e c} \sum_i [r_i, \Delta_i] \mid \Psi_g \rangle \qquad (19)$$

$\vec{\mu}$ and $\vec{m}$ may be assumed to be localized within one single chromophore (One-electron theory; static coupling) or not (polarizability theory; dynamic coupling). The 'one-electron theory' involves the motion of an electron during excitation in a chiral positive restfield of a nucleus not fully shielded by its electron cloud. In the 'polarizability theory' the movement of the excited electron polarizes all the bonds of the molecule[6]. In a molecule, if two chromophores with large $\vec{\mu}$ are located near each other, then W. Kuhn's model of

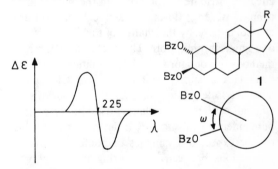

Fig. 3. CD-couplet of *vic*-dibenzoates
A negative torsion angle ω (defined in lit.[7]) as in 2a, 3β-dibenzoyloxy-5a-steroid (**1**) leads to a 'negative' couplet (branch at long wavelengths, negative, at short wavelengths, positive)[8].

[5] *A.F. Beecham, R.R. Sauers*, Tetrahedron Lett. 4761 (1970).
[6] s. *J.A. Schellman*, Accounts Chem. Res. *1*, 144 (1969).
[7] *W. Klyne, V. Prelog*, Experientia *16*, 521 (1960).
[8] *N. Harada, K. Nakanishi*, J. Am. Chem. Soc. *91*, 3989 (1969).

'coupled oscillators' can be applied successfully (the displacement of electron charges along the direction of $\vec{\mu}_1$ and $\vec{\mu}_2$ causes, by chiral rearrangement, a charge rotation and a magnetic transition moment $\vec{m}$. In such a case the CD-curve (Fig. 3) shows a 'couplet'[6] within an absorption band (i.e., two very intense CD-bands of opposite signs). This method has been applied in particular to the calculation of the CD-spectra of *biopolymers* with ordered secondary structures.

2 Polarimetry[9]

2.1 Measurements at a Single Wavelength

For many decades only rotations at a single wavelength (mostly NaD line, 589 nm) were measured instead of the various Cotton effects. Since $[\alpha]_D$ under reproducible conditions, is a constant for any substance, concentrations of optically active compounds in solution can be determined from it (*saccharimetry, sugar inversion,* etc.[9]). Originally, Whiffen[10] in the field of sugars, and later, Brewster[11], in general for chiral organic compounds, stated a rule which can be used to calculate $[M_D]$ from increments of the individual substituents. If the configuration is known, one can determine the preferred conformation from this.

2.2 Measurement of Differences

2.2.1 Basic Principles

According to the Drude equation (4) all Cotton effects contribute to rotation. On the other hand, changes of $[M_D]$ during a chemical reaction are caused, mostly, by the change of the Cotton effect of a single chromophore. This fact is used in many methods of configuration determination based on van't Hoff's 'superposition rule' and Chugaeff's 'distance rule'. According to the *superposition rule,* all centers (axis, planes) of chirality contribute independently to $[M_D]$; deviations from this can be expected if changes of conformations or interactions between neighboring chromophores occur ('vicinal effect' of Freudenberg). The *distance rule,* which has been reformulated by Klyne[12] for the Cotton effects states, that any chemical alteration will affect $[M_D]$ less as the distance from the chromophore increases.

2.2.2 Applications

Hudson's Isorotation Rules: $[M_D]$ of a sugar derivative is composed additively of contributions from C-1 and from the rest of the pyranose or furanose ring. From this the configuration at C-1 can be determined: For an α-D- or β-L-configuration

$[M_D]$ is more positive than for the diastereomer with β-D- or α-L-configuration

Pyrimidine nucleosides are exceptions to the rule, since, in these compounds, the heterocyclic ring is itself, a chromophore whose CD-bands appear at longer wavelengths than those of the acetal (or aminal) chromophore.

Rule of Bose – Chatterjee[13]:

is more dextrorotatory than its enantiomer. S and L stand for small and large substituents.

Mills' rule[14]:

An allylic alcohol of general formula A is more dextrorotatory than its isomer B.

Rules of shift (Levene, Hudson, Freudenberg et al.):
$\Delta\,[M_D] \equiv (M_D(\text{derivative})] - [M_D(\text{acid})]$ is positive for an α-hydroxy acid of configuration C at C-2, if the derivative is the alkali salt, the amide, the phenylhydrazide or the benzimidazole.

[9] s.a. *Houben-Weyl,* Methoden der Organischen Chemie, 4. Aufl., Bd. III/2, p. 425, Georg Thieme Verlag, Stuttgart 1955.
[10] *D.H. Whiffen,* Chem. Ind. [London] 964 (1956).
[11] *J.H. Brewster,* Topics in Stereochemistry *2,* 1 (1967).

[12] *W. Klyne,* Tetrahedron *13,* 29 (1961).
[13] *A.K. Bose, B.G. Chatterjee,* J. Org. Chem. *23,* 1425 (1958).
[14] *J.A. Mills,* J. Chem. Soc. 4976 (1952).

Lactone Rule[15-17]: $[M_D]$ of a γ- or δ-lactone of configuration D is more positive than that of the corresponding hydroxy acid, salt, or diol obtainable by $LiAlH_4$ reduction.

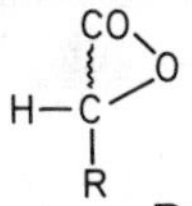

D

Rules of Barton: Barton[18] has published rotation increments for reactions such as acetylation, oxidation of OH-groups, hydrogenation of double bonds, epimerization of hydroxy groups, etc., in the steroid and terpenoid field; from these, the configuration at the respective chiral centers can be determined.

Lutz-Jirgensons' Rule: $[M_D]$ of an α-L-amino acid becomes more positive by going from neutral to acidic solution.

3 Application of CD and ORD to Low Molecular Weight Compounds

3.1 Introduction

Since about 1953 Djerassi has provided the basis for the many different applications of CD and ORD in stereochemistry by his intense investigations of the Cotton effect especially in the steroid and terpenoid field. Based upon thousands of measurements, some general rules can be stated. Thus, in an optically active compound the chromophore itself may be chiral ('inherently' or 'intrinsically chiral' or 'dissymmetric') leading, mostly to a very strong Cotton effect; or it may be locally symmetrical, but perturbed dissymmetrically by its chiral surrounding (classification according to Moscowitz). In the latter case in general, $|\Delta \epsilon|$, is less than 10. In the first case the electron absorption curve and the CD-curve are proportional to each other (i.e., g' is constant throughout the band); in the latter case they are not (the CD-maximum, in general, may be found at longer wavelengths by some nm than the absorption maximum) (Fig. 4 and 5).

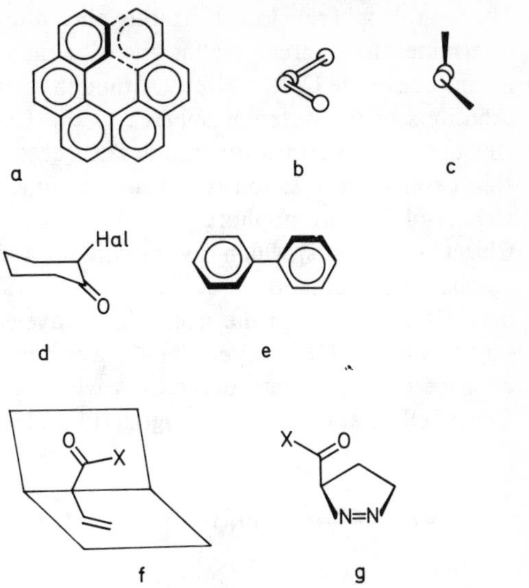

Fig. 4. Examples of inherently dissymmetric chromophores
a) Hexahelicene
b) $\pi \rightarrow \pi^*$-band of non-coplanar conjugated dienes C=C—C=C and enones C=C—C=O
c) Disulfides[19] (torsion angle $\neq 90°$)
d) axial α-haloketones
e) Bands of non-coplanar biphenyls whose corresponding electron excitations are not localized in one single ring
f) R-band of β,γ-unsaturated oxo-compounds (ketones, aldehydes, lactones...) with proper geometry (angle between the two planes appr. 100–120°)
g) n$\rightarrow \pi^*$-bands of the azochromophore in appropriately substituted (X=C or O) nonaromatic azo compounds[20]

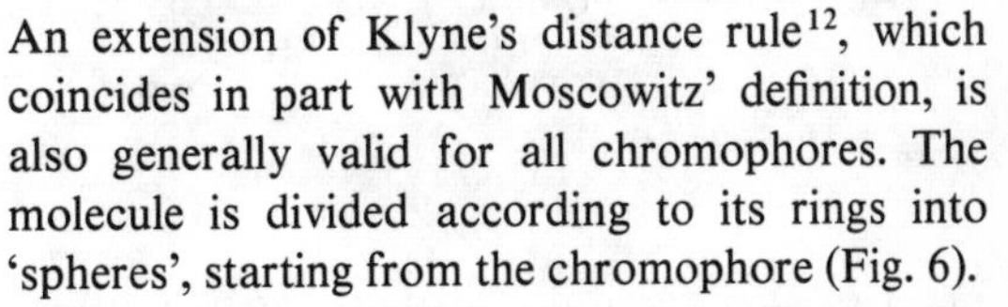

Fig. 5. Examples of inherently symmetric, but chirally perturbed chromophores.

An extension of Klyne's distance rule[12], which coincides in part with Moscowitz' definition, is also generally valid for all chromophores. The molecule is divided according to its rings into 'spheres', starting from the chromophore (Fig. 6).

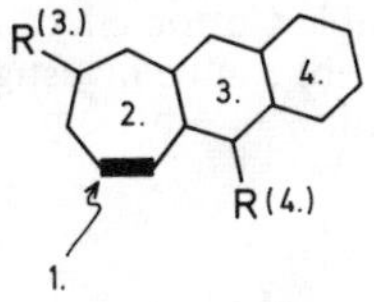

Fig. 6. Schematic division of a molecule into spheres. The chromophore itself (symbolized by the thick bar) forms the 1st sphere, the ring into which it is incorporated the second, rings or groups bonded to this latter form the third sphere, etc.

[15] *C.S. Hudson,* J. Am. Chem. Soc. *32,* 338 (1910).
[16] *W. Klyne,* Chem. Ind. [London] 1198 (1954); J. Chem. Soc. 7237 (1965).
[17] *M. Romaňuk, J. Křepinský,* Collect. Czech. Chem. Commun. *29,* 830 (1964).
[18] *D.H.R. Barton,* J. Chem. Soc. 813 (1945) und folgende Arbeiten.
[19] *J. Lindenberg, J. Michl,* J. Am. Chem. Soc. *92,* 2619 (1970).
[20] *G. Snatzke,* Riechst., Aromen, Körperpflegem. *19,* 1 (1969).

The chiral sphere closest to the chromophore determines to a great extent the sign and the magnitude of the Cotton effect. Although the contributions of the different spheres to the Cotton effect are additive, their magnitudes decrease if the chiral sphere in question is not in the immediate vicinity of the chromophore.

Chiral compounds, which do not contain a chromophore according to the definition of Witt, show only a ORD-curve in the accessable wavelength range, and no CD curve. They may, however, often be converted into derivatives which give a Cotton effect at longer wavelengths (Fig. 7).

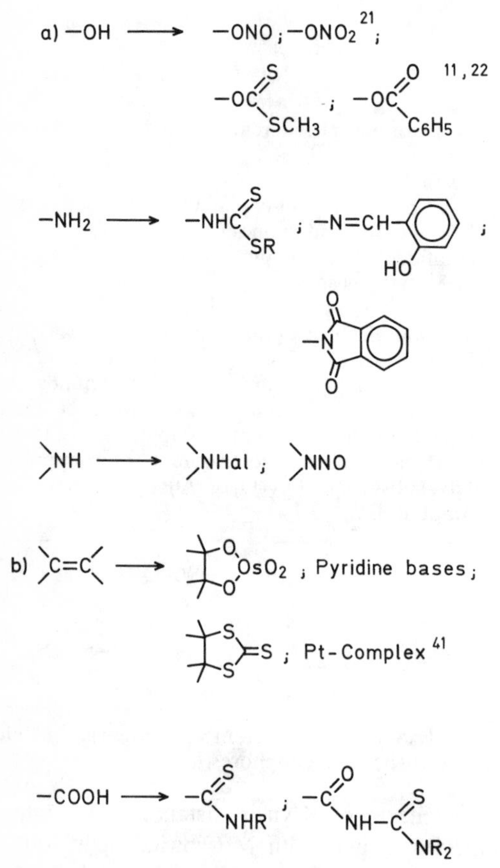

3.2 Oxo Compounds

Until now only the Cotton effect within the n→π* band has been investigated thoroughly, and many rules exist correlating its sign with the stereochemistry (schematically represented in Fig. 8),

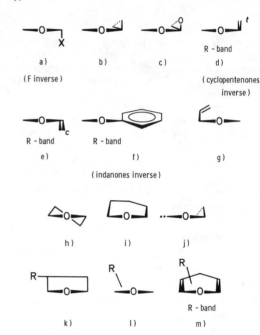

Fig. 8. Schematic representation of the more important rules for ketones (Projection from O to C of the carbonyl), which lead to a positive Cotton effect.

Chirality of the 1ˢᵗ sphere:	a) axial α-haloketones
	b) 'conjugated' cyclopropylketones
	c) 'conjugated' oxidoketones
	d) transoid non-coplanar conjugated enones
	e) cisoid non-coplanar conjugated enones
	f) non-coplanar aryl ketones
	g) β,γ-unsaturated oxocompounds with proper geometry (cf. Fig. 4f)
Chirality of the 2ⁿᵈ sphere:	h) cycloalkanones in twist conformation
	i) cycloalkanones of asymmetric conformation
	j) cyclohexenones with coplanar enone grouping
Chirality of the 3ʳᵈ (4ᵗʰ...) sphere:	k) cycloalkanone with C$_s$-symmetry
	l) acyclic oxocompounds
	m) crossed conjugated cyclohexadienones

Fig. 7. Examples of 'cottogenic derivatives' which have a clearly measurable Cotton effect.
The groups under b) can be investigated directly with modern equipment.

[21] *L.D. Hayward, S. Claesson*, Chem. Comm. 302 (1967). − *Y. Tsuzuki* et al., Bull. Chem. Soc. Jap. *38*, 274 (1965); *39*, 761, 1391, 2269 (1966). − *G. Snatzke, H. Laurent, R. Wiechert*, Tetrahedron *25*, 761 (1969).

[22] *N. Harada, M. Ohashi, K. Nakanishi*, J. Am. Chem. Soc. *90*, 7349 (1968). − *N. Harada, K. Nakanishi*, J. Am. Chem. Soc. *90*, 7351 (1968).

3.2.1 Saturated Oxo Compounds

Cyclohexanones in the chair conformation obey the octant rule (Figs. 8k, 9), according to which the contribution of any substituent (besides fluorine) to the Cotton effect is proportional to the product of

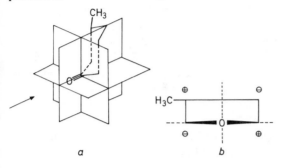

a

b

Fig. 9. Octant rule for oxocompounds with achiral 1ˢᵗ and 2ⁿᵈ sphere.
a) perspective view of the octants in case of (R)-3-methyl-cyclohexanone (chair conformation)
b) 'octant projection' from O to C of carbonyl. The signs given correspond to back octants

its coordinates. Thus, its sign changes if this substituent is mirrored by a nodal surface of the n- or π*-orbital, resp. Whether this is true also for the third nodal surface perpendicular to the C=O bond is still a matter of controversy; symmetry considerations[23] require at least a quadrant rule. According to calculations by Pao and Santry[24] a fourth nodal surface which is also perpendicular to the C=O bond (Fig. 10) must be added to the other three. The CD of monosubstituted adamantanones is in excellent agreement with this[25]. Also, according to group theory calculations by Ruch[26], a more complex dependence upon the coordinates is expected.

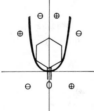

Fig. 10. Fourth nodal surface (thick) for ketones according to Pao and Santry[24] in a projection from the top on to the plane of the carbonyl group. The signs given correspond to the upper octants and sectors, resp.

[23] *J.A. Schellman*, J. Chem. Phys. *44*, 55 (1966).
[24] *Y.H. Pao, D.P. Santry*, J. Am. Chem. Soc. *88*, 4157 (1966).
[25] *G. Snatzke, G. Eckhardt*, Tetrahedron *24*, 4543 (1968); *26*, 1143 (1970); *G. Snatzke, B. Ehrig, H. Klein*, Tetrahedron *25*, 5601 (1969).

The octant rules has been applied in very many cases to the determination of constitution, configuration or conformation (cf. Tab. 1 for some keto steroids).

Table 1. Dependence of $\Delta\varepsilon$ for some keto steroids on the position and configuration at C-5 (keto group at C-1 through C-12) or C-14 (keto group at C-15 through C-17), resp.

Position of keto group	$\Delta\varepsilon_{max}$ α- configuration	β-	Position of keto group	$\Delta\varepsilon_{max}$ α- configuration	β-
1	−0.4	−3.4	11	+0.3	+0.4
2	+3.0	−0.8	12	+1.0	+0.6
3	+1.3	−0.5	15	+3.3	−2.5
4	−2.3	+0.1	16	−5.8	+3.4
6	−1.4	−4.2	17	+3.5	+1.1
7	−0.7	+0.7			

The application of the octant rule may be illustrated by the two 1-keto steroids isomeric at C-5 (Fig. 11). The first and second spheres are achiral in both isomers. For **2** methyl C-19 and methylene C-17 lie in nodal planes and, thus, do not contribute to the CD; the contribution of C-8 is compensated by that of C-6. All other C-atoms are in the lower left, i.e. negative octant, which leads to a medium strong negative Cotton effect (cf. Tab. 1). In case of **3**, however, extensive compensation occurs because some C-atoms lie in front octants. The CD is, thus, only weakly negative (cf. Tab. 1); highly characteristic, however, is the extraordinary strong positive background rotation as well as the change of sign of the CD by cooling the solution down to −190 °C.

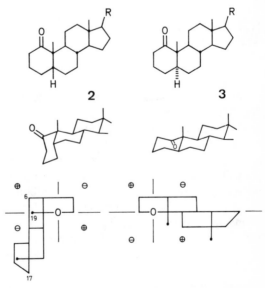

2

3

Fig. 11. Octant projections of 5β- (**2**) and 5α-cholestan-1-one (**3**). For **3** many atoms are already positioned in front octants.

Cl, Br, I, COOR and some other groups give an antioctant contribution to the CD if they are positioned in axial conformation at C-β to the carbonyl; F, methyl, and O-methyl at the same position have practically no influence upon the Cotton effect[25].

In the presence of a chiral second sphere the sign of the CD can be determined according to Fig. 8h, i or j. By this rule, the enantiomorphic CDs of A-nor-5α-cholestanone-2 (**4**) and the 16-keto steroid **5** follow from Fig. 12.

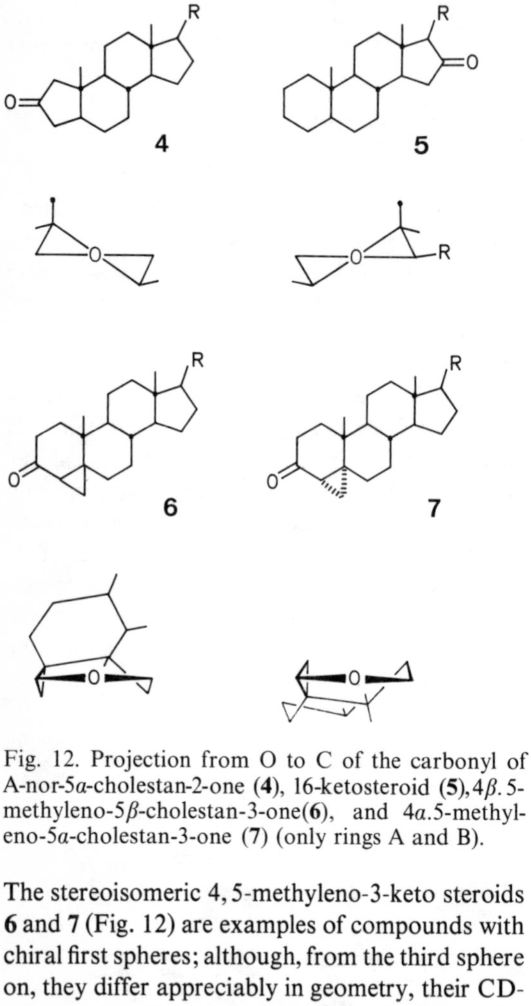

4   **5**

6   **7**

Fig. 12. Projection from O to C of the carbonyl of A-nor-5α-cholestan-2-one (**4**), 16-ketosteroid (**5**),4β.5-methyleno-5β-cholestan-3-one(**6**), and 4a.5-methyleno-5α-cholestan-3-one (**7**) (only rings A and B).

The stereomeric 4,5-methyleno-3-keto steroids **6** and **7** (Fig. 12) are examples of compounds with chiral first spheres; although, from the third sphere on, they differ appreciably in geometry, their CD-curves are exactly mirror images of each other; their CD is + 2.35 (**6**) and −2.39 (**7**), resp. (cf. Fig. 8b). An oxido ring 'conjugated' with a carbonyl acts in a way similar to the cyclopropane ring in the analogous position (Fig. 8c). In both cases a second Cotton effect around 200 nm can be observed[27]; in the case of cyclopropyl ketones, the

sign is, in general, opposite to, that of the CD within the n→π* band, in the case of oxido ketones it is the same. A few exceptions to these rules are known[27] and can be reasonably explained. If e.g. the chirality of the second (third,..) sphere is very strong, but that of the first very weak, then the second (third,..) sphere may determine the CD and thus lead to the 'wrong' sign. The octant rule was preceded by the 'axial α-halo keto rule' (Fig. 8a). NH$_2$ or S-groups act similar to halogens. The strong interaction of an axial halogen in α-position to a carbonyl is seen in the great bathochromic shift of the UV- and CD-band of the n→π* transition. Cl, Br, and I (as NH$_2$ and -S-) give positive CDs if the absolute configuration is as in Fig. 8a; F, on the contrary, gives a negative one. Exceptions are possible if the halogen (e.g. in a cyclopentanone) is not exactly axially positioned and the second or third sphere is strongly chiral.

The same rules can be applied to acyclic oxo compounds and, in favorable cases, the preferred conformation of the chain can be determined. Djerassi investigated in detail aldehydes (**8**) and methyl ketones (**9**) (Fig. 13). For the configuration

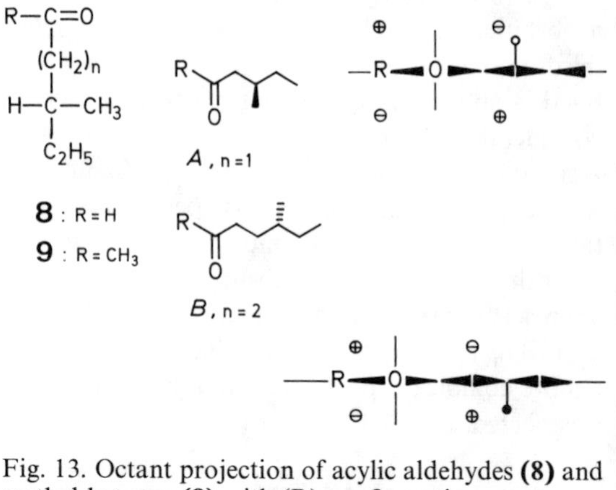

8 : R = H
9 : R = CH$_3$

Fig. 13. Octant projection of acyclic aldehydes (**8**) and methyl ketones (**9**) with (R)-configuration.
↑ symbolizes a methyl group in a back octant, ↑ one in a front octant. The signs given correspond to back octants, for the front octants they are opposite.

given, a negative Cotton effect is found in the case of n = 0, a positive one for all n ⩾ 1. If the carbonyl is bound directly to the center of chirality, then several conformations have the same energy, and the sign of the CD cannot be predicted directly for such a mixture of conformers. In the case of n ⩾ 1, that conformation is preferred, in which the longest chain has a zig-zag shape with the methyl group standing out from the plane. This is in agreement with the octant rule (first sphere achiral, no second sphere present, third sphere determines the CD); at the same time it is an indication of the fact that, indeed, an octant and not a quadrant rule is valid

[26] E. Ruch, A. Schönhofer, Theor. Chim. Acta *19*, 225 (1970).
[27] K. Kuriyama, H. Tada, Y.K. Sawa, S. Itŏ, I. Itoh, Tetrahedron Lett. 2539 (1968).

for oxo compounds. The Cotton effect of other similar ketones (e.g. of the 20-keto pregnanes **10–15** (Fig. 14), can be explained in the same way. Substituents which alter the conformation of the side chain appreciably, although the configuration remains the same, may, of course, influence strongly the CD (cf. e.g. **11**).

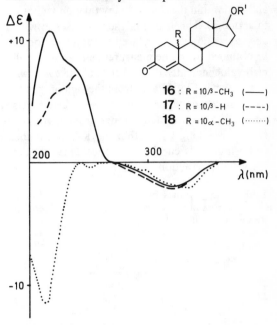

affects the R-band. For the absorption at lowest energy, mainly the helicity of the non-coplanar chromophore determines the CD band. In the transitions at higher energy, a slight hybridization with σ-orbitals already seems possible.

16 : R = 10β -CH$_3$ (——)
17 : R = 10β -H (----)
18 R = 10α -CH$_3$ (·······)

Fig. 15. CD-curves of some Δ^4-unsaturated 3-keto-steroids (**16–18**) (in ethanol).

Fig. 14. CD ($\Delta\varepsilon_{max}$) of some 20-ketopregnanes. **11** shows a double-peaked curve at the wavelengths given. All other data refer to the maximum at around 290–300 nm.

10   + 3.5
11   + 0.1 (326) − 0.2 (285)
12   + 4.8
13   + 1.5
14   − 2.6
15   − 1.4

3.2.2 Unsaturated Oxo Compounds

α-, β-*Unsaturated oxo compounds:* If the C=O group is not coplanar with the C=C group, as is usually the case in acyclic compounds, then the first sphere is chiral and the rules of Fig. 8d, e hold for the n→π*-CD-band. Since both the π and the π*-orbitals are chiral, the π→π* band of the C=C−C=O chromophore is an inherently dissymmetric system, and its sign is in general opposite to that of the n→π* CD-band. Furthermore, a third, frequently very intense, Cotton effect around 200 nm has been observed, which coincides with a minimum in the UV-absorption curve. The corresponding transition must, therefore, be forbidden, and may perhaps be the second n→π* band. Since both the rule mentioned above and correlation of the helicity of an enone system with the sign of its CD within the K-band often fail, another correlation with allylic axial subtituents[27a] has been suggested recently. Similarly the sign of the CD-band at shortest wavelengths has been correlated with the axial substituent next to the C=O group[27a]. Fig. 15 shows three examples of transoid enones; replacement of CH$_3$ (**16**) with H (**17**) scarcely

For the R-band of cyclopentenones, in general, an inverse rule to that of the above mentioned transoid enones has been found experimentally. This change of sign of the CD in spite of the same helicity can be explained theoretically[28].

For arylketones a rule (Fig. 8f) holds which is identical to that for corresponding transoid (Fig. 8d) and cisoid (Fig. 8e) enones. Its application to the determination of the configuration of some flavanone derivatives is shown in Table 2[29].

Table 2. CD values ($\Delta\varepsilon_{max}$) of the R-band of some flavanone derivatives at about 330 nm[29]

No.	Compound	R^1	R^2	R^3	$\Delta\varepsilon_{max}$
19	(−)-naringenin glucoside	glu-cose	H	H	+7.60
20	taxifolin	H	OH	OH	+4.90
21	astilbin	H	O-rham-nose	OH	+2.65

19–21

[27a] A.W. Burgstahler, R.C. Barkhurst, J. Am. Chem. Soc. *92*, 7601 (1970).

[28] W. Hug, G. Wagnière, Helv. Chim. Acta *54*, 633 (1971).

β,γ-Unsaturated oxo compounds: If the C=C bond is arranged in such a way relative to the C=O bond that a chiral interaction is possible with the n- as well as with the π-orbital (Fig. 4f and 8g), an inherently dissymmetric system is produced and $|\Delta\epsilon|$ within the R-band may take on values up to about 30. This is one of the few cases in which absolute configuration as well as conformation can be obtained from one measurement: an unusually high value indicates the presence of the *conformation* drawn in Fig. 4f, and, from the sign of the CD, the *absolute configuration* can be inferred. Also acyclic β,γ-unsaturated ketones or aldehydes can show this strong CD within the R-band, if their conformation is energetically favorable (cf. Tab. 3) for this interaction.

Table 3. $\Delta\epsilon_{max}$-values of the R-band of the 19-oxo-chromophore of some nonrigid β,γ-unsaturated oxo-steroids at $+20°$ and $-178°$, resp.[30]

Compound	$\Delta\epsilon_{max}$ at	
	20°	−178°
22	− 10.12	− 13.88
23	− 7.48	− 8.29
24	+ 9.64	+ 8.15

22 : R = H
23 : R = CH$_3$

24

Conjugated dienones: The rules of Fig. 8d, e, and j can be applied also to linear dienones. Cross conjugated cyclohexadienones, on the contrary, have, without the presence of perturbing polar groups, a symmetric first and second sphere. The octant rule may, therefore, be applied to their R-band CD (Fig. 8m). In addition to this, they show three more CD-bands at shorter wavelengths (Fig. 16).

3.3 Acid Derivatives

For the $n\rightarrow\pi^*$ CD band of *acid derivatives*, rules can be applied similar to those for the corresponding oxocompounds. In the case of simple acids and esters, as well as lactams or lactones

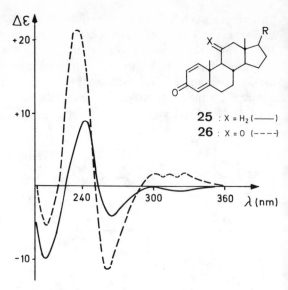

Fig. 16. CD of the cross conjugated dienones **25** and **26**.

25 : X = H$_2$ (———)
26 : X = O (- - - -)

with a coplanar ring, the first and (if present) second sphere is achiral; for the influence of more distant spheres, different rules have been proposed. Klyne uses a modified octant rule, Snatzke et al. an unsymmetric sector rule, and Schellman[31] a quadrant rule. Thus far, differentiation between these rules is not easily possible, since the relative magnitudes of the contributions in the different sectors, as well as the exact shapes of the nodal surfaces, are still uncertain.

3.3.1 Saturated Lactones

In general the $-C(=O)-O-C$ grouping of γ- or δ-*lactones* is coplanar; the ring, however, is not. According to Legrand[32] or Wolf[33] the torsion angle around the $C(=O)-C_\alpha$ bond determines, in this case of a chiral second sphere, the sign of the CD. Fig. 17 illustrates the rule and its application. In simple δ-lactones an equilibrium between half-chair and boat conformations is possible. Two CD-bands of opposite signs are, therefore, obtained, the one at shorter wavelength corresponding to the boat conformation[33]. The ratio of the rotational strengths of these two bands depends strongly upon the solvent used, since this influences the position of the equilibrium.

For a chiral first sphere, as e.g. in α,β-cyclopropyl[35] or α,β-oxido lactones, the rules of Fig. 8b, c can be applied analogous to the case of the corresponding ketones.

[29] *K.R. Markham, T.J. Mabry,* Tetrahedron *24,* 823 (1968).
[30] *G. Snatzke, K. Schaffner,* Helv. Chim. Acta *51,* 986 (1968).
[31] *B.J. Litman, J.A. Schellman,* J. Phys. Chem. *69,* 978 (1965).

[32] *M. Legrand, R. Bucourt,* Bull. Soc. Chim. Fr. 2241 (1967).
[33] *H. Wolf,* Tetrahedron Lett. 5151 (1966).

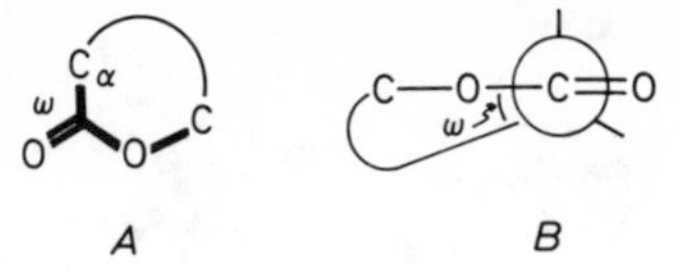

A

B

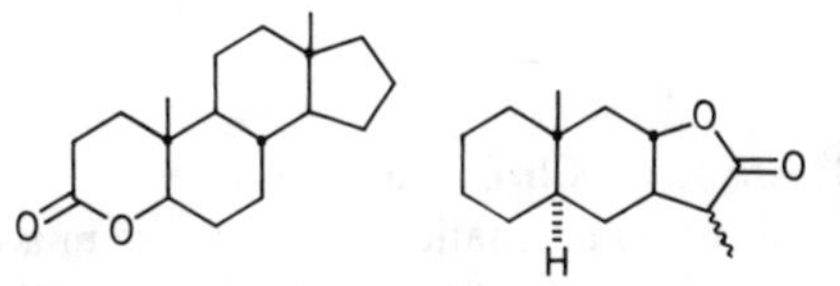

27   **28**

Fig. 17. Determination of the sign of the CD of saturated lactones with achiral first sphere.
A = Schematic representation of the ring. The bonds marked with bold-faced bars are coplanar.
B = Newman projection along the $(O=)C-C_\alpha$-bond. If the torsion angle ω is negative, as assumed here, the CD of the R-band is positive. This rule is applied to a δ- (**27**) and a γ-lactone (**28**)[34]

3.3.2 Unsaturated Acid Derivatives

If the first sphere is chiral, the corresponding ketone rules (Fig. 8d, g) are valid also for the derivatives of α,β- and β,γ-*unsaturated acids*. In applying them on the basis of molecular models one must be sure to keep the lactone grouping coplanar. Three examples are presented in Fig. 18. In the case of conjugated lactones which have their n→π* band around 260 nm frequently the π→π* CD-band can be measured also. Very often its sign is the same as for the R-band CD.

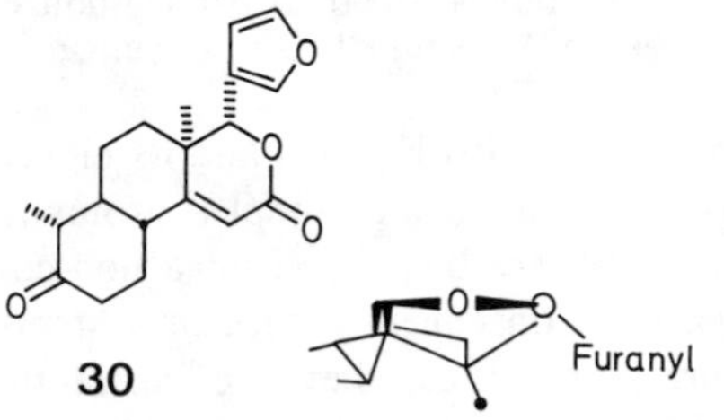

	$\lambda_{max.}$	$\Delta \varepsilon$
29	262	+2.25
30	264	+7.00
31	222	−14.2

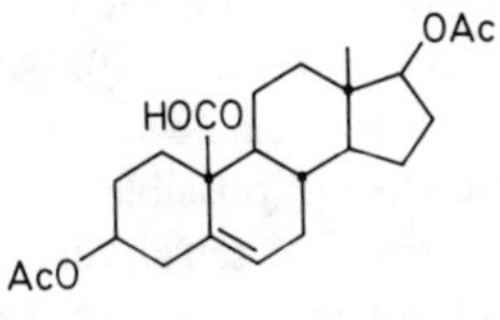

Fig. 18. CD of some unsaturated acid derivatives

[34] *J.P. Jennings, W. Klyne, P.M. Scopes*, J. Chem. Soc. 7211 (1965).
A.F. Beecham, Tetrahedron Lett. 3591 (1968).
[35] *G. Snatzke, E. Otto*, Tetrahedron **25**, 2041 (1969).

3.4 Olefins and Aromatic Compounds

3.4.1 Monoolefins

Only recently has the CD of monoolefins been measured (cf. Ref. [36–40]). The CD-band at longest wavelenght has presumably mainly $\pi\rightarrow\sigma^*$ character, but a contribution from other transitions (especially $\pi\rightarrow\pi^*$) cannot be excluded completely.

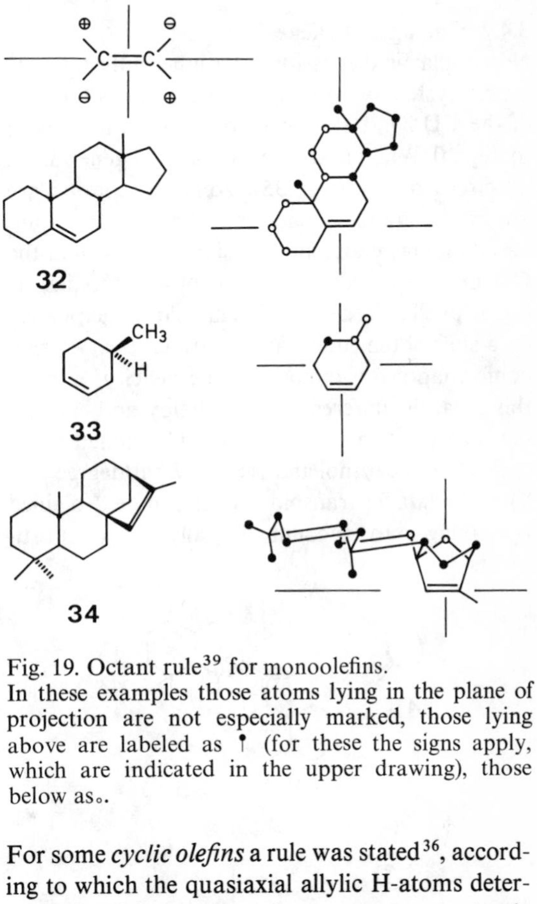

32

33

34

Fig. 19. Octant rule[39] for monoolefins.
In these examples those atoms lying in the plane of projection are not especially marked, those lying above are labeled as ● (for these the signs apply, which are indicated in the upper drawing), those below as ○.

For some *cyclic olefins* a rule was stated[36], according to which the quasiaxial allylic H-atoms determine the Cotton effect. Later on an octant rule was proposed (Fig. 19)[39], which, however, fails for some methylene steroids[40]. Since in *cyclohexenes* (halfchair conformation) the second sphere is chiral, the absolute configuration should determine the sign of the Cotton-effect. However, because of

[36] *A. Yogev, D. Amar, Y. Mazur*, Chem. Comm. 339 (1967).
[37] *M. Legrand, R. Viennet*, C.R. Acad. Sci., Paris **262 C**, 1290 (1966).
[38] *M. Yaris, A. Moscowitz, R.S. Berry*, J. Chem. Phys. **49**, 3150 (1968).
[39] *A.I. Scott, A.D. Wrixon*, Chem. Comm. 1182 (1969); Tetrahedron **26**, 3695 (1970).
[40] *M. Fétizon, I. Hanna*, Chem. Comm. 462 (1970); *M. Fétizon, I. Hanna, A.I. Scott, A.D. Wrixon, T.K. Devon*, Chem. Comm. 545 (1971).

the contribution from different states to the band at longest wavelength, most probably a single rule will not be sufficient to describe the Cotton effect.

Olefins form complexes with Pt II, whose d-d bands fall into an easily measured wavelength range. Recently a quadrant rule has been proposed which correlates the sign of the 440 nm CD-band with the stereochemistry of the olefin[41]. This rule has, however, many exceptions.

3.4.2 Conjugated Dienes

Non-coplanar dienes form an inherently dissymmetric system, whose helicity determines the sign of the CD-band at longest wavelength according to Fig. 20. Whereas this rule is valid, in general, for all cisoid dienes (e.g. **35**), some exceptions are known for transoid ones (e.g. **36**). Temperature variation may cause an inversion of the sign of the CD in compounds conformationally mobile, as e.g. α-phellandrene (**37**)[42]. This can be explained by a shift of the equilibrium between the two ring conformations of opposite diene helicity[43]. From this data, the differences of enthalpy and entropy between the two forms can be determined to be $\Delta H = 0.28$ kcal/mol and $\Delta S = 2.2$ cal/degree mol. The CD data for transoid enones can be explained by taking into account the allyl axial substituents[27a].

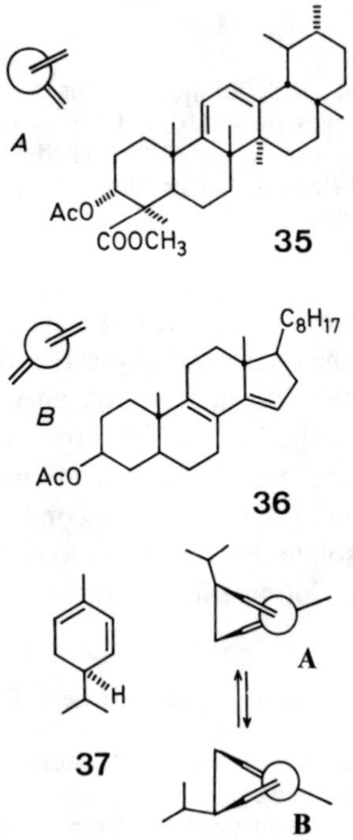

Fig. 20. Helicity rule for cisoid (A) and transoid (B) conjugated dienes.
Both Newman projections given lead to a positive CD. The sign of the CD of α-phellandrene (**37**) depends on the temperature.

Signs of the C=C—C=C Cotton effect

	predicted	found
35	+	+
36	−	−
37 A	−	− (+20°)
B	+	+ (−177°)

3.4.3 Aromatic Chromophores

All bands of an aromatic chromophore in a chiral compound are optically active. However, the CD of the ^1L_b-band is often very weak, especially in the absence of polar substituents. The reason is that $\vec{\mu}$ and $\vec{m}$ are perpendicular one to another, and, therefore, R will be zero without perturbation (hybridization, vibrational deformation etc.). According to their stronger $\vec{\mu}$, the bands at short wavelengths show more pronounced Cotton effects.

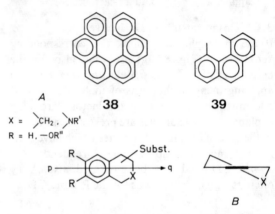

Fig. 21. Cotton effect of ^1L_b-band in tetralins (X=CH$_2$) or tetrahydroisoquinolines (X=NR′) with chiral 2nd sphere.
For a chirality as given in projection B (viewed from p towards q in projection A; the thick bar symbolizes the benzene ring) the ^1L_b-band CD is positive.

Hexahelicene (**38**, cf. also Fig. 4a) and its higher homologs, as well as some simpler aromatic compounds, as **39**, are built up from a non-coplanar σ-system. Consequently their π-systems must be chiral; and, thus, they are inherently dissymmetric chromophores. Their CD is the largest known for solutions ($|\Delta\epsilon|$ several 100). A simple correlation of the Cotton effects with the helicity of the aromatic system is not possible.

The non-aromatic ring of a tetralin or tetrahydroisoquinoline normally is present in the halfchair

[41] A.I. Scott, A.D. Wrixon, Chem. Comm. 1184 (1969); Tetrahedron 27, 2339 (1970).
[42] G. Snatzke, E. sz. Kováts, G. Ohloff, Tetrahedron Lett. 4551 (1966).
[43] A.W. Burgstahler, H. Ziffer, U. Weiss, J. Am. Chem. Soc. 83, 4660 (1961).

conformation and, therefore, is chiral; its chirality determines, according to Fig. 21, the sign of the CD of the ^1L_b-band[44]. For the ^1L_a-band CD no such straightforward rule is valid, as here – similar to the monoolefins – contributions from other transitions (e.g. $\pi \rightarrow \sigma^*$) cannot be completely excluded.

For simple benzene derivatives several rules have been proposed. However, the general principle (cf. chapter 5.7:3.1) explained in Figs. 21 and 22 can be applied.

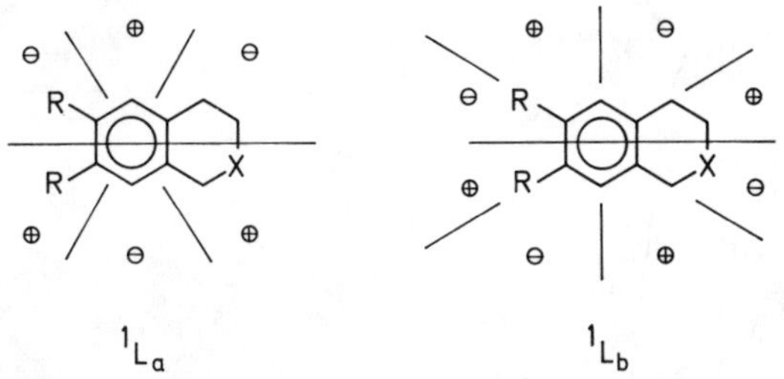

^1L_a ^1L_b

Fig. 22. Sector rule for benzene derivatives
Contributions of the 3rd (4th...) sphere to the Cotton effect. The signs given apply to the upper sectors, the plane of projection is also a nodal plane. The signs for the ^1L_a-band CD are not yet very certain. Hitherto only compounds with R = H or R = OR' have been investigated. Other substitutions may cause additional nodal planes as well as change of signs.

For the contributions of the 3th (4th...) sphere, sector rules hold. One of these[44] is cited in Fig. 22. The ^1L_b- and ^1L_a-transition have different nodal planes, which can be assumed to separate the different sectors, as in the case of ketones (cf. the octant rule). Additional planes (e.g. the plane of the benzene ring) must be added to meet the symmetry requirements of the ring system. This rule can explain why the CDs of hemanthamine (40) and crinamine (41) are of the same magnitude within the ^1L_b-band, but differ within the ^1L_a-band (Fig. 23)[45]. The positive sign of the CD of the first band, as well as the greatest part of its magnitude, is determined by the chirality of the second sphere

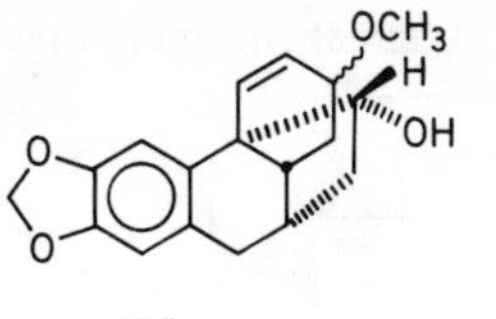

[44] *G. Snatzke, G. Wollenberg, J. Hrbek jr., F. Šantavý, K. Bláha, W. Klyne, R.J. Swan,* Tetrahedron *25,* 5059 (1969). *E. Dornhege, G. Snatzke,* Tetrahedron *26,* 3059 (1970); *G. Snatzke, J. Hrbek jr., L. Hruban, A. Horeau, F. Šantavý,* Tetrahedron *26,* 5013 (1970).
G. Snatzke, P.C. Ho, Tetrahedron *27,* 3645 (1971); *G. Snatzke, M. Kajtár, F. Werner-Zamojska,* XXIIIrd International Congress of Pure and Applied Chemistry, Special Lectures *7,* 117 (1971).

[45] *K. Kuriyama, T. Iwata, M. Moriyama, K. Kotera, Y. Hameda, R. Mitsui, K. Takeda,* J. Chem. Soc. (B) 46 (1967).

40 : 3β -OCH₃
41 : 3α -OCH₃

a   b
40   41
^1L_a -3.97 (244nm)   -2.85 (245nm)

c   d
e
40   41
^1L_b +3.42 (289nm)   +3.52 (290nm)

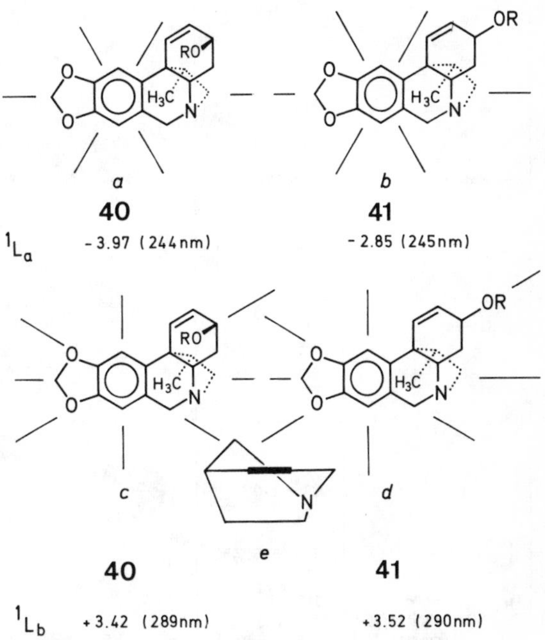

Fig. 23. CD of hemanthamine (40) and crinamine (41). The OR-group is located above the plane of projection for 40, below it for 41. Of the two 2nd spheres, one (six-membered ring) is much stronger chiral than the other (seven-membered ring) and, therefore, determines the CD(e).

(projection e in Fig. 23). Kuriyama et al.[45] and DeAngelis and Wildman[46] have proposed other rules, which are given in Fig. 24.

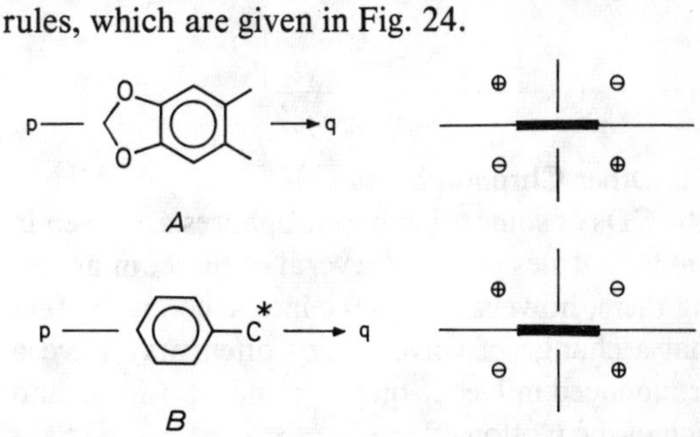

A

B

Fig. 24. Quadrant rule of Kuriyama et al.[45] (A) and of DeAngelis and Wildman[46] (B) for benzene derivatives
Projection is from p towards q, C* is a center of chirality. Rule A is valid mainly for the ^1L_b-band CD, for the ^1L_a-band CD the signs be reversed.
Rule B on the other hand is valid for the ^1L_a-band CD; the sign of the ^1L_b-band CD is mostly opposite

[46] *G.G. DeAngelis, W.C. Wildman,* Tetrahedron *25,* 5099 (1969).

Table 4. Cotton effects of different chromophores

Chromophore	λ_{max}
C=S	490
C=N	250
—C(S)(NR₂)	325 – 360
CH₃—C(O)(S—)	270
—NO	675
—NO₂	280 , 330
—N=N—	230 , 330
—N₃	290
—SCN	245
—S—S—	260
—Se—Se—	340
C—C (S)(S)	265
(O)(S)	240 – 250
(S)(S)	235 – 260
(S)(S) C=S	235 , 305 , 430
R,R' SO dialkyl- alkyl-aryl- diaryl-	210 / 220 , 250 / 245 , 285
PO	210
—O—NO	325 – 390
—O—C(S)(SCH₃)	350
—NH—C(S)(SR)	330
—N—NO	370
N—Hal	250 – 280

3.5 Other Chromophores

The CDs of some other chromophores are given in Table 4. Rules exist for several of these; in applying them, however, one take into account the fact that a change of solvent very often may have a pronounced influence on magnitude, position, and sign of the Cotton effects.

[47] H. Rosenkranz, W. Scholtan, Hoppe-Seyler's Z. Physiol. Chem. 352, 896 (1971).
[48] D.W. Miles, R.K. Robins, H. Eyring, Proc. Nat. Acad. Sci. U.S. 57, 1138 (1967). − D.W. Miles, M.J. Robins, R.K. Robins, M.W. Winkley, H. Eyring, J. Am. Chem. Soc. 91, 824, 831 (1969).
[49] s. W.C. Johnson jr., I. Tinoco jr., Biopolymers 7, 727 (1969).

4 Application of CD and ORD to Biopolymers

4.1 Polypeptides and Proteins

L-α-Amino acids show a positive CD around 203 nm (water). In the presence of other chromophores (e.g. -S-S-, aromatic systems) Cotton effects are found also within their bands. For polypeptides in ordered conformations, the individual amide chromophores cannot be excited independently, but only cooperatively.

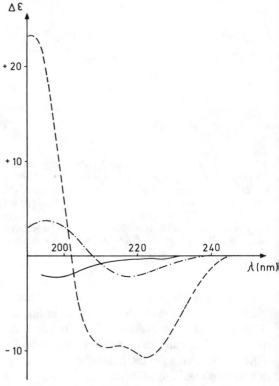

Fig. 25. Standard CD-curves for polypeptides and proteins[47]
α-Helix (----), β-structure (-·-·-) Random coil (——)

According to the exciton theory for an α-helix (righthanded screw), besides the negative n→π^* band, four π→π^* bands are predicted at shorter wavelengths (195 nm, negative; 193 nm, positive; 189 nm, positive; 185 nm, negative). Fig. 25 illustrates this for an α-helix. Similar «standard curves» are also shown for the β-form and the random coil[47]. The percentage of the three forms in the conformational equilibrium of a polypeptide or protein (without other interfering chromophores) can be determined with satisfactory accuracy by matching the experimental CD-curve with one synthesized from these three standard curves.

When CD-measurements at such short wavelengths were not possible directly, one had to resort to ORD-measurements outside the absorption bands of the amide chromophore (280–600 nm). Under the (restricting) assumption that only α-helix and random coil are present, Moffitt derived equation 20 for the molar rotation (corrected for the refractive index n of the solvent) per amino acid residue

$$[\varPhi] \cdot \frac{3}{n^2 + 2} = [m'(\lambda)] = a_0 \frac{\lambda_0^2}{\lambda^2 - \lambda_0^2} \qquad (20)$$
$$+ \, b_0 \frac{\lambda_0^4}{(\lambda^2 - \lambda_0^2)^2}$$

This has been modified by Yang to

$$[m'(\lambda)] \cdot \frac{\lambda^2 - \lambda_0^2}{\lambda_0^2} = a_0 + b_0 \frac{\lambda_0^2}{\lambda^2 - \lambda_0^2} \qquad (21)$$

Plotting the magnitude $[m'(\lambda)] \cdot \dfrac{\lambda^2 - \lambda_0^2}{\lambda_0^2}$

$$\text{vs.} \quad \frac{\lambda_0^2}{\lambda^2 - \lambda_0^2}$$

one obtains a straight line whose slope gives b_0 directly (−630 for 100% right handed helix, and 0 for the random coil). Shechter and Blout (Ref. [54]) give another equation which is mathematically equivalent to the Moffitt-Yang equation but more favorable for practical purposes.

4.2 Polynucleotides and Nucleic acids

Pyrimidine and purine bases are achiral. However, their glycosides show Cotton effects within their absorption bands[48]. This explains the fact that pyrimidine glycosides (but not the purine glycosides) do not follow Hudson's isorotation rule. In polynucleotides − as in polypeptides − only cooperative excitations are possible[49]. This leads to a great number of individual bands. In general, only the envelope of these can be measured. In neutral solution ribonucleic acids have a positive CD-band around 265 nm and two smaller negative ones ('nonconservative behavior') below 250 nm; whereas, deoxyribonucleic acids have a positive CD-maximum at 273 nm and a negative one of the same intensity at 243 nm ('conservative'). These bands change appreciably with temperature since, during heating, the helices 'melt'. This causes a decrease of the CD until finally the values of mononucleotides are reached.

5 Bibliography

5.1 Monographs

[50] C. Djerassi, Optical Rotatory Dispersion, McGraw-Hill, New York 1960.
[51] T.M. Lowry, Optical Rotatory Power, republication, Dover Publ., New York 1964.
[52] L. Velluz, M. Legrand, M. Grosjean, Optical Circular Dichroism, Verlag Chemie, Weinheim/Bergstr. 1965.
[53] P. Crabbé, Optical Rotatory Dispersion and Circular Dichroism in Organic Chemistry, Holden-Day, San Francisco 1965; − strongly extended French edn.: Gauthiers-Villars, Paris 1968.
[54] G. Snatzke, Optical Rotatory Dispersion and Circular Dichroism in Organic Chemistry, Heyden & Son, London 1967.
[55] R. Bonnett, J.G. Davis, Some Newer Physical Methods in Structural Chemistry: Mass Spectrometry, Optical Rotatory Dispersion and Circular Dichroism, United Trade Press, London 1967.
[56] B. Jirgensons, Optical Rotatory Dispersion of Proteins and Other Macromolecules, Springer Verlag, Berlin 1969.
[57] P. Crabbé, An Introduction to the Chiroptical Methods in Chemistry, Syntex, Mexico City 1971.

5.2 Review Articles

[58] W. Kuhn in K. Freudenberg, Stereochemie, Deuticke, Leipzig 1933.
[59] W. Heller in A. Weissberger, Physical Methods of Organic Chemistry, Vol. I, Part. 2, p. 1491, Interscience, New York 1949.
[60] W. Klyne, Adv. Org. Chem. 1, 239 (1960).
[61] H.G. Leemann, Chimia 14, 1 (1960).
[62] Symposium on Rotatory Dispersion: Theory and Application, Tetrahedron 13, 1 (1961).
[63] L. Velluz, M. Legrand, Angew. Chem. 73, 603 (1961).
[64] A. Moscowitz, Adv. Chem. Phys. 4, 67 (1962).
[65] I. Tinoco jr., Adv. Chem. Phys. 4, 113 (1962).
[66] H.G. Leemann, K. Stich, Chimia 17, 184 (1963).
[67] S.F. Mason, Quart. Rev. 17, 20 (1963).
[68] L. Velluz, M. Legrand, Angew. Chem. 77, 842 (1965).
[69] P. Crabbé in Encyclopedia of Industrial Chemical Analysis 2, 735 (1966).
[70] H. Ripperger, Z. Chem. 6, 161 (1966).
[71] A Discussion on Circular Dichroism: Electronic and Structural Principles, Proc. Roy. Soc. A 297, 1 (1967).
[72] P. Crabbé in N.L. Allinger, E.L. Eliel, Topics in Stereochemistry, Interscience, New York 1, 93 (1967).
[73] G. Snatzke, Z. Instrumentenkunde 75, 111 (1967).
[74] G. Snatzke, Z. Anal. Chem. 235, 1 (1968).
[75] S.F. Mason, Contemp. Phys. 9, 239 (1968).
[76] G. Snatzke, Angew. Chem. 80, 15 (1968).
[77] H. Eyring, H.-Ch. Liu, D. Caldwell, Chem. Rev. 68, 525 (1968).
[78] W. Klyne, P.M. Scopes, Il Farmaco 24, 5 (1969).
[79] C.W. Deutsche et al., Ann. Rev. Phys. Chem. 20, 407 (1969).
[80] L. Velluz, M. Legrand, Bull. Soc. Chim. Fr. 1785 (1970).
[81] R.D. Gillard, P.R. Mitchell, Structure and Bonding 7, 46 (1970).
[82] P.M. Scopes, Annu. Progr. Chem. 66B, 34 (1970).

5.8 Application of Linear Dichroism*

Amnon Yogew and *Yehuda Mazur*
Department of Chemistry, The Weizmann Institute of
Science, Rehovot, Israel

1 Interaction of Polarized Light with Oriented Molecules

According to both classical and quantum models, light absorbing molecules are optically anisotropic. The quantum model shows that the probability of electric dipole transition is proportional to the square of the transition dipole moment. The light intensity is represented in this model by the square of the electric field vector of the electromagnetic wave.

When the absorbing matter consists of an assembly of oriented molecules, and the incident light is plane polarized, the interaction between both will be proportional to the square of the scalar product of the electric field vector $\vec{E}$, and the vector of the transition dipole moment, $\vec{M}$, according to Equation (1).

$$\alpha \sim (\vec{M}, \vec{E})^2 \tag{1}$$

where α is the probability of light absorption.

This relation defines linear dichroism. Thus, linear dichroism is exhibited only by anisotropic media where the absorbing molecules have non-random distribution. It has been observed in *crystals, liquid crystals, fibers, biopolymers, oriented polymers, streaming liquids* and *monomolecular films*.

2 Measurement Techniques

The dichroic behavior of an anisotropic sample may be described by an ellipsoid whose radius vector corresponds to the square root of the optical density, and one of its axes superimposes the optical axis of the sample[1-3]. Thus, in order to define the dichroic behavior of an anisotropic sample, it suffices to measure its absorption in two directions only, one being in the direction of the optical axis, the second orthogonal to it. The ad-

vantage of such measurements is the elimination of complications due to phase retardations along the optical path.

For linear dichroism measurements of an anisotropic sample it is necessary to use a spectrophotometer equipped with a polarizer and to change either the direction of light polarization or the position of the sample with respect to the polarization of the incident light.

Generally, conventional spectrophotometers are used for these measurements. The polarizers may be Glan type calcite prisms, or Rochon prisms[4-10]. In the visible region, polaroid polarizers of polyvinyl alcohol and iodine have been used[6].

In most measurements it is preferable to rotate the sample rather than the plane of polarization[6]. For the reverse procedure the sample is retained and prisms with two orthogonal polarization directions are interchanged[10].

When the sample is rotated, it is possible that not all portions traversed by the light beam have the same absorption. On the other hand, turning of the polarizer is not always possible; and, when it is, photometric errors may occur due to different sensitivity of the photomultiplier to different states of polarization.

Measurements of linear dichroism in a *single crystal* are generally performed with spectrophotometers equipped with the above mentioned accessories and with a microscope. With the help of the microscope the crystal is adjusted so that one of its axes is parallel to the direction of polarization of the light[11-13].

To avoid difficulties due to high optical density of the absorbing crystal or any other absorbing media, an instrument measuring the *reflection spectrum* instead of the absorption spectrum has been developed[14].

[4] *K. Popov,* Opt. Spektrosk. *3,* 579 (1957).
[5] *R. Eckert, H. Kuhn,* Z. Elektrochem. Ber. Bunsenges. *64,* 356 (1960).
[6] *Y. Tanizaki,* Bull. Chem. Soc. Jap. *32,* 75 (1959).
[7] *Y. Tanizaki, H. Inoue, N. Ando,* J. Mol. Spectrosc. *17,* 156 (1965).
[8] *W.B. Gratzer, G.M. Holzwarth, P. Doty,* Proc. Nat. Acad. Sci. U.S. *47,* 1785 (1961).
[9] *L.V. Smirnov,* Dokl. Akad. Nauk SSSR *82,* 237 (1952).
[10] *D.M. Gray, I. Rubenstein,* Biopolymers *6,* 1605 (1968).
[11] *D.P. Craig, P.C. Hobbins,* J. Chem. Soc. 2309 (1955).
[12] *A.C. Albrecht, W.T. Simpson,* J. Chem. Phys. *23,* 1480 (1955).
[13] *J. Tanaka,* Bull. Chem. Soc. Jap. *36,* 833 (1963).

---

* This chapter deals only with ultraviolet absorption spectroscopy.

[1] *R. Zbinden,* Infrared Spectroscopy of High Polymers, Academic Press, p. 182, 213, New York 1964.
[2] *L.V. Smirnov,* Opt. Spektrosk. *3,* 123 (1957).
[3] *A. Yogev, L. Margulies, D. Amar, Y. Mazur,* J. Am. Chem. Soc. *91,* 4558 (1969).

The reflection spectrum, recorded by the instrument, can be mathematically transformed into an absorption spectrum using the Kroenig-Krammer equation.

A new device has been developed by which measurements of linear dichroism can be obtained in a single spectrum run by using two orthogonal modes of polarization emerging from the same polarizer[15].

The elements of this device are a Rochon prism, a quartz wave plate and a depolarizer. Two identical sets of these are placed in the sample and reference beams of the spectrophotometer with their wave plates set so that in one beam the angle between the fast axis and the plane of polarization is +45°, and in the other −45°. It follows that, as the wavelength is changed, the state of polarization of the light emerging from the waveplates alternates between horizontal and vertical, with all states of ellipticity in between.

Due to these successive alternations of the two orthogonal states of polarization, a non-continuous absorption spectrum is obtained, whose outer and inner envelopes correspond to the absorptions due to the two states of polarization of the incident light.

3 Quantitative Linear Dichroism Measurements

Linear dichroism measurements are conveniently expressed by a dimensionless parameter, the dichroic ratio, d_0, which describes the ratio of absorption coefficients in parallel ($\varepsilon_{\parallel}$), and perpendicular direction ($\varepsilon_{\perp}$), to the optical axis[3, 16, 17].

$$d_0 = \frac{\varepsilon_{\parallel}}{\varepsilon_{\perp}}$$

In some works a different expression[2, 4, 18] was preferred, namely:

$$d_0 = \frac{\varepsilon_{\parallel} - \varepsilon_{\perp}}{\varepsilon_{\parallel} + \varepsilon_{\perp}}$$

3.1 In Crystals

In order to evaluate quantitatively the linear dichroism measurements done on a *single crystal*, a previous knowledge of the crystal data, down to subcrystalline structure, is necessary. Once the position of the chromophore in a single crystal in relation to the electric vector of light propagation is found, the direction of the transition moment may be derived, using Equation (2)[11–13].

$$d_0 = \frac{a_z}{a_y} = \frac{(\vec{M}, \vec{E}_z)^2}{(\vec{M}, \vec{E}_y)^2} \tag{2}$$

where $\vec{M}$ is the vector of the transition dipole moment, $\vec{E}_y$ and $\vec{E}_z$ are electric field vectors along y and z axes respectively, and x is the direction of the light propagation vector. a_y and a_z are absorption coefficients along the y and z axes.

3.2 In Oriented Polymers

Unlike in crystals, the orientation of chain segments in polymers is not defined. Thus, in order to evaluate linear dichroism in polymers, it is necessary to determine the population of the repeating segments of polymers in relation to a definite direction. Once this orientation is known, the correlation between dichroic ratios and molecular properties, like the geometrical orientations of the chromophores, may be established with the aid of equation (2)[1, 16, 18–23].

Theoretical models (for IR linear dichroism, but also applicable for ultraviolet measurements) have been developed to define these partial orientations of the polymer segments in a three-dimensional model. If the polymer segments are partially oriented with an axial symmetry, the theoretically evaluated dichroic ratio is expressed by equation (3)[21].

$$d_0 = \frac{a_z}{a_y} = \frac{2\cos^2\phi + 2/3(1-f)/f}{\sin^2\phi + 2/3(1-f)/f} \tag{3}$$

where: a_y and a_z are the absorption coefficients in the y and z direction, x being the direction of light propagation vector; $\emptyset$ is the angle between the polymer axis and the transition moment vector of the oriented chromophores; f is the fraction of the oriented polymer segments and (1-f) their random fraction.

When the orientation about the symmetry axis is not uniform, it is possible to use the above equation after selecting appropriate values for $\emptyset$ and f.

Theoretical models have been developed to correlate the orientation of polymer segments with the stretch ratio, the ratio of the length of the stretched polymer to its original length.

The equation relating the stretch ratio, v, with orientation has the form:

$$F = 1 - \frac{v}{v^3 - 1} + \frac{v}{(v^3 - 1)^{3/2}} \cos^{-1} \frac{1}{v^{3/2}} \tag{4}$$

where: $F = 2/3 \ (1-f)$, f being the fraction of oriented polymer segments[1, 24, 25].

[14] *B.G. Anex*, Mol. Cryst. *1*, 1 (1966).
[15] *J.H. Jaffe, H. Jaffe, K. Rosenheck*, Rev. Sci. Instrum. *38*, 935 (1967).
[16] *R.D.B. Fraser*, J. Chem. Phys. *21*, 1511 (1953).
[17] *A. Yogev, J. Riboid, J. Marero, Y. Mazur*, J. Am. Chem. Soc. *91*, 4559 (1969).
[18] *F. Dörr*, Angew. Chem. *78*, 457, Internat. Edit. *5*, 478 (1966).
    R.D.B. Fraser, J. Chem. Phys. *24*, 89 (1956).

[19] *R.D.B. Fraser*, J. Chem. Phys. *28*, 1113 (1958).
[20] *R.D.B. Fraser*, J. Chem. Phys. *29*, 1428 (1958).
[21] *M. Beer*, Proc. Roy. Soc. Ser. *A236*, 136 (1956).
[22] *S. Krimm*, J. Chem. Phys. *32*, 313 (1960).

3.3 In Molecules Incorporated in Oriented Media

Stretched polymers: Non-polymeric molecules may be incorporated into stretched polymers and become oriented themselves[5, 26-29]. The polymers thus serve as a 'solvent' for the incorporated molecule. For evaluating the linear dichroism behavior of this mixed phase system, the same symmetry considerations are applicable as for stretched polymers.

Quantitative treatment: Popov[4] and Smirnov[2] developed a theoretical treatment relating the dichroic ratio of the incorporated molecules with the elongation (stretch ratio) of a polymer film. According to this treatment, the orientation of the incorporated molecules would be complete only on unlimited stretching. Since measurements of dichroic ratios are inconsistent with the above treatment, a new model was proposed which implies that incorporated molecules become only partially oriented. This model was used to determine the relative direction of transition moments in molecules possessing a *multitransition chromophore.*

According to this model, if the direction of one of the transition moments is known, linear dichroism measurement in the wavelength of this transition enables the calculation of the degree of orientation. The direction of other transition moments of this chromophore will be established after measuring their respective dichroic ratios, using an equation relating these values with the degree of orientation established beforehand.

According to a different approach[6], the incorporated material is represented as an imaginary sphere which, on stretching, changes into an ellipsoid, maintaining a constant volume. An equation has been derived which correlates the ratio of the axes of the ellipsoid, representing the stretch ratio of the film, with the measured dichroic ratio. It also correlates the ratio of the axes with the direction of the transition moment of the chromophore in relation to the direction of the stretching. Dichroic ratios of incorporated molecules in stretched polyethylene film[3, 17] show the validity of equation (3) for chromophores of incorporated molecules. In addition, it was found that the dichroic ratio of incorporated material in *stretched polyethylene films* is independent of the *amount of the stretching.*

When dichroic ratios were measured for molecules possessing chromophores at the same carbon skeleton, their degrees of orientation (f-values in Equation 3) were assumed to be the same. Thus, measurements of dichroic ratios of two compounds, having their chromophores in different geometrical positions, leed to a solution of Equation (3), and enabled determination of the direction of the transition moments of these chromophores.

A different quantitative approach to describe the distribution of incorporated molecules, which uses more than one orientational parameter, was suggested. Overlapping UV bands of highly symmetrical chromophores were separated according to their orthogonal polarizations using the above approach[29a, b, c].

Two different quantitative analyses of linear dichroic spectra were compared[29d].

4 Applications

4.1 UV-Spectroscopy

Crystals and *liquid crystals:* One of the standard procedures to determine the directions of transition moments is the measurement of the linear dichroism of a single crystal. By this method, the transition moments of large numbers of organic molecules, mainly *aromatic compounds* and *metal complexes,* were established[11-13, 18, 30, 31].

In addition, polarization absorption spectra of *molecules embedded on host crystals* were determined, and their respective transition moments were established.

Liquid crystals were used as hosts for a few organic molecules. The liquid crystal matrix was pre-oriented either by magnetic field or by deposition on uni-directional scratched quartz plates. The orientation of the liquid crystals was retained by supercooling[36-38].

[29a] *E.W. Thulstrup, J.H. Eggers,* Chem. Phys. Lett. *1,* 690 (1968).

[29b] *E.W. Thulstrup, J. Michl, J.H. Eggers,* J. Phys. Chem. *74,* 3868 (1970).

[29c] *J. Michl, E.W. Thulstrup, J.H. Eggers,* J. Phys. Chem. *74,* 3878 (1970).

[29d] *A. Yogev, L. Margulies, Y. Mazur,* Chem. Phys. Lett. *8,* 157 (1971).

[30] *F. Dörr, H. Held,* Angew. Chem. *72,* 287 (1960).

[31] *P. Heim, F. Dörr,* Tetrahedron Lett. *42,* 3095 (1964).

[32] *H.C. Wolf,* Advances in Solid State Physics, Vol. *9,* New York 1959.

[33] *G. Scheibe,* Angew. Chem. *52,* 631 (1939).

[34] *D.P. Craig, P.C. Hobbins, J.R. Walsh,* J. Chem. Phys. *22,* 1616 (1954).

[23] *A. Rich, M. Kasha,* J. Am. Chem. Soc. *82,* 6197 (1960).

[24] *O. Kratky,* Kolloid-Zh. *64,* 213 (1933).

[25] *W. Kuhn, F. Gruen,* Kolloid-Zh. *101,* 248 (1942).

[26] *A. Jablonsky,* Acta Physica Polon. *3,* 421 (1934).

[27] *J. Kern, F. Dörr,* Z. Naturforsch. A *16,* 363 (1961).

[28] *J. Kern,* Z. Naturforsch. A *17,* 271 (1962).

[29] *H. Jakobi, A. Novak, H. Kuhn,* Z. Elektrochem. Ber. Bunsenges. *66,* 863 (1962).

Measurements of polarized reflection spectra of *nickel dimethylglyoxime* and *β-ionylidene crotonic acid* were published, leading to the assignment of their various bands[14, 39, 40].

Biopolymers: Linear dichroism measurements of helical biopolymers were reported. *α*-Helical *poly-L-lysine hydrofluoride, poly-L-prolines I* and *II, poly-L-methyl-L-glutamate-poly-L-alanine* oriented by unidirectional stroking have been measured[8, 10, 15, 41, 42]. From these data, a spectroscopic assignment of the bands of the helices was made. Helix-coil transitions in *poly-L-glutamatic acid, poly-L-proline* and *poly-L-lysine* were studied with the aid of linear dichroism measurements[43].

Stretched films: The linear dichroism of many organic molecules incorporated in stretched films of *polyethylene, polyvinyl alcohol* and some other polymers was measured. Most of the organic molecules[2, 4, 5, 7, 18, 30, 44−51] were *organic dyes.* In addition, a number of polyenes like *vitamin B_{12}, β-carotene, 15,15'-cis-β-carotene* and aromatic hydrocarbons as pyrene, perylene, phenanthrene, fluoranthrene, p-dimethoxy benzene, 10a, 4a-boroazoro-phenanthrene were investigated[5, 29a, b, c, 52]. In most cases, it was assumed that the transition dipole of the main band of these compounds was parallel to the direction of stretching, allowing a qualitative determination of the direction of other transitions in these molecules.

Attempts were made to analyze quantitatively linear dichroic spectra of *aromatic hydrocarbons* like *naphthalene, naphthalenediols, anthracene, phenanthrene,* and their oxygen and aromatic azoderivatives incorporated in stretched polyvinyl alcohol or polyallyl alcohol[4, 47, 49, 51−55]. However, in many cases, the fact that hydrogen bonding between hydroxyl groups of polyvinyl alcohol and polar sites in the incorporated molecules may affect their orientation was not considered.

Quantitative evaluation of the direction of the transitions of *α,β-unsaturated ketones,* and the respective transitions of *unsubstituted and monosubstituted benzoic acids* and their esters have been described recently[17, 56, 57].

4.2 Structural and Conformational Analysis

Quantitative interpretation of the linear dichroic measurements of compounds embedded in stretched polyethylene film enables one to determine the orientation of its molecular skeleton. Knowing the orientation as well as the direction of the transition moment makes it possible to find the position of the absorbing chromophore in the skeleton of the molecule. Other applications emerging from linear dichroism measurements are the possibilities of distinguishing between *cis* and *trans* ring fusions in polycyclic compounds and establishing the conformation of substituents in cyclic compounds.

This method was used for *α,β-unsaturated ketones* within *cholestane* and *androstane* skeletons, and for *benzoic acid esters* of *hydroxy-cholestanes* and *hydroxy-androstane*[3, 17, 57, 58].

4.3 Molecular Association

The orientation of a compound incorporated in *stretched polyethylene films* depends on its geometrical dimensions. When the incorporated molecules are associated, their directions and orientations may change. This enables interpretation of the state of association of the incorporated molecules[52].

Thus, two states of association of the enol forms of *1,3-cyclohexanediones* were detected: one, the cyclic dimeric form of association in which the transition moment is perpendicular to the long axis of the associated molecule, and the second, a 'polymeric' form in which the transition moment is parallel to this axis. However, in polyvinyl alcohol films this association of the enol form of *1,3-cyclohexanediones* is prevented by hydrogen bonding with the polyvinyl alcohol; therefore, a different dichroic behavior is observed[52].

[35] *D.P. Craig, J.R. Walsh,* J. Chem. Soc. 1613 (1958).
[36] *G.P. Gaesar, H.B. Craig,* J. Am. Chem. Soc. *81,* 191 (1969).
[37] *E. Sackman, S. Meiboom, L.C. Syder, A.E. Meixrer, R.F. Dietz,* J. Am. Chem. Soc. *90,* 3567 (1968).
[38] *E. Sackman,* J. Am. Chem. Soc. *90,* 3569 (1968).
[39] *L.J. Paskhurst, B.G. Anex,* J. Chem. Phys. *45,* 862 (1966).
[40] *B.G. Anex, F.K. Krist,* J. Am. Chem. Soc. *89,* 6114 (1967).
[41] *K. Rosenheck, B. Sommer,* J. Chem. Phys. *46,* 532 (1967).
[42] *K. Rosenheck, H. Miller, A. Zakavia,* Biopolymers *7,* 614 (1969).
[43] *K. Rosenheck,* Molecular Associations in Biology, Academic Press, p. 517, New York 1968.

[44] *Y. Tanizaki, T. Kobayashi, N. Ando,* Bull. Chem. Soc. Jap. *32,* 1362 (1959).
[45] *Y. Tanizaki,* Bull. Chem. Soc. Jap. *33,* 979 (1960).
[46] *Y. Tanizaki, H. Ono,* Bull. Chem. Soc. Jap. *33,* 1207 (1960).
[47] *Y. Tanizaki,* Proc. Inter. Symp. Mol. Struct. Spectrosc. [Tokyo] 1962.
[48] *Y. Tanizaki,* Bull. Chem. Soc. Jap. *38,* 1798 (1965).
[49] *Y. Tanizaki, S. Kubodera,* J. Mol. Spectrosc. *24,* 1 (1967).
[50] *T. Tsunoda, T. Yamaoka,* J. Polym. Sci. *A3,* 3691 (1965).

[51] *K.R. Popov*, Opt. Spektrosk. *25*, 843 (1960).

[52] *E.W. Thulstrup, M. Vala, J.H. Eggers*, Chem. Phys. Lett. *7*, 31 (1970).

[53] *H. Inoue, Y. Tanizaki*, Z. Phys. Chem. [Frankfurt/Main] *73*, 48 (1970).

[54] *H. Inoue, T. Nakamura, T. Igarashi*, Bull. Chem. Soc. Jap. *44*, 1469 (1970).

[55] *K.R. Popov, L.V. Smirnov*, Opt. Spektrosk. *30*, 628 (1971).

[56] *A. Yogev, L. Margulies, Y. Mazur*, J. Am. Chem. Soc. *93*, 249 (1971).

[57] *L. Margulies*, M. Sc. Thesis, Feinberg Graduate School, Weizmann Institute of Science, Rehovoth (Israel) 1970.

[58] *A. Yogev, L. Margulies, Y. Mazur*, J. Am. Chem. Soc. *92*, 6059 (1970).

5.9 Applications of Faraday Effect Spectroscopy

Bernard Briat
EPCI, 10, rue Vauquelin, Paris 5è

This chapter contains:
Some basic definitions intended to describe the formulas and units in use. These cover magnetooptical rotatory dispersion (MORD) and magnetic circular dichroism (MCD) together with a comparison of the two techniques (5.9:1 and 5.9:2).
A brief account of the significance of the parameters derived from the experimental curves (5.9:4).
A comparison between natural and magnetic optical activity (5.9:5).
Several specific applications (5.9:6) including:
the resolution of overlapping bands (5.9:6.1) in e.g., vitamin B_{12} derivatives or nucleosides,
the study of biochemical problems (5.9:6.2) in e.g., hemcontaining molecules or vitamin B_2 derivatives, the study of carbonyl compounds.
A few comments on the study of spin degeneracy (5.9:7).

1 Introduction

The general term 'Faraday effect spectroscopy' refers to magnetically induced optical activity and covers both *magnetic circular dichroïsm* (MCD) and *magnetooptical rotatory dispersion* (MORD). These phenomena can be measured by means of a dichrometer (MCD) or a spectropolarimeter (MORD) provided a magnetic field is induced in the sample compartment parallel to the direction of the light beam. The origin of the Faraday effect lies basically in the helical symmetry of the magnetic field. The material under investigation responds differently to optical stimulation with right or left circularly polarized light[1]. The method is applicable − at least in principle − to any optically inactive or active molecule.

There has been a renewed interest in MORD and MCD techniques during the past five years. To a great extent, this interest has been motivated by instrumental progress as well as by the derivation of a general theory[2] which shed light on the results and stimulated further studies. A few reviews including a detailed history[2] have already been devoted to the subject[2-5].

2 Definitions

2.1 Phenomenological Approach and Units

Natural and/or magnetic optical activity can be defined in two ways according to the method of measurement employed. If circularly *polarized light* is used (left and right components), one can define circular dichroism as $\Delta\varepsilon = \varepsilon_L - \varepsilon_R$ and circular birefringence as $\Delta n = n_L - n_R$ where ε and n stand for the molar extinction coefficient and the refractive index of the medium respectively. If plane polarized light is employed, it becomes elliptically polarized after having travelled through the sample. It can be shown that the ellipicity θ and the angle α between the long axis of the ellipse and the initial direction of planar polarization constitute a means of measuring CD and ORD respectively.

In practice, commercial instruments provide $\Delta\varepsilon$ (or better ΔD which is a difference in optical density units) or θ on the one hand and α on the other hand. When a magnetic field H is added, since the phenomena are proportional to H, quantities should be standardized to one gauss. Following a number of authors, the following units are suggested:

$[\Delta D] = \Delta D/H$ magnetic dichroic optical density.

$[\Delta\varepsilon] = \Delta D/Hcl$ molar magnetic circular dichroism where c and l stand for the concentration (in mole l^{-1}) of the solution and the length of the cell (in cm).

$[\Phi]_M = 100 \dfrac{\alpha}{Hcl}$ molar magnetic rotation (α in degrees)

$[\theta]_M = 100 \dfrac{\theta}{Hcl}$ molar magnetic ellipticity (θ in degrees).

[1] *B. Briat, C. Djerassi*, Nature *217*, 918 (1968).

[2] *A.D. Buckingham, P.J. Stephens*, Rev. Phys. Chem. Jap. *17*, 399 (1966).

$[\Delta\varepsilon]$, $[\Phi]_M$, and $[\theta]_M$ are molar quantities and suitable for comparisons on a mole to mole basis. Furthermore, it is advantageous to switch easily from one system of units to the other. As in natural activity, $[\theta]_M = 3300[\Delta\varepsilon]$.

Finally, in all Faraday effect experiments, light should propagate from the south pole to the north pole of the magnet. With these conventions, the magnetic rotation of *water* is negative in the UV and visible regions. The MCD through the 500 nm band of *cobalt(II) salts* is also negative.

2.2 Dispersion Formulae

MCD has a non zero value only through absorption bands. Considering an isolated line, the MCD dependence on the energy ν (in cm^{-1}) or the wavelength has been shown to be[1,5]

$$F = [\Delta\varepsilon]/\varepsilon_m = a\frac{df}{d\nu} + bf \qquad (1)$$

where f is some mathematical function which describes the shape of the absorption curve (e.g., Lorentzian or Gaussian) and ε_m is the maximum molar extinction coefficient at the center of the band.

According to this equation, an MCD curve can be described as a linear combination of the absorption curve in zero magnetic field and its first derivative. The F parameter which is characteristic of magnetooptical activity is, in many respects, similar to Kuhn's factor[6] for natural optical activity. MORD and MCD are related through the Kramers-Kronig relationship:

$$[\Phi]_M/\varepsilon_m \sim a\mathrm{KK}\ \frac{df}{d\nu} + b\mathrm{KK}(f) \qquad (2)$$

where the simplified notation KK means 'take the Kramers-Kronig transform of'.

If a and b (Fig. 1) were negative, one would obtain curves symmetrical to the above ones with the wavelength axis as the symmetrical one.

It is apparent from the figure that an MCD or MORD curve is, in general, more complicated than an ordinary CD or ORD one. For current discussion of experimental results, it is, therefore, advisable to avoid the term Cotton effect and to emphasize the existence and sign of the a and b parameters. Finally, it should be noted that the a

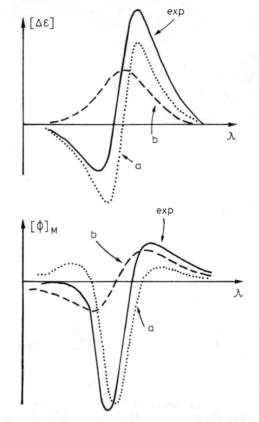

Fig. 1. Dependence of wave lengths of MCD and MORD bands on positive a and b values.

and b parameters proposed here are closely related to the A and B parameters employed by Buckingham and Stephens[2]. The two nomenclatures have been discussed in a recent paper[5]. The determination of the parameters from the data can be made according to Stephens et al.[7] who also provide the necessary explicit mathematical functions.

3 Comparison Between MCD and MORD

It follows from Equation (1) and (2) that MCD and MORD should, at least in principle, provide the same information, i.e. the value and sign of a and b. However, it is usually highly advantageous to work with MCD data.

It appears that the contribution of each band (Fig. 2) is reasonably well isolated in MCD whereas the MORD terms strongly overlap. This is so because MORD curves are more complicated than MCD ones and possess tails outside the region of absorption. For this reason it is necessary to add a background rotation (the so-called Drude term) deriving from the contribution of bands located outside the range of available data.

[3] *P.N. Schatz, A.J. McCaffery*, Quart. Rev. Chem. Soc. *4*, 552 (1969).

[4] *B. Briat*, Méthodes Physiques d'Analyse (GAMS) *6* (No 1), p. 19, 1970.

[5] *G. Barth, E. Bunnenberg, C. Djerassi, D. Elder, R. Records, V.E. Shashoua, J. Badoz, M. Billardon, A.C. Boccara, B. Briat* in Symposia of the Faraday Soc. *3*, 1969.

[6] *W. Kuhn*, Ann. Rev. Phys. Chem. *9*, 417 (1958).

[7] *P.J. Stephens, W. Suetaka, P.N. Schatz*, J. Chem. Phys. *44*, 4592 (1966).

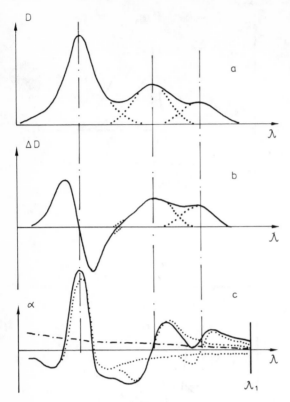

Fig. 2. Hypothetical MCD and MORD Curves.
(a) total absorption, (b) MCD, and (c) MORD curves for 3 different absorption bands.

Fig. 2 also shows the use of MORD measurements in the transparency region. Although this has not always been recognized in the past, the rotation (at λ_1) is generally the sum of many contributions of different origin. This severely restricts the possibility of a detailed interpretation of the spectra. Therefore, such measurements are of interest only when absorption bands are not within reach of the spectral range of the instrument. Much data has been accumulated in the literature and several additivity laws have been established.

4 Significance of the Parameters

Only MCD will be considered since MORD can be calculated from MCD. The spectra can be understood by considering the various perturbations caused by a static magnetic field. Absorption intensity in the absence of the field can be calculated when the orbitals and the perturbations caused by the electromagnetic fields of the light wave are known.

A static magnetic field causes other types of perturbations. For example, if the electrons have well defined orbital angular momentum, the field will cause normally degenerate states to have different energies (Zeeman effect). The implication of this effect in organic molecules is that two absorption components are allowed with left and right polarizations. The two components do not present their maximum for the same energy but are slightly separated by a quantity $s\beta H$ (in cm^{-1}) where s is the spectroscopic splitting factor (proportional to the so-called g or Lande factor) of the excited state and β stands for the Bohr magneton. Since MCD is defined as the difference between the two absorption curves, one obtains an S-shaped MCD spectrum and the peak-to-peak magnitude is proportional to the splitting $s\beta H$ (thus to H). Moreover, it can be shown that for any reasonable shape function for f, this splitting is inversely proportional to the width of the band; thus narrow-bands show large effects. It remains that the Faraday effect responds to the orbital Zeeman effect even for broad bands when no direct measurement of the Zeeman splitting is feasible. On theoretical grounds, degeneracy is expected for systems possessing at least a threefold axis of symmetry (e.g., *benzene*) and many MCD experiments have been performed on such systems. Examples of this type are *coronene* and *triphenylene*[8], *porphyrins*[9], *[18] annulene derivatives*[10] or *ferrocene derivatives*[11]. For these molecules MCD has provided an unambiguous assignment of degenerate excited states and has permitted the determination of their g factors. These, in turn, have provided theoretical chemists with some of the most stringent tests of calculated wave functions yet devised. Finally, it has been found that small perturbations to lower than three-fold symmetry do not alter the expected spectra to a considerable degree.

In general, the magnetic field will cause a perturbation of the orbitals, viz., the perturbed orbital

$$\Psi'_a = \Psi_a + \sum_{b \neq a} C_b \Psi_b$$

where Ψ_a and Ψ_b are unperturbed orbitals and the coefficient C_b is less than one.

This means that the shape of an orbital distorted by the field can be described by adding the proper amounts of the shapes of certain unperturbed orbitals of the set. In practice it is found that the a and b orbitals must have certain symmetry properties in order to overlap under the perturbation; moreover the orbitals with energies near that of ψ_a are used to a greater extent and this fact enters as an energy denominator into the expression of the Faraday effect. For an absorption between a ground state, o, and an excited state, a, and on the condition that the state a overlaps with one other state b,

$$\Delta D_{0a} \sim \frac{\vec{\mu}_{ab} \cdot (\vec{m}_{0a} \; \vec{m}_{b0})}{E_b - E_a}$$

The numerator involves the triple cross product of three vectors which are electric ($\vec{m}$) or magnetic ($\vec{\mu}$) moments for the various transitions. It can be shown from the above formula that $\Delta D_{0b} = -\Delta D_{0a}$ since only the energy difference changes its sign. Thus MCD is the same for the two bands even when they show different strengths as illustrated with *trans 15,16-dimethyl-dihydropyrene*[10] (Fig. 3). Here we have a favora-

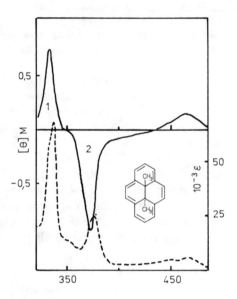

Fig. 3. Absorption (------) and MCD (——) spectra for trans 15,16-dimethyl dihydropyrene (Ref.[10])

ble situation since bands 1 and 2 are perpendicularly polarized in the plane of the molecule, whereas $\vec{\mu}$ is oriented along the third axis. It follows that MCD is a suitable technique to show the relative polarizations of transitions. Moreover, if a weak band is hidden under an intense one, it is likely to appear in an MCD spectrum if the bands have different polarizations.

5 Comparison of Natural and Magnetic Optical Activity

A similarity seems to exist between natural and magnetically induced optical activity. However, it should be kept in mind that the two phenomena are of completely distinct origin. Actually, they are additive. The MCD of the two anomers of *2'-Desoxy-5(trifluoromethyl)uridine* (Fig. 4) are essentially superimposable within experimental error, whereas the CD curves show the mirror image relationship expected[12]. This is why different quantum mechanical expressions are used to estimate the experimentally determined a and b parameters on the one hand (MCD) and the rotatory strength (CD) on the other hand.

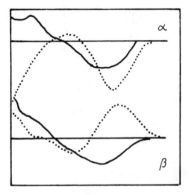

Fig. 4. CD and MCD spectra of α and β anomers of 2'-Desoxy-5 (trifluoromethyl) uridine (Ref.[12])

It is often difficult to make a-priori estimates of the relative magnitudes of CD (or ORD) and MCD (or MORD) Cotton effects for an optically active molecule. They are of the same order for *pyrimidine nucleosides*[12]; whereas, MCD is much stronger in the case of *chlorophyll-like molecules*[13] (Fig. 5).

From Fig. 5 it is clear that MCD signals of *pyro-methyl-pheophorbid a* are far more intense than CD ones. Moreover, it has been established that the magnitude of the MCD peak around 540 nm depends considerably upon the nature of the substituents attached to the skeleton. Instead of 269 units for the molecule, shown in the figure, one finds 407 units when the hydrogen atom on the additional ring is substituted by CO_2CH_3.

[8] *P.J. Stephens, P.N. Schatz, A.B. Ritchie, A.J. McCaffery*, J. Chem. Phys. *48*, 132 (1968).

[9] *E.A. Dratz, Ph.D. Thesis*, Univ. California, Berkeley (1966).

[9a] *P.S. Pershan, M. Gouterman, R.L. Fulton*, Mol. Phys. *10*, 397 (1966).

[10] *B. Briat, D.A. Schooley, R. Records, E. Bunnenberg, C. Djerassi*, J. Am. Chem. Soc. *89*, 7062 (1967).

[11] *H. Falk*, Monatsh. Chem. *100*, 411 (1969).

[12] *W. Voelter, R. Records, E. Bunnenberg, C. Djerassi*, J. Am. Chem. Soc. *90*, 6163 (1968).

[13] *B. Briat, D.A. Schooley, R. Records, E. Bunnenberg, C. Djerassi*, J. Am. Chem. Soc. *89*, 6170 (1967).

[14] *M.E. Carville, Ph.D. Thesis*, Iowa State University, Ames 1967.

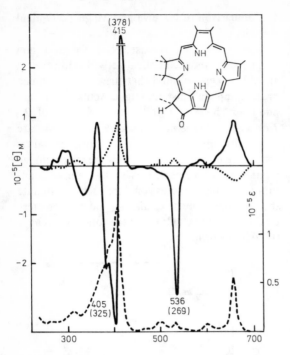

Fig. 5. Absorption (------) and CD spectrum of pyro-methyl-pheophorbid a in the presence (.......) or absence (——) of a 41.7-kG-field.

6 Specific Applications

6.1 Resolution of Overlapping Bands

One of the more obvious applications of MCD spectroscopy to organic systems is the study of accidental electronic degeneracy. Thus, in *p-aminobenzoic acid*[14], *indole,* and many *purines* it is possible to determine clearly the presence of two components in what might otherwise appear to be a single band. Furthermore, the fact that two bands interact in MCD (b terms) provides additional information since this means that they must have perpendicular components in their transition moments (in order that the vector cross product not vanish). Although the UV spectrum shows two bands in the 230–300 nm region for *guanosine,* only one band is observed for *adenosine* in the same region (see Fig. 6). It is clear from the MCD spectra, however, that two bands indeed exist for both guanosine and adenosine since two opposite and almost equal b terms are observed. This fact strongly supports the theoretically substantiated assignment of two perpendicularly polarized $\pi \rightarrow \pi^*$ transitions in the near UV. Thus MCD can have considerable advantage over other spectroscopic methods.

The fluorescence method is somewhat awkward for routine measurements since these must be made at liquid nitrogen temperature. On the other hand, linear dichroism measurements of *purines* require large crystals.

Finally, the very weak Cotton effects observed in the CD spectra of guanosine and adenosine, illustrated in Fig. 6, also emphasize the advantage of MCD for this series when small amounts of material are available.

For *corrine*[15] (Fig. 7) MCD provides a means of resolving overlapping absorption bands. In Fig. 7a, MCD peaks in the 380–420 nm region correspond to shoulders in the UV spectra. For the *nickel compound,* however, the situation is completely different since one observes (Fig. 7b) a

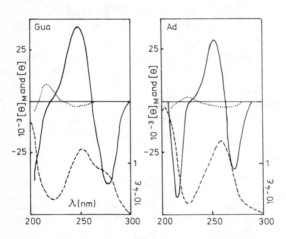

Fig. 6. Absorption (------), CD (.......) and MCD (——) spectra of guanosine and adenosine at 49.5 kG (Ref.[12]).

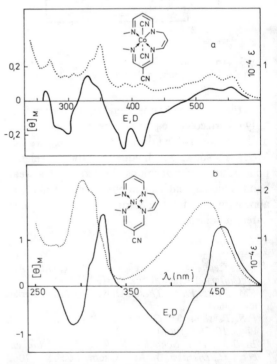

Fig. 7. Absorption (.......) and MCD (——) spectra of corrine (Ref.[15]).

broad and unresolved red absorption band between 350 and 500 nm. Comparison of the MCD spectra of the two molecules has been used to locate the E and D bands of the nickel compound around 400 nm with reasonable accuracy, as expected from the theory.

6.2 Use of the Faraday Effect in Biochemistry

MORD and MCD have been used mostly for the investigation of *porphyrins*. The *hemoglobin* chromophore (Fig. 8) is an iron complex with four nitrogen ligands from the planar porphyrin nucleus, a globin at the fifth position, and various substituent ligands at the sixth position about the central iron atom. Derivatives of *ferrohemoglobin* and *ferrihemoglobin (methemoglobin)* with H_2O, O_2, CO or CN as the ligands allow the observation[5, 16] of substantial magnetic rotation with amplitudes or/and dispersion features characteristic of the oxidation state of the central metal and of the nature of the substituents on the ring (Fig. 8).

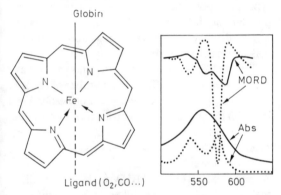

Globin

Fe

Ligand (O_2, CO...)   550   600

MORD

Abs

Fig. 8. Absorption and MCD spectra of hemoglobin (.......) and oxyhemoglobin (——) (Ref.[16]).

Quite similarly it was possible to follow the kinetics of the oxidation of *cytochrome c*[17] or *spinach ferredoxin*[18]. In the former case, it was also demonstrated that the magnetic rotation per optical density unit varied for samples from different sources.

Although no detailed mechanism has yet been offered to explain the spectra, some general concepts can be used to justify the large rotations (or MCD) observed, as well as the sensitivity of the technique to conformational changes. The most important features of the visible absorption spectrum of porphyrins can be explained[9] in terms of the promotion of a π electron of the porphyrin conjugated chain to an empty π^* orbital of the same chromophore[9a]. Thus the nature (i.e., the symmetry) of the excited states strong-

ly depends upon the mean symmetry of the conjugated chain. It has been shown in the case of metal and base-free *phthalocyanines* that this symmetry is approximately D4h (four fold axis) for the former and D2h (two-fold axis) for the latter. Then group theory predicts the existence of degenerate excited states for D4h symmetry and non degenerate but perpendicularly polarized excited states for D2h symmetry. As shown in the theoretical section, these are the conditions for the appearance of a terms and large b terms respectively. Furthermore, a terms should also be large since the absorption bands are sharp. This explains why most of the above experiments can be performed with the use of a permanent magnet providing a field of a few kilogauss. Actually, as suggested by Djerassi and Bunnenberg, MCD seems to be a very valuable tool for the qualitative and quantitative analysis of *porphyrins* if a superconducting magnet is used for the measurements. To conclude the preceding arguments, it is important to point out in Fig. 8 that *hemoglobin* shows an a term around 580 nm while only b terms are observed for *oxyhemoglobin;* this suggests that the symmetry of the chromophore is higher for the former molecule. Finally it should be understood that this symmetry will, in general, be dependent upon the nature of the substituents on the ring. This explains the sensitivity of the Faraday effect to conformational changes in porphyrins. The role of the central metal should also be considered for a refined treatment of the problem[16].

The MCD spectra of various *Vitamin B-2 derivatives*[19] (riboflavin analogs) have been used to elucidate the position of protonation of the *isoalloxazine* ring in solution. The best evidence comes from optical spectra of *flavins* (1) in which an alkyl bridge is placed between N-1 and N-10. These are similar to the flavin cation spectra, indicating that the protonation may occur at N-1.

1   2

15 *B. Briat, C. Djerassi*, Bull. Soc. Chim. Fr. 135 (1969).
16 *M.V. Volkenstein*, Nature *209*, 709 (1966).
    M.V. Volkenstein, Y.A. Sharonov, A.K. Shemelin, Mol. Biol. *1*, 395 (1967).
    M.V. Volkenstein, L.T. Metlyaev, I.S. Milevskaya, Mol. Biol. *3*, 75 (1969).
17 *V.E. Shashoua*, Nature *203*, 972 (1964).
18 *D.I. Marlborough, D.O. Hall, R. Cammack*, Biochem. Biophys. Res. Commun. *35*, 410 (1969).
19 *G. Tollin*, Biochemistry 7, 1720 (1968).

The first two rings (A and B) of the isoalloxazine structure are analogous to *quinoxaline* (2).

A protonated *quinoxaline* (3) would be electronically similar to the flavin cation, if the latter were protonated at N-1 (4). It turns out that the MCD of the quinoxaline cation is similar in shape to that observed for the flavin cation, although shifted to shorter wavelengths, thus confirming the assignment of the position of protonation.

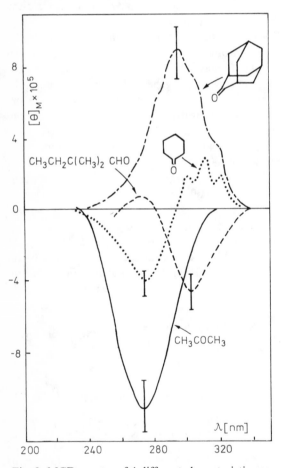

3   **4**

6.3 MCD Studies of Carbonyl Compounds

Finally, many MCD studies from Djerassi's group have been motivated, in part, by the possibility that the technique may provide information about the stereochemistry of organic compounds, particularly of molecules containing the carbonyl chromophore. Recently, a superconducting magnet in conjunction with a circular dichroism instru-

ment of increased sensitivity has produced reliable spectra for 53 carbonyl compounds[5,20] (Fig. 9). As a result of the signal to noise ratio (indicated by a vertical bar), the MCD effects corresponding to the $n \rightarrow \pi^*$ transition are always of very low intensity. The wavelength position of the MCD maximum usually does not coincide with the maximum of the UV absorption bands. Finally, the intensity, sign, and, when two bands have been observed, their relative intensities are highly dependent on the structure of the compounds. At present it is difficult to assess the level at which stereochemical information appears from these data. It can be imagined, however, that further investigations of conformationally well-defined systems are likely to point out some underlying patterns leading to a sector rule.

7 Concluding Remarks

A considerable amount of work has been done on substituted *quinolines* and *hydrocarbons* in an attempt to correlate the sign of the MCD with the type and position of the substituent[14,21]. The position of the substituted group was found to have little effect; what was much more important was its relative electron donating or withdrawing ability. MCD curves were also shown to have pronounced solvent effects in a number of cases[22]. As a rule, spin degeneracy could not be observed in MCD for singlet $\rightarrow$ triplet transitions or doublet $\rightarrow$ doublet transitions (free radicals). More experimental and theoretical work is needed to confirm and extend previous work.

Although MORD and MCD methods are still in their initial stages, they will soon be taught on a more general level. The information they provide usually cannot be obtained simply by any other method. Since the measurements require only the addition of a suitable magnet to existing instruments, it can be anticipated that the number of publications in this field will grow exponentialy during the next decade with a concomitant broadening of the scope of applications[23].

Fig. 9. MCD spectra of 4 different characteristic carbonyl compounds (Ref.[20]).

[20] *G. Barth, E. Bunnenberg, C. Djerassi*, Chem. Comm. 1246 (1969).

[21] *J.G. Foss, M.E. McCarville*, J. Am. Chem. Soc. *89*, 30 (1967).

[22] *B. Briat, D.A. Schooley, R. Records, E. Bunnenberg, C. Djerassi, E. Vogel*, J. Am. Chem. Soc. *90*, 4691 (1968).

[23] *C. Djerassi, E. Bunnenberg, D.L. Elder*, Pure Appl. Chem. *25*, 57 (1971).

5.10 Photoelectron Spectroscopy

Georg Hohlneicher
Institut für Physikalische Chemie und Elektrochemie
der Technischen Universität,
D-8 Munich

1 Introduction

Chemical applications of photoelectron spectroscopy are at present making such rapid strides that it is possible to give only a brief outline of the basic experimental principles. In addition, some of the concepts which can be used for the interpretation of photoelectron spectra will be discussed.

2 Definitions

In the past, the term *photoelectron spectroscopy* has not been used in a very specific way in the literature[1,2]. Used generally, it comprises all spectroscopic procedures which measure the energy distribution of electrons liberated from an atom or molecule by light quanta:

$$M + h\nu \rightarrow M_j^{\oplus} + e^{\ominus} \tag{1}$$

According to the energy of the exciting radiation the ion $M^{\oplus}$ is formed in its ground state $M_0^{\oplus}$ or in an excited state $M_j^{\oplus}$.

Although the term *electron spectroscopy* is used widely for all kinds of photoelectron spectroscopy, it is not adequate since it is used collectively to describe all absorption and emission processes in the visible and ultraviolet region.

When X-rays are used for excitation, highly excited ions are formed in which the detached electrons originate from an inner shell of one of the participating atoms. Siegbahn[7] refers to this method as *Electron Spectroscopy for Chemical Analysis (ESCA)*[3-7]. Other authors use the terms *Induced Electron Emission* (IEE, see Ref. [8]) and *X-Ray Photoelectron Spectroscopy*.

Excitation by light in the vacuum UV region instead of by X-rays liberates only *valence electrons*. This *low-energy photoelectron spectroscopy* generally is called simply photoelectron spectroscopy and is referred to in the abbreviated form PES.

Despite the substantially different energy of the exciting radiation, the respective spectrometer set-up (5.10:3) is very similar for ESCA and PES[4,7,12,14-17]. On the other hand, the wide differences in the chemical applications of ESCA and PES requires that they be discussed separately (5.10:4 and 5.10:5).

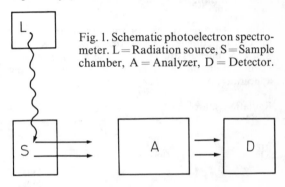

Fig. 1. Schematic photoelectron spectrometer. L = Radiation source, S = Sample chamber, A = Analyzer, D = Detector.

3 Principles of Measurement

The basic set-up of a photoelectron spectrometer is shown in Fig. 1.

In a target chamber, S, the radiation coming from the light source, L, strikes the substance being investigated. The photoelectrons generated are

[1] *D.M. Hercules*, Anal. Chem. *42*, 20A (1970).
[2] *D. Betteridge, A.D. Baker*, Anal. Chem. *42*, 43A (1970).
[3] *N. Svartholm, K. Siegbahn*, Ark. Mat. Astron. Fys. *33A*, 21 (1946).
[4] *K. Siegbahn, K. Edvarson*, Nucl. Phys. *1*, 137 (1956).
[5] *C. Nordling, S. Hagstrom, K. Siegbahn*, Ark. Fys. *13*, 483 (1958).
[6] *C. Nordling, S. Hagstrom, K. Siegbahn*, Z. Phys. *178*, 433, 439 (1964).
[7] *K. Siegbahn, C. Nordling, A. Fahlman, R. Nordberg, K. Hamrin, J. Hedman, G. Johansson, T. Bergmark, S.E. Karlsson, L. Lindgren, B. Lindberg*, ESCA «Atomic, Molecular and Solid State Structure Studied by Means of Electron Spectroscopy», Almquist and Wiksells, Uppsala 1967.
[8] Scientific Research, December 9, 1968, p. 51.
[9] *B.L. Kurbatov, F.I. Vilesov, A.N. Terenin*, Dokl. Akad. Nauk SSSR *140*, 797 (1961).
[10] *F.I. Vilesov, B.L. Kurbatov*, Dokl. Akad. Nauk SSSR *140*, 1361 (1961).
[11] *D.W. Turner, M.I. Al-Joboury*, J. Chem. Phys. *37*, 3007 (1962).
[12] *M.I. Al-Joboury, D.W. Turner*, J. Chem. Soc. 5141 (1963).
[13] *M.I. Al-Joboury, D.W. Turner*, J. Chem. Soc. 616 (1965).
[14] *D.W. Turner*, Proc. Roy. Soc. *A 307*, 15 (1968).
[15] *J.C. Helmer, N.H. Weichert*, Appl. Phys. Lett. *13*, 266 (1968).
[16] *D.W. Turner, D.P. May*, J. Chem. Phys. *45*, 47 (1966).
[17] *C.S. Fadley, C.E. Miner, J.M. Hollander*, Appl. Phys. Lett. *15*, 223 (1969).
L.N. Kramer, M.P. Klein, J. Chem. Phys. *51*, 3620 (1969).

separated according to their kinetic energy in an analyzer, A, and detected by the detector, D.

3.1 Radiation Sources

3.1.1 ESCA Excitation Sources

The preferred excitation sources in ESCA are the K_α lines of Al and Mg (Tab. 1). Since their half-widths lie just over 1 eV and narrower lines of sufficient energy are not known, the resolution attainable with ESCA is limited at present. However, with the use of a single-crystal monochromator one can obtain an excitation line half-width of 0.5 eV. The resulting loss in intensity is partly compensated for by elimination of the strong *bremsstrahlung*[18].

Table 1. Energies (in eV) of some radiation sources for use in photoelectron spectroscopy. The $K_{\alpha2}$ line is always much less intense than the $K_{\alpha1}$ line; in very intense photoionization bands, it may lead to the appearance of satellite bands.

Radiation source	Energies (in eV)	
	$K_{\alpha1}$	$K_{\alpha2}$
Mg	1250	
Al	1487.5	1487.0
Cr	5414	5404
Mn	5898	5886
Cu	8096	8035
	Resonance	
He	21.21	
He$^\oplus$	40.82	
Ne	16.67/16.84	
Ar	11.62/11.83	
Kr	10.03/10.64	
Xe	8.55/ 9.57	
Hg	4.89	

3.1.2 Low Energy Excitation (PES)

Here, the most prominent energy sources are the resonance lines of the noble gases[19]. Helium occupies a unique position since its resonance line does not display the doublet splitting common to the other noble gases (Tab. 1)[19, 20]. The resonance line of He$^\oplus$ is very important[21] for extending the

energy range. Continuous variation of the excitation energy is possible by using a vacuum monochromator, but the resulting spectral energy densities are much smaller than those obtained by resonance lines, except that it is possible to use synchrotron radiation as the initial source.

3.2 Analyzers

The analyzer works according to one of the following three principles:

Retarding field
Magnetic deflection
Electrostatic deflection

At the outset *retarding field* analyzers were used almost exclusively[9, 10, 12, 20]. Recently, Price[21] has improved them to the stage where the resolution obtained is almost equal to that first obtained by Turner[14] with electrostatic deflection (approx. 15 meV). In the retarding field method, because all electrons with a kinetic energy T greater than the retarding potential contribute to the measured current I, the actual photoelectron spectrum is obtained from dI/dT. The obvious advantage of retarding field analyzers is their high luminosity. Their main disadvantage is that at a given energy all of the background between the threshold value and this given energy contributes to the current, I. The signal to noise ratio thus decreases with increasing kinetic energy.

Most newer PES and ESCA instruments use *electrostatic deflection spectrometers* because magnetic shielding is much easier with them than with *magnetic deflection* instruments. Unlike the latter, where Helmholtz coils must be used to compensate for external magnetic fields — a difficult matter when dynamic magnetic perturbations arise side by side with the earth field[1] — paramagnetic screening is possible and is sufficient in most cases.

3.3 Detectors

The electrons emerging from the analyzer are detected either directly, or with a counter tube or a secondary electron multiplier (SEM). Direct detection requires very sensitive d.c. amplification and has the advantage that the output signal is proportional to the number of electrons leaving the analyzer. With the SEM the amplification is dependent on the electron energy, but its greater sensitivity generally outweighs this complication.

For very low counting rates scanning together with a small computer may be used to improve the signal to noise ratio.

[18] *R. Nordberg,* unpubl.

[19] *J.A.R. Samson,* Techniques of Vacuum Ultraviolet Spectroscopy, John Wiley & Sons, Inc., New York 1967.
R.B. Cairns, H. Harrison, R.I. Schoen, Applied Optics *9,* 605 (1970).

[20] *D.C. Frost, C.A. McDowell, D.A. Vroom,* Phys. Rev. Lett. *15,* 612 (1965).

[21] *W.C. Price* in *P. Hepple,* Molecular Spectroscopy, Proceedings of a Conference held at Brighton, England, 17.–19. April 1968, p. 221, Institute of Petroleum, Elsevier, London 1969.

4 High Energy Photoelectron Spectroscopy (ESCA)

Most ESCA measurements are performed on solids, but, as with PES, it is also fundamentally possible to investigate free molecules[22]. In a solid sample the photoelectrons generated originate from a surface layer not more than 100-Å thick[7]. Nonmetallic samples are deposited as a thin layer on a metallic carrier. As the measurement itself takes place under high vacuum conditions, the sample must have a sufficiently low vapor pressure and must not decompose under these conditions. In many cases strong cooling of the sample is necessary.

4.1 Theory

The kinetic energy of photoelectrons of solids is governed by the *energy level diagram* as shown in Fig. 2. There is considerably less direct interaction

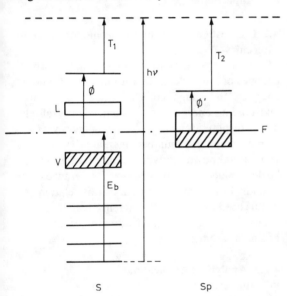

Fig. 2. Energy levels in high-energy photoelectron spectroscopy (ESCA). V = valence band, L = conduction band, S = sample, Sp = spectrometer, F = Fermi level.

between the inner electrons of different atoms than between valence electrons. The result is that the corresponding inner levels are not broadened as much as the valence band (V) or the conduction band (L). Very little is known presently about the actual width of the levels of the inner electrons in solids which determines the ultimate resolution obtainable in photoelectron spectroscopy.

4.1.1 Binding Energy

According to Siegbahn[7] the binding energy, E_b, is defined as the energy required to excite an electron from an occupied level to the Fermi level of the

solid. To remove this electron completely from the solid an additional amount of energy is required corresponding to the work function $\emptyset$. For a frequency, v, of the exciting radiation, the kinetic energy, T_1, with which the photoelectron leaves the solid is

$$T_1 = h v - E_b - \emptyset \qquad (2)$$

Strictly speaking, the energy balance should include the vibrational excitation, but, because the half-width of the exciting radiation is in the order of 1 eV, this contribution can be neglected. However, the fact that in most instances the energy $\emptyset$ is not identical with the work function $\emptyset$ of the spectrometer cannot be neglected. Thus, I, is usually not the exact kinetic energy of the electron when it enters the analyzer slit.

If sample S and spectrometer Sp are in electrical contact and in thermodynamic equilibrium, as should be the case under ESCA spectroscopy conditions[7], then the essential common reference level is not the energy level of the free electron but the common Fermi level[23].

Provided the sample is a *metal* or a *semiconductor*, the Fermi level and the contact potential, $\emptyset - \emptyset'$, are clearly defined. The contact potential causes the free electrons to be either braked or accelerated on their way to the inlet slit of the analyzer and to have a final kinetic energy, T_2. The following relationship is found between the measured variable, T_2, and the binding energy, E_b:

$$E_b = h \cdot v - T_2 - \emptyset' \qquad (3)$$

Where only *relative binding energies* are of interest, the quantity $\emptyset'$, which is generally not accurately known, is of no significance. It contributes approximately 3–5 eV in all measurements regardless of the nature of the sample.

Different conditions apply to *nonconductors*. Here, surface charges which are difficult to control may arise and contribute 1 or 2 eV to the difference $\emptyset - \emptyset'$. In this case, since the charges vary from sample to sample, the relative binding energies are affected.

---

[22] *K. Siegbahn, C. Nordling, G. Johansson, J. Hedman, P.F. Hedén, K. Hamrin, U. Gelius, T. Bergmark, L.O. Werme, R. Manne, Y. Baer,* ESCA Applied to Free Molecules, North-Holland Publishing Company, Amsterdam, London 1969.

[23] *A. Haug,* Theoretische Festkörperphysik I, p. 154, Franz Deuticke, Wien 1964.

4.1.2 Chemical Shifts

In principle, X-ray absorption measurements yield the same information as ESCA. The essential advantage of the latter is that in X-ray measurements the binding energies, E_b, can be observed merely as absorption edges superimposed on a strong continuous background. Photoelectron spectroscopy is, thus, superior in both sensitivity and resolution.

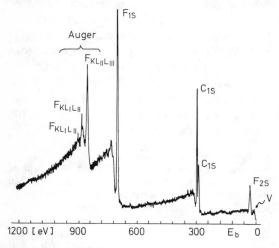

Fig. 3. ESCA spectrum of polytetrafluoroethylene (Teflon®). V = valence band. Recorded in the Development Laboratory of AEI Scientific Apparatus Limited, Manchester, England.

Fig. 3 shows the ESCA spectrum of *Teflon*® in the 0-1200 eV binding energy region. Before taking the spectrum a very thin layer of vacuum grease was applied to the Teflon sample. Together with the weak valence band, V, the ionizations from the 1s and 2s levels of fluorine and from the 1s level of carbon are clearly revealed. The carbon signal is split into two bands due to the C-atoms in the vacuum grease and the Teflon.

In a given atom, the varying chemical environments produce different binding energies of the inner electrons and *chemical shifts* are observed. Environmental influences on the inner electrons of an atom are mainly due to electrostatic interactions (via coulomb and exchange terms) with the valence electrons. There is an approximate correlation between this interaction and the effective charge of the atom under investigation[24]. In ionic crystals at least the first coordination sphere must be considered[7]. In organic compounds having several like atoms, usually the most positive atom, as determined by the molecular environment, displays the greatest binding energy[7, 25, 26]. However, this simple relationship does not always hold[26, 27] and the electrostatic interaction with charges at other atoms must be considered.

A much closer relationship exists between the binding energies and *orbital energies* of the inner electrons. Since the various shells in the inner electron region are clearly separated energetically, the single-particle picture gives a good approximation.

This fact is well known from X-ray absorption spectroscopy. The influence of correlation effects and relativistic corrections should be small especially if, instead of using absolute values, differences between orbital energies are compared with differences between binding energies[27]. As far as they are available, comparisons of this type yield good results even in those cases where there is no direct charge correlation[27].

4.2 Applications

4.2.1 Investigation of the Electronic Structure of Molecules

In addition to comparison of calculated orbital energies of inner electrons with measured binding energies, the rough correlation between charge and binding energy is significant. Nearly all elements display an increased binding energy as the formal oxidation number increases[1, 7]. In this way an unknown oxidation state of an atom can be determined by comparison with reference compounds. For organic compounds this relationship can be used together with reference compounds to determine the most positive atom out of a number of similar atoms.

4.2.2 Analytical Applications

Potential uses of ESCA are based on

Appearance of chemical shifts;
differences in binding energies of different elements.

The range of the chemical shift (about 15 eV) for a given level of an atom is small compared to the difference in binding energy between elements adjacent in the periodic table. For example, the binding energy of a 1s *carbon* electron is approximately 295 eV compared to 405 eV for a 1s *nitrogen* electron. Because the binding energies of various elements are so widely different, their signals do not overlap in the majority of cases. With ESCA it

[24] *U. Gelius, P.F. Hedén, J. Hedman, B.J. Lindberg, R. Manne, R. Nordberg, C. Nordling, K. Siegbahn, Physica Scripta, publ.*

[25] *J.M. Hollander, D.N. Hendrickson, W. Jolly*, J. Chem. Phys. *49*, 3315 (1968).

[26] *M. Barber, D.T. Clark*, Chem. Comm. 23, 24 (1970).

[27] *M. Barber, D.T. Clark*, Chem. Comm. 22 (1970).

is, therefore, possible to detect a trace amount of one element in the presence of others. However, the cross-section for the ionization from a given energy level is strongly dependent on the atomic number Z (proportional to Z^4). Therefore, the area of a photoionization band is no measure of the number of atoms present even if the energy dependence of the spectrometer sensitivity itself is known. For quantitative analysis it is possible to use a calibration technique with an accuracy of about 5% obtainable to date[7].

In many cases a correlation exists between chemical shift and charge of an atom. Except in strongly conjugated systems, the charge is determined mainly by the adjacent groups. Thus, the chemical shifts in ESCA techniques are also relatively specific for the immediate neighborhoold of an atom. For example, the ESCA spectrum of *ethyl trifluoroacetate* exhibits four partially resolved bands for the binding energy in the 1s carbon region corresponding to the four different C atoms[7]. Unfortunately, the resolution of 0.5 − 1 eV, which can be obtained only by means of a computerized deconvolution of the experimental spectrum, is often not sufficient to determine differences in binding energies. Consequently, in many instances, in tables of ESCA shifts, the ranges for differently bonded atoms overlap strongly.

4.2.3 Application to Surface Studies

Because the measured photoelectrons originate from a layer not thicker than 100 Å, the ESCA technique is really a *surface technique*. Where the ESCA spectrum of the bulk material is the subject of interest, the surface need not be absolutely clean. The spectra of molecules adsorbed on the surface are superimposed additively on the spectrum of the basic lattice. However, since most of the adsorbed molecules contain carbon and oxygen, it is advisable first to clean the surface by bombarding it with noble gas ions. In many cases the intensities of other signals, especially those due to higher transitions in heavier elements, increase remarkably if the surface is cleaned.

In surface studies coatings can be detected which, in favorable cases, amount to less than a monolayer[7]. Thus far chemical shifts produced by adsorption on a surface have not been reported. In general the ESCA technique is probably inferior to *Auger spectroscopy* (see 4.3) for the study of surface effects[28].

---

[28] *J.C. Rivière*, Phys. Bull. *20*, 85 (1969).

4.3 Secondary Effects in ESCA

Basically, ESCA yields the same information as X-ray fluorescence. However, in lighter atoms the latter is much less important than the Auger effect, in which the highly excited primary ion is stabilized by freeing a second electron, and not by emission of radiation.

$$M + h\nu \rightarrow M_j^{\oplus} + e^{\ominus} \rightarrow M_k^{\oplus\,\oplus} + 2e^{\ominus} \qquad (4)$$

The electrons emitted during this stabilization process once again have discrete energies characteristic of the atoms concerned leading to additional bands in the photoelectron spectrum (cf. *e.g.,* the bands above the F_{1s} signal in Fig. 3). In general, the Auger bands are broader than the photoelectron bands, and their position is independent of the excitation energy. Because, in some instances, Auger electron emission shortens the lifetime of the excited state, $M_j^{\oplus}$, appreciably, it can also introduce a broadening of the ionization band itself.

5 Low Energy Photoelectron Spectroscopy (PES)

Photoelectron spectra can be obtained either from gas-phase molecules or from solid surfaces. For *surface studies* the low-energy excitation has the advantage that the generated photoelectrons originate from a depth of only a few Å and can thus provide information about the surface it-

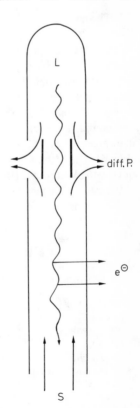

Fig. 4. Sample chamber arrangement for recording PES spectra of gases.

self[29, 30]. In addition, photoelectron spectroscopy seems to offer novel methods for studying adsorption processes at clean surfaces[31].

Here, the discussion is limited to the basic principles of low-energy photoelectron spectroscopy of free molecules. Detailed reviews are available in Refs. [2, 32, 33].

When *gaseous molecules* are excited with an energy greater than 10.5 eV, no highly transparent window is available to separate the resonance lamp L and the target chamber S illustrated in Fig. 4. Instead, differential pumping (diff. P) is used so that the gas from L cannot penetrate into the sample chamber and the sample S cannot penetrate into the lamp. The generated photoelectrons are separated preferentially parallel to the electric field, *i.e.*, perpendicular to the incident beam[34], and enter the analyzer through a narrow slit in the sample chamber wall. In the instruments now available a sample pressure of about 0.05–0.08 torr is required in the target chamber. Therefore, even where a heated inlet system is used, the application of PES is limited to compounds which reach a vapor pressure of about 0.1 torr at 200°C and vaporize without decomposition below this temperature. About 1 mg sample is required for the actual measurement, but to maintain a constant vapor pressure in the target chamber about 1 ml should be available.

5.1 Theory

With excitation energies less than 50 eV, only valence shell electrons are ionized. Because the valence orbitals are dependent on the total system, the information obtained from the low-energy photoelectron spectrum of a free molecule is different from that obtained from a solid sample of the same compound. The resulting ion, $M^{\oplus}$, is formed either in its ground state, $M_o^{\oplus}$, or in an excited state, $M_j^{\oplus}$ (Fig. 5). The resolution presently attainable is about 0.015 eV (approx. 100 cm^{-1}), so that, in favorable cases, it is possible to observe the excitations of vibrations together with the ionization processes (Fig. 5).

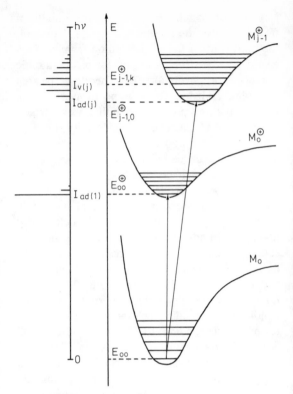

Fig. 5. Energy levels (right) to elucidate the photoelectron spectra (left).

Using a second subscript to denote the vibrational state, the adiabatic ionization potential is given by:

$$I_{ad(j)} = E^{\oplus}_{j-1,0} - E_{oo}$$

and the vertical ionization potential by:

$$I_{v(j)} = E^{\oplus}_{j-1, max,} - E_{oo}$$

for the jth ionization process. Here, 'j^{th} ionization process' denotes a process in which a monopositive ion is formed in an excited state. $E^{\oplus}_{j, max,}$ is that energy of the excited vibrational state, $M_j^{\oplus}$, which has the highest transition probability. Photoelectrons produced by an excitation radiation frequency, ν, have the kinetic energy, T:

$$T = h\nu - I_{ad(j)} - \Delta E_{vib} \qquad (5)$$

Fundamentally, repulsion effects, rotational excitation, and other perturbations, e.g., the velocity distribution of the molecules, also influence the kinetic energy of the photoelectrons. In discussing spectra these effects can be neglected because they are at least one order of magnitude smaller than the limit of resolution[14, 35].

From Fig. 5 it can be seen that the jth adiabatic ionization potential can be written in the following form:

$$I_{ad(j)} = I_{ad(1)} + \Delta E^{\oplus}_{o \to j-1} \qquad (6)$$

$\Delta E^{\oplus}_{o \to j}$ is the excitation energy for the transition $M^{\oplus}_{oo} \to M^{\oplus}_{jo}$ i.e., $I_{ad(j)} - I_{ad(1)}$ corresponds to an electronic excitation of $M^{\oplus}$ which may be observed also in the optical absorption spectrum of the ion itself. Admittedly, the selection rules for photoionization differ from those for optical transitions. For a given molecule the effective cross-section for ionization depends mainly on the

[29] *W.F. Krolikowski, W.E. Spicer*, Phys. Rev. *185*, 882 (1969).

[30] *W.F. Krolikowski, W.E. Spicer*, Phys. Rev. *B 1*, 478 (1970).

[31] *W.T. Bordass, J.W. Linnett*, Nature *222*, 660 (1969).

[32] *D.W. Turner*, Chem. Brit. *4*, 435 (1968).

[33] *G. Hohlneicher*, Angew. Chem. unpubl.

[34] *J.W. McGowan, D.A. Vroom, A.E. Comeaux*, J. Chem. Phys. *51*, 5626 (1969).

nature of the removed electron (σ, π, etc.) and, for the same kind of electrons, it depends on the degeneracy of the ionic state, $M_j^{\oplus}$[21].

From equation (6), T can be defined by:

$$T = h \cdot \nu - I_{ad(1)} - \Delta E_{o \to j-1}^{\oplus} - \Delta E_{vib}. \qquad (7)$$

The energy ΔE_{vib} corresponds to a vibrational excitation of the ion formed. The intensity distribution among different vibrations within the photoionization band is governed by the Franck-Condon factors which depend on the geometry of the ground state of the molecule, M_o, and on that of the excited ionic state, $M_j^{\oplus}$[36, 37] (cf. Fig. 5). Conversely, if the vibrational portion of a photoelectron band is resolved, the measured Franck-Condon factors provide information about the geometry of the excited ionic state[38].

The successful use of PES for investigating the electronic structure of molecules is due, above all, to the close relationship between the vertical ionization energy $I_{v(j)}$ of the ion and the single-particle energies of the neutral initial molecule.

Within the framework of a single-particle description (Fig. 6), the occupation B3 corresponds to an excited state relative to the ionic ground state B2 due to the excitation of an electron from orbital 3 to orbital 1. Alternatively, relative to the occupation in the ground state of the initial neutral molecule B1, it appears that an electron has been removed directly from orbital 3 to form B3.

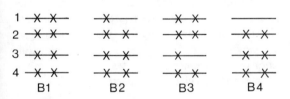

B1   B2   B3   B4

Fig. 6. Occupation schemes for some molecular states in a single-electron representation.

From this diagram it can be seen that the orbital energies of the neutral molecule represent a first order approximation to the various ionization energies (Koopmans' theorem)[39]. This means that

their differences represent a first order approximation to the particular excitation energies of the ion. By comparison to the ground state, excited states of the ion, as described by the occupation B4 in the single-particle picture, correspond to the separation of one electron and the simultaneous excitation of another one. Such two-particle processes at most yield very weak bands in the photoelectron spectrum. The orbital energies, however, represent only a first order approximation for the ionization energies[40]. They do not take into account rearrangement effects and changes in correlation contributions accompanying the transition from the initial neutral molecule to the final ionic state. Nevertheless, the measured vertical ionization potentials are in fairly good agreement with calculated orbital energies. The calculated orbital energies are invariably greater than the ionization energies, but in many cases the difference is nearly proportional to the orbital energy[41, 42]. Substantial deviations are observed, in general, only where the electron is detached from a degenerate orbital[41, 43, 44]. In this case a Jahn-Teller effect often occurs[45], which eliminates orbital degeneracy in the ion by decreasing molecular symmetry.

Possible uses of PES to 'visualize' orbitals extend also to their vibrational structure. For example, the geometry is little affected if an electron is removed from a nonbonding orbital. Therefore, the Franck-Condon factor for the 0–0 transition is very large, and transitions to higher vibrational levels become rapidly weaker with increasing excess energy. (cf. Fig. 5).

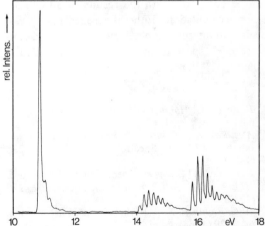

Fig. 7. Photoelectron spectrum of formaldehyde.

[35] D.W. Turner, Mol. Spectrosc. 209 (1969).

[36] M.E. Wacks, J. Chem. Phys. 41, 930 (1964).

[37] M. Halmann, I. Laulicht, J. Chem. Phys. 43, 1503 (1965).

[38] W.L. Smith, P.A. Warsop, Trans. Faraday Soc. 64, 1165 (1968).

[39] T. Koopmans, Physica 1, 104 (1933).

[40] W.G. Richards, Internat. J. Mass. Spectrosc. Ion Phys. 2, 419 (1969).

[41] H. Basch, M.B. Robin, N.A. Kuebler, C. Baker, D.W. Turner, J. Chem. Phys. 51, 52 (1969).

For example, in *formaldehyde* (Fig. 7), the ionization from the nonbonding oxygen orbital exhibits an intense line at 10.88 eV with two small vibrational satellites. If, instead, the ionization is from a bonding or antibonding orbital, then the geometry is altered more strongly and the greatest Franck-Condon factor is displayed by a transition higher than the 0–0 transition. By comparing the observed vibration frequencies with those of the neutral molecule, it may be possible to decide whether and in what region of the molecule the orbital is bonding or antibonding. Accordingly, the second ionization band of formaldehyde, which is assigned to the bonding π-orbital[43], exhibits a well-defined vibrational structure with $\tilde{v} = 1210 \pm 50$ cm^{-1} (0.15 eV). This figure must be compared with the C–O stretching vibration, $\tilde{v} = 1744$ cm^{-1} (0.22 eV), of the ground state of the neutral molecule[43]. The frequency is reduced, as one would predict when a bonding electron is removed.

5.2 Applications

5.2.1 Electronic Structure and Bonding Properties of Molecules

Due to the close relationship between ionization potentials and orbital energies, PES is a very useful technique for studying electronic structure and bonding properties of molecules. The helium resonance line produces photoelectrons from the σ-system as well as from the π-electron system. Consequently, PES is, in many instances, the only method that furnishes experimental data which can be compared with the results of all-valence electron calculations. Either theoretical results can be used to interpret photoelectron spectra or, conversely, PE-spectra can be used to judge calculations[41, 43, 46–51].

Simple perturbation treatments based substantially on a qualitative knowledge of molecular orbitals are at least as important as the comparison of measured ionization energies with calculated orbital energies. Such treatments often enable one to derive a relationship between the photoelectron spectra of related compounds and to deduce the electronic structure from the changes that occur. Interesting results have been obtained from the interaction between different *free electron pairs*[52, 53], between *conjugated double bonds*[54, 55], between *nonconjugated double bonds*[56] and between *halogen atoms* and *unsaturated systems*[57, 58]. In addition, studies of this type in many instances clearly demarcate *mesomeric effects* from *inductive effects* within a given model[55, 58].

5.2.2 Potential Analytical Applications of PES

To date very little information is available on the analytical uses of PES. However, from other applications, the following conclusions can be drawn about its capabilities.

For *qualitative* applications PES appears to be equivalent to electron spectroscopy in the visible and ultraviolet regions. Since most *saturated systems* exhibit more than one ionization band between the first ionization energy and the maximum upper limit governed by the exciting line (generally 21.21 eV), a photoelectron spectrum as a fingerprint of the compound is often more significant than an electronic spectrum.

The first ionization potential of an *unsaturated system* is usually 2–4 eV lower than that of a comparable *saturated system*. However, sometimes in unsaturated systems, too, the first ionization is displaced markedly towards higher energies by *heteroatoms*. The influence of alkyl substituents on photoelectron spectra is usually much greater than on electronic spectra.

[42] *F. Ecker, G. Hohlneicher*, Theor. Chim. Acta, unpubl.

[43] *A.D. Baker, C. Baker, C.R. Brundle, D.W. Turner*, Internat. J. Mass. Spectrosc. Ion Phys. *1*, 285 (1968).

[44] *C. Baker, D.W. Turner*, Chem. Comm. 480 (1969).

[45] *H.A. Jahn, E. Teller*, Proc. Roy. Soc. A *161*, 220 (1937).

[46] *W.G. Richards, R.C. Wilson*, Trans. Faraday Soc. *64*, 1729 (1968).

[47] *R. Manne*, Chem. Phys. Lett. *5*, 125 (1970).

[48] *E. Lindholm, B.Ö. Jonsson*, Chem. Phys. Lett. *1*, 501 (1967).

[49] *M.J.S. Dewar, S.D. Worley*, J. Chem. Phys. *50*, 654 (1969).

[50] *D.C. Frost, F.G. Herring, G.A. McDowell, I.A. Stenhouse*, Chem. Phys. Lett. *4*, 533 (1969); *5*, 291 (1970).

[51] *E. Heilbronner, K.A. Muszkat*, J. Am. Chem. Soc. *92*, 3818 (1970).

[52] *P. Bischof, J.A. Hashmall, E. Heilbronner, V. Hornung*, Tetrahedron Lett. 4025 (1969).

[53] *E. Haselbach, E. Heilbronner*, Helv. Chim. Acta *53*, 684 (1970).

[54] *C. Baker, D.W. Turner*, Proc. Roy. Soc. A *308*, 19 (1968).

[55] *R. Griebel, G. Hohlneicher*, unpubl.

[56] *P. Bischof, J.A. Hashmall, E. Heilbronner, V. Hornung*, Helv. Chim. Acta *52*, 173 (1969).

[57] *H.J. Haink, E. Heilbronner, V. Hornung, E. Kloster-Jensen*, Helv. Chim. Acta *53*, 1073 (1970).

[58] *E. Heilbronner, V. Hornung, E. Kloster-Jensen*, Helv. Chim. Acta *53*, 331 (1970).

Ionization of an electron from a lone-pair orbital of imines, aldehydes, ketones or halogens can be recognized easily in many molecules. In these cases, also, the ionization potential is not significant for the immediate neighborhood of the corresponding heteroatom. This is due to the sensitivity of the ionization potential to substituents far from the lone pair[55].

For *quantitative* analysis PES is probably of little use. Even where the sensitivity curve of the spectrometer and the relative cross-sections of the photoionization are known, the measured intensity gives information only on the partial pressures in the target chamber. These pressures are determined less by the original sample composition than by the vapor pressure and the effusion properties of the different compounds. By using a calibration, it is sometimes possible to reach an *analytical accuracy* of about 3% for some mixtures.

Smaller amounts of *impurities* can be detected if they have a lower ionization potential than the main component or if they have very sharp ionization bands, e.g., *cyclohexene* in *cyclohexane* or *water* in *benzonitrile* can be detected in concentrations less than 1%[59].

5.3 Comparison With Other Methods

The most accurate ionization potentials are obtained with the optical method[60], where the values are determined from the convergency limit of a Rydberg series[61]. However, this method is restricted to molecules displaying well resolved Rydberg bands and, usually furnishes only the first ionization potential. Except for differences in the selection rules, which make evaluation of absorption spectra more difficult, absorption spectra of free radical ions supply the same information as photoelectron spectra[40] of the corresponding neutral molecules. However, it is seldom possible to measure an absorption spectrum of a radical ion in the gas phase.

Since only the generated photoelectrons are recorded in PES, the result obtained is influenced by succeeding reactions only if the initially formed state of the cation is very short-lived[14, 35]. Unlike other methods used to determine ionization potentials, such as photoionization[62] or mass spectroscopic determination of appearance potentials[63], PES usually yields satisfactory results also in the determination of higher ionization energies. Autoionization processes occurring in highly excited neutral molecules can interfere in photoelectronic spectra due to the process

$$M_o + h\nu \rightarrow M_k \rightarrow M_i^\oplus + e^\ominus \qquad (8)$$

The electrons thus generated also contribute to the photoelectron spectrum. However, neutral molecules must absorb the whole quantum hν. Consequently, autoionization processes manifest themselves only in the vicinity of the energy used for excitation, *i.e.*, at the upper limit of the PE-spectrum[21].

Periodically the interaction of an autoionizing state with an excited state of an ion is observed. When continuous excitation is employed and those electrons are measured which are freed while possessing zero kinetic energy (photoionization resonance spectroscopy[64, 65]), these processes tend to dominate the photoelectron spectrum.

[59] *R. Griebel, G. Hohlneicher*, unpubl.

[60] Ionization Potentials, Appearance Potentials and Heats of Formation of Guseons Positive Ions, US Department of Commerce, National Bureau of Standards, 1969.

[61] *G. Herzberg*, Molecular Spectra and Molecular Structure I. Spectra of Diatomic Molecules, 2. Ed., p. 387, D. van Nostrand Co., Toronto, New York, London 1950.

[62] *F.I. Vilesov, M.E. Akopyan* in *S. Neporent*, Elementary Photoprocesses in Molecules, Engl. from the Russian, p. 22, Consultants Bureau, Plenum Press, New York 1968.

[63] *C.A. McDowell* in *C.A. McDowell*, Mass Spectrometry, p. 506, McGraw-Hill, New York 1963.

[64] *W.B. Peatman, T.B. Borne, E.W. Schlag*, Chem. Phys. Lett. *3*, 492 (1969).

[65] *T. Baer, W.B. Peatman, E.W. Schlag*, Chem. Phys. Lett. *4*, 243 (1969).

5.11 Advances in Photometry

J.K. Foreman
Ministry of Technology, Laboratory of the Government Chemist, London, S.E. 1

This chapter is concerned with ultraviolet and visible absorption and reflectance photometry, fluorescence and phosphorescence (5.11:4). The text describes recent developments on broadening the scope of the techniques and their application in less usual circumstances such as high and low temperatures, continuous measurements, flow through systems. Specific consideration is given to kinetic measurements (5.11:3) for which spectrophotometry is proving to be a particularly useful method of measurement.

1 Ultraviolet and Visible Absorption Spectrophotometry

1.1 Instrumentation

Commercial spectrophotometers offering a high standard of performance have been available for some time. Nevertheless improvements continue to be made, both in terms of quality of instrument performance and extension of application to new areas of analytical interest, such as flow-through and kinetic measurements (for automation s. Chapter 14.3).

1.1.1 Light Sources

The hydrogen discharge and tungsten lamps are well established for the ultraviolet and visible regions respectively; in each case an effectively continuous emission is achieved by pressure broadening of a line source.

In most modern light sources filling with deuterium results in a higher lamp output than with hydrogen. The spectral energy distribution for a deuterium lamp shows increasing intensity in the direction of low wavelength. The useful limit is set by the transmission characteristics of the window. Fused silica is conventionally used as the window material and this gives a useful output down to 180 nm. Extension of the limit to 160 nm is possible using synthetic silica windows. Measurements at wavelengths lower than 160 nm have been made in a vacuum spectrometer[1] using an efficient water-cooled lamp having calcium fluoride windows.

Wavelength calibration is achieved with the mercury-deuterium lamp[2]. The characteristic mercury lines are superimposed upon the continuous emission of the deuterium.

Continuous sources for the *vacuum ultraviolet* region have been extensively studied. These are provided by condensed discharges or microwave excited spectra in rare gases. Two of these, krypton and xenon, overlap the ultraviolet region in their wavelength coverage (Table 1) and may find increasing practical application. Indeed xenon lamps are used in several commercial ultraviolet spectrophotometers.

Table 1. Wavelength range for the continuum of Krypton and Xenon.

Element	condensed discharge	microwave discharge
Kr	125–185 nm	126–170 nm
Xe	147–225 nm	150–200 nm

Attempts to use mixed Kr-Xe discharges to provide a longer continuum for low wavelengths fail because the ionization potential of xenon (12.3 eV) is sufficiently lower than that of krypton (13.97 eV) that the exciting energy produces xenon emission almost entirely.

The rapid growth of atomic absorption spectrophotometry has led to the ready availability of sharp line sources of many elements based upon either the hollow cathode or the microwave-excited electrodeless discharge tube. Such line sources for absorption spectrophotometry have been used instead of continuous sources[3]. The approach is suggested as being of value for the examination of materials having sharp absorption maxima.

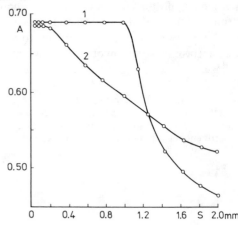

Fig. 1. Variation of absorbance with slit-width for anthracene in cyclohexane at 340.4 nm
1 Cadmium discharge lamp, 2 Hydrogen lamp;
A = Absorbance, S = Slit width
From Anal. Chim. Acta, *39*, 161 (1967).

[1] *B. Flood,* Photoelec. Spectrom. Group Bulletin *17*, 495 (1967).
[2] Manufacturers' Supply, Wickham Ltd., Southampton, England.
[3] *W.W. Harrison, K. Caufield,* Anal. Chim. Acta *39*, 161 (1967).

In such cases the use of continuous radiation can result in deviations from Beer's Law, giving rise to non-linear calibration curves and a reduction in sensitivity. Furthermore, instrumental slit-width settings must ideally be fixed to maintain a constant wavelength range of radiation and constant average intensity at the detector, which will record an intensity that is not necessarily related to the true absorbance. Fig. 1 illustrates the advantage of using a line source, absorbance is independent of slit-width until the latter is widened to a point where light from an adjacent line in the source spectrum enters the beam. Accurate instrumental wavelength setting, merely by adjusting to peak detector response at the line of interest, is an additional benefit.

Successful analytical applications of line sources for absorption measurements include *aniline*, *N-methylaniline* and *N,N-dimethylaniline*, each with broad peaks in the region of 300 nm. However, the small wavelength displacements between the three spectra are sufficient to enable the three compounds to be separately determined in a mixture.

A mercury arc source was used in conjunction with a conventional spectrophotometer, measurements were made at three wavelengths each corresponding to a mercury line (313.2 nm, 302.1 nm, 275.3 nm) and concentrations calculated from simultaneous equations[4]. Errors of 1% or better were achieved, for the *styrene* content of *ethylbenzene* samples in the range 0—10% utilizing the styrene band near 290 nm[5].

An essential prerequisite of the line source method is the availability of a source having an emission line at the appropriate wavelength for sample measurement. Although many element lamps are available offering wavelengths throughout most of the ultraviolet and visible spectrum (see Table 2) exact matching of emission line to the required absorption band may still prove difficult.

Again, for substances having broad absorption bands the line source method is not attractive; errors due to the use of polychromatic incident radiation are minimized in such instances and the limited improvement in sensitivity offered by line sources is unlikely to justify the practical inconvenience of lamp changing for different applications.

As evident from Table 2, the wavelengths in the ultra-violet range are well covered, the longer wavelength visible spectrum less so.

1.1.2 Optical System

For dispersion of the incident light many general purpose spectrophotometers are based upon a 30° prism in a Littrow mounting or, alternatively, a diffraction grating with associated filters to remove higher orders of diffraction. Occasionally both are provided in a single instrument. Most commercial spectrophotometers can be used over the wavelength range 190 to 850 nm and a number of prism instruments are operable to 2500—3000 nm using a purge of nitrogen to limit absorption by air.

In general prisms are fabricated of quartz or fused silica; their wavelength resolution is a function of wavelength (typically 0.2 nm at 200 nm, 2 nm at 800 nm). Grating spectrophotometers yield constant resolution throughout the wavelength range (Fig. 2).

Analytical determinations using light absorption comprise measurement of the extent of light absorption at a particular wavelength relative to that in a 'blank' or reference solution.

In recent years the incorporation of double beam facilities has become increasingly more widespread in commercial spectrophotometers.

Table 2. Emission wavelength (in increasing order of the main lines) of commercially available individual element lamps

Element	λ [nm]	Element	λ [nm]	Element	λ [nm]	Element	λ [nm]
As	193.7	Si	251.6	Al	309.3	La	391.2
Se	196.1, 204.0	Hg	253.7	Mo	313.3	Y	410.2
Zn	213.8	Ir	264.0	V	318.4	Nb	405.9
Te	214.3	Ge	265.1	Cu	324.7	Ca	422.7, 239.9
Sb	217.5, 231.1	Pt	265.9	Ag	328.1	Sm	429.7
Bi	223.1	Gd	268.4	Rh	343.5	Tb	432.6
Cd	228.8, 326.1	Ta	271.5	Re	346.0	Eu	459.4
Ni	232.0, 341.5	Tl	276.8	Ru	349.9	Sr	460.7
Be	234.8	Mn	279.5	U	351.5	Nd	463.4
Co	240.7	Pb	283.3	Cr	357.9	Pr	495.1
Au	242.8	Sn	286.3, 224.6	Zr	360.1	Ba	553.6
Pd	247.6	Mg	285.2	Ti	365.3	Na	589.0
Fe	248.3, 372.0	Hf	307.2	Sc	391.2	Li	670.8
B	249.7	In	304.0	W	400.9	K	766.5, 440.4

[4] *D.D. Tunnicliff*, Anal. Chem. *20*, 828 (1948).
[5] *N. Hadden, J.A. Perry*, Anal. Chem. *23*, 1337 (1951).

In optical double beam systems the light beam passes alternately through the sample and reference cells by means of a suitable device placed between the collima-

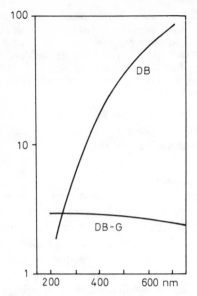

Fig. 2. Dispersion as a function of the nominal wavelength for the Beckman Model DB (Prism) and DB-G (Grating) Spectrophotometers from Beckman Instruments, Ltd.

tor and the cells. This can be rotating halfsector mirror (Hitachi Model 124), vibrating mirror (Beckman DK), rotating chopper-mirror (Cary Model 16) or a beam splitter (Shimadzu Model MPS-50L). The mechanical approach is utilized in the Unicam SP 3000 automatic spectrophotometer; reference and sample cells are moved alternately into the light path. The signal from the reference beam is first attenuated until it is equal in intensity to that provided by a modulated auxiliary light beam from a separate tungsten lamp. The intensity transmitted by the sample is measured as a ratio of the auxiliary beam. The resulting signal represents percentage transmittance of the sample relative to the reference solution.

The *kinetic study* of *enzyme* systems requires measurements of absorbance changes as little as 0.002 units. Frequently changes are monitored at two wavelengths simultaneously; one wavelength is typically an isosbestic point in relation to the two components of the system, e.g. oxidized and reduced forms of an enzyme, and the second is a wavelength at which changes in concentration of one of the components can be measured. Equipment for such studies is now available commercially (Cary Model 16); after selection of the two appropriate wavelengths the unit switches the monochromator successively between them.

The photometric accuracy claimed for this instrument, which has two prism monochromators mounted to yield additive dispersion, is $\pm$ 0.1% near unit absorbance or better. Since biochemical systems are frequently turbid, high performance measurement at high absorbances is essential.

Dual wavelength operation is achieved in the Aminco-Chance spectrophotometer by an optical arrangement which utilizes two diffraction gratings (Fig. 3). Incident radiation is divided into two beams by a mask and the intensity of the two beams equalized by a manual attenuator. The two beams are dispersed separately by the two gratings and focussed on to the exit slit by a single mirror. The two chopped beams are passed alternately through the sample cell by means of two mirrors. Signals from sample and reference solutions are sorted electrically after amplification by a photomultiplier tube. The sensitivity of absorbance difference measurement is such that full scale deflection is produced by a 1% change. Absorbancies of the order 2–3 are useable in favorable circumstances.

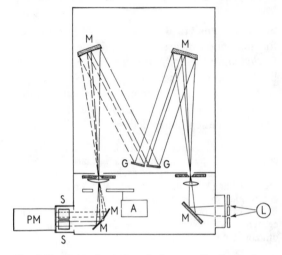

Fig. 3. Dual wavelength optical system for the Aminco-Chance Spectrophotometer

M Mirror	PM Photomultiplier
G Grating	S Sample cuvette
A Optical attenuator	L Light source
From Aminco-Chance	

Where absorption bands are poorly resolved the production of the first derivative of the absorption curve enhances the definition. An optical system capable of yielding a derivative spectrum is available for the Shimadzu MPS-50L spectrophotometer.

The light beam from the monochromator is split into two and each half is partially masked, one beam at the short wavelength end and the other at the long wavelength end. Consequently, there is a small difference between the mean wavelength of the two beams passing into the sample and reference cells. Scanning over the desired wavelength range with sample solution in both cells produces an output which approximates closely to a first derivative spectrum.

1.1.3 Detection of Transmitted Radiation

Two types of *detector* are used, photomultiplier tubes for the ultraviolet region and lead sulfide photocells for longer wavelengths. Several instruments are fitted with a single detector, either photomultiplier or photocell, covering the entire wavelength range. The performance of these detectors and associated electronics is such that photometric reproducibilities of the order 0.1—0.5% over the 0—100% transmittance range are obtainable; with high performance spectrophotometers this is improved by about one order of magnitude.

In the measurement of absorbancies in turbid or translucent samples the cell/detector geometry is important. In such circumstances radiation falling upon the detector, omitting reflected radiation, is of two types, parallel transmitted light I_p which has transversed the medium without striking the sample particles and diffuse transmitted light I_d which has interacted with the particles. The latter, due to scattering, is not necessarily parallel to the incident radiation. If I_0 denotes the intensity of the incident radiation the measured absorbance is given by[6]

$$\log \frac{I_0}{I_p + f\,I_d}$$

where f is the fraction of the diffused radiation collected by the photomultiplier. f clearly increases as the cell-detector distance decreases (Fig. 4).

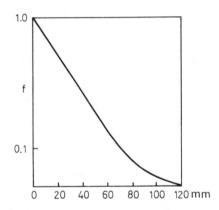

Fig. 4. Relationship between f and sample/detector distance L for the end-on photomultiplier orientation of the Shimadzu MPS-50 L spectrophotometer from Shimadzu

For the most meaningful results, therefore, the cell-detector separation should be minimal. This configuration is provided in the Shimadzu, Model MPS-50L and where the cell and detector are separated by a thin diffuser of high transmittance, either of opal glass or quartz. Under these conditions the value of f approximates to unity and maximum definition of absorption spectra is obtained.

1.1.4 Presentation of Photometric Data

Earlier spectrophotometers usually displayed transmitted light intensity on a meter with a linear transmittance scale and, as a result, the exponen-

tial absorption law, a logarithmic absorbance one. The error introduced in reading the logarithmic absorbance scale at high values is well known and it is becoming generally apparent in modern spectrometers that the presentation is arranged to read linearly in both transmittance and absorbance. The logarithmic absorbance output is linearized electronically. Advantage is taken of the fact that certain transistors, termed transductors, obey extremely closely an ideal logarithmic current to voltage relationship. Thus the logarithmic current output of a detector with reference to concentration of the absorbing species may be converted to a linear voltage relationship. The accuracy of such a conversion depends upon the closeness to logarithmic response of the selected transductor.

1.1.5 Cells

The application of spectrophotometry to systems at high or low temperatures, to flowing streams or to wide concentration ranges is made possible by the provision of suitable absorption cells and appropriate temperature controlling units. Almost all manufacturers of modern spectrophotometers offer a wide range of accessories such as long-path cells, flow-through cells, micro cells, constant temperature baths etc. (see the makers literature). In addition to commercial products there has, however, been widespread research by many authors into providing cells for particular applications and this chapter summarizes the outcome.

In a special flow-through microcell of 1 mm circular aperture the volume of sample in the light beam is approximately 8 μl[7]. The cell is fabricated from a plexiglass block, inlet and outlet for solutions are provided by stainless steel 1 mm hypodermic needles. An analytically useful property of this type of cell is that up to quite high flow rates the flow is laminar. Thus samples which have not undergone zonal mixing, e.g. effluents from chromatographic columns, gradient density solutions for electrophoresis, can be analyzed zonally. A flow rate of about 0.5 ml/min is suitable. A cell offering a 5 cm light path and being 2 mm in diamter has been fabricated from tygon tube[8]. When used for flow-through studies the inner wall is notched at the entrance and exit to guide air bubbles through. This cell has a higher length/volume ratio than most others which are readily available and is, therefore, advantageous for measurements on weakly absorbing species.

6 *K. Shibata*, Spectrophotometry of Translucent Biological Materials — Opal Glass Transmission Methods in *D. Glick*, Methods of Biochemical Analysis, p. 77—109, Interscience, New York 1959.
7 *R. Shapira, A.M. Wilson*, Anal. Chem. *38*, 1803 (1966).
8 *F.J. Kilzer, S.B. Martin*, J. Chromatogr. *31*, 204 (1967).

In making measurements on *samples of unknown absorbance* difficulties can arise in selecting the appropriate cell path-length or concentration at which the measurement is to be made. Recourse to concentration adjustment by dilution may be criticised on the grounds that, not only is it time-consuming, but Beer's Law may not be obeyed over the concentration range of interest and chemical equilibria may be disturbed in the dilution process. To overcome this problem double path length cells[9] (Fig. 5) allow the light beam to pass through 10 mm or through 4 mm. Thus, rotation of the cell through 90° provides a 2.5 fold reduction in path length. Alternately[10], a path length ratio of 10 to 1 can be achieved merely by inserting a silica spacer 9 mm × 9 mm into a 10 mm × 10 mm cell. To avoid overflow of sample solution the original cell is fitted with a reservoir at the top.

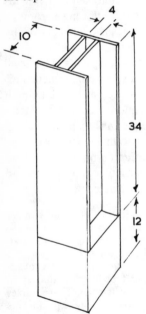

Fig. 5.
Double path quartz micro-cell[9] from Anal. Chem. *39*, 1679 (1967)

Difference spectrometry, i.e. spectral changes upon mixing two solutions is carried out in a special cell (Fig. 6). The two components are placed separately in the two compartments and, after balancing the spectrometer, the cell is inverted, the solutions allowed to mix and the spectrum remeasured. Two particular uses claimed for this cell are[11] zero-time absorbance measurement in spectrophotometric kinetic studies and improved precision in double-beam measurements on small quantitaties of sample.

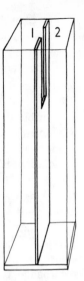

Fig. 6.
Split-compartment cell for difference spectrophotometry[11] from Anal. Biochem. *6*, 287 (1963)

Absorption spectra measured at reduced temperatures: In a double path absorption cell[12], 2 cm in optical path length, temperatures down to −196° can be reached with a suitable coolant (liquid nitrogen).

The cell is constructed of brass and comprises detachable inner and outer chambers separated by an evacuated region to provide insulation. Quartz windows are used, sealed in indium gaskets. The vacuum seal between the inner and outer sections is made with rubber or teflon rings. A stream of dry air is swept across the optical faces to avoid condensation of material on the windows. In use the interspace between the two compartments is first evacuated, then sample and reference are introduced into the two inner compartments and coolant into the outer compartment.

An alternate approach is to construct a Dewar flash which can be fitted into the cell compartment of a commercial spectrophotometer (Cary Model 14[13]).

The Dewar flask is of flattened oval section and the optical path passes through the region of minimum curvature. The cell can be lowered into the Dewar flask and has acrylic resin windows. The level of liquid nitrogen must be below the optical path before measurements are made.

Low temperature spectrophotometric measurements can also be made by depositing the material of interest as a thin film on a suitable window material such as sapphire. The maintenance of good thermal contact between the window and refrigerant is of prime importance. A simple method of ensuring this is to wind a strip of copper round the window, the ends of the strip are allowed to protrude a few cm upwards and are soldered into a slot in the base of a Kovar vessel which contains the coolant and forms the lower portion of an optical-type Dewar flask[14]. This arrangement can be used repetitively without frequent dismantling.

[9] *D.M. Abelson*, Anal. Chem. *39*, 1679 (1967).
[10] *M.H. Smith*, Nature *192*, 722 (1961).
[11] *J.A. Yankeelov jr.*, Anal. Biochem. *6*, 287 (1963).

[12] *F.J. Smith*, Rev. Sci. Instrum. *33*, 1367 (1962).
[13] *W.B. Elliott, W. Tanski*, Anal. Chem. *34*, 1672 (1963).

Absorption spectra measured in the vapor state:
Ultraviolet spectra of *steroids* recorded in conventional solvents show a single broad band over the wavelenght range 210–250 nm. Both definition and sensitivity are improved by recording spectra of the vapors. A cell and furnace assembly has been designed[15] which permits heating under reduced pressure (Fig. 7).

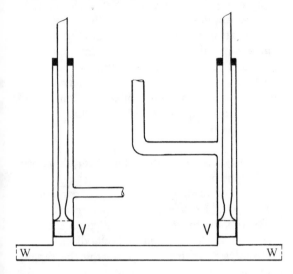

Fig. 7. Cell for measurement of vapor phase ultraviolet absorption spectra
V   Plug needle valve; quartz counting window

It is 10 cm by 1 cm in size and constructed of UV-transmitting silica. Two teflon plug needle valves are provided for sealing the unit and two side-arms for ingress and egress of sample. The furnace fits round the cell below the level of the side arms. In studying difficultly volatile materials such as steroids the sample is admitted as a chloroform solution $(10-20\mu$ l) by removing one of the plugs. The upturned side-arm is connected to a vacuum pump; after evaporation of the solvent the pumping system is isolated and the cell heated to a temperature at which the material vaporizes. The analytical value of absorption spectra obtained in this way is illustrated in Fig. 8 which shows the spectra of the vapors of *corticosterone acetate* and *deoxycorticosterone acetate*. These two compounds can be unequivocally differentiated by observation of the region below 220 nm. Purging of the spectrophotometer light path with dry nitrogen is advisable in order to reduce unwanted absorptions due to other vapors in the 220 nm region. This type of cell is inherently versatile; temperatures up to 250° are readily attainable, and, in addition, the design allows it to be linked to the outlet of a *gas chromatographic* column.

[14] *L.J. Schoer*, Rev. Sci. Instrum. *38*, 1531 (1967).
[15] *S. Natelson, J.E. Bonas*, Microchem. J. *9*, 68 (1965).
[16] *D.F. DeTar*, Anal. Chem. *38*, 1794 (1966).
[17] *R.B. Fraser, E. Suzuki*, Anal. Chem. *38*, 1773 (1966).

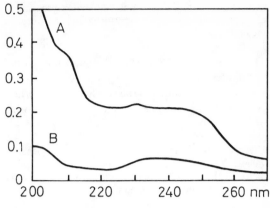

Fig. 8. Vapor phase ultraviolet spectra of
A Corticosterone acetate at 190°
B Deoxycorticosterone acetate at 180°
from Microchem. J. *9* 68 (1965)

1.2 Evaluation of Absorption Spectra. Application of Computer Techniques

For simple analysis the concentration of the component sought can be readily calculated from absorbancy measurements at an appropriate wavelength, either by reference to a calibration curve or by utilizing molar absorbance data.

Where the sample contains several components, each contributing to the total spectrum, the calculation of individual concentrations is still possible by making the appropriate number of absorbance measurements at different wavelengths and solving a series of simultaneous equations, provided molar absorbancies for each compound are known at each wavelength. The method is a time-consuming one and, in general, not used if alternatives are available. The advent of digital computer techniques now enables the calculations to be performed with minimal effort and hitherto laborious matrix calculations may now be contemplated and performed economically.

A computer programme called Analyz, written in Fortran IV, has been developed for Beer's Law calculations on multi-component samples[16]. The method involves calculation, from absorbance measurements on standard solutions, of the molar absorbancies of each component at a number of wavelengths. The latter must be at least equal to the number of components present. The concentration of each component may then be computed from an array involving the absorbance values and the molar absorbancy matrix. The accuracy of the method is improved if the least squares method is used whereby many more wavelengths and standard solutions of each component are used over and above those necessary for the basic matrix. Nonlinear Beer-Lambert relationships may be allowed for by fitting the data to a parabola, which gives an acceptable approximation. Use of this approximation requires three times the basic matrix number of molar absorbancy figures. Solution of the final concentration equations is not usually simple but the Analyz program is capable of solving them by perturbation and iterative methods.

The availability of high speed digital computing techniques now permits ready application of the least squares methods to the resolution of overlapping bands, a problem encountered in many forms of spectrometry[17].

Given the experimental parameters, e.g. absorbance values, Y_1, Y_2 .... Y_n related to the desired parameter X, the best fit results when the parameters are chosen to minimize the function

$$S = \sum_{i=1}^{n} W_i (F_i - Y_i)^2$$

In this equation

$$F_i = F(X_i, P_1 ..... P_m)$$

where the P values are parameters defining the component bands of the gross spectrum.

Since F_i is not necessarily a linear function of the m parameters the preferred method of solution is to estimate values P_1 ... P_m and assume F to be a linear function of the corrections $\Delta P_j = P_j - P_j^1$ etc, where $j = 1$ ... m. For solution purposes the function $F(X, P_1 .... P_m)$ is expressed as

$$F = \sum_{i=1}^{N} A_i + B$$

where $A_1 .... A_N$ are functions generating a single band and B generates a base line. Experimental data can often be fitted to Gaussian or Cauchy functions with the A_i values taking the appropriate form. The solution involves solving m simultaneous equations using the estimated parameters $P_1^1 .... P_m^1$ followed by matrix inversion to obtain the corrections ΔP_j. Because of the non-linearity between F and the parameters the solution is an iterative process.

In relating organic chemical structures to absorption spectra it is usually sufficient to observe or measure absorbancies at a few wavelengths. In the case of ultraviolet spectra of steroids, however, these are closely similar for certain groups of compounds regardless of variation of substituents elsewhere in the molecule. Thus the *3-oxo-Δ^4 steroids* all have markedly similar spectra. To detect subtle variations in spectra produced by the substituents a more refined approach is called for. For this purpose a function generator/analog computer system has been developed[18] which effects a comparison of two compounds over the whole spectrum. If A, E and C represent the

absorbance, molar absorbance and concentration, then for two compounds the following holds for every wavelength.

$$\log A_1/A_2 = \log E_1/E_2 + \log C_1/C_2$$

Hence a plot of $\log E_1/E_2$ against wavelength may be obtained merely by adjusting the ordinate to the value of $\log C_1/C_2$. Comparison of the plots so obtained reveals small spectral differences as displacements from the ordinate. Use of this method, which in practical terms involves plotting the two spectra of interest using conducting ink and placing the spectra in a curve-following function generator, has enabled a number of general conclusions to be drawn regarding the effect of substituents on steroid spectra.

1.3 Analytical Applications for Absorption Spectrophotometry

Applications to many specific types of compounds e.g. *resins, polymers, soaps, detergents, fatty acids*, organic functional groups etc. are conveniently reviewed in 'Standard Methods of Chemical Analysis'[19]. Also, applications in the rapidly growing field of *pesticide* and *pesticide residue* analysis[20-22] (cf. Chapter 11.1) are recommended. The contribution of ultraviolet spectrophotometry to the analysis of *drugs* (Chapter 11.2) has also been reviewed[23]. A comprehensive coverage of current literature is given in the biannual reviews of ultraviolet spectrometry and light absorption spectrometry in 'Analytical Chemistry'.

2 Reflectance Spectrometry

2.1 General Principles

The use of reflection techniques for the optical study of substances is less developed than those involving absorption. This may be attributed to the lack of commercial availability, until recently, of convenient measuring apparatus. However, it is now possible to obtain reflectance spectra[24], both manually and automatically, over the near ultraviolet and visible regions. Further developments

[18] *E.R. Garrett*, Anal. Chem. *34*, 1472 (1962).
[19] *F.J. Welcher*, Standard Methods of Chemical Analysis, Vol. 3, Instrumental Methods, D. van Nostrand, Princeton, New Jersey 1966.

[20] *U. Kiigemani*, J. Ass. Offic. Agr. Chem. *48*, 1001 (1965).
[21] *R.C. Blinn* in *F.A. Gunther*, Residue Reviews, Vol. 5, Academic Press, New York 1963.
[22] *H.F. Beckman* in *G. Zweig*, Analytical Methods for Pesticides, Plant Growth Regulators and Food Additives, Vol. 1, Academic Press, New York 1963.
[23] *I. Sunshine, S. Gerber*, Spectrophotometric Analysis of Drugs, C.C. Thomas, Springfield, Illinois 1963.
[24] *W.W. Wendlandt, H.G. Hecht*, Reflectance Spectroscopy, Interscience, New York 1966.

and applications in this field may be anticipated. Two types of reflection can occur when a substance is irradiated, specular (mirror) reflection which obeys the well-known laws of reflection, and diffuse reflection.

If the thickness of the sample is such that further increase in thickness does not alter the reflectance R then the Kubelka-Munk condition

$$(1-R_\infty)^2/2R_\infty = k/s$$

holds for diffuse reflectance. k is the absorption coefficient and has the same significance as in absorption spectrometry and s is termed the scattering coefficient. R_∞ is the limiting value of R for an infinitely thick sample.

In practice the reflectance R is usually measured comparatively by reference to a standard substance of known high reflectance ($R_\infty \sim 1$) and the ratio R_∞ (sample)/R_∞ (standard) is determined. A plot of this ratio over the wavelength range of interest constitutes a reflectance spectrum.

2.2 Reflectometers

Reflectance attachments are now available for almost all manual and recording spectrophotometers. The general approach to reflectance measurements will, therefore, be described together with a limited number of illustrative examples. Because reflectance measurements are overwhelmingly

made relative to a standard surface rather than absolutely, the attachments provide for illumination of both sample and a standard either consecutively or simultaneously. The standards in most common use are magnesium oxide, magnesium carbonate and certain opaque glasses. These provide a reproducible surface of high reflectivity, i.e. R_∞ is 0.98 to 0.99.

Considering that both specular and diffuse reflectance can occur, the system used for measurement must take into account whether total reflectance of one of the components is desired. The specular component may be eliminated either by absorption in a black surface in the reflected light path or by irradiating the sample at an angle of incidence of 0°, i.e. normal to the sample surface, and detecting only the off-axis diffuse reflectance, or conversely by illuminating at a finite angle of incidence and detecting reflection normal to the surface.

Attachments for measuring reflectance are commonly based on the integrating sphere principle. Reflected light from the sample, which is situated at the surface of the sphere, is integrated by the highly reflective inner surface. The detector is also located at a suitable point on the surface of the sphere.

A typical example of the use of the integrating sphere is the reflectance attachment for the Perkin-Elmer 4000A spectrophotometer (Fig. 9). The incident light beam is

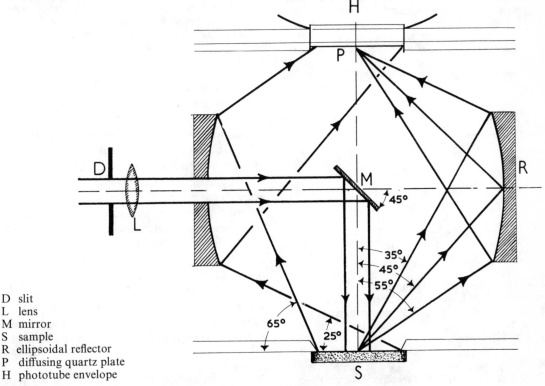

D slit
L lens
M mirror
S sample
R ellipsoidal reflector
P diffusing quartz plate
H phototube envelope

Fig. 10. Reflectance attachment for the Beckman Model DU Spectrophotometer.

normal to the sample surface in each case. Reflected light is integrated by the magnesium oxide inner surface of the sphere and is collected by the detector. The wavelength range 240 to 2500 nm can be studied using appropriate detectors. The specular component is eliminated by coating the interior of the sphere with black velveteen.

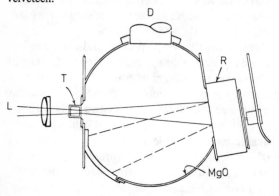

Fig. 9. Reflectance attachment for transparent and translucant samples (T) or reflecting samples (R) in monochromatic light (L) using a photomultiplying detector D of the Perkin-Elmer Model 4000 A spectrometer.

In the attachment for the Beckmann DU spectrophotometer (Fig. 10) an ellipsoidal mirror collects diffusely reflected light between angles of 35° and 55° to the normal angle of incidence and directs it into the phototube detector. The sample, either solid or liquid, is placed in a sliding drawer at the base of the attachment.

The Shimadzu MPS-50L spectrophotometer is provided with a reflectance attachment for opaque samples which separates the specular components by a series of mirrors (Fig. 11). A high angle of incidence coupled with a minimal sample-detector separation permits a large fraction of the diffuse reflectance to be measured. The specular component is collected in a light trap.

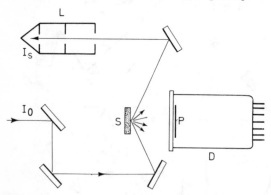

Fig. 11. Reflectance attachment for the Shimadzu Model MPS-50 L spectrophotometer
I_o light intensity
I_{dr} diffuse light
I_s specular light

D photomultiplying
    detector
S sample

All of these reflectance attachments utilize a collimated light beam from the spectrophotometer monochromator. Diffuse illumination of the sample may be accomplished by siting an auxiliary lamp inside the integrating sphere (Zeiss RA 3).

Evidently reflectance attachments vary widely in geometry and the fraction of reflected light collected at the detector is dependent on the geometric configuration. Thus, results are not directly comparable between one instrument and another and it is important to quote the type used when reporting results. It is an accepted convention to subscript angles of incidence and reflection; thus $R_{0,45}$ indicates incidence and viewing angles of 0° and 45°, $R_{O,D}$ implies normal incidence and diffuse viewing.

The measurement of reflectance spectra at low temperatures may be advantageous if the sample material is unstable at room temperature or if improved spectral resolution is needed. An attachment for the Unicam SP 500 spectrophotometer is capable of operation at liquid nitrogen temperature[25].

Sample and reference material are placed in recesses drilled in the top of two copper rods 1.25 inches in diameter. The rods are joined by copper tubing which passes from the base of the rods through the tray of the standard SP 540 reflectance attachment to lagged flexible tubing which carries the coolant. To prevent frosting of the spectrophotometer optics at low temperatures the instrument was enclosed in thin polyethylene sheet containing desiccant. Also a purge of dry nitrogen was passed into the mirror housing and allowed to flow into the polyethylene cover to maintain a slight positive pressure.

Diffuse reflectance measurements at elevated temperatures are particularly valuable for studying thermal transitions as a complementary technique to thermogravimetry and differential thermal analysis (cf. Chapter 9.3). The method has the advantage that results are independent of concomitant weight changes of the sample.

High temperature reflectance spectroscopy[26] consists of plotting the reflectance spectrum at a series of incremental temperatures; this enables wavelength ranges of transitions to be located and transition temperatures to be approximately deduced. More accurate transition temperatures are obtained by *dynamic reflectance spectroscopy*[26], which involves plotting reflectance against temperature at a fixed wavelength using a programmed continuous temperature rise. To date the principal

[25] *M.C.R. Symons, P.A. Trevalion*, Unicam Spectrovision *10*, 8 (1961).

[26] *W.W. Wendlandt*, Thermal Methods of Analysis, Interscience, New York 1964.

field of application has been solid-state transitions and dissociation of metal ion complexes with a variety of ligands. The apparatus[27] comprises an aluminum block with a recessed sample well. It is heated by a sheathed cartridge heater. Two thermocouples are employed, one in contact with the sample to measure its temperature, and one positioned in the aluminum block for temperature programming. Where the reaction involves fusion a quartz cell, inserted into the aluminum block, is satisfactory.

2.3 Applications

Reliable methods of comparative reflectance measurement were first demanded in fields where some index of surface whiteness or color was needed. Indeed, by far the most extensive studies to date refer to industries such as food (flour, opaque fruit juices etc.), paints, and building materials. Applications to qualitative and quantitative analysis are limited but advantageous in certain areas. They are most suitable for solid materials and they should prove valuable for insoluble materials or those unstable in solution. Interest is growing in evaluating analyses involving paper or thin layer chromatographic separations where reflectance analysis obviates the need to elute the separated materials from the substrate.

Reflectance in the visible region was applied to the analysis of mixtures of *dyes*[28]. *Aniline blue, eosine B, basic fuchsin, malachite green, naphthol yellow S* and *rhodamine B* were separated on plates of aluminum oxide or silica gel with n-butanol/-ethanol/water (80:20:10). Analysis of the plates, after drying at 110° for 15 minutes, was carried out by direct spectrometry of the plate or examination of the spots after scraping them off the plate.

In the former case the thin layer plate was covered by an identical clear glass plate and examined in the reflectance attachment of the Beckman DK 2 spectrophotometer. The reference standard was prepared by grinding some of the absorbent from the sample plate and packing it into the cell.

For material removed from the plate about 50 mg was compressed between a sheet of paper and a microscope cover glass to give a layer 0.3 mm thick. The paper and cover glass were fixed by tape and the whole examined in the reflectance attachment of the Beckman DU spectrophotometer. Again absorbent from the sample plate served as the reference.

The direct approach yielded a precision of ± 5%; analysis of individual portions was more precise and compared favorably with transmittance measurements. Satisfactory reflectance spectra could be obtained with as little as 0.04 μg of dye in some instances. Plotting the function 2–log% R against square root of concentration yielded a linear calibration up to 0.5 mg/ml whereas the same function plotted against concentration is linear only up to 0.05 mg/ml.

Alternately a linear relationship over the concentration range 0.2 to 3.5 mg/ml resulted if the Kubelka-Munk function $(1-R^2)/2R$ was plotted against logarithm of concentration. Provided that the reflectance spectra of the dyes are sufficiently different the analysis may be performed even where the chromatographic separation is incomplete[29].

Similarly, reflectance measurements have been used for determining amino acids separated by thin layer chromatography[30]. DL-*alanine*, L-*arginine*, L-*glutamic acid*, glycine, L-*leucine*, L-*lysine*, DL-*methionine*, DL-*phenylalanine*, DL-*serine* and DL-*valine* were separated in several solvent systems containing 0.2 to 0.4% of ninhydrin as a chromogenic reagent. The reflectance method proved somewhat less sensitive but more precise than transmittance. Depending upon the concentration of ninhydrin used in the developing solvent the sensitivity ranged from 5×10^{-9} to 10^{-8} mol for most amino acids. A similar reflectance method has been used for the direct determination of sugars separated by paper chromatography[31, 32].

The above methods rely on reflectance measurements in the visible spectrum and require colored samples (initially or after reaction with a chromogenic reagent). However, commercial reflectance spectrometers operate in the ultraviolet region. Thus, *thiamine hydrochloride, pyridoxine hydrochloride, nicotinic acid, nicotinamide,* and *p-aminobenzoic acid* were separated with glacial acetic acid/acetone/methanol/benzene (5:5:20:70) on thin layer plates of silica gel. For plates bearing 2% luminous pigment irradiation under an ultraviolet lamp revealed the spots as areas of diminished fluorescence. Plates free from pigment were scanned in the reflectance attachment of a Beckman DK2 spectrophotometer, selecting wavelengths of absorption maxima for each vitamin in turn. Of the five vitamins studied only nicotinamide and nicotinic acid have closely similar reflectance spectra, and these are readily resolved by R_f measurements.

[27] *W.W. Wendlandt*, Anal. Chem. *35*, 105 (1963).
[28] *M.M. Frodyma*, J. Chromatogr. *13*, 61 (1964).
[29] *R.W. Frei*, Can. J. Chem. *44*, 1945 (1966).
[30] *MüM. Frodyma, R.W. Frei*, J. Chromatogr. *17*, 131 (1965).
[31] *A. Benvenue, K.T. Williams*, J. Chromatogr. *2*, 199 (1959).

For quantitative analysis each spot was removed from the plate together with about 80 mg of the silica gel. This was ground under careful control, packed in a windowless reflectance cell and the reflectance measured at the absorption maximum of the vitamin. A near linear calibration curve of percent reflectance against logarithm of concentration was obtained for quantities in the range 0.3 to 3μ moles. The precision of measurement is 2 to 3% at 2μ moles of vitamin per 80 mg adsorbent.

Samples containing a few μ moles of *salicylic acid* and *acetylsalicylic acid (aspirin)*[33] were separated on silica gel thin layer plates using hexane/glacial acetic acid/chloroform (85:15:10), the R_f values being 0.35 and 0.5 respectively. On drying the plates at 90° for two hours aspirin converts to salicylic acid so that a single calibration may be used for both compounds. Quantitative analysis was performed by measuring the reflectance of the separated materials after removing them from the plate together with 70 mg of the accompanying adsorbent. Pure adsorbent was used as the reference material.

In addition to the analyses of solid specimens most reflectance attachments can also be used with liquid samples. Thus turbid or opaque solutions and suspensions are amenable to reflectance spectrometry. For example the oxidation and denaturation of hemoglobin carbonmonoxide may be followed by repetitive measurements of reflectance[34].

3 Kinetic Methods of Analysis

3.1 General

Most analytical methods are of the equilibrium type. The development, in recent years, of kinetic methods, where concentration is determined as a function of the rate of a reaction involving the species of interest, opens up a considerable new area for analytical chemists. Reactions which proceed at measurable rates may be utilized as may be the catalytic or inhibitory effect of substances upon a chemical reaction. One of the best known examples is the catalytic effect of iodine upon the slow reaction between cerium(IV) and arsenic(III). Parts per million quantities of iodine may be determined via this reaction.

The dependence of reaction rates upon concentration and temperature is well known. Thus, a reaction which proceeds too fast for convenient measurement may be slowed to an acceptable rate by reducing the concentration of one or both reactants by suitable orders of magnitude. This applies for trace analysis; indeed, there are examples of measurements down to 10^{-8} mole.

Meanwhile several well-defined areas of applicability of kinetic methods[35] have been established. These are redox reactions involving odd numbers of electrons, exchange reactions involving mono- and multidentate ligands, and enzyme reactions. Both uncatalyzed and catalyzed systems are utilized. In the latter case either the reactants or catalyst may be determined. Additionally, a number of types of organic compound are amenable to analysis by rate measurements.

3.2 Measurement of Reaction Rates

The principal techniques for measuring reaction rates have been electrometric and spectrophotometric; the present discussion is restricted to the latter.

The manner in which concentrations are derived from kinetic studies differs considerably from the better known equilibrium approach. Before discussing applications of kinetic methods it is advisable to outline the various alternate approaches. The preferred one will depend largely upon the physico-chemical nature of the reaction studied. Regardless of the approach used it is necessary to ensure that the substance being determined is the rate-limiting species.

Currently four approaches to rate measurements are recognised[36]. The most widely used methods are the constant time and variable time ones; in addition the so-called slope method and signal stat method have both found application.

3.2.1 Constant Time Method

The *constant time method,* as the name implies, involves measuring the extent of reaction over a fixed time for all samples in a batch regardless of concentration of the required constituent. The commencement may or may not be zero time with respect to the reaction, the important feature is that the chosen interval should be such that the concentrations of the reactants have not changed significantly from their initial value over the chosen period. If this condition exists, then the measured parameter change, e.g. absorbance, is a

[32] *M.M. Frodyma, V.T. Lieu,* Anal. Chem. *39,* 814 (1967).

[33] *M.M. Frodyma,* J. Chromatogr. *18,* 520 (1965).

[34] *R.E. Anacreon, R.H. Noble,* Appl. Spectrosc. *14,* 29 (1960).

[35] *G.A. Rechnitz, H.B. Mark,* Kinetics in Analytical Chemistry, Interscience, New York 1968.

[36] *H.L. Pardue* in *C.N. Reilly, F.W. McLafferty,* Advances in Analytical Chemistry and Instrumentation, p. 141, Wiley, New York 1969.

linear function of concentration. If during the time interval there is a significant build-up of products, then the linear relationship no longer applies.

At the practical level the constant time method possesses the inherent disadvantage that if any samples in a batch have an unduly low concentration compared with those for which the interval is optimized, then the absorbance change will be small and subject to increased error. Typical measurement times range from a few seconds to a few minutes.

3.2.2 Variable Time Method

In this approach the function which is maintained constant is the extent of reaction. Thus, for spectrophotometric measurement, the time required for the absorbance to change by a fixed pre-determined amount is observed. Immediately, it can be seen that this method overcomes the difficulty of poor precision for samples of low concentration noted above for the constant time method. While it is possible for the measurement to be started at zero time, it is in fact advisable to ensure thorough mixing of the reactants before starting the measurement.

Using the *variable time method,* the concentration sought is linearly related to the reciprocal time for the pre-determined signal change.

It has been demonstrated that, unlike the constant time method, this relationship holds regardless of whether the response curve for the reaction is linear or non-linear. Equally, it remains valid if the measurement system response is non-linear over the chosen interval. The latter feature represents a considerable advantage in setting up kinetic systems in that the choice of detection system is virtually unlimited. In consequence, the variable time method finds considerable use outside the limited spectrophotometric field considered here.

3.2.3 Slope Method

If the slope of the reaction response curve is measured at, or close to, zero-time, then it is proportional to the concentration of the rate determining species. In practice, the derivative of this curve is obtained using differentiation circuitry. The method is quick and applicable to both linear or non-linear responses; however, where non-linear curves are evaluated, all slopes must be determined at the same value of the parameter measured to maintain the same proportionality between slope and concentration over a range of sample concentrations.

[37] *P.A. Loach, R.J. Loyd,* Anal. Chem. *38,* 1709 (1966).

3.2.4 Signal Stat Method

This differs from the other three in being a continuous flow method. The measurement made is of the rate at which a reagent must be added to maintain constancy of the parameter (e.g. absorbance) measured. Obvious disadvantages of this method are the need to add continuously an additional solution and to correct for dilution. However, by maintaining the measured parameter constant, the concentration of the species being monitored is also held constant. This removes a potential source of error in a system where the rate constant is dependent upon the concentration of the species detected.

3.3 Spectrophotometers for Kinetic Measurements

The majority of kinetic methods in which spectrophotometry is used have been developed using commercially available instruments. Appropriate timing circuitry is added to permit initial and final absorbance measurements(seeSection5.11:3.4).

The desirability of making measurements close to zero-time often demands accurate measurements of small changes in absorbance. This has prompted the design of spectrophotometers specifically suited to the requirements of kinetic measurements.

A spectrophotometer of high sensitivity[37] enables changes as small as 5×10^{-5} absorbance unit to be quantitatively measured and reaction times between 10^{-4} sec. and several minutes to be determined with high precision.

The principal features contributing to the high performance are high spectral purity of the light beam, stabilization of the detecting light source by an optical feed-back system and enhancement of signal/noise ratio by means of a digital averaging unit. The instrument (Fig. 12) is based on the Cary 14R double-beam recording spectrophotometer. This incorporates a double monochromator (resolution 0.1 nm over the range of 260 to 1200 nm).

The spectrophotometer was designed to study light-induced reaction kinetics. The light source is either a 100 watt tungsten lamp or a 150 watt xenon lamp. The beam passes through condensing lenses and is focussed onto the sample cell. The beam is pulsed by a chopper and operated by a timing circuit providing time intervals of 14 msec. or 1.53 sec. to a reproducibility of 0.1%.

This high performance spectrophotometer was designed for studying reaction rates down to 10^{-4} sec., the lower limit being imposed by the performance of digital memory averaging device. Such rapid reactions have, to date, found

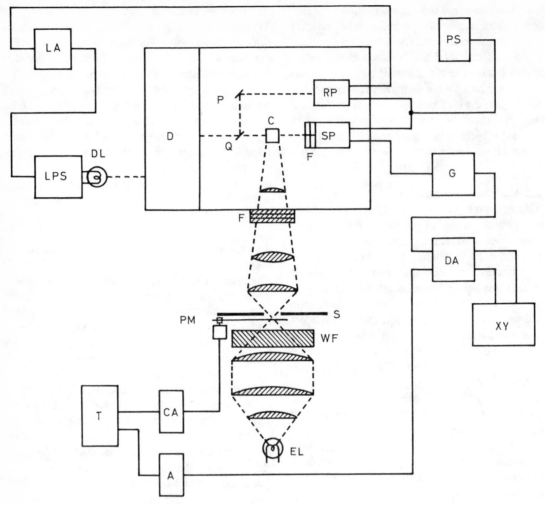

Fig. 12. Block diagram of kinetic spectrophotometer[37] from Anal. Chem. *38*, 1709 (1966)

LA	Lamp feedback amplifier	G	D.C. Amplifier and	Q	Quartz blank
LPS	Lamp power supply		Bias supply	C	Cuvette
DL	Detecting Lamp	DA	Digital memory averager	F	Filters
DM	Double monochromator	XY	Recorder	S	Slit
P	Parabolic mirror	A	Average scan trigger	WF	Water filter
RP	Reference	CA	Chopper drive amplifier	PM	Pen motor chopper
SP	Sample	EL	Exciting beam lamp	T	Timer
PS	Power supply				

very little analytical use. However, the optical feedback stabilizing circuitry provides a basis for designing sensitive kinetic photometers operating over an analytically useful time range of the order 5 to 500 sec.

High stability over longer periods[38] are achieved in a photometer with a transmittance stability of 0.02% over several hours, reproducibility of measurements to 0.01% over periods up to one hour and, for absorbancies between 0.2 and 0.3, errors in measured concentration of less than 0.2% relative. Since analytical applications of reaction rates rarely allow the use of repetitive observations[37], a digital averaging unit was not employed.

The photometer is double beam, a beam splitter directing light into both sample and reference paths. Stability of the light source is achieved using the optical feedback method, the programmable lamp supply being driven by the amplified signal from the reference phototube. The detector circuit for the measured light beam is a current-to-voltage converter the sensitivity of which is adjusted so that 100.00% transmittance gives 1.0000 volt. The wavelength of analytical interest is selected using appropriate interference and cut-off filters. Temperature control and rapid, complete mixing of reactants are of paramount importance in accurate kinetic analyses and the desired conditions were achieved using a simple thermostatted cell. It comprises a 1 cm^3 borosilicate glass cell sealed with epoxy resin into a Lucite outer vessel. Water controlled in temperature to $\pm 0.02°$ circulates through the interspace between the walls. The light path is circumscribed by 0.25 inch O-rings just below the vertical midpoint and the solution is stirred effectively, the stirrer being mounted above the light path. Complete mixing (99.5%) is achieved in less than 20 seconds. Based on a detailed evaluation of the photometer performance the authors conclude that the limiting factor in lamp output stability may well be inhomogeneities in the filament; they stress the importance of both phototubes viewing the same portion of the filament.

A continuous, semi-automatic spectrophotometer[39] has been designed primarily for kinetic analyses. Emphasis is placed on highly stable performance: drift stability 0.003 absorbance units per hour, photometric accuracy of 0.01 absorbance unit at 1.0 and 0.001 near zero absorbance. The instrument is modular in construction and a solid state digital logic system is provided to control the sequencing of sample handling, initiation of reaction, print out etc. The spectrophotometer light source is stabilized by circuitry which monitors both current and voltage outputs.

Convection currents generated by the heat from the lamp constitute a significant source of short-term instability if these cross the spectrophotometer light path; the effect is minimized by the use of baffles. Gas filled diodes were used for the detector as the preferred compromise between signal/noise ratio and useful wavelength range. The instrument may be used to measure absorbance and transmittance, and differentiating circuitry is incorporated to provide direct read-out of reaction rate.

In addition to the spectrophotometers described above, which have been designed specifically for rate measurements, several commercial spectrophotometers provide facilities for kinetic analysis.

The Cary Model 16 spectrophotometer is supplied with a variable timing sequence unit for rate measurement and also an automatic sample changer. The Technicon Autoanalyzer is suitable for rate measurement by the constant time method. The reaction time can be adjusted to the desired value by incorporating a suitable length of delay coil between the point of mixing of reagents and the colorimeter. The Durrum stopped-flow spectrophotometer is capable of mixing reactants to 99.5% completion in 0.002 seconds and enables reaction half-times to be determined down to 0.005 seconds. Its principal use is, therefore, for studying fast reactions, which to date have found little analytical use. The rapid mixing is achieved by forcing the reactants rapidly through a mixing jet into the measuring cell.

3.4 Applications

3.4.1 Enzyme Reactions
Enzymes have found considerable application in both chemical and clinical analysis; their remarkable specificity and high catalytic efficiency are often combined with easily measurable rates (5.11:3.4.2—3.4.6).

3.4.2 Glucose
There have been several studies of the determination of parts per million quantities of glucose in samples of clinical interest, particularly blood. The methods are based on the catalyzed oxidation of *glucose* to gluconic acid and hydrogen peroxide by glucose oxidase (Enzyme Classification 1.1.3.4). The hydrogen peroxide is determined by a rapid reaction to a colored product. The rate of change of absorbance in the early stages of the reaction is then proportional to the initial concentration of glucose. Two methods of producing a suitable colored product have been used, oxidation of a dye in the presence of horseradish peroxidase[40] (E.C. 1.11.1.7) and reaction with excess iodide in the presence of molybdate[41].

A method for continuous analysis of glucose in blood plasma samples was devised using the dye oxidation method[40]. The sample and composite reagent comprising glucose oxidase, peroxidase, and o-toluidine buffered to pH 4.2 were mixed continuously in a flow system controlled by peristaltic pumps. The mixed reactants pass through two cells in a sensitive photometer, a fixed delay of 30 seconds being introduced between the cells. Glucose concentrations are computed from the measured absorbance difference at 635 nm between the two cells by reference to a calibration line. A 30 second delay between mixing and the first cell is necessary to overcome an induction period in the reaction.

The molybdate-catalyzed oxidation of iodide[41] was used for determining *glucose* in serum, plasma and blood. The variable time method was employed, the time for a preselected absorbance change of about 0.07 being measured. The absorbance of the tri-iodide complex was measured at 365 nm (Sargent Spectro-Electro Titrator, Sargent Model Q concentration comparator and a cell thermostated to $40 + 0.1°$). The prepared sample (1.0 ml) was injected into 3.0 ml of composite glucose oxidase — iodide-molybdate reagent in phosphate buffer. The reaction rate is affected by glucose oxidase and iodide concentration, by pH, and by temperature, and these variables are held constant. Results are reproducible to 1 to 2%. Oxalate, citrate, fluoride, heparin and small amounts of ascorbic acid do not interfere.

3.4.3 Ethanol
A specific enzyme catalyzed reaction enables the *ethanol* content of blood to be rapidly determined without recourse to separation procedures[42]. Ethanol is selectively oxidized in the presence of alcohol dehydrogenase (ADH; E.C. 1.1.1.1) and

[38] *H.L. Pardue, P.L. Rodriguez*, Anal. Chem. *39*, 901 (1967).
[39] *T.E. Weichselbaum*, Anal. Chem. *41*, 725 (1969).
[40] *W.J. Blaedel, G.P. Hicks*, Anal. Chem. *34*, 388 (1962).
[41] *H.V. Malmstadt, S.J. Hadjiioannu*, Anal. Chem. *34*, 452 (1962).

nicotinamide-adenine dinucleotide (NAD) to yield the reduced form of NAD, denoted NADH. The reaction is

$$C_2H_5OH + NAD^{\oplus} \xrightarrow{ADH} CH_3CHO + NADH + H^{\oplus}$$

The rate of change of absorbance of NADH is measured at 350 nm by the variable time method at $25 \pm 0.1\%$. Each analysis requires a total time of 1—2 minutes compared with 90 minutes if the reaction is measured at completion. The method is applicable to ethanol concentrations ranging from 15 to 300 mg/100 ml of blood. Since an appreciable percentage of the ethanol (up to 10%) is consumed during the brief (1 to 25 second) reaction period, a linear relationship between concentration and reciprocal time is not obtained. Fairly frequent calibration is necessary due to the instability of the ADH solution.

3.4.4 Amino Acids
The enzyme L amino acid oxidase (E.C. 1.4.3.2) selectively catalyses the oxidation of Lα-*amino acids*[43]. The reaction may be written

$$RCH(NH_2)COOH + O_2 + H_2O \xrightarrow{\text{amino acid oxidase}}$$

$$RCO/COOH + NH_3 + H_2O_2$$

The rate of reaction is measured by the variable time method utilizing the colored products formed when the hydrogen peroxide formed is reacted at constant temperature with *o*-dianisidine in the presence of horseradish peroxidase. Quantities of the order $20—200\mu g$ of amino acid may be determined with a relative error of about 2%. Measured reaction times are in the range of a few seconds to two minutes. Response curves are linear except at the extreme lower and upper concentration limits. In the former case non-linearity results because a substantial part of the amino acid reacts in producing the predetermined absorbance change; in the latter the concentration of amino acid is sufficient to partially saturate the enzyme.

The specificity of L amino acid oxidase to acids having the L configuration enables L compounds to be determined in racemates or in the presence of the D acids. In the cases of *histidine* and *tyrosine* the presence of D configuration acids enhances the rate of reaction of the L form. The L forms of *isoleucine, α-amino-n-butyric acid, citrulline, leucine, methionine, norleucine, norvaline, phenylalanine* and *tryptophan* may be satisfactorily determined. A few L amino acids are not affected by the enzyme. This group includes alanine, arginine, aspartic acid, cysteine, glutamic acid, pro-

line, serine, threonine and valine. A study of interference revealed that certain *carboxylic acids,* including benzoic, mandelic, salicylic, iodoacetic and some sulfonic acids, inhibit the oxidation process.

3.4.5 Lactic Acid
In all the examples cited above, the enzyme-catalyzed reaction is used to determine the substrate. By adjusting the experimental conditions so that the enzyme activity is rate-limiting, the kinetic method can be used for the determination of enzyme activities. In clinical tests certain enzyme activities are diagnostically important. Consequently, there are several examples of kinetic methods for their determination.

Lactate dehydrogenase (LDH) can be kinetically determined using the following reaction sequence[44].

$$L\text{-lactic acid} + NAD^{\oplus} \xrightarrow[\text{(E.C.1.1.1.27)}]{LDH}$$

$$\text{pyruvic acid} + NADH + H^{\oplus}$$

$$NADH + H^{\oplus} + \text{dye}_{ox}\text{.(blue)} \xrightarrow{\text{diaphorase}}$$

$$NAD^{\oplus} + \text{dye}_{red}\text{.(colorless)}$$

where NAD represents nicotinamide-adenine-dinucleotide, NADH its reduced form, and dye_{ox} dye_{red} the reduced and oxidized forms of the dye 2,6-dichlorophenolindophenol. Concentrations of lactic acid, NAD, dye_{ox} and diaphorase are maintained sufficiently high for the LDH activity to be rate-limiting, whereupon, using the fixed time method, a linear relationship between LDH activity and absorbance change at 600 nm is obtained. Analyses were run continuously using the flow photometer outlined above for glucose determinations[40]. 0.2 ml *blood serum* could be processed at 20 samples per hour. For LDH activities between 300 and 1500 units the error is ± 20 units; below 300 LDH units, the error may reach ± 30 units. The Technicon Autoanalyzer has been successfully used for LDH analysis by this method[45] (Chapter 14.3).

Lactic dehydrogenase has also been determined in *urine*[46,47]. The reaction sequence utilized was

$$\text{lactate} + NAD \xrightarrow[\text{(E.C.1.1.1.27)}]{LDH} \text{pyruvate} + NADH_2$$

$$NADH_2 + \text{dye}_{ox}\text{.} \xrightarrow{PMS} NAD + \text{dye}_{red}\text{.}$$

Here NAD denotes nicotinamide adenine dinucleotide and NADH2 its reduced form, PMS is phenazino methosulfate. 2,6-dichlorophenolindophenol was used

[42] *H.V. Malmstadt, T.P. Hadjiioannu,* Anal. Chem. *34,* 455 (1962).

[43] *H.V. Malmstadt, T.P. Hadjiioannu,* Anal. Chem. *35,* 14 (1963).

[44] *W.J. Blaedel, G.P. Hicks,* Anal. Biochem. *4,* 476 (1962).

[45] *L. Brooks, H.G. Olken,* Clin. Chem. *11,* 748 (1965).

[46] *G.P. Hicks, S.J. Updike,* Anal. Biochem. *10,* 290 (1965).

as the dye. Analyses were run continuously by a parallel stream technique (Fig. 13). After mixing the sample with a composite reagent comprising non rate-limiting concentrations of components, the stream is divided into two lines of equal delay between the mixing vessel and the photometer cells. One stream is held at room temperature, the other at $40 \pm 0.01\%$. The absorbance difference between the two streams is proportional to the rate of reaction at 40°. Results revealed the presence of a 'blank' activity of non-enzymic origin. This was removed completely by prior treatment of the *urine* using gel filtration through a Sephadex column. Accordingly, the parallel stream analysis system was modified to accept chromatographic column effluents. Good recoveries were established for *malate dehydrogenase* (E.C. 1.1.1.37), *glutamic oxaloacetic transaminase* and *cysteine sulfinic acid*.

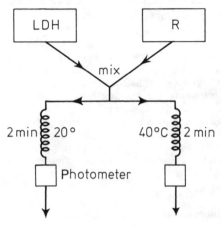

Fig. 13. Parallel stream technique for lactic dehydrogenase (LDH) determination with a suitable reagent (R); from Anal. Biochem. *10*, 290 (1965)

3.4.6 Transaminases
Glutamic oxaloacetic transaminase (GOT, aspartate aminotransferase, E.C. 2.6.1.1), may be determined using the following reaction sequence[48] in which KG and GDH are α-ketoglutarate and glutamic dehydrogenase respectively.

$$\text{Amino acid} + \text{KG} \xrightarrow{\text{GOT}} \text{Keto acid} + \text{glutamate}$$

$$\text{Glutamate} + \text{NAD} \xrightarrow{\text{GDH}} \text{KG} + \text{NADH}_2 + \text{NH}_4^\oplus$$

$$\text{NADH}_2 + \text{dye}_{ox.} \xrightarrow{\text{PMS}} \text{NAD} + \text{dye}_{red.}$$

The system is applicable to other transaminases provided a suitable amino acid is employed; aspartic acid is required for GOT and alanine for glutamic pyruvic transaminase (GPT, E.C. 2.6.1.2).

Enzymes are generally selective in catalyzing a particular reaction of a similar group of compounds, e.g. alcohols. Where the rates of reaction differ significantly for the individual compounds, these should be amenable to determination in admixture. Mixtures of ethanol and propanol were analyzed using alcohol dehydrogenase[49], the reaction being

$$R-CH_2OH + NAD \xrightarrow{\text{ADH}} RCHO + NADH_2$$

$NADH_2$ is a common product for each alcohol and the overall rate of reaction is obtained by monitoring the $NADH_2$ absorbance at 340 nm. The equation relating alcohol concentration to the composite reaction rate R is

$$R = K_1 C_1 + K_2 C_2$$

where C_1, C_2 are the initial alcohol concentrations and K_1, K_2 are constants which are determined separately for each alcohol. By performing the analysis at two ADH concentrations, C_1 and C_2 may be calculated. The method is capable of extension to probably four components.

3.4.7 Phenolic Compounds
Phenolic compounds have been determined over the concentration range 1 to $100 \mu g/ml$ by a kinetic method based on their reaction with N(arylsulfonyl)quinonimines to yield blue indophenols[50]. From an experimental study of reaction kinetics and residual blank values N(benzenesulfonyl)quinonimine proved to be the most suitable reagent; it was used as an alkaline solution in methyl cellosolve. Two methods of measurement were evaluated: the constant time method in which the absorbance was measured 3 minutes after initiation of the reaction, and the derivative method where the rate of change of absorbance with time was automatically recorded. *1-Naphthol* and *5-amino-1-naphthol* must be determined by the former method because the rate of indophenol formation is not proportional to concentration of the substrate. *Phenol, o-chlorophenol, 2,3-dimethylphenol, m-aminophenol* and *2-naphthol* may be determined by either approach. Suitable wavelengths for absorbance measurement depend on the phenol being determined, but all lie in the region 610–650 nm. The reaction rate is pH dependent; determinations are performed at pH 11.7 where the rate is maximal. The method is rapid, precise, and highly specific to phenolic compounds.

[47] *G.P. Hicks, G.N. Nalevac*, Anal. Biochem. *13*, 199 (1965).

[48] *G.P. Hicks, W.J. Blaedel*, Anal. Chem. *37*, 354 (1965).

[49] *H.B. Mark*, Anal. Chem. *36*, 1668 (1964).

[50] *G.G. Guilbault*, Anal. Chem. *38*, 1897 (1966).

Traces of *1-naphthol* in 2-naphthol may be determined by the reaction rate[51] of diazotized 2-naphthylamine-5,7-disulfonic acid which reacts more rapidly with 1-naphthol than with 2-naphthol. Nevertheless, for satisfactory results the bulk of the 2-naphthol must be removed by precipitation with acid. The supernate containing the 1-naphthol and the residual 2-naphthol is reacted in $0.5 N$ HCl with the diazotized reagent at 23–24°. The absorbance of the resulting coupled dye is read at 485 nm after 15 ± 1 minutes. The 1-naphthol content is found from a calibration graph. Over the concentration range 0.07 to 0.35% 1-naphthol, a standard deviation of ±0.004% was obtained.

3.4.8 Carboxylic Acids

The differing rates of reaction[52] of *carboxylic acids* with *diphenyldiazomethane* allow kinetic methods to be used for determining the composition of mixtures of carboxylic acids. In ethanol the reaction is second order. However, in addition to the formation of the benzyhydryl ester according to the equation

$$(C_6H_5)_2 CN_2 + RCOOH \xrightarrow[-N_2]{} (C_6H_5)_2 HC–OOCR$$

the acid catalyzes the formation of benzyhydryl ethyl ether through reaction with the solvent ethanol. To overcome the effect of this side reaction the carboxylic acid concentration is maintained in several fold excess over the diphenyldiazomethane and the pseudo unimolecular rate of disappearance of the latter is measured. The rate constants for the individual alcohols are first determined separately and the values are used to calculate the composition of mixtures from the overall rate of disappearance of diphenyldiazomethane. Binary mixtures of *acetic acid* and *benzoic acid* and *m*- and *p-methoxybenzoic acids* have been analyzed in this way. In each case the individual rate constants differ by less than a factor of two.

3.4.9 Reducing Compounds

The *fructose* and *glucose* contents of blood serum have been simultaneously determined by a kinetic procedure[53] dependent upon the different rates of reaction with ammonium molybdate in acid solution. The measured data for rates of molybdenum blue formation are used to calculate the composition of a binary mixture. A double-point method[54] was used in which the absorbance was measured at two separate times after initiating the reaction.

Alternately, a single-point method may be employed provided the total sugar concentration is determined independently. Mixtures of *sucrose* with *glucose* or *fructose* may be similarly analyzed.

Other reducing sugars represent a potential source of interference in this method, according to the rate at which the extraneous sugar reacts with ammonium molybdate. An alternative approach to the analysis of sugar mixtures has been the reaction with 2,3,5-triphenyl-$2H$-tetrazolium chloride to yield red formazan products. *Glucose, fructose, mannose, sorbose, galactose, xylose,* and *ribose,* all exhibit pseudo zero order kinetics if measured over a limited range of reaction. Thus the response curves are linear.

The double-point method was used. Absorbance of the formazan product was measured after 20 and 50 minutes reaction time. Several binary mixtures of the listed sugars were satisfactorily analyzed at millimolar concentrations, the reaction rates for the individual sugars having been determined first. The method has the advantage over the ammonium molybdate procedure that creatinine and glutathione, which are often present in blood serum, do not interfere. *Ascorbic acid,* if present at 1–2%, interferes seriously with both procedures.

4 Fluorescence and Phosphorescence

4.1 General

Transition from the first excited singlet state to the ground state constitutes *fluorescence;* the fraction of the total number of excited molecules undergoing this transition represents the *quantum efficiency* for the process, an important parameter in absolute fluorescence studies.

Alternately, the return to the ground state involves a conversion to the lowest triplet state of the molecule (singlet-triplet conversion). Return to the ground state from the triplet state with emission of a quantum of energy is termed *phosphorescence.* Thus, phosphorescence involves transitions between states of different multiplicity. Since they are forbidden transitions, they occur with extremely low probability.

Consequently, phosphorescence yields are many orders of magnitude smaller than fluorescence yields. Also phosphorescence lifetimes (in the range 10^{-4} to 10 seconds) are much longer than fluorescence lifetimes (less than 10^{-8} seconds). Therefore, phosphorescence is much more susceptible to *quenching* than fluorescence, and has been observed in relatively few instances.

Fluorescence measurements have proved of immense value in analytical chemistry, particularly

[51] *J.S. Parsons,* Anal. Chem. *27,* 21 (1955).

[52] *J.D. Roberts, C.M. Regan,* Anal. Chem. *24,* 360 (1952).

[53] *L.J. Papa,* Anal. Chem. *34,* 1443 (1962).

[54] *R.G. Garmon, C.N. Reilley,* Anal. Chem. *34,* 600 (1962).

for determining compounds at extremely low concentrations. The fluorescence intensity at any wavelength may be represented by

$$F = k_\lambda \ I_0 \ \Phi_f$$

I_0 is the incident light intensity, Φ the quantum efficiency, and $K\lambda$ a constant incorporating concentration, path length, molar absorption coefficient, detector efficiency, and instrument geometry.

Unlike absorptiometry, the sensitivity of which depends upon the ratio I/I_0, fluorescence intensity is directly proportional to the intensity of the light source. Within certain practical limits, such as susceptibility of the molecule to photo-decomposition, the sensitivity of fluorescence measurements can be increased by raising the intensity of the source. Since both the exciting and emission wavelengths may be varied according to the analytical requirement, a high degree of specificity can be achieved.

When small concentrations are being studied by fluorescence methods, the excitation spectrum is related to the absorption spectrum of the molecule. Indeed, they are identical if the exciting light intensity is constant for all wavelengths and the fluorescence efficiency is independent of the exciting wavelength. Provided the first and second absorption bands show no overlap, the fluorescence emission spectrum bears an approximately mirror-image relationship to the first absorption band and appears on the long wavelength side of the latter. Further, the structure of the fluorescence emission spectrum is independent of the exciting wavelength because fluorescence emission always originates from the first excited singlet state.

4.2 Instrumentation

4.2.1 Fluorescence
A fluoroscence spectrometer consists of the following essential units:

light source,
sample cell,
means of selecting the wavelength or waveband of exciting radiation,
means of selecting the desired emission wavelength or band, and
detector for measuring the fluorescent radiation.

Wavelength selection for both excitation and emission may be achieved using filters (in fluorimeters) or a monochromating system (in spectrofluorimeters). Commercial instruments of both types are available; the ultraviolet and visible ranges are usually covered; some instruments provide facilities for infrared measurement.

The geometric configuration of components is important in the design of instruments for measuring fluorescence. The optimum geometry is primarily dependent on the type of sample being examined. The most common case in analytical studies is that of very dilute solutions showing little light absorption. Here the preferred arrangement is to view the fluorescence at right-angles to the incident illumination. Then the acceptance by the detector of scattered light and background fluorescence due to the sample container and solvent is minimized. If, however, the fluorescence of concentrated solutions is involved, the method of frontal illumination is preferred as it limits light losses due to absorption. Geometry is also important in relation to sensitivity; fluorescence is emitted in all directions and sensitivity is related to the fraction of the radiation utilized at both excitation and emission stages.

Fluorescence spectra produced by a single beam spectrofluorimeter are not 'true' spectra. Several instrumental factors distort the curves. The excitation spectrum is affected by the intensity distribution of the exciting radiation as a function of wavelength. The emission spectrum is a function of the efficiency of the detector at each wavelength. Therefore, fluorescence spectra from different designs of spectrofluorimeter are not directly comparable.

This is of little consequence for analytical laboratory measurement. However, a true comparison is desirable for publication and for cases where absolute measurements of quantum efficiency are required. Several commercial spectrofluorimeters provide facilities for producing corrected spectra. A commonly used procedure is as follows. To correct the excitation spectrum, the incident light beam from the monochromator is divided; part excites the sample and the remainder falls on a thermopile. Since the sensitivity response of a thermopile is constant over the normal wavelength range, its voltage output reflects the variation of excitation intensity with wavelength. The corrected spectrum is obtained by dividing the emission detector voltage by the thermopile voltage at each wavelength. To correct the emission spectrum the wavelength response function of the emission detector is first determined. The correction function is then applied via the monochromator drive in a programmed manner to the emission signal. Other methods of correction are available [55-57].

4.2.2 Light Sources
Ideally the *light source* should produce a high intensity of even distribution over the wavelength range of interest, normally the UV/visible region.

[55] *H.B. Mark*, Talanta *12*, 27 (1964).
[56] *C.A. Parker, W.T. Rees*, Analyst *85*, 587 (1960).
[57] *G.K. Turner*, Science *146*, 183 (1964).

To date there is no source which fully satisfies this condition.

The tungsten lamp is satisfactory in the visible region but its intensity diminishes appreciably in the ultraviolet. Currently available spectrofluorimeters are usually equipped with a xenon arc source capable of operation between 100 and 150 watts. Although lamps of higher power, up to 500 watts, are available, their size and cost militate against their use under normal conditions.

While a source of even wavelength intensity distribution is desirable for measurement of fluorescence excitation spectra, the study of fluorescence emission is less dependent on wavelength discrimination. It is frequently preferable to isolate an intense line from the source spectrum for the purpose of excitation. The mercury spectrum is useful in this context, since it emits a number of analytically useful lines between 250 and 580 nm (enhanced by using a mercury-xenon lamp). Alternately, a mercury-cadmium lamp yields several intense cadmium lines in the range 320–650 nm in addition to the mercury spectrum.

The intense monochromatic radiation produced by *lasers* is likely to find increasing use in studying fluorescence. Lasers are now widely available commercially but their wavelength coverage is limited. However, where a laser line is suitable for exciting the molecule of interest, an intense fluorescence should result. The 257.9 nm emission from the frequency-doubled argon ion laser has been utilized[58] for studying fluorescence by the matrix isolation method wherein the sample is diluted by an inert gas and frozen on a suitable window material.

4.2.3 Wavelength Selection

For instruments based on filters a wide range of commercial filters are available, mostly of glass composition. Careful selection of filters is important; the secondary filter passing the fluorescence to be measured should exclude, as far as possible, other wavelengths transmitted by the primary filter to which the detector is sensitive.

For spectrofluorimeters grating monochromators appear to be preferred to prisms despite the inherently greater dispersion obtainable with a prism at the shorter wavelengths. Gratings possess the advantage of linear dispersion. However, at wavelengths longer than about 400 nm a glass filter is required to eliminate second order diffraction. Resolutions of the order 0.5 to 1 nm are attainable in commercial spectrofluorimeters.

4.2.4 Detectors

Photomultipliers are employed almost universally to detect fluorescence in commercial instruments. In some instances facilities are provided for blank subtraction. Careful design of the photomultiplier housing to eliminate light leakage is essential in high sensitivity instruments.

4.2.5 Sample Cell Compartment

Fused quartz cells or test tubes are normally employed as containers for liquid samples. For work of the highest sensitivity, careful selection of quartz or silica is essential to avoid fluorescent impurities. In addition, mountings for solid samples, microcells, thermostatted cells, flow-through systems etc. are now commonly available in commercial instruments.

A double-walled Dewar system, incorporating double-walled quartz windows[59], may be used for fluorescence measurements at selected temperatures down to that of liquid nitrogen.

4.2.6 Measurement of Phosphorescence

To minimize vibrational effects, phosphorescence is measured at low temperature or by maintaining the sample as a rigid glass. Advantage is taken of the long decay time of phosphorescence to measure it and separate it from accompanying fluorescence. The normal method of measurement is to surround the sample, suitably contained, by a rotating shutter having two slits so arranged that the sample is irradiated by exciting light and the phosphorescence is allowed to pass into the detector during the 'dark' period when the sample is not illuminated. By varying the speed of the rotating shutter decay times may be measured over a wide range. For decay times of less than about 1 sec. oscilloscope presentation is necessary. Since emission is not measured during the excitation process, wide wavelength or unfiltered light sources may be used. This enhances the sensitivity of the method considerably.

A solvent suitable for phosphorimetry[60] should set to a clear glass at liquid nitrogen temperature and be free from cracks. Many organic solvents fail to meet this criterion. Among the acceptable solvents are pentane, petroleum ether, diethyl ether, ethanol, propanol, butanol and 2-bromobutane. Several binary mixtures proved suitable, including ethanol/methanol, ethanol/glycerol, ethanol/water, n-pentane/n-heptane[37]. In addition diethyl ether/-

[58] *J.S. Shirk, A.M. Bass,* Anal. Chem. *41,* 103A (1969).

[59] *R.B. Nehrich, A.L. Lewis,* Applied Optics *4,* 1040 (1965).

[60] *J.D. Winefordner, P.A. St. John,* Anal. Chem. *35,* 2211 (1963).

isopentane/ethanol (5:5:2) is finding considerable application in phosphorescence measurements.

4.3 Analytical Applications

4.3.1 Fluorescence

Compared with absorption spectrophotometry fluorescence is capable of greater sensitivity ($10^{-9}-10^{-10}$ M in some instances) and can, in many organic molecules, be altered by varying parameters such as the solvent and pH[61]. The determination of one fluorescent compound in the presence of others is frequently possible by appropriate choice of exciting wavelength; both excitation and emission spectra are available to aid discrimination. A first-order choice of the most appropriate excitation wavelength can usually be inferred from the absorption spectrum of the compound of interest. Nevertheless, for optimum conditions certain factors must be taken into consideration. The choice of instrumental conditions should ideally eliminate potential interference from Raman spectra of the solvent, weak anti-Stokes radiation, and scattered light. In addition, the influence of quenching of fluorescence must be properly taken into account.

A very large number of organic molecules are fluorescent or may be converted chemically to fluorescent derivatives[65,66]. Extensive analytical applications are to be found in the fields of biology and medicine[62,63], including pathology[64]. *Amino acids, amines, vitamins, enzymes, nucleotides, nucleic acids, steroids,* and *hormones* are included in this category.

The intense fluorescence of *polynuclear hydrocarbons* has been utilized in relation to absolute measurements and quantum efficiencies. This group is also of particular analytical interest[67-69].

A substantially pure sample of *phenanthrene* was analyzed for small quantities of *fluorene* and *anthracene*. Examination of the absorption spectrum of fluorene indicates that no wavelength will excite either component alone. However, fluorene shows maxima in its emission spectrum at 302 and 309 nm which are free from phenanthrene interference. By examining phenanthrene at concentrations of 100µg/ml, absorption losses in the sample are minimized and fluorene contents to below 10 ppm could be determined. *Anthracene* can be excited by the 366 nm line of mercury, at which wavelength phenanthrene is not excited. However, a blank fluorescence, attributable to anti-Stokes transitions, had to be overcome by measurements at $-90°$. Contents down to 0.1 ppm could be determined. The fluorescence method proved helpful in monitoring the purity of fluorene, prepared by zone refining, as a function of distance along the zone.

Polycyclic hydrocarbons are of considerable importance in carcinogenic studies; general techniques for separating them from a range of smoked foods include paper and thin-layer *chromatography*. Spots on chromatograms were located by fluorescence and spectrofluorimetry was used to confirm the identity of the isolated material. For the particular carcinogen *benzo[a]pyrene* the spectrofluorimetric method enabled levels down to 1 part in 10^9 to be determined with recoveries of 73% and upwards in fish, cheese and frankfurters. The benzo[a]pyrene content of cigarette smoke has been determined[70] by spectrofluorimetry using perylene, also fluorescent, as internal standard.

Perylene: 427 nm (excitation) 438 nm (emission)
Benzo[a]pyrene: 376 nm (excitation) 405 nm (emission).

A novel system for displaying fluorescence data for polynuclear hydrocarbons[71] serves to improve discrimination between the rather similar spectra of these molecules. Termed triparametric recording, it enables a stereofluorograph to be obtained automatically. The variables measured are the excitation and emission wavelengths and the emission intensity.

The contribution of *polynuclear hydrocarbons* and their alkylated derivatives to atmospheric pollution has been examined[72] using gas chromatographic and thin-layer chromatographic clean-up procedures.

Fluorescence and phosphorescence measurements served to characterize the separated peaks or spots,

[61] *B.L. Van Duuren*, Chem. Rev. *63,* 325 (1963).
[62] *S. Udenfriend*, Fluorescence Assay in Biology and Medicine, Academic Press, New York 1964.
[63] *B.L. Van Duuren* in *D.M. Hercules*, Fluorescence and Phosphorescence Analysis, p. 195, Interscience Publishers Inc., New York 1966.
[64] *R.E. Phillips, F.E. Elevitch*, Progr. Clin. Pathol. *1,* 62 (1966).
[65] *E.L. Wehry, L.B. Rogers* in *D.M. Hercules*, Fluorescence and Phosphorescence Analysis, p. 81, Interscience, New York 1966.
[66] *W.J. McCarthy, J.D. Winefordner* in *G.G. Guilbault*, Fluorescence, Dekker, New York 1967.

[67] *C.A. Parker*, Soc. Anal. Chem. Conf. Nottingham, England, 1965, p. 208.
[68] *J.W. Howard*, J. Ass. Offic. Agr. Chem. *49,* 595 (1966).
[69] *J.W. Howard*, J. Ass. Offic. Agr. Chem. *49,* 611 (1966).
[70] *H.J. Davis*, Anal. Chem. *38,* 1752 (1966).
[71] *M.M. Schacter, E.O. Haenni*, Anal. Chem. *36,* 2045 (1965).

and in some cases revealed that compositions were more complex than examination of the chromatographs revealed. R_f or retention data alone are inadequate to characterize unequivocally these very similar compounds.

The combination of temperature-programmed gas chromatography on an SE 30 column with fluorescence and phosphorescence has been shown to be a powerful method of resolving *petroleum fractions* into their components. These fractions contain a large number of aromatic nitrogen and sulfur compounds[73] which can be characterized by fluorescence (excitation and emission), phosphorescence and phosphorescence decay curves.

The conversion of non- or weakly fluorescent compounds to highly fluorescent derivatives[74] is used for the determination of organic functional groups. *2-Hydroxy-, 2-amino-* and *2-nitro-1-alcohols, 17-ketosteroids* and *1-alkenes* are converted to formaldehyde by suitable oxidative reactions. Fluorescent dihydrolutidine derivatives are then formed by reaction of the aldehyde with a β-diketone in the presence of ammonia. For *nitriles, primary alkylamines* and *aliphatic aldehydes* conversion to dihydropyridine derivatives is satisfactory. Other reactions which proved valuable are the conversion of *phenols* to fluorescent coumarins and the conversion of *2-deoxysugars* and *aldehydes* containing an α-methylene group to quinaldines.

The combination of thin-layer chromatography with fluorescence analysis of separated spots has proved particularly valuable in determining pollutants in air. Several commercial fluorescence spectrometers offer attachments which enable a thin-layer plate to be scanned. Using a spectrofluorimeter both excitation and emission spectra may be examined, e.g. *dibenz[a, h]acridine, dibenz[a, j]acridine, benzo[l, m, n]phenanthridine,* and *acridine*[72]. The method is both simpler and more sensitive than procedures involving mechanical removal of separated spots from the chromatoplates. Two dimensional thin-layer chromatography followed by fluorescence measurements at various pH's was used to separate and determine *anthracene, benz[a]anthracene, benzo[a]pyrene, benzo[b]fluoranthene, benzo[k]fluoranthene, carbazole, chrysene* and *pyrene. 1-Phenalenone* and *7H-benz[d, e]anthracen-7-one* were similarly determined. *Cholesterol*[75] can be determined down to 0.005 μg as the fluorescent fluoranthenyl urane

derivative following isolation of the latter by thin-layer chromatography. Certain *carbohydrate* mixtures have been analyzed by thin-layer chromatographic separation followed by spraying with *p*-aminohippuric acid and scanning of the plates in a fluorimeter[76].

All of the examples quoted above are applications of prompt fluorescence. This is the normally encountered mode, having a life-time of 10^{-8} seconds or less. In certain circumstances sensitized delayed fluorescence, with life-times of a similar order to phosphorescence, is observed[77]. This requires the presence of a donor and an acceptor molecule. The incident radiation is absorbed by the donor which transfers it to the acceptor by triplet state interactions. Hence, the normal fluorescence spectrum of the acceptor is emitted because of the involvement of triplet states with a longer life time.

Traces of impurities in polynuclear aromatic hydrocarbons may be detected by examining delayed fluorescence produced by a high intensity light source. The presence of such an impurity can be masked in the examination of prompt fluorescence by the intense fluorescence of the major component. Results of delayed fluorescence studies must be interpreted with caution; a number of compounds, especially polynuclear hydrocarbons, exhibit delayed fluorescence due to 'excimer' (excited dimer) formation. Also the high exciting intensities required may induce significant photo-decomposition of the substrate.

4.3.2 Phosphorescence

Quantitative anlysis by phosphorimetry was first suggested in 1957[78], but the literature of applications is growing steadily.

Aromatic nitro compounds, which give rather poor absorption spectra, are readily determined in nanogram quantities or less by phosphorimetry. *p*-Nitroaniline and similar compounds for pesticide use have been determined by phosphorimetry[79]. Phosphorimetry was also applied to determine 22 compounds of pharmacological interest including *barbiturates,* a range of *cocaine derivatives, quinine, ephedrine* and *chlorotetracycline.* All the triplet states of the compounds have lifetime of a few seconds and detection limits between 0.004 and 0.2 μg/ml are quoted. Linear calibrations are

[72] E. Sawicki, Talanta *13*, 619 (1966).
[73] H.V. Drushel, A.L. Sommers, Anal. Chem. *38*, 10 (1966).
[74] M. Pesez, J. Bartos, Talanta *14*, 1097 (1967).

[75] H.T. Gordon, J. Chromatogr. *22*, 60 (1966).
[76] W.M. Connors, W.K. Boak, J. Chromatogr. *16*, 243 (1964).
[77] C.A. Parker, Photoluminescence of Solutions, p. 449, Elsevier, London 1968.
[78] R.J. Keirs, Anal. Chem. *28*, 202 (1957).
[79] E. Sawicki, J. Pfaff, Microchem. J. *12*, 7 (1967).
[80] J.D. Winefordner, M. Tin, Anal. Chim. Acta *31*, 239 (1964).

obtained over the concentration range 10^{-7} to 10^{-5} M.

Phosphorimetry can be applied to compounds separated by thin-layer chromatography. After separation of the tobacco alkaloids *nicotine, nornicotine* and *anabasine* on alumina thin-layers, each alkaloid was dissolved in ethanol for phosphorescence measurements. Direct phosphorimetry of compounds on cellulose or glass fiber paper has proved feasible but detection limits are poorer by several orders of magnitude.

For a number of organic compounds the sensitivity of phosphorescence determination can be improved by perturbing the system with a heavy atom[80-82] to increase the probability of transitions between states of different multiplicity. This is accomplished by incorporating the heavy element into the phosphorescent molecule or in the solvent, e.g. ethanol/ethyl iodide. Increased phosphorescence intensity resulted for *coumarin, fluorescein, N-N-dimethyl aniline* and several *dinitronaphthalenes*[83]. In 12 polynuclear hydrocarbons[84] the effect is of variable magnitude.

Using ethanol/ethyl iodide (5:1 v/v), enhancement factors over ethanol alone range from 25 for *benzo[b]fluorene* to 1.3 for *dibenz[a,h]anthracene*. The effect is not universal and marked decreases in sensitivity are found for *triphenylene* and *retene (7-isopropyl-1-methylphenanthrene)*.

Phosphorescence is akin to radioactivity in that its decay, with certain exceptions, is exponential. Hence, a mixture of phosphorescing compounds of differing life-times is amenable to analysis by resolution of the decay curve[85]. Modifications to a phosphorimeter permit the logarithm of composite intensity to be plotted against time. Using this technique of time-resolved phosphorimetry, decay times of milliseconds and longer may be measured. Mixtures of *benzoic acid* and *benzaldehyde* ($\sim 10^{-4}$ M) and *tryptophan* and *tyrosine* ($\sim 5 \times 10^{-6}$ M) were analyzed by this method with a relative error of up to $\pm 5\%$. The relative decay rates in these two instances are greater than 25 and 4.6 respectively.

[81] *J.D. Winefordner, H.A. Moye*, Anal. Chim. Acta *32*, 278 (1965).

[82] *S.P. McGlynn*, J. Chem. Phys. *39*, 675 (1963).

[83] *I.J. Graham-Bryce, J.M. Corkhill*, Nature *186*, 965 (1960).

[84] *L.V.S. Hood, J.D. Winefordner*, Anal. Chem. *38*, 1922 (1966).

[85] *P.A. St. John, J.D. Winefordner*, Anal. Chem. *39*, 500 (1967).

5.12 Mössbauer Spectroscopy

Ekkehard Fluck

Institut für Anorganische Chemie der Universität, D-7 Stuttgart

Recoil-free emission and resonance absorption of gamma quanta by atomic nuclei (Mössbauer Effect) allows:
determination of the electron density at the nucleus of suitable atoms (e.g., Fe or Sn, see Fig. 4) and information on the chemical bonding system (5.12:2.1);
determination of the symmetry of the electronic and thus, in general, the coordinative surrounding of the Mössbauer nuclides (5.12:2.2);
determination of Curie- or Néel temperatures (5.12:2.3).

1 Mössbauer Effect

The Mössbauer Effect is the phenomenon of recoil-free emission and resonance absorption of gamma quanta by atomic nuclei.

1.1 Basic Principles

An atom or molecule can be excited by absorption of a light quantum $h\nu$ of sufficient energy, i.e., it can be raised to a higher quantum state. On reverting to the ground state, the absorbed energy usually is re-emitted as light quanta, a process called fluorescence. A special form of this phenomenon, *resonance fluorescence,* occurs when the energy of incident quanta corresponds to the separation between the ground state and the first excited state of an atom or molecule. The process of reversion can lead only to the emission of quanta with a single energy equal to that of the quantum absorbed.

The main difference between the quantum systems of the electron shell of an atom or molecule and of a nucleus is that the *transition energies*, E_r, in the electron systems are smaller by several orders of magnitude compared with those in atomic nuclei.

On the other hand, the *natural line widths*, Γ, of the resonance lines are similar in both kinds of quantum systems.

$$\Gamma = \frac{h}{\tau} \qquad (1)$$

Accordingly, the sharpness of the resonance line, which is characterized by the ralation Γ/E_r, is much greater for atomic nuclei than for electron systems. Some characteristic data of a resonance transition in electron shells and in a nucleus are summarized in Table 1.

Table 1. Energy data (transitions) for resonance fluorescence

	Atomic shell (Na_D-line)	Nucleus (^{119}Sn-nucleus)
Transition energy E_r [eV]	2.1	23 800
Mean lifetime of the excited state τ [sec]	$1.5 \cdot 10^{-8}$	$2.7 \cdot 10^{-8}$
Natural resonance width [eV]	$4.4 \cdot 10^{-8}$	$2.4 \cdot 10^{-8}$
Resonance sharpness Γ/E_r	$2.1 \cdot 10^{-8}$	10^{-12}
Recoil energy R [eV]	10^{-10}	$2.5 \cdot 10^{-3}$
R/E_r	$\approx 4.8 \cdot 10^{-11}$	$\approx 10^{-7}$
R/Γ	$\approx 2.3 \cdot 10^{-3}$	$\approx 10^{5}$

If a free atomic or nuclear system reverts from the excited state to the ground state, a photon of the energy E_γ will be emitted (Fig. 1). However, the energy of the photon is smaller than E_r since, upon emission of the photon, the emitting system will suffer a *recoil* for which energy must be supplied.

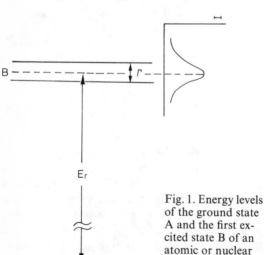

Fig. 1. Energy levels of the ground state A and the first excited state B of an atomic or nuclear system.

The *recoil energy* R of the atomic or nuclear system is then:

$$R = 1/2 mv^2 = \frac{p^2}{2m} = \frac{P^2}{2m} \qquad (2)$$

$$= \frac{E_\gamma^2}{2mc^2} \cong \frac{E_r^2}{2mc^2}$$

since $E_r = E_\gamma + R$
$E_r \gg R$

The energy of the emitted quanta, E_γ, is diminished by this amount, R, compared with the transition energy, E_r (see Fig. 2).

Since, on the other hand, the atoms or atomic nuclei which absorb quanta suffer a recoil too, the energy required for the excitation is increased by the amount of the recoil energy R (Fig. 2).

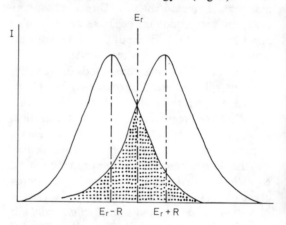

Fig. 2. Emission and absorption lines

Due to the short mean lifetime of the excited state, its energy according to Eq. (1) is not well defined, in contrast to that of the ground state which has a long mean lifetime:

the emitted quanta show a certain *energy distribution*;
the emitted line has a natural *line width*;
the energy distribution curve is of *Lorentzian shape* in the ideal case.

The *probability of emission* (and absorption) of quanta of energy E, which differs from energy E_0, is given by Eq. (3) (k = constant):

$$w(E) = \frac{k}{(E - E_0)^2 + 1/4\Gamma^2} \qquad (3)$$

In nuclear transitions the large transition energy, E_r, according to Eq. (2) causes a much larger recoil energy, and the ratio E_r/Γ can become very large. Therefore, the overlap between the line emitted by the radiation source and the line which is required for the excitation is negligibly small. A resonance phenomenon is no longer observed.

If, however, the emitting and absorbing atoms are not free but solidly built into a *crystal lattice,* the mass of the system which suffers the recoil becomes very large [m in Eq. (2)]. It corresponds to the mass of the whole crystal. The recoil energy is now much smaller than the line width.

The transition energy E_r can be distributed only between the emitted gamma quantum and possibly excited lattice vibrations.

If a transition occurs without the excitation of lattice vibrations, then the gamma quantum keeps the whole transition energy. Thus a 'recoil-free' emission of a gamma quantum occurs. The same is true for absorption.

With gamma quanta emitted and absorbed recoil-free resonance phenomena can be observed analogous to optical resonance fluorescence.

Experimental arrangements[1] for the observation of the Mössbauer effect are shown schematically in Fig. 3a and 3b. Generally the *resonance fluorescence* is observed by measuring absorption (Fig. 3a). For larger objects, however, the *scattered radiation* can be measured (Fig. 3b).

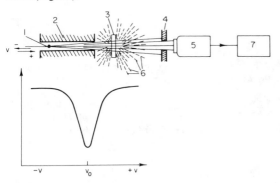

Fig. 3. Experimental arrangement for observing
a) the resonance absorption of gamma quanta:

1 radiation source which moves with velocity v relative to a stationary absorber	4 collimator
	5 radiation detector (scintillation counter or proportional counter)
2 source collimator	6 gamma quanta scattered by the absorber
3 absorber (material under investigation)	7 electronic recording

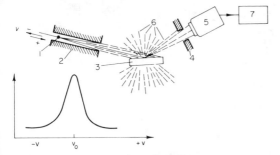

b) the resonance scattering of gamma quanta:
1, 2, 4–7 as in Fig. 3. 3 scattering material (from Chemical Applications of Mößbauer Spectroscopy, edited by V. I. Goldanskii and R. H. Herber, Academic Press, N. Y., 1968).

The easiest way to observe the *resonance condition* is to disturb the system and observe the influence upon measurable parameters such as the intensities of the transmitted or scattered radiation. The energy of the gamma quanta emitted by the radiation source can be changed by moving the radiation source and resonance absorber relative to each other. The velocity, v, which is necessary for the Doppler effect to shift the energy of the gamma quanta, E_γ, by an amount corresponding to the line width, is:

$$v = \frac{\Gamma \cdot c}{E_\gamma} \qquad (4)$$

Since the lines are very narrow, relatively small velocities are sufficient to destroy the resonance (Table 2).

Table 2. Velocities, corresponding to one line width, for the Doppler effect in some Mössbauer nuclides

Isotope	E_r	$T_{1/2}$	Γ	$v = \dfrac{\Gamma \cdot c}{E_\gamma}$
	[keV]	[sec]	[eV]	[mm/sec]
^{57}Fe	14.4	$1.0 \cdot 10^{-7}$	$4.55 \cdot 10^{-9}$	0.0945
^{119}Sn	23.8	$1.9 \cdot 10^{-8}$	$2.4 \cdot 10^{-8}$	0.303
^{191}Ir	129	$1.4 \cdot 10^{-10}$	$3.25 \cdot 10^{-6}$	7.53
^{197}Au	77	$1.9 \cdot 10^{-9}$	$2.4 \cdot 10^{-7}$	0.94

1.2 Mössbauer Nuclides

The Mössbauer effect has been observed with numerous elements. A limitation is that the energy difference between the ground state and the first excited state must not be too large. Especially with light nuclides the recoil energy is very large and, therefore, the Mössbauer effect becomes very small. Furthermore, the mean *lifetime* of the excited state should lie between 10^{-6} and 10^{-12} sec, since the observation of the resonance line beyond these limits becomes difficult due to line widths which are too narrow and too large, respectively. Mother nuclides, however, should have a relatively long mean life time (Table 3).

The *elements* with which the Mössbauer effect has been observed are indicated in Fig. 4 by shaded fields. With Fe, Zn, Sn, Sb, Te, I, Ta, W and Ir the effect occurs even at room temperature.

The experiment is generally carried out with nuclides which decay to produce the *first excited* state of the nucleus to be investigated. The decay schemes of some nuclides, which fulfill this condition are shown in Fig. 5.

2 Mössbauer Parameters

Chemists are most interested in the following parameters which can be determined from Mössbauer spectra and their temperature dependence:

Isomer shift
Quadrupole splitting
Magnetic hyperfine splitting
Resonance effect magnitude.

Table 3. Some Mössbauer Nuclides

Nuclide	Natural Abundance [%]	E [keV]	$T_{1/2}$ [sec]	Parent nuclei	half life	Nuclear reaction for the production
^{57}Fe	2.17	14.4	$1.0 \cdot 10^{-7}$	^{57}Co	267 d	^{56}Fe(d, n)
^{119}Sn	8.58	23.8	$1.9 \cdot 10^{-8}$	^{119}Sn	250 d	^{118}Sn(n, γ)
^{127}J	100	59				^{126}Te(n, γ)
^{129}J	0	26.8	$1.85 \cdot 10^{-8}$	^{129}Te	33 d or 70 m	^{128}Te(n, γ)
^{191}Ir	38.5	129	$1.4 \cdot 10^{-10}$	^{191}Os	14 h	^{190}Os(n, γ)
^{195}Pt	33.8	99	$1.4 \cdot 10^{-10}$	^{195}Au	192 d	^{194}Pt(n, γ)

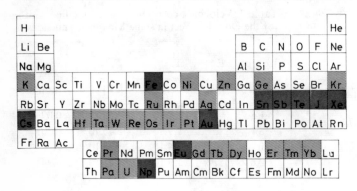

Fig. 4. A periodic chart in which heavy shades mark elements with which Mössbauer spectroscopy has been used for solving chemical problems; slight shades mark elements whith which the Mössbauer effect has been observed

Fig. 5. Some decay schemes of nuclei

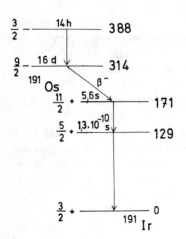

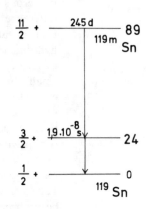

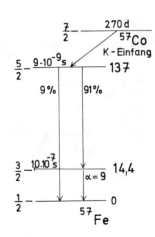

2.1 Isomer Shift

If Mössbauer nuclides in the radiation source and in the resonance absorber are in the same chemical bond state and at analogous lattice places, optimal absorption of radiation is observed when radiation source and resonance absorber have zero velocity relative to each other. If, however, the compared atoms in the radiation source and resonance absorber are in different environments (different bonds), generally, the energy of the gamma quanta emitted by the radiation source must be changed by a *Doppler effect* to obtain maximum absorption. This shift of the center of the resonance absorber spectrum relative to a reference substance (standard) is referred to as the *isomer shift*.

$$I.S. = \frac{4\pi}{5} Ze^2 R_g^2 \left(\frac{\delta R}{R_g}\right) \left\{ |\Psi(0)|_a^2 - |\Psi(0)|_s^2 \right\} \quad (5)$$

Z	atomic number
R_g	radius of the nucleus in the ground state
δR	difference between the nuclear radii in the excited state and in the ground state
$|\Psi(O)|^2$	total electron density at the nucleus of the atom
Index$_a$	in the resonance absorber
Index$_s$	in the radiation source

In many cases it is referred to the source of radiation used in the experiment. However, it is more useful to refer to a well defined substance

[1] *M. Kalvius*, Z. Naturforsch. *17A*, 494 (1962).

and to measure both the substance under investigation (resonance absorber) and the standard substance with the actual radiation source.

The nuclear term [1. term of Equation (5)] describes the relative change of the nuclear radius on excitation of the nucleus, the electronic term determined by the chemistry of the atom, the difference between the total electron densities at the nucleus in the resonance absorber and in the radiation source. If the sign of the nuclear term is known, the measurement of the isomer shift yields directly a quantitative comparison of the electron densities at the nuclei in the resonance absorber and in the radiation source.

The isomer shift is mainly a linear function of the *s electron* density at the nucleus since only s electrons do have a finite density at the nucleus (if small relativistic contributions by p electrons are neglected). At a given total s electron density of the atom, an increasing p and d electron density diminishes the electron density at the nucleus by screening the s electrons.

The observed shift of the resonance line is composed of the isomer shift which may be both temperature dependent and a temperature shift. The latter is determined by the mean kinetic energy of the nuclei in the lattice.

2.2 Quadrupole Splitting

The interaction between the electric quadrupole moment of the nucleus under investigation in the ground state and/or in the excited state and an electric field gradient causes the resonance line to be split into two components.

The so called *quadrupole splitting*, Δ , of the resonance line is proportional to the product of the electric field gradient and nuclear quadrupole moment.

$$\Delta = \frac{e^2qQ}{4I(2I-1)}\left|3m_I^2 - I(I+1)\right|\left(1+\frac{\eta^2}{3}\right)^{\frac{1}{2}} \qquad (6)$$

q   electrostatic field gradient
Q   nuclear quadrupole moment
I   nuclear spin
m_I magnetic quantum number of the nuclear level
η asymmetry parameter $= \dfrac{(\partial^2V/\partial x^2) - (\partial^2V/\partial y^2)}{\partial^2V/\partial z^2}$

$$\text{where} \quad \frac{\partial^2V}{\partial x^2} \quad \frac{\partial^2V}{\partial y^2} \text{ and } \frac{\partial^2V}{\partial z^2} \qquad (7)$$

(abbreviated V_{xx}, V_{yy} and V_{zz}) are the three components of the electric field gradient.

The coordinate system is chosen, so that

$$|V_{zz}| \geqslant |V_{yy}| \geqslant |V_{xx}| \text{ and consequently } -1 \leqslant \eta \leqslant +1$$

If the electronic environment of a nucleus has *cubic symmetry*, the electric field gradient disappears. The nuclear state with spin 1 or greater is degenerate and the Mössbauer spectrum consists of a singlet. If, on the other hand, an atom is at a non-cubic site of a lattice, the degeneracy of the nuclear level is removed and a quadrupole splitting of the resonance line is observed.

In many cases the components of a doublet caused by quadrupole splitting show different intensities. This effect can have various causes:

crystallographic orientation[1, 2],
anisotropic asymmetry: The probability of emission or absorption at a recoil-free transition may be dependent on the angle[3-5]; the latter also occurs with polycrystalline material (Goldanskii-Karyagin effect).

The Goldanskii-Karyagin effect allows conclusions to be drawn, from the anisotropy of the atomic vibrations in molecules or lattices, regarding structural properties of single crystals in polycrystalline material.

The correlation of isomer shifts and quadrupole splittings permits determination of the different contributions of s, p and d electrons to chemical bonding.

2.3 Magnetic Splitting

If a nucleus in the excited state has the nuclear spin number $I = 3/2$ and in the ground state the spin number $I = 1/2$ (cf. Chapter 5.5:1), as with the ^{57}Fe nucleus, then the excited state is split into 4 sublevels, and the ground state into 2 sublevels. In agreement with the selection $\Delta m = +1, 0, -1$, the six allowed transitions of the eight possible transitions between the levels of the excited state and the ground state give 6 resonance lines. The splitting of the levels can be obtained directly from the distances between the various lines (Fig. 6).

Since the splitting of the ground state, g_0, is larger in the case discussed here than that of the excited state, g_1, the energy of the six lines becomes larger in the sequence:

Line	Transition	Energy	rel. Intensity
1	$-3/2 \to -1/2$	E	$3(1+\cos^2\theta)$
2	$-1/2 \to -1/2$	$E + g_1$	$4\sin^2\theta$
3	$+1/2 \to -1/2$	$E + 2g_1$	$1+\cos^2\theta$
4	$-1/2 \to +1/2$	$E + g_1 + g_0$	$1+\cos^2\theta$
5	$+1/2 \to +1/2$	$E + 2g_1 + g_0$	$4\sin^2\theta$
6	$+3/2 \to +1/2$	$E + 3g_1 + g_0$	$3(1+\cos^2\theta)$

[2] *E. Fluck, W. Kerler, W. Neuwirth,* Angew. Chem. *75,* Engl. 277 (1963). J. Jap. Chem. *19,* 27 (1965).
[3] *V.I. Goldanskii,* Dokl. Akad. Nauk. SSSR, *147,* 127 (1962); C.A. *58,* 9755 (1963).
[4] *V.I. Goldanskii, E.F. Makarov, V.V. Chrapov, Zh. Eksp. Teor. Fiz.* 44, 752 (1963).
[5] *S.V. Karyagin,* Dokl. Akad. Nauk. SSSR, *148,* 1102 (1963); C.A. *59,* 1221 (1963).

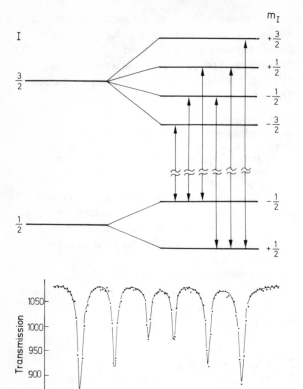

$$\varepsilon = \left(\frac{R_{\infty} - R_0}{R_{\infty}}\right) 100$$

R_0: counting rate at resonance maximum

R_{∞}: counting rate in the absence of a resonance effect

ε is primarily a function of the fraction of the recoil-free emission and absorption events, f and f' i.e. of the total number of transitions. These are dependent on the substance and temperature. Furthermore ε can vary strongly with the experimental arrangement.

3 Applications of Mössbauer Spectroscopy to Chemical Problems

3.1 Introduction

The *isomer shift* of a nucleus gives information about the *electron density* at the nucleus. The shift is a function of the *oxidation number* and of the state of *chemical bonding* of the atom. *Quadrupole splitting* of a resonance line is caused by an *electric field gradient* at the nucleus. It yields information about the *symmetry* of the electronic environment and thus also of the coordination sphere of the atom. However, the absence of a quadrupole splitting does not necessarily mean that the chemical symmetry is cubic (octahedral or tetrahedral). It means only that the electric field around the nucleus has cubic symmetry. For the chemist, Mössbauer spectroscopy is primarily a tool for *structural* research and for the study of the *chemical bond*.

The method is restricted to relatively few elements although in some cases, Mössbauer nuclides may be used as a probe for effects in other systems. From the changes of the values of the quadrupole splitting in the spectra of frozen solutions of *dibutyltin dichloride* and *triphenyl-tin chloride* in various aprotic solvents, a scale of *solvation capacity* was established[6].

The method is further restricted to samples in solid phases, though most liquid substances can be solidified by cooling. Frozen solutions can also be investigated. In certain cases the fraction of recoil-free absorption is high enough even in *liquids* to observe an effect (e.g. with very large organic iron or tin containing molecules).

Fig. 6. a) Energy level scheme for the magnetic splitting of the ground state and the first excited state of ^{57}Fe in metallic iron,
b) Mößbauer spectrum of metallic iron at room temperature. Ordinate: transmission (arbitrary units). Abscissa: relative velocity (mm/sec) between radiation source (^{57}Co in platinum at 25 C) and absorber.

The relative intensities of the six resonance lines depend on the angle between the magnetic field and the optical axis of observation (see above).

For an iron foil with statistically oriented magnetic areas an intensity ratio 3:2:1:1:2:3 as to be expected for the six lines.

A splitting into six resonance lines is observed also if there is an *electric quadrupole interaction* besides the magnetic interaction. The seperations of the lines depend on the magnitudes of the fields.

The study of magnetic interaction in the Mössbauer spectrum serves for the determination of transition temperatures such as the *Curie* or *Néel temperature*.

2.4 Magnitude of the Mössbauer Effect

As a measure of the magnitude of the resonance effect, the magnitude ε is generally used:

3.2 Iron Compounds

For the ^{57}Fe *nucleus* the magnitude $(\delta R/R)$ in equation (5) is negative. A shift of the resonance line or band to more negative velocities means an increase of electron density at the nucleus compared with the standard. It may be caused by an increased occupation of the 4s level and/or a decreased 3d electron density which causes a reduction in the screening of the s electron and

[6] *V.I. Goldanskii*, J. Organometal. Chem. *4*, 160 (1965).

herewith an increased s electron density at the nucleus. Typical ranges for the isomer shift of high-spin iron compounds using *sodium nitrosyl-prussiate-di-aquate* as a reference are:

	$Fe^{2\oplus}$	$Fe^{3\oplus}$	$Fe^{6\oplus}$
I.S. [mm/sec]	+1.0 to +1.8	+0.4 to +0.9	−0.6

Iron coordination compounds in which the bonds between the central iron atom and the ligands are mainly covalent have similar isomer shifts independent of the formal oxidation number of iron. The oxidation number of iron in these compounds is obtained from a possible *quadrupole splitting* of the resonance line and its temperature dependence. *Octahedral iron(II) coordination compounds* do not show quadrupole splitting of the resonance line if the six ligands are identical or have identical influence upon the electric field around the iron nucleus. However, if the resonance line is split, due to the non-equivalency of the ligands, this splitting is practically independent of the temperature. On the other hand the quadrupole splittings of *iron(III) coordination compounds* are highly temperature dependent.

Since the d electrons screen the s electrons, the isomer shift can give quantitative information on bond state in which d electrons are involved (for instance on the degrees of the π back-donation).

Changes of isomer shift and quadrupole splitting show in which way the electron density at the central iron atom is changed in a redox process. For instance, the thorough study of *iron(II) phthalocyanine* and its anions with 1–4 charges show that the added negative charge is distributed mainly over the complex as a whole[7].

Organo-iron compounds in some cases show only a slight Mössbauer effect at room temperature, generally due to low-lying Debye temperatures.

3.3 Tin Compounds

Organotin compounds have been investigated extensively. In general, *tin(II)-compounds* have a positive isomer shift relative to gray tin (α-tin), whereas *tin(IV)compounds* have a negative isomer shift.

Compounds of the type SnR_4 usually result in spectra showing singlet lines, even when the substituents are different, i.e. the electric field at the tin nucleus is influenced only by the nearest neigh-

bors. Only if the character of the substituents changes very strongly, *quadrupole splitting* occurs. Thus, for instance, in the spectrum of *pentafluorophenyl-triphenyl-tin* a relatively large quadrupole splitting of 1.1 mm/sec is observed. A similarly large quadrupole splitting in the spectrum of *triphenyl-phenylbarenyl-tin* (1) leads to the conclusion that the inductive effects of the electron acceptor properties of pentafluorophenyl- and phenylbarenyl groups are approximately equivalent[8].

Similarly the spectra of *trialkyl-* and *triaryltin halides* are characterized by remarkable *quadrupole splittings*. Their value in a series of organotin halides of the same type (R_iSnHal_{4-i}) is proportional to the total electron donor capacity of the organic substituents.

Kinetics and *Absorption mechanisms of totally symmetric organotin compounds:* The absorption of $(CH_3)_4Sn$ on the surface of γ-aluminum oxide, silica gel, and aluminum silicate has been investigated in this way. The shape of the Mössbauer spectrum of *absorbed tetramethyl-tin* depends highly on the absorption temperature. With increasing temperature, a surface complex is formed which probably contains *tin of coordination number six*[9].

3.4 Biochemical Problems

The ^{57}Fe nucleus plays an important role in biological systems[16, 27] (*hemoglobin*, numerous *enzymes, nucleic acids*). According to sections 5.12:3.1 and 5.12:3.2, information can be obtained from the Mössbauer spectrum about the oxidation number of iron and its changes in biochemical processes as well as about the structure of these coordination complexes.

The meet the difficulties which are connected with the low iron content of many biological materials, the sensitivity can be increased tremendously by the exchange of the isotope ^{56}Fe, which is mainly present, by ^{57}Fe, e.g. in *ferredoxin*[10] and in another *iron containing protein* from *Azotobacter vinelandii*[11].

[7] *R. Taube*, Z. Anorg. Allg. Chem. *364*, 297 (1969).

[8] *A. Yu. Aleksandrov*, Dokl. Akad. Nauk. SSSR *165*, 593 (1965); C.A. *64*, 9106 (1966).

[9] *A.N. Karasyov*, Kinet. Katal. *4*, 710 (1965).

[10] *D.C. Blomstrom*, Proc. Nat. Acad. Sci. U.S. *51*, 1085 (1965).

[11] *Y.I. Shethna*, Proc. Nat. Acad. Sci. U.S. *52*, 1263 (1964).

4 Bibliography

[12] *D.M.J. Compton, A.H. Schoen,* The Mößbauer Effect, John Wiley, New York 1962.

[13] *V.I. Goldanskii,* The Mößbauer Effect and Its Application to Chemistry, At. Energy Review *1,* 3 (1963), Consultants Bureau, New York.

[14] *E. Fluck,* Advan. Inorg. Chem. Radiochem. *6,* 433 (1964).

[15] *G.K. Wertheim,* Mößbauer Effect, Principles and Applications, Academic Press, New York 1964.

[16] *E. Fluck,* Applications to Iron Chemistry and to Biochemical Problems, Panel on Application of the Mößbauer Effect in Chemistry and Solid State Physics, I.A.E.A., Vienna 1965.

[17] *J. Duncan, R. Golding,* Quart. Rev. *19,* 36 (1965).

[18] *I.J. Gruverman,* Mößbauer Effect Methodology, Vol. 1–3, Plenum Press, New York 1965 ff.

[19] *R.H. Herber,* Introduction to Mößbauer Spectroscopy, J. Chem. Educ. *42,* 180 (1965).

[20] *H. Wegener,* Der Mößbauer-Effekt und seine Anwendung in Physik und Chemie, Bibliographisches Institut, Mannheim 1965.

[21] *E. Fluck,* Fortschritte der chem. Forschung, Springer Verlag, Berlin, Heidelberg, New York *5,* 395 (1966).

[22] *A.H. Muir, K.J. Ando, H.M. Coogan,* Mößbauer Effect Data Index, 1958–1965, Interscience Publ., New York 1966.

[23] Applications of the Mößbauer Effect in Chemistry and Solid-State Physics, Techn. Report Ser., Inter. At. Energy Agency *50,* 89 (1966).

[24] *V.I. Goldanskii,* Angew. Chem. *79,* 844; Engl.: 830 (1967).

[25] *R.H. Herber,* Progress in Inorg. Chem., Interscience Publ., New York, *8,* 1 (1967).

[26] *V.I. Goldanskii, R.H. Herber,* Chemical Applications of Mößbauer Spectroscopy, Academic Press, New York 1968.

[27] *Yu. Sh. Moshkovskii,* Applications of the Mößbauer Effect in Biology, in Ref. [26]. p. 524.

[28] *N.N. Greenwood,* Endeavour *27,* 33 (1968).

6 Fragmentation Methods

Contributions by

H. Batzer, Basel
H.-D. Beckey, Bonn
H.-W. Fehlhaber, Bonn
H. Schildknecht, Heidelberg
R. Schmid, Basel

6.1 Mass Spectrometry

Hans-Wolfram Fehlhaber
Organisch-Chemisches Institut der Universität,
D-53 Bonn

1 Fundamentals

1.1 Introduction

Mass spectrometry is based on the generation of gaseous ions and the determination of their relative masses and abundances. It enables the most accurate *molecular weight determinations* and, under certain circumstances, yields very precise information about the *elemental composition* of an organic compound. The excess energy transferred to the ions causes fragmentations which are generally highly structure-specific. Their interpretative value equals that of multi-stage chemical degradation reactions and may, therefore, afford detailed *structural assignments,* even in the case of complex compounds.

Several tenths of a milligram of substance are normally used to obtain a mass spectrum, but, with careful working, acceptable spectra can be obtained from much smaller amounts, — less than one microgram, under favorable conditions.

1.2 Measurement Principles

The measurement process in the mass spectrometer embraces the following steps: sample evaporation, ionization, mass separation, ion detection, and recording of the spectrum. High vacuum is maintained in the entire analyzer system ($< 10^{-6}$ torr) to suppress collisions between the ions and residual gas molecules (ion-molecule reactions, scattering of the ion beam).

A number of systems have been developed for sample evaporation, but all are based on two fundamental principles. In one method, the sample is evaporated in an evacuated and thermostatically controlled reservoir and passes via a pressure reduction device into the ion source as a continuous gas flow. The required vapor pressure of about 10^{-2} torr at a maximum temperature of 250° is feasible with gaseous, liquid, and sublimable solid substances. To minimize catalytic reactions, the inlet systems are made of glass as far as possible. In the second method, the sample is introduced directly into the ion source via a vacuum lock device and the optimum evaporation rate is adjusted by careful heating. As a vapor pressure of $10^{-5} - 10^{-7}$ torr is sufficient here, this method is particularly suitable for solid samples of low volatility, but is applied also to more volatile compounds if thermal decompositions occur at the temperature needed for working with a reservoir system.

The most important ionization method is that employing electron impact (for further special ionization methods s. Ref.[1]). In the ion source the sample vapor is bombarded by an electron beam emitted from a hot-cathode and accelerated by an electric field. In this way mainly positive ions are generated, negative ions being produced only to a very minor extent. Although the ionization potentials of organic compounds lie between 8 and 15 eV, much higher electron energies are usually employed (normally 70 eV) in order to improve the ion efficiency. Part of this excess energy is transferred to the ions and results in fragmentation reactions.

By means of a low-intensity electric field, the positive ions are extracted from the ion source and then pass through an acceleration region operating at several kV. For separation of the ions according to their mass (strictly their mass to charge ratio m/e), the deflection in a magnetic field (single focusing), the deflection in a combination of electrostatic and magnetic fields (double focusing), the mass-dependent time of flight in a field-free region (time-of-flight mass spectrometer), and various other methods are utilized[2].

The ions are detected by simple ion traps, secondary electron multipliers or, in special cases, photographic plates. After amplification, the signals are recorded by a potentiometer recorder, multichannel galvanometer recorder (Fig. 1), or cathode ray oscilloscope, on magnetic tape, or fed directly to an on-line computer.

As almost every nominal mass is represented by a peak in the mass spectrum (*spectrum background*, see Section 6.1:3.1), the mass scale can generally be established by counting the peaks. The beginning of the spectrum, (m/e 10 in the presence of boron, otherwise m/e 12), and the easily identified peaks of water (m/e 18) and air (m/e 28 and 32) from the residual gas in the apparatus serve as reference points. Due to the monotonic alteration in the mass spacing, small gaps can be bridged by interpolation, but in difficult cases suitable refer-

---

[1] *H.D. Beckey,* Methodicum Chimicum, 6.2.
[2] Ref.[39] [46].

ence compounds must be employed for calibration. As the half-width of the peaks is governed by the instrumental parameters and remains unaltered during the preparation of a mass spectrum, the peak height is proportional to the abundance of the trapped ions. For this reason mass spectra are conventionally presented as normalized bar graphs, the intensity of the highest peak (= base peak) set equal to 100 (as shown in Fig. 2, Section 6.1:2.1.2).

acid, loss of an alcohol molecule from *ethers, acetals* and *polyfunctional esters,* decarboxylation of *unsaturated carboxylic acids* and *cyclic anhydrides,* and removal of nitrogen from compounds containing an N=N group. In addition, hydrogenations and dehydrogenations (especially the quinone ⇌ hydroquinone conversion), double bond isomerizations and even skeletal rearrangements have been observed. The majority of these reactions can be attributed to surface reactions in

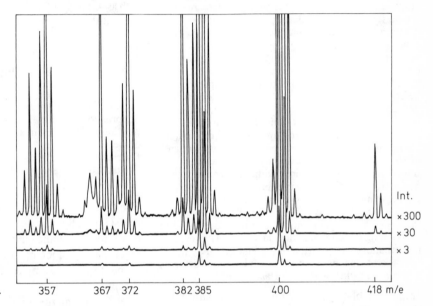

Fig. 1. Section of a mass spectrum taken with a galvanometer recorder.

1.3 Selection of the Sample

1.3.1 Volatility and Thermal Stability

In principle, the sample must be vaporizable at a temperature at which no appreciable thermal decomposition will occur. The minimum vapor pressure of about 10^{-7} torr required (using the direct insertion technique) is obtained with the majority of organic compounds, including a large number of metal complexes, and is, in fact, generally reached at temperatures well below the melting point. There are, in principle, no limitations with regard to the molecular weight, but experience has shown that volatility and stability problems are likely with molecular weights greater than 1000. Usable spectra from molecular weights above 1500 can be obtained in exceptional cases only.

Several types of compounds readily undergo thermal transformations[3]. The reactions involved are primarily *elimination reactions,* e.g. dehydration of *aliphatic alcohols,* cleavage of *aliphatic esters* (especially acetates) into olefin and carboxylic

the inlet system (e.g. with adsorbed water). They can be reduced considerably, but hardly ever completely suppressed, by employing direct evaporation in the ion source. Such reactions occasionally lead to false inferences in the interpretation of the mass spectrum. Therefore, whenever possible, more stable derivatives should be prepared (see Section 6.1:1.3.4).

The development of ionization techniques which bypass the gas phase will probably push back the limits of application of mass spectrometry set by insufficient volatility and thermal gas phase reactions. A first and important step in this direction is the field desorption technique (see Section 6.2:2). Of course, consideration must also be given to adverse effects due to ion-molecule reactions in the condensed phase.

It is fundamentally impossible to obtain mass spectra from ionic compounds (including betaines), except if specific thermal fissions result in analyzable compounds. For example, the hydrohalides of *amino compounds* are easily decomposed by heat to give the mass spectrum of the free base (and the hydrogen halide). Quaternary ammonium salts also break down into well-defined

---

[3] *H. Budzikiewicz,* Z. Anal. Chem. *244,* 1 (1969).

products[4,5]. In the case of polymers, certain deductions can be made concerning their structure by investigating the products of pyrolysis (see Chapter 6.4).

1.3.2 Purity Requirements

The permissible level of impurities in a sample depends mainly on their relative volatility and the molecular weight.

High molecular weight or non-volatile impurities (e.g. cellulose fibers or chromatographic carrier material residues) remain behind during evaporation; they may, of course, catalyze thermal decomposition reactions. Volatile low molecular weight constituents, particularly adherent solvent residues, similarly have no adverse effect in most cases, as, firstly, they can easily be identified and, secondly, can generally be pumped off before the spectrum is recorded. Solvents of crystallization are, however, highly undesirable because, when using the direct insertion technique, they may produce intermittent evaporation and splashing of the sample. Under these circumstances no constant ion current and, hence, no reproducible spectra can be obtained.

Impurities (of unknown kind) with a molecular weight similar to that of the sample present a considerable problem. In such cases it is often impossible to interpret the mass spectrum, as neither the molecular peak nor the fragments can be identified unambiguously. When evaporating the sample in the ion source, as a result of fractionation, the spectrum of the impurity may possibly become dominant, an effect which may exist even at a concentration of about 0.1%.

Impurities of any type interfere when automatic data acquisition and processing are employed and also when substances are being identified by (automatic) spectrum comparison.

1.3.3 Mixtures of Substances

The qualitative detection of given components in a mixture of substances generally does not present any problems, except where a mixture of isomers is involved. In the case of unknown compounds, the molecular ions can often be identified (see 6.1:4.1), but an allocation of the fragments is hardly feasible, with the exception of homologous compounds. When dealing with low-volatile substances which must be evaporated in the ion source, constituents may be overlooked because of the inevitable fractionation that occurs. On the other hand, sometimes one can utilize this fractionation in order to correlate the molecular ions and the corresponding fragments with the aid of the alteration of the peak intensities (measured at constant intervals of time).

Quantitative determinations are feasible only when reference spectra are available for all the components of the mixture and when the whole sample can be introduced into the mass spectrometer via a reservoir inlet system. They are thus limited to relatively volatile and thermally stable substances. Guidance on the evaluation of the spectra and the sources of error to be borne in mind is provided by Kienitz[46] and Benz[47] (see 6.1:5.1). The procedure requires much outlay and nowadays can usually be replaced by gas chromatography.

1.3.4 Conversion into Suitable Derivatives

The following effects can be achieved by selecting suitable derivatives:

Increase in volatility,
Suppression of thermal and mass spectrometric decomposition reactions as an aid to the definite identification of the molecular ion,
Differentiation of the constituents of a mixture,
Controlled formation of fragment ions (see Section 6.1:4.4).

The first two aims are coupled to each other, as a rule. Generally, it is desirable to reduce the polarity of the molecule and to suppress elimination reactions by blocking off the functional groups. Some typical examples follow (for details see Chapter 4).

Acetylation, methylation and, above all, the preparation of trimethylsilyl ethers have proved effective in raising the volatility of polyhydroxy compounds. In the case of trimethylsilyl ethers, a considerable increase of the molecular weight (72 mass units per hydroxyl group) must be accepted; the outstanding volatility and stability (except for sensitivity to hydrolysis), however, ensure a ready determination of molecular ions of up to about 2000 mass units. For molecular weight determinations of sugars, also acetonids[6] and diethyl dithioacetals[7] have been used. The dehydration which often predominates with simple aliphatic alcohols can be eliminated by oxidizing to the ketone or carboxylic acid. 2,4-dinitrophenylhydrazones have been proposed[8] for molecular weight determinations of unstable *ketones*. With *carboxylic*

[4] *M. Hesse, W. Vetter, H. Schmid*, Helv. Chim. Acta *48*, 674 (1965).

[5] *M. Hesse, H. Schmid*, Liebigs Ann. Chem. *696*, 85 (1966).

[6] *D.C. DeJongh, K. Biemann*, J. Am. Chem. Soc. *86*, 67 (1964);
    V. Kovačik, S. Bauer, P. Šipoš, Collect. Czech. Chem. Commun. *34*, 2409 (1969).

[7] *D.C. DeJongh, S. Hanessian*, J. Am. Chem. Soc. *88*, 3114 (1966).

[8] *R.J.C. Kleipool, J.T. Heins*, Nature *203*, 1280 (1964);
    C. Djerassi, S.D. Sample, Nature *208*, 1314 (1965).

acids, conversion to the methyl ester has been found to be advantageous. In the case of amino acids, the amino function frequently needs to be protected.

2 Determination of the Elemental Composition

2.1 Isotope Peaks

Most naturally occurring elements consist of an isotope mixture of practically constant composition (Table 1) — apart from very slight variations which depend on the origin of the sample.

The isotope combinations present in the molecules appear in the mass spectrum separated according to their mass numbers (sum of the integral nuclide weights). Usually, for molecular and fragment

Table 1. Natural isotope distribution[9] of the elements more frequently occuring in organic compounds

Element	Relative abundance (%) of the isotopes with relative masses X, X+1 and X+2				
	X		X+1		X+2
Hydrogen	^{1}H	99.985	^{2}H(D)	0.015	
Boron	^{10}B	19.61	^{11}B	80.39	
Carbon	^{12}C	98.893	^{13}C	1.107	
Nitrogen	^{14}N	99.634	^{15}N	0.366	
Oxygen	^{16}O	99.759	^{17}O	0.037	^{18}O 0.204
Fluorine	^{19}F	100			
Silicon	^{28}Si	92.21	^{29}Si	4.70	^{30}Si 3.09
Phosphorus	^{31}P	100			
Sulfur	^{32}S	95.0	^{33}S	0.76	^{34}S 4.22
Chlorine	^{35}Cl	75.53			^{37}Cl 24.47
Bromine	^{79}Br	50.537			^{81}Br 49.463
Iodine	^{127}I	100			

ions, only the mass number of the combination formed from the lightest isotopes (Column X in Table 1) is given. The neighboring mass lines, corresponding to the combinations containing the heavier isotopes, are referred to as 'isotope peaks'. There is, therefore, a difference between the mass spectrometrically determined molecular weight and the chemical molecular weight, since the latter includes all the isotope combinations present (in the case of CH_2Cl_2, for example, the mass-spectrometric molecular weight is 84, while the chemical molecular weight is 84.94).

To a certain extent, deductions on the elemental composition can be made from the relative intensities of the isotope peaks.

[9] Nuclear Data Tables, US Atomic Energy Commission, National Research Council, Washington 1959.

2.1.1 Analysis of the Isotope Peaks of Compounds Containing CHNO

The molecular peaks (M) of organic compounds containing only nitrogen and oxygen as hetero elements are accompanied by 'isotope peaks' (M + 1, M + 2, ...) of rapidly decreasing intensity at intervals of one mass unit (see Fig. 1: m/e 418, 419, 420). Their relative intensities, with reference to M, are closely related to the elemental composition. Nevertheless, they cannot be used directly to determine an empirical formula because the accuracy of measurement is insufficient for this.

However, the number of carbon atoms can be estimated with the aid of the M+1 peak. Neglecting the relatively small contribution by the nuclides D, ^{15}N and ^{17}O, the relative intensity of this peak approximately follows Equation 1, which by transformation gives Equation 2.

$$\frac{I_{M+1}}{I_M} \approx n_c \cdot \frac{1.107}{98.893} \qquad (1)$$

$$n_c \approx \frac{I_{M+1} \cdot 100}{I_M \cdot 1.12} \qquad (2)$$

I_M, I_{M+1} = Intensities of the peaks M and M + 1,
n_c = Number of carbon atoms.

The number of carbon atoms is obtained from Equation 2 to an accuracy of at least $\pm$ 2. As a rule, the values are rather on the high side, so they usually give the upper limit. The error increases at high nitrogen content, because the ^{15}N fraction of almost 0.4% then results in appreciable interference. Naturally, peaks of different origin (e.g. isotope peaks of M–H fragments) superimposed on the group of peaks concerned also lead to erroneous results. For this reason, Equation 2 can be applied to fragment peaks only when these appear completely isolated.

2.1.2 Analysis of Isotope Peaks in the Presence of Additional Hetero Elements

Monoisotopic elements like *fluorine, phosphorus* and *iodine* cannot be determined directly due to the lack of isotope peaks. Iodine, however, because of its high atomic weight (127), can be identified because compounds containing it and possessing high mass numbers exhibit abnormally small isotope peaks by comparison to CHNO compounds of similar molecular weight.

Sulfur and *silicon* both possess an isotope of relatively high abundance, that is two mass units heavier than the main isotope (see Table 1). Therefore, they can be recognized by the enhanced intensity of the M + 2 peak. The M + 1 peak enables the two elements to be distinguished as it is appreciably more intense in the presence of silicon (Fig. 2). The following equations have been successfully used to calculate the number of sulfur or silicon atoms:

$$n_s \approx \left(\frac{I_{M+2} \cdot 100}{I_M} - 1\right) : 4.4 \qquad (3)$$

$$n_{si} \approx \left(\frac{I_{M+2} \cdot 100}{I_M} - 1\right) : 3.4 \qquad (4)$$

I_{M+2} = Intensity of the M + 2 peak,
n_s, n_{si} = Number of sulfur and silicon atoms, respectively.

The accuracy of the values determined in this way is ± 1. With fragment peaks, the same limitations hold as in the determination of the number of carbon atoms.

To determine the number of carbon atoms from Equation 2 in the presence of sulfur, silicon or boron, it is necessary to subtract their contributions to the M+1 peak:

$$n_c \approx \left(\frac{I_{M+1} \cdot 100}{I_M} - 0.8 \cdot n_s - 5.1 \cdot n_{si} - 409.9 \cdot n_b\right) : 1.12 \qquad (6)$$

At a high proportion of boron ($n_b > 2$), this method is admittedly useless.

Many other elements, particularly *germanium, selenium, tellurium* and most *metals,* the majority of which consist of several isotopes with characteristic abundance distribution, can be identified without difficulty from their typical isotope peaks. Of course, satisfactory identification is possible only for one atom at a time. If several atoms or combinations of these elements with each other or with

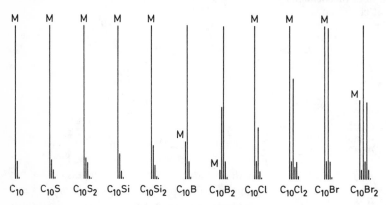

Fig. 2. Relative intensities of the molecular peaks (M) and the isotope peaks in a C_{10} compound containing various hetero elements.

Chlorine, bromine, and *boron* each consist of two isotopes whose abundances are of the same order of magnitude (Table 1). These elements, therefore, produce such characteristic peak groups (Fig. 2) that they can be positively identified by inspection up to relatively high atomic numbers. A quantitative determination is most simply carried out by comparison with the bar graphs published for numerous combinations of these elements (e.g. Beynon[34] and Kienitz[46]).

For *n* atoms of chlorine, bromine, or boron, the ratio of the isotope peak intensities can be calculated with the aid of the binomial theorem (Eq. 5), *a* being the relative abundance of the light isotope and *b* of the heavy isotope. It is sufficient to employ the following rounded-off values: $a=1$, $b=4$ for boron; $a=3$, $b=1$ for chlorine; $a=b=1$ for bromine.

$$(a+b)^n = a^n + n \cdot a^{n-1} \cdot b + \binom{n}{2} \cdot a^{n-2} \cdot b^2 + \ldots + b^n \qquad (5)$$

The individual terms on the right-hand side of Equation 5 correspond to the intensities of M, M + 1, M + 2 ... for boron and M, M + 2, M + 4 ... for chlorine or bromine.

other hetero elements are present, then the intensity distribution of the peaks must be calculated for a series of possible empirical formulas and compared with the measured values (attainable agreement is better than 5% of the peak height). A suitable procedure involving a matrix method of stepwise computation has been described by Riedel[10]. All elements displaying a typical isotope distribution can be assigned quite well even when smaller peaks of different origin are superimposed, i.e. also in the case of many fragment peaks.

2.2 High-resolution Mass Spectrometry

As a result of the different deviations of the nuclide masses from integral values (based on $^{12}C = 12.0$, for example, $^1H = 1.007825$, $^{14}N = 14.003074$ and $^{16}O = 15.994915$), ions of the same nominal mass but of different elemental composition have slightly different masses. The derivation of empirical formulas can, therefore, be carried out alternately by means of precise mass determinations, the prerequisite of which is a high resolving power of the mass spectrometer.

Table 2. Determination of the elemental composition of a CHNO compound by high-resolution mass spectrometry (for explanations see the text)

| Mass ($M_{exp.}$) | Found | | | Calculated | | | |
	$\dfrac{I_{M+1} \cdot 100}{I_M}$	$\dfrac{I_{M+2} \cdot 100}{I_M}$	Empirical formula	$M_{exp.} - M_{calc.}$	$\dfrac{I_{M+1} \cdot 100^a}{I_M}$	$\dfrac{I_{M+2} \cdot 100^a}{I_M}$
161.0477 ± 0.0008			$C_9H_7NO_2$	< 0.0001		
201.1030 ± 0.0010			$C_{12}H_{13}N_2O$	+ 0.0002		
582.2834 ± 0.0029	41.5 ± 2	8.2 ± 2	$C_{33}H_{42}O_9$	+ 0.0005	37.90	8.81
			$C_{36}H_{40}NO_6$	− 0.0022	41.48 X	9.60
			$C_{21}H_{46}N_2O_{16}$	− 0.0013	25.52	6.39
			$C_{31}H_{40}N_3O_8$	+ 0.0019	36.70	8.17
			$C_{34}H_{38}N_4O_5$	− 0.0008	40.28 X	8.92
			$C_{19}H_{44}N_5O_{15}$	< 0.0001	24.32	5.90
			$C_{22}H_{42}N_6O_{12}$	− 0.0027	27.90	6.20
			$C_{32}H_{36}N_7O_4$	+ 0.0005	39.08	8.25
			$C_{35}H_{34}N_8O$	− 0.0022	42.66 X	9.08
			$C_{17}H_{42}N_8O_{14}$	+ 0.0014	23.11	5.42
			$C_{20}H_{40}N_9O_{11}$	− 0.0013	26.70	5.68
			$C_{30}H_{34}N_{10}O_3$	+ 0.0019	37.87	7.59

[a] Calculated from the natural isotope abundances (Table 1) using the method of Beynon[34] (6.1:5.1)

When two neighboring peaks of equal height at masses M and $M + \Delta M$ are sufficiently distinct for the 'valley' between them to amount to 10% of the peak height, then the resolution of the instrument is $M/\Delta M$; the reciprocal $(\Delta M/M) \times 10^6$ [ppm] gives a measure of the peak width. To record normal mass spectra, $M/\Delta M$ is set to 2000 at most, whereas high-resolution work requires more than 10,000 (or < 100 ppm). Although this value can be obtained with modern single-focusing mass spectrometers, for precise mass determinations it is preferable to use double-focusing instruments specially developed for this purpose, which provide an appreciably higher resolving power. Using suitable reference compounds (e.g. perfluorokerosene), the masses of the ions can be determined to an accuracy of a few ppm. Conversion tables[11] or computers must then be employed to assign the empirical formulas.

The reliability with which the elemental composition can be determinal decreases as the number of hetero atoms and the mass number increase. Table 2 illustrates this for a *peptide alkaloid*[12] compound containing nitrogen and oxygen by means of two fragments (m/e 161 and 201) and the molecular ion (m/e 582) whose masses had been determined with a maximum error of ± 5 ppm.

Each of the two fragments (i.e. up to a mass number of about 200) fits only one possible elemental composition. However, at m/e 582 the absolute uncertainty of measurement and also the possible number of hetero atoms are considerably greater. No less than twelve empirical formulas are possible if a maximum of ten nitrogen atoms and twenty oxygen atoms is admitted. If, as in this case, the molecular ion is definitely concerned, a decision can be made with the aid of the laws of combination, the isotope peaks present, and the elemental composition of fragment ions. As an even molecular weight implies an even number of nitrogen atoms, five of the twelve formulas for m/e 582 can be excluded. A comparison of the relative intensities of the M + 1 and M + 2 peaks leaves only three possibilities (those marked by a cross in the last but one column in Table 2). Of these three, the first has only one nitrogen atom, the third only one oxygen atom, so both can be eliminated by considering the elemental composition of the fragment ions (m/e 201 contains two nitrogen atoms and m/e 161 two oxygen atoms). The formula that remains is $C_{34}H_{38}N_4O_5$.

This procedure does not always lead to an unambiguous result and in some cases must be supplemented by the analytical determination of at least one element.

When other hetero elements are present, in addition to nitrogen and oxygen, a prior selection following the rules laid down in the previous section is suitable.

[10] O. Riedel, Z. Anal. Chem. *271*, 1 (1966).
[11] J.H. Beynon, A.E. Williams, Mass and Abundance Tables for Use in Mass Spectrometry, Elsevier, Publ. Co., Amsterdam 1963;
J. Lederberg, Computation of Molecular Formulas for Mass Spectrometry, Holden-Day, San Francisco 1964;
D. Henneberg, K. Casper, Z. Anal. Chem. *227*, 241 (1967).
[12] R. Tschesche, L. Behrendt, H.-W. Fehlhaber, Chem. Ber. *102*, 50 (1969).

Table 3. Selection of some mass doublets

Difference in the elemental composition	Mass differences		
	ΔM (Mass units)	$(\Delta M/M) \cdot 10^6$ [ppm] at the nominal mass M	
		M = 200	M = 500
CH_4—O	0.0364	182	73
NH_2—O	0.0238	119	48
O_2—S	0.0178	89	36
CH_2—N	0.0126	63	25
N_2—CO	0.0112	56	22
C_7H_2—N_5O	0.0054	27	11
CH—^{13}C	0.0045	22	9
C_5—N_2O_2	0.0040	20	8
SH_4—C_3	0.0034	17	7
C_3N—H_2O_3	0.0027	12	5
C_2H_2O—N_3	0.0013	7	3
C_3—HFO	0.0011	6	2
CH_5O_5—N_7	0.000005	0.03	0.01

When ions of different elemental composition occur at the same mass number (e.g. with mixtures of substances or as a result of fragment overlapping), then these multiplets appear separated, provided the resolution exceeds the relative mass differences. With a resolving power of 50 ppm $(M/\Delta M = 20,000)$ which is applicable on a routine basis, from the doublets given in Table 3 only those can be resolved that lie within range *a*. Even at smaller mass differences, down to approximately 20 ppm (range *b*), doublets can be identified from their peak shape, as long as the two components exhibit similar intensities, and mass measurements can still be made. Doublets corresponding to mass differences of less than 15 ppm (range *c*) can no longer be identified, and mass measurements of such peaks may cause misinterpretations.

They include (above m/e 300) the relatively often occuring CH/^{13}C doublet which arises from overlapping by the ^{13}C satellite of a fragment lighter by one hydrogen atom. This may possibly lead to considerable difficulties in determining the empirical formula of compounds yielding intense M-1 peaks, e.g. acetals, amines and alkylsubstituted aromatic compounds (due to cleavages according to Equations 12, 15, or 16, see Section 6.1:3.2).

There are, however, already mass spectrometers with a maximum resolving power of 10 ppm, and in a few years instruments will probably be available that permit a resolving power of 5 ppm. Then mass measurements with most of the multiplets will be possible, even at high mass numbers.

3 Electron Impact Induced Decomposition Reactions

3.1 General

The *mass spectrometric decomposition reactions* are endothermic monomolecular, and kinetically controlled. Fragmentations taking place within about 10^{-6} seconds are detected, thus including, to a certain extent, also multi-stage breakdown processes.

Most structure-specific fragmentation reactions can be explained satisfactorily only by assuming that the positive charge in the molecular ion is preferably localized on hetero-atoms or π-electron systems, i.e. at the site of lowest ionization potential. This implies that, after ionization by 70 eV electrons which can remove an electron from any part of the molecule, a step-by-step decrease of energy of the electronic system (by transformation into vibrational energy) must take place prior to any fragmentation.

The probability of a fission occurring is governed by the relative binding forces, steric factors (formation of cyclic transition states) and, above all, the stability of the products formed. This is due to the fact that in an endothermic reaction the structure of the transition state closely resembles that of the end products; the activation enthalpy is approximately proportional to the reaction enthalpy (for a simple bond rupture both quantities are in any case practically equal). The virtual energy of the cation formed, which depends mainly on the stabilization of the positive charge, is generally of greater importance for the occurrence of a fragmentation than the stability of the ejected neutral particle.

In addition to the energetically favorable reactions, numerous high-energetic non-specific fragmentations occur with most compounds. They generally result in a rather uncharacteristic 'background spectrum', which, admittedly, is significantly different for aliphatic (maxima at m/e 43, 57, 71 or 55, 69 etc.) as against aromatic compounds (maxima at m/e 39, 50 or 51, 63–65, 74–77, 91).

The following sections contain a selection of the most important and adequately established fragmentation mechanisms. The symbols employed are: ($\oplus$) = radical ion, ($\frown$) = two-electron shift, ($\frown$) = one-electron shift.

3.2 Simple Bond Fissions

The basic reaction of aliphatic compounds consists in the cleavage of a C—C single bond:

$$\left[-\overset{|}{\underset{|}{C}}-\overset{|}{\underset{|}{C}}- \right]^{\oplus} \cdot \longrightarrow -\overset{|}{\underset{|}{C}}{}^{\oplus} + \cdot\overset{|}{\underset{|}{C}}- \qquad (7)$$

In *n-alkanes* all C—C bonds, except the terminal ones, are cleaved with equal probability. However, as the relative abundance of the low-mass fragments increases due to secondary fragmentations, the C_nH_{2n+1}, peaks appear in the mass spectrum (at intervals of 14 mass units) with intensities that increase progressively with decreasing mass number (with a maximum generally at m/e $57 = C_4H_9^{\oplus}$). Branched hydrocarbons are cleaved preferentially at the branching sites, since the stability of the carbonium ions formed increases with the degree of substitution; the corresponding peaks, therefore, display a significantly enhanced intensity.

Functional groups are ejected as radicals if these possess a certain degree of stability:

$$\left[-\overset{|}{\underset{|}{C}}-X \right]^{\oplus} \cdot \longrightarrow -\overset{|}{\underset{|}{C}}{}^{\oplus} + X^{\cdot} \qquad (8)$$

This reaction is found predominantly with *alkyl halides* (except for fluorides) and *nitro compounds*. If the carbon atom is tertiary, other substituents (OH, carbonyl derivatives) can be eliminated in this way, too. Corresponding fragments also occur with *dialkyl ethers* (X = OR), but here they are evidently not formed in a single-stage reaction. Carbonium ions formed in accordance with Equations (7) or (8) can undergo further decomposition through olefin elimination:

$$-\overset{|}{\underset{|}{C}}-\overset{|}{\underset{|}{C}}-\overset{|}{\underset{|}{C}}{}^{\oplus} \longrightarrow -\overset{|}{\underset{|}{C}}{}^{\oplus} + \overset{}{\underset{}{>}}C=C\overset{}{\underset{}{<}} \qquad (9)$$

Alkenes yield resonance-stabilized allylic cations as a result of *allylic fission:*

$$\left[>C=\overset{|}{\underset{|}{C}}-\overset{|}{\underset{|}{C}}-\overset{|}{\underset{|}{C}}- \right]^{\oplus} \cdot \longrightarrow$$

$$(10)$$

$$\left\{ >C=\overset{|}{\underset{|}{C}}-\overset{|}{\underset{|}{C}}{}^{\oplus} \longleftrightarrow {}^{\oplus}\overset{|}{\underset{|}{C}}-\overset{|}{\underset{|}{C}}=C< \right\} + \cdot\overset{|}{\underset{|}{C}}-$$

The diagnostic value of this fragmentation is admittedly small, as it is frequently preceded by double bond isomerizations. Cyclohexene deriva-

tives undergo double allylic fission, which is synonymous with the *retro-Diels-Alder reaction* (RDA)[13]:

$$\left[\text{(cyclohexene cation)} \right]^{\oplus} \cdot \longrightarrow \left[\text{(butadiene)} \right]^{\oplus} \cdot + \text{(ethylene)}$$

$$\text{or} \quad \text{(diene)} + \left[\text{(olefin)} \right]^{\oplus} \qquad (11)$$

Whether the positive charge remains preferentially in the diene or the olefin fragment depends on the relative stabilities of the ions (e.g. the degree of substitution). The RDA reaction can also be induced by double bonds belonging to a fused-on benzene ring (tetralin structure) or by those resulting from a preceding decomposition reaction (e.g. in accordance with Equation 18). The driving force of the reaction is not very great and the RDA is, therefore, frequently suppressed in the presence of other functionalities. Certain unsaturated pentacyclic triterpenes (especially Δ^{12} enes) present a notable exception in that the RDA reaction does have a high diagnostic value[14].

Hetero atoms which are able to stabilize a neighboring carbonium ion by means of their nonbonding electrons induce the α-cleavage:

$$R-\overset{|}{\underset{|}{C}}-\underset{\cdot}{\overset{\cdot\oplus}{X}}| \longrightarrow R^{\cdot} + \left\{ >C=\underline{\overline{X}}{}^{\oplus} \longleftrightarrow {}^{\oplus}\overset{|}{\underset{|}{C}}-\underline{\overline{X}}| \right\}$$

$$(12)$$

This is one of the most important primary reactions. It occurs in *alkyl halides, alcohols, mercaptans,* (thio)*ethers* and *amines* with greatly increasing reactivity in that order, in addition to being graded primary < secondary < tertiary for each type of compound. In the case of α-substituted cyclic ethers and amines, as well as *lactones* and *lactams,* characteristic fragment ions arise through the preferential cleavage of the exocyclic carbon-carbon bonds:

$$\overset{R^1}{\underset{R^2}{\diagup}}\overset{Z}{\underset{\overset{\oplus}{\underline{X}}}{\diagdown}}Y \longrightarrow \overset{R^1}{\diagup}\overset{Z}{\underset{\overset{\oplus}{\underline{X}}}{\diagdown}}Y + R^2 \qquad (12a)$$

$$X = O,\ NR^3\ ;\ Y = H_2,\ O$$

[13] *H. Budzikiewicz, J.I. Braumann, C. Djerassi,* Tetrahedron *21*, 1855 (1965).
[14] *H. Budzikiewicz, J.M. Wilson, C. Djerassi,* J. Am. Chem. Soc. *85*, 3688 (1963).

The α-cleavage is particularly favored in *acetals, ketals* and other geminally disubstituted compounds due to the outstanding resonance stabilization of the carbonium ion formed:

Carbonyl compounds such as *aldehydes, ketones, esters, amides* and other *carboxylic acid derivatives* also undergo the α-cleavage:

The resulting acyl ions liberate carbon monoxide in a characteristic secondary reaction:

The driving force for the *benzylic fission* of aromatic compounds is the resonance stabilization of the fragment ion which is considerably enhanced by electron-donating substituents (OR, NR₂) in the ortho- or para-position:

With respect to the group Y, there are practically no limitations. Analogous reactions take place with appropriately *substituted heterocyclic compounds*. The experimental findings for numerous benzyl derivatives, especially alkylbenzenes and benzyl halides — also in the presence of nuclear substituents *without* a marked mesomeric effect — indicate that the fragmentation occurs via a ring expansion leading to the formation of tropylium ions:

As a result of this reaction all the carbon atoms of the ring system become equivalent and positional isomers lose their identity. Corresponding skeletal rearrangements are assumed to take place with many *heterocycles*, especially five-membered ring compounds:

$$X = O, S, NR^2$$

Benzoyl derivatives behave in the same way as the aliphatic carbonyl compounds (Eq. 13 and 14). Aromatic halogeno, nitroso and nitro compounds tend to lose the substituent as a radical (by analogy with Eq. 8). This reaction is promoted by certain side chains in the ortho position. Since, in these cases, methyl and methoxy groups and even hydrogen radicals are removed from the aromatic nucleus, the reaction is assumed to follow a cyclization mechanism. A typical example of this effect is provided by cinnamoyl compounds, the decomposition of which leads to benzopyrylium ions:

The following general rule applies to the fragmentation reactions formulated in Equations 7, 12, 13 and 15: where splitting off of different alkyl groups can occur under energetically equivalent conditions, the relative intensities are greatest for those peaks that correspond to the loss of the largest group.

3.3 Fissions Accompanied by Hydrogen Rearrangement

A large number of mass spectrometric decomposition reactions are associated with a more or less specific *hydrogen transfer reaction*. They proceed preferably via six-membered cyclic transition states, but occasionally occur with five and seven-membered and, in exceptional cases, with eight or higher membered transition states.

Alcohols and *alkyl halides* easily undergo elimination of water or hydrogen halide. In contrast to the thermal and ionic mechanisms, the mass spectrometric reaction proceeds mainly through 1,3- or 1,4-elimination:

$$(CH_2)_n \quad \longrightarrow \quad (CH_2)_n \quad \longrightarrow \quad (CH_2)_n + HX \quad (17)$$
$$n = 1, 2$$

The radical ion formed can give rise to various secondary reactions that are generally induced by the radical electron. With certain *ortho*-disubstituted aromatic compounds an interesting variant of the reaction is found that enables other positional isomers to be distinguished:

$$\longrightarrow + HX^{(\cdot)} \quad (17a)$$

$$Z = CH_2, O, NH, NR$$

This mechanism is involved in the elimination of water from *ortho*-substituted benzylic alcohols (Y—X=CH$_2$—OH) and benzoic acids (Y—X=CO—OH), in the elimination of an alcohol molecule from their ethers and esters, respectively, and also in the loss of an OH radical from nitro compounds (Y—X=NO→O).

Apart from the α-cleavage (Eq. 12—13), the most important fragmentation reaction is the *McLafferty rearrangement* named after its discoverer. It can occur in practically all unsaturated systems where a hydrogen atom is available in the γ-position:

$$(18)$$

From the extensive experimental material available, the following characteristics can be derived: the steric relationships must allow the hydrogen atom to approach to within 1.8 Å of the acceptor atom; the acceptor X must carry the positive charge or a radical electron; the charge generally remains on the functional group and only in exceptional cases in the olefin fragment; if the requirements for several competing rearrangements are met, the reaction preferentially involves the hydrogen at the higher substituted carbon atom.

The McLafferty rearrangement is observed as a typical decomposition reaction with all *carbonyl compounds* (X=O, S), with *oximes, hydrazones* and *semicarbazones* (X=NR′) and with tri- and *tetra-substituted ethylenes*[15] (X=CR′$_2$). If the group Y also has hydrogen available in the γ-position, a second McLafferty rearrangement is subsequently possible. Corresponding reactions occur with *alkynes*, (iso)*nitriles, isothiocyanates* and *epoxy compounds* (in which $C\overset{O}{\diagup}\diagdown C$ replaces C=X). They also occur as a secondary fragmentation of *amino compounds* after the α-cleavage (X=$^\oplus$NR′$_2$), but here they are generally of minor importance. In the case of *alkylbenzenes* and *alkyl-substituted heterocyclic compounds* (Y—C=X being part of the aromatic system), the McLafferty rearrangement competes with the benzylic fission (Equation 15), and is promoted by electron donating substituents in the *m*-position[16].

Alkyl aryl ethers also undergo olefin elimination. In contrast to the McLafferty rearrangement, the hydrogen transfer is fairly unspecific here and probably moves to the oxygen atom with formation of a phenol ion. The same is true of *phenol* and *enol acetates* (and other esters) and of anilides which are cleaved by ketene elimination.

Allied to the McLafferty rearrangement is the so-called γ-cleavage, in which a double ('reciprocal') hydrogen transfer takes place:

$$(19)$$
$$n \geqslant 1$$

So far, this reaction has been proven in *fatty acid, methyl esters*[17], *ketones*[18], *aldehydes*[19], *oximes*[20]

[15] *M. Kraft, G. Spiteller*, Org. Mass Spectrom. *2*, 865 (1969).

[16] *M.I. Gorfinkel, L.Y. Nanovskaia, V.A. Koptyug*, Org. Mass Spectrom. *2*, 273 (1969).
H. Nakata, A. Tatematsu, Tetrahedron Lett. 4303 (1969).

[17] *G. Spiteller, M. Spiteller-Friedmann, R. Houriet*, Monatsh. Chem. *97*, 121 (1966).

[18] *M. Kraft, G. Spiteller*, Liebigs Ann. Chem. *712*, 28 (1968);
G. Eadon, C. Djerassi, J. Am. Chem. Soc. *91*, 2724 (1969).

[19] *R.J. Liedtke, C. Djerassi*, J. Am. Chem. Soc. *91*, 6814 (1969).

and *2-alkylquinolines*[21], but apparently it is generally feasible in compounds containing a polarized double bond.

In the case of *substituted cycloalkanes,* the α-cleavage (in accordance with Eq. 12 or 13) leads only to ring opening but not yet to the formation of fragments. The latter only takes place in a second reaction step associated with a hydrogen rearrangement:

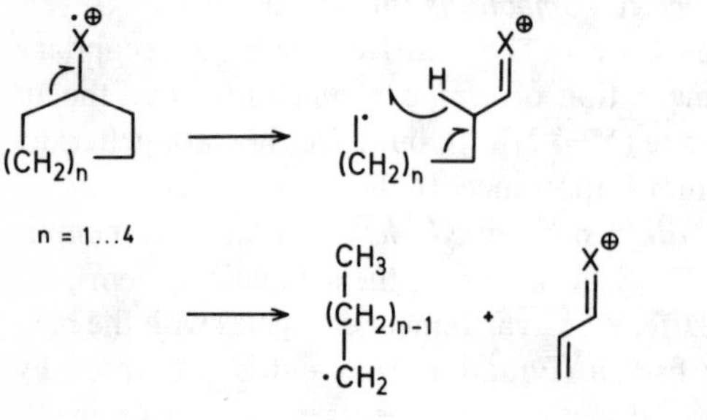

$$n = 1...4$$

$$(20)$$

This reaction sequence results in highly characteristic fragment ions in *steroid ethylene ketals* and *amines,* where several analogous steps may occur consecutively, as is demonstrated by the following example[22]:

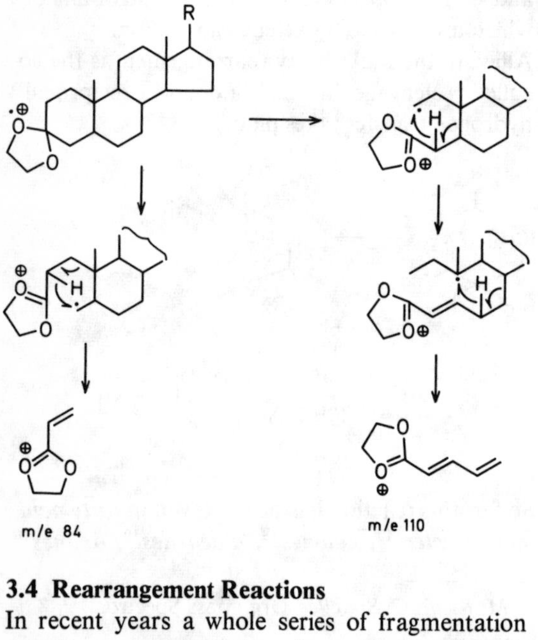

m/e 84   m/e 110

3.4 Rearrangement Reactions

In recent years a whole series of fragmentation reactions accompanied by the intramolecular migration of an alkyl or aryl moiety or by the trans-

fer of heteroatoms has been discovered[23]. But the majority of these rearrangements are connected to special structural prerequisites, and it is, therefore, difficult to derive generally applicable rules. These reactions can naturally cause severe problems during structure analyses. The relative abundance of the rearrangement fragments is, however, generally fairly small, except in pronounced escape reactions' like those appearing in highly unsaturated systems where simple bond ruptures require a very high energy of activation. Only in few cases the reaction mechanism has been fully established, but it appears that three- and four-membered cyclic transition states predominate.

Driving force of most rearrangements is the elimination of a stable neutral particle, e.g. in the following reactions:

CO expulsion from *quinones* (twofold), *pyrones* and related compounds, from *phenols* and *diaryl ethers* and from *aryl sulfones* and *aryl sulfoxides;*
CO_2 elimination in *fumarates* and *maleates,* in *α-cyano* and *acetylene carboxylic acid* esters and in *diaryl* and *alkyl aryl carbonates;*
SO and SO_2 ejection from *diaryl sulfones,* diaryl *sulfoxides* and *sulfonamides*
NO loss from *aromatic nitro compounds.*

Although 1,2-shifts of alkyl or aryl groups have been repeatedly formulated to explain unexpected fragmentation reactions, they have been proven in a few cases only, e.g. *substituted cyclohexenones* and *oxiranes*[23]:

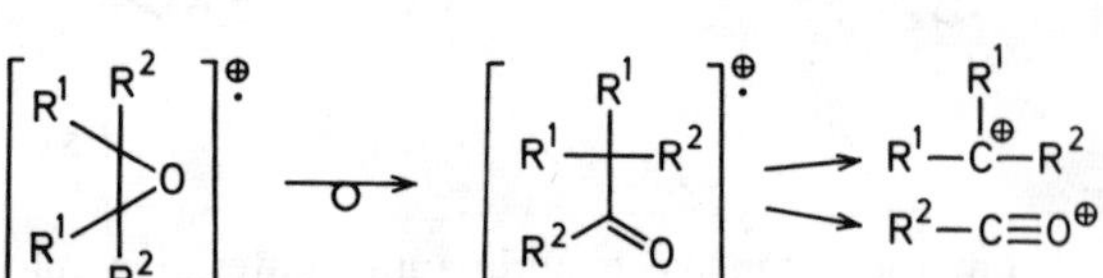

$$R^1 = CH_3, C_6H_5$$
$$R^2, R^3 = (Cyclo) Alkyl$$

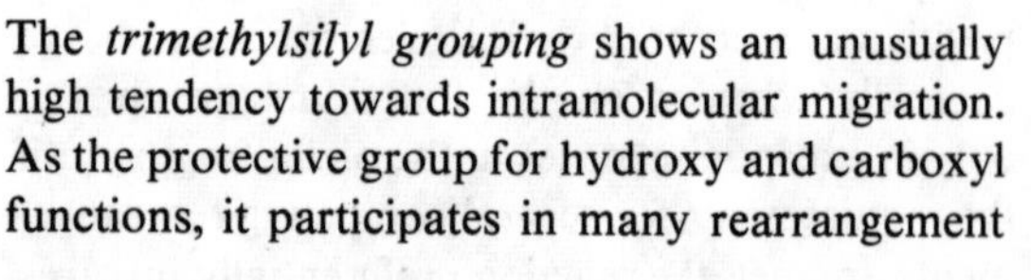

$$R^1 = CH_3, C_2H_5, C_6H_5$$
$$R^2 = H, CH_3, C_6H_5$$

The *trimethylsilyl grouping* shows an unusually high tendency towards intramolecular migration. As the protective group for hydroxy and carboxyl functions, it participates in many rearrangement

[20] *M. Kraft, G. Spiteller,* Org. Mass Spectrom. *2,* 541 (1969).
[21] *H.-W. Fehlhaber, W. Ochterbeck, W. Werner,* unpubl.
[22] *H. Audier, A. Diara, M. de J. Durazo, M. Fétizon, P. Foy, W. Vetter,* Bull. Soc. Chim. Fr. 2827 (1963); *G. von Mutzenbecher, Z. Pelah, D.H. Williams, H. Budzikiewicz, C. Djerassi,* Steroids 2, 475 (1963).
[23] *P. Brown, C. Djerassi,* Angew. Chem. *79,* 48 (1967); *R.G. Cooks,* Org. Mass Spectrom. *2,* 481 (1969).

reactions, even involving very large-sized cyclic transition states[24]. Intramolecular oxygen transfers have been observed in *N-methyl* and *N-phenyl acetanilides*[25], α-substituted *N-methyl benzylamides*[26] and in *o-nitro-* and *2,4-dinitrophenylhydrazones* of aromatic aldehydes and ketones[27]. *Aliphatic polyols* and their derivatives often yield fragments that are attributable to the 1,3-displacement of a hydroxy and methoxy or acetoxy group, respectively[28].

4 Interpretation of Mass Spectra for Structure Determination

4.1 The Molecular Ion Peak
The following criteria are suitable for checking whether the peak at the highest mass number (excluding the isotope peaks) corresponds to the molecular ion ($M^{\oplus}$). (i) The *molecular weight* of a compound containing the elements usual in organic chemistry (Table 1) can be odd only if an odd number of nitrogen atoms is present. (ii) The mass differences to the next peaks are restricted to certain values resulting from the elimination of defined neutral particles or radicals (e.g. in Fig. 1: m/e $418 = M^{\oplus}$, m/e $400 = M - 18 = M - H_2O$, m/e $385 = M - 33 = M - H_2O - CH_3$, m/e $382 = M - 36 = M - 2H_2O$); differences of 5–14 and 21–25 mass units between the molecular ion peak and a fragment peak definitely cannot arise. (iii) If the electron energy is reduced to about 10–12 eV, the relative intensity of the molecular ion peak is significantly increased.

The *abundance of the molecular ion peak* depends on the charge stabilization in the molecular ion and on the sensitivity to fragmentation. To a rough approximation, the abundance decreases in the following order, in accordance with the predominant type of structure: aromatic compounds, cycloalkanes, olefins, n-alkanes, amines, ketones, esters, ethers, alcohols, i-alkanes, carboxylic acids and their anhydrides, acetals, ketals, polyhydroxy compounds.

For the latter members of this series the molecular peak may be entirely absent and the molecular weight can be deduced only in favorable circumstances by a systematic analysis of the fragments. In these cases, however, a reliable determination of molecular weight can usually be obtained by means of field-ionization or electron-attachment mass spectrometry (see Chapter 6.2). If the molecular ion is absent as a result of a thermal decomposition reaction, then, as a rule, the mass spectrum alone does not provide any information about the original molecular weight.

With compounds containing polar groups, especially with alcohols, amines and caboxylic acids (occasionally even with esters), low-intensity M+1 peaks appear more often, as a result of hydrogen transfer from a neutral molecule. These ions are sometimes more stable (more abundant) than the molecular ion and can, therefore, lead to misinterpretations. As the reaction involved is bimolecular, the abundance is proportional to the square of the sample pressure in the ion source (the relationship is linear for all the other peaks). The reaction can thus be detected by varying the sample pressure or, alternately, by the introduction of an additional hydrogen donor (water, alcohol). An analogous transfer of larger groups has been observed in exceptional cases only.

4.2 Characteristic Fragment Peaks
The peaks arising from structure-specific fragmentation reactions are generally distinguished by a conspicuous mass number and/or high intensity by comparison with the fragments resulting from high-energy fissions appearing in the same mass range. The latter are almost completely suppressed at low ionization energies (< 15 eV) and, as most of the secondary reaction also diminish, this procedure allows a better identification of the important primary fragments. The most intense peaks appearing under these conditions are those which originate from reactions associated with low activation energies, i.e. reactions in which no net bond dissociation occurs, as e.g., in the retro-Diels-Alder reaction (Eq. 11) and the McLafferty rearrangement (Eq. 18).

Assigning the 'characteristic' peaks to specific fragmentation reactions and converting the data into structural information requires a great deal of experience and, naturally, cannot be laid down in rules. The interpretation of a mass spectrum always requires two aspects to be considered: (i) For peaks in the upper mass range the *mass difference* referred to the molecular ion peak is of

[24] *G.H. Draffarr, R.N. Stillwell, J.A. McCloskey,* Org. Mass Spectrom. *1,* 669 (1968);
J. Diekmann, J.B. Thomson, C. Djerassi, J. Org. Chem. *32,* 3904 (1967); *34,* 3147 (1969).

[25] *R.A.W. Johnstone, D.W. Payling, A. Prox,* Chem. Comm. 826 (1967);
H.-W. Fehlhaber, P. Welzel, Org. Mass Spectrom. *4,* 545 (1970).

[26] *A. Prox, J. Schmid,* Org. Mass Spectrom. *2,* 121 (1969).

[27] *J. Seibl, J. Völlmin,* Org. Mass Spectrom. *1,* 713 (1968).

[28] *H.F. Grützmacher, J. Winkler,* Org. Mass Spectrom. *1,* 295 (1968).

decisive importance, because the elimination of smaller neutral particles or radicals can provide information about functional groups, side-chains etc. (ii) In the middle and lower mass ranges, however, so-called *key fragments* of specific *mass number* can appear that are typical of complete structural units. A large number of classes of compounds are distinguished by such characteristic fragments, which thus provide primary data for structure elucidation. A selection has been collected in tabular form by Spiteller and Benz (see 6.1:5.1). Exploratory attempts have been made to employ computers for the interpretation of mass spectra[29].

4.3 Metastable Ions

Fragment ions which arise not within the ion source but on the way from the ion source to the magnetic field (due to relatively slow decomposition reactions) are recorded as fairly weak, broadened and diffuse peaks, known as metastable peaks (m*). They generally occur at a non-integral value on the mass scale, in good approximation to Equation 21 (for the sector field instruments normally employed). Two metastable peaks are evident in Fig. 1 slightly below m/e 365 and 371.

$$m^* = \frac{m_2^2}{m_1} \qquad \begin{array}{l} m_1 = \text{mass number of the parent ion,} \\ m_2 = \text{mass number of the daughter ion.} \end{array} \qquad (21)$$

Using a double-focusing mass spectrometer under special recording conditions (defocusing technique[30]), the metastable decompositions can be detected exclusively. In addition to improved sensitivity, this method offers the advantage that the mass numbers of the parent and daughter ions can be determined directly.

Metastable ions are of great importance as they provide evidence for the formation of a fragment from a specific ion, especially when multi-stage decomposition reactions are involved. For example, the two metastable peaks in Fig. 1 show that the fragments with mass numbers 382 and 385 arise from secondary reactions of the m/e 400 fragment (by loss of H_2O and CH_3 respectively), because the calculated values from Equation 21

for these transformations, $385^2/400 = 370.6$ and $382^2/400 = 364.8$, are in good agreement with the observed values. Such secondary decompositions accompanied by metastable ions can often assist in the identification of characteristic fragment ions. A typical example is provided by fragments of mass number 105, which in the presence of a peak at m/e 77 and a metastable peak at m/e 56.5 can easily be identified as benzoyl ions ($C_6H_5CO^{\oplus}$ → $C_6H_5^{\oplus} + CO$ in accordance with Eq. 14).

If a fragment ion is formed in several different pathways, however, the analysis of the metastable peaks can lead to erroneous conclusions because their intensities are not correlated in any way with the intensities of the parent or daughther ions. It may happen, therefore, that a relatively unimportant slow reaction can create metastable ions while the energetically most favored and thus rapidly occurring main reaction sequence does not. The same is accordingly true when fragment ions of different structures occur at the same mass number.

4.4 Influence of Chemical Transformations on Fragment Formation

If the mass spectrum of a compound indicates that specific cleavages occur, then the functional group can be localized in the fragmentation ion by preparing derivatives which do not alter the fragmentation reactions (peak-shift technique). The group introduced must exert the smallest possible directing effect on fragment formation. *Methylation* of hydroxy, carboxyl and amino groups is very often employed. Trifluoroacetic anhydride is preferred for the *acylation* of hydroxy and amino groups, because, in contrast to acetyl derivatives, the *trifluoroacetyl derivatives* cannot undergo a ketene elimination in the mass spectrometer. In the case of oxidation and reduction reactions or the hydrogenation of a C=C double bond, a change in the fragmentation behavior must be anticipated. Active hydrogen atoms are most simply identified by replacement with deuterium, which can be performed with an adequate yield (about 50%) directly in the ion source of the mass spectrometer (by the simultaneous introduction of D_2O). Labeling with deuterium, for example reactions with CD_3I or $LiAlD_4$ etc, in parallel with the analogous reactions with nondeuterated reagents, is naturally of particular advantage.

As far as the deliberate production of fragmentation reactions is concerned, which is often desirable particularly with compounds of complex structure, e.g. many *natural products*, relatively few suitable derivatives are available. First of all,

[29] K. Biemann, C. Cone, B.R. Webster, G.P. Arsenault, J. Am. Chem. Soc. 88, 5598 (1966);
R. Venkataraghavan, F.W. McLafferty, G.E. van Lear, Org. Mass Spectrom. 2, 1 (1969);
A.M. Duffield, A.V. Robertson, C. Djerassi, B.G. Buchanan, G.L. Sutherland, E.A. Feigenbaum, J. Lederberg, J. Am. Chem. Soc. 91, 2977 (1969).
[30] M. Barber, W.A. Wolstenholme, K.R. Jennings, Nature 214, 664 (1967).

acetals, ketals and amino functions must be considered for this purpose, because of their marked tendency to undergo specific α-cleavages (Eq. 12) and, if necessary, well-defined subsequent reactions (Eq. 20). Thus (poly)*cyclic ketones,* which themselves exhibit only a moderate directing effect, and hence yield characteristic fragment ions only in exceptional cases, are advantageously converted into *ethylene ketals* (cf. 6.1:3.3). In order to determine the amino acid sequence in oligopeptides, use has been made of the reduction of all the amide groups to a polyethylenediamine skeleton[31]. To localize isolated $C = C$ double bonds in open-chain compounds a whole range of chemical conversions have been proposed. The most promising of these is an osmium tetroxide oxidation to the glycol followed by the preparation of the isopropylidene derivative[32]. For cyclic olefins, especially unsaturated steroids, an oxidative cleavage to the dicarbonyl compound (most simply by means of RuO_4) has proved successful since the prerequisites are mostly fulfilled for a specific massspectrometric decomposition via the McLafferty rearrangement. (Eq. 18)[33].

[31] *K. Biemann, F. Gapp, J. Seibl,* J. Am. Chem. Soc. *81,* 2274 (1959).

[32] *J.A. McCloskey, M. McClelland,* J. Am. Chem. Soc. *87,* 5090 (1965).

[33] *W.R. Chan, D.R. Taylor, G. Snatzke, H.-W. Fehlhaber,* Chem. Comm. 548 (1967);
R. Tschesche, H.G. Berscheid, H.-W. Fehlhaber, G. Snatzke, Chem. Ber. *100,* 3289 (1967).

5 Bibliography

5.1 Monographs

[34] *J.H. Beynon,* Mass Spectrometry and Its Applications to Organic Chemistry, Elsevier Publ. Co., Amsterdam 1960.

[35] *K. Biemann,* Mass Spectrometry, Organic Chemical Applications, McGraw-Hill Book Co., New York 1962.

[36] *F.W. McLafferty,* Mass Spectrometry of Organic Ions, Academic Press, New York 1963.

[37] *H. Budzikiewicz, C. Djerassi, D.H. Williams,* Interpretation of Mass Spectra of Organic Compounds, Holden-Day, San Francisco 1964.

[38] *H. Budzikiewicz, C. Djerassi, D.H. Williams,* Structure Elucidation of Natural Products by Mass Spectrometry, Vol. I Alkaloids, Vol. II Steroids, Terpenoids, Sugars and Miscellaneous Natural Products, Holden-Day, San Francisco 1964.

[39] *C. Brunnée, H. Voshage,* Massenspektrometrie, Vieweg, Braunschweig 1964.

[40] *R.W. Kiser,* Introduction to Mass Spectrometry and Its Applications, Prentice-Hall, Englewood Cliffs, New Jersey 1965.

[41] *H.C. Hill,* Introduction to Mass Spectrometry, Heyden & Son, London 1966.

[42] *F.W. McLafferty,* Interpretation of Mass Spectra, W.A. Benjamin, New York 1966.

[43] *R.I. Reed,* Applications of Mass Spectrometry, Academic Press, New York 1966.

[44] *G. Spiteller,* Massenspektrometrische Strukturanalyse organischer Verbindungen, Verlag Chemie, Weinheim/Bergstr. 1966.

[45] *H. Budzikiewicz, C. Djerassi, D.H. Williams,* Mass Spectrometry of Organic Compounds, Holden-Day, San Francisco 1967.

[46] *H. Kienitz,* Massenspektrometrie, Verlag Chemie, Weinheim/Bergstr. 1968.

[47] *W. Benz,* Massenspektrometrie organischer Verbindungen, Akademische Verlagsgesellschaft, Frankfurt/M. 1969.

[48] *J. Seibl,* Massenspektrometrie, Akademische Verlagsgesellschaft, Frankfurt/M. 1970.

5.2 Spectra Compilations

[49] Catalogue of Mass Spectral Data, American Petroleum Institute, Research Project 44, Texas A+M University, College Station, Texas (ca. 3000 Spektren).

[50] *E. Stenhagen, S. Abrahamsson, F.W. McLafferty,* Atlas of Mass Spectral Data, John Wiley & Sons, New York 1969; Archives of Mass Spectral Data, John Wiley & Sons, New York 1970ff.

[51] Index of Mass Spectral Data, American Society for Testing and Materials (ca. 8000 Spektren), Philadelphia, Pa., 1969.

[52] *A. Cornu, R. Massot,* Compilation of Mass Spectral Data (ca. 6000 Spektren bis 1969), Heyden & Son, London 1966.

6.2 Field Ionization and Special Mass-Spectrometric Methods

Hans Dieter Beckey
Instiut für Physikalische Chemie der Universität,
D-53 Bonn

Mass spectrometric methods of investigation have acquired great significance in the elucidation of structures and in the quantitative analysis of organic compounds. The different ionization methods employed in the production of mass spectra vary in the extent of their degree of application. Their relative importance is also variable with time as the state of the art develops. In addition to the ionization of organic molecules by electron impact*, field ionization* (6.2:1), chemical ionization* (6.2:2) and ionization by electron capture* (6.2:3) are increasingly finding application in organic analysis. Photoionization is not dealt with here, although it may attain the same practical significance after development of commercial photoionization mass spectrometers.

Other important mass spectrometry ionization methods which are employed primarily in inorganic chemistry or in special investigations lie outside the scope of this chapter.

The choice of the optimum ionization method for a given mass spectrometric problem of analysis depends on the requirements. The electron impact method is usually used for organic structural analysis; with this method, as a consequence of the strong transfer of energy to the organic molecule, a variety of characteristic fragments are formed from which the molecular structure can be elucidated. A factor common to the three ionization methods described here is the fact that little energy is transferred to the ionized molecules, so that the undisrupted 'molecular ions' are more strongly marked in the mass spectrum. Frequently it is advantageous to combine one of these three methods with the electron impact method.

* Abbreviations: EI = electron impact
FI = field ionization
FD = field desorption
CI = chemical ionization
EC = electron capture

1 Field Ionization and Field Desorption

1.1 Theoretical Basis

Under the influence of an extremely high electrical field (about 10^8 V/cm) the atomic or molecular potentials are changed in such a way that one or more valence electrons can penetrate the potential wall. Positive ions are formed in this way.

The quantum mechanical tunnel effect forms the basis of this field ionization. Its first practical application was in the *field ion microscope*[1]. Field ion sources were developed later for mass spectrometric applications. In this case, the ionization of

an atom or molecule by a high electric field is required[2] (Fig. 1).

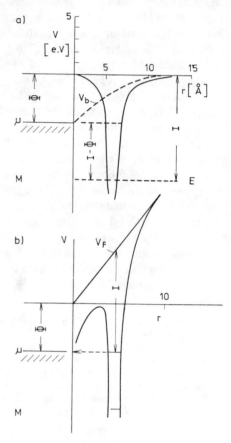

Fig. 1. Potential energy (V) of an atom near a metal surface
a) without field
b) with external field
I = ionization energy, Φ = work function of the metal
V_b = image force potential (------), V_f = external electrical potential = $e \cdot F \cdot r$, (F = field strength), μ = Fermi-level of the metal electron; E = electronic ground level.

Fig. 1a is a plot of the potential energy for the displacement of a valence electron relative to an atom considered to be fixed a few Å outside a metal surface. The metallic phase is shown to the left of the vertical coordinate axis, the gaseous phase to the right. The metal electrons occupy energy levels up to the Fermi level, μ. To release an electron the electronic work function ϕ must be expended. I is the ionization energy of the atom in the gaseous phase. V_B characterizes the 'image force

[1] *E.W. Müller*, Z. Physik, *131*, 136 (1951).
[2] *H.D. Beckey*, Angew. Chem. 81, 662; engl., 8, 623 (1969);
H.D. Beckey, Field Ionization Mass Sepctrometry, Pergamon Press, Oxford 1971.

potential', i.e., the interaction between the electron and the image charges generated by the electron and the ion in the field-generating metal surface.

Fig. 1b shows a potential diagram for the case where an atom is under the influence of a high electric field F. Let the field potential vary linearly, $V_F = eFr$ (e = electronic charge, r = distance from the metal surface). The resulting potential is made up of three components: the field potential; V_F, the image potential, V_B, and the atomic potential in Fig. 1a undisturbed by the field. Regarded from a classical electrodynamic point of view, the electron would have to surmount the potential barrier to penetrate the metal. According to quantum mechanics, however, there is a certain probability of the electron traversing the potential barrier of only a few Å thickness shown in Fig. 1b. The electron penetrates into the metal and the positively charged ion outside the metal surface can, after appropriate acceleration, be subjected to a mass spectrometric analysis.

Accordingly, because the lower levels of the metal are occupied, the valence electron must be raised at least to the Fermi level of the metal electrons by the external field potential so that the electron can penetrate into the metal. Only at a certain minimum distance d_{min} from the metal surface is the external potential sufficiently high to allow ionization to take place, at least with rare gas atoms like He or Ne.

Molecules can also be directly field-ionized from the adsorbed state. This is known as 'field desorption'[3]. FD is only a special version of FI.

The ion current generated by field ionization[2] is determined by two factors: the field ionization probability per particle and the transport of neutral particles into the ionization zone, which is increased by electrical polarization[2, 4].

The transport of neutral particles from the gas phase into the ionization zone can be calculated, whereas the particle supply by surface diffusion is theoretically very difficult to determine[5]; however, it constitutes a considerable fraction of the total particle supply.

1.2 Experimental Techniques

1.2.1 Field Ion Emitters

The high field strengths[6] (about 10^7 to several 10^8 V/cm) needed for field ionization are generated by strongly curved, electrically conductive electrode surfaces to which is applied a high positive potential (about 10 kV) opposite a negatively charged electrode. Fine *metal points*[7-9], thin wires (6.2:1.2.2), and sharp metal knife-edges (6.2:1.2.3) are used as field electrodes.

In organic analysis, however, the latter are preferred because of the higher ion currents and greatly reduced statistical ion current fluctuation as compared with emission points. Nevertheless, for special fields of application[10, 11] and especially for metallurgical investigation with the atom probe[12] the emission needles retain their importance. (For constructions of suitable FI emitter needles see Ref. [13].)

1.2.2 Thin Wires as Field Anodes

The high field strengths obtained with metal points cannot be generated with thin wires as FI emitters, since, with very high voltages, there is a risk of breaking the wire. However, the attainable field strength (10^7–10^8 V/cm) suffices completely for the field ionization of most *organic* substances. The advantage of using wires as field anodes lies in the substantially increased field ion current[14, 15]. The surface of a smooth wire 2×10^{-3} cm in diameter and 5 mm long is about 10^4 times greater than that of a single needle tip of a few 10^{-5} cm radius of curvature. The number of particles impinging on the emitter is increased correspondingly. Admittedly, the multiplication factor cannot be fully utilized, since the field strength at the smooth surface of a 2 μm wire to which a voltage of 10 kV is applied is insufficient for field ionization.

Electron micrographs show, however, that a very large number of microneedles grow on the wire emitter by *field polymerization* of organic molecules at room temperature[2, 16, 17] or by *high field pyrolysis*. Benzonitrile proved to be the most suitable substance for growing the microneedles. At the points of these microneedles locally increased field strengths are generated which suffice for the

[3] *E.W. Müller*, Phys. Rev. *102*, 618 (1956); *R. Gomer*, J. Chem. Phys. *31*, 341 (1959).

[4] *E.W. Müller*, Ergebn. exakt. Naturwiss. *27*, 290 (1953).

[5] *H.G. Metzinger*, *H.D. Beckey*, Z. Naturforschg. *22A*, 1020 (1967).

[6] *H.D. Beckey*, *H. Krone*, *F.W. Röllgen*, J. Sci. Instr. *1*, 118 (1968).

[7] *M.G. Inghram*, *R. Gomer*, Z. Naturforschg. *10A*, 863 (1955).

[8] *E.W. Müller*, *K.H. Bahadur*, Phys. Rev. *102*, 624 (1956).

[9] *H.D. Beckey*, Naturwiss. *45*, 259 (1958).

[10] *J. Block*, Z. Naturforschg. *18A*, 952 (1963).

[11] *W.A. Schmidt*, Z. Naturforschg. *19A*, 318 (1964).

[12] *E.W. Müller*, *J.A. Panitz*, *S.B. McLane*, Rev. Sci. Instr. *39*, 83 (1968).

[13] *E.W. Müller*, Advances in Electronics and Electron Physics, *13*, 83 (1960).

[14] *H.D. Beckey*, Z. Instrumentenkde. *71*, 51 (1963).

[15] *H.D. Beckey*, Z. anal. Chemie, *197*, 80 (1963).

[16] *H.D. Beckey*, *E. Hilt*, *A. Maas*, *M.D. Migahed*, *E. Ochterbeck*, Int. J. Mass Spectry, Ion Phys. *2*, 161 (1969).

[17] *H.D. Beckey*, *M.D. Migahed*, *F.W. Röllgen*, Adv. in Mass Spectrom. V, p. 622, The Institute of Petroleum, London 1971.

field ionization or field desorption of organic substances. Locally the field is increased by a maximum factor of about ten.

Tungsten wires of 10 μm diameter are now most commonly used as field anodes for analysis of organic samples. Their tensile strength is about 160 times larger than that of the previously used platinum-Wollaston wires. The wires are 'activated' at a high temperature (HT, about 1200 °C) with a vapor pressure of the benzonitrile of about 5×10^{-3} torr[31-33]. The needles produced in this way are resistent with respect to thermal and mechanical stress and to chemical attack by corrosive substances like perfluorokerosene which is commonly used for calibration of the mass scale.

1.2.3 Sharp Metal Knife-Edges as Field Anodes

As with thin wires, metal knife-edges were first introduced into FI mass spectrometry by Beckey[14, 15]. Robertson et al. have dealt exhaustively with the structure and field-strength distribution of such emitters[18-20]. Commercial razor blades, possibly degreased, or thin etched platinum or tungsten foils have been used in various laboratories.

In contrast to thin wires, metal blades have the advantage that they cannot be broken by strong field action. Admittedly, when using blades, electrical flashovers to the counter electrode (slotted cold cathode) only a few tenths of a millimeter away from the field anode (at potential differences of over 10 kV) often occur and damage the sharp part of the blade. Furthermore, because of electrostatic shielding effects, the ion emission with blades is somewhat smaller and the current fluctuations are somewhat greater than with thin wire field anodes.

Despite these disadvantages, sharp knife-edges are frequently used with success as FI emitters when only a qualitative detection of ions is necessary and when some loss of intensity is permitted, or for kinetic studies[35-37].

1.2.4 Field Ion Sources

While pure field ion-sources are used predominantly in fundamental field ionization research, combined electron impact/field ionization sources are used generally in quantitative analysis and structural determination of organic substances. Valuable additional information is often gained by combining both ionization methods. The transition from one ionization method to the other can be

effected by simple insertion or withdrawal of the FI emitter into or from the ion chamber by means of a sliding rod and by throwing a multiple switch for the different voltage supplies (Fig. 2).

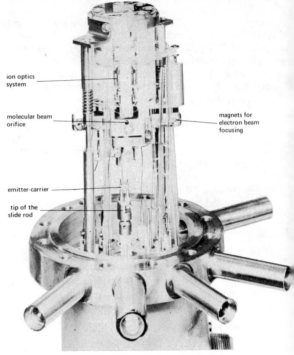

ion optics system

molecular beam orifice

magnets for electron beam focusing

emitter-carrier

tip of the slide rod

VARIAN MAT combined field desorption/field ionization/ electron impact ion source

Fig. 2. Combined field desorption/field ionization/ electron impact ion source (Varian MAT)

With FI the *total* ion intensity at the mass spectrometer detector is about 100 times smaller than that with EI operation. However, an increase in intensity is possible with very well activated emitters and with strongly polar substances. In addition, the ion optics of FI sources can be improved. Moreover, the *parent* ion intensity is much larger in many cases with FI or FD than with EI.

The energy inhomogeneity of the ions is somewhat greater with FI than with EI. With single focusing mass spectrometers a resolving power of $M/\Delta M = 800$ can be achieved with FI. With double focusing instruments the same resolving power is achieved with FI and EI. (A resolving power of 30.000 has been clearly demonstrated[21].)

1.2.5 Field Desorption Ion-Source

If solid organic substances are not allowed to evaporate from the micro-oven onto the FI emitter

[18] *C.M. Cross, A.J.B. Robertson*, J. Sci. Instr. *43*, 475 (1966).

[19] *D.F. Brailsford, A.B. Robertson*, Int. J. Mass Spectry. Ion Phys., *1*, 75 (1968).

[20] *P.A. Blenkinsop, B.E. Job, D.F. Brailsford, C.M. Cross, A.J.B. Robertson*, Int. J. Mass Spectr. Ion Phys., *1*, 421 (1968).

[21] *H.-R. Schulten, H.D. Beckey*, Mass Spectr. *7*, 861 (1973).

[22] *H.D. Beckey*, Int. J. Mass Spectr. Ion Phys. *2*, 500 (1969).

as a molecular beam in the usual way, but instead, the wire or knife-edge emitter is immersed directly into a solution of the substance to be analyzed (outside the ion source), a sufficiently large amount of substance for analysis is bound by adsorption to the strongly jagged surface of a benzonitrile-activated emitter (see 6.2:1.2.2). The solvent is allowed to evaporate and the thus-prepared emitter is inserted into the ion source[22] with the aid of a vacuum lock (see Fig. 3).

After a few minutes a background mass spectrum is taken with the wire at room temperature. With *samples of very low vapor pressure,* a mass line due to the solvent predominates in the background spectrum; with *samples of somewhat higher vapor pressure,* additional mass lines due to the substance are found. A current of a few mA is then passed through the emitter (a few A in the case of metal strips with sharp knife-edges); the emitter is heated in this way until a sufficiently intense mass spectrum results without the substance evaporating too rapidly.

The ionization from the adsorbed state is known as *field desorption.* The new method of sample insertion has the following advantages. When many organic substances are dissolved in a suitable solvent, no fragmentation occurs and, therefore, they can be adsorbed undecomposed on FD emitters; in contrast to this, thermal cleavage of small molecules such as H_2O, NH_3, etc., from organic substances frequently occurs with evaporation from a molecular beam oven. Furthermore, the minimum analyzable amount of substance can be kept smaller with the field desorption technique than with the oven evaporation technique. A droplet of a few microliters suffices for the immersion of the emitter. The smallest detectable amounts of substance lie far below the microgram range. In-

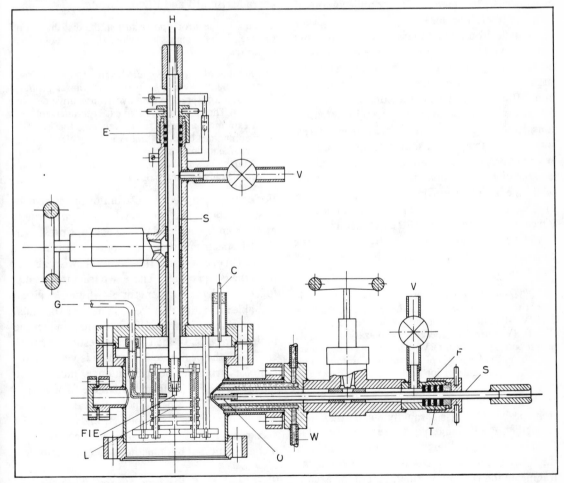

Fig. 3. Field desorption source with emitter (E) and solid lock (F)

FIE FI-emitter
L   (Fig. 2) Ion optic system
G   Gas inlet
W   Water cooler
T   Teflon joint
O   Sample heating oven
V   Valve
S   Sliding rod
C   Current
H   High voltage and heater for FIE

497

creased sensitivity also seems possible with this method. The field desorption technique is naturally restricted to substances of relatively low volatility.

1.3 Structural Analysis of Organic Molecules

In mass-spectrometric analyses of organic substances it is general first to apply electron impact ionization and to obtain as much structural information as possible from the spectra obtained. The following additional information can be gained from the FI or FD mass spectrum:

The molecular weight of the substance and its molecular formula,
a characteristic fragment spectrum which differs from the EI fragment spectrum,
a pyrolysis FD-mass spectrum in the case of polymers measured at high temperatures of the field anode.

1.3.1 Molecular Weight Determination

In many cases ion intensities of the undecomposed molecule (molecular ion) are missing or are very small in electron impact mass spectra of organic compounds. The reason for this is the instability of the molecular ions with respect to the relatively large internal energies taken up by the molecules from 70 eV electrons. Indeed, one can increase the ratio of molecular ion to fragment ion intensity considerably by decreasing the electron energy and using as low an ion-source and evaporation oven temperature as possible[23]. However, this increases only the relative molecular ion intensity, while the absolute ion intensity drops with decreasing electron energy. Thus, when a molecular ion cannot be detected with 70 eV electron impact, it will not be registered either by lowering the electron energy. In many cases, FI or FD is of help here.

The difference between the two ionization methods is that with electron impact the ionization probability $P = a(V-I)$ initially rises linearly with the excess energy $(V-I)$ (V = electron energy, I = ionization energy, a = proportionality factor). With increasing excess energy, the internal energy of the molecular ions, and, therefore, the fragmentation probability also rises.

On the other hand with field ionization, the ionization probability rises rapidly with the field strength[24], $P \simeq F^{30}$, and quickly reaches the value $P = 1$. However, with high ionization probabilities the internal energy excitation and, therefore, the fragmentation is small. Further, part of the excitation energy is transferred to the field anode[34].
In addition, the selection rules are changed in the presence of strong electric fields, and stable molecular ion states become accessible which cannot be reached by electron impact. For example, in the EI

mass spectrum of *neopentane (2, 2-dimethylpropane)*[2], the molecular ion is absent, while in the FI mass spectrum the relative molecular ion intensity amounts to several percent. It is possible even to increase the rel. intensity of the molecular ions of neopentane to 100% by cooling the field anode and applying the field impulse method[34]. By recording FI spectra, it was possible to detect molecular ions in microgram quantities missing in the EI spectrum, and in nanogram quantities with FD. Furthermore, a high resolution mass spectrometer enabled the elemental composition of a sample to be deduced from a precision mass determination[25, 26, 21].

1.3.2 FI and FD Fragments and Molecular Structure

The following conclusions may be drawn from the characteristic FI and FD fragments:

Because of the strong *polarization*, the distribution of the intensity of the mass lines differs from that obtained with EI.

The number of *reaction modes* leading to intense fragments is smaller with FI; in the gas phase, primary decompositions predominate over consecutive decompositions. The spectra are often easier to interpret.
With extremely *unstable molecular ions*, the EI spectrum sometimes lacks not only the molecular ions but also fragment ions which would have to appear in the spectrum somewhat below the molecular mass but are detectable only by means of FI or FD.

For example, in the FI spectrum of the extremely unstable compound *pentaerythritol tetranitrate*[26], apart from the molecular peak M (not present in the EI spectrum) there are found the fragment peaks $(M-122)^{\oplus}$ and $(M-76)^{\oplus}$ (likewise absent from the EI spectrum). These two fragment peaks, together with the fragments m/e = 46 and 76, give a valuable indication that the peak m/e = 316, which is very weak even in the FI spectrum, is in fact the molecular peak. In the case of a pesticide, *Temik sulfone*[27], the peak m/e = 79 which is very important for structural determination is much more intense in the FI spectrum than in the EI spectrum (Fig. 4a, b). The other important peaks m/e = 85 and 86 appear much more strongly in the EI spectrum. This example clearly demonstrates that a combination of both methods can be useful.

[23] *G. Remberg, E. Remberg, M. Spiteller-Friedmann, G. Spiteller*, Org. Mass Spectrom. *1*, 87 (1968).

[24] *M.J. Southon, D.G. Brandon*, Philos. Mag. *8*, 579 (1963).

[25] *E.M. Chait, F.W. Shannon, W.O. Perry, G.E. van Lear, F.W. McLafferty*, Int. J. Mass Spectry. Ion Phys. *2*, 141 (1969).

Inferences about *functional groups* can be drawn from the intensity and the characteristic line form of some FI mass lines. For example, the line $m/e = 29$ $(C_2H_5^\oplus)$ from n-paraffins and aliphatic amines is intense and relatively narrow. The mass line $m/e = 30$ from aliphatic amines, on the other hand, is spread strongly in the direction of smaller masses [28] and indicates the $CH_2NH_2^\oplus$ group (Fig. 4c).

1.4 Quantitative Analysis of Multicomponent Mixtures

Quantitative analyses of organic compounds are frequently carried out with EI mass spectrometers. A simplification of the analyses can be introduced in cases in which very many components are present in a mixture (e.g. 300 to 500 components in mineral oil or petroleum samples). With EI the fragment ions are dominant (Fig. 5a, b); with FI the molecular ions (Fig. 5c). The superimposition of fragment ions on the molecular ions in the FI spectrum (Fig. 5) is very small. Evaluation is very simple. Nevertheless, it must be remembered that the FI spectrum gives no information about the isomer distribution and that usable reproducibility of the spectra can be achieved only when very many parameters are held constant [29].

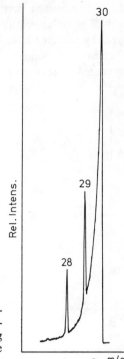

Fig. 4c) Section of the FI-mass spectrum of isopentyl-amine. Typical broadening of the CH_2NH_2-mass line $m/e = 30$.

Reproducible results are easier to achieve with high-boiling mineral oil fractions, with which the sensitivity coefficients for FI are almost independent of molecular weight [30].

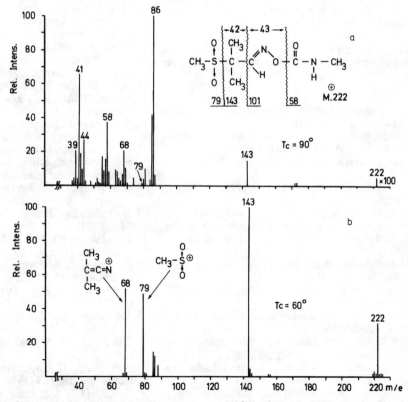

Fig. 4a) Electron impact (90 °C) and b) Field ionization mass spectrum (60 °C) of Temic-Sulfone[27]. M = 222

Recently, complex mixtures of the pyrolysis products of biopolymers have been successfully analyzed by Schulten et al. The pyrolysis products of bacteria were studied by FI[38], and DNA was pyrolized directly on the surface of an FD-emitter[39].

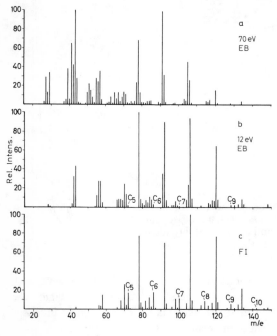

Fig. 5 a) 70 eV- b) 12 eV-electron impact-mass spectrum (EC) of a petroleum sample. T = 250°. c) Field-Ionization mass spectrum. T = 22° (Levsen and Hippe, Bonn). The paraffins are marked with C_5, C_6... Their intensity in the 12 eV-EI-spectrum appears very weak in comparison to the FI-spectrum.

This chapter has been written at the end of 1969. Rapid progress has been made since then with FI- and especially with FD-mass spectrometry of organic compounds. A few new references will be given, demonstrating the status of the subject at the beginning of 1974 (Ref. [31–55]).

Concerning the analysis of biomedical and natural products publications about the following classes of compounds have been published in the mean time: amino acids[40], peptides[41], nucleosides and nucleotides[21], glycosides[44], sugar phosphates[46], organic salts[47], pesticides[42] and pesticide metabolites[43], and drugs[45]. A survey on the application in biomedical research has been given in Ref. [48]. Recent technical developments are described in Ref. [49–51]. Examples for recent FI studies are given in Ref. [52–55].

[26] C. Brunée, G. Kappus, K.H. Maurer, Z. analyt. Chemie, 232, 17 (1967).

[27] J.N. Damico, R.P. Barron, J.A. Sphon, Int. J. Mass Spectry Ion Phys., 2, 161 (1969).

[28] H.D. Beckey, G. Wagner, Z. analyt. Chemie, 197, 58 (1963).

[29] K.G. Hippe, H.D. Beckey, Erdöl und Kohle, 24, 620 (1971).

[30] W.L. Mead, Analytic. Chem., 40, 743 (1968).

[31] H.D. Beckey, A. Heindrichs, E. Hilt, M.D. Migahed, H.-R. Schulten, H.U. Winkler, Meßtechnik 78, 196 (1971).

[32] H.-R. Schulten, H.D. Beckey, Org. Mass Spectrom. 6, 885 (1972).

[33] H.D. Beckey, E. Hilt, H.-R. Schulten, J. Phys. E: Sci. Instr. 6, 1043 (1973).

[34] F.W. Röllgen, H.D. Beckey, Int. J. Mass Spectr. Ion Phys. 12, 465 (1973).

[35] P.J. Derrick, A.M. Falick, A.L. Burlingame, J. Am. Chem. Soc. 94, 6794 (1972); ibid. 95, 437 (1973).

[36] P.J. Derrick, A.M. Falick, S. Lewis, A.L. Burlingame, Org. Mass Spectrom. 7, 887 (1973).

[37] P.J. Derrick, A.M. Falick, A.L. Burlingame, Int. J. Mass Spectr. Ion Phys. 12, 101 (1973).

[38] H.-R. Schulten, H.D. Beckey, A.J.H. Boerboom, H.L.C. Meuzelaar, Anal. Chem. 45, 191 (1973).

[39] H.-R. Schulten, H.D. Beckey, A.J.H. Boerboom, H.L.C. Meuzelaar, Anal. Chem. 45, 2358 (1973).

[40] H.U. Winkler, H.D. Beckey, Org. Mass Spectrom. 6, 655 (1972).

[41] H.U. Winkler, H.D. Beckey, Biophys. Biochem. Res. Commun. 46, 391 (1972).

[42] H.-R. Schulten, H.D. Beckey, J. Agr. Food Chem. 21, 372 (1973).

[43] H.-R. Schulten, H. Prinz, H.D. Beckey, W. Tomberg, W. Klein, F. Korte, Chemosphere 2, 23 (1973).

[44] W.D. Lehmann, H.-R. Schulten, H.D. Beckey, Org. Mass Spectrom. 7, 1103 (1973).

[45] H.-R. Schulten, H.D. Beckey, G. Eckhardt, S.H. Doss, Tetrahedron 29, 3861 (1973).

[46] H.-R. Schulten, H.D. Beckey, E.M. Bessel, A.B. Foster, M. Jarman, J.H. Westwood, J.C.S. Chem. Commun. 13, 416 (1973).

[47] D.A. Brent, D.J. Rouse, M.C. Sammons, M.M. Bursey, Tetrahedron Lett. 4127 (1973).

[48] H.-R. Schulten, H.D. Beckey in A.R. West, Advances in Mass Spectrometry, Vol. VI, p. 499–507, Applied Sciences Publ., London 1974.

[49] H.U. Winkler, H.D. Beckey, Org. Mass Spectrom. 7, 1007 (1973).

[50] H.-R. Schulten, H.D. Beckey, J. of Chromatogr. 83, 315 (1973).

[51] H.J. Heinen, Ch. Hötzel, H.D. Beckey, Int. J. Mass Spectr. Ion Phys. 13, 55 (1974).

[52] P. Brown, G.R. Pettit, R.K. Robins, Org. Mass Spectrom. 2, 521 (1969).

[53] P. Brown, G.R. Pettit, Org. Mass Spectrom. 3, 67 (1970).

[54] J.A. Sphon, J.N. Damico, Org. Mass Spectrom. 3, 51 (1970).

[55] J.B. Forehand, W.F. Kuhn, Anal. Chem. 42, 1839 (1970).

[56] F.H. Field, M.S.B. Munson, D.A. Becker, Adv. Chem. Ser. 58, 167 (1966).

[57] F.H. Field, M.S.B. Munson, J. Am. Chem. Soc. 87, 3289 (1965).

2 Chemical Ionization

2.1 Theoretical Basis

The technique of chemical ionization[56-59] has come in widespread use in practical organic analysis lately[60-69].

The fundamental principle of CI is as follows: A gas for reaction is admitted into the ion source of a mass spectrometer at a pressure of 1 torr, an unusually high value for conventional mass spectrometry. After the primary ionization of the reaction gas with about 500 eV electrons, positive secondary ion species with an abundance of 90% of the total ion current are generated by reaction of the primary ions with the reaction gas. Methane is very suitable. The secondary ions do not react further with the reaction gas.

Vapor of a sample to be analyzed at substantially lower pressure ($\sim 10^{-3}$ torr) is added to the reaction gas. The secondary ions then react with the sample substance and give rise to a characteristic mass spectrum of this substance.

This mass spectrum differs strongly from the conventional mass spectrum produced by a 70-eV-electron impact at pressures $\leq 10^{-4}$ torr. The relative fragment intensities are substantially smaller in the CI spectrum than in the EI spectrum. There is some resemblance between the CI and FI spectra but characteristic differences exist even here (see below).

When the reaction gas (e.g., methane) is introduced into the ion source at about 1 torr pressure, primary methane ions and not ions of the sample substance are generated predominantly by the impacting electrons ($p < 10^{-3}$ torr).

Primary reactions:
$$CH_4 + e \rightarrow CH_4^{\oplus}, CH_3^{\oplus}, CH_2^{\oplus}, CH^{\oplus}, \ldots + 2e \quad (a)$$

The primary ions react with the reaction gas.

Secondary reactions:
$$CH_4^{\oplus} + CH_4 \rightarrow CH_5^{\oplus} + CH_3 \quad (b)$$
$$CH_3^{\oplus} + CH_4 \rightarrow C_2H_5^{\oplus} + H_2 \quad (c)$$
$$CH_2^{\oplus} + CH_4 \rightarrow C_2H_4^{\oplus} + H_2 \quad (d)$$
$$\phantom{CH_2^{\oplus} + CH_4} \hookrightarrow C_2H_3^{\oplus} + H_2 + H$$
$$C_2H_3^{\oplus} + CH_4 \rightarrow C_3H_5^{\oplus} + H_2 \quad (e)$$

Further reactions are less significant. The $CH_5^{\oplus}$ and $C_2H_5^{\oplus}$ ions are by far the most intense secondary ions. They combine with the neutral methane only with difficulty but react strongly with the organic sample substance principally by proton- or hydride-transfer reactions.

$$CH_5^{\oplus} + RH \rightarrow RH_2^{\oplus} + CH_4 \quad (f)$$
$$RH_2^{\oplus} \rightarrow R^{\oplus} + H_2$$
$$\phantom{RH_2^{\oplus}} \hookrightarrow A^{\oplus} + B \quad (f')$$

where RH is the sample molecule and A and B are fragment ions. The $C_2H_5^{\oplus}$ ions can react in the following way:

$$C_2H_5^{\oplus} + RH \rightarrow RH_2^{\oplus} + C_2H_4 \quad (g)$$
$$\phantom{C_2H_5^{\oplus} + RH} \hookrightarrow R^{\oplus} + C_2H_6 \quad (g')$$

From reactions f to g' it may be inferred that with chemical ionization the $(M \pm H)^{\oplus}$ ions should be well-defined in the mass spectrum as is the case.

2.2 Mass Spectrometers

Conventional single focusing magnetic mass spectrometers or other instruments can be used for chemical ionization. However, an important difference from the normal pattern of such instruments is that the ion source, the ion-acceleration chamber, and the mass analyzer volume need to be evacuated by three separate, high-speed pumps (pumping rate > 300 l/sec). These maintain the large pressure drop between the ion source (~ 1 torr) and the analyzer ($\sim 10^{-5}$ torr). A well designed CI source has been described by Beggs et al.[70].

2.3 Applications

Chemical ionization can be used to supplement electron impact ionization for mass spectrometric structural analysis. The advantage of the method lies in the fact that the molecular ions of those

[58] *M.S.B. Munson, F.H. Field,* J. Am. Chem. Soc. 87, 3294 (1965); ibid. *88,* 221 (1966); ibid. *88,* 2621 (1966); ibid. *88,* 4337 (1966).

[59] *M.S.B. Munson,* Anal. Chem. *43,* 28A (1971).

[60] *F.H. Field,* J. Am. Chem. Soc. *91,* 2827 (1969).

[61] *F.H. Field, P. Hammet, W.F. Libby,* J. Am. Chem. Soc. *91,* 2839 (1969).

[62] *H.M. Fales, H.A. Lloyd, G.W.A. Milne,* J. Am. Chem. Soc. *92,* 1590 (1970).

[63] *H. Ziffer, H.M. Fales, G.W.A. Milne, F.H. Field,* J. Am. Chem. Soc. *92,* 1597 (1970).

[64] *G.W.A. Milne, T. Axenrod, H.F. Fales,* J. Am. Chem. Soc. *92,* 5170 (1970).

[65] *H.M. Fales, G.W.A. Milne, T. Axenrod,* Anal. Chem. *42,* 1483 (1970).

[66] *H.M. Fales, G.W.A. Milne, R.S. Nicholson,* Anal. Chem. *43,* 1785 (1971).

[67] *G.W.A. Milne, H.M. Fales, T. Axenrod,* Anal. Chem. *43,* 1815 (1971).

[68] *H.M. Fales, Y. Nagai, G.W.A. Milne, H.B. Brewer jr., T.J. Bronzert, J.J. Pisano,* Anal. Biochem. *43,* 288 (1971).

[69] *A.A. Kiryushkin, H.M. Fales, T. Axenrod, E.G. Gilbert, G.W.A. Milne,* Org. Mass Spectrom. *5,* 19 (1971).

compounds which yield only extremely weak $M^{\oplus}$ lines in the EI mass spectrum often yield very intense $(M+1)^{\oplus}$ or $(M-1)^{\oplus}$ lines in the CI mass spectrum. Furthermore, structural information can be derived from the fact that the interpretation of CI mass spectra can be based on a few simple postulates regarding the fundamental chemical reactions[56-59]. For example, the paraffin CI spectra can be explained quantitatively by the reacting ions (which result as secondary products from methane) attacking the paraffin molecules at statistically distributed locations, followed by localized decomposition leading to the observed ions (Fig. 6). The CI mass spectra of esters with methane as reaction gas can be explained by the preferred reaction of $CH_5^{\oplus}$, $C_2H_5^{\oplus}$ and $C_3H_5^{\oplus}$ with the carbonyl group. Although chemical ionization has been employed very little (1969) for quantitative mixture analysis, possibilities for future application are quite conceivable.

3 Electron Capture

3.1 Theoretical Basis

Capture of electrons by organic molecules can result in negatively charged molecular ions as well as positively and negatively charged fragment ions. The effective cross-section for electron capture is, however, in general several orders less than that for the formation of positive ions. Further, the reproducibility of mass spectra of negative ions is dependent on several parameters. Therefore, negative-ion mass spectrometry thus far has not reached the significance for practical analysis possessed by positive-ion mass spectrometry. Only electron capture mass spectrometry (EC-MS) with the aid of 'duoplasmatron' ion sources[71, 72] (see 6.2:3.2) has thus far been used for structural elucidation and quantitative analysis of organic molecules. Its significance can increase in the future with the development of corresponding commercial instruments, especially when this method, as with the FI, FD and CI methods, is combined with the EI method for positive ions.

Electron capture with the aid of monoenergetic electron beams is of significance for fundamental research, but will not be dealt with in further detail here.

[71] *M. v. Ardenne*, Kernenergie *1*, 1029 (1958).
[72] *M. v. Ardenne*, Z. Angew. Physik *11*, 121 (1959).

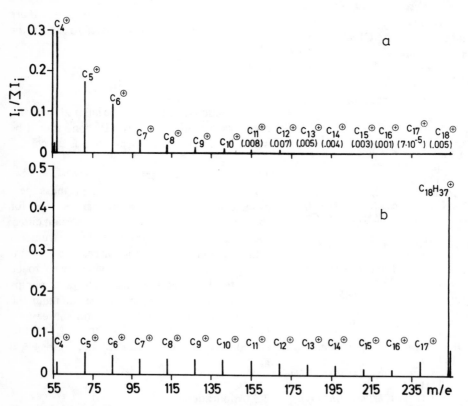

Fig. 6. Mass spectrum of n-$C_{18}H_{38}$ a) by electron impact b) by chemical ionization[56].

[70] *D. Beggs, L. Vestal, H.M. Fales, G.W.A. Milne,* Rev. Sci. Instr. *42*, 1578 (1971).

[73] *M. v. Ardenne, K. Steinfelder, R. Tümmler,* Z. Physikal. Chem. (Leipzig) *220*, 105 (1962).

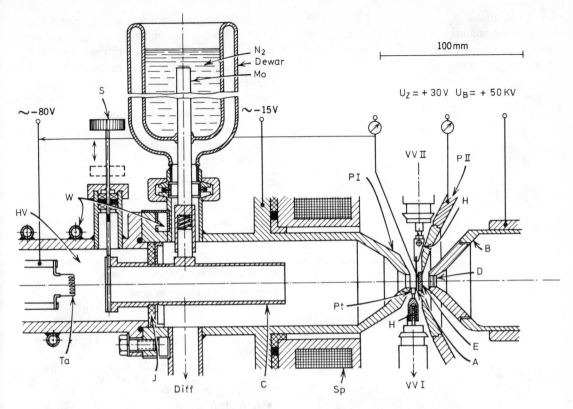

Fig. 7. Duoplasmatron-Ion source of v. Ardenne[78]
Dewar, filled with liquid nitrogen

W	Water cooler
Ta	Double cathode 0.5 mm W and Ta-wire
HV	High vacuum
J	Ceramic insulation
Diff	Hg-Diffusion pump
S	Rod to separate the cathode chamber after sample injection (shifted 90°)
C	Vapor condensation pipe (Cu, cooled)
Sp	Exciting coil

VVI, VVII	Prevacuum I and II
H	Heating bowl (W-wire 0.2 mm)
A	Insulated anode (Diam 1 mm) and pre-focusing of the slit adjustment
E	Moveable emission slit and scanning electrode (setting is 6 mm)
B	Acceleration electrode
D	Slit directed to B ($500 \times 1500 \ \mu^2$)
PI, PII	Pole piece
PI	also intermediate electrode

3.2 Duoplasmatron Ion-Source

One difficulty relating to the generation of electron capture mass spectra is the small effective cross-section for the corresponding ion-formation reactions. This difficulty can be overcome in principle by the use of intense electron beams, but this is difficult because the electron energy must be small, and beam-widening occurs in consequence of the space charge formed. This difficulty is circumvented[71, 72] by compensating the space charge of the electrons by an opposing positive charge. This can be achieved by means of quasi-neutral plasmas with a low-pressure discharge. Electrons of almost thermal energy are generated in great density in this manner.

In a duoplasmatron ion-source (represented schematically in Fig. 7) a hot tungsten cathode generates a current of 0.1–0.25 A in a low-pressure discharge in argon at about 10^{-2} torr or mercury at about 10^{-3} torr.

Recombination at the wall of the discharge tube is reduced by means of a strong magnetic field contracting the discharge. Charge carrier diffuses into the ionization space through a small aperture. The substances to be analyzed or reference samples (usually solids) are evaporated into the ionization space from small evaporation furnaces.

Since the energy of the ions generated in the duoplasmatron source is relatively inhomogeneous, in the von Ardenne spectrometer the ions are accelerated to 40 kV to achieve a sufficient mass resolving power. The resolution can likewise be increased by means of a double-focusing mass spectrometer.

The ionization processes taking place in the duoplasmatron source are very complex. In practice, however, the discharge can be controlled so that either the group of the molecular ions, $M^\oplus$, $(M \pm 1)^\oplus$, etc., or that of the fragment ions occurs more prominently in the spectrum.

3.3 Applications

Quantitative analyses of organic multicomponent mixtures have been attempted with the aid of the duoplasmatron ion-source[73]. However, the princi-

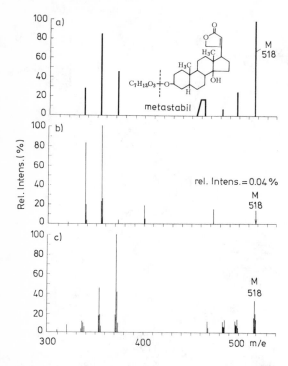

pal field of application lies in the structural analysis of organic compounds, especially in the comparison of the electron capture mass spectra[74, 75] with the corresponding positive-ion mass spectra. The EA mass spectrum of *Somalin*[76] (Fig. 8) shows a certain similarity to the FI mass spectrum. In the FI spectrum the molecular peak is the most intense; in the EC spectrum its intensity is likewise considerable, while in the EI spectrum it is negligibly small.

The FI spectrum was taken with low resolving power, so that the peaks $(M + nH)^{\oplus}$ etc. are represented by a single more strongly marked mass line. In both the FI and EC spectra fragment peaks appear which are not present in the EI spectra, and vice versa. This example shows the usefulness of the combination of different mass-spectrometric methods for the structural elucidation of organic substances.

More recently a review of electron capture mass spectrography was given by v. Ardenne et al.[79].

[74] *R. Tümmler,* Z. Physikal. Chem. (Leipzig) *229,* 58 (1965).
[75] *M. v. Ardenne, R. Tümmler, E. Weiss, T. Reichstein,* Helv. Chim. Acta *47,* 1032 (1964).
[76] *C. Brunée,* Z. Naturforschg. *22b,* 121 (1967).
[77] *M. v. Ardenne, K. Steinfelder, R. Tümmler,* Z. Chem. *5,* 287 (1965).
[78] *M. v. Ardenne, K. Steinfelder,* Kernenergie *3,* 717 (1960).
[79] *M. v. Ardenne, K. Steinfelder, R. Tümmler,* Elektronenanlagerungs-Massenspektrographie organischer Verbindungen, Springer Verlag, Berlin 1971.

Fig. 8. Mass spectrum of somalin (upper mass region). a) field ionization[2], b) electron impact ionization[76], c) electron capture ionization[71, 77].

6.3 Elektronenbrenzen

Hermann Schildknecht
Organisch-Chemisches Institut der Universität
D-69 Heidelberg

1 Reagents in Elektronenbrenzen

In Elektronenbrenzen of organic compounds in tritiated water, the electrons emitted on tritium decay are utilized in structure elucidation. All compounds examined so far in tritiated water afforded well-defined fragments when appropriate conditions, such as temperature and exposure time, were chosen. Electron pyrolysis possesses crucial advantages:

the same reagents are always being used, and
the smallest quantities of tritiated fragments can be detected by means of their radioactivity.

Consequently, electron pyrolysis may serve in structure elucidation of substances present in trace amounts.

The electrons formed on β-decay of tritium are directly or indirectly responsible for all the radiochemical reactions taking place in tritiated water. After the primary and secondary electrons have been slowed down within 10^{-11} sec to thermal energies of approximately 0.03 eV, they undergo hydration to hydrated electrons e_{aq}^{θ} which can be detected in alkaline tritiated ice by ESR[1].

Hydroxyl radicals are always formed in ^{3}HHO in the same yield as e_{aq}^{θ} and can be detected along with hydrogen atoms by ESR[2]. The G(H) values are smaller by a factor of four than $G(e_{aq}^{\theta})$ and G(HO·). For product formation both hydrogen and hydroxyl ions may become significant.

2 Mechanism of Elektronenbrenzen

The chemically reactive particles which attack molecules present in tritiated water are:

hydrated electrons which act as reducing agents and nucleophiles, and
hydroxyl radicals and hydrogen atoms which act as oxidizing agents and electrophiles.

The free radical fragments initially formed in the electron pyrolysis reaction combine at once with hydrogen atoms or hydroxyl radicals to yield stable fragments. Therefore, the formed fragments can be identified both qualitatively and quantitatively through tritium labeling.

2.1 The Solvated Electron as a Nucleophile

The first experiments with amino acids have already shown[3] that the amino group can easily be eliminated. In this case it is fair to assume that a nucleophilic attack by the hydrated electron takes place simultaneously with displacement of the $NH_3^\oplus$ group present in the aqueous solution (cf. Scheme 1).

Scheme 1.

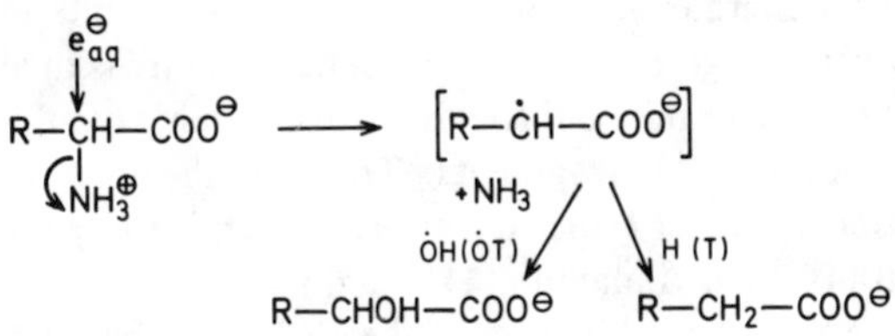

$$
\begin{array}{c}
e_{aq}^\ominus \\
\downarrow \\
R-\underset{NH_3^\oplus}{\overset{|}{C}H}-COO^\ominus \longrightarrow \left[R-\overset{\cdot}{C}H-COO^\ominus\right] \\
\quad\quad\quad \overset{+NH_3}{\underset{\dot{O}H(\dot{O}T)}{\swarrow}} \quad \overset{}{\underset{H\,(T)}{\searrow}} \\
R-CHOH-COO^\ominus \quad\quad R-CH_2-COO^\ominus
\end{array}
$$

The most important end products of the electron pyrolysis of *alanine,* namely propionic acid, lactic acid, and pyruvic acid, may still be analytically identified by this method although they are undetectable by chemical means. After incubation of the sample compound with a few microliters tritiated water of specific activity, e.g. 5 Ci/ml, the tube contents is applied to a radio-chromatogram, separated by thin-layer chromatography and the activity recorded along the thin-layer plate with a thin-layer scanner (cf. Fig. 1). The detection of the fragments is facilitated especially by the direct labeling. The electron pyrolysis product lactic acid is probably formed through combination of the propionic acid radical obtained by deamination and the OH radicals which are always present.

[1] *K. Eiben,* Dissertation, Universität Karlsruhe, 1964.
[2] *J. Kroh, B.C. Green, J.W.T. Spinks,* Can. J. Chem. 40, 413 (1962).
[3] *H. Schildknecht, O. Volkert,* Z. Anal. Chem. 216, 7 (1966).

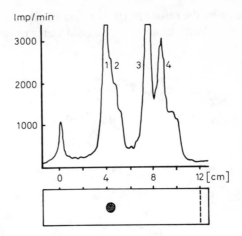

Fig. 1. Thin-layer radio-chromatogram of 5 μg alanine after electronenbrenzen.
1. alanine. 2. lactic acid; 3. propionic acid + pyruvic acid; 4. alanine ethyl ester

The lactic acid is then oxidized to pyruvic acid. This is analytically significant since it is more difficult to identify alcohols than ketones.
Ring nitrogen is also cleaved by nucleophilic attack as is demonstrated in the case of *proline* (1). In contrast to the mass spectral fragmentation pattern of this compound, no pyrrolidine is present

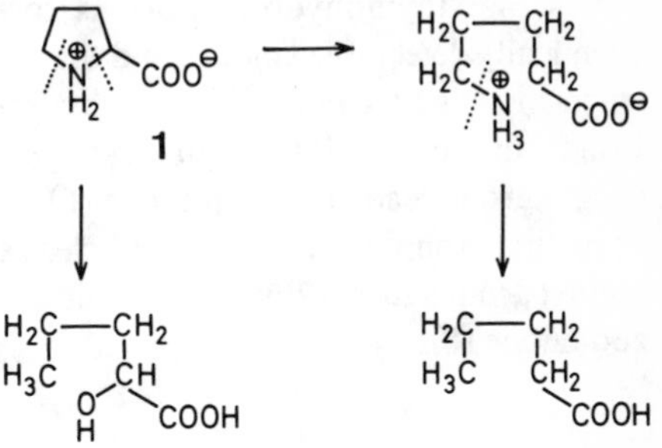

2.2 The Solvated Electron as a Base

That the hydrated electron e_{aq}^θ acting as a nucleophilic particle can, in special cases, behave like a base and abstract a proton is shown by the ring opening of the *purine* (2)[4]. First a purine anion (3) is formed, which is stabilized through ring opening (several resonance structures). The carbene (4) then reacts to yield 4-amino-5-formamidopyrimidine (5). The latter is also formed in the reaction of purine with inactive water; however, in substantially lower yield. It may be concluded from these results that the activation energy of the reaction purine $+ e_{aq}^\theta$ is much smaller than the activation energy of the reaction purine + hydroxyl ions, as the latter are generated about 10^4-fold faster than the hydrated electrons[5]. This conclusion is corroborated by the results of the reaction of esters and flavones[6] with tritiated and normal water respectively.

2.3 The Solvated Electron as a Reducing Agent

Benzoquinones are not only cleaved during their incubation with tritiated water, but are always hydrogenated to the more stable hydroquinones as well. As a reducing agent e_{aq}^{θ} possesses a reduction potential which is even larger than that of hydrogen atoms[7]. *Thymoquinone*[8] (**7**) may serve as a simple example. The electron pyrolysis spectrum shows that thymohydroquinone is formed in a chemically detectable amount and that the quinoid fragments also are converted into hydroquinones. Fragmentation of (**7**) through cleavage of the isopropyl group leads to toluquinone (**6**) which loses its methyl groups only reluctantly; cleavage of two methyl groups from (**7**) yields 2,6-dimethyl-*p*-benzoquinone (**8**).

2.4 The Solvated Electron as a Free Radical

The repeated observation of the cleavage of alkyl side chains, especially that of isopropyl groups, can be explained very well by the behaviour of e_{aq}^{θ} as a free radical.

2.5 The Solvated Electron as an Energy Carrier of Radiation-Induced Pyrolysis

The electrons are made effective solely through energy liberation. Energy-rich electrons liberate smaller amounts of energy, i.e. they possess a small LET (Linear Energy Transfer). On the other hand, the energy released by thermalized electrons is large. Substrate molecules which are hit by such electrons are apparently in an energy-rich resonance state. This leads to the fragmentation of the molecule and the formed fragments correspond to those obtained in a pyrolysis or to those generated in a mass spectrometer.

3 Structure Analysis by Elektronenbrenzen

3.1 Overlapping Combination of the Fragments

The presence of several bonds with similar dissociation energies in a molecule frequently results in several fragmentation modes. The structure is then simply determined through an 'overlapping combination' of the fragments. Thus, in the case of *aspartic acid* (**9**) the structure is determined by the α- (**10**) and β-alanine (**11**) fragments. The superimposable combination of the Elektronenbrenzen products tryptamine (**14**) and indolylpropionic acid (**12**) yields *tryptophan* (**13**); in this case, also, alanine and skatole (**15**) may be combined to give the correct formula.

3.2 Hydroxyl Groups as Labeling Groups

The position of substitution in a side chain is labeled more frequently by hydroxyl groups than by residual methyl groups. This will be illustrated in the case of the electron pyrolysis analysis of *dimethylmenthylamine*[9] (**16** in Scheme 2). Dimethylamine (**18**) and menthane (**17**) were detected by gas chromatography as radiolysis products. All the C_3 fragments originate from the isopropyl group. In cases where the identification of the cyclic fragments on the basis of their retention times is not unequivocally possible, they can be further degraded by electron pyrolysis. The degradation of *menthane* (**17**) and *menthol* shows that it

[4] *H. Jaggy*, Dissertation, Universität Heidelberg, 1968.

[5] *M. Eigen, de Mayer*, Naturwissenschaften *42*, 413 (1955).

[6] *A. v. Klaudy*, Dissertation, Universität Heidelberg, 1970.

[7] *J.H. Baxendale, R.S. Dixon*, Z. Phys. Chem. [Frankfurt am Main] *43*, 161 (1964).

[8] *F. Römer*, Dissertation, Universität Heidelberg, 1967.

[9] *K. Penzien*, Dissertation, Universität Heidelberg 1967.

is possible to degrade a molecule by fractional Elektronenbrenzen down to its basic constituents.

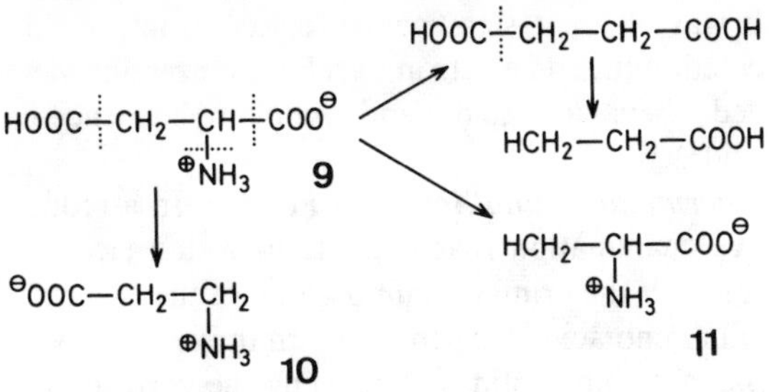

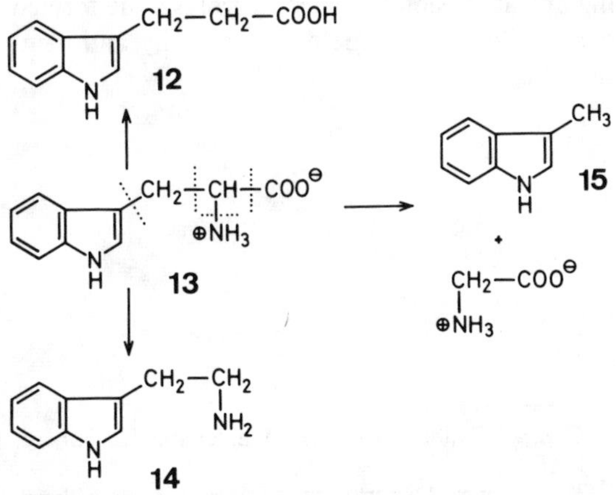

9

10

11

12

13

15

14

Scheme 2 illustrates how the structure is determined from the resulting fragments. Cyclohexanol indicates how methane and cyclohexane, as the simplest components found, may be combined to give methylcyclohexane. Methylcyclohexane and propane when combined yield menthane; here 4-methylcyclohexanol and isopropyl alcohol indicate that the locations of substitution are at C-atom 4 of the cyclohexane ring and at C-atom 2 of propane, respectively. In the case of the many possible combinations of menthane with dimethyl-

amine, the menthol detected in the analysis serves to indicate the correct structure.

4 Applications

4.1 Structure Elucidation of Synthetic Products

It was necessary to distinguish between thiadiazines and thiazoles, e.g. between **(19)** and **(20)**[10]. The fragmentation pattern obtained by electron pyrolysis indicated that the compounds were *thiazoles*[11].

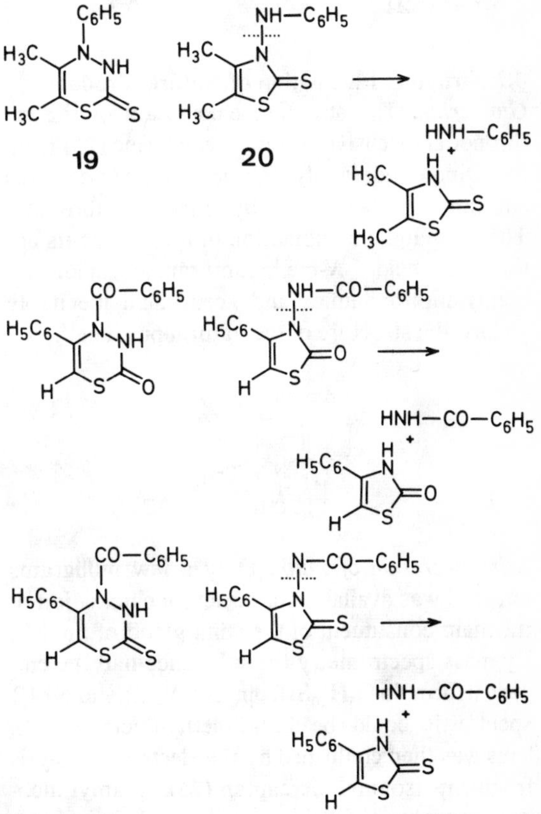

19

20

Scheme 2.

17

18

16

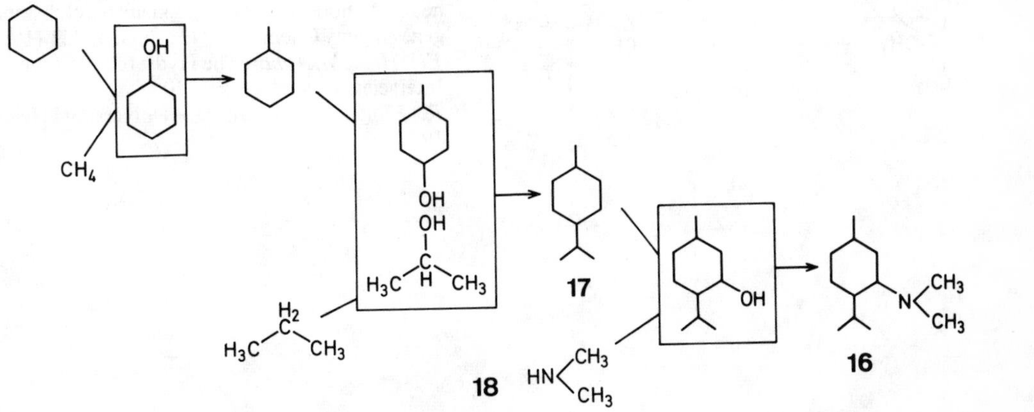

Thermal rearrangement of phenylazomethyl iso-cyanates led to heterocyclic compounds for which the structure could not be elucidated by physical methods alone. Electron pyrolysis furnished ethyl-amine and *N*-acetyl-*N'*-phenylurea (22). This in-dicated that, of all the possible isomers, the reaction product possessed structure (21)[12].

21   **22**

4.2 Structure Elucidation of Natural Products

Glomerine: The structure elucidation of the ar-thropodal defensive secretion glomerine (23) from the Dipolopod family *Glomeris marginata* was substantially facilitated by Elektronenbrenzen. The meaningful combination of the fragments an-thranilic acid, *N*-methylanthranilic acid, *N*-methylanthranilamide and acetic acid inevitably yielded the structure of the natural product[13].

23

Stink secretion of mink: Only a few milligrams material was available for structure elucidation of the main constituent of the stink gland of mink[14]. By mass spectrometry the substance had the em-pirical formula $C_5H_{10}S$; from its UV, IR and NMR spectra it could be 2,2-dimethylthietane (24). This was then confirmed by the electron pyrolysis fragments isoamyl mercaptan (25), isoamyl alco-hol, isopropyl alcohol, propane, ethyl mercaptan, ethyl alcohol and ethylene.

24   **25**

Mezereine: The empirical formula of the natural product mezereine $C_{38}H_{38}O_{10}$, the poison from *Daphne Mezereum,* indicated a complex structure. It was of great significance for its structure elu-cidation that on electron pyrolysis mezereine yiel-ded benzoic acid and cinnamylideneacetic acid[15, 16].

N-terminal Amino Acid in a Peptide: In a reduc-tive deamination reaction, the hydrated electron splits off the primary amino group of the N-termi-nal amino acid in a peptide (with the exception of S-containing peptides)[19], yielding the correspond-ing acylated peptide. After hydrolysis the formed carboxylic acid can be detected easily and iden-tified[17].

[10] *G. Ege,* Habilitationsschrift, Universität Heidelberg, 1968.
[11] *K. Gessner,* Dissertation, Universität Heidelberg, 1967.
[12] *H. Schildknecht, G. Hatzmann,* Angew. Chem. *81,* 469; Engl. *8,* 456 (1969).
[13] *H. Schildknecht, W.F. Wenneis,* Z. Naturforsch. *21B,* 552 (1966).
[14] *I. Wilz,* Dissertation, Universität Heidelberg, 1967.
[15] *H. Schildknecht, G. Edelmann, R. Maurer,* Chem,-Ztg. Chem. App. *94,* 347 (1970).
[16] *H. Schildknecht, R. Maurer,* Chem.-Ztg. Chem. App. *94,* 849 (1970).

5 Bibliography

[17] *H. Schildknecht, F. Enzmann, K. Gessner, K. Pen-zien, F. Römer, O. Volkert,* Elektronenbrenzen, eine neue Methode zur Analyse kleinster Substanzmen-gen, Angew. Chem. *78,* 841; Engl. *5,* 751 (1966).
[18] *E.J. Hart, M. Anbar,* The Hydrated Electron, Wiley-Interscience, New York 1970.
[19] *R. Küppers,* Dissertation, Universität Heidelberg 1971.

6.4 Pyrolysis of Polymers

Hans Batzer and *Rolf Schmid*
CIBA-GEIGY AG, CH-4002 Basel

1 Introduction

Pyrolysis is a general term for a chemical reaction which is brought about by the action of heat. This technique is of great importance, both in synthetic procedures (e.g. cracking processes) and as a tool for structural analysis. When a polymer is pyrolyzed, decomposition results as a rule, and the effect of temperature and time is to cause degradation of the polymer to compounds of relatively low molecular weight which can then be identified. Knowledge of the structure of the decomposition products provides information about both the structure of the macromolecule and the degradation mechanism. Secondary reactions can be expected to occur during and after the cracking process (rearrangements of intermediary radicals, etc.). The way in which the pyrolysis proceeds with time also provides information, both on the thermal stability of the polymers and on the kinetics of degradation reactions. In terms of apparatus and technique, pyrolysis involves two stages: firstly, the pyrolysis itself, and secondly, identification of the decomposition products. The latter can be carried out by any appropriate analytical technique (*spectroscopy, chromatography, fractionation, titration, formation of derivatives,* etc.).

2 Procedure

In principle, pyrolysis can be carried out in various ways, e.g. in an electric arc, or using a hot wire or a heated plate, or by simply warming (usually electrically, but also by immersion in a fused metal bath) in a suitable container consisting of a heat-resistant material, preferably quartz or platinum, but occasionally glass or Teflon[1,2].

Pyrolysis is generally carried out in vacuo, or in an inert carrier gas such as nitrogen, helium or argon. The apparatus is specially constructed to allow collection of the decomposition products, for example, in evacuated receivers[3]. The fission products are condensed as quickly as possible (sometimes being fractionated at the same time) and analyzed. For certain special purposes, pyrolysis may be carried out in oxygen or air (oxidative degradation, cf. Fig. 2) or in an atmosphere of hydrogen. A recent development combines the pyrolysis apparatus and the equipment used for identification of the products (e.g. chromatography) into one unit.

2.1 Combination of Pyrolysis with Spectroscopic Methods

The isolated fission products can be identified using spectroscopic methods such as *mass spectroscopy*[4,5], *IR* and *UV spectroscopy*[6], *nuclear magnetic resonance,* or *electron spin resonance.* A further possibility with *IR spectroscopy* (cf. Chapter 5.2) is to follow the pyrolysis of a polymer sample in the form of a film, infrared spectra of the film being measured after various heating times[7]. A recently developed apparatus permits pyrolysis of micro-samples to be carried out in a cell directly in the beam of an infrared spectrometer, in an inert or an oxidizing atmosphere, or in vacuo[8].

By use of *electron spin resonance* (cf. Section 5.5:4), very short-lived radicals can be detected[9,10]; this technique is a very useful complement to the other identification methods mentioned above.

2.2 Combined Pyrolysis/Gas Chromatography

Gas chromatography (cf. Chapter 2.10) is by far the most used chromatographic method for identification of the pyrolysis products from polymers[11-13].

Developments in instrumentation have led to equipment in which the pyrolysis vessel is coupled with the gas chromatograph. Condensation and secondary pyrolysis reactions are reduced to a minimum by using a rapid flow of a pre-heated

[1] *H.H.G. Jellinek*, Degradation of Vinyl Polymers, Academic Press, New York 1955.
[2] *S.L. Madorsky*, Thermal Degradation of Organic Polymers, J. Wiley & Sons, New York 1964.
[3] *H. Koch* in *Houben-Weyl*, Methoden der organischen Chemie, 4. Aufl., Bd. IV/2, p. 454, Georg Thieme Verlag, Stuttgart 1955.
[4] *S.L. Madorsky, S. Strauss*, J. Res. Nat. Bur. Stand. *40*, 417 (1948).
[5] *D.P. Maier*, Anal. Chem. *36*, 1678 (1964).
[6] *D. Hummel*, Kaut. Gummi, Kunstst. *11*, 185 (1958).
[7] *H. Kambe, Y. Shibazaki, T. Iwamoto*, Tokyo Daigaku Koku Kenkyusto Shuto *3*, 690 (1963).
[8] *M.P. Brash, T.S. Light*, Appl. Spectrosc. *19*, 114 (1965).
[9] *W. Holzmüller*, Plaste Kaut. *12*, 321 (1966).
[10] *A.A. Berlin, V.V. Yarkina, A.P. Firsov*, Vysokomol. Soedin, Ser. A *10*, 2157 (1968).
[11] *W.B. Swann, J.P. Dux*, Anal. Chem. *33*, 654 (1961).
[12] *K. Ettre, P.E. Varadi*, Anal. Chem. *35*, 69 (1963).

inert carrier gas which sweeps the fission products directly from the oven into the gas chromatograph[14-19]. Improved reproducibility can be achieved by conducting the pyrolysis at an exactly fixed temperature[20]. This can be done by using as the pyrolysis element a platinum wire which forms part of a Wheatstone bridge[21]. With specially constructed equipment, polymer specimens can be heated to 900 °C in 3–4 seconds; decomposition is then rapid and practically complete[22, 23].

For certain polymers, such as *proteins,* methods have been developed which allow micro-amounts to be used[24-26]. In some cases, the pyrolysis products are hydrogenated before they are analyzed in the gas chromatograph[27].

2.3 Combined Pyrolysis/Thermogravimetric Analysis (TGA)

The introduction of TGA represents an important advance in polymer pyrolysis (cf. Chapter 9.3 and 9.4). In this technique, the change in weight during the pyrolysis process is recorded (Fig. 1).

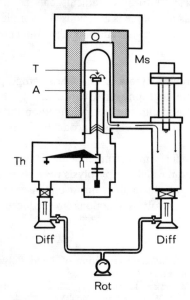

Fig. 1. Apparatus for combined pyrolysis/thermogravimetric analysis

Diff Diffusion pumps   MS Mass Spectrograph
Rot Rotation pumps   T crucible
A   Tube   O furnace

Th = thermobalance which must measure small weight changes rapidly and with high precision, and should be as insensitive as possible to temperature effects while the sample is heated.

Various highly developed thermobalances suitable for investigation of the pyrolysis of polymers[28] are commercially available. With some special equipment it is possible to separate the fission products into discrete fractions which can be collected and analyzed by known methods. By use of a special furnace, TGA may be coupled with a gas chromatograph or a mass spectrometer, as shown in Fig. 1[29].

In principle, there are two ways in which TGA can be carried out:

statically or isothermally: the change in weight is measured as a function of time;
dynamically or non isothermally: the weight change of the sample is followed as the temperature is steadily increased.

If the data (weight, weight change per unit time, pressure, temperature) are recorded automatically, characteristic curves are obtained for each polymer. As already mentioned, these allow significant conclusions to be drawn about the thermal decomposition processes and their kinetics, and about heat stability[30-36].

[17] *C.B. Honaker, A.D. Horton,* J. Gas Chromatogr. *3,* 396 (1965).

[18] *G. Edel,* Method Phys. Anal. 174 (1966).

[19] *S.S. Hirsch, M.R. Lilyquist,* J. Appl. Polym. Sci. *11,* 305 (1967).

[20] *D. Deur-Siftar, T. Bisticki, T. Tandi,* J. Chromatogr. *24,* 404 (1966).

[21] *M. Krejci, M. Deml,* Collect. Czech. Chem. Commun. *30,* 3071 (1965).

[22] *J. Zulacia, G. Guiochon,* Bull. Soc. Chim. Fr. 1343, 1351 (1966).

[23] *C.E.R. Jones, G.E.I. Reynold,* J. Gas Chromatogr. *5,* 25 (1967).

[24] *M.V. Stark,* J. Gas. Chromatogr. *5,* 22 (1967).

[25] *M. Dimbat, F.T. Eggertsen,* Microchem. J. *9,* 500 (1965).

[26] *G.J. Fleming,* J. Appl. Polym. Sci. *10,* 1813 (1966).

[27] *J. van Schooten, J.K. Evenhuis,* Polymer *6,* 343 (1965).

[28] *P.E. Slade jr., L.T. Jenkins,* Techniques and Methods of Polymer Evaluation, Vol. 1, Marcel Dekker, New York 1966.

[29] *Mettler,* Technical Bulletin T-107 (1969).

[30] *B. Vollmert,* Kunststoffe *56,* 680 (1966).

[31] *A.E. Newkirk,* Anal. Chem. *32,* 1558 (1960).

[32] *H.C. Anderson,* J. Polym. Sci. Part C *6,* 175 (1963).

[33] *H.T. Lee, L. Reich, D.W. Levi,* NASA Accession No. N 65–14462 Rept. No. AD 453629 Avail. CFSTI (1964).

[34] *J.M. Cox, B.A. Wright, W.W. Wright,* J. Appl. Polym. Sci. *9,* 513 (1965).

[35] *J.M. Lancaster, B.A. Wright, W.W. Wright,* J. Appl. Polym. Sci. *9,* 1955 (1965).

[36] *H.C. Anderson,* SPE Trans. *2,* 202 (1962).

[13] *P. Drienovskii, O. Kysel,* Chem. Zvesti *17,* 912 (1963).

[14] *W.M. Barbour,* J. Gas Chromatogr. *3,* 228 (1965).

[15] *A. Berton,* Chim. Anal. |Paris| *47,* 502 (1965).

[16] *R.A. Prosser, J.T. Stapler, W.E.C. Yelland,* Anal. Chem. *39,* 694 (1967).

Depending on the nature of the surrounding medium or carrier gas, additional reactions may occur between the polymer and the medium[37-40] (Fig. 2).

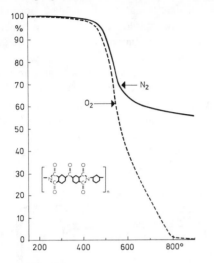

Fig. 2. Pyrolysis of a polyimide
In the presence of oxygen (O_2), and in an atmosphere of nitrogen (N_2)[37-40].

In certain cases — especially with highly cross-linked polymers — valuable results can be obtained by carrying out stepwise isothermal thermogravimetric analysis. Vaporization of relatively low molecular weight compounds is achieved by application of vacuum. In this way, it is possible to remove chemically unbound components before pyrolysis; samples should preferably be powdered. It is similarly possible to reverse equilibrium reactions involved in the degradation of a macromolecule before pyrolysis (in the strict sense of the word) occurs. This was demonstrated with a highly cross-linked polymer, obtained by reaction of phthalic anhydride with a diglycidyl ether of diphenylolpropane at 5×10^{-6} mmHg pressure[41] (Fig. 3).

3 Differential Thermal Analysis (DTA)

DTA (cf. Chapter 9.3 and 9.4) allows the measurement of thermal effects associated with chemical and physical processes. As with TGA, runs can be carried out either isothermally or dynamically. In most equipment, the temperature difference between sample and reference cells is recorded[28]. The corresponding heat changes can be measured directly by the recently developed technique, Differential Scanning Calorimetry (DSC). In DSC, the differential energy required to keep both sample and reference at the same temperature is measured throughout the analysis. The recorder pen deflection is directly proportional to the energy input or output, in millicalories per second, and the area under a peak gives the total energy of the transition involved (Fig. 4)[42]. Polymer decomposition is normally associated with a large heat of reaction; DTA thus allows one to obtain information about the energetics of the decomposition reactions[43]. In addition, information can be obtained about chemical and physical changes which do not involve weight changes.

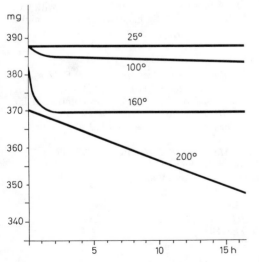

Fig. 3. Stepwise isothermal thermogravimetric analysis
Weight loss due to diffusion of low molecular weight products at 25°, 100° and 160° as compared with that due to decomposition of the polymer at 200 °C.

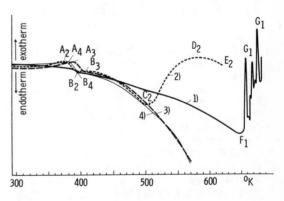

Fig. 4. Application of DSC-analysis
Curve 1: No transition or reaction with high energy change up to 640 °K is observed on low sensitivity, using the same material as with Fig. 3.
Curve 2: Higher sensitivity. The same process first shows a glass transition (A_2-B_2), then an exotherm (C_2) starting at 500 °K and with a maximum reaction rate at 580 °K (D_2).
Curve 3: If this process is interrupted at 620 °K (E_2) and re-started at 300 °K, the exotherm maximum disappears and the transition region shifts to slightly higher temperatures (A_3-B_3); the network structure is still intact.

The same diepoxide, cured with *hexahydrophthalic anhydride* instead of *phthalic anhydride* (Fig. 4, Curve 4), gives a pattern similar to Curve 3, suggesting that the exotherm maximum D_2 is caused by reformation of the anhydride from the half-ester, a reaction that does not occur in the case of hexahydrophthalic anhydride.

Interpretation of DTA diagrams, as shown in Fig. 4, may be complicated, because endothermic and exothermic processes may overlap; moreover, strongly endothermic evaporation nay occur, especially when open cells are used. DTA measurements are consequently especially meaningful when coupled with thermogravimetric analysis[44-48].

4 Special Methods

In certain cases, pyrolysis may be coupled with characteristic product identification methods. For example, interesting information on the degradation mechanism of *polyvinyl chloride*[49] can be obtained by thermogravimetric analysis and simultaneous measurement of the hydrogen chloride eliminated, by potentiometric titration. In a study on *methyl polymethacrylate* it proved useful to follow the change in pressure inside the closed quartz reaction vessel during pyrolysis[50,51]. With thermoplasts, the intrinsic *viscosity*[52], or *swelling* and *gel-formation*[53] can be determined. Useful information can also be obtained on cross-linked

resins by carrying out *dielectric measurements*[54]. Further data can be obtained by dynamic mechanical measurements on polymers (e.g. damping and modulus determinations)[38]. A new porous plug technique was employed for following the surface degradation of certain polymers caused by intense surface heating[55].

Information which is of special value in elucidation of degradation mechanisms can be obtained by marking specific groups in the molecule with radioactive elements[56]; for example ^{14}C can be used, and may be converted into $^{14}CO_2$. This *radiothermal analysis* technique can be coupled to thermogravimetric analysis (cf. Fig. 5).

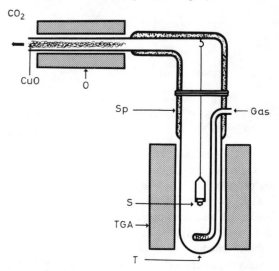

Fig. 5. Apparatus for combined radiothermal analysis/thermogravimetric analysis

Gas   carrier gas   S   sample in the crucible
TGA TGA furnace   O   oxidizing furnace with CuO
T   tube   Sp   insulated chromium-nickel spool

[37] *B. Groten*, Anal. Chem. *36*, 1206 (1964).
[38] *A.H. Frazer*, High Temperature Resistant Polymers Intersc. Publishers J. Wiley & Sons, New York 1968.
[39] *J.H. Freeman, L.W. Frost, G.M. Bower, E.J. Traynor, H.A. Burgman, C.R. Ruffing*, Polym. Eng. Sci. *9*, 56 (1969).
[40] *H.W. Lochte, E.L. Strauss, R.T. Conley*, J. Appl. Polym. Sci. *9*, 2799 (1965).
[41] *H. Batzer, R. Schmid* (unpubl.).
[42] *E.S. Watson, M.J. O'Neil, J. Justin, N. Brenner*, Anal. Chem. *36*, 1233 (1964).
[43] *L. Reich*, J. Appl. Polym. Sci. *10*, 813 (1966).
[44] *H.C. Anderson*, Nature *191*, 1088 (1961).
[45] *S. Kohn, G. Taguet*, Rech. Aeron. *88*, 27 (1962).
[46] *Shih-K'ang Wu, I. Ch'in Chou*, Ko Ken Tzu T'ung Hsun *6*, 153 (1964).
[47] *D.P. Bishop, D.A. Smith*, Ind. Eng. Chem. *59*, 32 (1967).
[48] *R.G. Neville, J.W. Makoncy, K.R. McDowall*, J. Appl. Polym. Sci. *12*, 607 (1968).
[49] *A. Guyot, P. Roux, Pham-Quang-Tho*, J. Appl. Polym. Sci. *9*, 1823 (1965).
[50] *J.E. Clark, H.H.G. Jellinek*, Proc. Battelle Symp. Thermal Stability Polymers, Columbus, Ohio *1963*, D 1–D 24.
[51] *H.H.G. Jellinek, H. Kachi*, Proc. Battelle Symp. Thermal Stability Polymers, Columbus, Ohio *1963*, C 1–C 14.

[52] *E. Kardash, A.N. Pravednikov, S.S. Medvedev*, Dokl. Akad. Nauk SSSR *156*, 658 (1964).
[53] *J.V. Zhuravleva, V.V. Rode*, Vysokomol. Soedin., Ser. A *10*, 569 (1968).
[54] *J.C. Patterson-Jones, D.A. Smith*, J. Polym. Sci. *12*, 1601 (1968).
[55] *R.F. McAlevy, J.G. Hansel*, AIAA (Am. Inst. Aeron. Astronaut.) J. *3*, 244 (1965).
[56] *D.O. Bowen*, Mod. Plast. 127 (1967).

5 Bibliography

H.G. Jellinek[1] (1955).
H. Koch[3] (1955).
S.L. Madorsky[2] (1964).
P.E. Slade jr., L.T. Jenkins[28] (1966).
A.H. Fraser[38] (1968).
D. Schultze, Differentialthermoanalyse, Verlag Chemie. Weinheim/Bergstr. 1969.
R.T. Conley, Thermol Stability of Polymers, Marcel Dekker, New York 1970.

7 Diffraction Methods

contributed by

J. Haase, Karlsruhe
A. Maas, Bonn
E. Schultze-Rhonhof, Bonn
G. Will, Bonn

7.1 X-ray Analysis

Ernst Schultze-Rhonhof

D-53 Bonn 1, Amsterdamer Str. 15

X-ray diffraction with single crystals permits complete crystal structure determinations on crystalline solids (7.1:5), i.e. determination of the space group symmetry of the crystal, the parameters of the unit cell, and the coordinates of all atoms in the crystal except hydrogen. (Localization of hydrogen atoms: cf. Chapter 7.3.) The following information may be obtained: atomic distances, bond angles, molecular structures, density, and, under some circumstances, chemical composition.

With X-ray diffraction, it is not possible to elucidate details of part of a structure, without complete determination of the entire structure. X-ray diffraction with powders (7.1:3.2.3) is, in addition, suitable for indentification of known substances (fingerprint method). Only in rare, very favorable cases are structure determinations with powder methods likely to succeed; for the most part determinations are possible only by analogy with known structures.

The utilization of anomalous scattering (7.2, see also 7.3:6) allows the absolute configuration of molecules to be determined. The wavelength used must be just below an absorption edge of one of the elements in the crystal. X-ray diffraction with non-crystalline materials, especially liquid and glassy samples (7.1:5.8), permits investigation of average inter-atomic distances and their frequencies, particle sizes of colloids and macromolecules by small angle diffraction, orientations of crystallites in polycrystalline materials, and filamentary molecules in fibers.

The literature listed in Section 7.1:6 should be consulted for derivations and evidence. This Chapter deals with structure determinations on crystals, especially with single crystal methods.

1 Introduction

X-ray structure analysis, formerly a domain of inorganic chemistry, has also gained increasing importance for determination of the structures of important organic compounds (Nobel prizes 1962–1964[1-6]). The advantage of X-ray structure analysis in contrast to most other physical methods is that it gives the entire configuration of a compound.

1.1 Fundamental Principles of Crystallography

1.1.1 Crystal, Lattice

From the structure of a crystalline chemical substance, a motif can always be picked out, which is repeated throughout the entire crystal, filling all the space without any gaps. In organic substances, this motif is normally the chemical molecule; in high polymers or similar substances, it can sometimes be a segment of a chain, or the like. These 'motifs' are related systematically by symmetry elements.

One kind of symmetry exists in all crystals, i.e. periodic repetitions or translations in three directions in space. Crystals, therefore, may be defined as three-dimensional, periodic arrangements of matter.

Each periodic translation in three dimensions of space may be expressed by three linear independent, i.e. noncoplanar, components of a vector, $\vec{a} = \vec{a}_1 + \vec{a}_2 + \vec{a}_3$. This vector generates from any point in space a so-called (point-) space-lattice[7, 8], by which it is represented uniquely (Fig. 1). Such a

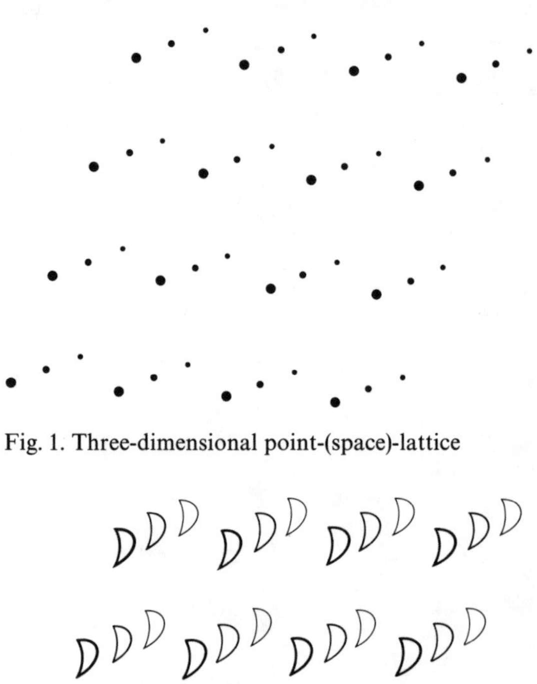

Fig. 1. Three-dimensional point-(space)-lattice

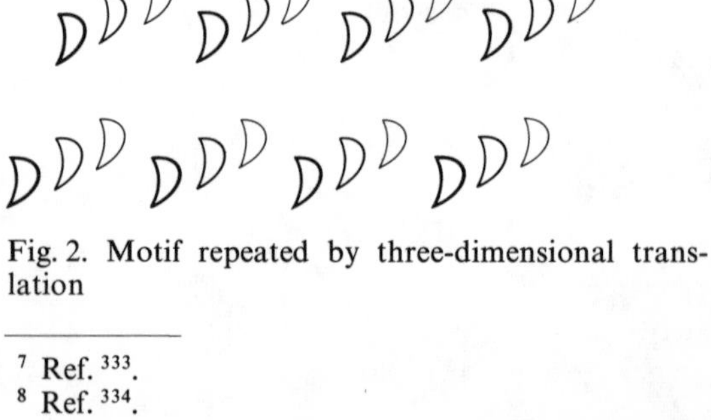

Fig. 2. Motif repeated by three-dimensional translation

[1] *F.H. Crick*, Angew. Chem. *75*, 425 (1963).
[2] *M.H.F. Wilkins*, Angew. Chem. *75*, 429 (1963).
[3] *J.D. Watson*, Angew. Chem. *75*, 439 (1963).
[4] *M.F. Perutz*, Angew. Chem. *75*, 589 (1963).
[5] *J.C. Kendrew*, Angew. Chem. *75*, 595 (1963).
[6] *D. Crowfoot-Hodgkin*, Angew. Chem. *77*, 954 (1965).

[7] Ref. 333.
[8] Ref. 334.

space-lattice may be placed in each crystal, one of its points being a definite point within a molecule or 'motif' of the crystal, and each other one an equivalent point within an equivalent 'motif'. The translation vector shifts not only a single point, but the motif as a whole parallel to itself, into identical positions (Fig. 2). A *net plane* may be positioned through every three points of a space lattice. All planes parallel to that net plane which also go through points of the space lattice form a *stack of net planes* (see Fig. 3). A simplified theory of X-ray diffraction by crystals uses those net planes as planes of reflection.

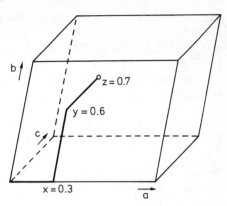

Fig. 4. Coordinates of the point $x = 0.3$; $y = 0.6$; $z = 0.7$ in a unit cell

Fig. 3. Stack of net planes in the point lattice of Fig. 1

1.1.2 The Coordinate System of Crystallography

The crystallographer uses the three components $\vec{a}_1, \vec{a}_2, \vec{a}_3$ of a suitably chosen translation vector $\vec{a}$ as the basis of a coordinate system, describing the structure of his crystal. These three components are then called the *lattice constants,* and in the literature are usually labeled a, b, c. The angles between pairs of axes are termed the *lattice angles* $\alpha_1, \alpha_2, \alpha_3$ or α, β, γ. The parallelepiped described by the three vector components is called the *unit cell*. It has a definite volume and within the crystal contains a definite quantity of matter, although it is only a mathematical quantity. The entire structure is composed of connected unit cells, each being identical with all the others.

Co-ordinates of a point in the unit cell, $(\vec{x}_1, \vec{x}_2, \vec{x}_3)$ or (x, y, z), are expressed as fractions of the lattice constants:

$$\vec{x}_k = \frac{\text{shift in direction } \vec{a}_k}{\text{lattice constant } \vec{a}_k} \quad (1)$$

The co-ordinates of equivalent points in different unit cells of the same structure thus always differ by integers (Fig. 4). To describe the entire structure, it is, therefore, sufficient to describe only one unit cell. Usually, x, y, z, are chosen to be between 0 and 1. It may sometimes be convenient to use coordinates between -1 and 0.

1.1.3 Crystal Systems, Bravais Lattice

In principle, every space lattice can be represented by an infinite number of translation vectors. In general, however, one chooses a possibility which describes the unit cell by the three shortest translations which do not lie in one plane. Sometimes it may be useful to refer to axes which are orthogonal to each other or, for reasons of symmetry, which are of equal size. (7.1:1.1.4). If this is done,

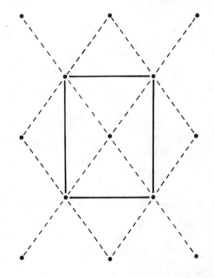

Fig. 5. Body-centered, rectangular cell (———) in two dimensions

The lattice may also be described by the primitive, but oblique cell (-----); but this cell does not show the existent symmetry sufficiently

one occasionally obtains not the smallest possible unit cell, but centered cells whose volumes are 2, 3 or 4 times that of the smallest possible, 'primitive' cell (Fig. 5).

The following 7 axial systems, the so-called crystal systems, can be distinguished[9-12]:

Triclinic crystal system: $a \neq b \neq c; \alpha \neq \beta \neq \gamma \neq 90°$
Monoclinic crystal system: $a \neq b \neq c; \alpha = \gamma = 90°;$ $\beta \neq 90°$
(sometimes also: $\alpha = \beta = 90°; \gamma \neq 90°$)
Orthorhombic crystal system: $a \neq b \neq c; \alpha = \beta = \gamma = 90°$
Hexagonal (trigonal) crystal system: $a = b \neq c;$ $\alpha = \beta = 90°; \gamma = 120°$.
Rhombohedral crystal system: $a = b = c; \alpha = \beta = \gamma \neq 90°$
Tetragonal crystal system: $a = b \neq c; \alpha = \beta = \gamma = 90°$
Cubic (isometric) crystal system: $a = b = c;$ $\alpha = \beta = \gamma = 90°$

With respect to the possible centerings, one finds 14 space-lattice types, the so-called *Bravais space lattices*[13-15].

1.1.4 Symmetry Elements
Besides the translations, there is a series of other symmetry operations, which can link together the 'motifs' in crystals:

Rotation axis,
Reflection plane (mirror),
Center of symmetry (inversion),
Rotoinversion axis (rotoreflection axis),
Screw axis, and
Glide plane.

The first four of these symmetry elements also express themselves in the morphology of the crystal.

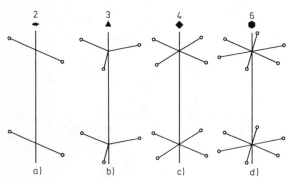

Fig. 6. Rotation axes (2-fold, 3-fold, 4-fold, 6-fold)

[9] Ref. [313], p. 10.
[10] Ref. [333], p. 101.
[11] Ref. [336], p. 25.
[12] Ref. [337], p. 119.
[13] Ref. [333], p. 84.
[14] Ref. [336], p. 26.
[15] Ref. [313], p. 6.
[16] Ref. [336], p. 49.

If a structure contains an n-fold (n = 1, 2, 3, 4, 6; generally X) *rotation axis,* it is repeated by rotation through an angle of $2\pi/n$ about this axis; n is called the multiplicity of the rotation axis. Multiplicities other than 1, 2, 3, 4, 6 cannot occur in crystals[16] (Fig. 6).

Symbol of a rotation axis is 1, 2, 3, 4, 6; in general X. Reflection by a *reflection plane* (symbol: m) obeys the known optical laws: a point in front of the plane is repeated at the same distance behind it (Fig. 10a).

Inversion (symbol): I) denotes reflection at a point, the center of symmetry (Fig. 9).

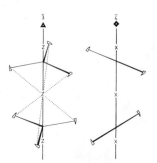

Fig. 7. 3- and 4-fold rotoinversion axes.

A *rotoinversion axis* couples rotation and inversion (or, leading to the same result, rotation and reflection), i.e. these two operations are executed in immediate succession. Only the 3- and 4-fold rotoinversion axes (symbols: $\bar{3}, \bar{4}$) (Fig. 7) are really new symmetry operations. The other ones are already obtained by combinations of other symmetry operations.

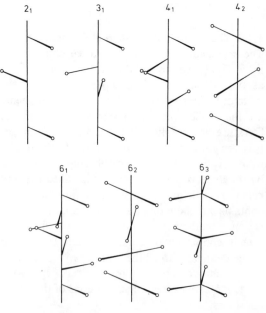

Fig. 8. Screw axes (The axes 3_2, 4_3, 6_5, 6_4 can be generated by reflection from 3_1, 4_1, 6_1, 6_2 and are not drawn.)

A *screw axis* is formed by coupling a rotation axis with a translation. From a given point, a further one is generated by rotation through $2\pi/n$ and simultaneous translation parallel to the direction of the axis. Here, the screw component is $\tau \cdot p/n$, τ being the period of the translation in the direction of the axis; $p = 1, 2,... (n-1)$. The symbol of a screw axis is X_p. The following can occur in crystal lattices; 2_1; 3_1; 3_2; 4_1; 4_2; 4_3; 6_1; 6_2; 6_3; 6_4; 6_5 (Fig. 8)[17].

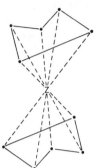

Fig. 9.
Inversion center

In executing a *glide reflection,* one displaces by half the translation length in the direction of gliding and then reflects immediately. Only certain directions, parallel to the lattice axes (symbols: a, b, c) and the planar and spatial diagonals (symbols: n, d), can be directions of glide reflection (Fig. 10b).

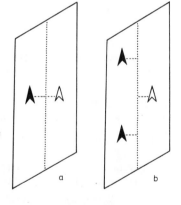

Fig. 10.
a) Reflection plane
b) Glide plane
(white: front;
black: back of the
figure.)

1.1.5 Further Crystallographic Basic Terms
The symmetry elements: rotation axis, reflection plane, center of symmetry and rotoinversion axis, which can be detected by the morphology of a crystal, can be combined unquestionably in 32 different ways. These 32 combinations are called *crystal classes*[18-20].

Since the presence or absence of a *center of symmetry* cannot be determined from the symmetry of an X-ray pattern, only 11 of these crystal classes, the *Laue classes,* can be distinguished by X-ray diffraction. The Laue symmetry of the remaining classes is obtained by addition of the center to the symmetry operations generating the classes[21, 22]. In Laue class $\bar{3}$ m, by comparison of intensities it is possible to distinguish two sub-classes $\bar{3}$ m (A) and $\bar{3}$m(B)[23, 24], which have a different orientation of the two-fold axis relative to the axial system.

For $\bar{3}$m (A) : $|F(hkl)| = |F(khl)| \neq |F(hk\bar{l})|$, whereas for $\bar{3}$m (B) : $|F(hkl)| \neq |F(khl)| = |F(hk\bar{l})|$.

If one adds the screw axes and glide planes, which are not recognizable from the morphology of the crystal, and the centers, one obtains 230 possible combinations, the so-called *space groups*[25]. A space group may be described either by listing the generating symmetry operations, or by statement of the so-called lattice complex. According to Niggli, the *lattice complex* is the totality of all points generated from one point xyz by all symmetry operations of a space group. It is of practical importance for the determination of a crystal structure. If the space group is known, one must determine only the co-ordinates xyz of one atom from each lattice complex; the co-ordinates of all the other atoms are then known automatically. The description of a certain part of the unit cell, called the *asymmetric unit,* is, therefore, sufficient for complete description of a structure. This asymmetric unit contains exactly one point of each lattice complex of the unit cell.

1.2 The Reciprocal Lattice
The coordinate system of a given crystal can be defined by the three components $\vec{a}_1, \vec{a}_2, \vec{a}_3$ (or a, b, c) of a vector $\vec{a}$. There is then a second vector $\vec{a}^*$ having the components $\vec{a}^1, \vec{a}^2, \vec{a}^3$ (or a*, b*, c*), linked with $\vec{a}_1, \vec{a}_2, \vec{a}_3$ in the following way:

$$(\vec{a}_i\vec{a}^k) = 0, \text{if i} \neq k \qquad (2a)$$
$$(\vec{a}_i\vec{a}^k) = 1, \text{if i} = k\,; (i, k = 1, 2, 3) \qquad (2b)$$

[17] Ref. [336], p. 87.
[18] Ref. [336], p. 61.
[19] Ref. [333], p. 46.
[20] Ref. [313], p. 32.

[21] Ref. [336], p. 336.
[22] Ref. [313], p. 30.
[23] *D.W.J. Cruickshank, E. Schultze-Rhonhof, J.G. Sime,* International Tables for X-ray Crystallography, Pilot Issues of Series A, part 2.4, International Union of Crystallography (1971), p. 35.
[24] Ref. [315], p. 464.
[25] Ref. [313-315].

From Equation (2a), it can be shown that if $i \neq k$, each $\vec{a}_i$ is perpendicular to each $\vec{a}^k$, and each $\vec{a}^k$ to each $\vec{a}_i$ because the scalar product then vanishes. Equation (2b) normalizes the $\vec{a}^k$; the projection of $\vec{a}^i$ on $\vec{a}_i$ should have length $1/\vec{a}_i$, $\vec{a}_i$ denoting the length of $\vec{a}_j$ (Fig. 11). The three vector components $\vec{a}^1$, $\vec{a}^2$, $\vec{a}^3$ now describe a second coordinate system which is called the *reciprocal lattice*[26–29], in contrast to the first, 'real' lattice. This reciprocal lattice is of considerable significance, for it permits very convenient description of the X-rays scattered by a crystal, and greatly simplifies many crystallographic computations.

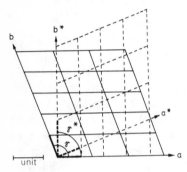

Fig. 11. Relationship between real lattice ———
$(a = 1.25; b = 0.75; \gamma = 110°)$ and reciprocal lattice-----
$(a^* = 0.85; b^* = 1.42; \gamma^* = 70°)$
The components of the basis vectors of both lattices are drawn more heavily

Each lattice plane of a crystal lattice (see 7.1:1.1.1) can be related to a certain point of the reciprocal lattice. This point lies along the direction of the normal to the net plane stack. Its distance from the origin of the reciprocal lattice is $1/d$, where d is the net plane distance, i.e. the distance of two neighboring planes of the same stack.

Usually, one denotes the lattice points of the reciprocal lattice, and also the related net plane stacks and the X-ray reflections scattered by them, with the letters h, k, l – the so-called *Miller's indices,* which originated in crystallography before the advent of X-rays[30, 31].

[26] *P.P. Ewald*, Z. Kristallogr. (A) *56*, 129 (1921).
[27] *P.P. Ewald*, Handbuch der Physik 24, p. 248. Geiger & Scheel Verlag, Berlin 1927.
[28] Ref. [329], p. 407.
[29] Ref. [338], p. 115.
[30] Ref. [333], p. 21.
[31] Ref. [336], p. 40.

2 Laws of X-ray Diffraction by Crystals

2.1 Braggs Law

X-rays of a certain wavelength λ which fall on a crystal stimulate its electrons to undergo oscillations and, therefore, to emit X-radiation of the same wavelength, generally in all directions in space. In most directions, however, this secondary radiation is extinguished through interference. A certain few directions remain, which obey the 'Bragg's Law'[32]. An exact theory of *X-ray interference* is given in the book by M. v. Laue[33]. For practical purposes, the simplified concept of net planes reflecting the X-rays (see Fig. 12) is sufficient.

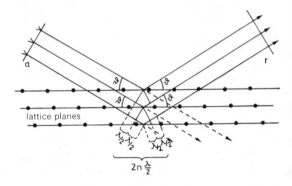

Fig. 12. Bragg's reflection
An X-ray beam (a) incident at an angle ϑ on a stack of planes of a crystal is not extinguished by interference only in the case where the difference in path length in the reflected beam (r) which occurs is an integral number of wavelengths λ.

Two coherent, parallel X-rays of the wavelength λ incident on two neighboring net planes of a stack of net planes having the interplanar spacing d are 'reflected' by these planes. They will mutually reinforce each other if the path length difference is λ or an integral number of wavelengths; in all other cases they will weaken each other. Since in a real crystal many rays interfere on many planes, they will extinguish each other completely. The difference between the path lengths of two X-ray beams is a function of the interplanar spacing d and the angle ϑ made by the incident beam and the reflecting plane (called the *Bragg's angle).* The relation between ϑ, d and λ for non-extinguished interferences is given by Bragg's Law

$$n\lambda = 2d \sin \vartheta \qquad (3)$$

n being any integral number[34].

[32] Ref. [334], p. 43.
[33] Ref. [330].

Bragg's Law describes the interference between beams scattered by equivalent points of different unit cells, but not the interference between beams scattered by different points of one single cell. The interplanar spacing d is a function of the lattice parameters a, b, c and α, β, γ. Each stack of net planes is related to one point of the reciprocal lattice. By Bragg's law, the lattice parameters are, therefore, related (for a given recording geometry) unambiguously to the position – but not to the intensity – of the reflections. The technique for determination of lattice parameters is discussed in Section 7.1:4.1. It should also be mentioned that Equation (3) indicates that reflections having $\sin \vartheta > 1$ cannot be measured with the given wavelength.

2.2 Diffraction by a Real Structure

Obviously, the X-rays scattered by different points of the same unit cell also interfere.

Each point p scatters with an amplitude ρ_p, which is the electron density in p. A wave starting from that point has a phase difference Φ_p compared with a wave scattered by the origin. This phase difference Φ_p for the point $p(x, y, z)$ and the reflection hkl is (see Refs. [35, 36]):

$$\phi_p = 2\pi \, (hx + ky + lz) \qquad (4)$$

The point $p(x, y, z)$ thus adds $f_p e^{i\Phi_p}$ to the total amplitude $|F(hkl)|$. $(i = \sqrt{-1})$. By integrating all the f_p over the unit cell (whose volume = V) one obtains the *structure factor* F(hkl):

$$F(hkl) = V\int_0^1\int_0^1\int_0^1 \rho(x, y, z)e^{i\phi_p}dxdydz \qquad (5)$$

In practice, the electron density distribution is seldom given by a function which can be integrated, but by atomic locations. In this case, one substitutes in (5)

$$f_n \, e^{i\phi_n} \qquad (6)$$

for the integration over the volume of an atom. Here, f_n is the *atomic scattering factor* of the nth atom of the structure, and Φ_n is the phase at the position of the atom center, as defined in Equation (4). The atomic scattering factors are monotonous decreasing functions of the type of atom and of the Braggs angle ϑ (3), which have been tabulated[37–39] and graphically drawn[40], and which for $\vartheta = 0$ are equal to the number of electrons of the atom. Equation (5) may thus be rewritten as

$$F(hkl) = \sum_n f_n \, e^{i\phi_n} \qquad (7)$$

As the graph in Fig. 13 shows for two atoms, Equation (7) can also be expressed by sine and cosine functions of the phase angle:

$$F(hkl) = \sum_n f_n \cos \phi_n + i\sum_n f_n \sin \phi_n \qquad (8)$$

If one takes into account the *temperature factor* T_n, a correction for thermal motion in the crystal, dealt with in section 7.1:3.3.4, one obtains

$$F(hkl) = \sum_n f_n \cos \phi_n \, T_n + i\sum_n f_n \sin \phi_n \, T_n \qquad (9)$$

by which form the structure factor F(hkl) is usually calculated.

If of the n atoms of a unit cell j belong to one lattice complex (see 7.1:1.1.5), and if k lattice complexes exist in the cell, then $j \cdot k = n$. The atoms of one lattice complex normally have the same values of f_n and T_n; (9) can thus be rewritten

$$F(hkl) = \sum_k f_k T_k \sum_j \cos\phi_{j,k} + i\sum_k f_k T_k \sum_j \sin\phi_{j,k} \qquad (10)$$

Since the lattice complexes are given for each of the space groups, the j trigonometric functions belonging to the partial sums

$$\sum_j \cos \phi_{j,k} \text{ and } \sum_j \sin \phi_{j,k}$$

may be summarized using the trigonometric addition theorems. Very simplified equations are obtained. For example, the lattice complex of space group No. 2: $P\bar{1} – C_i^1$ consists of the two points x, y, z and $\bar{x}, \bar{y}, \bar{z}$, (Crystallographers usually write negative signs over the quantity to which they belong to distinguish them from the subtraction sign). Equation (10) thus becomes

$$\sum_j \cos \phi_j = \cos 2\pi(hx + ky + lz) + \cos 2\pi(-hx - ky - lz)$$
$$= 2 \cos 2\pi(hx + ky + lz) \qquad (11)$$

and

$$\sum_j \sin \phi_j = \sin 2\pi(hx + ky + lz) + \sin 2\pi(-hx - ky - lz)$$
$$= 0 \qquad (12)$$

Equation (12) applies for all space groups having a center of symmetry at the origin.

The $\sum_j \cos \phi_j \equiv A$ and $\sum_j \sin \phi_j \equiv B$

have been tabulated as structure factor components A and B for all space groups[41], so that Equation (10) takes the form which is most convenient for practical computations:

$$F(hkl) = \sum_k f_k T_k A_k \, (hkl) + i\sum_k f_k T_k B_k \, (hkl) \qquad (13)$$

[34] Ref. [334], p. 43.

[35] Ref. [329], p. 259.

[36] Ref. [334], p. 47.

[37] *J.B. Forsyth, M. Wells*, Acta Crystallogr. *12*, 412 (1959).

[38] Ref. [318], p. 201

[39] *D.T. Cromer, J.T. Waber*, Acta Crystallogr. *18*, 104 (1965).

[40] Ref. [334], p. 52.

[41] Ref. [313–315], p. 367.

The absolute value of the structure factor, $|F(hkl)|$, is called the *structure amplitude*.

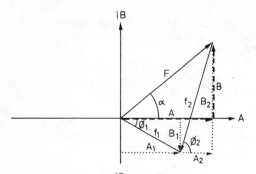

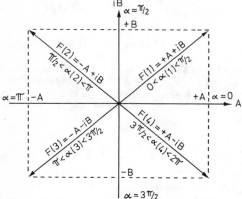

Fig. 13. a) Addition of the structure factor components of two atoms, giving the structure factor F in the complex plane
b) Relationship of A, B and α, giving the same $|F|$ in the four quadrants of the complex plane

$$|F(hkl)| = \left\{ \left(\sum_k f_k T_k A_k(hkl) \right)^2 + \left(\sum_k f_k T_k B_k(hkl) \right)^2 \right\}^{1/2} \tag{14}$$

As Equation (13) shows, the square of the structure amplitude may also be written as the product of the structure factor $F(hkl)$ and its complex conjugate $F^*(hkl)$:

$$|F(hkl)|^2 = F(hkl) \cdot F^*(hkl)$$

$$= \left\{ \sum_k f_k T_k A_k(hkl) + i \sum_k f_k T_k B_k(hkl) \right\}$$

$$\times \left\{ \sum_k f_k T_k A_k(hkl) - i \sum_k f_k T_k B_k(hkl) \right\} \tag{15}$$

This quantity $|F(hkl)|^2$, corrected as discussed in Section 7.1:3.3, is proportional to the intensity $I(hkl)$ of an X-ray beam, reflected by a net plane stack (hkl) as can be measured by different methods. Analogous to Equation (7), $F(hkl)$ also may be defined as

$$F(hkl) = |F(hkl)| \cdot e^{i\alpha(hkl)}$$

$$= |F(hkl)|\cos\alpha(hkl) + i|F(hkl)|\sin\alpha(hkl) \tag{16}$$

in which $\alpha(hkl)$ is the phase angle of the entire structure. Comparison with Fig. 13b shows that

$$\cos\alpha(hkl) = \text{const} \cdot A(hkl) \tag{17a}$$
$$\sin\alpha(hkl) = \text{const} \cdot B(hkl) \tag{17b}$$

Thus,

$$\text{tg }\alpha(hkl) = B(hkl)/A(hkl) \tag{18}$$

In the case of a centro-symmetric space group, for which $B = \sum_j \sin\alpha_j = 0$ (see Equation 12), (18) becomes

$$\text{tg }\alpha(hkl) = 0/A(hkl) = 0 \tag{19}$$

and

$$\alpha(hkl) = \begin{cases} 0 \text{ for positive } A(hkl) \\ 2\pi \text{ for negative } A(hkl) \end{cases} \tag{20}$$

(cf. Fig. 13b). In this case, one generally no longer speaks of the phase α, but of the sign:

$$F(hkl) = A(hkl) \tag{21}$$
$$= |F(hkl)| \cdot \text{sign of } A(hkl) \tag{22}$$

At this point it should be mentioned that the structure amplitude $|F(hkl)|$ is indeed proportional to $\sqrt{I(hkl)}$, the square root of the intensity, a quantity which can be measured. At the present state of knowledge, however, there is no means of direct access from any measurable values to the phase angle α (hkl), and thus to the structure factor. For Fourier synthesis (Section 7.1:2.3), therefore, which is dealt with in the next chapter, and which should permit direct calculation of an electron density function and thus of the structure itself, one half of the necessary information is in principle lacking.

Methods for stepwise approximation to the phase angles or signs are given in Sections 7.1:5 and 7.1:5.5. One can, of course, calculate the α (hkl) for a known or given structure, using Equations (4) to (18).

2.3 Electron Density Functions by Fourier Synthesis[42–44]

If the structure amplitudes $F(hkl)$ or the structure factors $|F(hkl)|$ are known, together with the phases α (hkl) or signs, one can invert Equation (5) and calculate the electron density ϱ (XYZ) of the point x, y, z in the unit cell (where V = volume of the unit cell):

$$\rho(XYZ) \tag{23}$$

$$= \left(\frac{1}{V}\right) \sum_{h=-\infty}^{+\infty} \sum_{k=-\infty}^{+\infty} \sum_{l=-\infty}^{+\infty} F(hkl) \cdot e^{-2\pi i(hX + kY + lZ)}$$

(Capitals X, Y, Z as in Ref. [45]). (23) is the fundamental equation for the Fourier synthesis[42–44], the so-called Fourier transform of (5). By Euler's relation[46], Equation (23) may be rewritten as:

$$\rho(XYZ) \tag{24}$$

$$= \left(\frac{1}{V}\right) \sum_{h=-\infty}^{+\infty} \sum_{k=-\infty}^{+\infty} \sum_{l=-\infty}^{+\infty} |F(hkl)|\cos(2\pi(hX + kY + lZ) - \alpha(hkl))$$

The Fourier synthesis is one of the most important tools for crystal structure determinations by X-ray methods (see 7.1:5 and 7.1:5.5.2).

Owing to the great practical importance of the method, many other forms of Equations (23) and (24) exist, which make these formulae, which are very cumbersome for numerical calculations, more convenient and less time-consuming. Older forms, which are sometimes still used, can be found in Refs. [47-50]; the equations used in Ref. [47] are discussed in full detail in Ref. [51]. Modern proposals, which are compatible with present-day electronic data-processing techniques, have been discussed[52]; in addition, they are incorporated in computer programs[53, 54].

3 Methods of Data Collection

3.1 Selection of Material for Crystal Structure Analysis

For single crystal structure determinations using X-rays, unfortunately, single crystals are by no means always used. In particular, rapidly grown 'crystals', which are usually formed in organic synthetic work, do not generally fulfil the minimum requirements for structure determination. Suitable material can be grown only by slow patient 'breeding' of crystals, sometimes over periods of months or years[55, 56].

3.1.1 Size and Shape of Crystal

The intensity of X-rays scattered by a crystal is inter alia proportional to the volume of the crystal. It should, therefore, be favorable to choose an especially big specimen. On the other hand, absorption (see 7.1:3.3.2) under certain circumstances destroys considerable parts of this intensity. Since the absorption is an exponential function of the path traversed by the beam through an absorbing medium, as small a crystal as possible might be more favorable. If the dimensions are chosen such that the intensity is attenuated by absorption by a factor of about $1/e$, for the mean diameter of the crystal, one has, in general, an appropriate compromise.

If absorption is low, a mean diamter of 0.1–0.2 mm should give useful measurements. If heavy atoms are present, the absorption will be greater. Crystals which are too large often can be detected on photographs, since their reflections have black edges (generated because of scattering by the edges of the crystal), but a white spot in the center.

When the intensities are recorded, for each of the reflections the crystal is orientated in a different way relative to the primary and reflected beams. If the shape of the crystal is very irregular, the path length of the beam – and, therefore, the absorption and other effects – will be very different from reflection to reflection. Consequently, the crystal to be investigated should be as regular in shape as possible. Some authors recommend the spherical form, but in many cases it is very difficult to grind spheres. The crystal should at least have approximately equal dimensions in all directions.

Some recording geometries have directions with little influence on absorption. With a Weissenberg goniometer one may, therefore, use needles oriented along the direction of the rotation axis; and with a precession camera, a leafshaped specimen parallel to the recording film, without adversely affecting the measurement results (see 7.1:3.2).

3.1.2 Crystal Imperfections

Crystals which at first sight appear to be single crystals, are often really aggregates of many little crystals grown from one common nucleus, but having slightly different orientations. If this effect increases, the X-ray reflections scattered by that crystal split in very different ways, and cannot be used for exact measurements (cf. Ref. [57]). If this crystal defect is not too great, it can sometimes be compensated for by integrated recording methods (see 7.1:3.2.5); all parts of the X-ray beam, scattered by all parts of the crystal, must be detected. In most cases, of course, one detects so much of the background radiation together with the reflections, that it would be better to seek a crystal of better quality. An effect connected with this crystal imperfection is *extinction* (see 7.1:3.3.3).

It is, of course, quite impossible to use material which is splintered during the preparation of the crystals, or which is contaminated by small crystallites which have grown with, or which adhere to, the single crystals, etc. Such crystals can be detected in most cases even under the microscope, or, if photographs are taken, by weaker scattering patterns which appear on the films in addition to the scattering pattern of the main crystal but which have diverging orientations. If *counter methods* are used, one cannot see a large part of the reciprocal lattice all at once, and it is, therefore, easy to fail to observe such errors. Consequently, a new crystal should always be checked first by a photographic method so as to avoid wasting time and labor on bad material.

3.1.3 Twinning

It very often happens that crystals of the same kind grow together following certain laws. In these cases, one speaks of *twins*[58-60] (also trillings, etc.).

[42] Ref. [323].
[43] Ref. [317], p. 317.
[44] Ref. [329], p. 352.
[45] Ref. [313-315], p. 373.
[46] Ref. [323], p. 16.
[47] Ref. [313], p. 374.
[48] Ref. [323], p. 37.
[49] Ref. [323], p. 178.
[50] Ref. [329], p. 370.
[51] Ref. [323], p. 21.

[52] Ref. [23].
[53] Ref. [319], No. 418–443.
[54] Ref. [319], No. 5063–5068.
[55] Ref. [329], Kap. 5, p. 53.
[56] Ref. [318], p. 5.
[57] Ref. [329], p. 69.

Some compounds nearly always form twins, and it is practically impossible to find a real single crystal. A crystal having re-entering angles is never a single crystal and often a twin, for a single crystal can only have salient angles. Twins can often also be detected with the *polarizing microscope*[61-63]. Unfortunately, neither of these methods is absolutely certain, since the effects to not occur with all twins. Because the two individuals of a twin crystal have the same lattice, their reciprocal lattices — i.e. the locations of their reflections — often coincide, in complete contrast to the case in which crystals accidentally grow together (see 7.1:3.1.2).

If the space group of the crystal has *systematic absences* (i.e. 'motifs' of systematically missing reflections in the reciprocal lattice originating from certain symmetry properties: see 7.1:4.2), which do not coincide within the partial lattices scattered by the single individuals of the twin, one sometimes observes virtual absences, which cannot be interpreted by any of the space group symmetries, and which are, therefore, not listed in the International Tables[64, 65]. In such a case, one should suspect twinning. However, sometimes cases exist in which, by chance, the atomic positions interfere in such a way that additional absences appear[66]. Care must, therefore, be exercised, when observing such absences.

It can be very unpleasant to discover that one has been working with a twin, only when the crystallographic computations are being worked out. This is sometimes discovered at a very late state of a structure determination, by obtaining results which cannot be interpreted in any other way. Recently, however, computer programs have been written[67], with which, under favorable circumstances, the structures of twin crystals may also be calculated[66, 68].

3.2 Measurement of Intensities and Methods of Recording

Intensities of reflected X-radiation are now recorded according to two different principles: with films or with counters. Films have the advantage that a larger part of the reciprocal lattice (i.e. many reflections together) can be seen simultaneously.

Films, therefore, offer a comparatively easy survey of symmetry, lattice parameters, diffuse regions between the reflections, and other similar features. Moreover, the *single crystal goniometers* used for film techniques are cheaper by a factor of about ten, than the automatic *counter diffractometers* which are used today.

When measuring with counters, each of the reflections must be recorded separately. Therefore, one can never see greater parts of the reciprocal lattice simultaneously. Some reflections or larger regions of the lattice may more easily be associated with the wrong indices, but, on the other hand, measurements are much more exact than when using film. The recording times may be of the same order, if very high accuracy is required, but, in most cases, they are shorter.

The best results are obtained with a combination of both methods. A general survey of the reciprocal lattice is obtained by a film procedure, and then the intensities are measured using a counter.

3.2.1 Primary Radiation

Monochromatic X-radiation is normally used for all recording techniques. The older Laue's method[69], using 'white' X-radiation, should be used only for special purposes.

An X-ray tube emits a series of discrete lines from the spectrum of its anodic material, but, unfortunately, this is partly obscured by a background of general radiation (see Fig. 14). A filter or a monochromator[70, 71] can be used to obtain the desired line of this spectrum monochromatically: the $K\alpha$-line (which is the strongest one) is chosen mostly.

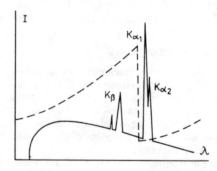

Fig. 14. X-ray spectrum (———) of an element of atomic number Z (schematic), with absorption edge of the element having atomic number Z−1 (------)

[58] Ref. [329], p. 53.
[59] Ref. [318], p. 6.
[60] Ref. [336], p. 109.
[61] J. Orcel, Bull. Soc. Fr. Mineral. Cristallogr. *51*, 197 (1928).
[62] Ref. [320].
[63] Ref. [322].
[64] Ref. [313], p. 75 ff.
[65] Ref. [323], supplement.
[66] G. Bergerhoff, E. Schultze-Rhonhof, Acta Crystallogr. *15*, 420 (1962).
[67] Ref. [319], No. 5068.
[68] G. Bergerhoff, H. Kasper, Acta Crystallogr. *B24*, 388 (1968).
[69] Ref. [336], p. 336.
[70] Ref. [329], p. 130 ff.
[71] Ref. [318], p. 73 ff.

The filter makes use of the fact that there are certain discontinuities in the behavior of the absorption coefficient as a function of the wavelength (these are known as absorption edges, see Fig. 14). If an element having an atomic number which is 1 or 2 less than that of the anodic material is chosen for the filter, absorption is relatively weak for the wavelength of the Kα-line, but very strong for wavelengths which are only a little shorter. By this method, one can achieve sufficient suppression, especially of the generally troublesome Kβ-line, together with a good deal of general radiation. Very pure monochromatic radiation is obtained by scattering the primary beam by a crystal which is orientated such that one of its strongest Bragg's reflections scatters. This arrangement is called a *monochromator*. Unfortunately, the yield of monochromatic X-rays, which serve as the primary beam for the real measurement, is very low, so that longer times or better detectors must be used for recording, than when using a filter.

3.2.2 Photographic Methods

Using photographic methods, one never measures the intensities of diffracted X-rays directly, but the blackening of a photographic film generated by them. Contrary to former practise, involving visual comparison of the reflections with a standard, a photometer is now normally used.

If a beam of light, of intensity L_0, penetrates a film which has been exposed to X-rays and developed, and which has the blackness S, the beam is attenuated in passing through the film, and has a lower intensity L. Equation (25) holds for L_0, L and S[72,73]:

$$S \cong \log (L_0/L) \qquad (25)$$

The relationship between S and the intensity I of the X-ray beam is linear to an upper limit of I, beyond which it is non-linear:

$$S \cong f(I) \cdot I \qquad (26)$$

In addition, for a good photometer the intensity L incident upon the cell should be proportional to the photo-voltage, which may be recorded on a galvanometer as scale deflection A. One may then write:

$$I \cong 1/f(I) \cdot \log(A_u/A_r) \qquad (27)$$

In this equation, f(I) is a constant to the upper limit of linearity mentioned above. A_r is the galvanometer deflection caused by a beam of light which has passed through the photographic image of an X-ray reflection. A_u is similarly the recorded value, produced by a beam which is weakened only by the general blackening of the film, i.e. the background. Since this general blackening also occurs at the position of the reflection, it is advisable to assume a value for L_0 which is the intensity of the light-beam after attenuation by the background. Not all relationships mentioned are always truly linear.

A standard curve should, therefore, be recorded (see Fig. 15) of the f(I) of Equation (27) for each film material, to correct for non-linear values.

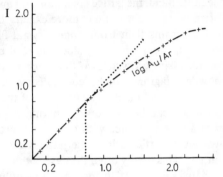

Fig. 15. Standard curve for an X-ray film, as in Ref.[74] (Linear part ········; I = Intensity)

3.2.3 Powder Methods

The simplest method of recording an X-ray diffraction diagram with crystalline material is the *powder method* published by Debije and Scherrer[75,76]. When a fine crystalline powder is placed in the path of an X-ray beam, for each of the net planes of the crystal lattice, some of the small crystallites are always orientated in such a way that Bragg's Equation is valid for that plane. In the usual 'Debije-Scherrer cameras' (see Ref.[77]), a cylindrical film encloses the specimen, which is in the form of a tiny cylinder. Rotation of the specimen during the exposure increases the probability of reflection at each of the net planes[78].

If a camera radius of 57.3 mm is chosen, as is the case for some commercial models, the conversion factor from millimeters distance on the film to Bragg's angle is $2 \vartheta = 1°$ per mm. The exposure times are of the order of 10 minutes to several hours, depending on the radiation used and the nature of the sample.
A modification of the Debije-Scherrer camera is the camera of A. Guinier[79]; this uses radiation which is rendered monochromatic and focused by a concave curved crystal arranged in the path of the primary radiation. The sample, therefore, need not be in the form of a small cylinder, but is simply spread on a planar plate.

Complete structure determination cannot be achieved with powder methods, except in a very few, extremely favorable cases, because correlation of the peaks on a film with distinct net planes is often uncertain; reflections of different net

---

[72] Ref.[329], p. 78 ff.
[73] *E. Schultze-Rhonhof,* Dissertation, p. 67, Universität Bonn 1964.

[74] Ref.[73], Fig. 22.
[75] *W. Kast* in Ref.[325], p. 551.
[76] Ref.[327].
[77] *W. Kast* in Ref.[325], p. 552, Fig. 5a.
[78] *W. Kast* in Ref.[325], p. 552, Fig. 5b.
[79] *W. Kast* in Ref.[325], p. 553, Fig. 6a+b.

planes often coincide, and the intensities of the reflections cannot normally be recorded sufficiently accurately.

On the other hand, the lattice constants and angles can often be determined much more exactly from powder diagrams than from single crystal films, especially if approximate values are already known from single crystals so that each peak of the powder diagram can be correlated with the correct net plane. Sometimes analogies of the lattice structure can be seen by comparing the diagram of a new substance with that of an already known one – particularly if the lattices are 'isomorphous'. This is very often the case with chemically related substances. In addition, often only the powder diagram of a substance is used for identification, since any kind of crystal can be uniquely characterized by the positions and relative intensities of its peaks. For this purpose, a voluminous catalog of such diagrams is available, the ASTM Powder Diffraction Data File[80] (cf. Ref. [81–85]). For a detailed description of the powder method, see Ref. [86]).

3.2.4 Single Crystal Diffractometers for Photographic Methods

To record data from *single crystals,* it must be possible to orient the crystal relative to the primary beam for each net plane in such a way that Bragg's Equation (3) is valid and the reflection appears at a defined position. The most important single crystal diffractometers are the Weissenberg camera and the Buerger precession goniometer.

On the *Weissenberg camera,* the single crystal can be rotated about one of its crystallographic axes. To record the diffracted radiation, the crystal is surrounded coaxially by a cylinder of film. The primary beam is incident on the crystal from the side; when the goniometer is in its starting position, the beam is perpendicular to the axis of rotation. If the crystal is rotated about its axis of rotation, Bragg's Equation will sometimes be satisfied for each of the net planes of the crystal; a diffracted beam, a reflection, therefore, results. These reflections appear as circular lines of intersection of cones with the cylindrical film. These curves are called 'layer lines' or 'levels'; each is an image of a plane of the reciprocal lattice, perpendicular to the axis of rotation (see Fig. 16, cf. Ref. [87]). The level in the middle of the film is assigned the number 0 and is generally called the *Zero level.* Its 'cone' is the plane

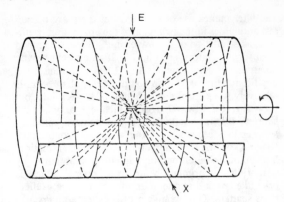

Fig. 16. Diffraction of an X-ray in the rotating-crystal method; schematic drawing. E = equator

which contains the primary beam and the image of the origin of the reciprocal lattice. The other levels are counted positive in one direction, and negative in the other direction. At the same time, the number of the level is a common Miller's index for all the reflections situated on it and indeed is that which corresponds to the rotation axis. For example, all reflections of a crystal which is rotated about b, have $k = 0$ for the zero level.

The distance a_n of the nth level from the zero level gives the diffraction angle μ_n, which corresponds to the lattice constant in the direction of the axis of rotation. If r is the radius of the camera, the equation holds:

$$\text{tg}\,\mu_n = a_n/r \tag{28}$$

and the lattice parameter p is given by:

$$p = \frac{n \cdot \lambda}{\sin\mu_n} \tag{29}$$

Unfortunately, this *rotating-crystal method*[88] yields very little information about the other two lattice constants and about the lattice angles. Indexing of single reflections within a level is also very difficult. Weissenberg, therefore, worked out an improvement of the method, which enables correct indexing of each single point. The other two lattice constants and one of the angles can also be obtained without difficulty. It is more difficult to determine the other two angles, unless they are exactly 90°.

Using the *Weissenberg method*[89–94], a single layer is selected by a slotted screen situated between film

[80] Ref. [311].
[81] Ref. [309].
[82] Ref. [312].
[83] Ref. [307].
[84] Ref. [308].
[85] Ref.[310].
[86] Ref. [327].

[87] *G. Habermehl,* Angew. Chem. *75,* p. 80, Fig. 4 (1963).
[88] Ref. [334], p. 92.
[89] Ref. [87], Fig. 5.
[90] Ref. [327], p. 275.
[91] Ref. [317], p. 185.
[92] Ref. [334], p. 252 ff.
[93] Ref. [334], p. 375 ff.
[94] Ref. [334], p. 133.

and crystal. At the same time, the film cassette is moved along the slot in the direction of the axis of rotation synchronously, as the crystal is rotated. The layer, which was formerly a line on the film, is thus spread to give a twodimensional, but distorted, image of a plane of the reciprocal lattice (see Fig. 17, Ref. [89]). The position of each single reflection on the film can, consequently, be correlated uniquely to the orientation of the crystal at the moment when the reflection appears.

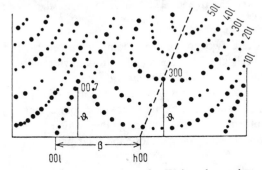

Fig. 17. Schematic drawing of a Weissenberg photograph
ϑ is drawn for the reflections (007) and (300); the distance which is equivalent to the angle β is also given

Reflections with a common index occur on characteristic curves (Fig. 18)[90, 91]. Each reflection lies on the point of intersection of two of these curves; its indexing is thus given. Lines of the reciprocal lattice, which contain the origin, on the Weissenberg photograph have images which are also straight lines. The angle between these lines can be determined very simply: on commercial cameras, a distance of 1 mm between two straight lines on the film is equivalent to an angle of 2° between the reciprocal lattice rows, i.e. 90° = 45 mm. In dealing with upper levels, one must consider, when taking the photographs, that the reflecting planes in the crystal are inclined with respect to the rotation axis.

In the *equi-inclination method*[92] the crystal is turned, together with film and slotted screen, through an *equi-inclination angle* ν with respect to the primary beam. By choosing for ν the μ_n from Equations (28) and (29), one obtains Weissenberg photographs whose interpretation is very easy. In the most common case, where the reciprocal axis belonging to the axis of rotation is perpendicular to the plane of the zero-layer, i.e. for which the reciprocal and the real axes coincide, the points of all layers lie exactly in the same positions if one superimposes the films. Such coinciding reflections differ only in the index which indicates the number of the layer. If the reciprocal axis belonging to the axis of rotation is not perpendicular to the zero plane, small shifts (*offsets*) occur, which can be used for the determination of lattice angles[93].

The Schiebold-Sauter method[95], in which a layer, separated by a slotted screen, is recorded on a plane film which rotates synchronously as the crystal rotates, is no longer of practical importance.

Fig. 18
Curves representing the reflections having an identical index in a Weissenberg photograph
Drawn by the IBM 7090 of the G.M.D., Bonn

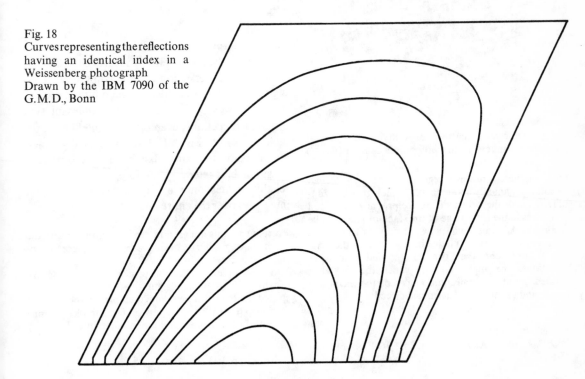

The *precession method* uses a very different principle. In this method, the normal of a plane of the reciprocal lattice (which is a lattice row of the real lattice), precesses around the primary beam. The beams generated by a reciprocal lattice plane then lie on a cone. Parallel planes of the reciprocal lattice, whose images also may be called 'layer lines' or 'levels', generate coaxial cones with different aperture angles. Each of these cones can be separated by a suitable circular slotted screen, so that each layer line can be registered on a single photographic film. The screen and the film are moved in such a way that both of them remain parallel to the reciprocal plane which is recorded. In contrast to the Weissenberg goniometer, the precession camera[96, 97] yields undistorted images of the reciprocal lattice (Fig. 19), so that its symmetry can be seen immediately. The range of recording of the precession camera, the number of reflections which can be observed, is, however, smaller than with the Weissenberg method.

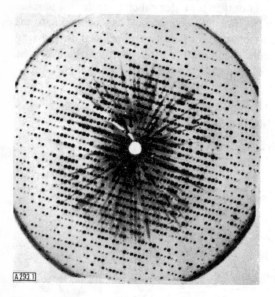

Fig. 19. Diffraction pattern of a myoglobin crystal photograph taken by the precession camera (Ref.[97])

On a commercial precession goniometer the crystal can be rotated manually about an axis, which goes horizontally through the origin of the reciprocal lattice, and which lies in the zero plane of the stack of planes to be recorded. All other planes which have this axis in common can thus be made the zero layer; the reciprocal lattice can, therefore, be recorded from other directions. The angles through which the crystal is rotated are at the same time angles of the reciprocal lattice and can thus be determined directly.

[95] Ref. [334], p. 312.
[96] *J.C. Kendrew*, Angew. Chem. *75*, 596, Fig. 1 (1963).
[97] Ref. [332].

Once a crystal is adjusted on a goniometer head on a precession camera, it can be transferred immediately onto a Weissenberg camera. There it is already correctly adjusted (except for small errors which are due to mechanical imperfections of the goniometer heads and their holders), for the axis about which the crystal may be turned by hand on the precession goniometer is identical with the rotational axis of the Weissenberg camera. On the other hand, transfer from the Weissenberg camera to the precession goniometer is less easy. On the precession goniometer, one has first to search for one of the planes of the reciprocal lattice by rotating about the rotational axis, and to adjust the crystal exactly.

These two goniometers complement each other excellently for a survey of an unknown reciprocal lattice. With the Weissenberg camera, one obtains a stack of planes normal to the axis of rotation. The precession camera provides complementary information about all the planes which have the rotation axis in common, and their parallels (Fig. 20).

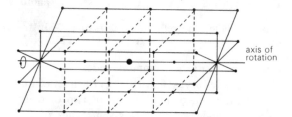

Fig. 20. Relationship of Weissenberg and precession photographs from an identically adjusted crystal ● Reflection 0, 0, 0; ------ Weissenberg photographs; ——— precession photographs

3.2.5 Integrating Methods

Apparently the most simple measure of the intensity of an X-ray reflection is the maximum blackening generated by the reflection on a film. Unfortunately, however, the blackening profiles of two different reflections, even on the same film, never look the same. Fig. 21 shows two peaks which actually have the same intensity, but very different maximal blackening. Therefore, one can obtain reliable values only by using a method which integrates over the entire intensity of a reflection[98].

Fortunately, this integration can be performed with the usual photographic recording techniques by use of very simple devices.
The area under a reflection profile can be found approximately by dividing it into a number of equally spaced vertical samples and multiplying the sum of their heights by the spacing of the samples. If one records a reflection several times, each time shifted by

[98] Ref. [329], p. 103.

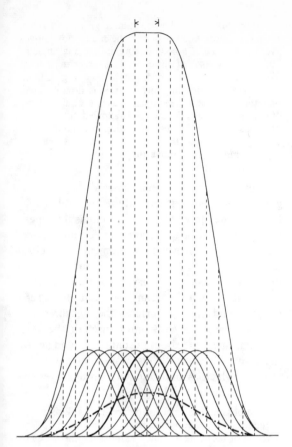

Fig. 21. Blackening profiles
———— and ·—·—·—·— circumscribe the same area. By summation of 10 single peaks, each of them shifted by a certain spacing, an integrated peak, having a plateau is obtained

such a spacing, all the peaks sum up to give a sum curve (Fig. 21), which has a nearly constant plateau, if one shifts often enough. The height of the plateau is then a measure for the desired *integrated intensity*. In practice, this principle is, of course, used in two dimensions; the spacing of the samples can be varied independently in both directions. The form of the plateau can be adapted to the requirements of the special recording problem.

In an integrating Weissenberg or precession camera, the film holder is shifted mechanically each time by one spacing, when all desired reflections are recorded; the single point of a reflection thus becomes a rectangle after a certain number of exposures. When the film holder has reached its starting position again, a 'cycle' has been completed. Integrated exposures for quantitative data collecting should be finished only after full cycles, since otherwise no constant plateau is obtained and the integrating effect is lost.

The intensities of the reflections from a single exposure often differ by several powers of ten, so that in most cases only part of the reflections have optimal conditions for observation. Therefore, one usually records from the same level different exposures which differ in the number of their cycles; each reflection will, thus, have optimal conditions for measurement on some film. Later, the different exposure times are corrected for by scale factors.

3.2.6 Counter Diffractometers

To measure the intensities of X-ray reflections with counters, for each single reflection one must arrange the primary beam, the crystal, and the counter in such a way, that:

the primary beam, the normal to the reflecting lattice plane stack, and the diffracted beam lie in one plane,
the diffracted beam passes through the axis of the counter, and
the angles of incidence and emergence equal the Bragg's angle of the recorded reflection.

Since the X-ray tube, together with the attached cables, water supply, monochromators etc., is a very inflexible object, the primary beam is given a fixed direction. The counter, which is likewise relatively heavy, in most instruments is rotated only about one of its axes: the socalled $2\,\theta$-axis. In this case, the counter and the primary beam lie in the plane of this rotation. On instruments with Weissenberg geometry, one has an additional rotation about the equi-inclination angle v, which leads the counter along the surface of a cone which may be imagined as being round the rotation axis (cf. Section 7.1:3.2.4 and Fig. 16), and which is set once for each level. In all cases, the angle between the primary and the secondary beam (=direction: X-ray tube − crystal/direction: crystal − counter) must be $180°-2\theta$. The crystal itself must be orientated in such a way, that the normal of the net plane stack, which is recorded, lies in the plane of the primary and the secondary beam and divides their angle into two halves.

The conditions mentioned above are thus realized. They are now no longer infringed by any rotations of the crystal about the normal to the lattice plane stack. In principle, an infinite number of possibilities exist for orientation of the crystal with respect to the primary beam and the counter.

Positioning can be done by hand for each single reflection, but this is very tedious and much more timeconsuming than photometric data collecting from the film. Manual operated counter diffractometers have, therefore, not achieved great success. On the other hand, a number of instruments have been developed during the last few years in which, once a crystal has been adjusted, positioning and data collecting are carried out automatically, more or less by electronic means. Simpler instruments use a normal electronic regulating device; others employ a regular processing computer. Within the foreseeable future, such computers may also adjust and correct themselves: hitherto, this has had to be done by hand.

At the moment, two different principles of construction are employed for counter goniometers: instruments having Weissenberg geometry (7.1:3.2.4), and those in which the crystal is orientated on an *Eulerian cradle*.

Using the Weissenberg geometry, one replaces the moving film-holder by the counter, which moves along the cone surfaces where the secondary radiation of the different levels appears (cf. Fig. 16).

The equi-inclination angle and the corresponding angle at the counter are normally adjusted manually, for the change from one level to another one is necessary only in intervals of hours or days. Thus, for the crystal itself, one single setting of an angle about the axis of rotation is sufficient. In addition, the counter must be moved into the recording position.

With one of the commercial instruments (PAILRED; Philips, Eindhoven), the rotation of the crystal about the axis of rotation is governed by a kind of analog computer. On a right-angled cross of two iron rods, two sliders (one of them connected to the second rod) are moved by step motors, proportional to the reciprocal lattice constants of the reflection to be recorded. The position of these sliders is transformed mechanically into a rotatory motion of the crystal, which positions the crystal correctly for the measurement[99].

The *Eulerian cradle* consists essentially of three axes, each connected with the others (see Fig. 22). Firstly, it can be rotated as a whole about the 2ϑ-axis of the counter, but it is independent of this. Its axis is therefore here called the ω-axis. Perpendicular to it is the χ-axis, accomplished by a circular arc or a full circle. On this axis, again perpendicular to χ, the Φ-axis moves.

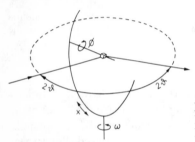

Fig. 22
Principle of the
Eulerian cradle

If these three axes can all be rotated about the full 360°, a crystal mounted at their common point of intersection can be brought into any desired position. For mechanical reasons, with some instruments the χ-circle can be constructed only for less than 180°. The number of measurable reflections is, however, always limited considerably. Some manufacturers try to remove this difficulty by incorporating additional circles.

In a newly developed instrument (CAD-4, Enraf-Nonius, Delft), the χ-axis, which is perpendicular to the ω-axis, is replaced by a so-called κ-axis, which is inclined by only 50° to the ω-axis. The Φ-axis is positioned on this κ-axis, likewise at an angle of 50°. When a rotation about κ is performed the Φ-axis now moves over the surface of a cone having an opening angle of 100°. The axis itself of this cone, the κ-axis can perform the full circular rotation about ω (see Fig. 23, cf. Ref.[100]). Since the hindering χ-circle is now omitted, only the goniometer head itself causes a small number of measurable reflections to be shadowed out. In addition, this principle has the advantage that cooling and heating devices and the like may be placed near the crystal, immediately above the top, without hindering any motion of the instrument.

The real measurement begins after all angles of the goniometer have been correctly set for a reflection. If one wishes to obtain correct intensities, measurements must be made about a certain angular region. This is achieved by movement of the crystal (ω-scan) or of the counter (ϑ-scan) about the theoretical position. The crystal or the counter is moved about half the 'scan-angle' out of the zero position, and then, performing the measurement, moved once or several times through the zero position, each time about the entire scan-angle. At the points of return of this movement, the background (which is unfortunately always present) must be measured.

Several possibilities exist for determining the sequence of recording reflections and background. Each has the purpose of reducing as much as possible the space of time during which the instrument is moving but not measuring. If a computer is connected with the diffractometer, the profile of the reflection can also be recorded in order to make the measurement more accurate by calculation.

The shape of a peak depends inter alia upon the mutual orientation of the normals to the net plane and to the rotation axis of the scan. If the two coincide, Bragg's equation is fulfilled, even for a scan of 360°. The reflection will become infinitely broad. The sharpest peaks are obtained when the two directions are mutually perpendicular because, in this case, Bragg's condition is realized only for a moment during the scan. Correction for this effect by calculation is very easy, however, if the orientation of all axes is known (see 7.1:3.3.1).

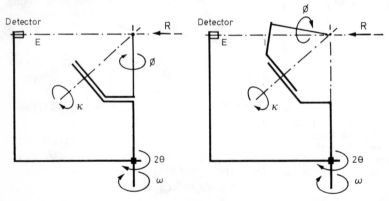

Fig. 23. The principle of the
κ-axis with the diffractometer
CAD-4 of Enraf-Nonius, Delft
(with kind permission of
Enraf-Nonius)

E = Equatorial plane
R = Direction of X-rays

3.3 Correction to the Intensities

To compare the intensities I(hkl) recorded by one of the methods described in the preceding section with the theoretically calculated structure factor F(hkl), one must consider a number of corrections for Equation (14). In this section, the proportionality constants f_k will be replaced by certain quantities, and the temperature factor T_k will be discussed.

3.3.1 The Lorentz and Polarization Factors

The observed intensity depends upon the geometry of the measurement under the special conditions of a certain reflection hkl (cf. 7.1:3.2.6). The correcting factor involved is called the *Lorentz factor* L(hkl)[101] (cf. Ref. [102], for theory, see Ref. [103]; formulae and tables are given in Ref. [104–106].)

As a result of a polarization effect during the diffraction of an X-ray beam by a crystal, one must apply a second correction, the so-called *polarization factor* P(hkl), which is a simple function of Bragg's angle ϑ:

$$P(hkl) = \frac{1 + \cos^2 2\vartheta}{2} \qquad (30)$$

and which does not depend upon the conditions of the experiment[107–113]. To simplify calculations and tables one usually reduces the Lorentz and polarization factors to:

$$LP(hkl) = L(hkl) \cdot P(hkl) \qquad (31)$$

Equation (14) therefore becomes

$$\frac{I(hkl)}{LP(hkl)} \sim |F(hkl)|^2 \qquad (32)$$

($\sim$ meaning proportionality.)

The computer programs which exist for data reduction of X-ray data usually contain a subroutine for LP-correction (see 7.1:5.7). However, one must pay strict attention to a subroutine, which contains the true L-factor for the recording technique used, as the results of the calculation may be completely wrong otherwise.

3.3.2 Absorption

The absorption of X-rays by matter is an exponential function of the thickness t of the absorbing layer and of the so-called linear absorption coefficient μ, which itself depends on the composition of the irradiated matter and on the wavelength of the X-rays. For monochromatic radiation, as is used for structure determination, the equation:

$$I = I_o \cdot e^{-\mu t} \qquad (33)$$

holds. In this equation, I_0 is the primary intensity of the X-rays before absorption, I the residual portion which was not absorbed. To a first approximation, μ can be calculated as the sum of the so-called atomic absorption coefficients μ_a[114]:

$$\mu = \frac{1}{V}\sum_i n_i(\mu_a)_i \qquad (34)$$

n is the number of atoms of element i in the volume V.

When a crystal scatters X-rays, each volume element dV scatters (see Fig. 24). The primary beam is weakened before scattering, by the factor $e^{-\mu t_1}$, and the leaving beam by the factor $e^{-\mu t_2}$ (t_1 = path before scattering; t_2 = path after scattering; $t = t_1 + t_2$ = total path of the X-ray beam in the crystal). The beam scattered by the volume element dV is thus attenuated in all by the factor $e^{-\mu(t_1 + t_2)} = e^{-\mu t}$. To calculate the attenuation of the total X-radiation scattered by the crystal, the *transmission factor* T_A, one must integrate over all volume elements dV, obtaining:

$$\frac{I}{I_o} = T_A = \frac{1}{V}\int_V e^{-\mu t}dV \qquad (35)$$

Fig. 24
Absorption
corrections:
The paths of the
X-ray beams in the
crystal

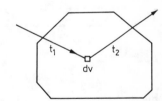

For very simply shaped crystals (spheres, cylinders) this integration can be performed quite easily. Tables of transmission factors are available[115, 116] for such crystal forms. For more complicated crystal shapes, these calculations soon become very difficult since the paths t_1 and t_2 leading to any volume element are not only different for each single reflection, but may have different values even for the same reflection in different orientations. With (35) one obtains from (14) and (32):

$$\frac{I(hkl)}{LP(hkl) \cdot T_A(hkl)} \sim |F(hkl)|^2 \qquad (36)$$

[99] Ref. [329], p. 118.

[100] Nonius Computer Controlled 4-circle Diffractometer CAD-4. Firmenprospekt der Firma Enraf-Nonius, Delft (1969).

[101] P. Debije, Ann. Phys. [Leipzig] 43, 93 (1914).

[102] H.A. Lorentz, unpubl.

[103] Ref. [329], p. 152, 156.

[104] Ref. [317], p. 266.

[105] Ref. [329], p. 178.

[106] Ref. [318], p. 136.

[107] P.P. Ewald, Phys. Z. 14, 471 (1913).

[108] H. Mark, L. Szilard, Z Phys. 35, 743 (1926).

[109] P. Kirkpatrick, Phys. Rev. 29, 632 (1927).

[110] R.W. James, Nature 121, 422 (1928).

[111] Ref. [329], p. 27.

[112] Ref. [329], p. 152.

[113] Ref. [329], p. 174.

[114] Ref. [318], p. 162.

[115] Ref. [329], p. 213.

[116] Ref. [318], p. 195.

Computer programs exist (see 7.1:5.7)[117-119] which can be used to calculate the transmission factors of each reflection, if the shape and orientation of the crystal are known. Some crystallographers, however, have doubts about the trustworthiness of such calculations (not of the programs themselves!).

The theory of absorption, and the tables necessary for calculation, can be found in Refs. [120-123], which also deal with various experiments for direct measurement of the transmission factors.

3.3.3 Extinction

Two different influences which can affect the intensity of a scattered X-ray are included in the general term 'extinction': 'primary' and 'secondary' extinction.

A beam which is scattered once by a plane of a crystal lattice obeys Bragg's Equation (3) exactly for all parallel planes through which it may pass as it exits. A part of this beam can thus be reflected several times so that, after 2n-fold scattering, it again travels in the direction of the primary beam, and after 2n + 1-fold scattering it travels in that of the secondary beam (see Fig. 25). Since, at each scattering, a phase lag of $\pi/2$, occurs, a twice reflected beam has a phase difference of exactly π with respect to the primary beam, which is therefore weakened by it.

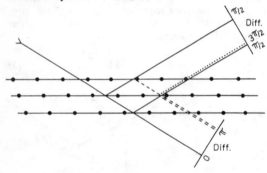

Fig. 25. Extinction by phase lags of $(2n+1)\pi/2$
-·-·-·-   lattice plane
————   primary beam; once-scattered beam
-------   twice-scattered beam
....   thrice-scattered beam

For a larger single crystal, which is of ideal structure through its entire lattice, i.e. an *ideally perfect crystal,* this effect may be of considerable magnitude especially for small angles ϑ. On the other hand, if the crystal consists only of small regions, which are twisted to each other at very small angles (such angles often cannot be detected by macroscopic methods), the multiple scattering occurs only within one such region. The exact phase relation with respect to the neighboring region is disturbed. For the *ideally imperfect crystal,* the primary extinction can, therefore, be neglected. The two limiting cases, the 'ideal perfect crystal' and the 'ideally imperfect crystal', can be fully dealt with theoretically[124-129]. Apart from a few exceptions, real crystals, unfortunately, have structures somewhere in between these two extreme cases, and calculation of the primary extinction is therefore very difficult in practice. Many authors, therefore, do not apply a correction for extinction, even though this often leads to serious errors.

When a primary beam is scattered by a crystal, it loses some of its energy at each plane through which it travels — the energy is taken up by the reflected beams. This attenuation of the intensity is termed 'secondary extinction'. Since this behaves much like ordinary absorption, it is customary when performing calculations to regard it as an additional term to be added to the absorption coefficient:

$$\mu' = \mu + g \cdot Q \qquad (37)$$

In this equation, g is a constant, and Q the energy of the reflected beam.

The correction becomes more complicated when primary and secondary extinction occur simultaneously. If E is written for the extinction corrections, Equations (14) and (36) give:

$$\frac{I(hkl)}{LP(hkl) \cdot T_A(hkl) \cdot E(hkl)} \sim |F(hkl)|^2 \qquad (38)$$

($\sim$ meaning proportionality)

3.3.4 The Temperature Factor

Hitherto, it has been assumed that the individual atoms remain motionless on defined lattice points. This is true only near 0 °K, however. Additional devices for low temperature measurements are commercially available for most goniometers. At higher temperatures the electrons of all the atoms undergo thermal motion. This causes the effective curves of the atomic scattering factors f to decrease more rapidly with $\sin \vartheta / \lambda$, than is indicated in Section 7.1:2.2 (Ref. [37-40]). This effect is corrected for by the 'temperature factor' T_k; cf. Equations (9) and (14).

In the simplest case, of an isometric (=cubic) structure, this correction has the form:

$$f_k = f_{k_0} \cdot T_k = f_{k_0} \cdot e^{-(B_k \sin^2 \vartheta / \lambda^2)} \qquad (39)$$

[117] Ref. [319], No. 362.
[118] Ref. [319], No. 5065.
[119] *B.J. Wuensch, C.T. Prewitt,* Z. Kristallogr. *122,* 24—59 (1965).
[120] Ref. [318], p. 157.
[121] Ref. [329], p. 204.
[122] *A. Classen,* Philosophical Magazine (VII) *9,* 57 (1930).
[123] *A.J. Bradley,* Proc. Phys. Soc. [London] *47,* 879 (1935).
[124] Ref. [329], p. 195.
[125] *C.G. Darwin,* Philosophical Magazine (VI) *27,* 675 (1914).

for any kind of atom, or even for any crystallographically different position of the same kind of atom. In this equation, B_k is the so-called *isotropic temperature coefficient;* in the literature this is often erroneously termed the 'temperature factor'. When demands for exactness are less, one often obtains a useful approximation by correcting the entire structure factor F(hkl) by a common mean temperature coefficient $\overline{B}$:

$$I(hkl) \sim |F(hkl)|^2 \cdot e^{-2(\overline{B}\, \sin^2\vartheta/\lambda^2)} \qquad (40)$$

Wilson's statistic gives such mean temperature coefficients $\overline{B}$, which may serve as starting values for subsequent refinements.

When the crystal under study is not isometric, its atoms normally undergo anisotropic motion. In this case, each atomic scattering factor is corrected by an 'anisotropic' temperature correction of the form:

$$T_k = e^{-(b_{11}h^2 + b_{22}k^2 + b_{33}l^2 + b_{12}hk + b_{23}kl + b_{31}lh)} \qquad (41)$$

Many authors however are content to use the simple isotropic temperature coefficient B_k as an approximation for anisotropic crystals also, since it is obviously more difficult to deal with six independent coefficients than with one.

3.3.5 Scale Factors; Wilson's Statistic

By introduction of a *scale factor s,* the proportionality (38) becomes an equation:

$$\frac{I(hkl)\cdot s}{LP(hkl)\cdot T_A(hkl)\cdot E(hkl)} = |F(hkl)|^2 \qquad (42)$$

which, together with Equation (14), describes completely the relationship between the measurable X-ray intensities I(hkl) and the parameters of a structure.

For some parts of a structure determination it is necessary to know these scale factors, especially for the quantitative evaluation of Patterson and Fourier syntheses, and for comparison of calculated with observed structure amplitudes $|F_c(hkl)|$ and $|F_o(hkl)|$ – which allows a conclusion for the quality of a proposed structure [cf. 7.1:5.5, Equation (7)]. Using data from different sources (e.g. from different films), it may be sufficient to know relative scale factors, which can often be calculated if a given reflection can be measured on more than one film. Approximate values of the absolute scale factors may, under certain circumstances, be obtained by statistical methods[139–142], the best known of which is *Wilson's statistic.*

For this, the total of all reflections is subdivided into zones having similar values of $\sin^2\vartheta$, e.g. $\sin^2\vartheta = 0$–0.2; 0.2–0.4 etc. Within each of these zones the mean value of the squares of the observed relative structure factors, $\overline{|F_r|^2}$ is formed. Under certain circumstances – in particular, the observed intensities must have a statistical distribution on the reciprocal lattice – the formula for the absolute structure should be valid:

$$\overline{|F_a|^2} \cong \sum_{j=1}^{N} \bar{f}_j^2 \qquad (43)$$

In this equation, the $\bar{f}_j$ are the values of the atomic scattering factors of the N atoms j for the mean of the zones. With (42) and (43), it follows that

$$|F_a|^2 = s' \cdot |F_r|^2 = s' \cdot I(hkl) \qquad (44)$$

and

$$\overline{|F_a|^2} \cong \sum_{j=1}^{N} \bar{f}_j^2 = s' \cdot \overline{I(hkl)} \qquad (45)$$

and thus that:

$$s' = \frac{\sum\limits_{j=1}^{N} \bar{f}_j^2}{\overline{I(hkl)}} \qquad (46)$$

Since the observed intensities include the influence of the temperature factor, s' may be subdivided into the form:

$$s' = \frac{s}{T_n} = s \cdot e^{+2(B\sin^2\vartheta/\lambda^2)} \qquad (47)$$

If ln s is plotted as a function of $\sin^2\vartheta/\lambda^2$, a straight line should be obtained. The extrapolated value of this graph for $\sin^2\vartheta/\lambda^2 = 0$ gives the scale factors s, and its slope gives a mean value for the temperature coefficient B.

[126] *W.H. Bragg,* Phil. Mag. (VI) 27, 881 (1914).
[127] *C.G. Darwin,* Phil. Mag. (VI) 43, 800 (1922).
[128] *K. Lonsdale,* Mineral. Mag. 28, 14 (1947).
[129] Ref. [317], p. 313.
[130] Ref. [329], p. 231.
[131] Ref. [318], p. 232.
[132] *H.G.J. Moseley, C.G. Darwin,* Phil. Mag. (VI) 26, 210 (1913).
[133] *M. v. Laue,* Ann. Phys. [Leipzig] 42, 1569 (1913).
[134] *P. Debije,* Verh. dtsch. physik. Ges. 15, 738 (1913).
[135] *P. Debije,* Ann. Phys. [Leipzig] 43, 49 (1914).
[136] *C.G. Darwin,* Phil. Mag. (VI) 27, 315 (1914).

[137] *W.H. Bragg,* Phil. Mag. (VI) 27, 881 (1914).
[138] *I. Waller,* Z. Phys. 17, 398 (1923).
[139] *A.J.C. Wilson,* Nature 150, 152 (1942).
[140] Ref. [323], p. 92.
[141] Ref. [329], p. 233.
[142] Ref. [335], p. 46.

4 Crystallographic Primary Information

4.1 Shape and Size of the Unit Cell

Three-dimensional measurements by one of the photographic single-crystal techniques always give an unambiguous description of the reciprocal lattice, i.e. three independent components of a vector and the angles between them, independent of the intensities of the reflections. In favorable cases, especially for cubic lattices, the parameters of the reciprocal lattice may be obtained also from a powder film. Details of interpretation of the measurements are given in the descriptions of methods for use in special cases.

Each reciprocal lattice has a one-to-one correlation to a real lattice. Unfortunately, a real lattice may be described by an infinite number of vector sets of three components each. A set of rules for the selection of a useful vector basis, i.e. of a coordinate system for a crystal, has, therefore, been established.

In the most common, triclinic, case, the following rule exists: the coordinate system must be chosen in such a way that the unit cell has the smallest possible volume, and its vector basis consists of the three shortest independent vectors. Certain rules for calculation exist, to find this so-called *reduced cell,* if an optional vector set is given[143, 144]. When the lattice has higher symmetry, the cell is chosen so that it repeats this symmetry. Such cells may also be centered (cf. Bravais lattices, 7.1:1.1.3); they then contain an integral multiple of the smallest possible cell.

4.2 Space Group Determination

Within each space group (see Section 7.1:1.1.5) except No. 1: P1–C$_1^1$, there are certain parts of the lattice within an unit cell which are equal to other ones and linked to them by symmetry. For certain reflections, therefore, the intensity can be extinguished by interference within the cell. The laws of these 'systematic absences'[145] are characteristic for certain symmetry elements and consequently for certain space groups. For example, the 'systematic absences' rule which states that:

Reflections are present only if $h + k + l = 2n$

always indicates a body-centered cell. If the absences are known and the space-group is sought, the tables published by Nowacki[146] are more convenient, as they are arranged by absences and give the possible space groups for each of them.

In addition, it is often possible to limit the number of space groups, if one knows the crystal class. For the Laue class D_{3d}–$\overline{3}$m, one can distinguish two sub-classes by the fact that certain pairs of reflections are equal or different. This was discussed in Section 7.1:1.1.5. Furthermore, in favorable cases it is possible to establish, with the aid of an intensity statistic[147], whether there is a center of inversion within the space group. With a few exceptions, all space groups should be distinguishable in this way. Only the so-called 'enantio-

morphous' space groups (which behave as image and reflected image) cannot be distinguished for fundamental reasons. This last can be achieved only by determination of the absolute configuration (Chapter 7.2) if chemical evidence cannot be used to establish whether the material has D or L configuration.

4.3 Contents of the Unit Cell

The volume V of a unit cell can be calculated from the cell parameters of the real lattice: cf. Refs. [148, 149]. The mass contained in the cell is calculated from:

$$m = n \cdot \frac{M}{N_L} = n \cdot \frac{M}{6.022 \cdot 10^{23}} \qquad (48)$$

where n = the number of molecules or formula units in the cell; M = the molecular weight or formula weight, and N_L = the Loschmidt number. From this, the theoretical density $\rho_{R\ddot{o}}$ can be calculated:

$$\rho_{R\ddot{o}} = \frac{m}{V} = n \cdot \frac{M}{N_L \cdot V} \qquad (49)$$

If the density is given (as usual) in g/cm³, but the cell constants in Å $\approx 10^{-8}$ cm and the cell volume in Å³ $\approx 10^{-24}$ cm³, one obtains:

$$\rho_{R\ddot{o}}\left[\frac{g}{cm^3}\right] \qquad (50)$$

$$= \frac{n \cdot M}{6,022 \cdot 10^{23} \cdot V \cdot 10^{-24}}\left[\frac{g}{Å^3}\right] = \frac{1,660\, n \cdot M}{V}\left[\frac{g}{Å^3}\right]$$

In most cases, the number n is unknown at the beginning of a structure determination; one, therefore, obtains it by comparing (50) with the actual, experimental density ρ_{obs} by

$$n \approx \rho_{obs.} \cdot \frac{V}{1.660 \cdot M} \qquad (51)$$

Normally, the observed value for the density ρ_{obs} is smaller than it should be because of gaps in the lattice, air included in a crystal powder, or similar reasons. Consequently, n also seems to be too small. Unless there are some extraordinary circumstances, n must of course agree with the symmetry of the space group.

5 Methods of Structure Determination

The decisive step in a crystal structure determination by X-ray methods, and at present the only step for which no general instructions exist regarding calculation, is the formulation of the correct structure model. This model must be sufficiently similar to reality, so that the exact parameters can be derived from it with the aid of the usual refining methods (see 7.1:5.5).

Unfortunately, although with normal X-ray diffraction methods the intensity I(hkl) of a reflection and thus the structure amplitude |F(hkl)| can be detected, this cannot be achieved with the structure factor F(hkl) itself, since the phases α(hkl) [cf. Equations (14), (18), (24)] cannot be observed. Only in the case of anomalous dispersion (cf.

---

[143] *M.J. Buerger,* Z. Kristallogr. *109,* 42 (1957).

[144] Ref. [334], p. 364.

[145] Ref. [313], p. 73.

[146] Ref. [323].

[147] Ref. [335], p. 46.

Chapter 7.2) can such information about the phases be obtained in favorable instances.

Except for a few structures, most of which are inorganic, in which the atoms occupy such special positions that a model for the structure is obvious (e.g. NaCl, diamond, Ref. [150]), there are today four practicable means of access to a structure model; naturally, these may also be used in combination:

the Patterson function (Section 7.1:5.1);

statistical direct methods for phase determination, combined with subsequent Fourier synthesis (7.1:5.2);

analogy methods (7.1:5.3), and

anomalous dispersion (7.1:5.4 and 7.2).

5.1 The Patterson Function

5.1.1 Fundamental principles

The *Patterson synthesis*[151-159] is a Fourier synthesis (see 7.1:2.3) in which, instead of the unknown $F(hkl)$, the $|F(hkl)|^2 \sim I(hkl)$ – which can be observed – are summed up. The Fourier Equation (23) is replaced by the Patterson Equation.

$$P(UVW) \tag{52}$$

$$= \left(\frac{1}{V^2}\right) \sum_{h=-\infty}^{+\infty} \sum_{k=-\infty}^{+\infty} \sum_{l=-\infty}^{+\infty} |F(hkl)|^2 \cdot e^{-2\pi i(hU+kV+lW)}$$

(The coordinates in the Patterson space are usually called U, V, W.) This function has the following characteristic quality[151]:

$$P(UVW) \tag{53}$$

$$= \int_0^1 \int_0^1 \int_0^1 \rho(X,Y,Z) \cdot \rho(X+U, Y+V, Z+W) dX dY dZ$$

Clearly, this means that: each pair of points within the crystal space (X,Y,Z) and $(X+U, Y+V, Z+W)$, having a mutual distance U, V, W, adds to the Patterson function at the point (U,V,W) the product of the two electron densities, $\varrho(X,Y,Z) \cdot \varrho(X+U, Y+V, Z+W)$.

This is integrated over the entire unit cell. When two points within the crystal, having the distance U, V, W, are both occupied by atoms, i.e. have high values of electron density, a maximum is generated at the point (U,V,W) of the Patterson function $P(U,V,W)$. Seen from another point of view, each maximum in a Patterson synthesis means a vector linking two atoms in the crystal. Fig. 26 shows schematically (in two dimensions) a crystal having $n = 3$ atoms, and its Patterson function. $n \cdot (n-1) = 3 \cdot 2 = 6$ maxima are generated, which are always arranged in a centrosymmetric way even when the crystal itself has no center. Furthermore, each Patterson function contains n (in this case $= 3$) maxima at its origin, corresponding to the distances of each of the n atoms 'to itself'. These maxima in the origin are trivial. The relationship between electron density distribution and Patterson function has been shown for *cholesteryl iodide*[152].

When interpreting a Patterson synthesis, one has to assign to each maximum, i.e. to each vector linking two atoms, its origin and end, so that a model of the structure free of any disagreement results. The model must interpret each vector of the synthesis, but must require no vectors which are not present in the synthesis. In practice, one must naturally reduce this strict requirement, in consequence of experimental errors and series-termination effects. In some, very rare cases, different, so-called 'homometric structures', may have identical Patterson functions.

Sections 7.1:5.1.2–5.1.6 deal with different methods for the interpretation of Patterson syntheses.

5.1.2 Harker Sections

Two atoms at points X_1, Y_1, Z_1 and X_2, Y_2, Z_2 generate Patterson maxima in U, V, W and $\bar{U}, \bar{V}, \bar{W}$ with $U = X_1 - X_2; V = Y_1 - Y_2; W = Z_1 - Z_2$ (7.1:5.1.1).

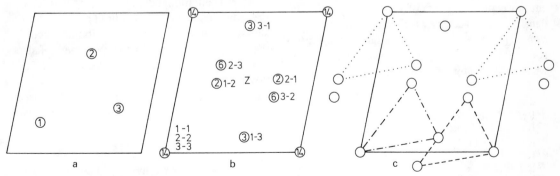

Fig. 26. a) Two-dimensional crystal structure with three atoms, having electron numbers 1:2:3
   b) The correlated Patterson synthesis
      Figures beside the circles: generating atoms. Figures in the circles: relative heights of the maxima
   c) 'Images' of the structure from Fig. 26a) in the Patterson synthesis of Fig. 26b)
      (The dotted image is drawn twice.)

When these two points are linked by a space group symmetry, U,V,W may assume special values. Some illustrative examples are:

A two-fold axis links the two points X,Y,Z and $\overline{X},\overline{Y},Z$. The corresponding Patterson maxima are then found at $\pm\{U=2X; V=2Y; W=0\}$.

The linking symmetry of the points X,Y,Z and X,Y,$\overline{Z}$ is a 'mirror' (reflecting plane). The coordinates of the corresponding Patterson maxima are $\pm\{U=0; V=0; W=2Z\}$.

When a screw axis links X,Y,Z to $\overline{X},\overline{Y}$, $1/2+Z$, the maxima lie on the plane $Z=1/2$ at the positions $\pm\{U=2X; V=2Y; W=1/2\}$.

Such 'Harker maxima', which link symmetry-connected points of the crystal space, can be distinguished in many cases from the vectors linking independent points, since they appear on quite distinct sections or lines of the Patterson space called *Harker sections*[160-163]. From them certain coordinates of the crystal space can be derived, although not always unambiguously. Thus, the value $W=\pm2Z$ from the second example above leads to $Z=\pm W/2$. Here, W can be seen immediately from the synthesis.

A systematic list of all differences between the points of a lattice complex (see 7.1:1.1.5) with general coordinates should always be set up when starting the interpretation of a Patterson synthesis. From this, a list of all Harker maxima of the corresponding space group will be obtained. By correct interpretation of the information which this list contains, and by intelligent combination of the possible coordinates derived from it, a correct model of a structure will often be obtained very rapidly.

A special method for interpreting Harker sections is Buerger's implication theory[160-163].

5.1.3 Heavy Atom Methods

If a crystal contains atoms which are very different in weight, i.e. points of very different electron density, those vectors which link only the *heavy atoms* (H–H vectors) will dominate in the Patterson synthesis[164, 165]. In addition, but clearly weaker, the vectors linking heavy and light atoms

(H–L vectors) also appear, but the vectors linking only light atoms (L–L vectors) generally disappear in the background. For example a crystal which contains Cl (36 e$^-$) as well as C(6 e$^-$) generates

C–C vectors of relative height 36 (1)
C–Cl vectors of relative height 216 (6)
and Cl–Cl vectors of relative height 1296 (36).

It is, therefore, desirable to introduce a heavy atom into organic molecules, for example Br, I, or a heavy metal atom.

The position of this heavy atom in the crystal can be determined in most cases with the aid of the few H–H vectors. In the most simple case, the cell contains only one single heavy atom at a center of symmetry. In this case, the H–L vectors of the Patterson synthesis immediately represent the structure of the crystal. With a cell which is not centrosymmetric, or which contains several heavy atoms, one can try to use these atoms as fixed points and to build up a model of the structure by intelligent combination of the H–L vectors.

In general, however, another method is chosen which uses the Patterson synthesis itself only for the determination of positions of the heavy atoms; phase angles α(hkl) are calculated using the known parts of the structure, i.e. initially using only the heavy atoms. From these phases and the observed structure amplitudes |F(hkl)|, a first Fourier synthesis can be calculated. Half of the information used for this synthesis, the |F(hkl)|, is correct except for experimental errors. The other half, the α(hkl), is certainly not totally incorrect, but is not completely correct. The synthesis will, therefore, probably begin to show not only the heavy atoms used for the first calculation, but also parts of the rest of the structure. If one now calculates a new α(hkl) using this additional information, the values should be better than the first ones. A second Fourier synthesis should, therefore, show additional details of the structure. This procedure can be continued until a final synthesis shows the entire structure. Additional details of the use of this method for the refinement of structure models are dealt with in Section 7.1:5.5.2: 'Fourier refinements'.

The heavy atom method, combined with consecutive cycles of Fourier refinement, has been parti-

[148] Ref. [316], p. 7.
[149] Ref. [317], p. 106.
[150] *W.L. Bragg,* Proc. Roy. Soc., Ser. A *89,* 248 (1914).
[151] Ref. [323], p. 111.
[152] Ref. [6], Fig. 1.
[153] Ref. [328].
[154] *A.L. Patterson.* Phys. Rev. (II) *46,* 372 (1934).
[155] *A.L. Patterson,* Z. Kristallogr. (A) *90,* 517 (1935).
[156] Ref. [323], p. 110.
[157] Ref. [317], p. 318.
[158] Ref. [329], p. 554.

[159] Ref. [335], p. 144.
[160] *D. Harker,* J. Chem. Phys. *4,* 381 (1936).
[161] *M.J. Buerger,* J. Appl. Phys. *17,* 579 (1946).
[162] Ref. [323], p. 142.
[163] Ref. [328], p. 132.
[164] Ref. [329], p. 509.
[165] Ref. [335], p. 171, 222.

cularly successful for structure determinations of *proteins* (e.g. *hemoglobin*[166], *myoglobin*[167], a *derivate of Vitamin B$_{12}$*[168]).

In unfavorable cases, however, the heavy atoms sometimes occupy very special positions and do not make a contribution to all of the reflections, but, considered by themselves, give systematic absences. Phases for these reflections cannot, of course, be calculated from a model containing only the heavy atoms. The heavy atom method must, therefore, fail in these cases.

If the substance under investigation itself contains no heavy atoms, so that one first must prepare a derivative from it, it is, of course, the structure of the derivative which is obtained by the heavy atom method, and not that of the substance. In this case, one has to ensure by some other method that the structure of the investigated molecule is not changed during the synthesis of the derivative.

5.1.4 Image-Seeking Functions

In any structure containing n atoms per unit cell, a (three-dimensional) polygon with n vertices (n-gon) can be drawn, whose vertices are the n atom positions. As Fig. 26c shows, this image can be found in the Patterson synthesis, and, in fact, n times in the real position, and n times, linked by a center, in the inverse position (not drawn in Fig. 26c). Here, each vertex of the n-gon is situated once in the origin maximum. These figures can be found with the aid of *image-seeking functions*[169, 170]. If one superimposes several Patterson syntheses $P_n(U_n, V_n, W_n)$, shifted by $-U_n, -V_n, -W_n$ in such a way that each of the vertices of one of the

images is superimposed once by the origin of one of the syntheses (Fig. 27), all of the superimposed syntheses have maximum values at the points of this image, while everywhere else at least one of the syntheses shows small values of the function. An image-seeking function is then such a function of the Patterson functions $P_n(U_n, V_n, W_n)$ which indicates these common maxima. The most powerful ones are the product function

$$\pi_n(X,Y,Z) = \prod_n [P_n(U_n+X, V_n+Y, W_n+Z)] \quad (54)$$

which gives the product of the superimposed Patterson functions at the point X, Y, Z[171], and the minimum function[172]

$$\mu_n(X,Y,Z) = \underset{n}{Min}[P_n(U_n+X, V_n+Y, W_n+Z)] \quad (55)$$

which is the minimum of the P_n at the point X, Y, Z, whilst the sum function

$$\sigma_n(X,Y,Z) = \sum_n [P_n(U_n+X, V_n+Y, W_n+Z)] \quad (56)$$

may (because of the ever-present background of a Patterson synthesis) simulate maxima of the function which do not correspond to any real positions of atoms in the crystal structure[173].

In practice, one proceeds stepwise, by first superimposing only two Patterson syntheses; any of the maxima of one synthesis is placed upon the origin of the other; one checks to see whether the image of a line so defined can be detected again. Harker maxima should not be used for this purpose. Third

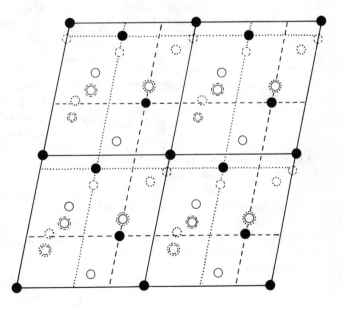

Fig. 27. Patterson synthesis from Fig. 26b thrice shifted, and superimposed
The full black circles are common in all three of the syntheses; the structure of Fig. 26a can be deduced from them

[166] Ref. [4].
[167] Ref. [5].
[168] Ref. [6], Fig. 4.
[169] Ref. [328], p. 181.

[170] Ref. [328], p. 218.
[171] Ref. [328], p. 233.
[172] Ref. [328], p. 239.
[173] Ref. [328], p. 247.

and fourth, etc. syntheses are then added to the image-seeking function. If the correct shifts have been chosen, an n-gon is obtained each time, which is repeated several times in the function, until finally the desired image of the structure appears.

5.1.5 The Convolution Molecule Method

For some structure determinations, the steric configuration of some parts of the structure (e.g. of an aromatic system) is already known at the beginning. Such partial structures generate defined arrangements of maxima in a Patterson synthesis, called *convolution molecules*[174–176] since, in the language of mathematics, this is 'convolution' or 'folding'[177].

The convolution molecules of the molecules themselves are their Patterson structures, which are invariant against translations and changes of orientation in their internal structure, and which, therefore, are very suitable for determination of orientations. In practice, these Patterson functions are calculated and then rotated about the origin of the Patterson synthesis which is to be interpreted, until the best possible agreement is found. In doing this, no maxima of the convolution molecules are allowed to fall on any zero values of the synthesis, since superposition of several convolution molecules can lead only to an addition of values of the function, and therefore to larger Patterson maxima, but never to a subtraction.

The orientation can also be determined with the 'equal indexed' convolution molecules which are parallel or linked by a center of inversion.

When the orientation of the molecules is known, the general 'mixed indexed' convolution molecules can be calculated. With these the translation parameters of the partial molecules can be determined. When such a part of the entire structure is known, a first set of phases $\alpha(hkl)$ for a Fourier synthesis may be calculated from it, in a manner similar to the heavy atom method; further parts of the total structure can thus be obtained. Subsequent procedure is as described in Sections 7.1:5.1.3 and 5.5.2.

As a special case, the Patterson synthesis itself is generated by 'convolution' of the total structure with its own inverse.

5.1.6 Method of Isomorphous Replacement

If one has two crystals, which differ only in one single atom, but have otherwise identical structures, one can determine the phases for a Fourier synthesis directly by the *method of isomorphous replacement*[178, 179]. If the two compounds have the compositions MR and NR, their structure factors may be written:

$$F^{MR}_{(hkl)} = F^{M}_{(hkl)} + F^{R}_{(hkl)} \qquad (57)$$

and

$$F^{NR}_{(hkl)} = F^{N}_{(hkl)} + F^{R}_{(hkl)}, \qquad (58)$$

and thus the difference of the two structure factors is:

$$F^{MR}_{(hkl)} - F^{NR}_{(hkl)} = F^{M}_{(hkl)} - F^{N}_{(hkl)} = \Delta F(hkl) \qquad (59)$$

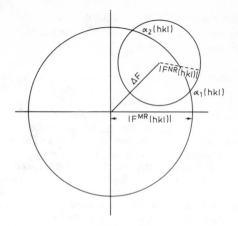

Fig. 28. Construction of the two possible phase angles $\alpha_1^{MR}(hkl)$ and $\alpha_2^{NR}(hkl)$ from the known values $|F^{MR}(hkl)|$, $|F^{NR}(hkl)|$ and ΔF
A circle of radius $|F^{MR}(hkl)|$, is drawn. The position of $F^{MR}(hkl)$ is on this circle; the vector ΔF is drawn, starting from the origin. If a second circle of radius $|F^{NR}(hkl)|$ is drawn about the end of ΔF, the required phase angle $\alpha^{MR}(hkl)$ should be given by one of the points of intersection of the two circles

The structure factors may be drawn as vectors in the complex plane (see Fig. 13b). If the exact parameters of M and N are known from any other method, the values and directions of the middle part of Equation (59) can be calculated. Only the absolute values can be observed of the $F^{MR}_{(hkl)}$ and $F^{NR}_{(hkl)}$; it is, therefore, known that they must lie on circles of radii $|F^{MR}_{(hkl)}|$ and $|F^{NR}_{(hkl)}|$. With this information, the vectors $F^{MR}_{(hkl)}$ and $F^{NR}_{(hkl)}$ can be graphically constructed, as Fig. 28 shows. There exist two solutions, of course, situated symmetrically with respect to the phase direction of $\Delta F(hkl)$, and coinciding in centrosymmetric structures. Often, this question can be decided for acentric crystals by two-fold isomorphous replacement, i.e. by comparing with a third isomorphous compound PR. The identical coordinates of M and N must be determined in advance by some method. If M and N are heavy atoms, it is not very difficult to determine their parameters by the heavy atom method. Otherwise, a difference Patterson synthesis can be calculated. This is a Patterson synthesis using $|F^{MR}_{(hkl)}|^2 - |F^{NR}_{(hkl)}|^2$ or $(|F^{MR}_{(hkl)}| - |F^{NR}_{(hkl)}|)^2$. The characteristics of these syntheses are discussed in Ref. [180, 182].

[174] *W. Hoppe*, Ber. Bunsenges. Phys. Chem. *61*, 1076 (1957).
[175] *W. Hoppe, E.F. Paulus*, Acta Crystallogr. *23*, 339 (1967).
[176] *W. Hoppe*, Angew. Chem. *78*, 292 (1966).
[177] Ref. [176], Fig. 3.
[178] Ref. [335], p. 228.
[179] Ref. [329], p. 521.

5.2 Direct Methods

During recent years, some methods have gained importance, which seek to determine the phases of the structure factors 'directly' from the scattering data by mathematical methods[183-185]. Some of them use inequalities, employing the relationships of the structure factors of different reflections; some of them are of a statistical nature.

For practical use of the direct methods it is customary and convenient to use, instead of the structure factors F(hkl) themselves, values which can be derived from them. Some authors use the *unitary structure factor* U(hkl)[186-189], while others prefer the *normalized structure factor* E(hkl)[190-193].

$$U(hkl) = \frac{F(hkl)}{\sum\limits_{j=1}^{N} f_j} \qquad (60)$$

with the atomic scattering factors f_j of the N atoms of the unit cell [see equation (6)]. Since $\sum\limits_{j=1}^{N} f_j$ is the greatest possible value for $|F(hkl)|$, U(hkl) can only have a value between $+1$ and -1. An example for the calculation of U(hkl) is given in Ref.[194].

Using the unitary atomic scattering factor

$$n_j = f_j / \sum\limits_{j=1}^{N} f_j \qquad (61)$$

one can formulate, as with equation (7):

$$U(hkl) = \sum\limits_{j=1}^{N} n_j e^{i\phi_j} \qquad (62)$$

In the centric case, this becomes:

$$U(hkl) = \sum\limits_{j=1}^{N} n_j \cos\phi_j \qquad (63)$$

[(ϕ_j is defined as for equation (4)]. From (61), it can be derived that:

$$\sum\limits_{j=1}^{N} n_j = 1 \qquad (64)$$

The normalized structure factor E(hkl) is so defined, that the mean of its squares becomes 1,

$$\overline{(E^2)} = 1. \qquad (65)$$

This can be achieved by defining:

$$E(hkl) = \frac{F(hkl)}{(\overline{F^2})^{1/2}} \qquad (66)$$

Moreover, in the centrosymmetric case the following relationships are valid:

$$\overline{|E^2 - 1|} = 4/\sqrt{2\pi e} = 0{,}968 \qquad (67)$$

and

$$\overline{|E|} \approx \sqrt{2/\pi} = 0{,}798. \qquad (68)$$

In the following text, F_h, U_h, E_h will always be written for F(hkl), U(hkl), E(hkl) etc., h meaning the vector h, k, l of the reciprocal lattice.

5.2.1 Harker-Kasper Inequalities

Harker and Kasper[195, 196] showed that correlations exist between certain F_h and $|F_h|^2$ for centrosymmetric crystals which can be formulated by inequalities[195-199]. These can be derived by application of Cauchy's inequality

$$\left| \sum\limits_{j=1}^{N} a_j b_j \right|^2 \leqq \left(\sum\limits_{j=1}^{N} |a_j|^2 \right) \left(\sum\limits_{j=1}^{N} |b_j|^2 \right) \qquad (69)$$

and of the inequality of Schwartz

$$\left| \int f(x)g(x)dx \right|^2 \leq \left(\int |f(x)|^2 dx \right) \left(\int |g(x)|^2 dx \right) \qquad (70)$$

upon the structure factors, which must be divided into factors in a suitable manner. In the most simple case, which can be taken as an example, (63) is divided, using

$$a_j = \sqrt{n_j} \qquad (71)$$

and

$$b_j = \sqrt{n_j} \cos\phi_j \qquad (72)$$

leading to

$$U_h = \sum\limits_{j=1}^{N} a_j b_j. \qquad (73)$$

From (64), further equations are valid:

$$\sum\limits_{j=1}^{N} a_j^2 = \sum\limits_{j=1}^{N} n_j = 1 \qquad (74)$$

and

$$\sum\limits_{j=1}^{N} b_j^2 = \sum\limits_{j=1}^{N} n_j \cos^2\phi_j$$

$$= \tfrac{1}{2} \sum\limits_{j=1}^{N} n_j(1 + \cos 2\phi_j) = \tfrac{1}{2}(1 + U_{2h}) \qquad (75)$$

[180] *P.M. Green, V.M. Ingram, M.F. Perutz*, Proc. Roy. Soc., Ser. A *225*, 287 (1954).
[181] *D.M. Blow*, Proc. Roy. Soc., Ser. A *257*, 302 (1958).
[182] Ref.[335], p. 189.
[183] Ref.[329], p. 551.
[184] Ref.[331].
[185] Ref.[335], p. 234.
[186] *D. Harker, J.S. Kasper*, J. Chem. Phys. *15*, 882 (1947).
[187] *D. Harker, J.S. Kasper*, Acta Crystallogr. *1*, 70 (1948).
[188] Ref.[329], p. 557.
[189] Ref.[331], p. 2.
[190] Ref.[324].
[191] *E.F. Bertraut*, Acta Crystallogr. *8*, 544 (1955).
[192] Ref.[329], p. 574.
[193] Ref.[331], p. 13.
[194] Ref.[331], p. 6.
[195] Ref.[186].
[196] Ref.[187].
[197] Ref.[329], p. 557.

Substituting (73), (74) and (75) into Equation (69), one obtains

$$U_h^2 \leqslant 1/2(1 + U_{2h}) \qquad (76)$$

If both U_h and U_{2h} have large values, from (76) it can be shown that U_{2h} is positive. For example, from (76) for

$$|U_h| = 0.7 \text{ and } |U_{2h}| = 0.8$$

the two solutions:

$$0.49 \leqslant 1/2(1 \pm 0.64),$$

result. From these, only the value $\frac{1}{2}(1 + 0.64) = 0.82$ matches the inequality. However, (76) fails for example for

$$|U_h| = 0.5 \text{ and } |U_{2h}| = 0.7,$$

since in this case (76) is already correct with negative signs:

$$0.25 \leqslant 1/2(1 \pm 0.49) \leqslant 1/2(1 - 0.49) = 0.255.$$

Equation (76) permits one to determine only the signs of reflections, all three of whose indices are even. However, inequalities also exist by which the signs of other reflections can be determined, as well as those which link more than two reflections. A centrosymmetric unit cell contains several centers each of which may be chosen as the origin. The change from one of these origins to another means a change of the signs for certain groups of reflections, each of them having all h, k or l even or odd. One can, therefore, assign an arbitrary sign to up to three structure factors, which must be chosen from three different of these groups. In doing this, one defines the origin of the cell.

5.2.2 Sign Relationships (Sayre's Equations)

Sayre[200] pointed out in 1952 the existence of precise equations linking the structure factors, for a structure which has equal, non-overlapping atoms[201-205]. With a factor g which is a function of $2\sin\vartheta/\lambda$, the following equation is valid:

$$F_h = 1/(g \cdot V) \sum_{h'} F_h \cdot F_{h+h'} \qquad (77)$$

If one of the terms in the sum on the right hand side of (77) is much greater than the other ones, it determines the sign of the total sum, so that one can derive from it, with s_x = sign of F_x:

$$s_h = s_{h'} s_{h+h'}. \qquad (78)$$

By comparison of a sufficient number of such relations, the signs of some of the strong structure factors can be determined. For less strong reflections (78) may be 'probably true', but not certain. Several papers have been published, concerning the probability $W_{h, h'}$ that (78) is 'true'[206, 208]. In practice, the most used formula may be that of Cochran and Woolfson[209]:

$$W_{h, h'} = 1/2 + 1/2 \tanh\left\{\left(\frac{\varepsilon_3}{\varepsilon^3}\right)|U_h U_{h'} U_{h+h'}|\right\} \qquad (79)$$

with

$$\varepsilon_3 = \sum_{j=1}^{N} n_j^3 \qquad (80)$$

and

$$\varepsilon = \sum_{j=1}^{N} n_j^2 \qquad (81)$$

Zachariasen[210] introduced a procedure which uses general numbers for some of the desired signs, and uses them for calculation until the correct values can be assigned later.

5.2.3 The Karle and Hauptman Procedure

In 1953, Hauptman and J. Karle[211-213] derived, by mathematical analysis, a number of relationships between structure factors, which led them to the conclusion that the phase problem must be soluble if sufficient data were available. The method was originally only for centrosymmetric structures but was subsequently also extended to acentric structures[214, 215]. It should be sufficient here to give as an example the formulae for the signs s_h of even reflections $(h = 2n; k = 2n; l = 2n)$ in space group $P\bar{1}$, as formulated in Ref. [216]:

with

$$s_h = \sum_1 + \sum_2 + \sum_3 + \sum_4 \qquad (82)$$

$$\sum_1 = \frac{\varepsilon_3}{4\varepsilon_2^{5/2}}(U_{1/2h}^2 - \varepsilon_2) \qquad (83)$$

$$\sum_2 = \frac{\varepsilon_3}{2\varepsilon_2^{5/2}}\sum_{h'} U_{h'} U_{h+h'} \qquad (84)$$

$$\sum_3 = \frac{\varepsilon_4}{4\varepsilon_2^{7/2}}\sum_{h'} U_{h'} (U_{1/2(h+h')}^2 - \varepsilon_2) \qquad (85)$$

$$\sum_4 = \frac{\varepsilon_5}{8\varepsilon_2^{9/2}}\sum_{h'} (U_{1/2h+h'}^2 - \varepsilon_2)(U_{h'}^2 - \varepsilon_2) \qquad (86)$$

and

$$\varepsilon_r = \sum_{j=1}^{N} n_j^r$$

[198] Ref. [331], p. 15.
[199] Ref. [335], p. 235.
[200] *D. Sayre*, Acta Crystallogr. *5*, 60 (1952).
[201] *W. Cochran*, Acta Crystallogr. *5*, 65 (1952).
[202] *W.H. Zachariasen*, Acta Crystallogr. *5*, 68 (1952).

[203] Ref. [329], p. 567.
[204] Ref. [331], p. 43.
[205] Ref. [335], p. 244.
[206] Ref. [324].

Similar formulae are valid for other groups of h. Initially, one knows only the U_h^2, so only Σ_1 and Σ_4 will contribute to s_h. When an increasing number of the signs are known, the Σ_2 and Σ_3 also become important. Because of the many numerical calculations involved in the Karle and Hauptman procedure, the method can be used only with a computer. Ref. [217] provides a good survey of the further development of the procedure, and a detailed bibliography.

5.2.4 Sign Permutations
In a centrosymmetric structure, each of the F_h can have positive or negative sign. It should, therefore, be theoretically possible to calculate with each of the sign combinations a Fourier synthesis, and to check the results for correctness by some other method. Unfortunately, however, one already needs 2^{n-d} syntheses for n reflections (d signs may be arbitrarily chosen for a d-dimensional synthesis), so that the limits of a computer are soon reached[218, 219]. Woolfson[220] has proposed a strategy which uses only a selected number of the theoretically possible combinations, leading to useful statements about a structure. He solved a problem having 16 reflections in 2 dimensions, by only 2^8 (instead of 2^{14}) syntheses. Later, this method was developed further by Good[221].

5.3 Analogy Methods
Sometimes a crystal structure which is to be determined is so similar to one already known that conclusions can be drawn, by analogy, from comparison of the two structures. In this case, an analogous model may be set up for the new structure without using the long and often tiresome methods dealt with in the preceding sections. This model, of course, must satisfy the same criteria as a model obtained by the usual methods. In particular, the observed and calculated structure factors must agree sufficiently [see Equation (87)]. Isotypous and isomorphous crystals, having similar lattice parameters and similar composition, can normally be treated by *analogy methods*. Often this 'similarity' does not need to be very great. In many cases, good results can also be obtained from powder data.

5.4 Anomalous dispersion
A method which under favorable circumstances allows even the direct determination of phases is *anomalous dispersion*. Because of its great practical importance for the determination of absolute configuration, which is a subject of very great interest to the organic chemist, it is discussed separately in the Chapter 7.2.

5.5 Refinement of a Structure Model
A model of a structure obtained by the methods described in the preceding sections will at the beginning be only a very rough image of the structure. In a second stage of the structure determination, the refinement' of the structure[222, 223], all the parameters describing the model are varied until the best possible agreement of the observed and calculated structure factors is reached. The *residual factors* R are taken as a measure of the progress of this process. Several forms of R are in common use. Most frequently, the definition is, with

$$F_o = F_{observed} \quad \text{and} \quad F_c = F_{calculated}:$$

$$R_1 = \frac{\sum \left\| F_o \right| - \left| F_c \right\|}{\sum \left| F_o \right|} \tag{87}$$

Sometimes this is also weighted[224, 225].

Wilson[226] has shown that an entirely wrong structure, an arbitrary arrangement of atoms (only the stoichiometry being correct), has $R = 0.828$ for centrosymmetric crystals and $R = 0.586$ for an acentric one. Models having $R < 0.4$ may, therefore, be worth an attempt of refinement. It is usual to regard a structure with $R < 0.20$ as correct. With modern methods of measurement, $R < 0.10$ can be achieved. Very good refined structures can have $R \approx 0.05$.

It is discussed whether the non-observed, weak reflections should be included in the calculation of the R-factor. If they are assigned the value $|F_0| = 0$, and if

[207] *E.F. Bertaut*, Acta Crystallogr. *8*, 544 (1955).

[208] *A. Klug*, Acta Crystallogr. *11*, 515 (1958).

[209] *W. Cochran, M.M. Woolfson*, Acta Crystallogr. *8*, 1 (1955).

[210] *W.H. Zachariasen*, Acta Crystallogr. *5*, 68 (1952).

[211] Ref. [324].

[212] Ref. [329], p. 574.

[213] Ref. [331], p. 88.

[214] *J. Karle, I.L. Karle*, Acta Crystallogr. *21*, 849 (1966).

[215] Ref. [324].

[216] Ref. [331], p. 88.

[217] Ref. [214].

[218] *M.M. Woolfson*, Acta Crystallogr. *7*, 65 (1954).

[219] Ref. [329], p. 576.

[220] Ref. [218].

[221] *I.J. Good*, Acta Crystallogr. *7*, 603 (1954).

[222] Ref. [329], p. 585.

[223] Ref. [335], p. 317.

[224] Ref. [317], p. 332.

[225] Ref. [329], p. 585.

[226] *A.C.J. Wilson*, Acta Crystallogr. *3*, 397 (1950).

$\|F_0\|-\|F_c\| \neq 0$, surely R will be too great. If, however, both of them are excluded from the calculation, R becomes too small. Some authors, therefore, give $|F_0|$ the value $|F_0| = |F_{min}|/2$, one half of the smallest, actually observed value. A precise statement about the way in which the R-factor was calculated should always be included in any publication about a crystal structure. Booth[227, 228] was able to show that another R-factor,

$$R_2 = \frac{\sum(|F_0| - |F_c|)^2}{\sum|F_0|^2} \qquad (88)$$

is a function of the mean deviation $\bar{\sigma}$ of the parameters of a structure from their true values.

Two methods for structure refinement which are different in principle but equally efficient are discussed below: 'Least-squares-refinement' (see 7.1:5.5.1) and refinement by successive Fourier syntheses (see 7.1:5.5.2).

5.5.1 Least-Squares-Refinement

With this method the parameters of a structure model should be varied in such a way that the calculated F_c agree with the observed F_0 as well as possible. The F_0 themselves are, however, subject to experimental errors; consequently, the F_c cannot be fitted exactly but only in an optimal manner. Legendre[235] was the first to study this problem[229-234]. He demonstrated that the optimal fit is reached when the sum of the squares of the errors reaches a minimum.

An observable quantity q may be a linear function of a set of variables $x_1, x_2, x_3, \dots x_j, \dots x_n$ and of a set of known constants $a_1, a_2, a_3, \dots a_j, \dots a_n$. Consequently:

$$q = a_1x_1 + a_2x_2 + a_3x_3 + \dots \dots + a_jx_j + \dots \dots a_nx_n \quad (89)$$

Let us also suppose that several similar experiments are possible, by which the constants a_j are varied. If the observations q were exact, it would be possible to determine all the parameters x_j from a system of n equations, i.e. from n observations. In reality, however, each observation q is associated with an error E; the x_j

thus become uncertain also. Furthermore, when a crystal structure determination is being carried out, where $q = |F_0|$, one normally has many more observations than there are parameters to be determined. The problem is, therefore, over-determined. If all m observations were theoretically correct, any set of n from these m equations should lead to exactly the same n parameters x_j. Because of the errors E of the observations, however, other results for the x_j are obtained from each set. If E is introduced into Equation (89), one obtains:

$$q + E = a_1x_1 + a_2x_2 + a_3x_3 + \dots \dots a_jx_j + \dots \dots a_nx_n \quad (90)$$

The error E is therefore

$$E = a_1x_1 + a_2x_2 + \dots \dots + a_jx_j + \dots \dots a_nx_n - q \quad (91)$$

From m observations q_k, a set of equations (91) can be written:

$$E_1 = a_{11}x_1 + a_{12}x_2 + \dots \dots + a_{1j}x_j + \dots \dots + a_{1n}x_n - q_1$$
$$E_2 = a_{21}x_1 + a_{22}x_2 + \dots \dots + a_{2j}x_j + \dots \dots + a_{2n}x_n - q_2$$
$$E_3 = a_{31}x_1 + a_{32}x_2 + \dots \dots + a_{3j}x_j + \dots \dots + a_{3n}x_n - q_3$$
$$\vdots \qquad \vdots \qquad \vdots \qquad \qquad \vdots \qquad \qquad \vdots \quad : \;(92)$$
$$E_k = a_{k1}x_1 + a_{k2}x_2 + \dots \dots + a_{kj}x_j + \dots \dots + a_{kn}x_n - q_k$$
$$\vdots \qquad \vdots \qquad \vdots \qquad \qquad \vdots \qquad \qquad \vdots \quad :$$
$$E_m = a_{m1}x_1 + a_{m2}x_2 + \dots \dots + a_{mj}x_j + \dots \dots + a_{mn}x_n - q_m$$

The best solution for (92) is obtained, according to Legendre[236], if

$$E_1^2 + E_2^2 + \dots \dots + E_k^2 + \dots \dots + E_m^2 = \sum_{k=1}^{m} E_k^2 = \text{Min.} \quad (93)$$

becomes a minimum. By addition of all the E_k^2 from Equation (92) one obtains for (93)

$$\sum_{k=1}^{m} E_k^2 = \sum_{k=1}^{m} (a_{k1}x_1 + a_{k2}x_2 + \dots \dots + a_{kj}x_j + \dots$$
$$\dots + a_{kn}x_n - q_k)^2 = \sum_{k=1}^{m}\left\{\left(\sum_{j=1}^{n} a_{kj}x_j\right) - q_k\right\}^2 \quad (94)$$

$\sum_{k=1}^{m} E_k^2$, the sum of the squares of the errors, is a minimum when its partial derivatives with respect to the x_j vanish, i.e. when

$$\frac{\partial \sum_{k=1}^{m} E_k^2}{\partial x_j} = 2\sum_{k=1}^{m}\left\{\left(\sum_{j=1}^{n} a_{kj}x_j\right)a_{kj}\right\} = 0 \quad (95)$$

By rearranging the n normal equations are obtained:

$$\left(\sum_{k=1}^{m} a_{k1}a_{k1}\right)x_1 + \left(\sum_{k=1}^{m} a_{k1}a_{k2}\right)x_2 + \dots + \left(\sum_{k=1}^{m} a_{k1}a_{kj}\right)x_j + \dots + \left(\sum_{k=1}^{m} a_{k1}a_{kn}\right)x_n = \sum_{k=1}^{m} a_{k1}q_k$$

$$\left(\sum_{k=1}^{m} a_{k2}a_{k1}\right)x_1 + \left(\sum_{k=1}^{m} a_{k2}a_{k2}\right)x_2 + \dots + \left(\sum_{k=1}^{m} a_{k2}a_{kj}\right)x_j + \dots + \left(\sum_{k=1}^{m} a_{k2}a_{kn}\right)x_n = \sum_{k=1}^{m} a_{k2}q_k$$

$$\vdots \qquad \vdots \qquad \vdots \qquad \vdots \qquad \vdots$$

$$\left(\sum_{k=1}^{m} a_{ki}a_{k1}\right)x_1 + \left(\sum_{k=1}^{m} a_{ki}a_{k2}\right)x_2 + \dots + \left(\sum_{k=1}^{m} a_{ki}a_{kj}\right)x_j + \dots + \left(\sum_{k=1}^{m} a_{ki}a_{kn}\right)x_n = \sum_{k=1}^{m} a_{ki}q_k \qquad (96)$$

$$\vdots \qquad \vdots \qquad \vdots \qquad \vdots \qquad \vdots$$

$$\left(\sum_{k=1}^{m} a_{kn}a_{k1}\right)x_1 + \left(\sum_{k=1}^{m} a_{kn}a_{k2}\right)x_2 + \dots + \left(\sum_{k=1}^{m} a_{kn}a_{kj}\right)x_j + \dots + \left(\sum_{k=1}^{m} a_{kn}a_{kn}\right)x_n = \sum_{k=1}^{m} a_{kn}q_k$$

or, more briefly, for the ith of them: $\qquad \sum_{j=1}^{n}\left(\sum_{k=1}^{m} a_{ki}a_{kj}\right)x_j = \sum_{k=1}^{m} a_{ki}q_k \qquad (97)$

The n unknowns x_j can be determined from these equations. Different reliabilities of the n observations may be taken into account by multiplying both sides of (97) by a weight w_i.

To use these fundamental equations for refinement of a crystal structure, one has to ask oneself which of the values appearing at the refinement procedure corresponds to the mathematical symbols of the Equations (89)–(97).

For observations, one can insert the differences

$$\Delta F = F_o - F_c \qquad (98)$$

Writing F_c as a function of its n parameters p_j:

$$F_c = f(p_1, p_2, \ldots \ldots p_j, \ldots \ldots p_n) \qquad (99)$$

and F_o as a function of the same parameters $(p_j + \varepsilon_j)$, corrected by small deviations ε_j:

$$F_o = f(p_1 + \varepsilon_1, p_2 + \varepsilon_2, \ldots p_j + \varepsilon_j, \ldots p_n + \varepsilon_n) \qquad (100)$$

one obtains

$$\Delta F = f(p_1 + \varepsilon_1, p_2 + \varepsilon_2, \ldots p_j + \varepsilon_j, \ldots p_n + \varepsilon_n)$$
$$- f(p_1, p_2, \ldots p_j, \ldots p_n) \qquad (101)$$

The structure factors are, of course, no linear functions of their parameters, as it is the basis of the least-squares method. If the ε_j are small, however, and, therefore, if the model which is to be refined already matches the reality well, one may develop according to Taylor[237].

$$f(p_1 + \varepsilon_1, \ p_2 + \varepsilon_2, \ldots p_j + \varepsilon_j, \ldots p_n + \varepsilon_n)$$
$$= f(p_1, p_2, \ldots p_j, \ldots p_n) \qquad (102)$$
$$+ \left(\varepsilon_1 \frac{\partial F_c}{\partial p_1} + \varepsilon_2 \frac{\partial F_c}{\partial p_2} + \ldots + \varepsilon_j \frac{\partial F_c}{\partial p_j} + \ldots + \varepsilon_n \frac{\partial F_c}{\partial p_n} \right) + R_1$$

With the assumptions mentioned, the residual term R_1 may be neglected: $R_1 = 0$. Thus:

$$\Delta F = F_o - F_c = F_c + \sum_{j=1}^{n} \varepsilon_j \frac{\partial F_c}{\partial p_j} - F_c = \sum_{j=1}^{n} \varepsilon_j \frac{\partial F_c}{\partial p_j} \qquad (103)$$

Taking into account the errors of the observations, which are contained in each of the ΔF, by comparison of (90) with (103), with a running index k assigned to each of the m observations, one obtains:

$$q_k + E_k = \sum_{j=1}^{n} a_{kj} x_j \qquad (90a)$$

and

$$\Delta F_k + E_k = \sum_{j=1}^{n} \varepsilon_j \frac{\partial F_{c_k}}{\partial p_j} \qquad (103a)$$

[227] A.D. Booth, Phil. Mag. (VII) 36, 609 (1945).
[228] Ref. [329], p. 588.
[229] Ref. [317], p. 326.
[230] Ref. [329], p. 609.
[231] Ch. Scheringer, Acta Crystallogr. 16, 546 (1963).
[232] Ch. Scheringer, Acta Crystallogr. 19, 504 (1965).
[233] Ch. Scheringer, Acta Crystallogr. 19, 513 (1965).
[234] Ref. [335], p. 340.
[235] A.M. Legendre, Nouvelles méthodes pour la détermination des orbites des comètes, p. 72, Courcier, Paris 1806.

yields:

$$x_j = \varepsilon_j \qquad (104)$$

and

$$a_{kj} = \frac{\partial F_{c_k}}{\partial p_j} \qquad (105)$$

i.e. the matrix elements a_{kj} of the normal Equations (96) and (97) are obtained by partial differentiation of all the structure factors F_{c_k} with respect to all parameters p_j. Thus, by substituting (104) and (105) in (97), and taking into consideration the weights w_i, the ith normal equation is:

$$\sum_{j=1}^{n} \left(\sum_{k=1}^{m} w_i \frac{\partial F_{c_k}}{\partial p_i} \cdot \frac{\partial F_{c_k}}{\partial p_j} \right) \varepsilon_j = \sum_{k=1}^{m} w_i \frac{\partial F_{c_k}}{\partial p_i} \cdot \Delta F_k \qquad (106)$$

The deviations ε_j of the parameters are the solutions of the system of normal equations. It is evident that such systems of equations can be established and solved only with the aid of electronic data processing systems.

Not only positional parameters x,y,z, temperature factors, and scale factors are parameters p_j, which can be refined, but also all the variables by which the structure factors can be described. For example, according to Scheringer[238, 239] one can put rigid bodies consisting of n atoms, such as *aromatic ring systems* or *inorganic complex ions,* as a whole into a *least-squares-refinement.* Consequently, one must refine not 3n positional parameters, but only 3, plus 3 additional angular parameters which describe the orientation of the rigid body in the crystal. In addition, with compounds in which there is variable occupation of some of the lattice positions (for example mixed crystals, or crystals having migrating ions in their structure[240]), it is possible to treat the densities of occupation of crystallographic lattice positions as parameters which can be refined.

When a calculated structure model is still very far from correct, so that the ε_j become too great, the R_1 in equation (102) can no longer be neglected. When solving the system of normal equations, one may obtain erroneous solutions ε_j, which converge to a satellite minimum of equation (93). The true minimum will, consequently, not be reached on continuation of the refinement. Also, it may sometimes happen, that, for some of the parameters, very large changes are suddenly calculated, i.e. that the parameters 'jump' or that the calculated standard deviations become unreasonably high.

[237] I.N. Bronstein, K.A. Semendjajew, Taschenbuch der Mathematik, 4. Aufl., p. 278, Verlag Harry Deutsch, Frankfurt/M., Zürich 1964.
[238] Ref. [231].
[239] Ch. Scheringer, Acta Crystallogr. 19, 513 (1965).
[240] E. Schultze-Rhonhof, Dissertation, Universität Bonn, 1964.

This phenomenon often occurs if one tries simultaneously to refine parameters which have very different influences on the structure factors, e.g. positional parameters of very heavy and very light atoms[241].

The least squares method has the advantage over Fourier refinement that the single observations can be 'weighted' according to their reliability, without spoiling the results. In particular, dubious reflections can be excluded entirely from the refining calculations.

5.5.2 Fourier Refinement

In the refinement of structure models[242-245] by Fourier synthesis (7.1:2.3), a centrosymmetric structure is the more simple case, since here only two possibilities, $\alpha = 0$ and $\alpha = \pi$ exist for the phases $\alpha(hkl)$. Equation (24) is therefore simplified to

$$\rho(X, Y, Z) \tag{107}$$
$$= 1/V \sum_{h=-\infty}^{+\infty} \sum_{k=-\infty}^{+\infty} \sum_{l=-\infty}^{+\infty} \pm |F(hkl)| \cos[2\pi(hX + kY + lZ)]$$

If a part of the structure is already known from any of the methods discussed in Section 7.1:5, or if one has a structure model from other sources, the structure amplitudes $F_c(hkl)$ for the known partial structure can be calculated in advance. If the R-factor of this data set is $R < 0.5$, the model may be worth an attempt at refinement. First of all, for all the reflections for which the values of both the F_c and the F_0 are large, it may be assumed that the calculated sign of the F_c also holds for the F_0. For the weaker structure factors one must recognize that the two signs may not be equal because of errors in the model. One can, therefore, first calculate a Fourier synthesis using only the strong F_0, which are fairly certain to have the sign of the F_c. Structure amplitudes having uncertain signs should be excluded, since the error in the electron density calculation generated by an F_0 having an incorrect sign is twice the error generated by simply omitting the reflection. Such a first synthesis should reproduce the starting model with corrected parameters. In addition, in most cases, further parts of the structure are indicated in the synthesis. The procedure may be repeated with this corrected model. With each refining cycle, the number of correct signs generally increases. The syntheses will each time show more details of the structure. When no more signs are altered by further cycles, and also the R-factor no longer decreases, the structure has been refined as far as is possible by this method. The synthesis should then give the correct chemical composition of the compound as far as the numbers of the atoms, and the heights and distances of the peaks are concerned. In the ideal case, the peaks should be round. The space between the atoms should contain no large negative regions, since the electron density in a crystal is a positive function.

The refinement of acentric structures is more difficult and converges more slowly, since the phases $\alpha(hkl)$ in equation (24) can take any arbitrary value between 0 and 2π, without any gap. In principle, however, it is possible, using the procedure described above for the refinement of centric structures.

Difference Fourier synthesis is the difference of two syntheses, of which one is calculated from the F_0, and the other from the F_c. The F_0-synthesis should show the positions where the atoms are really located, whereas the F_c-synthesis shows where they are situated in the model. The difference of the two syntheses will, therefore, be zero at points where both of the atomic positions are identical, thus indicating the correctness of these positions. On the other hand, if an atom occupies an incorrect position in the model, the difference synthesis will show a 'hole' at this point, and instead a maximum appears at the correct position (see Fig. 29). The model can be corrected very simply by shifting the atom in the direction of the maximum.

It is not necessary actually to compute two Fourier syntheses with F_0 and F_c and to subtract the results point by point. It is sufficient, to insert into Equation

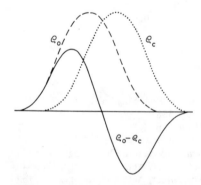

Fig. 29. Difference Fourier synthesis
The full curve is generated as the difference of two Fourier syntheses, calculated from F_0 and F_c and having maxima which are not exactly identical

[241] Ref. [232].
[242] Ref. [323].
[243] Ref. [317], p. 317.
[244] Ref. [329], p. 590.
[245] Ref. [335], p. 318.

(23) the differences $F_0(hkl) - F_c(hkl)$ instead of the $F(hkl)$, and to calculate using that one single synthesis. This will immediately show $\rho_0(X,Y,Z) - \rho_c(X,Y,Z)$.

Differential synthesis[246-249] starts from the fact that at the point of a maximum in a Fourier synthesis, the first derivatives of the electron density must vanish:

$$\frac{\partial \rho}{\partial x} = \frac{\partial \rho}{\partial y} = \frac{\partial \rho}{\partial z} = 0 \qquad (108)$$

Suppose that one coordinate of an atom before refinement is x, and after it with a shift ε

$$x_0 = x + \varepsilon \qquad (109)$$

it can then be shown that three equations of the type

$$\frac{\partial \rho}{\partial x} + \varepsilon_x \frac{\partial^2 \rho}{\partial x^2} + \varepsilon_y \frac{\partial^2 \rho}{\partial x \partial y} + \varepsilon_z \frac{\partial^2 \rho}{\partial x \partial z} = 0 \qquad (110)$$

are valid, from which the shifts ε may be determined.

5.6 Evaluation of the Results

By the methods dealt with above, for each of the atoms of a structure a set of coordinates is obtained, given in a coordinate system which is defined by the edges of the unit cell. The chemist, however, is interested in the distances and angles between the atoms. Using the formulae given below, these distances and angles may be calculated from the crystallographic parameters[250].

In general, one can calculate the distances from a given atom to all its neighbors within a given maximum distance. When doing this, it is often forgotten that every unit cell has 26 neighboring cells, and also that atoms located in these cells may belong to the immediate environment of a definite atom. In most cases, those neighboring cells are forgotten which have only one common point with the cell concerned in the direction of its space diagonal. Furthermore, all crystallographic calculations are carried out with only one representative from each lattice complex (see 7.1:1.1.5). It is also sufficient, always to examine only the neighborhood of these representatives, since all atoms linked by symmetry have an identical environment. However, for a certain atom, the members of its own lattice complex lying within the given maximum distance must be taken into con-

sideration as neighbors since their distances from the atom concerned may all be different.

The distance s_{12} of two atoms at the positions x_1, y_1, z_1 and x_2, y_2, z_2, in a lattice having the cell parameters a, b, c, α, β, γ is:

$$\begin{aligned} s_{12} = \{ &(x_1 - x_2)^2 a^2 + (y_1 - y_2)^2 b^2 + (z_1 - z_2)^2 c^2 \\ &+ 2(x_1 - x_2)(y_1 - y_2)\, ab \cos \gamma \\ &+ 2(y_1 - y_2)(z_1 - z_2)\, bc \cos \alpha \\ &+ 2(z_1 - z_2)(x_1 - x_2)\, ca \cos \beta \}^{1/2} \end{aligned} \qquad (111)$$

This equation can, of course, be simplified, when the symmetry of the unit cell is higher than triclinic and their parameters assume special values[251]. The bond angle ψ between three atoms, x_0, y_0, z_0 at the vertex of the angle, and x_1, y_1, z_1 and x_2, y_2, z_2 at its ends, can be calculated, with the distances s_{01} and s_{02}, from equation (111), by the equation:

$$\cos \Psi = \frac{p}{s_{01} s_{02}} \qquad (112)$$

with

$$\begin{aligned} p = &(x_0 - x_1)(x_0 - x_2)a^2 + (y_0 - y_1)(y_0 - y_2)b^2 \\ &+ (z_0 - z_1)(z_0 - z_2)c^2 \\ &+ \{(x_0 - x_1)(y_0 - y_2) + (y_0 - y_1)(x_0 - x_2)\}\, ab \cos \gamma \\ &+ \{(y_0 - y_1)(z_0 - z_2) + (z_0 - z_1)(y_0 - y_2)\}\, bc \cos \alpha \\ &+ \{(z_0 - z_1)(x_0 - x_2) + (x_0 - x_1)(z_0 - z_2)\}\, ca \cos \beta \end{aligned} \qquad (113)$$

This formula may also be simplified for higher symmetry.

The following procedure is used for determination of the *standard deviations*[252-254]: When the measurement of a value x is repeated a number of times, the single observations x_i will vary about a mean value $\bar{x}$. If p_i is the probability of obtaining a certain observation x_i, then

$$\bar{x} = \sum_i p_i x_i \qquad (114)$$

with

$$\sum_i p_i = 1 \qquad (115)$$

The scatter of the observations about $\bar{x}$ may be characterized by the variance

$$\sigma^2 = \sum_i p_i (x_i - \bar{x})^2 \qquad (116)$$

whose square root, σ, is called the standard deviation. With the least-squares method, σ can be calculated from the matrix (A_{ij}) of the normal equation elements. According to (106), the matrix element a_{ij} of (A_{ij}) is:

$$a_{ij} = \left(\sum_{k=1}^{m} w \frac{\partial F_{c_k}}{\partial p_i} \cdot \frac{\partial F_{c_k}}{\partial p_j} \right) \qquad (117)$$

If (A^{ij}) is the inverse matrix of (A_{ij}), and one of its elements is a^{ij}, the variance of the parameter p_i is then:

$$\sigma^2(p_i) = a^{ii} \qquad (118)$$

[246] *A.D. Booth*, Trans. Faraday Soc. *42*, 444 (1946).
[247] *A.D. Booth*, Trans. Faraday Soc. *42*, 617 (1946).
[248] *A.D. Booth*, Fourier technique in X-ray organic structure analysis, p. 46. Cambridge University Press, Cambridge 1948.
[249] Ref. [329], p. 600.
[250] Ref. [329], p. 629.

[251] Ref. [329], p. 631.
[252] Ref. [317], p. 330.
[253] Ref. [329], p. 616.
[254] Ref. [329], p. 643.

The standard deviation of the electron density $\varrho(XYZ)$ at the point X,Y,Z of a Fourier synthesis is given by

$$\sigma(\rho(XYZ)) \tag{119}$$

$$= \frac{1}{V}\left\{\sum_h \sum_k \sum_l \sigma^2[F(hkl)]\cos^2[2\pi(hX+kY+lZ)\right.$$

$$\left. -\alpha(hkl)]\right\}^{\frac{1}{2}}$$

For a reasonably large set of reflections, (119) reduces to

$$\sigma(\rho(XYZ)) = \frac{1}{V}\left\{\sum_h \sum_k \sum_l \sigma^2[F(hkl)]\right\}^{\frac{1}{2}} \tag{120}$$

The equation below has proved useful as an estimate for the $\sigma[F(hkl)]$ which appears in (120),

$$\sigma|F(hkl)| \approx \Delta F(hkl) = ||F_o(hkl)| - |F_c(hkl)|| \tag{121}$$

The standard deviations of bond lengths and angles may be derived from those of the atomic parameters. In each case, one must consider whether the related parameters are independent or are linked by symmetry[255, 256]. Detailed research on this topic has been published by Cruickshank and Robertson[257].

5.7 Computer Programs

With early crystal structure determinations on compounds of simple constitution, it was possible to perform the calculations by hand or with desk calculators. It would also be possible today, with sufficient expense of time and patience, to process all the data by pencil and paper, and to correct them with the aid of the available tables for Lorentz and polarization factors, absorption, etc., to obtain finally a useful list of the $|F_0(hkl)|$. Even Patterson and Fourier syntheses might be calculated, at least in projections, by simple means. However, a single calculation of such a kind would take weeks of hard labor, which could cause occasional errors to appear. Various strip techniques have proved suitable[258-262] as classical methods for this purpose.

Three-dimensional syntheses and least-squares refinements cannot be performed at all without the aid of high-speed digital computers, since the number of single steps necessary for such computations is simply too large. Computer programs exist for nearly all steps occuring in a crystal structure determination, discussed in this chapter. The 'Commission on Crystallographic Computing' of the 'International Union of Crystallography' always compiles a 'World List of Crystallographic Computer Programs'[263]. The last issue of this list contains 697 programs. Some groups of authors have developed program systems which link the partial programs necessary for a structure determination to program chains. In these chains, one link passes the data on to the next one[264-267].

5.8 Non-crystalline Materials

Non-crystalline materials do not produce sharp reflection patterns such as crystals generate, since the regular correlations between the atomic positions do not exist here, or only short-range order exists (in contrast to the long-range order in crystals). Nevertheless, here, too, one can derive conclusions about structural features of the substance under examination from the X-ray diagrams. In particular, conclusions about mean atomic distances and their relative frequencies, and about sizes and orientations of particles are possible.

Although, in principle, gaseous molecules also can be examined in regard to their intramolecular distances by X-ray methods, electron diffraction (see Chapter 7.5), which needs much less time for recording, has achieved success.

5.8.1 Determination of Mean Atomic Distances.
In liquids and glasses, the atoms are packed so closely that their immediate environment is ordered. Apart from the intramolecular distances, certain other intermolecular distances may also occur quite often which correspond to direct contact of molecules. These distances generate some interference effects at certain scattering angles[269-271]. Because of the small number of the shared atoms, the scattering maxima are very broad and flat. Therefore, one cannot simply use Bragg's law to determine the distances in the substance under study from the positions of the peaks. One has rather to determine the periods appearing in the entire scattering curve by a Fourier analysis.

[255] Ref. [317], p. 331.

[256] Ref. [329], p. 634.

[257] D.W.J. Cruickshank, A.P. Robertson, Acta Crystallogr. 6, 698 (1953).

[258] C.A. Beevers, H. Lipson, Phil. Mag. (VII) 17, 855 (1934).

[259] C.A. Beevers, H. Lipson, Nature 137, 825 (1936).

[260] J.M. Robertson, Phil. Mag. (VII) 21, 176 (1936).

[261] A.L. Patterson, Phil. Mag. (VII) 22, 753 (1936).

[262] A.L. Patterson, G. Tunell, Amer. Mineral. 27, 655 (1942).

[263] Ref. [319].

[264] M.G.B. Drew, ATSYS — Atlas Crystallographic System. In Ref. [319], No. 3068.

[265] E. Schultze-Rhonhof, BN-X-64, Programmsystem für Kristallstrukturuntersuchungen. In Ref. [319], No. 5063–5068.

[266] J.M. Steward, D.F. High, J.R. Holden, R.V. Chastain, C.W. Dickinson, Brown, M.O. Dayhoff, B. Morosin, E.C. Lingafelter, L.H. Jensen, H. Takeda, S. Wax, W. Keefe, X-Ray-63. In Ref. [319], No. 413–445.

[267] J.M. Steward et al.[266], X-Ray-67.

[268] Ref. [325], p. 577.

[269] A. Prietzschk, Z. Phys. 177, 482 (1941).

5.8.2 Determination of Particle Sizes: Small Angle Scattering

When colloidal particles or macromolecules are distributed in a very dilute solution, it can be assumed that the X-ray radiation diffracted by different particles practically does not interfere. On the other hand, if the wavelength of the radiation is small compared with the size of the particles, at small scattering angles interference may occur with the radiation which was scattered by different points of the same particle. This may be observed as 'small angle scattering'[272-286]. With nearly spherical particles, a Gaussian-shaped scattering curve is obtained (see Fig. 30, Ref.[285]), which is identical with that of a single particle. Deviations from the Gaussian form indicate that the particle has a different shape.

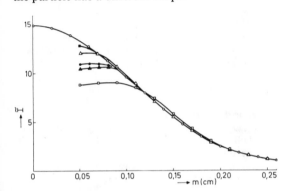

Fig. 30. Small angle scattering curves of apoferritin (a protein of molecular weight 462000)
Uncorrected curves from solutions of different concentrations. With kind permission of Anton Paar, Graz

[270] W. Kast, A. Prietzschk, Z. Elektrochem. Ber. Bunsenges. 47, 112 (1941).
[271] Ref.[325], p. 578, Fig. 27.
[272] O. Kratky, Naturwissenschaften 26, 94 (1938).
[273] O. Kratky, A. Sekora, R. Treer, Z. Elektrochem. Ber. Bunsenges. 48, 587 (1942).
[274] H. Kiessig, Kolloid-Z. Z. Polym. 98, 213 (1942).
[275] O. Kratky, Naturwissenschaften 30, 542 (1942).
[276] O. Kratky, A. Sekora, Naturwissenschaften 31, 31 (1943).
[277] O. Kratky, Monatsh. Chem. 76, 325 (1947).
[278] D. Heikens, P.H. Hermans, P.F. van Felden, A. Weidinger, J. Polym. Sci. 11, 433 (1953).
[279] O. Kratky, Öst. Chem.-Ztg. 54, 193 (1953).
[280] O. Kratky, A. Sekora, R. Treer, Z. Elektrochem. Ber. Bunsenges. 52, 49 (1954).
[281] Ref.[326].
[282] Ref.[325], p. 583.
[283] A. Guinier, Comt. rend. 204, 1115 (1937).
[284] A. Guinier, Thésis de l'Université Paris, Ser. A., 1854 (1939).
[285] H.J. Bielig, O. Kratky, G. Rohns, H. Wawra, Biochim. Biophys. Acta 112, 110 (1966).
[286] O. Kratky, Angew. Chem. 72, 467 (1960).
[287] Ref.[325], p. 587.

An especially interesting area of application of small angle scattering exists with solutions of fiber molecules. At the center their curve appears Gaussian; at the outer regions, it is shaped like $1/\vartheta^2$ if the fiber molecules are statistically coiled, and at the outermost portion it takes the form of $1/\vartheta$. The mean diameter of the ball may be calculated from the width of the Gaussian curve. In the $1/\vartheta$-part of the curve, the single statistical fiber elements scatter independently. From the angle at which this curve form begins, the length of the fiber molecules can, therefore, be derived.

5.8.3 Crystal Orientations

If, in the patterns of a polycrystalline material, closed interference circles no longer occur, but only sickle-shaped bows, a preferred orientation of the single crystallites[287] can be presumed. If the intensity is measured along the bow of such a sickle, a curve of the shape of the error function is found: an *azimutale blackening curve*[288]. The reciprocal of the width of this curve at one half the peak height is a relative measure of the orientation of the crystallites in the material under examination.

With small angle scattering, preferred orientations of extended crystallites can also be indicated. According to Heyn[289, 290], a number of vegetable fibers generate diagrams showing two crossed lines. These diagrams may be interpreted in terms of a helical structure for these fibers.

[288] Ref.[325], p. 590, Fig. 37.
[289] A.N.J. Heyn, Textile Res. J. 19, 163 (1949).
[290] Ref.[325], p. 592, Fig. 38.

6 Bibliography

6.1 Indices

[307] Crystal Data. 1st Ed. New York 1954.
[308] Crystal Data. 2nd Ed. Pittsburgh 1963.
[309] Crystal Structures. 1st Ed. John Wiley & Sons, New York 1948—1960.
[310] Crystal Structures 2nd Ed. John Wiley & Sons, New York 1963—1965.
[311] Powder Diffraction Data File (PD-1) (sog. «ASTM-Kartei»). American Society for Testing and Materials, Philadelphia 1957.
[312] The Barker Index of Crystals. Vol. I 1951, Vol. II 1956, Vol. III 1964, W. Heffer and Sons, Cambridge, England.

6.2 Tables

[313] International Tables for X-ray Crystallography. Vol. I. N.F.M. Henry, K. Lonsdale, Symmetry groups. The Kynoch Press, Birmingham 1952.
[314] Ref.[313], improved edition 1965.
[315] Ref.[313], improved and supplemented edition 1969.

[316] *K. Sagel*, Tabellen zur Röntgenstrukturanalyse. Springer Verlag, Berlin, Göttingen, Heidelberg 1958.
[317] International Tables for X-ray Crystallography. Vol. II. *J.S. Kasper, K. Lonsdale*, Mathematical Tables. The Kynoch Press, Birmingham 1959.
[318] International Tables for X-ray Crystallography. Vol. III. *C.H. MacGillavry, G.D. Rieck*, Physical and Chemical Tables. The Kynoch Press, Brimingham 1962.
[319] International Union of Crystallography, World List of Crystallographic Computer Programs. 2nd Ed. A. Oosthoek's Uitgevers, Utrecht 1966.

6.3 Textbooks and Monographs

[320] *A.N. Winchell*, Elements of optical mineralogy, part 1. 5th Ed. John Wiley & Sons, New York 1937.
[321] *R. Glocker*, Materialprüfung mit Röntgenstrahlen. 3. Aufl. Springer Verlag, Berlin, Göttingen, Heidelberg 1949.
[322] *E.E. Wahlstrom*, Optical Crystallography, 2nd Ed. John Wiley & Sons, New York 1951.
[323] *W. Nowacki*, Fouriersynthesen von Kristallen, Birkhäuser, Basel 1952.
[324] *H. Hauptman, J. Karle*, The solution of the phase problem. I. The centrosymmetric crystal. American Crystallographic Association, Monograph No. 3. Edwards Bros., Ann Arbor, Michigan 1953.
[325] *Houben-Weyl*, Methoden der organischen Chemie, 4. Aufl., Bd. III/2, p. 543. Georg Thieme Verlag, Stuttgart 1955.
[326] *A. Guinier, G. Fournet*, Small angle scattering of X-rays. John Wiley & Sons, New York, Chapman & Hall, London 1955.

[327] *L.V. Azároff, M.J. Buerger*, The powder method in X-ray crystallography. McGraw-Hill Book Company, New York, Toronto, London 1958.
[328] *M.J. Buerger*, Vector space and its application in crystal-structure investigation. John Wiley & Sons, New York 1959.
[329] *M.J. Buerger*, Crystal-structure analysis, 1st Ed. John Wiley & Sons, New York 1960.
[330] *M. v. Laue*, Röntgenstrahlinterferenzen. 3. Aufl. Akademische Verlagsgesellschaft, Frankfurt/M. 1960.
[331] *M.M. Woolfson*, Direct Methods in Crystallography. Clarendon Press, Oxford 1963.
[331a] *A. Guinier*, X-ray Diffraction in Crystals, Imperfect Crystals, and Amorphous Bodies. W.H. Freeman, San Francisco 1963.
[332] *M.J. Buerger*, The precession method in X-ray crystallography. John Wiley & Sons, New York, Sydney 1964.
[333] *M.J. Buerger*, Elementary crystallography, 3rd Ed. John Wiley & Sons, New York, London, Sydney 1965.
[334] *M.J. Buerger*, X-ray crystallography, 7th Ed. John Wiley & Sons, New York, London, Sydney 1966.
[335] *H. Lipson, W. Cochran*, The Determination of Crystal Structures, 3rd Ed. Bell & Sons, London 1966.
[336] *W. Kleber*, Einführung in die Kristallographie. 10. Aufl. VEB Verlag Technik, Berlin 1967.
[337] *H. Krebs*, Grundzüge der anorganischen Kristallchemie. Ferdinand Enke Verlag, Stuttgart 1968.
[338] *J.M. Bijvoet, W.G. Burgers, G. Hägg*, Early Papers on Diffraction of X-rays by Crystals. A. Oosthoek's Uitgevers, Utrecht 1969.
[339] *G.H. Stout, L.H. Jensen*, X-ray Structure Determination. Macmillan, London 1969.

7.2 Determination of Absolute Configuration (Anomalous Scattering of X-rays)

Ernst Schultze-Rhonhof
Amsterdamer Straße 15, D-53 Bonn

1 Principles

Using the methods described in Chapter 7.1, the structures of organic molecules can, of course, be determined. However, none of the procedures discussed there provides the answer to the question of the *absolute configuration* of asymmetric molecules[1-6]. The reason for this is that two X-ray reflections, $F(hkl)$ and $F(\overline{hkl})$, linked by a center in the reciprocal lattice, cannot normally be distinguished by observation techniques, even when the scattering crystal itself has no center.

According to Equation (16) of Section 7.1:2.2:

$$F(hkl) = |F(hkl)| \cdot e^{+i\alpha(hkl)} \tag{1}$$

and

$$F(\overline{hkl}) = |F(hkl)| \cdot e^{-i\alpha(hkl)} \tag{2}$$

Since, however, the phases $a\,(hkl)$ cannot be measured, but only the intensities of the reflections

$$I(hkl) \sim |F(hkl)|^2 \tag{3}$$

[cf. after (15) in Section 7.1:2.2], one obtains:

$$I(hkl) = I(\overline{hkl}). \tag{4}$$

---

[1] *J.M. Bijvoet*, Endeavour *14*, 161 (1955).
[2] *M. v. Laue*, Röntgenstrahlinterferenzen, 3. Aufl., p. 52. Akademische Verlagsgesellschaft, Frankfurt/M. 1960.
[3] *M.J. Buerger*, Crystal Structure Analysis, 1st Ed., p. 542. John Wiley & Sons, New York 1960.
[4] *W. Schlenk*, Angew. Chem. *77*, 161 (1965).
[5] *H. Lipson, W. Cochran*, The Determination of Crystal Structures, 3rd Ed., p. 382, Bell & Sons, London 1966.

Equation (4) is known as Friedel's law.

Equations (1)–(4) are based on the assumption that, on scattering of X-rays by each of the electrons of a crystal, the same phase lag, theoretically 180°, always occurs. If, however, the wavelength of the primary radiation is just below that corresponding to an absorption edge (see Fig. 14 in (7.1:3.2.2; cf. 7.1:3.3.2) of one of the atoms in the crystal under study, i.e. near a resonance scattering of the atom, so-called anomalous dispersion occurs. The phase lag at the scattering differs slightly from 180°. For calculation, one assigns to the anomalous scattering atom, instead of the usual atomic scattering factor f_o, a scattering factor, which is corrected by a complex term:

$$f = f_o + \Delta f' + i\Delta f''. \tag{5}$$

Because of this anomalous scattering, Friedel's law (4) becomes invalid; instead, one obtains

$$I(hkl) \neq I(\overline{hkl}). \tag{6}$$

If one divides the structure factor $F(hkl)$ into one portion which is generated by the anomalous scattering atom, $F_A(hkl)$, and another from the residue of the structure, $F_R(hkl)$:

$$F(hkl) = F_A(hkl) + F_R(hkl), \tag{7}$$

and further divides $F_A(hkl)$ into a real part ${}^rF_A(hkl)$ and an imaginary ${}^iF_A(hkl)$ part, one can then write:

$$F(hkl) = F_R(hkl) + {}^rF_A(hkl) + {}^iF_A(hkl) \tag{8}$$

and

$$F(\overline{hkl}) = F_R(\overline{hkl}) + {}^rF_A(\overline{hkl}) + {}^iF_A(\overline{hkl}). \tag{9}$$

For the correlated phases of the real parts one obtains, as usual,

$$\alpha_R(hkl) = -\alpha_R(\overline{hkl}) \tag{10}$$

and

$${}^r\alpha_A(hkl) = -{}^r\alpha_A(\overline{hkl}). \tag{11}$$

In contrast, the phase of an imaginary part exceeds that of the correlated real part by $+\pi/2$:

$${}^i\alpha_A(hkl) = {}^r\alpha_A(hkl) + \pi/2 \tag{12}$$

and

$${}^i\alpha_A(\overline{hkl}) = {}^r\alpha_A(\overline{hkl}) + \pi/2; \tag{13}$$

thus,

$${}^i\alpha_A(\overline{hkl}) = -{}^i\alpha_A(hkl) + \pi. \tag{14}$$

Fig. 1 shows equations (8) and (9) in the complex plane. It can now immediately be seen that:

$$|F(hkl)| \neq |F(\overline{hkl})| \tag{15}$$

The corrections $\Delta f'$ and $\Delta f''$ from (5) depend upon the wavelength λ of the primary radiation, and upon Bragg's angle θ. They were first calculated by Dauben and Templeton[7,8]. An enlarged table is included in Ref.[9].

2 Practical Procedure and Applications

2.1 Determination of the Absolute Configuration for an Already Known Crystal Structure

The classical experiment on the invalidity of Friedel's law (4), when anomalous dispersion occurs, was carried out with *zinc blende*[10]. On natural crystals of zinc blende, one of the faces (to which the indices (111) were arbitrarily assigned) is always well developed and shiny, whereas the opposite $(\overline{1}\overline{1}\overline{1})$-face appears dull and less well developed. Regarding the structure of zinc blende, it was known that layers of zinc and sulphur alternate in the [111]-direction, separated by distinct distances. Zinc has an absorption edge (which is the K-edge) at 1.284 Å. One might, therefore, expect normal dispersion for the $L\alpha_2$-line of gold ($\lambda\alpha_2 = 1.288$ Å), whereas the $L\alpha_1$-line ($\lambda\alpha_1 = 1.276$ Å) should be expected to give anomalous dispersion. Zn should scatter under these conditions with a phase advance of 10.5° relative to S. The authors, in fact, succeeded by this procedure in showing that the (111)-plane is occupied by the sulphur, and the $(\overline{1}\overline{1}\overline{1})$-plane by the zinc.

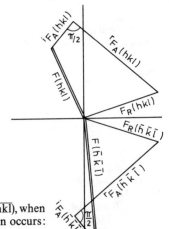

Fig. 1. $F(hkl)$ and $F(\overline{hkl})$, when anomalous dispersion occurs: Equations (8) and (9), drawn in vector form

[6] *J.M. Bijvoet, W.G. Burgers, G. Hägg*, Early Papers on Diffraction of X-rays by Crystals, p. 350, A. Oosthoek's Uitgevers, Utrecht 1969.

[7] *C.H. Dauben, D.H. Templeton*, Acta Crystallogr. *8*, 729 (1955).

[8] Ref.[3], p. 546.

[9] *C.H. MacGillavry, G.D. Rieck*, International Tables for X-ray Crystallography, Vol. III., p. 213. The Kynoch Press, Birmingham 1962.

[10] *D. Coster, K.S. Knol, J.A. Prins*, Z. Phys. *63*, 345 (1930).

The absolute configuration of optically active, symmetric organic compounds can be determined[11] with the aid of anomalous dispersion. In 1951, Bijvoet examined the configuration of *sodium rubidium tartrate*[12, 13]; that of D-*isoleucine* followed in 1954[14]. By this means, it proved possible to establish the configurations of all substances related to tartaric acid and isoleucine; previously, determinations of the relative configurations had required many years' work.

The methods used are very well exemplified by the determination of the configuration of L*(+)-tartaric acid* from that of $NaRbC_4H_4O_6 \cdot 2H_2O$. Rb has an absorption edge at $\lambda = 0.8155$ Å. One can, therefore, use ZrK_α-radiation ($\lambda_{\alpha_1} = 0.7901$ Å; $\lambda_{\alpha_2} = 0.7859$ Å) for stimulation of anomalous dispersion. From a series of reflections, the I(hkl) and I($\overline{hkl}$) were measured each time. In addition, the expected intensities for the two possible enantiomorphs A (which corresponds to the L-configuration, according to Emil Fischer's convention) and B were calculated from the already known structure of tartaric acid. As Table 1 (cf. Ref. [12]) shows, the configuration of L(+)-tartaric acid corresponds to the model A, which is drawn in Fig. 2. The correlation of Emil Fischer (1891), which was originally chosen arbitrarily, was thus proved correct.

Anomalous scattering factors are calculated according to the equation:

$$F(hkl) = \sum_j (f'_j + {}^if''_j) \cdot e^{-2\pi i(hX_j + kY_j + lZ_j)} \cdot T_j \quad (16)$$

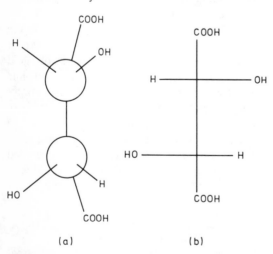

(a)   (b)

Fig. 2. Absolute configuration of L(+)-tartaric acid = model A

In this equation, the formula is simplified by

$$f'_j = f_o + \Delta f' \quad (17)$$

and

$$f''_j \equiv \Delta f'' \quad (18)$$

from equation (5). T_j is the temperature factor (see 7.1:3.3.4). As Fig. 1 shows, the real parts may be calculated in the usual way (see 7.1:2.2); in principle, the imaginary parts also are calculated in the same manner, but their phases are shifted by $\Delta \alpha = + \pi/2$.

2.2 Direct Structure Determination with the Aid of the Anomalous Dispersion

2.2.1 Combination with the Patterson Method

If a crystal contains only one or a small number of anomalous scattering atoms in its unit cell, the *Patterson synthesis*[15, 16] can be interpreted very easily, if one succeeds in recording the differences $|F(hkl)|^2 - |F(\overline{hkl})|^2 \sim |I(hkl)| - |I(\overline{hkl})|$ with sufficient precision. The Patterson function

$$P(UVW) \quad (19)$$
$$= \left(\frac{1}{V^2}\right) \sum_{h=-\infty}^{+\infty} \sum_{k=-\infty}^{+\infty} \sum_{l=-\infty}^{+\infty} |F(hkl)|^2 \cdot e^{-2\pi i(hU + kV + lW)}$$

[see 7.1:5.1.1, equation (52)], can be divided into a real and an imaginary part:

$$P(UVW) = {}^rP(UVW) - i \cdot {}^iP(UVW) \quad (20)$$

with

$${}^rP(UVW) \quad (21)$$
$$= \left(\frac{1}{V^2}\right) \sum_{h=-\infty}^{+\infty} \sum_{k=-\infty}^{+\infty} \sum_{l=-\infty}^{+\infty} |F(hkl)|^2 \cdot \cos 2\pi(hU + kV + lW)$$

and

$${}^iP(UVW) \quad (22)$$
$$= \left(\frac{1}{V^2}\right) \sum_{h=-\infty}^{+\infty} \sum_{k=-\infty}^{+\infty} \sum_{l=-\infty}^{+\infty} |F(hkl)|^2 \cdot \sin 2\pi(hU + kV + lW)$$

As is derived in complete detail in Ref.[17], the imaginary part (22) may be rearranged to:

$${}^iP(UVW) = \frac{1}{V^2} \sum_{h=0}^{+\infty} \sum_{k=-\infty}^{+\infty} \sum_{l=-\infty}^{+\infty} \left\{ |F(hkl)|^2 \right. \quad (23)$$
$$\left. - |F(\overline{hkl})|^2 \right\} \sin 2\pi(hU + kV + lW)$$

[11] C. Bokhoven, J.C. Schoone, J.M. Bijvoet, Proc. Koninkl. Ned. Akad. Wetenschap. (B) 52, No. 2 (1949).

[12] A.F. Peerdeman, A.J. van Bommel, J.M. Bijvoet, Proc. Koninkl. Ned. Akad. Wetenschap. (B) 54, 16 (1951).

[13] J.M. Bijvoet, A.F. Peerdeman, A.J. van Bommel, Nature 168, 271 (1951).

[14] J. Trommel, J.M. Bijvoet, Acta Crystallogr. 7, 703 (1954).

In a regular Patterson synthesis (see 7.1:5.1), each vector V_{AB} linking two atoms A and B generates a positive peak. The $^iP(UVW)$-synthesis according to equation (23) differs essentially from this. Suppose that A is an anomalous scattering atom, and N a normal scattering one. A vector V_{AN} then generates a positive peak, and a vector A_{NA}, having the inverse direction, generates a negative one; the vectors V_{NN} linking the normal scattering atoms vanish. When a regular Patterson synthesis contains N^2 peaks from N atoms per unit cell, there then appear $n.N-n^2$ positive peaks and the same number of negative ones in the $^iP(UVW)$-synthesis with n anomalous scattering atoms. Such a synthesis is much more simple than a normal Patterson synthesis, even in the case where the unit cell contains several anomalous scattering atoms. When the cell contains only one anomalous scattering atom, the positive peaks immediately give an image of the structure, and the correct configuration included. With several anomalous scattering atoms, the application of image-seeking functions (see 7.1:5.1.4) may promise success, if one neglects the negative peaks and superimposes the origin of the $_iP(UVW)$-function successively over the positions of the anomalous atoms.

2.2.2 Combination with the Method of Isomorphous Replacement

For the structure factors F(hkl) of two compounds MR and NR having identical structures except for the atoms M and N, and in which the positions of M and N are known, the phases for the structure factors F^{MR} (hkl) and F^{NR} (hkl) can be determined by the method of isomorphous replacement (see 7.1:5.1.6). Unfortunately, these phases have two

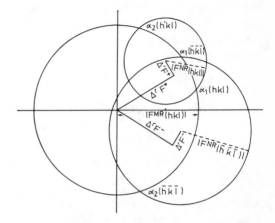

Fig. 3. Construction of the possible phase angles α(hkl) with anomalous dispersion (cf. also Fig. 28, 7.1:5.1.6)

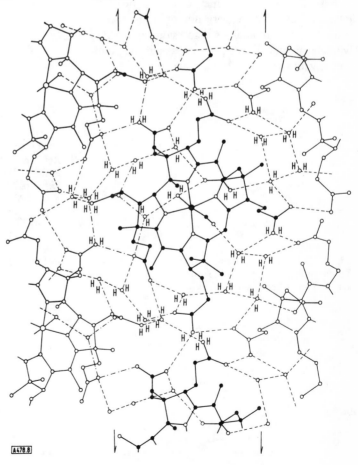

Fig. 4.[22] Crystal structure of cobyric acid in a projection in the direction of the a-axis. Molecules with center of gravity at x are drawn in heavy print; molecules with center of gravity at x̄ are drawn more faintly. Dotted lines indicate hydrogen bonds.

different meanings if one uses simple isomorphous replacement for their determination.

When the conditions of the experiment can be chosen in such a way that one of the two atoms M or N scatters anomalously, ΔF [see 7.1:5.1.6, equation (59)] may be split into two components, perpendicular to each other. The construction drawn in Fig. 28, Chapter 7.1:5.1.6, can now be performed with $F(hkl)$ and also with $F(\overline{hkl})$ (see Fig. 3; here, N is assumed to be the anomalous scattering atom). Since for the normal scattering compound MR, the usual condition (10) must hold which is realized for the pair

$$\alpha_2^{MR}(hkl) = -\alpha_2^{MR}(\overline{hkl}) \qquad (24)$$

in Fig. 3, but not for the other pair

$$\alpha_1^{MR}(hkl) \neq -\alpha_1^{MR}(\overline{hkl}) \qquad (25)$$

$\alpha_2^{MR}(hkl)$ is the true phase of the structure factor $F^{MR}(hkl)$.

An example of the use of this method is the determination of the structure of factor V_{1a} in vitamin B_{12}, *cobyric acid* (1), as in Ref. [21], D. Crowfoot-Hodgkin [22] (see Fig. 4, as Ref. [23]), Fig. 5, as Ref. [24], shows the asymmetric distribution of the intensities in a Weissenberg photograph which is caused by the anomalous dispersion, and which without anomalous dispersion would be symmetric with respect to the line 001.

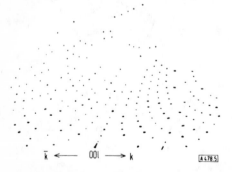

Fig. 5. 0kl-reflections from a sample of factor V_{1a}. The asymmetry on both sides of the line marked by arrows can be seen easily.

(1)

Table 1. Anomalous dispersion with sodium rubidium tartrate

Indices	Observed ratios of intensities		Calculated intensities			
			Model A		Model B	
hkl	hkl	$\overline{hkl}$	$I(hkl)$	$I(\overline{hkl})$	$I(hkl)$	$I(\overline{hkl})$
1 4 1	(?)		361	377	377	361
1 5 1	(?)		337	313	313	337
1 6 1	>		313	241	241	313
1 7 1	<		65	78	78	65
1 8 1	>		185	148	148	185
1 9 1	>		65	46	46	65
1 10 1	>		248	208	208	248
1 11 1	<		27	41	41	27
2 6 1	>		828	817	817	828
2 7 1	>		18	8	8	18
2 8 1	>		763	716	716	763
2 9 1	(?)		170	166	166	170
2 10 1	<		200	239	239	200
2 11 1	(?)		159	149	149	159
2 12 1	<		324	353	353	324

[15] *Y. Okaya, Y. Saito, R. Pepinsky,* Phys. Rev. *98,* 1857 (1955).

[16] Ref. [5], p. 389−394.

[17] Ref. [5], p. 389−392.

[18] Ref. [5], p. 392−394.

[19] *A.C.T. North,* Acta Crystallogr. *18,* 212 (1965).

[20] *J. Trommel, J.M. Bijvoet,* Acta Crystallogr. *7,* 703 (1954).

[21] *D. Crowfoot-Hodgkin,* Angew. Chem. *77,* 958 (1965).

[22] Ref. [21], p. 954−962.

[23] Ref. [21], p. 959, Fig. 8.

[24] Ref. [21], p. 957, Fig. 5.

7.3 Neutron Diffraction

Georg Will
Abteilung für Kristallstrukturlehre und Neutronenbeugung Mineralogisches Institut der Universität,
D-53, Bonn

Neutron diffraction has two aspects: one characterized by nuclear interaction, and one by magnetic interaction.

Neutron diffraction based on nuclear interaction permits absolute determination of the location of atoms and molecules in a crystal, and of *molecular configurations* and study of the following individual topics:

Precise location of hydrogen atoms in a molecule and in a crystal, especially at low temperatures (see Section 7.3:3.2) – for example, with benzene.

Determination of the crystal structure of large organic molecules such as Vitamin B_{12} (Section 7.3:3).

Differentiation between H and D in molecules (Section 7.3:3).

Position of hydrogen (or elements with low atomic numbers) in the presence of heavy elements (7.3:3) e.g. in uranium hydride.

Determination of the absolute configuration of molecules by the method of anomalous dispersion with absorbing nuclei (7.3:6).

Investigation of oscillations and vibrations of atoms and molecules in crystals (7.3:3.3), for example, in benzene.

Differentiation between neighboring elements in the Periodic Table in molecules and crystals — for example, C and N, or $Mn^{2\oplus}$ and $Fe^{3\oplus}$ (7.3:3.4).

Labelling of atoms by isotopes (7.3:3.4).

The magnetic moment of the neutron is the basis of *magnetic structure analysis*. The following can be studied:

The transition into a ferro- or anti-ferromagnetic state (7.3:4.2); for example, with MnO.

The existence and properties of helical spin structures (7.3:4.3), e.g. in $MnSO_4$ or the rare earth elements.

The magnitude and direction of the magnetic moments (spins) in a crystal (7.3:4.3).

The determination of covalency contributions in a bond (7.3:4.4), for example in $LaCrO_3$ and $LaFeO_3$.

Determination of spin densities in crystals (7.3:4.5), for example in Fe or in Ni.

Determination of the density of delocalized bonding electrons, as in covalent bonds (7.3:4.5); for example, in MnF_2 or in $ZrZn_2$.

Further problems of fundamental interest exist, which can be studied best with neutron diffraction experiments:

Whether the thermal vibrations of the nucleus and the electron cloud are synchronous and of the same amplitude (cf. Section 7.3:5); for example, $Co_{0.92}Fe_{0.08}$.

Anisotropic vibrations of individual atoms, even in crystals with cubic symmetry (7.3:5), for example in CaF_2, UO_2, ThO_2.

Atomic polarization in molecules and atoms (7.3:7) e.g. triazine or Mn_2Sb.

Determination of the *molecular configuration in the gas or liquid state* (7.3:8), for example N_2, or O_4.

1 General Principles and Relationships to X-ray Diffraction

X-ray and neutron diffraction are today the most important and informative methods for the determination of crystal structures and molecular configurations. *Crystal structure* analysis occupies a unique position in determination of the electron density spatial distribution in a molecule forming part of a crystal. At present such investigations are based mainly on the various X-ray diffraction techniques. However, neutrons can be used in a very similar way, and for many problems offer certain advantages over X-rays; thus, rapid development can be foreseen for neutron diffraction.

A neutron is an elementary particle with finite mass and spin, and without electric charge. The wavelength λ follows from the neutron energy E [in eV], or the neutron velocity v [in cm/sec], and the neutron mass m by the de Broglie relation (h = Planck's quantum):

$$\lambda = \frac{h}{m \cdot v} = \frac{h}{\sqrt{2 m \cdot E}} \, \Delta \, \frac{3{,}956 \cdot 10^5}{v} \, \Delta \, \frac{0{,}286}{E} \qquad (1)$$

$$1 \text{ Å} \; \Delta \; 0{,}082 \text{ eV} \; \Delta \; 3956 \text{ m/sec}$$

Neutrons with the desired wavelengths are thus found in the thermal neutron spectra of nuclear reactors with energies of about 0.1 eV. These energies are about 10,000 times smaller than the energy of X-rays; they are comparable to the kinetic energy of atoms in matter. When a beam of thermal neutrons strikes a crystal, elastic Bragg scattering without energy transfer will be observed, as well as inelastic scattering involving energy exchange with the atomic or crystalline vibrations. Since the energies of the neutrons and the transferred energies are comparable, the gain or loss in energy is quite easy to measure, and this forms the basis of inelastic neutron diffraction or neutron spectroscopy, which is described in Chapter 5.4.

The general fundamental principles, crystallographic expressions, interference conditions, reciprocal lattice, space group extinctions, and different methods for phase determination and least squares refinement of the parameters are the same for both X-ray and neutron diffraction. More details are given in Chapter 7.1 Table 1 summarizes the major properties and differences of the two types of radiation.

Table 1. Comparison between X-ray and neutron diffraction

Feature	X-rays	Neutrons
Mass	0	1.008922 atomic mass units
Spin	0 (Bose particle)	$\frac{1}{2}$ (Fermi particle)
Magnetic moment	No moment	$\mu_n = -1.91319$ nuclear magnetons
Wavelength	Characteristic line spectrum from quantum transitions in the atomic "shell" e.g. Cu-Ka = 1.54 Å; in addition, continuous soft "Bremsspektrum", e.g. for Laue photos.	Filtration of a narrow wavelength band, e.g. 1.1 ± 0.05 Å from the continuous Maxwell spectrum of a reactor
Energy	$h \cdot v = \dfrac{h \cdot c}{\lambda}$	$\dfrac{1}{2} mv^2 = \dfrac{h \cdot v}{2\lambda}$
Energy for 1 Å	3×10^{18} h	1.5×10^{13} h (comparable with the energy of crystal vibrations)
Speed at 1 Å	3×10^{10} cm/sec	4×10^5 cm/sec
Type of radiation	Electromagnetic radiation	Particles (de Broglie wavelengths)
Type of interaction	Electromagnetic interaction of the electrons of the atom with the electromagnetic field of the X-radiation	Two types of interaction: 1) Nuclear interaction between neutrons and atomic nucleus. 2) Magnetic dipole interaction $=0.539 \times 10^{-12} \times S$ cm (S =spin quantum number)
Isotope effect	No difference between isotopes	Isotopes have different scattering powers
Phase change on scattering	180°	Normally 180°; for some exceptions (e.g. H, Mn) 0°, i.e. negative scattering lengths.
Absorption	Very high $\mu \sim 10^2$–10^3 cm^{-1}	Generally low $\mu \sim 10^{-1}$ cm^{-1}; some exceptions, e.g. Gd (19000 cm^{-1}), Cd (2700 cm^{-1}), Eu (1600 cm^{-1}), Er (125 cm^{-1}), B (440 cm^{-1})[a]
Detection methods	Direct interaction between radiation and detector: 1) Film 2) Geiger counter, Prop. counter, Scintillation counter	Indirect interaction with the detector via mechanical collision processes: BF$_3$-counter, H^3-counter, U^{235}-fission chamber
Inelastic scattering	Energy change negligible; insignificant.	Energy change considerable (inelastic neutron diffraction for study of lattice dynamics)

[a] The values given are the total cross-sections for 0.073 eV neutrons (λ=1.055 Å)

2 Diffraction Technique and Physical Principles

2.1 Diffraction by Crystals

A crystal can be described as a three-dimensional periodic array of atoms or molecules. This three-dimensional lattice acts as a three-dimensional grating for an incoming neutron beam, and inter-ference effects are observed as described by the v. Laue equations, or by the Bragg equation [see Equation (6), and Chapter 7.1]. The locations of the diffraction maxima in space are described by purely geometrical relationships between the dimensions of the grating, e.g. the dimensions of the unit cell, and the wavelength λ of the radiation.

It is also possible to calculate the unit dimension, the lattice constants a, b, c and the angles α, β, γ solely from the observed diffraction angles:

Scheme 1:   Position of $\rightleftharpoons$ Dimensions of
diffraction   the unit cell
maxima   a, b, c, α, β, γ

The intensity of the diffracted beam is determined by the contents of the unit cell, i.e. by the spatial distributions of the atoms throughout the unit cell. This information then contains the molecular orientation and molecular configuration – the crystal structure. In order to describe the diffracted, observable intensities, a structure factor F must be introduced, which describes the scattering power of one single unit cell:

$$F_{hkl} = \sum_j \varphi_j \cdot e^{2\pi i \cdot (hx_j + ky_j + lz_j)} \qquad (2)$$

The summation is over all atoms j of the unit cell. ϕ_j is the Fourier-transform of the scatterer; it depends on the interaction between radiation and matter, and will be discussed in Section 7.3:2.4.

The intensity of the coherently diffracted wave is the sum of the contributions from all the unit cells which form the crystal. The resulting amplitude q_{hkl} of a wave diffracted at the atomic planes of one mosaic block is one half of the scattering within the first Fresnel zone:

$$q_{hkl} = N_c \cdot d_{hkl} \cdot \frac{\lambda}{\sin\theta} \cdot F_{hkl} = 2\,N_c \cdot d_{hkl}^2 \cdot F_{hkl} \qquad (3)$$

N_c is the number of cells per unit volume. For neutrons, q is of the order of 10^{-4} to 10^{-5}; q is independent of the wavelength.

There is no coherence between the waves diffracted from the individual mosaic blocks of a real crystal. Diffraction experiments with real crystals in a monochromatic beam, therefore, use a rocking technique, whereby the crystal is turned through a small angular interval. This yields an integrated intensity I:

$$I = Q \cdot \Delta v \qquad (4)$$

with

$$Q = \frac{\lambda^3 \cdot N_c^2}{\sin 2\theta} \cdot F_{hkl}^2 \qquad (5)$$

Δv is the volume of one crystallite. The observed intensity diffracted from a real crystal is, therefore, proportional to F^2.

[1] Acta Crystallogr. A 25, 391 (1969).
[2] G.E. Bacon, N.A. Curry, S.A. Wilson, Proc. Roy. Soc., Ser. A 279, 98 (1964).

From the observed intensities I_{hkl}, the crystal structure can be calculated (the fact that the phase of the diffracted wave is lost in the experiment is neglected; the methods for phase determination are discussed in Section 7.1:5).

Scheme 2:
Intensity of $\rightleftharpoons$ Spatial arrangement
interference   of atoms in the
unit cell =
crystal structure

Therefore, X-ray and neutron diffraction experiments must be set up in such a way that the positions of the interference spots (in relation to the crystal orientation), and also the intensities, can be determined accurately, reliably and fast.

2.2 Experimental Techniques for Diffraction Measurements

The basis for elastic neutron diffraction is Bragg's law. The lattice spacing d in the crystal is fixed by the unit cell dimensions of the crystal. The wavelength λ and the scattering angles 2θ are linearly interdependant, however. Therefore, one quantity can be chosen as a variable by the experimenter. This means that it is possible to make an analysis either of the Bragg-angle θ (with λ kept constant), or of the energy of the diffracted beam, with the scattering angle 2θ fixed. Both methods are in use today.

Scheme 3: $\lambda = 2\,d\,\sin\theta$ \qquad (6)

crystal diffractometer	Time of flight spectrometer
λ = constant	λ = variable
θ = variable	θ = constant
Analysis of scattering angles at fixed wavelength	Analysis of the energy of the diffracted neutrons at fixed scattering angle

2.3 Physical Principles of Neutron Diffraction

The elastic scattering of thermal neutrons when a beam of neutrons is incident on a crystal is determined essentially by two processes, of approximately equal magnitude:

interaction between the neutron and the atomic nuclei, through the nuclear forces;

magnetic interaction (classical dipole interaction) between the magnetic moment of the neutron and the magnetic moments of the unpaired electrons (in paramagnetic ions or atoms).

Both these interactions are of the order of 10^{-12} to 10^{-13} cm.

The nuclear scattering interaction between neutrons and matter is described by the nuclear scattering length b (measured in fermis; 1f= 10^{-13} cm).

Whereas the scattering of X-rays increases proportional to the atomic number Z, no such simple rule holds for the scattering of neutrons by nuclei, which can be represented by two processes. One is the scattering of a plane wave at a rigid sphere (hard sphere or *potential scattering*), which increases with volume and thus with $\sqrt[3]{A}$ (where A is the mass number of the nucleus; cf. Fig. 1). More important than potential scattering is *resonance scattering*, in which the neutrons are captured and re-emitted by excited nuclei.

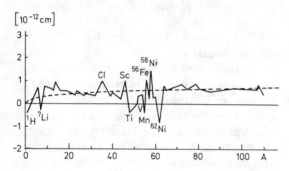

Fig. 1. Nuclear scattering amplitudes b as a function of atomic weight A.
The monotonic curve for 'potential scattering' of neutrons by the nuclei is shown by the dotted line

This results in a random distribution of nuclear scattering amplitudes through the periodic system, depending on the nuclear energy levels (Fig. 1). Because of this resonance scattering, different isotopes of the same element may exhibit marked differences in scattering power (isotope-specific scattering amplitudes:) e.g. ^{58}Ni: b= 14.4 f, ^{60}Ni: b= 3.0 f, ^{62}Ni: b=−8.7 f; or ^{1}H: b=−3.78 f, ^{2}H: b=6.5 f.

A positive sign of b indicates a phase change of 180°, analogous to X-ray scattering; a negative sign for the scattering length b means that there is no phase difference between the incident and scattered waves. The Fourier peaks for such nuclei with negative b values are thus negative in the Fourier diagrams. Important nuclei with negative scattering lengths are ^{1}H: −3.72 f, Li (natural mixture): −1.9 f, ^{55}Mn: −3.6 f, ^{51}V: −0.5 f. The nuclear scattering lengths, which are necessary for evaluation of diagrams, must be determined by diffraction experiments on crystals with a simple structure, because the energy levels of nuclei are not known with sufficient accuracy for theoretical calculations.

In contrast to X-ray scattering, which is strongly angle-dependent (atomic diameter $\approx 10^{-8}$ cm), the *nuclear scattering* of neutrons is isotropic (nuclear diameter $\approx 10^{-13}$ cm). The atomic form factor used in X-ray structure analysis is unknown in neutron nuclear scattering. The *magnetic scattering* of neutrons is angle-dependent however, and is described by a magnetic form factor.

2.4 Evaluation of Data

The intensity of a neutron wave diffracted at a lattice plane (hkl) is proportional to the square of the crystallographic structure factor F_{hkl} for this plane. The diffraction is by nuclear interaction and magnetic dipole interaction, and the observed intensity of the diffracted neutron beam is additively composed of the two components:

$$I = I_{nukl} + I_{magn} \qquad (8)$$

Since the intensity is proportional to the square of the crystallographic structure factor, the following relation holds for unpolarized neutrons:

$$|F|^2 = |F|^2_{nukl} + q^2 \cdot |F|^2_{magn} \qquad (9)$$

q is known as the magnetic interaction vector (see Equation (14)), and F is the structure factor − with the following meanings:

neutrons − nuclear scattering:

$$F_{nukl} = \sum_j b_j \cdot e^{2\pi i(hx_j + ky_j + lz_j)} \qquad (10a)$$

neutrons − magnetic scattering:

$$F_{magn} = \sum_j p_j \cdot e^{2\pi i(hx_j + ky_j + lz_j)} \qquad (10b)$$

For comparison, the structure factor for X-ray scattering

$$F_{X-rays} = \sum_j f_j^R \cdot e^{2\pi i(hx_j + ky_j + lz_j)} \qquad (10c)$$

$f_{\frac{x}{j}}$ are the *scattering amplitudes for X-rays,* which are the Fourier transforms, and describe the spatial distribution of *all* electrons surrounding the nuclei. b_j are the nuclear scattering lengths (not dependent on the scattering angle 2θ), and p_j are the angle-dependent magnetic scattering amplitudes (see Section 7.3:4). The summation is over all atoms of one unit cell.

Fourier inversion of the experimental structure factor data yields a density distribution $\varrho_{(xyz)}$ in the unit cell, assuming that the correct phases of the F-values are known.

$$\rho(xyz) = \frac{1}{V} \sum_{h=-\infty}^{+\infty} \sum_{k=-\infty}^{+\infty} \sum_{l=-\infty}^{+\infty} F_{hkl} \cdot e^{2\pi i(hx + ky + lz)} \quad (11)$$

The structure factor F_{hkl} contains a phase factor $e^{i\alpha}$, which reduces to +1 or −1 for centro-symmetric crystals. The phases will be assumed to be known; for a detailed discussion of the phase problems, see Section 7.1:5.

Table 2. Nuclear Scattering lengths (in units of 10^{-12} cm) of elements and isotopes[1]

Nr.	Isotope	Scattering length	Nr.	Isotope	Scattering length	Nr.	Isotope	Scattering length
1	^{1}H	—0.372		^{58}Ni	1.44	54	Xe	0.47
	^{2}H	0.621		^{60}Ni	0.30	55	^{133}Cs	0.75
	^{3}H	0.47		^{61}Ni	0.76	56	Ba	0.52
2	^{4}He	0.30		^{62}Ni	—0.87	57	^{139}La	0.83
3	Li	—0.194		^{64}Ni	—0.037	58	Ce	0.46
	^{6}Li	0.18 +0.025i	29	Cu	0.76		^{140}Ce	0.47
	^{7}Li	—0.21		^{63}Cu	0.67		^{142}Ce	0.45
4	^{9}Be	0.774		^{65}Cu	1.11	59	^{141}Pr	0.44
5	B	0.54 +0.021i	30	Zn	0.57	60	Nd	0.72
	^{11}B	0.60		^{64}Zn	0.55		^{142}Nd	0.77
6	^{12}C	0.665		^{66}Zn	0.63		^{144}Nd	0.28
	^{13}C	0.60		^{68}Zn	0.67		^{146}Nd	0.87
7	^{14}N	0.94	31	Ga	0.72	62	^{152}Sm	—0.5
8	^{16}O	0.577	32	Ge	0.84		^{154}Sm	0.8
	^{17}O	0.578	33	As	0.64	63	Eu	0.55
	^{18}O	0.600	34	Se	0.78	64	Gd	1.5
9	^{19}F	0.55	35	Br	0.67	65	Tb	0.76
10	Ne	0.46	36	Kr	0.74	66	Dy	1.69
11	^{23}Na	0.351	37	Rb	0.85		^{160}Dy	0.67
12	Mg	0.52		^{85}Rb	0.83		^{161}Dy	1.03
13	^{27}Al	0.35	38	Sr	0.656		^{162}Dy	—0.14
14	Si	0.42	39	^{89}Y	0.79		^{163}Dy	0.50
15	^{31}P	0.51	40	Zr	0.69		^{164}Dy	4.94
16	^{32}S	0.28	41	^{93}Nb	0.69	67	^{165}Ho	0.85
17	Cl	0.96	42	Mo	0.66	68	Er	0.79
	^{35}Cl	1.18	43	Tc	0.68	69	^{169}Tm	0.69
	^{37}Cl	0.26	44	Ru	0.73	70	Yb	1.26
18	^{40}A	0.20	45	Rh	0.59	71	Lu	0.73
19	K	0.37	46	Pd	0.60	72	Hf	0.78
	^{39}K	0.37	47	Ag	0.61	73	^{181}Ta	0.70
20	Ca	0.49		^{107}Ag	0.83	74	W	0.466
	^{40}Ca	0.49		^{109}Ag	0.43	75	Re	0.92
	^{44}Ca	0.18	48	Cd	0.37 +0.16i	76	Os	1.07
21	^{45}Sc	1.18		^{113}Cd	—1.5 +1.2i		^{188}Os	0.78
22	Ti	—0.34	49	In	0.39		^{189}Os	1.10
	^{46}Ti	0.48	50	Sn	0.61		^{190}Os	1.14
	^{47}Ti	0.33		^{116}Sn	0.58		^{192}Os	1.19
	^{48}Ti	—0.58		^{117}Sn	0.64	77	Ir	1.06
	^{49}Ti	0.08		^{118}Sn	0.58	78	Pt	0.95
	^{50}Ti	0.55		^{119}Sn	0.60	79	Au	0.76
23	V	—0.05		^{120}Sn	0.64	80	Hg	1.27
24	Cr	0.352		^{122}Sn	0.55	81	Tl	0.89
	^{52}Cr	0.490		^{124}Sn	0.59	82	Pb	0.96
25	^{55}Mn	—0.36	51	Sb	0.54	83	^{209}Bi	0.864
26	Fe	0.95	52	Te	0.56	90	^{232}Th	0.99
	^{54}Fe	0.42		^{120}Te	0.52	92	U	0.84
	^{56}Fe	1.01		^{123}Te	0.57		^{235}U	0.98
	^{57}Fe	0.23		^{124}Te	0.55		^{238}U	0.85
27	^{59}Co	0.25		^{125}Te	0.56	93	Np	1.055
28	Ni	1.03	53	^{127}I	0.52	94	Pu	0.75

Since any diffraction experiment is a *Fourier inversion* itself, we can formulate *scheme 4* (see page 556).

The Fourier inversion Equation (11) of the experimental structure factors (with correct phases) yields Fourier diagrams analogous to electron density diagrams in X-ray diffraction. When the nuclear diffraction data are used, the diagrams are representative of the nuclear scattering density and describe the position of the atomic nuclei in the unit cell.

In general, $\varrho_{(xyz)}$ is a function of the radiation and of the structure factors used in Equation (10):

X-rays: ρ (xyz) = *electron density*

neutrons:
nuclear scattering ρ (xyz) = *nuclear scattering density*

neutrons:
magnetic scattering ρ (xyz) = *magnetic moment density.*
(In 3d-transition elements this is, to good approximation, the spin density, and, therefore, the density of unpaired outer 3d-electrons).

Scheme 4

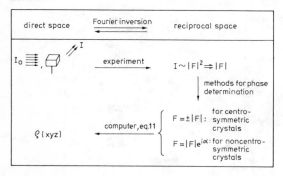

3 Diffraction of Neutrons at the Nucleus

3.1 Introduction

In general, a neutron diffraction study is carried out as the continuation of an X-ray diffraction investigation, since the heavy expenditure involved justifies such experiments only when other methods (especially X-ray diffraction) fail to give satisfactory results. Neutron diffraction is especially effective for studying the three following kinds of problems:

Determination of the coordinates of light elements in the presence of heavy elements, particularly in locating hydrogen atoms: $Z_1 \gg Z_2; b_1 \approx b_2.$

Differentiation between elements of neighboring atomic numbers, or even between isotopes: $Z_1 \approx Z_2; b_1 \neq b_2.$

Determination of the structures of crystals which absorb X-rays very strongly, for example *lead* or *mercury* compounds, or crystals which contain *lanthanides* or *uranium:* $\mu_{\text{Neutrons}} < \mu_{\text{X rays}}$

3.2 Localization of Hydrogen Atoms

For localization and investigation of hydrogen atoms in crystals, neutron diffraction experiments give far better results than X-ray diffraction methods. With neutrons, the attainable accuracy is some 10 times higher. There are two reasons.

First, the scattering power of hydrogen for X-rays is intrinsically very weak, because of the proportionality with Z, the number of electrons. For neutrons, there is no great difference in scattering length between H and the other elements of the periodic table (cf. Table 2):

$$b_{1_H} = -3.72 \ f; \qquad b_{2_H} = 6.21 f.$$

Secondly, the scattering power of X-rays decreases, in the case of hydrogen, much more rapidly with increasing scattering angle 2θ than for the other elements ($f_{(H)} = 0.1$ at $\sin\theta/\lambda = 0.45 \ \text{Å}^{-1}$). This unfavorable situation cannot be improved drastically even with better X-ray techniques, since there are not sufficient reflections at high angles containing singificant contributions from the hydrogen atoms. On the other hand, the resolution and accuracy of the atomic coordinates depends greatly on the measurements at higher angles. The scattering of neutrons at hydrogen nuclei (diameter 10^{-12} to 10^{-13} cm), however, is independent of the angle θ.

In order to reveal finer details of the structure and of the hydrogen atoms, the technique of difference Fourier diagrams is used, in which the scattering contribution of the skeleton molecule (without hydrogen atoms) is subtracted. Figure 2 illustrates the difference Fourier projection[2] of benzene on the (001)-plane at $-135\,°C$. After subtraction of the C_6-skeleton, only the hydrogen atoms remain in the Fourier diagram. Because of the negative scattering length of H, these peaks are drawn with dotted (negative) lines. The C—H distance was determined as 1.09 Å, and the C—C-distance as 1.398 Å. Difference diagrams can be calculated

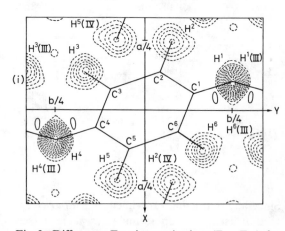

Fig. 2. Difference Fourier projection $(F_{\text{obs}} - F_C)$ for benzene on the (001) plane at $-135°$.
After substraction of the C_6 skeleton, the diagram contains only the H atoms. These are indicated by dotted contours (negative regions) because of the negative nuclear scattering length of hydrogen.

very simply by replacing the structure factor amplitudes F_{hkl} in Equation (11) by the difference values.

$$\Delta F_{hkl} = (F_{obs} - F_{skeleton}) \qquad (12)$$

In the case of benzene (Fig. 2), $F_{skeleton}$ is the calculated structure factor for the C_6 carbon ring. In neutron diffraction, such difference Fourier diagrams — where the major part of the structure has been subtracted — are of much greater importance than in X-ray diffraction. Because of the isotropy of the nuclear scattering, the effect of series termination is a considerable problem. As a result, the normal diagrams show major fluctuations of density. Instead of using mathematical procedures to correct this error, the difference Fourier technique described above is applied.

The finite dimension of the observed maxima in the Fourier diagram, which in a way represent a picture of the nuclei, is primarily a consequence of the series termination effects in Equation (11), and is only secondarily due to the thermal vibration of

Fig. 4. Three-dimensional superimposed neutron scattering density of the corrin unit in vitamin B_{12}
The central Co atom appears in the neutron Fourier diagram (resolution 1.3 Å) as a 'light' atom because of its relatively low nuclear scattering length. H nuclei are negative (dotted contour lines)

the nuclei, which causes some additional blurring. As a second example, Fig. 3 shows part of a Fourier diagram of a Vitamin B_{12}-derivative[3], $C_{63}H_{87}O_{15}N_{13}PCo \cdot 16H_2O$. This compound forms monoclinic crystals with a = 14.915 Å, b = 17.486 Å, c = 16.409 Å and β = 104.11°, space group $P2_1$. As in X-ray structure analysis of proteins, measurements were first made on this *vitamin B_{12} complex* at a reduced resolution of 1.3 Å, and the results were subsequently improved at an increased resolution of 1.0 Å. The improvement achieved by the increased resolution is shown in Fig. 3. Fig. 4 shows the three-dimensional superimposed neutron scattering density of the *corrin* unit as part of the structure of Vitamin B_{12}. The central cobalt atom appears in the neutron Fourier diagram (resolution 1.3 Å) as a 'light' atom because of its relatively short nuclear scattering length. The H nuclei are again negative (dotted contour lines). A total of 4400 independent reflections were measured; the crystal weighed 11 mg. The measurement speed was 30 min/reflection.

The high incoherent background scattering from hydrogen poses particular difficulties in structure analyses of compounds containing hydrogen. This is due essentially to the nuclear-spin incoherence, resulting from the different possible orientations of the neutron spin ($I_n = 1/2$) relative to the nuclear spin ($I_{1H} = 1/2$), and is particularly large in the case of hydrogen (about 34 barns in crystals). By comparison, the coherent scattering cross-section of hydrogen is only 1.79 barns. The

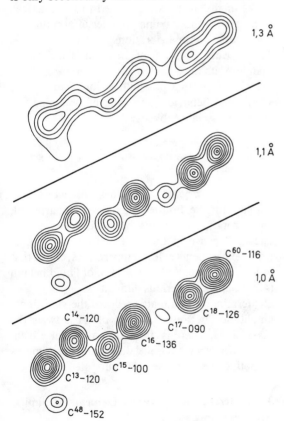

Fig. 3. Improvement in the resolution in a Fourier section of vitamin B_{12} at y = 0.120b (part of diagram). It is not possible to distinguish individual atoms at a resolution of 1.3 Å; complete separation is achieved at 1.0 Å. The numbers next to the assignment of the atoms' identity give the y-coordinates in units of $10^3 y/b$

[3] *F.M. Moore, B.T.M. Willis, D.C. Hodgkin*, Nature *214*, 130 (1967);
D.C. Hodgkin, F.H. Moore, B.H. O'Connor, unpubl.

situation is much more favorable with deuterium ($I_{2H} = 1$), which has an incoherent scattering cross-section of about 2.2 barns. Therefore, the investigation of deuterated crystals is preferred wherever possible.

3.3 Hydrogen Bonds and Ferroelectricity

Special advantages result when neutrons are used for the crystal structure analysis of ferroelectric crystals. Such crystals exhibit a spontaneous electric polarization on cooling through a characteristic temperature, the ferroelectric Curie point. This effect is associated with small displacements of individual atoms in the crystal structure. Generally, light atoms (often, hydrogen atoms) change their positions in the neighborhood of much heavier atoms, and move from a special position in the unit cell (without free parameters) into a general position with the general coordinates x, y, z. The ferroelectric transition is, therefore, always associated with a crystallographic phase transition to a crystal class with pyroelectric symmetry. For determination of the position coordinates x, y, z, neutron diffraction has the advantages that all elements have comparable neutron scattering powers, and that the neutron scattering power does not depend on the scattering angle; there is no form factor. Reflections with large indices, which are especially sensitive to small changes in the parameters, are not, as in X-ray diffraction, already weakened by the form factor fall-off and are thus not yet at the limit of detection.

One of the most important *ferroelectric crystals* is *potassium dihydrogen phosphate*, with a transition temperature of $-123\,°C$. The ferroelectric polarization can be recognized clearly in the Fourier and difference Fourier diagrams taken at $20\,°C$ and $-180\,°C$: above and below the Curie temperature. The potential curve of the H atoms has two minima, which are symmetrically situated between two oxygen atoms. In the paraelectric phase, the two potential minima are statistically equally occupied: therefore, the Fourier maxima appear as blurred ellipsoids. In the ferroelectric phase, the hydrogen atoms become ordered, and now only one potential minimum is occupied: hydrogen bonds O—H...O now exist, and the hydrogen Fourier peaks are spherical and concentrated near one oxygen atom.

The position of the hydrogen in the hydrogen bridge is dependant on an external electrical field. If the field is reversed, the polarization in the hydrogen bridge is reversed, resulting in a considerable change of the intensity of sensitive reflection pairs, like (400) and (040).

3.4 Effect of Thermal Vibration

Since, in neutron diffraction, there is no form factor at nuclei, neutron diffraction experiments are especially suitable for the investigation of atomic or molecular vibrations (see also Section 7.3:6). The thermal motion of the atoms in the crystal causes a blurring of the nuclear positions, and, therefore, introduces (via Fourier transformation) an artificial form factor, e.g. a decrease of the otherwise isotropic scattering length with increasing angle. In X-ray diffraction, this additional decrease is superimposed on the atomic form factor and may cause errors in interpretation of the observed X-ray diffraction data. In neutron diffraction the decrease in scattering is caused solely by the thermal motion of the atoms and molecules. This allows accurate and reliable determination of the temperature factors, and, hence, calculation of oscillation parameters. Analysis of the diffraction data of benzene indicates rotary oscillations of the molecule in the plane of the molecule, around the 6-fold axis. Since the Fourier peaks of the nuclei are expected to be spherical, inspection of the maxima provides information on anisotropic thermal vibrations.

The thermal vibration of an atom is a function mainly of its mass and of the temperature. There are, therefore, two possibilities for changing the vibrational behavior, and thus of obtaining answers to special problems:

cooling to low temperatures;
changing the mass, e.g. substitution of hydrogen by deuterium.

Cooling the crystal diminishes the vibrational amplitudes; the Fourier maxima are sharpened, and errors in calculating bond distances are reduced. Because of the high penetration of neutrons, experiments at low temperatures, e.g. at that of liquid nitrogen ($77\,°K$) or even of liquid helium ($4.2\,°K$), pose no serious difficulties.

Deuterating a crystal also reduces the vibrational amplitudes, because the mass of ^2H is twice the mass of ^1H. There is, therefore, a special advantage in performing parallel diffraction experiments on both deuterated and non-deuterated crystals.

3.5 Differentiation Between Elements of Similar Atomic Number

Difficulties are always encountered with X-rays when the atoms involved are neighbors in the periodic table, e.g. C and N, Mg and Al, or Mn and Fe. There is little difference between the X-ray scattering power of such elements, and sometimes it is impossible to differentiate between two elements, e.g. $Mn^{2+} \triangle Fe^{3+}$, or $Mg^{2+} \triangle Al^{3+}$ (the latter, as in natural spinel). With neutrons, the situation is more favorable because the scattering lengths are determined by the energy terms of the nuclei, without linear dependence on the atomic number Z. For example, the elements C, N and O (Z = 6, 7, 8) have scattering lengths b = 6.65, 9.4 and 5.77 f. As a rule, elements of similar Z can be distinguished without difficulty with the aid of neutron diffraction. In some rare cases, where the differences in b are small, recourse can be had to isotopes, and the isotope-specific scattering length can be used (cf. Table 2). By appropriate selection, it is always possible to localize the elements in the structure. The advantage of this isotope specificity of nuclear scattering has been used in studies[4] on the short range order in β-brass (β-CuZn), by using a crystal enriched with ^{65}Cu.

Application of this isotope specificity of nuclear scattering offers the possibility of labelling individual atoms in a molecule. Replacement of ^1H by ^2H is again of particular importance in the study of special molecular constitution problems.

3.6 Absorption of Neutrons

Except for a few nuclei, neutrons are absorbed about 10^3 to 10^4 times less by crystals than are X-rays. This means that it is possible to study crystal structures containing elements like *lead, mercury, rare earth elements* or *uranium*, which absorb X-rays strongly.

4 Analysis of Magnetic Scattering

4.1 Physical Principles

The neutron has a magnetic moment of -1.91319 nuclear magnetons, and, therefore, can enter into magnetic *dipole-dipole interaction* with the magnetic moments of atoms. Interaction with the magnetic moments of valence electrons which are situated in incomplete shells, and, therefore, produce a resultant magnetic moment, is particularly important. Interaction with the magnetic moments of the atomic nuclei is weaker by a factor of about 2000, and can be detected only under certain conditions; it can be disregarded in the present discussion.

The magnetic scattering of neutrons was discussed by Halpern and Johnson[4a], and later in an extended form by Trammel[4b] and Blume[4c]. According to these authors, the magnetic scattering amplitude p_j is given by

---

[4] *C.B. Walker, D.T. Keating*, Phys. Rev. *130*, 1726 (1963).

$$p_j = p_j(k) = \frac{e^2 \cdot \mu_N}{m \cdot c^2} \cdot S_j \cdot f_j(k) \qquad (13)$$

$$= 0.539 \cdot 10^{-12} \cdot S_j \cdot f_j(k) \text{ cm}$$

e, m = charge and mass of the electron
μ_N = magnetic moment of the neutron (in Bohr magnetons)
S_j = spin quantum number of the atom j
$f_j(k)$ = corresponding magnetic form factor

$$\underline{k} = \frac{4\pi \sin \theta}{\lambda} \cdot \underline{\varepsilon}$$

$\underline{\varepsilon}$ = scattering vector (unit vector)
2θ = scattering angle (θ = Bragg angle)

$f_j(k)$ are now *magnetic form factors*, which essentially describe the spatial distribution of the *unpaired electrons*.

For magnetic scattering, a structure factor was defined in Equation (10b) which is very similar to the structure factor used in X-ray diffraction and also to that used in nuclear neutron diffraction. Since magnetic scattering is the result of an interaction between two *vectors*, which are defined by length as well as direction in space, the diffracted intensity must contain a dependence on the orientation of these two vectors relative to each other. This dependence on the orientation is given by the so-called 'magnetic interaction vector' *q*, which represents a relation between the direction of the magnetic moment, described by a unit vector κ, and the orientation of the reflecting crystal plane (hkl), described by the scattering vector ε (unit vector), which is normal to the plane (hkl) (cf. Fig. 5):

$$\underline{q} = \underline{\varepsilon}(\underline{\varepsilon} \cdot \underline{\kappa}) - \underline{\kappa} \qquad (14a)$$

Multiplication gives

$$q^2 = 1 - (\underline{\varepsilon} \cdot \underline{\kappa})^2 \qquad (14b)$$

$$= \sin^2 a$$

where a is the angle between $\underline{\varepsilon}$ and $\underline{\kappa}$.

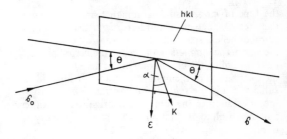

Fig. 5. Vector relation between incident neutron beam N, diffracted beam R, orientation of the reflecting plane and the magnetization direction κ. The scattering vector ε is proportional to $\sin^2\alpha$, i.e. the intensity has a maximum when the magnetic moments lie in the reflecting plane hkl.

Since the incident beam, the diffracted beam and the scattering vector ε are in the same plane (normally, the horizontal plane), the diffracted intensity reaches a maximum when the magnetic moment is perpendicular to that plane, and the magnetic intensity will be zero when the spin-vector is perpendicular to the reflection plane (i.e. parallel to ε). These conditions can often be realized by application of an external magnetic field. With single crystals, the direction of the magnetic moment vectors in the crystal (with 3d-elements this is the direction of the spins) can be determined uniquely from this relation.

Since the orientation of the various crystallites in a powder is random, there are limits to the determination of the angle α from powder measurements.

It can be shown that for *cubic crystal symmetry*, $\langle q^2 \rangle = \text{constant} = 2/3$, and, consequently, the diffracted intensity is independent of the orientation of the spins. In a cubic crystal therefore, the orientation of the spins cannot be determined from powder measurements. In crystals with *rhombohedral, tetragonal, or hexagonal symmetry*, the orientation of the spins can be given in relation to the unique axis of the crystal system, i.e. in relation to [111] in rhombohedral crystals and in relation to the c-axis in tetragonal or hexagonal crystals. In the other crystal systems, the orientation of the spins can be determined from powder data, provided that the resolution is high enough to separate the individual powder diffraction lines.

4.2 Determination of Spin Configurations

The magnetic scattering of neutrons is particularly important, when the spins (or the magnetic moments) undergo a phase transition into a magnetically ordered state, i.e. on the appearance of collective magnetism.

Neutron diffraction experiments provide information about:

The occurrence of *collective magnetism;*
the transition temperature (*Curie* or *Neél temperature*, T_C or T_N);
the type of *magnetism:* paramagnetism, ferromagnetism, ferrimagnetism or anti-ferromagnetism, including spiral or helical spin configurations;
the size of the *magnetic unit cell,* or the periodicity of a helix (including the orientation of the helix);
the distribution of the magnetic atoms (or ions) in the sublattice, and the *sublattice magnetization;*
the orientation of the moments relative to each other, and the absolute orientation with respect to the crystal axes;

[4a] *O. Halpern, M.H. Johnson,* Phys. Rev. *55,* 898 (1939).
[4b] *G.T. Trammell,* Phys. Rev. *92,* 1387 (1953).
[4c] *M. Blume,* Phys. Rev. *130,* 1670 (1963); 133, A 1366 (1964).

the magnitude of the *magnetic moments* of the individual ions, with the possibility of distinction between ions like Fe^{3+} and Fe^{2+};
the magnitude and direction of *internal fields;*
the *temperature-dependence* of magnetization; the local distribution of the magnetic moment density or spin density in the crystal;
the determination of *covalency parameters.*

Magnetic order in a crystal becomes apparent by the sudden appearance of coherent magnetic diffraction on transition from the paramagnetic state to a state of collective magnetism. The transition can be of first or second order. The onset of ordering can be recognized immediately in the diffraction diagrams by the observation of additional reflections (the so-called magnetic reflections), or by a sudden increase of certain nuclear reflections. The magnetic lines can generally be indexed with Miller indices in the usual way. The Miller indices directly describe the periodicity of the spin configuration, and, therefore, the magnetic unit cell.

The interference maxima are described by the Bragg equation: $\lambda = 2d \sin \theta$. When the periodicity of the spins is identical with the atomic or molecular period of the chemical or crystallographic unit cell, these maxima correspond to lattice points in the reciprocal lattice, which are described by a reciprocal lattice vector d*. For crystals of orthorhombic symmetry, as an example, d* is of the form:

$$(d^*)^2 = \frac{h^2}{a^2} + \frac{k^2}{b^2} + \frac{l^2}{c^2} \qquad (15)$$

In the case where the magnetic and chemical unit cells are the same sizes, the nuclear and magnetic scattering coincide on the same reciprocal lattice point, and the intensity is diffracted into the same diffraction line. Magnetic long range order can then be seen in the diffraction diagram by the appearance of the coherent magnetic diffraction as an intensity increase when cooling through the *Curie* or *Neel temperature.*

A classical example is the anti-ferromagnetic structure of MnO, which crystallizes with the face-centered cubic rock salt structure. Below $T_N = 122\ °K$, anti-ferromagnetic order is observed, with a magnetic unit cell of twice the dimensions of the chemical unit cell: $A = 2a$ (Fig. 6).

Because nuclear and magnetic diffraction are observed superimposed in the diagram, magnetic and nuclear contributions must be separated for analysis of the magnetic structure:

$$I_{magn} = I_{(magn + nucl)} - I_{nucl} \qquad (16)$$

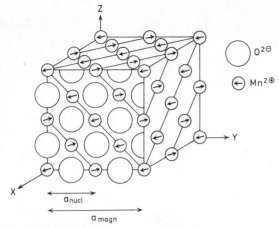

Fig. 6. Antiferromagnetic structure MnO
The spins lie in the (111) planes, and all the spins are parallel within the same (111) plane. Proceeding along [111], e.g. from one (111) plane to the next gives an antiferromagnetic order $+ - + -$. The same magnetic structure is also found in NiO and EuTe.

Two methods of achieving this separation are possible:

A *temperature difference diagram:* Two measurements (above and below the transition temperature) are made, and the difference between the two diagrams with $I\frac{T > T_N}{nucl}$ and $I\frac{T < T_N}{(magn+nucl)}$ from Equation (16) directly gives the magnetic scattering, I_{magn}.

Magnetic field dependence: In most ferro- and ferrimagnetic crystals, the magnetic moments can be aligned by application of an external *magnetic field*. At the correct geometry, the moments can be aligned parallel and perpendicular to the scattering vector ε (see Fig. 7) and Equation (14)b $q^2 = \sin^2 \alpha$ will then be 1 and 0 respectively; and the magnetic contribution can be determined by constructing the difference diagram and using Equation (16).

4.3 Magnetic Spiral Structures
In some magnetic neutron diffraction diagrams, reflections are observed which cannot be indexed in terms of the underlying chemical unit cell or a single multiple of it. Such reflections are termed satellite reflections, because they are observed in pairs to the right and left of nuclear reflections. In the reciprocal lattice, the satellites are at equal distances $+ \tau^*$ and $- \tau^*$ from a central fundamental lattice point.

Satellite reflections are the interference lines caused by magnetic spiral structures, where the magnetic moments are rotated by a general angle α, which is different from 2π (ferromagnetism) or π (colinear antiferromagnetism), when proceeding along a crystal axis. τ^* is the propagation vector in reciprocal space. The direction of τ is the propagation direction of the spiral in the crystal. The

(a)

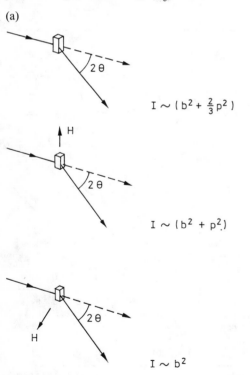

(b)

(c)

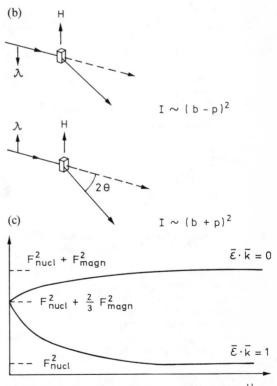

Fig. 7. Comparision of the magnetic diffraction intensity with field direction for
a) polarized and
b) unpolarized neutrons
c) variation with field strength of the diffracted intensity of the reflection from a crystal with cubic symmetry

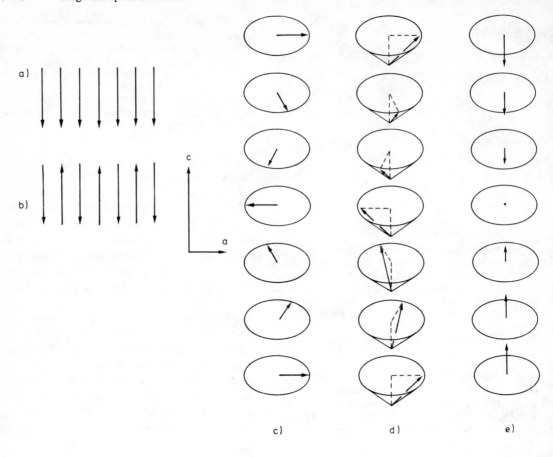

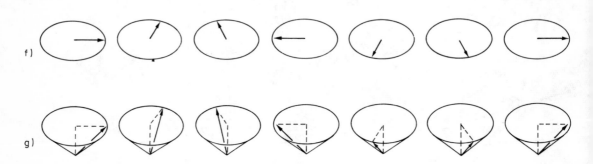

Fig. 8. Schematic representation of the important spin configurations found in crystals.

a) ferromagnetism
b) linear antiferromagnetic ordering
c) normal helical structure (spiral)
d) open cone spiral (umbrella)
e) oscillating spin modulation
f) zycloidal spiral
g) open cone with zycloidal spiral

τ: propagation vector
n: normal to the plane of the rotating moment vectors
γ: moment vector

	a)	b)	c)	d)	e)	f)	g)
τ: parallel	–	–	c	c	c	a	a
n: parallel	–	–	c	c	c	c	c
γ: perpendicular	–	–	c	c^a	$-^b$	c	c^a
	a)	b)	c)	d)	e)	f)	g)
examples	Fe	MnO	Tb	Ho	Tm		MnSO$_4$
	Ni	NiO	Dy	Er			
	EuO	EuTe	Ho				

a) additional ferromagnetic component parallel c
b) parallel c

length of τ ($= 1/\tau^*$) gives the periodicity of the helix. This periodicity is not necessarily in a rational proportion to a lattice constant; it can be a function of temperature, as is observed with the *rare earth elements*.

4.4 Determination of Covalent Bond Components

Neutron diffraction experiments can be performed with such precision that the effect of partly covalent bonding in magnetic salts can be seen. This enables one to determine experimentally, from the neutron diffraction data, the admixture of wave functions from the magnetic ions, i.e. from ions with unpaired electrons, to the diamagnetic ligand ions (Fig. 9). In the chosen anti-ferromagnetic orientation, the tails of the magnetic form

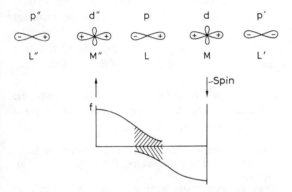

Fig. 9. Schematic diagram of the orbitals of an antiferromagnetic chain and the overlap of the form factors of 3d electrons (M, M″) and ligands (L, L′, L″). In the case of the shown antiferromagnetic orientation of the spins, interference extinction occurs in the neutron diffraction experiment in the region of the ligands.

factors may induce through the ligand a partial overlap with opposite sign. As a consequence, part of the diffracted intensity is of opposite phase, interference extinction results, and the full magnetic moment μ_0 connected with the ion M cannot be observed. The moment actually observed is

$$\mu_{(r)\,obs} = \int \mu_{(r)} \cdot dr = 1 - 2\,A^2 \cdot \qquad (17)$$

The total integrated moment should be $\mu_0 = \int \mu_{(r)}\, dr = 1$. A is an admixture parameter which can be calculated from the antibonding orbitals. If the actual magnetic moment at the ion M is μ_0, the observed value μ is too small by $2A^2$. An extension of this theory to a 3-dimensional

[5] J. Hubbard, W. Marshall, Proc. Phys. Soc. [London], 86, 561 (1965).

[6] R. Nathans, H.A. Alperin, S.J. Pickart, P.J. Brown, J. Appl. Phys. 34, 1182 (1963).

crystal with 6 anti-ferromagnetic neighbors and including interference extinction yields for Bragg scattering[5]:

$$\text{at } e_g\text{-orbitals: } 1 - 3\,A_\sigma^2 - 3\,A_s^2 \qquad (18)$$

$$\text{at } t_{2\pi}\text{-orbitals: } 1 - 4\,A_\pi^2 \qquad (19)$$

Application to the 3d-ions gives the following expression:

$$\mu_{obs}/\mu_0 = 1 - 3\,(A_\sigma^2 + A_s^2) \qquad \text{Ni}^{2\oplus}\ (e_g) \quad (20)$$
$$1 - 4\,A_\pi^2 \qquad \text{Cr}^{3\oplus}\ (t_{2g})$$
$$1 - 1.2\,(A_\sigma^2 + 2A_\pi^2 + A_s^2) \qquad \begin{matrix}\text{Mn}^{2\oplus}\\ \text{Fe}^{3\oplus}\end{matrix}\ (e_g \text{ and } t_{2g})$$

Experiments of this kind have been carried out with polarized neutrons on MnF_2 single crystals[6] and with unpolarized neutrons on NiO single crystals[7]. Integration of the spin density in the Fourier diagram in the case of NiO gave only 82% of the expected moment: $\mu_{obs}/\mu_0 = 0.82$. Further experiments were performed with polycrystalline samples[8]. For comparison, the results derived from NMR measurements[9] are included in Table 3. Figure 10 shows a neutron diffraction diagram used for the determination.

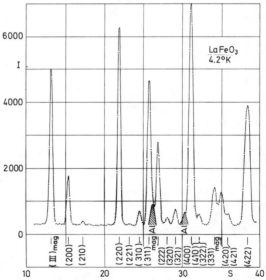

Fig. 10. Neutron diffraction diagram of $LaFeO_3$ recorded at 4.2°K in the antiferromagnetic state I = Intensity (neutrons); S = Scattering angle 2θ

4.5 Determination of Spin Densities in Crystals

As is well known, the X-ray form factor is the Fourier transform of the spatial distribution of all the electrons of an atom, or of the total electron density. By analogy, the magnetic form factor describes, through Fourier transformation, the spatial distribution of the magnetic moment densi-

Table 3. Covalent bonding components in antiferromagnetic salts of 3d-elements.

Ion	From nuclear resonance measurements		From neutron diffraction measurements	
	Compound	Covalency coefficients [%]	Compound	Covalency coefficients [%]
$Ni^{2\oplus}$	$KNiF_3$	$A_\sigma^2 + A_s^2 = 4.32$	NiO	$A_\sigma^2 + A_s^2 = 6.0$
$Cr^{3\oplus}$	K_2NaCrF_6	$A_\pi^2 \quad = 4.7$	$LaCrO_3$	$A_\pi^2 \quad = 4.6$
$Fe^{3\oplus}$	Fe^{3+} in	$A_\sigma^2 + A_\pi^2 = 3.3$	$LaFeO_3$	$A_\sigma^2 + 2A_\pi^2 + A_s^2 = 10.0$
	$KMgF_3$	$A_s^2 = 0.76$		
$Mn^{2\oplus}$	$KMnF_3$	$A_\sigma^2 - A_\pi^2 = 0.18$	MnF_2	$A_\sigma^2 + 2A_\pi^2 + A_s^2 = 3.3$
		$A_s^2 = 0.52$	MnO	$= 5.5$
$Mn^{3\oplus}$			$LaMnO_3$	No effect found

ty in a crystal. For 3d-elements, this is to a good approximation the spin density. By *Fourier inversion* using Equation (11), we can, therefore, derive the spatial distribution of the magnetic moments in the unit cell from the experimental diffraction data. The result is a diagram of the density of the magnetic moments around and between the nuclear centers, which is in its appearance quite similar to the electron density diagrams derived from the X-ray diffraction data. In crystals containing 3d-transition elements, the crystal fields are large compared with the LS coupling and the orbital moments are quenched. The magnetic moment density as derived from the neutron diffraction data is then (to a good approximation) the spin density, and thus the density of the unpaired (outer) electrons.

From neutron form factor measurements, and the subsequent spin density calculations, three main problems can be investigated:

The asymmetric distribution of the outer electrons around the nucleus (deviation from spherical symmetry);
Delocalization of the electron density, for example, as covalent bonding contributions;
Continuous spread of electrons through the whole crystal, for example, as a consequence of electron polarization in the conduction band.

In X-ray crystal structure analysis, the electron distribution around the nucleus is considered to be spherically symmetrical. With neutrons, on the other hand, only the few outer electrons are seen, and these are not necessarily spherically symmetrical, especially when involved in directed chemical bonding. Also, the individual wave functions (except for S-electrons) do not have spherical symmetry.

The magnetic form factors for $Mn^{2\oplus}$ and $Gd^{3\oplus}$, which are shown in Fig. 11, are the Fourier transforms of $3d^5$ and $4f^7$ electron configurations. These ions are in S-states, and the electron distri-

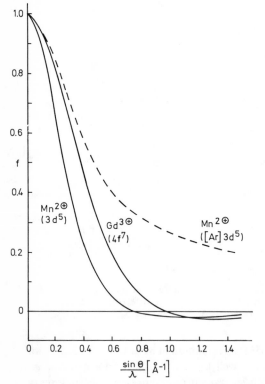

Fig. 11. Form factor curves of the 3d electrons in $Mn^{2\oplus}$ and of the 4f electrons in $Gd^{3\oplus}$ for the magnetic scattering of neutrons
The form factor curve of $Mn^{2\oplus}$ (with all electrons) for the diffraction of X-rays is also given for comparison

[7] *H.A. Alperin*, Phys. Rev. Lett. *6*, 55 (1961).
[8] *R. Nathans, G. Will, D.E. Cox*, Prov. Int. Conf. Magnetism, p. 327, Nottingham 1964.
[9] *R.G. Shulman, S. Sugano*, Phys. Rev. *130*, 506 (1963);
R.G. Shulman, K. Knox, Phys. Rev. Lett. *4*, 603 (1969);
T.P.P. Hall, W. Hayes, R.W. Stevenson, J. Wilkens, J. Chem. Phys. *38*, 1977 (1963).

bution is of spherical symmetry. In general, the single orbitals do not have spherical symmetry, and the form factors cannot be represented by isotropic smooth curves. From the neutron diffraction experiments, form factor values are derived, which are scattered around the spherical symmetric form factor curves. Fig. 12 shows the results of such an experiment with polarized neutrons on an iron crystal in its ferromagnetically ordered state, together with theoretical spherically symmetrical electron configurations. The deviations of the experimental points from the spherically symmetrical curve are real, and are a direct measure of the asymmetry of the distribution of the unpaired electrons. (The size of the circles in this figure corresponds to the experimental error.)

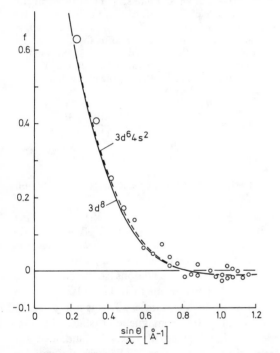

Fig. 12.a) Experimental magnetic form factor values f (o) found for elementary iron with polarized neutrons, compared with theoretical curves for various electronic configurations

Complete delocalization of electrons or magnetic moments into regions of bonding between nuclei leads to such large deviations of the experimental points from the general form factor curve that this curve is no longer a mean curve for the observed values. Such extreme delocalizations have also been observed with polarized neutrons in intermetallic $ZrZn_2$[10] and MnF_2[6]. In the case of MnF_2, the delocalized electron density can be explained by a covalent bonding contribution between $Mn^{2\oplus}$ and $F^{\ominus}$.

---

[10] S. J. Pickart et al., Phys. Rev. Letters *12*, 444 (1964).

Fourier inversion of the experimental data by Equation (11) yields a moment density or *spin density distribution*. In experiments of this type we are interested in the deviation of the spin density from spherical symmetry, or the spin density between atoms. Therefore, difference Fourier diagrams give the best results and reveal any deviation very clearly and in great detail. The Fourier coefficients are the differences between the observed and the theoretical values for spherical symmetry.

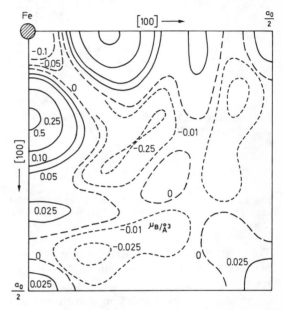

Fig. 12.b) Deviation of the spin density distribution from spherical symmetry in an iron crystal (difference Fourier diagram)

The extended negative regions are due to the negatively polarized 4s electrons which form a uniform antiferromagnetic background in the otherwise ferromagnetic crystal (3d electrons). The maxima, therefore, essentially denote an additional 3d electron density; a distinct concentration of the 3d electrons along the crystal axes is evident. This distribution is characteristic of the e_g symmetry of the 3d electrons.

The number of reflections available, and, therefore, the number of Fourier coefficients, are usually very limited. There is, consequently, always some danger that the diagrams may contain errors due to series termination effects. In order to achieve an artificial and faster convergence of the series, the Fourier coefficients are modified by the following expression:

$$F'^{m}_{hkl} = \left(\frac{\sin 2\pi h\delta}{2\pi h\delta}\right) \cdot \left(\frac{\sin 2\pi k\delta}{2\pi k\delta}\right) \cdot \left(\frac{\sin 2\pi l\delta}{2\pi l\delta}\right) \cdot F^{m}_{hkl}$$

$$(21)$$

This modification takes as the density at xyz the average density in the surrounding cube of dimension δ; δ is of the order of 0.3 Å.

4.6 Measurements with Polarized Neutrons

Extremely high accuracy in the determination of the magnetic scattering contribution can be obtained by the technique involving diffraction of polarized neutrons at the crystal.

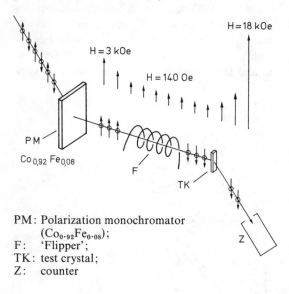

PM : Polarization monochromator
      ($Co_{0.92}Fe_{0.08}$);
F :   'Flipper';
TK : test crystal;
Z :   counter

Fig. 13. Schematic representation of the experimental arrangement for measurements with polarized neutrons

The neutron is a Fermi particle of spin 1/2, and can occur in the energetically different spin states $+1/2$ and $-1/2$. Both spin states normally occur with equal probability in a neutron beam (unpolarized neutrons). With a suitable experimental arrangement (Fig. 13), however, it is possible to obtain a neutron beam that contains only neutrons in a single spin state (polarized neutrons). A polarized neutron beam of this type is obtained by Bragg reflection from a 'polarization monochromator' (nowadays, a single crystal having the composition $Co_{0.92}Fe_{0.08}$ which is magnetized in an external field is nearly always used). The direction of the spin vector is described by a unit vector $\pm\lambda$, which is related to the external magnetic field. The neutron beam can be converted from one spin state into the other by superposition of a low frequency field on the polarized beam (flipper).

The conditions for the polarization of the neutrons are obtained by insertion of Equation (10a) and (10b) into Equation (9). The *structure factor* F can then be expressed in the general form

$$F_{hkl} = \sum_j \left\{ b_j + (q \cdot \lambda)\, p_j \right\} \cdot e^{2\pi i\,(hx_j + ky_j + lz_j)} \quad (22)$$

The vectors $+\lambda$ and $-\lambda$ are present in equal proportions in a beam of *unpolarized neutrons*. The diffracted intensity is proportional to F^2, and thus proportional to $(b^2 + p^2)$; the cross term $(q \cdot \lambda)$ becomes zero [Equation

(9)]. For polarized neutrons, on the other hand, we have $+\lambda = 1$ and $-\lambda = 0$. Equation (22) now leads to phase dependence of the nuclear scattering on the magnetic scattering.

$$F_{hkl}^{\pm} = \sum_j (b_j \pm p_j) \cdot e^{2\pi i(hx_j + ky_j + lz_j)}. \quad (23)$$

The resultant intensity is now proportional to $(b + p)^2$, and the general relation is:

> *polarized neutrons:* $I \sim (b + p)^2$
> *unpolarized neutrons:* $I \sim b^2 + p^2$.

Another advantage of polarized neutrons lies in the measuring technique itself. For the accurate determination of p one needs only to determine the ratio $\gamma = p/b$ accurately; the method is, therefore, relative and is, thus, independent of systematic errors (such as the determination of an absolute scale or of the temperature factor, which is known to cause a decrease in the scattering intensity with increasing scattering angle, similar to the decrease in the form factor itself). In measurements with polarized neutrons, one simply determines the polarization ratio P by recording first $I^{\oplus}$ with neutrons having $\lambda = +1$, and then $I^{\ominus}$ with neutrons having $\lambda = -1$.

$$P = \frac{I^+}{I^-} = \frac{(b + p)^2}{(b - p)^2} = \frac{(1 + \gamma)^2}{(1 - \gamma)^2} \quad (24)$$

with $\gamma = p/b$

This leads to very accurate values of p, and hence of f.

The limits of the method can also be found very readily from Equation (24). To obtain a difference between $I^{\oplus}$ and $I^{\ominus}$, coherent overlap of neutron scattering and magnetic scattering must occur in the reflections. Investigations so far have been concentrated on ferromagnetic crystals, where this condition is always satisfied. This is not generally the case in anti-ferromagnetic crystals. Measurement with polarized neutrons is fundamentally impossible for MnO or NiO where the neutron scattering reflections and the magnetic reflections are rigorously separated because the magnetic unit cell is twice as large as the chemical unit cell, and the crystallographic extinction rules separate the two sets of reflections exactly.

5 Investigation of the Thermal Motion of Atoms in Crystals

As with X-ray diffraction, the intensity in neutron diffraction diagrams decreases with increasing scattering angle 2θ, owing to the thermal vibration of the atoms.

The observed intensity is given by the Equation (25)

$$I = I_o \cdot K \cdot L \cdot A \cdot F^2 \cdot e^{-2B\left(\frac{\sin\theta}{\lambda}\right)^2} \quad (25)$$

where I_0 is the intensity of the incident primary beam. K is an apparatus constant. L the Lorentz factor, A the absorption factor, and F the structure factor from Equation (10).

It can be shown that the temperature factor B is related to the mean thermal amplitude of the atoms, u:

$$B = \frac{8}{3} \cdot \pi^2 \cdot \langle u^2 \rangle \qquad (26)$$

so that knowledge of B provides information about the thermal vibrations of the atoms.

Determination of the temperature factor B from X-ray diffraction diagrams is difficult and unreliable because in this case there is an additional drop in intensity due to the angle-dependent form factor f, so that the reflections in an X-ray diagram always become small at large values of $2\,\theta$. A precise determination of B from X-ray diffraction data requires separation of these two effects, and this means that the electron distribution around an atom in the crystal must be known. (The form factor is the Fourier transform of the electron distribution of an atom.) In neutron diffraction diagrams, on the other hand, apart from the well-known effect of the Lorentz factor (for measurements on single crystals, $L = 1/\sin 2\,\theta$), the drop in intensity is due entirely to the thermal vibrations of the atoms and ions, and $2B\ (\sin\theta/\lambda)^2$ varies linearly with ln I.

Anisotropic thermal motion of the atoms and molecules leads to different temperature factor values for different crystal directions. The appearance of the Fourier maxima is then no longer spherical. The same effect is encountered in X-ray diffraction when the electron density is not spherically symmetrical. Neutron diffraction experiments are, therefore, especially suitable for differentiation of an anisotropic electron distribution from anisotropic atomic vibrations. Exact results can be obtained by studying the dependence of temperature on the temperature factors. This may be done by measuring several reflections at the same Bragg angle but at different temperatures.

Especially surprising results were found in some crystals with the CaF_2 structure, in CaF_2[11] itself, UO_2 and ThO_2[12]. These compounds crystallize with cubic symmetry, space group Fm3m. Measurement of the temperature dependence of the temperature factors of reflections at the same Bragg angle, e.g. $f(T) = (I_{755} - I_{771})$, revealed different thermal motion in different crystal directions; this was not expected for cubic symmetry. The reason for this effect is assumed to be an anharmonic interaction between neighboring ions.

[11] *B.T.M. Willis*, Acta Crystallogr. *18*, 75 (1965).
[12] *B.T.M. Willis*, Proc. Roy. Soc., Ser. A *274*, 122, 134 (1963).

The question as to whether the electron cloud and the nucleus experience synchronous thermal vibration with the same amplitudes is of fundamental interest. In principle this question should be answerable by systematic comparison of the temperature factors derived from neutron and X-ray diffraction experiments. In practice, however, the temperature factor values derived from X-ray data are regarded as dubious, as mentioned previously. This method is, therefore, not reliable. A very precise, independent determination of the thermal movements of nuclei and of charge clouds can be achieved in some favorable cases, by a combined measurement of nuclear and magnetic neutron scattering.

For the diffraction of polarized neutrons, a polarization ratio P was defined in Equation (24). If a different temperature behavior is assumed for the nucleus and for the electron cloud, individual temperature factors B_{nucl} and B_{magn} must be introduced; from Equation (24), then:

$$P = \left\{ \frac{b \cdot e^{-B_{nucl} \cdot \left(\frac{\sin\theta}{\lambda}\right)^2} + p \cdot e^{-B_{magn} \cdot \left(\frac{\sin\theta}{\lambda}\right)^2}}{b \cdot e^{-B_{nucl} \cdot \left(\frac{\sin\theta}{\lambda}\right)^2} - p \cdot e^{-B_{magn} \cdot \left(\frac{\sin\theta}{\lambda}\right)^2}} \right\}^2 \qquad (27)$$

A difference in the temperature-dependence of B_{nucl} and B_{magn} must, therefore, indicate a temperature dependence of the normally constant polarization ratio P. In order to trace any such effect unambigously to thermal motion, proof has first to be given that b and p are not dependent on temperature. This is certainly so for b, but not necessarily for p, which contains, according to Equation (13), through the magnetic form factor f, the spatial distribution of the 3d electrons. In experiments on a $(Co_{0,92} Fe_{0,08})$ single crystal[13], between 20° and 600°, such an effect was observed. It remains, however, to be shown that the spatial distribution of the 3d-electrons remains unchanged when the Curie temperature is approached.

6 Anomalous Dispersion

In X-ray diffraction, the effect of anomalous dispersion can be observed; the same effect is known in neutron diffraction. Anomalous dispersion causes a violation of Friedel's law, and one observes:

$$I_{hkl} \neq I_{\bar{h}\bar{k}\bar{l}} \qquad (28)$$

The physical reason for this effect is the high absorption of the neutron in the nucleus, which gives re-emission with a phase factor different from 0° (−b) or 180° (+b). In neutron diffraction this effect occurs when a resonance line is in the region of the neutron energy. The nuclear scattering amplitude must then be written:

$$b = b_0 + \Delta b' + \Delta b'' \qquad (29)$$

$\Delta b'$ and $\Delta b''$ are correction terms for the real and imaginary component. The imaginary part, $\Delta b''$, is related to the total scattering cross section: $\Delta b'' = \sigma/2\lambda$. The effect is therefore especially large for strong absorbers like ^{113}Cd, ^{149}Sm or ^{157}Gd. For these elements, $\Delta b''$ is larger than $47f$ in resonance. Since $\Delta b'$ and $\Delta b''$ are strongly energy-dependent, the scattering lengths of such elements are wavelength-dependent. The most important resonance nuclei for *anomalous dispersion* are listed in Table 4.

As in X-ray diffraction, the method of anomalous dispersion is of principal importance in determinations of the absolute configuration of polar compounds by measurements of Bijvoet pairs of reflections. However, the resonance lines of most nuclei are well below 1 Å, and practical application of this method will not be possible until neutron beams with wavelengths of the order of 0.1 Å and sufficiently high intensity are available.

Table 4. Imaginary parts of the Scattering Amplitude (in fermi $= 10^{-13}$ cm)

Element	$\Delta b''$ ($\lambda = 1.0$Å)	$\Delta b''$ at resonance
Cd	1.4	5.8
^{113}Cd	11.4	47.0
Sm	7.0	8.8
^{149}Sm	51.0	63.0
Eu	0.7	13.1
^{151}Eu	1.5	27.4
Gd	8.5	12.6
^{157}Gd		66.0
^{6}Li	0.25	
^{10}B	1.1	

7 Investigation of the Chemical Bond

7.1 Combination of X-ray and Neutron Diffraction

Comparative simultaneous investigations with X-rays and neutrons are gaining interest steadily, because the information from diffraction data on the nuclei and the electron clouds provide answers to special problems concerning the chemical bond. In classical X-ray structure analysis, the electron distribution is assumed to be of spherical symmetry, and, therefore, the scattering angle-dependent form factors are independent of the orientation in the crystal. With increasing accuracy in X-ray diffraction and in electron density determina-

tion by X-rays, this assumption is certainly no longer valid. It is well established by the quantum mechanical theories of chemical bonding, for example, that π-orbitals (and also the individual atomic p- and d-orbitals) are non-spherical. Likewise, an electron density concentration can be expected at lone electron pairs, or in certain chemical bonds between atoms. Highly accurate X-ray structure determination has, in fact, made such asymmetric charge distributions visible in electron density Fourier diagrams[14]. As a consequence, the center of the electron cloud will be shifted from the nucleus, and X-rays and neutrons will yield different atomic positional parameters. Furthermore, the non-spherical electron density distribution is the main reason for the difficulties in determining anisotropic temperature factors and also anisotropic oscillations derived from X-ray data. Since both effects yield a dependence on direction in the crystal, in general it is more likely that a directional concentration of the charge into the chemical bond will be found than an anisotropic vibration.

From very precise measurements, differences of the order of 0.01 Å for *carbon atoms,* and about 0.1 Å for *hydrogen atoms,* between the positions of the nuclei and the center of the electron density around the nuclei can be observed.

Increased accuracy and reliability can be achieved when the measurements are performed at low temperatures, for example, 77 °K (liquid nitrogen), a temperature which can be employed without much difficulty. At low temperatures thermal vibrations are reduced greatly.

For reliable analysis of asymmetric electron density distribution, the following approach is recommended: The experiments are performed at as low a temperature as possible. From the neutron diffraction data, the positions of the nuclei in the crystal and the thermal, anisotropic parameters are determined by least squares refinement. All sources of error, especially absorption, extinction or 'Umweganregung', must be corrected. In the least squares analysis of the X-ray diffraction data, these parameters are used without further variation and the refinement is based solely on localized molecular orbitals, their occupation density, the molecular exponents and similar parameters. At present no algorithm has been developed which can be generally recommended, and no computer programs for such refinements have been written.

7.2 Investigation of Atomic Polarization by Combination of Nuclear Scattering and Magnetic Scattering

In those relatively rare cases where the atomic orbitals are singly occupied, the magnetic moments of such ions allow simultaneous measurement of the nuclear position through the *nuclear interaction* and the spin density through the *mag-*

netic interaction, with the same crystal, in one experiment. The crystal must be in an ordered magnetic state: ferro, antiferro or ferrimagnetic. When nuclear and magnetic diffraction are superimposed in the same diffraction spots ($I_{hkl}^{obs} = I_{hkl}^{nucl} + I_{hkl}^{magn}$) the technique with polarized neutrons can be used (see Section 7.3:4.6). In this case, there is coherence between nuclear and magnetic scattering, and very accurate determination of the density of the unpaired electrons is possible.

Divalent *nickel* ($Ni^{2\oplus}$) in a crystal field of octahedral symmetry can be considered as an example. The five degenerate 3d-orbitals are divided in to a doublet, with e_g-symmetry, and a triplet with t_{2g}-symmetry, the triplet being lower in energy. The orbitals are occupied by the eight (3d)-electrons, according to Hund's rule. In a crystal field of octahedral symmetry, the energetically lower t_{2g}-orbitals will be occupied first, and the following configuration is observed ([Ar] = argon-electron configuration):

$$Ni^{2+} = [Ar] + (3d)^8$$
$$= [Ar] + (3d)^6_{t_{2g}} + (3d)^2_{e_g}$$

The six electrons in t_{2g} occupy the orbitals in pairs with no magnetic moment and make no contribution to the magnetic neutron diffraction. The remaining two electrons in e_g are unpaired, however, and interact with the magnetic moment of the neutrons. Analysis of the magnetic diffraction lines, consequently, yields the density of the e_g-electrons directly.

In ferrimagnetic Mn_2Sb, the atomic polarization has been measured with polarized neutrons[15]. Fourier inversion of the experimental data shows the center of density of the 3d-electrons to be shifted by 0.02 Å along the c-axis from the position of the nucleus (in $Mn^{2\oplus}$ there are five 3d-electrons).

[13] *F. Menzinger, A. Paoletti*, Phys. Rev. Letters *10*, 290 (1963).
[14] *P. Coppens*, Science *158*, 1577 (1967).
[15] *H.A. Alperin, P.J. Brown, R. Nathans*, J. Appl. Phys. *34*, 1201 (1963).

7.4 Methods Using Electron Microscopy

Albrecht Maas
Arbeitsgemeinschaft Strukturen und physikalisch-chemische Eigenschaften von Festkörperoberflächen der Universität, D-53 Bonn

In recent years electron microscopy has become an increasingly popular tool for investigating the physics and chemistry of inorganic solids. By contrast, it has been used little in organic chemistry. One reason is that some organic substances suffer irreversible changes or cannot be investigated because of vaporization under the physical conditions of direct microscopic examination (10^{-5} torr pressure, temperature rise, ionization due to electron bombardment).

Nevertheless, the newest instruments and techniques developed, when suitably combined, are likely to make their contribution to the solution of numerous problems in organic chemistry. The present volume describes several fields of work in which electron microscopy provides morphological data and, in combination with high resolution electron diffraction, direct qualitative analytical information. In particular they are the following:

Electron diffraction analysis of solids (Chapter 7.5)
Electron beam microanalysis (10.3)
X-ray structure analysis (7.1)
X-ray fluorescence analysis (10.3)
Crystallization (2.1)
Zone melting and column crystallization (2.2)
Methods for measuring very small amounts of substances in technical products (10.7)
Determination of catalytically active surfaces (9.7)
Biotransformation of chemicals (13)
Solid and liquid phase polymer studies (2.7.9.2)
Development of ultramicroanalytical methods (14.4)
Trace analysis (10.1)

Within certain limitations, electron microscopy serves also as an auxiliary technique during the analysis of pesticides, pharmaceuticals, foodstuff additives, dyestuffs, petroleum products, fats, oils and waxes (11).

Here, it is intended primarily to discuss the possibilities and limitations of the use of electron microscopy for investigating organic substances. Principal items will be image and contrast formation. A knowledge of these features is a requisite, firstly, for selecting suitable methods of investigation and preparation and, secondly, for interpreting the image obtained.

In respect to the physical basis of electron microscopy and the very extensive and complex details of preparation techniques, the reader is referred ot the bibliography (7.4:4).

1 Introduction

The first electron micrograph of an organic molecule was published in 1970[1]; in it sites of heavy metal atoms incorporated at definite points were visible as local blackenings. A novel microscopic technique, scanning transmission electron microscopy, combined with electron spectroscopy, was employed to obtain the micrographs. As early as 1968, Müller[2], using a special field ion microscope with a mass spectrometer connected in series (atom probe field ion microscope), observed metal atoms during their diffusion on a tungsten surface and detected them individually in a time-of-flight mass spectrometer after they had left the surface.

However, these outstanding successes do not justify the expectation that it is possible to image the molecular and atomic make-up of objects with the aid of electron microscopy at will. The complex interaction between the image-forming corpuscular radiation and the object being investigated leads to appreciable instrumental and procedural difficulties.

One specific problem encountered during the study of organic objects is that the generally energy-rich corpuscular beam produces radiation damage in the object which can lead to the complete breakdown of an organic molecule. On the other hand, setting aside the ideal objective of a direct observation of molecules and their atoms, there are numerous problems in organic chemistry in which damaging an object can be avoided by adopting special techniques.

The very rapid strides made by electron microscopy – and by corpuscular beam optics in general – have led to instrumental methods which, in addition to micromorphological information, also furnish quantitative and qualitative analytical data. Particularly noteworthy instruments of this type are the *electron beam probe microanalyzers* and *ion beam probe microanalyzers* which allow one to perform an elemental and an isotopic analysis of the investigated object. These instruments, sometimes combined directly with the microscope, represent the necessary supplementary devices for the analytical interpretation of electron micrographs.

An electron micrograph of an object in no way represents its 'natural image' in the usual sense of the word. Imaging of an object in the electron microscope is fundamentally modified by the specific interaction between beam and object, the nature of the imaging system, the behaviour of the object under the instrumental conditions (vacuum, temperature rise, ionization, residual gas effects) and the changes occurring in the object during preparation. Before examination, it is generally necessary to convert the object into a definite state, which is substantially stable during imaging, so as to yield an interpretable image as unambiguous as possible. A very careful match between the analytical aim, the physical and chemical properties of the object, and the envisaged technique must be made.

Therefore, it is not possible to state standard procedures for solving particular problems that will inevitably produce success.

2 Radiation Damage to Organic Substances

Irradiation of the objects in the electron microscope takes place at a charge density j.t (coulombs/cm²), j being the number of electrons impinging per square centimeter of object and t the duration of irradiation in seconds. When transmission electron microscopy is used for medium magnification (about $10,000 \times$ primary magnification equivalent to a max. final $100,000 \times$ photographic magnification), a beam current density of the order of 10^{-2} coulomb/cm² is required to reproduce the object on the fluorescent screen. Under these conditions, according to Lippert[3], the object receives a dose of 4.6×10^9 roentgens (i.e. the energy absorbed by ionization) with an electron acceleration voltage of 110 kV, and 9.4×10^9 roentgens with 40 kV. To illustrate these values in practical terms, bacteria and viruses are killed by a dose of 10^6 roentgens. It is only with high acceleration voltages of the order of 1 million volts used in developing very high voltage electron microscopy that the probability of ionization and, hence, the possible radiation damage is diminished.

From chemical irradiation studies[4,5] it follows that at this ionization level individual H atoms and also entire molecule groups are split off. During irradiation of thin layers (a few 10 nm thick) of *high polymers* – polymethacrylate (Plexiglas), polyvinyl formal (Formvar), polyester (Vestopal), epoxy resin (Araldite) – in the transmission electron microscope, it was found that both crosslinking and splitting of adjacent chains can occur as initial radiation damage. On further irradiation the layer is broken down to a residue consisting essentially of carbon.

[1] *A.V. Crewe, J. Wall, J. Langmore,* Science 168, 1338 (1970).

[2] *W. Müller,* An atom probe field ion microscope. Europ. Reg. Conf. on Electron Microscopy, Rome 1968, 135.

[3] *W. Lippert,* Optik *15,* 293 (1958).

[4] *K. Little,* The action of electrons on high polymers. Proc. 3. Internat. Conf. Electron Microscopy London 1954, 165.

[5] *L. Reimer,* Lab. Invest. *14,* 1082 (1965).

The weight loss and the residual carbon content following irradiation can provide information about the extent of the damage (cf. Refs.[6-10] and Reimer and Bittner[11]; also illustrations in Reimer[12], in particular Table 9.3).

At object irradiation current densities below 10^{-4} A/cm^2 the damage to Vestopal, Araldite and Plexiglas in the transmission microscope (at 60 kV) is dependent essentially on the irradiation dose j·t. Where the current density exceeds 10^{-4} A/cm^2 (e.g. with the normal current density of around 10^{-2} A/cm^2 in the transmission electron microscope), heating of the object leading to further breakdown evidently becomes more apparent.

A change in, and the destruction of, the crystalline structure of organic substances is revealed very sensitively by the changing electron diffraction diagrams and their ultimate disappearance[13, 14, 16]; measurements of the change in image contrast and infrared spectroscopy can also be used for this purpose[7, 8, 15]. The general conclusion from these investigations is that aromatic compounds are less sensitive to radiation damage than aliphatic ones.

Reimer[12] names several organic substances for whose complete destruction doses between 10^{-3} coulomb/cm^2 (*glycine, leucine, stearic acid, paraffin wax* and *polymethacrylate*) and 1 to 3 coulombs/cm^2 (for *phthalocyanine* and copper phthalocyanine) were found. The values for *anthracene, tetracene, pentacene, fuchsine, neutral red* and *indigo* lie in between. Predictably, at higher object temperatures the damage-producing dose is less.

2.1 Object-Linked Limitations

Because of these possible changes in the object due to heating and ionization, special precautions (object cooling, minimum beam current density, use of an image intensifier combined with a TV-display system) are necessary for a purposeful direct study of organic objects by transmission electron microscopy. Such limitations which restrict or even prevent investigation of an object are called *object-linked limitations*.

Among the remaining object-linked limitations, that of vacuum (pressure limit about 10^{-4}torr) means that liquid and gaseous objects can be investigated only if it is possible to transform them into the solid aggregate state having a sufficiently low intrinsic vapor pressure. A preliminary trial must determine whether the analytical problem can be elucidated under the altered conditions of state. If is possible, for example, to solidify emulsion systems containing about 90% water by very rapid freezing (in about 10^{-2}sec) to about $-140°$ and to reproduce their fine structure by electron microscopy using freeze-etching and surface replication techniques (cf. Maas et al.[17, 18]).

A further object-linked limitation is the object contrast required for imaging; this will be discussed later.

2.2 Instrumental Limitations

The most important instrumental limitation is the point-to-point resolution limit. In electron microscopy this is understood to be the minimum distance between two object points which can still be separated in the final image. The resolution limit achieved in practice is dependent, firstly, on the resolving power of the instrument used (under optimum object and operating conditions); secondly, on the particular interaction between the object being studied and the electrons. In general, the image resolution limit attainable with normal preparation expenditure is often substantially greater (several times) than the resolution limit of the electron microscope used.

3 Imaging Techniques of Electron Microscopy

As in light microscopy, every imaging method used in electron microscopy presupposes certain fundamental properties of the objects. For ex-

[6] A. Brockes, M. Knoch, H. König, Z. Wiss. Mikr. 62, 450 (1955).

[7] A. Brockes, Z. Phys. 149, 353 (1957).

[8] L. Reimer, Z. Naturforsch. 14b, 566 (1959).

[9] A. Cosslett, The effect of the electron beam on thin sections. Europ. Reg. Conf. Electron Microscopy Delft, Vol. II, 678 (1960).

[10] W. Lippert, Optik 15, 293 (1958); 19, 145 (1962).

[11] L. Reimer, X. Bittner unpubl.

[12] L. Reimer, Elektronenmikroskopische Untersuchungs- und Präparationsmethoden 2. Aufl. Springer Verlag, Berlin, Heidelberg, New York 1967.

[13] L. Reimer, Z. Naturforsch. 15a, 405 (1960). Veränderungen organischer Substanzen im Elektronenmikroskop. Proc. Europ. Reg. Conf. EM Delft, Vol. II, 668 (1960).

[14] K. Kobayashi, K. Sakaoku, Lab. Invest 14, 1047 (1965).

[15] G.F. Bahr, F.B. Johnson, E. Zeitler, Lab. Invest. 14, 1115 (1965).

[16] H. Orth, E.W. Fischer, Makromol. Chem. 88, 188 (1965).

[17] A. Maas, Über spezielle Meß- und Untersuchungsmethoden bei der elektronenmikroskopischen Gefrierätztechnik. Europ. Reg. Conf. on Electron Microscopy, Rome 1968, Vol. II, 35.

[18] F. Gstirner, D. Kottenberg, A. Maas, Arch. Pharm. 302, 340 (1969).

ample, transmission electron microscopy requires thin transmitting object layers several 10 nm to a maximum of 100 nm thick dependent on the electron energy and object material. Reflection and emission electron microscopy enable nontransmitting objects to be investigated but, in turn, generally presuppose electrical conductivity or certain other physical properties of the object surface.

To give a brief survey of the many available imaging techniques, the following is a list of the principal instruments in use:

Transmission electron microscopes
Scanning transmission electron microscopes
Emission electron microscopes [utilizing secondary electron emission from object surfaces caused by primary electrons, ions or photons (UV radiation)]
Reflection electron microscopes
Scanning reflection electron microscopes
Electron-mirror emission microscopes
Scanning mirror electron microscopes
Field electron microscopes.

(The ion emission microscope, ion scanning microscope and field ion microscope complete the list.)

Various of these instruments have now achieved great importance for investigating solid surfaces in physics and chemistry (e.g. emission microscopy combined with spectrometry) but most cannot be employed for studying organic objects. The scanning transmission electron microscope, the scanning reflection electron microscope, the transmission electron microscope and perhaps the scanning mirror electron microscope might be suitable in particular for the investigation of organic substances.

Of these the scanning transmission electron microscope has very recently received particular attention because it combines the necessary small radiation exposure with sensitive objects, good image contrast, and a high resolution power. However, further development of this instrument for practical use is likely to take some years.

The trend that modern high-resolution microscopy and physical microanalytical methods are following shows clearly that universal procedures and instruments for stringent requirements in respect to reliability and accuracy of analytical data will not be available in the near future. In general, to solve a particular fundamental problem, it will be necessary to estimate the possible information obtainable from various methods allowing for the special conditions present (object properties, experimental conditions) and to deduce a combined optimum method. In this respect analytical microscopy is probably only in its infancy today.

At present there are generally only two types of instrument available for practical use: the transmission electron microscope and the scanning reflection electron microscope. For investigating organic objects the first-named instrument is limited by the high degree of exposure the object receives (heating, ionization), the second because of its relatively poor resolving power. However, the two methods complement each other in a beneficial manner, especially because of their different method of image and contrast generation.

3.1 Transmission Electron Microscopy

3.1.1 Image Formation

The reproduction of the object is performed along an optical path which, seen in a simplified manner, is comparable to that of a light microscope (the generation of the image *contrast* differs fundamentally from that in the light microscope). The energy of the electron beam is normally between 20 and 100 keV and can penetrate up to about 100-nm-thick layers (of light elements). Using very high voltage instruments, it has recently become possible to penetrate aluminum layers a few microns thick with a resolution limit of about 1 nm $=$ Å.

The wavelength λ of the beam is obtained from the de Broglie relationship

$$\lambda = \frac{h}{m \cdot v}$$

with h $=$ Planck's constant, m the mass and v the velocity of the electrons. For an acceleration voltage of 100 kV the relativistically corrected wavelength is 0.037 Å. This very short wavelength of the electron beam compared to light rays (4000–7000 Å) and x-rays (order of magnitude 1 Å) is one reason for the high resolving power (3 Å for the transmission electron microscope compared to 3000 Å for the light microscope) and for the high sensitivity of electron diffraction analysis methods of very small crystallites (smallest diameter about 10 Å compared to about 1000 Å for x-ray diffraction).

The optical imaging is performed with axially symmetrical electrostatic or electromagnetic fields (electron lenses).

Analogous to the construction of a light microscope, a transmission electron microscope consists of an electron beam source, a condenser system, the objective lens and generally a multilens projection system. The projector system images the intermediate image of the object generated by the objective lens in the final image plane. A fluorescent screen located in the final image plane is used for observing the final image. It can be replaced by a photographic plate if the image is to be stored, or, in more up-to-date manner, by an electronic image store. An unfavorable characteristic of electron lenses is their large aberration, so that the objective lens, which determines the optical resolution limit, must make use only of the electron beam near the axis (optimum aperture $= 10^{-3}$).

3.1.2 Image Contrast

The *image contrast* K is defined by $K = \log I_o/I$ with I_o the absent-object beam current density in Å/cm^2 and I that with the object present, both measured in the image screen plane.

How the image contrast in the transmission electron microscope is generated can be explained clearly by postulating that an aperture in the rear focal plane of the objective (diffraction image plane) retains all electrons diffracted at the object at an angle greater than the aperture angle. Only elastically scattered electrons in the nuclear field of the object atoms, i.e., those that lose no energy, contribute to the image; inelastically scattered, slowed down electrons produce a 'scattering background' − background noise − in the final image because of the electron-velocity-dependent refractive index of the electron lenses. For a given aperture angle and electron acceleration voltage, the contrast generated in the image plane is proportional to the 'mass thickness' $x = \rho D$ ($\rho =$ density, $D =$ layer thickness in the transmission direction). The functional relationship can be written $K = c\,(\alpha, U_o, Z)\rho D$, with the aperture angle α, the acceleration voltage U_o, and the atomic number Z of the scattering object atom acting as contrast-influencing factors in the factor c. Contrast defined in this way is also denoted as the *scattering-absorption-contrast*, but the definition applies strictly only to *amorphous* layers of uniform thickness (here $c(\alpha, U_o, Z) = \frac{\log e}{x_k}$, with $x_k =$ 'contrast thickness' = mass thickness allowing only $1/e$ (37%) of the intensity to pass through the aperture, thus enabling contrast measurements to be used for qualitative analyses in such layers).

3.1.3 Diffraction Contrast

In crystalline objects an appreciable additional contrast effect acts on crystallites at greater than about 50 Å; this is the *diffraction contrast*. It is observed in crystalline substances and is due to scattering of the electrons at the three-dimensional atom arrangement of the crystal lattice instead of at the randomly distributed atoms of amorphous substances. Interference effects at the electrons scattered at this space lattice produce an electron scattering into discrete spatial directions in correspondence with the Bragg condition ($n\lambda = 2d \sin\theta$,

with $\lambda =$ wavelength, $d =$ network plane spacing, $2\theta =$ diffraction angle, $n =$ an integer) termed the diffraction at the crystal lattice. In general, the diffraction angle exceeds the objective aperture angle except for very large lattice constants. The primary beam producing the electron optical imaging of a crystal thus loses additional electrons by diffraction at the crystal lattice. If now the Bragg condition is not uniformly fulfilled throughout an observed crystal, e.g., because of crystal structure faults or bending of the crystal, then the lattice diffraction induced intensity loss of the primary beam also varies locally. The consequent intensity fluctuations in the image are called *diffraction contrasts*.

Diffraction contrasts can often be used for analyzing crystal defects corresponding to their cause of formation. However, it is necessary to remember that certain diffraction contrast phenomena may be due to conditions during preparation and observation. Examples are moiré pattern resulting from orientation and structure differences of superposed single crystals, contrast strips of 'same inclination' due to bent crystals, contrast strips of 'same thickness' due to a wedge-shaped object thickness, and 'schlieren effects' in adjacent single crystalline regions of slightly different crystallographic orientation.

3.1.4 Phase Contrast

A further very significant feature is the *phase contrast* arising near the resolution limit of very highly magnified images. It is produced by an interaction between the scattered and the nonscattered electron waves in conjunction with the spherical aberration in the objective lens. Since the phase contrast is greater than the scattering contrast, and can be positive or negative (contrast reversal, i.e. white turns black) according to the objective defocusing, the interpretation of image contrast in the vicinity of the resolution limit is very difficult. For a given defocusing only part of the information describing the object appears in the image. To analyze an object structure having the order of magnitude of the resolution limit, it is, therefore, necessary to combine stepwise focusing while recording images with a complex evaluation procedure.

3.1.5 Increasing the Contrast by Preparation Methods

The organic-chemical objects to be investigated may be regarded as almost pure phase objects because their scattering absorption is negligibly small due to the predominance of elements of low atomic number. However, intensification of the phase contrast in transmission-electron microscopy cannot be executed easily yet, due to considerable experimental difficulties. An increase in the image contrast is possible therefore only by an increase of the scattering absorption contrast.

The scattering absorption contrast permits local contrast variations of the image elements only if there are considerable local differences of the mass thickness $x = \rho D$. To obtain a sufficient contrast of the image elements, which is also necessary to achieve an optimal resolution, in regard to the scattering absorption contrast, either considerable local differences of the density or of the transmitted layer thickness or both are necessary. Details of an object lying within the optical resolution limit become an obvious image element only if the image contrast of this element can be differentiated sufficiently from the image contrast of its surroundings within the detection sensitivity (photoplate and/or fluorescent screen, or phototube). By means of a preparative alteration of the locally transmitted layer thickness or of the locally present density of a detail of an object, it is possible to increase the scattering absorption contrast. The possibility of changing the transmitted layer thickness of an object is used e.g. in chemical etching of layers that can be transmitted, and, in a more complicated form, for the surface replica-methods in order to recognize the morphology of surfaces.

An increase of the scattering absorption contrast by alteration of the present local density of the object can be achieved, if, in the object to be investigated, local variations of thickness are produced by appropriate preparation techniques. These variations are related definitely to the local properties of the object.

For this purpose heavy metal atoms are incorporated in the object; either a chemical incorporation into molecules or physical absorption on internal or external interfaces of the object may be used. By virtue of the high specific gravity of the heavy metal atoms, even a small local concentration produces a sharp increase in the local density and a corresponding increase in the scattering contrast. However, the enhanced scattering power of the heavy metal due to its high atomic number does not come into play until high electron beam acceleration voltages (100 kV) and large objective apertures are reached. In respect to the heavy metal induced contrast increase, Reimer[21] noted, that in a biological object (mitochondria membrane) the incorporation of one osmium atom for 30 carbon atoms led to an approximately 50% increase in local density as recorded in contrast measurements. The selective incorporation of atoms of high atomic number into certain molecules or molecule groups can be used to perform a specific contrasting of object constituents. Experiments of this kind, however, are still in their infancy.

A knowledge of contrast and image formation in the transmission electron microscope is essential for selecting suitable preparation methods and for interpreting the image.

In the foreseeable future most high resolution electron microscopy studies, including organic chemical objects, will likely be carried out with existing, refined transmission microscopy techniques.

3.2 Scanning Reflection Electron Microscope

The opaque object is scanned with an electron beam probe of small beam cross-section (beam diameter at incidence = about 50–100 Å) using rastered line-by-line scanning. Most of the primary electrons scattered by the object and the secondary electrons detached from the object surface reach an electron detector (scintillation detector plus photomultiplier). The amplified electrical signal generated by this detector is used to control the brightness of the image beam of a cathode ray tube; the sequential movement of this image beam and that of the electron beam spot scanning the object are synchronized. Thus the image built up in the cathode ray tube reproduces the locally varying loss of electrons from the object surface measured by the detector.

According to desired experimental conditions (e.g. the depth of penetration of the electrons) the acceleration voltage of the primary ray may be selected to lie between 1 and 30 kV. The magnification obtained is given by the ratio of the size of the raster spot on the object to that on the cathode tube screen.

Because of the almost parallelly bound primary electron beam, the effective aperture is considerably smaller than in the case of the transmission electron microscope; a substantially greater depth of focus is thus obtained.

At present the optimum *resolution limit* is about 150 Å; under normal routine conditions it is a few 100 Å.

The *image contrast* is related in a complex manner to the angle of incidence and the energy of the primary electrons, to the secondary electron yield (which varies with the material, the crystallographic orientation, and the surface structure), to the geometric arrangement of the electron detector, to the ratio of primary and secondary electrons registered by the detector, and to the detector amplification. These manifold factors affecting the object contrast generally alter any universally applicable conclusions regarding their individual influence. Just as in the case of the transmission electron microscope, the interpretation of the image must be considered for the totality of the contrast-determining factors involved.

The advantage of the scanning reflection electron microscope is that the exposure of the object to heat and ionization is appreciably less than in the case of the transmission electron microscope. In addition, generally object imaging without preliminary preparation is feasible. Evaporation on a thin metal (generally Al or Au) or carbon layer is necessary only for increasing the contrast and with electrically insulating object surfaces.

Consequently, it is possible to carry out a combined investigation of the same object site with the scanning microscope and the transmission microscope by depositing the layer to be evaporated as a high resolution surface replica film. Thus, a surface replica film of sites selected by scanning microscopy using diameters up to 20 mm can be examined by transmission electron microscopy down to resolution limits of 30 to 50 Å.

Similarly, the use of the microanalyzer can be combined with scanning reflection electron microscopy.

The reflection scanning electron microscope supplements the transmission electron microscope very well: by virtue of its very large focus depth, the large viewing field (e.g. starting with a 5-mm object field and a $20 \times$ image magnification) and the possibility of observing objects directly without preliminary preparation.

In addition, this technique allows the direct observation of conducting physical and chemical examinations. Corresponding studies often cannot be carried out in the transmission electron microscope because of the limitation of the object space to a few millimeters or because exclusively objects are available which cannot be transmitted by the electron beam.

Bibliography

[19] *B. von Borries,* Die Übermikroskopie. Einführung, Untersuchung ihrer Grenzen und Abriß ihrer Ergebnisse. Saenger, Berlin 1949.

[20] *S. Leisegang,* Elektronenmikroskope. Hdb. Physik Bd. 33. Springer Verlag, Berlin 1956.

[21] *M.E. Haine, V.E. Cosslett,* The Electron Microscope. Spon, London 1961.

[22] *H. Müller,* Präparation von technisch-physikalischen Objekten für die elektronenmikroskopische Untersuchung. Akad. Verlagsges. Geest & Portig, Leipzig 1962.

[23] *M. von Ardenne,* Tabellen zur Angewandten Physik, 2. Aufl. Bd. I: Elektronenphysik, Übermikroskopie und Ionenphysik. Deutscher Verlag der Wissenschaften, Berlin 1962.

[24] *L. Reimer,* Elektronenmikroskopische Untersuchungs- und Präparationsmethoden. 2. Aufl. Springer Verlag, Berlin, Heidelberg, New York 1967.

[25] *H. Seiler,* Abbildung von Oberflächen mit Elektronen, Ionen und Röntgenstrahlen. Bibliographisches Institut, Mannheim/Zürich 1968.

[26] *G. Schimmel,* Elektronenmikroskopische Methodik. Springer Verlag, Berlin, Heidelberg, New York 1969.

[27] *G. Schimmel, W. Vogell,* Methodensammlung der Elektronenmikroskopie, Deutsche Gesellschaft für Elektronenmikroskopie. Wissenschaft. Verlagsges. [fortlauf. Ergänzung].

[28] *A. Maas,* Beitr. El. Mikr. Direkt-Abb. Oberfl. *5* (1972), unpubl.

7.5 Electron Diffraction Analysis

Joachim Haase
Section Elektronen- und Röntgenbeugung der Universität Ulm, D-75 Karlsruhe

This survey deals with structural determinations of free molecules by electron diffraction of gases. As with microwave spectroscopy (cf. Chapter 5.6) and Raman spectroscopy (5.3), this method has become increasingly important. Electron diffraction permits investigation of molecules which do not have too complex a structure. The other condition is that the sample should have a sufficiently high vapor pressure of ca. 5–20 mm Hg (7.5:2 and 7.5:3).

If one wishes to investigate the influence of substituents on molecular structures, it is best to start with molecules of the highest possible symmetry. Generally these have no dipole moment and structural determination can be achieved only by electron diffraction in the gas phase.

The present article gives a short summary of the most important formulae (7.5:1) and a survey of the experimental techniques. The interpretation of electron diffraction patterns is described in more detail (7.5:2). The advantages of the method will be demonstrated by four selected examples (7.5:4).

1 Theory of Electron Diffraction of Gases

The theory and formulae necessary for description of experimental techniques for electron diffraction have been given by various authors [1–12].

In structural analysis of gaseous samples by electron diffraction, a monochromatic electron beam strikes the sample of molecules under investigation and is scattered. It is necessary to find a solution of the Schrödinger equation for the electron beam — gas molecule system, i.e. to calculate the probability of finding electrons in a certain direction after the scattering process. In practice, this probability is measured as an angular-dependent intensity distribution of the electrons scattered by the molecules.

No exact solution of the Schrödinger equation for the system mentioned above exists because it is a many particle problem. For this reason some simplifying assumptions must be made. Firstly, only a small sample volume is considered in which the electron beam interacts with the molecules. This assumption excludes multiple scattering[13-16], *i.e.* independent scattering of electrons by all the molecules.

The problem is now reduced to finding a solution of the Schrödinger equation for the scattering of electrons by one molecule only; however, in this case, also, no exact solution exists. The next simplification is to neglect the influence of chemical bonding on the scattering[17]. This means that one considers the electrostatic potential of the molecule as being built up from the potentials of the atoms in the molecule (independent atom model)[18].

The electrostatic potential of the molecule is the scattering potential in the Schrödinger equation. With this simplification, it is possible to treat the Schrödinger equation describing the scattering of electrons by atoms with a spherically symmetrical charge distribution. Various methods are known by which the solution of the wave equation in this case can be found to a good degree of approximation[19, 20, 89, 90]. This approximation is allowed because, for the angular range which is most accessible experimentally, the electrons are scattered from the inner part of the potential which comes mainly from the nucleus[11]. This region is hardly

affected by chemical bonding. Thus far an effect by chemical bonding on the scattering of electrons by molecules has been treated theoretically only for certain small molecules[21-24]. This is to be expected at small scattering angles if electrons having energies in the region of 40–60 keV are used for structural determination.

With the approximations described above, the angular distribution of the scattered electrons can be calculated. After averaging all atomic coordinates over all possible orientations, Equation (1) is obtained.

The diffraction pattern shows radial symmetry.

$$I_T(s) = I_A(s) + I_n(s) = K \left\{ \sum_{i=1}^{N} \left(|f_i(s)|^2 + \frac{S_i(s)}{s^4} \right) \right. \quad (1)$$

$$\left. + \sum_{\substack{i,j=1 \\ i<j}}^{N} |f_i(s)||f_j(s)| \cos\left(\eta_i(s) - \eta_j(s)\right) e^{-l_{ij}^2 s^2/2} \cdot \frac{\sin s \cdot r_{ij}}{s \cdot r_{ij}} \right\}$$

The symbols in Equation (1) have the following meaning:

N is the number of atoms in the molecule, K is a constant for static electron energy, $|f_j(s)|$ the absolute value and $n_j(s)$ the phase of the complex scattering factor $f_j(s) = |f_j(s)| \exp[-i_j(s)]$ of the j'th atom; $S_j(s)$ is the incoherent scattering intensity of the j'th atom. Heisenberg[25] deduced the formulae for the calculation of incoherent scattering factors; Bewilogua[26] and, more recently, Tavard[27] calculated and tabulated values for $S_j(s)$[28]. The angular variable s is given by

$$S = 4\pi/\lambda \times \sin(\theta/2)$$

where λ is the de-Broglie wavelength of the electrons which is used instead of the scattering angle θ; r_{ij} signifies the distance between the atoms i and j in the molecule, and l_{ij}^2 denotes the mean square amplitude of the vibration of the pair (i, j) of atoms.

[1] *J. Karle, I.L. Karle* in Ref. [2], p. 427–461.
[2] *Braude-Nachod,* Determination of Organic Structures by Physical Methods, Academic Press, New York 1955.
[3] *O. Bastiansen, P.N. Skancke,* Adv. Chem. Phys., Vol. III, 323 (1960).
[4] *A. Almenningen, O. Bastiansen, A. Haaland, H.M. Seip,* Angew. Chem. 77, 877; Engl. 4, 819 (1965).
[5] *P. Debye,* Ann. Phys. 46, 809 (1915).
[6] *P. Debye,* J. Chem. Phys. 9, 55 (1941).
[7] *P. Ehrenfest,* Amsterdam. Acad. 23, 132 (1915).
[8] *N.F. Mott,* Proc. Camb. Phil. Soc. 25, 304 (1929); Proc. Roy. Soc. Ser. A127, 658 (1930).
[9] *H. Viervoll,* Acta Chem. Scand. 1, 120 (1947).
[10] *N.G. Rambidi,* Zh. Strukt. Khim. 3, 347 (1962).
[11] *V.P. Spiridinov, N.G. Rambidi, N.V. Alekseev,* Zh. Strukt. Khim. 4, 779; Engl. 717 (1963).
[12] *T. Iijima, R.A. Bonham, T. Ando,* J. Phys. Chem. 67, 1472 (1963).
[13] *I.L. Karle, J. Karle,* Chem. Phys. 18, 963 (1950).
[14] *R.A. Bonham,* J. Chem. Phys. 43, 1103 (1965).

[15] *Y. Morino,* Bull. Chem. Soc. Jap. 38, 1796 (1965).
[16] *G. Moliere,* Z. Naturforsch. A 2, 137 (1947); A 3, 78 (1948).
[17] Ref. [11], 797.
[18] *L.O. Brockway,* Rev. Mod. Phys. 8, 271 (1936).
[19] *M. Born,* Z. Phys. 37, 863 (1926); 38, 803 (1926).
[20] *G.F. Drukarew,* Zh. Eksp. Teor. Fiz. 19, 247 (1949).
[21] *T. Iijima, R.A. Bonham,* J. Phys. Chem. 67, 2796 (1963); 67, 2266 (1963); J. Chem. Phys. 42, 2612 (1965).
[22] *T. Iijima,* Bull. Chem. Soc. Jap. 38, 1757 (1965).
[23] *C. Tavard,* Cah. Phys. 17, 165 (1963).
[24] *D.A. Kohl, L.S. Bartell,* J. Chem. Phys. 51, 2891 (1969); 51, 2898 (1969).
[25] *W. Heisenberg,* Phys. Z. 32, 740 (1931).
[26] *L. Bewilogua,* Phys. Z. 32, 737 (1931).
[27] *C. Tavard,* J. Chim. Phys. 64, 540 (1967). *Cromer,* J. Chem. Phys. 47, 1892 (1967); 50, 4857 (1969).
[28] *R.A. Bonham,* J. Chem. Phys. 43, 1460 (1965); 43, 1933 (1965).

The intensity function consists of two parts. The first term $I_A(s)$ is the atomic scattering intensity which is a superposition of the scattering intensities of the N atoms in the molecule; these are treated as independent scattering centers ('atomic background scattering'). The second term, the 'molecular scattering intensity', $I_M(s)$, contains the essential information about the geometrical arrangement of the atoms in the molecule. The two terms of Equation (1) differ very much in their dependence on s. The atomic constituent is a function which decreases very rapidly as s values increase. The molecular contribution is a function which oscillates as $\sin(sr_{ij})$ and which is damped by $(sr_{ij})^{-1}$. This part of the intensity function appears as a small variation in the atomic scattering background. The effect of the vibrations of the atoms in the molecule is to bring about additional damping $[\exp{-l_{ij}^2 s^2/2}]$ of the molecular portion of the intensity function which is also called the 'temperature factor' (derivation see Ref.[29-32]). Equation (1) was obtained by assuming a harmonic interatomic potential function. To arrive at an improved expression for the intensity function and to take account of the anharmonicity of the interatomic potential, a Morse function[33] for the potential should be used as starting point, and then the quantity P(r) calculated, which is the probability that the distance between two atoms lies between r and r + dr. From this probability, now asymmetric, there follows an equation similar to (1)[34, 35]:

$$I_T(s) = I_A(s) + K \sum_{\substack{i,j=1 \\ i<j}}^{N} |f_i(s)||f_j(s)| \cdot \cos\left(\eta_i(s) - \eta_j(s)\right)$$

$$e^{-l_{ij}^2 s^2/2} \cdot \frac{\sin s\, (r_g\,(1)_{ij} - \Phi(s))}{s \cdot r_{eij}} \qquad (2)$$

where r_e is the minimum of the potential function V(r), $r_g(1)$ is the center of gravity of the function P(r)/r, and $\Phi(s)$ is a phase shift which takes account of the dependence of the intensity function on the anharmonicity of the potential.

Glauber and Schomaker[36] were the first to show the necessity of the complex scattering factors f(s) in Equation (1), since it was found impossible to determine unambiguously the molecular structure of *uranium hexafluoride* from the electron diffraction diagram[37]. Since then, the complex scattering factors have been calculated and tabulated[38-45]. The bigger the difference between the atomic numbers of the atoms in the molecule under study (*i.e.* the more the atoms differ in size)[5], the bigger is the influence of the phases η on the scattering intensity. If the molecule contains only atoms of nearly equal atomic numbers, the phase factor $\cos[\eta_i(s) - \eta_j(s)]$ can be neglected, and the scattering factors in the first Born approximation are used[19] instead of f(s). Scattering factors in Born approximation are related to the form factors of X-rays $F(s)$[46] by the expression:

$f^B(s) = K\,[Z - F(s)]/s^2$. (Z signifies the atomic number of the corresponding atom, s is as defined above, and K is a constant.)

2 The Analysis of Electron Diffraction Patterns

2.1 Determination of Relative Intensities

The information about the intensity distribution of the scattered electrons is fixed on photographic plates. Visual analysis[47] is seldom used; instead, the sector-photometer method is usually employed to obtain more accurate data[48]. The purpose of this sector[49] (which rotates just above the photographic plate during exposure) is to suppress the steep decrease in the atomic background scattering during the recording of the diffraction image. To achieve this, the sector has an opening which is proportional to $s^3 - s^{3,8}$ (s is the variable explained in section 7.5:1). It is thus possible to record on the photographic plate in the region where the image density on the plate is linearly proportional to the intensity over the whole angular range experimentally accessible[50]. By use of a rotating sector one of the main difficulties of the electron diffraction method can be overcome, *i.e.* that of detecting the weak oscillations of the molecular portion of the intensity function in contrast with the intensity function determined via the steeply decreasing atomic scattering background.

[29] R.W. Kames, Phys. Z. 33, 737 (1932).

[30] D. Debye, J. Chem. Phys. 9, 55 (1941).

[31] J. Karle, I.L. Karle, J. Chem. Phys. 18, 957 (1950).

[32] Y. Morino, J. Chem. Phys. 20, 726 (1952).

[33] P.K. Morse, Phys. Rev. 34, 57 (1929).

[34] L.S. Bartell, J. Chem. Phys. 23, 1219 (1955).

[35] K. Kuchitsu, L.S. Bartell, J. Chem. Phys. 35, 1945 (1961).

[36] V. Schomaker, R. Glauber, Phys. Rev. 89, 667 (1953); Nature 170, 290 (1952).

[37] S.H. Bauer, J. Chem. Phys. 18, 27 (1950).

[38] J.A. Hoerni, J.A. Ibers, Phys. Rev. 91, 1182 (1953); Acta Crystallogr. 7, 405 (1954).

[39] J. Karle, R.A. Bonham, J. Chem. Phys. 40, 1396 (1964).

[40] J. Haase, Z. Naturforsch. A21, 187 (1966); A23, 1000 (1968); A25, 936, 1219 (1970).

[41] T.G. Strand, J. Chem. Phys. 44, 1611, 2426 (1966).

[42] S. Konaka, Bull. Chem. Soc. Jap. 39, 1146 (1966); 39, 1134 (1966).

[43] K. Kimura, J. Chem. Phys. 46, 2599 (1967).

[44] J.L. Peacher, J.C. Wills, J. Chem. Phys. 46, 4807 (1967).

[45] H.C. Cox, R.A. Bonham, J. Chem. Phys. 47, 2599 (1966).
R.A. Bonham, J. Chem. Phys. 50, 1056 (1969).

[46] International Tables for X-Ray Crystallography, Vol. 3. The Kynoch Press, Birmingham, Engl. 1968.

[47] H. Mark, R. Wierl, Naturwissenschaften 18, 205 (1930).

[48] I.L. Karle, J. Karle, J. Chem. Phys. 17, 1052 (1949).

[49] C. Finback, Avnandl. Norske Videnskap. Acad. Oslo, Mat. Naturv. Kl. 1937, No. 3.
P.P. Debye, Phys. Z. 40, 66 (1939).

[50] L.S. Bartell, J. Chem. Phys. 35, 1211 (1961).

[51] J. Haase, W. Zeil, Z. Phys. Chem. N.F. 45, 202 (1965).

The oscillations now become clearly visible and can be measured by a microphotometer. A digital voltmeter and a printer are connected to the microphotometer to collect the photometrically measured light transmittance data[51].

There is another difficulty, however. In spite of the use of a rotating sector, the oscillations of the molecular scattering contributions are very weak at scattering angles corresponding to $s > 20$ and are distorted because of the grainy nature of the photographic plate. There are various ways of measuring the transmittance of the plates in order to suppress this effect:

A number (e.g. 10) of radial directions on the plate are chosen, and the transmittance is measured along them. Corresponding values from individual series of measurements are averaged, and thus the light transmittance curve to be used is determined[52].
The photographic plate is rotated around its center while the light transmittance values are recorded[53].
The photographic plate is made to oscillate about an axis of symmetry of the diffraction pattern. In this case, light transmittance values are measured in a preselected angular region of the plate[5, 54].

After the transmittance values have been found in one of the above ways, they are converted into optical densities by use of the relation: $S = \log(I_0/I(r))$[88]. Here, I_0 denotes the transmittance of the plate at an unexposed part of the plate and $I(r)$ is the transmittance at radius r.

The values for optical density are converted into relative intensities by a calibration procedure[31], after the radii of the individual density values have been converted into the appropriate s-values by use of the electron wavelength λ. λ is determined either from a diffraction pattern of a crystalline powder (e.g. *zinc oxide*) with exactly-known lattice constants[55], or from very precise voltage measurements. Additional intensity values are calculated by an interpolation procedure with equidistant s-values. Finally, the intensity function can be smoothed to remove remaining irregularities in this function[56].

The intensity values finally obtained are plotted against the appropriate s-values. They contain a residual background which originates from the difference between the sector shape and the real form of the atomic scattering background. From Equation (1), the atomic scattering background is a function of the form factors of the individual atoms in the molecule under investigation. A smooth background line is drawn through the oscillations of the intensity function in such a way that the areas above and below this line, between the molecular intensity function and the background line, are approximately equal[48]. The residual background is eliminated, using Equation (3):

$$M(s) = \frac{I_T(s)}{I_A(s)} - 1 \qquad (3)$$

$M(s)$ is termed the reduced molecular intensity function. From this function $M(s)$, the radial distribution function can be calculated by use of a Fourier-sine-transform. The radial distribution function $F(r)$ contains all the information about the structure of the molecule:

$$F(r) = \int_0^\infty s \cdot M(s) \cdot \sin(sr) ds. \qquad (4)$$

To calculate the radial distribution function, the integration is replaced by a summation from $s = 0$ to $s = s_{max}$, where s_{max} denotes the value of the scattering angle for which intensity measurements are possible. In the angular range $0 < s < 2$, experimental data are often either unobtainable or else very inaccurate depending on the experimental technique used. Consequently, in this s-range one uses a theoretical intensity curve which is calculated for an assumed model of the molecule using Equation (1). A damping factor $\exp(-bs^2)$ is introduced to ensure that the intensity curve, which also exists for s-values $s > s_{max}$, does not make a significant contribution from this region to the radial distribution function. This factor is termed the artificial temperature factor and it ensures sufficiently rapid convergence of the integrand of Equation (4). The value of b is chosen such that $\exp(-bs_{max}^2) \leqslant 0,1$. The final formula for calculating the experimental radial distribution function has the following form:

$$F(r) \sim \sum_{s=0}^{s_{max}} s\, M(s) e^{-bs^2} \cdot \sin(s \cdot r)\, \Delta s \qquad (5)$$

2.2 Determination and Evaluation of the Radial Distribution Function

The maxima of the radial distribution function (Fig. 1 for *carbon tetrachloride*) directly give the probability that a certain interatomic distance will be found in the molecule at a specific r-value. In Equation (1), for whose derivation harmonic interatomic potentials were used, the radial distribution function[57, 58] can be represented as a superposition of Gaussian functions of the form $A \exp(-B(r-r_0)^2)$.

[52] *B. Beagly*, Trans. Faraday Soc. *65*, 1219 (1969).
[53] *J.F. Chiang, S.H. Bauer*, Trans. Faraday Soc. *64*, 2247 (1968).
[54] *O. Bastiansen, P.N. Skancke*, Adv. Chem. Phys. *3*, 323 (1969).
[55] *C. Lu, E.W. Malmberg*, Rev. Sci. Instrum. *14*, 271 (1943).
[56] *P.G. Guest*, Numerical Methods of Curve Fitting, Chap. 10, p. 322, University Press, Cambridge 1961.

[57] *J. Karle, H. Hauptmann*, J. Chem. Phys. *18*, 875 (1950).
[58] *J. Karle*, J. Chem. Phys. *45*, 4149 (1966).

The maxima of the Gaussians occur at the positions of the most probable interatomic distances r_0. The half-width B of the curves is determined by the vibrational amplitude[59, 60] which is related to B by the expression: $l^2 = (1-4bB)/2B$. It should be noted that the half-widths of the Gaussian curves are increased by the artificial temperature factor; this can be seen from the equation given for B.

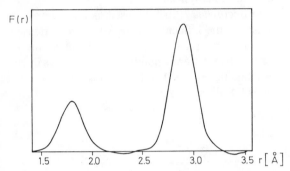

Fig. 1. Radial distribution function for carbon tetrachloride

The maxima of the radial distribution curve are related to the interatomic distances in the following manner. The area under a peak is proportional to the product of the atomic numbers and the corresponding pair of atoms, multiplied by a number n, and divided by the interatomic distance; n gives the number of the identical interatomic distances in the molecule[30, 48].

Theoretical intensity curves and radial distribution functions can be used to improve the background line of the experimental intensity function and the zero line of the radial distribution function[48, 61–65].

To refine the molecular parameters with a least squares procedure, starting parameters are taken from the radial distribution function to calculate the theoretical intensity curve using Equation (3). This theoretically calculated reduced molecular intensity function is then compared with the experimentally-obtained intensity function. It should be remarked here that now only the experimentally

obtained s-range is used[66–68]. A least squares procedure on the radial distribution function is described by Bonham[69].

In some special cases, it is also possible to estimate anharmonicity constants of the interatomic potential from asymmetries of the peaks of the radial distribution function. These values can be refined in a least squares calculation if a model for the anharmonic interatomic potential is used (for details of Fourier analysis see Ref.[9, 70]).

3 Experimental Technique

From an electron source, S, i.e. a hot filament, an electron beam with an energy of 30 to 60 keV is directed towards a photographic plate PL which acts as receiver (Fig. 2). Two centering devices Z_1 and Z_2 are

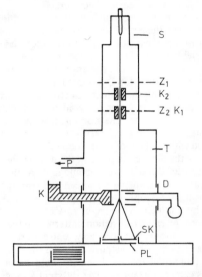

Fig. 2. Schematic diagram of an electron diffraction apparatus for structural determination on gaseous samples[71]

provided to center the electron beam mechanically. K_1 and K_2 denote two magnetic lenses which focus the electron beam on the photographic plate PL. The diffraction chamber T (which is evacuated by a pumping system P) has several openings D where the gas inlet system, a nozzle, can be mounted. The sample under investigation is introduced through this nozzle. The electron beam strikes the gas beam close to the nozzle opening. A cooling system K is mounted opposite the nozzle in order to prevent the pressure in the diffraction chamber from rising sharply. There, the substance is

[59] R.A. Bonham, T. Ukaji, J. Chem. Phys. 36, 72 (1962).

[60] M. Traetteberg, R.A. Bonham, J. Chem. Phys. 42, 587 (1965).

[61] L.S. Bartell, J. Chem. Phys. 23, 1854 (1955).

[62] A. Almenningen, Kgl. Norske Vidensk. Selsk. Skr. No. 3, 1958.

[63] T.G. Strand, Thesis, The Technical University of Norway, Trondheim 1961.

[64] R.A. Bonham, F.A. Momany, J. Phys. Chem. 67, 2474 (1963).

[65] T. Iijima, R.A. Bonham, J. Phys. Chem. 68, 3146 (1964).

[66] K. Hedberg, M. Iwasaki, Acta Crystallogr. 17, 529 (1964).

[67] M. Iwasaki, Acta Crystallogr. 17, 533 (1964).

[68] O. Bastiansen, Acta Crystallogr. 17, 538 (1964). H.M. Seip, Chem. Phys. Lett. 3, 617 (1969).

[69] R.A. Bonham, L.S. Bartell, J. Chem. Phys. 31, 702 (1959).

[70] J. Waser, V. Schomaker, Rev. Mod. Phys. 25, 671 (1953).

frozen out after it has passed the electron beam. The cooling system also has the purpose of reducing the scattering volume, *i.e.* diminishing the pressure just in front of the nozzle so that multiple scattering is minimized according to the requirements in Section 7.5:1. During the scattering experiment, the pressure inside the chamber is $1–3 \times 10^{-5}$ mm Hg.

The rotating sector SK is shown just above the photographic plate PL. The purpose of the sector is to annul the steep decrease of the atomic background. It must, therefore, have an opening which is inversely proportional to $I_A(s)$ cf. Equation (1). Such a sector has a very narrow opening, however, in the region of small radii. Machining of a sector of this kind with the necessary high precision is extremely difficult. Most of the sectors currently used in electron diffraction work, therefore, have an opening proportional to $s^3–s^{3,8}$. In this case also there are mechanical irregularities, and the result is that the intensity function in the scatter-angle region $0 < s < 2$ cannot be measured very precisely.

To determine the residual background mentioned in Section 7.5:2, the actual sector shape is determined by comparison of the diffraction pattern of a rare gas, *e.g.* argon, with the theoretical intensity function of the same gas calculated from the appropriate form factor. The so-called beam stop is mounted in the center of the sector to shield off the primary electron beam from the diffraction image in order to avoid over-irradiation of the center of the plate. The stop has a small opening in its center so that the primary electron beam can be centered with the help of the systems Z_1 and Z_2. Before recording the diffraction pattern, the primary beam can be observed through this opening, with the sector stationary, via a transparent screen and with the aid of an optical microscope. During exposure, this opening is closed by a shutter as the sector rotates.

To record the entire diffraction pattern with the highest possible accuracy, it is necessary to use more than one camera distance (the distance between the nozzle and the photographic plate). Intensity data recorded at various camera distances cover different s-ranges. The intensity function for the entire s-range can be calculated by adjusting the different part-functions with scale factors. The scale factors are found from the overlap regions of individual diffraction patterns from the various camera distances.

The camera distance can be changed in several ways. One possibility is to move the plane of the photographic plate up and down keeping the nozzle in a fixed position. This arrangement has the disadvantage that the plate box, which contains cassettes holding several plates, must be moved with very high accuracy through considerable distances.

Another solution, which is used in commercial apparatus, is to keep the plate box and the entire sector mechanism in a fixed position at the bottom of the diffraction chamber. The camera distance is altered by moving the nozzle along the direction of the primary electron beam[71]. In addition, the s-range can be varied by changing the accelerating voltage. The photographic plates can be moved in turn from the cassette to the exposure position; during this process, the electron beam is directed by a deflector into a receiver. The nozzle and the whole inlet system can be heated so that it is possible to investigate the temperature dependence of the structure of the substance under study[72–74]. In addition, use of a heatable nozzle system makes it

possible to study samples whose vapor pressure is too low at room temperature for diffraction patterns to be obtainable.

The exposure times range from 15 sec to 4 min depending on the vapor pressure of the sample.

4 Examples

4.1 Tetramethyllead

Tetramethyllead[75] will be used as an example to illustrate the influence of the complex scattering factors. This molecule is composed of atoms of very different atomic numbers. Fig. 3 shows intensity functions (A–C) and radial distribution functions (D–F) of tetramethyllead.

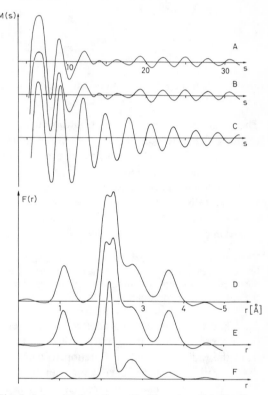

Fig. 3. Intensity functions (A–C) and radial distribution function (D–F) for tetramethyllead
D calculated from curve A, E and F calculated from curves B and C respectively.

Curve B was calculated using complex scattering factors; curve C was calculated with scattering factors only in first Born approximation. The in-

[71] *W. Zeil*, Z. Instrumentkd. *74*, 84 (1966).
[72] *K. Hedberg, M. Iwasaki*, J. Chem. Phys. *36*, 589 (1962).
[73] *R.A. Bonham, J.L. Peacher*, J. Chem. Phys. *38*, 2319 (1963).
[74] *A. Reitan*, Acta Chem. Scand. *12*, 131 (1958).
[75] *J. Haase, W. Zeil*, Ber. Bunsenges. Phys. Chem. *72*, 1066 (1968).

fluence of the term cos $\Delta\eta$ in Equation (1) is noticeable: the intensity function B is damped very strongly and, beyond a certain s-value, the sign of this function is changed with respect to curve C. The absolute values of the maxima and minima increase after this s-value. The reason for the change of the sign in the intensity function is that the phases of the complex scattering factors for lead on the one hand, and for carbon and hydrogen on the other, are so different that the difference $\Delta\eta$ will reach the value $\pi/2$ within the experimental angular range. The result is an extra zero position in the intensity function in addition to the zero positions arising from $\sin(sr_{ij})$. The radial distribution functions are calculated from the corresponding intensity functions (curves D–F).

The experimental radial distribution function could be interpreted in terms of two different Pb–C distances for tetramethyllead, *i.e.* it seems there is no tetrahedral symmetry in $Pb(CH_3)_4$. The same holds for the Pb...H distance. The use of the complex scattering factors shows that the theoretical radial distribution function E is in good agreement with the experimental one, using the same model, *i.e.* distances and vibrational amplitudes, as for calculation of curve F.

It can now be seen that the molecule has tetrahedral symmetry. From the term $\Delta\eta \cdot \frac{\sin s \cdot r_{ij}}{s \cdot r_{ij}}$ in Equation (1), it can be deduced that in the case where the phase difference $\Delta\eta$ for a combination of atoms is nonvanishing, two different distances $r + \Delta r$ and $r - \Delta r$ appear in the intensity function. The value of Δr is given by $\Delta r = \pi/s^0$, where s^0 denotes the s-value for which $\Delta\eta$ becomes $\pi/2$ for the pair of atoms concerned[76]. If the difference between the atomic numbers is not too large, $|Z_i - Z_j| \leq 10$, the effect of the phase factor cos Δs is only to broaden the peaks of the radial distribution function. This causes an apparent increase in the values of the vibrational amplitudes. Ref. [59] should be consulted regarding corrections for failure of the Born approximation.

4.2 Sulfur Dioxide

In a structural determination on *sulfur dioxide*[77], not only structural parameters but also the anharmonicity constant for the S–O bond were determined successfully using a Morse potential. The two interatomic distances in SO_2 are well separated in the radial distribution function. As can be seen from the relative areas of the radial distribution function, the S–O distance (which appears twice) has a greater influence than the O...O distance. Therefore, this distance (1.4309 Å) determines mainly the character of the whole intensity function. The following values for the other structure parameters were obtained: r(O...O) = 2.463 Å. From these values, the calculated bond angle $\angle OSO = 118° 59'$. These values are in good agreement with the results, obtained by microwave spectroscopy[78]: r(S–O) = 1.4308 Å, $\angle OSO = 119° 19'$.

From electron diffraction measurements, the anharmonicity constant was found to be a = 2.6 ± 0.4 $Å^{-1}$. Neglecting all interactions, the force constant for the S–O bond was calculated from this value using the dissociation constant of SO[87]. The calculated value was $k_{SO} = 9.9$ mdyn/Å (cf. 10.41 mdyn/Å[78], 10.03 mdyn/Å[79], 10.05 mdyn/Å[80]; the last two values from R-spectroscopy).

4.3 Tert-butylchloroacetylene

From microwave spectroscopy, it was found for *tert-butylchloroacetylene*[81, 82] that the C–C bond adjacent to the acetylenic bond is considerably shorter than the value for the corresponding bond in tert-butylacetylene[83]. Because of the position of the center of mass in tert-butylchloroacetylene, the coordinates of the acetylenic carbon atom adjacent to the tertiary carbon atom could not be

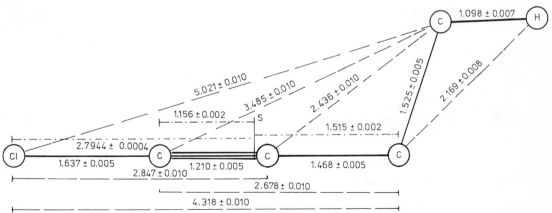

Fig. 4. Structure of *tert*-butylchloroacetylene. a = ----- Parameters obtained from electron diffraction data; b = -·-·-·- Parameters obtained from rotational spectra; S = center of mass

determined from the rotational constants. A total determination of the molecular structure was achieved with the aid of electron diffraction (Fig. 4).

The results actually obtained by the respective methods are given, *i.e.* the interatomic distances as determined from electron diffraction, and the separations of mass centers as obtained from microwave spectroscopy. Agreement between the two sets of results is good.

4.4 Disulfane

Disulfane[84] is a further example for the combination of two different methods for structural analysis. In evaluation of the radial distribution function of H_2S_2, it was found that only three of the four interatomic distances can be determined from the data. The missing distance H...H ≈ 2.9 Å could not be determined, since the contribution of this distance to the intensity function is very small because of the low scattering intensities of the hydrogen atoms. The corresponding peak in the radial distribution function is likewise very weak and is masked by residual fluctuations in the zero line. Consequently, the dihedral angle ε, which is an important parameter for the structural determination, cannot be obtained from electron diffraction data alone. This parameter can be determined from the value C–B = 3,12 MHz (difference between the rotational constants C and B) as obtained from microwave spectroscopy[85]. The value calculated for the dihedral angle is $\varepsilon = 90° 37'$ (Fig. 5).

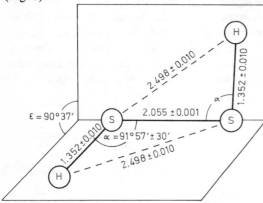

Fig. 5. Structure of disulfane as obtained from electron diffraction data and rotational constants.

[76] *C.H. Wong, V. Schomaker*, J. Chem. Phys. *28*, 1007 (1958).
[77] *J. Haase, M. Winnewisser*, Z. Naturforsch. *23a*, 61 (1968).
[78] *Y. Morino*, J. Mol. Spectrosc. *13*, 95 (1965).
[79] *D. Kivelson*, J. Chem. Phys. *22*, 904 (1954).
[80] *C.R. Polo, M.K. Wilson*, J. Chem. Phys. *22*, 900 (1954).

5 Accuracy of Results

Systematic experimental errors include: errors in determinations of the wavelength and of the camera distance, irregularities in the s-scale which can be caused by the microphotometer, and errors in the background function. Theoretical errors include: errors caused by the failure of the Born approximation, and uncertainties in the elastic and inelastic scattering factors deriving from different calculation procedures. Finally, the influence of unknown interatomic potentials also falls in this category. Some of these errors have an especially large effect on the determination of interatomic distances. The main error results from uncertainties in the determination of the wavelength which is needed for determination of the s-scale. Errors in the background line have an especially large effect on the values of vibration amplitudes[91]. For this reason, the values of the mean square amplitudes as determined by electron diffraction may, under certain circumstances, deviate appreciably from the values calculated from vibrational spectra. Uncertainties in the background line can also influence geometric parameters if there is a strong correlation between geometric parameters and mean square amplitudes. This is always the case if the peaks in the radial distribution function overlap to a large extent, *i.e.* if some interatomic distances in the molecule are close together. The values of the standard deviations as obtained from a least squares analysis of the molecular intensity function are usually too small to represent the true errors. It has been agreed that the error limit is taken as 2.5–3 times the standard deviations. This error limit provides a confidence interval of 99%. This procedure gives values for the error limit of a few thousandths of an Angstrom unit for the geometric parameters[5, 86].

[81] *J. Haase*, Z. Naturforsch. *22a*, 195 (1967).
[82] *H.K. Bodenseh*, Z. Naturforsch. *22a*, 523 (1967).
[83] *L.J. Nugent*, J. Chem. Phys. *36*, 965 (1962).
[84] *M. Winnewisser, J. Haase*, Z. Naturforsch. *23a*, 56 (1968).
[85] *G. Winnewisser*, J. Chem. Phys. *49*, 3465 (1968).
[86] *Y. Morino*, Acta Crystallogr. *18*, 549 (1965).

6 Bibliography

[87] *G. Herzberg*, Molecular Spectra and Molecular Structure, D. van Norstrand Co., Princetown 1950.
[88] *L. Reimers*, Elektronenoptische Untersuchungs- und Präparationsmethoden, Springer Verlag, Berlin, Göttingen, Heidelberg 1959.
[89] *T.Y. Wu, T. Ohmura*, Quantum Theory of Scattering, Prentice Hall, Englewood Cliffs, New Jersey 1962.
[90] *N.F. Mott, H.S.W. Massey*, The Theory of Atomic Collisions, 3. Ed., Clarendon Press, Oxford 1965.
[91] *S.J. Cyvin*, Molecular Vibrations and Mean Square Amplitudes, Elsevier Publ. Co., Amsterdam 1968.

8 Equilibrium and Kinetic Methods

Contributions from

J.C. Haartz, Cincinnati
F. Kaplan, Cincinnati
B. Kastening, Jülich
H.W. Nürnberg, Jülich
H. Wamhoff, Bonn

8.1 Polarographic and Voltammetric Techniques

Hans Wolfgang Nürnberg and *Bertel Kastening*
Zentralinstitut für Analytische Chemie der Kernforschungsanlage Jülich (KFA), D–517 Jülich

1 Introduction

A large number of inorganic and organic substances in solution can be reduced or oxidized by
electron transfer at an electrode, according to the
general equation

$$Ox + ne^{\ominus} \rightleftharpoons Red \tag{1}$$

The range of electrode potentials within which the
reaction proceeds in the forward (cathodic) or
backward (anodic) direction is specific for the
particular substance. The overall reaction can be
studied by the corresponding *current-voltage-
curve* (i-E-curve). It may involve homogeneous
and heterogeneous chemical steps as well as adsorption and desorption processes and may be
subject to inhibiting or catalysing effects.

Polarographic and voltammetric techniques aim at the
complete or partial determination of these i-E-curves or
their first and second derivative, respectively, under
stationary or non-stationary conditions. Certain general working conditions are associated with all these techniques: small concentrations ($\leqslant 10^{-3}$ M) of the reacting substances Ox and Red in solution, the presence of
an excess of inert supporting electrolyte (usually 0.1 to
1 M), use of small volumes of solution (between ca.
30 ml and some μl), well defined mass-transfer conditions to the working electrode, application of microelectrodes as working electrodes and counter-elec-
trodes sufficiently large so that polarization of these
latter is negligible at the low currents.
The vast potentialities for quantitative determination of
even the smallest amounts of substance in solution
(down to the ppb-range) become evident from the equivalence of one mole of substance with the large electri-
cal charge of $n \cdot F \equiv n \cdot 96,500$ coulombs (F = Faraday's
constant). Moreover, due to the negligibly small
amounts of substance necessary for recording an i-E-
curve, these methods are almost ideal indicating techniques for the investigation of a large number of chemical and physical-chemical problems.

In organic chemistry and biochemistry, polarographic and voltammetric techniques are important tools in the following fields:

Determination of the level of inorganic trace elements in
organic substances. This is becoming increasingly important in pollution and environmental research, cf.
Chapter 10.6.
Analysis of organic substances. The frequency and
breadth of application are certainly more restricted
than in the determination of inorganic substances, espe-

cially since gas chromatography, mass-spectroscopy
and spectrophotometric methods are available as alternatives. Nevertheless, polarographic and voltammetric
techniques serve for series analyses and control determinations especially because, in principle, there are no
problems of standardization. The quantity of transferred electric charge, corresponding to the amount of
substance according to the equivalence shown above,
always serves as a reference.
The methods under discussion are actually much more
important for the determination of characteristic physical-chemical parameters and their correlation with the
structure and constitution of organic and biochemical
substances in solution as well as at the electrode/solu-
tion interface (cf. 8.1:5). This also concerns the study of
electrode processes, including the development and
optimization of technical electrosyntheses of organic
compounds.
A particularly successful and promising field of application is the study of reaction kinetics and the determination of elementary steps of involved reaction sequences
in homogeneous solution as well as for heterogeneous
reactions at the electrode/solution interface. The range
of kinetic determinations is extraordinarily broad; time
constants may range from some hours to the order of
nanoseconds. In addition, there is a possibility of studying chemical reactions within the electrical double-layer
of the working electrode. This region constitutes to
some extent an easily controllable analog of living cell
membranes which are also electrically charged interfaces.

Frequently the terms *polarography* and
voltammetry[1-24] are still applied only to conventional dc polarography. While this was essentially justified 15 to 20 years ago, classical polarography nowadays constitutes only one method,
though important, among a variety of techniques
more recently available even in assembled form in
multipurpose instruments[4].

In classical *dc polarography* stationary current
voltage curves are recorded, the dropping mercury
electrode receives preference because of its spe-

[1] *P. Zuman*, Bibliography of Publications Dealing
with the Polarographic Method ('Heyrovský-Biblio-
graphy'), Academia, Prague 1951–1966.

[2] *L. Jellici, L. Griggio*, Bibliografia Polarografia,
Vol. 1–20, Consiglio Nazionale delle Ricerche,
Roma 1948–1968.

[3] *G. Milazzo*, Electroanalytical Abstracts, Vol. I–IX,
Birkhäuser Verlag, Basel 1963–1971.

[4] *D.N. Hume*, Anal. Chem. *38*, 261 R (1966); *40*,
174 R (1968).
R.S. Nicholson, Anal. Chem. *42*, 130 R (1970).

[5] *D.J. Pietrzyk*, Anal. Chem. *38*, 278 R (1966); *40*,
194 R (1968); *42*, 139 R (1970).

[6] *I.S. Longmuir*, Advances in Polarography,
Vol. I–III, Pergamon Press, Oxford 1960.

cial advantages. However, microelectrodes from other materials, have become important because they extend considerably the range of potentials and thus (as well as for other reasons) enlarge the number of substances which can be investigated. Properly, the term *polarography* should be restricted to the application of the dropping mercury electrode; whereas, the more general term *voltammetry* applies to other types of microelectrodes.

Modern techniques have resulted in a considerable increase in sensitivity and precision and in a clearer understanding of reaction mechanisms. In reaction kinetics they allow advances into previously inaccessible ranges. They use pulsed and periodic changes of electrode potentials from low to high frequencies, decreasing considerably the polarization time t_p which determines the range of kinetic accessibility.

2 The d.c. Polarographic Method

2.1 Principle

Polarography and the related voltammetric techniques, in principle, consist of an electrolysis in which molecules or ions present in an electrolyte are reduced or oxidized at a measuring electrode by electron transfer. The reactions proceeding at the counter electrode are essentially without significance. Peculiarities in comparison with the more general conception of electrolysis are given by the two following characteristics.

Due to the small surface of the working electrode, the amount of substance reacting electrochemically is small and avoids a change of bulk concentrations in the solution during the measurement.

The potential of the working electrode can be changed at will within certain limits by applying an external voltage because the layout of the external measuring circuit and the electrolyte solution avoids a marked voltage drop across them even under current load and because the counter electrode is 'non-polarizable'; that is, the potential drop across it remains constant even under current load.

The working electrode is a typical indicator electrode. From the relation between the external applied voltage and the measured current ('current-voltage curve', i-E-curve), sensitive and reproducible conclusions with respect to the composition of the solution can be drawn without causing any perceptible change in the system under investigation. Outstanding reproducibility is achieved especially with the dropping mercury electrode due to the continuous renewal of its surface. The *halfwave potential* $E_{1/2}$, at which the current is half the height of the step-shaped dc polarogram, is more or less specific to the reacting substance and may be used as a qualitative criterion. The *limiting current* at the plateau of the wave is proportional to the concentration of the respective substance and, therefore, a quantitative measure; from these limiting currents diffusion coefficients can also be evaluated and the course of low- and medium-rate homogeneous reactions in solution can be followed. Moreover, with relaxation techniques, very fast chemical reactions in solution can be investigated, if an equilibrium established in the solution is disturbed in a layer surrounding the working electrode by the process occuring at the electrode surface. Finally, the electrochemical behavior and

[7] *P. Zuman, I.M. Kolthoff*, Progress in Polarography, Vol. I, II, Interscience Publishers Inc., New York 1962;
P. Zuman, I.M. Kolthoff, L. Meites, Vol. III, IV unpubl.

[8] *G.J. Hills*, Polarography 1964, Vol. I, II, Mac Millan, London 1966.

[9] *J.O'M. Bockris, B.E. Conway*, Modern Aspects of Electrochemistry, Vol. I—V, Butterworths, London 1954—1969.

[10] *P. Delahay, C.W. Tobias*, Advances in Electrochemistry and Electrochemical Engineering, Vol. I—VII, Interscience Publishers Inc., New York 1961—1970.

[11] *A.J. Bard*, Electroanalytical Chemistry, Vol. I—V, Marcel Dekker, New York 1966—1971.

[12] *J. Heyrovský*, Angew. Chem. *72*, 427 (1960).

[13] *P. Delahay*, New Instrumental Methods in Electrochemistry, Interscience Publishers Inc., New York 1954.

[14] *M. v. Stackelberg*, Polarographie organischer Stoffe in *Houben-Weyl*, Methoden der organischen Chemie, 4. Aufl., Vol. III/2, p. 295, Georg Thieme Verlag, Stuttgart 1955.

[15] *J. Heyrovský, P. Zuman*, Einführung in die praktische Polarographie, Verlag Technik, Berlin 1959.

[16] *G. Charlot, J. Badoz-Lambling, B. Trémillon*, Electrochemical Reactions, Elsevier Publ. Co., Amsterdam 1962.

[17] *H. Strehlow*, Electrochemical Methods in *A. Weissberger*, Technique of Organic Chemistry, Vol. VIII, Part II, p. 799, Interscience Publishers Inc., New York 1963.

[18] *T.A. Krjukowa, S.I. Sinjakowa, T.V. Arefjewa*, Polarographische Analyse, VEB Deutscher Verlag f. Grundstoffindustrie, Leipzig 1964.

[19] *L. Meites*, Polarographic Techniques, 2nd Ed., Interscience Publishers Inc., New York 1965.

[20] *J. Heyrovský, J. Kůta*, Grundlagen der Polarographie, Akademie-Verlag, Berlin 1965.

[21] *H.W. Nürnberg, M. v. Stackelberg*, J. Electroanal. Chem. Interfacial Elektrochem. *2*, 181, 350 (1961); *4*, 1 (1962).

[22] *H.W. Nürnberg*, Z. Anal. Chem. *186*, 1 (1962).

[23] *H.W. Nürnberg, G. Wolff*, Chem.-Ing.-Tech. *37*, 977 (1965).

[24] *D.R. Crow, J.V. Westwood*, Polarography (Monographs on Chemical Subjects, series), Methuen & Co., London 1968.

the reduction and oxidation mechanisms, respectively, of chemical compounds can be studied in detail in an experimentally simple way. Redox potentials of reversible electrode processes are closely correlated to thermodynamic functions, to equilibrium constants of coupled chemical reactions, and to quantum chemical parameters.

2.2 Electrodes, Cells, Instrumentation

During measurement, a dc voltage supply of low internal resistance with variable voltage output applies the voltage U to the circuit consisting of the electrolysis cell and the current recorder in series. The voltage U corresponds to the sum of the ohmic drops $i \cdot R_M$ at the current recorder and $i \cdot R_L$ across the electrolyte solution and the difference of the potential drops $\Delta\phi_M$ and $\Delta\phi_G$ at the working and counter electrodes:

$$U = i \cdot R_M + i \cdot R_L + \Delta\phi_M - \Delta\phi_G \qquad (2)$$

The ohmic drop $i \cdot R_M$ including that at the leads can be neglected due to the small currents and the low resistance of the recording instrument. By addition of an inert supporting electrolyte (cf. 8.1:2.3) the solution is given a conductivity sufficiently large so that the term $i \cdot R_L$ is negligible. (Under unfavorable conditions it may be necessary to correct for the ohmic drop or to eliminate it with the aid of commercially available 'iR compensators', or to use an electronic potentiostat, which then requires the introduction of a third, unloaded, reference electrode.

The counter electrode should have a sufficiently large area to avoid noticeable deviations under current load from its 'rest potential'. The potential drop $\Delta\phi_G$ then remains constant so that any change of the external voltage U produces a corresponding change of the potential drop $\Delta\phi_M$ at the working electrode. Hence, U corresponds to the potential E of the working electrode with respect to the constant potential of the counter electrode:

$$U = \Delta\phi_M - \Delta\phi_G = E \qquad (3)$$
(vs. counter electrode)

In classical polarography the working electrode is a *dropping mercury electrode*. Every new drop grows into an (almost) fresh solution and, at a given value of E, every drop establishes the same reactant concentration profile around it, thus establishing quasi-stationary conditions. Due to the drop growth, a periodic change of the current takes place. This is reflected, due to the critical damping of the recorder, only by the characteristic 'notches' of the recorded curves. Other liquid

metals (gallium, amalgams) as well as suspensions of solids of sufficient conductivity ('carbon paste electrodes') are used for special purposes. For some applications, use is made occasionally of the advantages of special constructions and techniques (tapered capillaries, bent capillaries according to Smoler, control of the drop time by mechanical knocking).

If the working *electrode* is made of solid materials[25], the continuous supply of fresh solution and the stationary mass transfer conditions are usually achieved by rotating the electrode in the solution. Due to the mathematical difficulties in solving the hydrodynamic problems, rigorous theoretical derivations similar to the case of the dropping electrode have been developed only for special constructions (e.g. the *rotating disc electrode*[26, 27]). The *rotating ring disc electrode*[28, 29] serves especially for the investigation of short-lived intermediates. After the electrochemical reaction at the first working electrode (disc), the solution passes a second working electrode (ring) concentric to and isolated from the disc. This allows for a ring potential different from that of the disc and, thus, for the selective observation at the ring of intermediates of the disc electrode reaction.

Apart from the study of specific influences of the *electrode material,* its selection depends upon the *range of potentials* in which investigations are to be made. *Mercury* offers the advantage of a large *hydrogen overvoltage* which extends the range of accessible potentials to very negative values. However, its range is rather limited at positive potentials due to the ease of mercury oxidation (cf. 8.1:2.3). The majority of investigations with mercury, therefore, are concerned with reduction processes; whereas, platinum as well as graphite have frequently been used for the study of oxidations.

Numerous constructions have been suggested as *polarographic cells*[19]. Either counter electrode is situated in a separated electrolyte solution (connection through a saltbridge or a diaphragm) and

[25] *R.N. Adams,* Electrochemistry at Solid Electrodes, Marcel Dekker, New York 1969.
[26] *V.G. Levich,* Physicochemical, Hydrodynamics, Prentice-Hall, Englewood Cliffs, New Jersey 1962.
[27] *A.C. Riddiford,* The Rotating Disk System, in Ref. [10], Vol. IV, p. 47 (1966).
[28] *A.N. Frumkin, L.N. Nekrasov, V.G. Levich, Yu.B. Ivanov,* J. Electroanal. Chem. Interfacial Elektrochem. *1,* 84 (1959/60).
[29] *W.J. Albery,* Trans. Faraday Soc. *62,* 1915 (1966); *63,* 1771 (1967).
W.J. Albery, S. Bruckenstein, Trans. Faraday Soc. *62,* 2584 (1966).

serves simultaneously as reference electrode of known potential (e.g. the calomel electrode), or a mercury pool in the sample solution acts as the counter electrode, the potential of which can be determined with a separated reference electrode. Prior to the measurement, an inert gas (nitrogen or hydrogen, preferably of special purity grade) is bubbled through the solution for 'deaeration' to avoid interference by the reduction of oxygen. Special flow cells have been developed for continuous measurements, special micro-cells for investigations of very small volumes.

Commercially available 'polarographic instruments', (cf. Ref. [23]) which automatically generate a linearly changing polarization voltage, U, are applied almost exclusively for polarographic investigations. The immediate recording of current-voltage-curves (i-E-curves, 'polarograms') is achieved by synchronizing the paper speed of a current recorder with the voltage scan.

2.3 Solutions, Supporting Electrolytes, Buffer Systems

The solutions are composed of

the solvent,
the supporting electrolyte, which often consists of a pH buffer system, especially in aqueous solutions,
the substance under investigation ('depolarizer'),
occasionally special additives (maximum suppressors, inhibitors, catalysts).

While water served almost exclusively as a *solvent* previously, *aprotic organic solvents*[30-33] have gained importance more recently. The main reason for the application of these follows. While in water or in solvents of similar proton-availability, e.g. alcohols, most organic electrode reactions proceed irreversibly with the transfer of two or more electrons and protons, many compounds are reduced or oxidized reversibly with the transfer of a single electron only in aprotic solvents. This simplifies the reactions and often provides more unequivocal information. In water, protons are almost always involved in the reactions; this causes redox potentials and mechanisms to be strongly pH-dependent. Hence, in water and similar solvents, the addition of a buffer in considerable excess is usually indicated. Due to the required conductivity, only polar solvents with sufficiently high dielectric constants are useful for nonaqueous work. Besides alcohols, dimethylformamide, dimethylsulfoxide, acetonitrile and, more recently, propylene carbonate have been used.

The *supporting electrolyte* eliminates the voltage drop $i \cdot R_L$ and prevents mass transfer by migration of charged depolarizers. It must be present in considerable excess compared to the depolarizer. It determines the ionic strength of the solution and the thickness of the electric double-layer of the working electrode. Of course, the components of the supporting electrolyte should not undergo reduction or oxidation in the potential range under investigation. In the negative range, the accessible potentials are limited by hydrogen evolution or by the reduction of the supporting electrolyte cation. At positive potentials, the range is limited by the dissolution of the electrode metal or by oxidation of the supporting electrolyte anions; with mercury, this limit is determined by the solubility of the mercury (I) salt of the supporting electrolyte anion. Generally, an additional supporting electrolyte is not necessary if a pH-buffer system is applied. However, comparison of results often requires control of the ionic strength.

The most favorable concentration range of the *depolarizer* is between 10^{-4} and 10^{-3} moles/liter. At higher concentrations the voltage drop $i \cdot R_L$ may become appreciable due to relatively large currents. At lower concentrations the 'capacity current' (cf. 8.1:3.7) exerts a strong influence at the dropping electrode; it can be eliminated to some extent by suitable techniques or devices (compensation, 'Tast-polarography'). Special techniques (cf. 8.1:4) permit investigations at depolarizer concentrations several orders of magnitude lower.

At the dropping electrode, a forced flow of the solution occasionally interferes with pure radial diffusion leading to enhanced currents, known as *polarographic maxima*[34], by increased mass transfer. This effect can be eliminated by small additions of surfactants (e.g. gelatin, Triton-X-100, at about 10^{-3} percent). Increased concentrations of such additives may cause special effects (cf. 8.1:3.6).

[30] *G. LeGuillanton*, Bull. Soc. Chim. Fr. 2359 (1963).

[31] *C.K. Mann* in *A.J. Bard*[11], Vol. III, p. 57 (1966).

[32] *S. Wawzonek*, Talanta *12*, 1229 (1965).

[33a] Colloquium on Electrochemistry in Non-Aqueous Media, Paris 1970, J. Electroanal. Chem. Interfacial Electrochem. *29*, No. 1 (1971).

[33b] *J.N. Butler*, Reference Electrodes in Aprotic Organic Solvents, in *P. Delahay, C.W. Tobias*[10], Vol. VII, p. 77, 1970.

[33c] *J. Badoz-Lambling, G. Cauquis*, Analytical Aspect of Voltammetry in Non-Aqueous Solvents and Melts, in *C.N. Reilley*, Advances in Analytical Chemistry and Instrumentation, Vol.: *H.W. Nürnberg*, Electroanalytical Techniques and Applications, John Wiley Inc., New York (unpubl.).

[34] *M. v. Stackelberg*, Fortschr. Chem. Forsch. *2*, 229 (1951).

3 Theoretical Principles

3.1 Current-voltage Curves

The basic principle of dc polarography[13, 17, 20, 35] (and of related techniques like cyclic voltammetry, cf. 8.1:4.3) is the relation between the current i flowing through the cell and the potential E of the working electrode determined by the external voltage U and expressed relative to the respective reference electrode (cf. 8.1:2.2).

The shape of such c.v. curves may be illustrated by a simple example. The solution may contain the molecule M, which can be reversibly reduced by transfer of one electron to an anion-radical $M^{\ominus}$:

$$M + e^{\ominus} \rightleftharpoons M^{\ominus} \qquad (4)$$

(Such systems are frequently realized in aprotic solvents.) Thermodynamics requires that the concentrations of M and $M^{\ominus}$ at the electrode obey Equation (5)

$$E = E_0 + \frac{RT}{F} \cdot \ln \frac{[M]}{[M^{\ominus}]} \qquad (5)$$

where E, the potential of the working electrode, and E_0, the standard redox potential of Equation (4), are given with respect to the same reference electrode.

If only M is present in the solution, then a certain amount of $M^{\ominus}$ must always be generated. This amount is immeasurably small if E is considerably more positive than E_0. On the other hand, M must disappear completely at the electrode/solution interface if E is considerably more negative than E_0; then, each molecule M approaching the electrode is immediately transferred to $M^{\ominus}$. The number of molecules M, which can approach the electrode per second, limits the current at these potentials and is controlled by diffusion and convection (cf. 8.1.3.2). Even with charged depolarizers (ions) mass-transfer does not take place by ion migration, because the addition of the supporting electrolytes eliminates any noticeable potential gradient outside the electrochemical double-layer.

At quasi-stationary mass transfer to the working electrode, the c.v. curve inhibits the wave-shaped solid curve of Fig. 1. It increases with negative potential in the potential range about E_0 until it reaches, for $E \ll E_0$, the *limiting current* i_d, controlled by mass transfer only. In the reversible case, the half wave potential $E_{1/2}$ corresponds to E_0 (provided $D_{ox} = D_{Red}$, cf. Equation 9).

Besides the simple c.v. curve of a reversible one-electron process, other types of c.v. curves are frequently observed if the electrode reaction does not proceed at a sufficiently high rate (curve a) or

[35] *W.M. Reinmuth*, Theory of Electrode Processes, in *C.N. Reilley*, Advances in Analytical Chemistry and Instrumentation, Vol. I, p. 241, John Wiley Inc., New York 1960.

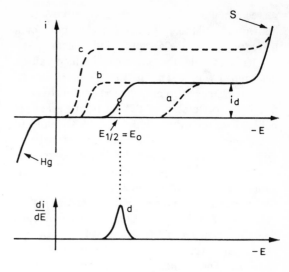

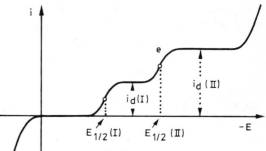

Fig. 1. Examples of current-voltage curves
Solid line: reversible one-electron reduction
a: slow irreversible electron transfer
b: irreversible reduction (e.g. by rapid dimerization)
c: irreversible 2-electron reduction (e.g. with intermediate protonation)
d: 'derivative' current-voltage curve
e: current-voltage curve with two waves (two one-electron processes)
S: supporting electrolyte reduction
Hg: mercury dissolution

the overall reaction does not proceed reversibly (curve b; cf. 8.1:3.3) and are shown in Fig. 1. Frequently the transfer of more than one electron takes place (curve c). The limiting current will then be a multiple of the one-electron current, and $E_{1/2}$ generally is more positive than E_0 of the primary one-electron process, though usually more negative than the standard redox potential of the overall reaction. This transfer of more than one electron at the same potential is frequently connected with a proton transfer to $M^{\ominus}$ and, therefore, often observed in aqueous solutions.

Polarographic c.v. curves often consist of more than one wave (cf. curve e in Fig. 1) if the solution contains either several depolarizers reduced at different potentials or a single depolarizer reduced to different products at different potentials. Expressing this latter case by the general Equation (6)

$$M \xrightarrow[E_{1/2}(I)]{+\ n_1 e^{\ominus}} M' \xrightarrow[E_{1/2}(II)]{+\ n_2 e^{\ominus}} M'' \qquad (6)$$

$E_{1/2}(II)$ usually corresponds to the half-wave potential observed if M' is present as the depolarizer.

Occasionally *derivative curves* (curve d in Fig. 1; di/dE vs. E) are recorded as a variant of ordinary c.v. curves. These curves are not as sensitive as common polarograms, but may facilitate the separate observation of two waves situated close to each other.

3.2 Mass Transfer

The *limiting current* i_d is controlled by diffusion and convection (cf. 8.1:3.5 for limiting currents controlled by the kinetics of chemical reactions). Table 1 shows some important relations derived from the laws of mass transfer.

In Fig. 2 are shown the time-dependent concentration profiles for the depolarizer and the reaction product within the *diffusion layer* in the vicinity of the electrode.

With the usual dropping capillaries, the limiting current amounts to about 2.5 to 3.5 μamps at natural drop times, if $n = 1$ and $^0c = 10^{-3}$ moles/liter.

Because the diffusion coefficient D depends on the viscosity and solvation, i_d varies for the same depolarizer in different media even if n remains constant. Short drop lives (e.g. 0.1 s) controlled by mechanical knocking are applied in the so-called *rapid polarography*. They permit considerably more rapid voltage scanning while the values of i_d are somewhat smaller.

The dependence of the currents on the height h of the mercury column (measured from the top of the mercury reservoir to the lower end of the capillary), $i_d \sim \sqrt{h}$, is used frequently as a proof for mass transfer control by diffusion at the dropping electrode. Correspondingly, $i_d \sim \sqrt{\omega}$ applies to the rotating disc-electrode for this case.

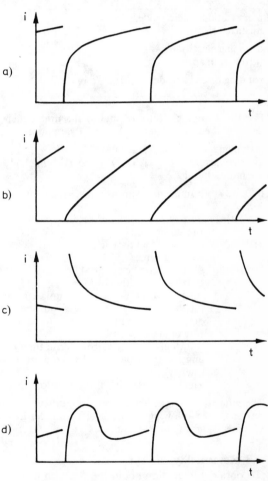

Fig. 3. Examples for current-time curves at single drops
a: diffusion control, $i \sim t^{1/6}$
b: current controlled by reaction kinetics, $i \sim t^{2/3}$
c: current controlled by the adsorption of the product, $i \sim t^{-1/3}$
d: inhibition of the electrode process by a surfactant the adsorption of which is retarded by diffusion control

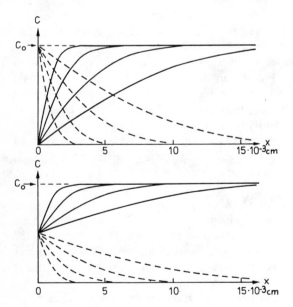

Fig. 2. Concentration profile in the diffusion layer
———— : depolarizer
- - - - - : reaction product
above: at the limiting current
below: at the half-wave potential
from left to right: 0.05; 0.2; 0.8; 3.2 seconds after beginning of drop life (calculated with $D = 10^{-5} \, cm^2 s^{-1}$)

Recording *current-time curves* during the life of individual drops with an oscilloscope may serve as another useful proof for mass transfer control as well as for the elucidation of reaction mechanisms and for kinetic investigations. Some examples of such i-t-curves are shown in Fig. 3. This technique corresponds to 'chrono-amperometry' (cf. 8.1:4.2).

Table 1. Relations for limiting currents controlled by mass transfer

General relation:

$$i_d = n \cdot F \cdot \left(\frac{dN}{dt}\right)_{x=0} = n \cdot F \cdot q \cdot D \cdot \left(\frac{dc}{dx}\right)_{x=0}$$

Type of electrode	Concentration gradient $(dc/dx)_{x=0}$
plane stationary electrode, unstirred electrolyte	$^0c/\sqrt{\pi Dt}$
dropping electrode (growing drop)	$^0c/\sqrt{\frac{3}{7}\pi Dt}$
rotating disc electrode	$0,62 \cdot {}^0c \cdot \omega^{1/2} \cdot D^{-1/3} \cdot \nu^{-1/6}$

For the dropping electrode, due to the time dependence of the electrode area ($A \sim t^{2/3}$) the relations

instantaneous current: $i_d = 706 \cdot n \cdot {}^0c \cdot D^{1/2} \cdot m^{2/3} \cdot t^{1/6}$
average current: $\bar{i}_d = 607 \cdot n \cdot {}^0c \cdot D^{1/2} \cdot m^{2/3} \cdot \tau^{1/6}$
are more suitable.

n: number of electrons transferred per molecule
$(dN/dt)_{x=0}$ number of moles per second reacting at the surface (flux)
x: distance from the surface
A: electrode area
D: diffusion coefficient of M
^0C: bulk concentration of M
t: time from the beginning of polarization (at the dropping electrode: from the beginning of drop life)
m: rate of mercury flow
τ: duration of a single drop ('drop time')
ω: angular frequency ($= 2\pi \cdot$ rotation frequency)
ν: kinematic viscosity of the solution

The numerical factors 706 and 607, respectively, correspond to the following units: i: μA; ^0c: mmoles/liter; D: cm^2s^{-1}; m: $mg \cdot s^{-1}$; t and τ: s.

3.3 Reversibility

To obtain information about the reaction mechanism and to judge whether $E_{1/2}$ is a thermodynamic quantity, the *degree of reversibility* of a wave must be known. Organic electrode reactions seldom proceed reversibly in aqueous solutions, and not always even in aprotic solutions. The shape of the wave itself provides the first indication. Relations applying to simple cases are listed in Table 2. Accordingly, a plot of log $(i/i_d - i)$ vs.

E would yield straight lines, with reciprocal slope of $59/n$ mv per log unit for reversible and $59/\alpha n_\alpha$ mv for irreversible waves at 25 °C.

Table 2. Shape of reversible and irreversible waves

Reversible wave: $i = i_d / 1 + \exp\left[\dfrac{nF}{RT}(E - E_0)\right]$

(change of sign of the exponent for anodic waves)

Irreversible wave (rate-determining electron transfer, written for reduction):

$$i = i_d / 1 + \frac{\rho}{k_0}\exp\left[\frac{\alpha n_\alpha F}{RT}(E - E_0^*)\right]$$

dropping electrode: $\rho = 1,13 \cdot \sqrt{D/\tau}$
rotating disc electrode: $\rho = 0,62 \cdot D^{2/3} \cdot \omega^{1/2} \cdot \nu^{-1/6}$

n: number of electrons transferred in the overall reaction
n_a: number of electrons transferred in the rate-determining step ($n_a \leq n$, frequently $n_a = 1$)
E_0:   standard redox potential of the overall reaction
E_0^*:   standard redox potential of the rate-determining step
k_0:   rate constant of electron transfer at the potential E_0^*
a: 'transfer coefficient' ($0 < a < 1$, in general $a \simeq 0.5$)

If the electron transfer proper is slow, then
$$k_0 < \rho; \; E_0 \geqslant E_0^* > E_{1/2}$$

If the product of the first electron transfer is removed by fast secondary reactions, then
$$k_0 > \rho; \; E_0 > E_{1/2} > E_0^*$$

Because protons generally participate in the overall reaction in aqueous solution, reversible potentials as well as $E_{1/2}$ values for irreversible processes *depend* strongly *on pH*[36] (cf. example in Fig. 4). In considering the value of $dE_{1/2}/dpH$ for irreversible reactions one must keep in mind that proton transfer often takes place heterogeneously with the absorbed species.

Auxiliary techniques are frequently used as proof of reversibility, e.g. cyclic voltammetry (cf. 8.1:4.3) and a.c. polarography (cf. 8.1:4.4)

3.4 Double Layer Effects

Any influence of the supporting electrolyte on reversible $E_{1/2}$-values is due to a change of the activity coefficients of the reacting species; particularly in organic solvents there may be a strong effect due to ion pairing.

For irreversible reactions the kinetics of individual steps play an important role and these can be influenced to a considerable extent by the *electrochemical double layer*[37-40] due to the following effects:

[36] *P.J. Elving*, Pure Appl. Chem. 7, 423 (1963).

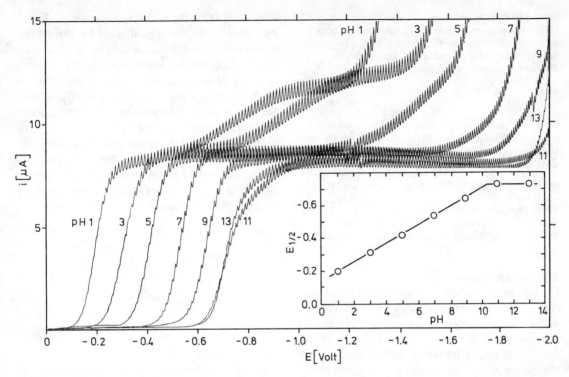

Fig. 4. pH-dependence of current-voltage curves: 10^{-3} M nitrobenzene with 0.01 percent gelatin. (The first wave corresponds to the 4-electron reduction to phenylhydroxylamine, the second wave seen at pH < 5 corresponds to the reduction to aniline.) Potentials vs. SCE.

The difference $(E — \psi_1)$ enters the rate equation of electron transfer instead of the potential E, where ψ_1 is the potential of the 'outer Helmholtz plane' with respect to the bulk of the solution.

Even under equilibrium conditions, the concentrations of ions in the double layer differ from those in the bulk of the solution.

The considerable electric field strength in the diffuse double layer affects the rate and equilibrium of coupled chemical reactions through the dissociation field effect[41].

Therefore, to compare half-wave potentials it is necessary to keep constant the *ionic strength* of the solution, which determines the structure and thickness of the double layer.

3.5 Coupled Chemical Reactions

The change in solvation at charge transfer, as well as the equilibrium constants of the chemical reactions participating in the overall reaction, contribute to the value of reversible redox potentials; therefore, these depend upon the medium.

For irreversible reactions, halfwave potentials and limiting currents are determined frequently by the rate of *coupled chemical reactions*[42]. Because of the large variety of mechanisms, only some characteristic examples are mentioned here:

Preceding acid recombination reaction:

$$M^{\ominus} + H^{\oplus} \; \rightleftharpoons \; MH \xrightarrow{\text{Reduction}} Product$$

The equilibrium is shifted in favour of $M^{\ominus}$; within the potential range under consideration only MH is reducible. Under certain conditions the limiting current can be controlled by the rate of formation of MH instead of mass transfer.

Subsequent dimerization:

$$M + e^{\ominus} \; \underset{E_0^*}{\rightleftarrows} \; M^{\ominus} \begin{cases} \longrightarrow \frac{1}{2}{}^{\ominus}M\!-\!M^{\ominus} \\ \xrightarrow{+H^{\oplus}} MH \xrightarrow{+e^{\ominus}} MH^{\ominus} \end{cases}$$

[37] *R. Parsons*, The Structure of the Electrical Double Layer and Its Influence on the Rates of Electrode Reactions, in Ref. [10], Vol. I, p. 1, 1961.

[38] *P. Delahay*, Double Layer and Electrode Kinetics, Interscience Publishers Inc., New York 1965.

[39] *D.M. Mohilner*, The Electrical Double Layer in *A.J. Bard*[11], Vol. I, p. 241, 1966.

[39a] *R. Payne*, The Electrical Double Layer in Non-Aqueous Solutions, in *P. Delahay, C.W. Tobias*[10], Vol. VII, p. 1, 1970.

[40] *L. Gierst, L. Vandenberghen, E. Nicolas, A. Fraboni*, J. Electrochem. Soc. *113*, 1025 (1966).

[41] *H.W. Nürnberg, G. Wolff*, J. Electroanal. Chem. Interfacial Electrochem. *21*, 99 (1969).

[42] *H.W. Nürnberg, M. v. Stackelberg*[21], *2*, 350 (1961). *L. Holleck*, Öst. Chem.-Ztg. *63*, 241 (1962). *R. Brdička, V. Hanuš, J. Koutecký*, General Theoretical Treatment of the Polarographic Kinetic Currents, in *P. Zuman, I.M. Kolthoff*[47], Vol. I, p. 145, 1962.

Since $M^\ominus$ is removed from the equilibrium, $E_{1/2} > E_0^*$. If the two indicated branches of the reaction compete with each other, the limiting current may change with the pH value and the depolarizer concentration.
Subsequent disproportionation:

$$M + e^\ominus \;\underset{E_0^*}{\rightleftharpoons}\; M^\ominus \longrightarrow \tfrac{1}{2}M + \tfrac{1}{2}M^{2\ominus} \underbrace{\qquad\qquad}_{\text{subsequent reactions}}$$

Since $M^\ominus$ is again removed from the equilibrium, $E_{1/2} > E_0^*$. The limiting current depends on the rate of disproportionation, since M is regenerated.
Catalytic hydrogen evolution[43, 44]:

$$M + H^\oplus \longrightarrow MH^\oplus \xrightarrow{+e^\ominus} MH \longrightarrow M + \tfrac{1}{2}H_2$$

These reactions proceed preferably with an adsorbed catalyzer M (especially N-bases) and may decrease the hydrogen overvoltage at mercury to a considerable extent.

3.6 Adsorption and Inhibition

During charge transfer organic molecules are generally in immediate contact with the metal without an intermediate layer of solvent molecules: the reaction proceeds in the adsorbed state[45-51]. *Adsorption* is caused by the lyophobic character of the molecules and, occasionally, by a specific (e.g. π-electron) interaction with the metal. The extent of adsorption depends on the metal, on the potential, on the kind of molecules, and on the solvent. As a rule, adsorption from organic solvents is considerably weaker than from water.

The *adsorption prewaves* of polarography are thermo-dynamically determined. Due to preferred adsorption of the reaction product, as many molecules can be reduced (oxidized) at a potential more positive (negative) than E_0 as adsorb at the surface. The shift of $E_{1/2}$ corresponds to the gain in free energy of adsorption.

Kinetic effects of adsorption are more frequently observed. The charge transfer and, in many cases, also proton transfer reactions take place with the adsorbed species; the corresponding rates are determined by the surface concentration which in turn depends on the solution composition. Kinetics may also be strongly affected by surfactants acting as *inhibitors* by displacing the more weakly adsorbed depolarizer molecules, e.g. if maximum suppressors are applied at higher concentrations (cf. 8.1:2.3); occasionally, *catalytic* effects may also occur. By the elimination of heterogeneous subsequent reactions, the mechanism may be changed and, under certain conditions, one observes the behaviour characteristic for aprotic conditions in aqueous solution in the presence of inhibitors (Fig. 5)[52]. Occasionally, adsorption of the reaction products causes an *autoinhibition*.

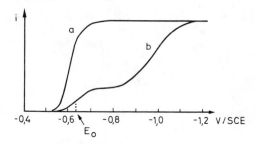

Fig. 5. Influence of surfactants (inhibitors) on the electrode mechanism. m-chloronitrobenzene, pH 13
a: without additive
b: in presence of 0.05 percent camphor
E_0: standard redox potential of the reversible one-electron process (cf. Holleck et al.[52])

3.7 The Electrocapillary Curve and Double-layer Capacity

The *surface tension* σ at an electrode depends on the potential; σ has a maximum value at the *potential of zero charge*[53]. The curve σ vs. E, the *electrocapillary curve,* has an approximately parabolic shape; in the presence of specific adsorption the curve is asymmetric and the potential of zero charge is shifted (Fig. 6).

The natural drop time (not controlled by mechanical knocking) of a dropping electrode is proportional to the surface-tension, $\tau \sim \sigma$. At very negative potentials, therefore, τ decreases strongly, and a decrease of limiting currents may be observed prior to the final supporting electrolyte reduction. For the investigation of adsorption phenomena, a precise recording of the

[43] *S.G. Mairanovskii*, Catalytic and Kinetic Waves in Polarography, Plenum Press, New York 1968.

[44] *H.W. Nürnberg* in *I.S. Longmuir*[6], Vol. II, p. 694, 1960.

[45] *L. Gierst, D. Bermane, P. Corbusier*, Ric. Sci. *29*, 75 (1959).

[46] *H.W. Nürnberg, M. v. Stackelberg*, J. Electroanal. Chem. Interfacial Elektrochem. *4*, 1 (1962).

[47] *C.N. Reilley, W. Stumm*, Ref.[7], Vol. I, p. 81 (1962).

[48] *A.N. Frumkin, B.B. Damaskin*, Adsorption of Organic Compounds in *J.O'M. Bockris, B.E. Conway*[9], Vol. III, p. 149, 1964.

[49] *A.N. Frumkin*, J. Chim. Phys. 785 (1966).

[50] *B. Kastening, L. Holleck*, Talanta *12*, 1259 (1965).

[51] *B. Kastening, L. Holleck*, J. Electroanal. Chem. Interfacial Elektrochem. *27*, 355 (1970).

[52] *L. Holleck, H.J. Exner*, Z. Elektrochem. Ber. Bunsenges. *56*, 46 (1952).
L. Holleck, B. Kastening, Über den Mechanismus der polarographischen Reduktion aromatischer Nitroverbindungen in *T. Kambara*, Modern Aspects of Polarography, p. 129, Plenum Press, New York 1966.

[53] *L. Campanella*, J. Electroanal. Chem. Interfacial Electrochem. *28*, 228 (1970).

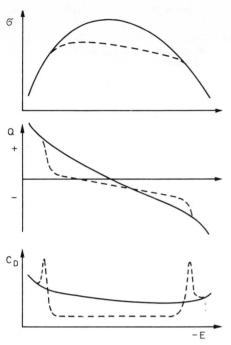

Fig. 6. Parameters of the phase boundary as a function of the electrode potential.
———— : without specific adsorption
- - - - : in presence of an adsorbed organic compound
σ: surface tension
Q: electrode charge
C_D :   differential double layer capacity

electrocapillary curve can be obtained by measuring τ with special devices[54-56], even simultaneously with the polarogram. A capillary electrometer with a stationary mercury electrode has often been applied as an alternate approach with special merits but also disavantages.

For electrochemical investigations other important surface parameters are the *charge of the electrode*, Q, and the *electrode capacity (differential double layer capacity*[38]*)*, C_D, which are correlated to σ as follows

$$Q = -\frac{d\sigma}{dE}; \quad C_D = \frac{dQ}{dE} = -\frac{d^2\sigma}{dE^2} \quad (7)$$

Q corresponds to the excess charge at the metal side and — with inverse sign — that at the electrolyte side of the double layer; the electrical analog of this is a condenser of capacity C_D. Due to thermal motion, the 'plate' situated at the electrolyte side is degenerated to a diffuse layer.

[54] B. Kastening, Ber. Bunsenges. Phys. Chem. 68, 979 (1964).

[55] B. Nygård, E. Johansson, J. Olofsson, J. Electroanal. Chem. Interfacial Elektrochem. 12, 564 (1966).

[56] H.W. Nürnberg, G. Wolff, Collect. Czech. Chem. Commun. 30, 3397 (1965).

If the surface area, the potential of the electrode, or the dielectric in the double layer are changed, then the potential-dependent surface charge Q requires a *capacity charging current* i_c. At the dropping electrode (where the area increases, $q \sim t^{2/3}$), this current ($i_c \sim t^{-1/3}$) can interfere considerably with the observation of the faradaic current i_F which is the current corresponding to the electron transfer in the electrode process and thus usually the component of interest in the total current $i = i_c + i_F$. Therefore, it is often preferable to record the current only during the last part of the drop life ('Tast polarography', c.f. the application of this principle with other techniques, 8.1:4). At stationary electrodes appreciable charging currents flow and must be considered if the potential is changed rapidly (cf. 8.1:4).

In the case of specific adsorption σ, Q and C_D are subject to strong changes (Fig. 6). If a surfactant desorbs within a narrow potential range then, as the potential moves across this range, a change in the dielectric of the double-layer condenser is produced and a sudden change in the charge takes place which is reflected in sharp peaks on the capacity-voltage curve (C_D vs. E curve). The measurement of C_D with a.c. techniques (e.g. *'tensammetry'*, cf. 8.1:4.4) is, therefore, a suitable tool for the study of the adsorption behaviour of organic compounds, even in connection with the mechanism of reduction or oxidation processes (cf. 8.1:3.6).

4 Special Polarographic and Voltammetric Techniques

In any polarographic measurement, whether the purpose is analysis or the evaluation of physicochemical data, one or several of the following fundamental problems are involved to some extent.

High *sensitivity* to obtain a low limit of determination; the highest possible *accuracy;*
high *resolution* and *separation* of the polarographic signals of different compounds present in solution occasionally in greatly different amounts;
extension of the range of kinetic accessibility to very fast reactions with *half lives* down to micro- and even nanoseconds.

Certain methodological limitations are inherent to classical dc polarography and these can be overcome only by the application of special polarographic and voltammetric techniques[57-61].

[57] P. Delahay, The Study of Fast Electrode Processes by Relaxation Methods in P. Delahay, C.W. Tobias[10], Vol. I, p. 233, 1961.

The range of application of these special advanced polarographic techniques is frequently more restricted than that of classical dc polarography, but within these limits they are superior to the latter; hence, the application is not alternative but complementary to classical dc polarography. For a survey on commercially available instruments, cf. Ref. [60].

interest. Hence, the sensitivity can be expressed in terms of the attainable ratio i_F/i_c.

Techniques which efficiently eliminate capacity currents offer essential advantages with regard to the three other problems listed above.

Among the special polarographic techniques, there is a variety of methods, the most advanced of which are still being tested and the number of

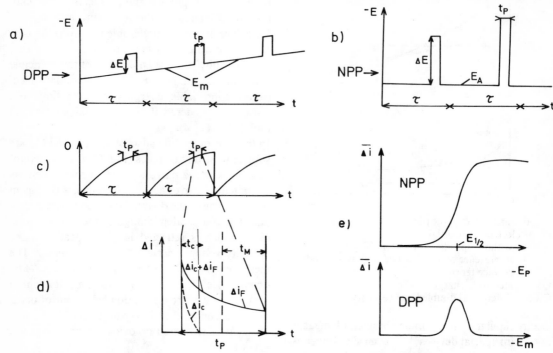

Fig. 7. Principle of polarization and current recording of pulse polarography (Barker type)

a and b: change of potential E of the dropping electrode with time for derivative (DPP) and normal (NPP) pulse polarography. τ: drop time; t_p: pulse duration; E: pulse amplitude; E_m: mean electrode potential for DPP; E_A: given starting potential for NPP

c: change of electrode area, A, during droptime τ with pulse interval t_p indicated

d: change of current during pulse interval t_p. Δi_c: capacity current; Δi_F: faradaic current; t_c: charging time of the double layer capacity; t_M: measuring interval within t_p

e: schematic representation of the resulting polarograms for NPP and DPP. For the peak potential, E_p, of the pulses with NPP: $E_p = E_A + \Delta E = E_m$

With respect to *sensitivity*, it must be kept in mind that the electrical signal resulting during the course of an electrode reaction consists of two components, only one of which is of interest. Thus, for the current which is frequently the measured quantity, one has $i = i_F + i_c$. Here, i_c denotes the current necessary for charging the double layer of the working electrode up to the charge density, Q, corresponding to the applied electrode potential, E. The faradaic current, i_F, connected with the transfer of electrons during the course of the electrode reaction is, except in tensammetry, the only quantity of

which increases each year due to the enormous possibilities and advances of modern electronics. Therefore, only some of those techniques are discussed which have turned out remarkably successful in solving important problems.

4.1 Pulse Polarography

According to the shape of the recorded curves, two versions are distinguished: derivative and normal pulse polarography[62]. The working principle of both is essentially the same (cf. Fig. 7).

[58] *H. Schmidt, M. v. Stackelberg*, Neuartige polarographische Methoden, Verlag Chemie, Weinheim/Bergstr. 1962.

[59] *R. Neeb*, Fortschr. Chem. Forschg. *4*, 335 (1963).

[60] *H.W. Nürnberg*, Chem.-Ing.-Tech. *38*, 160 (1966).

[61] *W.H. Reinmuth*, Anal. Chem. *36*, 211 R (1964); *40*, 185 R (1966).

[62] *G.C. Barker, A.W. Gardner*, Z. Anal. Chem. *186*, 73 (1960).

At the same instant during the life of each drop, the electrode is polarized with rectangular pulses of 4 to 100 msec duration. During polarization, the electrode area is, therefore, practically constant. In the *derivative version* small pulses, with amplitudes ΔE between 2 and 100 mv., are superimposed at the respective working point being given, as in dc polarography, by slow and linear increase of the applied dc voltage in the negative or positive direction. The *normal version* employs pulses of increasing amplitude from drop to drop, superimposed on a fixed starting potential. Only the change in current, Δi, caused by the voltage pulse, ΔE, is recorded. The extraordinarily high resolution of pulse polarography is based on this principle. Since the electrode area is almost constant during the short pulse interval, t_p, the capacity current component due to the change of the electrode area, which interferes in dc polarography, is practically eliminated. On the other hand, a considerable capacity current component, Δi_C, is caused by the pulsed polarization. This component, however, decays rapidly to negligible amounts, according to

$$\Delta i_c = \Delta i_{c_o} \exp\left(-\frac{t}{R_L C_D}\right) \qquad (8)$$

After a charging time, $t_c = 3\, R_L C_D$, Δi_c has decreased to less than 5 percent of its original value, Δic_o. For instance, at ohmic resistances, R_L, of the solution of 100 Ω and 1000 Ω, respectively, and for a typical value of the double-layer capacity, C_D, of 0.4 μF at an electrode area of $2 \cdot 10^{-2}$ cm^2, t_c-periods of 0.12 and 1.2 msec., respectively, are sufficient.

On the other hand, the faradaic current component, i_F, which usually is the component of interest, depends on time in a different way, viz. $i_F \sim t^{-b}$ with $b \le 0.5$; b attains its maximum value of 0.5 at pure diffusion control without any contributions of kinetic complications.

These considerations apply to all pulse techniques in general. Due to the different dependences on time of Δi_c and Δi_F, very efficient elimination of capacity currents can be achieved in pulse polarography by restricting electronically the recording of the current to the period t_M of the pulse duration t_p. The choice of t_M is such as to let Δi_c decay to negligible amounts and to record, over t_M, the average value Δi, being practically identical with Δi_F.

The considerable increase of sensitivity and accuracy is based mainly on this elimination of capacity current. The lower limits of determination are reduced to 10^{-7}M or 10^{-8}M with a typical maximum standard deviation of $\pm$ 10 percent at 10^{-7}M. This error decreases to only $\pm$ 1 percent or less for the range between 10^{-6}M and 10^{-5}M, which is the limiting range of reasonable application of classical dc polarography. Due to the restriction to the current change, Δi, resulting from the pulsed voltage change, ΔE, during t_p an extraordinarily high resolution is obtained. Thus, if $E_{1/2}$ values differ by ca. 200 mv., at least 100-fold increase of compounds reduced at more positive potentials is possible; at larger differences of $E_{1/2}$ up to 10^4 and more.

For quantitative analysis, the *derivative method* is usually preferable. Bell-shaped polarograms are obtained corresponding to the first derivative of the i-E characteristic. Width and symmetry of the waves are influenced by kinetic effects[63, 64]. However, normal step-shaped polarograms are obtained with the normal version. It is applied especially for the evaluation of adsorption and kinetic effects in studies of reaction mechanisms[63, 65]. The wide application range of pulse polarography which, contrary to ac techniques, has approximately equal capabilities for reversible and irreversible electrode processes with respect to sensitivity accuracy, and resolution, should be emphasized.

The above advanced version of pulse polarography, according to Barker, is realized in the commercially available instrument of Southern Analytical Ltd., Camberley, Surrey, England. This is, to a somewhat lesser extent, also true for the multipurpose instrument PAR 170 of the Princeton Applied Research Corp., Princeton, N.J., USA. There are also simpler versions of pulse polarography which, of course, meet more modest requirements.

The principle of current recording at quasi-constant electrode area has already been applied in *Tast-Polarography,* though without a pulsed polarization. By eliminating to a considerable extent the capacity current component due to the increase of surface area, an increase of sensitivity i_F/i_c is obtained compared with dc-polarography.

4.2 Chrono-Techniques

With these techniques, a certain working point of the i-E-characteristic of a solid or mercury drop electrode in an unstirred electrolyte solution is established by a step or pulse change in polarization whereupon the resulting change with time of the *current,* the *potential,* or the *charge* is recorded with a recorder or, for fast processes, with an oscilloscope. According to the measured quantity one distinguishes between *chronoamperometry, chronopotentiometry,* and *chronocoulometry*[66, 67]. Rather limited application is made of these techniques in quantitative analysis. They are of con-

[63] G. Wolff, H.W. Nürnberg, Z. Anal. Chem. *216,* 169 (1966).

[64] G.C. Barker, J.A. Bolzan, Z. Anal. Chem. *216,* 215 (1966).

[65] E. Paleček, J. Electroanal. Chem. Interfacial Electrochem. *22,* 347 (1969).

siderable importance in the evaluation of kinetic parameters, for the elucidation of mechanisms of electrode reactions and/or coupled homogeneous chemical reaction sequences as well as for the investigation of adsorption, inhibition, and catalysis effects.

Chronoamperometry[25, 68–73], also denoted as the *potentiostatic* step or pulse method, usually applies electronic control of polarization with a potentiostat of short rise time. For the evaluation of kinetic effects it is necessary to polarize quickly enough to make the capacity current (decreasing exponentially with time according to Equation 8) negligible, before the exponent b in the time function of the faradaic current, $i_F \sim t^{-b}$, increases to 0.5 due to the diffusion rate's decrease with $\sqrt{t}$, and thus before the overall rate of the electrode process and the corresponding current have become completely diffusion controlled. For precise kinetic measurements, the diffusion control of i_F should not exceed 75 percent; rough information can be obtained at up to 90 percent diffusion control[74]. In this way the rate constants of fast electrode reactions as well as of fast preceding or intermediate chemical steps (CE and ECE mechanisms, respectively) can be evaluated. Pulsed polarization must be applied in the study of chemical reactions following a reversible electrode reaction (EC mechanism). From the amount of intermediate that is re-oxidizable after a certain pulse duration, t_p, one can calculate that portion which has not been removed by the chemical reaction from the electrode/solution interface. Hence, the rate constant of the chemical step can be evaluated.

Analogous considerations apply to the other chronotechniques. For *chronopotentio-metry*[25, 75–77], also denoted the galvanostatic method, however, it must be remembered that, since the polarization is performed by a given constant current, an unequivocal determination of the working point is not possible in those ranges where the i-E characteristic exhibits a minimum. Only the rising part of the i-E-curve is amenable to this technique within the minimum range.

4.3 Voltammetric Sweep Techniques

As a rule, these techniques[25, 78–83] apply, elevated voltage scan rates, $v = dE/dt$, at stationary or quasi stationary electrodes in unstirred solutions. For dropping electrodes, the condition of quasi stationary electrode area means the application of sufficiently large v values (e.g. 0.3 to 5000 volts/sec) to achieve the registration of the entire i-E-characteristic of the electrode process under investigation during a section of a single drop life. This requires observation with an oscilloscope. The use of a recorder is possible at true stationary solid electrodes (noble metals, graphite, or carbon paste electrodes) at small values of v. Peak-shaped polarograms are obtained (cf. Fig. 8a), the peak current, i_p, being proportional to the concentration. From the peak (i_p) onwards, the depolarizer concentration at the electrode is zero analogous to the range of limiting currents for techniques with a quasi stationary diffusion layer (e.g. classical dc polarography). The drop of the current is due to the continuous decrease of the concentration gradient, $\partial c/\partial x$, and, therefore, the diffusion rate due to the spreading of the diffusion layer, $\delta = \sqrt{\pi Dt}$.

[66] *H. Christie, G. Lauer, R.A. Osteryoung,* Anal. Chem. *35,* 1979 (1963); J. Electroanal. Chem. Interfacial Electrochem. *7,* 60 (1964).
J.H. Christie, J. Electroanal. Chem. Interfacial Electrochem. *13,* 79 (1967).
J.H. Christie, R.A. Osteryoung, F.C. Anson, J. Electroanal. Chem. Interfacial Electrochem. *13,* 236 (1967).
[67] *H.B. Herman, H.N. Blount,* J. Phys. Chem. *73,* 1406 (1969).
[68] *H. Gerischer, W. Vielstich,* Z. Phys. Chem. NF. *3,* 16 (1955).
[69] *Y. Okinaka, S. Toshima, H. Okaniwa,* Talanta *11,* 203 (1964).
[70] *P.J. Lingane, H. Christie,* J. Electroanal. Chem. Interfacial Electrochem. *10,* 284 (1965).
[71] *W.M. Schwarz, I. Shain,* J. Phys. Chem. *69,* 30 (1965).
[72] *K.B. Oldham, R.A. Osteryoung,* J. Electroanal. Chem. Interfacial Electrochem. *11,* 397 (1966).
[73] *R.F. Nelson, S.W. Feldberg,* J. Phys. Chem. *73,* 2623 (1969).

[74] *H.W. Nürnberg,* Fortschr. Chem. Forsch. *8,* 241 (1967); Ber. Kernforschungsanlage Jülich, Jül-475-CA (1967).
[75] *D.G. Davis,* Applications of Chronopotentiometry to Problems in Analytical Chemistry in *A.J. Bard*[11], Vol. I, p. 157, 1966.
M. Paunovic, J. Electroanal. Chem. Interfacial Electrochem. *14,* 447 (1967).
[76] *C. Furlani, G. Morpurgo,* J. Electroanal. Chem. Interfacial Electrochem. *1,* 351 (1959/60).
[77] *A.C. Testa, W.H. Reinmuth,* Liebigs Ann. Chem. *32,* 1552 (1960).
[78] *H.I. Shalgosky, I. Watling,* Anal. Chim. Acta *26,* 66 (1962).
[79] *H.M. Davis,* Chem.-Ing.-Tech. *37,* 715 (1965).
[80] *R.S. Nicholson, I. Shain,* Liebigs Ann. Chem. *36,* 706 (1965): *37,* 178 (1965).
[81] *R.S. Nicholson,* Liebigs Ann. Chem. *37,* 1351 (1965).
[82] *D.S. Polcyn, I. Shain,* Liebigs Ann. Chem. *38,* 370 (1966); *39,* 376 (1966).
[83] *J.M. Savéant,* Electrochim. Acta *12,* 999 (1967).

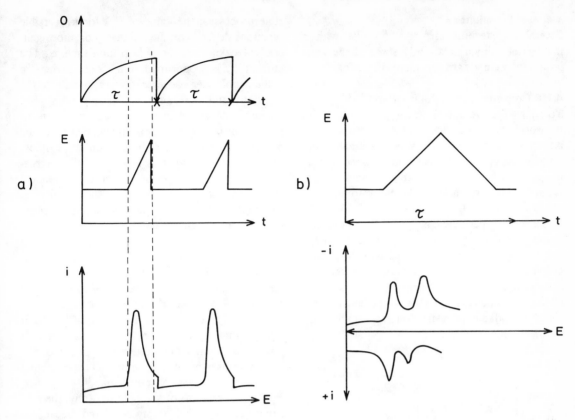

Fig. 8. Principles of sweep techniques

a: single sweep method: change of the electrode area, A, during drop time τ, sawtooth-change of electrode potential, E, at the rate v during the polarization interval within each drop life τ; resulting i-E-curve (polarogram)

b: cyclic voltammetry: change of the electrode potential, E, during the polarization interval within each drop life τ, resulting i-E-curve (polarogram) for a two-step electrode process of the type:

$A + n_1 e^{\ominus} \leftrightarrows B + n_2 e^{\ominus} \leftrightarrows C$. First step reversible; reduction of B to C rather irreversible.

Essentially, two varieties have been applied: the *single-sweep technique* with sawtooth-shaped change of voltage, and *cyclic voltammetry* (CV) with triangular voltage scan (cf. Fig. 8b). Both methods are suitable for quantitative analysis, particularly if extended series of control analyses of the same system must be performed rapidly. A considerable decrease of standard deviation to ± 0.1 to ± 0.5 percent can be achieved by making use of differential signals obtained from comparison with a standard solution in a double cell *(comparative differential cathode-ray polarography)*[78, 79]. In this way the separation capabilities in presence of an excess component can also be improved considerably (to ratios of $1:10^4$ to $1:10^5$ with differences of the peak potentials exceeding 100 mv.).

The *C.V. technique* supplies the complete i-E-characteristic for the reduction and the re-oxidation of a compound. Ample application of this method has been made particularly for the investigation of electrode reactions with respect to the degree of

reversibility as well as for the detection of unstable intermediates of involved reaction sequences.

The C.V. technique, especially its repetitive version, may properly be listed among the a.c. techniques, since a periodic (triangular) polarization of the working electrode takes place.

A special galvanostatic version is used in Czechoslovakia, i.e. *oscillographic polarography*[84]. A *sine-wave current* of *constant amplitude* (adjustable between 0.03 and 1 milliamps) is applied, usually at a frequency of 50 Hz. The functional relationship of electrode potential or its first derivative, vs. time is recorded with an oscilloscope.

[84] *J. Heyrovský, R. Kalvoda,* Oszillographische Polarographie mit Wechselstrom, Akademie-Verlag, Berlin 1960.

[85] *B. Breyer, H.H. Bauer,* Alternating Current Polarography and Tensammetry, Interscience Publishers, New York 1963.

[86] *D.E. Smith,* AC Polarography and Related Techniques: Theory and Practice in *A.J. Bard*[11], Vol. I, p. 1, 1966.

4.4 A.c. Techniques

Several a.c. techniques[58-61, 85-96] exist with a number of sub-variants. Only some of the most important methods are considered in the following.

4.4.1 Convenional a.c. Polarography[85, 86]

To adjust the average electrode potential, E_m, of the working electrode and, thus, the working point on the i-E-characteristic, a slow, linear and continuous d.c. voltage scan is applied. Superimposed upon it is a sine-wave a.c. voltage of small amplitude ($E_{ac} \simeq 10$ mv$_{pp}$) and usually of low frequency ν (30 to sometimes 100 Hz), (cf. Fig. 9). This causes corresponding periodic changes of the concentrations c_{ox} and c_{red} at the electrode surface; only the resulting a.c. current, i_{ac}, is recorded by the polarograph through capacitive coupling to the cell. For a noticeable signal i_{ac}, the degree of irreversibility of the electrode reaction should not be too high. This is a fundamental restriction in the application of a.c. techniques.

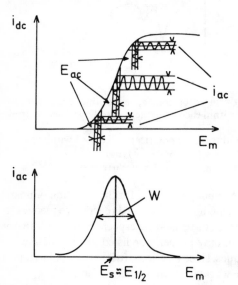

Fig. 9.

top: polarization principle of ac polarography with sine wave alternating voltage

bottom: resulting ac polarogram (schematically). The half width W amounts to $W = 90.6/n$ mv for reversible and to $W = 90.6/\alpha n_\alpha$ for irreversible electrode reactions at 25°C.

Because, at any working point E_m, one observes the change of the current, $\Delta i = i_{ac}$, produced by the a.c. voltage polarization, $E_{ac} = \Delta E$, it can easily be seen that a bell-shaped polarogram with a peak at $E_{1/2}$ will result analogous to the derivative techniques. Strictly speaking, this is true only for rever-

sible processes; the value of $E_{1/2}$ for irreversible reactions depends on the frequency or, more generally speaking, on the polarization time, t_p. The current, i_{ac}, is proportional to $n^2 \cdot {}^0c \cdot \omega^b$, where ${}^0c$ is the depolarizer concentration in the bulk of the solution and $\omega = 2\pi\nu$ is the angular frequency. The exponent $b \leq 0.5$ attains its maximum value 0.5 at complete diffusion control of the overall rate of the electrode reaction and exhibits smaller values as long as kinetics of the charge-transfer reaction proper or of preceding chemical steps are involved in the control of the overall rate. The voltage drop, $i_{ac} \cdot R$, across the ohmic resistance, R, of the circuit (sum of the resistances of the solution, the electrode, the cell diaphragm, and circuit resistance of the measuring net work) affects considerably and in a complicated manner the shape and height of the a.c. polarogram, since the effective a.c. voltage across the interface electrode/solution depends on the value of $i_{ac} \cdot R$. It follows that $i_{ac} \cdot R$ should be kept as small as possible and that a quantitative comparison of a.c. polarograms is reasonable only if all experimental parameters are kept constant.

For simple versions of a.c. polarography, the sensitivity is also considerably restricted, because a capacitive a.c. component directly proportional to the frequency, ω, resulting from the charging and discharging of the double-layer due to the periodic changes of potentials, is always present besides the faradaic component. (Determination limit at $\nu = 50$ Hz. ca. $5 \cdot 10^{-5}$M.) A more efficient elimination of the capacitive a.c. current must be achieved for more precise analytical requirements or for covering a wide range of frequencies necessary for kinetic investigations. This can be done by *phase-sensitive rectification* of i_{ac} prior to its registration, making use of the phase shift between the capacitive and faradaic components of i_{ac}. The phase angle between these components in the ideal case amounts to 45° for completely reversible processes. Actually, changes of the *double layer capacity* during the growth of the drop as well as the interference of the ohmic drop $i_{ac} \cdot R$ cause some deviation from this ideal value of the phase angle. Modern advanced a.c. polarographs are, therefore, fitted with an arrangement for continuous adjustment of the selected phase angle thus allowing for rather efficient elimination of the capacitive a.c. current or, on the other hand, a complete suppression of the faradaic component of i_{ac} (cf. tensammetry). Since these instruments also allow a rather

[87] H. Jehring, J. Electroanal. Chem. Interfacial Electrochem. 21, 77 (1969).

[88] G.C. Barker, J.L. Jenkins, Analyst 77, 685 (1952).

[89] G.C. Barker, Anal. Chim. Acta 18, 118 (1958).

precise measurement of the phase angle between i_{ac} and E_{ac} (the value of which is characteristic for the degree of reversibility as well as for the type, the rate, and sometimes for the mechanisms of coupled chemical reactions), a.c. polarography with advanced phase sensitive rectification has become an important and promising tool for the study of the kinetics and mechanisms of electrode reactions. At the same time, it is suitable for sensitive analytical determinations.

Tensammetry[85, 87] constitutes an important special technique of particular interest to the organic chemist. It permits the investigation of surfactants which are neither reducible nor oxidizable in the range of potentials within which they are adsorbed at the electrode. The appreciable changes of the dielectric of the double layer connected with their adsorption or desorption cause considerable *capacitive a.c. currents* (cf. Fig. 10) which can be used for the analysis of these compounds as well as for the study of problems of the phase boundary, e.g., with respect to the steric orientation of the adsorbed surfactant molecules. Such measurements can frequently be performed even with very simple a.c. polarographs.

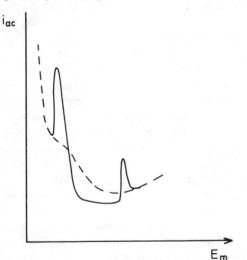

Fig. 10. Tensammetric curves (schematically)
- - - - -: i_{ac} vs. E_m curve of supporting electrolyte without surfactant and at complete desorption of the surfactant
————: i_{ac} vs. E_m curve in presence of surfactant

4.4.2 Square Wave Polarography

Instead of a sine wave a.c. voltage, this advanced a.c. technique[88, 89] applies a rectangular voltage of 225 Hz with amplitudes between 2 and 32 mV_{pp} superimposed on the average electrode potential, E_m (cf. Fig. 11). Moreover, the version of Barker, which considered here, carries out the registration at quasi constant area of the drop by restricting the

respective square wave polarization to a small period (ca. 2 sec.) after the beginning of drop life. Very efficient elimination of capacity current permits current registration by the same principle as in pulse polarography. Due to the restriction of the measuring time, t_M, to the last twelfth of each half period of the square-wave pulse (cf. Fig. 11), an

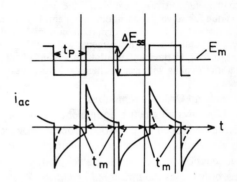

Fig. 11. Principle of rectangular ac polarization (top) and resulting alternating current components (below) for square wave polarography (Barker type)
- - - - -: i_c
————: i_F

almost pure faradaic current is recorded, since the capacity current resulting from the pulsed polarization has then become negligibly small because of its exponential decay. However, supporting electrolyte concentrations of at least 0.1 M, more favorably 1 M, are necessary to ensure short rise-times of the potential compared to the small pulse periods, t_p, with a half period of 1/450 sec. For substances having a reversible electrode reaction, the sensitivity of the method, with a determination limit of $5 \cdot 10^{-8}$ M to 10^{-7} M, is remarkably high. However, the restrictions inherent in all a.c. techniques with small amplitudes apply to irreversible electrode reactions.

4.5 Multipurpose Instruments

The extensive application of transistors and of integrated and miniaturized circuits in modern electronic instruments have stimulated the development of voltammetric multipurpose instruments which are now commercially available. Thus, the instrument PAR 170 (Princeton Applied Research Corp., Princeton, N.J., USA) offers all polarographic and voltammetric techniques discussed so far except square-wave polarography.

[90] *W.A. Brocke, H.W. Nürnberg,* Z. Instrumentenkunde 75, 291, 315, 355 (1967).
[91] *G.C. Barker,* Faradaic Rectification in *E. Yeager,* Trans. Sympos. Electrode Processes, Philadelphia 1959, Chapt. 18; John Wiley Inc., New York 1961.

However, not all the advanced techniques (e.g. pulse polarography) included in these instruments show the same sensitivity and experimental flexibility as those instruments specializing in one certain technique. On the other hand, the price of such multipurpose instruments is of the same order as that of a powerful special instrument. The importance of multipurpose instruments is obvious. The investigator can study a certain system with a wide spectrum of techniques, allowing for a convenient and rapid choice of the most suitable technique for solving a particular problem. Due to the large amount of diverse information obtained in this way, such multipurpose instruments will, within a few years, belong to the fundamental equipment of all laboratories dealing with electrochemical investigations.

4.6 'Second Order' Techniques

Fig. 12 shows an electric equivalent circuit of an electrode at which an electrode process proceeds. The electrical equivalent component of the electrode reaction proper, including coupled chemical steps, constitutes the *faradaic impedance*, Z_F. It is primarily a non-linear quantity and, therefore, the i-E-characteristic of an electrode process, strictly speaking, also exhibits non-linear behaviour. For the 'low level'-techniques discussed so far, which apply small changes ΔE_{pp} of the electrode potential, a *linear* approximation of the respective section of the i-E-characteristic around the working point, E_m, is permissible: i.e. qualitatively for $\Delta E_{pp} < 50$ mv; for more precise calculations only for $\Delta E_{pp} < 10$ mv or even < 5 mv. Although, for small values of ΔE_{pp}, those signals (e.g. current components for techniques where the potential is the adjusted entity) resulting from the *second derivative* of the i-E-characteristic are small, they nevertheless have finite values and may be recorded. The techniques specializing in the measurement of those signal components of higher order are generally called *second order techniques*[57-61, 90, 96]. Considering that the *double layer capacity,* C_D, always exhibits essentially linear behaviour compared with the degree of non-linearity of the faradaic impedance Z_F, it is evident that these techniques offer the possibility of very

efficient elimination of the *capacity current*. Obviously, second order techniques will have increasing importance for electrochemical kinetics as well as for electrochemical trace- and microanalysis. The corresponding theory is somewhat involved and instruments are not yet commercially available for all methods.

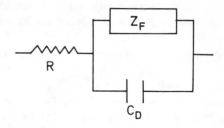

Fig. 12. Equivalent circuit of the working electrode
R: ohmic resistance of solution and electrode
C_D: double layer capacity
Z_F: faradaic impedance

According to the type of the non-linear effect of the faradaic impedance used, *faradaic rectification*[57, 90-96], *second harmonic*[59, 86, 96], *intermodulation*[59, 92, 96] and *demodulation*[95] techniques are distinguished. For a selective determination of the corresponding signal component, *low pass, high pass* and *band pass,* filters are applied. The quality and suitable design of these are decisive for the efficiency of the instruments.

The recently developed method of *High Level Faradaic Rectification* (HLFR) has gained special significance for studies on the kinetics of fast electrode reactions and especially for the investigation of very rapid coupled chemical reactions proceeding homogeneously or heterogeneously[74, 90, 94]. The working electrode (if a dropping mercury electrode, during a 40 msec interval of drop life) is periodically polarized with a train of rectangular voltage pulses of large amplitude (up to 1 Volt) and a pulse duration t_p adjustable down to 0.7 μsec. The resulting faradaic rectification current i_{FR} is the current component filtered out by a low-pass filter and is the recorded signal.

For a considerable number of aliphatic and aromatic *carboxylic acids* the rate constants of dissociation (k_d) and recombination (k_r) have been determined in aqueous electrolyte media for the first time[74]. Correlations of this kinetic data with structure[97], pK and thermodynamic functions of the acids and with double layer parameters have revealed fundamental information on the mechanism of dissociation and recombination[98]. These correlations are also useful to study the influences on reactivity of local water structure around reactants, salt effects, and the high electric field

[92] R. Neeb, Z. Anal. Chem. *208,* 168 (1965).
[93] G. Wolff, H.W. Nürnberg, Z. Anal. Chem. *224,* 332 (1967).
[94] G.C. Barker, H.W. Nürnberg, Naturwissenschaften *51,* 191 (1964).
[95] W.A. Brocke, J. Electroanal. Chem. Interfacial Electrochem. *30,* 237 (1971).
[96] G.C. Barker, Non-Linear Relaxation Methods for the Study of very fast Electrode Processes in *G.J. Hills*[8], Vol. I, p. 25, 1966.
[97] H.W. Nürnberg, H.W. Dürbeck, Z. Anal. Chem. *205,* 217 (1964); H.W. Nürnberg, H.W. Dürbeck, G. Wolff, Z. Phys. Chem. NF *52,* 144 (1967).

strength[41] in the double layer at electrically charged interfaces.

The HLFR-technique is at present the most powerful voltammetric approach[74] for kinetic resolution ($3 \cdot 10^8$ sec^{-1} for homogeneous 1st order rate constants; 5000 cm sec^{-1} for heterogeneous rate constants). This method also offers remarkably high potentialities for the study of adsorption effects[64].

4.7 Coupling With Other Techniques

Joint applications of polarographic or other voltammetric techniques with various physicochemical methods have become frequent and important.

Obviously, dropping mercury or solid indicator electrodes are suitable probes for following the course of electrolysis at electrodes of large area by indicating the changes of concentrations of reactants as well as of intermediates and products[99-101].

Coupling with non-electrochemical methods has proved successful for practical applications, e.g. for routine analysis, and for scientific investigations by providing mutually complementary information and enhanced potentialities.

In *chromato-polarography*[102] a dropping electrode serves for continuous analysis of the eluate of a chromatographic column. The eluate flows slowly through a polarographic cell, permitting either the recording of the current at a controlled potential of the indicator electrode and as a function of the volume flowing out of the column, or the repeated recording of complete i-E-curves. Besides compounds acting as depolarizers, it is possible also to follow the concentration in the eluate of such compounds exerting electrochemical effects through their catalytic or surface active properties.

Increased application has been made of the coupling of voltammetry techniques with *electron spin resonance* (ESR)[103-106]. This is due to the great importance of free radicals in the course of electrode reactions. The life time of such radicals (especially anion and cation radicals) may be very short. They may also be considerably long especially in aprotic or sometimes even aqueous solutions. The ESR technique is based on the fact that radicals are paramagnetic and will, in an external magnetic field, absorb electromagnetic energy of the corresponding resonant frequency (in the micro-wave range). A highly informative resonance spectrum is obtained by interaction with the nuclear spins in the molecule (especially of H and N atoms), allowing conclusions about molecular structure of the radicals, their conformation, the distribution of electron density, the interaction with the solvent and with other species in solution (e.g. ion-pairing). By applying MO-theories, electrochemical and ESR parameters can be correlated closely. Moreover, ESR measurements can serve for the determination of radical concentrations, e.g. in kinetic investigations of stabilization reactions.

In some cases electrochemistry has been used to generate radicals for investigation with ESR if the generation by chemical means presented difficulties. In general, however, the electrochemical behaviour has been investigated simultaneously. For 'external generation' a macro-scale electrolysis is carried out in a separate electrolysis cell at a large electrode area. The electrolyte is then transferred into the microwave cavity placed in the magnetic field. The internal ('intra muros') technique uses small polarizable electrodes situated directly in sample cells within the ESR cavity. The high sensitivity of ESR permits the investigation of the small amounts of radicals being generated at such electrodes; small mercury pools or solid electrodes are generally used.

Radioactive isotopes have been applied repeatedly for electrochemical investigations[107]. These are especially suitable for the study of adsorption at the electrode surface. The activity is determined either from the back of the electrode (made appropriately thin) or from the solution side. In the latter case, the information necessary to determine the degree of coverage is obtained by measuring the activity for a range of distances between the electrode and the sensors.

Polarographic and voltammetric techniques are excellent means for the investigation of radicals

[98] *H.W. Nürnberg* in *G.J. Hills*[8], Vol. I, p. 149; *H.W. Nürnberg*, Cronache di Chimica [Milano], n. 31, Marzo 1971.

[99] *B. Kastening*, Collect. Czech. Chem. Commun. *30*, 4033 (1965).

[100] *P. Zuman*, J. Polarogr. Soc. [London], *13*, 53 (1967).

[101] *H. Lund*, Öst. Chem.-Ztg. *68*, 152 (1967).

[102] *W. Kemula* in *P. Zuman, I.M. Kolthoff*[7], Vol. II, p. 397, 1962.

[103] *R.N. Adams*, J. Electroanal. Chem. Interfacial Electrochem. *8*, 151 (1964).

[104] *G. Cauquis*, Bull. Soc. Chim. Fr. 1618 (1968).

[105] *B. Kastening*, Chem.-Ing.-Tech. *42*, 190 (1970).

[106] *B. Kastening*, Joint Application of Electrochemical and ESR-Techniques in *C.N. Reilley*, Advances in Analytical Chemistry and Instrumentation, *H.W. Nürnberg*, Electroanalytical Techniques and Applications, John Wiley Inc., New York, unpubl.

and other short-lived products generated by non-electrochemical methods, e.g. through *hydrogenation* by suspended catalysts[108], through photolysis *(photo-polarography)*[109-111] and through *pulse radiolysis*[112] with high energy electrons. Here, the indicator electrode is placed directly in the solution in which the corresponding reaction is initiated.

Several varieties of spectrophotometric investigations[114] use the interaction with *light*[113] to study electrode reactions. To investigate processes occuring at the electrode, or in the immediate vicinity, either light has been passed through optically transparent electrodes[115] or changes of the reflectance of the surface have been studied[116].

Finally, research[117-122] has been done in which the influence of light on electrode processes is studied. According to the results obtained so far such interactions are due partly to the optical activation of molecules (especially adsorbed molecules) involved in the electrode reaction, and partly to the optically stimulated generation of *solvated electrons* which are re-oxidizable or react with the substrate molecules. Particular interest has been devoted further to *luminescence* induced by electrode processes[113, 123].

5 Applications

5.1 Determination of Physico-Chemical Quantities

The polarographic and voltammetric techniques discussed above furnish a large amount of information about the chemical and physico-chemical behaviour of organic substances in solution and at electrodes[124, 137].

The quantity determined most frequently is the *concentration* ^0c of compounds either for analytical purposes[128, 129], for the elucidation of *reaction mechanisms*[131, 135-138] or as a function of time during kinetic investigations for the determination of *rate constants* of chemical reactions[134].

The *diffusion coefficient,* D, of compounds acting as a depolarizer or generating the depolarizer in a chemical reaction preceding electron transfer, is also an easily accessible quantity (limits of error $\pm$ 1 to 2 percent).

Knowledge of the i-E-characteristics and the influence of various conditions on it is essential for elucidation of the electrochemical behaviour of a compound. Such information serves as the basis for the development[139-141] of electrosyntheses. It is also helpful for the elucidation of molecular structures, of the position and the kind of substituents, of the possibility of fission of rings and other bonds, of the position of double bonds. In addition, it enables the detection of isomer formation and of mesomeric behaviour. Specific electrochemical parameters here are the number of transferred electrons, n, which also allows conclusions about the various oxidation stages; the transfer coefficient, α, the value of which may be affected considerably by chemical reactions preceding

[107] *N.A. Balashova, V.E. Kazarinov*, Use of the Radioactive Tracer Method for the Investigation of the Electric Double Layer Structure in *A.J. Bard*[11], Vol. 3, p. 135, 1969.

[108] *H. Berg*, Z. Phys. Chem. *229*, 138 (1965).

[109] *H. Berg*, Z. Anal. Chem. *216*, 151 (1966).

[110] *H. Berg*, Grundlagen und Möglichkeiten der Photo-Polarographie in *T. Kambara*, Modern Aspects of Polarography, p. 29, Plenum Press, New York 1966.

[111] *H. Berg*, J. Electroanal. Chem. Interfacial Electrochem. *15*, 415 (1967).

[112] *J. Lilie, G. Beck, A. Henglein*, Ber. Bunsenges. Phys. Chem. *75*, 458 (1971).

[113] *T. Kuwana*, Photoelectrochemistry and Electroluminescence in Ref. [11], Vol. I, p. 197, 1966.

[114] *I. Bergmann* in *G.J. Hills*[8], Vol. II, p. 925, 1966.

[115] *T. Osa, T. Kuwana*, J. Electroanal. Chem. Interfacial Electrochem. *22*, 389 (1969).

[116] *T. Takamura, K. Takamura, E. Yeager*, J. Electroanal. Chem. Interfacial Electrochem. *29*, 279 (1971).

[117] *G.C. Barker, A.W. Gardner, D.C. Sammon*, J. Electrochem. Soc. *113*, 1182 (1966); *G.C. Barker*, Ber. Bunsenges. Phys. Chem. *75*, 728 (1971).

[118] *M. Heyrovský*, Z. Phys. Chem. NF *52*, 1 (1967); Proc. Roy. Soc., Ser. A *301*, 411 (1967).

[119] *H. Berg*, J. Electroanal. Chem. Interfacial Electrochem. *14*, 351 (1967); Electrochim. Acta *13*, 1249 (1968).

[120] *G.C. Barker*, Electrochim. Acta *13*, 1221 (1968).

[121] *Z.A. Rotenberg, V.I. Lakomov, Y.V. Pleskov*, J. Electroanal. Chem. Interfacial Electrochem. *27*, 403 (1970).

[122] Solvated Electrons in Liquid and Solid Solution, Discussion Meeting, Deutsche Bunsengesellschaft, Herrenalb 1971, Ber. Bunsenges. Phys. Chem. *75*, No. 4 (1971).

[123] *D.M. Hercules*, Accounts of Chemical Research *2*, 301 (1969). *A.J. Bard* in *G. Guilbault*, Fluorescence, Chapter 14, Marcel Dekker, New York 1967.

[124] *M. Březina, P. Zuman*, Die Polarographie in der Medizin, Biochemie und Pharmazie, Akadem. Verlagsgesellschaft, Leipzig 1956; Engl., Interscience Publishers Inc., New York 1958.

[125] *G.W.C. Milner*, The Principles and Applications of Polarography and other Electroanalytical Processes, Part III, p. 469, Longmans, Green and Co., London 1957.

[126] *K. Schwabe*, Polarographie und chemische Konstitution organischer Verbindungen, Akademie Verlag, Berlin 1957.

[127] *H.W. Nürnberg*, Angew. Chem. *72*, 433 (1960).

charge transfer and/or by adsorption in an opposite manner; the rate constant k_{ct} of charge-transfer which depends exponentially on the electrode potential E.

The parameters of the double layer — the double layer capacity C_D, the surface charge Q, the potential drop ψ_1 across the diffuse part of the double layer — are important in elucidation of the mechanism of electrode processes, especially since the latter may be influenced significantly, via preferential or suppressive effects exerted to a different degree, by double layer parameters on parallel reactions. In addition there is a general effect through the shift of the potential range in which the electrode process occurs.

The study of the *surface activity* of organic compounds by the determination of surface tension σ, the degree of coverage θ, the interactions within the adsorbed layer, and the orientation of the adsorbed molecules, together with the study of the inhibition and catalysis effects of such an adsorbed layer on electrode reactions and on coupled heterogeneous and homogeneous chemical processes, constitute an important special field of investigation.

The whole spectrum of polarographic and voltammetric techniques with their different characteristic polarization times, t_p, may be applied according to the required time resolution to obtain the quantities and information mentioned above. However, the measurement of $E_{1/2}$-values and the calculation of parameters from them are a domain of conventional d.c. polarography which must be performed very accurately for these purposes.

For *reversible electrode reactions*, the *standard redox-potential*, E_0, can be derived from the half-wave potential, $E_{1/2}$. If, as is often the case, homogenous or heterogeneous protonation steps are involved, then corresponding pK-values may be determined from the pH-dependence of $E_{1/2}$ for reversible as well as for irreversible electrode processes. In an analogous way, pK-values of coupled chemical equilibria (e.g. stability constants of charge transfer complexes of aromatics with solvent molecules or of ion pairs with the supporting electrolyte[142]) can be derived from the dependence of $E_{1/2}$ on the concentration of the complexing excess component. The determination of E_o and pK, respectively, from $E_{1/2}$ is of particular importance if a species participating in some redox equilibrium or a protonation equilibrium (which is sometimes detected only in this way) cannot be isolated but exists only as an unstable intermediate at the electrolyte/solution interface.

It becomes obvious from the close relation to thermodynamic equilibrium or activation functions of a reversible or irreversible electrode reaction, respectively, that the *half-wave potential* $E_{1/2}$ is the fundamental parameter for the electrochemical behaviour of a substance.

In the case of reversible electrode processes, less frequently occuring with organic compounds, one has

$$E_{1/2_{rev}} = E_0' + \frac{RT}{nF} \ln \sqrt{\frac{D_{Red}}{D_{Ox}}} \qquad (9)$$

(cf. Table 2 where, for simplification, the assumptions $D_{red} = D_{ox}$, $f_{red} = f_{ox}$ have been made). Here, E_o denotes the value of the standard redox potential at the given supporting electrolyte concentration.

$$E_0' = E_0 + \frac{RT}{nF} \ln \frac{f_{Ox}}{f_{Red}} \qquad (10)$$

It contains the constant term involving the ratio of activity coefficients of Ox and Red, which deviates usually from 1. Because the concentrations of Ox and

[128] *P. Zuman*, Organic Polarographic Analysis, Pergamon Press, Oxford 1964; Z. Anal. Chem. *216*, 151 (1966).

[129] *P.J. Elving*, Voltammetry in Organic Analysis in *C.N. Reilley*, Advances in Analytical Chemistry and Instrumentation, *H.W. Nürnberg*, Electroanalytical Techniques and Applications, John Wiley Inc., New York, unpubl.

[130] *J. Proszt, V. Cieleszky, K. Györbiró*, Polarographie, Teil III, p. 375, Akadémiai Kiadó, Budapest 1967.

[131] *Ch. L. Perrin*, Mechanisms of Organic Polarography in *S.G. Cohen, A. Streitwieser jr., R.W. Taft*, Progress in Physical Organic Chemistry, Vol. 3, p. 165, Interscience Publishers Inc., New York 1965.

[132] *P. Zuman*, Physical Organic Polarography in *A. Streitwieser jr., R.W. Taft*, Progress in Physical Organic Chemistry, Vol. 5, p. 81, Interscience Publishers Inc., New York 1967.

[133] *P.J. Elving*, Pure Appl. Chem. *15*, 297 (1967).

[134] *P. Zuman*, Polarography and Reaction Kinetics in *V. Gold*, Advances in Physical Organic Chemistry, Vol. 5, p. 1, Academic Press, New York 1967.

[135] *P. Zuman*, Substituent Effects in Organic Polarography, Plenum Press, New York 1967; Fortschr. Chem. Forsch. *12*, 1 (1969).

[135a] *B. Kastening*, Radicals in Organic Polarography in Ref. [7], Vol. III/IV, unpubl.

[136] Discussions of the Faraday Society, *45*, (1968), Electrode Reactions of Organic Compounds.

[137] *G. Semerano, L. Griggio*, Ric. Sci. Suppl. *27*, 243 (1957).

[138] *P.J. Elving*, Ric. Sci. *30*, 205 (1960).

[139] *S. Wawzonek*, Science *155*, 39 (1967).

[140] *M.M. Baizer*, Naturwiss. *56*, 405 (1969).

[141] *F. Beck*, Chem.-Ing.-Tech. *42*, 153 (1970).

[142] *M.E. Peover, J.D. Davies* in *G.J. Hills*[8], Vol. II, p. 1003, 1966.

Red are small ($<10^{-3}$ M), they have only a negligible influence on the activity coefficients. The thermodynamic parameters mentioned hereafter are related to a constant E'_0-value resulting from the kind and concentration of the supporting electrolyte, the respective solvent, and the buffer components often present in large excess as compared with c_{ox} and c_{red}.

In general the assumption $D_{ox} = D_{red}$ can be made; hence, $E_{1/2rev} = E'_0$. Otherwise, D_{ox} and D_{red} must be determined and E_0 obtained from Equation 10 must be inserted instead of $E_{1/2rev}$ in the following relations. If $E_{1/2rev} = E'_0$, then

$$E_{1/2_{rev}} = -nF \Delta G^0 \tag{11}$$

where ΔG^0 is the standard free enthalpy of the electrode reaction $Ox + ne^{\ominus} \rightleftarrows Red$.

The free enthalpy ΔG is composed of the enthalpy and the entropy term according to the Gibbs-Helmholtz equation.

$$\Delta G = \Delta H - T \Delta S \tag{12}$$

For organic systems the electrode process, e.g. $Ox + ne^{\ominus} \longrightarrow Red$, frequently occurs irreversibly. Then analogous consideration applies correlating $E_{1/2_{irr}}$ with the thermodynamic activation functions corresponding to the rate constant of the electrode process at the potential $E = 0$.

In Equation 13

$$E_{1/2_{irr}} = \frac{RT}{\alpha n_\alpha F} \ln \left\{ 0,886\, k_{E=0} \sqrt{\frac{\tau}{D}} \right\} \tag{13}$$

α is the transfer coefficient and n_α the number of electrons transferred in the rate determining step of the electrode reaction; hence, in general $n_\alpha \leq n$.

From

$$\ln k_{E=0} \sim -\frac{\Delta G^{\#}}{RT} \quad \text{it follows} \atop \text{that} \quad E_{1/2_{irr}} \sim \frac{\Delta S^{\#}}{R} - \frac{\Delta H^{\#}}{RT} \tag{14}$$

it follows that

where $\Delta G^{\#}$, $\Delta S^{\#}$ and $\Delta H^{\#}$ are the free enthalpy, the entropy, and the enthalpy of activation of the irreversible electrode reaction related to the potential $E = O$.

The thermodynamic functions determing the electrochemical reactivity of a compound reflect in their values the readiness with which an electrochemically active group or position of an organic molecule accepts or delivers an electron upon reduction or oxidation, respectively. Ultimately, the determining quantities for this electron transfer are the electronic configuration and the electron density at the reactive site of the molecule. This will depend on the structure of the molecule, the resulting strength and kind of bonds, as well as the local solvation structure and other special steric effects (e.g. the orientation of the molecule at the electrode). In physical organic chemistry these differ-

ent influences on the electronic configuration at a certain position or group of a molecule are usually described by the inductive (I), mesomeric (M), and steric (S) effects. The half-wave potential offers, in principle, a well defined and convenient experimental approach to these manifold problems connected with the structure of electrochemically active compounds.

Due to the variety of superimposed effects and interactions contained in the quantity $E_{1/2}$, it is frequently necessary in the study of the different types of compounds and different mechanisms of electrode reactions to restrict oneself to *qualitative* predictions. Thus, the introduction of an *electrophilic substituent,* in principle, results in a positive shift of $E_{1/2}$ via its (—)I-and perhaps also (—)M-effect on the electron density at the electroactive position in the molecule. Accordingly, *nucleophilic substituents* cause a negative shift of $E_{1/2}$ due to their (+)I- and (+)M-effects. As a rule the effects of several substituents sum up linearly. Steric effects as well as intramolecular hydrogen bonding may superimpose in a dominant manner.

Quantitative correlations, however, result from the study of relative differences due to structural effects of compounds showing the same mechanism of the electrode reaction, e.g. at systematic variation of substituents of a certain type of compounds, where the electron transfer occurs always at the same group or position in the molecule. Here one must be sure that αn_α or n, respectively, as well as D and, if there is pH dependence, $dE_{1/2}/dpH$ remain constant for all the members of the homologous series examined. The relative differences $\Delta E_{1/2}$ corresponding to the relative differences $\Delta\Delta G^0$ and $\Delta\Delta G^{\#}$, respectively, then obey *linear relationships* of the Hammet-type equation [132,135]

$$\Delta E_{1/2} = \rho \sigma \tag{15}$$

In this equation the slope ρ of the $\Delta E_{1/2}$ vs. σ line depends only on the particular type and mechanism of the electrode reaction, while σ reflects the respective *substituent effect*. Here, for the linear correlations described by equation 15, the term 'Hammett-type' has been used. The σ values given by Hammett are valid only for meta and para substituted aromatics, e.g. benzoic acid derivatives, while different values of σ apply to other homologous series of compounds like aliphatics or heterocyclics. The reader is referred to the relevant literature for details.

The sign of ρ indicates whether the electrode process of the respective type of compounds has nucleophilic $(+\rho)$ or electrophilic $(-\rho)$ character.

The substituent constant σ is predominantly a function of *inductive effects*, while marked *mesomeric* effects may cause deviations from linearity and require additional terms. Inductive and mesomeric effects shift the enthalpy and entropy terms in parallel or opposite directions. Special *steric effects*[143], especially those by substituents which disturb the coplanarity of the molecule, e.g. ortho-substituents of aromatics and heterocyclics, affect considerably $\Delta\Delta S^0$ and $\Delta\Delta S^{\neq}$, respectively. Under well defined conditions, the deviations from linearity resulting hereby usually induce a special negative shift of $E_{1/2}$; its extent may be quantitatively correlated with the strength of the respective steric effect. In principle, therefore, it must be noted that $\Delta E_{1/2}$ is a function of the inductive, mesomeric, and steric effects, though one of these effects is frequently dominant.

By careful experiments, $\Delta E_{1/2}$-values can be determined at an error limit of ± 10 to ± 2 mV; hence, if for instance $\rho = 0.3$ V, then differences in σ of 0.033 and 0.0066, respectively, are still significant. Correspondingly, for larger values of ρ smaller differences of σ and, therefore, of the substituent effect can be determined.

Although, several thousand organic compounds have been investigated successfully in this way, the extensive possibilities for the determination of $E_{1/2}$ have not yet been applied extensively in this field.

Successful and promising calculations have been made for the correlation of $E_{1/2}$ with quantum-chemical parameters of the electronic states of the molecules. Thus, reversible $E_{1/2}$-values for uniform types of compounds (e.g. alternating hydrocarbons) can be correlated by linear relationships with the MO-energies of the lowest vacant (LVMO, for reduction) and the highest occupied (HOMO, for oxidation) orbitals[144-147]. Moreover, linear relationships exist between $E_{1/2}$ and characteristic spectroscopic frequencies providing information on the difference of *solvation energies* for Red and Ox in the respective solvent from a comparision of the ionization potentials and $E_{1/2}$ (cf. Ref. [148, 149]).

Numerous and various applications of polarography and voltammetry are possible in pure and technical organic chemistry. Many of these have not yet been explored. This applies also to biochemistry, medicine, pharmacy[124, 150], biology, agricultural and food chemistry[151].

5.2 Determination of Important Types of Compounds

For details, reference is made to the relevant monographs, reviews[152-156] partly being published in series[7, 9-11], and to the original publications accessible from bibliographies[1-5] and partially collected in conference reports[6, 8].

Within the potential range accessible with the mercury electrode (about $+0.5$ to -2.8 V (SCE)), the following types of compounds can be studied.

Unsaturated hydrocarbons[152]: For *aliphatics* the presence of conjugated or cumulated double or triple bonds is a prerequisite; for *aromatics*[154] condensed rings or conjugation with double bonds in aliphatic or aromatic side chains must be present.

Carboxylic acids: Apart from the $H^{\oplus}$-reduction after dissociation, carboxylic acids show electrochemical activity if conjugation of double bonds with a carboxylic group exists. Generally, the anions $A^{\ominus}$ are reduced at more negative potentials than the acid HA.

Halogen compounds: The reduction of

$$\text{C-X} \quad ; \quad \text{C=C-X} \quad ; \quad \text{-C(=O)-C-X}$$

or Het-X is facilitated according to the decrease of electronegativity in the series: (F) < Cl < Br < I. Several simple saturated aliphatic halogen compounds are also reducible.

Oxo compounds: Reducible are: aliphatic and aromatic aldehydes and ketones and, sometimes more easily, their derivatives like hydrazones, dinitrophenyl-

[143] *A.J. Bard, K.S.V. Santhanam, J.T. Maloy, J. Phelps, L.O. Wheeler*, Discuss. Faraday Soc. *45*, 167 (1968).

[144] *J. Koutecký*, Z. Phys. Chem. NF *52*, 1 (1967).

[145] *G.J. Hoijtink*, The Electrochemical Reduction of Aromatic Hydrocarbons in *P. Delahay, C.W. Tobias*[10], Vol. VII, p. 221, 1970.

[146] «Relationships between polarographic constants and molecular structure», Ric. Sci. *30*, 177 (1957).

[147] *R. Zahradník, C. Párkányi*, Talanta *12*, 1289 (1965).

[148] *M.E. Peover*, Electrochemistry of Aromatic hydrocarbons and Related Substances in *A.J. Bard*[11], Vol. 2, p. 1 (1967).

[149] *B. Case, N.S. Hush, R. Parsons, M.E. Peover*, J. Electroanal. Chem. Interfacial Electrochem. *10*, 360 (1965).

[150] *E. Hoffmann, J. Volke*, Polarographic Analysis in *C.N. Reilley*, Advances in Analytical Chemistry and Instrumentation, Vol.: *H.W. Nürnberg*, Electroanalytical Techniques and Applications, John Wiley Inc., New York, unpubl.

[151] *P. Nangniot*, La polarographie en agronomie et en biologie, Editions J. Duculot, Gembloux 1970.

[152] *H. Lund*, Talanta *12*, 1065 (1965).

[153] *J. Volke*, Talanta *12*, 1081 (1965).

[154] *J. Tirouflet, E. Laviron*, Talanta *12*, 1105 (1965).

[155] *H. Berg, K. Kramarczyk*, Talanta *12*, 1127 (1965).

[156] *P. Zuman*, Talanta *12*, 1337 (1965).

hydrazones, aldimines and ketimines, semi-carbazones, azomethines. The betainylhydrazone of saturated ketones resulting from reaction with Girard-T reagent (trimethylammoniumaceto-hydrazide chloride) is of particular importance, especially for 17-ketosteroids. In this group, electrochemical activity is exhibited also by the series of tropones and tropolones; terpenealdehydes and ketones; aldoses and ketoses; diketones and endiols; quinones including their oximes, many quinones exhibiting reversible electrode processes; and steroids where the CO group is conjugated to a C=C bond of the steroid framework, viz. Δ^4–3–; Δ^1–3–; $\Delta^{1,4}$–3–; and Δ^9–12–ketosteroids, to which numerous important hormones belong.

Peroxides
Nitrogen compounds: Aliphatic and aromatic nitro compounds; nitroso compounds (especially aromatic); certain hydroxylamines; oximes and hydroxamic acids; amine oxides; azo and azoxy compounds.

Sulfur compounds (including certain *selenic* and *telluric compounds*): Thiols like mercaptans, thio acids, dithiocarbamic acids, thiocarbamides, and xanthates can be oxidized. Disulfides, diselenides, ditellurides, and thioethers; sulfonium salts; sulfoxides; sulfinic acids; sulfones and sulfonic acids if conjugated with a double bond, show reducibility. Thiobenzophenones and thiobenzamides; aromatic thiocyanates; metal complexes of thiols, disulfides, and diselenides.

Oxygen heterocyclics: Flavones and isoflavones; anthocyanes; coumarines; coumaranes and chromanes; xanthones; isobenzpyrylium salts; penicillic acid, Angelica lactones, digitoxins.

Nitrogen heterocyclics: Indole derivatives; pyridines; quinolines and isoquinolines; acridines; pyrazines; quinoxalines; phenazines; carbamide compounds like pyrimidines, purines, barbituric acid derivatives; pterines and pteridines; isalloxazines; riboflavines; tetrazolium salts.

Oxygen-nitrogen heterocyclics: Oxazines, gallocyanines, actinomycines.

Sulfur-nitrogen heterocyclics: Benzthiazoles and thiadiazoles including penicillin and its derivatives; saccharin; methylene blue.

Organometallics:

Here, the group $\diagdown \!\!\! C\!-\!x$

with X = Hg, Pb, Sn, Sb, As, P, Se, Te, should be listed in particular. For the generation of optically active tertiary phosphines and arsines, the cathodic elimination of ligands from *quaternary phosphonium* and *arsonium salts* has recently become of interest[157].

Classes of substances of interest in biochemistry and physiology: Many applications have been made and further applications are to be expected in this field[158]. The polarographic determination of all important antibiotics[159], of most of the vitamins, of hormones with a

ketosteroid framework, of adrenalins, of heart glycosides as well as of nucleosides and nucleotides derived from purines, pyrimidines, and flavines[160] should be mentioned. Polarographic techniques also offer very interesting aspects in the investigation of nucleic acids[65, 161, 162] because of the electrochemical activity of denatured DNA and RNA and their components. In biochemistry and medicine, the catalytic hydrogen waves of alkaloids[43] and of blood serum proteins if complexed with $Co^{2\oplus}$ ions[163] are of diagnostic importance. By making use of the oxygen reduction-wave, the metabolism of the lens and the aqueous humour of the eye[164–166] as well as the respiration of microorganisms and corresponding enzymes can be investigated.

Upon the application of *solid electrode materials*[25] (metals like platinum, gold etc. as well as graphite, carbon paste, and semiconductor electrodes), the accessible range of potentials as compared with that on mercury is considerably extended to more positive values, in part by the formation of electron-conducting surface layers preventing further anodic dissolution of the metal. Even in aqueous solutions, potentials more positive than +1 V (SCE) are accesible; in non aqueous media +2 V (SCE). On the other hand, the range towards negative potential is more restricted in aqueous solution (especially because of the lower hydrogen overvoltage) as compared with mercury.

It is obvious, therefore, that the number of reducible compounds is more restricted at such electrodes while the investigation of the oxidation[167, 168] of numerous compounds, concealed from observation at mercury, becomes possible. Almost all types of compounds of which the investigation at mercury has been restricted to their reduction, can be oxidized at correspondingly positive potentials: unsaturated hydrocarbons (olefins, aromatics) halogen compounds, oxygen compounds (aldehydes, ketones, peroxides, hy-

[157] *L. Horner, J. Haufe*, Chem. Ber. *101*, 2903 (1968); J. Electroanal. Chem. Interfacial Electrochem. *20*, 245 (1969).

[158] *H. Berg*, Elektrochemische Methoden und Prinzipien in der Molekularbiologie, Akademie-Verlag, Berlin 1966.

[159] *H. Berg*, Ric. Sci. *27*, 184 (1957).

[160] *B. Janik, P.J. Elving*, Chem. Rev. *68*, 295 (1968).

[161] *E. Paleček*, Polarographic Techniques in Nucleic Acid Research in *J.N. Davidson, W.E. Cohn*, Progress in Nucleic Acid Research and Molecular Biology, Vol. 9, p. 31, Academic Press, New York 1969.

[162] *H. Berg, K. Eckardt*, Z. Naturforsch. *25b*, 362 (1970).

[163] *R. Brdička*, Z. Phys. Chem. 1958, Sonderheft Internat. Polarograph. Koll., Dresden 1957.
R. Brdička, M. Březina, V. Kalous, Talanta *12*, 1149 (1965).

[164] *W. Hans, O. Hockwin, O. Kleifeld*, Graefes Arch. Ophthalm. *157*, 72; 140 (1955).

[165] *O. Hockwin, H. Bergeder*, Naturwissenschaften *18*, 491 (1957), Graefes Arch. Ophthalm. *160*, 1 (1958).

[166] *O. Hockwin*, Z. Anal. Chem. *216*, 255 (1966).

[167] *N.L. Weinberg, H.R. Weinberg*, Chem. Rev. *68*, 449 (1968).

[168] *K. Sasaki, W.J. Newby*, J. Electroanal. Chem. Interfacial Electrochem. *20*, 137 (1969).

droquinones), nitrogen compounds (nitroso compounds, hydroxylamines), heterocyclics, metalorganic compounds.

Naturally, oxidation not only yields other products but also the oxidation mechanism of the compound differs from its mechanism reduction. These mechanisms, which depend not only on the type of compound but also on the conditions to a great extent, cannot be discussed here in detail; besides hydrogenation for reduction and dehydrogenation for oxidation, these mechanisms comprise elimination, dimerization, dismutation, substitution reactions etc.

Some types of compounds, the reduction of which cannot be observed at any electrode, can be oxidized at solid electrodes, e.g. saturated aliphatics in the group of hydrocarbons; simple carboxylic

acids in the group of oxygen compounds (Kolbe-electrolysis), esters, alcohols and phenols (including alkoxy phenols); amines, hydrazines, amides, amino acids, and lactams in the group of nitrogen compounds. The oxidation of metal porphyrins may be mentioned as an example of metal-organic compounds, the oxidation of analogs of the hydroforms of coenzymes (nicotinamide-adenine-dinucleotide)[169] as an example of compounds of interest in biology.

General reference is again made to the respective quoted review articles for more detailed information.

[169] *W.J. Blaedel, R.G. Haas*, Anal. Chem. *42*, 918 (1970).

8.2 Organic Applications of Ion Resonance Spectroscopy

Fred Kaplan and *Janet C. Haartz*
Department of Chemistry University of Cincinnati
Cincinnati, Ohio 45221 U.S.A.

Ion cyclotron resonance (ICR) spectroscopy represents a valuable addition to the spectroscopic tools available for the study of the gas phase ion chemistry. The technique can be used for detailed investigation of either positive or negative ion-molecule reactions at low kinetic energies and relatively low pressures (10^{-6} to 10^{-4} torr). Reaction pathways may be elucidated even in systems where more than one set of reaction products can be produced. Possible studies of gaseous ion properties include structure, unimolecular decomposition, autoionization, thermodynamic stability, electron affinity and collision induced fragmentation. More importantly, the chemical reactions of gaseous ions with neutral molecules can be explored free of the usual interference of solvation effects. In addition, knowledge of energy levels within molecules may be obtained from ICR studies utilizing scavenger techniques to detect interaction of electrons with molecules[1-3].

1 Basic Principles and Instrumentation
The principles and instrumentation involved in ICR spectroscopy have been described in detail[4,5]. The path of a charged particle in a uniform mag-

netic field, B, is a circular orbit in a plane normal to B. The angular frequency of the circular component of motion perpendicular to B is called the natural cyclotron frequency, ω_c:

$$\omega_c = v/r = eB/m \text{ (radians sec}^{-1}) \qquad (1)$$

where e is the charge on the ion, m its mass, v its velocity and r the radius of its path.

Although not all ions of identical e/m will have the same velocity (and therefore the same orbital radius), the ratios, v/r, will still characterize a sharp cyclotron frequency, ω_c, which depends only on e, B, and m. If an alternating electric field E_1 (t) of frequency ω_1 is applied normal to the magnetic field and if $\omega_1 = \omega_c$, the ions will absorb energy from E_1 (t) and will be accelerated to greater velocities and orbital radii. This absorption of energy can be readily detected by using a marginal oscillator and phase sensitive detector similar to those used in NMR instrumentation.

[1] *D. Ridge, J.L. Beauchamp*, J. Chem. Phys. *51*, 470 (1969).

[2] *R. O'Malley, K. Jennings*, J. Mass Spectrosc. Ion Phys. 2. App. 1 (1969).

[3] *P. Kreimler, S.W. Buttrill jr.*, J. Am. Chem. Soc. *92*, 1123 (1970).

[4] *J.D. Baldeschwieler*, Science *159*, 263 (1968). S.a. Ref. [29,30].

[5] *J.L. Beauchamp*, Ph. D. Thesis Harvard University, 1968.

The ICR 'single resonance' experiment is normally performed by sweeping the magnetic field while observing the power absorption of the oscillator operating at a fixed radio frequency. Thus, a plot of the power absorption from the marginal oscillator versus the magnetic field B (and thus m/e) is essentially a mass spectrum of the ions present in the ICR cell. However, in contrast to the method of detection utilized in traditional mass spectrometry, the ions detected in the ICR spectrometer need not arrive at a specific point to be detected. (This is an important feature since it enables concurrent observation of both reactant and product ions.) At a fixed frequency of, $\omega_1/2\pi = 153.57$ kHz and a magnetic field sweep of 100 to 15,000 gauss, ions of mass 1 to 150 may be observed. The ICR spectrometer thus utilizes the same principles as the omegatron mass spectrometer described by Hipple, Sommer and Thomas[6].

The ICR cell is divided into a source and an analyzer region. (See Fig. 1). Primary ions are generally produced by electron impact in the source region and are drifted away from their point of origin by application of a static electric field normal to the magnetic field. The drift velocities imparted to ions (~ 500 cm/sec) are generally less than thermal velocities and, therefore, the ions have essentially unperturbed kinetic energies if the ionizing energy of the electron beam is only a few electron volts above the ionization potential of the molecule.

Trapping voltages are applied to the side plates of the cell to constrain the motion of the ions in the direction of the magnetic field; selection of positive or negative ions is possible by applying the appropriate polarity for the trapping voltages. Ions are collected by an electrode at the end of the cell and this gives a measure of the ion current.

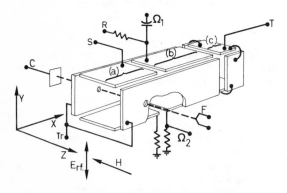

Fig. 1. Diagram of ICR cell. (a) = Ion source; (b) = Ion analyzer; (c) = Ion collector; F = Filament; C = Collector; Tr = Trapping 0.1 V; T = Total ion current; w_1 = Observing oscillator; w_2 = Irradiating oscillator; R = Region drift and modulation 0.1 V.

The ICR spectrometer used in the single resonance mode of operation does not have any significant advantage over other mass spectrometers (See Chapter 6) for obtaining information about ion-molecule reactions. In fact, it is more limited with respect to mass range and resolution than most mass spectrometers. However, an additional mode of operation, that of 'double resonance' makes this instrument a very powerful and convenient tool in the study of ion-molecule reactions. At relatively higher pressures (10^{-6} to 10^{-4} torr) there is ample opportunity for ion-neutral molecule collisions which may result in reaction. For the process:

$$A^{\oplus} + M \longrightarrow B^{\oplus} + N \qquad (2)$$

if the rate constant, k_i, is energy dependent (i.e. if $dk_i^{\circ}/dE_A \neq 0$), then changing the energy of $A^{\oplus}$ will result in a change in the number of ions $B^{\oplus}$ in the cell. In a typical double resonance experiment, the change in power absorption from ω_1 due to $B^{\oplus}$ as one varies the kinetic energy of $A^{\oplus}$ is monitored. The 'heating' of $A^{\oplus}$ is achieved by irradiating with a separate HF electric field E_2 (t) at frequency ω_2 which can be tuned so that $\omega_2 = \omega_c$ $_{(A^{\oplus})}$. By sweeping through the cyclotron frequencies of all ions present while observing the intensity of the product ion, $B^{\oplus}$, only those ions which give rise to $B^{\oplus}$ will cause a change in the intensity of $B^{\oplus}$. By observing whether the intensity of $B^{\oplus}$ increases or decreases when $A^{\oplus}$ is heated, one can draw inferences about exothermicity or endothermicity of the reaction. Thus the term 'double resonance' arises because two different frequencies are used to bring ions of two different m/e values into resonance simultaneously. The term 'single resonance' is used to describe the experiment when only one HF signal is applied and ions of only one m/e are brought into resonance at a given field strength. The relative ease of the ICR experiments has made this instrument increasingly popular for the study of ionmolecule reactions.

Modulation of the HF power absorption and amplification via phase sensitive detection results in quite high sensitivity in ICR spectrometers. The common mode of operation for single resonance involves modulation of the magnetic field during the field sweep and results in the derivative shape signals shown in the spectrum of $(CH_3)_3CH$ (Fig. 2).

[6] J. Hipple. H. Sommer. H. Thomas. Phys. Rev. 76. 1877 (1949).

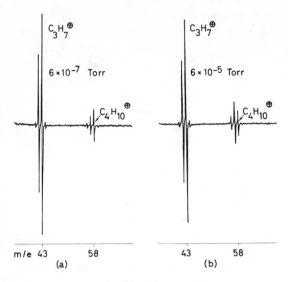

Fig. 2. Single resonance spectra of $(CH_3)_3CH$ at (a) low and (b) high pressure using field modulation of 15 eV. Note the changes in relative intensities of peaks with change in pressure.

In the double resonance experiment, *modulation* of ω_2 is the usual and most convenient method. Unfortunately, due to artifacts arising from the instrumentation, the sign of dk/dE_{ion} reported from observation of the double resonance signal may occasionally be in error[7]. Accordingly, for thermochemical studies, it is necessary that conclusions not be based on an isolated double resonance observation but supplemented by internal checks on the system, including theoretical calculations, pressure-abundance studies and the use of other modulation modes. Fig. 3 shows a double resonance spectrum.

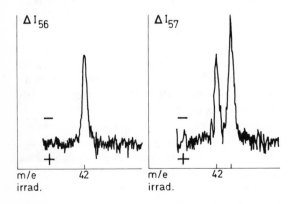

Fig. 3. Pulsed double resonance spectra of m/e = 56, m/e = 57 (observed arising from $(CH_3)_3CH$ at P = 6×10^{-5} torr indicating (a) $42^{\oplus} \to 56^{\oplus}$ (ΔI_{56}) and (b) $42^{\oplus} \to 57^{\oplus}$; $43^{\oplus} \to 57^{\oplus}$ (ΔI_{57}).

The basic instrument described above is that manufactured by Varian[8] the V-5900 ICR spectrometer. Various adaptations have been devised to extend and enhance the capabilities of the instrument[9–12]. These include several different modifications of the mode of ion production, such as electron energy modulation, pulsed grid modulation and photoionization, as well as alternate double resonance techniques.

2 Applications

2.1 Ion Thermochemistry.

The relationships between structure and reactivity have long been a matter of considerable importance to chemists. *'Acidity* and *basicity'* of molecules are often the *crux* of these correlations. ICR studies promise to be very fruitful in determining the inherent acidities or basicities and related properties of molecules, free of solvation effects.

The proton affinity (PA) of a molecule is defined as the negative of the enthalpy change for the reaction:

$$M + H^{\oplus} \to MH^{\oplus} \qquad (3)$$

In thermochemical terms,

$$PA(M) = \Delta H_f(M) + \Delta H_f(H^{\oplus}) - \Delta H_f(MH^{\oplus}) \quad (4)$$

Beauchamp[5] developed an ICR technique for determining the limits of PA of a molecule from competition experiments for ion-molecule reactions:

$$M + AH^{\oplus} \rightleftarrows A + MH^{\oplus} \qquad (5)$$

$$M + AH^{\oplus} \;\overrightarrow{\tiny{\not\leftarrow}}\; + A + MH^{\oplus} \qquad (5a)$$

$$B + MH^{\oplus} \;\overrightarrow{\tiny{\not\leftarrow}}\; BH^{\oplus} + M \qquad (5b)$$

If the reactions (5a) and (5b) are observed, they must be exothermic. Therefore,

$$\Delta H_f(AH^{\oplus} - A) > \Delta H_f(MH^{\oplus} - M) > \Delta H_f(BH^{\oplus} - B).$$

If M is the substance under investigation and if A and B are reagents with a known

$$\Delta H_f(AH^{\oplus} - A) \text{ and } \Delta H_f(BH^{\oplus} - B)$$

the limitation for $\Delta H_f (MH^{\oplus} - M)$ can be determined. The determination of this value allows a calculation of the proton affinity, thus giving the basicity of M.

[7] *G.C. Goode, A.J. Ferrer-Correia, K.R. Jennings*, J. Mass. Spectrosc. Ion Phys. *5*, 229 (1970).

[8] Varian Associates, Palo Alto, Calif., U.S.A.
[9] *J.M.S. Henis. W. Frasure.* Rev. Sci. Instrum. *39.* 1772 (1968).
[10] *J.L. Beauchamp. J.T. Armstrong.* Rev. Sci. Instrum. *40.* 123 (1969).
[11] *L.R. Anders.* J. Phys. Chem. *73.* 469 (1969).
[12] *R.T. McIvor*, Rev. Sci. Instrum. *41*, 126, 555 (1970).

Kaplan, Cross, and Prinstein[13] utilized these techniques in the hope of determining the structure of the *norbornyl cation*. Center of a long standing controversy, the norbornyl cation. (Fig. 1) may be represented as two rapidly equilibrating classical ions or as an intermediate, the so-called 'nonclassical carbonium ion', which is lower in energy than either classical ion.

1

ICR studies of proton affinities enabled the calculation of heats of formation of a set of bicyclic cations, including the olefins and ketones. The results show the bicyclo [2.2.1] heptyl cation compared to its hydrocarbon to be approximately 6 kcal/mole more stable than the bicyclo [2.2.2] octyl cation compared to its hydrocarbon. The corresponding olefin systems show an even greater difference, 11 kcal/mole.

The results are interpreted in terms of changes in angle strain, nonbonded interactions, torsional strain and charge delocalization in going from the hydrocarbon to the ion; or as supporting the non-classical structure. By contrast, the 'classical' ion structure should exhibit greater stability for the bicyclo [2.2.2]-octyl cation. As a check against any unforeseen effects, the 2-bicyclo[2.2.1] heptanone and 2-bicyclo[2.2.2] octanone compounds were also examined. The experiment showed only small differences in the stabilities of the two compounds as would be expected since the charge is localized on the oxygen atom.

Applying similar techniques to negative ions, Brauman and Blair[14] have studied the relative gas phase acidities of a number of organic compounds, including a series of *alcohols*. Consider the reaction:

$$HA \rightarrow A^{\ominus} + H^{\oplus} \qquad (6)$$

wherein the enthalpy of the reaction may be calculated from the relationship:

$$\Delta H^0 = PA(A^{\ominus}) = D^0(HA) - EA(A\cdot) + IP(H\cdot) \qquad (7)$$

The lack of data for electron affinities of radicals makes it difficult to calculate ΔH^0 for the gas phase reaction. The relative acidities determined by ICR

experiments for alcohols in the gas phase was $OH>(CH_3)_3CCH_2OH>n\text{-}C_5H_{11}OH \approx t\text{-}BuOH \approx n\text{-}C_4H_9OH > (CH_3)_2CHOH > C_2H_5OH > CH_3OH>H_2O$. The gas phase order is the reverse not only of the solution order but also contradicts the widely accepted belief that alkyl groups are electron-releasing and, hence, should be destabilizing toward negative charge. That alkyl groups can stabilize anions in simple saturated systems in the gas phase is evident from the data. Brauman and Blair conclude that the solution order is probably an artifact and does not represent any intrinsic property of the molecules but rather results from steric hindrance to solvation in the larger alkoxides. They postulate that the stabilization arises from the polarization of the alkyl group.

Since the O-H bond strength is approximately the same for all these alcohols, it would be of considerable interest to know the *electron affinities of the alkoxyl radicals*. The alcohol acidity order is such that the electron affinity in general must increase with the size of the alkyl group. Although good estimates of these electron affinities are not now available, Brauman and Smyth[15] recently utilized photoelectron detachment techniques coupled with the ICR spectrometer to obtain electron affinities for OH and SH. Extension of the technique to the alkoxyl radicals would alleviate the difficulties of calculating ΔH^0 for acid dissociation in the gas phase.

2.2 Ion Structure

Mass spectral fragmentation processes have been studied extensively in the last ten years and much information is available concerning the details of the occurrence of these processes. However, assignment of structure to an observed ion of known composition is a difficult and recurring problem and has been attacked by a variety of techniques including deuterium labeling, appearance potential measurements, and molecular orbital calculations. Studies of ion-molecule reactions by ICR have shown that structural differences in isomeric ions could be detected by observing the differences in the reactivity of various fragment ions with neutral molecules.

2.2.1 McLafferty Rearrangement

One mass spectral fragmentation process which has been studied extensively, the *McLafferty rearrangement,* occurs in many compounds in-

[13] *F. Kaplan. P. Cross. R. Prinstein.* J. Am. Chem. Soc. *92.* 1445 (1970).

[14] *J.I. Brauman. L.K. Blair.* J. Am. Chem. Soc. *92.* 5986 (1970), *90.* 5636 (1968), *90.* 6551 (1968), *91.* 2126 (1969).

[15] *J.I. Brauman. K. Smyth.* J. Am. Chem. Soc. *91.* 7778 (1969).

cluding simple *ketones, esters, carboxylic acids* and *amides*. In this fragmentation, transfer of a γ-hydrogen atom to the carbonyl group is followed by elimination of an olefin. For example, in 2-hexanone, the following reaction occurs:

2

The structure of the product ion, $C_3H_6O^\oplus$, remained in question until recently and included three possibilities: the keto ion (**a**), the enol ion (**b**), and the symmetrical oxonium ion (**c**), proposed as the product for the 'double McLafferty' rearrangement.

a   b   c

Diekman *et al*[16] have utilized ICR spectroscopy to study the ion-molecule reactions of the keto species (**a**), generated from *acetone;* the enol species (**b**), generated from *2-hexanone* and *1-methylcyclobutanol;* and the 'double McLafferty' species (**c**), generated from *5-nonanone*. In all reactions studied, the enol ion from 2-hexanone and that from 1-methylcyclobutanol behaved identically, an indication of equivalent structure. However, the authors found seven ion-molecule reactions which served to distinguish between the keto and enol isomers and attributed these 'distinguishing reactions' to structural differences. The ion resulting from the 'double McLafferty' rearrangement was shown to react identically to the enol ion in all systems studied and the authors concluded that the enol ion and the 'double McLafferty' ions probably have identical structures.

This conclusion is in discord with the earlier analysis and interpretation[17] of metastable peaks observed in conventional mass spectrometry which led to the proposal of the symmetrical double McLafferty oxonium ion. If the enol ion and the product ion from the double McLafferty rearrangement are not the same, then either this is the first instance of ions of different structure exhibiting identical reactivity or, since the lifetimes in the ICR cell are significantly longer than in a conventional mass spectrometer, the symmetrical double McLafferty ion exists, but isomerizes prior to undergoing ion-molecule reactions.

2.2.2 Deuterium Labelling

The usefulness of *deuterium labeling* coupled with ICDR is well illustrated by investigations[18] in which it was established that the enol ion is a *direct* product of the double McLafferty rearrangement and that at least a substantial part of the double McLafferty ion must consist of ions in the enolic form. From *4-nonanone-1,1,1,-d_3*, the following three possible $C_3H_6O^\oplus$ ions could arise, depending on the mechanism of the rearrangement:

d   e   f

Since independent studies had established that protonated ketones, which serve as gas phase acids, transfer only H atoms attached directly to oxygen, the three possible intermediates **d, e,** and **f** could serve as a source of both H and D in a 1:1 ratio (**d**), deuterium only (**e**) or H only (**f**). ICDR experiments indicated that proton transfer to a base, the parent ketone, contained a 4:1 ratio of H to D. That this result was not due to a kinetic deuterium isotope effect was established through the use of 4-nonanone-7,7-D_2 where the H/D ratio possibilities are reversed and the D/H ratio is now 4:1 for proton transfer. These data strongly support **f** as the product ion from the double McLafferty rearrangement.

2.3 Mechanisms of Reactions

Ions in the gaseous phase may or may not react by the same mechanism as in solutions. Nevertheless, knowledge of the reaction pathway in the gas phase can be of help in postulating mechanisms for solution reactions. Properly designed ICR studies make it possible to determine reasonable reaction pathways for many gas phase reactions. Nucleophilic[19] and electrophilic substitutions[20], which proceed in analogy to those in solution, also belong to such reactions.

[16] *J. Diekman. J. MacLeod. C. Djerassi. J. Baldeschwieler*. J. Am. Chem. Soc. *89*. 5953 (1967).

[17] *F.W. McLafferty. W.T. Pike*. J. Am. Chem. Soc. *89*. 5953 (1967).

[18] *G. Eadon. J. Diekman. C. Djerassi*. J. Am. Chem. Soc. *91*. 3486 (1969), *92*. 6205 (1970).

[19] *D. Holtz, J.L. Beauchamp, S. Woodgate*, J. Am. Chem. Soc. *92*, 7484 (1970).

[20] *S.A. Benezra, M.K. Hoffman* and *M.M. Bursey*, J. Am. Chem. Soc., *92*, 7501 (1920).

2.3.1 Carbon Compounds

Kaplan *et al.*[19] have investigated the ion-molecule reactions of *hexadeuterioacetone* (m/e 64, d) which produce the deuterated molecular ion (m/e 66) and the deuterated dimer (m/e 130) as well as an ion at m/e 110 (**e**)[21]. The proposed mechanism for the reaction producing ion **e** involves condensation of ionized acetone with a molecule of neutral acetone and subsequent loss of a methyl radical:

a m/e 58

b   **c** m/e 101

(8)

Corroboration of this mechanism and structure was obtained from the double resonance studies of the ion-molecule reactions of mixtures of deuterioacetone and acetone:

d m/e 64

e m/e 110   **f** m/e 104

(9)

g m/e 107   **c**

(10)

Double resonance results indicate a coupling between ions **d** and **e** as well as between **c** and **g**. Significantly, there is no coupling between **c** and **e** nor between **e** and **g**. If the postulated intermediate **b** is correct, then **a** should give rise only to (**c**) and (**g**). Further, double resonance experiments should indicate coupling between **e** and **f** as well as between **c** and **g** but not between **c** and **e** nor between **e** and **g**. The double resonance experiments support these conditions and lead to the conclusion that **b** is the reaction intermediate.

Bursey *et al.*[20] have recently investigated acetylation by using *butanedione ions the acylation* of a variety of compounds, including ketones, alcohols, ethers and esters[22]. It was shown that the acetyl ion, $CH_3CO^{\oplus}$ (m/e 43), was not the acylating agent but rather that the precursors of the acylated products were the molecular ion m/e 86 **h** and the m/e 129 **i** ion formed in the reaction:

h m/e 86   +

i m/e 129   + $CH_3CO\bullet$ (11)

Further experiments[23] utilizing *deuterium labelling* show that the acetyl group that is added to the neutral molecule is the group that is lost when **i** reacts to acetylate some other, neutral species. That is, m/e 138 (**j**) and m/e 132 (**k**) transfer only $CD_3CO^{\oplus}$, while m/e 129 (**i**) and 135 (**l**) transfer $CH_3CO^{\oplus}$:

j m/e 138   **k** m/e 132

l m/e 135   +   ROR   ⟶

(12)

---

[21] *F. Kaplan. J.L. Beauchamp. J. Diekman. C. Djerassi.* unpubl.

[22] *M. Bursey. T. Elwood. M. Hoffman. T. Lehman. J. Tesarek.* Anal. Chem. *42.* 1370 (1970).

[23] *M. Hoffman, T. Elwood, T. Lehman, M. Bursey,* Tetrahedron Lett. 4021 (1970).

[24] *R.C. Dunbar,* J. Am. Chem. Soc. *90,* 5676 (1968).

2.3.2 Boron Compounds

Many *boron*-containing negative ions are isoelectronic with organic compounds and it should be possible to utilize these compounds as charge-labeled prototypes for the exploration of organic reactions. Dunbar[24] has investigated some negative ion-molecule reactions of *diborane* and found that $BH_4^\ominus$ (m/e 14, 15) and $B_2H_7^\ominus$ (m/e 27, 28, 29) predominate in the gas phase at higher pressures. The reaction:

$$BH_4^\ominus + B_2H_6 \rightarrow B_2H_7^\ominus + BH_3 \qquad (13)$$

has been studied by ICDR techniques and was postulated to proceed through the short-lived intermediate $B_3H_{10}^\ominus$. A priori, one could not say whether a hydride transfer is involved with a concurrent opening of one of the bridged bonds of B_2H_6 or whether a more intimate intermediate might be involved in which case several possibilities of boron scrambling would occur. Analysis of the boron hydride ICR data was complicated due to the possibilities of overlapping double resonance spectra resulting from the presence of the ^{10}B and ^{11}B isotopes. A method for the reduction of such data was developed in this study and should be useful in cases where similar isotopic complications exist. The result in the present case indicated complete scrambling of the boron atoms in an intermediate $B_3H_{10}^\ominus$ which is too short-lived to be detected in an ICR spectrometer:

$$\rightleftharpoons \quad etc. \qquad (14)$$

2.4 Reaction Rates

The use of ICR spectrometry for the measurement of *reaction rates* and *rate constants* holds great promise for application to organic reactions in the gas phase. At this writing, the techniques have been applied mainly to simple systems, including the reactions of diatomic molecules such as N_2 and H_2[25]. However, Buttrill[26] has developed a method utilizing linewidths to calculate ion-molecule reaction rate constants and applied it to several ion-molecule reactions, including the well-studied system[27]:

$$CH_4^\oplus + CH_4 \rightarrow CH_5^\oplus + CH_3 \qquad (15)$$

The rate constants derived agreed well with values obtained using other techniques.

Single resonance signal intensities[28] (i.e. peak heights) have been used in a more general method for obtaining ion-molecule reaction rate constants in systems which include primary, secondary and stable tertiary ions. Applied to the methyl fluoride system, the method accurately predicts relative single resonance signal intensities from the calculated rate constants for the reactions:

$$CH_3F^\oplus + CH_3F \rightarrow CH_3FH^\oplus + CH_2F\cdot \quad (16)$$

$$CH_3F^\oplus + CH_3F \rightarrow C_2H_4F^\oplus + HF + H \quad (17)$$

$$CH_3FH^\oplus + CH_3F \rightarrow C_2H_6F^\oplus + HF \quad (18)$$

Clearly, more work must be done in this area before it is possible to use it for correlations and predictions or comparisons of solution and gas-phase reactions.

The full potential of ICR spectrometry is yet to be realized. Although it is still a relatively new technique, its application to a wide range of chemical problems in the gas phase has already shown its versatility. Future developments will include not only refinements in the instrumentation, but also application to a wider range of problems. The high sensitivity of the instrument and ease of conversion from observation of positive to observation of negative ions make it an especially promising technique for the field of negative ion studies, in contrast to traditional mass spectrometry.

[25] *M.T. Bowers. D.D. Elleman. J. King jr..* J. Chem. Phys. *50.* 1840, 4787 (1969).

[26] *S.E. Buttrill jr..* J. Chem. Phys. *50.* 4125 (1969).

[27] Ref. [29], p. 306 ff.

[28] *A. Marshall. S.E. Buttrill jr..* J. Chem. Phys. *52.* 2752 (1970).

[29] *E.W. McDaniel, V. Cermak, A. Dalgarno, E.E. Ferguson, L. Friedman,* 'Ion-Molecule Reactions', Interscience Publishing Co., New York, 1970.

[30] *J.M.S. Henis,* Ion Cyclotron Resonance Spectroscopy, Wiley Intersci. Publ., New York, 1971.

8.3 Determination of Tautomeric Equilibria

Heinrich Wamhoff
Organisch-Chemisches Institut der Universität,
D-53 Bonn

Tautomerism is defined as the phenomenon in which isomeric compounds can be interconverted by rapid and reversible prototropic rearrangements. The proton migrates between two or more basic and conjugated structures of an organic molecule. The various structural isomers which result are called tautomers.

Tautomerism is most marked with carbonyl compounds, and among these the β-dicarbonyl compounds have been investigated most intensively. Recently, however, interest has been concentrated on heterocyclic compounds, many of which show tautomeric behaviour. In particular, numerous natural products such as alkaloids and nucleic acids contain groups potentially capable of tautomerism. It has been suggested that these groups have great biological importance, e.g. in the proposed mechanism of spontaneous mutation of genes[1, 2].

The classical methods (still used to some extent) available for quantitative determination of tautomerism are supplemented nowadays by a range of extremely powerful physical techniques.

Chemical determination (8.3:1–8.3:2) of tautomeric equilibria was used shortly after the discovery and formulation of the principle of tautomerism. Comparison of these chemical methods with the *physical techniques* which have recently been applied successfully (8.3:3–8.3:4) leads to the conclusion that the chemical methods must be regarded very critically, since they do not always provide unambiguous results (see Section 8.3:1).

For all chemical methods, the rate of interconversion of the individual tautomers must be very slow compared with the rate of the intended chemical reaction with one of the tautomeric forms.

The chemical methods for determination of tautomeric equilibria are of two classes:

Determination of the reactivity of a specific structural group present in only one of the tautomeric forms.
Indirect determination of a keto-enol equilibrium from the nature of the reaction products.

1 Reactivity of a Specific Structural Group Present in only one of the Tautomeric Forms

(e.g. $C=O_{keto}$, $C=C_{enol}$, $-OH_{enol}$).

In this class, the most intensely investigated method is the K.H. Meyer titration technique[3–5]. Here, the double bond of the enol form present in the equilibrium mixture adds bromine instantaneously on titration with an alcoholic solution of bromine. Both direct titration and a modified back-titration[6–9] are widely employed.

The enthalpies of the keto-enol rearrangements of substituted *ethylacetoacetates* and *1,3-diketones* have been determined both by K.H. Meyer titrations[10–12], and with the physical methods[13, 14] described in Section 8.3:2. Values of 1.0–2.8 kcal/mole were found. From this rather small value, it can be concluded that the keto-enol equilibrium is easily shifted – as is often observed when chemical methods are employed.

Titration methods do not give reliable results in every case[15, 16]; for example, with *fluoro-β-keto-esters*, the results from bromine titration and from physical methods are significantly different[9, 17, 18].

The bromine titration method has also been applied to several heterocyclic compounds which potentially can exist in tautomeric forms[19]. Kohler and Blatt[20] concluded from titration results that *3,4-diphenyl-5-isoxazolone* in ethanol exists predominantly (i.e. almost 90%) as the OH-form:

(1)

[3] *K.H. Meyer*, Liebigs Ann. Chem. *380*, 212 (1911).
[4] Ref. [124], p. 388 ff.
[5] *H. Henecka*, Chemie der β-Dicarbonylverbindungen, p. 4, 41, Springer Verlag Berlin 1950.
[6] *K.H. Meyer*, Ber. *45*, 2843 (1912); *47*, 826 (1914).
[7] *G. Schwarzenbach, C. Wittwer*, Helv. Chim. Acta *30*, 657, 669 (1947).
[8] *A.B. Ness, S.M. McElvain*, J. Am. Chem. Soc. *60*, 2213 (1938).
[9] *J.D. Park, H.A. Brown, J.R. Lacher*, J. Am. Chem. Soc. *75*, 4753 (1953).
[10] *K.H. Meyer, P. Kappelmeyer*, Ber. *45*, 2852 (1912).
[11] *J.B. Conant, A.F. Thompson jr.*, J. Am. Chem. Soc. *54*, 4039 (1932).
[12] *G. Briegleb, W. Strohmeier, I. Höhne*, Ber. Bunsenges. Phys. Chem. *56*, 240 (1952).

[1] *J.D. Watson, F.H.C. Crick*, Cold Spring Harbor Symposia Quant. Biol. *18*, 123 (1953);
M.H.F. Wilkins, Angew. Chem. *75*, 429 (1963).
[2] *E. Freese*, J. Mol. Biol. *1*, 87 (1959).

However, this method fails with *3-phenyl-5-isox-azolone*, since the bromine reacts in an additional step with the 5-isoxazolone nucleus[21] and an apparent value of 190% is obtained.

2 Indirect Determination of a Keto-Enol Equilibrium from the Nature of the Reaction Products

After a reaction has been carried out, if the structure of the final product is compared with those of the possible tautomeric forms of the starting material, it may be possible to reach conclusions about the position of equilibrium. As mentioned in Section 8.3:1, however, this method must be used very critically, and the rate of the reaction used for the determination must be taken into consideration.

Arndt[22] attempted to investigate the lactam-lactim tautomerism of amides and heterocyclic compounds, using methylation with diazomethane:

$$\begin{array}{c} \underset{\substack{|\\ \text{NH}\\|\\ \text{C}=\text{O}\\|}}{} \rightleftharpoons \underset{\substack{|\\ \text{N}\\\|\\ \text{C}-\text{OH}\\|}}{} \end{array} \qquad (2)$$

$$(3)$$

Subsequently, however, Gompper[23, 24] showed that the transition state (i.e. whether S_N1- or S_N2 type reaction is preferred) is of decisive importance, as also is the electron distribution of the mesomeric anion, which dictates the nucleophilicity (electron density), and thus determines the final position of methylation.

On consideration of the tautomeric 'CH', 'OH' and 'NH' forms, which are normally in equilibrium in the case of the 5-isoxazolones[25], one might

expect *dihydrofuro[2,3-d]isoxazoles* to result from intramolecular condensation of *4(2-hydroxyethyl)-5-isoxazolones:*

However, the product obtained is *7-methyl 5-oxa-6-azaspiro[2,4]hept-6-en-4-one*[26], which is formed from the 'NH' form (this being the predominant tautomer in polar solvents). Substituted *5-pyrazolones* and *glutaconimides* react in the same manner to form spiro-fused heterocyclic products[26–28].

3 Basicity and Ionization Constants

3.1 Theory

Ionization constants of acids and bases are good indicators of the tendency for proton abstraction or addition in solution. This method is generally

[13] *L.W. Reeves,* Can. J. Chem. *35,* 1351 (1957).

[14] *G. Allen, R.A. Dwek,* J. Chem. Soc. B 161 (1966).

[15] *H. Suhr,* Anwendungen der kernmagnetischen Resonanz in der organischen Chemie, p. 301, Springer Verlag, Berlin 1965.

[16] *H.S. Jarret, M. Sadler, J.N. Shoolery,* J. Chem. Phys. *21,* 2092 (1953).

[17] *R. Filler, S.M. Naqui,* J. Org. Chem. *26,* 2571 (1961).

[18] *J.C. Reid, M. Calvin,* J. Am. Chem. Soc. *72,* 2952 (1950).

[19] Ref. [135], Vol. I, 321.

[20] *E.P. Kohler, A.H. Blatt,* J. Am. Chem. Soc. *50,* 504 (1928).

[21] *C.L. Angyal, R.J.W. Le Fèvre,* J. Chem. Soc. 2181 (1953).

[22] *F. Arndt,* Angew. Chem. *61,* 397 (1949).

[23] *R. Gompper,* Chem. Ber. *93,* 187, 198 (1960).

[24] *F. Arndt, B. Eistert, R. Gompper, W. Walter,* Chem. Ber. *94,* 2125 (1961).

[25] *A.J. Boulton, A.R. Katritzky,* Tetrahedron *12,* 41 (1961).

[26] *H. Wamhoff, F. Korte,* Chem. Ber. *99,* 2962 (1966).

[27] *E. Magnien, W. Tom,* J. Org. Chem. *32,* 1229 (1967).

[28] *D.W. Jones,* J. Chem. Soc. C 1678 (1969).

used in heterocyclic chemistry to investigate tautomeric equilibria.

For a compound which can undergo tautomerism, there are two empirical dissociation constants K_1 and K_2, which are derived from the actual dissociation constants K_A, K_B, K_C and K_D of the individual tautomeric forms:

$$K_1 = K_A + K_B$$

$$\frac{1}{K_2} = \frac{1}{K_C} + \frac{1}{K_D}$$

For the tautomeric constant K_T, then:

$$K_T = \frac{K_A}{K_B} = \frac{K_C}{K_D}$$

If a hydrogen atom is replaced by a methyl group (the influence on the ionization constant is normally very small), the tautomerism constant K_T follows from K_1 and the ionization constants of the individual alkyl derivatives:

$$K_T = \frac{K_{(HXCH_3^{\oplus})}}{K_1} - 1 = \frac{K_1}{K_{(CH_3XH^{\oplus})} - K_1}$$

3.2 Measurements

Ionization constants are determined most easily by potentiometric titration, using glass and calomel electrodes, with carbonate-free potassium hydroxide solution as titrant. Using these experimental conditions, pK ranges between 1.25 and 11.0 can be determined; pK values above 11 are measured using a hydrogen electrode.

Albert and Phillips[29] have investigated the acid and base ionization constants of 87 hydroxy derivatives of six-membered heterocyclic nitrogen compounds.

Table 1 shows values experimentally determined for some 2-, 3- and 4-hydroxy-pyridines and -quinolines and the corresponding methoxy- and N-methyl-derivatives.

From the values listed in Table 1, the equilibrium ratio of the lactam and hydroxy forms can be determined[30, 31] from the Equation:

$$\log R = pK_{(OCH_3)} - pK_{OH}$$

Here, R is the ratio of the amide tautomer to the enol tautomer. $pK_{(OCH_3)}$ and pK_{OH} are the values determined experimentally for addition of a proton to the methoxy compound and to the hydroxy compound respectively.

Proportion of amide and enol tautomers (measured in neutral aqueous solution 20°)

	pyridine	quinoline
2-Hydroxy	340	3000
4-Hydroxy	2200	24000

From the calculated ratios of amide and enol tautomers, it is found that the 2- and 4-hydroxy-pyridines and quinolines exist predominantly in the lactam form in aqueous solution. These findings have been confirmed by spectroscopic investigations[32].

The tautomerism of amino-pyridines[33, 34] has also been studied. The following scheme of tautomeric structures can be formulated for the *4-amino-pyridine*:

Table 1. p$'K_a$-values of hydroxy derivatives of pyridine and quinoline (measured in H_2O at 20°; concentration 0.02–0.0001 M).

Substituent groups	p$'K_a$-values of protonated form (Amide/enol ratio)	
	pyridine derivatives	quinoline derivatives
2-Hydroxy	0.75 (±0.01) (340 : 1)	−0.31 (±0.06) (3000 : 1)
2-Methoxy	8.28 (±0.06)	3.17 (±0.03)
2-Oxo-1-methyl	0.32 (±0.02)	−0.71 (±0.10)
3-Hydroxy	4.86 (±0.01)	4.30 (±0.05)
3-Methoxy	4.88 (±0.03)	
4-Hydroxy	3.27 (±0.02) (2200 : 1)	2.77 (±0.02) (24 000 : 1)
4-Methoxy	6.62 (±0.02)	6.65 (±0.02)
4-Oxo 1-methyl	3.33 (±0.02)	2.46 (±0.02)

[29] *A. Albert, J.N. Phillips*, J. Chem. Soc. 1294 (1956).
[30] *J.T. Edsall, M.H. Blanchard*, J. Am. Chem. Soc. *55*, 2337 (1933).
[31] *G.F. Tucker, J.L. Irvin*, J. Am. Chem. Soc. *73*, 1923 (1951).
[32] *S.F. Mason*, J. Chem. Soc. 5010 (1957).

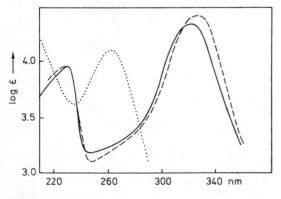

From the Equation:

$$K = [amino] / [imino]$$
$$= K_{a(amino)} / K_{a(imino)}$$

the tautomeric equilibrium constant K_T is obtained by titrimetric determination of the dissociation constants $K_{a(amino)}$ and $K_{a(imino)}$. Measurements of the pK$_a$-values showed that the ratio of amino to imino forms is 10^3:1 and more, i.e. these compounds exist predominately in the amino form.

On the basis of the ionization constants of *2-pyridone/2-hydroxypyridine* tautomers, it has been shown by Spinner and Yeoh[35] that the size of a ring which is fused to the pyridine nucleus has a considerable effect on the position of the equilibrium (Mills-Nixon-effect[36]). On changing from a 5-membered fused ring to a 6-membered fused ring, the ratio of the 2-hydroxy-pyridine form to the 2-pyridone form drops by a factor of 20–30.

4 Spectroscopic methods

Spectroscopic methods (IR, UV, NMR) are being used increasingly and with great success to study tautomeric equilibria. In contrast to the chemical methods, the spectroscopic techniques do not affect the equilibria in any way.

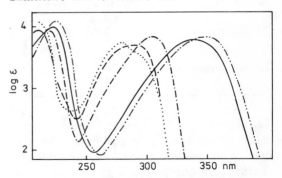

Fig. 1. UV spectra of pyridine-4-thione (———), 1-methyl-pyridine-4-thione(------)and 4-benzylthiopyridine (·····); cf. Ref.[37]

4.1 UV Spectroscopy

Determination of tautomeric equilibria by means of UV spectroscopy involves comparing the UV spectrum of a potentially tautomeric compound with the spectra of the two possible alkylated forms.

From comparison of the UV spectra of *4-mercaptopyridine, 1-methyl-4-mercaptopyridine* and *4-benzylmercaptopyridine*[37] (Fig. 1), it has been concluded that 4-mercaptopyridine exists predominantly as pyridine-4-thione.

Similarly, 2-mercapto-pyridine exists in the thione form and not in the mercapto form. Whereas N-alkylation has only a small effect, S-alkylation causes a slight bathochromic UV absorption shift, of ca. 15 nm.

Other good examples of UV spectroscopy for determination of tautomeric equilibria are the measurements made by Mason[38] on aqueous solutions of the cationic, anionic and zwitterionic forms of *2-hydroxypyridine* derivatives.

It was shown experimentally that a zwitterionic or amide-type structure absorbs at higher wavelengths than the corresponding enolic hydroxy-form[39]. The absorption bands with longest wavelengths of the several tautomeric species can be arranged in the following sequence:

zwitterion > anion > cation > enol

Fig. 2. UV spectra of 2-methoxypyrazine in water at pH 7 (------) and in cyclohexane (·····), 2-methoxypyrazine in 5 N H$_2$SO$_4$ (-·--·-·), 2-hydroxypyrazine in 10 N H$_2$SO$_4$ (———) and 1-methyl-2-pyrazone in 10 N H$_2$SO$_4$ (-··--··--··); cf. Ref.[38]

[33] *A. Albert, R. Goldacre, J.N. Phillips*, J. Chem. Soc. 2240 (1948).

[34] *S.J. Angyal, C.L. Angyal*, J. Chem. Soc. 1461 (1952).

[35] *E. Spinner, G.B. Yeoh*, Tetrahedron Lett. 5691 (1968).

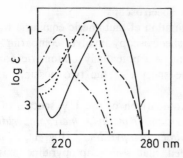

Fig. 3. UV spectra of 4-methoxy-pyridine in water at pH 9 (-·-·-··-), in 1 N HCl (·····), at pH 13 (------), and of 4-hydroxypyridine at pH 7 (——); cf. Ref. [38]

From Figures 2 and 3 and Table 2, the 2- and 4-*hydroxypyridines* and *2-hydroxypyrazine* exist predominantly as the 2- and 4-pyridones and 2-pyrazinone respectively, while with *3-hydroxypyridine* the enolic structure is favored.

The equilibrium constant of *3-thionaphthenones* was determined from UV-spectra and pK-values using the equation:

$$K_T = [Enol]/[Keto]$$

The values obtained were:
in H_2O: 0.18 ± 0.01; in C_2H_5OH: 2.02 ± 0.12[41]. Here, the keto form is the preferred one in aqueous solution whereas the enol structure predominates in ethanol.

4.2 IR Spectroscopy

IR spectroscopy is also very suitable for determination of tautomeric rearrangements. It supplements the methods described above in that it offers the following possibilities:

Table 2. UV-Data for the 2- and 4-hydroxy-pyridines and 2-hydroxy-pyrazine, and their derivatives. $K^{\oplus}$: cationic form; $A^{\ominus}$: anionic form; E: enolic OH (or OR) form; ZW $\pm$: zwitterionic form.

Compound	pH	Species	λ_{max} [nm]	ε	
2-Hydroxy-pyridine	6	ZW±	293. 224	5 890	7 230
	13	A$^{\ominus}$	298. 230	5 070	9 000
	<0	K$^{\oplus}$	277. 209	6 950	3 600
2-Methoxy-pyridine	7	E	269. 205	3 230	5 300
	1	K$^{\oplus}$	279. 210	6 920	3 550
1-Methyl-pyridone-(2)	5	ZW±	297. 226	5 700	6 100
	<0	K$^{\oplus}$	279. 210	6 250	3 500
4-Hydroxy-pyridine	7	ZW±	253	14 800	
	13	A$^{\ominus}$	260. 239	2 200	14 150
	0	K$^{\oplus}$	234	9 800	
4-Methoxy-pyridine	9	E	235. 222	2 000	9 300
	4	K$^{\oplus}$	235	9 500	
1-Methyl-pyridone-(4)	7	ZW±	260	18 900	
	1	K$^{\oplus}$	239	11 800	
2-Hydroxy-pyrazine	5	ZW±	317. 221	5 520	8 820
	10.5	A$^{\ominus}$	316. 227	5 640	11 240
	<0	K$^{\oplus}$	342. 222	6 220	10 400
2-Methoxy-pyrazine	C_6H_{12}	E	277. 210	5 760	11 500
	7	E	290. 209	5 240	9 600
	<0	K$^{\oplus}$	304. 218	6 900	9 220
1-Methyl-pyrazone-(2)	7	ZW±	321. 224	5 830	9 100
	<0	K$^{\oplus}$	345. 225	6 410	10 900

The hydroxypyridine/pyridone equilibrium is shifted clearly towards the enol form when positions on the pyridine nucleus are substituted by electron-attracting groups, e.g. chlorine atoms. Here too, UV spectroscopy[40] gave unambiguous results.

Tautomeric equilibria and their approximate positions (in some cases, the existence of only one tautomeric form) can be determined from the characteristic absorption bands of specific functional groups (for example: >C=O, —NH$_2$, —OH, >NH, C=C, etc.).

[36] W.H. Mills, I.G. Nixon, J. Chem. Soc. 2510 (1930).
[37] A.R. Katritzky, R.A. Jones, J. Chem. Soc. 3610 (1958).
    A. Albert, G.B. Barlin, J. Chem. Soc. 2384 (1959).
[38] S.F. Mason, J. Chem. Soc. 1253 (1959).
[39] S.F. Mason, J. Chem. Soc. 5010 (1957).
[40] A.R. Katritzky, J.D. Rowe, S.K. Roy, J. Chem. Soc. B 758 (1967).
[41] W. Rubaszewska, Z.R. Grabowski, Tetrahedron 25, 2807 (1969).

Suitable absorption bands are used for quantitative determination, if they can be assigned to one of the potential tautomeric structures, and their extinction can be measured: C=O, C=C, C=N and ring-vibration bands (1900–1500 cm⁻¹).

Chelate bridges[42], which often occur in tautomeric systems, as a rule cause great changes and shifts of individual adsorption bands[43].

4.2.1 β-Dicarbonyl Compounds

With enolized and chelated *β-dicarbonyl compounds*, the ν_{OH}-vibration is either totally absent or appears as a very broad peak[44–46], bathochromically shifted to 2700 cm⁻¹. Carbonyl absorptions[47, 48] of chelated β-dicarbonyl compounds occur as very broad bands at 1538–1639 cm⁻¹, about 100 times more intense than the normal C=O absorption bands. *o-Hydroxyacetophenone* and *salicylaldehyde* also show bathochromically shifted carbonyl absorption bands at 1639–1613 cm⁻¹.

4.2.2 Lactam-like Compounds

The NH vibrations of secondary amides and of lactam-type structures (e.g. heterocyclic compounds such as *hydantoin, 5-pyrazolones, quinazolones* and *pyridiminones*) are found at lower wave-numbers[49] because of *hydrogenbridged association*. In many cases these absorptions are below 3000 cm⁻¹ as shown in work by Sohár[50–54].

Besides the 1500–1900 cm⁻¹ range, absorption bands in other regions can be used for determination of tautomeric structures. *2-Aminopyridine* has the following amine absorption bands: ν_{as} 3500 cm⁻¹ and ν_{sym} 3410 cm⁻¹. ν_{sym} is calculated[55] from the equation

$$\nu_{sym} = 345.53 + 0.876\nu_{as} = 4312 \text{ cm}^{-1}.$$

From these facts, it follows that this compound exists predominantly in the amine form:

4.2.3 β-Keto-carboxylic Esters

Le Fèvre and Welsh[56] have determined the enol content of ethyl acetoacetate quantitatively by means of IR spectroscopy. By measurement of the extinction of the C=C double bond of the enol form in ethylene bromide at 18°, an enolic constant was obtained:

$$K = [\text{enol}]/[\text{keto}] = 0.08$$

which is in good agreement with the value determined by bromometric titration[57].

The tautomeric equilibria in the cases of *ethyl cyclopentanone-2-carboxylate* and *methyl 4-methyl-1,2-cyclopentadione-3,4,5-tricarboxylate*[58, 59] are especially suitable for study (Fig. 4).

[42] *N.J. Leonard, H.S. Gutowsky, W.J. Middleton, E.M. Petersen*, J. Am. Chem. Soc. 74, 4070 (1952).

[43] *L.J. Bellamy, L. Beecher*, J. Chem. Soc. 4487 (1954).

[44] *R.S. Rasmussen, P.D. Tunicliff, R.R. Brattain*, J. Am. Chem. Soc. 71, 1068 (1949).

[45] *R.S. Rasmussen, R.R. Brattain*, J. Am. Chem. Soc. 71, 1073 (1949).

[46] *J.D. Park, H.A. Brown, I.R. Lacher*, J. Am. Chem. Soc. 75, 4753 (1953).

[47] *W. Gordy*, J. Chem. Phys. 8, 516 (1940).

[48] *I.M. Hunsberger*, J. Am. Chem. Soc. 72, 5626 (1950).

[49] *R.E. Richards, H.W. Thompson*, J. Chem. Soc. 1248 (1947).

[50] *P. Sohár*, Acta Chim. Budapest 54, 91 (1967); 57, 425 (1968).

[51] *P. Sohár, I. Kosa*, Acta Chim. Budapest 57, 411 (1968).

[52] *P. Sohár, A. Nemes*, Acta Chim. Budapest 56, 25 (1968).

[53] *P. Sohár, J. Nyitrai, K. Zauer, K. Lempert*, Acta Chim. Budapest 58, 31 (1968).

[54] *P. Sohár, G. Varsanyi*, Acta Chim. Budapest 58, 189 (1968); Spectrochim. Acta 23 A, 1947 (1967).

[55] *L.J. Bellamy, R.L. Williams*, Spectrochim. Acta 9, 341 (1957).

[56] *R.J.W. Le Fèvre, H. Welsh*, J. Chem. Soc. 2230 (1949).

[57] *K.H. Meyer, P. Kappelmeier*, Ber. 44, 2718 (1911).

[58] *A.S.N. Murthy, A. Balasubramanian, C.N.R. Rao, T.R. Kasturi*, Can. J. Chem. 40, 2267 (1962).

[59] Ref. [137], p. 212.

In CCl$_4$, **A** shows the following absorption bands (Fig. 4):

1750 [cm^{-1}] ring ketone
1730 β-keto-ester
1670   enolized β-keto-ester
1620   C=C vibration

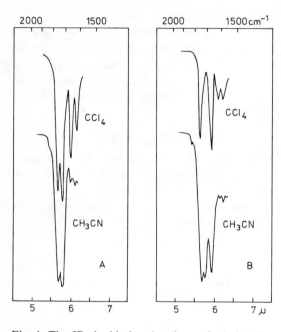

Fig. 4. The IR double bond regions of ethyl cyclopentanone-2-carboxylate (A) and methyl 4-methyl-cyclopentane-1,2-dione-3,4,5-tricarboxylate (B) in CCl$_4$ and CH$_3$CN; cf. Ref.[58]

4.2.4 Heterocyclic Keto-compounds

The *5-hydroxyisoxazole/5-isoxazolone* system is another tautomeric equilibrium[60, 61]; 'OH', 'CH' and 'NH' forms may exist (Scheme 1 and 3; Sections 1 and 2). Infrared spectra indicate that 3,4-dimethyl-5-isoxazolone exists predominantly in the 'OH' form in aqueous solution, in chloroform solution, and in the solid state. *3-Phenyl-4-bromo-3-phenyl-*, and *4-methyl-3-phenyl-5-isoxazolones* form mixtures of 'CH'- and 'NH'-forms; the proportion of the 'NH'-form is distinctly increased in solvents of greater polarity. This can be seen from the C=O absorptions:

Solvent:	CHCl$_3$	CCl$_4$	KBr
C=O absorption:	1740–1725	1760	1680 [cm^{-1}]

Some of the results obtained[60] are summarized in Tables 3 and 4.

4.3 NMR Spectroscopy

Only with NMR spectroscopy is it possible to obtain the absolute values of equilibrium constants directly by integration of the signals from the structural elements of particular tautomeric species. The rate of keto-enol tautomerism at room temperature is sufficiently slow compared with the speed of NMR processes to permit the exact determination of the equilibrium constant.

Table 3. Tautomeric forms of substituted 5-isoxazolones
(OH = OH-form; CH = CH-form; NH = NH-form)

Substituent in position		Tetrachloroethylene	CHCl$_3$	solid
3	4			
CH$_3$	CH$_3$	—	98% OH	OH
C$_6$H$_5$	H	—	98% OH	CH
C$_6$H$_5$	CH$_3$	90% CH + 10% NH	70% CH + 30% NH	NH
C$_6$H$_5$	Br	95% CH	90% CH + 10% NH	NH

Extinction measurements in CCl$_4$ gave an enol value of ca. 22%. In acetonitrile solution, there is negligible (<1%) enolization. Besides the absorptions at 1750 and 1730 cm^{-1} only very weak vibrations of the enolic form (at 1670 and 1620 cm^{-1}) are found.

In CCl$_4$ solution, the tricarboxylic ester **B** exists predominantly in the enol form and sharp absorption bands are found in the double bond region at 1750, 1680, 1618 and 1580 cm^{-1}. In this case, too, the keto form is favored in acetonitrile.

For the equilibrium:

the different resonance signals of the protons of both the keto and enol tautomers can be observed.

[60] *A.J. Boulton, A.R. Katritzky*, Tetrahedron *12*, 41 (1961).

[61] *A.R. Katritzky, A.J. Boulton*, Spectrochim. Acta *17*, 238 (1961).

[62] *S.F. Mason*, J. Chem. Soc. 4874 (1957).

[63] *Y.N. Sheinker, V.V. Kushkin, I.Y. Postovskii*, J. Phys. Chem. [USSR] *31*, 214 (1957); C.A. *51*, 17455 (1957).

Table 4. Examples of the use of IR spectroscopy to determine tautomeric forms.

Tautomeric equilibrium	Ref.	Tautomeric equilibrium	Ref.
	62		70–73
	63 64		74,75
$H_5C_6 - \overset{S}{\underset{\parallel}{C}} - CH_2 - COOC_2H_5$	65 66		
$H_3C - \overset{O}{\underset{\parallel}{C}} - (CH_2)_2 - CH_2OH$	67		
H_2N isoxazole	68		76
$H_5C_6 - \overset{S}{\underset{\parallel}{C}} - CH_2 - \overset{NH_2}{\underset{\parallel}{C}}=NH$ (OR)			
$H_5C_6 - \overset{S}{\underset{\parallel}{C}} - CH = \overset{NH_2}{\underset{\parallel}{C}} - NH_2$ (OR)	69		77

4.3.1 Open-Chain β-Dicarbonyl Compounds

The tautomerism of open-chain β-*dicarbonyl compounds* was the first to be investigated by means of NMR spectroscopy[78–80] (Fig. 5).

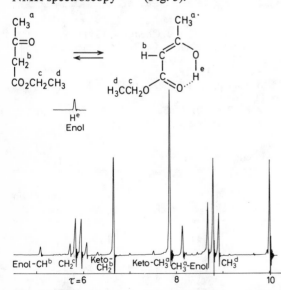

Fig. 5. NMR spectra of ethyl acetoacetate, α-methyl-acetylacetone and diethylacetonedicarboxylate, in cyclohexane-d$_{12}$ as solvent, 60 MHz (TMS:$\tau = 10$)

[64] *Y.N. Sheinker, I.Y. Postovskii, N.M. Voronina*, J. Phys. Chem. [USSR] *33*, 302 (1959); C.A. *54*, 4147 (1960).

[65] *S.K. Mitra*, J. Indian. Chem. Soc. *15*, 205 (1938).

[66] *Z. Reyes, R.M. Silverstein*, J. Am. Chem. Soc. *80*, 6367 (1958).

[67] *W. Lüttke*, Chem. Ber. *83*, 571 (1950).
C.D. Hurd, W.H. Saunders, J. Am. Chem. Soc. *74*, 5324 (1952).
J.E. Whiting, J.T. Edward, Can. J. Chem. *49*, 3799 (1971).

[68] *A.J. Boulton, A.R. Katritzky*, Tetrahedron *12*, 51 (1961).

[69] *J. Goerdeler, W. Mittler*, Chem. Ber. *96*, 944 (1963).

[70] *H.M. Randall, R.G. Fowler, N. Fuson, J.R. Dangl*, Infrared Determination of Structures, Van Nostrand, New York 1949.

[71] *C.A. Coulson*, Research [London] *10*, 149 (1957).

[72] *C.G. Cannon*, Spectrochim. Acta *10*, 341 (1958).

[73] *S. Refn*, Spectrochim. Acta *17*, 40 (1961).

[74] *K. Lempert, J. Nyitrai, P. Sohár, K. Zauer*, Tetrahedron Lett. 2679 (1964).

[75] *P. Sohár, J. Nyitrai, K. Zauer, K. Lempert*, Acta Chim. Budapest *58*, 165 (1968).

[76] *P. Sohár, K. Körmendy, A. Pfisztner-Freud, F. Ruff*, Acta Chim. Budapest *60*, 273 (1969).

[77] *N. Akitada, K. Shozo*, Chem. Pharm. Bull. [Tokyo] *17*, 425 (1969); C.A. *71*, 21452e (1969).

[78] *H.S. Jarret, M.S. Sadler, J.N Shoolery*, J. Chem. Phys. *21*, 2092 (1953).

The following characteristic signals from Fig. 5 are summarized in Table 5.

The enol forms of unsubstituted β-dicarbonyl compounds give spectra containing a signal due to the methine protone. The enol proton which is usually involved in intramolecular chelation causes a paramagnetically shifted signal, and, therefore, is found at low field ($\tau < 0$), as the electron density in the vicinity of this proton is diminished by the chelate bridge[81]. With β-dicarbonyl compounds, a linear relation can be established between the paramagnetic shift of the enolic OH proton and the bathochromically-shifted IR carbonyl frequency[82, 83].

Burdett and Rogers[84, 85] made a quantitative study of the tautomeric equilibria of β-dicarbonyl compounds and, from the signal intensities, obtained the equilibrium constants given in Table 6.

$$K_e = [\text{enol}]/[\text{keto}]$$

The accuracy is ± 3–5%.

Table 6. Enol values and equilibrium constants of some β-dicarbonyl compounds (measured at $33 \pm 2°$; samples neat, without solvent).

compound	% Enol	K_e	signals
Acetylacetone	81	4.3	CH_{3E} /CH_{3K}
α-Bromo-acetylacetone	46	0.85	CH_{3E} /CH_{3K}
Ethyl acetoacetate	8	0.09	CH_{3E} /CH_{3K}
Butyl acetoacetate	15	0.18	CH_{3K} /$CH_{2(Ethyl)}$
Ethyl benzoyl-acetate	22	0.28	$=CH_E/CH_{3 E+K}$
Ethyl α-cyano-acetate	93	13	OH_E /$CH_{2 E+K}$

[79] L. W. Reeves, Can. J. Chem. 35, 131 (1957).
[80] C. Giessner-Prettre, C. R. Acad. Sci. 250, 2547 (1960).
[81] Ref. [138], p. 322 ff.
[82] S. Forsen, M. Nilsson, Acta Chem. Scand. 13, 1383 (1959).
[83] A. L. Porte, H. S. Gutowsky, I. M. Hunsberger, J. Am. Chem. Soc. 82, 5057 (1960).

Table 5. Determination of keto/enol proportions from the intensitres of NMR signals.

Substance		R——CH₂——CO——	COOCH₂————CH₃ / CH	CH₃		Proportion (approx.)
Ethylacetoacetate	keto	7.84	6.65 } 5.88	—	} 8.78	80
	enol	8.13	5.08	—		20
3-Methyl-2,4-pentanedione	keto	7.95	6.41	8.78	—	51
	enol	8.00	—	8.21	—	49
Diethyl-acetonedicarboxylate	keto	6.48 } 5.85	6.48 } 5.85		} 8.76	62
	enol	6.88	6.88			38

Using high resolution NMR spectroscopy, the ratio of the enolic structures E I⇌E II in substituted *benzoylacetones* can be calculated approximately, by considering long range proton-proton couplings[86]. In E I, the methine proton couples with the methyl group through a C–C bond, whereas in E II allylic coupling should be expected; in the former case, coupling is too small to be resolved, whereas in the latter, J = 0.5–0.7 Hz. In practice, very small coupling constants (0–0.13 Hz) are observed, which appear as line broadening (e.g. of the methyl signal). The amount of E II in the equilibrium E I⇌E II can be calculated as amounting to ca. 5–10%.

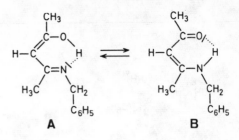

A   B

the NMR spectrum (Fig. 6) clearly shows that, of the three possible tautomeric forms, the resonance-stabilized and chelated ketamine form **B** is predominant.

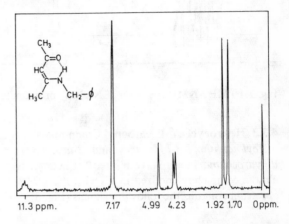

Fig. 6. 60 MHz-NMR spectrum of neat 4-benzyl-amino-3-penten-2-one (TMS: τ = 10); cf. Ref.[88]

Similarly, from observation of the coupling constants between the methyl protons and the chelating proton (J = 0.2–0.3 Hz), Klose and co-workers[87] concluded that at room temperature *β-thioxo-ketones* exist exclusively in the enol form **B**. At higher temperatures, form **A** exists in equilibrium with **B**.

This k etamine form is revealed unambiguously by a characteristic coupling of the NH-proton with the benzylic protons; the latter ones are accordingly found as a doublet (J = 6.3 Hz), while the chelate proton exhibits a triplet (J = 6 Hz) at 11.3 ppm.

H₃C–C=C–C–C₆H₅ T≫RT° H₃C–C=C–C–C₆H₅
                T=RT°

A   B

4.3.2 β-Keto-amines

Dudek and Holm[88] have studied keto-enol equilibria of *α,β-unsaturated β-keto-amines*. With *4-benzylamino-3-penten-2-one*, for example:

$$CH_3CH_2-C \rightleftharpoons CH_3CH_2-C$$

The NMR spectrum of *N-phenylpropionamidine* in CDCl₃ (Fig. 7) shows that the equilibrium is well over on the side of the phenylimine form[89]. Besides the triplet and the quartet due to the ethyl group, the aromatic protons are found as a broad multiplet because of the electron-releasing imine structure. The amino-protons give a sharp signal at τ = 5.19.

[84] *J.L. Burdett, M.T. Rogers*, J. Am. Chem. Soc. *86*, 2105 (1964).

[85] *M.T. Rodgers, J.L. Burdett*, Can. J. Chem. *43*, 1516 (1965).

[86] *D.J. Sardella, D.H. Heinert, B.L. Shapiro*, J. Org. Chem. *34*, 2817 (1969).

[87] *G. Klose, P. Thomas, E. Uhlemann, J. Märki*, Tetrahedron *22*, 2695 (1966).

[88] *G.O. Dudek, R.H. Holm*, J. Am. Chem. Soc. *84*, 2691 (1962).

[89] *J. Neuffer, J. Goerdeler*, unpubl.

[90] *H. Wamhoff, G. Schorn, F. Korte*, Chem. Ber. *100*, 1296 (1967).

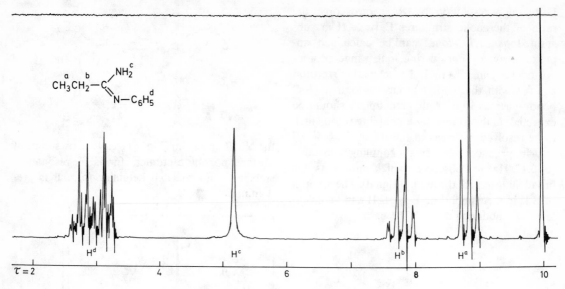

Fig. 7. 60 MHz-NMR spectrum of N-phenylpropionamidine in CDCl₃ (TMS τ = 10)

4.3.3 Heterocyclic β-Dicarbonyl Compounds

3-Acyl-naphtho[2,3-b]pyranes and *3-acetyl-3,4-dihydrocoumarin* were investigated[90]; the latter is almost completely enolized in CDCl₃; the NMR spectrum contains the following signals (Fig. 8):

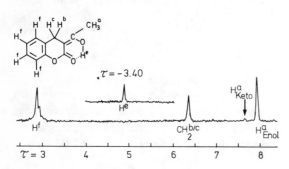

Fig. 8. 60 MHz-NMR spectrum of 3-acetyl-3,4-dihydrocoumarin in CDCl₃ (TMS:τ = 10)

CH₃ᵃ₍E₎	7.90	CH₂ᵇ/ᶜ	6.38
aromatic Hᶠ	2.6–3.1	Hᵉ₍E₎	−3.4

The small signal at ca. τ = 7.6 represents the very small amount of the keto form (<2%).

3-Benzoyl-3,4-dihydrocoumarin in CDCl₃ exists 100% in the keto form (Fig. 9).

The protons Hᵇ, Hᶜ and Hᵈ form an ABM pattern, with 8 lines for the AB portion and 4 lines for the M part:

Hᵇ/ᶜ	6.90	6.47	(J_{bc} = 16.3 Hz)
Hᵈ	5.29		(J_{db} = 7.0 Hz; J_{dc} = 9.0 Hz)
Hᶠ	1.9–3.0		

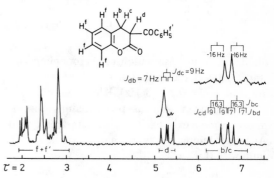

Fig. 9. 60 MHz-NMR spectrum of 3-benzoyl-3,4-dihydrocoumarin in CDCl₃ (TMS:τ = 10). When a spin decoupler is employed, the couplings of protons Hᵇ with Hᵈ and of Hᶜ with Hᵈ are annulled, and Hᵈ gives a singlet; in the inverse case, Hᵇ and Hᶜ each give a doublet.

In DMSO-d₆ as solvent, *3-acetyl-3,4-dihydrocoumarin* exists as an 81:19 mixture of keto and enol forms as determined from the intensities of the signals (Fig. 10).

To establish equilibrium, the substances are left in solution for five days at constant temperature[3]. The enol content of *α-acetyl-δ-thiovaleroacetone* changes from 44% immediately after solution in CDCl₃ to the final value of 75% after five days[90]. The signals to be used for determination of keto-enol equilibrium should preferably be integrated 10–20 times, and the average intensity calculated.

With the exception of those β-dicarbonyl compounds which are fixed sterically as *trans* forms[91],

[91] B. Eistert, W. Reiss, Chem. Ber. 87, 92, 108 (1954).

[92] D.J. Sardella, D.H. Heinert, B.L. Shapiro, J. Org. Chem. 34, 2817 (1969).

[93] S.T. Yoffe, E.I. Fedin, P.V. Petrovskii, M. Kabachnik, Tetrahedron Lett. 2661 (1966).

	CH₃[a]	CH₂[b/c]	H[d]		H[e]	H[f]
Keto form	7.70	6.95	5.86		—	2.5–3.2
		6.63 (J_{bc} = 16.2)	(J_{db} = J_{dc} = 7.6)			
Enol form	7.90	6.3	—		–2.37	2.5–3.2

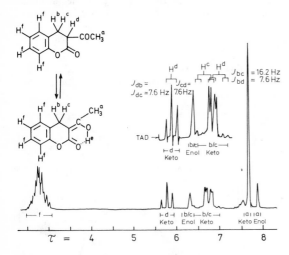

Fig. 10. 60 MHz-NMR spectrum of 3-acetyl-3,4-di-hydrocoumarin in DMSO-d₆ (TMS:τ = 10)

NMR measurements on *open chain β-diketones*[92], *β-keto-esters* and *α-acyl-lactones*, *-thio-lactones* and *-lactams*[90] provide no evidence for the existence of *trans*-enols. These compounds exist predominantly in the chelated *cis*-enol form. If bulky substituent groups cause steric hindrance of *cis*-enolization, these compounds do not form *trans*-enols; instead, there is an increase in the amount of the energetically more stable keto form[93].

Kinetic NMR investigations of the trans-cis-enol-rearrangement of *ethyl 2-formyl-phenylacetate* showed that *trans*-enols are highly unstable in solvents (e.g. CDCl₃), and undergo rapid conversion into *cis*-enols[94]. Only in the case of *β-thioketo-thiol esters*[95] does a rather small amount of the *trans*-enethiol form enter into discussion. Recently it has been shown that for 3-thioacyl-lactones and -3,4-dihydrocoumarins both *cis* and *trans* enol forms can co-exist[96].

The inductive effect of alkyl chain lengths on the keto-enol equilibria of 1-alkyl-3-acetyl-2-piper-idinones and open chain β-dicarbonyl compounds can be studied by NMR spectroscopy[97]. NMR spectroscopy is also useful for measuring the thermodynamic functions of keto-enol equilibria[10–14, 98]. The equilibrium constants K_e and their temperature dependency can be calculated from the NMR spectra. Assuming ideal behavior, the enthalpy of the change, ΔH, is calculated from the Gibbs equation

$$\Delta G° = -RT \ln K_e$$

by the least squares method. Table 7 shows some values of ΔH and ΔS for substituted β-diketones[98].

Table 7. Thermodynamic functions for the enolization of substituted β-diketones.

R¹—CO—CHR²—COR³

R¹	R²	R³	$-\Delta H$	$-\Delta S$	temperature range
CH₃	H	CH₃	2.8	7.3	38–170°
C₆H₅	H	CH₃	2.8	3.9	90–170°
C₆H₅	H	C₆H₅	3.2	3.4	90–180°
CH₃	CH₃	CH₃	1.3	6.1	38–170°
CH₃	C₃H₅	CH₃	2.1	7.3	25–125°
CH₃	CH₂C₆H₅	CH₃	2.5	8.4	38–170°

The values show that the keto-enol equilibrium is easily shifted, even by very small effects. For this reason, spectroscopic methods, which exclude chemical influences, should be used, wherever possible, to determine equilibria.

4.3.4 Pyrimidinones

NMR spectroscopy can be used also for investigation of the tautomerism (and the effects of substituent groups) of *pyrimidines*[99–100] and related *nucleosides*[101]. Being very polar and capable of tautomerism, pyrimidines are insoluble in most nonpolar solvents. Dimethyl sulfoxide-d₆ is a very suitable solvent for this class of compounds since there is no exchange of the type normally observed when protic solvents are used.

[94] *S.T. Yoffe, P.V. Petrovskii, E.I. Fedin, K.V. Vatsuro, P.S. Burenko, M.I. Kabachnik,* Tetrahedron Lett. 4525 (1967).

[95] *F. Duus, P. Jakobsen, S.O. Lawesson,* Tetrahedron 22, 5323 (1968).

[96] *F. Duus, E.B. Pedersen, S.O. Lawesson,* Tetrahedron 25, 5703 (1969).

[97] *H. Wamhoff, G. Höffer, H. Lander, F. Korte,* Liebigs Ann. Chem. 722, 12 (1969).

[98] *G. Allen, R.A. Dwek,* J. Chem. Soc. B 161 (1966).

[99] *C.D. Jardetzky, O. Jardetzky,* J. Am. Chem. Soc. 82, 222 (1960).

[100] *J.P. Kokko, J.H. Goldstein, L. Mandell,* J. Am. Chem. Soc. 83, 2909 (1961).

[101] *L. Gatlin, J.C. Davis jr.,* J. Am. Chem. Soc. 84, 4464 (1962).

The NMR values in Table 8 show, that 'hydroxy'- and amino-pyrimidines exist predominantly in the lactam and amino forms respectively.

Table 8. NMR data for some substituted pyrimidines in DMSO —d_6 as solvent (TMS: $\tau = 10$).

pyrimidine	position				
	2	3	4	5	6
2-Oxo-1,2-dihydro-	—	—	1.68	3.58	1.68
2-Amino-[a]	4.28	—	1.72	3.42	1.72
2,4-Dioxo-1,2,3,4-tetrahydro-	—	–1.31	—	4.18	2.26
4-Amino-2-oxo-1,2-dihydro-[b]	—	—	—	3.90	2.20
5-Brom-2,4-dioxo-1,2,3,4-tetrahydro-	—	–1.81	—	—	1.81
5-Nitro-2,4-dioxo-1,2,3,4-tetrahydro-	—	—	—	—	0.80
5-Methyl-2,4-dioxo-1,2,3,4-tetrahydro-	—	–1.34	—	7.94	2.41

[a] In $CDCl_3 \cdot$ — [b] $J_{5/6} = 8$ Hz

In the NMR spectra of *2-aminopyrimidine*[102] and of *1H-2-pyrimidinone*[103], besides the broad signal of the amine protons and the intermolecularly chelated NH proton, there is a characteristic A_2B pattern due to the remaining three ring protons: a doublet (H^4, H^6) and a triplet (H^5), whose intensities are in the ratio 2:1. *Cytosine*[104] and *Uracil*[105] give a weakly deshielded signal for the chelated lactam proton (or for the amino group) and two AB doublets for the protons H^5 and H^6. *5-Methyl-5-bromo-* and *5-nitro-uracils*[105] give a singlet for the proton in the 6-position, which is paramagnetically shifted by electron-attracting substituent groups in the 5-position.

4.3.5 Benzotriazole

The tautomerism of *benzotriazole* can be studied by low temperature measurements[106].

At 0° the 60 MHz NMR spectrum contains a broad resonance signal caused by rapid proton exchange (Fig. 11). The four aromatic protons form the two symmet-

rical multiplets of a typical AA′BB′ system. At a lower temperature, the symmetry of this signal group changes sharply. At −100° the NH proton gives a sharp signal (width: 2 Hz), while the aromatic protons now yield two asymmetric signal groups. These observations can be explained only in terms of the asymmetric tautomeric form, in which the shielding of the aromatic protons is different. Using a modified Arrhenius Equation:

$$v = v_0\ e^{-\frac{\Delta H}{RT}}$$

ΔH for this prototropic transition, calculated from the temperature dependency of the line width, has a value of ca. 2 Kcal.

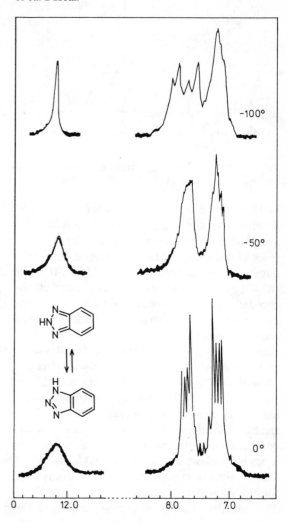

Fig. 11. NMR spectrum of benzotriazole in THF at various temperatures[106]. (TMS: $\tau = 10$)

Further examples of the use of NMR spectroscopy for determination of tautomeric equilibria are summarized in Table 9.

[102] High Resolution NMR Spectra Catalogue (2) Varian Palo Alto.

[103] *S. Gronowitz, R.A. Hoffmann*, Ark. Kemi *16*, 459 (1960).

[104] *A.R. Katritzky, A.J. Waring*, J. Chem. Soc. 3046 (1963).

[105] *J.P. Kokko, L. Mandell, J.H. Goldstein*, J. Am. Chem. Soc. *84*, 1042 (1962).

[106] *A.N. Nesmeyanov, V.N. Babin, L.A. Fedorov, M.I. Rybinskaya, E.I. Fedin*, Tetrahedron *25*, 4667 (1969).

[107] *E.W. Garbisch jr.*, J. Am. Chem. Soc. *85*, 1696 (1963).

[108] *F. Merenyi, M. Nilsson*, Acta Chem. Scand. *18*, 1208 (1964).

Table 9. Examples of the determination of tautomeric forms using NMR spectroscopy.

Tautomeric Equilibrium	Ref.	Tautomeric Equilibrium	Ref.
	107		113
	108		114
	109		116
	110		117
	111		
	112		118

[109] R. Mondelli, L. Merlini, Tetrahedron 22, 3253 (1966).

[110] C.A. Grob, H.J. Wilkens, Helv. Chim. Acta 50, 725 (1967).

[111] L.A. Paquette, R.W. Begland, J. Am. Chem. Soc. 88, 4685 (1966).

[112] H. Yasuda, H. Midorikawa, J. Org. Chem. 31, 1722 (1966).

[113] F.A. Snavely, C.H. Yoder, J. Org. Chem. 33, 513 (1968).

[114] J. Elguero, R. Jacquier, G. Tarrago, Bull. Soc. Chim. Fr. 3780 (1967).

[115] G. Dudek, E.P. Dudek, Tetrahedron 23, 3245 (1967).

[116] A. Yogev, Y. Mazur, J. Org. Chem. 32, 2162 (1967).

[117] J.W. Schulenberg, J. Am. Chem. Soc. 90, 7008 (1968).

[118] S. Hünig, H. Hoch, Liebigs Ann. Chem. 716, 68 (1968).

[119] D.C. Nonhebel, J. Chem. Soc. C 676 (1968).

Table 9 (cont.)

Tautomeric Equilibrium	Ref.
	119
	120
	121
	123
	123a

[120] H. Brockmann, T. Reschke, Naturwissenschaften 55, 544 (1968).

[121] J. Adams, R.G. Shepherd, Tetrahedron Lett. 2747 (1968).

[122] H. Sterk, Monatsh. Chem. 100, 916 (1969).

[123] W. Steglich, G. Höfle, L. Wilschowitz, G.C. Barret, Tetrahedron Lett. 169 (1970).

[123a] V.G. Granik, B.M. Pyatiu, J.U. Persianova, E.M. Peresleni, N.P. Kostyaichenko, R.G. Gluskov, Y.N. Sheinker, Tetrahedron 26, 4367 (1970).

5 Bibliography

[124] H. Henecka, B. Eistert, in Houben-Weyl, Methoden der organischen Chemie, 4. Aufl. Bd. II, p. 380, Georg Thieme Verlag, Stuttgart 1953.

[125] L.J. Bellamy, W. Brügel, Ultrarot-Spektrum und chemische Konstitution, 2. Aufl., Steinkopff, Darmstadt 1966.

[126] R.N. Jones, C. Sandorfy, The Application of Infrared and Raman Spectroscopy to the Elucidation of Molecular Structure in W. West, Chemical Applications of Spectroscopy Vol. 9, p. 247, Interscience Publishers Inc., New York 1956.

[127] L.M. Jackman, Applications of NMR Spectroscopy in Organic Chemistry, Pergamon Press, London 1959.

[128] J.A. Pople, W.G. Schneider, H.J. Bernstein, High Resolution Nuclear Magnetic Resonance, McGraw Hill Book Co., New York 1959.

[129] K.B. Wiberg, B.J. Nist, The Interpretation of NMR-Spectra, Benjamin, New York 1962.

[130] K. Nakanishi, Infrared Absorption Spectroscopy (Practical) Holden-Day, San Francisco 1962.

[131] H.H. Jaffé, M. Orchin, Theory and Applications of Ultraviolet Spectroscopy, p. 572, Wiley, New York 1962.

[132] A. Albert, E. Serjeant, Ionisation Constants, a Laboratory Manual, Methuen & Co., London 1962.

[133] A. Albert in A.R. Katritzky, Physical Methods in Heterocyclic Chemistry, Vol. I, p. 2, Academic Press, New York 1962.

[134] A.R. Katritzky, A.P. Ambler in A.R. Katritzky, Physical Methods in Heterocyclic Chemistry, Vol. II, p. 165, Academic Press, New York 1962.

[135] A.R. Katritzky, J.M. Lagowski in A.R. Katritzky, Advances in Heterocyclic Chemistry, Vol. I, p. 311, Vol. II, p. 1, Academic Press, New York 1963.

[136] S.F. Mason in A.R. Katritzky, Physical Methods in Heterocyclic Chemistry, Vol. II, p. 7, Academic Press, New York 1963.

[137] C.N.R. Rao, Chemical Applications of Infrared Spectroscopy, Academic Press, New York 1963.

[138] H. Suhr, Anwendungen der kernmagnetischen Resonanz in der organischen Chemie, Springer Verlag, Berlin 1965.

[139] D.W. Mathieson, Nuclear Magnetic Resonance for Organic Chemists, Academic Press, New York 1967.

[140] D.H. Williams, I. Fleming, Spektroskopische Methoden in der organischen Chemie, Georg Thieme Verlag, Stuttgart 1971.

[141] A. Albert in A.R. Katritzky, Physical Methods in Heterocyclic Chemistry, Vol. III, p. 1. Academic Press, New York 1971.